Single-Ion Solvation

Experimental and Theoretical Approaches to Elusive Thermodynamic Quantities

RSC Theoretical and Computational Chemistry Series

Titles in the Series:

1. Knowledge-based Expert Systems in Chemistry: Not Counting on Computers
2. Non-Covalent Interactions: Theory and Experiment
3. Single-Ion Solvation: Experimental and Theoretical Approaches to Elusive Thermodynamic Quantities

Single-Ion Solvation
Experimental and Theoretical Approaches to Elusive Thermodynamic Quantities

Philippe Hünenberger and Maria Reif
Laboratory of Physical Chemistry, ETH Zürich (ETHZ),
Hönggerberg, HCI, Zürich, Switzerland

RSC Publishing

RSC Theoretical and Computational Chemistry Series No. 3

ISBN: 978-1-84755-187-0
ISSN: 2041-3181

A catalogue record for this book is available from the British Library

Published by The Royal Society of Chemistry,
Thomas Graham House, Science Park, Milton Road,
Cambridge CB4 0WF, UK

Registered Charity Number 207890

For further information see our website at www.rsc.org

Preface

Ions are fascinating chemical species. They are ubiquitous in nature and play a fundamental role in many essential ecological, biochemical, physiological and technological processes. Under the conditions of pressure and temperature prevailing on earth, ions would probably not play such an important role if their presence was not associated with that of large amounts of water. In the absence of water and at room temperature and ambient pressure, the long-ranged and extremely strong Coulombic interactions existing between oppositely-charged ions promote their aggregation into solid salts (exceptionally ionic liquids). The magnitude of these interactions is such that gas-phase ions are essentially non-existing species at equilibrium. Yet, in polar solvents (prominently water), dielectric screening of these direct Coulombic interactions by the dipolar solvent molecules represents an almost equally strong opposing force that permits, in synergy with entropic effects, the existence of solvated ions as dissociated entities. The properties of ions in solution crucially depend on the nature and magnitude of these solvation forces, and their experimental and theoretical characterization is thus an essential step towards the understanding of ionic properties. The key quantity accounting for the magnitude of ion-solvent interactions, in the simplest context of an individual ion at infinite dilution in a given solvent, is the corresponding *single-ion solvation free energy*. Knowledge of the pressure and temperature dependence of this property also gives access to all other derivative thermodynamic solvation parameters, *e.g.* single-ion solvation enthalpy, entropy, volume, heat capacity, compressibility or expansivity, as well as corresponding partial molar variables in solution. However, this fundamental quantity can be defined in two distinct ways.

The *real* single-ion solvation free energy corresponds to the reversible work associated with the hypothetical process of transferring the ion from the gas phase to the solution environment. Although this process is hypothetical in the sense that it cannot be carried out directly in an experiment involving a single ion, the corresponding free energy can still be deduced from the results

RSC Theoretical and Computational Chemistry Series No. 3
Single-Ion Solvation
Experimental and Theoretical Approaches to Elusive Thermodynamic Quantities
By Philippe Hünenberger and Maria Reif

Published by the Royal Society of Chemistry, www.rsc.org

of appropriate electrochemical and spectroscopic measurements. However, the real solvation free energy encompasses two physically very distinct contributions: a *bulk* contribution, arising from the interaction of the ion with the surrounding solvent molecules, and a *surface* contribution, arising from the electric potential jump upon entering the solvent. In contrast, the *intrinsic* single-ion solvation free energy only accounts for the bulk component. This free energy is in principle of greater theoretical value compared to the corresponding real quantity, because it does not intermingle surface and bulk effects. Unfortunately, the intrinsic solvation free energy cannot be determined on the sole basis of experimental data. Its evaluation requires the introduction of some *extra-thermodynamic assumption*, namely an apparently reasonable but not strictly provable postulate concerning the physics of the bulk solvation process or of the solvent surface properties. Note that the above issues only affect *single-ion* solvation parameters. The need to distinguish between real and intrinsic quantities disappears when corresponding sums over overall neutral sets of ions (salts) are considered instead. These sums are readily available by means of thermodynamic cycles relying on an appropriate combination of calorimetric, electrochemical or spectroscopic measurements.

The problem of the experimental or theoretical evaluation of intrinsic single-ion solvation free energies, as well as of corresponding derivative thermodynamic solvation parameters, has represented a challenge to the physical chemistry community for more than one century, and cannot be claimed to have found a satisfactory solution even nowadays.

From the *experimental* point of view, the problem amounts to determining *conventional* (relative, *e.g.* taking the proton as a reference ion) or *real* single-ion solvation free energies, both accessible experimentally, along with either of three solvent-dependent quantities: the *air-liquid interfacial potential* χ_{svt} *of the pure solvent*, the *intrinsic absolute potential* $\mathcal{V}_{H,svt}$ *of the reference hydrogen electrode in this solvent*, or the *intrinsic solvation free energy* $\mathcal{G}_{H,svt}$ *of the proton in this solvent*. Similarly, the pressure or temperature derivatives of these quantities, along with corresponding conventional or real single-ion solvation parameters, would give access to all other intrinsic derivative thermodynamic solvation properties. However, due to the constraints imposed by the electroneutrality of macroscopic matter at equilibrium and by the unavoidable presence of polarization effects at interfaces, these three quantities are experimentally *elusive*, *i.e.* neither of them can be evaluated without the help of some extra-thermodynamic assumption. Unfortunately, considering the case of water at atmospheric pressure and room temperature, the use of different experimental approaches along with distinct (apparently equally reasonable) assumptions leads to a very large uncertainty range on the order of $0.5 - 1.0$ V (potentials) or $50 - 100$ kJ·mol^{-1} (free energy) in the estimated values for the above elusive quantities.

From the *theoretical* point of view, three main approaches have been employed to evaluate intrinsic single-ion solvation free energies, as well as corresponding derivative thermodynamic solvation parameters: continuum-electrostatics calculations, classical atomistic simulations, and quantum-mechanical computations (calculations or simulations).

On the low-resolution end, *continuum-electrostatics calculations* (Born model and modified versions thereof) typically represent the ion as a rigid non-polarizable sphere and the solvent as a continuous (structureless) medium characterized by an infinite extent and a linear dielectric response. These approaches have a long history and the merit of providing a simple (often analytical) and qualitatively correct framework for intuitive reasoning, as well as a solid baseline for the discussion of more complex effects. However, their major shortcomings are to neglect the microscopic structure of the solvent molecules and the specific details of ion-solvent interactions, and to rely on the ill-defined concept of an ionic radius. As a result, these methods fail, even after application of various corrections, to provide a quantitative description of ionic solvation.

In the middle-resolution range, *classical atomistic simulations* represent the ion in solution as a system of classical point particles (atoms) interacting according to an empirically designed and calibrated potential energy function (force field). These methods have been applied both to the direct calculation of intrinsic single-ion solvation free energies and to the evaluation of the air-liquid interfacial potential of pure solvents. They should in principle be more accurate than continuum-electrostatics approaches, by accounting for the microscopic structure of the solvent molecules. However, in contrast to the latter methods, the averaging over solvent configurations must be carried out explicitly and the considered system is now of finite extent (*e.g.* liquid droplet or periodic computational box). The first difference does not represent a serious problem in the context of ionic solvation. Given the computational resources available at present, the sampling of a statistically representative number of configurations can easily be achieved, and simulations are able to provide a direct and accurate connection between molecular and thermodynamic properties *via* statistical mechanics. However, the second difference introduces serious methodological issues regarding the choice of boundary conditions, the approximate treatment of electrostatic interactions, and the evaluation of electrostatic potentials based on the sampled configurations. These issues have, up to recently, prevented the obtension of consistent results from atomistic simulations. Fortunately, the situation has changed in the past few years with the realization that the corresponding errors could be corrected *ex post*, so as to achieve methodological independence in the simulation results (*i.e.* consistent results across various choices regarding the three above issues). This achievement, which takes the atomistic approach to the same level of consistency as the Born model, opens new perspectives for the reliable application of atomistic simulation techniques to the investigation of problems related to ionic solvation. Still, the outcome of these calculations remains affected by three major sources of error: the mean-field treatment of electronic polarizability (in most calculations), the approximate representation of van der Waals interactions (functional form and combination rules) and the dependence of the results on the ion-solvent van der Waals interaction parameters. The latter parameters can be viewed as representing the atomistic analog of the ionic radius used in continuum-electrostatics calculations, and their calibration involves a similar ambiguity. Nevertheless, it is possible that the results of atomistic simulations concerning a large spectrum of ionic properties, including but extending beyond single-ion solvation free energies, provides in the near future a reliable approach for the accurate evaluation of the above elusive quantities. The corresponding extra-thermodynamic assumption will be termed in this book the *atomistic-consistency assumption.*

Finally, on the high-resolution end, *quantum-mechanical computations* represent the ion in solution as a many-particle system characterized by a wavefunction obeying the Schrödinger equation, given a Hamiltonian encompassing Coulombic interactions between all the elementary particles involved. These methods have also been applied both to the direct calculation of intrinsic single ion solvation free energies and to the evaluation of the air-liquid interfacial potential of the pure solvent. They have a bright future and the potential to definitely settle the problem by providing the framework for formulating what will be referred to as the *ultimate extra-thermodynamic assumption.* As first-principles (at least in principle) approaches, they promise to enable the calculation of the three above elusive quantities based on the most accurate physical model available nowadays, and without relying on the specification of ambiguous quantities such as ionic radii or ion-solvent van der Waals interaction parameters. This calculation nevertheless relies on the acceptance of a definition for partitioning the charge density (nuclei and electrons) into bulk and surface components. That such a definition is required can be viewed as the ultimate cause for the need of extra-thermodynamic assumptions in the determination of the three elusive quantities. Unfortunately, these approaches have the major shortcoming of being computationally expensive. This high computational cost results in practice nowadays in severe restrictions concerning the system size, configuration sampling, basis-set size and treatment of electron correlation. It may be reduced by interfacing a quantum-mechanical treatment of the ion and its first solvation shells with a classical atomistic or/and continuum-electrostatics approach for the surrounding bulk solvent. However, even in this case, the computation of single-ion solvation free energies represents a considerable challenge, and the necessary approximations may still for a while compromise the gain in accuracy expected from a more fundamental treatment of the problem.

The purpose of this book is to provide the reader with: (*i*) a clear picture of the concepts, definitions and difficulties underlying the problem of evaluating single-ion solvation free energies, as well as corresponding derivative thermodynamic parameters, *via* experiment or theory; (*ii*) an overview of the various experimental approaches employed to date; (*iii*) a discussion of the various theoretical approaches employed to date, with a particular focus on classical atomistic simulations; (*iv*) an extensive and consistent compilation of experimental data regarding ionic solvation, in the restricted context of alkali and halide hydration. As a compact account (experimental and theoretical issues, physical concepts, methods, results, data and literature) of a relatively complex field with both a very long history and a significant share of subsisting ambiguities and controversies, it is the hope of the authors that this book will provide the reader with a useful and solid basis for further investigations in the area. Towards this goal, a particular attention has been paid throughout this text to: (*i*) the clarity and consistency of the employed terminology and mathematical notation, including the introduction of new terms, acronyms or symbols, whenever necessary; (*ii*) the critical examination of the physics underlying key concepts, semantic, terminological and symbolic ambiguity representing here also an important source of complexity, inconsistencies and mistakes in the field; (*iii*) the consistent application of a clearly defined thermodynamic standard-state convention, alternative standards representing another important source of confusion and errors; (*iv*) the extensive citation of relevant literature sources, including over 2300 references spanning more than 100 years of research; (*v*) the establishment of appropriate connections between experimental and theoretical research, encompassing both historical and front-edge components.

Since the presented material was initially planned for a review article rather than for a book (see below), the text is intended for a linear rather than modular reading. However, readers who are primarily concerned with experimental or theoretical aspects, respectively, might decide to focus on Chapters 2 and 5 or on Chapters 3 and 6, respectively, in first priority, Chapter 4 providing common background information. Readers who are exclusively interested in an experimental data compilation concerning alkali, halide and proton hydration properties (including estimates for the three above elusive quantities) will find their happiness in Chapter 5, and in particular in Section 5.19, where recommended values are provided based on an extensive account and analysis of the available literature.

One thing should be stated once (and never mentioned again): we never really intended to write this book, we rather let ourselves be "sucked" into this task. As many (good and bad) things in life, all started "accidentally", when we were asked in February 2007 by the editors of the *Proceedings of the Royal Society of Chemistry* (RSC) to write a review about the theoretical calculation of ionic solvation properties. In March 2009, the planned date for the manuscript delivery, the "review" was already 400 (double-spaced) pages long, and there was no way it could be published as a single (even long) article. Thus, we asked RSC if they would consider publishing it as a book instead, which they kindly agreed to do. With this sort of birth, it is no surprise that the book turned into a (multi-)recidivist in terms of broken deadlines. Yet, almost two years (and quite some work) later, the book was ready to be published in the present state. In addition to the initially planned theory overview, it also included a significant experimental component. In fact, the simultaneous discussion of experiment and theory appears to be essential in this area, considering that the two aspects are so intimately intertwined.

Writing a scientific book of this kind is a massive effort, and the authors would like to express their deep gratitude to a number of persons for their help and support, before and during the writing process. The theoretical part of this text concerning classical atomistic simulations relies largely on the recent achievement of calculating methodology-independent single-ion solvation free energies. This success is a result from the remarkable scientific work of (in chronological order) Nathan Baker, Wolfgang Weber, Christine Peter, Michael Bergdorf and Mika Kastenholz (in addition to the authors' own contributions). A wealth of information concerning this theoretical part has also been gathered from insightful discussions with (in chronological order) Wilfred van Gunsteren, Alan Mark, Paul Smith, Andrew Torda, Brock Luty, Andy McCammon, Adrian Elcock, Gerhard Hummer, Herman Berendsen, Thomas Hofer, Nico van der Vegt, Benoit Roux, Joost van de Vondele and Chris Oostenbrink. The authors are also very grateful to Wilfred van Gunsteren, head of the group for Informatikgestützte Chemie (IGC) at ETHZ, for being both a leading and inspiring scientist in the area of molecular simulation, as well as the laureate of the Golden Tricycle Award 2009. Thanks to Wilfred, the IGC group is at the same time a highly stimulating and a socially extremely pleasant (family-like) working environment. This wonderful atmosphere was and is also the fact of all past and present IGC group members (including Jolande, Prisca, Daniela, Ana and Carmen), but they are too numerous to be thanked here individually. A special mention, still, for the subgroup of "Phil's angels", Cristina Pereira, Lovorka Perić-Hassler, Halvor Hansen, Bruno Horta and Monika Laner. Many thanks also go to Peter Chen, Markus Reiher, Beat Meier and Wilfred van Gunsteren for being the very first readers of (an early version of) this book, and for their insightful feedback on these pages. The editorial team at RSC also deserves our acknowledgments for their kindness and professionalism. First Graham Webb and Robert Eagling, for their patience with our never-to-appear review article. Then Janet Freshwater, Alice Toby-Brant and Katrina Harding for their support in designing the soon-to-appear book.

Maria Reif would like to dedicate this book to her parents. She is thankful to all the people who showed continuous and motivating interest in the progress of the book. She thanks Cristina, Lovorka, Halvor and Bruno for the unforgettable time in IGC. She would also like to express her gratitude to the MMS group of Chris Oostenbrink at BOKU University Vienna, where she was during the finalization of the manuscript, for the very nice and helpful atmosphere.

Philippe Hünenberger would like to dedicate this book to his four spiritual fathers (for one, also biological) in the various areas of Natural Sciences, Léopold Hünenberger (Gymnase de Chamblandes, Lausanne) for Chemistry, Marco Nicollerat (Gymnase du Belvédère, Lausanne) for Mathematics, Christian Gruber (EPFL) for Physics, and Wilfred van Gunsteren (ETHZ) for Computational Chemistry. This book is dedicated as well to his "little" family, Valérie, Achille and Gaspard. Without them, even the best science would have no sense, and life would be dangerously imbalanced towards the intrinsic rather than the real side. Gratitude also extends to his "larger" family (Léo, Gisèle, Fabien, Anne-Catherine, Babette, Germain, Suzanne, Marie-Line, Martine and Brigitte, and their own families), and to his friends of always (Marc-Aurèle, Sylvie, Zoltan & Agnès, Xavier & Judith, Roberto & Thereza, Wolfgang & Jelka, Nicolas & Sylvie, Eric, Vincent & Karin, Chantal & Catherine, François & Gabrielle, Franca & Jens, Daniel & Karin, Kathia & Dieter, Karin & Sarah, Jozi and Zrinka, and their own families) for their love, care and steady support in life.

And, when they are at it, the two authors of this book would also like to express their gratitude to each other for what has been a beautiful and inspiring four-years scientific collaboration, often as thrilling as a detective investigation on a particularly intricate case, and sometimes as frustrating as a crazy race through an endless maze.

Finally, research does not work nowadays without financial support, for which the Swiss National Science Foundation and ETH Internal Grants (ETH Forschungskomission) are also gratefully acknowledged. Philippe Hünenberger is also thankful for the award of the Ruzicka price 2008 concerning his work on the very topic of this book.

Last but not least, the authors would like to thank Ana Nedić-Eigenmann for designing the cover picture of this book and providing the funny ion drawings appearing at the end of each chapter. These drawings are meant to summarize the reader's (expected) state of mind after reading the preceding text, and as an invitation to turn the page. With a wink. Just like the one on the next page...

A single ion has decided to jump from the gas-phase into a swimming pool of pure liquid. It is eager to experience solvation! This ion is also about to discover a fundamental and painful fact of nature: it is impossible to reach the bulk of a liquid without crossing its surface.

CONTENTS

Table of symbols

Table I *Symbols, acronyms and notations (most important or/and frequently encountered only) used in this book.* The section column refers to the text section where the symbol or notation is first introduced or/and defined ("all" indicates that no specific definition was required).

Symbol	Meaning	Section
Physical constants and units		
ϵ_o	Permittivity of vacuum $\epsilon_o = 8.854187817 \cdot 10^{-12}$ C·V^{-1}·m^{-1}	2.1
h	Planck's constant $h = 6.62606896 \cdot 10^{-34}$ J·s ($\hbar = (2\pi)^{-1}h$)	3.3.1
e	Elementary charge $e = 1.602176487 \cdot 10^{-19}$ C	2.1
D	Debye unit for dipole moments 1 D = $3.33564 \cdot 10^{-30}$ C·m	2.1
bar	Usual thermodynamic unit for pressure 1 bar = 10^5 J·m^{-3}	all
N_A	Avogadro's constant $N_A = 6.02214179 \cdot 10^{23}$ mol^{-1}	2.1
k_B	Boltzmann's constant $k_B = 1.3806504 \cdot 10^{-23}$ J·K^{-1}	2.1
R	Ideal-gas constant $R = N_A k_B = 8.314472$ J·K^{-1}·mol^{-1}	2.1
F	Faraday's constant $F = N_A e = 96485.3399$ C·mol^{-1}	2.2
Standard functions		
$h(r)$	Heaviside step function	all
$\delta(r)$	Dirac delta function ($\delta(r) = dh(r)/dr$)	all
Physical quantities (generic)		

(continues on the next page)

Table I *(continued)*

Symbol	Meaning	Section
m	Mass	all
M	Molar mass	all
q	Charge	2.1
z	Signed integer charge of an ion ($q = ze$)	2.1
μ_d	Dipole moment	2.1
α_{el}	Static dipolar polarizability	3.3.1
n_{el}	Effective number of valence electrons	3.3.1
$\tilde{\alpha}_{el}$	Static dipolar polarizability volume ($\tilde{\alpha}_{el} = (4\pi\epsilon_o)^{-1}\alpha_{el}$)	3.3.1
$\mathbf{x}$	Cartesian coordinate vector (point)	all
r	Interatomic distance	2.1
E	Energy (per mole basis)	2.1
$\tilde{E}$	Energy (per particle basis)	2.1
ρ	Mass density	all
γ	Macroscopic surface tension coefficient	3.2
$\tilde{\gamma}$	Microscopic surface tension coefficient	3.2
ϵ	Static relative dielectric permittivity	2.1
$\tilde{\epsilon}$	Infinite-frequency relative dielectric permittivity	2.1,3.2
R_I	Effective ionic radius (in-crystal, in-solution or gas-phase)	5.4
Acronyms		
CE	Continuum-electrostatics calculation	3.2
BO	Born model (charging free energy)	3.2
ST	Surface tension model (cavitation free energy)	3.2
MSA	Mean-spherical approximation	6.1.2
AT	Classical atomistic simulation	3.3
MD	Molecular dynamics simulation	3.3.1
MC	Monte Carlo sampling	3.3.1
FBC	Fixed boundary conditions	3.3.1
PBC	Periodic boundary conditions	3.3.1
HBC	Hyperspherical boundary conditions	3.3.1
CB	Coulombic electrostatics (FBC/CB)	3.3.1
LS	Lattice-sum electrostatics (PBC/LS)	3.3.1
CT	Cutoff-based electrostatics (PBC/CT)	3.3.1
SC	Straight-cutoff truncation (PBC/CT/SC)	3.3.1
CA	Atom-based straight-cutoff truncation	6.2.1
CM	Molecule-based straight-cutoff truncation	6.2.1
RF	Cutoff truncation with reaction-field correction (PBC/CT/RF)	3.3.1
BA	Atom-based cutoff truncation with reaction-field correction	6.2.1
BM	Molecule-based cutoff truncation with reaction-field correction	6.2.1
MT	Cutoff truncation with *ad hoc* modification (PBC/CT/MT)	3.3.1
GM	Geometric-mean combination rule	3.3.2
LB	Lorentz-Berthelot combination rule	3.3.2
WH	Waldman-Hagler combination rule	3.3.2
TI	Thermodynamic integration	3.3.3
ODL	Orientational disorder limit	3.3.3

(continues on the next page)

Table I *(continued)*

Symbol	Meaning	Section
IQ	Isotropic quadrupole	3.3.3
ACA	Atomistic-consistency assumption	6.2.3
QM	Quantum-mechanical computation	3.4
CPMD	Car-Parrinello molecular dynamics	6.3.3
QM/MM	Hybrid quantum-mechanical molecular-mechanical computations	6.3.2
IG	Ideal-gas	4.2.3
IM	Ideal-mixture	4.2.3
IS	Ideal-solution	4.2.3
IB	Ideal-molal-solution	4.2.3
RXC	Reaction calorimetry	4.2.18
TLC	Third-law cycle	4.2.15
TCI	Temperature-scaled heat capacity integration	4.2.11
SBS	Spectroscopy-based statistical mechanics	4.2.12
FDR	Formation-dissolution-reformation cycle	4.2.15
BFH	Born-Fajans-Haber cycle	4.2.15
ISR	Ion-formation-solvation-reformation cycle	4.2.15
IHSP	Isolated phase with homogeneous surface polarization	4.3.3
NPIE	Non-polarizable interface at equilibrium	4.3.3
PIZC	Polarizable interface at zero charge	4.3.3
PICE	Polarizable interface after charge equilibration	4.3.3
AGZF	Air gap at zero field	4.3.3
POT	Potentiostat	4.3.3
EIM	Electrostatic interface model	4.3.2
fcc	Face-centered cubic crystal lattice	5.2
bcc	Body-centered cubic crystal lattice	5.2
Phases and interfaces		
Ω	System of phases	4.3.1
$\overline{\Omega}$	Exterior of system of phases	4.3.1
Σ	Interfacial region of system of phases	4.3.1
$\overline{\Sigma}$	Interfacial region of exterior of system of phases	4.3.1
Ω'	Extended system of phases $\Omega' = \Omega + \overline{\Omega}$	4.3.1
Σ'	Interfacial region of extended system of phases $\Sigma' = \Sigma + \overline{\Sigma}$	4.3.1
P_i	Phase i	4.3.1
$\tilde{\mathcal{E}}$	Electrostatic interface characteristics	4.3.1
$\mathcal{E}'$	Electrostatic interface characteristics concerning bound interfacial charge distribution	4.3.1
$\mathcal{C}$	Physical interface characteristics	4.3.1
$\mathcal{L}$	Lateral interface characteristics	4.3.2
$\mathcal{F}$	Frontal interface characteristics	4.3.2
N_{P_i}	Nature of phase i	4.3.1
F_{P_i}	Type of crystallographic face exposed at the interface by phase i (monocrystalline solids only, implying a planar interface)	4.3.1
C	Interface curvature	4.3.1
$\mathcal{P}$	Potentiostat setting	4.3.1
$\tilde{S}$	Surface charge	4.3.1

(continues on the next page)

Table I *(continued)*

Symbol	Meaning	Section
$\tilde{D}$	Surface polarization	4.3.1
$\tilde{\rho}_F$	Free interfacial charge distribution	4.3.1
$\tilde{\rho}_S$	Surface charge contribution to $\tilde{\rho}_F$	4.3.1
$\tilde{\rho}_D$	Surface polarization contribution to $\tilde{\rho}_F$	4.3.1
ρ'_B	Bound interfacial charge distribution	4.3.1
ρ'_S	Surface charge contribution to ρ'_B	4.3.1
ρ'_D	Surface polarization contribution to ρ'_B	4.3.1
ρ_T	Total interfacial charge distribution $\tilde{\rho}_F + \rho'_B$	4.3.1
ρ_S	Surface charge contribution to ρ_T	4.3.1
ρ_D	Surface polarization contribution to ρ_T	4.3.1
κ_{DH}	Inverse Debye screening length	4.3.1
Thermodynamics: phases (subscripts)		
g	Gas phase	4.1
ℓ	Liquid phase	4.1
s	Solid phase	4.1
d	Dissolved, *i.e.* in solution	4.1
aq	In aqueous solution	4.1
svt	Solvent (pure liquid or solution component)	4.1
wat	Water (pure liquid or solution component)	4.1
el	Element (one atom) in its standard compound and phase	4.1
Thermodynamics: species		
Σ	Arbitrary species (generic)	4.1
I	Ion-forming species (others: J, K, ...)	4.1
I^z	Ionic species (integer charge z)	4.1
$I^{z_I}_{n_I} J^{z_J}_{n_J} K^{z_K}_{n_K} \cdots$	Neutral combination of ions ($n_I z_I + n_J z_J + n_K z_K = 0$)	4.1
$I^{z_I}_{n_I} J^{z_J}_{n_J}$	Neutral binary combination of ions ($n_I z_I + n_J z_J = 0$)	4.1
IJ	Pair of (isovalent) ionic species (I^z and J^{-z}; crystal: associated; gas-phase, solution: dissociated)	4.1
$I_g + J_g$	Atom pair in the gas phase (I_g and J_g; not to be confused with the ion pair IJ_g)	4.1
$[I^z \cdot J^{-z}]$	Associated ion pair (I^z bound to J^{-z}; gas-phase or solution)	4.1
M	Alkali ion-forming species (ion: M^+)	4.1
X	Halide ion-forming species (ion: X^-)	4.1
e^-	Electron	4.1
H^+	Proton	4.1
Thermodynamics: variables		
P	Pressure	4.2.1
T	Absolute temperature	4.2.1

(continues on the next page)

Table I *(continued)*

Symbol	Meaning	Section
Y	Extensive thermodynamic variable (*e.g.* U, G, H, S, C_P, V, K_T or A_P)	all
y	(Partial) molar variable derived from Y (*e.g.* u, μ, h, s, c_P, v, k_T or a_P)	all
U	Internal energy	4.2.1
G	(Gibbs) Free energy (free enthalpy)	4.2.1
H	Enthalpy	4.2.1
S	Absolute (third-law) entropy	4.2.1
C_P	Isobaric heat capacity	4.2.1
V	Volume	4.2.1
K_T	Isothermal volume-compressibility	4.2.1
A_P	Isobaric volume-expansivity	4.2.1
κ_T	Isothermal compressibility	4.2.1
α_P	Isobaric expansivity (thermal expansion coefficient)	4.2.1
$\hat{O}_Y$	Thermodynamic-derivative operator	4.2.1
Thermodynamics: standard states		
$bbme_T$	Standard state convention used in this book (bar[s,ℓ,svt]/bar[g]/molal[sol]/"warm" electron[e])	4.2.4
P°	Reference pressure of 1 bar	4.2.4
b°	Reference molality of 1 mol·kg^{-1}	4.2.4
T^-	Reference temperature of 298.15 K	4.2.4
P^∇	Alternative (non-$bbme_T$) reference pressure	4.2.17
c^∇	Alternative (non-$bbme_T$) reference (molar) concentration	4.2.17
x^∇	Alternative (non-$bbme_T$) reference mole fraction	4.2.17
$\circ$	Standard quantity according to $bbme_T$ (superscript), standard variant	4.2.4
$\circ *$	Standard quantity according to $bbme_T$ (superscript), density-corrected variant	4.2.4
$-$	Quantity at temperature T^- (superscript)	4.2.4
$\ominus$	Standard quantity at temperature T^- according to $bbme_T$ (superscript), standard variant	4.2.4
$\ominus *$	Standard quantity at temperature T^- according to $bbme_T$ (superscript), density-corrected variant	4.2.4
$(\ominus)$	Quantity that is meant to be standard, but is approximate	4.2.13
∇	Quantity based on alternative (non-$bbme_T$) standard-state convention at temperature T	4.2.4
$\overline{\nabla}$	Quantity based on alternative (non-$bbme_T$) standard-state convention at temperature T^- (superscript)	4.2.4
$\bullet$	Conventional quantity (relative to the proton) based on standard states at temperature T according to $bbme_T$ convention (superscript)	4.5.1
$-\!\bullet$	Conventional quantity (relative to the proton) based on standard states at temperature T^- according to $bbme_T$ convention (superscript)	4.5.1
$\Delta_c^\circ y$	Conversion factor between alternative convention and $bbme_T$ convention ("∇" $\rightarrow$ "$\circ$")	4.2.17

(continues on the next page)

Table I *(continued)*

Symbol	Meaning	Section
$\Delta_c^{\ominus} y$	Conversion factor between alternative convention and $bbme_T$ convention at temperature T^- ("$\nabla\!\!\!-$" → "$\ominus$")	4.2.17
$\square$	Semi-standard point-to-point quantity at liquid pressure P°	4.2.16
$\boxminus$	Semi-standard point-to-point quantity at liquid pressure P° and temperature T^-	4.2.16
Thermodynamics: derivatives		
$\partial_T Y$	$(\partial Y/\partial T)$ at constant P, system extent and system composition	4.2.1
$\partial_P Y$	$(\partial Y/\partial P)$ at constant T, system extent and system composition	4.2.1
$\partial_T Y^{\ominus}$	$\partial_T Y$, evaluated under standard conditions	4.2.1
$\partial_P Y^{\ominus}$	$\partial_P Y$, evaluated under standard conditions	4.2.1
Thermodynamics: reaction and formation parameters		
$\Delta_r Y$	Thermodynamic reaction parameter of a specified reaction	4.2.7
$\Delta_m Y[\Sigma_s]$	Melting parameter of solid species Σ_s at $T_m(P)$ ($\Delta_m Y^\circ[\Sigma_s]$ at P° and T_m°)	4.2.7
$\Delta_b Y[\Sigma_\ell]$	Vaporization (boiling) parameter of liquid species Σ_ℓ at $T_b(P)$ ($\Delta_b Y^\circ[\Sigma_\ell]$ at P° and T_b°)	4.2.7
$\Delta_h Y[\Sigma_g]$	Heat-up parameter of gas-phase species Σ_g (gas-phase standard state; from the absolute zero temperature to T)	4.2.10
$E_A[\Sigma]$	Electron affinity of species Σ	4.2.12
$E_I[\Sigma]$	Ionization potential of species Σ	4.2.12
$\Delta_f Y[\Sigma]$	Formation parameter of species Σ (solvated ions: intrinsic)	4.2.10
$\Delta_f Y[I_d^z]$	Intrinsic formation parameter of solvated ion I_d^z	4.5.1
$\Delta_f \hat{Y}[I_d^z]$	Real formation parameter of solvated ion I_d^z	2.2,4.5.1
$\Delta_f Y^{\bullet}[I_d^z]$	Conventional (relative) formation parameter of solvated ion I_d^z	4.5.1
$\Delta_d Y[IJ_s]$	Dissolution parameter of salt IJ_s (assumed binary; default: in water)	4.2.14
$\Delta_a Y[I_{el}]$	Atomization parameter of ion-forming species I_{el} (*i.e.* one atom of the element in its standard state)	4.2.14
$\Delta_i Y[I_g]$	Ionization parameter of ion-forming species I_g (cation-forming: e_g^- abstraction; anion-forming: e_g^- capture)	4.2.14
$\Delta_l Y[IJ_g]$	Reticulation (lattice-formation) parameter of gas-phase ion pair IJ_g (*i.e.* gas-phase I_g^z plus gas-phase J_g^{-z})	4.2.14
$\Delta_s Y[IJ_g]$	Solvation parameter of gas-phase ion pair IJ_g (*i.e.* gas-phase I_g^z plus gas-phase J_g^{-z}; default: in water)	4.2.14
$\Delta_s Y^{\bullet}[I_g^z]$	Conventional (relative) single-ion solvation parameter of ion I_g^z ($\Delta_s Y^{\bullet}[H_g^+] = 0$; default: in water)	4.5.1
$\Delta_s \hat{Y}_{(svt)}[I_g^z]$	Real solvation parameter of ion I_g^z	2.2,4.2.14
$\Delta_s Y_{(svt)}[I_g^z]$	Intrinsic solvation parameter of ion I_g^z	2.2,4.2.14

(continues on the next page)

Table I *(continued)*

Symbol	Meaning	Section
Thermodynamics: standard solvation parameters and related quantities		
$\Delta_s \hat{Y}^{\ominus}_{svt}[I^z_g]$	Standard real solvation parameter, standard variant	4.2.5
$\Delta_s \hat{Y}^{\ominus *}_{svt}[I^z_g]$	Standard real solvation parameter, density-corrected variant	4.2.5
$\Delta_s Y^{\ominus}_{svt}[I^z_g]$	Standard intrinsic solvation parameter, standard variant	4.2.5
$\Delta_s Y^{\ominus *}_{svt}[I^z_g]$	Standard intrinsic solvation parameter, density-corrected variant	4.2.5
$\Delta_s Y^{\ominus}_{srf}[I^z_g]$	Surface term	2.2
$\Delta_s Y^{\boxminus}_{svt}[I^z_g]$	Intrinsic point-to-point solvation parameter	4.2.16
$\Delta_s Y^{\ominus}_{std}$	Standard-state correction term, standard variant	4.2.5
$\Delta_s Y^{\ominus *}_{std}$	Standard-state correction term, density-corrected variant	4.2.5
$\Delta_f Y^{\ominus}[I^z_d]$	Standard intrinsic formation parameter of dissolved ion, standard variant	4.2.5,4.2.14
$y^{\ominus}[I^z_d]$	Intrinsic partial molar variable of dissolved ion, standard variant	4.2.5,4.2.14
$\mathcal{Y}^{\ominus}_{H,svt}$	$= \Delta_s Y^{\ominus}[H^+_g]$ (*e.g.* $\mathcal{Y} = \mathcal{G}, \mathcal{H}, \mathcal{S}$ for $Y = G, H, S$)	4.5.3
$\hat{\mathcal{Y}}^{\ominus}_{H,svt}$	$= \Delta_s \hat{Y}^{\ominus}[H^+_g]$ (*e.g.* $\mathcal{Y} = \mathcal{G}, \mathcal{H}, \mathcal{S}$ for $Y = G, H, S$)	4.5.2
$y^{\ominus}_{H,svt}$	$= y^{\ominus}[H^+_d]$ (*e.g.* $y = s, c_P, v, k_T, a_P$, for $y = s, c_P, v, k_T, a_P$)	4.5.3
$\hat{y}^{\ominus}_{H,svt}$	$= \hat{y}^{\ominus}[H^+_d]$ (*e.g.* $y = s, c_P, v, k_T, a_P$, for $y = s, c_P, v, k_T, a_P$)	4.5.2
$\Delta_s Y_{chg}[I^z_g]$	Charging parameter of ion I^z (intrinsic)	3.2
$\Delta_s Y_{cav}[I^z_g]$	Cavitation parameter of ion I^z (intrinsic)	3.2
$\Delta_s Y^{\ominus}_{BO+ST}[I^z_g]$	Solvation free energy of ion I^z (BO+ST model; intrinsic; standard)	3.2
$\Delta_s G^{AT}_{chg}[I^z]$	Charging free energy of ion I^z evaluated using an atomistic (AT) model (recalculated with P-truncation: $\Delta_s G^{AT,P}_{chg}[I^z]$; M-truncation: $\Delta_s G^{AT,M}_{chg}[I^z]$)	3.3.3
$\Delta_s G^{AT}_{cav}[I^z]$	Cavitation free energy of ion I^z evaluated using a classical atomistic (AT) model	3.3.3
$\Delta_s G^{\ominus}_{AT}[I^z_g]$	Standard intrinsic single-ion solvation free energy of ion I^z evaluated using a classical atomistic (AT) model (recalculated with P-truncation: $\Delta_s G^{\ominus}_{AT,P}[I^z_g]$; M-truncation: $\Delta_s G^{\ominus}_{AT,M}[I^z_g]$)	3.3.3
$\Delta_s G^{\ominus}_{AT,cor}$	Correction term	3.3.3
Thermodynamics: molar and partial molar variables		
$y[\Sigma_d]$, y_i	Intrinsic partial molar thermodynamic variable of solvated species Σ_d or Σ_i (u, g, h, s, c_P, v, k_T, a_P for U, G, H, S, C_P, V, K_T, A_P)	4.2.2

(continues on the next page)

Table I *(continued)*

Symbol	Meaning	Section
$y[\Sigma]$, y_i	Molar thermodynamic variable of species Σ or Σ_i (pure species; *e.g.* u, g, h, s, c_P, v, k_T, a_P for U, G, H, S, C_P, V, K_T, A_P)	4.2.2
y, y_{svt}	Binary mixture, infinitely dilute regime (y refers to solute, y_{svt} to solvent)	4.2.2
$\hat{y}[\Sigma_d]$, $\hat{y}_i$	Real partial molar thermodynamic variable of solvated species Σ_d or Σ_i ($\hat{u}$, $\hat{g}$, $\hat{h}$, $\hat{s}$, $\hat{c}_P$, $\hat{v}$, $\hat{k}_T$, $\hat{a}_P$ for U, G, H, S, C_P, V, K_T, A_P)	2.2,4.2.2
$y^{\bullet}[\Sigma_d]$, $y_i^{\bullet}$	Conventional (relative) partial molar thermodynamic variable of solvated species Σ_d or Σ_i ($u^{\bullet}$, $g^{\bullet}$, $h^{\bullet}$, $s^{\bullet}$, $c_P^{\bullet}$, $v^{\bullet}$, $k_T^{\bullet}$, $a_P^{\bullet}$ for U, G, H, S, C_P, V, K_T, A_P)	4.5.1
$\hat{y}$, y_{svt}	Binary mixture, infinitely dilute regime ($\hat{y}$ refers to solute, y_{svt} to solvent)	4.2.2
$K_E^{\ominus}$	Equilibrium constant of a reaction	4.2.6,4.4.3
K_R	Reaction quotient of a reaction	4.2.6,4.4.3
P_t, T_t	Triple-point pressure and temperature	all
P_c, T_c	Critical pressure and temperature	5.17
P_{vap}	Vapor pressure (P_{vap}^- at T^-)	all
T_m	Melting temperature ($T_m^\circ[S_s]$ at P°)	4.2.7
T_b	Boiling temperature ($T_b^\circ[S_\ell]$ at P°)	4.2.7
$\Delta\mathcal{V}_{ext}$	Potential difference in circuit coupled to the system	4.2.7
Thermodynamics: important elusive quantities		
$\mathcal{Y}_{H,svt}$	Intrinsic solvation parameter of the gas-phase proton for the solvent	2.3,7
$y_{H,svt}$	Intrinsic partial molar variable of the proton dissolved in the solvent	2.3,7
χ_{svt}	Air-liquid interfacial potential of the solvent	2.3,7
$\mathcal{V}_{H,svt}$	Intrinsic absolute potential of the reference hydrogen electrode in the solvent	2.3,7
Potentials: phase characterization		
w	Spectroscopic work function	2.1
ω	Standard work function	4.3.5
a_i	Thermodynamic activity of species Σ_i ($bbme_T$ convention)	4.2.6
γ_i	Thermodynamic activity coefficient of species Σ_i ($bbme_T$ convention)	4.2.6
μ	(Intrinsic) chemical potential (μ_i: species Σ_i; $\mu_{\Sigma_i:P}$: species Σ_i present in phase P)	4.3.4
$\tilde{\mu}$	Electrochemical potential ($\tilde{\mu}_i$: species Σ_i; $\tilde{\mu}_{\Sigma_i:P}$: species Σ_i present in phase P)	4.3.4
$\hat{\mu}$	Real chemical potential ($\hat{\mu}_i$: species Σ_i; $\hat{\mu}_{\Sigma_i:P}$: species Σ_i present in phase P)	4.3.4

(continues on the next page)

Table I *(continued)*

Symbol	Meaning	Section
ψ	Volta (or outer) potential	4.3.2
ψ	Lateral Volta potential	4.3.2
$\tilde{\psi}$	Frontal Volta potential	4.3.2
ϕ	Galvani (or inner potential), external	4.3.2
$\breve{\phi}$	Galvani (or inner potential), internal	4.3.8
χ	Surface potential (out-to-in direction), external	4.3.2
$\breve{\chi}$	Surface potential (out-to-in direction), internal	4.3.8
χ	Lateral surface potential	4.3.2
$\tilde{\chi}$	Frontal surface potential	4.3.2
ξ	Exclusion potential of a phase	4.3.8
ξ_{AT}	Classical exclusion potential of a phase	4.3.8
ξ_{QM}	Quantum-mechanical exclusion potential of a phase	4.3.8
$\mathcal{V}$	Electrode potential	4.4
$\Delta\mathcal{V}$	Potential difference (probed or imposed between two metal electrodes)	4.4
$\Delta_H\mathcal{V}^{\ominus}[I_{el}, I_d^z]$	Redox potential of the I_{el}-I_d^z element-ion couple with respect to the reference hydrogen electrode	4.4.3
$\Delta_H\overline{\mathcal{V}}_{M,S}$	Potential of zero charge of the M-S interface with respect to the reference hydrogen electrode	4.4.5
$\Delta\overline{\psi}_{M,S}$	Lateral Volta potential difference at the M-S interface at the potential of zero charge of the M-S interface with respect to the reference hydrogen electrode	4.4.5
Potentials: solvent characterization		
$\chi_{svt}^{\ominus}$	Standard air-solvent interfacial potential (air $\rightarrow$ liquid; external)	2.2
$\tilde{\chi}_{svt}^{\ominus}(R)$	Air-solvent interfacial potential (curved interface; external; standard)	3.3.5
$\tilde{\chi}_{svt,+}^{\ominus}$, $\tilde{\chi}'_{S+}$	Curvature dependence of $\chi_{svt}^{\ominus}$ for a convex interface ($R > 0$)	6.1.1,6.2.1
$\tilde{\chi}_{svt,-}^{\ominus}$, $\tilde{\chi}'_{S-}$	Curvature dependence of $\chi_{svt}^{\ominus}$ for a concave interface ($R < 0$)	6.1.1,6.2.1
$\breve{\chi}_{svt}^{\ominus}$	Air-solvent interfacial potential (air $\rightarrow$ liquid; internal; standard)	4.3.8
$\mathcal{V}_{H,svt}^{\ominus}$	Standard intrinsic absolute potential of the hydrogen electrode in the solvent	4.3.6
$\hat{\mathcal{V}}_{H,svt}^{\ominus}$	Standard real absolute potential of the hydrogen electrode in the solvent	4.3.6
$\tilde{\mathcal{V}}_{H,svt}^{\ominus}$	Standard vacuum absolute potential of the hydrogen electrode in the solvent	4.3.6
Electrochemical cell schemes		
$\vdots$	Generic interface between two phases	4.4
$\|\|$	Polarizable interface between two phases	4.4
$\|$	Non-polarizable interface between two phases	4.4
$\square$	Air gap (insulant space) between two phases	4.4

(continues on the next page)

Table I *(continued)*

Symbol	Meaning	Section
$\sim$	Phase is in common between two half-cells	4.4
$P_1 \vdots [P_2]P_3$	Phases P_2 and P_3 contact phase P_1	4.4
$P_1 \vdots P_2(P_3) \vdots P_4$	Phase P_3 solely contacts phase P_2	4.4
Solvation free-energy contributions		
$\Delta_s G^{\bullet}[I_g^z]$	Conventional (relative) single-ion solvation free energy of ion I_g^z ($\Delta_s G^{\bullet}[\mathrm{H}_g^+] = 0$; default: in water)	4.5.1
$\Delta_s \hat{G}[I_g^z]$	Real single-ion solvation free energy of ion I_g^z	2.2
$\Delta_s G[I_g^z]$	Intrinsic single-ion solvation free energy of ion I_g^z	2.2
$\Delta_s G_{srf}[I_g^z]$	Surface contribution to $\Delta_s \hat{G}[I_g^z]$	2.2
$\Delta_s G^{\ominus}_{std}$	Standard-state correction to $\Delta_s G[I_g^z]$ and $\Delta_s \hat{G}[I_g^z]$	2.2
$\Delta_s G_{chg}[I_g^z]$	Intrinsic charging free energy of ion I^z	3.2
$\Delta_s G_{cav}[I_g^z]$	Cavitation free energy of ion I^z	3.2
Atomistic simulations		
$\mathbf{r}$	$3N$-dimensional Cartesian coordinate vector (N particles)	3.3.1
$V(\mathbf{r})$	Classical potential-energy function (force field; per particle basis)	3.3.1
$V_{exc}(\mathbf{r})$	Excluded-sites contribution to the classical potential-energy function (force field; per particle basis)	3.3.1
$V_{srf}(\mathbf{r})$	Surface term contribution to the classical potential-energy function	3.3.1
V_{slf}	Self-term contribution to the classical potential-energy function	3.3.1
$\mathbf{r}_{ij}$	Vector connecting site i to site j ($\mathbf{r}_j - \mathbf{r}_i$; norm r_{ij})	3.3.1
$\overline{\mathbf{r}}_{ij}$	Corresponding minimum-image vector under PBC (norm $\overline{r}_{ij}$)	3.3.1
$\mathbf{R}_{ij}$	Vector connecting the centers of the molecules containing sites i and j ($\mathbf{R}_j - \mathbf{R}_i$; norm R_{ij})	3.3.1
$\overline{\mathbf{R}}_{ij}$	Corresponding minimum-image vector under PBC (norm $\overline{R}_{ij}$)	3.3.1
q_i,q_j	Charges of sites i and j	3.3.1
$V_{LJ,pair}$	Lennard-Jones interaction function	3.3.1
C_{12}, C_6	Lennard-Jones C_{12} and C_6 parameter (for a given pair of sites)	3.3.1
R_{zer}	Lennard-Jones distance at the interaction zero (for a given pair of sites)	3.3.1
R_{min}	Lennard-Jones distance at the interaction minimum (for a given pair of sites)	3.3.1
E_{min}	Lennard-Jones energy at the interaction minimum (for a given pair of sites)	3.3.1
ii, ii', is, ss	Value used for ion-ion (same ion type), ion-ion (different ion type), ion-solvent or solvent-solvent interactions, respectively (superscript for C_{12}, C_6, R_{min} or E_{min})	3.3.2
C_6^{is}, C_{12}^{is}	Ion-solvent Lennard-Jones interaction parameters C_6 and C_{12}	3.2
C_6^{ii}, C_{12}^{ii}	Ion-ion Lennard-Jones interaction parameters C_6 and C_{12}	3.2

(continues on the next page)

Table I *(continued)*

Symbol	Meaning	Section
λ	Coupling parameter	3.3.2
$\psi_{EL,oi}$	Ion-solvent electrostatic interaction function used in the simulation	3.3.3
$\psi'_{EL,oi}$	Idem, used to recalculate the potential based on given configurations	3.3.3
ζ_q^{AT}	Average solvent-generated potential at the center of the ionic cavity with charge q (using $\psi_{EL,oi}$)	3.3.3
$\zeta_q^{\prime AT}$	ζ_q^{AT} recalculated based on simulated configurations (integration over a sphere of radius R centered at the cavity; Coulombic pairwise interaction function; P-truncation: $\zeta_q^{\prime AT,P}$; M-truncation: $\zeta_q^{\prime AT,M}$); using $\psi'_{EL,oi}$	3.3.3
ζ_{AT}	Short notation for $\zeta_0^{\prime AT}$ in the limit of large R (uncharged cavity; P-truncation: $\zeta_{AT,P}$; M-truncation: $\zeta_{AT,M}$)	3.3.3
χ_{AT}	Air-solvent interfacial potential for planar interface in equilibrium evaluated using a classical atomistic (AT) model (P-integration: $\chi_{AT,P}$; M-integration: $\chi_{AT,M}$)	3.3.5
χ'_{AT}	Interfacial potential at a cavity	3.3.4
η_{AT}	Average external potential of the pure solvent (P-truncation: $\eta_{AT,P}$; M-truncation: $\eta_{AT,M}$)	3.3.4
$\check{\eta}_{AT}$	Average internal potential of the pure solvent (P-truncation: $\check{\eta}_{AT,P}$; M-truncation: $\check{\eta}_{AT,M}$)	3.3.4
$\mathcal{Q}$	Quadrupole-moment trace of a solvent model	4.3.8
$\overline{\mathcal{Q}}$	Effective quadrupole-moment trace of a solvent model	6.2.1
$\mathcal{Q}_{AT}$	Quadrupole-moment trace of an AT solvent model	4.3.8
$\mathcal{Q}_{QM}$	Quadrupole-moment trace representative for a QM description of the solvent	4.3.8
ξ_{AT}	Classical exclusion potential of an atomistic solvent model	4.3.8
$\psi_{LJ,12,ij}$	Pairwise Lennard-Jones repulsion energy between sites i and j (scaled by $C_{12,ij}$)	3.3.1
$\psi_{LJ,6,ij}$	Pairwise Lennard-Jones dispersion energy between sites i and j (scaled by $C_{6,ij}$)	3.3.1
$\psi_{EL,ij}$	Pairwise electrostatic interaction energy between sites i and j (scaled by $(4\pi\epsilon_o)^{-1}q_iq_j$)	3.3.1
R_C	Cutoff distance (CT electrostatics)	3.3.1
H_{ij}	Cutoff truncation function for sites i and j (CT electrostatics; $H_{P,ij}$: P-truncation; $H_{M,ij}$: M-truncation)	3.3.1
τ_{CT}	Effective pairwise electrostatic interaction function (CT electrostatics; τ_{SC}, τ_{RF} or τ_{MT})	3.3.1
ϵ_{RF}	Reaction-field permittivity (RF electrostatics)	3.3.1

Chapter 1
Introduction

Ions are ubiquitous in nature and play a fundamental role in numerous ecological (*e.g.* geology[1–4], atmospheric chemistry[5–8], interstellar chemistry[7,9–11]), technological (*e.g.* electrolysis[12–14], batteries[15–17], materials engineering[18–20]), chemical (*e.g.* reaction intermediates[21], ionic liquids[22–27], catalysis[28–30]), biochemical (*e.g.* stabilization and modulation of biomolecular structure[31–41], mediation of biomolecular interactions[42–47], intracellular signal transduction[48,49], enzyme[50–52] and nucleic acid[51–55] catalysis), physiological (*e.g.* membrane potentials[56], nerve action potentials[56], muscle contraction[56], biomineralization[57]), pathological (*e.g.* neurodegenerative[58,59] and allergic[60] diseases) and medicinal (*e.g.* radiation therapy[61,62], therapeutic agents[63,64], diagnostic agents[64]) processes.

An *ion* is an atomic or molecular species characterized by a deficiency or an excess of electrons compared to its proton content, resulting in a net positive or negative charge. Because of the very large magnitude and long-ranged nature of Coulombic interactions in free space, isolated ions are not found as free entities in the gas phase at equilibrium and under usual conditions of pressure and temperature. They tend to aggregate so as to form electroneutral ionic salts[65–68] or, in special cases, ionic liquids[22,23,27]. However, the situation is different in polar solvents, prominently water, which are capable of dissolving ions, *i.e.* of stabilizing them as free dissociated entities, at least up to a certain concentration corresponding to saturation. This stabilization finds its origin in the Coulombic interactions between the isolated ions and the solvent dipoles, *i.e.* in specific solvation forces. These forces counteract the direct Coulombic interactions between ions of opposite charges and, along with entropic effects and up to the saturation concentration, prevent the aggregation of the ions. In the simplest context of an individual ionic species at infinite dilution in a given solvent, the central quantity to characterize the magnitude of the solvation forces at constant pressure and temperature is the

RSC Theoretical and Computational Chemistry Series No. 3
Single-Ion Solvation
Experimental and Theoretical Approaches to Elusive Thermodynamic Quantities
By Philippe Hünenberger and Maria Reif

Published by the Royal Society of Chemistry, www.rsc.org

corresponding *single-ion solvation free energy*.

The determination of single-ion solvation free energies *via* experimental measurements or theoretical calculations is a problem that has preoccupied the physical chemistry community (in a direct or indirect form, including the determination of *e.g.* Volta potential differences and absolute electrode potentials) for more than one century[69–84]. It is a very basic one, that could well be viewed as the "hydrogen-atom problem" of ion thermodynamics and electrochemistry. Yet, this apparently simple problem hides a significant and unexpected amount of complexity, and cannot be claimed to have found a satisfactory solution even nowadays.

The goal of this book is to provide the reader with: (*i*) a clear picture of the *concepts*, *definitions* and *difficulties* underlying the problem of determining single-ion solvation free energies *via* experiment or theory; (*ii*) an overview of the various *experimental approaches* employed to date; (*iii*) a discussion of the various *theoretical approaches* employed to date, with a particular focus on classical atomistic simulations. *Derivative thermodynamic properties* such as solvation enthalpies, entropies, heat capacities, volumes, compressibilities and expansivities are also discussed. However, the scope of the book does not extend beyond *bulk solvation thermodynamics* in the *infinitely dilute regime*. In particular, structural (*e.g.* solvent distribution functions, solvation structure, coordination numbers and geometries, Jahn-Teller distortions) and dynamic (*e.g.* ionic diffusion, rate of solvent exchange between solvation shells and with the bulk, ion-water and first-shell water vibrational properties) aspects, as well as finite-concentration effects (*e.g.* direct or solvent-mediated ion-ion interactions, ion-ion association, activity coefficients, osmotic coefficients, colligative properties and, in general, any concentration dependence of the bulk and surface properties of electrolyte solutions), interface effects (*e.g.* air-solution or metal-solution interfaces), and interactions in gas-phase ion-solvent clusters, ionic crystals, ionic liquids or ionic melts are not considered. Reviews concerning the latter topics (*i.e.* properties related to structure[85–109], dynamics[88, 89, 92–94, 96, 98, 100–103, 110–112], finite-concentration effects[98, 112–134], interface effects[87,99,106,135–158], gas-phase ion-solvent clusters[159–178], ionic crystals[65, 67, 68, 179–197], ionic liquids[22–27, 198–206] and ionic melts[207–220]) can be found elsewhere.

This book also provides an extensive compilation of experimental data regarding ionic solvation. Obviously, however, such a compilation has to be restricted in terms of its scope. The ions considered are those of the *alkali* (Li^+, Na^+, K^+, Rb^+ and Cs^+) and *halide* (F^-, Cl^-, Br^- and I^-) series. The ions Fr^+ and At^- predominantly derive from the radioactive elements ^{223}Fr and ^{210}At, with short lifetimes of 22 minutes and 8.1 hours, respectively[221], and are left out due to limited relevance and lack of experimental data (see, however, Refs.[222–224]). Similarly, alkali anions and halide cations are relatively exotic species, and are not considered in this text (see, however, Refs.[224–231]). The compilation is also restricted to *water* as a solvent, because hydration properties of ions are not only the most practically relevant but also the most thoroughly investigated. Since the ions of the alkali and halide series are monoatomic and monovalent, the physics involved in their hydration is expected to be comparatively simpler than that of polyatomic or/and polyvalent ions. Unlike polyatomic ions, their (nuclear and electronic) charge densities and average hydration structures are spherically symmetric. In comparison to polyvalent ions, they also exert a more moderate electric field on the surrounding water molecules, resulting in a more labile and shorter-ranged solvation structure, a predominantly electrostatic (rather than partially covalent, as *e.g.* for heavy metals) water binding mode, a more limited electronic polarization of the surrounding water molecules, and negligible water deprotonation effects in their neighborhood. However, even in the limited context of these two "simplest" ion series, important variations in the ionic size, effective charge density and polarizability result in relatively large qualitative and quantitative differences in terms of hydration, at the thermodynamic (this book), as well as structural[91, 94, 97, 104, 105, 109, 130, 232–250] and dynamic[94, 130, 233–236, 238, 239, 241, 243, 244, 246, 249, 251–256] levels. For this reason, these ions permit to illustrate the key issues related to single-ion solvation thermodynamics while remaining in the most simple possible context.

The *properties considered* for the alkali and halide ions (or alkali-halide ion pairs) with water as a solvent, which will be defined precisely in Chapter 4, include: (*i*) pair formation and (partial) molar parameters in different states (crystal, solution); (*ii*) pair reaction parameters for various processes (reticulation, dissolution, solvation); (*iii*) single-ion formation and (partial) molar parameters in different states (element standard state, gas-phase atom, gas-phase ion, dissolved ion on real or conventional scales); (*iv*) single-ion reaction parameters for various processes (atomization, ionization, solvation on real or conventional scales); (*v*) available estimates for the pure solvent properties involved in the determination of intrinsic single-ion solvation parameters, as well as formation and partial molar parameters in solution (intrinsic solvation free energy of the proton, air-liquid interfacial potential of the solvent, intrinsic absolute potential of the reference hydrogen electrode in the solvent, as well as temperature or/and pressure derivatives thereof); (*vi*) data of theoretical relevance (*e.g.* estimates for the ionic radii or ion-solvent van der Waals interaction parameters). The reported formation and reaction parameters are those deriving from the (Gibbs) free energy G, the enthalpy H, and the entropy S. The quantities for which (partial) molar parameters are reported are those deriving from the (absolute) entropy S, the (isobaric) heat capacity C_P, the volume V, the (isothermal) volume-compressibility K_T, and the (isobaric) volume-expansivity A_P.

Further references are provided whenever possible concerning ions, solvents and properties not explicitly included in the present compilation. General reviews about ionic solvation can also be found elsewhere[85, 93, 98, 109, 257–285].

This book is organized as follows. Chapters 2 and 3 discuss the fundamental problems encountered in the determination of single-ion solvation free energies, as well as corresponding derivative thermodynamic parameters, *via* either experimental measurements (Chapter 2) or theoretical calculations (Chapter 3). Chapter 4 clarifies the concepts and definitions required to formulate the problem in a well-defined framework within thermodynamics and electrochemistry. Chapters 5 and 6 describe the various approaches that have been employed to date for the characterization of the thermodynamics of processes involving ions, *via* either experimental measurements (Chapter 5) or theoretical calculations (Chapter 6). They also provide data compilations and illustrative results pertinent to these measurements and calculations, given the restricted scope of the book in terms of ions, solvent and properties considered. Based on the careful investigation of literature data, a set of recommended values for properties related to alkali, halide and proton hydration is ultimately provided in Section 5.19. Finally, Chapter 7 provides concluding remarks, as well as a perspective on still unsettled issues and progresses expected to result from on-going research in the area.

A detailed *list of the symbols, acronyms and notations* employed in this book is provided in Table I. These are also explained in the text when first encountered, with the following exceptions, defined precisely in Chapter 4 but introduced earlier in the text: the *reference pressure* $P^\circ = 1$ bar, the *reference molality* $b^\circ = 1$ mol·kg^{-1}, the *reference temperature* $T^- = 298.15$ K, the superscript "∘" indicating a *standard* (but temperature-dependent) quantity appropriate for P° and b°, the superscript "-" indicating a (pressure- and, possibly, molality-dependent) quantity evaluated at T^-, the combined superscript "⊖" indicating a standard quantity appropriate for P°, b° and T^-, and the notations $\partial_P Y^\ominus$ and $\partial_T Y^\ominus$ indicating the first *isothermal pressure derivative* and *isobaric temperature derivative*, respectively, of a thermodynamic variable Y evaluated at constant system composition and under standard conditions (similar notations ∂_P^2 and ∂_T^2 being used for self-second derivatives, and $\partial_{P,T}^2$ for cross-second derivatives). The latter derivatives of standard (thus formally pressure- and temperature-independent) quantities are short notations actually implying derivation with respect to P° and T^- considered as variables, followed by evaluation at the reference values of these variables. For convenience, a summary of the properties of water relevant in the context of ionic solvation is also provided in Table 1.1.

Table 1.1 *Standard properties of water relevant in the context of ionic solvation.* The reported data includes the molar mass M, the mass density $\rho^{\ominus}$, the macroscopic surface tension coefficient $\gamma^{\ominus}$, the static relative dielectric permittivity $\epsilon^{\ominus}$, the infinite-frequency relative dielectric permittivity $\tilde{\epsilon}^{\ominus}$, the gas-phase molecular dipole moment $\mu_{d,g}^{\ominus}$, the liquid-phase (average) molecular dipole moment $\mu_{d,\ell}^{\ominus}$, the solid-phase (average; hexagonal ice I_h) molecular dipole moment $\mu_{d,s}^{\ominus}$, the triple point pressure P_t and temperature T_t, the critical point pressure P_c and temperature T_c, the melting temperature T_m°, the boiling temperature T_b°, the melting enthalpy $\Delta_m H^{\circ}$ at T_m°, the vaporization enthalpy $\Delta_b H^{\circ}$ at T_b°, the vapor pressure P_{vap}^{-}, the formation free energy $\Delta_f G^{\ominus}$, the formation enthalpy $\Delta_f H^{\ominus}$, the formation entropy $\Delta_f S^{\ominus}$, the absolute molar entropy $s^{\ominus}$, the molar isobaric heat capacity $c_P^{\ominus}$, the molar volume $v^{\ominus}$, the isothermal compressibility $\kappa_T^{\ominus}$, the isobaric expansivity $\alpha_P^{\ominus}$ (thermal expansion coefficient), the molar volume-compressibility $k_T^{\ominus}$ and the molar volume-expansivity $a_P^{\ominus}$. The first pressure (∂_P) and temperature (∂_T) derivatives, as well as the corresponding second derivatives (∂_P^2, ∂_T^2, $\partial_{P,T}^2$), are also reported for some properties. The standard states are according to the $bbme_T$ convention (Section 4.2.4; reference pressure $P^{\circ} = 1$ bar and reference temperature $T^{-} = 298.15$ K).

Property			Ref.	Notes
M	[kg·mol^{-1}]	0.018015	221	
$\rho^{\ominus}$	[kg·m^{-3}]	997.0480	221	a
$\partial_P \rho^{\ominus}$	[10^{-3} kg·m^{-3}·bar^{-1}]	45.1		b
$\partial_P^2 \rho^{\ominus}$	[10^{-6} kg·m^{-3}·bar^{-2}]	-9.41	286	c
$\partial_T \rho^{\ominus}$	[kg·m^{-3}·K^{-1}]	-0.259		d
$\partial_T^2 \rho^{\ominus}$	[10^{-4} kg·m^{-3}·K^{-2}]	-95.2	287	e
$\partial_{P,T}^2 \rho^{\ominus}$	[10^{-4} kg·m^{-3}·K^{-1}·bar^{-1}]	-1.23	287	f
$\gamma^{\ominus}$	[10^{-3} J·m^{-2}]	71.99	221	a
$\partial_P \gamma^{\ominus}$	[10^{-5} J·m^{-2}·bar^{-1}]	7.02	288	
$\partial_T \gamma^{\circ}$	[10^{-3}J·m^{-2}·K^{-1}]	-0.154	289	g
$\epsilon^{\ominus}$		78.36	221	
$\partial_P \epsilon^{\ominus}$	[10^{-4} bar^{-1}]	37.4	290	a
$\partial_P^2 \epsilon^{\ominus}$	[10^{-6} bar^{-2}]	-0.566	290	a
$\partial_T \epsilon^{\ominus}$	[K^{-1}]	-0.359	290	a
$\partial_T^2 \epsilon^{\ominus}$	[10^{-2} K^{-2}]	0.159	290	a,h
$\partial_{P,T}^2 \epsilon^{\ominus}$	[10^{-5} K^{-1}·bar^{-1}]	-1.17	290	i
$\tilde{\epsilon}^{\ominus}$		5.2	221	
$\mu_{d,g}^{\ominus}$	[D]	1.8546 ± 0.0040	221	
$\mu_{d,\ell}^{\ominus}$	[D]	2.9 ± 0.6	291	
$\mu_{d,s}^{\ominus}$	[D]	2.9	292	

(continues on the next page)

Table 1.1 *(continued)*

Property			Ref.	Notes
P_t	$[10^{-5}$ bar$]$	611.7	221	
T_t	[K]	273.16	221	
P_c	[bar]	220.6	221	
T_c	[K]	647.14	221	
T_m°	[K]	273.15	221	a
T_b°	[K]	373.15	221	a
$\Delta_m H^\circ$	$[\text{kJ}\cdot\text{mol}^{-1}]$	6.01	221	
$\Delta_b H^\circ$	$[\text{kJ}\cdot\text{mol}^{-1}]$	40.65	221	a,j
P_{vap}^-	[bar]	0.031691	293	
$\Delta_f G^\ominus$	$[\text{kJ}\cdot\text{mol}^{-1}]$	-237.129	294	
$\Delta_f H^\ominus$	$[\text{kJ}\cdot\text{mol}^{-1}]$	-285.830	294	
$\Delta_f S^\ominus$	$[\text{J}\cdot\text{mol}^{-1}\cdot\text{K}^{-1}]$	-163.344	-	k
$s^\ominus$	$[\text{J}\cdot\text{mol}^{-1}\cdot\text{K}^{-1}]$	69.95	221	
$c_P^\ominus$	$[\text{J}\cdot\text{mol}^{-1}\cdot\text{K}^{-1}]$	75.291	294	
$v^\ominus$	$[10^{-6}\ \text{m}^3\cdot\text{mol}^{-1}]$	18.07	-	l
$\kappa_T^\ominus$	$[10^{-6}\ \text{bar}^{-1}]$	45.2	295	a
$\alpha_P^\ominus$	$[10^{-4}\ \text{K}^{-1}]$	2.6	295	a
$k_T^\ominus$	$[10^{-10}\ \text{m}^3\cdot\text{mol}^{-1}\cdot\text{bar}^{-1}]$	8.17	-	m
$a_P^\ominus$	$[10^{-8}\ \text{m}^3\cdot\text{mol}^{-1}\cdot\text{K}^{-1}]$	0.47	-	n

(a) the value reported for the pressure $P^\circ = 1$ bar is approximated here by a value reported for a pressure $P = 1$ atm; (b) calculated as $\partial_P \rho^\ominus = \rho^\ominus \kappa_T^\ominus$; (c) calculated based on an empirical analytical equation for $\rho^-(P)$; (d) calculated as $\partial_T \rho^\ominus = -\rho^\ominus \alpha_P^\ominus$; (e) calculated based on an empirical analytical equation for $\rho^\circ(T)$; (f) calculated from the temperature derivative of the isothermal compressibility of water, based on an empirical analytical equation for $\kappa_T^\circ(T)$, as $\partial_{P,T}\rho^\ominus = \rho^\ominus \partial_T \kappa_T^\ominus + (\rho^\ominus)^{-1}\partial_P \rho^\ominus \partial_T \rho^\ominus$; (g) calculated based on an empirical analytical equation for $\gamma^\circ(T)$; (h) an alternative value based on an empirical analytical equation[296] for $\epsilon^\circ(T)$ is $0.164\cdot10^{-2}$ K^{-2}; (i) a (probably less accurate) value estimated by a finite temperature difference of the pressure derivatives, based on two empirical analytical equations for $\epsilon_S(P)$ of Ref.[286] at $T = 273$ and 323 K is $-2.81\cdot10^{-5}$ $\text{K}\cdot\text{bar}^{-1}$; (j) the reported value corresponds to $\Delta_b H^\circ$ at T_b° (both approximated here by values at 1 atm rather than 1 bar), whereas the value $\Delta_b H^\ominus$ (*i.e.* the standard value at P° and T^-, which is not on the boiling curve of the phase diagram) is[221] 43.98 $\text{kJ}\cdot\text{mol}^{-1}$; (k) calculated using the Gibbs equation ($\Delta_f G^\ominus = \Delta_f H^\ominus - T^- \Delta_f S^\ominus$); (l) calculated from M and $\rho^\ominus$ as $v^\ominus = (\rho^\ominus)^{-1} M$; (m) calculated from $\kappa_T^\ominus$ and $v^\ominus$ as $k_T^\ominus = v^\ominus \kappa_T^\ominus$; (n) calculated from $\alpha_P^\ominus$ and $v^\ominus$ as $a_P^\ominus = v^\ominus \alpha_P^\ominus$.

Stimulated by the reading of this opening chapter, a single ion (under the board) proudly introduces its ionic family (the alkali-halides) before the great dive into water. As evident in the drawing, even apparently simple ions like the alkali and halides have very different characters.

Chapter 2
Fundamental experimental problems

The *experimental determination* of single-ion solvation free energies, as well as of corresponding derivative thermodynamic solvation parameters, is complicated by two fundamental problems: (*i*) the local *electroneutrality* of macroscopic matter at equilibrium (and its corollary, the *absence of free gas-phase ions*); (*ii*) the presence of a *surface polarization* at air-liquid interfaces. These two fundamental problems are discussed in sequence in Sections 2.1 and 2.2, and jointly summarized in Section 2.3. For completeness, it should also be mentioned that the ambiguity in the thermodynamic conventions regarding the definition of *standard states*, *i.e.* the existence of alternative conventions and the absent or incomplete specification in many experimental studies of the convention employed, also represents a serious (although not formally "fundamental") problem in the experimental literature and a significant factor of "uncertainty" in practice. The discussion of this third problem is deferred to Chapter 4. The description of practical experimental approaches, and the presentation of corresponding results in the context of alkali and halide hydration will be the concern of Chapter 5.

2.1 Electroneutrality

The first fundamental experimental problem is due to the local *electroneutrality* of macroscopic matter at equilibrium and its corollary, the absence of free gas-phase ions.

The electrostatic interaction energy $\tilde{E}$ between two point charges q_1 and q_2 at a distance r_{12} in free space is given by *Coulomb's law*[297],

$$\tilde{E}_{qq} = \frac{q_1 q_2}{4\pi\epsilon_o r_{12}} \quad , \qquad (2.1)$$

RSC Theoretical and Computational Chemistry Series No. 3
Single-Ion Solvation
Experimental and Theoretical Approaches to Elusive Thermodynamic Quantities
By Philippe Hünenberger and Maria Reif

Published by the Royal Society of Chemistry, www.rsc.org

where ϵ_o is the permittivity of vacuum (Table I). Since ionic charges are integer multiples of the elementary charge e (Table I), Eq. 2.1 can be rewritten (on a per mole basis) in the context of the interaction between two ions as $E_{qq} = N_A \tilde{E}_{qq} = C_{qq} z_1 z_2 r_{12}^{-1}$, where z_1 and z_2 are the integer charges of the two ions (signed charges in units of e) and $C_{qq} = N_A \left(4\pi\epsilon_o\right)^{-1} e^2 = 138.9$ kJ·nm·mol^{-1}, N_A being Avogadro's constant (Table I). This interaction is long-ranged, decaying as the inverse of the charge-charge distance, and of very large magnitude, because $C_{qq} r_{12}^{-1}$ for r_{12} of atomic dimensions is about two to three orders of magnitude larger than the thermal energy $RT^- = 2.48$ kJ·mol^{-1} at room temperature (where $R = N_A k_B$ is the ideal-gas constant, k_B being Boltzmann's constant; Table I). For example, the energy necessary to separate a $[Na^+ \cdot Cl^-]_g$ ion pair at equilibrium distance[298, 299] (≈ 0.24 nm; Table 5.3) in the gas phase into two free Na_g^+ and Cl_g^- ions at infinite distance as predicted by Eq. 2.1 (*i.e.* approximating the ions as point charges) is as high as 579 kJ·mol^{-1} ($\approx$233 RT^-).

Similarly, the electrostatic interaction energy $\tilde{E}_{q\mu}$ between a point charge q_1 and a point dipole of magnitude $\mu_{d,2}$ in its most favorable orientation (charge situated along the dipole line and on the oppositely-charged side of the dipole) at a distance r_{12} in free space is given by

$$\tilde{E}_{q\mu} = \frac{q_1 \mu_{d,2}}{4\pi\epsilon_o r_{12}^2} \quad . \tag{2.2}$$

Since molecular dipole moments are typically expressed in units of Debye D (Table I), Eq. 2.2 can be rewritten (on a per mole basis) in the context of the interaction between an ion and a dipolar molecule as $E_{q\mu} = N_A \tilde{E}_{q\mu} = C_{q\mu} z_1 \mathrm{D}^{-1} \mu_{d,2} r_{12}^{-2}$, where z_1 is the integer charge of the ion, $\mathrm{D}^{-1}\mu_{d,2}$ the molecular dipole in Debye and $C_{q\mu} = N_A \left(4\pi\epsilon_o\right)^{-1} e\mathrm{D} = 2.892$ kJ·nm^2·mol^{-1}. This interaction is shorter-ranged compared to charge-charge interactions, decaying as the inverse square of the charge-dipole distance, but still of a quite large magnitude, because $C_{q\mu} r_{12}^{-2}$ for r_{12} of atomic dimensions is still about one to two orders of magnitude larger than the thermal energy $RT^- = 2.48$ kJ·mol^{-1} at room temperature. For example, the energy necessary to separate a $[Na^+ \cdot H_2O]_g$ ion-water pair at equilibrium distance[300] (≈ 0.23 nm) in the gas phase into a free Na_g^+ ion and a free water molecule H_2O_g ($\mu_d = 1.85$ D; Table 1.1) at infinite distance as predicted by Eq. 2.2 (*i.e.* approximating the ion as a point charge and the water molecule as a point dipole in optimal orientation), is still as high as 101 kJ·mol^{-1} ($\approx$41 RT^-). Similarly, separating a $[Cl^- \cdot H_2O]_g$ ion-water pair at equilibrium distance[301] (≈ 0.31 nm) into a free Cl_g^- ion and a free water molecule H_2O_g requires 56 kJ·mol^{-1} ($\approx$23 RT^-).

An immediate consequence of the long range and large magnitude of electrostatic interactions (charge-charge as well as charge-dipole) in free space is that *free ions in the gas phase* are essentially *non-existent species* under most experimental conditions, *i.e.* when considering systems at equilibrium under usual conditions of pressure and temperature. The reason is that hypothetical free gaseous ions would always spontaneously aggregate with oppositely charged ions (*e.g.* Na_g^+ and Cl_g^- to form an overall neutral salt crystal $NaCl_s$) and, whenever present, with dipolar solvent molecules (*e.g.* Na_g^+ and Cl_g^-, in the presence of liquid water H_2O_ℓ, to form an overall neutral aqueous solution $NaCl_{aq}$, possibly in equilibrium with solid salt if the amount of ions is above the saturation level). The latter aggregated states (crystal or solution) are inevitably electroneutral, for otherwise, they would strongly attract further oppositely-charged gaseous ions. Note that the situation of a gas of ions with identical charges at equilibrium is experimentally equally improbable. For example, a sample of 1 mm^3 of a pure sodium ion (real) gas (no counter-ions) at a density of 40.3 $mol{\cdot}m^{-3}$ (the density of an ideal gas at pressure P° and temperature T^-) would have a self-energy of about $1.6{\cdot}10^{17}$ J, *i.e.* about 2000 times the energy released by a 0.02 megaton[302] nuclear explosion[303] (Hiroshima, 1945).

In contrast to the gas-phase situation, solvated ions can exist as free dissociated entities up to a certain concentration corresponding to saturation. To understand the dissolution process of an ionic crystal at constant pressure and temperature, one needs to examine the detailed enthalpic and entropic balance resulting from ion-ion, ion-solvent and solvent-solvent interactions in the crystal, pure solvent and solution systems. However, from a broad perspective, this process can be viewed as resulting from a combination of three main effects: (*i*) the reduction of the enthalpic cost for separating oppositely-charged ions resulting from the *dielectric screening* of ion-ion interactions by the solvent; (*ii*) the *entropy increase* resulting from the *free-volume increase*, *i.e.* upon changing the state of the ions from restricted oscillations within limited effective volumes around fixed lattice positions to free diffusion in a given (typically much larger) solution volume; (*iii*) the *entropy decrease* resulting from the (generally observed) *solvent structuring* in the neighborhood of the ions. Note that the volume-related entropy change actually depends on a specified reference concentration in solution. This change is typically positive and large when considering water and a reference molality b°. Similarly, the sign and magnitude of the solvent-structuring entropy change actually depend on the ion and solvent considered. For water, this change is typically negative for monoatomic ions (*e.g.* alkali and halides), its magnitude generally increasing with the ion charge and decreasing with the ion size, but can possibly become positive for large, hydrophobic ions and weakly charged ions. The three above contributions to the driving force of the dissolution process are discussed in turn below.

The electrostatic interaction energy $\tilde{E}_{qq}^{scr}$ between two point charges q_1 and q_2 at a distance r_{12} in a continuous medium of homogeneous static relative dielectric permittivity ϵ_{svt} representing the solvent is given by the screened version of Coulomb's law,

$$\tilde{E}_{qq}^{scr} = \frac{q_1 q_2}{4\pi\epsilon_o\epsilon_{svt} r_{12}} \quad . \tag{2.3}$$

This equation can be rewritten (on a per mole basis) in terms of the integer ion charges z_1 and z_2 as $E^{scr}_{qq} = N_A \tilde{E}^{scr}_{qq} = C_{qq} z_1 z_2 \epsilon^{-1}_{svt} r^{-1}_{12}$. The scaling factor ϵ^{-1}_{svt} characterizes the ability of the solvent molecules to respond to the influence of the dissolved charges, thereby reducing the magnitude of the effective electrostatic force between them. Note that Eq. 2.3 is only strictly valid for point charges and represents an approximation in the context of finite-sized ions, which is most accurate in the long-distance range.

As illustrated in Figure 2.1, different mechanisms may be involved in the *dielectric response* of a solvent at the molecular level[304–310], and are thus encompassed within the quantity ϵ_{svt}. For dipolar solvent molecules (*e.g.* water), the most efficient response mechanism arises from the reorientation of the solvent dipoles under the effect of the ionic electric field. Weaker mechanisms may also be active, including the alteration of the solvent molecular geometry and the polarization of its electron density. Since they are faster than molecular reorientation, these mechanisms become predominant when the dielectric response is probed using electric fields oscillating at high frequencies[311–313]. Electronic polarization also represents the leading mechanism for non-dipolar solvent molecules. For conducting media such as liquid metals or electrolyte solutions, the static permittivity is in principle infinite, because the free charge-carrying species contained in these media can redistribute so as to entirely screen any applied electric field within a very short distance

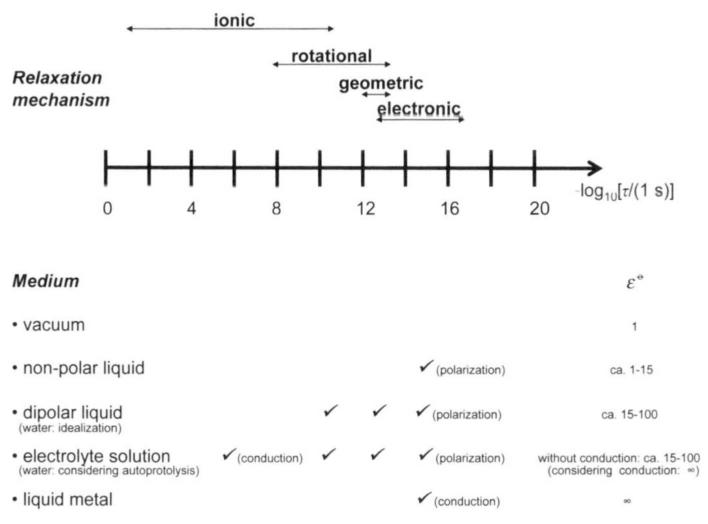

Figure 2.1 *Relaxation mechanisms contributing to the static relative dielectric permittivity of liquid media along with corresponding indicative characteristic timescales.* The different relaxation mechanisms are ionic conduction, solvent rotational reorientation, solvent geometric distortion, solvent electronic polarization and electronic conduction. Indicative characteristic timescales τ at ambient pressure and room temperature are also provided. The mechanisms active in the different types of liquid media are indicated by check marks, along with indicative ranges for the resulting static relative dielectric permittivity $\epsilon^{\ominus}$.

range from its source. For liquid metals, the dielectric response mechanism is electronic conduction, which is extremely fast. For electrolyte solutions, the mechanism is ionic conduction, which is typically much slower than the solvent (rotational, geometric and electronic) response. Thus, it is still possible in this case to define an effective static permittivity of the solvent by extrapolating the intermediate-frequency response (*i.e.* probing a timescale where solvent relaxation is active, but ionic conduction inactive) to zero frequency. Since this effective quantity is the one relevant in the context of ion-ion interactions, it will be referred to here as the static permittivity of the solvent in the electrolyte solution, and noted ϵ_{svt} in the limit of infinite dilution. Note that the above distinction between true and extrapolated permittivities is important even for pure water at neutral pH which is formally a conductor due to autoprotolysis (H^+_{aq} and OH^-_{aq} species), with a screening length on the order of 1 μm (Section 4.3.1).

The static relative dielectric permittivity of a pure non-conducting (or weakly conducting, *e.g.* water) liquid can be determined using an alternating-current plate capacitor[296,314]. In a first measurement, the capacitance is determined when the plates are separated by vacuum. In a second measurement, the plates are separated by the liquid and the capacitance measured at intermediate frequencies (see above) is extrapolated to zero frequency. The ratio of the two capacitances (liquid divided by vacuum) corresponds to ϵ_{svt}. For conducting (*e.g.* electrolyte, possibly non-aqueous[315]) solutions, the contribution of conductance to the impedance of the liquid phase becomes non-negligible even at high frequencies. In this case, the most appropriate measurement method is dielectric relaxation spectroscopy[316–322]. In this approach, the impedance of the solution is measured as a function of the potential difference associated with an alternating electric field. After removal of the ionic conductance contribution, the static dielectric permittivity can be deduced from an analysis of the components of the complex permittivity (representing dispersion and absorption processes) as a function of frequency[316] (Argand diagram). The frequencies involved are on the order of 0.9 – 400 GHz for aqueous solutions[316], the Debye model[304] being valid up to about 40 GHz for pure water[98,323]. The interpretation of the resulting dielectric spectra[323] may present different levels of complexity, depending on whether a simple Debye model[304] can be used and whether ion-pairing plays a significant role[131,132,318,324].

Water as a solvent is characterized by a very high screening ability, with a static relative dielectric permittivity $\epsilon^{\ominus}_{svt} = 78.36$ at pressure P° and temperature T^{-} (Table 1.1). For this reason, the interaction between ions in water, although still long-ranged, is about two orders of magnitude weaker than in the gas phase. In this sense, water is much closer to a conducting medium like a liquid metal, than to a non-polar medium like an organic solvent or vacuum. For example, the energy necessary to separate a $[Na^+{\cdot}Cl^-]_{aq}$ ion pair at contact distance[248] ($\approx$ 0.275 nm) in water into free hydrated Na^+_{aq} and Cl^-_{aq} ions at infinite distance in solution as predicted by Eq. 2.3 (*i.e.*

approximating the ions as point charges and the solvent as a homogeneous dielectric medium) is as low as 6.4 kJ·mol^{-1} ($\approx$ 2.6 RT^{-}). Although this estimate may not be very accurate for finite-sized ions, where a more exact continuum-electrostatics treatment should account for the lower dielectric permittivity within the ionic cavities[325–336] (see also corresponding investigations by atomistic simulations in Refs.[328,337–376] and by quantum-mechanical simulations in Ref.[377]), it is sufficient to illustrate the point. Note in passing that liquid metals (*e.g.* mercury or molten metals), that are characterized by an infinite permittivity value, are also able to dissolve ions[378–386]. For example, an estimate for the solvation enthalpy of $NaCl_g$ (*i.e.* the sum of the values for Na_g^+ and Cl_g^-) in liquid sodium, extrapolated to P° and T^-, is -695.5 kJ·mol^{-1} (derived from experimental solvation enthalpies of halogen gases in liquid metals[383], using data from Table 5.9), *i.e.* comparable in magnitude with the corresponding value of -728.0 kJ·mol^{-1} in water (Table 5.11, Set 1G).

Because dielectric screening reduces the enthalpic driving force towards ion aggregation in solution (dramatically for water or liquid metals, more moderately for less polar solvents), this driving force can be compensated by *entropic effects*. According to statistical mechanics, the entropy gain upon releasing ions from a crystal into a solution increases approximately linearly with the logarithm of the effective solution volume accessible to each ion, *i.e.* with the negative logarithm of the solution concentration. This approximation becomes exact in the infinite-dilution limit, *i.e.* when direct or solvent-mediated ion-ion interactions can be neglected. For this reason, ions can only dissolve in a solvent up to a finite concentration corresponding to saturation, the extra ions remaining in the crystalline form. This volume-related entropy gain is generally counteracted by an entropy decrease resulting from the structuring of the solvent in the neighborhood of the ions. Nevertheless, even when ions are dissolved in a solvent as dissociated species, the magnitude of electrostatic interactions remains such that no violations of electroneutrality are possible at the macroscopic level, *i.e.* only volume elements of truly microscopic sizes can contain an excess of positive or negative ions at a given instant.

It is important to realize that the constraint of *electroneutrality of macroscopic matter at equilibrium* not only implies a vanishing macroscopic charge, but even vanishing microscopic fluctuations of this charge. This point can be illustrated by a simple (yet striking) calculation of the vapor pressure of Na_g^+ and Cl_g^- in equilibrium with a salt crystal or a 1 molal (ideal) aqueous salt solution. Using estimates $\Delta_l G^{\ominus}[NaCl_g] = -719.0$ kJ·mol^{-1} (Table 5.10) for the reticulation free energy of $NaCl_g$ ($Na_g^+ + Cl_g^- \rightarrow NaCl_s$) and $\Delta_s G^{\ominus}[NaCl_g] = -728.0$ kJ·mol^{-1} (Table 5.11, Set 1G) for the hydration free energy of $NaCl_g$ ($Na_g^+ + Cl_g^- \rightarrow NaCl_{aq}$), the corresponding equilibrium vapor pressures of free ions at room temperature, $P^{-}_{vap,\mathrm{Na}^+}$ and $P^{-}_{vap,\mathrm{Cl}^-}$, are found to be on the order of 10^{-64} bar in both cases. This implies that a volume about 17 orders of magnitude larger than that of the entire atmosphere (total atmosphere volume $6 \cdot 10^{20}$ m^3, assuming an atmosphere height of 1000 km and an earth radius[221] of 6378 km) would be required to find a single gaseous ion in equilibrium with the salt crystal or solution.

Although the present discussion is mostly concerned with single-ion solvation free energies, the constraints related to the absence of free gas-phase ions and to the electroneutrality of macroscopic matter at equilibrium clearly apply to all extensive thermodynamic parameters, *e.g.* free energy G, enthalpy H, entropy S, heat capacity C_P, volume V, volume-compressibility K_T and volume-expansivity A_P. The consequence of the above observations in the context of the determination of thermodynamic parameters for ionic processes are quite drastic and can be summarized as follows[85,93,98,109,257–285]:

Equilibrium experiments performed at usual pressures and temperatures (or equivalent calorimetric procedures) can provide information on thermodynamic parameters for processes involving ions, but (i) only in the form of sums over overall neutral sets of ions, and (ii) only for processes that do not involve free gas-phase ions.

For instance, such experiments alone permit the determination of parameters for salt formation (*e.g.* $Na_s + {}^1\!/\!_2\, Cl_{2,g} \rightarrow NaCl_s$), dissolved salt formation (*e.g.* $Na_s + {}^1\!/\!_2\, Cl_{2,g} \rightarrow NaCl_{aq}$), salt dissolution (*e.g.* $NaCl_s \rightarrow NaCl_{aq}$) or atomization (*e.g.* $Na_s \rightarrow Na_g$ and ${}^1\!/\!_2\, Cl_{2,g} \rightarrow Cl_g$), as will be discussed in more detail in Chapter 5. But they do not permit the determination of parameters for ionization (*e.g.* $Na_g \rightarrow Na_g^+ + e_g^-$ and $Cl_g + e_g^- \rightarrow Cl_g^-$; even in the form of a sum), reticulation (*e.g.* $Na_g^+ + Cl_g^- \rightarrow NaCl_s$) or solvation (*e.g.* $Na_g^+ \rightarrow Na_{aq}^+$ and $Cl_g^- \rightarrow Cl_{aq}^-$; even in the form of a sum). Note that although salt formation, dissolved salt formation and salt dissolution parameters can be measured by independent experiments, they are not independent, *i.e.* they form the three legs of a thermodynamic cycle that will be termed the formation-dissolution-reformation (FDR) cycle (Section 4.2.15). In the above statement, the expression *equivalent calorimetric procedure* refers to the indirect determination of equilibrium thermodynamic parameters *via* calorimetry, which may involve non-equilibrium measurements as well as pressures and temperatures differing from usual conditions. This includes the determination of reaction enthalpies *via* reaction calorimetry (RXC) measurements (Section 4.2.18), which are generally not performed under equilibrium conditions (*e.g.* combustion or explosive process), and of reaction entropies *via* third-law cycle (TLC) procedures (Section 4.2.15), the latter involving temperature-scaled heat capacity integration (TCI) to the absolute zero temperature (Section 4.2.11), *i.e.* low-temperature measurements. The restrictive statement formulated above deserves three important comments.

First, the absence of free gaseous ions at equilibrium only applies to the *usual range of pressures and temperatures.* At very high temperatures, *i.e.* when the average kinetic energy of neutral species in the gas phase exceeds their ionization potentials, a mixture of ions and electrons in equilibrium can be formed. The mixture remains, however, macroscopically electroneutral. This particular state is referred to as a plasma state. Plasma states actually account for about 99% of the known matter in the universe, including stars and parts of the interstellar medium, and are involved in natural phenomena such as lightnings and aurorae boreales[387]. Artificial plasmas can also be generated by arc, radiofrequency or microwave discharges, or using laser beams[388]. At extremely low pressures, free gaseous ions could also in principle remain stable at room temperature due to entropic effects, the large volume accessible to each ion shifting the aggregation equilibrium towards the free ionic species. However, as illustrated above by the vapor pressure calculation for $NaCl_s$ or $NaCl_{aq}$, the pressure should be much lower than any naturally (*e.g.* about 10^{-13} bar, the pressure in the interplanetary space, based on data from the Rosetta spacecraft[389]) or experimentally (about 10^{-11} bar, using a cryopump or turbomolecular pump[390]) achievable value, or the pressure decrease applied in conjunction with a temperature increase. Very slow equilibration would represent a serious issue at the corresponding extremely low densities.

Second, the absence of net charge in macroscopic samples at usual pressures and temperatures only applies to *equilibrium situations*. In non-equilibrium situations, a net charge separation between macroscopic samples can actually be observed, as is the case in static electricity phenomena[391]. For example, such a charge separation can be promoted artificially by means of a van de Graaff generator or naturally by the circulation of air masses in the atmosphere. In both cases, electrons are extracted from one system under the action of macroscopic forces (friction) and conveyed to the other system. Although the resulting relaxation to equilibrium (electric discharge or lightning bolt) may be quite spectacular, these situations are definitely not the ones most frequently encountered by the chemist. Of course, one may argue that the distinction between what is considered to be an equilibrium (stable) and a non-equilibrium (metastable) situation is somewhat arbitrary. For example, a macroscopic charge separation between two conductors can be maintained for very long times if the medium separating them is highly insulating. To avoid this ambiguity, an equilibrium situation will be *defined* in this book as a situation of *homogeneous effective pressure* (mechanical equilibrium), *homogeneous temperature* (thermal equilibrium), and *homogeneous effective chemical potentials* for all species (chemical equilibrium) throughout the system, with the additional constraint of *local electroneutrality*. The word effective is used here to encompass the possible influence of interfacial effects, *i.e.* the alteration of the pressure resulting from interfacial tension and the alteration of the chemical potential of charged species resulting from the electric potential generated by interfaces (this effective chemical potential will be subsequently termed the electrochemical potential; Section 4.3.4). The concept of local electroneutrality implies that only volume elements of truly microscopic sizes within the system, including at interfaces, may contain an excess of positive or negative charges at a given instant. These conditions automatically imply that the system is characterized by *time-independent* intensive properties. Note, however, that the restriction of local electroneutrality is only imposed to the considered system and not necessarily to its surroundings. For example, a potential difference applied between two components of the system from the outside (as done in electrochemistry by means of a potentiostat), although it implies a macroscopic charge separation outside the system, is not viewed as incompatible with an equilibrium situation, as long as the local electroneutrality constraint is satisfied within the system itself. The above definition and concepts will be further developed when examining interfacial effects in Chapter 4 (Section 4.3.1).

Third, it should be stressed that equilibrium *electrochemical measurements* involving either Galvanic (Section 4.4.3) or Voltaic (Section 4.4.4) cells represent *no exception* to the above restrictive statement, *i.e.* taken alone, they only provide information on thermodynamic parameters for processes involving ions in the form of sums over overall neutral sets. An equilibrium Galvanic cell measurement probes the potential difference (electromotive force) between two electrodes (half-cells) within a cell encompassing only non-polarizable interfaces (*i.e.* allowing for an exchange of charges between two phases in contact), given the constraint that the number of electrons released at one electrode and absorbed at the other are equal. The electromotive force is directly connected to the free energy of the overall reaction (sum of the two half-cell reactions). However, the corresponding stoechiometric equation must satisfy the electroneutrality balance in terms of the ions only, since no electron can appear on either side of this equation. Therefore, such a measurement alone can only characterize a process involving overall neutral sets of ions. An equilibrium Voltaic cell measurement probes the potential difference (compensating potential difference) between two metal electrodes within a cell encompassing a polarizable interface (*i.e.* capable of presenting a net charge separation), under the constraint that this interface is actually not polarized (*i.e.* presents no charge separation given the selected magnitude of the compensating potential difference). In this situation, the compensating potential difference is directly connected to the Volta potential difference across another interface within the cell. When this second interface separates a metal electrode from a solution containing its ions, the measured Volta potential difference is in turn related to the formation free energy of the corresponding metal ion in solution *via* the "solvation" free energy of the electron in the isolated metal (*i.e.* its so-called work function). In the absence of information concerning the latter quantity, combinations of compensating potential differences can only provide useful information on reactions involving no net change in the number of electrons in the metal. The corresponding stoechiometric equation must again satisfy the electroneutrality balance in terms of the ions only. As a result, such measurements alone can also only characterize processes involving overall neutral sets of ions.

In view of the above restrictions concerning the absence of free gas-phase ions and the electroneutrality of macroscopic matter at equilibrium, it may seem that the endeavor of obtaining single-ion thermodynamic parameters, as well as relevant parameters involving gas-phase ions, even in the form of sums over overall neutral sets, is bound to fail. Fortunately, there still exist two main entrance points to access these quantities, namely *spectroscopy* and *theory*, or a combination of both. Spectroscopic measurements escape the above restrictions, because they do not represent equilibrium measurements. They directly probe quantum energy levels, without requiring the corresponding excited states to be present in equilibrium. On the other hand, the obtained information is purely energetic, *i.e.* adiabatic (or possibly semi-adiabatic; Sections 4.2.12 and 4.3.5), and must be post-processed into thermodynamic quantities *via* an appropriate statistical-mechanical analysis. This approach will be referred to as the spectroscopy-based statistical mechanics (SBS) approach (Section 4.2.12). Theory also escapes the above restrictions, because it is obviously not bound to experimental constraints, *i.e.* experimentally-inaccessible conditions as well as unphysical processes can be considered. On the other hand, any theoretical approach relies on an underlying (simplified) model and, therefore, only leads to approximate predictions. Theoretical methods will be the concern of Chapters 3 and 6. As mentioned above, spectroscopy and theory give access to a number of thermodynamic parameters that are inaccessible to equilibrium measurements or equivalent calorimetric procedures. The corresponding approaches are detailed in Chapters 4 and 5, and only a brief summary is provided below.

Using optical, (photo-)electron or threshold spectroscopy experiments, it is possible to measure the ionization potentials[392] E_I (cation-forming species) or electron affinities[393] E_A (anion-forming species) of gas-phase atoms. An appropriate statistical-mechanical analysis of this data leads to the thermodynamic parameters of ionization (*e.g.* $Na_g \rightarrow Na_g^+ + e_g^-$ and $Cl_g + e_g^- \rightarrow Cl_g^-$), ionization referring in this context to the reaction forming the ion, *via* either addition or abstraction of an electron.

Using Madelung-type theoretical calculations[394–398], it is possible to evaluate reticulation (lattice formation) enthalpies of ionic crystals. If ionization parameters are experimentally available (SBS), experimental (and typically more accurate) values for these enthalpies may be obtained from a thermodynamic cycle also involving atomization and salt formation parameters (equilibrium measurements or equivalent calorimetric procedures), namely the Born-Fajans-Haber[79–81] (BFH) cycle (Section 4.2.15). In special cases where ionization parameters are not experimentally available, the BFH cycle may also be used to construct corresponding estimates based on theoretically evaluated reticulation parameters. The lattice entropies may also in principle be estimated from (calculated or experimental) vibration frequencies using (quasi-)harmonic models[196,399–401]. Alternatively, experimental (and typically more accurate) values for these entropies may be deduced from the absolute entropies of the gas-phase ions (SBS) and of the salt (TCI). The above estimates provide access to the thermodynamic parameters of reticulation (*e.g.* $Na_g^+ + Cl_g^- \rightarrow NaCl_s$).

Once ionization parameters are available (SBS or BFH), salt solvation parameters (*e.g.* $Na_g^+ + Cl_g^- \rightarrow NaCl_{aq}$) may be deduced from a thermodynamic cycle also involving atomization and dissolved salt formation parameters (equilibrium measurements or equivalent calorimetric procedures), that will be termed the ion-formation-solvation-reformation (ISR) cycle (Section 4.2.15). Note, however, that single-ion solvation parameters (*e.g.* $Na^+_g \rightarrow Na^+_{aq}$ and $Cl^-_g \rightarrow Cl^-_{aq}$), of which the salt solvation parameters are the sum, remain inaccessible at this point.

Finally, using photo-electron, thermionic or field-emission spectroscopy[402,403] on metal samples with highly clean surfaces, it is possible to measure the work function of a metal (Section 4.3.5). This quantity represents the "solvation" free energy of the electron in the isolated metal. The use of an appropriate thermodynamic cycle involving this quantity, atomization parameters (equilibrium measurements or equivalent calorimetric procedures), ionization parameters (SBS or BFH), as well as the result of an equilibrium Voltaic cell measurement (Section 4.4.4), it is possible to derive thermodynamic parameters for single-ion solvation (*e.g.* $Na^+_g \rightarrow Na^+_{aq}$ and $Cl^-_g \rightarrow Cl^-_{aq}$). The single-ion solvation parameters accessed *via* this type of approach are called *real* solvation parameters, and will be written $\Delta_s \hat{Y}[I^z_g]$, where Y is a generic notation for an extensive thermodynamic variable (*e.g.* G, H, S, C_P, V, K_T or A_P) and I^z is a generic notation for an ion with signed integer charge z. The notation $\Delta_s \hat{Y}[I^z_g]$ does not encompass any reference to the considered solvent, which is generally obvious from the context. Whenever such an explicit reference is desired, the notation $\Delta_s \hat{Y}_{svt}[I^z_g]$ will sometimes be used instead, or $\Delta_s \hat{Y}_{wat}[I^z_g]$ in the specific case of hydration.

The above restrictive statement can now be reformulated in a somewhat more optimistic fashion:

The combination of equilibrium experiments (or equivalent calorimetric procedures), spectroscopic (non-equilibrium) measurements (post-processed using statistical mechanics) and (possibly) theoretical calculations enables the determination of single-ion thermodynamic parameters, including those of processes involving gas-phase ions. In the case of solvation, the parameters that can be accessed in this way pertain to the real solvation process.

The restriction implied by the last sentence of this statement is related to a second fundamental experimental problem in the determination of single-ion solvation parameters, namely the presence of surface effects. These effects are discussed in the next section.

2.2 Surface polarization

The second fundamental experimental problem can be viewed as resulting from the unavoidable presence of a *surface polarization* at vacuum-liquid (air-liquid) interfaces.

The real single-ion solvation free energy $\Delta_s \hat{G}_{svt}[I^z_g]$ of an ion I^z in a given solvent and in the limit of infinite dilution, written here $\Delta_s \hat{G}[I^z_g]$ for simplicity, can be defined as the reversible work (expressed on a per mole basis) associated with either of two equivalent *hypothetical processes* (Figure 2.2).

The first hypothetical process will be referred to here as the *physical pro-*

cess. This process involves: (*i*) taking the ion I^z from some fixed location in vacuum (*i.e.* infinitely far away from any other particle); (*ii*) bringing it towards an infinite (spherical, uncharged) sample of the pure solvent and crossing the (quasi-planar) vacuum-liquid interface; (*iii*) dropping it at some fixed location within the bulk solvent (*i.e.* infinitely far away from the vacuum-liquid interface).

The second hypothetical process will be referred to here as the *unphysical process*. This process involves: (*i*) progressively growing an uncharged ion-sized cavity at some fixed location within the bulk (*i.e.* infinitely far away from the vacuum-liquid interface) of an infinite (spherical, uncharged) sample of the pure solvent; (*ii*) progressively transferring the charge of a single ion I^z at some fixed location in vacuum (*i.e.* infinitely far away from any other particle) to the grown cavity, across the (quasi-planar) vacuum-liquid interface and the liquid-cavity interface; (*iii*) progressively ungrowing the uncharged ion in vacuum. Obviously, the third subprocess is associated with no work (*i.e.* free-energy) contribution since no forces act on the ion in vacuum.

The two above hypothetical processes must be carried out *reversibly*, *i.e.* infinitely slowly. This ensures that the system remains at equilibrium along

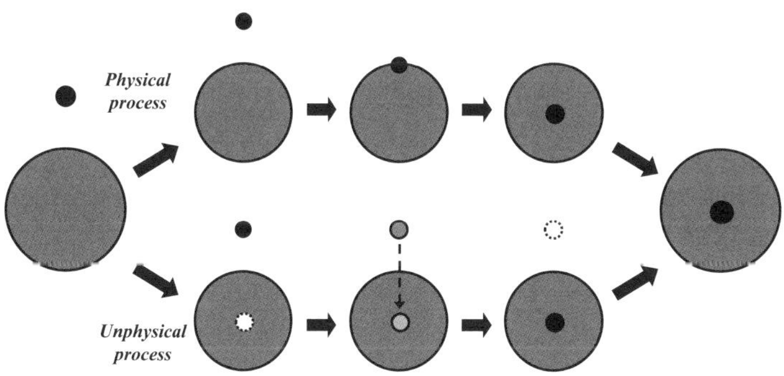

Figure 2.2 *Hypothetical processes defining the real solvation free energy of an ion.* The physical process (top) involves taking the ion from a fixed location in vacuum to a fixed location in the bulk solvent, passing through the vacuum-liquid interface. The unphysical process (bottom) involves progressively growing an uncharged ion-sized cavity at a fixed location within the bulk solvent, and then progressively transferring the ion charge from the ion at a fixed location in vacuum into the cavity, passing through the vacuum-liquid and liquid-cavity interfaces. It is assumed that the solvent sample is pure, uncharged, spherical and of infinite size, resulting in a quasi-planar vacuum-liquid interface, and that the fixed locations in vacuum and in the bulk are infinitely far from this interface. To account for pressure effects, the vacuum region is actually filled by an ideal gas (no interaction with the gas-phase ion), representing vapor-saturated air at a pressure P. The reversible work associated with either of the two processes for an ion I^z in a given solvent defines (when expressed on a per mole basis) the real single-ion solvation free energy $\Delta_s \hat{G}[I_g^z]$ of the ion in this solvent. The corresponding standard value $\Delta_s \hat{G}^{\ominus}[I_g^z]$ is obtained by considering the process at pressure P° and temperature T^-, and adding an appropriate (purely entropic) standard-state correction term $\Delta_s G^{\ominus}_{std}$ (Eq. 2.9). The corresponding standard intrinsic value $\Delta_s G^{\ominus}[I_g^z]$ is obtained by further subtracting the surface term $\Delta_s G^{\ominus}_{srf}[I_g^z]$ arising from the air-liquid interfacial potential $\chi^{\ominus}_{svt}$ of the pure solvent, also evaluated at P° and T^- for a planar interface in equilibrium (Eqs. 2.6 and 2.8).

the entire process, *i.e.* that all possible solvent configurations are sampled with an appropriate equilibrium distribution at each new location, growth-state or charge-state of the ion. This requirement poses two little conceptual problems.

First, a liquid sample cannot be at equilibrium when it is surrounded by vacuum. Without further restraints, the liquid should partly evaporate within a specified finite total volume, until it reaches equilibrium with its own vapor. In this case, the equilibrium pressure is determined by the vapor pressure of the liquid at the given temperature, and cannot be chosen freely. This difficulty can be overcome by assuming that: (*i*) the region outside the liquid sample is actually filled by an ideal gas (non-interacting particles) at pressure P; (*ii*) this gas exerts a pressure P on the liquid but has no interaction with the gas-phase ion; (*iii*) this gas encompasses evaporated solvent molecules at a partial pressure equal to the vapor pressure of the liquid given the total pressure P. Under these assumptions, the situation considered approximates that of a macroscopic liquid sample surrounded by vapor-saturated air at a pressure P, and the word air (rather than vacuum) will be retained in the following discussion. Second, even in the absence of external forces (*e.g.* gravitation) beyond the pressure imposed by the surrounding air, the liquid sample will undergo equilibrium fluctuations away from an ideal spherical shape. However, the sphericality of the sample is not essential in the present considerations, as long as this sample is of macroscopic size and its surface quasi-planar. In a theoretical context, these two difficulties may be remedied by introducing artificial confinement forces that maintain a given external pressure, prevent solvent evaporation and preclude deviations from the spherical shape. In this case, a possible influence of the artificial forces on the solvation process, including *via* a perturbation of the surface structure of the sample, must be considered. Although these considerations become very relevant in the case of atomistic simulations or quantum-mechanical computations (Chapter 3), they are not essential to describe the hypothetical solvation processes from a broad perspective, as done in the present section.

Since the physical and unphysical solvation processes share identical initial and final states (Figure 2.2), while free energy is a state function, these processes are associated with the same free-energy change $\Delta_s \hat{G}[I_g^z]$. The same consideration also holds for the corresponding derivative thermodynamic properties. These two processes have been termed hypothetical, because they correspond to thought experiments and cannot be carried out in practice, due to: (*i*) various idealizations (*e.g.* infinite spherical sample of the absolutely pure solvent surrounded by air viewed as an ideal gas, single ion located at fixed points in the initial and final states, process carried out infinitely slowly); (*ii*) the experimental impossibility of observing free gas-phase ions at equilibrium (Section 2.1); (*iii*) for the unphysical process, the artificial nature of the process itself (cavity growth and charge transfer). While the unphysical process is typically more convenient for theoretical calculations (Chapter 3), the *physical process* is the one closest to the experimental situation, and is the one to be kept in mind in the following discussion. The key point in the experimental context is that although the hypothetical physical process cannot be probed directly in a single experiment (at least not using currently available experimental techniques), this process is implicitly encompassed within any experimental measurement providing information on the solvation of an ion. As will be seen below, this observation results in a severe restriction on the nature of the single-ion solvation parameters that are experimentally accessible.

Since the initial and final ion locations of the hypothetical solvation process (Figure 2.2) are both situated infinitely far from the air-liquid interface, while interfacial effects are expected to be finite-ranged, the solvation free energy $\Delta_s \hat{G}[I_g^z]$ can be formally decomposed into a *bulk* and a *surface* contribution[264,272,277,404–420], as

$$\Delta_s \hat{G}[I_g^z] = \Delta_s G[I_g^z] + \Delta_s G_{srf}[I_g^z] \quad . \tag{2.4}$$

The quantity $\Delta_s \hat{G}[I_g^z]$ is called the *real* single-ion solvation free energy, because it is the one encompassing the totality of the reversible work associated with the physical solvation process. The bulk contribution $\Delta_s G[I_g^z]$ is called the *intrinsic* single-ion solvation free energy, because it exclusively arises from the interaction between the ion and its local solution environment (*i.e.*, at infinite dilution, with the neighboring solvent molecules). Finally, the contribution $\Delta_s G_{srf}[I_g^z]$ is called the *surface* term, because it arises from the reversible work associated with the ion crossing the air-liquid interface, but excluding the interactions characteristic of bulk solvation, which are already encompassed in $\Delta_s G[I_g^z]$. Obviously, the process involving a finite-sized ion going through an air-liquid interface is a very complex one, *i.e.* its detailed description at the microscopic level would be extremely complicated. Fortunately, it is not necessary to know the details of this process for giving an interpretation of the surface term. This is because: (*i*) the ion is infinitely far from the interface at the beginning and at the end of the solvation process, so that the state of the interface remains unaffected by the process; (*ii*) the solvent sample is overall neutral in the initial state of the process and only infinitesimally charged in the final state, so that net-charge effects can be ruled out. Consequently, the only property of the interface that can possibly affect $\Delta_s \hat{G}[I_g^z]$ through $\Delta_s G_{srf}[I_g^z]$ is the presence of a *surface polarization* (surface dipole-moment density), leading to a discontinuity of magnitude χ_{svt} (in the air-to-liquid direction) in the electric potential. One may thus write

$$\Delta_s G_{srf}[I_g^z] = zF\chi_{svt} \ , \tag{2.5}$$

where $F = N_A \cdot e$ is Faraday's constant (Table I).

Physically, the presence of a polarization at the air-liquid interface of a macroscopic solvent sample arises from the fact that the solvent molecules in this region are subject to an anisotropic environment. For this reason, the spatial and (whenever relevant) orientational distributions of these molecules depend on the distance to the interface, additional perturbations involving the distortion of molecular geometries and electronic clouds. As a result, the local time-averaged solvent charge density, as contributed by the nuclei and electrons of the solvent molecules, is also a function of this distance. The leading surface multipole moment of this charge distribution is the surface polarization, which results in an interfacial jump of magnitude χ_{svt} in the electric potential. The surface charge distribution is in general not purely dipolar, *i.e.* it is also characterized by non-vanishing higher-order multipole moments. These higher-order moments contribute to local variations of the electric potential in the interface region. However, it is easily seen based on arguments from electrostatics [421] that they result in no net potential jump across this region, *i.e.* they do not contribute to χ_{svt}. Obviously, there is no zeroth-order moment (surface charge) at the interface between an uncharged liquid sample and air, considering that the presence of such a charge would violate the local electroneutrality constraint (Section 2.1). The case of a liquid sample bearing a net charge does not need to be considered here, because it involves a macroscopic charge separation and would thus only be relevant in the context of non-equilibrium phenomena. Note, finally, that the surface polarization at a planar air-liquid interface in equilibrium and the resulting interfacial potential χ_{svt} are properties of the pure solvent at a given thermodynamic state point, *i.e.* they are independent of the ion.

The above considerations are by no means limited to the situation of dipolar solvent molecules. Even if the solvent molecule (or atom) bears no net dipole (*e.g.* carbon tetrachloride or mercury), electronic polarization and geometric distortion effects induced by the anisotropic interfacial environment may still contribute to χ_{svt}. These considerations can also be extended to electrolyte solutions. In this case, the preferential affinity or exclusion of specific ions with respect to the interfacial region [135, 141, 149, 240, 242, 422–440], *i.e.* the distinct ionic charge distribution profiles in this region, will induce an additional (concentration-dependent) contribution to the interfacial potential [441–447], as well as an indirect alteration of the solvent contribution to this quantity. However, even in this case, both the bulk and the surface regions remain electroneutral, so that the overall effect can still be interpreted in terms of a surface polarization. A similar (concentration-dependent) effect of solute molecules on χ_{svt} is also observed for solutions of uncharged organic molecules [448–454]. In this case, the interfacial potential will encompass contributions related to the orientational preferences of both solute and solvent dipoles close to the surface. In the Helmholtz model [455], commonly used to interpret the results of corresponding experiments, these contributions are assumed to be additive with weights determined by the surface fractions of the solute and solvent, themselves typically evaluated based on the Gibbs adsorption equation [456]. This approach amounts to neglecting mutual correlations in the corresponding orientational distributions.

It is interesting to note the close parallel between the mechanisms contributing to the dielectric response of a liquid medium (Section 2.1 and Figure

2.1) and those contributing to the corresponding air-liquid interfacial potential, the associated characteristic timescales also being comparable. In the case of pure water, four mechanisms contribute to χ_{svt}: (*i*) altered orientational distribution of the molecular dipoles close to the surface; (*ii*) preferential affinity or exclusion of the H^+_{aq} and OH^-_{aq} ions generated by autoprotolysis; (*iii*) alteration of electronic polarization properties at the surface; (*iv*) alteration of molecular geometric properties at the surface. The relative magnitudes of these contributions are difficult to assess, although the two former ones are likely to be predominant. This issue will be further considered in Chapter 5 (Section 5.17). Interfacial effects in general are discussed in detail in Chapter 4 (Section 4.3).

To obtain standard values $\Delta_s \hat{G}^\ominus[I_g^z]$ and $\Delta_s G^\ominus[I_g^z]$ for the real and intrinsic single-ion solvation free energies at the reference temperature T^-, the above values $\Delta_s \hat{G}[I_g^z]$ and $\Delta_s G[I_g^z]$, must be evaluated at the reference liquid pressure P° and at temperature T^-, and adjusted to match a specified definition of thermodynamic standard states. Given the standard-state convention adopted in this book ($bbme_T$ convention, reference pressure P° and molality b°, described in detail in Section 4.2.4), this results in

$$\Delta_s \hat{G}^\ominus[I_g^z] = \Delta_s G^\ominus[I_g^z] + \Delta_s G^\ominus_{srf}[I_g^z] \tag{2.6}$$

with

$$\Delta_s G^\ominus[I_g^z] = \Delta_s G^{\boxminus}[I_g^z] + \Delta_s G^\ominus_{std} \tag{2.7}$$

and

$$\Delta_s G^\ominus_{srf}[I_g^z] = zF\chi^\ominus_{svt} \ , \tag{2.8}$$

where $\chi^\ominus_{svt}$ is the air-liquid interfacial potential of the pure solvent at pressure P° and temperature T^- for a planar interface at equilibrium, and the notation $\Delta_s G^{\boxminus}[I_g^z]$ has been introduced to denote the intrinsic free-energy change of the hypothetical physical solvation process appropriate for a liquid pressure P° and a temperature T^-. The *standard-state correction* $\Delta_s G^\ominus_{std}$ is calculated as (equation derived in Section 4.2.16, Eq. 4.178).

$$\Delta_s G^\ominus_{std} = RT^- \ln \frac{RT^- b^\circ \rho^\ominus_{svt}}{P^\circ} \ , \tag{2.9}$$

where $\rho^\ominus_{svt}$ is the density of the pure solvent at pressure P° and temperature T^-. For water, $\rho^\ominus_{svt} = 997.0$ kg·m^{-3} (Table 1.1), resulting in $\Delta_s G^\ominus_{std} = 7.95$ kJ·mol^{-1}. The standard-state correction accounts for the fact that the idealized solvation process (Figure 2.2) involves the transfer of the ion from a fixed point in vacuum to a fixed point in the bulk liquid (expressed on a per mole basis), while thermodynamics considers the transfer of the ion from and to a random location within a specified effective (molar) volume (gas: volume accessible to a particle in an ideal gas at pressure P° and temperature T^-; solution: volume accessible to a particle in an ideal solution at pressure P°, molality b°, temperature T^- and solvent density $\rho^\ominus_{svt}$)[457,458]. The quantity

$\Delta_s G^{\boxminus}[I_g^z]$ will be further referred to as the semi-standard *intrinsic point-to-point* solvation free energy of the ion. It is semi-standard in the sense that it depends on the choice of a reference temperature T^- and on the choice of a reference pressure P° only to determine the state of the solvent, whereas P° is irrelevant for the gas-phase state of the ion and b° irrelevant for the corresponding dissolved state. The standard-state correction is attributed here to the bulk (intrinsic) term rather than to the surface term, as appropriate for a correction related to volume effects. In terms of standard values, the intrinsic solvation free energy $\Delta_s G^{\ominus}[I_g^z]$ thus encompasses two effects according to Eq. 2.7: (*i*) the contribution due to the interaction between the ion and the neighboring solvent molecules at infinite dilution; (*ii*) the contribution due to the change in ideal (partial) molar entropy of the ion, given the reference values prescribed by the standard-state convention for the effective molar volumes accessible to the gas-phase and solvated ions. According to Eq. 2.6, the corresponding real quantity $\Delta_s \hat{G}^{\ominus}[I_g^z]$ encompasses in addition the effect of the interface polarization on the electrostatic potential within the bulk solution relative to the corresponding potential outside the sample.

Illustrative (experimental and theoretical) estimates for $\chi_{svt}^{\ominus}$ for various non-aqueous solvents are reported in Table 2.1. The corresponding estimates for water are relevant in Chapter 5, and are therefore reported later in Table 5.18. It should be stressed that there is a considerable uncertainty in these values, especially apparent in the case of water, which arises from serious methodological issues in both the experimental (requirement for an extra-thermodynamic assumption, as discussed in Sections 2.3 and 5.17) and theoretical (ambiguity in the potential-evaluation scheme and parameter dependence of the results, as discussed in Sections 3.3.5 and 4.3.8) approaches. Nevertheless, it can be observed that: (*i*) for the non-aqueous solvents, $\chi_{svt}^{\ominus}$ does not appear to be systematically correlated with the polarity of the solvent, as estimated by its static relative dielectric permittivity $\epsilon_{svt}^{\ominus}$, except for the fact that non-dipolar solvents are characterized by much lower values of both $\epsilon_{svt}^{\ominus}$ and $\chi_{svt}^{\ominus}$; (*ii*) $\chi_{svt}^{\ominus}$ appears to be negative for all non-aqueous solvents considered, suggesting, for dipolar solvents, a preferential orientation of the positive end of the solvent dipoles in the surface region away from the liquid phase, and on the order of −0.5 to 0.0 V, resulting in $z^{-1}\Delta_s G_{srf}^{\ominus}[I_g^z]$ values in the approximate range −50 to 0 kJ·mol^{-1}; (*iii*) even the sign of $\chi_{wat}^{\ominus}$ is not known with certainty, the reported values for water ranging from −1.15 to +1.4 V, resulting in $z^{-1}\Delta_s G_{srf}^{\ominus}[I_g^z]$ values in the approximate range −115 to +140 kJ·mol^{-1}. The determination methods and the results for water will be discussed in more detail in Section 5.17. For reasons provided there, the authors of this book consider a positive value of $\chi_{wat}^{\ominus}$ on the order of +0.13 V with an estimated uncertainty of ±0.10 V as the most realistic estimate at present, resulting in a $z^{-1}\Delta_s G_{srf}^{\ominus}[I_g^z]$ value of about +13 kJ·mol^{-1}.

As has been stated above, any experimental measurement providing information on the solvation of an ion implicitly encompasses the hypothetical physical process of Figure 2.2. Indeed, as explained in Section 2.1, real single-ion solvation free energies can be determined by means of an appropriate thermodynamic cycle involving an equilibrium Voltaic cell measurement, along with the work function of a metal (SBS) and the corresponding atomization (equilibrium measurement or equivalent calorimetric procedure) and ionization (SBS) parameters. However, in this process, bulk (intrinsic) solvation and surface effects are intermingled, *i.e.* they are both simultaneously active in the second step, and cannot be unambiguously separated. In other words, the partitioning of Eq. 2.6, although appealing, cannot be unambiguously defined on the sole basis of experimental data. The question whether the same restriction applies to the unphysical process will be discussed in Chapter 3.

Although the above discussion was mostly concerned with single-ion solvation free energies, the distinction between real and intrinsic single-ion solvation properties also exists for *derivative thermodynamic solvation parameters*, *e.g.* free energy G, enthalpy H, entropy S, heat capacity C_P, volume V, volume-compressibility K_T and volume-expansivity A_P. This is obvious from the fact that all the above variables can be expressed as functions of G and of its first and second derivatives with respect to pressure and temperature (Section 4.2.1), while χ_{svt} itself depends on pressure and temperature. Therefore, the notations $\Delta_s\hat{Y}[I_g^z]$ and $\Delta_s Y[I_g^z]$, or $\Delta_s\hat{Y}_{svt}[I_g^z]$ and $\Delta_s Y_{svt}[I_g^z]$ when the identity of the solvent is not obvious from the context (*e.g.* subscript "*wat*" for water), will be used systematically to distinguish between real and intrinsic single-ion solvation parameters, respectively, where Y is a generic notation for any extensive thermodynamic parameter (G, H, S, C_P, V, K_T or A_P). Eqs. 2.6-2.8 can then be generalized to

$$\Delta_s\hat{Y}^{\ominus}[I_g^z] = \Delta_s Y^{\ominus}[I_g^z] + \Delta_s Y_{srf}^{\ominus}[I_g^z] \tag{2.10}$$

with

$$\Delta_s Y^{\ominus}[I_g^z] = \Delta_s Y^{\oplus}[I_g^z] + \Delta_s Y_{std}^{\ominus} \tag{2.11}$$

and

$$\Delta_s Y_{srf}^{\ominus}[I_g^z] = zF\hat{O}_Y\chi_{svt}^{\ominus} \quad , \tag{2.12}$$

where $\hat{O}_Y$ is the thermodynamic-derivative operator relating a given extensive thermodynamic variable Y to the free energy and its first and second derivatives with respect to pressure and temperature, *i.e.* (equation introduced in Section 4.2.1, Eq. 4.28)

$$Y = \hat{O}_Y G \tag{2.13}$$

and $\Delta_s Y_{std}^{\ominus}$ is a standard-state correction term appropriate for the quantity Y.

Note that there exist two variants of standard-state definitions for derivative thermodynamic properties, concerning processes involving species in solution, referred to in this book as the *standard variant* and the *density-corrected variant* (indicated by an additional "$*$" superscript), as will be discussed in Sections 4.2.5 and 4.2.16. Eqs. 2.10 and 2.11 are formulated in terms of the

Table 2.1 *Estimates for the air-liquid interfacial potential and temperature derivative for non-aqueous solvents.* The reported data includes estimates for the air-liquid interfacial potential $\chi^{\ominus}_{svt}$ of the pure solvent (planar surface at equilibrium) and for its temperature derivative $\partial_T \chi^{\ominus}_{svt}$, under standard conditions (reference pressure $P^{\circ} = 1$ bar and reference temperature $T^{-} = 298.15$ K). The static relative dielectric permittivity $\epsilon^{\ominus}_{svt}$ of the solvent (from Ref.[471]) is also indicated. Corresponding estimates for water can be found in Table 5.18. The methods (and involved extra-thermodynamic assumptions) refer to the literal enumeration of Section 5.17. The estimates are listed in chronological order of publication, separately for each solvent. The values calculated using atomistic simulations (method G_{13}) correspond to the external surface potential $\chi^{\ominus}_{svt}$ calculated based on the M-scheme ($\chi^{\ominus}_{AT,M}$; Sections 3.3.5 and 4.3.8). The corresponding estimates for the internal surface potential $\check{\chi}^{\ominus}_{svt}$ calculated based on the P-scheme ($\chi^{\ominus}_{AT,P}$) are reported between parentheses (note that these values depend strongly on the classical solvent model employed). The value in boldface ($\chi^{\ominus}_{AT,M}$ or $\chi^{\ominus}_{AT,P}$) is the one reported in the quoted reference, while a value in italics was calculated from the latter using Eq. 3.86 based on the exclusion potential $\xi^{\ominus}_{AT}$ appropriate for the solvent model employed and the experimental density of the pure solvent (Eq. 4.345). This interconversion could not always be performed, an analytical expression for $\xi^{\ominus}_{AT}$ only being available for non-polarizable solvent models with a single Lennard-Jones interaction site.

Method	$\chi^{\ominus}_{svt}$ [V]	$\partial_T \chi^{\ominus}_{svt}$ [10^{-3} V·K^{-1}]	$\epsilon^{\ominus}_{svt}$	Ref.	Notes
		Mercury			
A_{13}	2.2	-	∞	98,459	
		Methanol			
B_{13}	-0.2±0.13	1.85	33.1	460	a
B_{13}	-0.23	1.85	*id.*	272	
F_{13}	-0.18	-	*id.*	461	
C_{13}	-0.09	-	*id.*	462	
G_{13}	**-0.155**(*-0.432*)	0.08	*id.*	463,464	b
A_{13}	-0.13	-	*id.*	418	c
G_{13}	**-0.26**(*-0.537*)	1.5	*id.*	465	d,e
B_{13}	-0.15±0.05	-	*id.*	465	f,g
D_{13}	-0.184	-	*id.*	466	f
A_{13}	-0.13	-	*id.*	420	h
		Ethanol			
B_{13}	-0.3	1.0	24.8	272	
F_{13}	-0.12	-	*id.*	461	
A_{13}	-0.29	-	*id.*	418	c
D_{13}	-0.26	-	*id.*	466	f
A_{13}	-0.21	-	*id.*	420	h
		n-Propanol			
D_{13}	-0.267	-	20.3	466	f
		iso-Propanol			
D_{13}	-0.275	-	19.4	466	f
		n-Butanol			
D_{13}	-0.277	-	17.5	466	f
		Acetonitrile			
A_{13}	-0.1±0.006	-	36.6	467	i
A_{13}	-0.41	-	*id.*	413	j
F_{13}	-0.1	-	*id.*	461	
D_{13}	-0.108	-	*id.*	466	f
A_{13}	-0.1	-	*id.*	420	h

(continues on the next page)

Table 2.1 *(continued)*

Method	$\chi^{\ominus}_{svt}$ [V]	$\partial_T \chi^{\ominus}_{svt}$ [10^{-3} V·K^{-1}]	$\epsilon^{\ominus}_{svt}$	Ref.	Notes
		Dimethylsulfoxide			
F_{13}	-0.29	-	46.5	461	
A_{13}	-0.21	-	*id.*	418	c
D_{13}	-0.238	-	*id.*	466	f
A_{13}	-0.19	-	*id.*	420	h
		Dimethylformamide			
F_{13}	-0.26	-	36.7	461	
A_{13}	-0.46	-	*id.*	418	c
D_{13}	-0.434	-	*id.*	466	f
A_{13}	-0.39	-	*id.*	420	h
		Acetone			
F_{13}	-0.34	-	20.6	461	
A_{13}	-0.4	-	*id.*	418	c
D_{13}	-0.337	-	*id.*	466	f
A_{13}	-0.32	-	*id.*	420	h
		Formic Acid			
A_{13}	-0.46	-	56.1	413	j
		Methane			
G_{13}	*0.001* (**-0.22**)	-	1.7	468	k
		n-Hexane			
G_{13}	- (**-0.17871**)	-	1.9	469	l,m
G_{13}	*0.007* (**-0.4338**)	-	*id.*	469	n
		Bromine			
G_{13}	*-0.048* (**-0.65**)	-	3.1	470	o

(a) based on a differential measurement (solvent-water) along with the $\chi^{\ominus}_{svt}$ value for water estimated in Ref.[448] (Table 5.18); (b) based on the TIPS methanol model[472], temperature derivative by finite difference between 250 and 350 K, value of $\chi^{\ominus}_{AT,P}$ calculated based on $\xi^{\ominus}_{AT} = 0.277$ V estimated using $\rho^{\ominus}_{svt} = 791.4$ kg·m^{-3} (Ref.[221]) and $\mathcal{Q} = 0.006185$ e·nm^2 (Ref.[472]; oxygen atom as molecular center); (c) real values from the indicated reference, intrinsic values based on the Born model; (d) value at 287 K, based on the TIPS methanol model[472]; (e) value for $\chi^{\ominus}_{AT,P}$ calculated using $\xi^{\ominus}_{AT} = -277 \cdot 10^{-3}$ V, deduced from the estimates for $\chi^{\ominus}_{AT,P}$ and $\chi^{\ominus}_{AT,M}$ of Refs.[463, 464] (corresponding to a temperature of 278 K) reported in this table; (f) based on a differential measurement (solvent-water) along with the $\chi^{\ominus}_{svt}$ value for water estimated by the same authors (Table 5.18); (g) based on consideration of previous experiments[272,413,448,473]; (h) real values from experimental literature data[474–477], intrinsic values based on the Born model; (i) real values from the indicated reference, intrinsic values based on an assumption of equal intrinsic solvation free energies for large hydrophobic ions; (j) real values from the indicated reference, intrinsic values from Ref.[478]; (k) value at 100 K based on the IP-METH methane model[468], value of $\chi^{\ominus}_{AT,M}$ calculated based on $\xi^{\ominus}_{AT} = 0.221$ V estimated using $\rho^{\ominus}_{svt} = 440$ kg·m^{-3} (Ref.[479]; 1 bar, 100 K) and $\mathcal{Q} = 0.004444$ e·nm^2 (Ref.[468]; carbon atom as molecular center); (l) based on a modified FQ hexane model[480]; (m) interconversion between $\chi^{\ominus}_{AT,M}$ and $\chi^{\ominus}_{AT,P}$ could not be performed easily (polarizable solvent model); (n) based on the C27r hexane model[481], value of $\chi^{\ominus}_{AT,M}$ calculated based on $\xi^{\ominus}_{AT} = 0.441$ V estimated using $\rho^{\ominus}_{svt} = 660.6$ kg·m^{-3} (Ref.[221]) and $\mathcal{Q} = 0.031687$ e·nm^2 (Ref.[481]; midpoint of the C3-C4 bond as molecular center); (o) based on the bromine model described in Ref.[470], value of $\chi^{\ominus}_{AT,M}$ calculated based on $\xi^{\ominus}_{AT} = 0.602$ V estimated using $\rho^{\ominus}_{svt} = 3102.8$ kg·m^{-3} (Ref.[221]) and $\mathcal{Q} = 0.017077$ e·nm^2 (Ref.[470]; midpoint of the Br-Br bond as molecular center).

standard variant, where one has (Eq. 4.185)

$$\Delta_s Y^{\ominus}[I_g^z] = \hat{O}_Y \Delta_s G^{\ominus}[I_g^z] + \Delta_s Y_{std}^{\ominus} \quad . \tag{2.14}$$

Equations similar to Eqs. 2.10 and 2.11 also hold for the density-corrected variant upon replacement of $\Delta_s \hat{Y}^{\ominus}$, $\Delta_s Y^{\ominus}$ and $\Delta_s Y_{std}^{\ominus}$ by the corresponding quantities $\Delta_s \hat{Y}^{\ominus *}$, $\Delta_s Y^{\ominus *}$ and $\Delta_s Y_{std}^{\ominus *}$ appropriate for this variant. However, in the density-corrected variant, Eq. 2.14 is replaced by (Eq. 4.186)

$$\Delta_s Y^{\ominus *}[I_g^z] = \hat{O}_Y (\Delta_s G^{\ominus}[I_g^z] - \Delta_s G_{std}^{\ominus}) + \Delta Y_{std}^{\ominus *} \quad . \tag{2.15}$$

Note that the solvation free energy is the same in both variants. The two variants only differ in the way they handle the standard-state correction term, which depends on the solvent density, upon differentiation with respect to pressure and temperature. They represent the natural standardized outcome of different types of experimental determinations. These relatively intricate issues will be discussed in detail in Chapter 4. For the moment, it is sufficient to note that derivative thermodynamic parameters for the intrinsic point-to-point solvation process are always related to the corresponding free energy as (Eq. 4.179)

$$\Delta_s Y^{\boxminus}[I_g^z] = \hat{O}_Y \Delta_s G^{\boxminus}[I_g^z] \quad , \tag{2.16}$$

but can be standardized to the two distinct variants as

$$\Delta_s Y^{\ominus}[I_g^z] = \Delta_s Y^{\boxminus}[I_g^z] + \Delta_s Y_{std}^{\ominus} \quad , \tag{2.17}$$

or (Eq. 4.180)

$$\Delta_s Y^{\ominus *}[I_g^z] = \Delta_s Y^{\boxminus}[I_g^z] + \Delta_s Y_{std}^{\ominus *} \quad . \tag{2.18}$$

Expressions for the two types of standard-state correction terms $\Delta_s Y_{std}^{\ominus}$ and $\Delta_s Y_{std}^{\ominus *}$ can be deduced from Eqs. 4.183 and 4.184, along with expressions listed in Table 4.1.

Similar considerations apply to all corresponding single-ion formation parameters and partial molar variables in solution. In this case, the notations $\Delta_f \hat{Y}[I_d^z]$ and $\Delta_f Y[I_d^z]$ will be used to distinguish between real and intrinsic single-ion formation parameters, respectively, and the notations $\hat{y}[I_d^z]$ and $y[I_d^z]$ to distinguish between real and intrinsic single-ion partial molar variables in solution (where y refers to Y, *i.e.* μ, h, s, c_P, v, k_T and a_P for G, H, S, C_P, V, K_T and A_P, respectively). Here, I_d^z is a generic notation for a dissolved ion, and the subscript may be modified to specify the identity of the solvent when not obvious from the context (*e.g.* "*aq*" for aqueous). In terms of these quantities, Eq. 2.10 becomes

$$\Delta_f \hat{Y}^{\ominus}[I_d^z] = \Delta_f Y^{\ominus}[I_d^z] + \Delta_s Y_{srf}^{\ominus}[I_g^z] \tag{2.19}$$

or

$$\hat{y}^{\ominus}[I_d^z] = y^{\ominus}[I_d^z] + \Delta_s Y_{srf}^{\ominus}[I_g^z] \quad . \tag{2.20}$$

Here also, Eqs. 2.19 and 2.20 are formulated in terms of the standard variant, similar equations holding for the density-corrected variant upon replacement of $\Delta_f \hat{Y}^{\ominus}$, $\Delta_f Y^{\ominus}$, $\hat{y}^{\ominus}$ and $y^{\ominus}$ by the corresponding quantities $\Delta_f \hat{Y}^{\ominus *}$, $\Delta_f Y^{\ominus *}$, $\hat{y}^{\ominus *}$ and $y^{\ominus *}$ appropriate for this variant. The single-ion formation parameters in solution are directly related to the corresponding solvation parameters by addition of the corresponding standard formation parameters $\Delta_f Y^{\ominus}[I_g^z]$ of the gas-phase ion, *i.e.* in terms of the intrinsic point-to-point process

$$\Delta_f Y^{\boxminus}[I_d^z] = \Delta_f Y^{\ominus}[I_g^z] + \Delta_s Y^{\boxminus}[I_g^z] \quad . \tag{2.21}$$

The single-ion partial molar variables in solution are also directly related to the corresponding solvation parameters by addition of the corresponding standard molar parameters $y^{\ominus}[I_g^z]$ of the gas-phase ion, *i.e.* in terms of the intrinsic point-to-point process

$$y^{\boxminus}[I_d^z] = y^{\ominus}[I_g^z] + \Delta_s Y^{\boxminus}[I_g^z] \quad . \tag{2.22}$$

Following from Eqs. 2.17 and 2.18, these quantities can be standardized to the two distinct variants in an analogous way to $\Delta_s Y^{\boxminus}[I_g^z]$ by addition of either $\Delta_s Y_{std}^{\ominus}$ or $\Delta_s Y_{std}^{\ominus *}$.

The consequence of the above considerations in the context of the determination of single-ion solvation parameters, as well as single-ion formation parameters and partial molar variables in solution, are quite drastic and can be summarized as follows[82,83,264,404,405,407–409,411,415]:

Single-ion solvation parameters, as well as single-ion formation parameters and partial molar variables in solution, can be determined experimentally, but only in the form of real quantities, i.e. as a sum of contributions from bulk (intrinsic) solvation and surface effects.

Since the surface term $\Delta_s G_{srf}^{\ominus}[I_g^z]$ is linear in the ionic charge (Eq. 2.8), this term cancels out when considering a sum of solvation free energies for overall neutral sets of ions (salt solvation parameters), *i.e.* the distinction between intrinsic and real values becomes irrelevant in these sums. As seen in Section 2.1, the salt solvation parameters can be determined directly by application of the ISR cycle. Taking the example of an alkali-halide ion pair MX with water as a solvent, one may thus write

$$\begin{aligned} \Delta_s G^{\ominus}[MX_g] &= \Delta_s \hat{G}^{\ominus}[M_g^+] + \Delta_s \hat{G}^{\ominus}[X_g^-] \\ &= \Delta_s G^{\ominus}[M_g^+] + \Delta_s G^{\ominus}[X_g^-] \quad , \end{aligned} \tag{2.23}$$

where the pair solvation process is $MX_g \rightarrow MX_{aq}$, MX_g representing $M_g^+ + X_g^-$ (gas-phase ions, non-interacting in their standard state) and MX_{aq} representing $M_{aq}^+ + X_{aq}^-$ (dissociated ions in solution, non-interacting in their standard state). Here, the quantity $\Delta_s G^{\ominus}[MX_g]$ can be measured directly (ISR) or indirectly (by summation of $\Delta_s \hat{G}^{\ominus}[M_g^+]$ and $\Delta_s \hat{G}^{\ominus}[X_g^-]$, as determined based on Voltaic cell experiments and spectroscopic work function measurements), which allows for a consistency checking between the two types of determinations. In contrast, $\Delta_s G^\ominus[M_g^+]$ and $\Delta_s G^\ominus[X_g^-]$ are not accessible on the sole basis of experimental data. Since Eq. 2.12 is also linear in the ionic charge, similar considerations apply to all derivative thermodynamic solvation parameters, as well as to all corresponding formation parameters and partial molar variables in solution. For example, for an alkali-halide ion pair MX in water, one may generalize Eq. 2.23 to

$$\begin{aligned}\Delta_s Y^\ominus[MX_g] &= \Delta_s \hat{Y}^\ominus[M_g^+] + \Delta_s \hat{Y}^\ominus[X_g^-] \\ &= \Delta_s Y^\ominus[M_g^+] + \Delta_s Y^\ominus[X_g^-] \quad ,\end{aligned} \tag{2.24}$$

and also write

$$\begin{aligned}\Delta_f Y^\ominus[MX_{aq}] &= \Delta_f \hat{Y}^\ominus[M_{aq}^+] + \Delta_f \hat{Y}^\ominus[X_{aq}^-] \\ &= \Delta_f Y^\ominus[M_{aq}^+] + \Delta_f Y^\ominus[X_{aq}^-] \quad ,\end{aligned} \tag{2.25}$$

and

$$\begin{aligned}y^\ominus[MX_{aq}] &= \hat{y}^\ominus[M_{aq}^+] + \hat{y}^\ominus[X_{aq}^-] \\ &= y^\ominus[M_{aq}^+] + y^\ominus[X_{aq}^-] \quad .\end{aligned} \tag{2.26}$$

It is important to realize that the above restriction concerning the experimental inaccessibility of intrinsic single-ion solvation parameters amounts in practice to the inaccessibility of *a single unknown quantity*, for a given solvent and including its functional dependence on pressure and temperature. If the intrinsic single-ion solvation free energy of a single special ion (say, a cation) in a given solvent was available, this would enable the determination of corresponding intrinsic single-ion solvation free energies for all ions in this solvent. This is because the sum of intrinsic values is in principle experimentally accessible for all neutral ion pairs, so that the value for the specific cation would fix the values for all anions, and the resulting value for a given anion would in turn fix the values for all other cations. If the intrinsic single-ion solvation free energy of this special ion in the given solvent was further known in its functional dependence on pressure and temperature, all the corresponding intrinsic single-ion solvation parameters, as well as single-ion formation parameters and partial molar variables in solution, could be accessed.

The usual choice for this special ion is the *proton* H^+, *i.e.* the simplest possible ion in terms of internal structure (no electrons at all). Note, however, that in spite of this apparent simplicity, H^+ solvation typically involves complex electronic effects, which are actually far less significant in *e.g.* alkali and halide solvation (Section 4.3.8). In this case, the single experimentally inaccessible quantity for a given solvent is the *intrinsic solvation free energy* $\Delta_s G_{svt}^\ominus[H_g^+]$ *of the proton in this solvent*, including its functional dependence on pressure and temperature. Since this quantity plays a central role in this book, it will be attributed a special symbol $\mathcal{G}_{H,svt}^\ominus$, where the intrinsic quantity $\mathcal{G}_{H,svt}^\ominus(= \Delta_s G_{svt}^\ominus[H_g^+])$ is to be clearly distinguished from the corresponding real quantity $\hat{\mathcal{G}}^{\ominus}_{H,svt}(=\Delta_s\hat{G}^{\ominus}_{svt}[\mathrm{H}^+_g])$, which encompasses the surface term and is experimentally accessible, *i.e.* (Eqs. 2.6 and 2.8)

$$\hat{\mathcal{G}}^{\ominus}_{H,svt} = \mathcal{G}^{\ominus}_{H,svt} + F\chi^{\ominus}_{svt} \quad . \tag{2.27}$$

Some of the corresponding derivative thermodynamic solvation properties and partial molar variables in solution also play a central role in this text and will be given special symbols. These are the solvation enthalpy $\mathcal{H}^{\ominus}_{H,svt}$ $(=\Delta_s H^{\ominus}_{svt}[\mathrm{H}^+_g])$, entropy $\mathcal{S}^{\ominus}_{H,svt}$ $(=\Delta_s S^{\ominus}_{svt}[\mathrm{H}^+_g])$, as well as the partial molar entropy $s^{\ominus}_{H,svt}$ $(=s^{\ominus}[\mathrm{H}^+_d])$, heat capacity $c^{\ominus}_{P,H,svt}$ $(=c^{\ominus}_P[\mathrm{H}^+_d])$, volume $v^{\ominus}_{H,svt}$ $(=v^{\ominus}[\mathrm{H}^+_d])$, volume-compressibility $k^{\ominus}_{T,H,svt}$ $(=k^{\ominus}_T[\mathrm{H}^+_d])$ and volume-expansivity $a^{\ominus}_{P,H,svt}$ $(=a^{\ominus}_P[\mathrm{H}^+_d])$. Here also, the intrinsic quantities ($\mathcal{H}^{\ominus}_{H,svt}$, $\mathcal{S}^{\ominus}_{H,svt}$, $s^{\ominus}_{H,svt}$, $c^{\ominus}_{P,H,svt}$, $v^{\ominus}_{H,svt}$, $k^{\ominus}_{T,H,svt}$ and $a^{\ominus}_{P,H,svt}$) are to be clearly distinguished from the corresponding real quantities ($\hat{\mathcal{H}}^{\ominus}_{H,svt}$, $\hat{\mathcal{S}}^{\ominus}_{H,svt}$, $\hat{s}^{\ominus}_{H,svt}$, $\hat{c}^{\ominus}_{P,H,svt}$, $\hat{v}^{\ominus}_{H,svt}$, $\hat{k}^{\ominus}_{T,H,svt}$ and $\hat{a}^{\ominus}_{P,H,svt}$). For all the above quantities, the solvent subscript will sometimes be omitted when the identity of the solvent is clear from the context. Considering Eqs. 2.10, 2.12 and 2.20, the corresponding real and intrinsic quantities are connected by the relationships

$$\hat{\mathcal{Y}}^{\ominus}_{H,svt} = \mathcal{Y}^{\ominus}_{H,svt} + F\hat{O}_Y\chi^{\ominus}_{svt} \tag{2.28}$$

for $Y = H$ or S with $\mathcal{Y} = \mathcal{H}$ or $\mathcal{S}$, respectively, and

$$\hat{y}^{\ominus}_{H,svt} = y^{\ominus}_{H,svt} + F\hat{O}_Y\chi^{\ominus}_{svt} \tag{2.29}$$

for $Y = S$, C_P, V, K_T or A_P, with $y = s$, c_P, v, k_T or a_P, respectively. Here again, Eqs. 2.28 and 2.29 are formulated in terms of the standard variant, similar equations holding for the density-corrected variant upon replacement of $\mathcal{Y}^{\ominus}$ and y° by the corresponding quantities $\mathcal{Y}^{\ominus*}$ and $y^{\ominus*}$ appropriate for this variant. In the special case of the proton, Eq. 2.22 can be written

$$y^{\boxminus}_{H,svt} = y^{\ominus}[\mathrm{H}^+_g] + \mathcal{Y}^{\boxminus}_{H,svt} \quad , \tag{2.30}$$

and the intrinsic point-to-point quantities $\mathcal{Y}^{\boxminus}_{H,svt}$ and $y^{\boxminus}_{H,svt}$ can be standardized to the two distinct variants by addition of either $\Delta_s Y^{\ominus}_{std}$ or $\Delta_s Y^{\ominus*}_{std}$ in analogy with Eqs. 2.17 and 2.18.

Another point of view, however, is to regard the *air-liquid interfacial potential* $\chi^{\ominus}_{svt}$ *of a given solvent* as the experimentally-inaccessible quantity, including its functional dependence on pressure and temperature. If this quantity was known, real single-ion solvation free energies could be immediately converted to intrinsic values (Eqs. 2.6 and 2.8), including the proton as a special case (Eq. 2.27), a similar consideration holding for all other intrinsic single-ion solvation parameters (Eqs. 2.10, 2.12, 2.19 and 2.20), including the proton as a special case (Eqs. 2.28 and 2.29).

As will be seen in Chapter 4, (Sections 4.3.6 and 4.3.7), a third point of view is to regard the *intrinsic absolute potential* $\mathcal{V}^{\ominus}_{H,svt}$ *of the reference hydrogen electrode in a given solvent* as the experimentally-inaccessible quantity, including its functional dependence on pressure and temperature. This quantity turns out to be directly connected to $\mathcal{G}^{\ominus}_{H,svt}$ and $\chi^{\ominus}_{svt}$ *via* (Eqs. 4.315 and 4.317)

$$\mathcal{G}^{\ominus}_{H,svt} = F\mathcal{V}^{\ominus}_{H,svt} - \Delta_f G^{\ominus}[\mathrm{H}^+_g] \quad , \tag{2.31}$$

and

$$\chi^{\ominus}_{svt} = -\mathcal{V}^{\ominus}_{H,svt} + F^{-1}\left(\hat{\mathcal{G}}^{\ominus}_{H,svt} + \Delta_f G^{\ominus}[\mathrm{H}^+_g]\right) \quad , \tag{2.32}$$

where $\Delta_f G^{\ominus}[\mathrm{H}^+_g]$ is the formation free energy of the gas-phase proton. Here again, the intrinsic absolute potential $\mathcal{V}^{\ominus}_{H,svt}$ is to be clearly distinguished from the corresponding real quantity $\hat{\mathcal{V}}^{\ominus}_{H,svt}$, which encompasses the interfacial potential $\chi^{\ominus}_{svt}$ and is experimentally accessible, namely (Eq. 4.317)

$$\hat{\mathcal{V}}^{\ominus}_{H,svt} = \mathcal{V}^{\ominus}_{H,svt} + \chi^{\ominus}_{svt} \quad , \tag{2.33}$$

from which it follows *via* Eqs. 2.31 and 2.32

$$\hat{\mathcal{G}}^{\ominus}_{H,svt} = F\hat{\mathcal{V}}^{\ominus}_{H,svt} - \Delta_f G^{\ominus}[\mathrm{H}^+_g] \quad . \tag{2.34}$$

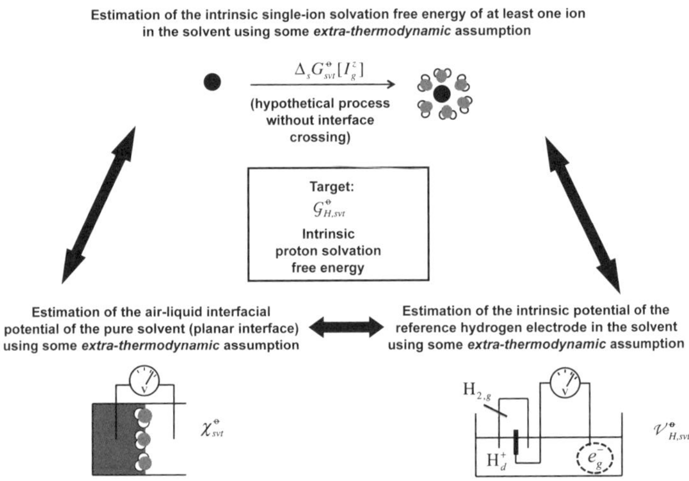

Figure 2.3 *Experimentally-elusive quantities required for the evaluation of intrinsic single-ion solvation free energies.* These quantities are, for a given solvent, the intrinsic solvation free energy $\Delta_s G^{\ominus}_{svt}[I^z_g]$ of at least one ion I^z_g in the solvent, the air-liquid interfacial potential $\chi^{\ominus}_{svt}$ of the pure solvent, or the intrinsic potential $\mathcal{V}^{\ominus}_{H,svt}$ of the reference hydrogen electrode in the solvent. These three properties are directly related to the intrinsic solvation free energy $\mathcal{G}^{\ominus}_{H,svt}$ of the proton in the solvent, commonly taken as the central (target) elusive parameter. The determination of any of these quantities is impossible on the sole basis of experimental data and requires the introduction of an extra-thermodynamic assumption. Knowledge of the pressure and temperature dependence of any of these quantities would also give access to all other intrinsic single-ion solvation parameters, as well as to corresponding intrinsic single-ion formation parameters and partial molar variables in solution.

Eqs. 2.31 and 2.34 can be used to formulate a simple and intuitive definition of $\hat{\mathcal{V}}^{\ominus}_{H,svt}$ and $\mathcal{V}^{\ominus}_{H,svt}$. These quantities represent the real and intrinsic formation free energies, respectively, of the dissolved proton in the given solvent, divided by F.

In the literature, the adjective "absolute" is commonly employed to refer to single-ion or single-electrode quantities, often without clearly specifying whether these quantities are intrinsic or real. In the present text, this ambiguity is avoided by the systematic use of the adjectives "intrinsic" or "real", instead of (or in addition to) "absolute", and of "no-hat" *vs.* "hat" symbols, respectively, for the two types of quantities. The adjective "absolute" is nevertheless retained, in addition to "intrinsic" or "real", for single-electrode potentials to avoid a possible confusion with corresponding relative (redox) potentials, measured against the reference hydrogen electrode (Section 4.4.3).

The relationship between the equivalent problems of determining $\Delta_s G^{\ominus}_{svt}[I^z_g]$ for an arbitrary ion I^z in a given solvent, or simply $\mathcal{G}^{\ominus}_{H,svt}$, $\chi^{\ominus}_{svt}$ or $\mathcal{V}^{\ominus}_{H,svt}$ for the given solvent, is illustrated schematically in Figure 2.3, considering $\mathcal{G}^{\ominus}_{H,svt}$ as the central target property. The fact that these interrelated quantities are inaccessible experimentally has not prevented scientists to make numerous attempts at their evaluation over more than one century[69–84]. This is because there still exists an entrance point to access these quantities, namely the introduction of an *extra-thermodynamic assumption*. Such an assumption consists of a hypothesis, typically concerning: (*i*) the partitioning of a set of salt parameters into single-ion intrinsic values, guided by intuition, by a theoretical analysis, or by experiments probing observables assumed correlated with single-ion properties, and leading to an estimate for $\mathcal{G}^{\ominus}_{H,svt}$ (or for the intrinsic proton value of some derivative thermodynamic parameter); (*ii*) the expected hypothetical limiting situation of a given experiment aiming at the determination of $\chi^{\ominus}_{svt}$ (or of the corresponding pressure or temperature dependence); (*iii*) the partitioning of bulk and surface effects in a theoretical model assumed to account appropriately for the solvation physics of an ion or the surface physics of a solvent. To our knowledge, no experimental approach, and associated extra-thermodynamic assumption, has targeted the direct determination of $\mathcal{V}^{\ominus}_{H,svt}$ to date. This statement is based in particular on a careful analysis of the experimental literature concerning "absolute" electrode potentials, which revealed that all existing measurements actually concern the real value $\hat{\mathcal{V}}^{\ominus}_{H,svt}$ (see, however, Refs.[482,483]; Section 5.14).

Since the quantities $\mathcal{G}^{\ominus}_{H,svt}$, $\chi^{\ominus}_{svt}$ and $\mathcal{V}^{\ominus}_{H,svt}$ are experimentally inaccessible, yet *a priori* well defined and amenable to a determination given the introduction of an extra-thermodynamic assumption, they will be referred to here as *experimentally elusive*, rather than indeterminable. A surprisingly large number of alternative extra-thermodynamic assumptions have been proposed over the years, many of which will be presented and discussed in Chapter 5. Their common point is that they always sound intuitively reasonable, at least to their author, but can be neither proved nor disproved on the sole basis of experimental data. Note that although the expression is usually employed referring to a heuristic assumption concerning the partitioning of pair properties or some limiting experimental situation (see above), theory can also be viewed as a form of extra-thermodynamic assumption, justified by an approximate physical representation of the microscopic world. The problems inherent to theoretical approaches will be discussed in Chapter 3.

2.3 Fundamental experimental problems: summary

The general conclusions from the two previous sections regarding the fundamental problems involved in the experimental determination of single-ion solvation properties are summarized in Figure 2.4, taking an alkali-halide ion pair as example. The most relevant observations, formulated in terms of the free energy for simplicity, are the following.

The real single-ion solvation free energies, $\Delta_s\hat{G}^\ominus_{svt}[M_g^+]$ for the alkali cation and $\Delta_s\hat{G}^\ominus_{svt}[X_g^-]$ for the halide anion in a given solvent, can be obtained from experimental data. Their determination relies on a thermodynamic cycle involving equilibrium (atomization parameters, Voltaic cell experiment) and spectroscopic (ionization parameters, metal work function) measurements. The real potential $\hat{\mathcal{V}}^\ominus_{H,svt}$ of the reference hydrogen electrode in the given solvent is also accessible through these measurements, as well as the cor-

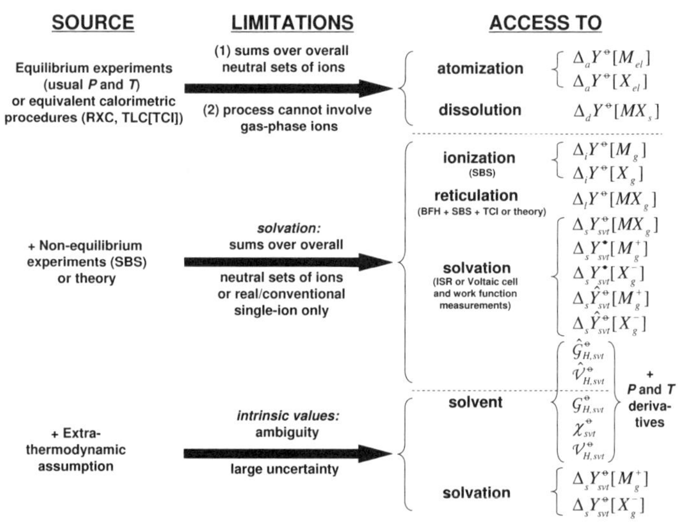

Figure 2.4 *Fundamental experimental problems in the determination of thermodynamic parameters for processes involving ions.* The limitations involved in equilibrium experiments performed at usual temperatures and pressures, or equivalent calorimetric procedures, are detailed in Section 2.1. The use of non-equilibrium experiments (spectroscopy) or theory relaxes some of these limitations, as also detailed in Section 2.1. Finally, the determination of some quantities is only possible using extra-thermodynamic assumptions, as detailed in Section 2.2. The acronyms and notations employed are defined in Table I (see also Figure 4.1).

responding real proton solvation free energy $\hat{\mathcal{G}}^{\ominus}_{H,svt}$. The ion-pair solvation free energy $\Delta_s G^{\ominus}_{svt}[MX_g]$, for which the distinction between real and intrinsic values becomes irrelevant, can also be obtained from experimental data. The determination of this sum can be performed in two different ways: (*i*) by summing up the two real single-ion solvation free energies $\Delta_s \hat{G}^{\ominus}_{svt}[M^+_g]$ and $\Delta_s \hat{G}^{\ominus}_{svt}[X^-_g]$ obtained above; or (*ii*) using a thermodynamic cycle (ISR), involving equilibrium (atomization parameters, dissolved ion-pair formation parameters) and spectroscopic (ionization parameters) measurements. The quantities $\Delta_s G^{\ominus}_{svt}[MX_g]$, $\Delta_s \hat{G}^{\ominus}_{svt}[M^+_g]$, $\Delta_s \hat{G}^{\ominus}_{svt}[X^-_g]$, $\hat{\mathcal{V}}^{\ominus}_{H,svt}$ and $\hat{\mathcal{G}}^{\ominus}_{H,svt}$ are of course affected by experimental errors, including those arising from possible ambiguities in the choice of standard states. However, these errors remain relatively limited in practice, and the results reported by different investigators, possibly using different experimental approaches, are essentially consistent with each other. Similar considerations apply in principle to the corresponding real derivative thermodynamic solvation parameters of the proton and partial molar properties of the solvated proton (*e.g.* $\hat{\mathcal{H}}^{\ominus}_{H,svt}$, $\hat{\mathcal{S}}^{\ominus}_{H,svt}$, $\hat{s}^{\ominus}_{H,svt}$, $\hat{c}^{\ominus}_{P,H,svt}$, $\hat{v}^{\ominus}_{H,svt}$, $\hat{k}^{\ominus}_{T,H,svt}$ and $\hat{a}^{\ominus}_{P,H,svt}$), which are related to the derivatives of $\hat{\mathcal{G}}^{\ominus}_{H,svt}$ with respect to temperature and pressure. Available experimental data concerning the above quantities in the context of hydration will be presented and discussed in Chapter 5 (Sections 5.12 and 5.14-5.16), including the suggestion of corresponding recommended values (Section 5.19).

In contrast, the corresponding intrinsic single-ion solvation free energies, $\Delta_s G^{\ominus}_{svt}[M^+_g]$ for the alkali cation and $\Delta_s G^{\ominus}_{svt}[X^-_g]$ for the halide anion in the given solvent, cannot be obtained on the sole basis of experimental data. Their evaluation, *i.e.* the partitioning of $\Delta_s G^{\ominus}_{svt}[MX_g]$ into two intrinsic rather than two real contributions, always requires the formulation of some experimentally unprovable extra-thermodynamic assumption. The use of such an assumption ultimately permits to fix the values of the intrinsic proton solvation free energy $\mathcal{G}^{\ominus}_{H,svt}$, of the air-liquid interfacial potential $\chi^{\ominus}_{svt}$, and of the intrinsic potential of the reference hydrogen electrode $\mathcal{V}^{\ominus}_{H,svt}$ for the given solvent. Unfortunately, the use of distinct types of experiments and extra-thermodynamic assumptions typically leads to very different results in terms of these quantities. At present, for water, the uncertainty range in the value of $\mathcal{G}^{\ominus}_{H,svt}$ (and thus, in the derived values of intrinsic single-ion solvation free energies for monovalent ions) is probably as broad as $50 - 100$ kJ·mol^{-1}, the corresponding uncertainty range in the values of $\chi^{\ominus}_{svt}$ and $\mathcal{V}^{\ominus}_{H,svt}$ being on the order of $0.5 - 1.0$ V. Very large uncertainty ranges are also associated with corresponding intrinsic derivative thermodynamic parameters of the proton and partial molar properties of the hydrated proton (*e.g.* $\mathcal{G}^{\ominus}_{H,svt}$, $\mathcal{H}^{\ominus}_{H,svt}$, $\mathcal{S}^{\ominus}_{H,svt}$, $s^{\ominus}_{H,svt}$, $c^{\ominus}_{P,H,svt}$, $v^{\ominus}_{H,svt}$, $k^{\ominus}_{T,H,svt}$ and $a^{\ominus}_{P,H,svt}$), which are related to the derivatives of $\mathcal{G}^{\ominus}_{H,svt}$ with respect to pressure and temperature. Experimentally-derived data (bound to specific extra-thermodynamic assumptions) concerning the above quantities in the context of hydration will be presented and discussed in Chapter 5 (Sections 5.17 and 5.18), including the suggestion of corresponding recommended values (Sections 5.19 and 5.20).

If the quantities $\mathcal{G}^{\ominus}_{H,svt}$, $\chi^{\ominus}_{svt}$ and $\mathcal{V}^{\ominus}_{H,svt}$ (and thus, intrinsic single-ion solvation free energies) are experimentally elusive, one is entitled to wonder if there is any interest at all in trying to evaluate them. Obviously, if these values fail to appear in the outcome of any experimental measurement, they possess no practical usefulness or predictive power. In other words, a description of ion thermodynamics and electrochemistry relying on a real or relative scale, *e.g.* a scale where one of these three quantities is arbitrarily set to zero, is sufficient to account for the entire body of experimental results. Yet, a thorough (though surely not exhaustive) scanning of the scientific literature reveals that about 160 articles have directly or indirectly aimed over the past 90 years at the determination of one of these quantities using some extra-thermodynamic assumption. The main reasons for which this (arguably practically irrelevant) problem has attracted so much attention are probably the following.

First, the quantities $\mathcal{G}^{\ominus}_{H,svt}$, $\chi^{\ominus}_{svt}$ and $\mathcal{V}^{\ominus}_{H,svt}$ (and thus, intrinsic single-ion solvation free energies) appear at first sight to be physically well defined. If the time-dependent Schrödinger equation could be integrated numerically without approximations, for a macroscopic system and on a long enough timescale, which will not be computationally feasible in any foreseeable future, the situation of the unphysical solvation process (Figure 2.2) could be realized exactly, and these quantities could in principle be determined from first principles. For this reason, their determination seems to represent a considerable challenge, but not a hopeless quest (an example of a hopeless quest that has also in its time attracted much attention is the design of a *perpetuum mobile* violating the first law of thermodynamics). Note that in spite of its apparent solidity, this argument is not entirely bulletproof, as will be discussed in Chapter 7.

Second, the description of ion thermodynamics and electrochemistry based on a real (intermingling bulk and surface effects) or relative (rather than absolute) scale, although practically satisfactory, is intellectually frustrating if one thinks that an absolute scale is in principle accessible. In addition, the choice of an anchoring point is subjective, different choices presenting specific advantages in specific contexts, so that a number of different scales have been introduced (*e.g.* real, and different variants of conventional scales, namely standard, gaseous-proton, aqueous-proton and redox conventional scales; Sections 4.5.1 and 4.5.2). As a result, possible ambiguities in the scale definitions, as well as the need for interconversions, complicate the field unnecessarily.

Finally, intrinsic solvation free energies have a higher "theoretical value" compared to real ones. Here, higher theoretical value is not meant in the sense of increased predictive power. The ability of an intrinsic scale to account for experimentally accessible quantities would be just the same as that of the real scale or of any arbitrary relative scale. It is meant in the sense of being more insightful, *i.e.* amenable to intuitive reasoning and direct comparison with theoretical models. The reason is that real values simultaneously characterize two physically very distinct phenomena: solvation forces within the bulk and polarization at the air-liquid interface. In contrast, intrinsic values specifically characterize the former phenomenon, as is the case for intuitive reasoning and theoretical calculations (Chapter 3). For instance, one of the most powerful (yet simplistic, *i.e.* qualitative) models to understand the key (electrostatic) component of ionic solvation is the Born model (Section 3.2). This model suggests in particular that anions and cations of identical sizes have equal solvation free energies, and that at identical sizes, an ion should be four times more strongly solvated when its charge is doubled. These predictions, which turn out to be quantitatively inaccurate in the case of hydration, can only be assessed in terms of intrinsic single-ion solvation free energies, but not in terms of real or relative ones. A similar problem arises in classical atomistic simulations (Section 3.3) where a parameterization of the ion-solvent van der Waals interactions against experimental single-ion solvation free energies (Section 6.2.2) requires an estimate for $\mathcal{G}^{\ominus}_{H,svt}$. The quality of the resulting model, *i.e.* its ability to reproduce experimentally accessible properties of ionic systems, will then crucially depend on the accuracy of the estimate employed.

At this point, it is worth stating four important and still controversial questions regarding the determination of the above experimentally-elusive quantities, namely whether: (*i*) these quantities truly escape any purely experimental (*i.e.* assumption-free) determination (or whether this restriction only results from limitations in currently available experimental techniques that could possibly be lifted in a far future); (*ii*) this restriction really also applies to the derivatives of these quantities with respect to intensive system parameters (*e.g.* pressure, temperature or composition); (*iii*) this restriction really also applies to the differences between these quantities for different solvents (or whether the knowledge of the values for one reference solvent, *e.g.* water, would be sufficient to deduce corresponding values for any other solvent solely based on experimental data); (*iv*) these experimentally-elusive quantities are also theoretically elusive (or whether they could in principle be calculated based on a first-principles approach in an assumption-free fashion, even if only in a far future). These questions are of fundamental importance in the present context, yet surprisingly difficult to answer appropriately. Elements of answers will be provided throughout this book, and these questions will be reconsidered at the very end of the text (Chapter 7).

A single ion has lost itself in the surprising complexity of this chapter and feels dramatically dehydrated. Is is also starting to wonder whether the vision ahead is the oasis of single-ion intrinsic properties, or whether it is a mere mirage.

Chapter 3
Fundamental theoretical problems

This chapter is concerned with the problems inherent to the *theoretical determination* of single-ion solvation free energies, as well as of corresponding derivative thermodynamic solvation parameters. Considering the two alternative hypothetical processes describing the phenomenon of ionic solvation (Figure 2.2), the experimental situation is best discussed with the physical process in mind (Section 2.2). In contrast, the calculation of single-ion solvation free energies *via* theoretical methods is typically more conveniently performed according to the *unphysical process*. The reason is that describing the transfer of a finite-sized ion through an interface with a sufficient accuracy is far more complicated, *i.e.* would require much more complex analytical models or the sampling of a much larger number of solvent configurations compared to the processes of growing a cavity within the bulk solvent and transferring infinitesimal charges from vacuum to this cavity. There is, however, a second advantage. In the physical process, bulk solvation and surface effects are intermingled, *i.e.* they are both simultaneously active in the second step and cannot be unambiguously separated. This is in principle no longer the case in the unphysical process. In the second step of the latter process, infinitesimal charges are progressively transferred from the ion in vacuum to the ion-sized cavity within the bulk. The corresponding infinitesimal work, at a given charge state of the ion, consists of a *solvent contribution* and a *self-energy contribution*. The former is immediately related to the solvent-generated electric potential in the cavity relative to a zero-point in vacuum, and includes bulk and surface effects. The latter is related to the interaction of the transferred infinitesimal charge with the charge density still present in the vacuum ion and the charge density already present in the cavity. However, the specific contribution of the surface molecules to the solvent term can in principle be separated out or omitted in the calculation. Given such a partitioning, an intrinsic value can be obtained by integrating the sum of

RSC Theoretical and Computational Chemistry Series No. 3
Single-Ion Solvation
Experimental and Theoretical Approaches to Elusive Thermodynamic Quantities
By Philippe Hünenberger and Maria Reif

Published by the Royal Society of Chemistry, www.rsc.org

the bulk-solvent and self-energy contributions to the reversible work of charge transfer. Indeed, the value that is sought after in the first place in a theoretical calculation is the intrinsic (rather than real) solvation free energy, as well as corresponding derivative thermodynamic solvation parameters, because this number characterizes exclusively the physics of bulk solvation, without contamination from the pure solvent interfacial properties (Section 2.2).

At this point, however, a very important proviso should be kept in mind. In the above discussion, it has been assumed that contributions of bulk and surface solvent molecules to the reversible work of charging can be unambiguously separated. In practice, this implies that such an unambiguous partitioning is achievable in terms of the solvent-generated electric potential. If this statement is correct, theoretical methods permit the estimation of intrinsic single-ion solvation parameters with an accuracy that is solely limited by the approximate nature of the underlying molecular model. There is then hope that a "perfect" first-principles model ultimately leads to the assumption-free evaluation of these parameters. In contrast, if the above partitioning between bulk and surface components to the solvent-generated electric potential is ambiguous, the choice of a specific partitioning scheme must be introduced as an additional assumption that will not disappear even in the case of a "perfect" first-principles model. This relatively intricate issue will be encountered first in Sections 3.3.3-3.3.5, in the form of puzzling inconsistencies affecting the results of theoretical calculations. The cause for these inconsistencies will be analyzed in more detail in Section 4.3.8. Finally, the point of view of the authors on this question, namely that a specific definition for this partitioning is indispensable and represents the root of all extra-thermodynamic assumptions, will be presented in Chapter 7.

Compared to experiment, theoretical methods require the specification of additional information concerning the ion-solvent system that defines the physical nature of the underlying model. For example, one generally needs to formulate a (classical or quantum-mechanical) representation for: (*i*) the charge distribution within the ion in vacuum, *e.g.* point charge, homogeneous charge density, or nuclear point charge plus electronic charge density; (*ii*) the corresponding charge distribution within the ion in solution; (*iii*) the non-polar component of solute-solvent interactions that determines, in balance with electrostatic interactions, the effective ion size and solvation structure in solution, *e.g.* hard wall (rigid solute-solvent boundary implying constraint forces), classical (soft) interactions (*e.g.* van der Waals repulsion and dispersion forces), or quantum-mechanical interactions (electronic exchange and correlation).

3.1 Types of theoretical models

The basic choices defining any model for the theoretical investigation of a molecular system are illustrated in Figure 3.1. There are four basic choices[484–542]:

(A$_1$) The choice of a *model resolution*[484–489], *i.e.* of the degrees of freedom or "elementary particles" of the model, *e.g.* subatomic particles, individual atoms, coarse-grained beads representing atom groups, or quantities associated with volume elements tiling a continuous medium.

(B$_1$) The choice of a representation for the *interaction*[484–531] between these degrees of freedom or "elementary particles", *e.g.* quantum-mechanical Hamiltonian operator for subatomic particles, classical potential-energy function for atoms, effective interaction function for beads, or operators involved in conservation, response or transport equations for continuous media.

(C$_1$) The choice of a method for the *generation of configurations*[496,532–541] in the space accessible to the these degrees of freedom or "elementary particles". Except for the simplest analytical models, the number of generated configurations for a given system is in general dictated by a tradeoff between the cost of the interaction evaluation and the available computing resources. One may distinguish the following categories of methods to generate configurations: (*i*) *searching*, *i.e.* the generation of a configurational *set* of local minimum-energy configurations with the broadest possible distribution in configuration space; (*ii*) *sampling*, *i.e.* the generation of a configurational *ensemble* obeying a well-defined probability distribution in terms of energy, so that thermodynamic information can be evaluated; (*iii*) *simulating*, *i.e.* the generation of a configurational *trajectory* obeying a well-defined probability distribution in terms of energy and in a sequence dictated by a physically-motivated equation of motion, so that thermodynamic as well as dynamic information can be

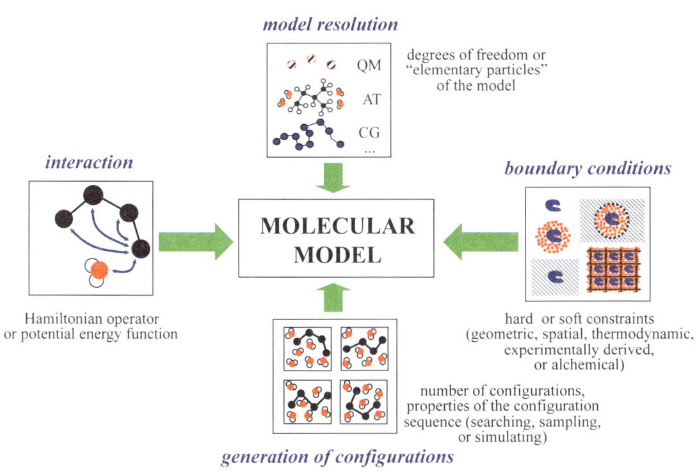

Figure 3.1 *Four basic choices defining a molecular model.* The acronyms and notations employed are defined in Table I (CG stands here for coarse-grained model).

evaluated. Common methods include *e.g.* simulations using quantum-mechanical equations of motion for subatomic particles[487,543,544], classical equations of motion for atoms or beads (molecular dynamics[487,490,492]; MD), and conservation, response or transport equations for continuous media[487,545], as well as searching or sampling approaches using other "unphysical" deterministic or stochastic algorithms (*e.g.* Monte Carlo sampling[504,546–552]; MC).

(D_1) The choice of *boundary conditions*[542], namely *hard* (enforced in all configurations) or *soft* (enforced in terms of configurational averages) constraints imposed on the generated system configurations. One may distinguish the following types of boundary conditions: (*i*) *geometric*, corresponding to rigid generalized coordinates, *e.g.* bond-length[553–559] or artificial[560–563] constraints (typically hard); (*ii*) *spatial*[490,492,564–566], *e.g.* finite *vs.* infinite system, non-periodic *vs.* periodic system, system shape, size and confinement method for finite systems (typically hard); (*iii*) *thermodynamic*[490,492,542,567], *e.g.* constant energy *vs.* temperature for MD[568–576] or MC[577–579], volume *vs.* pressure for MD[567,571,580–584] or MC[585–591], number of particles *vs.* chemical potential for MD[592–600] or MC[355,592,600–603] (in the context of microscopic systems, should be hard for independent extensive variables and soft for independent intensive variables); (*iv*) *experimentally-derived*[604,605], used to enforce agreement with measured observables, *e.g.* structure factors from X-ray crystallography[606–610] or interatomic distances and J-coupling constants from NMR measurements[606,609,611–618] (should be soft); (*v*) *alchemical*, concerning the coupling parameters defining an unphysical superposition of physical states in free-energy calculations[619–627] (hard, as in thermodynamic integration[628] and free-energy perturbation[629], or soft, as in λ-dynamics[630,631] and, from a certain point of view, enveloping distribution sampling[632–635] or Hamiltonian replica exchange[636–642]).

As illustrated in Figure 3.2, three main types of theoretical models, encompassing different combinations of the basic choices of Points A_1-D_1 can be applied to the calculation of single-ion solvation free energies, as well as of corresponding derivative thermodynamic solvation parameters. These three main types of models are the following:

(A_2) *Continuum-electrostatics calculations* (CE) are analytical (*e.g.* Born plus surface tension) or numerical (inclusion of various types of corrections) models relying on an implicit description of the solvent, typically as a continuous dielectric medium with linear response, and a hard-sphere description of the ion. The basic parameters involved in such a calculation are the integer ionic charge z and an effective ionic radius R_I, along with the static relative dielectric permittivity ϵ_{svt} and macroscopic surface tension coefficient γ_{svt} (or its microscopic analog $\tilde{\gamma}_{svt}$) of the pure solvent. The sampling over solvent configurations is implicit, the electrostatic interactions are normally exactly Coulombic, and the spatial boundary conditions correspond to those of a macroscopic non-periodic system (of formally infinite extent).

(B_2) *Classical atomistic simulations* (AT) are numerical (computational) models typically relying on an all-atom description of the ion-solvent system, possibly along with CE approximations or corrections for long-range effects. The basic parameters involved in such a calculation are the integer ionic charge z and the van der Waals interaction parameters for ion-solvent interactions (*e.g.* Lennard-Jones interaction parameters C_6^{is} and C_{12}^{is}), along with the parameters of a solvent model (geometry, atomic charges, van der Waals interaction parameters for solvent-solvent interactions). In these models, the sampling over solvent configurations must be carried out explicitly (*e.g. via* MD or MC; in the former case, atomic masses are also required as parameters), the electrostatic interactions may differ from Coulombic (*e.g.* cutoff-based schemes), and the spatial boundary conditions correspond to those of a microscopic non-periodic or periodic system (finite extent, possibly replicated to form a pseudo-crystal). These methods go beyond the continuous-medium and linear-response approximations typically used in CE (solvent microscopic description and non-linear response effects

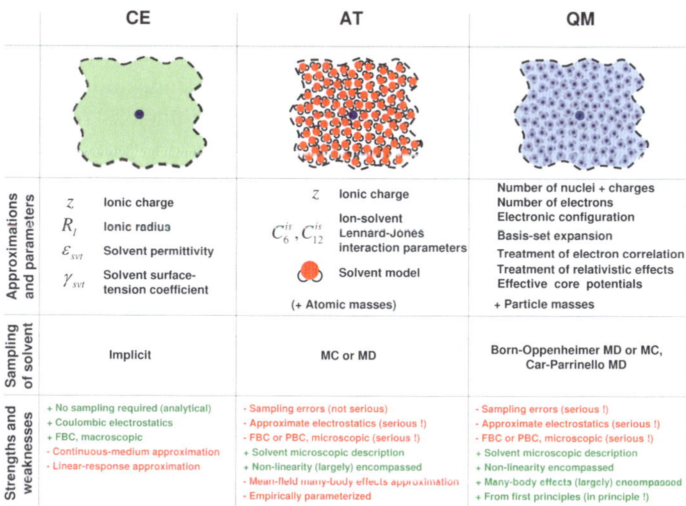

Figure 3.2 *Three main types of theoretical models applicable to the evaluation of single-ion solvation free energies, as well as derivative thermodynamic solvation parameters.* These models are summarized in terms of the involved approximations, parameters, solvent sampling method, strengths and weaknesses. The types of models are continuum-electrostatics calculations (CE), classical atomistic simulations (AT) and quantum-mechanical computations (QM). Note that the application of AT and QM approaches to the simulation of a pure-solvent system can also be used to evaluate the air-liquid interfacial potential of this solvent. The acronyms and notations employed are defined in Table I.

are largely encompassed), but at the cost of introducing errors related to finite sampling (nowadays limited), approximate electrostatics (in most cases), and finite (microscopic) system size (always). AT methods applied to pure liquid samples may also be used to evaluate the air-liquid interfacial potential. Note that the use of AT approaches has been preceded historically by the design of simple molecular-mechanics (plus continuum-electrostatics) models[643–646]. These approaches, at the time insight-providing, are rather approximate and will not be discussed in this book.

(C_2) *Quantum-mechanical computations* (QM; calculations or simulations) are numerical (computational) models relying on a quantum-mechanical description of the ion-solvent system, involving nuclei and electrons, or pseudo-nuclei and valence electrons, possibly along with AT or/and CE approximations or corrections for long-range effects. Most of these models rely on the Born-Oppenheimer approximation[647] (separability of the nuclear and electronic Schrödinger equations) and on a classical treatment of the nuclear motions. In principle, when all electrons, including core electrons, are treated explicitly, the physical parameters involved in such a calculation are restricted to nuclear charges, number of electrons and electronic configuration, along with particle masses and physical constants. However, further numerical parameters are introduced in practice by a number of unavoidable approximations concerning the expansion of the wavefunction or electron density onto a finite basis set, the approximate treatment of electron correlation, and the possible introduction of effective core potentials (pseudo-nucleus representing the nucleus plus the core electrons as a single interaction site). In these models, the sampling over solvent configurations must be carried out explicitly[487,490,544,648,649] (*e.g. via* Born-Oppenheimer MC or MD, or *via* Car-Parrinello MD), the electrostatic interactions may formally differ from Coulombic (in the quantum-mechanical sense, *i.e.* at the level of the nuclei and electrons; *e.g.* approximate treatment of electron correlation, use of effective core potentials) and the spatial boundary conditions correspond to those of a microscopic non-periodic or periodic system (finite extent; possibly periodically replicated to form a pseudo-crystal). These methods go beyond the AT description by encompassing quantum and many-body effects (*e.g.* electronic polarization; typically only included in a mean-field fashion in the AT description), but at the cost of introducing potentially larger errors caused by finite sampling and finite basis-set expansion. Furthermore, the problems related to approximate electrostatics (partial neglect of electron correlation, use of core potentials) and finite (microscopic) system size remain at least as serious as in the AT case. Due to their very high inherent computational cost, QM approaches are often hybridized with classical simulations, as in the QM-MM scheme[649,650], where the ion and its first solvation shells are treated quantum-mechanically, while the solvent molecules at larger distances are treated classically. QM methods applied to pure liquid samples may also be used to evaluate the air-liquid interfacial potential.

Just as was the case for their experimental determination (Section 2), the theoretical evaluation of single-ion solvation free energies, as well as of all corresponding derivative thermodynamic solvation parameters, is complicated by a number of fundamental problems, which depend on the type of model employed. The problems associated with the types of approaches of Points A_2-C_2 are discussed in sequence in Sections 3.2-3.4 and summarized in Section 3.5.

For completeness, it should also be mentioned that the ambiguity in the thermodynamic conventions regarding the definition of *standard states*, *i.e.* the existence of alternative conventions and the absent or incomplete specification in many theoretical studies of the convention employed, also represents a serious (although not formally "fundamental") problem in the theoretical literature and a significant factor of "uncertainty" in practice, just as it was the case for experimental studies (Chapter 2). The discussion of this problem is deferred to Chapter 4. The description of theoretical approaches as used in practice and the presentation of corresponding illustrative results will be the concern of Chapter 6.

3.2 Continuum-electrostatics calculations

The basic model for the CE analysis of single-ion solvation is the *Born model*[84] (BO). In this model, the solvated ion I^z is represented as a rigid non-polarizable (internal relative dielectric permittivity of one) sphere of effective radius R_I bearing a spherically-symmetric charge distribution of magnitude ze, and embedded in a medium of infinite extent and of homogeneous static relative dielectric permittivity ϵ_{svt} equal to that of the pure solvent. It is also assumed that the spherically-symmetric charge distribution within the solvated ion is identical to the corresponding distribution within the isolated ion in vacuum. For simplicity, this charge distribution can be represented by a single point charge of magnitude ze at the ion center, the final result being independent of a specific choice.

Continuum electrostatics results from the application of Maxwell's equations[651–653] for classical electrodynamics in continuous media[421], restricted to the case of zero electric current and magnetic field (electrostatic limit). In this situation, these equations are conveniently combined into *Poisson's equation*[654]

$$\nabla \cdot [\epsilon(\mathbf{x}) \nabla \Phi(\mathbf{x})] = -\epsilon_o^{-1} \tilde{\rho}_F(\mathbf{x}) \ , \tag{3.1}$$

where $\tilde{\rho}_F(\mathbf{x})$ is the free charge density ($\mathbf{x}$ denoting position in space), $\Phi(\mathbf{x})$ the (scalar) electrostatic potential and $\epsilon(\mathbf{x})$ the (local) relative dielectric permittivity. Note that unlike Maxwell's equations in free space, Poisson's equation is an approximate form where the contribution of the bound charges to the charge density is omitted from $\tilde{\rho}_F$ and treated implicitly within ϵ. This amounts to a *continuous-medium* along with a *linear-response* approximation, which are in principle only valid at low fields and on a semi-macroscopic scale. Given the indicated problem geometry (spherical cavity of radius R_I, permittivity of one inside the ion, permittivity ϵ_{svt} outside the ion, free charge density equivalent to a point charge q at the cavity center), and setting the zero of the potential at infinity, the solution of Eq. 3.1 reads

$$\Phi(x;q) = \frac{q}{4\pi\epsilon_o}\left[\frac{1}{\epsilon_{svt}x}h(x-R_I) + \left(\frac{1}{x} - \frac{1}{R_I} + \frac{1}{\epsilon_{svt}R_I}\right)h(R_I - x)\right] \quad , \tag{3.2}$$

where x is the distance from the cavity center and h the Heaviside step function, *i.e.* $h(x) = 0$ for $x \leq 0$ and $h(x) = 1$ for $x > 0$. The reaction-field potential Φ_{RF} at the cavity center, obtained by subtracting the vacuum contribution of the charge from this electric potential and setting $x = 0$, is thus

$$\Phi_{RF}(q) = -\frac{q}{4\pi\epsilon_o}\frac{\epsilon_{svt}-1}{\epsilon_{svt}}\frac{1}{R_I} \quad . \tag{3.3}$$

The corresponding (intrinsic; see below) *charging free energy* $\Delta_s G_{chg}[I_g^z]$ may be calculated by integrating Φ_{RF} from $q = 0$ to ze and expressing the result on a per mole basis, leading to the *Born equation*

$$\Delta_s G_{chg}^{BO}[I_g^z] = -\frac{N_A e^2 z^2}{8\pi\epsilon_o}\frac{\epsilon_{svt}-1}{\epsilon_{svt}}\frac{1}{R_I} \quad . \tag{3.4}$$

Note that the same result would be obtained considering any other spherically-symmetric charge distribution of magnitude ze for the isolated and solvated ion, provided that this distribution is entirely confined within the sphere representing the ion. Eq. 3.4 can equivalently be written as

$$\Delta_s G_{chg}^{BO}[I_g^z] = \frac{1}{2}N_A ez\Phi_{RF}(ze) = N_A ez\Phi_{RF}(ze) - \frac{1}{2}N_A ez\Phi_{RF}(ze) \quad . \tag{3.5}$$

The first term on the right-hand side represents the Coulombic interaction energy of the ion with the continuum solvent, while the second term accounts for a penalty incurred by the ion-induced perturbation of this continuum, *i.e.* the response to the ionic field causes an unfavorable alignment of the solvent dipoles within the continuum solvent. However, although $\Delta_s G_{chg}^{BO}[I_g^z]$ is clearly a free energy (reversible work of charging), it would be incorrect to interpret these two terms as an enthalpy and an entropy contribution (the correct enthalpy-entropy partitioning within the Born model is discussed further below). These terms actually represent pseudo-enthalpy and pseudo-entropy contributions in the sense of Refs.[655–659] (Section 3.3.3).

The quantity $\Delta_s G_{chg}[I_g^z]$, of which $\Delta_s G_{chg}^{BO}[I_g^z]$ is the estimate according to the BO model, corresponds to the second step of the unphysical solvation process (Figure 2.2), *i.e.* it accounts for the reversible work of infinitely slowly transferring the charge from the ion in vacuum to the empty ion-sized cavity within the bulk of the solvent. Note that this only holds under the assumption that the ionic charge distribution is identical in the free and solvated ion, in which case the self-energy of this distribution is identical in the two states and cancels out. Because the Born model does not account for the possible presence of a surface polarization at the air-liquid interface (Section 2.2), the single-ion solvation free-energy derived from this model will exclude the surface contribution $\Delta_s G_{srf}[I_g^z]$ to this free energy (Eqs. 2.4 and 2.5), and therefore represent an estimate for the intrinsic (rather than real) solvation free energy of the ion.

To obtain an estimate for the total intrinsic solvation free energy, the charging free energy must be increased by the *cavitation free energy*[259, 261, 660–668] $\Delta_s G_{cav}[I_g^z]$. A simple model for the evaluation of this term is the *surface tension* (ST) model[669], where the free energy associated with the formation of a microscopic cavity in a solvent is assumed to consist of a surface term, depending linearly on the surface area of this cavity *via* a microscopic (air-liquid) surface tension coefficient $\tilde{\gamma}_{svt}$ of the pure solvent, and a pressure-volume term, accounting for the work performed against the external pressure P, *i.e.*

$$\Delta_s G_{cav}^{ST}[I_g^z] = 4\pi N_A \left(\tilde{\gamma}_{svt} R_I^2 + \frac{1}{3} P R_I^3 \right) \quad . \tag{3.6}$$

The quantity $\Delta_s G_{cav}[I_g^z]$, of which $\Delta_s G_{cav}^{ST}[I_g^z]$ is the estimate according to the ST model, corresponds to the first step of the unphysical solvation process (Figure 2.2), *i.e.* it accounts for the reversible work of infinitely slowly growing an uncharged ion-sized cavity within the bulk of the solvent.

Finally, to obtain an estimate $\Delta_s G^{\ominus}[I_g^z]$ for the *standard intrinsic single-ion solvation free energy* at the reference temperature T^-, $\Delta_s G[I_g^z]$ must be evaluated using parameters appropriate for the reference liquid pressure P° and at temperature T^-, resulting in the corresponding semi-standard intrinsic point-to-point solvation free energy $\Delta_s G_{BO+ST}^{\boxminus}[I_g^z]$, and increased by an appropriate *standard-state correction term* $\Delta_s G_{std}^{\ominus}$ (Section 2.2, Eqs. 2.7 and 2.9). Combining Eqs. 2.7, 3.4 and 3.6 finally leads to a theoretical estimate (BO+ST model) for $\Delta_s G^{\ominus}[I_g^z]$, as

$$\begin{aligned} \Delta_s G_{BO+ST}^{\ominus}[I_g^z] &= \Delta_s G_{chg}^{BO}[I_g^z] + \Delta_s G_{cav}^{ST}[I_g^z] + \Delta_s G_{std}^{\ominus} \\ &= -\frac{N_A e^2 z^2}{8\pi\epsilon_o} \frac{\epsilon_{svt}^{\ominus} - 1}{\epsilon_{svt}^{\ominus}} \frac{1}{R_I^{\ominus}} \\ &\quad + 4\pi N_A \left[\tilde{\gamma}_{svt}^{\ominus} \left(R_I^{\ominus}\right)^2 + \frac{1}{3} P^\circ \left(R_I^{\ominus}\right)^3 \right] + \Delta_s G_{std}^{\ominus} \ , \end{aligned} \tag{3.7}$$

where $\epsilon_{svt}^{\ominus}$ and $\tilde{\gamma}_{svt}^{\ominus}$ are the static relative dielectric permittivity and microscopic surface tension coefficient of the pure solvent at P° and T^-, $R_I^{\ominus}$ denotes the effective radius of the ion in this solvent at P° and T^-, and $\Delta_s G_{std}^{\ominus}$ is given by Eq. 2.9 ($bbme_T$ standard-state convention; Section 4.2.4). For water, the value of the relative permittivity $\epsilon_{svt}^{\ominus}$ is 78.36 and that of the macroscopic (air-liquid) surface tension coefficient $\gamma^{\ominus}_{svt}$, which can be used as a first estimate for $\tilde{\gamma}^{\ominus}_{svt}$, is $71.99 \cdot 10^{-3}$ J·m^{-2} (Table 1.1). Finally, if a real solvation free energy $\Delta_s G^{\ominus}_{BO+ST}[I^z_g]$ is sought after, the intrinsic estimate $\Delta_s G^{\ominus}_{BO+ST}[I^z_g]$ of Eq. 3.7 must be further increased by the surface term $\Delta_s G^{\ominus}_{srf}[I^z_g]$ (Eqs. 2.6 and 2.8).

The charging term in Eq. 3.7 is always negative, while the cavitation term, in both its surface and pressure-volume components, as well as the standard-state correction term, are always positive. In the context of polar solvents (*e.g.* water), the charging term is typically by far the largest in magnitude. To give a simple numerical example, the different terms in the case of Na^+ hydration (assuming $R^{\ominus}_I = 0.17$ nm, $\tilde{\gamma}^{\ominus}_{wat} \approx \gamma^{\ominus}_{wat}$, and using data from Table 1.1) evaluate to -403.4 (charging), +15.7 (surface contribution to cavity formation), +0.001 (pressure-volume contribution to cavity formation) and +7.95 (standard-state correction) kJ·mol^{-1}, respectively, leading to an estimated solvation free energy of -379.7 kJ·mol^{-1}.

The BO+ST model as introduced above is restricted to monoatomic (spherical) ions. If polyatomic ions are to be included, one possibility is to consider an approximate non-spherical charge distribution[670–676] in a spherical cavity with an appropriate effective radius, or in a non-spherical cavity of simple (*e.g.* ellipsoidal[677,678] or spheroidal[672,675,679,680]) shape[494,514]. In this first case, additional reaction-field terms associated with the solvent dielectric response to the higher-order moments of the ionic charge distribution (*e.g.* Onsager model[671,681] for the dipolar contribution) must be included. In the second case, the CE analysis must be adapted to the specified (non-spherical) symmetry of the problem. Alternatively, the simple spherical, ellipsoidal or spheroidal cavity may be replaced by a non-spherical cavity adapted to the shape of the polyatomic ion, *e.g.* constructed as the union of spherical cavities associated with the individual atoms. More complex definitions based on the contact or solvent-accessible surface have also been formulated and applied to macromolecules[497,513,682–703] (see also the so-called generalized Born model[704–706]). Except for the simplest cavity shapes, the charging free energy generally cannot be given an analytical form analogous to Eq. 3.4 for the spherical cavity, and Poisson's equation (Eq. 3.1) has to be solved numerically, *e.g.* using a finite-difference approach[497,513,682–703].

The application of the BO+ST model for the evaluation of single-ion solvation free energies involves numerous *sources of error*:

(A_3) Neglect of the *solvent polarization at the ion-solvent interface*[707]: as stated above, the BO estimate for the charging free energy omits the contribution from the solvent polarization at the air-liquid interface, leading to a solvation free energy that is intrinsic. However, it also neglects the contribution from a possible polarization at the interface between the solvent and the ionic cavity. The results of AT simulations[707] (Section 6.2.1) suggest that: (*i*) such a polarization indeed exists and is responsible for a corresponding interfacial potential at the ion-solvent interface[707]; (*ii*) due to the high curvature of this interface, the interfacial potential is of relatively small (but non-negligible) magnitude[707] compared to the corresponding potential at a quasi-planar air-liquid interface; (*iii*) the magnitude of this interfacial potential is essentially independent of the charge state of the ion during the charging process[708]. Owing to the latter property, this potential already exists when the ion-sized cavity is uncharged, and leads to a corresponding solvation free-energy contribution that is essentially linear in the ionic charge. This effect also increases in magnitude with increasing ionic radius, due to the corresponding decrease in the interface curvature. When the solvent is water, the results of AT simulations[707] suggest that this contribution, which is overlooked by the BO+ST model and, to our knowledge, in the entire literature considering possible corrections to this model (Section 6.1.1), should increase the magnitude of the solvation free energy for cations and decrease it for anions. This contribution is also expected to be directly proportional to the ionic charge and to increase in magnitude with increasing ion size.

(B_3) Neglect of the *ion-compression work*[660,667,668,709–714]: as stated above, the BO estimate for the charging free energy assumes that the spherically-symmetric charge distributions within the free ion in vacuum and within the solvated ion are identical, so that the self-energies of the two distributions cancel out. However, when an ion is transferred from vacuum to a condensed phase (solution or crystal), its electron cloud is expected to be compressed by the denser environment, resulting in a decrease of its effective radius and an increase in its self-energy. Note, however, that this suggested compression is not entirely obvious when considering available estimates for gas-phase, in-crystal and in-solution radii (Section 5.4). This contribution, which is overlooked by the BO+ST model, is called the ion-compression work, and should decrease the magnitude of the solvation free energy for both cations and anions. Based on a simple model involving a homogeneous charge density within the ion (Section 6.1.1), this contribution is also expected to increase quadratically with the ion charge and to decrease in magnitude with increasing ion size.

(C_3) Neglect of the *ion polarizability*[715,716]: the BO model for the calculation of the charging free energy considers a non-polarizable ion in solution, *i.e.* characterized by an internal dielectric permittivity of one. However, the effect of ionic polarizability cannot be appropriately reintroduced by simply increasing the value of the ionic permittivity above one in the model (Section 6.1.1). This is because the Born model enforces spherical symmetry, while an electronic polarization of the ion can only result from instantaneous solvent configurations that precisely break this symmetry. For this reason, the effect of the ionic polarizability is best regarded as a contribution to specific ion-solvent interactions, also neglected in the BO+ST model (Point G_3 below). The only distortion of the ionic electron charge density that could result from a spherically-symmetric solvent-generated field is a compression or an expansion, also neglected in the BO+ST model (Point B_3 above).

(D_3) Neglect of the *ion-solvent dispersion interactions*[712, 713, 716–718]: besides the hard wall (implicit constraint force) excluding the continuum solvent from the ionic volume, and the interactions contributing to the cavitation term, the BO+ST model considers ion-solvent interactions to be exclusively of (classical) electrostatic nature. However, in reality, a second (minor) contribution arises from dispersion (London) interactions [719–721]. From a semi-classical perspective, these interactions can be interpreted as resulting from the favorable interaction between "instantaneous dipoles", caused by fluctuations in the electronic cloud of one species, and "induced dipoles", caused by the response of the electronic cloud of neighbor species to these instantaneous dipoles. A more appropriate description of dispersion requires, however, the quantum-mechanical analysis of electron correlation effects involving two closed-shell species. A simplified version of this analysis is provided by the Drude model[722–724], while approximate expressions in the context of atoms are given by the London[719–721,725] (Eq. 3.14) and Slater-Kirkwood[726] (Eq. 3.15) formulae. Such a quantum-mechanical analysis shows that the leading term of dispersion interactions is proportional to the inverse-sixth power of the distance between the interacting species and to the product of their polarizabilities (Eqs. 3.14 and 3.15). This contribution, which is overlooked by the BO+ST model, should increase the magnitude of the solvation free energy for both cations and anions. Because it is related to the polarizability of the ion, this effect is also expected to decrease in importance with increasing ionic charge magnitude, to increase in importance with increasing ion size, and to be comparatively larger for anions than for cations at comparable size (see *e.g.* the progressively decreasing polarizability volumes $\tilde{\alpha}_{el}$ of Cl^-, Ar, K^+ and Ca^{2+}, namely[221,727] $3.69 \cdot 10^{-30}$, $1.64 \cdot 10^{-30}$, $0.84 \cdot 10^{-30}$ and $0.55 \cdot 10^{-30}$ m^3, for the first effect; see *e.g.* alkali and halide polarizability volumes $\tilde{\alpha}_{el}$ in Table 5.9 for the two latter effects).

(E_3) Neglect of *electrostriction*[666, 714, 728–746]: since the ion exerts strong attractive forces on the surrounding solvent molecules, it is expected to increase the local density of the solvent in its neighborhood above the corresponding bulk value. This compression of the solvent around the ion is called electrostriction. It is expected to result in an increase of the local solvent permittivity above the corresponding bulk value ϵ_{svt} through an increase in the local density of the solvent dipoles (see *e.g.* the positive value $\partial_P \epsilon^{\ominus}_{wat} = 37.4 \cdot 10^{-4}$ bar^{-1} for water; Table 1.1), but also to induce an energetic cost caused by an increased overlap in the ion-solvent and solvent-solvent electron clouds. Model calculations[730, 731, 742] (Section 6.1.1) suggest that the first effect dominates, so that the overall contribution of electrostriction, which is overlooked by the BO+ST model, should increase the magnitude of the solvation free energy for both cations and anions. Because it is directly related to the field strength around the ion, this effect is also expected to increase in importance with increasing ion charge magnitude and to decrease in importance with increasing ion size[730,731,742] (see, however, Ref.[728]). These model calculations[730,731,742] also suggest that the favorable contribution of electrostriction to the solvation free energy is very small, at least for monovalent cations in water, compared to the unfavorable contribution of dielectric saturation (Point F_3 below).

(F_3) Neglect of *dielectric saturation*[259,261,325,660–666,675,676,714,717,729–732,734,743–774]: the CE approach involves a linear-response approximation, *i.e.* it assumes that the polarization of an infinitesimal solvent volume element is proportional to the local electric field. This approximation is only accurate at low fields. At higher fields, the dielectric response of the solvent will become sublinear, because the polarization of dipoles of fixed magnitude at a given density cannot exceed a maximum value, reached when all dipoles are perfectly aligned with the field. This reduction of the solvent dielectric response at high fields, which is intimately related to the discrete nature of the solvent molecules, is called dielectric saturation. Saturation may also induce an energetic cost, the alignment of solvent dipoles causing a significant perturbation of the solvent-solvent interactions (*e.g.* a partial disruption of the hydrogen-bonding network in the case of water). Beyond the saturation limit, there is still a contribution to the dielectric response associated with the electronic polarizability (and, possibly, geometric distortion) of the solvent molecules (Section 2.1 and Figure 2.1), *i.e.* the magnitude of the solvent dipoles is not fixed, but increases slightly upon increasing the applied field. However, this contribution is typically much lower than the one arising from dipole reorientation below the saturation limit, and also itself undergoes saturation at very high fields. The static permittivity (*e.g.* $\epsilon_{wat}^{\ominus} = 78.36$ for water; Table 1.1) only characterizes the solvent response at low field. A value more characteristic for the solvent response beyond the saturation of dipolar reorientation is the corresponding infinite-frequency value (*e.g.* $\tilde{\epsilon}_{wat}^{\ominus} = 5.2$ for water; Table 1.1). The latter value may be viewed as a coarse estimate for the local solvent permittivity in the immediate neighborhood of an ion. For large monovalent ions, it may represent a lower bound (only partial saturation of the dipolar reorientation). For small or/and polyvalent ions, it may even represent an upper bound (saturation of the solvent electronic polarizability itself[259,660]). An obvious way to tentatively include saturation effects into the CE approach is to formulate an expression for the local permittivity as a function of the partial ionic charge along the charging process and of the distance to the ion surface[661,666,717,730,731,744,748,752,754,757–759,762,764–766,769] (Section 6.1.1). Qualitatively speaking, dielectric saturation always results in a decrease of the local solvent permittivity below the corresponding bulk value. This contribution, which is overlooked by the BO+ST model, should thus decrease the magnitude of the solvation free energy for both cations and anions. Because it is directly related to the field strength around the ion, this effect is also expected to increase in importance with increasing ionic charge magnitude and to decrease in importance with increasing ion size.

(G_3) Neglect of the *solvation structure* and of *specific ion-solvent interactions*[260, 325, 712–714, 717, 718, 758, 759, 775–781]: the CE approach involves a continuous-medium approximation, *i.e.* it models the solvent as a homogeneous distribution of point dipoles at a given density and neglects all the microscopic details of the solvation structure, as well as the possibility of specific (other than purely charge-dipole) ion-solvent interactions. In reality, ions have a structuring effect on the surrounding solvent molecules[85, 86, 88–93, 98, 101, 104, 105, 108, 109, 237, 241, 243, 245, 247–249, 775, 782–784], leading to inhomogeneous radial and orientational distributions in successive solvation shells. Furthermore, ion-solvent interactions may include components that are not strictly of a charge-dipole nature. These specific effects may be related to *e.g.* electronic polarization of the ion under the effect of the instantaneously anisotropic solvent-generated field (Point C_3 above), anisotropy of the solvent molecular shape and charge distribution[785–789] (see also Refs. [94, 260, 426–429, 717, 780, 790–803] for theoretical investigations of this aspect), anion-solvent hydrogen-bonding in protic solvents[101, 172, 804, 805] (and, exceptionally, cation-solvent hydrogen bonding; *e.g.* NH_4^+), possible partial deprotonation of water molecules in the tight hydration shell of highly charged cations[806, 807] (*e.g.* partial deprotonation of $[Al(H_2O)_6]^{3+}$ in water[808], water acidity enhancement[807] in aqueous solutions of Mg^{2+} and Zn^{2+}), possible partial protonation of an anion[172, 809] (*e.g.* quantitative protonation of O^{2-} to OH^- and partial protonation[172,804,810,811] of F^- to HF in water), or partially covalent character (charge transfer) of the ion-solvent interaction[804, 805, 812]. The contribution of all these effects, which is overlooked by the BO+ST model, to the solvation free energy strongly depends on the ion charge (sign and magnitude), size, and electronic structure (polarizability), as well as on the nature of the solvent. For this reason, it is extremely difficult to provide a generic description of its influence. Solvent structuring, which is also somewhat related to dielectric saturation (Point F_3 above), is also expected to significantly increase the magnitudes of both the solvation entropies and enthalpies, *e.g.* due to the presence of quasi-immobilized solvent molecules in the first (or first few) solvation shells of the ion, sometimes referred to as irrotationally-bound solvent molecules[734], or to the structuring or destructuring action of the ion on the solvent (Sections 2.1 and 5.20). However, due to entropy-enthalpy compensation effects, solvent structuring appears to have a much more limited influence on the solvation free energy (see below).

(H$_3$) Ambiguity in the *choice of the ionic radius*[661, 714, 716, 746, 748, 754, 759, 765, 769, 770, 772, 776, 778–781, 790, 791, 813–819]: the key quantity characterizing a given ion within the BO+ST model, besides its charge, is the effective ionic radius $R_I^{\ominus}$. Unfortunately, this parameter, which should characterize the extent of the ionic cavity in the solution environment, is not an experimental observable, *i.e.* it does not possess an unambiguous physical definition. There exist numerous ways to derive sets of ionic radii using experimental or theoretical data (Section 5.4). Historically, ionic radii have been most commonly derived based on crystallographic structures. However, even if the lattice parameters can be accurately measured (Section 5.2), the fitting of single-ion radii for cations and anions based on the lattice parameters of a corresponding set of mixed salts can be performed in many different ways, resulting in different sets of so-called in-crystal (or crystallographic) radii. These radii were then used as estimates for the corresponding in-solution and gas-phase radii, assumed identical in the BO+ST model, thereby introducing a possible additional source of error related to the different environments. Many other approaches have been followed, leading to estimates for in-crystal but also in-solution and gas-phase radii. The resulting sets are generally qualitatively consistent in terms of trends along the ion series considered, but quantitatively quite diverse (Table 5.4). The ambiguity in the choice of ionic radii is a particularly serious issue, because the results of the BO+ST model are very sensitive to the value of $R_I^{\ominus}$, small differences in this parameter giving rise to large changes in the predicted solvation free energies. For example, the BO+ST estimate for the hydration free energy of the Na^+ cation varies by as much as 50 kJ·mol^{-1} upon changing $R_I^{\ominus}$ by ±0.01 nm (assuming $R_I^{\ominus} = 0.17$ nm, $\tilde{\gamma}_{wat}^{\ominus} \approx \gamma_{wat}^{\ominus}$ and using data from Table 1.1). Another common approach is to treat the ionic radii as empirical parameters of the model. This is typically done by introducing one or more adjustable correction terms to the in-crystal radii of a selected set, *e.g.* common additive corrections for the cations and the anions separately[408, 748, 776, 814, 816, 818, 820–822] (Section 6.1.1). The optimization of the correction terms follows from comparison of the BO+ST predictions with experimental data, typically in the form of estimated intrinsic single-ion solvation free energies. However, this choice has the disadvantages that: (*i*) the model loses predictive ability to rather become a fitting tool; (*ii*) the correction terms encompass in an effective way all the sources of errors discussed in the present enumeration, so that their physical significance becomes rather unclear; (*iii*) the results depend on the selection of a given set of ionic radii; (*iv*) these correction terms also encompass the error on the set of intrinsic single-ion solvation free energies employed in the fitting, which depend on the use of a specific extra-thermodynamic assumption and can be extremely large (Section 2.3).

(I$_3$) Ambiguity in the choice of the *microscopic surface tension coefficient*[823–843]: the ST model for the calculation of the cavitation free energy assumes a linear dependence of the surface contribution to this free energy on the surface area of the ion-sized cavity. This relationship is based on thermodynamic considerations for macroscopic interfaces, and may not be accurate for microscopic ones, where *e.g.* a volume term arising from interactions, *i.e.* distinct from the plain pressure-volume term, might have to be considered in addition[844]. Even accepting the linear dependence on the surface area, the microscopic surface tension coefficient $\tilde{\gamma}^{\ominus}_{wat}$ need not be immediately related to the corresponding macroscopic coefficient $\gamma^{\ominus}_{wat}$, although the latter value can be used as a first estimate in the absence of a better choice. For water, the value of $\gamma^{\ominus}_{wat}$ is $71.99 \cdot 10^{-3}$ J·m^{-2} (Table 1.1). A tentative experimental estimate for the corresponding microscopic value $\tilde{\gamma}^{\ominus}_{wat}$ based on positronium-induced nanobubble formation[845] is $35 \cdot 10^{-3}$ J·m^{-2}. A more indirect estimate based on a linear fit of the experimental hydration free energies of small alkane molecules (gas-phase to aqueous solution) against calculated solvent-accessible surface areas suggests[692] a much smaller value $\tilde{\gamma}^{\ominus}_{wat} = (3.5 \pm 0.4) \cdot 10^{-3}$ J·m^{-2}. However, it should be stressed that: (*i*) the solvent-accessible surface area, obtained by rolling a probe of 0.14 nm radius over the van der Waals envelope of the molecule, is an upper-bound estimate of the molecular surface area; (*ii*) experimental alkane solvation free energies also include non-polar solute-solvent interactions (*e.g.* dispersion); (*iii*) the linear fit also involves an offset of 1.72 nm^2 to be added to the calculated surface area prior to multiplication by $\tilde{\gamma}^{\ominus}_{wat}$. Another indirect estimate based on a linear fit of the experimental transfer free energies of ten small alkane molecules (non-polar solvent to aqueous solution) against calculated solvent-accessible surface areas suggests[846] a value $\tilde{\gamma}^{\ominus}_{wat} = 32.7 \cdot 10^{-3}$ J·m^{-2}. The contribution of dispersion interactions is presumably weaker along the transfer of an alkane molecule from a non-polar solvent (rather from the gas-phase) to water. However, this fit still appears to involve an offset (not reported in the original article) and the solvent-accessible surface area, apparently also obtained by rolling a probe of 0.14 nm radius over the van der Waals envelope of the molecule, remains an upper-bound estimate of the molecular surface area. AT simulations have been used[847–860] to evaluate $\gamma^{\ominus}_{wat}$, but seldom[834,835] $\tilde{\gamma}^{\ominus}_{wat}$. For example, one study reported[834] an estimate of $47.5 \cdot 10^{-3}$ J·m^{-2} for the surface tension of a small water cluster (64 molecules) using the TIP4P model[861] (note that this estimate pertains to a convex rather than to a concave interface). In summary, considering water and atom-sized cavities, most estimates suggest that $\tilde{\gamma}^{\ominus}_{wat}$ is smaller than $\gamma^{\ominus}_{wat}$, possibly by about a factor two. Note that an alternative to the ST model for estimating the reversible work of cavitation relies on scaled-particle theory[862–871] (Section 6.1.2). In any case, the cavitation free energy is typically much smaller than

> the charging free energy, which is why it was typically neglected in early work. For example, in the case of Na^+ hydration (assuming $R_I^{\ominus} = 0.17$ nm, $\gamma_{wat}^{\ominus} = \tilde{\gamma}_{wat}^{\ominus}$ and using data from Table 1.1), the charging term is about 25 times larger in magnitude than the cavitation term, and assuming *e.g.* $\tilde{\gamma}_{wat}^{\ominus} = (1/2)\gamma_{wat}^{\ominus}$ would change the calculated solvation free energy by 7.9 kJ·mol^{-1} only. For this reason, the ambiguity in the choice of the microscopic surface tension coefficient remains a minor (although non-negligible) source of error in the calculation of single-ion solvation free energies using the BO+ST model.

Note, finally, that although the BO+ST model neglects the effect of the ion polarizability (Point C_3), it formally does not neglect the effect of the solvent polarizability, which is included as a contribution to the solvent dielectric permittivity. On the other hand, the model overlooks the possible saturation of the solvent polarizability, as well as of the solvent reorientational response, at high electric fields (Point F_3).

Most of the above errors find their origin in the fact that CE is a low-field macroscopic theory, based on Maxwell's equations for continuous media[421], which is applied in the BO+ST model at the microscopic level and at high local field strengths, *i.e.* formally outside its scope of validity. In particular, the concept of dielectric permittivity is a semi-macroscopic concept which loses any meaning at the microscopic level. The representation of the bulk solvent reasonably far away from the ion as a dielectric continuum certainly makes physical sense. Rather, it is the description of the ion surface as a sharp dielectric boundary that is questionable. Similar considerations apply to the surface tension component of the model, but are less relevant in terms of incurred errors because the cavitation term is typically much smaller in magnitude than the charging term.

Given the above list of potential sources of errors in the BO+ST model, it is not surprising that this model fails to provide estimates for intrinsic single-ion solvation free energies in *quantitative* agreement with experiment[84,661,667,716,759,766,772]. The validity of this comparison is of course limited in view of the very large range of uncertainty in the experimental estimates to compare with (Section 2.3). In fact, one should rather marvel at the fact that this model still provides results in remarkable *qualitative* agreement with experiment. The reasons are that the BO+ST model: (*i*) captures the *dominant component of solvation*, namely the electrostatic interaction of the ion charge with its polarized solvent environment; (*ii*) benefits from a significant *enthalpy-entropy compensation effect.*

The latter point is clearly illustrated by the calculation of ionic solvation entropies[261,663,733,754,758,776,777,872] and enthalpies[717,742,746,758,790,791,813,816,819,873]. The BO+ST formula for the solvation entropy of an ion can be obtained by taking the negative derivative of the free energy (Eq. 3.7) with respect to temperature, namely

$$\begin{aligned}\Delta_s S^{\ominus *}_{BO+ST}[I^z_g] &= -\partial_T \left(\Delta_s G^{\ominus}_{BO+ST}[I^z_g] - \Delta_s G^{\ominus}_{std}\right) + \Delta_s S^{\ominus *}_{std} \\ &= \frac{N_A e^2 z^2}{8\pi\epsilon_o} \frac{1}{R^{\ominus}_I} \frac{\partial_T \epsilon^{\ominus}_{svt}}{(\epsilon^{\ominus}_{svt})^2} \\ &\quad -4\pi N_A (R^{\ominus}_I)^2 \partial_T \tilde{\gamma}^{\ominus}_{svt} - (T^-)^{-1} \Delta_s G^{\ominus}_{std} - R \ ,\end{aligned} \tag{3.8}$$

where $\Delta_s S^{\ominus *}_{std}$ is given by Eq. 4.183 and it is recalled that ∂_P and ∂_T applied to a standard quantity imply derivation with respect to P° and T^- considered as variables, followed by evaluation at the reference values of these variables. The expression for the charging component (excluding standard-state correction) of Eq. 3.9 is sometimes referred to as the Born-Bjerrum equation[874]. From the Gibbs equation, the corresponding enthalpy is given by

$$\begin{aligned}\Delta_s H^{\ominus *}_{BO+ST}[I^z_g] &= (1 - T^- \partial_T)(\Delta_s G^{\ominus}_{BO+ST}[I^z_g] - \Delta_s G^{\ominus}_{std}) + \Delta_s H^{\ominus *}_{std} \\ &= -\frac{N_A e^2 z^2}{8\pi\epsilon_o} \frac{1}{R^{\ominus}_I} \frac{(\epsilon^{\ominus}_{svt})^2 - \epsilon^{\ominus}_{svt} - T^- \partial_T \epsilon^{\ominus}_{svt}}{(\epsilon^{\ominus}_{svt})^2} \\ &+ 4\pi N_A (R^{\ominus}_I)^2 \left(\tilde{\gamma}^{\ominus}_{svt} - T^- \partial_T \tilde{\gamma}^{\ominus}_{svt} + \frac{1}{3} P^\circ R^{\ominus}_I\right) - RT^- \ ,\end{aligned} \tag{3.9}$$

where $\Delta_s H^{\ominus *}_{std}$ is given by Eq. 4.183. It has also been assumed here that $\partial_T R^{\ominus}_I = 0$, *i.e.* that the effective ionic radius is temperature independent (see, however, Refs. [791, 819, 875, 876] for the consideration of temperature-dependent ionic radii). The additional "$*$" superscript in these equations arises from the use of a density-corrected solute standard-state variant, already briefly mentioned in Section 2.2 and explained in detail in Sections 4.2.5 and 4.2.16. In the present case, this choice implies that the differentiation leading to Eqs. 3.8 and 3.9 is performed according to Eq. 2.15 rather than Eq. 2.14. As a result, the standard-state correction terms to the intrinsic point-to-point solvation parameters $\Delta_s S^{\oplus}[I^z_g] = -\partial_T \Delta_s G^{\oplus}[I^z_g]$ and $\Delta_s H^{\oplus}[I^z_g] = (1 - T^- \partial_T)\Delta_s G^{\oplus}[I^z_g]$ (Eqs. 2.16-2.18) are not equal to $\Delta_s S^{\ominus}_{std} = -\partial_T \Delta_s G^{\ominus}_{std}$ and $\Delta_s H^{\ominus}_{std} = (1 - T^- \partial_T)\Delta_s G^{\ominus}_{std}$, respectively, as would be the case in the standard variant, but to $\Delta S^{\ominus *}_{std} = -(T^-)^{-1} \Delta_s G^{\ominus}_{std} - R$ and $\Delta H^{\ominus *}_{std} = -RT^-$, respectively (Eq. 4.183). For water, the derivative $\partial_T \gamma^{\ominus}_{wat}$ corresponding to the macroscopic surface tension coefficient, which can be used as a first estimate for $\partial_T \tilde{\gamma}^{\ominus}_{wat}$, is $-0.154 \cdot 10^{-3}$ J·m^{-2}·K^{-1}, while the derivative $\partial_T \epsilon^{\ominus}_{wat}$ is -0.359 K^{-1} (Table 1.1). The derivatives $\partial_T \epsilon^{\ominus}_{wat}$ and $\partial_T \gamma^{\ominus}_{wat}$ are actually negative for most solvents (there are, however, exceptions[471]), due to the perturbation of the interface structuring and of the bulk dielectric responsiveness caused by increased thermal motion. As a result, the charging and standard-state correction contributions to the solvation entropy in Eq. 3.8 are typically both negative, while the corresponding cavitation contribution (surface component only) is normally positive. This is intuitively reasonable, considering that the presence of an ion is likely to promote solvent ordering in terms of the charging component (alignment of the solvent dipoles) and disordering in terms of the cavitation component (creation of an interface), while the third term accounts for an entropy reduction upon reducing the effective volume accessible to the ion, as determined by the standard-state reference pressure P° for the gas-phase ion and concentration b° for the ion in solution. In the context of polar solvents (*e.g.* water), the solvation free energy is typically dominated by the charging term (see above). However, the three terms typically provide contributions of comparable magnitudes to the solvation entropy. For example, the different terms in the case of Na^+ hydration (assuming $R_I^\ominus = 0.17$ nm, $\tilde{\gamma}_{wat}^\ominus \approx \gamma_{wat}^\ominus$, $\partial_T \tilde{\gamma}_{wat}^\ominus \approx \partial_T \gamma_{wat}^\ominus$, and using data from Table 1.1) evaluate to -23.9 (charging), 33.7 (cavitation, surface contribution only) and -34.98 (standard-state correction) J·mol^{-1}·K^{-1}, leading to an estimated solvation entropy of -25.2 J·mol^{-1}·K^{-1}.

In contrast to the free energies, the solvation entropies calculated from the BO+ST model are not even in qualitative agreement with experiment[754,791,872]. A similar observation applies to the solvation enthalpies, although the relative differences are typically smaller. For example, in water, assuming an ionic radius $R_I^\ominus = 0.155(336)$ nm for the Na^+ ion leads to $\Delta_s G_{BO+ST}^\ominus[Na_g^+] = -419.5$ kJ·mol^{-1}, an "experimental" value based on the assumption $\mathcal{G}_{H,wat}^\ominus = -1100$ kJ·mol^{-1} (Section 5.20, Table 5.26), but $\Delta_s S_{BO+ST}^{\ominus *}[Na_g^+] = -33.0$ J·mol^{-1}·K^{-1} and $\Delta_s H_{BO+ST}^{\ominus *}[Na_g^+] = -429.3$ kJ·mol^{-1}, which are much lower in magnitude than corresponding "experimental" values of -110.7 J·mol^{-1}·K^{-1} and -452.5 kJ·mol^{-1} assuming $\mathcal{H}_{H,wat}^\ominus = -1139.0$ kJ·mol^{-1} and $\mathcal{S}_{H,wat}^\ominus = -131.0$ J·mol^{-1}·K^{-1} (Section 5.20, Table 5.26). The reason for this failure is that both solvation entropies and enthalpies are extremely sensitive to the representation of the short-range solvation structure, *i.e.* to the microscopic nature of the solvent. This structure is, however, essentially overlooked by the BO+ST model. More precisely, it is entirely encompassed within the relatively inaccurate cavitation contribution, resulting in too low magnitudes for both solvation entropies and enthalpies. Fortunately, the errors committed in the two terms largely cancel upon summation (Gibbs equation), resulting in a more limited (although still significant) error on the solvation free energy. Due to enthalpy-entropy compensation effects, the accuracy of the BO+ST solvation free energy may still be acceptable, if not quantitative, even though the actual partitioning is entirely inaccurate. A similar feature holds in AT simulations, where solvation enthalpies and entropies are much more difficult to reproduce accurately compared to the corresponding free energies[877] (see, however, Refs.[793,878]), because of their more pronounced dependence on the description of the solvation structure (Section 6.2.3). Note, finally, that if enthalpy-entropy compensation effects reduce the error on the BO+ST estimate for the free energy close to room temperature, this may not necessarily be the case at a different temperature. This is easily seen by observing that even if one assumes $\Delta H_{err}^\ominus - T^- \Delta S_{err}^\ominus \approx 0$, where ΔH_{err} and ΔS_{err} denote the BO+ST errors in the enthalpy and entropy, respectively, one still has $\partial_T[\Delta H_{err}^\circ(T) - T\Delta S_{err}^\circ(T)] = -\Delta S_{err}^\circ(T)$, *i.e.* the extent of the compensation depends on temperature.

Although less commonly done (see, however, Refs.[409,730,735–737,879] for the evaluation of electrostriction corrections to ionic volumes), the BO+ST model can also be assessed in terms of volume properties by considering its pressure dependence. Taking the derivative of the free energy (Eq. 3.7) with respect to pressure, one obtains

$$\begin{aligned}\Delta_s V^{\ominus *}_{BO+ST}[I^z_g] &= \partial_P(\Delta_s G^{\ominus}_{BO+ST}[I^z_g] - \Delta_s G^{\ominus}_{std}) + \Delta_s V^{\ominus *}_{std} \\ &= -\frac{N_A e^2 z^2}{8\pi\epsilon_o}\frac{1}{R^{\ominus}_I}\frac{\partial_P \epsilon^{\ominus}_{svt}}{(\epsilon^{\ominus}_{svt})^2} \\ &\quad +4\pi N_A\left[(R^{\ominus}_I)^2\partial_P\tilde{\gamma}^{\ominus}_{svt} + \frac{1}{3}(R^{\ominus}_I)^3\right] + \Delta_s V^{\ominus *}_{std}\ ,\end{aligned} \tag{3.10}$$

where $\Delta_s V^{\ominus *}_{std}$ is given by Eq. 4.183. It has been assumed here that $\partial_P R^{\ominus}_I = 0$, *i.e.* that the effective ionic radius is pressure independent (the possibility of pressure-dependent ionic radii has, to our knowledge, never been considered in the literature). Here again, the additional "$*$" superscript arises from the use of a density-corrected solute standard-state definition. In the present case, this choice implies that the differentiation leading to Eq. 3.10 is performed according to Eq. 2.15 rather than Eq. 2.14. As a result, the standard-state correction term to the intrinsic point-to-point solvation volume $\Delta_s V^{\boxminus}[I^z_g] = \partial_P \Delta_s G^{\boxminus}[I^z_g]$ (Eqs. 2.16-2.18) is not equal to $\Delta_s V^{\ominus}[I^z_g] = \partial_P \Delta_s G^{\ominus}[I^z_g]$, as would be the case in the standard variant, but to $\Delta_s V^{\ominus *}_{std} = -RT^-(P^\circ)^{-1}$ (Eq. 4.183). The corresponding partial molar volume $v^{\ominus *}_{BO+ST}$ of the solvated ion is trivially related to the solvation volume $\Delta_s V^{\ominus *}_{BO+ST}$, by addition of the ideal-gas molar volume $RT^-(P^\circ)^{-1}$. This term cancels out with the standard-state correction term $\Delta_s V^{\ominus *}_{std}$, resulting in

$$\begin{aligned}v^{\ominus *}_{BO+ST}[I^z_d] &= -\frac{N_A e^2 z^2}{8\pi\epsilon_o}\frac{1}{R^{\ominus}_I}\frac{\partial_P \epsilon^{\ominus}_{svt}}{(\epsilon^{\ominus}_{svt})^2} \\ &\quad +4\pi N_A\left[(R^{\ominus}_I)^2\partial_P\tilde{\gamma}^{\ominus}_{svt} + \frac{1}{3}(R^{\ominus}_I)^3\right]\ .\end{aligned} \tag{3.11}$$

Note that $v^{\ominus *}_{BO+ST}[I^z_d]$ has no dependence on the gas-phase reference pressure P° and solution molality b°, as expected. For water, the derivative $\partial_P \gamma^{\ominus}_{wat}$ corresponding to the macroscopic surface tension coefficient, which can be used as a first estimate for $\partial_P \tilde{\gamma}^{\ominus}_{wat}$, is $7.02 \cdot 10^{-5}$ J·m^{-2}·bar^{-1}, while the derivative $\partial_P \epsilon^{\ominus}_{wat}$ is $37.4 \cdot 10^{-4}$ bar^{-1} (Table 1.1). The derivatives $\partial_P \epsilon^{\ominus}_{wat}$ and $\partial_P \tilde{\gamma}^{\ominus}_{wat}$ are actually positive for most solvents, due to the enhancement of the surface structuring and of the bulk dielectric responsiveness caused by the increased solvent density. As a result, the charging contribution to the partial molar volume in Eq. 3.11 is typically negative, while the cavitation contribution (both in its surface and pressure-volume components) is positive. The positive pressure-volume component of the cavity term is simply equal to the excluded volume of the ion viewed as a hard sphere. The positive surface component of the cavitation term can be viewed as an expansion of the effective ionic volume resulting from the structuring of its first solvation shells. Finally, the negative charging term can be interpreted as a volume contraction resulting from electrostriction (Point E_3). Note that there is no contradiction between the occurrence of this term and the above statement that the BO+ST model neglects the effects of electrostriction on the solvation free energy. Electrostrictive effects are present in a relative sense within the pressure dependence of the free energy, but omitted in an absolute sense (relaxation of the solvent density around the ion upon solvation) within the free energy itself, *i.e.* there is a lacking integration constant. To give a simple numerical example, the different terms in the case of Na^+ hydration (assuming $R_I^{\ominus} = 0.17$ nm, $\tilde{\gamma}_{wat}^{\ominus} \approx \gamma_{wat}^{\ominus}$, $\partial_P \tilde{\gamma}_{wat}^{\ominus} \approx \partial_P \gamma_{wat}^{\ominus}$, and using data from Table 1.1) evaluate to $-2.5 \cdot 10^{-6}$ (charging), $153.5 \cdot 10^{-6}$ (surface contribution to cavity formation) and $12.4 \cdot 10^{-6}$ (pressure-volume contribution to cavity formation), $m^3 \cdot mol^{-1}$, leading to an estimated partial molar volume of $163.5 \cdot 10^{-6}$ $m^3 \cdot mol^{-1}$.

Just as was the case for the entropy-enthalpy partitioning, although interesting from a conceptual point of view, the partial molar volumes evaluated using the BO+ST model lack quantitative accuracy. For example, in water, assuming an ionic radius $R_I^{\ominus} = 0.155(336)$ nm for the Na^+ ion leads to $v_{BO+ST}^{\ominus *}[Na_{aq}^+] = 134.9 \cdot 10^{-6}$ $m^3 \cdot mol^{-1}$, to be compared with an "experimental" value of $-5.9 \cdot 10^{-6}$ $m^3 \cdot mol^{-1}$ assuming $v_{H,wat}^{\ominus} = -4.5 \cdot 10^{-6}$ $m^3 \cdot mol^{-1}$ (Section 5.20, Table 5.26). The calculated value suggests an effective hard-sphere radius of 0.38 nm, which appears unrealistically large for a Na^+ ion. Here again, the reason for this failure is that solvation volumes are extremely sensitive to the representation of the short-range solvation structure, *i.e.* to the microscopic nature of the solvent. This structure is, however, essentially overlooked by the BO+ST model or, rather, it is entirely encompassed within the relatively inaccurate cavitation contribution. In particular, the surface contribution to the cavitation term ($128.2 \cdot 10^{-6}$ $m^3 \cdot mol^{-1}$) is extremely large, likely due to the macroscopic $\partial_P \gamma_{wat}^{\ominus}$ being a poor approximation for the corresponding microscopic value. The magnitude of the charging term is also probably underestimated, considering that this term is the only one that could justify a negative partial molar volume for Na^+ due to electrostriction, as suggested by experiment. Similarly to the observations made in the context of entropy-enthalpy compensation, the inaccurate partial molar volume has essentially no influence on the BO+ST estimate for the free energy at atmospheric pressure, because the corresponding pressure-volume term is extremely small at P° for solute species of atomic size. The situation might, however, be different at elevated pressures.

In view of the quantitative inaccuracy of the BO+ST model in terms of solvation entropy and partial molar volume, analysis of the higher-order pressure or/and temperature derivatives, namely of the partial molar heat capacity, volume-compressibility and volume-expansivity, although in principle possible, is of limited interest.

Another main failure of the BO+ST model in the context of water, besides the inaccurate entropy-enthalpy partitioning and partial molar volume description, is its inability to capture the differences in hydration free energy between anions and cations of identical (or similar) size, namely *asymmetric solvation effects*[803]. Since Eq. 3.7 depends quadratically on the integer ion charge z, the result is insensitive to the sign of this charge for given values of $R_I^{\ominus}$, $\epsilon_{wat}^{\ominus}$ and $\tilde{\gamma}_{wat}^{\ominus}$. In reality, at comparable sizes, water solvates an anion significantly better than a cation[94, 260, 426–429, 717, 780, 790–803], the corresponding solvation enthalpies and entropies also being of larger magnitudes. Note that this statement, which is certainly qualitatively correct, is actually very difficult to recast in quantitative terms. From the experimental point of view, a quantitative assessment requires estimates for the intrinsic proton solvation free energy, enthalpy and entropy, which are elusive quantities (Section 2.3). Furthermore, a pair of equisized cation and anion formally does not exist, and the notion of comparable size relies on the acceptance of a scale of ionic radii, which are also elusive (Section 5.4). From the theoretical point of view, AT simulations (Section 3.3) may consider hypothetical equisized cation-anion pairs, and clearly evidence anisotropic solvation effects[803]. However, their quantitative evaluation in terms of thermodynamic solvation parameters requires corrections for a number of artifacts, which have been recognized only recently (Section 6.2.1). In spite of these difficulties, the preferential solvation of anions over cations at comparable sizes can be viewed as an established qualitative property of water.

The reason for the discrepancy between the BO+ST model and experiment in terms of anisotropic hydration is generally attributed to the neglect of solvation structure and specific ion-solvent interactions in this model (Point G_3). The usual explanation is that the center of positive charge in a water molecule (hydrogen atoms) is closer to the molecular surface and can thus approach an ion more closely compared to the center of negative charge (oxygen atom and lone pairs). Alternatively, this can be viewed as a consequence of the (traceless) quadrupole moment of the water molecule relative to the center of its molecular "envelope" or to the formation of hydrogen-bonds with anions but not with cations (unless these are hydrogen-bond donors, *e.g.* NH_4^+). Note, however, that the preferential solvent polarization at a cavity-solvent interface (Point A_3) follows the opposite trend, favoring cation over anion solvation, suggesting that the latter effect is weaker compared to the former one[803].

The flaws of the BO+ST model, preventing its predictions from reaching quantitative agreement with experiment, are essentially identified (Points A_3-I_3). There is a considerable amount of literature suggesting empirical (physically-motivated or *ad hoc*) approaches to remedy one or the other of these problems. These will be discussed in more detail in Section 6.1.1. However, although interesting from a conceptual point of view, none of the resulting "improved" approaches can be regarded as entirely satisfactory. This is mainly because: (*i*) the derivation of physically-motivated correction terms for the different neglected effects (Points A_3-G_3) relies on additional assumptions and approximations, that are also themselves subject to large errors; (*ii*) the introduction of *ad hoc* correction terms for these effects relies on additional adjustable parameters, which are optimized for agreement with "experimental" data, and it is then unclear whether the "improvement" comes from a more accurate physical description, or merely from the inclusion of additional fitting parameters; (*iii*) correction terms estimated separately for the different effects cannot be assumed to be exactly additive, due to a significant extent of correlation among them; (*iv*) in the absence of unambiguous values for the ionic radii (Point H_3), for the microscopic surface tension coefficient (Point I_3), and for the experimental intrinsic values to compare with (Section 2.3), even the resultant of all these effects cannot be characterized unambiguously. Note that other models inspired from the CE approach also attempt to provide a semi-atomistic description of the of ion-solvent system. These approaches will be discussed in Section 6.1.2.

In summary, the BO+ST model is a *powerful qualitative and conceptual tool*. It captures the dominant component of solvation, namely the electrostatic interaction of the ion charge with its polarized solvent environment, enables a correct first-order reasoning on the effect of ionic charge and radius on the solvation free energy, and provides a solid reference point for the discussion of more complex secondary effects. However, although a considerable amount of work has been dedicated in the past to tentatively make this model quantitative (Sections 6.1.1 and 6.1.2), this is probably an unreachable goal. The reason is that the main flaw of the BO+ST model is the *neglect of the solvent microscopic structure*, the description of which is simply incompatible with the very principle of the CE approach

3.3 Classical atomistic simulations

In the context of AT simulations, intrinsic single-ion solvation free energies can be evaluated in two distinct ways: (*i*) *direct calculation of the intrinsic single-ion solvation free energy* $\Delta_s G^{\ominus}_{svt}[I^z_g]$ of an ion I^z in a given solvent *via* determination of the reversible work associated with the unphysical solvation process (Figure 2.2), with omission or artificial exclusion of the surface term (Eq. 2.7); (*ii*) *evaluation of the air-liquid interfacial potential* $\chi^{\ominus}_{svt}$ of the pure solvent at a planar interface in equilibrium (Section 2.2), and combination of the resulting estimate with experimentally-accessible real single-ion solvation free energies $\Delta_s \hat{G}^{\ominus}_{svt}[I^z_g]$ (Eqs. 2.6 and 2.8). The first approach can be viewed as an indirect estimation of the elusive quantity $\mathcal{G}^{\ominus}_{H,svt}$, while the second one represents a direct evaluation of the elusive quantity $\chi^{\ominus}_{svt}$. Both approaches can in principle be extended to the estimation of derivative thermodynamic solvation properties, *via* the pressure and temperature dependences of the calculated quantity. The direct evaluation of the intrinsic potential $\mathcal{V}^{\ominus}_{H,svt}$ of the hydrogen electrode, the third experimentally-elusive quantity intimately connected to $\mathcal{G}^{\ominus}_{H,svt}$ and $\chi^{\ominus}_{svt}$ (Section 2.3), is beyond the scope of purely classical AT simulations, due to the involvement of electronic processes.

Although the present part of this book (Chapters 2 and 3) is dedicated to the fundamental problems encountered in the determination of thermodynamic parameters for single-ion solvation, it is difficult to discuss these problems in the context of AT simulations without a minimal knowledge of the involved methodology. For this reason, background information on the AT approach is provided here rather than in the methods and results part of the book concerned with theoretical approaches (Chapter 6), the latter providing more advanced considerations (Section 6.2).

The *basic methodology* is essentially the same for both above types of calculations ($\mathcal{G}^{\ominus}_{H,svt}$ *via* $\Delta_s G^{\ominus}_{svt}[I^z_g]$, and $\chi^{\ominus}_{svt}$) and is summarized in Section 3.3.1. The four following sections, Sections 3.3.2-3.3.5, describe more specific aspects, namely: (*i*) the *parameters* (solvent, ions) involved in these calculations; (*ii*) the specific procedures for the calculation of an *intrinsic single-ion solvation free energy*; (*iii*) the specific procedures for the calculation of the *intrinsic bulk electric potential* in a pure liquid; (*iv*) the specific procedures for the evaluation of the *air-liquid interfacial potential* of a solvent. Finally, Section 3.3.6 summarizes the fundamental problems encountered by AT simulations in the context of ionic solvation properties, considering both types of approaches.

3.3.1 Basic methodology

In this section, the basic principles underlying the AT description of molecular systems are summarized, broadly following the general scheme of Figure 3.1. This description is exclusively concerned with the methodological aspects relevant for the calculation of an intrinsic single-ion solvation free energy (Section 3.3.3), of the bulk electric potential in a pure liquid (Section 3.3.4), or of the air-liquid interfacial potential of a solvent (Section 3.3.5). Furthermore, the discussion only encompasses the features encountered in the *most common types of models*, *e.g.* rigid solvent molecule, mean-field treatment of electronic polarization and Lennard-Jones representation of van der Waals interactions. Possible (less common) alternatives, *e.g.* flexible solvent molecules, explicit treatment of electronic polarization, or other representations of van der Waals interactions, are merely mentioned without additional details. The explicit-polarization alternative and the possible use of different representations for van der Waals interactions, along with various possible combination rules (Section 3.3.2), being the most relevant variants in the present context, will be discussed in Section 6.2.4. For simplicity, it is also assumed that the systems considered consist of *pure solvent samples*, for the determination of the intrinsic bulk electric potential or of the air-liquid interfacial potential of a solvent, along with *at most one ion*, for the calculation of an intrinsic single-ion solvation free energy. In the latter case, this single ion is also assumed to be *monoatomic*.

The AT description typically relies on *atoms* as "elementary particles" of the model, *i.e.* the individual atoms constituting the ion and the solvent molecules are considered to be point particles associated with a mass and a unique interaction site, and exempt of any internal structure. Some common solvent models also involve off-atom interaction sites, called *virtual sites.*

The interaction between the sites (ion and solvent atoms plus, possibly, solvent virtual sites) is described by a classical *potential-energy function* $V(\mathbf{r})$, where $\mathbf{r}$ is a $3N$-dimensional vector containing the Cartesian coordinates of the N sites in the system. This function $V(\mathbf{r})$ is also commonly called a *force field.* The representation of $V(\mathbf{r})$ in the present context of a pure solvent system possibly including a single monoatomic ion, typically involves the following approximations: (*i*) covalent interactions within a solvent molecule are represented by constraints enforcing the full rigidity of the molecule, possible virtual sites being held at fixed positions relative to this rigid molecular framework; (*ii*) non-bonded interactions are represented as a sum of pairwise (site-site) interaction terms, except for the possible presence of self and surface terms (see further below), each involving a van der Waals and an electrostatic component, the direct non-bonded interaction between sites within the same solvent molecule being omitted (intramolecular non-bonded exclusion); (*iii*) the van der Waals contribution (Pauli repulsion[880] and London dispersion[719–721]) to the pairwise non-bonded interaction is represented by means of a Lennard-Jones (12-6) function[881–883]; (*iv*) the electrostatic contribution to the pairwise non-bonded interaction is ideally represented using Coulomb's law (Eq. 2.1). Although the four above approximations are the most commonly encountered, many variants can also be found, including flexible solvent models[884–900], inclusion of many-body interaction terms (*i.e.* beyond pairwise additive; *e.g.* three body terms[901–913] or explicit polarization[141, 242, 354, 359, 366, 373, 439, 440, 798, 799, 905, 908–910, 914–980]), or alternative van der Waals functions (*i.e.* differing from the Lennard-Jones 12-6 function; *e.g.* 9-6 function[981–984], Morse function[985], Kihara function[986], Born-Mayer-Huggins function[180, 987, 988], Buckingham exp-6 function[989, 990], exp-6-8 function[991], or buffered-14-7 function[992]). With the exception of explicit polarization and alternative van der Waals representations (Section 6.2.4), these alternatives do not represent critical changes in the present context, and the four common force-field choices listed above will be assumed to hold in the following discussion.

The *rigidity of the solvent molecules*, including the fixed positioning of possible virtual sites, can formally be thought of as belonging to the potential-energy function $V(\mathbf{r})$ in the form of a sum of Dirac delta functions, each enforcing a specific site-site distance constraint, weighted by Lagrange multipliers. The values of the multipliers are determined on the flight during the simulation by a specific constraint-enforcement procedure[554–559].

The *Lennard-Jones function*[881–883] $V_{LJ,pair}$ used to represent van der Waals interactions can be written in different ways, involving different reference parameters, as

$$
\begin{aligned}
V_{LJ,pair}(r) &= C_{12}r^{-12} - C_6 r^{-6} \\
&= 4E_{min}\left[(r/R_{zer})^{-12} - (r/R_{zer})^{-6}\right] \\
&= E_{min}\left[(r/R_{min})^{-12} - 2(r/R_{min})^{-6}\right] \ ,
\end{aligned}
\tag{3.12}
$$

where r is the site-site distance, C_{12} the repulsion coefficient (accounting for Pauli repulsion[880]), C_6 the dispersion coefficient (accounting for London dispersion[719–721]), R_{zer} the distance at zero interaction (sometimes also called the collision diameter and often written σ), R_{min} the distance at the interaction minimum, and E_{min} the energy at the interaction minimum (sometimes also called the well depth and often written ϵ). These quantities are interconnected *via*

$$
\begin{aligned}
R_{zer} &= C_{12}^{1/6} C_6^{-1/6} \\
R_{min} &= (2C_{12})^{1/6} C_6^{-1/6} \\
E_{min} &= (4C_{12})^{-1} C_6^2 \quad .
\end{aligned}
\tag{3.13}
$$

The inverse-sixth power dependence of the Lennard-Jones (London) dispersion term is physically motivated. This dependence corresponds to the leading term in the quantum-mechanical analysis of electron correlation effects involving two closed-shell species[719–721], a simplified version of this analysis being provided by the Drude model[722–724]. Approximate expressions in the context of atoms are given by the London[719–721, 725] and Slater-Kirkwood[726] formulae. The *London equation* approximates the dispersion energy (on a per mole basis) between two closed-shell atoms i and j as[719–721, 725]

$$
E_{ij}(r) \approx -\frac{3N_A \tilde{E}_i \tilde{E}_j}{2(\tilde{E}_i + \tilde{E}_j)} \frac{\tilde{\alpha}_{el,i}\tilde{\alpha}_{el,j}}{r_{ij}^6} \ ,
\tag{3.14}
$$

where r_{ij} is the internuclear distance, while $\tilde{\alpha}_{el,i}$ and $\tilde{\alpha}_{el,j}$ are the static dipolar polarizability volumes, related to the static polarizability α_{el} as $\tilde{\alpha}_{el} = (4\pi\epsilon_o)^{-1}\alpha_{el}$, and $\tilde{E}_i$ and $\tilde{E}_j$ the average excitation energies (ground to exited electronic state; average energy approximation originally introduced by Unsöld[725]) of the two interacting atoms. In the approach of London, the quantities $\tilde{E}_i$ and $\tilde{E}_j$ are further approximated by the first ionization potentials of the two atoms. In the approach of Slater and Kirkwood[726], these quantities are approximated instead as $\tilde{E}_i \approx \hbar e\,(m_e\,\alpha_{el,i}/n_i)^{-1/2}$ and $\tilde{E}_j \approx \hbar e\,(m_e\,\alpha_{el,j}/n_j)^{-1/2}$, where $\hbar = (2\pi)^{-1}h$, h being Planck's constant (Table I), m_e is the electron rest mass (Table 4.2), and n_i and n_j are effective numbers of valence electrons, *i.e.* the numbers of electrons effectively contributing to the atomic polarizability (see Table 5.9 for examples of numerical values of $\tilde{\alpha}_{el,i}$ and $n_{el,i}$ in the context of alkali and halide gas-phase atoms and ions), resulting in the *Slater-Kirkwood equation*

$$E_{ij}(r) \approx -C_{SK} \frac{2\tilde{\alpha}_{el,i}\tilde{\alpha}_{el,j} n_{el,i}^{1/2} n_{el,j}^{1/2}}{(n_{el,i}\tilde{\alpha}_{el,j})^{1/2} + (n_{el,j}\tilde{\alpha}_{el,i})^{1/2}} \frac{1}{r_{ij}^6} \quad , \tag{3.15}$$

where C_{SK} evaluates to 23.97 kJ·mol^{-1}·nm$^{3/2}$.

In contrast, the corresponding inverse-twelfth power dependence of the (Pauli) repulsion term[880] is entirely *ad hoc.* Although not explicitly stated in the original article[993], this exponent was probably originally chosen for convenience in numerical calculations, because it is the square of the dispersion power law. The use of an exponential repulsion function[180, 987–991] would in principle be more appropriate. Note that the original articles of Lennard-Jones[882,883], published under the name of Jones before his marriage with Kathleen Lennard in 1926, actually refer to a 14⅓-5 functional form[882]. However, it is noted[882] that any inverse n^{th}-power law for the repulsive term would lead to similar results for the temperature dependence of gas-phase argon viscosity, provided that $n > 9$. The actual introduction of the "true" 12-6 Lennard-Jones function dates from the article[993] of 1937, a study of the equation of state of real gases where any n between 9 and 12 gave satisfactory results in comparison to experimental observations, and a value of 12 was selected arbitrarily. The very idea of representing van der Waals interactions as a sum of two terms with opposite signs, each involving a power law with arbitrary exponent, is actually much older[881] and due to Mie in 1903. The Lennard-Jones function could thus be referred to as a 12-6 Mie potential.

It should be noted that the accuracy of the Lennard-Jones function, or of any variant including a more physically appropriate representation of Pauli repulsion, is also limited by: (*i*) the neglect of higher-order dispersion terms[723,724] (all negative, in r^{-2n} with $n \geq 4$); (*ii*) the application of the function to sites that in general represent atoms within molecules rather than isolated closed-shell atoms, and that may in addition carry partial charges within the force-field representation.

Using the above approximations, and omitting the delta function terms enforcing the rigidity of the solvent molecules and the fixed positioning of possible virtual sites, the *potential-energy function* (force-field) for the system of N sites can be written

$$\begin{aligned} V(\mathbf{r}) \;=\; & \sum_{i=1}^{N} \sum_{j=i+1}^{N}{}' \Big[C_{12,ij}\psi_{LJ,12,ij}(\mathbf{r}) - C_{6,ij}\psi_{LJ,6,ij}(\mathbf{r}) \\ & + (4\pi\epsilon_o)^{-1} q_i q_j \psi_{EL,ij}(\mathbf{r}) \Big] + V_{exc}(\mathbf{r}) + V_{srf}(\mathbf{r}) + V_{slf} \;, \end{aligned} \tag{3.16}$$

where q_i and q_j are the partial charges of sites i and j, $C_{12,ij}$ and $C_{6,ij}$ the Lennard-Jones interaction parameters (repulsion and dispersion) for the site pair ij, and the prime after the second summation symbol indicates that pairs ij within a common solvent molecule should be omitted (intramolecular non-bonded exclusions). The functions $\psi_{LJ,12,ij}$, $\psi_{LJ,6,ij}$ and $\psi_{EL,ij}$ are *pairwise interaction functions* defining the form of the interaction between sites i and j. The simplest choices are

$$\psi_{LJ,12,ij}(\mathbf{r}) = r_{ij}^{-12} \quad , \quad \psi_{LJ,6,ij}(\mathbf{r}) = r_{ij}^{-6} \tag{3.17}$$

and

$$\psi_{EL,ij}(\mathbf{r}) = r_{ij}^{-1} \ , \tag{3.18}$$

where r_{ij} denotes the distance between sites i and j. As will be discussed further below, these choices, along with the choices $V_{exc} = V_{srf} = V_{slf} = 0$, lead to a force-field expression appropriate for the simulation of a finite (non-periodic) and microscopic (due to computational costs) system in vacuum, without explicit confinement and mean-field correction for the effect of an implicit surrounding solvent. They must be adjusted (boundary conditions, approximate pairwise interaction functions) for most practical purposes (see further below).

In addition to the pairwise terms, Eq. 3.16 may also encompass an *excluded-sites term* V_{exc}, a *surface term* V_{srf}, and a *self-term* V_{slf}. The former two are configuration-dependent, while the latter one is configuration-independent. These terms are absent in the simple case of a finite system in vacuum (also relying on Eqs. 3.17 and 3.18), but must be clearly specified in all other cases, *e.g.* for a periodic system with exact (lattice-sum) or approximate pairwise interaction functions, or for a finite system with mean-field correction for a surrounding implicit solvent. Broadly speaking, these terms may arise when a contribution to the non-bonded interaction (*e.g.* long-range tail) or to the solvation volume (*e.g.* far-range solvent) is treated implicitly within $V(\mathbf{r})$. In this case, the term V_{exc} accounts for a possible residual pairwise interaction between excluded atom pairs within the same solvent molecule, beyond the deleted direct Coulombic interaction, *via* the implicit component of the model. The term V_{slf} accounts for a possible corresponding interaction of an atom with itself. Finally, the term V_{srf} accounts for a possible corresponding interaction of the dipole moment of the sample (finite system or computational box) with itself. These terms are explicited further below in the context of the different boundary conditions and non-bonded interaction schemes considered.

The generation of configurations, for a molecular system represented at atomic resolution and given a classical potential-energy function of the form described above, is typically performed according to either of two main procedures: (*i*) *molecular dynamics* (MD) simulation[487,490,492]; (*ii*) *Monte Carlo* (MC) sampling[504,546–552].

MD simulation relies on the integration of the classical (Newtonian) equations of motion in time (t), *i.e.*

$$\frac{d^2\mathbf{r}}{dt^2} = \underline{\mathbf{M}}^{-1}\mathbf{F}(\mathbf{r}) \ \text{ with } \ \mathbf{F}(\mathbf{r}) = -\nabla_{\mathbf{r}}V(\mathbf{r}) \ , \tag{3.19}$$

where $\underline{\mathbf{M}}$ is the mass matrix of the system, a diagonal matrix containing the masses of the N sites by triplets along its diagonal, the masses of possible solvent virtual sites being formally set to zero (see below), and $\mathbf{F}$ the $3N$-dimensional force vector derived from $V(\mathbf{r})$ in configuration $\mathbf{r}$. In practice, this integration requires the simultaneous propagation of a $3N$-dimensional vector containing the Cartesian velocities of the N sites, and must be coupled to the constraint enforcement procedure guaranteeing the rigidity of the solvent molecules (see above). In the presence of virtual sites, Eq. 3.19 must be slightly modified to alleviate the singularity problem arising from formally setting the masses of the virtual sites to zero. In practice, the corresponding contribution to the force is redistributed onto the atomic sites involved in the definition of the virtual site prior to integration. The MD approach is a simulation method (Section 3.1), *i.e.* it generates configurations with a well-defined (microcanonical; NVE) probability distribution in a physically-meaningful sequence. For this reason, MD provides thermodynamic as well as dynamic information on the considered molecular system. In practice, Eq. 3.19 is commonly modified by introduction of a thermostatting[568–576,580,994–996] (see Ref.[542] for a review) or/and barostatting[571,580–584] (see introduction of Ref.[567] for an overview) term, permitting the generation of a canonical (NVT), isoenthalpic-isobaric (NPH), or isothermal-isobaric (NPT) probability distribution instead of the microcanonical one. Modifications for the simulation of open ensembles (*e.g.* grand canonical) are also possible[593–599,602,603].

MC sampling[504,546–552] relies on the successive application of random trial moves ($\mathbf{r} \rightarrow \mathbf{r} + \Delta\mathbf{r}$) followed by their conditional acceptance according to a probability depending on a reference temperature T, and given by

$$p_{acc}(\mathbf{r} \rightarrow \mathbf{r} + \Delta\mathbf{r}) = \min\{1;\ \exp[-(V(\mathbf{r} + \Delta\mathbf{r}) - V(\mathbf{r}))/(k_B T)]\} \quad . \quad (3.20)$$

Note that the moves must be generated in such a way that any trial configuration $\mathbf{r} + \Delta\mathbf{r}$ satisfies the constraints enforcing the rigidity of the solvent molecules and the fixed positioning of possible virtual sites within these molecules. The MC approach is a sampling method (Section 3.1), *i.e.* it generates configurations with a well-defined (canonical; NVT) probability distribution, counting all rejected moves as reoccurrences of the same configuration, in an arbitrary sequence. For this reason, MC provides thermodynamic but no dynamic information on the considered molecular system. In practice, Eq. 3.20 may also be modified by introduction of an ergostatting[577–579] or/and barostatting[585–591] term, permitting the generation of a microcanonical (NVE), isothermal-isobaric (NPT), or isoenthalpic-isobaric (NPH) probability distribution instead of the canonical one. Modifications for the sampling of open ensembles (*e.g.* grand canonical) are also possible[355,592,600,601].

The outcome of a thermostatted MD simulation or of a plain MC sampling (or of the variants of the two methods involving barostatting) in terms of configurational probability distributions should in principle be identical in the limit of an infinite ensemble of generated configurations. There may, however, be differences in practice in either of two situations: (*i*) in the presence of metric-tensor effects[997–1007], *i.e.* of a mass-dependent correlation between the configuration and momentum probability distributions arising from the use of constraints in the simulation (see also Refs.[1005,1006,1008–1011] for the contribution of constraints to free-energy changes); (*ii*) in the presence of discontinuities in the potential-energy function, *e.g.* arising from cutoff truncation of the non-bonded interactions. The former effects are present in MD, which samples both the coordinate and the momentum spaces, but not in MC, which samples only the coordinate space. However, in the context of MD, they are absent in systems consisting of monoatomic ions and fully rigid or fully flexible solvent molecules. The latter effects are relevant in MD, due to the omission of corresponding Dirac delta force terms[1012], but not in MC, which only relies on energy differences[1013]. They can be removed by not applying a cutoff, by applying smooth cutoff truncation, or by using a lattice-sum scheme for periodic systems (see further below).

The choice of spatial boundary conditions in an AT simulation is closely connected to that of a method for representing the *electrostatic interactions* in the system (see Refs[1014–1031] for reviews on the topic). The most common types of combinations in the present context of a pure solvent system possibly including a single monoatomic ion are summarized in Figure 3.3. Formally speaking, similar considerations apply to the treatment of van der Waals interactions in the system. However, as will be discussed further below, this treatment typically has a much more limited impact on the simulated properties.

Ideally, one would like the boundary conditions and electrostatic interaction scheme in the simulated system to match the experimental situation as closely as possible. This would correspond to a non-periodic system with exactly Coulombic interactions, compatible with the simple force-field representation suggested by Eqs. 3.16-3.18. However, to match the experimental situation, the system should also be of macroscopic extent. More precisely, in the context of the calculation of a single-ion solvation free energy, it should be sufficiently large so that: (*i*) the entire distance range where ion-solvent interactions contribute significantly to the solvation free energy is encompassed within the sample; (*ii*) surface effects at the liquid-vacuum interface are clearly separated from bulk effects in the neighborhood of the ion.

As an illustration, one may try to estimate the minimal system size required for these conditions to be fulfilled in the simple case of a sodium ion at the center of a spherical water droplet. According to the Born model[84], or, more specifically, to a straightforward extension of Eq. 3.4 to finite systems, the fraction f of the bulk (infinite droplet) charging free energy (leading component of the solvation free energy) accounted for when considering a finite droplet of radius S is $1 - R_I^{\ominus}/S$. Using an estimate $R_I^{\ominus} = 0.17$ nm for the in-solution radius of Na^+, and a fraction $f = 0.995$ (*i.e.* 99.5%, mapping to an error on the order of RT^- in the calculated solvation free energy) leads to a minimal required droplet radius $S = 34$ nm. Assuming that the surface perturbation at the liquid-vacuum interface extends over 10 molecular diameters (*i.e.* about 2.8 nm for water), the droplet radius should be further increased to about 37 nm, so as to ensure that the polarized solvent within the sphere of 34 nm radius is not perturbed by surface effects. This corresponds to a system containing about $7 \cdot 10^6$ water molecules, given a water number density $N_A M_{wat}^{-1} \rho_{wat}^{\ominus}$ of 33.32 nm^{-3} (Table 1.1). In practice, the largest systems that can be simulated nowadays contain on the order of 10^4 water molecules, given a single-processor calculation time of about a month and the sampling of a sufficient number of configurations to obtain a precise estimate for an ionic solvation free energy, *i.e.* about 10^6 configurations. Assuming exact Coulombic electrostatics, implying an $\mathcal{O}[N^2]$ scaling of the computational cost with the number N of interaction sites, the calculation of an ionic solvation free energy for such a system would require about 40000 years of single-processor calculation time at present. Using Moore's law[1032], one may estimate that this time will be reduced to a month around year 2045. This simple model calculation illustrates the fact that the system sizes currently accessible to AT simulations are still very far from reaching the macroscopic (experimental) regime. As a result, they are unavoidably affected by both *finite-size effects* (system too small to encompass the entire range where ion-solvent interactions contribute significantly to the solvation properties) and surface effects (liquid-vacuum interface too close to the ion and its polarized solvent environment to permit a clear-cut separation between bulk and surface influences on

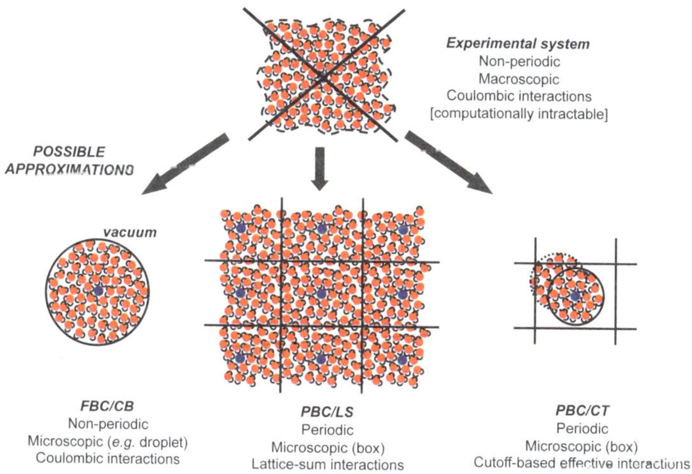

Figure 3.3 *Types of boundary conditions and electrostatic schemes commonly employed in classical atomistic simulations.* The figure displays the experimental situation (top; computationally intractable) and common approximations for the boundary conditions and electrostatic scheme (bottom) employed in the context of the determination of the intrinsic bulk electric potential or of the air-liquid interfacial potential of a liquid (no ion) or the calculation of an intrinsic single-ion solvation free energy (single monoatomic ion; situation illustrated in the figure). The ion is represented as a blue sphere and the solvent is depicted by water molecules. The acronyms and notations employed are defined in Table I.

the solvation properties). Recent overviews of these problems can be found in Refs.[1028, 1031]. In turn, the analysis of these simulations has to deal with these effects, *e.g.* by means of appropriate correction terms (Section 6.2.1), if reliable results are to be obtained.

The computationally tractable choices of boundary conditions and approximate electrostatic schemes used in the present context of a pure solvent possibly including a single monoatomic ion are the following (Figure 3.3; see Refs.[492, 503, 515, 1014–1017, 1019–1027, 1029–1031, 1033–1037] and in particular Ref.[1031] for recent overviews): (*i*) *fixed boundary conditions*[565, 566, 1038–1067] (FBC) with *Coulombic* (CB) electrostatics; (*ii*) *periodic boundary conditions*[490, 564, 1068] (PBC) with *lattice-sum* (LS) electrostatics; (*iii*) *periodic boundary conditions*[490, 564, 1068] (PBC) with *cutoff-based* (CT) electrostatics.

In FBC/CB[565, 566, 1038–1067], the system consists of a *fixed solvent volume* (*e.g.* a spherical droplet) surrounded by vacuum. The interactions are then handled using Eqs. 3.16-3.18, *i.e.* including in particular an *exact Coulombic sum* over all site pairs for the electrostatic interactions, except for intramolecular non-bonded exclusions, along with $V_{exc} = V_{srf} = V_{slf} = 0$. A confinement potential (*e.g.* Lennard-Jones wall) is typically also introduced to prevent solvent evaporation at the sample surface. Because the computational cost of such a simulation scales as $\mathcal{O}[N^2]$ (summation over all unique pairs among the N sites, excluding pairs within the same solvent molecule), tractable systems remains truly microscopic (see above), and FBC/CB simulations are affected by strong finite-size and surface effects. On the other hand, these simulations are expected to provide a reasonable description of ionic solvation within microscopic droplets, unfortunately largely irrelevant in the context of bulk single-ion solvation properties. Note, however, that numerous modifications of the FBC scheme have been designed with the goal of reproducing bulk-like rather than finite-droplet behavior[565, 1040, 1042, 1044, 1051, 1054, 1055, 1058, 1061, 1067, 1069–1071]. This requires the calibration of appropriate confinement and reaction-field potentials, that not only prevent solvent evaporation, but also ideally correct for the perturbation of the solvent structure at the interface, reintroduce missing long-range solvation forces in a mean-field fashion, and compensate for surface tension effects and the resulting pressure increase in the droplet. The introduction of such potentials will generally imply non-vanishing V_{exc}, V_{srf} and V_{slf} contributions in Eq. 3.16. Unfortunately, the calibration of these potentials is a difficult task, and their transferability from one system to another is often limited. The computational cost of a FBC calculation may also be reduced by application of cutoff truncation (see below), possibly along with a modification of the functional form of the interaction, in order to afford larger system sizes. However, such modifications are probably not recommended in the present context of the calculation of an intrinsic single-ion solvation free energy or of the air-liquid interfacial potential of a solvent, where it is certainly preferable to use a smaller system size but exactly Coulombic interactions, together with the application of appropriate correction terms (Section 6.2.1). As mentioned above, the computational cost of a

FBC/CB calculation scales as $\mathcal{O}[N^2]$. As a result, tractable systems remain truly microscopic and these simulations are affected by *strong finite-size and surface effects.*

In PBC[490,564,1068], the system consists of a *computational box* (space-filling shape, *e.g.* a cube) that is surrounded by an infinite lattice of *periodic replicas* of itself, so as to define an infinite periodic system. The interactions are then handled using a potential-energy function of the form of Eq. 3.16, in which N represents the number of particles in the box, along with suitably modified (periodic) Lennard-Jones and electrostatic pairwise interaction functions, as well as the possible inclusion of non-vanishing excluded-sites, surface and self-terms.

For the Lennard-Jones interactions under PBC, one usually applies minimum-image *spherical cutoff truncation*, and the corresponding pairwise interaction functions to be used in Eq. 3.16 are written

$$\psi_{LJ,12,ij}(\mathbf{r}) = H_{ij}(\mathbf{r}; R_C)\overline{r}_{ij}^{-12} \tag{3.21}$$

and

$$\psi_{LJ,6,ij}(\mathbf{r}) = H_{ij}(\mathbf{r}; R_C)\overline{r}_{ij}^{-6} \quad , \tag{3.22}$$

where $\overline{r}_{ij}$ denotes the minimum-image distance between sites i and j, *i.e.* the distance between site i within the computational box and the closest periodic copy of site j after application of lattice translations within the periodic system, and R_C represents a cutoff distance, assumed here to be smaller than half the shortest edge of the computational box. The *truncation function* H_{ij} may be defined in either of two ways, denoted P (by site) or M (by molecule), and further referred to as *P-truncation* (or atomic cutoff truncation) and *M-truncation* (or group-based cutoff truncation). In P-truncation, the cutoff criterion is applied to the actual minimum-image distance $\overline{r}_{ij}$ between the sites i and j, *i.e.* H_{ij} in Eqs 3.21 and 3.22 is written

$$H_{P,ij}(\mathbf{r}; R_C) = h(R_C - \overline{r}_{ij}) \ , \tag{3.23}$$

where h is the Heaviside step function, *i.e.* $h(x) = 0$ for $x \leq 0$ and $h(x) = 1$ for $x > 0$. In contrast, M-truncation relies on the definition of a reference point, further referred to here as a *molecular center*, for the solvent molecule (for water, this center is normally taken to be the oxygen site). A monoatomic ion is also considered as a "molecule" in this context, the "molecular" center coinciding in this case with the single interaction site. The cutoff criterion is then applied to the minimum-image distance $\overline{R}_{ij}(\mathbf{r})$ between the centers of the molecules to which sites i and j belong, *i.e.*

$$H_{M,ij}(\mathbf{r}; R_C) = h(R_C - \overline{R}_{ij}(\mathbf{r})) \ . \tag{3.24}$$

Note that when P-truncation is applied for the Lennard-Jones interactions, the pairwise interaction functions $\psi_{LJ,12,ij}(\mathbf{r})$ and $\psi_{LJ,6,ij}(\mathbf{r})$ may be rewritten as ij-independent functions of $\overline{r}_{ij}$, *i.e.* as $\psi_{LJ,12}(\overline{r}_{ij})$ and $\psi_{LJ,6}(\overline{r}_{ij})$.

This is, however, no longer possible when M-truncation is applied, because $H_{M,ij}(\mathbf{r}; R_C)$ depends on the positions of the molecular centers of the molecules to which i and j belong (Eq. 3.24). In practice, the choice of P-truncation or M-truncation makes little difference in the context of Lennard-Jones interactions given the cutoff distances typically used in simulations. Often, it makes no difference at all for systems involving water and monoatomic ions, because the solvent molecular center is generally taken to be the oxygen atom while most water models do not include Lennard-Jones interaction sites at the hydrogen atoms. However, the distinction becomes crucial in the context of electrostatic interactions (see further below). Typical cutoff distances R_C employed in simulations are on the order of $1.0 - 1.5$ nm. Because the Lennard-Jones forces are generally very small beyond such distances and tend to cancel out upon spherical averaging around a given atom, the error caused by the truncation of the Lennard-Jones interactions on the generated system configurations is typically very small. This may, however, not be the case for the corresponding error on the calculated energy and virial (pressure), which do not benefit from such a cancellation by spherical averaging. An unusually large impact of the Lennard-Jones cutoff distance on the simulated properties may also be encountered in strongly anisotropic systems, where error cancellation in the forces upon spherical averaging is less effective. Note, finally, that Lennard-Jones interactions are generally associated with no contribution to the excluded-sites, surface and self-terms in Eq. 3.16. Possible exceptions will be mentioned at the end of this section.

For the electrostatic interactions under PBC, one distinguishes (Figure 3.3) between PCB/LS, *i.e.* lattice-sum schemes and PBC/CT, *i.e.* cutoff-based schemes. These two types of approaches are described in turn below.

In PBC/LS[330,1072–1078], the electrostatic term of Eq. 3.16 is reformulated as a *lattice-sum*, including the pairwise interactions between all sites i within the computational box and all distinct sites j in the entire periodic system. The corresponding pairwise interaction function to be used in Eq. 3.16 is written in the form of a reciprocal-space sum, as

$$\psi_{EL,ij}(\mathbf{r}) = 4\pi \, |\underline{\mathbf{L}}|^{-1} \sum_{\mathbf{l}, l \neq 0} k^{-2} \exp[i\mathbf{k} \cdot \mathbf{r}_{ij}] \;, \tag{3.25}$$

where $\underline{\mathbf{L}}$ is a 3×3-matrix containing the Cartesian components of the box-edge vectors in its columns (triclinic computational box), $\mathbf{l}$ is a vector with (positive or negative) integer components, $\mathbf{k} = 2\pi\underline{\mathbf{L}}^{-1}\mathbf{l}$ is the associated reciprocal-lattice vector, and $\mathbf{r}_{ij} = \mathbf{r}_j - \mathbf{r}_i$ is the vector connecting sites i and j in the computational box. The resulting pairwise interaction is *formally Coulombic* in the sense that it becomes exactly Coulombic for pairwise distances within a finite range when the size of the computational box becomes infinite[329–331, 1079–1081]. However, there are some important differences in terms of the single-charge electric potential it derives from, namely[330, 1082]: (i) this potential is a particular solution of a Poisson equation involving all periodic copies of the charge as well as a homogeneous neutralizing background charge density, to be compared with a single charge and no background charge for the Coulomb potential; (*ii*) this potential satisfies the boundary conditions of a vanishing average and a vanishing derivative (field) average over the computational box, to be compared with a vanishing value and derivative at infinite distance from the charge for the Coulomb potential.

Another difference is that an excluded-sites term V_{exc} and a self-term V_{slf} are commonly introduced in Eq. 3.16. The term V_{exc} accounts for the pairwise interaction of a site i within the computational box with an excluded site j within the same solvent molecule, as well as with its periodic copies, from which only the direct (minimum-image) Coulombic interaction has been subtracted, namely[1082, 1083]

$$V_{exc}(\mathbf{r}) = (4\pi\epsilon_o)^{-1} \sum_{i=1}^{N} \sum_{j=i+1}^{N}{}'' q_i q_j \left[\psi_{EL,ij}(\mathbf{r}) - \overline{r}_{ij}^{-1}\right] \quad , \tag{3.26}$$

where the double-prime after the second summation symbol indicates that only excluded atom pairs should be considered *i.e.*, in the present context, all atom pairs within a common solvent molecule. This term is introduced to account for the fact that excluded sites should only be excluded in their direct Coulombic interactions, but not in terms of their formally intermolecular interactions across periodic boundaries. The term V_{slf} accounts for the interaction of a site i within the computational box with its own periodic copies (but not with itself) and with the homogeneous neutralizing background charge density, namely[330, 1082]

$$V_{slf} = (8\pi\epsilon_o)^{-1} \tilde{\alpha}_{LS}(\underline{\mathbf{L}}) \sum_{i=1}^{N} q_i^2 \tag{3.27}$$

where

$$\tilde{\alpha}_{LS}(\mathbf{L}) = \lim_{r\to 0} \left\{ \left(4\pi \left|\underline{\mathbf{L}}\right|^{-1} \sum_{\mathbf{l}, l\neq 0} k^{-2} \exp[i\mathbf{k}\cdot\mathbf{r}] \right) - r^{-1} \right\} . \tag{3.28}$$

This term is introduced to account for the fact that an atom presents formally intermolecular interactions with its own copies across periodic boundaries. This interaction energy is finite due to the presence of the neutralizing background charge density. To our knowledge, $\tilde{\alpha}_{LS}$ is only known in closed form for the special case of a cubic computational box[330,792,1082,1084,1085], where $\tilde{\alpha}_{LS} = \alpha_{LS}/L$ with $\alpha_{LS} = -2.837297$, L being the box-edge length. However, it is easily evaluated numerically for other box shapes. Most implementations of the LS scheme in current simulation programs include the terms V_{exc} and V_{slf} of Eqs. 3.26 and 3.27 in Eq. 3.16. The latter term is actually automatically introduced in computationally efficient implementations of this method (see further below), although it could be subsequently subtracted if desired.

Finally, a surface term V_{srf} is also sometimes introduced in Eq. 3.16 to account for the interaction between the charges within the computational box and a non-conducting dielectric medium assumed to encompass the infinite periodic system. Three possible choices have been proposed for this term in the literature, referred to as vacuum[1086] (external permittivity of one), adjusted[331,1080,1087,1088] (external permittivity equal to that of the solvent) and tinfoil[330,1028,1031,1073,1084] (infinite external permittivity) boundary conditions. Assuming that the infinite periodic system is constructed as a spherical assembly of cubic computational boxes, the term V_{srf}, corresponding to the limit of an infinite sphere radius, can be written

$$V_{srf} = \frac{\mathbf{M}^2}{\epsilon_o V(2\epsilon_{srf}+1)} \quad , \tag{3.29}$$

where $\mathbf{M}$ is the box dipole moment, V the box volume and ϵ_{srf} the permittivity or the external medium assumed to surround the periodic system. Besides the choice of ϵ_{srf}, the application of Eq. 3.29 involves a number of other intricate issues concerning: (*i*) the inclusion or omission of the ion contribution to $\mathbf{M}$, considering that this contribution depends on the arbitrary choice of a box origin; (*ii*) the possible consideration of an itinerant[1089] rather than box-defined dipole moment $\mathbf{M}$, which would in particular offer a way to solve the latter problem; (*iii*) the justification for the arbitrary choice of a growing spherical assembly (rather than any other shape) of computational boxes; (*iv*) the generalization to non-cubic computational boxes; (*v*) the possible additional inclusion of a term depending on the quadrupole-moment trace of the box[1082]. Although the proper choice for the surface term is still matter of debate[330,331,1028,1031,1080,1082,1084,1086–1098] (see Refs.[330,1028,1031,1082] for recent discussions), its effect in the context of pure water simulations and ionic solvation free-energy calculations appears to be essentially negligible[1081] (see, however, Refs.[1086,1095]). Most implementations of the LS scheme in current simulation programs exclude the term V_{srf}, *i.e.* rely on tinfoil boundary conditions. Based on considerations developed in Section 4.3.1 (see *e.g.* discussion of Figure 4.4), the authors of the present book think that tinfoil boundary conditions are the appropriate choice if a simulation is to represent a bulk situation within a conducting phase (including aqueous solutions and even pure water considering autoprotolysis), because the electric field in such a phase must vanish as a result of screening by compensating charges at its surface (Faraday cage). Tinfoil boundary conditions nevertheless have their detractors[331,1073,1080,1099,1100], the main argument being that this choice may artificially enhance dipole alignment within the system. Note, finally, that the medium assumed to surround the periodic system may also be thought to encompass a surface polarization, leading to a corresponding offset in the electric potential within the system[1082]. This contribution is also typically omitted, in which case the potential is determined by a constraint of vanishing average over the computational box.

The electrostatic term resulting from the insertion of Eq. 3.25 into Eq. 3.16

consists of a triple sum over primary sites i, secondary sites j, and lattice vectors $\mathbf{l}$. However, in practice, it can be rewritten as a double sum over single sites i and lattice vectors $\mathbf{l}$, provided that it also includes the self-term of Eq. 3.27. Many numerical methods have been developed to further increase the evaluation efficiency of this double sum, including the Ewald[1072] or related lattice-summation[1101–1111], particle-mesh (P^3M[330,1112], PME[1113] or SPME[1114]), fast-multipole (FMM[1115–1130] under PBC), multigrid (MG[1131–1133] under PBC), fast-Fourier Poisson (FFP[1134,1135] under PBC) or Maggs[1136–1140] approaches. The computational cost of a PBC/LS calculation scales at best as $\mathcal{O}[N \log N]$, which is more favorable than the $\mathcal{O}[N^2]$ of FBC/CB. Nevertheless, tractable systems (size of the computational box) remain truly microscopic, so that PBC/LS simulations are also affected by *strong finite-size effects*, just as FBC/CB simulations. In particular, the *periodicity* and *anisotropy* of the LS electrostatic interactions are entirely *artificial features* when simulating solutions, which are inherently non-periodic and isotropic media. On the other hand, because a periodic system is formally of infinite extent, *surface effects* have been *entirely eliminated.*

In PBC/CT[490], the electrostatic term of Eq. 3.16 is reformulated as a truncated sum, only including the pairwise interactions between sites i and j within the computational box for which the minimum-image distance is below a specified *cutoff distance.* Furthermore, the functional form of the interaction may either *remain Coulombic* or be *further modified.* The corresponding pairwise interaction function to be used in Eq. 3.16 is accordingly written as

$$\psi_{EL,ij}(\mathbf{r}) = H_{ij}(\mathbf{r}; R_c)\tau_{CT}(\overline{r}_{ij}) \quad , \tag{3.30}$$

where R_C is the cutoff distance, assumed here to be smaller than half the shortest edge of the computational box and usually taken equal to the cutoff distance also applied for Lennard-Jones interactions, and τ_{CT} is a Coulombic or modified *effective pairwise interaction function* (see further below). Here also, the truncation function H_{ij} may be defined in either of two ways, corresponding to *P-truncation* (Eq. 3.23) or *M-truncation* (Eq. 3.24). Note that when P-truncation is applied for the CT electrostatic interactions, the pairwise interaction function $\psi_{EL,ij}(\mathbf{r})$ may be rewritten as an ij-independent function of $\overline{r}_{ij}$, *i.e.* as $\psi_{EL}(\overline{r}_{ij})$. This is, however, no longer possible when M-truncation is applied, because $H_{M,ij}(\mathbf{r}; R_C)$ depends on the positions of the molecular centers of the molecules to which i and j belong (Eq. 3.24). In contrast to the choice of a truncation scheme for the Lennard-Jones interactions (see above), the choice of P-truncation or M-truncation for the CT electrostatic interactions may have a very strong impact on all simulated properties (see *e.g.* Section 3.3.6 and Figure 3.8) (see below). This impact depends on the effective pairwise interaction function τ_{CT} employed and on the value selected for the cutoff distance. Typical cutoff distances R_C employed in simulations are on the order of 1.0 - 1.5 nm. Because the electrostatic interactions are long-ranged, the error on the forces, energy and virial (pressure) caused by their truncation is typically very large, and the application of a cutoff represents a major source of error in AT simulations, irrespective of the choice of the effective pairwise interaction function.

Some CT schemes also introduce an excluded-sites term V_{exc} and a self-term V_{slf} in Eq. 3.16. Whenever included, these terms, analogous to those defined by Eqs. 3.26-3.28 in the PBC/LS case, are given by[1082,1083]

$$V_{exc}(\mathbf{r}) = (4\pi\epsilon_o)^{-1} \sum_{i=1}^{N} \sum_{j=i+1}^{N}{}'' q_i q_j \left[\tau_{CT}(\overline{r}_{ij}) - \overline{r}_{ij}^{-1}\right] \tag{3.31}$$

and

$$V_{slf} = (8\pi\epsilon_o)^{-1} \tilde{\alpha}_{CT} \sum_{i=1}^{N} q_i^2 \tag{3.32}$$

where

$$\tilde{\alpha}_{CT} = \lim_{r \to 0} \left[\tau_{CT}(r) - r^{-1}\right] \quad . \tag{3.33}$$

Although the introduction of a surface term V_{srf} is uncommon, this possibility has also been considered[1082].

The computational cost of a PBC/CT calculation scales at best as $\mathcal{O}[N]$, which is significantly more favorable than the $\mathcal{O}[N^2]$ of FBC/CB, and somewhat more favorable than the $\mathcal{O}[N \log N]$ of PBC/LS. Note that a linear scaling can only be achieved by means of grid-based pairlist methods[1141–1164] (see the introduction of Refs.[1163,1164] for recent overviews). Nevertheless, tractable systems (size of the computational box) remain truly microscopic, so that PBC/CT simulations are also affected by *strong finite-size effects*, just as FBC/CB and PBC/LS simulations. On the other hand, because a periodic system is formally of infinite extent, *surface effects* have been *entirely eliminated*, just as in PBC/LS simulations. In addition, given the cutoff distances typically employed in AT simulations (see above), neglecting the contribution of electrostatic interactions beyond the cutoff distance R_C represents an approximation that is difficult to justify, considering the long-ranged nature of these interactions. As a result, PBC/CT methods also generally suffer from *severe cutoff-induced artifacts*.

The PBC/CT methods can be further classified as: (*i*) *straight cutoff* (SC) methods; (*ii*) *reaction-field* (RF) methods; (*iii*) *modified truncation* (MT) methods. A recent overview of these approaches (as well as PBC/LS) can be found in Ref.[1031]. These three types of approaches are described in turn below. Note that the corresponding equations are formulated with the constraint $\tau_{CT}(R_C) = 0$, to ensure the continuity of the electrostatic interaction function at the cutoff distance. This constraint is not absolutely necessary in MC, which can sample discontinuous potential-energy functions, but it is in MD, which cannot take into account such discontinuities (unless a corresponding Dirac delta function force term is explicitly included[1012]).

PBC/CT/SC methods[490,492] rely on truncating electrostatic interactions while preserving their *Coulombic form*, *i.e.* using

$$\tau_{SC}(r) = r^{-1} - R_C^{-1} \tag{3.34}$$

in Eq. 3.30. When applying the SC scheme to the simulation of systems involving a dipolar solvent, M-truncation (Eq. 3.24) is the only viable option. The reason is that the cutoff noise incurred by the truncation of unmodified Coulombic interactions in a system of dipolar molecules is of order $\mathcal{O}[R_C^{-3}]$ (cutoff effectively applied to dipole-dipole interactions) for M-truncation, but of order $\mathcal{O}[R_C^{-1}]$ for P-truncation (cutoff effectively applied to charge-charge interactions). As a result, simulations of dipolar fluids (*e.g.* water) evidence dramatic cutoff artifacts when the SC scheme is applied with P-truncation[1013]. Obviously, however, the error is still of order $\mathcal{O}[R_C^{-1}]$ in both cases when multiple ions are present, or of order $\mathcal{O}[R_C^{-2}]$ for a single solvated ion. For this reason, simulations of systems involving charged species (*e.g.* electrolyte solutions, ionic crystals, melts, or liquids, macromolecules with charged functional groups) are also affected by very large cutoff artifacts even when using M-truncation[328,331,333,334,357,1071,1081,1086,1088,1165–1212]. No excluded-sites term, self-term or surface term is generally introduced in Eq. 3.16 for the SC scheme, although a self-term of the form of Eq. 3.37 with $\epsilon_{RF} = 1$ could be introduced for consistency with the RF scheme (see below).

PBC/CT/RF methods[1013,1050,1082,1185,1213–1215] rely on truncating electrostatic interactions and altering their functional form by using the so-called *reaction-field scheme*, namely

$$\tau_{RF}(r) = r^{-1} + \frac{\epsilon_{RF} - 1}{2\epsilon_{RF} + 1}\frac{r^2}{R_C^3} - \frac{3\epsilon_{RF}}{2\epsilon_{RF} + 1}\frac{1}{R_C} \tag{3.35}$$

in Eq. 3.30. This modification is physically-motivated in the context of dipolar systems[1013,1185,1213,1214]. The modified interaction function of Eq. 3.35 includes, in an effective pairwise form, the mean effect of the omitted electrostatic interactions beyond the cutoff distance, by assuming that the medium surrounding the cutoff sphere of each particle is a dielectric continuum of static relative dielectric permittivity ϵ_{RF}, rather than vacuum in the SC scheme. In the case of microscopically-homogeneous solutions (on the cutoff scale), a reasonable choice is to set ϵ_{RF} to the mean permittivity of the simulated system, *e.g.* to the pure solvent permittivity for dilute solutions. However, if the solution is microscopically heterogeneous (*e.g.* solvated macromolecule), the choice of this parameter and the validity of the RF approach may be questioned[1082], because the medium outside the cutoff sphere differs from particle to particle and may be of heterogeneous permittivity. Furthermore, the RF approach was derived in the context of dipolar fluids, and there is substantial evidence for important cutoff artifacts affecting systems that are rich in charged species[333,334,357,1202,1205,1216–1219]. The RF scheme is usually applied using M-truncation, but P-truncation is also possible here[707,1013,1031,1081,1082,1216,1220]. The latter implementation may offer some advantages in the case of polar solvents (large ϵ_{RF} value), because the effective interaction of Eq. 3.35 is characterized by a quasi-vanishing value and derivative at R_C in this case. In contrast to P-truncation, M-truncation may involve the occurrence of charge-charge distances r larger than R_C in Eq. 3.35, for which the interaction is unphysically attractive for like charges and repulsive for opposite charges. Most implementations of the RF scheme in current simulation programs exclude the terms V_{exc} and V_{slf} in Eq. 3.16. However, some recent programs do include them for consistency[1082,1083]. In this case, Eqs. 3.31-3.33 and 3.35 imply

$$\begin{aligned} V_{exc}(\mathbf{r}) &= (4\pi\epsilon_o)^{-1}\left[\frac{\epsilon_{RF}-1}{2\epsilon_{RF}+1}\frac{1}{R_C^3}\left(\sum_{i=1}^{N}\sum_{j=i+1}^{N}{}'' q_i q_j \bar{r}_{ij}^2\right)\right. \\ &\left. -\frac{3\epsilon_{RF}}{2\epsilon_{RF}+1}\frac{1}{R_C}\left(\sum_{i=1}^{N}\sum_{j=i+1}^{N}{}'' q_i q_j\right)\right] \end{aligned} \tag{3.36}$$

and

$$V_{slf}(\mathbf{r}) = -(8\pi\epsilon_o)^{-1}\frac{3\epsilon_{RF}}{2\epsilon_{RF}+1}\frac{1}{R_C}\sum_i q_i^2 \quad . \tag{3.37}$$

Considering a single rigid dipolar solvent molecule, it is easily seen from Eqs. 3.16, 3.36 and 3.37 that $V_{exc} + V_{slf}$ is configuration-independent and evaluates to

$$V_{exc} + V_{slf} = (8\pi\epsilon_o)^{-1}\frac{2(\epsilon_{RF}-1)}{2\epsilon_{RF}+1}\frac{\mu_d^2}{R_C^3} \quad , \tag{3.38}$$

where μ_d is the molecular dipole moment. This term is thus equivalent to the solvation free energy of the solvent molecule within a cavity of radius R_C surrounded by a dielectric continuum of permittivity ϵ_{RF} as given by the Onsager model[671], and is therefore physically reasonable. On the other hand, for a single monoatomic ion (charge q_I), $V_{exc} = 0$ and one has

$$V_{slf} = -(8\pi\epsilon_o)^{-1}\frac{3\epsilon_{RF}}{2\epsilon_{RF}+1}\frac{q_I^2}{R_C} \quad , \tag{3.39}$$

which has no clear physical interpretation and is in particular not equivalent to the corresponding Born expression[84] for a continuous solvent outside the cutoff sphere of the ion. Usually, no surface term is introduced in Eq. 3.16 for the RF scheme, although the possibility of including such a term has been considered[1082] (note, however, that the self-term expression proposed therein slightly differs from Eq. 3.37).

Finally, the PBC/CT/MT methods rely on truncating electrostatic interactions and altering their functional form in an *ad hoc* fashion. This modification is generally performed so as to enforce vanishing values of the effective interaction function τ_{CT} and of its derivative at R_C. Variants of the MT approach include the use of *switching*[1019,1166,1173,1182,1197,1221–1224], *shifting*[792,1019,1071,1166,1173,1182,1197,1208,1221,1222,1225–1232] or *damping*[1211,1233–1235] schemes. A few examples are: (*i*) the CHARMM[1221] switching function[1182]

$$\tau_{MT,CSW}(r) = \begin{cases} r^{-1} & r \leq R_B \\ h(R_C - r) r^{-1} \frac{(R_C^2 - r^2)^2 (R_C^2 + 2r^2 - 3R_B^2)}{(R_C^2 - R_B^2)^3} & r > R_B \end{cases} , \tag{3.40}$$

where R_B (with $0 < R_B < R_C$) and R_C determine the switching range; (*ii*) the CHARMM[1221] shifting function[1182]

$$\tau_{MT,CSH}(r) = h(R_C - r) \left(r^{-1} - \frac{2r}{R_C^2} + \frac{r^3}{R_C^4} \right) \quad ; \tag{3.41}$$

(*iii*) the generalized force shifting function[1182]

$$\tau_{MT,GSH}(r) = h(R_C - r) \left(r^{-1} + \frac{r^\beta}{\beta R_C^{\beta+1}} - \frac{1}{R_C} - \frac{1}{\beta R_C} \right) \quad , \tag{3.42}$$

where β (with $\beta > 0$) determines the rate of the force damping ($\beta = 1$ corresponds to what is commonly referred to as force shifting[1182]; note the presence of a sign error in Eq. 66 of Ref.[1194]); (*iv*) the generalized reaction field (GRF) shifting function[792,1227,1228]

$$\tau_{MT,GRF}(r) = h(R_C - r) \left(r^{-1} - \frac{12}{5R_C} + \frac{4r^2}{R_C^3} - \frac{3r^3}{R_C^4} + \frac{2r^5}{5R_C^6} \right) \quad ; \tag{3.43}$$

(*v*) the charge-neutralized Ewald shifting function[1233] (also referred to as Wolf damping[1211,1233–1235], WDP)

$$\tau_{MT,WDP}(r) = h(R_C - r) \left[r^{-1} \mathrm{erfc}(a^{-1} r) - R_C^{-1} \mathrm{erfc}(a^{-1} R_C) \right] \quad , \tag{3.44}$$

where erfc is the complementary error function and a is the width of an effective Gaussian charge distribution replacing the point charge. Both M-truncation and P-truncation have been used in MT schemes[707,1081,1182,1197,1236–1239]. Most implementations of MT schemes in current simulation programs exclude the terms V_{exc} and V_{srf} in Eq. 3.16. See, however, Ref.[792] for the GRF and Refs.[1211,1233–1235] for the WDP schemes. Usually, no surface term is introduced in Eq. 3.16 for the MT schemes.

Note, finally, that the SC scheme can be viewed as a special case of MT scheme with no modification after truncation (except for the addition of an offset $-R_C^{-1}$, Eq. 3.34), and that the RF scheme becomes equivalent to the SC scheme in the limit $\epsilon_{RF} \to 1$ (no solvent) and to a MT (shifting) scheme in the limit $\epsilon_{RF} \to \infty$ (conducting solvent; almost satisfied in practice for highly polar solvents such as water).

For completeness, it should be added that a number of approaches have been proposed to alleviate the artificial periodicity and anisotropy induced by the use of LS methods under PBC. Among these, one may mention: (*i*) orientational averaging of the LS potential[1225, 1229–1232], effectively leading to a MT scheme; (*ii*) hybridization of LS with FBC or RF methods[335, 1050, 1082, 1172, 1177, 1240]; (*iii*) use of hyperspherical boundary conditions[1241–1251] (HBC). However, these methods have only found limited use in practical applications, compared to the most popular LS, SC, RF and MT methods. Note also that if the truncation of the Lennard-Jones interactions generally has a minor effect on the forces, given the cutoff distances typically used in simulations, it can have a significant impact on the energy and virial (pressure). These artifacts can be largely remedied by using LS-type schemes[856, 1082, 1117, 1252–1261] or tail corrections[490, 856, 1258, 1261–1269] for these interactions. The tail corrections are generally designed to be configuration-independent, so that they effectively represent a self-term V_{slf} in Eq. 3.16.

As a final (but important) note, current simulation programs may present different implementations of the various electrostatic schemes concerning the excluded-sites term V_{exc}, the self-term V_{slf} and the surface term V_{srf} (PBC), as well as concerning the choice of P- or M-truncation (PBC/CT). For this reason, these choices should be specified very clearly when reporting the results of AT simulations. Of particular importance in the context of single-ion solvation free energies is the choice of V_{slf}, because this term appears directly in the calculated charging free energies as an offset constant (Section 3.3.3) and is essentially artificial (*e.g.* Eq. 3.27 for PBC/LS or Eq. 3.32 for PBC/RF).

3.3.2 Parameters

Since they rely on empirical parameters that must be calibrated based on experimental and, possibly, quantum-mechanical data, AT simulations (Section 3.3.1) are not first-principles approaches. Considering a potential-energy function of the form of Eq. 3.16, the following parameters need to be specified: (*i*) the geometric parameters of the rigid solvent molecule, including the fixed locations of possible virtual sites; (*ii*) the partial charges of the different sites in the solvent molecule, including atoms and possible virtual sites; (*iii*) the pairwise Lennard-Jones parameters for the interaction between the different sites within two solvent molecules, including atoms and possible virtual sites; (*iv*) the ionic charge; (*v*) the pairwise Lennard-Jones parameters for the interaction between the ion and the different sites within a solvent molecule, including atoms and possible virtual sites; (*vi*) the pairwise Lennard-Jones parameters for the interaction between two ions of the same or of different types, if ionic crystals, melts or solutions at finite concentrations are to be considered. The first three types of parameters exclusively belong to the definition of a solvent model. The fourth parameter, namely the ionic charge, is unambiguous. The fifth type of parameters concerns the ion in its interaction with the selected solvent model. Finally, the last type of parameters concerns ions and their mutual interactions. For simplicity, the discussion will be restricted to the case of water as a solvent.

Nearly all commonly used *water* models are characterized by a single Lennard-Jones interaction site located at the oxygen atom (see, however Refs.[900,1270–1273]). The corresponding single pair of *solvent-solvent Lennard-Jones interaction parameters* characterizing the interaction between the oxygen sites of two water molecules will be noted C_6^{ss} and C_{12}^{ss}, referring to the coefficients introduced in Eq. 3.12. Similarly, for a given monoatomic ion, the single pair of *ion-solvent Lennard-Jones interaction parameters* characterizing the interaction between the ion and the oxygen site of a water molecule will be noted C_6^{is} and C_{12}^{is}, and the single pair of *homoionic ion-ion Lennard-Jones interaction parameters* characterizing the interaction between two ions of this type, will be noted C_6^{ii} and C_{12}^{ii}. Finally, pairs of *heteroionic ion-ion Lennard-Jones interaction parameters* characterizing the interaction between two ions of different (given) types will be noted $C_6^{ii'}$ and $C_{12}^{ii'}$. Alternatively, all these parameters may also be written in terms of the distance at zero interaction R_{zer} or of the distance at the interaction minimum R_{min}, along with the energy at the interaction minimum E_{min}, the interconversion between C_6, C_{12} and these quantities being given by Eq. 3.13. The same superscripts will be used for the corresponding parameters R_{zer}, R_{min} and E_{min} as defined above in the context of C_6 and C_{12}. Finally, in the case of a solvent model also involving Lennard-Jones interaction sites at the hydrogen atoms, the corresponding additional Lennard-Jones interaction parameters may be written C_6^{sh} and C_{12}^{sh} (solvent-solvent, interaction between hydrogen and oxygen atoms), C_6^{hh} and C_{12}^{hh} (solvent-solvent, interaction between hydrogen atoms), and C_6^{ih} and C_{12}^{ih} (ion-solvent, interaction of the ion with hydrogen atoms). These considerations are easily further generalized to other solvents involving an arbitrary number of Lennard-Jones interaction sites, as well as to polyatomic ions.

The geometry, charges and Lennard-Jones interaction parameters characterizing some of the most commonly employed water models are summarized in Figure 3.4 and Table 3.1. The associated molecular dipole moments, as well as characteristic values for the simulated density and vaporization enthalpy

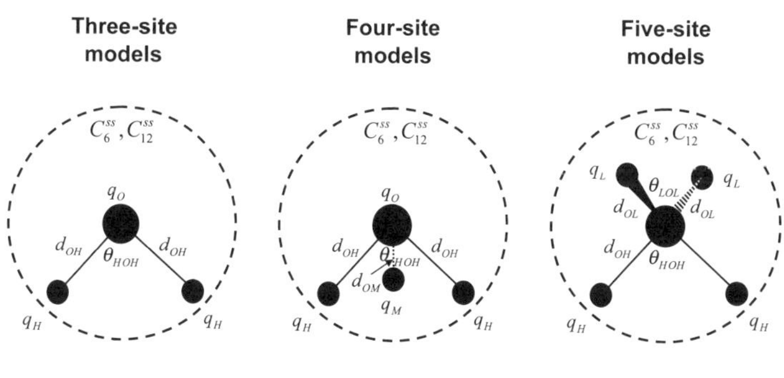

Figure 3.4 *Common types of non-polarizable water models used in atomistic simulations.* Corresponding sets of parameters, along with other characteristics, are reported in Table 3.1 for a few widely-used water models.

are also reported. The latter quantities depend sometimes significantly on the simulation protocol employed, which is also indicated. Water models (geometry, charges, C_6^{ss} and C_{12}^{ss}) are typically calibrated or/and validated by comparison of simulated properties with experimental thermodynamic, transport, structural, dielectric or dynamic properties of liquid water. All the reported models essentially reproduce the key properties within reasonable errors, indicating that parameterizing a water model based on experimental data is an *underdetermined problem*, *i.e.* it admits multiple solutions of similar quality. In particular, appropriate parameter choices may be significantly correlated with the methodology to be employed in simulations relying on a specific water model, especially concerning the approximate treatment of electrostatic interactions (Section 3.1). It is also of interest to observe that all these models are characterized by *artificially-enhanced charges* and overestimate the dipole moment of the water molecule compared to its experimental gas-phase value of 1.8546 D (Table 1.1), being much closer to the corresponding effective value in liquid water or ice, *i.e.* about 2.9 D (Table 1.1). The reason is that simple force fields of the kind described in Section 3.3.1 rely on an implicit *mean-field description of many-body terms*, and in particular of *electronic polarizability*, through the use of effective pairwise interaction parameters, and in particular atomic partial charges. One effect of electronic polarizability in bulk water is to enhance the average molecular dipole moment, so that the effective charges appropriate for a bulk-phase water model with implicit polarizability must be artificially enhanced compared to values appropriate for the gas phase. Force fields with explicit inclusion of electronic polarizability will be mentioned in Section 6.2.4.

Ion-water Lennard-Jones interaction parameter sets (C_6^{is} and C_{12}^{is}) are typically calibrated or/and validated against[1289] experimental thermodynamic, transport, structural, dielectric or dynamic data concerning ionic hydration in the infinitely dilute regime, ion-water clusters in the gas phase (possibly also including results of quantum-mechanical calculations), finite-concentration effects in aqueous solutions (related to direct or water-mediated ion-ion interactions), as well as properties of ionic crystals and melts. The inclusion of the latter type of information requires simulations of electroneutral ionic systems involving different types of ions (at least a cation and an anion) and thus, the introduction of additional Lennard-Jones parameters for ion-ion interactions between like (C_6^{ii} and C_{12}^{ii}) as well as unlike ($C_6^{ii'}$ and $C_{12}^{ii'}$) ions. For simplicity, it is common practice to relate these parameters to the corresponding ion-solvent (C_6^{is} and C_{12}^{is}) and solvent-solvent (C_6^{ss} and C_{12}^{ss}) interaction parameters by means of *combination rules*[719–721, 726, 992, 1290–1299]. Combination rules are also used to simplify the calibration task in the more general situation of water models involving Lennard-Jones interaction sites at the hydrogen atoms, arbitrary solvents with multiple Lennard-Jones interaction sites, or polyatomic ions.

Table 3.1 *Parameters and properties of a few non-polarizable water models commonly used in atomistic simulations.* Corresponding generic representations and parameter definitions are provided in Figure 3.4. The reported data includes the number of sites n, the oxygen-hydrogen distance d_{OH}, the oxygen-virtual(M) distance d_{OM} ($n = 4$), the oxygen-virtual(L) distance d_{OL} ($n = 5$), the hydrogen-oxygen-hydrogen angle θ_{HOH}, the virtual(L)-oxygen-virtual(L) angle θ_{LOL} ($n = 5$), the oxygen charge q_O, the hydrogen charge q_H, the virtual(M) charge q_M ($n = 4$), the virtual(L) charge q_L ($n = 5$), the distance at the solvent-solvent Lennard-Jones interaction minimum R^{ss}_{min}, the energy at this interaction minimum E^{ss}_{min}, the molecular dipole moment μ_d, the quadrupole-moment trace $\mathcal{Q}$, the classical exclusion potential $\xi^{\ominus}_{AT}$, the simulated density $\rho^{\ominus}_{AT}$ and the simulated vaporization enthalpy $\Delta_b H^{\ominus}_{AT}$ at T^-. The exclusion potential $\xi^{\ominus}_{AT}$ of the water model is calculated according to Eq. 4.345 from the quadrupole-moment trace $\mathcal{Q}$ of this model and the experimental density of water $\rho^{\ominus}_{wat} = 997.048$ kg·m^{-3} (Table 1.1). The electrostatic scheme used for the determination of $\rho^{\ominus}_{AT}$ and $\Delta_b H^{\ominus}_{AT}$ by simulation (MD, unless otherwise specified) is also indicated, with reference to the nomenclature of Section 3.3.1 (P or M stand for P-truncation or M-truncation, respectively; the cutoff distance R_C in nm is indicated between parentheses whenever relevant). Unless otherwise specified, the reported simulation data is from the same reference as the one reporting the model parameterization. In the opposite case, the indicated electrostatic scheme may differ from the one employed during the parameterization of the model. Note also that some of these simulations were carried out at 1 atm rather than 1 bar (negligible difference). The corresponding experimental values are $\rho^{\ominus}_{wat} = 997.048$ kg·m^{-3} and $\Delta_b H^{\ominus} = 43.98$ kJ·mol^{-1} (Table 1.1; see footnote j therein). Note that the latter quantity is distinct from the vaporization enthalpy $\Delta_b H^{\circ}$ at the boiling point T°_b, namely 40.65 kJ·mol^{-1}. If required, the R^{ss}_{min} and E^{ss}_{min} values can be converted to R^{ss}_{zer}, C^{ss}_{12} or C^{ss}_{6} values using Eq. 3.13. The standard states are according to the $bbme_T$ convention (Section 4.2.4; reference pressure $P^{\circ} = 1$ bar and reference temperature $T^- = 298.15$ K).

Model	n	d_{OH} [nm]	d_{OM} [nm]	d_{OL} [nm]	θ_{HOH} [deg.]	θ_{LOL} [deg.]	q_O [e]	q_H [e]	q_M [e]	q_L [e]	R^{ss}_{min} [nm]	E^{ss}_{min} [kJ·mol^{-1}]
SPC	3	0.1	-	-	109.47	-	-0.82	0.41	-	-	0.355332	0.650167
SPC/E	3	0.1	-	-	109.47	-	-0.8476	0.4238	-	-	0.355332	0.650167
TIPS	3	0.09572	-	-	104.52	-	-0.8	0.4	-	-	0.360897	0.497075
TIPSmod	3	0.09572	-	-	104.5	-	-0.8	0.4	-	-	0.360897	0.497075
TIPS2	4	0.09572	0.015	-	104.52	-	-	0.535	-1.07	-	0.363756	0.541813
TIP3P	3	0.09572	-	-	104.52	-	-0.834	0.417	-	-	0.353649	0.636272
TIP4P	4	0.09572	0.015	-	104.52	-	-	0.52	-1.04	-	0.353985	0.648520
TIP4P/EW	4	0.09572	0.0125	-	104.52	-	-	0.52422	-1.04844	-	0.355186	0.680946
TIP5P	5	0.09572	-	0.07	104.52	109.47	-	0.241	-	-0.241	0.350208	0.669440
TIP5P/EW	5	0.09572	-	0.07	104.52	109.47	-	0.241	-	-0.241	0.347626	0.744752
ROW	5	0.096	-	0.02539	105.0	180.0	-	0.3278	-	-0.3278	0.305871	2.958088
ST2	5	0.1	-	0.08	109.47	109.47	-	0.2357	-	-0.2357	0.347963	0.316938
CC	4	0.09572	0.025	-	104.52	-	-	0.65789	-1.31578	-	-	-
CFF	-	-	-	-	-	-	-0.8	0.4	-	-	0.3558	1.146416

(continues on the next page)

Table 3.1 *(continued)*

Model	μ_d [D]	$\mathcal{Q}$ [$10^{-3}e\cdot$nm^2]	$\xi^{\ominus}_{AT}$ [V]	$\rho^{\ominus}_{AT}$ [kg·m^{-3}]	$\Delta_b H^{\ominus}_{AT}$ [kJ·mol^{-1}]	Elec.	Ref.	Notes
SPC	2.27	8.20	0.82	972	43.7	RF/M(1.4)	1274	a
SPC/E	2.35	8.48	0.85	994	43.6	RF/M(1.4)	1275	a,b
TIPS	2.25	7.33	0.74	1000	37.2	SC/M(0.75)	1276	c,d
TIPSmod	2.25	7.33	0.74	-	-	-	1270	e
TIPS2	2.24	9.56	0.96	997	44.48	SC/M(0.75)	1277	c
TIP3P	2.35	7.64	0.77	1002	43.56	SC/M(0.9)	861	c,f
TIP4P	2.18	9.30	0.93	1001	44.56	SC/M(0.9)	861	c,f
TIP4P/EW	2.32	9.44	0.95	995.4	44.28	LS/P	1278	b
TIP5P	2.29	2.05	0.21	999	43.76	SC/M(0.9)	1279	c
TIP5P/EW	2.29	2.05	0.21	1000	43.42	LS/P	1280	
ROW	1.84	5.62	0.57	-	-	-	1281	g
ST2	2.35	1.70	0.17	-	-	-	1282	
CC	2.12	11.23	1.13	-	-	-	1283	
CFF	-	-	-	-	-	-	-	h

(a) simulation data from Ref.[1272]; (b) simulation data including a correction to $\Delta_b H^{\ominus}_{AT}$ for the self-polarization energy, as suggested in the indicated reference; (c) simulation data from MC instead of MD; (d) simulation data from Ref.[1277] (density; simulation protocol unspecified) and Ref.[1276] (vaporization enthalpy); (e) differs from the TIPS water model[1276] by introduction of a non-zero repulsive Lennard-Jones interaction between the oxygen and hydrogen atoms, with a corresponding C_{12}^{hs} parameter of $9.4215 \cdot 10^{-10}$ kJ·mol^{-1}·nm^{12}; (f) simulation data from Ref.[1284]; (g) see also Refs.[1282,1285] for other simulated properties; (h) the CFF water model is referred to in Ref.[1286] as "Thacher, T. Unpublished results", and the information given in this reference is restricted to atomic partial charges and Lennard-Jones interaction parameters R_{min}^{ss} and E_{min}^{ss} for use with a 6-9 potential in the CFF force field[981–984,1287,1288].

For example, one may apply a *geometric-mean* (GM) combination rule[1297,1300,1301], defined by

$$R_{zer}^{is} = (R_{zer}^{ii} R_{zer}^{ss})^{1/2} \quad \text{and} \quad E_{min}^{is} = (E_{min}^{ii} E_{min}^{ss})^{1/2} \tag{3.45}$$

along with

$$R_{zer}^{ii'} = (R_{zer}^{ii} R_{zer}^{i'i'})^{1/2} \quad \text{and} \quad E_{min}^{ii'} = (E_{min}^{ii} E_{min}^{i'i'})^{1/2} \quad , \tag{3.46}$$

or, equivalently,

$$C_6^{is} = (C_6^{ii} C_6^{ss})^{1/2} \quad \text{and} \quad C_{12}^{is} = (C_{12}^{ii} C_{12}^{ss})^{1/2} \tag{3.47}$$

along with

$$C_6^{ii'} = (C_6^{ii} C_6^{i'i'})^{1/2} \quad \text{and} \quad C_{12}^{ii'} = (C_{12}^{ii} C_{12}^{i'i'})^{1/2} \quad . \tag{3.48}$$

There exist many other combination rules[992, 1299–1319] among which one may mention two, that are also very common. The *Lorentz-Berthelot* (LB) combination rule[1303,1305] reads

$$R_{zer}^{is} = \frac{1}{2}\left(R_{zer}^{ii} + R_{zer}^{ss}\right) \quad \text{and} \quad E_{min}^{is} = \left(E_{min}^{ii} E_{min}^{ss}\right)^{1/2} \tag{3.49}$$

and similarly for $R_{zer}^{ii'}$ and $E_{min}^{ii'}$, or, equivalently,

$$\begin{aligned} C_6^{is} &= \frac{1}{64} C_6^{ii} C_6^{ss} \left(C_{12}^{ii} C_{12}^{ss}\right)^{-1/2} \left[\left(\frac{C_{12}^{ii}}{C_6^{ii}}\right)^{1/6} + \left(\frac{C_{12}^{ss}}{C_6^{ss}}\right)^{1/6}\right]^6 \\ &\text{and} \\ C_{12}^{is} &= \frac{1}{4096} C_6^{ii} C_6^{ss} \left(C_{12}^{ii} C_{12}^{ss}\right)^{-1/2} \left[\left(\frac{C_{12}^{ii}}{C_6^{ii}}\right)^{1/6} + \left(\frac{C_{12}^{ss}}{C_6^{ss}}\right)^{1/6}\right]^{12} \end{aligned} \tag{3.50}$$

and similarly for $C_6^{ii'}$ and $C_{12}^{ii'}$. The *Waldman-Hagler* (WH) combination rule[1319] reads

$$\begin{aligned} R_{zer}^{is} &= \left(\frac{(R_{zer}^{ii})^6 + (R_{zer}^{ss})^6}{2}\right)^{1/6} \\ &\text{and} \\ E_{min}^{is} &= \left(E_{min}^{ii} E_{min}^{ss}\right)^{1/2} \frac{2(R_{zer}^{ii})^3 (R_{zer}^{ss})^3}{(R_{zer}^{ii})^6 + (R_{zer}^{ss})^6} \quad , \end{aligned} \tag{3.51}$$

and similarly for $R_{zer}^{ii'}$ and $E_{min}^{ii'}$, or, equivalently,

$$\begin{aligned} C_6^{is} &= (C_6^{ii} C_6^{ss})^{1/2} \\ &\text{and} \\ C_{12}^{is} &= \frac{1}{2}\left(C_6^{ii} C_6^{ss}\right)^{-1/2} \left(C_{12}^{ss} C_6^{ii} + C_{12}^{ii} C_6^{ss}\right) \quad , \end{aligned} \tag{3.52}$$

and similarly for $C_6^{ii'}$ and $C_{12}^{ii'}$. Although the present discussion has been restricted to water as a solvent for simplicity, the same combination rules can also be applied to evaluate ion-solvent Lennard-Jones interaction parameters for a given non-aqueous solvent, based on the corresponding ion-water, water-water and solvent-solvent interaction parameters.

Combination rules are essentially *ad hoc* devices that do not rest on any solid physical ground, with the possible exception of the GM rule for the C_6 coefficients, partly justified by the presence of a polarizability product in Eqs. 3.14 or 3.15. They can be compared based on experimental data in the simple context of the interaction between rare-gas atoms[992,1290,1291,1293,1294,1298,1299,1309–1320], and, possibly, gas-phase ions[1295], ions in crystals[1289] or ions in molten salts[1306]. However, the validity of these tests is limited by the fact that the practical application of combination rules in AT simulations involves sites that in general represent atoms within molecules rather than isolated closed-shell species, and that these sites generally also carry partial charges.

Once a water model has been selected and a specific type of combination rule adopted, the Lennard-Jones interaction properties of a set of ions can be entirely characterized by a *single pair of parameters per ion*. A particularly convenient choice for this ion-specific parameter pair is C_6^{ii} and C_{12}^{ii}, because it exclusively characterizes the ion, and can in principle be subsequently used in the combination rules with arbitrary solvent models (water or non-aqueous solvent), as well as with arbitrary ions. Alternatively, one may use R_{zer}^{ii} or R_{min}^{ii} along with E_{min}^{ii}, where the former values can be interpreted as a measure for an effective ionic radius. There is, however, a fundamental qualitative difference between the *homoionic dispersion parameter* C_6^{ii} and the *homoionic repulsion parameter* C_{12}^{ii}.

As discussed in Section 3.3.1, the inverse-sixth power dependence of the Lennard-Jones (London) dispersion term is physically motivated, while the inverse-twelfth power dependence of the (Pauli) repulsion term is entirely arbitrary. Consequently, the parameter C_6^{ii} has some physical significance in terms of single-ion electronic properties, while the parameter C_{12}^{ii} has none. In fact, C_6^{ii} can in principle be related to the static *electronic polarizability* of the ion *via* the (approximate) Slater-Kirkwood formula[726] (Eq. 3.15), namely

$$C_6^{ii} \approx C_{SK} \tilde{\alpha}_{el,i}^{3/2} n_{el,i}^{1/2} \,, \tag{3.53}$$

where $\tilde{\alpha}_{el,i}$ is the static dipolar polarizability volume of the ion, related to the static polarizability α_{el} as $\tilde{\alpha}_{el} = (4\pi\epsilon_o)^{-1}\alpha_{el}$, $n_{el,i}$ is the effective number of valence electrons, *i.e.* the number of electrons effectively contributing the atomic polarizability, and $C_{SK} = 23.97$ kJ·mol^{-1}·nm$^{3/2}$. For example, for alkali and halide atoms or ions, experimental values are available for the gas-phase polarizability volumes[727], and are reported in Table 5.9. Corresponding effective electron numbers can be deduced from the polarizabilities of the isoelectronic noble gases[1321,1322] and from quantum-mechanical[1323] or experimental[1321] estimates of their dispersion coefficients. For the alkali and halide gas-phase ions, using values appropriate for the isoelectronic noble gases[1322,1323] leads[1289] to n_i values of 1.92 (He→Li^+), 3.31 (Ne→Na^+, F^-), 5.36 (Ar→K^+, Cl^-), 6.78 (Kr→Rb^+, Br^-) and 8.46 (Xe→Cs^+, I^-). An alternative set reported in Ref.[1321] based on different data sources is 1.42, 3.81, 5.40, 6.31 and 7.25. These values are also reported in Table 5.9.

In contrast, the parameter C_{12}^{ii} of AT simulations must be calibrated by comparison of simulation results with experimental data. In this sense, this

Table 3.2 *Overview of the available sets of ion-solvent (water) Lennard-Jones interaction parameters for the alkali and halide ions.* The sets are listed in chronological order of publication. The reported information includes the authors of the set and the corresponding original references, the water model(s) considered, and the combination rule (c.r.) compatible with the set (if specified; GM: geometric-mean combination rule; LB: Lorentz-Berthelot[1303, 1305] combination rule; WH: Waldman-Hagler[1319] combination rule). The corresponding parameters are listed in Tables 3.3-3.4.

Code	Authors	Ref.	Water models	Ref.	Ions	c.r.	Notes
Ch	Chandrasekhar *et al.*	1324	TIP4P	861	Li^+, Na^+, F^-, Cl^-	GM	
Pt	Pettitt & Rossky	1325	TIPSmod	1270	Li^+, Na^+, K^+, F^-, Cl^-	-	a
St	Straatsma & Berendsen	1171	SPC	1274	Na^+, K^+, F^-, Cl^-, Br^-	-	b
Aq	Åqvist	1047	SPC	1274	Li^+, Na^+, K^+, Rb^+, Cs^+	GM	
Da	Dang *et al.*	344,923,924,	SPC/E	1275	Na^+, K^+, Rb^+, Cs^+,	LB	c
		351, 235			F^-, Cl^-, I^-		
Ro	Roux	1326	TIP3P	861	K^+, Cl^-	LB	
Pn	Peng *et al.*	1286	CFF	982,983,	Li^+, Na^+, K^+, Rb^+, Cs^+,	WH	d
				981,984	F^-, Cl^-, Br^-, I^-		
Ps	Peslherbe *et al.*	939	TIP4P	861	I^-	-	
Bo	Borodin *et al.*	1327	SPC/E, TIP4P	861,1275	K^+	-	
We	Weerasinghe & Smith	1328	SPC/E	1275	Na^+, Cl^-	-	
Je	Jensen & Jorgensen	1329	TIP4P	861	Li^+, Na^+, K^+, Rb^+, Cs^+,	GM	
					F^-, Cl^-, Br^-, I^-		
Al	Alejandre & Hansen	1330	SPC/E	1275	Cl^-	LB	
Jo	Joung & Cheatham III	1331	SPC/E, TIP3P,	861,1275,	Li^+, Na^+, K^+, Rb^+, Cs^+,	LB	
			TIP4P/EW	1278	F^-, Cl^-, Br^-, I^-		
Ho	Horinek *et al.*	877	SPC/E	1275	Li^+, Na^+, K^+, Cs^+,	-	e
					F^-, Cl^-, Br^-, I^-		
Re	Reif & Hünenberger	1289	SPC, SPC/E	1274,1275	Li^+, Na^+, K^+, Rb^+, Cs^+,	GM	f
					F^-, Cl^-, Br^-, I^-		

(a) ion-ion interactions described by a Born-Mayer-Huggins potential[180, 987, 988] (this set represents a generalized version of ion-ion interaction parameters by Tosi & Fumi[67, 1332]); (b) parameters for Na^+, K^+, F^- by Straatsma & Berendsen[1171], complemented by van Gunsteren *et al.*[1333] for Cl^- and Br^-; (c) parameters for Na^+, K^+, Rb^+, Cs^+, F^-, Cl^-, I^- F^- by Dang *et al.*[344,351,923,924], complemented by Lee & Rasaiah[235] for Li^+ and Br^- (based on earlier work of Dang[921] and Lybrand *et al.*[1334]); (d) using a 6-9 van der Waals potential; (e) three different parameter sets relying on different Lennard-Jones well depths are provided for the cations; (f) three different parameter sets relying on different estimates for the standard intrinsic hydration free energy of the proton are provided, namely $\mathcal{G}^{\ominus}_{H,wat} = -1100$ kJ·mol^{-1} (L, L_E), -1075 kJ·mol^{-1} (M, M_E) or -1050 kJ·mol^{-1} (H, H_E), where sets L, M and H were calibrated for the SPC water model and sets L_E, M_E and H_E for the SPC/E water model.

Table 3.3 *Available Lennard-Jones interaction parameter sets for the alkali cations.* The reported data includes the homoionic ion-ion Lennard-Jones dispersion coefficient C_6^{ii} and repulsion coefficient C_{12}^{ii}, and the ion-solvent distance R_{min}^{is} at the Lennard-Jones interaction minimum and corresponding interaction energy E_{min}^{is} (based on the indicated water model). The different sets correspond to values quoted in different literature sources (Table 3.2), listed in chronological order of publication for each ion separately. Corresponding values for the halide anions can be found in Table 3.4. The water model and type of combination rule (c.r.; see caption of Table 3.2 for the naming) used in the parameterization of the ion is also indicated. These determine R_{min}^{is} and E_{min}^{is} which, unlike C_6^{ii} and C_{12}^{ii}, characterize the solvent as well as the ion. The values of the solvent-solvent Lennard-Jones interaction parameters C_6^{ss} and C_{12}^{ss} can be found in Table 3.1.

Set	C_6^{ii} [10^{-4} kJ·mol^{-1}·nm^6]	C_{12}^{ii} [10^{-8} kJ·mol^{-1}·nm^{12}]	R_{min}^{is} [nm]	E_{min}^{is} [kJ·mol^{-1}]	Water	c.r.	Ref.	Notes
			Li^+					
Ch	4.184000000	$1.673600000 \cdot 10^{-1}$	0.223740764	4.118659590	TIP4P	GM	1324	
Pt	$1.722973925 \cdot 10^{-1}$	$1.017274631 \cdot 10^{-2}$	0.255900000	0.602200000	TIPSmod	(LB)	1325	a,b,c
Aq	$2.828384000 \cdot 10^{-1}$	$2.615000000 \cdot 10^{-1}$	0.291226417	0.222989879	SPC	GM	1047	
Da	$3.247906996 \cdot 10^{-1}$	$3.819643387 \cdot 10^{-2}$	0.262300000	0.670000000	SPC/E	LB	921, 235	d
Pn	$4.077831743 \cdot 10^{-1}$	$(3.758129734 \cdot 10^{1})$	0.326000000	0.154808000	CFF	WH	1286	e
Je	$4.676415076 \cdot 10^{-2}$	$2.613391248 \cdot 10^{-1}$	0.337662615	0.036838411	TIP4P	GM	1329	
Jo	$1.736412873 \cdot 10^{-1}$	$6.436616528 \cdot 10^{-2}$	0.279324596	0.272970364	TIP3P	LB	1331	

(continues on the next page)

Table 3.3 *(continued)*

Set	C_6^{ii} [10^{-4} kJ·mol^{-1}·nm^6]	C_{12}^{ii} [10^{-8} kJ·mol^{-1}·nm^{12}]	R_{min}^{is} [nm]	E_{min}^{is} [kJ·mol^{-1}]	Water	c.r.	Ref.	Notes
Jo	$1.551713417·10^{-1}$	$1.383523269·10^{-2}$	0.258393139	0.544307881	TIP4P/EW	LB	1331	
Jo	$4.409660061·10^{-1}$	$3.450412908·10^{-2}$	0.256765998	0.957088732	SPC/E	LB	1331	
Ho	$1.388673514·10^{-2}$	$7.836200998·10^{-2}$	0.339000000	0.020000000	SPC/E	(LB)	877	b,f
Ho	$4.146292181·10^{-1}$	$2.794377392·10^{-2}$	0.254800000	1.000000000	SPC/E	(LB)	877	b,g
Ho	$2.667568761·10^{-1}$	$2.737598702·10^{-2}$	0.260400000	0.650000000	SPC/E	(LB)	877	b,h
Re	$5.114174717·10^{-2}$	$1.407983494·10^{-1}$	0.318949181	0.054948951	SPC	GM	1289	i
Re	$5.114174577·10^{-2}$	$3.354429537·10^{-1}$	0.342878235	0.035599905	SPC	GM	1289	j
Re	$5.114174716·10^{-2}$	$8.428213071·10^{-1}$	0.370239701	0.022459001	SPC	GM	1289	k
Re	$5.114174542·10^{-2}$	$1.892223194·10^{-1}$	0.326903366	0.047399307	SPC/E	GM	1289	l
Re	$5.114174507·10^{-2}$	$4.346408156·10^{-1}$	0.350361119	0.031274683	SPC/E	GM	1289	m
Re	$5.114174560·10^{-2}$	$9.877150971·10^{-1}$	0.375166746	0.020746380	SPC/E	GM	1289	n
			Na^+					

(continues on the next page)

Table 3.3 *(continued)*

Set	C_6^{ii} [10^{-4} kJ·mol^{-1}·nm^6]	C_{12}^{ii} [10^{-8} kJ·mol^{-1}·nm^{12}]	R_{min}^{is} [nm]	E_{min}^{is} [kJ·mol^{-1}]	Water	c.r.	Ref.	Notes
Ch	$1.255200000 \cdot 10^{1}$	5.857600000	0.274572697	2.088541606	TIP4P	GM	1324	
Pt	3.059137692	3.707771460	0.305300000	0.560046000	TIPS(mod)	(LB)	1325	a,b,c
St	$7.059868578 \cdot 10^{-1}$	2.014334563	0.319900000	0.200545740	SPC	(GM)	1171	o
Aq	$6.331270640 \cdot 10^{-1}$	8.639830000	0.364463620	0.086840219	SPC	GM	1047	
Da	5.005816319	$1.497062417 \cdot 10^{1}$	0.322800000	0.521600000	SPC/E	LB	351, 235	
Pn	5.050938089	$(9.367175815 \cdot 10^{2})$	0.338000000	0.439320000	CFF	WH	1286	e
We	2.768257440	5.986913481	0.312596064	0.342096692	SPC/E	GM	1328	p
Je	$3.803541217 \cdot 10^{-1}$	$1.728839124 \cdot 10^{1}$	0.402115362	0.036838411	TIP4P	GM	1329	
Jo	3.080556234	6.484850119	0.313724596	0.482470195	TIP3P	LB	1331	
Jo	3.059188643	3.319877526	0.300193139	0.692742078	TIP4P/EW	LB	1331	
Jo	5.993878892	6.087380805	0.298865998	0.979434404	SPC/E	LB	1331	
Ho	$7.570792515 \cdot 10^{-2}$	2.329099008	0.391700000	0.020000000	SPC/E	(LB)	877	b,f
Ho	5.829953595	5.524531440	0.297500000	1.000000000	SPC/E	(LB)	877	b,g
Ho	3.239687379	4.037799485	0.303100000	0.650000000	SPC/E	(LB)	877	b,h

(continues on the next page)

Table 3.3 *(continued)*

Set	C_6^{ii} [10^{-4} kJ·mol^{-1}·nm^6]	C_{12}^{ii} [10^{-8} kJ·mol^{-1}·nm^{12}]	R_{min}^{is} [nm]	E_{min}^{is} [kJ·mol^{-1}]	Water	c.r.	Ref.	Notes
Re	$7.884022461 \cdot 10^{-1}$	7.289667269	0.352830656	0.117727149	SPC	GM	1289	i
Re	$7.884022565 \cdot 10^{-1}$	$1.648670224 \cdot 10^{1}$	0.377660722	0.078282302	SPC	GM	1289	j
Re	$7.884022371 \cdot 10^{-1}$	$4.011891507 \cdot 10^{1}$	0.406711484	0.050182915	SPC	GM	1289	k
Re	$7.884022546 \cdot 10^{-1}$	8.565450078	0.357604673	0.108606393	SPC/E	GM	1289	l
Re	$7.884022451 \cdot 10^{-1}$	$1.944576315 \cdot 10^{1}$	0.382891818	0.072080532	SPC/E	GM	1289	m
Re	$7.884022444 \cdot 10^{-1}$	$4.493872499 \cdot 10^{1}$	0.410574903	0.047415475	SPC/E	GM	1289	n
			K^+					
Pt	$2.164771642 \cdot 10^{1}$	$1.939198202 \cdot 10^{2}$	0.354700000	0.548002000	TIPS(mod)	(LB)	1325	a,b,c
St	$1.639541460 \cdot 10^{-1}$	$1.186020143 \cdot 10^{2}$	0.507400000	0.006069585	SPC	(GM)	1171	o
Aq	$7.917174000 \cdot 10^{-1}$	$1.143132770 \cdot 10^{2}$	0.443639491	0.029854130	SPC	GM	1047	
Da	$2.300255714 \cdot 10^{1}$	$3.161129768 \cdot 10^{2}$	0.364800000	0.521600000	SPC/E	LB	351, 235	
Ro	$1.403580112 \cdot 10^{1}$	$1.353017744 \cdot 10^{2}$	0.353199596	0.481256690	TIP3P	LB	1326	

(continues on the next page)

Table 3.3 *(continued)*

Set	C_6^{ii} [10^{-4} kJ·mol^{-1}·nm^6]	C_{12}^{ii} [10^{-8} kJ·mol^{-1}·nm^{12}]	R_{min}^{is} [nm]	E_{min}^{is} [kJ·mol^{-1}]	Water	c.r.	Ref.	Notes
Pn	$4.726874135{\cdot}10^1$	$(1.470246931{\cdot}10^4)$	0.360000000	0.912112000	CFF	WH	1286	e
Bo	$3.092715944{\cdot}10^1$	1.589926848	0.228905424	9.888583579	SPC/E	(GM)	1327	o
Bo	$3.171609232{\cdot}10^1$	$3.990954748{\cdot}10^1$	0.298233582	2.021768045	TIP4P	(GM)	1327	o
Je	1.597956417	$3.051463564{\cdot}10^2$	0.453225047	0.036838411	TIP4P	GM	1329	
Jo	$2.548411541{\cdot}10^1$	$2.003531525{\cdot}10^2$	0.347324596	0.718063403	TIP3P	LB	1331	
Jo	$2.418013351{\cdot}10^1$	$1.250080963{\cdot}10^2$	0.336593139	0.892310420	TIP4P/EW	LB	1331	
Jo	$3.757481278{\cdot}10^1$	$1.963229956{\cdot}10^2$	0.336965998	1.081169700	SPC/E	LB	1331	
Ho	$2.136464136{\cdot}10^{-1}$	$1.854797390{\cdot}10^1$	0.432100000	0.020000000	SPC/E	(LB)	877	b,f
Ho	$2.808570850{\cdot}10^1$	$1.282141689{\cdot}10^2$	0.333400000	1.000000000	SPC/E	(LB)	877	b,g
Ho	$1.527823197{\cdot}10^1$	$8.980173753{\cdot}10^1$	0.340100000	0.650000000	SPC/E	(LB)	877	b,h
Re	$1.296027955{\cdot}10^1$	$1.865925722{\cdot}10^2$	0.366094319	0.382516189	SPC	GM	1289	i
Re	$1.296027935{\cdot}10^1$	$4.642524638{\cdot}10^2$	0.394985645	0.242504544	SPC	GM	1289	j
Re	$1.296027966{\cdot}10^1$	$1.180343735{\cdot}10^3$	0.426925899	0.152087230	SPC	GM	1289	k
Re	$1.296027966{\cdot}10^1$	$2.150388288{\cdot}10^2$	0.370448807	0.356318652	SPC/E	GM	1289	l

(continues on the next page)

Table 3.3 *(continued)*

Set	C_6^{ii} [10^{-4} kJ·mol^{-1}·nm^6]	C_{12}^{ii} [10^{-8} kJ·mol^{-1}·nm^{12}]	R_{min}^{is} [nm]	E_{min}^{is} [kJ·mol^{-1}]	Water	c.r.	Ref.	Notes
Re	$1.296027959 \cdot 10^1$	$5.169109134 \cdot 10^2$	0.398538029	0.229820702	SPC/E	GM	1289	m
Re	$1.296027955 \cdot 10^1$	$1.317023598 \cdot 10^3$	0.430841891	0.143979353	SPC/E	GM	1289	n
			Rb^+					
Aq	$9.007984640 \cdot 10^{-1}$	$2.843594000 \cdot 10^2$	0.473521447	0.021536565	SPC	GM	1047	
Da	$3.240514404 \cdot 10^1$	$6.273616983 \cdot 10^2$	0.375800000	0.521600000	SPC/E	LB	351, 235	
Pn	$9.981803402 \cdot 10^1$	$(3.947403973 \cdot 10^4)$	0.377000000	1.012530000	CFF	WH	1286	e
Je	2.580773161	$7.959357204 \cdot 10^2$	0.471709699	0.036838411	TIP4P	GM	1329	
Jo	$6.230239822 \cdot 10^1$	$7.074893364 \cdot 10^2$	0.358124596	0.934191981	TIP3P	LB	1331	
Jo	$5.778474036 \cdot 10^1$	$4.606138029 \cdot 10^2$	0.348493139	1.110889938	TIP4P/EW	LB	1331	
Jo	$6.547517192 \cdot 10^1$	$5.754936919 \cdot 10^2$	0.351365998	1.100370631	SPC/E	LB	1331	
Re	$3.095999651 \cdot 10^1$	$6.419026469 \cdot 10^2$	0.377391412	0.492661865	SPC	GM	1289	i
Re	$3.095999697 \cdot 10^1$	$1.550117727 \cdot 10^3$	0.406162639	0.317030609	SPC	GM	1289	j

(continues on the next page)

Table 3.3 *(continued)*

Set	C_6^{ii} [10^{-4} kJ·mol^{-1}·nm^6]	C_{12}^{ii} [10^{-8} kJ·mol^{-1}·nm^{12}]	R_{min}^{is} [nm]	E_{min}^{is} [kJ·mol^{-1}]	Water	c.r.	Ref.	Notes
Re	$3.095999677 \cdot 10^1$	$4.201403986 \cdot 10^3$	0.441352698	0.192568972	SPC	GM	1289	k
Re	$3.095999715 \cdot 10^1$	$7.019398261 \cdot 10^2$	0.380213826	0.471122230	SPC/E	GM	1289	l
Re	$3.095999718 \cdot 10^1$	$1.755419015 \cdot 10^3$	0.410394295	0.297915528	SPC/E	GM	1289	m
Re	$3.095999642 \cdot 10^1$	$4.694318230 \cdot 10^3$	0.445451706	0.182178560	SPC/E	GM	1289	n
			Cs^+					
Aq	1.238196224	$1.136196000 \cdot 10^3$	0.517557142	0.014809666	SPC	GM	1047	
Da	$5.770322546 \cdot 10^1$	$1.989254127 \cdot 10^3$	0.395800000	0.521600000	SPC/E	LB	347, 235	
Pn	$2.475587369 \cdot 10^2$	$(1.312176833 \cdot 10^5)$	0.403000000	1.071100000	CFF	WH	1286	e
Je	4.753043714	$2.699740027 \cdot 10^3$	0.496319154	0.036838411	TIP4P	GM	1329	
Jo	$1.296434141 \cdot 10^2$	$2.470282433 \cdot 10^3$	0.374424596	1.040323657	TIP3P	LB	1331	
Jo	$9.566641761 \cdot 10^1$	$1.386421977 \cdot 10^3$	0.366393139	1.060078765	TIP4P/EW	LB	1331	
Jo	$3.278988502 \cdot 10^1$	$7.149548589 \cdot 10^2$	0.379765998	0.494405405	SPC/E	LB	1331	

(continues on the next page)

Table 3.3 *(continued)*

Set	C_6^{ii} [10^{-4} kJ·mol^{-1}·nm^6]	C_{12}^{ii} [10^{-8} kJ·mol^{-1}·nm^{12}]	R_{min}^{is} [nm]	E_{min}^{is} [kJ·mol^{-1}]	Water	c.r.	Ref.	Notes
Ho	$4.726389059 \cdot 10^{-1}$	$9.077456972 \cdot 10^{1}$	0.468100000	0.020000000	SPC/E	(LB)	877	b,f
Ho	$8.454753549 \cdot 10^{1}$	$1.161895739 \cdot 10^{3}$	0.364800000	1.000000000	SPC/E	(LB)	877	b,g
Ho	$4.263452903 \cdot 10^{1}$	$6.992967019 \cdot 10^{2}$	0.370400000	0.650000000	SPC/E	(LB)	877	b,h
Re	$7.936269667 \cdot 10^{1}$	$2.672310223 \cdot 10^{3}$	0.392954436	0.618950056	SPC	GM	1289	i
Re	$7.936269716 \cdot 10^{1}$	$6.356696981 \cdot 10^{3}$	0.422380878	0.401313058	SPC	GM	1289	j
Re	$7.936269676 \cdot 10^{1}$	$1.772852023 \cdot 10^{4}$	0.460070440	0.240304996	SPC	GM	1289	k
Re	$7.936269754 \cdot 10^{1}$	$2.807759232 \cdot 10^{3}$	0.394576859	0.603836159	SPC/E	GM	1289	l
Re	$7.936269659 \cdot 10^{1}$	$7.021675709 \cdot 10^{3}$	0.425897431	0.381837571	SPC/E	GM	1289	m
Re	$7.936269707 \cdot 10^{1}$	$1.904842151 \cdot 10^{4}$	0.462831816	0.231829954	SPC/E	GM	1289	n

(a) uses a modified[1270] TIPS water model; (b) combination rule not specified in the original reference (LB was used by default); (c) original study uses a Born-Mayer-Huggins potential[180,987,988] for ion-ion interactions; (d) parameters used in Ref.[235] with the SPC/E water model[1275], although they were originally derived in Ref.[921] for the RPOL water model[920,1335]; (e) original study uses a 6-9 van der Waals potential (the reported C_{12}^{ii} is actually C_9^{ii}); (f) set (1) in Ref.[877]; (g) set (2) in Ref.[877]; (h) set (3) in Ref.[877]; (i) set L in Ref.[1289]; (j) set M in Ref.[1289]; (k) set H in Ref.[1289]; (l) set L_E in Ref.[1289]; (m) set M_E in Ref.[1289]; (n) set H_E in Ref.[1289]; (o) combination rule not specified in the original reference (GM was used by default); (p) the parameter E_{min}^{is} was deduced from the reported ion-ion and solvent-solvent parameters (GM) and scaled[1328] by a factor of 0.75.

Table 3.4 *Available Lennard-Jones interaction parameter sets for the halide anions.* The reported data includes the homoionic ion-ion Lennard-Jones dispersion coefficient C_6^{ii} and repulsion coefficient C_{12}^{ii}, and the ion-solvent distance R_{min}^{is} at the Lennard-Jones interaction minimum and corresponding interaction energy E_{min}^{is} (based on the indicated water model). The different sets correspond to values quoted in different literature sources (Table 3.2), listed in chronological order of publication for each ion separately. Corresponding values for the alkali cations can be found in Table 3.3. The water model and type of combination rule (c.r.; see caption of Table 3.2 for the naming) used in the parameterization of the ion is also indicated. These determine R_{min}^{is} and E_{min}^{is} which, unlike C_6^{ii} and C_{12}^{ii}, characterize the solvent as well as the ion. The values of the solvent-solvent Lennard-Jones interaction parameters C_6^{ss} and C_{12}^{ss} can be found in Table 3.1.

Set	C_6^{ii} [10^{-4} kJ·mol^{-1}·nm^6]	C_{12}^{ii} [10^{-8} kJ·mol^{-1}·nm^{12}]	R_{min}^{is} [nm]	E_{min}^{is} [kJ·mol^{-1}]	Water	c.r.	Ref.	Notes
				F^-				
Ch	$5.020800000 \cdot 10^1$	$2.092000000 \cdot 10^{-1}$	0.185305872	44.206093531	TIP4P	GM	1324	
Pt	$4.366878007 \cdot 10^1$	$1.633664727 \cdot 10^2$	0.331100000	1.204400000	TIPS(mod)	(LB)	1325	a,b,c
St	$1.217224685 \cdot 10^1$	$7.851270318 \cdot 10^1$	0.342400000	0.553838845	SPC	(GM)	1171	d
Da	$2.781989123 \cdot 10^1$	$2.568787815 \cdot 10^2$	0.352800000	0.699800000	SPC/E	LB	344, 235	
Pn	$2.021663369 \cdot 10^1$	$(1.311870838 \cdot 10^4)$	0.424000000	0.221752000	CFF	WH	1286	e
Je	$9.565530306 \cdot 10^1$	$7.700307849 \cdot 10^2$	0.348140457	1.388177703	TIP4P	GM	1329	
Jo	2.686066473	$1.281521386 \cdot 10^2$	0.407124596	0.094633553	TIP3P	LB	1331	

(continues on the next page)

Table 3.4 *(continued)*

Set	C_6^{ii} [10^{-4} kJ·mol^{-1}·nm^6]	C_{12}^{ii} [10^{-8} kJ·mol^{-1}·nm^{12}]	R_{min}^{is} [nm]	E_{min}^{is} [kJ·mol^{-1}]	Water	c.r.	Ref.	Notes
Jo	2.254087229	$1.927321057 \cdot 10^2$	0.431393139	0.066991550	TIP4P/EW	LB	1331	
Jo	5.242822176	$2.219307727 \cdot 10^2$	0.403365998	0.141885819	SPC/E	LB	1331	
Ho	$3.052531368 \cdot 10^1$	$5.006798872 \cdot 10^2$	0.370400000	0.550000000	SPC/E	(LB)	877	b
Re	$2.055034009 \cdot 10^1$	$2.511336407 \cdot 10^2$	0.361127156	0.522816390	SPC	GM	1289	f
Re	$2.055033980 \cdot 10^1$	$1.690160163 \cdot 10^2$	0.349404689	0.637290880	SPC	GM	1289	g
Re	$2.055033973 \cdot 10^1$	$1.121798102 \cdot 10^2$	0.337671416	0.782247677	SPC	GM	1289	h
Re	$2.055034000 \cdot 10^1$	$2.706359479 \cdot 10^2$	0.363384889	0.503626875	SPC/E	GM	1289	i
Re	$2.055034001 \cdot 10^1$	$1.781065390 \cdot 10^2$	0.350933420	0.620814292	SPC/E	GM	1289	j
Re	$2.055034033 \cdot 10^1$	$1.257493501 \cdot 10^2$	0.340899909	0.738837229	SPC/E	GM	1289	k
			Cl^-					
Ch	$1.464400000 \cdot 10^2$	$1.087840000 \cdot 10^4$	0.418937168	0.565414920	TIP4P	GM	1324	
Pt	$6.273421179 \cdot 10^2$	$2.157792495 \cdot 10^4$	0.398500000	1.505500000	TIPS(mod)	(LB)	1325	a,b,c

(continues on the next page)

Table 3.4 *(continued)*

Set	C_6^{ii} [10^{-4} kJ·mol^{-1}·nm^6]	C_{12}^{ii} [10^{-8} kJ·mol^{-1}·nm^{12}]	R_{min}^{is} [nm]	E_{min}^{is} [kJ·mol^{-1}]	Water	c.r.	Ref.	Notes
St	$1.366760880 \cdot 10^2$	$1.049551505 \cdot 10^4$	0.420900000	0.537865005	SPC	(GM)	1171	d
Da	$1.223232694 \cdot 10^2$	$8.939397486 \cdot 10^3$	0.424900000	0.521600000	SPC/E	LB	923, 235	
Ro	$1.099648996 \cdot 10^2$	$4.816873462 \cdot 10^3$	0.403824596	0.631920993	TIP3P	LB	1326	
Pn	$1.574445522 \cdot 10^2$	$(1.492959024 \cdot 10^5)$	0.473000000	0.326352000	CFF	WH	1286	e
We	$1.364187005 \cdot 10^2$	$9.898969066 \cdot 10^3$	0.418925379	0.552791748	SPC/E	GM	1328	
Je	$5.014946848 \cdot 10^2$	$2.116521346 \cdot 10^4$	0.399642956	1.388177703	TIP4P	GM	1329	
Al	$7.233668671 \cdot 10^1$	$3.126551293 \cdot 10^3$	0.404165998	0.521565024	SPC/E	LB	1330	
Jo	$4.802820915 \cdot 10^1$	$3.872584730 \cdot 10^3$	0.428124596	0.307813206	TIP3P	LB	1331	
Jo	$2.761450293 \cdot 10^1$	$3.907224600 \cdot 10^3$	0.453593139	0.182275955	TIP4P/EW	LB	1331	
Jo	$2.716659348 \cdot 10^1$	$3.449196753 \cdot 10^3$	0.448765998	0.186491407	SPC/E	LB	1331	
Ho	$1.198144253 \cdot 10^2$	$8.629324174 \cdot 10^3$	0.424300000	0.520000000	SPC/E	(LB)	877	b
Re	$1.280993692 \cdot 10^2$	$6.046188274 \cdot 10^3$	0.404170130	0.664184574	SPC	GM	1289	f
Re	$1.280993709 \cdot 10^2$	$3.588179831 \cdot 10^3$	0.386972638	0.862169238	SPC	GM	1289	g
Re	$1.280993710 \cdot 10^2$	$2.268412106 \cdot 10^3$	0.372463944	1.084347908	SPC	GM	1289	h

(continues on the next page)

Table 3.4 *(continued)*

Set	C_6^{ii} [10^{-4} kJ·mol^{-1}·nm^6]	C_{12}^{ii} [10^{-8} kJ·mol^{-1}·nm^{12}]	R_{min}^{is} [nm]	E_{min}^{is} [kJ·mol^{-1}]	Water	c.r.	Ref.	Notes
Re	$1.280993692 \cdot 10^2$	$6.046188274 \cdot 10^3$	0.404170130	0.664184574	SPC/E	GM	1289	i
Re	$1.280993716 \cdot 10^2$	$3.683285927 \cdot 10^3$	0.387817165	0.850965414	SPC/E	GM	1289	j
Re	$1.280993697 \cdot 10^2$	$2.327482819 \cdot 10^3$	0.373262719	1.070499276	SPC/E	GM	1289	k
				Br^-				
Da	$1.476432203 \cdot 10^2$	$1.447214449 \cdot 10^4$	0.437300000	0.494800000	SPC/E	LB	1334, 235	l
St	$1.488937085 \cdot 10^2$	$1.473838155 \cdot 10^4$	0.429900000	0.494463315	SPC	(GM)	1171	d
Pn	$2.790711750 \cdot 10^2$	($3.044988809 \cdot 10^5$)	0.494000000	0.334720000	CFF	WH	1286	e
Je	$7.304199557 \cdot 10^2$	$4.489885275 \cdot 10^4$	0.412371951	1.388177703	TIP4P	GM	1329	
Jo	$9.885433110 \cdot 10^1$	$9.954781606 \cdot 10^3$	0.437624596	0.395158341	TIP3P	LB	1331	
Jo	$7.317104277 \cdot 10^1$	$1.053119494 \cdot 10^4$	0.454393139	0.294189223	TIP4P/EW	LB	1331	
Jo	$6.260209136 \cdot 10^1$	$8.686177174 \cdot 10^3$	0.452765998	0.270805280	SPC/E	LB	1331	
Ho	$1.075507622 \cdot 10^2$	$1.373373902 \cdot 10^4$	0.449000000	0.370000000	SPC/E	(LB)	877	b

(continues on the next page)

Table 3.4 *(continued)*

Set	C_6^{ii} [10^{-4} kJ·mol^{-1}·nm^6]	C_{12}^{ii} [10^{-8} kJ·mol^{-1}·nm^{12}]	R_{min}^{is} [nm]	E_{min}^{is} [kJ·mol^{-1}]	Water	c.r.	Ref.	Notes
Re	$2.251976244 \cdot 10^2$	$1.515492979 \cdot 10^4$	0.416295933	0.737512639	SPC	GM	1289	f
Re	$2.251976269 \cdot 10^2$	$8.805677514 \cdot 10^3$	0.397880729	0.967531355	SPC	GM	1289	g
Re	$2.251976212 \cdot 10^2$	$5.386818189 \cdot 10^3$	0.381915280	1.237030098	SPC	GM	1289	h
Re	$2.251976245 \cdot 10^2$	$1.439112893 \cdot 10^4$	0.414505769	0.756831152	SPC/E	GM	1289	i
Re	$2.251976269 \cdot 10^2$	$8.805677514 \cdot 10^3$	0.397880729	0.967531355	SPC/E	GM	1289	j
Re	$2.251976212 \cdot 10^2$	$5.386818189 \cdot 10^3$	0.381915280	1.237030098	SPC/E	GM	1289	k
			I^-					
Da	$3.194858724 \cdot 10^2$	$6.098084042 \cdot 10^4$	0.467800000	0.521600000	SPC/E	LB	924, 235	
Pn	$5.697350276 \cdot 10^2$	($7.296407841 \cdot 10^5$)	0.519000000	0.355640000	CFF	WH	1286	e
Ps	$6.568483466 \cdot 10^2$	$7.895178385 \cdot 10^4$	0.445600000	0.941400000	TIP4P	(LB)	939	b
Je	$1.471568470 \cdot 10^3$	$1.822430321 \cdot 10^5$	0.437163890	1.388177703	TIP4P	GM	1329	
Jo	$1.573452490 \cdot 10^2$	$2.755688666 \cdot 10^4$	0.462824596	0.378033145	TIP3P	LB	1331	

(continues on the next page)

Table 3.4 *(continued)*

Set	C_6^{ii} [10^{-4} kJ·mol^{-1}·nm^6]	C_{12}^{ii} [10^{-8} kJ·mol^{-1}·nm^{12}]	R_{min}^{is} [nm]	E_{min}^{is} [kJ·mol^{-1}]	Water	c.r.	Ref.	Notes
Jo	$1.478385102 \cdot 10^2$	$3.131136800 \cdot 10^4$	0.472793139	0.344717155	TIP4P/EW	LB	1331	
Jo	$1.417291723 \cdot 10^2$	$2.805306858 \cdot 10^4$	0.469565998	0.341154961	SPC/E	LB	1331	
Ho	$1.450176927 \cdot 10^2$	$3.338161003 \cdot 10^4$	0.477000000	0.320000000	SPC/E	(LB)	877	b
Re	$4.445206073 \cdot 10^2$	$5.628558757 \cdot 10^4$	0.438811937	0.755398162	SPC	GM	1289	f
Re	$4.445206170 \cdot 10^2$	$3.166338523 \cdot 10^4$	0.418271664	1.007153956	SPC	GM	1289	g
Re	$4.445206136 \cdot 10^2$	$1.886179835 \cdot 10^4$	0.400599673	1.304917283	SPC	GM	1289	h
Re	$4.445206167 \cdot 10^2$	$5.211219802 \cdot 10^4$	0.436003807	0.785063596	SPC/E	GM	1289	i
Re	$4.445206086 \cdot 10^2$	$2.997613190 \cdot 10^4$	0.416367317	1.035110542	SPC/E	GM	1289	j
Re	$4.445206152 \cdot 10^2$	$1.771211674 \cdot 10^4$	0.398505700	1.346602141	SPC/E	GM	1289	k

(a) uses a modified[1270] TIPS water model; (b) combination rule not specified in the original reference (LB was used by default); (c) original study uses a Born-Mayer-Huggins potential[180,987,988] for ion-ion interactions; (d) combination rule not specified in the original reference (GM was used by default); (e) original study uses a 6-9 van der Waals potential (the reported C_{12}^{ii} is actually C_9^{ii}); (f) set L in Ref.[1289]; (g) set M in Ref.[1289]; (h) set H in Ref.[1289]; (i) set L_E in Ref.[1289]; (j) set M_E in Ref.[1289]; (k) set H_E in Ref.[1289]; (l) parameters used in Ref.[235] with the SPC/E water model[1275], although they were originally derived in Ref.[1334] for the SPC water model[1274].

parameter is affected by a similar kind of ambiguity as the effective ionic radius R_I in CE calculations (Point H_3 in Section 3.2). However, there is an important difference. The parameter R_I can only be connected to a solvation free energy and, possibly, to crystallographic lattice parameters, considering that derivative thermodynamic solvation properties cannot be reproduced accurately within CE approaches. In contrast, the parameter C_{12}^{ii}, in combination with C_6^{ii} and along with the choices of a water model and of combination rules, can in principle be tested in terms of many other simulated properties of pure ionic systems and ionic solutions (Section 6.2.3).

The C_6^{ii} and C_{12}^{ii} Lennard-Jones interaction parameters previously developed for alkali and halide ions (tentatively exhaustive list) are reported in Tables 3.2-3.4 (see also supplementary material of Ref.[1289] for a detailed overview of the different sets and of their calibration procedures). Note that various sets have also been derived for explicitly polarizable force fields[798,921,923,924,928,939,946,950,970,1327,1336].

As a final note concerning force-field parameters, it should be stressed that these are by no means physical constants. They represent effective quantities that encompass in an average sense corrections for the neglect of second-order effects (*e.g.* many-body interactions and quantum effects), as well as methodological approximations (*e.g.* approximate electrostatics, *ad hoc* combination rules) through careful calibration against experimental and, possibly, quantum-mechanical data. For this reason, force-field parameters are typically appropriate for: (*i*) a specific type of phase, *e.g.* gas or condensed phase; (*ii*) a specific simulation methodology, *e.g.* electrostatic scheme, cut-off distance if relevant, combination rules; (*iii*) a specific state point, *i.e.* a given choice of pressure and temperature (or, at least, a corresponding limited range), through calibration against data appropriate for this specific state point. Considering the latter dependence, one might actually consider adding a "$\ominus$" superscript to force-field parameters (*e.g.* solvent model parameters, C_6^{ii} and C_{12}^{ii}) appropriate for pressure P° and T^-.

3.3.3 Calculation of ionic solvation free energies

The calculation of a single-ion solvation free energy on the basis of AT simulations (Section 3.3.1) and given force-field parameters (Section 3.3.2) relies on a *free-energy calculation.* Reviews on these types of calculations can be found elsewhere[492,515,619–623,1021,1337–1346], and only the essential points are summarized below.

Free-energy calculations can be applied to the three following types of state changes: (*i*) *thermodynamic*, corresponding to the free-energy change upon altering a state parameter such as the pressure or the temperature at which a molecular system is considered; (*ii*) *conformational*, corresponding to the free-energy difference between two conformational states of the system; (*iii*) *alchemical*, corresponding to the free-energy change upon altering the atom-type content or the covalent structure of one or more species within the system. For reasons detailed previously (Section 2.2), the calculation of a single-ion solvation free energy *via* theoretical approaches, including AT simulations, is most conveniently performed based on the unphysical solvation process (Figure 2.2). In this case, the free-energy change that must be evaluated is of alchemical type, namely the one corresponding to the sum of the cavitation and charging processes.

The free-energy change associated with an *alchemical modification* has in most cases no physical meaning *per se*. This is obvious when the atom-type content, the net mass or the net charge of a species is altered, assuming that a single species is concerned by the alchemical change, because such a process has no experimental counterpart, excluding electron transfer and radioactive processes, which are in any case beyond the scope of classical models. It is in most cases also true when only the covalent structure of the species is altered (constitutional isomerization), because most of the force fields designed for the investigation of condensed-phase properties cannot accurately account for electronic effects within molecules (unlike so-called spectroscopic force fields, specifically designed for the description of gas-phase molecular properties[981–984,1287,1288,1347–1360]). Generally, the physically-relevant quantity is actually a *difference in free-energy changes* upon performing the same alchemical modification in two *distinct environments*, *e.g.* in solution *vs.* in vacuum or in the bulk solvent *vs.* within a molecular complex. Such a difference is related, *via* a thermodynamic cycle, to the difference in free-energy changes upon transferring the original and alchemically modified species from one environment to the other, *e.g.* to the difference in solvation or in binding free energy between the two species. Another point to keep in mind is that alchemical modifications can interconvert atom types within a species, but may not add or delete atoms, because the dimensionality of the classical potential-energy function must be preserved. However, when considering the difference in free-energy changes upon performing the same alchemical modification in two different environments, the mutation of an atom into a non-interacting or solely covalently interacting (but not massless) site within the species (dummy atom) is generally sufficient to achieve the same purpose. On the other hand, the covalent dissociation of a species or the covalent binding of two species in an alchemical change is in principle allowed, although not recommended because it leads in practice to a difficult configurational sampling problem.

In an alchemical modification, the initial state (A) and the final state (B) are defined by distinct potential energy functions, *e.g.* of the form of Eq. 3.16 in the context of the calculation of a single-ion solvation free energy, namely $V_A(\mathbf{r})$ and $V_B(\mathbf{r})$. The free-energy calculation relies on the definition of a hybrid (unphysical) potential-energy function $V(\mathbf{r};\lambda)$, depending on a *coupling parameter* λ in such a way that

$$V(\mathbf{r};0) = V_A(\mathbf{r}) \quad \text{and} \quad V(\mathbf{r};1) = V_B(\mathbf{r}) \; . \tag{3.54}$$

In the case of MD simulations, one rather uses a corresponding equation in terms of the Hamiltonian functions (potential plus kinetic energy) of the two states, so as to also account for possible alchemical mass changes. In the present context, however, the coupling is still formulated in terms of the potential energy because: (*i*) Hamiltonian coupling is not applicable in MC; (*ii*) mass changes have no influence on the outcome of the calculation, given sufficient sampling (although they may influence its convergence properties). The second observation holds as long as a difference in free-energy changes upon performing the same alchemical modification in two distinct environments is considered, and provided that the situation is not complicated by metric-tensor effects[997–1007] arising from the use of geometric constraints. As mentioned in Section 3.3.1, however, the latter effects are absent in systems consisting of monoatomic ions and fully rigid solvent molecules. In any case, no mass changes are required in the present context of ionic solvation. Note that Eq. 3.54 only specifies boundary conditions for an appropriate coupling scheme. In practice, the additional freedom in the design of a specific coupling scheme can be exploited to optimize the accuracy and convergence properties of the free-energy calculation[642,1345,1361–1367].

Assuming sampling in the isothermal-isobaric (NPT) ensemble, there exist many different approaches for the evaluation of the Gibbs free-energy change ΔG_{AB} associated with the alchemical state transformation $A \rightarrow B$, *i.e.* in the context of Eq. 3.54, with the λ-parameter change $0 \rightarrow 1$. Only the simplest one will be described here, namely *thermodynamic integration*[540,628,1366,1368–1373] (TI). Possible alternatives will be mentioned at the end of the discussion. In the TI approach, the free-energy change (expressed on a per mole basis) is evaluated based on the (statistical-mechanically exact) formula

$$\Delta G_{AB} = N_A \int_0^1 d\lambda \left\langle \frac{\partial V(\mathbf{r};\lambda')}{\partial \lambda'} \right\rangle_\lambda , \tag{3.55}$$

where $\langle ... \rangle_\lambda$ denotes (isothermal-isobaric) ensemble averaging based on the potential-energy function $V(\mathbf{r};\lambda)$ at a specified pressure P, temperature T and λ-value λ, *i.e.* averaging over configurations $\mathbf{x}$ and instantaneous volume values $\mathcal{V}$ occurring with a probability proportional to the Boltzmann factor $\exp\{-[V(\mathbf{r};\lambda)+P\mathcal{V}]/(k_B T)\}$. Note that canonical (NVT) ensemble averaging, involving a fixed volume rather than a constant reference pressure, would lead instead to the corresponding Helmholtz free-energy change $\Delta F_{A'B'}$, where the primes underline that although the initial and final states correspond to the same potential-energy function, they generally represent distinct thermodynamic state points of the system, *i.e.* even if the initial pressure or volume can in principle be chosen to make A and A' coincide, the final states B and B' will still generally be distinct (for this reason, the interconversion between $\Delta F_{A'B'}$ and ΔG_{AB} is in general not trivial).

In practice, two types of approximations are required to convert Eq. 3.55 to a form suitable for computer-based calculations: (*i*) replacement of the integral by a weighted sum over a finite number N_λ of λ-points, using some *quadrature scheme*; (*ii*) replacement of the infinite ensemble averages by *finite trajectory averages*, evaluated through independent equilibrium simulations of N_s steps (the number of steps may also be chosen differently for each of the N_λ points, whenever useful). Note that thermostatted and barostatted MD as well as barostatted MC directly produce trajectories corresponding to Boltzmann-weighted isothermal-isobaric ensembles, so that the ensemble average in Eq. 3.55 actually corresponds to an unweighted average over sampled configurations in both cases. The resulting approximate TI expression is thus

$$\Delta G_{AB} \approx N_A \sum_{n=1}^{N_\lambda} w_n N_s^{-1} \sum_{i=1}^{N_s} \frac{\partial V(\mathbf{r}_{n,i};\lambda')}{\partial \lambda'} \quad , \tag{3.56}$$

where w_n is the quadrature weight of the nth λ-point in the numerical integration, which depends on the spacing between the N_λ successive λ-points, and $\mathbf{r}_{n,i}$ the ith configuration of the simulation performed at the nth λ-point λ_n, *i.e.* using the potential-energy function $V(\mathbf{r};\lambda_n)$.

Of course, TI is by no means the only method to calculate an alchemical free-energy difference based on AT simulations. Other possible methods include: (*i*) slow growth[1369, 1374–1379], an approximate version of TI where the λ parameter is progressively increased from 0 to 1 in the course of a single simulation (not recommended due to its formally inexact nature[1369,1376–1378]); (*ii*) fast growth[1379–1383], relying on an ensemble of slow-growth simulations initiated from a Boltzmann-weighted distribution of configurations at $\lambda = 0$; (*iii*) free-energy perturbation[560, 629, 1369, 1373, 1384–1386], relying on a free-energy extrapolation from one λ point to the next or/and previous point (see also particle insertion[1387]); (*iv*) TI with cumulant expansion[792, 1081, 1289, 1388–1390], relying on the use of higher-order derivatives of the potential energy with respect to λ at each λ-point, (*v*) λ-MC[1391] or λ-dynamics[630,631,1392–1395], relying on the evolution of the λ variable simultaneously with the configurational sampling, according to a coupled trial-acceptance prescription or equation of motion (should be applied in conjunction with a suitable umbrella-sampling biasing potential in λ to ensure the sampling of all λ values[1346]); (*vi*) replica exchange in the λ-variable[636–642], relying on parallel simulations at the different λ-points along with a trial-acceptance prescription for the exchange of configurations across these simulations (should be applied in conjunction with a suitable biasing potential in λ to ensure the occurrence of exchanges between all parallel simulations); (*vii*) enveloping distribution sampling[632–634,1396–1399], relying on a single simulation involving an optimally-designed artificial potential-energy function, and free-energy extrapolations to the initial and final states based on the simulated ensemble (no coupling parameter is used in this approach).

There also exist a number of different approaches to calculate the *entropy difference* along an alchemical process, including: (*i*) application of the Gibbs equation[657, 877, 1400–1406], using the result of a free-energy calculation in terms of ΔG_{AB} along with an estimate for the enthalpy change ΔH_{AB} based on the simulations of the two end points; (*ii*) application of methods analogous to those involved in the evaluation of the free-energy change ΔG_{AB} (see above), but using statistical-mechanical formulae appropriate for the corresponding entropy change[656,1403–1405,1407–1409]; (*iii*) calculation of the free-energy change ΔG_{AB} at two different temperatures and use of a finite-difference approximation[1403–1405,1410] (see below; see also Refs.[1167,1404,1411] for a combination between this and the previous approach). For completeness, one may also mention the partitioning of the free-energy change ΔG_{AB} in terms of pseudo-enthalpy $\Delta H'_{AB}$ and pseudo-entropy $\Delta S'_{AB}$ components[655–659]. These components satisfy the Gibbs equation but omit the canceling and usually large contributions of interactions that are not directly involved in the alchemical perturbation, *i.e.* not active in the λ-derivative of $V(\mathbf{r};\lambda)$ (*e.g.* all solvent-solvent interactions in an alchemical change solely involving the solute). It should be stressed, however, that $\Delta H'_{AB}$ and $\Delta S'_{AB}$ cannot be interpreted in the same way as the true enthalpy and entropy contributions ΔH_{AB} and ΔS_{AB} of the Gibbs equation.

The method that appears generally to give the most accurate and reliable results in entropy calculations is *finite-difference in temperature*. In this approach, the free-energy change ΔG_{AB} is computed at two different temperatures $T - \Delta T/2$ and $T + \Delta T/2$ using one of the above free-energy calculation methods (*e.g.* TI). The entropy change is then evaluated through the exact thermodynamic relationship $S = -\partial_T G$ (Section 4.2.1), approximated as

$$\Delta S_{AB}(T) \approx -[\Delta G_{AB}(T + \Delta T/2) - \Delta G_{AB}(T - \Delta T/2)]/\Delta T \ . \qquad (3.57)$$

The temperature interval ΔT should be chosen sufficiently small, so that the first-order finite-difference approximation is accurate (*i.e.* heat-capacity effects are negligible), but also sufficiently large, so that errors in the ΔG_{AB} values at the two temperatures have a limited impact on the calculated ΔS_{AB}. Note that the *Gibbs equation method* has also been applied with success for the evaluation of single-ion solvation enthalpies[351,1324,1327] and entropies[877,1406]. In principle, the finite-difference approach could be applied as well to the evaluation of *other derivative thermodynamic solvation properties*, *e.g.* the solvation heat capacities, volumes, volume-compressibilities and volume-expansivities[1412], considering that all these quantities are related to the solvation free energy and its first or/and second pressure or/and temperature derivatives (Sections 2.2 and 4.2.1). However, for the volume and its derivatives, it is generally more convenient to analyze the volume change upon inserting the ion in a sufficiently large quantity of the pure solvent at constant pressure and temperature, along with the pressure and temperature derivatives of this quantity. This approach has also been applied with success for the evaluation of single-ion partial molar volumes in solution[1289,1324].

As will be discussed in Sections 4.2.5 and 4.2.16, the natural standardized outcome of the different approaches to evaluate *derivative thermodynamic solvation properties* may differ in terms of the *solute standard-state variant* involved. Finite-difference approaches based on the analysis of the first or/and second pressure or/and temperature dependence of the solvation free energy will lead to results in the standard variant. The same applies to heat capacities evaluated from calculated solvation enthalpies *via* finite-difference in temperature, irrespective of the enthalpy calculation scheme. However, solvation enthalpies or volumes calculated as differences between the corresponding properties of a large quantity of solvent in the presence of absence of the ion, which corresponds to the evaluation of isobaric and isothermal geometric volume variations or heats of mixing, respectively, will lead to results in the density-corrected variant. The same applies to volume-compressibilities and volume-expansivities evaluated *via* finite-difference in temperature or pressure based on solvation volumes calculated using the above geometric-variation approach.

The calculation of the *standard single-ion solvation free energy* $\Delta_s G^{\ominus}[I^z_g]$ of an ion I^z (assumed intrinsic; see below) *via* a free-energy calculation based on AT simulations is performed, considering the unphysical solvation process (Figure 2.2), according to the equation

$$\Delta_s G^{\ominus}_{AT}[I^z_g] = \Delta_s G^{AT}_{chg}[I^z_g] + \Delta_s G^{AT}_{cav}[I^z_g] + \Delta_s G^{\ominus}_{std} \; . \tag{3.58}$$

The three terms on the right-hand side are similar to those encountered in the discussion of the BO+ST model (Section 3.2 and Eq. 3.7), the two former ones now being calculated using an AT model instead of a CE model. The quantity $\Delta_s G_{chg}[I^z_g]$, of which $\Delta_s G^{AT}_{chg}$ is the estimate according to the AT model, corresponds to the second step of the unphysical solvation process, namely the *charging process*. The quantity $\Delta_s G_{cav}[I^z_g]$, of which $\Delta_s G^{AT}_{cav}[I^z_g]$ is the estimate according to the AT model, corresponds to the first step of the unphysical solvation process, namely the cavitation process. In order for $\Delta_s G^{\ominus}_{AT}[I^z_g]$ to represent an appropriate estimate for the standard intrinsic single ion solvation free energy at the reference temperature T^-, $\Delta_s G^{AT}_{chg}[I^z_g]$ and $\Delta_s G^{AT}_{cav}[I^z_g]$ must be evaluated using isothermal-isobaric simulations (Section 3.3.1) at the reference liquid pressure P° and at temperature T^-, resulting in the corresponding semi-standard intrinsic point-to-point solvation free energy $\Delta_s G^{\boxminus}_{AT}[I^z_g]$. The effective condensed-phase force-field parameters employed must also have been calibrated for these conditions (Section 3.3.2). The value of $\Delta_s G^{\boxminus}_{AT}[I^z_g]$ must then be increased by an appropriate standard-state correction term $\Delta_s G^{\ominus}_{std}$ (Eq. 2.7; see Refs.[1066,1412–1419] for more general discussions concerning the importance of standard-state corrections in free-energy calculations).

The changes involved in the calculation of $\Delta_s G^{AT}_{chg}$ and $\Delta_s G^{AT}_{cav}$ are both of alchemical type. It has been stated (see above) that free-energy changes associated with alchemical modifications have in most cases no physical meaning *per se*, unless they are compared with corresponding changes in a different environment. The present calculation is no exception and the free-energy changes $\Delta_s G^{AT}_{chg}$ and $\Delta_s G^{AT}_{cav}$ are actually implicitly compared with corresponding changes in vacuum. However, these vacuum contributions turn out to vanish in the present situation. For $\Delta_s G^{AT}_{cav}$, the growth of a cavity in the liquid is compared to the growth of a "cavity" in vacuum, the latter involving no work in the absence of neighbor particles for Lennard-Jones interactions. For $\Delta_s G^{AT}_{chg}$, the charging of the cavity in the liquid is compared to the charging of the "cavity" in vacuum, the latter involving no work since the electronic self-energy contribution to the charging process is omitted in classical models.

Note that the latter statement is only strictly valid for a potential-energy function exempt of self-term V_{slf} in Eq 3.16. In the presence of such a term, the subtraction of the vacuum contribution amounts to the *removal of the self-energy term* from the free-energy change calculated in solution. For example, in the PBC/LS scheme (Section 3.3.1), a self-term of the form of Eq. 3.27 may be included in the charging free energy of the ion in solution[330,792]. However, the same term will account for the charging free energy in vacuum, where the vacuum situation is interpreted here as a periodic system consisting of the ion and its periodic copies only, along with the corresponding homogeneous neutralizing background charge density. In other words, the calculation in vacuum can be omitted provided that the term V_{slf} is removed from the potential energy function when performing the charging process in solution. The same discussion applies to the PBC/RF scheme (Section 3.3.1), considering a self-term of the form of Eq. 3.37 or of any other form[1082,1083]. There, the vacuum situation is to be interpreted as that of a single ion inside a cutoff sphere, the term V_{slf} being based on the same ϵ_{RF} value as in solution and accounting for the interaction with an effective neutralizing background charge density within the sphere[1082]. Similar considerations also apply to a possible corresponding excluded-atoms term V_{exc} in Eq. 3.16, *e.g.* based on Eq. 3.26 for the PBC/LS scheme or Eq. 3.36 for the PBC/RF scheme, but are only relevant for polyatomic ions. Finally, the proper treatment of a possible surface term V_{srf} in Eq. 3.16 is not entirely obvious, considering that the nature and relevance of this term is itself already matter of debate (Section 3.3.1). Note, finally, that some of the above comments may also apply to the *cavitation term* when the description of the Lennard-Jones interactions involves a lattice-sum treatment or tail correction (Section 3.3.1).

In the case of the cavitation term $\Delta_s G^{AT}_{cav}$, the potential energies $V_A(\mathbf{r})$ and $V_B(\mathbf{r})$ associated with the initial and final states of the alchemical change differ in the ion-solvent Lennard-Jones interaction parameters, $C^{is}_{12,A} = C^{is}_{6,A} = 0$ in state A, as appropriate for a dummy particle and $C^{is}_{12,B} = C^{is}_{12}$ and $C^{is}_{6,B} = C^{is}_{6}$ in state B, as appropriate for an ion-sized cavity. The ionic charge is zero in both states, *i.e.* $z_A = z_B = 0$. Some care should be taken during the corresponding free-energy calculation to choose an appropriate coupling scheme (Eq. 3.54). For example, a coupling scheme linear in the Lennard-Jones interaction parameters is not recommended, since the ensemble average involved in Eq. 3.55 diverges at $\lambda = 0$ in this case. This can be remedied by the use of a so-called *soft-core coupling scheme*[642,1361–1365,1420,1421]. Besides this issue, the calculation of $\Delta_s G^{AT}_{cav}$ poses no special methodological problems, and gives essentially consistent results for all (reasonable) choices of boundary conditions and approximate electrostatic scheme (Section 3.3.1), given a water model and a set of ion-solvent Lennard-Jones interaction parameters (Section 3.3.2).

In the case of the *charging term* $\Delta_s G^{AT}_{chg}$, the potential energies $V_A(\mathbf{r})$ and $V_B(\mathbf{r})$ associated with the initial and final states of the alchemical change differ in the ionic charge, with $z_A = 0$ in state A, as appropriate for an uncharged ion-sized cavity and $z_B = z$ in state B, as appropriate for the fully-charged ion. The Lennard-Jones parameters are in both states C_6^{is} and C_{12}^{is}. In general, a simple *linear coupling scheme* compatible with Eq. 3.54 is adopted, *i.e.*

$$V(\mathbf{r};\lambda) = (1-\lambda)V_A(\mathbf{r}) + \lambda V_B(\mathbf{r}) \ . \tag{3.59}$$

In other words, the charge of the ion increases linearly with λ, as λze. In this case, the ensemble averages involved in Eq. 3.55 read

$$\left\langle \frac{\partial V(\mathbf{r};\lambda')}{\partial \lambda'} \right\rangle_\lambda = \langle V_B(\mathbf{r}) - V_A(\mathbf{r}) \rangle_\lambda = ze\zeta^{AT}_{\lambda ze} \quad , \tag{3.60}$$

where $\zeta^{AT}_{\lambda ze}$ is the *solvent-generated electric potential* potential at the center of the ionic cavity, ensemble-averaged over configurations sampled at a fractional charge λze of the ion. The corresponding expression for the charging free energy is then

$$\Delta_s G^{AT}_{chg}[I^z_g] = N_A ze \int_0^1 d\lambda \, \zeta^{AT}_{\lambda ze} \quad . \tag{3.61}$$

Assuming that the response of the solvent is *linear*, *i.e.*

$$\zeta^{AT}_{\lambda ze} \approx C_{LR} \lambda ze \quad , \tag{3.62}$$

where C_{LR} is a constant characterizing the magnitude of this response, for an ion of a given size, leads after substitution into Eq. 3.61 for the charging process to the approximate result

$$\Delta_s G^{AT}_{chg}[I^z_g] \approx N_A C_{LR} \frac{z^2 e^2}{2} \ . \tag{3.63}$$

This expression is identical with the corresponding Born expression from CE (Eq. 3.4), provided that one identifies

$$C_{LR} = -\frac{\epsilon_{svt} - 1}{4\pi\epsilon_o \epsilon_{svt} R_I} \quad . \tag{3.64}$$

However, AT simulations suggest that this approximation is not very accurate[707, 708, 792, 1422–1424]. A more appropriate linear approximation involves an *offset constant* ζ^{AT}_0 for the solvent-generated electric potential at the center of the uncharged ionic cavity, *i.e.*

$$\zeta^{AT}_{\lambda ze} \approx C_{LR} \lambda ze + \zeta^{AT}_0 \quad , \tag{3.65}$$

leading to

$$\Delta_s G_{chg}^{AT}[I_g^z] \approx N_A C_{LR} \frac{z^2 e^2}{2} + N_A z e \zeta_0^{AT} \ . \qquad (3.66)$$

The absence of the second term in the original Born expression is in large part related to the neglect of the *solvent polarization at the ion-solvent interface* (Point A_3 in Section 3.2). However, ζ_0^{AT} as evaluated from AT simulations may also contain a spurious contribution related to a methodological problem in the procedure used to calculate the solvent-generated potential at the cavity center (see further below). Although Eq. 3.66 typically represents a reasonable approximation in the case of monovalent ions[707, 708, 792, 1422–1424], $\zeta_{\lambda ze}^{AT}$ may evidence significant deviations from linearity at large λ for polyvalent ions, more specifically, a *sublinear behavior*. In this case, $\Delta_s G_{chg}^{AT}$ will exhibit a dependence on z that is no longer the sum of a quadratic and a linear term (Eq. 3.66). The rigorously quadratic dependence of the Born expression on the ionic charge (Eq. 3.4) is predominantly a consequence of the neglect of *dielectric saturation* (Point F_3 in Section 3.2) and, to a minor extent, *electrostriction* (Point E_3 in Section 3.2). Approximate expressions such as Eqs. 3.63 and 3.66 are useful in establishing connections between the AT and CE descriptions. However, in practice, no such approximations are involved when calculating a single-ion solvation free energy from AT simulations based *e.g.* on the TI approach (see, however, Refs.[1425, 1426]).

In principle, the solvent-generated potential $\zeta_{\lambda ze}^{AT}$, to be used in Eq. 3.55 *via* Eq. 3.60, should be obtained by averaging the electric potential generated by the solvent at the center of the ionic cavity over trajectory configurations corresponding to the given λ-point (*i.e.* based on a simulation at charge state λze), given the choice of boundary conditions and approximate electrostatic scheme used in the simulation (Section 3.3.1). Based on Eq. 3.16, assumed exempt of self-term V_{slf} (see above), one may write in the specific context of a monoatomic ion,

$$\zeta_{\lambda ze}^{AT} = \frac{1}{4\pi\epsilon_o} \left\langle \sum_i^{N-1} q_i \psi_{EL,oi}(\mathbf{r}) \right\rangle_\lambda \ , \qquad (3.67)$$

where the summation includes all solvent sites i, while o indicates the single ion site. The ion-solvent pairwise electrostatic interaction function $\psi_{EL,oi}$, which is a special case of the functions $\psi_{EL,ij}$ introduced in Section 3.3.1, is given by Eq. 3.18 (FBC/CB), Eq. 3.25 (PBC/LS) or Eq. 3.30 (PBC/CT). In the latter case, the ion-solvent truncation function H_{oi} in Eq. 3.30, which is a special case of the functions H_{ij} introduced in Section 3.3.1, is given by Eq. 3.23 (P-truncation) or Eq. 3.24 (M-truncation), and the effective pairwise interaction function τ_{CT} is given by Eq. 3.34 (PBC/CT/SC), Eq. 3.35 (PBC/CT/RF), or *e.g.* Eqs. 3.40-3.44 (PBC/CT/MT).

Unfortunately, since the first free-energy calculations of this kind in the eighties[1334, 1427, 1428], it has appeared that the use of Eqs. 3.55, 3.60 and 3.67, based on classical potential-energy functions of the form of Eq. 3.16 leads to results that are *essentially inconsistent*[350,708,792,1056,1057,1081,1167,1169,1171,1187,1191,1289,1412,1422–1424,1429–1436]. Depending on the *boundary conditions* and *electrostatic scheme* employed in the simulation (FBC/CB, PBC/LS, PBC/CT/SC, PBC/CT/RF or PBC/CT/MT, choice of P- *vs.* M-truncation for PBC/CT), and on the *simulation parameters* used when applying these schemes (droplet radius for FBC, box size for PBC, cutoff distance for PBC/CT), extremely large variations are observed in the calculated solvation free energies for a given ion, solvent model and set of ion-solvent Lennard-Jones interaction parameters. The situation for the PBC/LS scheme is further complicated by the fact that the numerical implementation of this scheme generally involves the inclusion of a self-term[792,1081,1423,1430,1431,1437] V_{slf} in Eq. 3.16 (Eq. 3.27), also sometimes suggested[1082,1083] in the PBC/RF scheme (*e.g.* Eq. 3.37). When this term is not appropriately removed by consideration of the charging process in vacuum, it induces an additional and typically very large contribution to the calculated solvation free energy. As an illustration of these problems, the application of 72 parameter combinations within 7 different choices of boundary conditions and electrostatic schemes (the self-term V_{slf} being excluded in all PBC schemes) revealed a standard deviation of 83 kJ·mol^{-1} in the raw hydration free energies calculated for the sodium ion[1081] based on the SPC water model[1274] along with the Straatsma and Berendsen ion-solvent Lennard-Jones interaction parameters[1171]. It can be shown that this unacceptable spread in the calculated values is caused almost exclusively by the charging free-energy contribution. The cavitation term, which is relatively small in magnitude, is only weakly affected by the choice of boundary conditions and electrostatic scheme[1081,1289] within reasonable bounds.

In order to investigate the causes for the discrepancies in the calculated charging free energies, it is important to disentangle the error caused by the choice of boundary conditions and electrostatic scheme during the *generation of the configurations* from the corresponding error in the *analysis of the configurations* in terms of the solvent-generated potential at the cavity center[708,1423,1429]. This can be done by recalculating a solvent-generated potential at the cavity center $\zeta'^{AT}_{\lambda ze}$ for the same set of configurations, but using different choices $\psi'_{EL,oi}$ for the effective ion-solvent electrostatic interaction function in Eq. 3.67, *i.e.* defining a *recalculated solvent-generated electric potential* $\zeta'^{AT}_{\lambda ze}$ at the cavity center based on the configurations generated at a charge state λze as

$$\zeta'^{AT}_{\lambda ze} = \frac{1}{4\pi\epsilon_o} \left\langle \sum_i^{N-1} q_i \psi'_{EL,oi}(\mathbf{r}) \right\rangle_\lambda . \tag{3.68}$$

Two particularly interesting choices of pairwise interaction functions for this recalculation are given by

$$\psi'_{EL,oi}(\mathbf{r}) = H_{oi}(\mathbf{r}; R)\bar{r}_{oi}^{-1} , \tag{3.69}$$

where the influence of a solvent charge is of a truncated Coulombic form, while H_{oi} is given by either Eq. 3.23 (P-truncation) or Eq. 3.24 (M-truncation).

The two distinct recalculation procedures will be further referred to as *P-summation* and *M-summation*, respectively. Note that $\overline{r}_{oi}$ should be replaced by r_{oi} if the configurations have been generated using the FBC/CB scheme.

The recalculated potentials $\zeta'^{AT}_{\lambda ze}$ depend on a summation range defined by the parameter R in Eq. 3.69. The corresponding P-truncated function of R will be written $\zeta'^{AT,P}_{\lambda ze}$ and the corresponding M-truncated function of R will be written $\zeta'^{AT,M}_{\lambda ze}$, keeping the dependence on R implicit for the simplicity of notation. Using these recalculated values in Eq. 3.61, and substituting the result into Eq. 3.58, one obtains two alternative definitions for the *recalculated intrinsic single-ion solvation free energy*, given a selected value of R. For the P-summation scheme, one has

$$\Delta_s G^{\ominus}_{AT,P}[I^z_g] = \Delta_s G^{AT,P}_{chg}[I^z_g] + \Delta_s G^{AT}_{cav}[I^z_g] + \Delta_s G^{\ominus}_{std} \tag{3.70}$$

with

$$\Delta_s G^{AT,P}_{chg}[I^z_g] = N_A ze \int_0^1 d\lambda\, \zeta'^{AT,P}_{\lambda ze} \ . \tag{3.71}$$

For the M-summation scheme, one has

$$\Delta_s G^{\ominus}_{AT,M}[I^z_g] = \Delta_s G^{AT,M}_{chg}[I^z_g] + \Delta_s G^{AT}_{cav}[I^z_g] + \Delta_s G^{\ominus}_{std} \tag{3.72}$$

with

$$\Delta_s G^{AT,M}_{chg}[I^z_g] = zeN_A \int_0^1 \zeta'^{AT,M}_{\lambda ze} \ . \tag{3.73}$$

The notations $\Delta_s G^{\ominus}_{AT}[I^z_g]$ and $\Delta_s G^{AT}_{chg}[I^z_g]$ will be reserved to the solvation and charging free energies evaluated directly from the simulation without recalculation of the potential, *i.e.* defined by Eqs. 3.58, 3.61 and 3.67 in their original forms.

For both $\Delta_s G^{\ominus}_{AT,P}$ and $\Delta_s G^{\ominus}_{AT,M}$ taken individually, and given a selected value of R, the results from simulations employing different boundary conditions and electrostatic schemes still differ. However, the differences are typically much more limited compared to those observed in the values $\Delta_s G^{\ominus}_{AT}$ calculated using the corresponding original electrostatic interaction functions. Furthermore, both quantities evidence expected convergence properties with respect to the simulation parameters, *i.e.* they converge to two well-defined values upon increasing the droplet radius in FBC, the box size in PBC, and the cutoff distance in PBC/CT. But there still remains an important problem. The P- and M-summation results $\Delta_s G^{\ominus}_{AT,P}$ and $\Delta_s G^{\ominus}_{AT,M}$ for a given R *do not converge to the same value.*

Irrespective of the boundary conditions and electrostatic scheme used to generate the configurational ensemble, the recalculated potentials $\zeta'^{AT,P}_{\lambda ze}$ and $\zeta'^{AT,M}_{\lambda ze}$ always differ by an offset[707, 708, 1429, 1437] ξ_{AT}, that is essentially independent of the value of R above a certain minimal distance (plateau value) and of the charge state λ. This difference ξ_{AT} will be called here the *classical exclusion potential of the solvent model* (this quantity is noted $\Delta\Phi_{ODL}$ in

Ref.[707]). More specifically, for any sufficiently large R, one has

$$\zeta'^{AT,P}_{\lambda ze} = \zeta'^{AT,M}_{\lambda ze} + \xi_{AT} \quad \forall \lambda \; . \tag{3.74}$$

The exact significance of the exclusion potential will be discussed later (Section 4.3.8). For the moment, it is just noted that the exclusion potential is a unique property of a given classical solvent model at a specific solvent density and including the choice of a molecular center for M-summation. For solvent models characterized by a single Lennard-Jones interaction site, also taken to be the molecular center (Section 3.3.1), ξ_{AT} is immediately related to the quadrupole-moment trace $\mathcal{Q}$ of the solvent model relative to this site *via* the solvent density (Eq. 4.345). For solvent models with multiple Lennard-Jones interaction sites, ξ_{AT} must be evaluated numerically[707,803] for a given solvent density. Values of $\mathcal{Q}$ and $\xi^{\ominus}_{AT}$, the latter corresponding to ξ_{AT} at the (experimental) solvent density $\rho^{\ominus}_{svt}$ appropriate for P° and T^{-}, are reported in Table 3.1 for some common water models involving a single Lennard-Jones interaction site at the oxygen atom.

Eq. 3.74 implies in particular that the potentials $\zeta'^{AT,P}_{\lambda ze}$ and $\zeta'^{AT,M}_{\lambda ze}$ already differ by ξ_{AT} when the ionic charge is zero, a situation that is depicted in Figure 3.5. The short notations $\zeta_{AT,P}$ and $\zeta_{AT,M}$ will be used in the

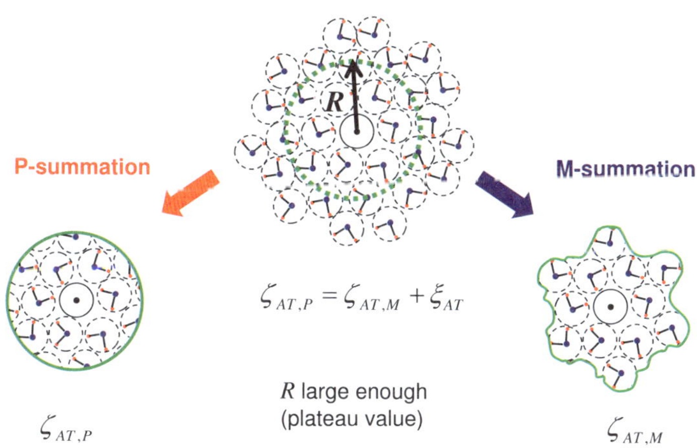

Figure 3.5 *Intrinsic electric potential at the center of an uncharged cavity within a pure liquid recalculated using P- or M-summation.* The electric potential is recalculated based on configurations issued from an AT simulation. The solvent is symbolized by water molecules. The recalculation relies on Eqs. 3.68 and 3.69, along with a truncation function defined either by Eq. 3.23 (P-summation; summation of the Coulombic potential contributions from the solvent by charges up to a distance R) or Eq. 3.24 (M-summation; summation of the Coulombic potential contributions from the solvent by molecules, based on a specified molecular center, up to a distance R). For sufficiently large sphere radii R (still smaller than the size of the droplet or computational box), the solvent-generated potential at the cavity center converges in both cases to a well-defined value ($\zeta_{AT,P}$ and $\zeta_{AT,M}$, respectively). However, these values differ by the classical exclusion potential ξ_{AT} of the solvent model (Eq. 3.75).

following to denote the values $\zeta_0'^{AT,P}$ and $\zeta_0'^{AT,M}$ for any sufficiently large R (plateau value), *i.e.* the two corresponding estimates for the *intrinsic electric potential at a cavity within the pure liquid.* Here, the word intrinsic refers to the fact that the solvent at the boundary of the spherical sample of radius R does not present the inhomogeneity and anisotropy features (Section 2.2) characteristic of an interface to vacuum (air), *i.e.* its configurational properties are still bulk-like. However, P-summation terminates the bulk sample on the basis of individual charges, while M-summation terminates it on the basis of intact molecules. In both cases, the resulting potential is meant to represent the potential at the cavity relative to a point outside the liquid at an infinite distance. In this special situation of an uncharged cavity, Eq. 3.74 reads

$$\zeta_{AT,P} = \zeta_{AT,M} + \xi_{AT} \; . \tag{3.75}$$

Note that the convergence of $\zeta_{\lambda ze}'^{AT,P}$ and $\zeta_{\lambda ze}'^{AT,M}$ as a function of R is generally not achievable in practice for $\lambda \neq 0$ given accessible sizes for simulated systems, due to the long-range nature of the electrostatic ion-solvent interactions. In contrast, distances R for which $\zeta_{\lambda ze}'^{AT,P} - \zeta_{\lambda ze}'^{AT,M}$ (Eq. 3.74), and, as a special case, $\zeta_0'^{AT,P}$ and $\zeta_0'^{AT,M}$ (uncharged cavity) converge (to $\zeta_{AT,P}$ and $\zeta_{AT,M}$, respectively, in the latter case) can easily be achieved. Thus, $\zeta_{AT,P}$ and $\zeta_{AT,M}$, as well as their difference ξ_{AT}, can be evaluated based on simulated trajectories.

The situation of Eq. 3.75 can be illustrated numerically[707] based on simulations of an uncharged sodium-sized cavity[1171] and the SPC water model[1274] at 1 bar and 300 K, the corresponding data being reported in Table 3.5. The following considerations apply irrespective of the boundary conditions and electrostatic scheme originally used to generate the configurational ensemble. When the potential at the center of the cavity is recalculated using P-summation (Eqs. 3.68 and 3.69 along with Eq. 3.23), the cavity potential $\zeta_{AT,P}$ evaluates to about +0.70 V (plateau value upon increasing the radius R of the solvent range considered for summation of the potential). However, when the value is recalculated using M-summation instead (Eqs. 3.68 and 3.69 along with Eq. 3.24; solvent molecular center at the oxygen atom), the cavity potential $\zeta_{AT,M}$ evaluates to about −0.12 V. Although the actual values of $\zeta_{AT,P}$ and $\zeta_{AT,M}$ depend on the cavity size and on the pressure and temperature conditions, the difference ξ_{AT} does not, except for a small dependence on pressure and temperature *via* the solvent density. The classical exclusion potential of the SPC water model at P° and T^- is thus $\xi_{AT}^{\ominus} = +0.82$ V (Table 3.1). In this case, the P-summation convention will lead to a recalculated solvation free energy that is about 80 kJ·mol^{-1} less negative for a monovalent cation and about 80 kJ·mol^{-1} more negative for a monovalent anion compared to the M-summation convention. Note that the value of $\zeta_{AT,M}$ for a sodium-sized cavity in SPC water at 1 bar and 300 K is slightly negative (-0.12 V). This accounts for a slight orientational preference of the water molecules close to the cavity surface, favoring orientations where the hydrogen atoms point

Table 3.5 *Average electric potentials calculated by atomistic simulations for the SPC water model.* The different entries correspond to the intrinsic electric potential ζ at a cavity within the pure liquid (Figure 3.5), the external intrinsic bulk electric potential η of the pure liquid (Figure 3.6), the internal intrinsic bulk electric potential $\breve{\eta}$ (Figure 3.6), the interfacial potential χ' at the cavity surface based on ζ and η (air-to-cavity direction), the interfacial potential $\breve{\chi}'$ at the cavity surface based on ζ and $\breve{\eta}$ (air-to-cavity direction) and the surface potential χ of the pure liquid (air-to-liquid direction) (Figure 3.7), as evaluated by atomistic (AT) simulations with the SPC water model[1274] at 1 bar, using either the M-scheme (M) or the P-scheme (P), at either 300 K or in the orientational disorder limit (ODL; idealized situation of an infinite rotational temperature for the solvent molecules, realized in practice here by zeroing the atomic partial charges of the solvent during the configuration generation). The cavity potential ζ corresponds to a sodium-sized cavity based on the Straatsma & Berendsen ion-solvent Lennard-Jones interaction parameters[1171]. The external potential η corresponds to an average over points distant by at least 0.1 nm from the oxygen atom of any water molecule.

	300 K	ODL
	Potential [V]	
$\zeta_{AT,M}$	−0.12	0.00
$\zeta_{AT,P}$	+0.70	+0.82
$\eta_{AT,M}$	−0.03	0.00
$\eta_{AT,P}$	+0.79	+0.82
$\breve{\eta}_{AT,M}$	−0.82	−0.82
$\breve{\eta}_{AT,P}$	0.00	0.00
$\chi'_{AT,M}$	+0.09	-
$\chi'_{AT,P}$	+0.09	-
$\breve{\chi}'_{AT,M}$	−0.70	-
$\breve{\chi}'_{AT,P}$	−0.70	-
$\chi_{AT,M}$	+0.72	0.00
$\chi_{AT,P}$	−0.10	−0.82

towards the bulk. In the idealized situation where the water molecules are characterized by an infinite rotational temperature, a situation referred to as the *orientational disorder limit*[707,1412] (ODL), no such preference is observed, resulting in $\zeta_{AT,M} = 0$ and $\zeta_{AT,P} = \xi_{AT}$.

Although the difference between P- and M-summation results in terms of calculated single-ion solvation free energies may be extremely large, the ambiguity affecting the experimentally deduced values for single-ion solvation free energies (Section 2.3) and the arbitrariness involved in the choice of ion-solvent Lennard-Jones interaction parameters (Section 3.3.2) has prevented the resolution of this problem on the basis of purely experimental arguments. As a result, a *hot debate* has taken place in the literature between the advocates of the P-summation[708,1056,1423,1429,1430,1433,1438] scheme and those of the

M-summation[707,1081,1082,1289,1289,1412,1432,1434,1436,1439] scheme for the calculation of single-ion solvation free energies, a debate which is, at least in the opinion of the authors, at present settled *in favor of the M-summation scheme*[707], based on arguments that will be provided in Section 4.3.8 and further developed in Chapter 7.

Accepting the M-summation result $\Delta_s G^{\ominus}_{AT,M}$ as the correct reference in the context of the recalculated solvation free energy, it is possible to design consistent correction terms to the raw solvation free energy $\Delta_s G^{\ominus}_{AT}$ calculated directly from the AT simulation (Section 6.2.1), *i.e.* considering the electrostatic interaction function actually used in this simulation. As mentioned above, the recalculated quantity $\Delta_s G^{\ominus}_{AT,M}$ itself would be very difficult to evaluate accurately, due to the long-range nature of the Coulomb potential. However, the corrected result for $\Delta_s G^{\ominus}_{AT}$ provides in most cases a well-converged estimate for this quantity, provided that an additional issue is addressed appropriately by the correction scheme. By construction, $\Delta_s G^{\ominus}_{AT,M}$ is an intrinsic solvation free energy, because it is based on an extrapolation to a large solvation range (distance R in the above discussion) excluding the presence of an air-liquid interface. In contrast, $\Delta_s G^{\ominus}_{AT}$ may in some cases encompass interfacial effects, which must be removed by the correction scheme if an intrinsic solvation free energy is sought after. For example, the result of a PBC/CT calculation considering a periodic system exempt of air-liquid interface (besides the ionic cavity) will represent an intrinsic value, due to the vanishing nature of the CT electrostatic interaction function in the long distance range. In contrast, the result of a FBC/CB calculation will represent a real-like value, but including a surface contribution appropriate for a microscopic system, involving *e.g.* the curved surface of a simulated droplet, rather than for a macroscopic one. Finally, the results of a PBC/LS calculation considering a periodic system exempt of air-liquid interface (besides the ionic cavity) will represent a value that obeys yet another convention, because the zero of the potential is defined here by a vanishing average over the computational box. Only when the latter differences are also taken into account in the correction scheme (Section 6.2.1) can one obtain *methodology-independent* intrinsic single-ion solvation free energies from AT simulations. By methodological independence, it is meant that a one-to-one correspondence is achieved between the choices of a given solvent model and of a pair of ion-solvent Lennard-Jones interaction parameters, and a resulting estimate for the intrinsic solvation free energy of the ion.

After evaluation of the appropriate correction term $\Delta_s G_{cor}[I_g^z]$ according to a procedure that will be described in detail in Section 6.2.1, a value for $\Delta_s G^{AT}_{chg}[I_g^z]$ calculated directly from an AT simulation at P° and T^- using Eqs. 3.61 and 3.67 in their original forms, *i.e.* based on the $\psi_{EL,oi}$ function actually used in the simulation, along with an estimate for $\Delta_s G^{AT}_{cav}[I_g^z]$, can be used to calculate the corresponding methodology-independent intrinsic single-ion solvation free energy $\Delta_s G^{\ominus}_{AT,cor}[I^z_g]$ as

$$\Delta_s G^{\ominus}_{AT,cor}[I^z_g] = \Delta_s G^{AT}_{chg}[I^z_g] + \Delta_s G_{cor}[I^z_g] + \Delta_s G^{AT}_{cav}[I^z_g] + \Delta_s G^{\ominus}_{std} \quad , \tag{3.76}$$

which substitutes the methodology-dependent expression of Eq. 3.58. The correction term $\Delta_s G_{cor}[I^z_g]$ depends on the solvent model, on the ion-solvent Lennard-Jones interaction parameters, on the ionic charge, on the boundary conditions and electrostatic scheme used in the simulation including the corresponding parameters, as well as on P° and T^{-} through pressure- and temperature-dependent properties of the real and model solvent.

3.3.4 Calculation of intrinsic bulk electric potentials

The discussion of the end of Section 3.3.3 concerning the intrinsic electric potential at cavities within liquids is naturally complemented by the investigation of the *intrinsic bulk electric potential in liquids*[707], *i.e.* the potential in the bulk excluding the air-liquid interfacial potential contribution. The calculation of this intrinsic bulk potential *via* AT simulations involves the averaging of the solvent-generated Coulombic potential over time and over a certain region of the simulated system[707] exempt of and sufficiently far from any interface. This analysis can be applied to a simulation of the pure solvent performed under any type of boundary conditions and using any electrostatic scheme, and is found to give approximately the same result irrespective of these choices (within reasonable bounds) for a given solvent model and under specified pressure and temperature conditions, provided that the environment of the solvent molecules can be considered to be homogeneous and isotropic (*i.e.* bulk-like) throughout the considered region. However, this analysis can be performed in four different ways, as illustrated in Figure 3.6 in the case of a spherical region.

These four ways involve the combination of two choices. First, the potential may be averaged over the *entire considered region*, leading to what will be referred to as the *internal* intrinsic bulk electric potential, *i.e.* the potential felt by a *test charge without excluded volume*, or over the part of the considered region that is *outside the molecular envelopes of the solvent molecules* in a given system configuration, leading to what will be referred to as the *external* intrinsic bulk electric potential, *i.e.* the potential felt by a *test charge with an excluded volume*. Second, the potential averaging may be performed on the basis of the *solvent charges* enclosed within the region, a procedure that will be referred to as *P-averaging*, or on the basis of the *charges of the solvent molecules* enclosed within the region, *i.e.* considering all solvent molecules with their *molecular centers* within the region, a procedure that will be referred to as *M-averaging*. Note that when an internal potential is determined, care should be taken to still exclude small volumes in the immediate vicinity of the solvent partial charges for a given system configuration[707]. If these volumes are small, the error incurred on the averaging is negligible, but the corresponding integrable singularities are avoided and the noise in the calculation is considerably reduced.

The resulting intrinsic bulk electric potentials will be noted $\eta_{AT,P}$ (external, P-truncation), $\eta_{AT,M}$ (external, M-truncation), $\check{\eta}_{AT,P}$ (internal, P-truncation) and $\check{\eta}_{AT,M}$ (internal, M-truncation). The differences between these quantities are again found to be connected to the *classical exclusion potential of the solvent model* ξ_{AT}, namely as[707]

$$\eta_{AT,P} = \eta_{AT,M} + \xi_{AT} \ , \tag{3.77}$$

$$\check{\eta}_{AT,P} = \check{\eta}_{AT,M} + \xi_{AT} \tag{3.78}$$

and

$$\check{\eta}_{AT,M} \approx \eta_{AT,M} - \xi_{AT} \ . \tag{3.79}$$

The exact significance of the exclusion potential, which was already encountered in Section 3.3.3 in the context of the intrinsic potential at an uncharged

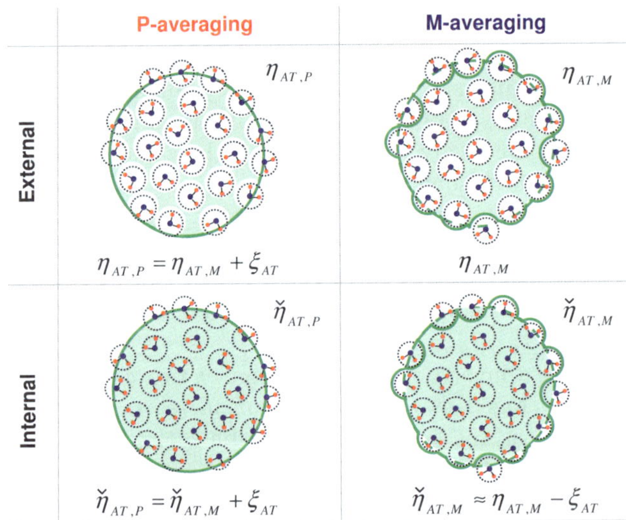

Figure 3.6 *External or internal intrinsic bulk electric potentials calculated based on a spherical region using P- or M-averaging.* The average electric potential is calculated for a spherical sample region within the simulated system based on configurations issued from an AT simulation of a larger solvent volume. The solvent is symbolized by water molecules, the averaging region corresponds to the green area, and the solvent charges considered in the average potential evaluation are encompassed within the thick green line. The external potential (η_{AT}) corresponds to the electric potential averaged outside the solvent molecules (test charge with excluded volume). The internal potential ($\check{\eta}_{AT}$) corresponds to the electric potential averaged over the entire sample (test charge without excluded volume). This calculation relies on considering all solvent charges within the sample region (P-averaging; $\eta_{AT,P}$ and $\check{\eta}_{AT,P}$) or the charges of all solvent molecules with their molecular centers within the sample region (M-averaging; $\eta_{AT,M}$ and $\check{\eta}_{AT,M}$). The calculated values are related to each other *via* the exclusion potential ξ_{AT} of the solvent model (Eqs. 3.77-3.79).

cavity within a pure liquid, will be discussed later (Section 4.3.8). Values of $\xi_{AT}^{\ominus}$ at the (experimental) solvent density $\rho_{svt}^{\ominus}$ appropriate for P° and T^{-} are reported in Table 3.1 for some common water models involving a single Lennard-Jones interaction site at the oxygen atom.

The situation of Eqs. 3.77-3.79 can be illustrated numerically[707] based on PBC simulations of water boxes using the SPC water model[1274] at 1 bar and 300 K, considering an averaging region set to the entire box, which is licit considering that PBC systems are exempt of any air-liquid interface. The corresponding data is reported in Table 3.5. In this case, the values of the different estimates for the intrinsic bulk electric potential are found to be $\eta_{AT,P} = +0.79$ V (external, P-truncation), $\eta_{AT,M} = -0.03$ V (external, M-truncation; solvent molecular center at the oxygen atom), $\check{\eta}_{AT,P} = 0.00$ V (internal, P-truncation) and $\check{\eta}_{AT,M} = -0.82$ V (internal, M-truncation; solvent molecular center at the oxygen atom). Although these values depend slightly on the electrostatic scheme used to generate the solvent configurations, and on the pressure and temperature conditions, the differences $\pm\xi_{AT}$ do not, except for a small dependence on pressure and temperature *via* the solvent density. The classical exclusion potential of the SPC water model at P° and T^{-} is thus about $\xi_{AT}^{\ominus} = +0.82$ V (Table 3.1). Note that the value of $\eta_{AT,M}$ for SPC water at 1 bar and 300 K is slightly negative (-0.03 V). This accounts for a marginal orientational preference of the water molecules around intermolecular voids, favoring orientations where the hydrogen atoms point away from these voids, consistent with the slightly negative value of $\zeta_{AT,M}$ for a sodium-sized cavity, namely -0.12 V (Section 3.3.3). In the ODL situation (Section 3.3.3), no such preference is observed, resulting in $\eta_{AT,M} = 0$ and $\eta_{AT,P} = \xi_{AT}$. In contrast, $\check{\eta}_{AT,P}$ is always rigorously zero because the P-averaged charge density within the bulk of a liquid is exactly zero. As a result, $\check{\eta}_{AT,P} = 0$ and $\check{\eta}_{AT,M} = -\xi_{AT}$ are always valid, even outside the ODL situation. This slight difference explains why Eq. 3.79 only holds exactly in the ODL situation. Here also, although the four variants may seem justifiable for the calculation of the intrinsic bulk electric potential of a pure liquid, only one is, at least in the opinion of the authors, reasonable, namely *M-averaging of the external potential*[707], based on arguments that will be provided in Section 4.3.8 and further developed in Section 7.

The intrinsic bulk electric potential $\check{\eta}$ of the pure solvent is not immediately relevant in the context of the calculation of an intrinsic single-ion solvation free energy (Section 3.3.3), where the key quantity is the intrinsic electric potential ζ at a cavity within the solvent. The latter potential will induce a free-energy contribution linear the ionic charge, that adds up to the quadratic contribution resulting from the dielectric response of the solvent. Although the latter quadratic contribution presents no fundamental ambiguity, the value of ζ as estimated *via* AT simulation depends on a specific convention for its evaluation, application of P- or M-summation (Figure 3.5) leading to distinct results $\zeta_{AT,P}$ and $\zeta_{AT,M}$, respectively, related by Eq. 3.75. The difference $\chi' = \eta - \zeta$ can be interpreted as representing the *interfacial potential at the cavity-solvent boundary*, measured in the cavity-to-liquid direction. As seen above, the value of η as estimated *via* AT simulation also depends on a specific convention for its evaluation, application of P- or M-summation for an internal or external potential (Figure 3.6) leading to the four distinct results $\eta_{AT,P}$, $\eta_{AT,M}$, $\check{\eta}_{AT,P}$ and $\check{\eta}_{AT,M}$, respectively, related by Eqs. 3.77-3.79. Based on consistent combinations of ζ and η, which involve a common choice concerning the way an interface-free bulk solvent region should be terminated (*i.e.* at individual solvent charges *vs.* molecules) and a decision concerning the way test charges probe the bulk potential (*i.e.* test charge with *vs.* without excluded volume), one has

$$\chi'_{AT,M} = \eta_{AT,M} - \zeta_{AT,M} \quad , \tag{3.80}$$

$$\check{\chi}'_{AT,M} = \check{\eta}_{AT,M} - \zeta_{AT,M} \approx \eta_{AT,M} - \xi_{AT} \quad , \tag{3.81}$$

$$\chi'_{AT,P} = \eta_{AT,P} - \zeta_{AT,P} = \chi'_{AT,M} \quad , \tag{3.82}$$

and

$$\check{\chi}'_{AT,P} = \check{\eta}_{AT,P} - \zeta_{AT,P} = \check{\chi}'_{AT,M} \quad , \tag{3.83}$$

the second equalities in these expressions following from Eqs. 3.75 and 3.77-3.79. The corresponding values concerning the numerical example of SPC water at 1 bar and 300 K (see above) are also reported in Table 3.5. It is seen that when P-summation and P-averaging or M-summation and M-averaging are applied consistently to evaluate the interfacial potential at the cavity-solvent boundary, the result is the same and only depends on whether an internal or an external definition is adopted for the intrinsic bulk potential of the liquid. In other words, it is the latter choice that indirectly causes the differences observed between P- and M-summation results in the calculation of intrinsic single-ion solvation free energies (Section 3.3.3).

3.3.5 Calculation of air-liquid interfacial potentials

The second approach to obtain intrinsic single-ion solvation free energies based on AT simulations takes an indirect route. It relies on computing the *air-liquid interfacial potential of the pure solvent* for a planar interface in equilibrium, $\chi^{\ominus}_{svt}$, and combining the resulting value with experimentally-accessible real single-ion solvation free energies $\Delta_s\hat{G}[I^z_g]$ (Section 2.2; Eqs. 2.6 and 2.8). To evaluate $\chi^{\ominus}_{svt}$, the simulation must involve a pure liquid sample surrounded by vacuum (air), and be performed at pressure P° and temperature T^-. For a given solvent model, simulations of this kind typically differ in four main aspects: (*i*) the interface geometry; (*ii*) the spatial boundary conditions; (*iii*) the nature of the confinement potential at the interface; (*iv*) the approach used for the interfacial potential calculation based on the simulated configurations. These choices are discussed in turn below.

The *interface geometry* is expected to affect the value of the interfacial potential. For simplicity, only spherical interfaces of radius $|R|$ will be considered, with $R < 0$ for concave interfaces (*e.g.* cavity), and $R > 0$ for convex interfaces (*e.g.* droplet), the limiting situation $R \to \pm\infty$ corresponding to a planar interface. In this case, the function $\tilde{\chi}^{\ominus}_{svt}(R)$ may be introduced to account for the interface curvature[707], with $\tilde{\chi}^{\ominus}_{svt}(\pm\infty) = \chi^{\ominus}_{svt}$. Most calculations to date have been performed for planar interface geometries[430,463–465,468,469,707,850,947,973,1440–1450] and only a few have considered spherical droplets or cavities[470,707,833,1451].

Different types of *boundary conditions* (Section 3.3.1) may be adopted, depending on the interface curvature. For a cavity[707], the most obvious choice is PBC. The use of FBC is also possible, but leads to the introduction of a second interface at the droplet surface. For a droplet[470,707,833,1451], the most obvious choice is FBC. The use of PBC, *i.e.* the periodic replication of the droplet, would also be possible, but does not appear to present significant advantages. For a planar slab[430,463–465,468,469,707,850,947,973,1440–1448,1450,1452], the most obvious choice is FBC orthogonally to the slab, and two-dimensional PBC along the slab plane[1440,1441]. In this case, the use of three-dimensional PBC, *i.e.* the periodic replication of the slab also orthogonally to its plane is possible and computationally simpler[463–465,468,850,947,1442–1446,1448,1449]. Most calculations to date have been performed for periodic planar slabs.

The stabilization of the interface in a given geometry typically involves the application of some artificial *confinement potential*. In principle, such a potential should not be required provided that: (*i*) the desired interface geometry is a stable (equilibrium) geometry, which is in principle the case for spherical interfaces due to surface tension effects, but not necessarily for planar slabs (the latter may in some cases spontaneously convert to *e.g.* rows of droplets or alternative periodic geometries under PBC); (*ii*) the system has a finite total volume, which is the case under PBC (volume of the computational box) but not under FBC (unspecified free volume encompassing the system); (*iii*) the simulation can be carried out over a sufficient timescale to appropriately sample interface fluctuations around the equilibrium geometry, as well as the vaporization equilibrium of the liquid. In this case, as previously discussed in Section 2.2 in the context of the hypothetical process of single-ion solvation, a simulation without confinement potential would sample an interface between the liquid and its vapor at the corresponding equilibrium vapor pressure, including equilibrium fluctuations in the interface geometry. To adjust to the desired situation of a pressure P°, one would have to add an inert gas in the system[583,584], so that the total pressure of the evaporated solvent and the inert gas equals P°. In practice, however, the three above conditions can seldom be satisfied, *i.e.* the system may not have a finite volume, the selected interface geometry may not be the most stable one or/and the interface fluctuations and the vaporization equilibrium may be too slow to be appropriately sampled. For this reason, it is necessary or simply easier to introduce an artificial confinement potential that fixes the interface geometry and prevents solvent evaporation.

The most common confinement potential is the Lennard-Jones wall, analogous to Eq. 3.12, but with the interparticle distance replaced by the distance to the interface[565, 707, 1052]. In the case of water, this term is typically only applied to the water oxygen atoms, using parameters C_6^{ss} and C_{12}^{ss} appropriate for oxygen-oxygen interactions based on the given water model (Section 3.3.2). A potential acting on the hydrogen atoms would be likely to perturb more significantly the surface structure of the sample. Note that a wall potential excluding the dispersion component would probably be more appropriate to mimic an interface to vacuum (air). Many other types of confinement potentials have also been used in AT simulations, including 3-9 function[1441], purely repulsive[565,1049,1058,1063,1067], harmonic[566,1043,1044,1048,1053,1055,1060,1069–1071], hard[1051,1440], or mean-field[1040,1042] potentials. Alternative approaches involve a shell of positionally-constrained[1039] or harmonically-restrained atoms[1069,1070] or molecules[1044, 1071], or molecules confined by a hard potential along with effective[1051] or stochastic[1044,1069–1071] representations for the effect of the surrounding medium.

One should be aware, however, that the choice of a specific form for the confinement potential may have a non-negligible influence on the positional and orientational distributions of the liquid molecules near the surface of the sample and thus, on the resulting polarization and interfacial potential[566,1043,1049, 1052, 1053, 1058, 1071]. Furthermore, the situation of a confined system with an unphysical confinement potential poses problems in terms of the pressure. Although it would in principle be possible to apply some pressure control mechanism on the sample (barostat[567,571,580–584]), *e.g.* by altering the sample volume based on the instantaneous pressure it exerts against the confinement wall, this is seldom done in practice. As a result, it is generally difficult to assess whether the calculated interfacial potential corresponds to realistic pressure conditions. Note that under standard conditions, P° represents the pressure of a hypothetical gas phase surrounding the system, so that the pressure in the liquid phase may be higher due to surface tension effects, *e.g.* $P^\circ + 2R^{-1}\gamma_{svt}$ for a liquid droplet of radius R, where γ_{svt} is the surface-tension coefficient of the liquid.

Finally, there are different approaches for the *interfacial potential calculation*, namely *via* integration of the surface polarization, a procedure that will be referred to as M-integration, or of the surface charge density, a procedure that will be referred to as P-integration. These will be discussed in the specific context of a planar slab, periodic along the x- and y-axes (periodic square of corresponding edges X and Y), and non-periodic along the z-axis (slab width Z, slab center at $z = 0$), as illustrated in Figure 3.7. The formalism is easily adjusted to the case of a cavity or droplet if needed.

The *M-integration scheme* relies on the principle that an infinitesimal solvent layer of width dz at z characterized by a homogeneous *polarization* $P(z)$ along the z-axis induces a *potential jump*[421] of magnitude $\epsilon_o^{-1} P(z) dz$. Knowing the form of $P(z)$, the potential profile $\Phi(z)$ may be calculated by integration, setting $\Phi(z) = 0$ outside the slab. If Z is sufficiently large, the liquid close to the interface midplane ($z = 0$) will not show any orientational preferences, *i.e.* it will behave as bulk liquid, with an average polarization $P(z) = 0$ in the z-direction. In this case, the potential is about constant (converged) close to $z = 0$ and the M-integration estimate for the air-liquid interfacial potential may be written

$$\chi_{AT,M} = -\epsilon_o^{-1} \int_0^{Z/2} dz\, P(z) \ . \tag{3.84}$$

In practice, $P(z)$ is estimated in the form a histogram of finite bin width Δz, and the integral in Eq. 3.84 approximated by a discrete sum. The value of $P(z)$ in each bin is obtained by calculating the z-component of the total solvent dipole moment of the corresponding layer, *i.e.* considering all solvent molecules with their molecular center within the layer, dividing by the layer volume $XY\Delta z$, and averaging over trajectory configurations. By symmetry, the corresponding average x- and y-components are expected to vanish given sufficient sampling, as well as variations of the solvent polarization along the x- and y-axes. The accuracy of the calculation can also be improved by averaging over the two sides of the slab, an equation similar to Eq. 3.84 also holding for $z < 0$.

The *P-integration scheme* relies on the principle that an infinitesimal solvent layer of width dz at z characterized by a homogeneous *charge density* $\rho_T(z)$ generates a *discontinuity in the electric field* along the z-axis of magnitude $\epsilon_o^{-1}\rho_T(z)dz$ (Gauss' law[421,1453]). Knowing the form of $\rho_T(z)$, the potential

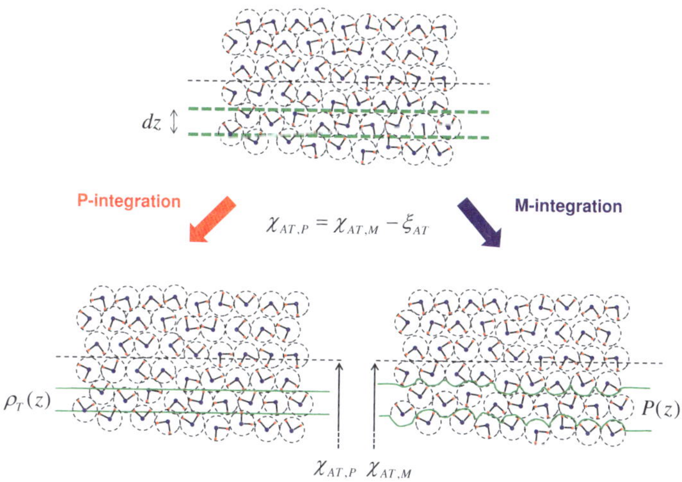

Figure 3.7 *Air-liquid interfacial potential of a planar slab of a pure liquid calculated using P- or M-integration.* The interfacial potential (interface midplane relative to a reference point outside the liquid at an infinite distance) is calculated based on an AT simulation. The solvent is symbolized by water molecules. The calculation relies on M-integration (Eq. 3.84), corresponding to the integration of the polarization $P(z)$, or on P-integration (Eq. 3.85), corresponding to the integration of the charge density $\rho_T(z)$. The calculated values ($\chi_{AT,P}$ and $\chi_{AT,M}$) differ by the exclusion potential ξ_{AT} of the solvent model (Eq. 3.86).

profile $\Phi(z)$ may be calculated by integration of the field, setting $\Phi(z) = 0$ outside the slab. If Z is sufficiently large, the liquid close to the interface mid-plane ($z = 0$) will not show any orientational preferences, *i.e.* it will behave as bulk liquid, with an average charge density $\rho_T(z) = 0$. In this case, the potential is about constant (converged) close to $z = 0$ and the P-integration estimate for the air-liquid interfacial potential may be written

$$\chi_{AT,P} = -\epsilon_o^{-1} \int_0^{Z/2} dz' \int_{z'}^{Z/2} dz'' \rho_T(z') \tag{3.85}$$

In practice, $\rho_T(z)$ is estimated in the form a histogram of finite bin width Δz, and the integral in Eq. 3.85 replaced by a discrete sum. The value of $\rho_T(z)$ in each bin is obtained by calculating the total solvent charge of the corresponding layer, *i.e.* considering all solvent sites within the layer, dividing by the layer volume $XY\Delta z$, and averaging over trajectory configurations. By symmetry, variations of the solvent charge density along the x- and y-axes are expected to vanish given sufficient sampling. The accuracy of the calculation can also be improved by averaging over the two sides of the slab, an equation similar to Eq. 3.85 also holding for $z < 0$.

When the results from P- and M-integration are compared, it is found that the values always differ by a constant offset, again related to the *classical exclusion potential of the solvent model* ξ_{AT}, namely as[707]

$$\chi_{AT,P} = \chi_{AT,M} - \xi_{AT} \; . \tag{3.86}$$

The exact significance of the exclusion potential, which was already encountered in Sections 3.3.3 and 3.3.4 in the context of the intrinsic potential at a cavity within a pure liquid and of the intrinsic bulk electric potential of a pure liquid, respectively, will be discussed later (Section 4.3.8). Values of $\xi_{AT}^{\ominus}$ at the (experimental) solvent density $\rho_{svt}^{\ominus}$ appropriate for P° and T^{-} are reported in Table 3.1 for some common water models involving a single Lennard-Jones interaction site.

The situation of Eq. 3.86 can be illustrated numerically[465, 707] based on FBC/CB simulations of water slabs confined by a Lennard-Jones wall using the SPC water model[1274] at 300 K and at a density appropriate for a liquid pressure of 1 bar, the corresponding data being reported in Table 3.5. When the air-liquid interfacial potential is calculated using P-integration (Eq. 3.85), an estimate $\chi_{AT,P} = -0.10$ V is obtained. When this interfacial potential is calculated using M-integration (Eq. 3.84; solvent molecular center at the oxygen atom), an estimate $\chi_{AT,M} = +0.72$ V is obtained instead. Although these values depend on the interface geometry, on the type of confinement potential and on the pressure and temperature conditions, the difference ξ_{AT} does not, except for a small dependence on pressure and temperature *via* the solvent density. The classical exclusion potential of the SPC water model at P° and T^{-} is thus about $\xi_{AT}^{\ominus} = +0.82$ V (Table 3.1). When these two different estimates for the air-liquid interfacial potential are combined with experimentally-accessible values of the real single-ion solvation free energies $\Delta_s \hat{G}[I_g^z]$ (Section 2.2; Eqs. 2.6 and 2.8) so as to obtain intrinsic values, the P-integration convention will lead to a solvation free energy that is about 80 kJ·mol^{-1} less negative for a monovalent cation and about 80 kJ·mol^{-1} more negative for a monovalent anion compared to the M-integration convention. This difference is exactly identical to the corresponding difference between intrinsic single-ion solvation free energies evaluated using AT simulations based on P- and M-summation. Since the sign of ξ_{AT} in Eq. 3.86, is opposite to the corresponding sign in Eq. 3.75, one has

$$\zeta_{AT,P} + \chi_{AT,P} = \zeta_{AT,M} + \chi_{AT,M} \quad . \tag{3.87}$$

This means in particular that the P-scheme applied consistently (P-summation for ionic solvation free energies and P-integration for air-liquid interfacial potential calculations) will lead to *the same real solvation free energies* as the M-scheme applied consistently (M-summation and M-integration for these two types of calculations, respectively). However, these schemes *disagree on the partitioning* of these values into bulk (intrinsic) and surface contributions. In contrast, inconsistent combinations (P-summation with M-integration or M-summation with P-integration) will generally lead to different results, even in terms of real solvation free energies.

Note that the value of $\chi_{AT,M}$ for SPC water at 300 K and a solvent density appropriate for a liquid pressure of 1 bar is positive and of relatively large magnitude (+0.72 V). This accounts for the preferential orientation of water molecules at an interface to vacuum (air), favoring orientations where the hydrogen atoms point towards the liquid, consistent with the negative value of $\zeta_{AT,M}$ for a sodium-sized cavity, namely -0.12 V (Section 3.3.3), and the negative value of $\eta_{AT,M}$ corresponding to intermolecular voids, namely -0.03 V (Section 3.3.4). In the ODL situation (Section 3.3.3), no such preference is observed, resulting in $\chi_{AT,M} = 0$ and $\chi_{AT,P} = -\xi_{AT}$. The numerical similarity between $\chi_{AT,M}$ (+0.72 V) and ξ_{AT} (+0.82 V) is coincidental and specific to the SPC water model under the given pressure and temperature conditions. Due to this coincidence, an "intrinsic" single-ion solvation free energy evaluated directly using the P-summation scheme (Section 3.3.3) or indirectly using the P-averaging scheme (this section) actually almost represents an estimate for the corresponding real solvation free energy.

The fact that the above observations concerning the difference between the P- and M-integration schemes parallel the ones made previously concerning the difference between the P- and M-summation schemes in the calculation of intrinsic solvation free energies *via* AT simulations (Section 3.3.3), as well as corresponding differences between the P- and M-averaging schemes, or internal and external averaging schemes in the calculation of the intrinsic bulk electric potential of a liquid (Section 3.3.4) is not fortuitous. Just like the P-summation scheme for recalculating the solvent-generated potential at the center of a cavity, the P-integration and P-averaging schemes regard the *solvent charge density as the source of potential.* In terms of multipole expansion, this means that not only the effect of the solvent molecular dipole moment is considered, but also the effect of higher order moments. In contrast, just like the M-summation scheme for recalculating the solvent-generated potential at the center of a cavity, the M-integration and M-averaging schemes regard the *solvent dipoles as the only source of potential*, thereby omitting the effect of the solvent higher-order multipole moments. As will be seen later, the higher-order moment responsible for the difference between the *P-scheme* (P-summation, P-averaging or P-integration) and the *M-scheme* (M-summation, M-averaging or M-integration) is the trace of the solvent quadrupole moment (Section 4.3.8).

Although the difference between P- and M-integration results in terms of the air-liquid interfacial potential may be extremely large, the ambiguity affecting the experimentally deduced values for this quantity (Section 2.3) has prevented the resolution of this problem on the basis of purely experimental arguments. As a result, a hot debate has taken place in the literature between the advocates of the P-integration[430,468–470,833,850,947,973,1442,1443,1445,1447–1449,1451,1454–1456] scheme and those of the M-integration[454,463–465,707,1170,1176,1440,1441,1444,1446,1450,1457] scheme for the calculation of air-liquid interfacial potentials, a debate which is, at least in the opinion of the authors, at present settled *in favor of the M-integration scheme*[707], based on arguments that will be provided in Section 4.3.8 and further developed in Chapter 7.

3.3.6 Fundamental problems

The following discussion of the *fundamental problems* encountered in AT simulations refers in turn to the direct calculation of intrinsic single-ion solvation free energies (Section 3.3.3) and to their indirect calculation *via* evaluation of the air-liquid interfacial potential of the pure solvent (Section 3.3.5).

Considering first the calculation of an *intrinsic single-ion solvation free energy* $\Delta_s G^{\ominus}_{AT}[I^z_g]$ using free-energy calculations based on AT simulations, it is useful to perform a *comparison with the CE approach* (Section 3.2), where a number of effects (Points A_3-I_3) were neglected. At the AT level, a representation of the microscopic solvation structure, rather than a continuous-medium linear-response approximation, is now included. This approach should thus account in an appropriate way for ion-solvent interface polarization (Point A_3), ion-solvent dispersion interactions (Point D_3), electrostriction (Point E_3), dielectric saturation (Point F_3) and solvation structure (Point G_3), while the issue concerning the microscopic surface tension coefficient (Point I_3) has become irrelevant. The ion-compression work (Point B_3) is in some way present, in the form of the repulsive component of the ion-solvent Lennard-Jones interaction contribution to the calculated solvation free energy, although this representation may not be very accurate. The specific ion-solvent interactions (Point G_3) are also in part accounted for, although the partially covalent character of these interactions, as well as possible protonation equilibria, involving the ion or closest solvent molecules, whenever relevant, are still neglected. On the other hand, the AT approach in its basic form (Section 3.3.1; see, however, Section 6.2.4) still neglects the (presumably weak) contribution of ionic polarization (Point C_3) and treats the electronic polarization of the solvent molecules in a mean-field fashion. Because the effective partial charges selected for the solvent model are calibrated to encompass the mean effect of polarization in the bulk liquid environment (Section 3.3.2), they may fail to account for the additional polarization caused by the high electric field in the immediate vicinity of an ion (see *e.g.* Table 6.5). Finally, the ambiguity in the choice of the ionic radius (Point H_3) persists in the form of a similar ambiguity concerning the choice of the ion-solvent Lennard-Jones interaction parameters C_6^{is} and C_{12}^{is} and, more specifically, of the latter coefficient, which is purely empirical and not connected to a physically-grounded description of ion-solvent interactions (Section 3.3.2; see, however, Section 6.2.3).

In spite of the point concerning electronic polarization, the inclusion of atomistic details into the model is expected to represent *a significant improvement* over the CE approach. However, it should be realized that the AT approach also introduces *new fundamental problems* that were avoided in CE calculations, namely (Figure 3.2): (*i*) solvent configurations have to be sampled explicitly (*e.g. via* MD or MC), while the averaging was implicit in the analytical BO+ST model; (*ii*) the considered system is microscopic and spatial boundary conditions have to be introduced, while the solvent continuum was of infinite extent in the BO+ST model; (*iii*) approximate electrostatic schemes must generally be introduced (except for FBC/CB) and a methodological ambiguity appears concerning the appropriate summation scheme (P- or M-summation) in the potential recalculation, while the interactions were exactly Coulombic and the potential unambiguously defined in the BO+ST model; (*iv*) a number of additional empirical parameters must be introduced (solvent geometry and partial charges, solvent-solvent and ion-solvent Lennard-Jones interaction parameters), while the basic BO+ST model solely involved the ionic radius, the two other parameters $\epsilon_{svt}^{\ominus}$ and $\tilde{\gamma}_{svt}^{\ominus}$, being in principle experimentally accessible. With respect to the third point, it can be noted that the potential considered in the BO+ST model, which considers the dipole moment as the only source of electric potential for the solvent, corresponds to M-summation in AT simulations. This point will be further considered in Section 4.3.8 and Chapter 7. As a result of these additional issues, the AT approach has, up to recently, failed to provide estimates for intrinsic single-ion solvation free energies in quantitative agreement with experiment, just as was said previously to be the case for the comparatively much simpler BO+ST model (Section 3.2).

Atomistic studies to evaluate intrinsic single-ion solvation free energies in water are numerous, including: (*i*) AT simulations under FBC[878,1047,1056,1057,1081,1329,1424,1458–1461]; (*ii*) AT simulations under PBC[693,708,792,800,877,1056,1081,1167,1171,1191–1193,1289,1331,1334,1422–1424,1430,1431,1439,1459,1462–1466]; (*iii*) AT simulations with explicit polarization[440,798,908,910,946,961,969,970,1336,1439,1467]. Following from the above discussion, the application of AT simulations for the evaluation of intrinsic single-ion solvation free energies is affected, predominantly in the charging component, by the following *sources of errors*, considered with reference to the ideal case of an ion in a macroscopic non-periodic solution within an exact quantum-mechanical description:

(A_4) *Approximate representation of electrostatic interactions*[333,334,792,1056,1081,1088,1166,1167,1169–1171,1191–1194,1199,1289,1422–1424,1468]: even in the limit of very large system sizes (see, however, Point B_4 below for finite-size effects), the approximations generally made in the representation of electrostatic interactions during the simulation (Section 3.3.1) have two consequences: (*i*) the solvent polarization around the solvated ion may not be identical to the polarization corresponding to an ideal Coulombic system; (*ii*) the potential evaluated at the center of the ionic cavity during the charging process, and thus the charging free energy itself, is typically based on a non-Coulombic interaction function, and may be incorrect even when the polarization around the ion is correctly described. In the limit of very large system sizes, the first problem only affects the PBC/CT scheme, *i.e.* non-Coulombic interactions involving cutoff truncation in the form of PBC/CT/SC, PBC/CT/RF or PBC/CT/MT schemes along with either P- or M-truncation, but not the FBC/CB and PBC/LS schemes, the latter being formally Coulombic in the limit of infinite box sizes[329–331,333,334,1079–1081,1289]. The second problem affects all schemes except FBC/CB. This problem is further complicated if a self-term V_{slf} is introduced[792,1081,1423,1430,1431,1437] in Eq. 3.16, as commonly done in PBC/LS (Eq. 3.27) and sometimes also[1082,1083] in PBC/RF (*e.g.* Eq. 3.37), because this term directly translates into a corresponding charging free-energy contribution, unless it is explicitly removed by consideration of the canceling vacuum contribution (Section 3.3.3). In the limit of very large system sizes, the two sources of error can in principle be disentangled by recalculation of the charging free energy based on the simulated trajectory using exact Coulombic interactions (Section 3.3.3; Eqs. 3.70 and 3.72 considering a very large distance range R). However, in practice, the results of this recalculation for a given force-field description are still affected by errors in the polarization related to finite-size effects (Point B_4 below) and by the ambiguity in the recalculation scheme with respect to the choice of a summation procedure (Point C_4 below).

(B_4) *Finite (microscopic) system size*[329,330,350,792,1056,1057,1081,1289,1422–1424,1430,1431,1439]: due to the consideration of a finite system, *e.g.* microscopic droplet under FBC or computational box under PBC (Section 3.3.1), the solvent polarization around the solvated ion is not identical to the polarization corresponding to an ideal macroscopic and non-periodic system, even considering microscopic interactions dictated by the specific (generally approximate) electrostatic interaction scheme used in the simulation (see Point A_4 above for the effect of approximate electrostatics). In the FBC/CB scheme, there is a lack of solvent polarization beyond the surface of the finite system[1047, 1056, 1057, 1329, 1424]. In PBC schemes, the solvent molecules in the computational boxes neighboring the reference one are perturbed by the periodic copies of the ion, so that they do not solvate the reference ion[329, 330, 333, 334, 350, 792, 1056, 1081, 1171, 1194, 1199, 1422–1424, 1430, 1431, 1468] (the net dipole moment of such a neighboring box is on average zero by symmetry). The associated error is typically much larger for the FBC/CB and PBC/LS schemes compared to the PBC/CT scheme[1081, 1289] (the self-term V_{slf} being excluded in all PBC schemes), because the application of a cutoff dramatically reduces the effective range of electrostatic interactions within the periodic system. As a result, the incurred error predominantly appears in this case as an effect related to approximate electrostatics rather than to finite system size (Point A_4 above).

(C_4) *Ambiguity in the definition of the potential at the ion site*[707,708,1056,1081,1187,1289,1424,1429,1432–1439,1469]: even if a sufficiently large (macroscopic) system could be simulated with exact Coulombic interactions, thereby alleviating the problems of Points A_4 and B_4 above, there would remain a fundamental ambiguity concerning the appropriate way to calculate the solvent-generated potential at the ionic site during the charging process based on the sampled configurations, namely *via* P- *vs.* M-summation (Section 3.3.3). As will be seen later (Section 4.3.8), the appropriate convention in terms of a recalculated Coulombic potential at the ion site is M-summation. In terms of the original potential generally involving an approximate electrostatic scheme (Point A_5 above), the discussion is more intricate (Section 6.2.1). The situation is further complicated in FBC/CB and PBC/LS simulations, because the calculated solvation free energy may also include a contribution from the air-liquid interfacial potential of the solvent at the surface of the droplet (FBC/CB) or of the ionic cavity (PBC/LS), leading to an ambiguity concerning the nature (real, intrinsic, or neither of the two) of the calculated solvation free energy.

(D_4) *Approximate force-field representation*[440, 937, 946, 1424, 1439, 1470]: even in the hypothetical absence of any of the previous sources of error (Points A_4, B_4 and C_4 above), which can be effectively achieved in practice by means of appropriate correction terms to the raw simulation results (Section 6.2.1), the accuracy of the calculated charging free energies for a given choice of ion-solvent Lennard-Jones interaction parameters (see, however, Point E_4 below for the ambiguity of this choice) is ultimately limited by the accuracy of the force-field representation employed, *i.e.* by the physical adequacy of the functional form chosen for the potential-energy function in Eq. 3.16. The main issue here is probably the mean-field treatment of the solvent electronic polarizability (Section 6.2.4). However, there are many other shortcomings in the force-field approximation, including: (*i*) the neglect of the ionic polarizability; (*ii*) the neglect of other quantum effects, *e.g.* implicit representation of electronic degrees of freedom, representation of water as a rigid molecule, exclusion of protonation equilibria; (*iii*) the treatment of electrostatic interactions using a monopole-based approximation, *i.e.* by means of partial charges located at discrete sites; (*iv*) the approximate nature of the Lennard-Jones van der Waals function, involving an *ad hoc* representation of the Pauli repulsion term, the neglect of higher-order dispersion coefficients, the application to water molecules based on a single interaction site at the oxygen atom and the use of *ad hoc* combination rules (Section 3.3.1); (*v*) the choice of a specific solvent model (Section 3.3.2). These approximations will lead in particular to an inaccurate or absent description of ionic polarization effects, of the ion-compression work, of specific ion-solvent interactions (*e.g.* partially covalent character) and of possible protonation equilibria (see above, as well as Points B_3, C_3, D_3 and G_3 in Section 3.2). In particular, it has been noted previously (Section 3.3.2) that the charges of solvent models with implicit polarization for AT simulations are artificially enhanced, so as to account in a mean-field fashion for electronic polarization effects in the bulk. This charge enhancement might, however, be insufficient for water molecules in the first solvation shells of an ion, which are subject to the very intense ionic electric field.

(E_4) *Ambiguity in the choice of the ion-solvent Lennard-Jones interaction parameters*[800,801,1081,1192,1289,1424,1439,1459,1463,1465,1470–1472]: just as was the case for the ionic radius to be employed in the BO+ST model (Point H_3 in Section 3.2), the intrinsic solvation free energy of an ion evaluated using AT simulations, assuming appropriate corrections for Points A_4, B_4 and C_4 above, and for a given approximate potential-energy function according to Point D_4 above, will still be bound to a specific choice of the ion-solvent Lennard-Jones interaction parameters C_6^{is} and C_{12}^{is} or, given a solvent model and combination rules, the corresponding homoionic ion-ion interaction parameters C_6^{ii} and C_{12}^{ii} (Section 3.3.2). However, these parameters, and more specifically C_{12}^{ii}, cannot be determined directly and must be inferred by comparison with experimental data. If these parameters are calibrated against experimental single-ion solvation free energies, assuming a set of conventional parameters and a given value for the intrinsic solvation free energy $\mathcal{G}_{H,svt}^{\ominus}$ of the proton (Section 6.2.2), they will lose any predictive power with respect to these properties. However, unlike the effective ionic radii of the BO+ST model, they may still in principle be cross-validated against other experimental properties (Section 6.2.3).

(F_4) *Finite sampling*[535,537,538,540,541]: since free-energy calculations based on AT simulations, unlike the BO+ST model, require the explicit sampling of solvent configurations, they may be affected by finite sampling errors. However, given the computing resources available nowadays, this problem is no longer an issue[1081,1289] for microscopic system sizes, yet sufficiently large to permit the obtension of reliable results given appropriate correction terms for Points A_4, B_4 and C_4 above (Section 6.2.1).

An illustration of the problems of Points A_4 and B_4 above is provided in Figure 3.8. In this figure, the structural solvation properties of a positive or negative monovalent ion in water (identical radii) as evaluated from AT simulations under PBC using different approximate electrostatic schemes are compared[1289]. The radial and orientational structuring of the solvation shells is seen to be significantly affected by the choice of a specific scheme. When the resulting solvent polarization around the ion is compared to the curve predicted by CE (Born model), the following observations can be made: (*i*) in contrast to the CE representation, the AT simulations correctly account for the effect of solvent structuring close to the ion (successive solvation shells); (*ii*) compared to the CE representation, the AT simulations tend to present a general underpolarization of the solvent (application of PBC, see Point B_4 above); (*iii*) the introduction of a cutoff (PBC/CT) leads to a large perturbation of the solvent polarization around the ion (approximate electrostatics, see Point A_4 above; cutoff artifacts and, in the case of PBC/CT/SC, strong overpolarization of the solvent within the cutoff sphere of the ion). When ionic solvation free energies are calculated based on the four schemes of Figure 3.8 in the absence of corrections for the effects of Points A_4 and B_4, the results differ by as much as 130 and 30 kJ·mol^{-1} for the positive and negative ions, respectively[1289].

As will be discussed later, accurate correction terms can be designed to remedy the problems of Points A_4, B_4 and C_4 (Section 6.2.1). Furthermore, simulations relying on force fields including an explicit treatment of electronic polarization (ion, solvent) have become feasible if not standard (Section 6.2.4), which may partially remedy a further major source of error (Point D_4). Since sampling errors (Point F_4) are no longer an issue in the present context nowadays, carefully-performed free-energy calculations based on AT simulations have in principle at present the capacity of associating a given solvent model and a given single C_{12}^{ii} value, assuming that C_6^{ii} is obtained from experimental ionic polarizabilities, to a unique value for the intrinsic solvation free energy in an essentially *methodology-independent* way[1081,1289] (at least, for the moment, in the context of force fields with a mean-field representation of electronic polarization). This represents an important achievement in the area, considering that the obtension of reliable numbers was not possible at all even a few years ago. The mean-field treatment of electronic polarizability (included in Point D_4 above) is then likely to represent the next major source of error in this context. Since the charges of solvent models with implicit polarization are artificially enhanced to account for the effective molecular dipole within the bulk (pure liquid) environment, while the first solvation shells of an ion represent a significantly more polar environment (high ionic electric field), the ionic solvation free energies are likely to be underestimated by these models.

However, if the ion-solvent Lennard-Jones interaction parameters have been calibrated against single-ion solvation free energies, this error will be compensated for by the parameterization, resulting in a model that may be inaccurate in the description of other ionic properties.

The problems related to the calculation of the *air-liquid interfacial potential* $\chi_{svt}^{\ominus}$ of a solvent (typically water) based on AT simulations are considered

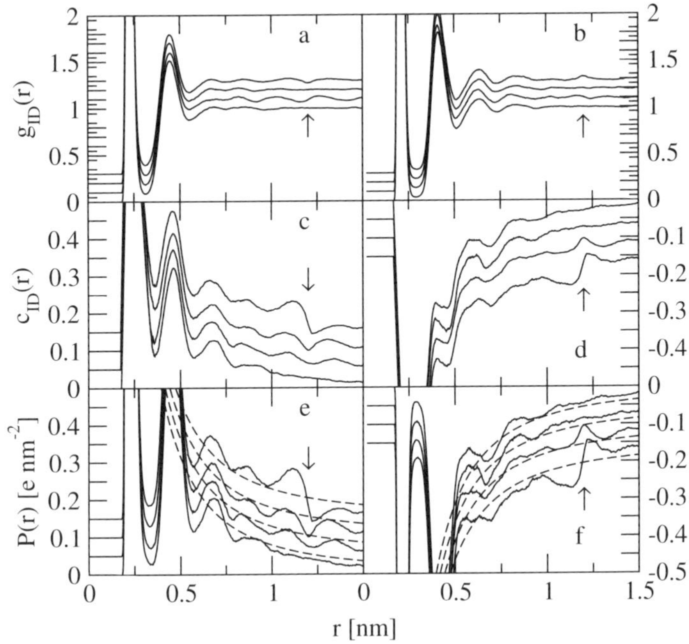

Figure 3.8 *Solvent structuring and polarization around an ion as obtained from atomistic simulations.* Structural solvation properties of a monovalent ion in water are evaluated from AT simulations (MD, 1 ns) using different electrostatic schemes. The properties displayed are the ion-dipole radial distribution function $g_{ID}(r)$, *i.e.* the local average density of water oxygen atoms at a distance r from the ion, relative to the bulk, the ion-dipole orientation correlation function $c_{ID}(r)$, *i.e.* the local average angle-cosine between the dipole of a water molecule at ion-oxygen distance r and a radial vector from the ion center, and the radial polarization $P(r)$, *i.e.* the local average dipole moment density of the solvent at ion-oxygen distance r which is proportional to $g_{ID}(r) \cdot c_{ID}(r)$. The simulations were performed using the SPC water model[1274] and involve a single monovalent ion with positive (a,c,e) or negative (b,d,f) charge (Lennard-Jones parameters of Straatsma and Berendsen[1171] for Na^+ in both cases). They were performed under NVT conditions (298.15 K) with one ion and $N_W = 1024$ water molecules in cubic computational boxes of edge length $L = 3.13$ nm (solvent density $\rho = 1$ g·cm^{-3}). The successive curves correspond to different electrostatic schemes, namely PBC/LS, PBC/CT/RF(P), PBC/CT/RF(M) and PBC/CT/SC(M), where the letter between parentheses refers to P-truncation or M-truncation. The successive curves are shifted upwards on the vertical axis by 0.10 (a and b) or 0.05 (c and e), or downwards by 0.05 (d and f) for readability. In the PBC/CT schemes the cutoff distance was set to $R_C = 1.2$ nm (indicated by an arrow). In the PBC/CT/RF schemes, the reaction-field permittivity was set to $\epsilon_{RF} = 66.6$. The dashed curves (e and f) depict the Born polarization corresponding to a CE representation (Section 3.2), calculated using $\epsilon_{svt} = 66.6$, as appropriate[1272] for the SPC water model.

next. Atomistic studies to evaluate the surface potential of water are numerous, including: (*i*) AT simulations of droplets[470,707,833,1451]; (*ii*) AT simulations of cavities[707]; (*iii*) AT simulations of slabs[430, 463–465, 468, 469, 850, 947, 973, 1440–1449]; (*iv*) AT simulations with explicit polarization[430, 947, 973]. The application of AT simulations for the evaluation of the air-liquid interfacial potential of a solvent is affected by the following *sources of errors*:

(A_5) *Approximate representation of electrostatic interactions*: this is generally not a major issue in this case, since FBC/CB simulations (*e.g.* droplet simulation with Coulombic interactions) are not affected by these types of errors, while PBC/LS or PBC/CT (*e.g.* cavity simulation or slab simulation with two- or three-dimensional periodicity) are affected by them only to a limited extent (purely dipolar system, large enough box or/and long enough cutoff distances affordable).

(B_5) *Finite (microscopic) system size*: this is generally not a major issue in this case, besides the problems related to sample geometry and confinement potential (see Points F_5 and G_5 below), because the transition from surface to bulk properties is relatively rapid in these purely dipolar systems, and can be assessed explicitly, so that relatively small system sizes are sufficient to reach convergence of the surface properties.

(C_5) *Ambiguity in the definition of the interfacial potential based on the sampled configurations*[470,707,833,1439,1445,1446]: as discussed in Section 3.3.5, the surface potential can be evaluated by surface integration in two distinct ways (M- or P-integration), leading to two distinct estimates, differing among each other by the classical exclusion potential ξ_{AT} of the solvent model. As will be seen later (Section 4.3.8), the appropriate convention is M-integration.

(D_5) *Approximate force-field representation*[470, 1446, 1447, 1473]: here also, the accuracy of the calculated interfacial potential is ultimately limited by the accuracy of the force-field representation employed, *i.e.* by the physical adequacy of the functional form chosen for the potential-energy function in Eq. 3.16 (see Point D_4 above). Of special relevance in this context, the mean-field representation of electronic polarization, the approximate nature of the Lennard-Jones van der Waals function, and the choice of a specific solvent model may have a significant influence on the results of the calculation. In particular, it has been noted previously (Section 3.3.2) that the charges of solvent models with implicit polarization for AT simulations are artificially enhanced, so as to account in a mean-field fashion for electronic polarization effects in the bulk. This charge enhancement might, however, be unrealistically high for surface water molecules in contact with a highly non-polar environment, namely vacuum (air).

(E_5) *Finite sampling*[535,537,538,540,541]: given the computing resources available nowadays, this problem is no longer an issue.

(F_5) *Dependence on the interface geometry*[707]: as discussed in Section 3.3.5, the calculated surface potential will depend on the geometry of the interface. Only values evaluated for planar interfaces (slab geometry) qualify as measures of the standard air-liquid interfacial potential $\chi_{svt}^{\ominus}$ of the solvent (Section 2.2).

(G_5) *Confinement potential*[566, 1043, 1049, 1052, 1053, 1058, 1071]: as discussed in Section 3.3.5, the calculated surface potential may be significantly affected by the choice of a specific confinement method used to prevent solvent evaporation at the sample surface and to preserve the interface geometry. This influence includes in particular an alteration of the pressure within the sample and a possible perturbation of its surface properties. These effects should be assessed, *e.g.* by repeating the calculation using confinement potentials of different forms.

The mean-field treatment of electronic polarizability (included in Point D_5 above) is likely to represent a major source of error in this context. Since the charges of solvent models with implicit polarization are artificially enhanced to account for the effective molecular dipole within the bulk (pure liquid) environment, while the surface region of a liquid is significantly less polar (interface to air), the air-liquid interfacial potential is likely to be significantly overestimated by these models. For example, the surface potential of the implicit-polarization SPC water model[1274] at 300 K and a density appropriate for a liquid pressure of 1 bar is about +0.72 V (Section 3.3.5 and Table 3.5, based on M-integration), which appears very high considering available "experimental" estimates and the value of about +0.13 V recommended by the authors of this book (Section 5.17).

3.4 Quantum-mechanical computations

In principle, QM computations (calculations or simulations) treat all nuclei and electrons constituting a molecular system *according to the laws of (relativistic) quantum mechanics.* For this reason, the QM approach is expected to present a number of key advantages over AT simulations (Figure 3.2), in particular: (*i*) it is at least in principle a first-principles approach, *i.e.* firmly grounded in microscopic physics, except for the treatment of nuclei as single particles; (*ii*) it does not rely on empirical parameters, a QM system being in principle entirely specified by the number of nuclei, their charges and masses, the number of electrons, the spin configuration, and physical constants; (*iii*) it accounts for specific quantum effects[1474–1479], *e.g.* isotope effects[1480–1483], tunneling[1484–1488] and low-temperature[1489–1495] or high-frequency statistical mechanics[1496]; (*iv*) it accounts for many-body effects[907, 913, 1497–1499] beyond the usual classical (*e.g.* mean-field, pairwise) approximation, *e.g.* electronic polarization[141, 1467, 1500], Pauli repulsion[880] and London dispersion[719–721]; (*v*) it accounts for covalent bond formation or breaking in contrast to the usual classical (*e.g.* harmonic) approximation, *e.g.* covalent binding[66] and charge transfer[1501–1504]; (*vi*) it accounts for processes involving electrons and light[1505–1508], *e.g.* electronic excitation[1509, 1510], laser effect[1511], ionization[1512–1514], chemical reactions[528, 1515–1525] and protonation equilibria[1526–1529].

Among all these advantages, the most relevant in the context of ionic solvation (compare with Points B_3, C_3, D_3 and G_3 of Section 3.2 and Point D_4 of Section 3.3.6) are certainly: (*i*) the improved description of *many-body effects* (ion and solvent electronic polarization); (*ii*) the improved description of *bonding and charge-transfer interactions* (specific ion-solvent interactions); (*iii*) the description of *protonation equilibria* (involving the ion or the closest solvent molecules). The QM approach is also expected to provide a much better description of the ionic solvation structure[102, 103] (*e.g.* solvent radial and orientational distribution functions, coordination numbers and geometries, Jahn-Teller distortions, altered ion concentration at the surface of water clusters) and dynamics (*e.g.* ionic diffusion, rate of solvent exchange between solvation shells and with the bulk, ion-water and first-shell water infra-red spectroscopic properties). However, maybe the most significant qualitative advantage of QM computations for the calculation of single-ion solvation free energies is that they require *no input parameter defining the ion size*, such as the ionic radius R_I of the CE approach (Point H_3 in Section 3.2) or the ion-ion repulsion parameter C_{12}^{ii} of the AT approach (Point E_4 in Section 3.3.6). As a result, these methods are the only ones that can formally be called *predictive* in this context.

Unfortunately, all these potential advantages are counter-balanced by a major drawback: QM calculations are *computationally extremely demanding*. Approaches to *reduce the computational cost* can be broadly classified according to the basic choices of Figure 3.1, as:

(A_6) *Use of hybrid methods*[649,650]: in these approaches, a high-resolution QM description of the most relevant region of the system, *e.g.* ion alone or ion plus its first (or first few) solvation shells, is coupled to a lower-resolution description of its surroundings, *e.g.* AT or CE representation of the solvent or of the surrounding solvent beyond the first solvation shells, or possibly both representations in two successive layers.

(B_6) *Use of effective core potentials*[1530]: here, the resolution of the model is decreased by removing the explicit description of the core electrons which are grouped with the corresponding nuclei, thereby defining *pseudo-nuclei* and associated effective core potentials, while only the explicit description of the valence electrons is retained. This also offers the advantage that relativistic effects, most relevant for the core electrons and for elements with atomic number $Z \geq 18$, need no longer be handled explicitly in the system description[1531, 1532], since they are implicitly included in the effective core potentials.

(C_6) *Approximations in the interaction description*: the first key approximation involved in most QM computations is the *Born-Oppenheimer approximation*[647], which assumes the separability of the nuclear and electronic Schrödinger equations. A second common approximation is the *neglect of relativistic effects* in the motion of the (valence) electrons[1531,1532]. A third common approximation is the treatment of the motions of the nuclei (or pseudo-nuclei) using *a classical approximation*, typically based on the hypersurface defined by the ground-state electronic energy level (Born-Oppenheimer pseudo-classical MD or MC[400,649], Car-Parrinello MD[648,1533], path-integral sampling[1534,1535]). A *basis-set expansion* of the wavefunction or electron density is also in any case necessary to make the (analytical) QM equations amenable to a numerical (computational) treatment. Furthermore, an exact *treatment of electron correlation*[1536,1537] (*e.g.* full configuration-interaction, exact cluster operator[521,1538], or exact perturbation expansion) are impossible for all but the smallest systems, so that a finite level of theory (*e.g.* limited configuration-interaction, truncated cluster operator expansion, or finite perturbation expansion) must be selected in practice. These two choices can be adjusted to achieve a trade-off between the desired accuracy of the interaction description and the incurred computational cost. An alternative is to dramatically simplify the form of the interaction Hamiltonian and compensate for its deficiencies through parameterization against experimental data (semi-empirical methods[1539–1561]).

(D_6) *Limited system sizes*: since the computational cost of QM calculations increases very rapidly with the number of electrons involved, especially for the most accurate methods (see Point C_6 above), an obvious way to reduce this cost is to restrict their application to small system sizes. This limitation can be partly relaxed by the use of hybrid methods (Point A_6 above).

(E_6) *Limited configurational sampling*: since the computational cost of QM calculations scales linearly with the number of system configurations to be evaluated, an obvious way to reduce this cost is to restrict their application to a small set of configurations. This limitation can be partly relaxed by application of hybrid methods (Point A_6 above).

These various approaches can be used in combination if required. Note that the approximations in the level of resolution and interaction description (Points A_6, B_6 and C_6) introduce adjustable (physical or numerical) parameters into the model. In this case, the QM approach is only a first-principles approach in principle, and no longer in practice. It should also be stressed that the approximations of Point C_6 do not represent fundamental limitations of the QM approach. Extremely accurate QM calculations including explicit coupling of the nuclear and electronic motions, relativistic corrections, extrapolations to infinite basis-set size and full configuration interaction have become possible nowadays[521,1531,1532,1537,1538,1562–1564]. These approximations only become unavoidable when large systems and a high number of system configurations need to be considered, as is the case for the characterization of ionic solvation properties.

As a result of the above approximations (Points A_6-E_6), the *fundamental problems* in QM computations related to ionic solvation can be broadly classified as follows:

(A_7) *Approximate interaction description*: this includes the errors incurred by: (*i*) the possible use of effective core potentials (usually limited); (*ii*) the Born-Oppenheimer approximation (usually very limited); (*iii*) the approximate treatment of relativistic effects (usually very limited); (*iv*) the classical description of the motion of the nuclei or pseudo-nuclei (usually limited, except for the description of high-frequency vibrations and proton transfer); (*v*) the finite basis-set size used for the wavefunction or electron density expansion (usually important); (*vi*) the finite level of theory applied to electron correlation (usually important). Of particular relevance in the present context is the difficulty to properly account for dispersion interactions (an electron correlation effect) at the lowest levels of theory (*e.g.* Hartree-Fock[1536,1565–1568] and density functional theory[1569,1570]).

(B_7) *Finite (microscopic) system size*: this includes the errors incurred by: (*i*) the extremely small tractable system sizes (pure QM description); (*ii*) the approximate AT or CE representation of the surroundings, and the effective coupling between the two subsystems (hybrid description).

(C_7) *Ambiguity in the definition of the potential at the ion site of the air-liquid interfacial potential based on the sampled configurations*[1448–1450, 1571–1574]: Computations relying on a QM approach have been recently confronted with the very same issue affecting AT simulations (see Points C_4 and C_5 in Section 3.3.6) regarding the appropriate calculation of potentials (Sections 3.3.3 and 3.3.5). For example, a cavity potential can also be calculated by integration over the solvent raw charge density (analogous to P-summation) or the solvent charge density sorted by molecules (analogous to M-summation). Similarly, an air-liquid interfacial potential can be calculated by integration of the surface charge density (analogous to P-integration) or of the surface polarization (analogous to M-integration). These distinctions are again also related to the definition of the intrinsic bulk potential of the pure liquid (Section 3.3.4), and in particular whether this potential is viewed as an internal (test charge without excluded volume) or an external (test charge with excluded volume) potential. Here also, the authors of this book recommend the use of the M-scheme and the adoption of an external potential definition, for reasons that will be provided in Section 4.3.8 and further developed in Chapter 7.

To give an example, Car-Parrinello simulations of liquids performed under PBC are the QM analogs of AT simulations using the PBC/LS scheme. These simulations thus also enforce a constraint of vanishing average internal potential over the computational box, and as such, effectively rely on a P-scheme (*e.g.* the potential evaluated at a cavity within the liquid will correspond to a P-summation result). Therefore, the calculated solvation free energies[1571,1572] will require a correction term for improper potential summation that is of a similar nature as that required in corresponding AT simulations using the PBC/LS scheme (Section 6.2.1; type-C_1 correction).

(D_7) *Finite sampling*: this includes the error incurred on the calculated properties (*e.g.* thermodynamic properties *via* statistical mechanics) by insufficient averaging over system configurations.

In the context of the calculation of ionic solvation free energies, the errors associated with the above types of problems may be so large that, in many cases, an AT description may ultimately be preferable to a QM one. However, if a QM approach is retained, the issues related to finite system size and finite sampling can probably only be overcome satisfactorily at present by using a hybrid description. Although many QM studies have been concerned with the structure and dynamics of ionic solvation, *e.g.* Car-Parrinello[141,808,810,811,1467,1575–1604,1604–1612] or hybrid[236,239,244,250,1478,1613–1671] studies, calculations of single-ion solvation free energies using these approaches are still extremely scarce, including *e.g.* hybrid methods involving solely the ion as QM object[1614,1618] or Car-Parrinello free-energy calculations of single-ion solvation[1571,1572]; (see also Ref.[377] for an ion-ion potential of mean force *via* Car-Parrinello simulations). In this specific context, it should also be mentioned that free-energy calculations for alchemical changes, which are relatively straightforward based on AT simulations (Section 3.3.3), raise a number of methodological issues in the context of QM calculations. Considering the unphysical solvation process (Figure 2.2), the concepts of a progressively growing cavity or of an ion-sized cavity with continuously increasing charge makes little sense within a QM description (see, however, Refs.[602,1672,1673]).

To our knowledge, there are only two available estimates for the air-liquid interfacial potential $\chi^{\ominus}_{svt}$ of a solvent, based on the Car-Parrinello method applied to for water[1448–1450], and none for the absolute intrinsic potential $\mathcal{V}^{\ominus}_{H,svt}$ of the reference hydrogen electrode in a given solvent. More details on the available QM methods will be provided in Section 6.3.

3.5 Fundamental theoretical problems: summary

The general conclusions from the previous sections regarding the *fundamental problems* involved in the *theoretical determination* of intrinsic single-ion solvation parameters can be summarized as follows (see also Figure 3.2).

Continuum-electrostatics (CE) calculations (Section 3.2) represent a powerful qualitative and conceptual tool to account for the leading electrostatic contribution to ionic solvation, and provide a solid basis for the discussion of more subtle effects. In particular, they represent the only methods that are, at least in appearance, exempt of any ambiguity concerning the calculation of electric potentials (see Point C_4 in Section 3.3.6 and Point C_7 in Section

3.4), and can thus serve as a reference point for microscopic approaches. As will be seen in Section 4.3.8 and further discussed in Chapter 7, this is because the convention relevant for the potential calculation is actually implicit in Maxwell's equations for electrodynamics in continuous media (and corresponds to the M-scheme in the context of AT and QM approaches). However, because they overlook the microscopic description of the solvent, quantitative results cannot be expected from these models, even by inclusion of correction terms (Section 6.1.1) or extension to semi-atomistic approaches (Section 6.1.2).

Classical atomistic (AT) simulations (Section 3.3) include a detailed microscopic description of the solvent, which represents a significant improvement over CE models. However, the price to pay is that the sampling over solvent configurations must be carried out explicitly on a computer, which is only achievable in practice for systems of finite (microscopic) size and usually based on an approximate (non-Coulombic) representation of electrostatic interactions. As a result, the gain of an improved solvent description is partly counter-balanced by a number of methodological issues including the need to account for approximate electrostatics, finite-size effects and the ambiguity in the evaluation of electric potentials. Finite-sampling problems are no longer a serious issue here nowadays. Furthermore, in its basic implementation relying on an effective pairwise treatment of many-body terms and, in particular, on a mean-field representation of electronic polarization, this approach still suffers from the neglect of an important component of ionic solvation. The issues affecting the indirect calculation of intrinsic single-ion solvation properties *via* the calculation of the air-liquid interfacial potential of the solvent are of a similar nature, including the influence of the sample geometry and confinement potential, and encompassing as well the ambiguity in the evaluation of the electric potential and the mean-field treatment of many-body terms. Finite-sampling problems are also no longer a serious issue here nowadays.

Finally, *quantum-mechanical (QM) computations* (Section 3.4) should in principle provide the most accurate representation of the ionic solvation phenomenon and of interfacial effects. In particular, in contrast to the CE and AT descriptions, the QM approach does not require any input parameter defining the ion size, such as the ionic radius R_I in CE calculations or the homoionic ion-ion repulsion parameter C_{12}^{ii} in AT simulations. It is thus the only approach that can formally be called predictive in this context. However, QM computations suffer from their high intrinsic computational cost. This results in very severe restrictions concerning the accessible system sizes and extent of configurational sampling, and in the need for a number of often drastic (physical or numerical) approximations. Consequently, the gain of an improved description of the molecular system is largely counter-balanced by a number of methodological issues, including the partial account of electron correlation effects, finite-size effects and finite-sampling errors, the latter two now representing a very serious issue. Furthermore, the ambiguity in the evaluation of electric potentials represents a major issue in QM computations, just as was the case in AT simulations.

In the present state, it therefore appears that none of the above theoretical approaches is able to provide quantitative predictions concerning intrinsic single-ion solvation free energies. In the case of an ideal exact and parameter-free QM approach, such a prediction (*via* either $\mathcal{G}^{\ominus}_{H,svt}$, $\chi^{\ominus}_{svt}$ or $\mathcal{V}^{\ominus}_{H,svt}$; Section 2.3) would be qualitatively more powerful, due to its independence on empirical parameters. However, a truly first-principles treatment of the problem, for a system of sufficient size and given a sufficient configurational sampling, will not become computationally feasible in any foreseeable future. Nevertheless, the idea that an exact QM calculation of these experimentally-elusive quantities could at least in principle be performed suggests that these quantities are well defined and not a mere illusion (Section 2.3). There is, however, an important proviso to be made concerning the partitioning of bulk and surface effects in these calculations (see further below). In the case of the CE and AT models, such a prediction would only be useful in conjunction with a cross-validation against other properties of ionic systems, *i.e.* the empirical parameters (R_I in CE calculations, C^{ii}_{12} in AT simulations) should be adjusted in such a way as to account for a large body of experimental properties, including single-ion solvation parameters but also extending beyond these. These empirical parameters may thus lose their arbitrariness and the model becomes truly predictive as soon as it is able to cross-link multiple aspects of ion and solvent thermodynamics and dynamics, *e.g.* pure solvent properties, solvation properties at infinite dilution and at finite concentration and properties of ionic solids and melts. This statement is moderately relevant for the CE approach, which can only cross-link very few properties of ionic systems (solvation free energy, crystal lattice parameters). It is more relevant for the AT approach, which has the ability to capture a much wider range of properties. The corresponding self-consistent refinement of ambiguous size parameters in the context of AT simulations will be later termed the atomistic-consistency assumption (Section 6.2.3). Again, there is an important proviso to be made here concerning the results of such a self-consistent modeling method with respect to the partitioning of bulk and surface effects (see further below). Irrespective of these issues, CE and AT models also provide *per se* a considerable insight into the nature of the solvation process. Note that in some sense, even a hypothetical "exact" QM description could also be viewed as representing an idealization, considering that any physical model, as accurate as one may think it is at a given time in history, still remains only a model. Finally, the discussion of the problems inherent to theoretical approaches opens up an interesting and intricate issue already briefly raised in Section 2.3, namely whether the above *experimentally-elusive quantities* are also *theoretically elusive.* As seen above, both AT and QM approaches are affected by an inherent ambiguity in the calculation of electric potentials. The CE approach does not appear to suffer from this problem, but it is solely because a specific convention is built into the model *via* Maxwell's equations. Ultimately, this ambiguity arises from the requirement for an arbitrary *defi-*

nition permitting the separation of the contributions of the solvent-generated electric potential due to bulk and surface region. As discussed in detail in Section 4.3.8 and Chapter 7, this definition is at the root of the need for an extra-thermodynamic assumption in the evaluation of the above elusive quantities, and the most reasonable choice relies on aligning the explicit definition to be made in AT and QM calculations onto the implicit definition involved in the CE analysis, which corresponds to application of the M-scheme. This choice is not compulsory, but is the only reasonable one if the resulting intrinsic solvation free energies are to be amenable to intuitive reasoning, given that this reasoning is itself so much anchored into the description provided by the Born model.

After a thorough reading of this chapter, a single ion has become more skeptical concerning the outcome of theoretical investigations. It would probably consider heading back to the lab and switching to experimental studies, but it has also read Chapter 2 before. Clearly, it is in desperate need for more fundamental knowledge concerning thermodynamics, interfacial effects, electrochemistry and single-ion properties, which it will gain in the next chapter.

Chapter 4
Concepts and definitions

The problem of the experimental and theoretical determination of single-ion solvation free energies, as well as of corresponding derivative thermodynamic solvation properties, is a relatively complex one. However, it is often further complicated by imprecisions, alternatives or ambiguities in the definitions of a number of *fundamental concepts*. In order to subsequently discuss the problem in a well-defined framework, this part of the text aims at clarifying these concepts and definitions, in the context of: (*i*) *notations*, as used throughout this book (see also Table I); (*ii*) *thermodynamics* (variables, standard states, processes and cycles); (*iii*) *electrostatics* (interfacial effects, electric potentials); (*iv*) *electrochemistry* (types of measurements, relationship to thermodynamic quantities and electric potentials); (*v*) *single-ion properties* (conventional, real and intrinsic scales).

4.1 Notations

Due to the constraint of electroneutrality in macroscopic matter at equilibrium (Section 2.1), *overall neutral sets of ions* play a special role in ion thermodynamics. In the most general case, such a set should be written $I_{n_I}^{z_I} J_{n_J}^{z_J} K_{n_K}^{z_K} \ldots$ with $n_I z_I + n_J z_J + n_K z_K + \ldots = 0$, where I, J, K, ... represent ion types. However, to avoid making the notation too cumbersome, the discussion will typically only refer to *binary combinations* of the form $I_{n_I}^{z_I} J_{n_J}^{z_J}$ with $n_I z_I + n_J z_J = 0$, the generalization to ternary or higher-order combinations being straightforward. It will often be further assumed that the two ion types are *isovalent*, leading to $I^z J^{-z}$, the generalization to combinations involving ions with different valences also being straightforward. Finally, a large part of the discussion is dedicated to the *alkali* and *halide* ions, and the specific notation M^+X^- will be used in this context, where M^+ denotes the alkali cation and X^- the halide anion.

RSC Theoretical and Computational Chemistry Series No. 3
Single-Ion Solvation
Experimental and Theoretical Approaches to Elusive Thermodynamic Quantities
By Philippe Hünenberger and Maria Reif

Published by the Royal Society of Chemistry, www.rsc.org

The *phase* in which a given species is considered, whenever relevant, will be indicated by a subscript, namely: g (gas phase), ℓ (liquid phase), s (solid phase), d (dissolved, *i.e.* in solution), aq (in aqueous solution), svt (solvent; pure liquid or component of a solution, depending on the context), wat (water; pure liquid or component of a solution, depending on the context), and el (one atom of an element in its standard-state compound and phase).

Given the above subscript convention for the indication of the phase, the following notations will also be employed for various *chemical species*: Σ (arbitrary species; generic), e^- (electron species), H^+ (proton species), I (ion-forming species; other ion-forming species: J, K, ...), I^z (ionic species; z: signed integer ion charge), IJ (pair of ionic species I^z and J^{-z}; crystal: associated; gas phase, solution: dissociated), $I_g + J_g$ (atom pair I_g and J_g in the gas phase; not to be confused with the pair of gas-phase ions IJ_g), and $[I^z \cdot J^{-z}]$ (associated ion pair; I^z bound to J^{-z}; gas phase or solution). Thus, for example, M_g (or X_g) denotes a gas-phase alkali (or halide) atom, M^+_{aq} (or X^-_{aq}) an aqueous alkali (or halide) ion, MX_s an alkali-halide salt, MX_g (or MX_{aq}) the sum (in terms of thermodynamic parameters) of an alkali and a halide gas-phase (or aqueous) ion, $M_g + X_g$ the sum (in terms of thermodynamic parameters) of an alkali and a halide gas-phase atom, and $[M^+ \cdot X^-]_g$ (or $[M^+ \cdot X^-]_{aq}$) an associated gas-phase (or aqueous) alkali-halide ion pair. All these notations are summarized in Table I for the ease of reference.

Further notations will be introduced in the course of this chapter concerning thermodynamic variables, standard states, processes and cycles (Section 4.2), as well as electric potentials (Section 4.3), electrochemical cells and measurements (Section 4.4), and single-ion properties (Section 4.5) relevant in the context of ionic solvation. During the discussion of electrostatics and electrochemistry (Sections 4.3 and 4.4), the notation $\Sigma : P$ will also sometimes be used to denote a chemical species Σ present in a given phase P of a multiple phase system, with the variants $\Sigma : M$, $\Sigma : S$ or $\Sigma : G$ to indicate a metal, solution or gas phase. These expand the more generic phase notations (subscripts) defined above. When there is no ambiguity concerning the identity of the phase, notations like *e.g.* $M : M$, $\Sigma : S$ and $\Sigma : G$ can be translated to M_s (or M_ℓ for a liquid metal), Σ_d (or Σ_{aq} if the solvent is water) and Σ_g, respectively, in the latter notation.

4.2 Thermodynamics

In this section, the key concepts providing the framework for the *thermodynamic description* of properties related to ionic solvation are summarized. These include: (*i*) thermodynamic variables (definitions and basic relationships); (*ii*) molar and partial molar variables; (*iii*) ideal behaviors; (*iv*) standard states (convention used in this book); (*v*) solute standard-state variants; (*vi*) activity; (*vii*) reaction parameters; (*viii*) phase-transition reactions; (*ix*) heat-up reactions; (*x*) formation parameters; (*xi*) temperature-scaled heat capacity integration (TCI); (*xii*) spectroscopy-based statistical mechanics (SBS); (*xiii*) thermodynamic parameters for the gas-phase electron and proton; (*xiv*) thermodynamic reactions relevant to ionic solvation; (*xv*) thermodynamic cycles relevant to ionic solvation; (*xvi*) standard-state corrections to the hypothetical solvation process; (*xvii*) standard states (alternative conventions and interconversion formulae); (*xviii*) experimental determination of thermodynamic parameters.

4.2.1 Thermodynamic variables

The basic *extensive* thermodynamic variables considered in this book are the Gibbs *free energy* (or free enthalpy) G, the *enthalpy* H, the (absolute) *entropy* S, the (isobaric) *heat capacity* C_P, the *volume* V, the (isothermal) *volume-compressibility* K_T and the (isobaric) *volume-expansivity* A_P. The *energy* (or internal energy) U may also be added to this list. This quantity is central in thermodynamics, but the consideration of H instead of U is typically more convenient for practical applications. To express equations valid for any of the seven above variables, the generic notation Y will be used in the present text (*i.e.* $Y = G$, H, S, C_P, V, K_T or A_P). The basic *intensive* thermodynamic variables are the *pressure* P and the (absolute) *temperature* T.

Unless otherwise specified, all macroscopic states considered here for the evaluation of the above variables (extensive as well as intensive) are assumed to correspond to *closed* (no exchange of particles between system and environment), *diaerg* (free exchange of volume work between system and environment), and *diatherm* (free exchange of heat between system and environment) systems *at equilibrium*. This implies in particular that in the considered systems: (*i*) the values of all the above variables are constant in time, (*ii*) P and T are single-valued (mechanical and thermal equilibrium), *i.e.* the corresponding local observables are homogeneous in space; (*iii*) the values of P and T in the system are identical to those in its environment, *i.e.* to values that can be freely chosen in a given experimental setup.

According to the *Gibbs phase rule*[456], the macroscopic state of a system at equilibrium involving N_P *phases* (subsystems with different microscopic organization and intensive properties other than P and T, *e.g.* density), N_S *species* (types of substances present in the system, counting identical substances occurring in multiple phases as distinct species), and N_C *constraints* (linearly-independent chemical equilibria or fixed ratios of species in some phases), can be entirely specified using

$$N_V = N_S - N_C - N_P + 2 \tag{4.1}$$

intensive variables along with N_P extensive ones. The number N_V is referred to as the *variance* of the system. In the present context (closed, diaerg and diatherm systems), a natural choice of variables for a monophasic system ($N_P = 1$) exempt of constraints ($N_C = 0$) is to take P and T along with $N_S - 1$ intensive *composition variables* (*e.g.* mole fractions of $N_S - 1$ species or concentrations of $N_S - 1$ species in a reference species) and one extensive *extent variable* (*e.g.* total number of moles of all species or number of moles of the reference species). The variables forming the selected set are referred to as the *state variables* of the system. Any thermodynamic variable that only depends on the current macroscopic state of the system, irrespective of the way this state was arrived at, can be written as a unique function of the state variables, and is thus referred to as a *state function.*

The notation $\partial_P Y$ will be used to indicate the *partial derivative* of a state function with respect to P (at constant T, system composition and system extent), $\partial_P^2 Y$ representing the corresponding second derivative. The notation $\partial_T Y$ will be used to indicate the partial derivative of a state function with respect to T (at constant P, system composition and system extent), $\partial_T^2 Y$ representing the corresponding second derivative. The notation $\partial_{P,T}^2$ will be used for the corresponding mixed second derivative (with respect to P at constant T and with respect to T at constant P, the order of the two differentiations being irrelevant).

The seven extensive variables noted Y, as well as the energy U, are defined as follows. The *volume* V of a system is a purely geometric property, and obviously a state function. The *(internal) energy* U of a system is defined, *within a constant*, by the *first law of thermodynamics*[1674–1681]. This law contains: (i) a differential definition of U, namely that the variation of U over any infinitesimal process in a closed system is equal to the sum of the work and heat supplied to the system during this process; (ii) the postulate, based on the absence of reported experimental contradictions, that U is a state function.

The*entropy* S of a system is defined, within an arbitrary constant, by the *second law of thermodynamics*[1682] (see also Refs.[1681, 1683–1685] for early work having led to this law and Ref.[1686] for a historical overview). This law contains: (i) a differential definition of S, namely that the variation of S over any reversible infinitesimal process in a closed system is equal to the ratio of the heat supplied to the system during the process to the absolute temperature of the system; (ii) the postulate, based on the absence of reported experimental contradictions, that S is a state function; (iii) a spontaneity condition, also based on the absence of reported experimental contradictions, stating that the heat supplied to a system during any spontaneous (irreversible) infinitesimal process in a closed system is always smaller than the product of the corresponding variation of S with the absolute temperature.

In contrast to the energy U, however, the entropy S can be attributed a *zero point* owing to the *third law of thermodynamics*[1687, 1688]. This law postulates, based on the extrapolation of low-temperature experiments, that at the absolute zero temperature, the entropy of a compound in a perfect crystal (any molecule, isomorph and pressure) is equal to the sum of the entropies of the constituting elements in perfect crystals (any molecule, isomorph and pressure). As a result, setting to zero the entropies of all 118 currently known elements[1689] in perfect crystals at the absolute zero (most stable molecule and isomorph, and for a given reference pressure) is sufficient to construct a system of entropies that is consistent throughout chemistry. This system, which merely results from a convention within phenomenological thermodynamics, turns out to be the correct one when considering the quantum statistical mechanical interpretation of entropy. Entropies based on this convention are referred to as *absolute* or *third-law* entropies. Owing to the third law and to the latter convention, the entropies of all elements and compounds in perfect crystals (any molecule, isomorph and pressure) are zero at the absolute zero temperature, and the corresponding absolute entropies at finite temperatures can be determined by temperature integration of the temperature-scaled heat capacity (TCI; Section 4.2.11). Alternatively, they can in principle also be determined using spectroscopy-based statistical mechanics (SBS; Section 4.2.12). In practice, this second approach is only experimentally applicable to the determination of absolute entropies for gas-phase species. The absolute entropy is a *positive quantity* at any finite temperature. This follows from the positiveness of the heat capacity of any system (see below).

Note that the first and third laws are formulated here in a chemical context. The corresponding statements must be refined in the presence of mass-energy interconversions (first law) and non-conservation of the nuclear spins (third law) occurring *e.g.* in the presence of nuclear processes.

The five other variables, namely the enthalpy H, the free energy G, the (isobaric) heat capacity C_P, the (isothermal) volume-compressibility K_T, and the (isobaric) volume-expansivity A_P are defined based on the three former variables V, U and S, as

$$H = U + PV \quad , \tag{4.2}$$

$$G = H - TS \quad , \tag{4.3}$$

$$C_P = \partial_T H \quad , \tag{4.4}$$

$$K_T = -\partial_P V \tag{4.5}$$

and

$$A_P = \partial_T V \quad . \tag{4.6}$$

These five variables are also state functions, because they are defined exclusively in terms of state functions. Thus, the seven extensive variables noted Y, namely G, H, S, C_P, V, K_T and A_P, are all state functions. Note, however, that G and H, just like U, are only defined *within a constant*, because U is only defined in a differential form by the first law of thermodynamics. In contrast, S, C_P, V, K_T and A_P are *absolute quantities*, *i.e.* with a well-defined zero point.

The *volume-compressibility* and *volume-expansivity* are introduced here as extensive variables characterizing the negative isothermal volume change of a system in response to a pressure increase or the corresponding isobaric volume change in response to a temperature increase, respectively. For pure substances, the variables K_T and A_P are commonly expressed in the form of corresponding *intensive* variables, which are also state functions, namely the *isothermal compressibility* κ_T and the *isobaric expansivity* (or *thermal expansion coefficient*) α_P, defined as

$$\kappa_T = V^{-1} K_T = -V^{-1} \partial_P V \tag{4.7}$$

and

$$\alpha_P = V^{-1} A_P = V^{-1} \partial_T V \quad . \tag{4.8}$$

However, as will be discussed later (Section 4.2.2), the generalization of these expressions to mixtures leaves room for ambiguity concerning the choice of the scaling volume. Consequently, this book will almost exclusively refer to the well-defined extensive quantities K_T and A_P, termed here volume-compressibility and volume-expansivity, respectively, where the word "volume" emphasizes the extensive (volume-dependent) nature of these quantities. Note that K_T is always a *positive quantity*, which is a consequence of the second law of thermodynamics. In contrast, A_P is commonly, but not systematically, positive.

Finally, another quantity may be of interest, namely the *isochoric heat capacity* C_V, defined as

$$C_V = \left(\frac{\partial U}{\partial T} \right)_{\{n_i\},V} \quad , \tag{4.9}$$

where n_i is the number of moles of species i in the system. Note that the temperature derivative is performed here at constant volume rather than at constant pressure, which is why the symbol ∂_T is not used. This quantity is related to previously defined variables by the equation

$$C_V = C_P - T K_T^{-1} A_P^2 \quad . \tag{4.10}$$

For this reason, C_V will not be considered in detail in this text, and heat capacity will always refer by default to the isobaric quantity C_P.

Note that C_P and C_V are always *positive quantities*, which is a consequence of the second law of thermodynamics, C_P being always larger than C_V, following from Eq. 4.10 along with the positiveness of the volume-compressibility (see above).

The three laws of thermodynamics are actually not sufficient to formulate chemical thermodynamics in a complete fashion, including a description of open systems and of chemical reactions. They must be complemented by a definition of the *chemical potential* of a species i, namely

$$\mu_i = \left(\frac{\partial U}{\partial n_i} \right)_{\{n_j, j \neq i\},V,S} \quad . \tag{4.11}$$

Introducing this definition, the first and second laws can be recast in the form of Gibbs' *fundamental equation chemical thermodynamics*[456,1690–1692] (see also Refs.[1693–1695] for historical overviews) together with its associated *spontaneity condition.* For any infinitesimal process, the corresponding expressions read

$$dU = -PdV + TdS + \sum_{i=1}^{N_S} \mu_i dn_i \tag{4.12}$$

and

$$\delta W_v + \delta W_n + \delta D \geq dU - TdS \quad , \tag{4.13}$$

where δW_v, δW_n and δD are the volume work, non-volume work (*e.g.* gravitational or electric), and chemical energy transfer (through exchange of substances between system and surroundings; for open systems only) supplied to the system during the infinitesimal process, while the equality (inequality) in the second expression applies to reversible (spontaneous) processes. The above quantities are defined by $\delta W_v = -P_{ext}dV$, $\delta W_n = -\Delta \mathcal{V}_{ext} dq$ (assuming electric work only) and $\delta D = -\sum \mu_{i,ext} \delta_d n_i$, where P_{ext} is the pressure in the surroundings, $\mu_{i,ext}$ the chemical potential of species i in the surroundings, $\Delta \mathcal{V}_{ext}$ an *external potential difference* in the surroundings, dq the charge transferred across the system along the potential difference $\Delta \mathcal{V}_{ext}$ (measured in the direction low potential to high potential, *i.e.* negative to positive pole of the external electric circuit) and $\delta_d n_i$ the number of moles of species i transferred to the system. In the present context, the fundamental equation and its associated spontaneity condition are more conveniently reformulated in terms of the free energy G (using the definitions of Eqs. 4.2 and 4.3), as

$$dG = VdP - SdT + \sum_{i=1}^{N_S} \mu_i dn_i \tag{4.14}$$

and

$$\delta W_v + \delta W_n + \delta D \geq dG - PdV - VdP + SdT \quad . \tag{4.15}$$

Eq. 4.14 permits in particular to replace the chemical potential definition of Eq. 4.11 by the more convenient expression

$$\mu_i = \left(\frac{\partial G}{\partial n_i} \right)_{\{n_j, j \neq i\}, P, T} \quad . \tag{4.16}$$

By application of the *Euler theorem*, Eq. 4.14 may be integrated to

$$G = \sum_{i=1}^{N_S} \mu_i n_i \ . \tag{4.17}$$

The compatibility between Eqs. 4.14 and 4.17 represents a constraint on the relative changes of all intensive variables in a system at equilibrium, known as the *Gibbs-Duhem equation*[456,1692,1696,1697] (see also Refs.[1694,1695] for historical overviews),

$$\sum_{i=1}^{N_S} n_i d\mu_i = VdP - SdT \ . \tag{4.18}$$

This equation shows in particular that the chemical potential of a pure substance in a given phase is a unique function of temperature and pressure, and can be used to derive the Gibbs phase rule[456] (Eq. 4.1).

The variables G, H and S play key roles in chemical thermodynamics for closed, diaerg and diatherm systems at equilibrium, and it is therefore important to characterize their dependences on P and T. Up to second order in G (first order in H, S and V), it can be shown that these dependences are described by the relationships

$$\partial_P G = V \ , \tag{4.19}$$

$$\partial_T G = -S \ , \tag{4.20}$$

$$\partial_P H = V - TA_P \ , \tag{4.21}$$

$$\partial_T H = C_P \ , \tag{4.22}$$

$$\partial_P S = -\partial_T V = -\partial^2_{P,T} G = -A_P \ , \tag{4.23}$$

$$\partial_T S = -\partial^2_T G = T^{-1} C_P \ , \tag{4.24}$$

and

$$\partial^2_P G = \partial_P V = -K_T \quad . \tag{4.25}$$

These important equations have no specific names, except for Eq. 4.22, which can be viewed as a basic form of the *Kirchhoff equation*[1698], more usually expressed in terms of reaction enthalpies (Section 4.2.7). Note that this equation has been introduced here as a definition (Eq. 4.4). Eq. 4.20 is also commonly combined with the definition of Eq. 4.3, leading to

$$\partial_T (T^{-1} G) = -T^{-2} H \ , \tag{4.26}$$

which can be viewed as a basic form of the *Gibbs-Helmholtz equation*[1699–1708] (see also Ref.[1709] for a historical overview), more usually expressed in terms of reaction enthalpies (Section 4.2.7). Differentiating this equation with respect to P, one might also write

$$\partial^2_{P,T} (T^{-1} G) = -T^{-2} \partial_P H \quad . \tag{4.27}$$

Eqs. 4.19, 4.20 and 4.23-4.26, along with the trivial equation $G = G$, can be collectively written in the generic form[1412]

$$Y = \hat{O}_Y G \quad , \tag{4.28}$$

where $\hat{O}_Y$ is the *thermodynamic-derivative operator* associated with variable Y, namely

$$\begin{aligned} \hat{O}_G &= 1 & \hat{O}_{C_P} &= -T\partial_T^2 \\ \hat{O}_H &= -T^2\partial_T T^{-1} & &= \partial_T(1-T\partial_T) \\ &= 1-T\partial_T & \hat{O}_V &= \partial_P \\ \hat{O}_S &= -\partial_T & \hat{O}_{K_T} &= -\partial_P^2 \\ & & \hat{O}_{A_P} &= \partial_{P,T}^2 \quad . \end{aligned} \tag{4.29}$$

Further introducing the notations $H'_P = \partial_P H$, $H'_T = \partial_T H$, $S'_P = \partial_P S$ and $S'_T = \partial_T S$, Eq. 4.28 can also be applied together with

$$\begin{aligned} \hat{O}_{H'_P} &= \hat{O}_V - T\hat{O}_{A_P} & \hat{O}_{S'_P} &= -\hat{O}_{A_P} \\ &= -T^2\partial_{P,T}^2 T^{-1} & & \\ \hat{O}_{H'_T} &= \hat{O}_{C_P} & \hat{O}_{S'_T} &= T^{-1}\hat{O}_{C_P} \quad . \end{aligned} \tag{4.30}$$

Note that any set of derived thermodynamic quantities f_Y satisfying a corresponding version of Eq. 4.28, *i.e.* the *set of thermodynamic-derivative relationships*

$$f_Y = \hat{O}_Y f_G \quad , \tag{4.31}$$

also automatically satisfies a *form of the Gibbs equation* (see further below), namely

$$f_G = f_H - T f_S \quad . \tag{4.32}$$

For $f_Y = Y$, this equation is Eq. 4.3. A number of versions of Eqs. 4.31 and 4.32 will be encountered throughout this chapter concerning *e.g.* changes of thermodynamic parameters for processes with $\Delta P = 0$ and $\Delta T = 0$ (this section), molar and partial molar variables (Section 4.2.2), standard molar and partial molar variables (Section 4.2.4), reaction parameters (Section 4.2.7) and formation parameters (Section 4.2.10).

For a given system extent and composition, G, H and S are *discontinuous functions* of P and T at *first-order phase transitions*, while C_P, K_T and A_P are generalized functions of P and T at these transitions, *i.e.* they include Dirac delta function terms. For example, for a system consisting of one mole of a pure substance, the first-order phase transition contribution to $C_P(P,T)$ at the melting temperature $T_m(P)$ for a pressure P may be written $\Delta_m H(P)\delta(T - T_m(P))$, where $\Delta_m H(P)$ is the melting enthalpy at pressure P. The variables C_P, K_T and A_P may also themselves be discontinuous at second-order phase transitions.

For finite processes connecting two equilibrium states of a given system, Eqs. 4.19-4.27 can be immediately rewritten into corresponding equations involving the differences ΔY instead of Y, provided that the corresponding differences ΔP for pressure derivatives or ΔT for temperature derivatives are zero. In the generic notation of Eq. 4.28, this can be written

$$\Delta Y = \hat{O}_Y \Delta G \quad \text{if} \quad Y = \begin{cases} G & \\ V, K_T & \text{and} \quad \Delta P = 0 \\ H, S, C_P, H'_T, S'_T & \text{and} \quad \Delta T = 0 \\ A_P, H'_P, S'_P & \text{and} \quad \Delta P = \Delta T = 0 \quad , \end{cases} \tag{4.33}$$

to which one may add

$$\partial_P \Delta H = \Delta V - \Delta[T A_P] \quad \text{and} \quad \partial_P \Delta S = -\Delta A_P \quad \text{if} \quad \Delta P = 0 \tag{4.34}$$

and

$$\partial_T \Delta H = \Delta C_P \quad \text{and} \quad \partial_T \Delta S = T^{-1} \Delta C_P \quad \text{if} \quad \Delta T = 0 \quad . \tag{4.35}$$

The above equations can be written in the generic form of Eq. 4.31 as

$$\Delta Y = \hat{O}_Y \Delta G \qquad (\Delta P = 0, \ \Delta T = 0) \quad , \tag{4.36}$$

and imply in particular (Eq. 4.32)

$$\Delta G = \Delta H - T \Delta S \qquad (\Delta P = 0, \ \Delta T = 0) \quad . \tag{4.37}$$

Note, however, that Eqs. 4.36 and 4.37 are less general than Eqs. 4.33-4.34 and the Gibbs equation (see further below, Eq. 4.41), respectively.

The relevance of the *free energy* G resides in its connection with the *driving force* associated with chemical processes carried out under experimentally common boundary conditions. More specifically, for any infinitesimal closed ($\delta D = 0$), isobaric (diaerg, $dP = 0$, $\delta W_v = -P dV$) and isothermal (diatherm, $dT = 0$) process where the only possible exchange of non-volume work is electric ($\delta W_n = -\Delta \mathcal{V}_{ext} dq$), the spontaneity condition (Eq. 4.15) becomes

$$dG \leq -\Delta \mathcal{V}_{ext} dq \ \ (\text{closed}, \ dP = 0, \ dT = 0) \quad . \tag{4.38}$$

Note that this equation implicitly requires that the system is in mechanical and thermal equilibrium (internal, as well as with its surroundings) in both the initial and final states of the infinitesimal process, so that P and T are single-valued in both states. Otherwise, the terms isobaric and isothermal are undefined. However, the system does not need to be in (internal) chemical equilibrium. Because G is a state function, so that dG is solely determined by the nature of the process, reversibility can only be achieved in the general case by adjusting $\Delta \mathcal{V}_{ext}$. This is exactly the situation encountered in *equilibrium electrochemistry measurements*, *i.e.* closed, isothermal, isobaric and at zero electric current (Section 4.4). If the process is carried out in an *uncoupled* fashion, *i.e.* without exchange of electric work between system and surroundings, it will be spontaneous when $dG < 0$ and unnatural when $dG > 0$, *i.e.*

not occurring experimentally and spontaneous in the reverse direction. The reversible situation $dG = 0$ for a given system will then only be encountered for specific processes. An uncoupled system for which all possible infinitesimal (closed, isobaric and isothermal) processes satisfy $dG = 0$ is a system at *chemical equilibrium.*

Integration of Eq. 4.38 leads to a corresponding equation for any finite closed, isobaric ($P = cst$ over the entire process) and isothermal ($T = cst$ over the entire process) process where the only possible exchange of non-volume work is electric, namely

$$\Delta G \leq W_{el} \quad (\text{closed, } P = cst, \, T = cst) \quad , \tag{4.39}$$

where W_{el} is the integral of $-\Delta\mathcal{V}_{ext}dq$ over the entire process, *i.e.* the total electric work supplied to the system. Reversibility can only be achieved in the general case by progressively adjusting $\Delta\mathcal{V}_{ext}$ along the entire process so that $dG = -\Delta\mathcal{V}_{ext}dq$ for all infinitesimal subprocesses. This situation can be achieved through *quasi-equilibrium electrochemistry measurements, i.e.* closed, isothermal, isobaric and in the limit of infinitesimal electric current (Section 4.4). In this case, the minimal amount of electric work will be supplied to ($\Delta G > 0$) or the maximal amount of electric work collected from ($\Delta G < 0$) the system along the process. For an uncoupled irreversible process, isobaric and isothermal conditions are often difficult to achieve in practice, because they imply that mechanical and thermal equilibrium can be maintained along the entire process, which requires some control on the reaction rate. For example, an uncontrolled chemical reaction (*e.g.* combustion or explosion) should be carried out instead using a very small amount of catalyst, so that the local observables associated with P and T can be assumed to remain at any time homogeneous, constant and equal to their values in the surroundings. This is of course only possible if the reaction is an activated reaction, so that its rate can be controlled through the amount of catalyst.

Eq. 4.39 cannot be rigorously generalized to any (possibly non-isothermal or/and non-isobaric) process connecting two states at the same pressure P ($\Delta P = 0$) and temperature T ($\Delta T = 0$). Mechanical and thermal irreversibilities will always lead to an increase in $W_{el} + W_n$ balanced by a decrease in Q for a given process. In the common situation where both W_n and W_{el} increase, this still results in $\Delta G \leq W_{el}$. But there are also situations where a system may partially convert supplied volume work into exported non-volume work, in which case one may find $\Delta G \geq W_{el}$. On the other hand, if the process is uncoupled ($W_{el} = 0$), Eq. 4.39 reads $\Delta G \leq 0$. Since G is a state function, ΔG is path-independent and the conditions $\Delta P = 0$ and $\Delta T = 0$ (rather than $P = cst$ and $T = cst$ over the entire process) are sufficient to guarantee the validity of this equation, *i.e.*

$$\Delta G \leq 0 \quad (\text{closed, uncoupled, } \Delta P = 0, \, \Delta T = 0) \quad . \tag{4.40}$$

It follows that all the finite uncoupled processes in a closed system sharing identical initial and final states, themselves characterized by identical pressures and temperatures (at equilibrium), will be spontaneous when $\Delta G < 0$ and unnatural when $\Delta G > 0$. In other words, at constant system extent and composition, a state with high free energy under given P, T conditions will always spontaneously convert to a state of lower free energy under the same P, T conditions by the action of spontaneous uncoupled chemical processes, irrespective of the pathway followed in this conversion. The description of the rates at which various possible spontaneous processes actually occur is, however, beyond the scope of thermodynamics. Fortunately, many metastable states of a closed system (with a free energy higher than that of the most stable state under the given P, T conditions) are sufficiently inert (*i.e.* slowly transforming on the experimental timescale) that they can be characterized experimentally. Other states that are more reactive can also sometimes be characterized through a combination of spectroscopy (non-equilibrium measurements probing directly the quantum energy levels of the system) and statistical mechanics (SBS; Section 4.2.12).

The relevance of the enthalpy H and of the entropy S resides in: (i) their relationship with the *temperature dependence* of the driving force associated with chemical processes carried out under the considered (experimentally common) boundary conditions (ii) the insight they provide on the *physical nature* of this driving force, in terms of enthalpic and entropic contributions to ΔG. The former point results from Eqs. 4.20 and 4.26 reexpressed in terms of ΔG and assuming $\Delta T = 0$, namely $\partial_T \Delta G = -\Delta S$ and $\partial_T (T^{-1} \Delta G) = -T^{-2} \Delta H$, see also Eq. 4.33. The latter point results from two equations, discussed in turn in the following paragraphs (Eqs. 4.41 and 4.42).

The first equation is the *Gibbs equation*[456,1692]. It states that for any finite process in a given system connecting two states at the same temperature T ($\Delta T = 0$), one has

$$\Delta G = \Delta H - T \Delta S \qquad (\Delta T = 0) \quad . \tag{4.41}$$

This equation is easily derived by writing the definition of Eq. 4.3 in differential form for an infinitesimal isothermal process ($dT = 0$), integrating the result for a finite isothermal process ($T = cst$ over the entire process) in a closed system, and observing that G, H and S are state functions, so that the same equation must hold for any other path connecting the two states, even if the system is not closed or if T is not constant along this path. Note that Eq. 4.41 is more general than Eq. 4.37 as deduced from Eqs. 4.32 and 4.36, since it is actually exempt of any constraint on ΔP.

The second equation connects the *enthalpy change* to the *heat plus electric work* (assuming no other form of non-volume work) supplied to the system during a process. It states that for any finite isobaric process ($P = cst$ over the entire process) in a closed system

$$\Delta H = Q + W_{el} \;\; (\text{closed}, \;\; P = cst) \quad , \tag{4.42}$$

where Q is the integral of δQ over the entire process, *i.e.* the total heat supplied to the system. This equation is easily derived by writing the definition of Eq. 4.2 in differential form for an infinitesimal isobaric process ($dP = 0$) in a closed system, inserting the first-law expression for dU and the expression $-PdV$ for the isobaric volume work, and integrating the result for a finite isobaric process ($P = cst$ over the entire process) in a closed system.

Eq. 4.42 cannot be rigorously generalized to any (possibly non-isobaric) process connecting two states at the same pressure P ($\Delta P = 0$). Mechanical irreversibility will always lead to an increase in $W_{el} + W_n$ balanced by a decrease in Q for a given process. In the common situation where both W_n and W_{el} increase, W_{el} will increase less than Q decreases, resulting in $\Delta H \geq Q + W_{el}$. But there are also situations where a system may partially convert supplied non-volume work into exported volume work, in which case one may find $\Delta H \leq Q + W_{el}$. On the other hand, if the process is uncoupled ($W_{el} = 0$), irreversibility always increases W_n and decreases Q, and one may write

$$\Delta H \geq Q \quad \text{(closed, uncoupled, } \Delta P = 0) \quad , \tag{4.43}$$

where the equality (inequality) applies to isobaric (non-isobaric) processes. In most practical cases, mechanical irreversibility has a limited effect and one may use $\Delta H \approx Q$ as a good approximation. These considerations show that for closed uncoupled processes with $\Delta P = 0$, ΔH can be approximately (exactly if the process is isobaric) identified with the heat absorbed by the process ($\Delta H > 0$) or minus the heat released by the process ($\Delta H < 0$). Finally, for coupled as well as uncoupled processes, Eqs. 4.40, 4.41 and 4.42 can also be combined into

$$Q \leq T\Delta S \quad \text{(closed, } \Delta P = 0, \Delta T = 0) \quad , \tag{4.44}$$

where the equality (inequality) applies to a reversible (spontaneous) process. This equation is first derived for an isobaric process (requirement for Eq. 4.42), and generalized to the $\Delta P = 0$ situation by noting that the right-hand side is path-independent, while Q on the left-hand side will be further decreased by the presence of mechanical irreversibility along the process.

These considerations lead to the *interpretation of the Gibbs equation* (Eq. 4.41) in terms of enthalpic and entropic components to the chemical driving force ΔG of a closed uncoupled process with identical initial and final P and T. As seen above, an uncoupled process will be spontaneous if $-T^{-1}\Delta G > 0$ or unnatural if $-T^{-1}\Delta G < 0$. Owing to the Gibbs equation, $-T^{-1}\Delta G$ can be partitioned into an enthalpic ($-T^{-1}\Delta H$) and an entropic (ΔS) contribution. The entropic contribution ΔS obviously corresponds to the entropy production incurred by the process within the system itself. Following from Eq. 4.43 and neglecting the possible effect of mechanical irreversibility ($\Delta H \approx Q$), $-\Delta H$ corresponds to the heat release from the system into the surroundings. According to the second law of thermodynamics, $-T^{-1}\Delta H$ is thus the entropy increase incurred by the process within the surroundings. The enthalpic contribution ΔH and the entropic contribution $-T\Delta S$ to the free-energy change ΔG are very distinct in nature, so that the partitioning afforded by the Gibbs equation provides insight into the physical nature of the chemical driving force associated with the process. For example, a negative entropic contribution $-T\Delta S$ reveals an increase in the extent of disorder within the system, while a negative enthalpic contribution ΔH reveals a decrease in the average energy of the system (neglecting volume effects) and the concomitant release of heat into the surroundings. The fact that $\Delta G < 0$ for a spontaneous process is then seen as an immediate consequence of the second law of thermodynamics, namely that the sum of the direct entropy production inside the system and the indirect entropy production in the surroundings (due a heat transfer from the system) must be positive. This observation is also immediately visible from Eq. 4.44 which shows, for example, that no spontaneous process decreasing the entropy of a system ($\Delta S < 0$) can release less heat $-Q$ than an amount $-T\Delta S$.

As already stated above, another main point of relevance of the *entropy difference* ΔS resides in its relationship with the *temperature dependence* of the driving force associated with chemical processes carried out under the considered (experimentally common) boundary conditions, resulting from Eq. 4.20 reexpressed in terms of ΔG and assuming $\Delta T = 0$, namely $\partial_T \Delta G = -\Delta S$. In turn, the changes in *volume-expansivity* and *heat capacity* control the dependence of ΔS itself on P and T, respectively, assuming $\Delta P = 0$ and $\Delta T = 0$, respectively (Eqs. 4.23 and 4.24). Analogously, the *volume* change ΔV is connected with the *pressure dependence* of the driving force associated with chemical processes carried out under the considered boundary conditions, resulting from Eq. 4.19 reexpressed in terms of ΔG and assuming $\Delta P = 0$, namely $\partial_P \Delta G = \Delta V$. In turn, the changes in *volume-compressibility* and *volume-expansivity* control the dependence of ΔV itself on P and T, respectively, assuming $\Delta P = 0$ and $\Delta T = 0$, respectively (Eqs. 4.23 and 4.25).

4.2.2 Molar and partial molar variables

For a system (or phase) consisting of a pure substance i (species Σ_i), an extensive quantity Y expressed on a per mole basis defines the corresponding intensive *molar variable* $y[\Sigma_i]$ (or simply y_i) of the pure substance i, with $y = g_i,\ h_i,\ s_i,\ c_{P,i},\ v_i,\ k_{T,i}$ or $a_{P,i}$ for $Y = G,\ H,\ S,\ C_P,\ V,\ K_T$ or A_P, respectively, namely

$$y_i = \left(\frac{\partial Y}{\partial n_i}\right)_{P,T} = \frac{Y}{n_i} \quad . \tag{4.45}$$

Owing to Eq. 4.16, the *molar free energy* of a pure substance i can be identified with its *chemical potential*, and the usual notation μ_i will be used in the following instead of g_i. Considering Eqs. 4.7 and 4.8, the *molar volume-compressibility* $k_{T,i}$ and the *molar volume-expansivity* $a_{P,i}$ are also related to the more usual compressibility $\kappa_{T,i}$ and expansivity (thermal expansion coefficient) $\alpha_{P,i}$, respectively, as

$$\kappa_{T,i} = v_i^{-1} k_{T,i} = -v_i^{-1} \left(\frac{\partial v_i}{\partial P} \right)_{n_i,T} \tag{4.46}$$

and

$$\alpha_{P,i} = v_i^{-1} a_{P,i} = v_i^{-1} \left(\frac{\partial v_i}{\partial T} \right)_{n_i,P} \quad . \tag{4.47}$$

These four quantities are intensive properties of the pure substance at a given state point, but have different units, $k_{T,i}$ (minus volume variation per mole and pressure unit) and $a_{P,i}$ (volume variation per mole and temperature unit) referring to a molar scaling, and $\kappa_{T,i}$ (minus unitless relative volume variation per pressure unit) and $\alpha_{P,i}$ (unitless relative volume variation per temperature unit) referring to a volume scaling.

For a given substance i, the variables y_i are functions of P and T only for a given phase. Usually, this phase is the most stable phase under the given P, T conditions. When a metastable phase is considered instead, or at phase transitions, the phase must be specified explicitly. Note that μ_i and h_i are only defined within an unknown (species-specific) constant, as was the case for G and H (Section 4.2.1), while s_i, $c_{P,i}$, v_i, $k_{T,i}$ and $a_{P,i}$ are absolute quantities, as was the case for S, C_P, V, K_T and A_P. Obviously, for a system consisting of n_i moles of a pure substance i in a single phase, one has

$$Y = n_i y_i \quad . \tag{4.48}$$

The concept of molar variable can be generalized to a system (or phase) consisting of a homogeneous mixture of different substances. In this case, the intensive (intrinsic, see below) *partial molar variable* $y[\Sigma_i]$ (or simply y_i) associated with an extensive quantity Y for the substance i within the mixture is defined as the partial derivative of Y with respect to the number of moles of i, keeping the pressure, temperature and number of moles of all other substances constant, namely

$$y_i = \left(\frac{\partial Y}{\partial n_i} \right)_{\{n_j, j \neq i\}, P, T} \quad . \tag{4.49}$$

Owing to Eq. 4.16, the *partial molar free energy* can be identified with the *chemical potential* μ_i of i in the mixture, and the usual notation μ_i will again be used for this quantity instead of g_i. The definition of partial molar compressibilities $\kappa_{T,i}$ and expansivities (thermal expansion coefficients) $\alpha_{P,i}$ is ambiguous, these quantities being defined by different authors as[1710]

$$\kappa_{T,i} = v_i^{-1} k_{T,i} \qquad \text{and} \qquad \alpha_{P,i} = v_i^{-1} a_{P,i} \quad , \tag{4.50}$$

as

$$\kappa'_{T,i} = |v_i|^{-1} k_{T,i} \qquad \text{and} \qquad \alpha'_{P,i} = |v_i|^{-1} a_{P,i} \quad , \tag{4.51}$$

or as

$$\kappa''_{T,i} = k_{T,i} \qquad \text{and} \qquad \alpha''_{P,i} = a_{P,i} \quad , \tag{4.52}$$

the latter set of equations leading to quantities with different units compared to Eqs. 4.46 and 4.47. To avoid this ambiguity, the quantities defined by Eqs. 4.50 and 4.51 will not be used, and the discussion of volume derivatives in mixtures will exclusively rely on the unambiguous *partial molar volume-compressibilities* $k_{T,i}$ and *partial molar volume-expansivities* $a_{P,i}$, as defined by Eq. 4.49 with $Y = K_T$ or A_P (as in Eq. 4.52, but avoiding the ambiguous notation involving the ambiguous symbols κ and α for these quantities). For consistency, these quantities will also be generally used for pure substances, the conversion to $\kappa_{T,i}$ and $\alpha_{P,i}$ being trivial and unambiguous in this case (Eqs. 4.46 and 4.47).

The variables y_i for a given substance i in a mixture are functions of P, T and system composition (but not system extent) for a specific type of mixture in a given phase. Here also, μ_i and h_i are only defined within an unknown (species-specific) constant, the same that enters into the molar variables of the corresponding pure substance, while s_i, $c_{P,i}$, v_i, $k_{T,i}$ and $a_{P,i}$ are absolute quantities. Owing to the *Euler theorem*, for a system consisting of a homogeneous mixture of N_S substances i with associated numbers of moles n_i, one has

$$Y = \sum_{i=1}^{N_S} n_i y_i \quad . \tag{4.53}$$

The simplest way to define the composition of a mixture relies on the specification of mole fractions. The *mole fraction* x_i of species i in the mixture is defined as

$$x_i = n_{tot}^{-1} n_i \qquad \text{with} \qquad n_{tot} = \sum_{i=1}^{N_S} n_i \quad . \tag{4.54}$$

Note that only $N_S - 1$ mole fractions need to be provided, since mole fractions obey the constraint $\sum_{i=1}^{N_S} x_i = 1$.

A common alternative is to distinguish one species, typically the one dominating the composition of the mixture, referred to as a *solvent* (subscript "*svt*"), and define the composition of the mixture in terms of the concentrations of the other species, referred to as *solutes*, in this solvent. The concentration unit most commonly used in thermodynamics is the *molality* (molal concentration), defined as

$$b_i = m_{svt}^{-1} n_i = M_{svt}^{-1} n_{svt}^{-1} n_i \quad , \tag{4.55}$$

where m_{svt} is the mass of solvent in the mixture, and M_{svt} is the molar mass of the solvent. Here also, only the $N_S - 1$ molalities of the solutes need to be specified, since the molality of the solvent is constant (formally, $b_{svt} = M_{svt}^{-1}$).

For the solutes (indices $i = 1..N_S - 1$), molalities and mole fractions can easily be interconverted as

$$b_i = M_{svt}^{-1} \left(1 - \sum_{j=1}^{N_S - 1} x_j \right)^{-1} x_i \tag{4.56}$$

and

$$x_i = \left(M_{svt}^{-1} + \sum_{j=1}^{N_S - 1} b_j \right)^{-1} b_i \quad . \tag{4.57}$$

Molalities are the concentration units used systematically throughout this book. Other concentration units will only be discussed in Section 4.2.17, in the context of standard-state interconversion formulae.

Since the main concern of this book is ionic solvation thermodynamics in the infinitely dilute regime, *i.e.* in the absence of ion-ion interactions, *homogeneous binary mixtures* involving a single solute in a solvent will mainly be considered. In this case, the notation y (without index i) will refer to the partial molar variables of the solute, while those of the solvent, if required, will be noted y_{svt}. When the solute considered is a *binary salt*, the species in solution are *dissociated ions* and should in principle be considered as two separate solutes. However, due to the electroneutrality constraint (Section 2.1), there is in effect only one composition variable and the dissolved salt may still be considered as a single "effective solute". The same holds for solutions of ternary or higher-order salts if they are produced by dissolution of a salt with a given stoechiometric ratio. Otherwise, the solution containing more than two ion types should be considered as a mixture of multiple salts, with as many composition variables. When the solute considered is a *single ion*, the solution refers to a hypothetical (unphysical) situation violating electroneutrality. In this case, the corresponding single-ion partial molar variables in solution can be defined either as *intrinsic* or as *real* quantities (Section 2.2). Only the intrinsic quantities will be noted $y[\Sigma_i]$ (or simply y_i), while the corresponding real quantities will be noted $\hat{y}[\Sigma_i]$ (or simply $\hat{y}_i$). For the chemical potential, μ_i and $\hat{\mu}_i$ will be used rather than g_i and $\hat{g}_i$ to denote the intrinsic and real chemical potentials, respectively. For overall neutral species, where the distinction between intrinsic and real values is irrelevant, the notation y_i will be retained. Obviously, the partial molar variables of a *dissolved salt* are equal to sums of (either intrinsic or real) partial molar variables for the dissolved ions (at any P, T and composition). Taking the example of an alkali-halide salt dissolved in water (MX_{aq}), one may thus write

$$y[MX_{aq}] = y[M_{aq}^{+}] + y[X_{aq}^{-}] = \hat{y}[M_{aq}^{+}] + \hat{y}[X_{aq}^{-}] \quad . \tag{4.58}$$

Considering the definitions of Eqs. 4.45 and 4.49, it is easily seen that Eqs. 4.19-4.25 are immediately transposable to (partial) molar variables. In the generic notation of Eq. 4.28, this reads for intrinsic quantities

$$y_i = \hat{O}_Y \mu_i \quad , \tag{4.59}$$

a corresponding set of equations holding for real quantities. Eq. 4.59 is of the form of Eq. 4.31 and implies in particular a corresponding form of the Gibbs equation (Eq. 4.32), namely for intrinsic quantities $\mu_i = h_i - Ts_i$, a corresponding equation holding for real quantities.

4.2.3 Ideal behaviors

Ideal behaviors represent *idealizations* concerning the pressure or/and composition dependence of (partial) molar variables (Section 4.2.2) at a specified temperature, that can be used as approximations within a certain domain of validity and become exact in a given limiting situation. They are required for the definition of standard states (Sections 4.2.4 and 4.2.17).

The *ideal-gas* (IG) behavior is defined as the idealized situation of a pure gas where the pressure dependence of the molar variables at constant T is of the form

$$y_i = y_i^{IG} \quad \text{for} \quad y = h \text{ and } c_P \quad , \tag{4.60}$$

$$s_i = s_{i,ref}^{IG} - R\ln(P/P_{ref}) \quad , \tag{4.61}$$

$$\mu_i = \mu_{i,ref}^{IG} + RT\ln(P/P_{ref}) \quad , \tag{4.62}$$

$$v_i = RTP^{-1} \qquad \text{(IG)} \quad , \tag{4.63}$$

$$k_{T,i} = RTP^{-2} \qquad \text{(IG)} \quad , \tag{4.64}$$

and

$$a_{P,i} = RP^{-1} \qquad \text{(IG)} \quad , \tag{4.65}$$

where the parameters y_i^{IG} are constants for a given choice of T, the parameters $y_{i,ref}^{IG}$ relying in addition on the choice of a reference pressure P_{ref}. Note that Eqs. 4.64 and 4.65 imply $\kappa_{T,i} = P^{-1}$ and $\alpha_{P,i} = T^{-1}$ *via* Eqs. 4.46, 4.47 and 4.63.

Ideal-gas behavior is typically a good approximation for pure gases at sufficiently low pressures and high enough temperatures compared to the boiling point. This corresponds to a situation of high molar volume (Eq. 4.63), where the gas molecules are essentially isolated from each other except for occasional collisions. In this case, the *ideal-gas equation of state*

$$PV = nRT \qquad \text{(IG)} \tag{4.66}$$

holds, and can be used, along with the absence of intermolecular interactions, to derive Eqs. 4.60-4.65. The ideal-gas approximation becomes rigorously exact, at any temperature and irrespective of P_{ref}, in the limit of infinitely low pressures. The parameters y_i^{IG} and $y_{i,ref}^{IG}$ are defined as those appropriate in this limiting situation. Note that the parameters $s_{i,ref}^{IG}$ and $\mu_{i,ref}^{IG}$ represent values extrapolated from an infinitesimal pressure to P_{ref} assuming the validity of Eq. 4.66 at any intermediate pressure, *i.e.* they generally differ from the corresponding molar parameters of the real gas at pressure P_{ref}.

The *ideal-mixture* (IM) behavior is defined as the idealized situation where the concentration dependence of the partial molar variables of a species i in a mixture at constant P and T is of the form

$$y_i = y_i^{IM} \quad \text{for} \quad y = h,\ c_P,\ v,\ k_T \text{ and } a_P \quad , \tag{4.67}$$

$$s_i = s_i^{IM} - R \ln x_i \quad , \tag{4.68}$$

and

$$\mu_i = \mu_i^{IM} + RT \ln x_i \quad , \tag{4.69}$$

where the parameters y_i^{IM} are constants for given choices of P and T. The occurrence of the logarithmic terms in Eqs. 4.68 and 4.69 accounts for the entropy increase (free-energy decrease) upon mixing species, *i.e.* for the so-called *cratic entropy*[257]. Note that the presence of the cratic terms in Eqs. 4.68 and 4.69 causes the divergence of s_i and μ_i to $+\infty$ and $-\infty$, respectively, in the limit $x_i \to 0$. However, the values of S and G for the mixture remain finite in this limit, as easily seen from Eq. 4.53, considering that $\lim_{x \to 0} x \ln x = 0$.

Ideal-mixture behavior is typically a good approximation for *gas mixtures* at sufficiently low pressures, high enough temperatures compared to the boiling point of all individual species, and in the absence of strong interactions between the different species. In this situation, the parameters y_i^{IM} are equal to the corresponding molar parameters for the pure gases. The above conditions actually encompass those of the ideal-gas approximation for all individual gases. As a result, in this case, the pressure dependence of the molar variables for the individual gases are given by Eqs. 4.60-4.65, and Eqs. 4.67-4.69 can be rewritten at constant T

$$y_i = y_i^{IG,IM} \quad \text{for} \quad y = h,\ c_P,\ v,\ k_T \text{ and } a_P \quad , \tag{4.70}$$

$$s_i = s_{i,ref}^{IG,IM} - R \ln(P_i/P_{ref}) \quad , \tag{4.71}$$

and

$$\mu_i = \mu_{i,ref}^{IG,IM} + RT \ln(P_i/P_{ref}) \quad , \tag{4.72}$$

where the parameters $y_i^{IG,IM}$ and $y_{i,ref}^{IG,IM}$ are identical to the parameters y_i^{IG} and $y_{i,ref}^{IG}$ involved in Eqs. 4.60-4.65 for the pure gas and for the same choice of T (the parameters $y_{i,ref}^{IG,IM}$ also relying on the choice of a reference pressure P_{ref}), and the *partial pressure* P_i of a gas in a mixture is defined as

$$P_i = x_i P \quad . \tag{4.73}$$

Here also, only $N_S - 1$ partial pressures need to be provided, because the partial pressures obey the constraint $\sum_{i=1}^{N_S} P_i = P$. Note that the latter equation is a constraint imposed by the definition of partial pressures and is always satisfied, even if the gas mixture is not ideal. The so-called *law of partial pressures*[1711] (Dalton's law) has a different content, namely

$$\sum_{i=1}^{N_S} P_i^{pure}(V,T) = P^{mixed}(V,T) \qquad \text{(IM)} \quad , \tag{4.74}$$

where P_i^{pure} is the pressure of n_i moles of gas i alone in a container of volume V at temperature T, and P^{mixed} the pressure of the mixture in the same volume and at the same temperature. The latter equation need not be satisfied for non-ideal mixtures. For gases, the ideal-mixture approximation becomes rigorously exact over the whole composition range and at any temperature in the limit of infinitely low pressures. The parameters $y_i^{IG,IM}$ and $y_{i,ref}^{IG,IM}$ are defined as those appropriate in this limiting situation, and are identical to the corresponding parameters y_i^{IG} and $y_{i,ref}^{IG}$, respectively.

For *solutions*, ideal-mixture behavior is a good approximation for mixtures where one species (solvent) is present in much larger amount compared to all other species (solutes). In this limit, for the solvent, $x_i \rightarrow 1$ and the appropriate parameters y_i^{IM} in Eqs. 4.67-4.69 are equal to the corresponding molar parameters of the pure liquid solvent. For the solutes, $x_i \rightarrow 0$ and the appropriate parameters y_i^{IM} are solvent-dependent effective parameters characterizing the single solute at infinite dilution in the given solvent. These parameters refer to a hypothetical extrapolation to $x_i = 1$ and are not identical to the corresponding molar parameters for the solute in a pure (solid, liquid or gas) state. The situation defined by the ideal-mixture behavior in the context of a solution can be referred to as that of a *"dissolved ideal gas"*, *i.e.* the solute molecules are isolated from each other and surrounded by a pure solvent environment. For solutions, the ideal-mixture approximation becomes rigorously exact at any temperature where the solvent remains liquid in the limit of infinitely low solute concentrations. For solvent as well as solutes, the parameters y_i^{IM} are defined as those appropriate in this limiting situation.

The *ideal-solution* (IS) behavior is a slight variant of the ideal-mixture behavior in the context of solutions. In this case, the dominant species (solvent) is singled out (subscript "*svt*") and Eqs. 4.67-4.69 are replaced for this species only by

$$y_{svt} = y_{svt}^{IS} \quad \text{for} \quad y = \mu,\ h,\ s,\ c_P,\ v,\ k_T \text{ and } a_P \quad , \tag{4.75}$$

with parameters y_{svt}^{IS} equal to the corresponding molar parameters of the pure liquid solvent at the given temperature T. The corresponding equations for the solute are still Eqs. 4.67-4.69 (with the superscripts "IM" replaced by "IS"). In other words, the ideal-solution approximation, unlike the ideal-mixture one, neglects the cratic entropy contribution of the solvent. Obviously, the ideal-solution approximation is nearly equal to the ideal-mixture one for dilute solutions ($x_{svt} \rightarrow 1$) and becomes identical to the latter, as well as rigorously exact at any temperature where the solvent remains liquid, in the limit of infinitely low solute concentrations. For solvent as well as solutes, the parameters y_i^{IS} are identical to the corresponding parameters y_i^{IM}.

The *ideal-molal-solution* (IB) behavior is a variant of the ideal-solution behavior, where Eq. 4.75 (with the superscript "IS" replaced by "IB") is used for the solvent, along with Eq. 4.67 (with the superscript "IM" replaced by "IB") for the solutes, namely

$$y_i = y_i^{IB} \quad \text{for} \quad y = h,\ c_P,\ v,\ k_T \text{ and } a_P \quad , \tag{4.76}$$

but with Eqs. 4.68 and 4.69 replaced for these species by

$$s_i = s_{i,ref}^{IB} - R \ln(b_i/b_{ref}) \tag{4.77}$$

and

$$\mu_i = \mu_{i,ref}^{IB} + RT \ln(b_i/b_{ref}) \quad , \tag{4.78}$$

where the solute parameters $s_{i,ref}^{IB}$ and $\mu_{i,ref}^{IB}$ rely on the choice of a reference molality b_{ref}. Owing to Eqs. 4.56 and 4.57, the ideal-molal-solution approximation is nearly equal to the ideal-solution one for dilute solutions (x_i, $b_i \to 0$ for all solutes) and becomes identical to the latter, as well as rigorously exact at any temperature where the solvent remains liquid, in the limit of infinitely low solute concentrations. For solvent as well as solutes, the parameters y_i^{IB} are defined as those appropriate in this limiting situation, and are identical to the corresponding parameters y_i^{IS} (and y_i^{IM}), except for $s_{i,ref}^{IB}$ and $\mu_{i,ref}^{IB}$ in the case of the solutes. For these parameters, the relationship reads

$$s_{i,ref}^{IB} = s_i^{IS} - R \ln(M_{svt} b_{ref}) \tag{4.79}$$

and

$$\mu_{i,ref}^{IB} = \mu_i^{IS} + RT \ln(M_{svt} b_{ref}) \quad . \tag{4.80}$$

In the case of water at P° and T^-, the second terms in Eqs. 4.79 and 4.80 evaluate to 33.39 J·mol^{-1}·K^{-1} and -9.96 kJ·mol^{-1}, respectively (based on data from Table 1.1). Note that the parameters $s_{i,ref}^{IB}$ and $\mu_{i,ref}^{IB}$ represent values extrapolated from infinitesimal molality to b_{ref}, assuming the validity of the ideal-molal-solution approximation at any intermediate molality, *i.e.* they generally differ from the corresponding partial molar parameters of the given solute in a real solution at molality b_{ref}.

4.2.4 Standard states (this book)

The purpose of defining *standard states* is to permit the tabulation of thermodynamic variables (Section 4.2.2) characterizing substances, either pure or within mixtures, under *precisely defined conditions.* Unlike the variables themselves, which are real functions of the intensive state variables defining the pressure, temperature and composition of a system and, therefore, characterized by an infinite number of values, the limited set of standard values can be measured in a finite amount of time and stored in a finite amount of space. The resulting tables can be subsequently used to evaluate (partial) molar thermodynamic variables corresponding to different experimental conditions, by means of thermodynamic extrapolation formulae, *e.g.* assuming ideal behaviors as defined in Section 4.2.3, or using more accurate expressions. The tabulated standard quantities are also often used as such to characterize and compare the thermodynamics of specific processes assumed to occur under the idealized standard-state conditions. Alternatively or additionally, the functional dependence of a (partial) molar thermodynamic variable with respect to pressure, temperature or/and composition can also be approximated by an analytical function with fitted parameters, in which case the tabulation contains the optimized parameters.

Unfortunately, there is considerable room for *ambiguity* in the definition of standard states, due to the existence of *alternative conventions* and the *absent or incomplete specification* in many experimental or theoretical studies of the convention employed, including those pertaining to ionic solvation properties. This results in a significant factor of "uncertainty" in practice[259,262,278,457,1571,1712–1729]. To avoid this problem, the present section provides a precise definition of the standard-state conventions employed throughout this book. The connection with common alternative conventions will be discussed in Section 4.2.17.

The *reference pressure* P°, *reference molality* b° and *reference temperature* T^- are set to the values $P^\circ = 1$ bar, $b^\circ = 1$ mol·kg^{-1} (1 molal) and $T^- = 298.15$ K (25°C), respectively. The following standard states are then applied for all quantities marked with a "∘" superscript, considering a given temperature T: (*i*) the *gas-phase* (subscript "*g*") standard state is a virtual state referring to one mole of the pure substance at temperature T and infinitesimal pressure, with properties extrapolated to pressure P° assuming ideal-gas behavior; (*ii*) the *liquid* (subscript "ℓ") standard state is a real state referring to one mole of the pure substance in the liquid phase at temperature T and pressure P° (in the presence of polymorphic forms, *e.g.* liquid crystals, the specific isomorph must also be indicated); (*iii*) the *solid* (subscript "*s*") standard state is a real state referring to one mole of the pure substance in the solid phase at temperature T and pressure P° (in the presence of polymorphic forms, *e.g.* crystalline solids, the specific isomorph must also be indicated); (*iv*) the *dissolved* (subscript "*d*", or "*aq*" for aqueous solutions) or *solute* standard state for a solute in a solution is a virtual state referring to one mole of the pure substance dissolved in the pure solvent at temperature T, pressure P°, and at infinitesimal concentration, with properties extrapolated to the molal concentration b° assuming ideal-molal-solution behavior; (*v*) the *solvent* (subscript "*svt*", or "*wat*" for water) standard state for the solvent in a solution is identical to the *liquid* standard state as defined above; (*vi*) the *element* (subscript "*el*") standard state is the one appropriate for one mole of the element (atom) within the most stable compound of this element in its most stable phase (gas, solid or liquid, most stable isomorph) at temperature T and pressure P°, according to the standard state for the given phase as defined above; (*vii*) the *electron* (e_g^-) standard state is a virtual state referring to one mole of electrons in the gas phase at temperature T and at a prescribed reference electric potential, according to the gas-phase standard state as defined above.

No specific definition is included for the standard state of an *ion*, which follows directly from the above definitions, depending on the phase in which the ion is considered. The only peculiarity is that the element standard state I_{el}^z of an anion ($z < 0$) is viewed as the "sum" of the neutral element standard state I_{el} plus $-z$ times the electron standard state e_g^-, while the element standard state I_{el}^z of a cation ($z > 0$) is viewed as the corresponding "difference" of I_{el} minus z times e_g^- (see Section 4.2.10 for more details).

At this point, the standard-state definition is nearly complete and unambiguous except for four points: (*i*) a convention setting the zero point of the absolute molar entropies of the elements based on the third law of thermodynamics; (*ii*) a clear specification of what is meant by extrapolations assuming ideal-gas or ideal-molal-solution behaviors; (*iii*) a choice of the reference potential for the electron standard state; (*iv*) the discussion of an ambiguity affecting partial molar properties corresponding to derivative thermodynamic parameters in the dissolved (solute) standard state, namely the existence of distinct variants for the solute standard state in terms of these parameters. These issues are considered in turn below.

According to the third law of thermodynamics (Section 4.2.1), the simplest convention to define a consistent system of absolute molar entropies is to set the *standard molar entropies of the elements to zero at the absolute zero temperature, i.e.* to impose $s^\circ[\Sigma_{el}] = 0$ at $T = 0$ K for perfect crystals of the most stable compound and isomorph, for all 118 currently known elements[1689]. Owing to the third law, this choice implies that the *standard molar entropies of all solid compounds are zero at the absolute zero temperature, i.e.* $s^\circ[\Sigma_s] = 0$ at $T = 0$ K for perfect crystals of any compound in any isomorph. This choice also turns out to be the one consistent with quantum statistical mechanics. A similar definition is applied for the standard molar entropy of the electron, namely $s^\circ[e_g^-] = 0$ at $T = 0$ K, except that it still refers to a gas-phase electron. It can be shown based on quantum-statistical mechanical arguments that the third law of thermodynamics also extends to the ideal-gas standard state. More precisely, setting $s^\circ[\Sigma_{el}] = 0$ at $T = 0$ K also implies that the *standard gas-phase molar entropies of all compounds are zero at the absolute zero temperature, i.e.* $s^\circ[\Sigma_g] = 0$ at $T = 0$ K for all compounds. Since the same property applies by definition to the electron (see above), it follows that the property $s^\circ[\Sigma_g]$ at $T = 0$ K also applies to this species as well as to any ion. Note that the above statements only strictly hold in the absence of nuclear processes, implying conservation of the nuclear spin. In the presence of such processes, the anchoring of the absolute entropy scale should be modified so as to account for the distinct nuclear spin degeneracies of the different elements.

The concept of *extrapolations* assuming *ideal-gas* or *ideal-molal-solution* behavior is based on ideas developed in Section 4.2.3. The molar properties of the gas-phase standard state correspond to the coefficients y^{IG} and y^{IG}_{ref} involved in Eqs. 4.60-4.65 with $P_{ref} = P^\circ$, and are thus defined by

$$
\begin{aligned}
u^\circ(T) &= \lim_{P\to 0} u(P,T) \quad , && (4.81)\\
h^\circ(T) &= u^\circ(T) + RT \quad , && (4.82)\\
s^\circ(T) &= \lim_{P\to 0} [s(P,T) + R\ln(P/P^\circ)] \quad , && (4.83)\\
\mu^\circ(T) &= h^\circ(T) - Ts^\circ(T) \quad , && (4.84)\\
c_P^\circ(T) &= \partial_T h^\circ(T) = \partial_T u^\circ(T) + R \quad , && (4.85)\\
v^\circ(T) &= RT(P^\circ)^{-1} \quad , && (4.86)\\
k_T^\circ(T) &= RT(P^\circ)^{-2} \quad , && (4.87)
\end{aligned}
$$

and

$$a_P^\circ(T) = R(P^\circ)^{-1} \quad . \tag{4.88}$$

These extrapolations concern a system consisting of a pure substance Σ at pressure P and temperature T. The resulting standard-state molar variables y° are functions of the temperature T only. For a given temperature, they only require the specification of Σ along with the state under consideration, and can thus be noted $y^\circ[\Sigma_g]$. The partial molar properties of the dissolved (solute) standard state correspond to the coefficients y^{IB} and y^{IB}_{ref} in Eqs. 4.76-4.78 with $b_{ref} = b^\circ$ and $P = P^\circ$, and are thus defined as

$$
\begin{aligned}
u^\circ(T) &= \lim_{b\to 0} u(b,P^\circ,T) \quad , && (4.89)\\
h^\circ(T) &= u^\circ(T) + P^\circ v^\circ(T) \quad , && (4.90)\\
s^\circ(T) &= \lim_{b\to 0} [s(b,P^\circ,T) + R\ln(b/b^\circ)] \quad , && (4.91)\\
\mu^\circ(T) &= h^\circ(T) - Ts^\circ(T) \quad , && (4.92)\\
c_P^\circ(T) &= \partial_T h^\circ(T) = \partial_T u^\circ(T) + P^\circ a_P^\circ(T) \quad , && (4.93)\\
v^\circ(T) &= \lim_{b\to 0} v(b,P^\circ,T) \quad , && (4.94)\\
k_T^\circ(T) &= \lim_{b\to 0} k_T(b,P^\circ,T) \quad , && (4.95)
\end{aligned}
$$

and

$$a_P^\circ(T) = \lim_{b\to 0} a_P(b,P^\circ,T) \quad . \tag{4.96}$$

These extrapolations concern a system consisting of the single solute Σ at molality b in a given solvent, at pressure P° and temperature T. The resulting standard-state partial molar variables y° are functions of the temperature T only. For a given temperature and choice of solvent, they only require the specification of Σ along with the state under consideration, and can thus be written $y^\circ[\Sigma_g]$. From the point of view of the solvent, an ideal molal solution is characterized by the absence of perturbation of the solvent-solvent interactions by the solute and by the independence of the solvent partial molar entropy on the solute molality. Thus, the standard partial molar properties of the solvent in a solution are simply equal to the corresponding standard molar properties of the pure liquid solvent. As discussed in Section 4.2.3, the ideal-molal-solution behavior (that distinguishes between solute and solvent and is based on molality) is an approximation of the ideal-solution behavior (that also distinguishes between solute and solvent, but is based on mole fractions), which is itself an approximation of the ideal-mixture behavior (that treats all species on an equal basis and is also based on mole fractions). In the context of dilute solutions, the three types of idealizations describe thermodynamic properties with similar accuracies, while they typically fail for concentrated solutions. The three approximations lead to identical values for the standard partial molar parameters of the solvent. The two former approximations also lead to identical standard values for the solute. However, these values differ from those derived based on extrapolations assuming ideal-mixing behavior (Eqs. 4.79 and 4.80).

The importance of including a clear definition for the *standard state of the electron* in any system meant to encompass ion thermodynamics, a practice that is not very widespread, has been underlined previously[278, 1719, 1720, 1723, 1727, 1728] and also appears essential to the authors of this book. As will be discussed later (Sections 4.2.12, 4.2.13 and 4.2.17), the choice made here to define this standard state as an ideal electron gas at temperature T, a definition that will be referred to as the *warm-electron convention*, is common[294, 1723], but not unanimously accepted. The other common convention refers to an ideal electron gas at the absolute zero temperature[1723,1730], a definition that will be referred to as the *cold-electron convention*. However, even within the warm-electron convention, the standard-state definition remains incomplete because it lacks the choice of a *reference electric potential* for the standard-state ideal electron gas at temperature T. This choice has no relevance *per se* within ionic (even single-ion) thermodynamics. However, it becomes crucial in the context of electrochemistry, namely in the definition of *absolute electrode potentials*. For this reason, the issue will be discussed later in Section 4.3.6, where definitions for these quantities are introduced.

Finally, the above standard-state convention still leaves room for some *ambiguity* in the definition of partial molar properties corresponding to *derivative thermodynamic parameters* in the *dissolved (solute) standard state*. This ambiguity resides in the possible inclusion or omission of terms related to the pressure or/and temperature derivatives of the solvent density into the standard derivative thermodynamic properties of the solute. This issue, which is not widely recognized in thermodynamics (and even possibly discussed for the first time in this book), is addressed in Section 4.2.5 (see also Section 4.2.16). Since it is relatively intricate and possibly controversial, no corresponding choice is included into the present standard-state definition. Instead, two specific choices are introduced and referred to as the *standard variant* (no additional superscript) and the *density-corrected variant* (additional "*" superscript).

At this point, the standard-state definition is complete and unambiguous. It is important to stress, however, that this definition is still bound to the *choice of a specific temperature* T. The standard states for the solid and liquid phases, the latter including the solvent in a solution, are *real states*. This implies in particular that they are only definable within a certain range of temperatures T, where the specific phase is thermodynamically stable at pressure P°. In contrast, the standard states for the gas phase and for the solute in a solution are *virtual states*, *i.e.* without experimental counterpart. A gas at pressure P° or a solution at concentration b° is never experimentally ideal. The gas-phase standard state is defined at all temperatures T for which the molecule of the given substance is stable, because any substance becomes gaseous at infinitesimal pressure. The dissolved standard state is only definable within a certain range of temperatures T where the liquid phase of the solvent is thermodynamically stable at pressure P°, but is then definable for any substance, even if its real solubility is below b°. The standard state of an element is defined at all temperatures for which the atom of the element is stable, because an element always exists in some most stable form. Note that the standard state of an element refers to one mole of the element (atom) within its most stable compound and phase at P° and T. Even though this compound may be a polyatomic molecule, the definition refers to one mole of this compound *in terms of atoms* and not to one mole of molecules. For example, the standard state of the element bromine (Br_{el}) at room temperature refers to one mole of bromine atoms within the bromine diatomic liquid in its standard state, *i.e.* to half a mole of bromine liquid in its standard state ($1/2\mathrm{Br}_{2,\ell}$). To avoid any ambiguities, the phase of the substance should always be specified explicitly when referring to a standard quantity, along with the specific polymorphic form (solid and, possibly, liquid phase) whenever relevant. This is essential, because a given species can generally be associated to multiple possible standard states. Taking again the example of bromine, a standard (partial) molar variable $y^\circ(T)$ at $T = T^- = 298.15$ K can be associated to *e.g.* Br_{el}, $\mathrm{Br}_{2,\ell}$, $\mathrm{Br}_{2,g}$, $\mathrm{Br}_{2,aq}$ or Br_g. These correspond, however, to different states at P° and T^-, referring to a bromine atom within real liquid bromine ($1/2\mathrm{Br}_{2,\ell}$), a bromine molecule within real liquid bromine (*i.e.* a twice larger quantity), a bromine molecule within ideal gas-phase bromine, a bromine molecule within an ideal aqueous bromine solution, and a bromine atom within an ideal gas of bromine atoms, respectively. For $T > 331.91$ K (boiling point of bromine at P°), $y^\circ(T)$ can no longer refer to $\mathrm{Br}_{2,\ell}$, while Br_{el} now stands for $1/2\mathrm{Br}_{2,g}$. For $T > 373.15$ K (boiling point of water at P°) or $T < 273.15$ K (melting point of water at P°), $\mathrm{Br}_{2,aq}$ also becomes undefined. Finally, for $T < 265.95$ K (melting point of bromine at P°), $y^\circ(T)$ can no longer refer to $\mathrm{Br}_{2,\ell}$, but can refer to $\mathrm{Br}_{2,s}$, *i.e.* a bromine molecule within real solid bromine, while Br_{el} now stands for $1/2\mathrm{Br}_{2,s}$.

As illustrated in the above discussion, the standard state of a species in a given phase is associated with a selected temperature T. For elements, the standard compound and phase also depend on T. In the special case where the temperature T is chosen equal to the *reference temperature* T^-, the corresponding standard-state quantity will be marked with a "$\ominus$" superscript rather than a "$\circ$" superscript. Finally the superscript "$-$" may be used for (non-standard) quantities evaluated at the reference temperature T^-.

As was the case for *single-ion* partial molar variables in solution (Section 4.2.2), standard single-ion partial molar variables in the dissolved state can be defined either as *intrinsic* or as *real* quantities (Section 2.2). Only the intrinsic quantities will be noted $y^\circ[\Sigma_{i,d}]$ and $y^\ominus[\Sigma_{i,d}]$ (or simply y_i° and $y_i^\ominus$), while the corresponding real quantities will be noted $\hat{y}^\circ[\Sigma_{i,d}]$ and $\hat{y}^\ominus[\Sigma_{i,d}]$ (or simply $\hat{y}_i^\circ$ and $\hat{y}_i^\ominus$). For the chemical potential, μ_i° and $\mu_i^\ominus$ or $\hat{\mu}_i^\circ$ and $\hat{\mu}_i^\ominus$ will be used rather than g_i°, $g_i^\ominus$, $\hat{g}_i^\circ$ and $\hat{g}_i^\ominus$ to denote the standard intrinsic and real chemical potentials, respectively. For overall neutral species, where the distinction between intrinsic and real variables is irrelevant, the notations μ_i° and $\mu_i^\ominus$ will be retained. Obviously, the standard partial molar variables of a *dissolved salt* are equal to the sums of (either intrinsic or real) standard partial molar variables for the dissolved ions. Taking the example of an alkali-halide salt dissolved in water (MX_{aq}), one may thus write (for any temperature T)

$$y^\circ[MX_{aq}] = y^\circ[M_{aq}^+] + y^\circ[X_{aq}^-] = \hat{y}^\circ[M_{aq}^+] + \hat{y}^\circ[X_{aq}^-] \tag{4.97}$$

and

$$y^\ominus[MX_{aq}] = y^\ominus[M_{aq}^+] + y^\ominus[X_{aq}^-] = \hat{y}^\ominus[M_{aq}^+] + \hat{y}^\ominus[X_{aq}^-] \quad . \tag{4.98}$$

Note that both the real and intrinsic single-ion quantities $\hat{y}^\circ$ and y° (or $\hat{y}^\ominus$ and $y^\ominus$ at T^-) are referred to as standard. These are just two different types of standard variables. As briefly mentioned above and discussed in detail in Section 4.2.5, there are also distinct variants for the solute standard state. The above symbols refer by default to the standard variant. Corresponding quantities in the density-corrected variant will be noted $\hat{y}^{\circ *}$ and $y^{\circ *}$ (or $\hat{y}^{\ominus *}$ and $y^{\circ *}$ at T^-).

Eq. 4.59, stated for (partial) molar variables, is in principle immediately generalizable to standard (partial) molar variables, *i.e.* in the generic notation of Eq. 4.28 and for intrinsic quantities in the standard variant

$$y_i^\ominus = \hat{O}_Y \mu_i^\ominus \quad , \tag{4.99}$$

a corresponding set of equations holding for real quantities, where it is recalled that the *derivative symbols* ∂_P and ∂_T applied to a *standard* (thus formally pressure- and temperature-independent) quantity imply derivation with respect to P° and T^- considered as variables, followed by evaluation at the reference values of these variables (Chapter 1). As will be discussed in Section 4.2.5, however, Eq. 4.99 is not valid for intrinsic (or real) quantities corresponding to derivative thermodynamic parameters in the dissolved standard state when applying the density-corrected variant. In this specific case, the dependence of $\mu_i^{\ominus}$ (or $\hat{\mu}_i^{\ominus}$) on the solvent density $\rho_{svt}^{\ominus}$ is excluded from the differentiation, resulting in the replacement of Eq. 4.99 by a slightly different expression (Eq. 4.104). Eq. 4.99 is of the form of Eq. 4.31 and implies in particular a corresponding variant of the Gibbs equation (Eq. 4.32), namely for intrinsic quantities in the standard variant

$$\mu_i^{\ominus} = h_i^{\ominus} - T^{-} s_i^{\ominus} \quad , \tag{4.100}$$

corresponding equations holding for real quantities in the standard variant, as well as intrinsic and real quantities in the density-corrected variant (including in the dissolved standard state for this variant, even though Eq. 4.99 itself does not hold in this case).

The standard-state convention defined above will be referred to as the *bbme*$_T$ convention. The first letter (*b* for bar) refers to the choice of a reference pressure $P^{\circ} = 1$ bar for the solid, liquid (including solvent) and dissolved standard states. The second letter (*b* for bar) refers to the choice of a reference pressure $P^{\circ} = 1$ bar, along with an ideal-gas extrapolation, for the gas-phase (including electron) standard states. The third letter (*m* for molal) refers to the choice of a reference molality $b^{\circ} = 1$ mol·kg^{-1}, along with an ideal-molal-solution extrapolation, for the dissolved (solute) standard state. Finally, the fourth letter and subscript e_T refers the choice of the gas-phase standard state at temperature T for the electron, *i.e.* of a warm-electron convention. The choice of the corresponding reference electric potential remains unspecified for now, and will be discussed in Section 4.3.6. The standard encompasses two possible solute standard-state variants as discussed in Section 4.2.5, quantities referring to the density-corrected variant being identified by an additional "$*$" superscript. Standard single-ion properties may be real or intrinsic (Section 2.2), quantities of the former type being identified by a "$\wedge$" overscript. Finally, the convention for absolute entropies is not explicitly indicated in the *bbme*$_T$ symbol because the adopted choice is unanimously accepted in chemical thermodynamics.

The *bbme*$_T$ convention appears to the authors as the most reasonable one, because: (*i*) it is (arguably) the most widely used in physical chemistry, with the possible exception of electrochemistry, where a reference pressure of 1 atm and a reference concentration of 1 mol·dm^{-3} (molarity) are also commonly employed; (*ii*) the bar is a multiple of the International System (SI) unit for pressure, the Pascal (1 Pa = 1 kg·m^{-1}·s^{-2}, 1 bar = 10^5 Pa), while 1 bar is also very close to the average atmospheric pressure at sea level (1 atm = 1.013 bar); (*iii*) the molality solely characterizes the composition of a solution (*i.e.* the number of moles of all species in solution, irrespective of P and T, see Eqs. 4.56 and 4.57), while it is also numerically very close to the commonly used molarity (1 M = 1 mol·dm^{-3}) for dilute aqueous solutions (in contrast to the molality, however, the molarity of a solution depends not only on its composition, but also on P and T through the solution density); (*iv*) the choice of the gas-phase standard state at temperature T for the electron (warm-electron convention) treats this species on a similar footing as the elements, the sole distinction being the phase involved in the definition of the absolute entropy at $T = 0$ K. The $bbme_T$ convention is essentially the one recommended by IUPAC[1731], except for the electron standard state, which appears to refer[1718] to an electron gas at the absolute zero temperature (cold-electron convention) rather than at temperature T. Unfortunately, alternative standard-state conventions are also very commonly used. These alternative conventions will be discussed in Section 4.2.17 and interconversion formulae for standard quantities will be provided.

To conclude this discussion, a clear statement should be made concerning the standard-state convention used in this book: *only standard quantities defined according to the $bbme_T$ convention are indicated with a "$\circ$" superscript (or a "$\ominus$" superscript when $T = T^-$) in the present text, figures and tables.*

4.2.5 Solute standard-state variants

The standard-state definitions provided in Section 4.2.4 hide an unexpected *ambiguity* concerning partial molar properties corresponding to *derivative thermodynamic parameters* in the *dissolved (solute) standard state.* As already mentioned in Section 4.2.4, Eq. 4.59 is in principle immediately generalizable to standard (partial) molar variables, *i.e.* in the generic notation of Eq. 4.28 and for intrinsic quantities (Eq. 4.99),

$$y_i^\ominus = \hat{O}_Y \mu_i^\ominus \quad , \tag{4.101}$$

a corresponding set of equations holding for real quantities, where it is recalled that ∂_P and ∂_T applied to a standard quantity imply derivation with respect to P° and T^- considered as variables, followed by evaluation at the reference values of these variables.

For the dissolved standard state, however, the application of Eq. 4.101 leads to somewhat *counter-intuitive results.* These are most easily seen by considering the hypothetical case of a non-interacting and volume-free single-particle solute. The standard-state chemical potential of the particle in solution can be inferred from statistical mechanics (Section 4.2.12) and is given, within the particle self-energy, by

$$\mu_i^\ominus = RT^- \ln \left[c(RT^-)^{-3/2} b^\circ \rho_{svt}^\ominus \right] \quad , \tag{4.102}$$

where c is a constant related to the particle mass and $(b^\circ \rho_{svt}^\ominus)^{-1}$ represents the effective molar volume accessible to the particle in solution at the standard-state reference molality b°. Application of Eq. 4.101 to Eq. 4.102 leads in particular to the results

$$\begin{aligned} s_i^{\ominus} &= R\left\{\frac{3}{2} - \ln\left[c(RT^-)^{-3/2}b^\circ\rho_{svt}^{\ominus}\right]\right\} - RT^-\partial_T \ln\rho_{svt}^{\ominus} \\ h_i^{\ominus} &= \frac{3}{2}RT^- - R(T^-)^2\partial_T \ln\rho_{svt}^{\ominus} \\ v_i^{\ominus} &= RT^-\partial_P \ln\rho_{svt}^{\ominus} \quad . \end{aligned} \tag{4.103}$$

Note that $\partial_T \ln\rho_{svt}^{\ominus} = \rho^{-1}\partial_T\rho_{svt}^{\ominus}$ and $\partial_P \ln\rho_{svt}^{\ominus} = \rho^{-1}\partial_P\rho_{svt}^{\ominus}$, so that the units of these derivatives are of inverse pressure and inverse temperature, respectively, and independent of the units selected for $\rho_{svt}^{\ominus}$. These equations suggest that the particle has a partial molar enthalpy differing from its mere thermal (kinetic) energy, although it presents no interaction with itself and with the solvent, and that it has a non-vanishing partial molar volume, although it has been assumed to be volume-free. Similarly, the partial molar entropy differs from that expected for a free particle within an effective molar volume $(b^\circ\rho_{svt}^{\ominus})^{-1}$. The unexpected entropy, enthalpy and volume contributions, namely the last terms in the above equations for s and h, and the single term in the equation for v, are related to the expansivity and compressibility of the pure solvent, two properties that appear to be entirely irrelevant in terms of solvation under isobaric and isothermal conditions. Their occurrence is even more surprising considering that the particle deos not "feel" the solvent, and should thus be insensitive to its nature beyond the role of $\rho_{svt}^{\ominus}$ in determining the effective molar volume associated with the dissolved standard state. The reason for the appearance of these terms is the following. In its standard state, the solute represents a "dissolved ideal gas". However, in contrast to the gas-phase standard state, the effective pressure P_{eff} of this ideal gas is not P°. Rather, it is dictated by the density of the solvent as $P_{eff}^{\ominus} = RT^-b^\circ\rho_{svt}^{\ominus}$. For this reason, the expected ideal-gas results in terms of derivative properties are increased by contributions related to the pressure and temperature derivatives of $P_{eff}^{\ominus}$ *via* $\rho_{svt}^{\ominus}$, leading to the unexpected terms in Eq. 4.103

These counter-intuitive features can be alleviated by introducing a slightly different definition for the dissolved (solute) standard state, referred to here as the *density-corrected variant*, as opposed to a standard strictly obeying Eq. 4.101, that will be referred to as the *standard variant*. Standard-state properties corresponding to the former variant will be marked with an additional "$*$" superscript. The distinction only affects derivative thermodynamic properties. Standard free energies and chemical potentials are identical in the two variants. It also only affects standard properties concerning systems and processes involving species in solution. The two variants are also equivalent otherwise.

In the density-corrected variant, Eq. 4.101 is replaced by the alternative expression, for the dissolved standard state only,

$$y_i^{\ominus *} - y_{std}^{\ominus *} = \hat{O}_Y\left(\mu_i^{\ominus} - \mu_{std}^{\ominus}\right) \quad , \tag{4.104}$$

corresponding sets of equations holding for the real quantities. In both the intrinsic and the real cases, the quantities $y_{std}^{\ominus *}$ in this equation are defined as

$$\mu_{std}^{\ominus *} = -T^{-} s_{std}^{\ominus *} \tag{4.105}$$

with

$$s_{std}^{\ominus *} = -R \ln \left(v_{ref} b^{\circ} \rho_{svt}^{\ominus} \right) \quad , \tag{4.106}$$

and

$$y_{std}^{\ominus *} = 0 \quad \text{for} \quad y = h,\ c_P,\ v,\ k_T \text{ and } a_P \quad . \tag{4.107}$$

Eq. 4.101 is still applied in this variant just as in the standard variant, to all other standard states, namely solid, liquid, gas and solvent. Note that the arbitrary choice of a reference volume v_{ref}, which has been introduced in Eq. 4.106 for unit consistency, has no influence on the results of Eq. 4.104.

Application of Eq. 4.104 to Eq. 4.102 in the case of the hypothetical non-interacting and volume-free single-particle solute (see above) leads in particular to the results

$$\begin{aligned} \mu_i^{\ominus *} &= \mu_i^{\ominus} \\ s_i^{\ominus *} &= R \left\{ \frac{3}{2} - \ln \left[c (RT^{-})^{-3/2} b^{\circ} \rho_{svt}^{\ominus} \right] \right\} \\ h_i^{\ominus *} &= \frac{3}{2} RT^{-} \\ v_i^{\ominus *} &= 0 \quad . \end{aligned} \tag{4.108}$$

The above expressions appear much more intuitive compared to those of Eq. 4.103 in this context. The chemical potential is the same as in the standard variant, but the partial molar enthalpy corresponds to that expected from the mere thermal (kinetic) energy of a non-interacting particle, the partial molar entropy to that expected for a free particle within an effective molar volume $(b^{\circ} \rho_{svt}^{\ominus})^{-1}$, and the partial molar volume to that expected for a volume-free particle.

Both variants are internally consistent and thus in principle acceptable. They also both satisfy the Gibbs equation, *i.e.*

$$\mu_i^{\ominus} = h_i^{\ominus} - T^{-} s_i^{\ominus} = \mu_i^{\ominus *} = h_i^{\ominus *} - T^{-} s_i^{\ominus *} \quad , \tag{4.109}$$

similar equations holding in terms of real quantities. The standard variant is less pleasant in terms of interpretation (see, *e.g.* Eq. 4.103 *vs.* Eq. 4.108). However, the density-corrected variant may also appear confusing in view of the rather unusual form of Eq. 4.104, as opposed to Eq. 4.101, for the dissolved standard state. Interconversion formulae between the two variants are summarized in Table 4.1 in the form of differences

$$\Delta_c y^{\ominus *} = y_i^{\ominus *} - y_i^{\ominus} = y_{std}^{\ominus *} - \hat{O}_Y \mu_{std}^{\ominus *} \quad , \tag{4.110}$$

Table 4.1 *Interconversion formulae between the standard and the density-corrected standard-state variants.* The correction terms refer to a standard partial molar thermodynamic variable y_i of an arbitrary species i in the dissolved state. The difference $\Delta_c y^{\ominus *} = y_i^{\ominus *} - y_i^{\ominus}$ between the density-corrected ($y_i^{\ominus *}$) and the standard ($y_i^{\ominus}$) variant is evaluated for $y = \mu$, h, s, c_P, v, k_T and a_P and water as a solvent (Eq. 4.110) and can be viewed as an additive correction to convert standard partial molar properties of solutes from the standard to the density-corrected variant. The standard water density $\rho_{svt}^{\ominus}$ and the corresponding pressure and temperature derivatives are taken from Table 1.1. The reference pressure, temperature and molality are set to $P^{\circ} = 1$ bar, $T^{-} = 298.15$ K and $b^{\circ} = 1$ mol·kg^{-1}. Note that the value of $-\Delta_c c_P^{\ominus *}$ for water reported in Ref.[1412] is 8.40 instead of 8.39 J·mol^{-1}·K^{-1}, due to the use of a solvent density rounded to 997.0 kg·m^{-3} in this reference.

y	$\Delta_c y^{\ominus *}$	water
μ	0	0
h	$+R(T^{-})^2 \partial_T \ln \rho_{svt}^{\ominus}$	-0.19 kJ·mol^{-1}
s	$+RT^{-} \partial_T \ln \rho_{svt}^{\ominus}$	-0.64 J·mol^{-1}·K^{-1}
c_P	$+RT^{-}[2\partial_T \ln \rho_{svt}^{\ominus} + T^{-} \partial_T^2 \ln \rho_{svt}^{\ominus}]$	-8.39 J·mol^{-1}·K^{-1}
v	$-RT^{-} \partial_P \ln \rho_{svt}^{\ominus}$	-1.12·10^{-6} m^3·mol^{-1}
k_T	$RT^{-} \partial_P^2 \ln \rho_{svt}^{\ominus}$	-2.85·10^{-10} m^3·mol^{-1}·bar^{-1}
a_P	$-[R\partial_P \ln \rho_{svt}^{\ominus} + RT^{-} \partial_{P,T}^2 \ln \rho_{svt}^{\ominus}]$	-0.0994·10^{-8} m^3·mol^{-1}·K^{-1}

that can be used as additive corrections to convert standard partial molar properties of solutes from the standard to the density-corrected variant, along with numerical values appropriate for water (based on data from Table 1.1).

The *existence of the two above variants* for the dissolved standard state is *not widely recognized in thermodynamics*, and even possibly discussed for the first time in this book. However, it is a potential source of inconsistencies between experimental or theoretical data concerning derivative thermodynamic properties for systems or processes involving dissolved species reported as "standard" in different literature sources. For example, experimental "standard" data encompassing solute entropies, enthalpies and heat capacities is typically determined by investigating the temperature dependence of corresponding standard free energies, *via* Eq. 4.101. Thus, the results obtained pertain to the standard variant. In contrast, experimental "standard" data encompassing solute volumes, volume-compressibilities and volume-expansivities is typically determined by investigating the concentration dependence of the corresponding solution properties (geometric volume of the solution and its pressure and temperature derivative) and extrapolation to infinite dilution. In such experiments, a hypothetical non-interacting and volume-free single-particle solute will be characterized by partial molar variables of the form of Eq. 4.108 and not of Eq. 4.103, in agreement with Eq. 4.104 and not with Eq. 4.101. Thus, the corresponding results pertain to the density-corrected variant. Similarly, the "standard" results of most theoretical calculations (Chapters 3 and 6) are formulated in the density-corrected variant, which is more intuitive in terms of interpretation. However, the above statements only concern typical situations. For example, "standard" partial molar enthalpies can also be determined (within a constant) by investigating the concentration dependence of the heat of mixing and extrapolation to infinite dilution, in which case the results will pertain to the density-corrected variant. Conversely, "standard" partial molar volumes can also be determined by investigating the pressure dependence of the solvation free energy, in which case the results will pertain to the standard variant.

The assignment of a specific variant to the results of a given type of experiment is not always entirely obvious, especially for *second-derivative properties*. Considering Eqs. 4.101 and 4.104, there are only two options for evaluating first-derivative properties (enthalpy, entropy, volume), namely standard or density-corrected. However, there are in principle four options for evaluating self-second derivatives (heat capacity, volume-compressibility), depending on the choices made for the first and the second differentiation. For the heat capacity, one may also proceed *via* the entropy, *i.e.* using $\hat{O}_{C_P} = -T\partial_T\partial_T$, or *via* the enthalpy, *i.e.* using $\hat{O}_{C_P} = \partial_T(1 - T\partial T)$, see Eq. 4.29, resulting in eight possible options. For the cross-second derivative (volume-expansivity), one may consider two different orders for the pressure and temperature differentiation, *i.e.* $\hat{O}_{A_P} = \partial_P\partial_T$ or $\partial_T\partial_P$, again resulting in eight possible options. These different routes potentially lead to as many different expressions. However, many combinations actually produce the same result, and only a few combinations appear relevant in the context of practical determination methods. In short, experimental determination of partial molar first derivatives (enthalpy, entropy, volume) based on isothermal measurements (*e.g.* heat of mixing or geometric volume) and extrapolation to infinite dilution lead to results in the density-corrected variant. Experimental determinations of partial molar second derivatives (heat capacity, volume-compressibility, volume-expansivity) based on further pressure or temperature analysis of the above first derivatives imply a density-corrected derivative followed by a standard derivative (*via* the enthalpy for the heat capacity, and pressure-first for the volume-expansivity), and lead to results in the density-corrected variant. All determinations of first or second partial molar derivative thermodynamic properties that are directly based on the pressure or temperature analysis of the free energy without relying on such an intermediate step (*via* heat of mixing or geometric volume measurements) will lead to results in the standard variant.

There is a close parallel between the above discussion and the question whether specific experimental approaches access real or intrinsic partial molar derivative thermodynamic properties of solutes. If the natural standard outcome of a specific experimental approach is in the density-corrected variant, it is in essence because the design of the experiment renders it insensitive to the pressure and temperature derivatives of the solvent density $\rho^{\ominus}_{svt}$. For example, the determinations of a partial molar enthalpy *via* the heat of mixing or of a partial molar volume *via* the geometric volume change upon mixing cannot be sensitive to $\partial_P \rho^{\ominus}_{svt}$ or $\partial_T \rho^{\ominus}_{svt}$ because they correspond to isobaric and isothermal measurements. The same reasoning can be applied in terms of the sensitivity of a given experiment to the pressure and temperature derivatives $\partial_P \chi^{\ominus}_{svt}$ and $\partial_T \chi^{\ominus}_{svt}$ of the surface potential $\chi^{\ominus}_{svt}$ of the pure liquid (Section 2.2). It follows that experiments targeting non-elusive quantities (*i.e.* not relying on an interpretation *via* some extra-thermodynamic assumption) and leading to standard results in the standard variant also automatically lead to real values $\hat{y}^{\ominus}$ for the partial molar derivative thermodynamic parameters. In contrast, experiments of this kind leading to standard results in the density-corrected variant automatically lead to corresponding intrinsic rather than real values $y^{\ominus *}$.

In Chapter 5, where standard experimental data concerning alkali, halide and proton hydration is reported, this intricate issue is only partially addressed. Rather than investigating the results of individual experimental studies on a case-to-case basis, it is globally assumed that: (*i*) all experimental determinations encompassing information on $\mu^{\ominus}$, $h^{\ominus}$, $s^{\ominus}$ and $c_P^{\ominus}$ for ions in the dissolved state pertain to the standard variant and deliver real values, from which corresponding intrinsic values can possibly be deduced by means of extra-thermodynamic assumptions; (*ii*) all experimental determinations encompassing information on $v^{\ominus}$, $k_T^{\ominus}$ and $a_P^{\ominus}$ for ions in the dissolved state pertain to the density-corrected variant, and are therefore noted $v^{\ominus *}$, $k_T^{\ominus *}$ and $a_P^{\ominus *}$, and correspond to intrinsic values. These assumptions certainly pertain to most (if not all) the experimental determinations considered in Chapter 5.

Considering Table 4.1, the magnitude of the difference between the two variants in the case of water is not very large for h, s and a_P, but quite significant for c_P, v and k_T. A more precise idea of the significance of these differences can be gained by comparison with suggested single-ion solvation and partial molar parameters in solution for the alkali and halide ions, as well as the proton in water (Table 5.26). Clearly, this represents a non-negligible and unpleasant source of inconsistency within the standard-state definition (Section 4.2.4). The ambiguity affects derivative thermodynamic properties, not the chemical potentials or free energies themselves, and concerns: (*i*) partial molar variables for dissolved species (this section); (*ii*) formation parameters for dissolved species (Section 4.2.10); (*iii*) reaction parameters for any reaction involving a different number of dissolved species in a given solvent on the left- and right-hand side of its stoechiometric equation, including solvation parameters as an important special case (Section 4.2.7). To our knowledge, this issue has never been documented in thermodynamic textbooks (with the possible exception of Ref.[458]) and has only recently been raised by the authors of this book in the literature[1412]. In the present text, no attempt was made to convert all data and equations into a single variant. Whenever data or equations clearly (equations) or probably (data) pertain to the density-corrected variant, this is indicated by the "$*$" superscript. Whenever they pertain to the standard variant, no "*" superscript is added. The rest of this chapter is mostly (see, however, Section 4.2.16) formulated within the standard variant for simplicity. Interconversions can be performed using Table 4.1, if desired.

4.2.6 Activity

The *activity* a_i of a species i relates its actual chemical potential μ_i to the corresponding standard value μ_i°, both at the same temperature T, as

$$\mu_i = \mu_i^\circ + RT \ln a_i \quad . \tag{4.111}$$

The activity is a function of P, T and system composition, but not system extent. This unitless quantity is bound to a specific definition of the standard states (Sections 4.2.4 and 4.2.17), and the $bbme_T$ convention is applied here. Note that the gas-phase standard state is *not* equivalent to a pure real gas at unit activity, as unfortunately suggested in many textbooks. Although the chemical potentials are identical (Eq. 4.111), the physical situations are very different. The standard state is a virtual state exempt of intermolecular interactions. The unit-activity state is a real state involving intermolecular interactions and coincidentally characterized by the same chemical potential. In particular, these two states will generally differ in terms of their molar derivative thermodynamic properties, *e.g.* enthalpy or entropy. Similarly, the dissolved standard state is *not* equivalent to the solute alone in solution at unit activity, as also often suggested. The chemical potentials are here also identical (Eq. 4.111), but the physical situations are distinct. The dissolved standard state is a virtual state exempt of direct or solvent mediated solute-solute interactions, as well as of any solute-induced perturbation of the solvent-solvent interactions. The unit-activity state is a real state involving solute-solute interactions and coincidentally characterized by the same chemical potential. Here again, the partial molar derivative thermodynamic properties will generally differ between the two situations.

Approximate expressions for the activity are easily derived for ideal gases, solids, liquids or solvents with negligible molar volumes, and solutes in ideal molal solutions (Section 4.2.3). Here, the expression negligible molar volume implies that the integral of $v(P)\,dP$ from P° to the selected pressure P is very small, which can loosely be written $v^\circ(P - P^\circ) \approx 0$, *i.e.* that it corresponds to a very small change in chemical potential. For more accurate calculations, these approximate expressions must be corrected by the introduction of *activity coefficients* γ_i. The resulting equations are

$$a_i = \begin{cases} \gamma_i (P_i/P^\circ) & \text{(gas)} \\ \gamma_i & \text{(solid, liquid or solvent)} \\ \gamma_i (b_i/b^\circ) & \text{(solute)} \quad , \end{cases} \tag{4.112}$$

where P_i is the partial pressure of a gaseous species i in a gas-phase mixture (Eq. 4.73), the quantity $\gamma_i P_i$ also sometimes being referred to as the fugacity, and b_i is the molality of a solute species i in a solution (Eq. 4.55). The unitless activity coefficients are functions of P, T and system composition, but not system extent, and bound to a specific definition of the standard states (Sections 4.2.4 and 4.2.17). They should be close to one in situations where the above approximations (ideal gas, solid, liquid or solvent with negligible molar volume, ideal molal solution) are valid. In practice, they are often neglected (*i.e.* set to one) in cases where simple approximate results are sufficient.

In the present context of *single-ion solvation properties*, it is necessary to distinguish between intrinsic and real chemical potentials (Section 4.2.2) for solvated ions. Eq. 4.111 has been written in terms of the intrinsic chemical potentials μ_i and μ_i°. However, the same equation holds in terms of the corresponding real chemical potentials $\hat{\mu}_i$ and $\hat{\mu}_i^\circ$, which differ from the former by the same constant offset, for a given charged species, related to the air-liquid interfacial potential of the pure solvent (Section 2.2). Consequently, there is no distinction to be made between real and intrinsic activities or activity coefficients for ionic species. However, solutions do not escape the constraint of electroneutrality of macroscopic matter at equilibrium (Section 2.1). Since equilibrium experiments or equivalent calorimetric procedures can only probe properties related to the sum of chemical potentials for overall neutral sets of ions, the distinction between intrinsic and real quantities being irrelevant in these sums, they only provide information on the corresponding *products* of activities or activity coefficients. For example, for an electroneutral solution of the binary salt $I_{n_I} J_{n_J}$ containing the ions I^{z_I} and J^{z_J}, with $n_I z_I + n_J z_J = 0$, the experimentally relevant quantity is the combined activity

$$\begin{aligned} a_{[I_{n_I} J_{n_J}]_d} &= (a_{I_d^{z_I}})^{n_I} (a_{J_d^{z_J}})^{n_J} \\ &= \gamma_{[I_{n_I} J_{n_J}]_d} \left(\frac{b_{I_d^{z_I}}}{b^\circ} \right)^{n_I} \left(\frac{b_{J_d^{z_J}}}{b^\circ} \right)^{n_J} \quad , \end{aligned} \tag{4.113}$$

where $\gamma_{[I_{n_I} J_{n_J}]_d}$ is a combined activity coefficient. In the special case of an alkali-halide MX dissolved in water, the combined activity and activity coefficient read $a_{MX_{aq}} = a_{M_{aq}^+} a_{X_{aq}^-}$ and $\gamma_{MX_{aq}} = \gamma_{M_{aq}^+} \gamma_{X_{aq}^-}$, respectively.

An approximate estimate for combined activity coefficients, valid in the dilute regime but still representing a significant improvement over the straight neglect of these coefficients, is provided by the *Debye-Hückel theory*[113]. The resulting expression in the low-concentration limit is called the *Debye-Hückel limiting law* and reads

$$\gamma_{[I_{n_I} J_{n_J}]_{aq}} = \exp\left[-A_{DH,svt} \left| z_I z_J \right| I_s^{1/2} \right] \quad , \tag{4.114}$$

where the solvent-dependent constant $A_{DH,svt}$ is defined, under standard conditions, as

$$A^{\ominus}_{DH,svt} = \left(2\pi N_A \rho^{\ominus}_{svt} b^{\circ}\right)^{1/2} e^3 (4\pi\epsilon_o \epsilon^{\ominus}_{svt} k_B T^{-})^{-3/2} \quad , \tag{4.115}$$

and the unitless *ionic strength* I_s is introduced as

$$I_s = \frac{1}{2b^{\circ}} \sum_{i=1}^{N_i} b_i z_i^2 \ , \tag{4.116}$$

the summation including the N_i ionic species (molality b_i, integer charge z_i) present in solution. The constant $A^{\ominus}_{DH,svt}$ evaluates to 1.175 for water (using data from Table 1.1).

Expressions analogous to Eq. 4.111 connect the actual values of all (partial) molar variables to the corresponding standard values, both at the same temperature T, *via* the activity along with its pressure or/and temperature derivatives. Considering the standard variant (Section 4.2.5), these can be easily derived by applying Eq. 4.101 to Eq. 4.111, which gives,

$$y_i = y_i^{\circ} + R\hat{O}_Y(T \ln a_i) \quad . \tag{4.117}$$

For example, for the (partial) molar enthalpies, entropies and volumes, one finds

$$h_i = h_i^{\circ} - RT^2 \partial_T \ln a_i \quad , \tag{4.118}$$

$$s_i = s_i^{\circ} - R(\ln a_i + T\partial_T \ln a_i) \quad , \tag{4.119}$$

and

$$v_i = RT\partial_P \ln a_i \quad . \tag{4.120}$$

Upon inserting Eq. 4.112, these can be translated into corresponding relationships involving the activity coefficients. When the activity coefficients are neglected, *i.e.* set to one for any P and T (ideal gas, solid, liquid or solvent with negligible molar volume, ideal molal solution), this leads to the expected approximations $s_i = s_i^{\circ} - R\ln(P_i/P^{\circ})$ and $v_i = RT(P^{\circ})^{-1}$ (gas), $s_i = s_i^{\circ} - R\ln(b_i/b^{\circ})$ and $v_i = 0$ (dissolved species), or $s_i = s_i^{\circ}$ and $v_i = 0$ (solid, liquid or solvent), along with $h_i = h_i^{\circ}$.

In the generic notation of Eq. 4.28, the corresponding expressions for the density-corrected variant (Section 4.2.5) are easily formulated by considering Eq. 4.104 instead of Eq. 4.101 for the solute species. In this case, Eq. 4.117 is replaced for these species by

$$y_i = y_i^{\circ *} + R\hat{O}_T(T \ln a_i) - \Delta_c y^{\ominus *} \quad , \tag{4.121}$$

where the conversion term $\Delta_c y^{\ominus *}$ is defined by Eq. 4.110 and corresponding expressions and numerical values for water summarized in Table 1.1. Alternatively, one could also consider adjusting Eq. 4.59 to the density-corrected variant in analogy with Eq. 4.101, in which case the activity coefficients would become insensitive to a selected variant. This possibility is not explored further here, considering that the main topic of the book is ionic solvation in the infinitely dilute regime, where activity is an irrelevant quantity.

4.2.7 Reaction parameters

Given the *stoechiometric equation* of a reaction, written in the form

$$\sum_{i=1}^{N'_S} \nu_i \Sigma_i = 0 \quad , \tag{4.122}$$

where N'_S denotes the number of species participating in the reaction, Σ_i a species i, and ν_i the associated stoechiometric coefficient, negative for reactants and positive for products, the isobaric and isothermal *reaction parameter* $\Delta_r Y$ associated with the thermodynamic variable Y is defined as

$$\Delta_r Y = \sum_{i=1}^{N'_S} \nu_i y_i \quad , \tag{4.123}$$

where it is assumed that all reactants and products (pure or in mixtures) are present at the same pressure P and temperature T. Because it is a linear combination of (partial) molar variables, a reaction parameter is a function of P, T and system composition, but not system extent. In the context of *ions*, a distinction between *intrinsic* ($\Delta_r Y$) and *real* ($\Delta_r \hat{Y}$) reaction parameters is only required when considering any reaction involving a different net charge between reactants and products in terms of solvated species in each solvent considered. Particularly important examples include single-ion solvation reactions (*e.g.* $Na^+_g \rightarrow Na^+_{aq}$), dissolved single-ion formation reactions (*e.g.* $Na_{el} \rightarrow Na^+_{aq} + e^-_g$), and transfer solvent-solvent single ion reactions (*e.g.* $Na^+_{aq} \rightarrow Na^+_d$). For simplicity, however, the following discussion will only refer to intrinsic parameters, corresponding expressions also holding for real ones.

Since $\Delta_r Y$ characterizes an isobaric and isothermal process (common P and T for all reactants and products), Eq. 4.59, stated for (partial) molar variables is immediately generalizable to reaction parameters, *i.e.* in the generic notation of Eq. 4.28, one has for intrinsic quantities

$$\Delta_r Y = \hat{O}_Y \Delta_r G \quad , \tag{4.124}$$

a corresponding set of equations holding for real quantities. Eq. 4.124 is of the form of Eq. 4.31 and implies in particular a corresponding form of the *Gibbs equation* (Eq. 4.32), namely for intrinsic quantities

$$\Delta_r G = \Delta_r H - T \Delta_r S \quad , \tag{4.125}$$

a corresponding equation holding for real quantities.

By analogy with Eq. 4.123, a *standard reaction parameter*, assuming a temperature T^- for simplicity, is defined as

$$\Delta_r Y^\ominus = \sum_{i=1}^{N'_S} \nu_i y_i^\ominus \ . \tag{4.126}$$

This quantity is a constant for a given stoechiometric equation and thermodynamic parameter Y. In terms of standard reaction parameters, a distinction between quantities in the standard ($\Delta_r Y^\ominus$) and density-corrected ($\Delta_r Y^{\ominus *}$) variants (Section 4.2.5) is only required for reactions involving a different number of dissolved species in a given solvent on the left- and right-hand side of the stoechiometric equation. Particularly important examples include again solvation reactions, formation reactions of dissolved species and solvent-solvent transfer reactions for neutral as well as charged species. For simplicity, however, the following discussion will only refer to parameters formulated in the standard variant, analogous expressions being easily derived for the density-corrected variant upon substituting Eq. 4.101 by Eq. 4.104 for the dissolved species. This amounts to adding to expressions formulated in the standard variant a conversion term $\Delta_c y^{\ominus *}$ (Eq. 4.110 and Table 4.1) for all standard partial molar variables corresponding to solute species involved in the reaction. Note that, due to Eq. 4.108, the standard reaction free energy is independent of the selected variant, *i.e.* $\Delta_r G^{\ominus *} = \Delta_r G^\ominus$.

Note also that although the thermodynamic variable Y is written in upper case in $\Delta_r Y$ and $\Delta_r Y^\ominus$ to respect usual thermodynamic notation, (standard) reaction parameters are molar quantities just like (partial) molar variables, and the notations $\Delta_r y$ and $\Delta_r y^\ominus$ would actually be more consistent.

Within the standard variant for dissolved species, Eq. 4.101, stated for standard (partial) molar variables, is immediately generalizable to standard reaction parameters, *i.e.* in the generic notation of Eq. 4.28 one has for intrinsic quantities

$$\Delta_r Y^\ominus = \hat{O}_Y \Delta_r G^\ominus \quad , \tag{4.127}$$

a corresponding set of equations holding for real quantities. Eq. 4.127 is of the form of Eq. 4.31 and implies in particular a corresponding form of the Gibbs equation (Eq. 4.32), referred to in this case as *the Gibbs equation*[456, 1692] (standardized form of Eq. 4.41, expressed on a per mole basis), namely

$$\Delta_r G^\ominus = \Delta_r H^\ominus - T^- \Delta_r S^\ominus \quad . \tag{4.128}$$

A number of equations implied by Eq. 4.127 are also important enough to be formulated and named explicitly. These are the *Gibbs-Helmholtz equation*[1699–1708] (see Eq. 4.26)

$$-(T^-)^2 \partial_T \left[(T^-)^{-1} \Delta_r G^\ominus \right] = (1 - \partial_T) \Delta_r G^\ominus = \Delta_r H^\ominus \ , \tag{4.129}$$

the *temperature dependence* (see Eq. 4.20)

$$\partial_T \Delta_r G^\ominus = -\Delta_r S^\ominus \ , \tag{4.130}$$

the *Kirchhoff equation*[1698] (see Eq. 4.22)

$$\partial_T \Delta_r H^\ominus = T^- \partial_T \Delta_r S^\ominus = -T^- \partial_T^2 \Delta_r G^\ominus = \Delta_r C_P^\ominus \ , \tag{4.131}$$

the *pressure dependence* (see Eq. 4.19)

$$\partial_P \Delta_r G^\ominus = \Delta_r V^\ominus \ , \tag{4.132}$$

the *second pressure dependence* (see Eq. 4.25)

$$\partial_P^2 \Delta_r G^\ominus = \partial_P \Delta_r V^\ominus = -\Delta_r K_T^\ominus \quad , \tag{4.133}$$

the *pressure-temperature dependence* (see Eq. 4.23)

$$\partial_{P,T}^2 \Delta_r G^\ominus = \partial_T \Delta_r V^\ominus = -\partial_P \Delta_r S^\ominus = \Delta_r A_P^\ominus \quad , \tag{4.134}$$

and the *enthalpy pressure dependence* (see Eq. 4.21)

$$\partial_P \Delta_r H^\ominus = \Delta_r V^\ominus - T^- \Delta_r A_P^\ominus \quad . \tag{4.135}$$

As usual, in the above equations, partial differentiation of a standard quantity with respect to P or T refers to a partial derivative with respect to P° and T^- considered as variables, followed by evaluation at the reference values of these variables. Note that Eqs. 4.128, 4.129 and 4.130 contain a redundant equation.

Another important property of standard reaction parameters is that they obey the *Hess law*[1674]. This law states that if the stoechiometric equation of a reaction (Eq. 4.122) can be written as a linear combination of corresponding equations for a set of other reactions, with given coefficients, any standard reaction parameter of the combined reaction can also be written as a linear combination of the corresponding standard reaction parameters for these other reactions, with the same coefficients. This property follows immediately from the definition of standard reaction parameters by Eq. 4.126. Its application is very convenient to obtain parameters for reactions that cannot be carried out experimentally, or for which the parameters cannot be determined with a sufficient accuracy, from parameters available for other reactions.

Considering the free energy as a special case of variable Y, the connection between the reaction parameter $\Delta_r G$ at P° and T^- for a given system composition and the corresponding standard reaction parameter $\Delta_r G^\ominus$ involves activities (Section 4.2.6). Comparing Eqs. 4.111, 4.123 and 4.126, this connection is given by

$$\Delta_r G = RT^- \ln(K_R / K_E^\ominus) \ , \tag{4.136}$$

where the unitless *equilibrium constant* $K_E^\ominus$ is defined as

$$K_E^\ominus = \exp[-\Delta_r G^\ominus/(RT^-)] \tag{4.137}$$

and the unitless *reaction quotient* is defined as

$$K_R = \prod_{i=1}^{N'_S} a_i^{\nu_i} \ . \tag{4.138}$$

Note that no distinction is required here concerning standard *vs.* density-corrected solute standard-state variants (Section 4.2.5), which only differ in terms of derivative thermodynamic properties. Inserting Eq. 4.112 into Eq. 4.138, the reaction quotient can be rewritten

$$K_R = \Gamma \prod_{i=1}^{N'_S} \left(\frac{P_i}{P^\circ}\right)_{i\in g}^{\nu_i} \left(\frac{b_i}{b^\circ}\right)_{i\in d}^{\nu_i} \quad , \tag{4.139}$$

where

$$\Gamma = \prod_{i=1}^{N'_S} \gamma_i^{\nu_i} \quad , \tag{4.140}$$

and the notations $i \in g$ and $i \in d$ indicate that the product only encompasses corresponding terms for gaseous or dissolved species, respectively, pure solids and liquids as well as solvents being entirely omitted. When idealized conditions (ideal gases, solids, liquids and solvents with negligible molar volumes, ideal molal solutions) represent a good approximation, Γ is commonly set to one for simplicity.

It is important to realize, however, that if $\Delta_r G^\ominus$ is the *exact* free-energy change associated with one reaction cycle (on a per mole basis) converting standard-state reactants to standard-state products, $\Delta_r G$ at P° and T^- only represents an *approximation* to the corresponding free-energy change at the given system composition. The reason is that Eq. 4.126 only involves summation over the N'_S species participating in the reaction, and overlooks possible changes in the chemical potentials of the $N_S - N'_S$ *non-participating species* (*e.g.* solvent, salt buffer, inert gas) among the N_S species present in the system.

If $\Delta_r G$ in Eq. 4.136 is to represent the exact free-energy change at a given composition Eq. 4.140 should be modified to

$$\Gamma_{tot} = \Gamma \prod_{i=1}^{N_S - N'_S} a_i \quad , \tag{4.141}$$

so as to also include the activities of the non-participating species. As a simple example, the solubility of a salt in an aqueous solution of a different salt (no common ion) depends on the concentration of this second salt in the solution *via* the activities of the corresponding ions and of water in the solution, although these species do not enter into the stoechiometric equation for the dissolution reaction. Setting Γ to one in Eq. 4.139 thus involves an additional approximation besides the validity of idealized behaviors (see above) for all participating species, namely that of using unit activities for all non-participating species.

In the context of *reactions involving ions*, no distinction between intrinsic and real quantities is required for the reaction quotient, because it is defined in terms of activities, for which the distinction does not exist (Section 4.2.6). For the equilibrium constant, a distinction is only required for reactions involving a different net charge between reactants and products in terms of solvated species in each solvent considered. Particularly important examples include single0ion solvation reactions, dissolved single-ion formation reactions and solvent-solvent single-ion transfer reactions. In these situations, Eqs. 4.136 and 4.137 can be reexpressed in terms of real quantities as

$$\Delta_r \hat{G} = RT^- \ln(K_R/\hat{K}_E^\ominus) \tag{4.142}$$

with

$$\hat{K}_E^\ominus = \exp[-\Delta_r \hat{G}^\ominus/(RT^-)] \quad , \tag{4.143}$$

where $\Delta_r \hat{G}$ now differs from the corresponding intrinsic free energy change $\Delta_r G$. Since the crossing of the interface by charges is physical, $\Delta_r \hat{G}$ rather than $\Delta_r G$, and $\hat{K}_E^\ominus$ rather $K_E^\ominus$, should be interpreted here as the true free energy and equilibrium constant of the reaction, respectively.

For a closed system where a single reaction may take place, the advance of the reaction according to the stoechiometric equation (Eq. 4.122), *i.e.* the composition of the system at any time, may be characterized by a single *advance parameter* ξ, with units of moles. This parameter is defined in such a way that the infinitesimal changes in the numbers of moles of all participating species Σ_i in the system associated with an infinitesimal change of $d\xi$ of ξ are given by $\nu_i d\xi$ (no change for non-participating species). This condition only determines the value of ξ within an arbitrary constant. The constant may be set *e.g.* by assigning the value $\xi = 0$ to the initial system composition. Note that the advance parameter ξ is confined to a specific range, between ξ_{min} and ξ_{max}, depending on the initial conditions due to the presence of *limiting factors*, *i.e.* ξ_{min} and ξ_{max} are determined by situations where a product or a reactant, respectively, has entirely disappeared. It is easily seen from the definitions of the (partial) molar variables (Eqs. 4.45 and 4.49) and of the advance parameters (Eq. 4.123) that an infinitesimal isobaric ($dP = 0$) and isothermal ($dT = 0$) process associated with a variation $d\xi$ of ξ leads to a corresponding infinitesimal change $dY = \Delta_r Y d\xi$ of Y, neglecting the contribution of non-participating species (see above).

As seen previously (Section 4.2.1), for closed and *uncoupled* systems, spontaneous isobaric and isothermal changes are characterized by $dG < 0$ (Eq. 4.38

with $dq = 0$). Since $dG = \Delta_r G d\xi$, it follows that a reaction will evolve spontaneously forward if $\Delta_r G < 0$ or spontaneously backward if $\Delta_r G > 0$. Considering that $\lim_{\xi \to \xi_{min}} \Delta_r G(\xi) = -\infty$ and $\lim_{\xi \to \xi_{max}} \Delta_r G(\xi) = +\infty$, because a limiting factor disappears from the system at these two extreme values of ξ so that its chemical potential goes to infinity, a situation where $\Delta_r G(\xi) = 0$ will always be reached, which corresponds to *chemical equilibrium*. Considering Eq. 4.136, one may equivalently say the reaction will evolve spontaneously forward if $K_R < K_E^{\ominus}$ or backward if $K_R > K_E^{\ominus}$, to reach chemical equilibrium when $K_R = K_E^{\ominus}$. Note that the above arguments do not rule out the possible existence of multiple zeros for $\Delta_r G(\xi)$, *i.e.* of multiple equilibrium compositions for a closed reaction system under isobaric and isothermal conditions. Assuming idealized behaviors for all participating species, the monotonic dependence of activities on composition (Eq. 4.112 with $\gamma_i = 1$), and thus of the reaction quotient, implies the uniqueness of this point. However, one could imagine pathological situations where the effect of activity coefficients or non-participating species causes the occurrence of (an odd number of) multiple zeros (the authors are not aware of any example of this kind).

For a closed and electrochemically *coupled* system at equilibrium, reversible isobaric and isothermal infinitesimal processes are characterized by $dG = -\Delta\mathcal{V}_{ext} dq$ (Eq. 4.38). The charge dq transferred across the system along the potential difference $\Delta\mathcal{V}_{ext}$ (measured in the direction low potential to high potential, *i.e.* negative to positive pole of the external electric circuit) is also related to the infinitesimal change $d\xi$ of the advance parameter as $dq = nFd\xi$, where n is the number of electrons transferred across the potential difference for one cycle of the stoechiometric equation. Using $dG = \Delta_r G d\xi$ and Eq. 4.136, the *electrochemical equilibrium* condition can be written in the form of the *Nernst equation*[1732],

$$\Delta\mathcal{V} = -\frac{RT^-}{nF} \ln \frac{K_R}{K_E^{\ominus}} = \Delta\mathcal{V}^{\ominus} - \frac{RT^-}{nF} \ln K_R \tag{4.144}$$

with

$$\Delta\mathcal{V}^{\ominus} = \frac{RT^-}{nF} \ln K_E^{\ominus} = -\frac{1}{nF} \Delta_r G^{\ominus} \quad , \tag{4.145}$$

where $\Delta\mathcal{V}$ is used instead of $\Delta\mathcal{V}_{ext}$ to indicate a zero-current (equilibrium) situation.

Important special cases of reactions, and the associated (standard) reaction parameters will be described subsequently, namely phase transition reactions (Section 4.2.8), heat-up reactions (Section 4.2.9), formation reactions (Section 4.2.10), and reactions relevant to ionic solvation (Section 4.2.14).

4.2.8 Phase-transition reactions

The *phase-transition reactions* form a special type of reactions (Section 4.2.7) where reactant and product consist of the same species in two different phases. Remember that these are formally treated as distinct species in the context of thermodynamics (Section 4.2.1). According to the Gibbs phase rule[456] (Eq. 4.1), a system consisting of a pure substance can only present two phases in equilibrium at a unique temperature for a given pressure P. For solid and liquid phases, this temperature is the *melting temperature* $T_m[\Sigma_s]$ of the solid species Σ_s, or $T_m^\circ[\Sigma_s]$ at $P = P^\circ$. The corresponding reaction parameters, which are functions of P only since T_m is determined by P, are called *melting parameters* and noted here $\Delta_m Y[\Sigma_s]$, or $\Delta_m Y^\circ[\Sigma_s]$ at P° and $T_m^\circ[\Sigma_s]$. For liquid and gas phases, this temperature is the *boiling temperature* $T_b[\Sigma_\ell]$ of the liquid species Σ_ℓ, or $T_b^\circ[\Sigma_\ell]$ at $P = P^\circ$. The corresponding reaction parameters, which are functions of P only since T_b is determined by P, are called *boiling (or vaporization) parameters* and noted here $\Delta_b Y[\Sigma_\ell]$, or $\Delta_b Y^\circ[\Sigma_\ell]$ at P° and $T_b^\circ[\Sigma_s]$. Another common phase transition involves the solid and gas phases (sublimation).

Since the system is in equilibrium at the transition temperature for a given pressure, the associated reaction free energy is zero. As a consequence, melting and boiling parameters satisfy the relationships

$$\Delta_m H[\Sigma_s] = T_m[\Sigma_s]\Delta_m S[\Sigma_s] \quad \text{and} \quad \Delta_b H[\Sigma_\ell] = T_b[\Sigma_\ell]\Delta_b S[\Sigma_\ell] \qquad (4.146)$$

at any pressure P, following from the Gibbs equation at zero free-energy change. Note that the phase-transition enthalpies and entropies can be viewed as describing the magnitude of discontinuities in the molar enthalpy and entropy of a pure substance, as a function of temperature at constant pressure, mapping to a corresponding singularity (Dirac delta function term of magnitude equal to the phase-transition enthalpy) in the heat capacity (Section 4.2.1).

In the context of phase-transition parameters involving gas-phase species (vaporization, sublimation), thermodynamic tables and textbooks often leave room for some ambiguity concerning whether the considered parameters are appropriate *for a real or for an ideal gas phase*. In principle, standard parameters such as T_b° and $\Delta_b H^\circ$ should refer to an ideal gas phase, as required by the standard-state definition (Section 4.2.4). However, this convention is virtually never applied for the boiling temperature in the context of tabulated data and phase diagrams, because it would imply that T_b° may differ from the temperature at which a liquid actually boils at pressure P°. For $\Delta_b H^\circ$, an ideal gas phase is generally considered, but the evaluation refers to the real boiling temperature rather than to the formally standard T_b°. As a result, Eq. 4.146 may be slightly violated. In both cases, the error is likely to be very small in most practical situations and can probably be safely disregarded.

Note, finally, that phase-transition parameters exclusively involving phases corresponding to *real standard states* (Section 4.2.4), namely liquids and solid phases, can only be evaluated at the corresponding transition temperature for a given pressure, *i.e.* along the corresponding transition line in a P,T-phase diagram. This is not the case when a phase with a *virtual standard state* is involved, such as the gas phase. As a result, it is possible to generalize *e.g.*

the concept of vaporization enthalpy to arbitrary pressures and temperatures where the liquid phase is stable, *i.e.* to determine $\Delta_b H(P,T)$ for $T \neq T_b(P)$, or $\Delta_b H^\circ(T)$ for $T \neq T_b^\circ(P)$. The latter quantities then unambiguously refer to an ideal gas, the ambiguity only pertaining to the situation $T = T_b$. A simple example of the above considerations in the context of water can be found in Table 1.1, where the vaporization enthalpy $\Delta_b H^\circ$ at P° and T_b° is 40.65 kJ·mol^{-1}, which differs from the vaporization enthalpy $\Delta_b H^\ominus$ at P° at T^-, namely $\Delta_b H^\ominus$ = 43.98 kJ·mol^{-1} (see footnote j in this table). In the parameterization of water models for AT simulations (Section 3.3.2), the latter quantity rather than the former one must be compared to simulation results at P° and T^- (Table 3.1).

4.2.9 Heat-up reactions

The *heat-up reaction* of a pure species in the gas-phase will be defined here as the (virtual) reaction taking the pure species in its gas-phase standard state from the absolute zero temperature to a finite temperature T. The corresponding standard reaction parameters will be noted for a gas-phase species Σ_g as $\Delta_h Y^\circ[\Sigma_g]$ for an arbitrary temperature T, or $\Delta_h Y^\ominus[\Sigma_g]$ for $T = T^-$. In practice, the only standard heat-up parameter relevant in the present context is the standard heat-up enthalpy $\Delta_h H^\ominus[\Sigma_g]$. Obviously, heat-up reactions do no represent isothermal processes. For this reason, one should be cautious in applying results derived previously for isobaric and isothermal reaction parameters (Section 4.2.7) to these types of reactions.

Following from Eq. 4.22, the *standard heat-up enthalpy* of a species Σ_g can be written

$$\Delta_h H^\ominus[\Sigma_g] = \int_0^{T^-} dT' c^\circ_{P,g}(T') \quad , \tag{4.147}$$

where $c^\circ_{P,g}(T')$ is the standard molar isobaric heat capacity of the gas-phase species Σ_g at temperature T'. Based on considerations made in Section 4.2.4 concerning the extension of the third law of thermodynamics to the gas-phase standard state at $T = 0$ K, the *standard heat-up entropy* of a gas-phase species is equal to its *absolute molar entropy*, *i.e.*

$$\Delta_h S^\ominus[\Sigma_g] = s^\ominus[\Sigma_g] \quad . \tag{4.148}$$

Given this trivial identity, the symbol $\Delta_h S^\ominus[\Sigma_g]$ will not be used further in this text.

The standard heat-up parameters should not automatically be expected to satisfy the Gibbs equation (Eq. 4.128), because they are not associated with an isothermal process (see above). They nevertheless satisfy this equation due to the absolute zero initial temperature. Thus, the *heat-up free energy* is

$$\Delta_h G^\ominus[\Sigma_g] = \Delta_h H^\ominus[\Sigma_g] - T^- s^\ominus[\Sigma_g] \quad . \tag{4.149}$$

The heat-up parameters are in principle not directly accessible experimentally, because all gases will undergo transitions to liquid or/and solid phases upon being cooled down to the absolute zero temperature. However, these parameters can be calculated using SBS (Sections 4.2.12 and 4.2.13). Since the heat-up entropy is equivalent to the absolute molar entropy, it can also be accessed for uncharged species *via* TCI (Section 4.2.11).

4.2.10 Formation reactions

The standard *formation reaction* of a species (atom, molecule, ion) in a given phase is defined as the reaction that combines elements in their standard states to form one unit of this species in its standard state for the specified phase. Since the standard state of an element depends on a selected temperature T (Section 4.2.4), the same holds for the reactants involved in a formation reaction. If ions are to be considered, this definition has to be extended by allowing electrons in their ideal gas-phase standard state to be absorbed or produced along the formation reaction (Section 4.2.4). The standard reaction parameters (Section 4.2.7) associated with the formation reaction are referred to as the standard *formation parameters* of the species (atom, molecule, ion) in the given phase. These parameters will be noted $\Delta_f Y^\circ[\Sigma]$, or $\Delta_f Y^\ominus[\Sigma]$ at temperature T^-, where the symbol between the square brackets represents the species Σ that is formed by the reaction, including its phase. Obviously, the standard formation parameters of an element in its standard state are zero by definition at any temperature. In this book, the electron is treated on the same footing as the elements, and the standard formation parameters of the ideal gas-phase electron are also zero at any temperature. This convention is, however, not unanimously employed (Section 4.2.17). Note that although the thermodynamic variable Y is written in upper case in $\Delta_f Y^\circ$ and $\Delta_f Y^\ominus$ (just like in all reaction parameters; Section 4.2.7) to respect usual thermodynamic notation, standard formation parameters are molar quantities, and the notations $\Delta_f y^\circ$ and $\Delta_f y^\ominus$ would actually be more consistent.

Assuming the temperature T^- for simplicity, the standard formation parameters are of interest, because they allow for an *alternative formulation* of the standard parameters of an arbitrary reaction, as

$$\Delta_r Y^\ominus = \sum_{i=1}^{N'_S} \nu_i \Delta_f Y^\ominus[\Sigma_i] \quad , \tag{4.150}$$

where N'_S denotes the number of species participating in the reaction, Σ_i a species i, and ν_i the associated stoechiometric coefficient, negative for reactants and positive for products (Eq. 4.122). This equation is easily derived from Eq. 4.126 by application of the Hess law (Section 4.2.7). This reformulation is convenient in the context of thermodynamic variables with an undefined zero point, such as G and H (Section 4.2.1), because the corresponding partial molar quantities $y_i^\ominus$ to be used in Eq. 4.126 also lack a zero point in this case (Section 4.2.2). In contrast, standard formation parameters are unambiguously defined, and can be tabulated for all relevant species (atoms, molecules, ions) and phases. These parameters may also be used for thermodynamic variables with a well-defined zero point, such as S, C_P, V, K_T and A_P. However, in this case, the tabulation of (partial) molar variables and the application of Eq. 4.126 are usually preferred.

In the context of *ions*, a distinction between *intrinsic* ($\Delta_f Y^\circ$ or $\Delta_f Y^\ominus$) and *real* ($\Delta_f \hat{Y}^\circ$ or $\Delta_f \hat{Y}^\ominus$) formation parameters is only required for charged species in solution, just as was the case for the corresponding standard partial molar variables in solution (Section 4.2.2). Obviously, the formation parameters of a *dissolved salt* are equal to sums of (either intrinsic or real) formation parameters for the dissolved ions. Taking the example of an alkali-halide salt dissolved in water (MX_{aq}), one may thus write (for any temperature T)

$$\begin{aligned} \Delta_f Y^\circ[MX_{aq}] &= \Delta_f Y^\circ[M^+_{aq}] + \Delta_f Y^\circ[X^-_{aq}] \\ &= \Delta_f \hat{Y}^\circ[M^+_{aq}] + \Delta_f \hat{Y}^\circ[X^-_{aq}] \end{aligned} \tag{4.151}$$

and

$$\begin{aligned} \Delta_f Y^\ominus[MX_{aq}] &= \Delta_f Y^\ominus[M^+_{aq}] + \Delta_f Y^\ominus[X^-_{aq}] \\ &= \Delta_f \hat{Y}^\ominus[M^+_{aq}] + \Delta_f \hat{Y}^\ominus[X^-_{aq}] \quad . \end{aligned} \tag{4.152}$$

As a particular case of standard reaction parameters, standard formation parameters obey an equation analogous to Eq. 4.127, namely

$$\Delta_f Y^\ominus[\Sigma] = \hat{O}_Y \Delta_f G^\ominus[\Sigma] \quad , \tag{4.153}$$

implying in particular a corresponding form of the *Gibbs equation* (Eq. 4.128), *i.e.*

$$\Delta_f G^\ominus[\Sigma] = \Delta_f H^\ominus[\Sigma] - T^- \Delta_f S^\ominus[\Sigma] \quad , \tag{4.154}$$

as well as equations analogous to Eqs. 4.129-4.135. For dissolved species, these equations have been written here in terms of intrinsic parameters, similar equations holding for real ones.

In terms of standard formation parameters, a distinction between quantities in the standard ($\Delta_f Y^\ominus[\Sigma]$) and density-corrected ($\Delta_f Y^{\ominus *}[\Sigma]$) variants (Section 4.2.5) is only required for dissolved species Σ_d. In this case, the two quantities are related by

$$\Delta_f Y^{\ominus *}[\Sigma_d] = \Delta_f Y^\ominus[\Sigma_d] + \Delta_c y^{\ominus *} \quad , \tag{4.155}$$

where the conversion term $\Delta_c y^{\ominus *}$ is defined in Eq. 4.110, corresponding expressions and numerical values for water being summarized in Table 4.1.

4.2.11 Temperature-scaled heat capacity integration (TCI)

The third law of thermodynamics (Sections 4.2.1 and 4.2.2) permits to consistently set to zero the molar entropies of all pure species in perfect crystals (any molecule and isomorph, and at any pressure) as well as in ideal gases (at any pressure) at the absolute zero temperature[1733]. Starting from the perfect-crystal variant of this law and owing to Eq. 4.24, the corresponding standard *absolute molar entropies* can then be determined by integrating the temperature-scaled molar isobaric heat capacity at P° from the absolute zero temperature to T^-, including integration over phase transitions, namely

$$s^{\ominus}[\Sigma] = \int_0^{T^-} dT'(T')^{-1} c_P^\circ(T') \quad , \qquad (4.156)$$

where $c_P^\circ(T')$ is the molar isobaric heat capacity of species Σ in its most stable phase at pressure P° and temperature T', including transition enthalpies in the form of Dirac delta functions at phase transitions (Section 4.2.8). The heat-capacity of the species is typically determined as a function of temperature *via calorimetry*. In practice, however, extrapolations based on *e.g.* the *Debye approximation*[1734] or the *Einstein model*[1735,1736] must be used at very low temperatures. For the ease of further reference, this approach to determine the absolute entropies of pure substances will be termed *temperature-scaled heat capacity integration* (TCI). This method is the one most commonly applied in the case of *solids* and *liquids*. For gaseous species, however, it is typically more convenient (neutral species) or only possible (charged species) to start from the ideal-gas variant of the third law and to obtain the standard absolute molar entropies using a heat-up principle (Section 4.2.9). This is done in practice *via* SBS (Sections 4.2.12 and 4.2.13). Note that the two approaches could also be combined to obtain absolute molar entropies for liquids and solids. More precisely, the entropy of the gas could be determined at a temperature T *via* SBS, and Eq. 4.156 applied backwards by integrating from T to T^-, including contributions from condensation (liquids and solids) and solidification (solids).

4.2.12 Spectroscopy-based statistical mechanics (SBS)

The purpose of *spectroscopy-based statistical mechanics* (SBS) is the evaluation of thermodynamic properties on the basis of information concerning the energy levels of a species under consideration, as probed *via spectroscopic techniques*. In the present context of ion thermodynamics, SBS has two main applications. First, for pure gaseous species, the determination of *absolute molar entropies* (and molar heat capacities) is often more conveniently performed (neutral species) than TCI (Section 4.2.11), or only possible (charged species), using quantum statistical mechanics based on spectroscopic data[1737,1738], *i.e.* on the knowledge of the electronic, vibrational and rotational energy levels of the isolated species. Second, it enables the determination of thermodynamic parameters associated with the *ionization* of gas-phase species, by addition or abstraction of an electron, or with the *ejection of an electron* from the conduction band of a solid or liquid metal (metal work function). These two types of applications are discussed in turn below. For completeness, one should mention a third relevant application of SBS in the present context, namely the determination of *atomization* parameters for gas-phase molecular species, corresponding *e.g.* to the dissociation of a polyatomic molecule into individual atoms. This specific application will be discussed later in Section 5.9.

The statistical-mechanical equation used for the determination of the standard absolute entropy of a gas-phase species is a generalization of the *Sackur-Tetrode equation*[1739–1741] originally proposed for monoatomic[1739–1741] or rigid polyatomic[1739] gas-phase species, to include the contribution of intramolecular and electronic degrees of freedom, namely

$$s^{\ominus}[\Sigma_g] = R\left\{\frac{5}{2} + \ln\left[\frac{1}{N_A^4 h^3}\left(2\pi M R T^-\right)^{3/2}\frac{RT^-}{P^\circ}q_{in}^-\right]\right\} \quad , \tag{4.157}$$

where M is the molar mass of the species and q_{in}^- its intramolecular partition function at temperature T^-. The latter quantity is defined, according to *Boltzmann statistics*, as

$$q_{in}^- = \sum_{i=0}^{\infty} g_i \exp\left[-\frac{\epsilon_i}{k_B T^-}\right] \quad , \tag{4.158}$$

where the sum runs over all energy levels i of the species (electronic, vibrational and rotational levels of the isolated molecule), ϵ_i is the energy of level i, and g_i the corresponding degeneracy. Strictly speaking, Eq. 4.157 is an approximation assuming that the spacing between the translational energy levels of the species is very small compared to the thermal energy at T^-. In practice, the corresponding error is entirely negligible for any species with a mass on the atomic scale. However, when considering s° as a function of T, this expression becomes incorrect at extremely low temperatures, which explains its divergence to $-\infty$ in the limit $T \to 0$, instead of the expected zero value.

Eq. 4.157 is applicable to both uncharged and charged species. It is also applicable to polyatomic as well as monoatomic species. In the latter case, the partition function q_{in}^- reduces to the electronic partition function q_{el}^-. This quantity evaluates to $2J_0 + 1$ when the species has no thermally-accessible excited electronic states at room temperature, where $2J_0 + 1$ is the angular-momentum degeneracy of the electronic ground state (total angular-momentum quantum number J_0). As will be seen in Chapter 5, this represents the most common situation for alkali and halide atoms and ions (see, however, Section 5.9 for the Cl and F atoms). In this case, Eq. 4.157 can be simplified to

$$s^{\ominus}[\Sigma_g] = a^{\ominus} + b^{\ominus} \ln(M/M^{\circ}) \tag{4.159}$$

with $M^{\circ} = 10^{-3}$ kg·mol^{-1}, $a^{\ominus} = 108.856$ J·mol^{-1}· K^{-1} and $b^{\ominus} = 12.4718$ J·mol^{-1}· K^{-1}. Note that the contribution from the nuclear partition function (spin degeneracy) is not included in Eq. 4.157, which is consistent with the anchoring of the absolute entropy scale defined in Section 4.2.4.

The corresponding equation for the *molar isobaric heat capacity* is easily derived based on Eq. 4.24, namely

$$c_P^{\ominus}[\Sigma_g] = \frac{5}{2}R + RT^{-}\partial_T \ln q_{in}^{-} \quad . \tag{4.160}$$

Most of the considerations made in the context of Eq. 4.157 for the entropy also apply to the heat capacity. Here again, when considering c_P° as a function of T, this expression becomes incorrect at extremely low temperatures, which explains its incorrect convergence to $(5/2)R$, instead of the expected value R. For monoatomic species, q_{in}^{-} reduces to q_{el}^{-}, which is a constant evaluating to $2J_0 + 1$ at any temperature T when the species has no thermally-accessible excited electronic states at room temperature. In this case, q_{el}^{-} is also characterized by a vanishing temperature derivative at T^{-}, and Eq. 4.160 can be simplified to

$$c_P^{\ominus}[\Sigma_g] = \frac{5}{2}R \quad . \tag{4.161}$$

The corresponding isochoric heat capacity is $(3/2)R$, as appropriate for three thermally-active translational degrees of freedom, the additional contribution R arising from the temperature derivative of the pressure-volume contribution to the enthalpy of an ideal gas.

The above equation for the heat capacity also gives access to the *heat-up enthalpy* of a gas-phase species (Section 4.2.9 and Eq. 4.147), as

$$\Delta_h H^{\ominus}[\Sigma_g] = \frac{5}{2}RT^{-} + R\int_0^{T^{-}} dT'\, T'\, \partial_T \ln q_{in}(T') \quad . \tag{4.162}$$

For monoatomic species with no thermally-accessible excited electronic states at room temperature, this leads to

$$\Delta_h H^{\ominus}[\Sigma_g] = \frac{5}{2}RT^{-} \quad . \tag{4.163}$$

As previously stated in Section 4.2.9 (Eq. 4.148), the heat-up entropy $\Delta_h S^{\ominus}[\Sigma_g]$ is equal to the absolute molar entropy $s^{\ominus}[\Sigma_g]$, and the heat-up free energy $\Delta_h G^{\ominus}[\Sigma_g]$ can be obtained by application of the Gibbs equation (Eq. 4.149).

It should be stressed that Eqs. 4.157, 4.160 and 4.162, or 4.159, 4.161 and 4.163 in the special case of monoatomic species without thermally-accessible electronically-excited states at room temperature, are in principle not applicable to the *electron* and the *proton*, for which *Fermi-Dirac statistics* should be used in instead of Boltzmann statistics (Section 4.2.13). Note, finally, that the standard molar volume, volume-compressibility and volume-expansivity of a gas are analytical (Section 4.2.3), namely

$$v^{\ominus}[\Sigma_g] = RT^{-}(P^{\circ})^{-1} \quad , \tag{4.164}$$

$$k_T^{\ominus}[\Sigma_g] = RT^{-}(P^{\circ})^{-2} \quad , \tag{4.165}$$

and

$$a_P^{\ominus}[\Sigma_g] = R(P^{\circ})^{-1} \quad , \tag{4.166}$$

irrespective of the statistics employed.

Besides the calculation of molar and heat-up parameters for gas-phase species, the second main application of SBS relevant in the context of this book is the determination of thermodynamic parameters associated with the *ionization* of a gaseous species or the *ejection of an electron* from the conduction band of a solid or liquid metal (the latter encompassing a contribution associated with the crossing of the metal surface). The basic principle of such a measurement is to evaluate an electron absorption or emission spectrum, indicative of the probability with which electrons of a specific kinetic energy can be added to or abstracted from the gaseous species or metal at a temperature T^{-}. If such a spectrum is processed using statistical mechanics, an estimate may be obtained for the semi-adiabatic (Helmholtz) free-energy change $\Delta\tilde{A}$ associated with the process. This estimate is semi-adiabatic in the sense that it pertains to the species in equilibrium at temperature T^{-}, but to an electron at rest (zero kinetic energy, *i.e.* at the absolute zero temperature). In the special case of a gas-phase monoatomic species without thermally-accessible electronically-excited states at the selected experimental temperature, the spectrum will consist of single lines, corresponding to a set of adiabatic energy changes $\Delta\tilde{U}$, the lowest of which correspond to the relevant electron addition or abstraction processes. The semi-adiabatic free-energy change $\Delta\tilde{A}$ (or the adiabatic energy change $\Delta\tilde{U}$) is associated to the cold-electron convention, and must be converted to a standard parameter appropriate for the warm-electron convention used in this book (Section 4.2.4). This conversion is based on the heat-up parameters (Section 4.2.9) of the electron, which is the main reason for the importance of the latter parameters. Since the electron is a fermion, their accurate calculation requires the application of Fermi-Dirac statistics (Section 4.2.13). Two important examples of this approach will be encountered in the context of ion thermodynamics, namely the evaluation of the *ionization parameters* of gas-phase atoms (Sections 4.2.14 and 5.10) and the evaluation of the *work functions* of solid or liquid metals (Sections 4.3.5 and 5.13). In the first case and considering monoatomic species without thermally-accessible electronically-excited states, the adiabatic energy changes $\Delta\tilde{U}$ in the cold-electron convention will be termed the (first) *ionization potential* (E_I) for cation-forming species and the *electron affinity* (E_A) for anion-forming species, both referring to an electron abstraction process. The corresponding standardized quantities in the warm-electron convention will be termed the *standard ionization parameters* ($\Delta_i Y^{\ominus}$), always referring to the ionization process, *i.e.* to an electron abstraction process for cation-forming species but to an electron addition process for anion-forming species. In the second case, the semi-adiabatic free-energy change $\Delta\tilde{A}$ in the cold-electron convention, multiplied by F^{-1}, will be termed the *spectroscopic work function* (w) of the metal, referring to the electron abstraction (ejection) process. The corresponding standardized quantity in the warm-electron convention will be termed the *standard work function* ($\omega^{\ominus}$).

4.2.13 Gas-phase electron and proton parameters

The treatment of the electron in the $bbme_T$ standard-state definition (Section 4.2.4) relies on what is usually referred to as the *electron convention*[294, 1723], although a more precise expression (introduced in this book and adopted throughout) is the *warm-electron convention*, denoted here by e_T. In this convention, the standard state of the electron is the ideal gas-phase electron at temperature T. The alternative convention, usually referred to as the *ion convention*[1723,1730], although a more precise expression (also introduced here and adopted throughout) is the *cold-electron convention*, denoted by e_0 instead of e_T, and will be discussed in Section 4.2.17. In this alternative convention, the standard state of the electron is the ideal gas-phase electron at the absolute zero temperature $T = 0$ K, which corresponds to the idealized situation of a motionless electron in vacuum. This convention arises naturally from the outcome of SBS measurements concerning gas-phase ions as well as solid or liquid metals (Section 4.2.12), which are of relevance for the determination of the ionization parameters of gas-phase atoms (Sections 4.2.14 and 5.10) and of the work functions of metals (Sections 4.3.5 and 5.13), and of absolute electrode potentials (Sections 4.3.6 and 5.14). However, as discussed in Section 4.2.17, the cold-electron convention does not permit an entirely consistent formulation of thermodynamics, and is therefore not recommended. The conversion of the raw SBS results from the cold-electron to the warm-electron convention requires knowledge of the heat-up parameters (Section 4.2.9) of the electron.

The standard *heat-up parameters* $\Delta_h H^{\ominus}[e_g^-]$ and $\Delta_h S^{\ominus}[e_g^-] = s^{\ominus}[e_g^-]$ of the *electron* are not experimentally accessible, but can be calculated[1742, 1743] using quantum statistical mechanics. However, because the electron is a light particle and a fermion, *Boltzmann statistics* (Section 4.2.12) is inappropriate in this case and *Fermi-Dirac statistics* must be applied instead. The resulting values[1723], as well as the corresponding standard molar heat capacity $c_P^{\ominus}[e_g^-]$, volume $v^{\ominus}[e_g^-] = RT^-(P^\circ)^{-1}$, volume-compressibility $k_T[e_g^-] = RT^-(P^\circ)^{-2}$ and volume-expansivity $a_P[e_g^-] = R(P^\circ)^{-1}$

are reported in Table 4.2. These quantities are also sometimes incorrectly calculated using Boltzmann statistics, resulting in different values $\Delta_h H^{(\ominus)}[e_g^-]$, $\Delta_h S^{(\ominus)}[e_g^-] = s^{(\ominus)}[e_g^-]$ and $c_P^{(\ominus)}[e_g^-] = (5/2)R$ for the former three parameters, the "$(\ominus)$" superscript indicating a value that is meant to be standard but is inaccurate. These estimates are also reported in Table 4.2 for com-

Table 4.2 *Standard thermodynamic parameters of the gas-phase electron and proton.* The reported data includes the formation free energy $\Delta_f G^\ominus$, the formation enthalpy $\Delta_f H^\ominus$, the formation entropy $\Delta_f S^\ominus$, the molar entropy $s^\ominus$ (equivalent to the heat-up entropy $\Delta_h S^\ominus$), the molar heat capacity $c_P^\ominus$, the molar volume $v^\ominus$, the molar volume-compressibility $k_T^\ominus$, the molar volume-expansivity $a_P^\ominus$, the compressibility $\kappa_T^\ominus$, the expansivity (thermal expansion coefficient) $\alpha_P^\ominus$, the heat-up enthalpy $\Delta_h H^\ominus$, the heat-up free energy $\Delta_h G^\ominus$, the rest mass m and the rest energy u. The reported values were determined according to Fermi-Dirac statistics, as appropriate for fermions. Corresponding approximate values $\Delta_f G^{(\ominus)}$, $\Delta_f H^{(\ominus)}$, $\Delta_f S^{(\ominus)}$, $s^{(\ominus)}$, $c_P^{(\ominus)}$ and $\Delta_h H^{(\ominus)}$ determined according to Boltzmann statistics are also reported between parentheses. The values of $v^\ominus$, $k_T^\ominus$, $a_P^\ominus$, $\kappa_T^\ominus$ and $\alpha_P^\ominus$ do not differ between the two statistics. The notations e_g^- and H_g^+ refer to the gas-phase electron and proton, respectively. The standard states are according to the *bbme*$_T$ convention (Section 4.2.4; reference pressure $P^\circ = 1$ bar and reference temperature $T^- = 298.15$ K; including a warm-electron convention).

	e_g^-		Ref.	Notes
$\Delta_f G^\ominus$	[kJ·mol^{-1}]	0 (0)	1723	
$\Delta_f H^\ominus$	[kJ·mol^{-1}]	0 (0)	1723	
$\Delta_f S^\ominus$	[J·mol^{-1}·K^{-1}]	0 (0)		a
$s^\ominus (= \Delta_h S^\ominus)$	[J·mol^{-1}·K^{-1}]	22.734 (20.979)	1723	
$c_P^\ominus$	[J·mol^{-1}·K^{-1}]	17.129 (20.786)	1723	b
$v^\ominus$	[10^{-3} m^3·mol^{-1}]	24.79		c
$k_T^\ominus$	[10^{-2} m^3·mol^{-1}·bar^{-1}]	2.48		d
$a_P^\ominus$	[10^{-5} m^3·mol^{-1}·K^{-1}]	8.31		e
$\kappa_T^\ominus$	[bar^{-1}]	1.00		f
$\alpha_P^\ominus$	[10^{-3} K^{-1}]	3.35		g
$\Delta_h H^\ominus$	[kJ·mol^{-1}]	3.145 (6.197)	1723	
$\Delta_h G^\ominus$	[kJ·mol^{-1}]	-3.63 (-0.058)		h
m	[10^{-31} kg]	9.10938	221	
u	[10^{6} kJ·mol^{-1}]	49.30		i
	H_g^+		Ref.	Notes
$\Delta_f G^\ominus$	[kJ·mol^{-1}]	1513.324 (1516.959)	1723	
$\Delta_f H^\ominus$	[kJ·mol^{-1}]	1533.101 (1536.197)	1723	
$\Delta_f S^\ominus$	[J·mol^{-1}·K^{-1}]	66.34 (64.52)		a
$s^\ominus (= \Delta_h S^\ominus)$	[J·mol^{-1}·K^{-1}]	108.946 (108.947)	1723	
$c_P^\ominus$	[J·mol^{-1}·K^{-1}]	20.782 (20.785)	1723	b
$v^\ominus$	[10^{-3} m^3·mol^{-1}]	24.79		c
$k_T^\ominus$	[10^{-2} m^3·mol^{-1}·bar^{-1}]	2.48		d
$a_P^\ominus$	[10^{-5} m^3·mol^{-1}·K^{-1}]	8.31		e
$\kappa_T^\ominus$	[bar^{-1}]	1.00		f
$\alpha_P^\ominus$	[10^{-3} K^{-1}]	3.35		g
$\Delta_h H^\ominus$	[kJ·mol^{-1}]	6.199 (6.140)	1723	
$\Delta_h G^\ominus$	[kJ·mol^{-1}]	-26.3 (-26.3)		h
m	[10^{-27} kg]	1.67262	221	
u	[10^{10} kJ·mol^{-1}]	9.053		i

(a) calculated using the Gibbs equation ($\Delta_f G^\ominus = \Delta_f H^\ominus - T^- \Delta_f S^\ominus$); (b) incorrectly reported with units of kJ·mol^{-1} in the quoted reference (assumed to be J·mol^{-1}·K^{-1}); (c) calculated as $RT^-(P^\circ)^{-1}$; (d) calculated as $k_T^\ominus = v^\ominus \kappa_T^\ominus$; (e) calculated as $a_P^\ominus = v^\ominus \alpha_P^\ominus$; (f) calculated as $\kappa_T^\ominus = (P^\circ)^{-1}$; (g) calculated as $\alpha_P^\ominus = (T^-)^{-1}$; (h) calculated using the Gibbs equation ($\Delta_h G^\ominus = \Delta_h H^\ominus - T^- \Delta_h S^\ominus$); (i) calculated as $N_A mc^2$ with $c = 2.99792458 \cdot 10^8$ m·s^{-1} (Ref.[221]).

parison. They are used in Chapter 5 to correct literature data that has been derived based on these inaccurate values. The differences between the results of the two types of statistics are clearly non-negligible.

Since the *proton* is also a fermion, the properties of the gas-phase reference state for this particle must also be calculated using Fermi-Dirac statistics. The resulting values[1723] of $\Delta_h H^{\ominus}[\mathrm{H}_g^+]$, $\Delta_h S^{\ominus}[\mathrm{H}_g^+] = s^{\ominus}[\mathrm{H}_g^+]$, $c_P^{\ominus}[\mathrm{H}_g^+]$, $v^{\ominus}[\mathrm{H}_g^+] = RT^-(P^\circ)^{-1}$, $k_T^{\ominus}[\mathrm{H}_g^+] = RT^-(P^\circ)^{-2}$ and $a_P^{\ominus}[\mathrm{H}_g^+] = R(P^\circ)^{-1}$ are reported in Table 4.2. These quantities are also sometimes incorrectly calculated using Boltzmann statistics, resulting in different values $\Delta_h H^{(\ominus)}[\mathrm{H}_g^+]$, $\Delta_h S^{(\ominus)}[\mathrm{H}_g^+] = s^{(\ominus)}[\mathrm{H}_g^+]$ and $c_P^{(\ominus)}[\mathrm{H}_g^+] = (5/2)R$ for the three former parameters, also reported in Table 4.2 for comparison. However, since the proton is much heavier than the electron, the error committed is more limited in this case.

The standard *formation parameters* of the electron in its standard state are zero by definition, irrespective of the choice of a cold- or warm-electron convention. In the warm-electron convention adopted here, these correspond to the formation parameters of the gas-phase electron e_g^-. In the cold-electron convention, however, the standard formation parameters of the gas-phase electron are formally equal to the corresponding heat-up parameters and thus non-vanishing. The standard formation parameters of the gas-phase proton can be evaluated, as for any other ion, according to a procedure described in Section 5.10. These values are also reported in Table 4.2.

4.2.14 Thermodynamic reactions relevant to ionic solvation

The standard formation and reaction parameters relevant to ionic solvation, as well as the corresponding standard (partial) molar variables, are summarized in Figure 4.1, in the specific context of an alkali-halide ion pair (MX) and water as a solvent. They are discussed below in the slightly more general context of an arbitrary ion pair IJ, of possibly polyvalent ions, and a given solvent, possibly non-aqueous. If necessary, the corresponding definitions can easily be generalized to ion sets involving cations and anions of distinct valences, *e.g.* $I^{2z}J_2^{-z}$, and to ternary or higher-order sets, *e.g.* $I^z+J^z+K^{-2z}$ (Section 4.1).

The different standard *reaction parameters* (Section 4.2.7) are noted $\Delta_x Y^{\ominus}[\Sigma]$, where the symbol between the square brackets represents the species Σ that *undergoes* the reaction, *i.e.* the reactant species including its phase, and the subscript x characterizes one specific type of reaction, with $x = a, i, l, s, d$ (see below). For a given ion-forming species I, capable of forming an ion I^z, and, possibly, a second ion-forming species J, capable of forming an ion J^{-z}, these include: (*i*) the *atomization* parameters $\Delta_a Y^{\ominus}[I_{el}]$, associated with the reaction $I_{el} \rightarrow I_g$; (*ii*) the *ionization* parameters $\Delta_i Y^{\ominus}[I_g]$, associated with the reaction $I_g \rightarrow I_g^z + ze_g^-$ (cation, $z > 0$) or $I_g + (-z)e_g^- \rightarrow I_g^z$ (anion, $z < 0$); (*iii*) the *reticulation* (lattice formation) parameters $\Delta_l Y^{\ominus}[IJ_g]$, associated with the reaction $IJ_g \to IJ_s$, where IJ_g represents the pair of gas-phase ions I_g^z plus J_g^{-z}; (*iv*) the *salt solvation* parameters $\Delta_s Y^{\ominus}[IJ_g]$, associated with the reaction $IJ_g \to IJ_d$ (where IJ_d represents the pair of dissolved ions I_d^z plus J_d^{-z}); (*v*) the *dissolution* parameters $\Delta_d Y^{\ominus}[IJ_s]$, associated with the reaction $IJ_s \to IJ_d$. Recall that the notations IJ_g and IJ_d are not meant to represent a bound ion-pair species, *i.e.* a complex (which would be noted $[I^z \cdot J^{-z}]_g$ or $[I^z \cdot J^{-z}]_d$, respectively), but a sum of thermodynamic parameters appropriate separately for I_g^z and J_g^{-z} or I_d^z and J_d^{-z}, respectively (Section 4.1). Note also that ionization parameters refer here indifferently to the transformation of a neutral atom I_g into a cation I_g^z ($z > 0$) by abstraction of z electrons, or into an anion I_g^z ($z < 0$) by addition of $-z$ electrons. These parameters should be distinguished from the corresponding ionization potential (E_I) and electron affinity (E_A) as determined by spectroscopy, which represent (semi-)adiabatic quantities in the cold-electron convention (Sections 4.2.12 and 5.10). Note that the latter quantities both refer to electron abstraction. In the case of anion formation, the electron affinity thus characterizes a process in the opposite direction compared to the ionization process as defined above.

The different standard *formation parameters* (Section 4.2.10) are noted $\Delta_f Y^{\ominus}[\Sigma]$, where Σ now represents the species that is *formed* by the reaction, *i.e.* the product species, including its phase. These include: (*i*) the *gas-phase atom formation* parameters $\Delta_f Y^{\ominus}[I_g]$, equivalent to the atomization parame-

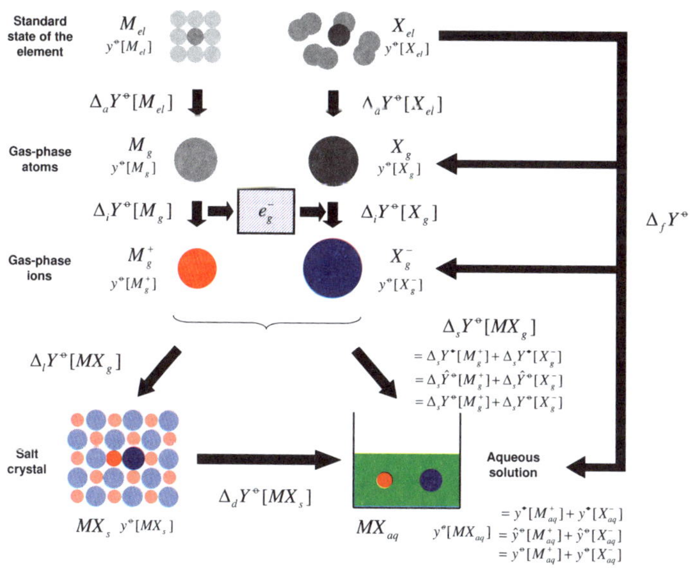

Figure 4.1 *Thermodynamic reactions relevant to ionic solvation.* The different reactions are illustrated in the specific context of an alkali-halide ion pair (MX) and water as a solvent. The symbols and notations employed are summarized in Table I.

ters $\Delta_a Y^{\ominus}[I_{el}]$; (*ii*) the *gas-phase ion formation* parameters $\Delta_f Y^{\ominus}[I_g^z]$, associated with the reaction $I_{el} \to I_g^z + ze_g^-$ (cation, $z > 0$) or $I_{el} + (-z)e_g^- \to I_g^z$ (anion, $z < 0$); (*iii*) the *salt formation* parameters $\Delta_f Y^{\ominus}[IJ_s]$, associated with the reaction $I_{el} + J_{el} \to IJ_s$; (*iv*) the *dissolved salt formation* parameters $\Delta_f Y^{\ominus}[IJ_d]$, associated with the reaction $I_{el} + J_{el} \to IJ_d$.

The different standard (partial) molar variables (Sections 4.2.2 and 4.2.4) are noted $y^{\ominus}[\Sigma]$, where Σ represents the considered species including its phase. Here, only variables derived from $Y = S$, C_P, V, K_T and A_P are relevant. Variables derived from G and H are defined within an unknown constant, and formation parameters are considered instead (see above). These standard variables include: (*i*) the *element* molar parameters $y^{\ominus}[I_{el}]$; (*ii*) the *gas-phase atom* molar parameters $y^{\ominus}[I_g]$; (*iii*) the *gas-phase ion* molar parameters $y^{\ominus}[I_g^z]$; (*iv*) the *salt* molar parameters $y^{\ominus}[IJ_s]$; (*v*) the *dissolved salt* partial molar parameters $y^{\ominus}[IJ_d]$.

In the above description, reticulation, solvation, dissolution, salt formation, dissolved salt formation, salt molar and dissolved salt partial molar parameters have been formulated on an *ion-pair* (salt) basis. The partitioning of the solvation ($\Delta_s Y^{\ominus}[IJ_g]$), dissolved salt formation ($\Delta_f Y^{\ominus}[IJ_d]$) and dissolved salt partial molar ($y^{\ominus}[IJ_d]$) parameters into *single-ion contributions* is, of course, one of the main themes of this book. As discussed in detail later (Section 4.5), this partitioning can be performed in at least three distinct ways, leading to *conventional* ($\Delta_s Y^{\bullet}[I_g^z]$, $\Delta_f Y^{\bullet}[I_d^z]$, $y^{\bullet}[I_d^z]$), *real* ($\Delta_s \hat{Y}^{\bullet}[I_g^z]$, $\Delta_f \hat{Y}^{\bullet}[I_d^z]$, $\hat{y}^{\bullet}[I_d^z]$) or *intrinsic* ($\Delta_s Y^{\ominus}[I_g^z]$, $\Delta_f Y^{\ominus}[I_d^z]$, $y^{\ominus}[I_d^z]$) single-ion parameters. In the context of the three above types of thermodynamic parameters and when considering derivative properties, a distinction must also be made between the standard and density-corrected variants (the latter indicated by an additional "$*$" superscript) of the solute standard-state definition (Section 4.2.5), at the ion-pair level as well as at the single-ion level irrespective of the partitioning employed.

Reticulation parameters $\Delta_l Y^{\ominus}[IJ_g]$ could also possibly be partitioned into *single-ion reticulation parameters* ($\Delta_l Y^{\ominus}[I_g^z]$), enabling a similar partitioning for salt dissolution ($\Delta_d Y^{\ominus}[I_s^z]$), salt formation ($\Delta_f Y^{\ominus}[I_s^z]$) and salt molar ($y^{\ominus}[I_s^z]$) parameters[1744,1745]. One might intuitively think that a natural partitioning should be symmetric for a given binary salt, *i.e.* assign one half to the cation and one half to the anion, because electrostatic interactions within a crystal appear at first sight to be pairwise symmetric. However, taking the example of the free energy and considering an alkali-halide salt, it is easily seen that this is not the case. The quantity $\Delta_l G^{\ominus}[M_g^+]$ can be defined by the reversible work of transferring an M_g^+ ion from the gas phase to an empty M_s^+ site within a MX_s crystal. This definition is analogous to the hypothetical solvation process of Section 2.2 (Figure 2.2), except that the "solvent" is a crystal with a specified empty site. Clearly, the negative electric potential generated by the surrounding ions on a void M_s^+ site will differ in magnitude from the corresponding positive electric potential generated on a void X_s^- site. This is obvious if the anionic and cationic sites have different crystallographic symmetries. But it is also the case if the crystallographic symmetry is the same, when interactions beyond pairwise point-charge electrostatics (*e.g.* including polarization effects) are considered. Besides this expected asymmetry, which would render it non-trivial, a partitioning of reticulation parameters into single-ion contributions would be largely irrelevant, because the assigned single-ion values would be specific to a given salt, just as single-ion solvation values are specific to a given solvent. Furthermore, while the solvation of ions is of extreme practical importance, the transfer of single ions between a crystal and the gas phase has no experimental relevance. For these reasons, the possibility of defining single-ion reticulation, dissolution, salt formation and salt molar parameters will not be further considered in this book.

The ionization ($\Delta_i Y^{\ominus}[I_g]$), gas-phase ion formation ($\Delta_f Y^{\ominus}[I_g^z]$) and dissolved single-ion formation ($\Delta_f Y^{\ominus}[I_d^z]$, written here as intrinsic and in the standard variant) processes share the common peculiarity that they involve *electrons*. When expressing the associated reaction parameters in terms of corresponding (partial) molar variables, as commonly done for S, V, C_P, K_T and A_P, it is important to remember to include the corresponding contribution for the electrons. The resulting expressions are

$$\Delta_i Y^{\ominus}[I_g] = y^{\ominus}[I_g^z] + z y^{\ominus}[e_g^-] - y^{\ominus}[I_g] \quad , \tag{4.167}$$

$$\Delta_f Y^{\ominus}[I_g^z] = y^{\ominus}[I_g^z] + z y^{\ominus}[e_g^-] - y^{\ominus}[I_{el}] \tag{4.168}$$

and

$$\Delta_f Y^{\ominus}[I_d^z] = y^{\ominus}[I_d^z] + z y^{\ominus}[e_g^-] - y^{\ominus}[I_{el}] \quad . \tag{4.169}$$

However, when expressing ionization and single-ion dissolution parameters in terms of formation parameters of the reactants and products, as commonly done for G and H, no contribution is involved for the electron, the standard-state formation parameters of which are zero by definition. The resulting expressions are

$$\Delta_i Y^{\ominus}[I_g^z] = \Delta_f Y^{\ominus}[I_g^z] - \Delta_f Y^{\ominus}[I_g] \tag{4.170}$$

and

$$\Delta_s Y^{\ominus}[I_d^z] = \Delta_f Y^{\ominus}[I_d^z] - \Delta_f Y^{\ominus}[I_g^z] \quad . \tag{4.171}$$

Note that the choice of a convention regarding the standard state of the electron, *e.g.* warm- or cold-electron convention (Sections 4.2.12 and 4.2.17), defines $y^{\ominus}[e_g^-]$ in Eqs. 4.167-4.169, but does not alter the form of Eqs. 4.170 and 4.171. Thus, this choice only affects ionization, gas-phase ion formation and dissolved single-ion formation parameters, without any influence on (partial) molar variables besides those of the electron itself, on other formation parameters and on other reaction parameters.

4.2.15 Thermodynamic cycles relevant to ionic solvation

A *thermodynamic cycle* is a process that starts and ends in the same state, *i.e.* in the present context, with the same species in the same phases. According to the Hess law (Section 4.2.7), the sums of the standard reaction parameters associated with the reactions constituting the different legs of a cycle should all be zero. Thermodynamic cycles are extremely useful in two respects: (*i*) when reaction parameters have been determined for all legs of the cycle through independent measurements or calculations, their consistency may be assessed in the form of a *cycle-closure assessment*; (*ii*) when reaction parameters have been determined for all but one leg of the cycle, parameters may be *deduced* for this remaining leg without any further measurement or calculation. The first option has to be preferred whenever possible, since the use of a cycle for the prediction of a parameter removes the possibility of a useful consistency checking.

The thermodynamic cycles relevant to ionic solvation are summarized in Figure 4.2 in the specific context of an alkali-halide ion pair (MX) and water as a solvent. They are discussed below in the slightly more general context of an arbitrary ion pair IJ, of isovalent but possibly polyvalent ions, and a given solvent, possibly non-aqueous. If necessary, the corresponding definitions can easily be generalized to ion sets involving cations and anions of distinct valences, *e.g.* $I^{2z}J_2^{-z}$, and to ternary or higher-order sets, *e.g.* I^z+J^z+K^{-2z} (Section 4.1).

There are three important thermodynamic cycles in the context of ionic solvation: (*i*) the three-legged *formation-dissolution-reformation cycle* (FDR) corresponds to the sequence of reactions $I_{el} + J_{el} \rightarrow IJ_s \rightarrow IJ_d \rightarrow I_{el} + J_{el}$; (*ii*) the four-legged *Born-Fajans-Haber cycle*[79–81] (BFH) corresponds to the sequence of reactions $I_{el} + J_{el} \rightarrow I_g + J_g \rightarrow IJ_g \rightarrow IJ_s \rightarrow I_{el} + J_{el}$; (*iii*) the

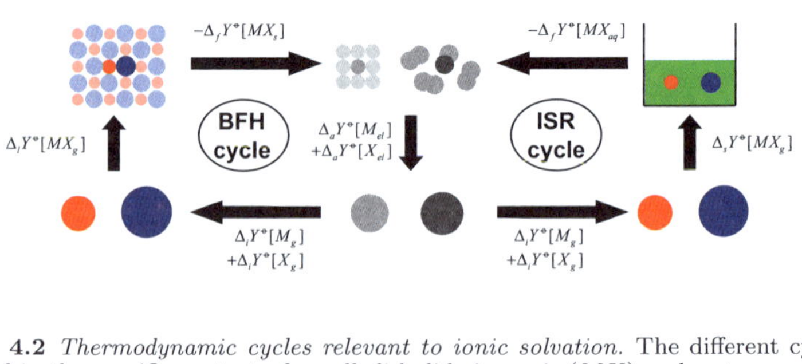

Figure 4.2 *Thermodynamic cycles relevant to ionic solvation.* The different cycles are illustrated in the specific context of an alkali-halide ion pair (MX) and water as a solvent. The individual processes forming the legs of the cycles are illustrated in Figure 4.1. The symbols and notations employed are summarized in Table I.

four-legged *ion-formation-solvation-reformation cycle* (ISR) corresponds to the sequence of reactions $I_{el} + J_{el} \rightarrow I_g + J_g \rightarrow IJ_g \rightarrow IJ_d \rightarrow I_{el} + J_{el}$. Recall that the notations IJ_g and IJ_d are not meant to represent a bound ion-pair species, *i.e.* a complex (which would be noted $[I^z \cdot J^{-z}]_g$ or $[I^z \cdot J^{-z}]_d$, respectively), but a sum of thermodynamic parameters appropriate separately for I_g^z and J_g^{-z} or I_d^z and J_d^{-z}, respectively (Section 4.1). The notations $I_{el} + J_{el}$ and $I_g + J_g$ denote corresponding sums for the two species in their element standard states (Section 4.2.4) and for the two gas-phase atoms, respectively. The "invention" of the BFH cycle in 1919, independently by Max Born, Kasimir Fajans and Fritz Haber, who agreed on a joint publication date (see Ref.[1746] for a historical overview), has represented an important step in the development of ion thermodynamics. The names used here for the two other cycles (FDR, ISR) do not reflect standard terminology and were introduced by the authors of this book for convenience of reference.

The equations resulting from the application of the Hess law to these cycles are

$$\Delta_f Y^{\ominus}[IJ_s] + \Delta_d Y^{\ominus}[IJ_s] - \Delta_f Y^{\ominus}[IJ_d] = 0 \tag{4.172}$$

for the FDR cycle,

$$\begin{aligned} & \Delta_a Y^{\ominus}[I_{el}] + \Delta_a Y^{\ominus}[J_{el}] + \Delta_i Y^{\ominus}[I_g] + \Delta_i Y^{\ominus}[J_g] \\ + \quad & \Delta_l Y^{\ominus}[IJ_g] - \Delta_f Y^{\ominus}[IJ_s] = 0 \end{aligned} \tag{4.173}$$

for the BFH cycle, and

$$\begin{aligned} & \Delta_a Y^{\ominus}[I_{el}] + \Delta_a Y^{\ominus}[J_{el}] + \Delta_i Y^{\ominus}[I_g] + \Delta_i Y^{\ominus}[J_g] \\ + \quad & \Delta_s Y^{\ominus}[IJ_g] - \Delta_f Y^{\ominus}[IJ_d] = 0 \end{aligned} \tag{4.174}$$

for the ISR cycle. Note that the three cycles are redundant, *i.e.* each of the three above equations can be expressed as a linear combination of the two others, also considering an additional (trivial) cycle $IJ_g \rightarrow IJ_s \rightarrow IJ_d \rightarrow IJ_g$ resulting in

$$\Delta_d Y^{\ominus}[IJ_s] = \Delta_s Y^{\ominus}[IJ_g] - \Delta_l Y^{\ominus}[IJ_g] \quad . \tag{4.175}$$

As will be discussed in detail in Chapter 5, the FDR cycle is typically used in either of two ways: (*i*) to verify the consistency between available salt formation, dissolution and dissolved salt formation parameters; (*ii*) to estimate unavailable dissolution parameters based on available salt formation and dissolved salt formation parameters. The BFH cycle is typically used in either of three ways: (*i*) to evaluate reticulation parameters from available atomization, ionization and salt formation parameters; (*ii*) to verify the consistency of calculated reticulation parameters with available atomization, ionization and salt formation parameters; (*iii*) to estimate unavailable ionization parameters based on available atomization and salt formation parameters, along with a theoretical estimate for the reticulation parameters. The ISR cycle is typically used to evaluate solvation parameters from available atomization, ionization and dissolved salt formation parameters.

Finally, there is a type of thermodynamic cycle of general relevance (not only to ionic solvation), which will be termed *third-law cycle* (TLC). A TLC, which relies on the third-law of thermodynamics (Section 4.2.1), can be applied to determine the reaction entropy associated with any reaction involving *pure species*. It involves performing the reaction at a temperature T, cooling down the products to the absolute zero temperature, performing the backward reaction, and heating up the reactants (of the original reaction) back to the temperature T. In simplified notation, with the reactant species collectively noted Σ and the product species collectively noted Σ', including specification of their phases, this cycle results in an equation of the form $\Delta_r S - s[\Sigma'] + 0 + s[\Sigma] = 0$, where it was noted that the entropy of the reaction at the absolute zero temperature vanishes owing to the third law of thermodynamics. Considering standard conditions, this leads to

$$\Delta_r S^{\ominus}[\Sigma \rightarrow \Sigma'] = s^{\ominus}[\Sigma'] - s^{\ominus}[\Sigma] \quad , \tag{4.176}$$

where the absolute entropies can be determined *via* TCI for solids or liquids, and possibly gases (Section 4.2.11), or *via* SBS for gases (Sections 4.2.12 and 4.2.13). Since the third law is only directly applicable to pure species, a TLC cannot include any state corresponding to a mixture (*e.g.* dissolved salt). The corresponding reaction entropies must therefore be determined by other means, *e.g.* temperature analysis of the reaction free energy (Section 4.2.18).

4.2.16 Standard-state corrections to the solvation process

In Section 2.2, two *hypothetical processes* describing the phenomenon of ionic solvation were introduced (Figure 2.2). The reversible work associated with both processes, expressed on a per mole basis, can be identified with the ionic solvation free energy. Considering the situation of a liquid pressure P° and temperature T^{-}, this free energy was referred to as a semi-standard *point-to-point* single-ion solvation free energy, because it refers to the transfer of the ion from a fixed point in vacuum (air) to a fixed point within the bulk of the solvent. If the surface contribution to this reversible work is omitted, the resulting point-to-point solvation free energy $\Delta_s G^{\boxminus}[I_g^z]$ is intrinsic. However, standard-state thermodynamics considers the transfer of the ion from and to a random location within a specified effective (molar) volume (gas: volume accessible to a particle in an ideal gas at pressure P° and temperature T^{-}; solution: volume accessible to a particle in an ideal solution at pressure P°, molality b°, temperature T^{-} and density $\rho_{svt}^{\ominus}$). As a result, the *standardization* is performed as (Eqs. 2.7 and 2.9)

$$\Delta_s G^{\ominus}[I_g^z] = \Delta_s G^{\boxminus}[I_g^z] + \Delta_s G_{std}^{\ominus} \quad , \tag{4.177}$$

with

$$\Delta_s G^{\ominus}_{std} = RT^- \ln \frac{RT^- b^\circ \rho^{\ominus}_{svt}}{P^\circ} \quad . \tag{4.178}$$

The quantity $\Delta_s G^{\ominus}_{std}$ corresponds to a correction for the process of "isothermal ideal-gas compression" from a molar-volume appropriate for $P^\circ = 1$ bar for the gaseous ion to an effective molar volume appropriate for $b^\circ = 1$ mol·kg^{-1} (molality) for the solvated ion, the latter given the solvent density $\rho^{\ominus}_{svt}$. More precisely, $\Delta_s G^{\ominus}_{std}$ represents the reversible work associated with the compression of an ideal gas from a molar volume $v_g = RT^-(P^\circ)^{-1}$ corresponding to the pressure P°, to a molar volume $v_{sol} = (b^\circ \rho^{\ominus}_{svt})^{-1}$, corresponding to a pressure $RT^- b^\circ \rho^{\ominus}_{svt}$, at constant temperature T^-.

In the present section, analogous expressions are provided for the corresponding *derivative thermodynamic properties*. Since it is more common in theoretical calculations, the *density-corrected* standard-state variant (Section 4.2.5) will be adopted here for the dissolved ion, *i.e.* Eq. 4.104 will be applied rather than Eq. 4.101 for this species. Corresponding expressions for the *standard variant* will be provided subsequently. Considering the derivative quantity

$$\Delta_s Y^{\oplus}[I^z_g] = \hat{O}_Y \Delta_s G^{\oplus}[I^z_g] \tag{4.179}$$

characterizing the hypothetical point-to-point solvation process, one may write

$$\Delta_s Y^{\ominus *}[I^z_g] = \Delta_s Y^{\oplus}[I^z_g] + \Delta_s Y^{\ominus *}_{std} \quad . \tag{4.180}$$

Using Eq. 4.101 for the gas-phase ion and Eq. 4.104 for the dissolved ion, one has

$$\Delta_s Y^{\ominus *}[I^z_g] - y^{\ominus *}_{std} = \hat{O}_Y \left(\Delta_s G^{\ominus}[I^z_g] - \mu^{\ominus}_{std} \right) \quad . \tag{4.181}$$

Inserting Eqs. 4.178-4.180, as well as Eqs. 4.105 and 4.106 leads to

$$\Delta_s Y^{\ominus *}_{std} = \hat{O}_Y \left[RT^- \ln \left(\frac{RT^-}{v_{ref} P^\circ} \right) \right] + y^{\ominus *}_{std} \quad . \tag{4.182}$$

Considering the form of the different $\hat{O}_Y$ operators (Eq. 4.29), along with Eqs. 4.105-4.107 and 4.178, one obtains the standard-state correction terms for derivative properties within the density-corrected variant, namely

$$\begin{aligned}
\Delta_s G^{\ominus *}_{std} &= \Delta_s G^{\ominus}_{std} \\
\Delta_s H^{\ominus *}_{std} &= -RT^- \\
\Delta_s S^{\ominus *}_{std} &= -(T^-)^{-1} \Delta_s G^{\ominus}_{std} - R \\
\Delta_s C^{\ominus *}_{P,std} &= -R \\
\Delta_s V^{\ominus *}_{std} &= -RT^-(P^\circ)^{-1} \\
\Delta_s K^{\ominus *}_{T,std} &= -RT^-(P^\circ)^{-2} \\
\Delta_s A^{\ominus *}_{P,std} &= -R(P^\circ)^{-1} \quad ,
\end{aligned} \tag{4.183}$$

where the free-energy correction term is also listed for completeness.

In contrast to $\Delta_s G^{\ominus}_{std}$, the above correction terms for the derivative properties do not correspond to the results expected for the isothermal compression of an ideal gas, even assuming an incompressible and inexpansive solvent, as effectively implied by the density-corrected standard-state variant (Section 4.2.5). The reason is that upon being dissolved, a particle loses its ideal-gas influence on the system volume, which is now controlled by the solvent density. Thus, for example, a non-interacting and volume-free single-particle species has a molar enthalpy of $(5/2)RT^-$ in the gas phase, corresponding to the thermal (kinetic) energy $(3/2)RT^-$ plus a pressure-volume contribution RT^-, but a partial molar enthalpy of only $(3/2)RT^-$ in the dissolved state, corresponding to the thermal energy $(3/2)RT^-$ only, since it has no influence on the solution volume. These considerations justify the above results for $\Delta_s H^{\ominus *}_{std}$ and $\Delta_s C^{\ominus *}_{P,std}$, as well as $\Delta_s S^{\ominus *}_{std}$ *via* the Gibbs equation. Similarly, such a particle has a molar volume $RT^-(P^\circ)^{-1}$ in the gas phase, as appropriate for an ideal gas, but a vanishing partial molar volume in the dissolved state, as appropriate for a volume-free particle that has no influence on the solution volume. These considerations justify the above results for $\Delta_s V^{\ominus *}_{std}$, $\Delta_s K^{\ominus *}_{T,std}$ and $\Delta_s A^{\ominus *}_{P,std}$.

Finally, correction terms $\Delta_s Y^{\ominus}_{std}$ appropriate for the *standard variant* of the solute standard state (Section 4.2.5) are easily derived as

$$\Delta_s Y^{\ominus}_{std} = \Delta_s Y^{\ominus *}_{std} - \Delta_c y^{\ominus *} \quad , \tag{4.184}$$

with $\Delta_s Y^{\ominus *}_{std}$ as given in Eq. 4.183, where the conversion term $\Delta_c y^{\ominus *}$ is defined by Eq. 4.110, corresponding expressions and numerical values for water being summarized in Table 4.1. In summary, the standard derivative thermodynamic solvation properties are related to the corresponding standard solvation free energy as

$$\Delta_s Y^{\ominus} = \hat{O}_Y \Delta_s G^{\ominus} + \Delta_s Y^{\ominus}_{std} \tag{4.185}$$

in the standard variant, or as

$$\Delta_s Y^{\ominus *} = \hat{O}_Y (\Delta_s G^{\ominus} - \Delta_s G^{\ominus}_{std}) + \Delta_s Y^{\ominus *}_{std} \tag{4.186}$$

in the density-corrected variant, where the correction terms involved are defined by Eqs. 4.178, 4.183 and 4.184 (along with the expressions provided in Table 4.1). Note again that the standard solvation free energy is identical in both variants. Finally, it is interesting to observe that for V, K_T and A_P, the correction terms in Eq. 4.183 are exactly equal to the negative of corresponding gas-phase molar properties. As a result, one may write

$$y^{\ominus *}[\Sigma_d] = \Delta_s Y^{\oplus}[\Sigma_g] \quad , \tag{4.187}$$

i.e. for these three variables, dissolved state partial molar variables in the density-corrected variant are equal to point-to-point solvation parameters.

4.2.17 Standard states (alternative conventions)

The standard states used throughout this book ($bbme_T$ convention) are precisely described in Section 4.2.4. However, in the literature, *different conventions* are also commonly employed. These alternative standards will be indicated with the "$\triangledown$" and "$\underline{\triangledown}$" superscripts instead of the "$\circ$" and "$\ominus$" superscripts, respectively. They include (with the change from the $bbme_T$ symbol indicated between parentheses): (i) the use of $P^\triangledown = 1$ atm rather than $P^\circ = 1$ bar as the reference pressure (first and second letters $b \rightarrow a$); (ii) the use of a $c^\triangledown = 1$ mol·dm^{-3} (molar) reference concentration rather than a $P^\circ = 1$ bar reference pressure for the gas-phase standard state (second letter $b \rightarrow M$); (iii) the use of $c^\triangledown = 1$ mol·dm^{-3} (molar) rather than b° =1 mol·kg^{-1} (molal) for the reference concentration in solution (third letter $m \rightarrow M$); (iv) the use of $c^\triangledown = (RT)^{-1}P^\triangledown$, where $P^\triangledown$ is the gas-phase reference pressure, *i.e.* of a gas-phase equivalent (molar) concentration, for the reference concentration in solution (third letter equal to second letter); (v) the use of a mole fraction $x^\triangledown = 1$ rather than a b° =1 mol·kg^{-1} (molal) concentration for the reference concentration in solution (third letter $m \rightarrow x$); (vi) the use of a gas-phase standard-state at the absolute zero temperature, rather than at T, for the electron, *i.e.* of a cold-electron instead of a warm-electron convention (fourth letter and subscript $e_T \rightarrow e_0$). The corresponding conversion formulae for (partial) molar enthalpies (h) and entropies (s) are sufficient to interconvert the most relevant data from one standard to another. The corresponding equations, assuming a given temperature T, are summarized below in the form of conversion terms $\Delta_c^\circ y$ in the direction "$\triangledown$" to "$\circ$", *i.e.*

$$y^\circ = y^\triangledown + \Delta_c^\circ y \tag{4.188}$$

for $y = h$ or s. Whenever required, conversion terms for the chemical potential may be obtained from the Gibbs equation, *i.e.*

$$\Delta_c^\circ \mu = \Delta_c^\circ h - T\Delta_c^\circ s \quad . \tag{4.189}$$

The conversion terms are calculated separately for each of the six above changes. Note, however, that multiple changes do not automatically result in additive conversion terms. These terms are summarized in Table 4.3 for convenience.

First, consider the use of a different *reference pressure* $P^\triangledown = 1$ atm, as opposed to $P^\circ = 1$ bar in the $bbme_T$ convention, *i.e.* the conversion $aame_T \rightarrow bbme_T$. Using Eqs. 4.23 and 4.25, one has

$$\Delta_c^\circ h = \int_{P^\triangledown}^{P^\circ} dP \, [v(P,T) - T a_P(P,T)] \quad (aame_T \rightarrow bbme_T) \tag{4.190}$$

and

$$\Delta_c^{\circ} s = -\int_{P^{\nabla}}^{P^{\circ}} dP\, a_P(P,T) \quad (aame_T \rightarrow bbme_T) \quad , \tag{4.191}$$

where $a_P = v\alpha_P$ in the case of pure species. The exact evaluation of these conversion terms is difficult for the liquid, solid, solvent and solute standard states. However, the corresponding corrections are usually negligible. The evaluation can be performed analytically for the gas-phase standard state (ideal-gas behavior, $a_P = RP^{-1}$), where the correction is expected to be most significant, leading to

$$\Delta_c^{\circ} h_g = 0 \quad (aame_T \rightarrow bbme_T) \tag{4.192}$$

and

$$\Delta_c^{\circ} s_g = -R \ln(P^{\circ}/P^{\nabla}) \quad (aame_T \rightarrow bbme_T) \quad . \tag{4.193}$$

For an alternative reference pressure $P^{\nabla} = 1$ atm, $\Delta_c^{\circ} s_g$ evaluates to 0.11 J·mol^{-1}·K^{-1} irrespective of T. The corresponding value of $\Delta_c^{\ominus} \mu_g$ is -0.03 kJ·mol^{-1} at T^{-}. These correction terms are thus also very small, *i.e.* essentially negligible. The $aame_T$ convention, or rather a $aaMe_T$ or $aaMe_0$ convention, is the one most commonly used in electrochemistry[221].

Second, consider the use of a $c^{\nabla} = 1$ mol·dm^{-3} *molar reference concentration* for the *gas-phase* reference state, as opposed to a $P^{\circ} = 1$ bar reference pressure in the $bbme_T$ convention, *i.e.* the conversion $bMme_T \rightarrow bbme_T$. Here, the conversion only concerns the gas-phase standard state and may be evaluated from Eqs 4.192 and 4.193 with $P^{\nabla} = RTc^{\nabla}$, leading to

$$\Delta_c^{\circ} h_g = 0 \quad (bMme_T \rightarrow bbme_T) \tag{4.194}$$

and

$$\Delta_c^{\circ} s_g = -R \ln[P^{\circ}/(RTc^{\nabla})] \quad (bMme_T \rightarrow bbme_T) \quad . \tag{4.195}$$

For an alternative gas-phase reference concentration $c^{\nabla} = 1$ mol·dm^{-3}, $\Delta_c^{\ominus} s_g$ evaluates to 26.69 J·mol^{-1}·K^{-1} at T^{-}. The corresponding value of $\Delta_c^{\ominus} \mu_g$ is -7.96 kJ·mol^{-1} at T^{-}. This value is very close to the negative of $\Delta_s G^{\ominus}_{std} = 7.95$ kJ·mol^{-1} previously encountered in Eq. 2.9 (or Eq. 4.178) considering water as a solvent. The slight difference arises from the fact that $\Delta_s G^{\ominus}_{std}$ corrects for a value based on a molal-equivalent rather than molar reference concentration for the gas phase. The latter convention could be termed $bmme_T$ and is only similar to $bMme_T$ when considering water as a solvent and a reference temperature close to T^{-}. The $bMme_T$ convention, or rather a $bMMe_T$ convention, has been used by some authors in the context of ionic solvation[733, 814, 819, 821, 1722, 1747–1749]. It is convenient because the effective reference concentration is (nearly in $bMme_T$ when considering water as a solvent, or exactly for $bmme_T$ or $bMMe_T$ considering any solvent) the same in the gas phase as in solution, which implies that standard-state correction terms like $\Delta_s G^{\ominus}_{std}$ in Eq. 2.9 (or 4.178) can be omitted. However, it also has the unpleasant feature that the gas-phase reference pressure becomes temperature-dependent, *via* $P^\nabla = RTc^\nabla$, and generally differs from the condensed-phase reference pressure, namely $P^\nabla = 24.79$ bar for gases at T^- *vs.* $P^\circ = 1$ bar for liquids, solids and solutions. In other words, the standard-state definition assembles systems with different reference pressures, which seems somewhat clumsy when thinking of real equilibrium systems, although it remains perfectly acceptable in terms of standard states. Similarly, considering the use of a $c^\nabla = b^\circ \rho^\circ_{svt}$ *molal reference concentration* for the *gas-phase*, as opposed to a $P^\circ = 1$ bar reference pressure in the $bbme_T$ convention, *i.e.* the conversion $bmme_T \rightarrow bbme_T$, $\Delta_c^\ominus s_g$ and $\Delta_c^\ominus \mu_g$ evaluate to 26.67 J·mol^{-1}·K^{-1} and -7.95 kJ·mol^{-1} at T^-, respectively.

Third, consider the use of a $c^\nabla = 1$ mol·dm^{-3} *molar reference concentration* for *solutions*, as opposed to a $b^\circ = 1$ mol·kg^{-1} molal reference concentration in the $bbme_T$ convention, *i.e.* the conversion $bbMe_T \rightarrow bbme_T$. Here, the conversion only concerns the solution standard state. It can be viewed as a change of the reference molality from $b^\nabla = c^\nabla (\rho^\circ_{svt})^{-1}$ to b°, leading to

$$\Delta_c^\circ h_d = \int_{b^\nabla}^{b^\circ} db\, \partial_b h_d(b, P^\circ, T) \quad (bbMe_T \rightarrow bbme_T) \tag{4.196}$$

and

$$\Delta_c^\circ s_d = \int_{b^\nabla}^{b^\circ} db\, \partial_b s_d(b, P^\circ, T) \quad (bbMe_T \rightarrow bbme_T) \quad . \tag{4.197}$$

The evaluation can be performed for the solution standard state assuming ideal-molal-solution behavior (Section 4.2.3), which implies $\partial_b h_d(b, P, T) = 0$ and $\partial_b s_d(b, P, T) = -Rb^{-1}$, leading to

$$\Delta_c^\circ h_d = 0 \quad (bbMe_T \rightarrow bbme_T) \tag{4.198}$$

and

$$\Delta_c^\circ s_d = -R \ln(b^\circ \rho^\circ_{svt}/c^\nabla) \quad (bbMe_T \rightarrow bbme_T) \quad . \tag{4.199}$$

For an alternative solution reference (molar) concentration $c^\nabla = 1$ mol·dm^{-3} and in the case of water as a solvent, $\Delta_c^\ominus s_d$ evaluates to 0.03 J·mol^{-1}·K^{-1} at T^- (using $\rho^\ominus_{wat} = 997$ kg·m^{-3} from Table 1.1). The corresponding value of $\Delta_c^\ominus \mu_d$ is -0.01 kJ·mol^{-1} at T^-. These correction terms are thus essentially negligible for water. Of course, the situation could be very different in the case of another solvent or for water at temperatures differing significantly from T^-. This convention, or rather a $aaMe_T$ or $aaMe_0$ convention, is the one most commonly used in electrochemistry[221].

Fourth, consider the case of a $c^\nabla = (RT)^{-1} P^\nabla$ *gas-phase equivalent molar reference concentration* for *solutions*, where P^∇ is the gas-phase reference pressure, as opposed to a $b^\circ = 1$ mol·kg^{-1} reference concentration in the $bbme_T$ convention, *i.e.* the conversion $bbbe_T \rightarrow bbme_T$. It can again be viewed as a change of reference molality from $b^\nabla = c^\nabla (\rho^\circ_{svt})^{-1} = (RT\rho^\circ_{svt})^{-1} P^\nabla$ to b°, and Eqs. 4.196-4.199 can also be applied in this case (the change considered being

$bbbe_T \rightarrow bbme_T$ instead of $bbMe_T \rightarrow bbme_T$ in this case). For an alternative solution reference (molar) concentration $c^\nabla = 0.040$ mol·dm^{-3} (corresponding to $P^\nabla = 1$ bar) and in the case of water as a solvent, $\Delta_c^\ominus s_d$ evaluates to -26.67 J·mol^{-1}·K^{-1} at T^- (using $\rho_{wat}^\ominus = 997$ kg·m^{-3} from Table 1.1). The corresponding value of $\Delta_c^\ominus \mu_d$ is 7.95 kJ·mol^{-1} at T^-. This convention, or rather a $aaae_T$ convention, was *e.g.* used in Refs.[733, 821, 1748], the corresponding correction terms (neglecting the influence of the reference pressure change *via* Eqs. 4.190 and 4.191) for water at T^- being $\Delta_c s_d^\ominus = -26.56$ J·mol^{-1}·K^{-1} and $\Delta_c \mu_d^\ominus = 7.92$ kJ·mol^{-1} (a corresponding correction is also required for the gas-phase properties *via* Eqs. 4.192 and 4.193). This convention is convenient because, just like for *e.g.* $bmme_T$ or $bMMe_T$ (see above), the effective reference concentration is the same in the gas phase as in solution, which implies that standard-state correction terms like $\Delta_s G_{std}^\ominus$ in Eq. 2.9 (or 4.178) can be omitted. Furthermore, unlike for *e.g.* $bmme_T$ or $bMMe_T$ (see above), the gas-phase reference pressure is equal to the condensed-phase reference pressure. However, this comes at the cost of introducing a very unusual and temperature-dependent concentration unit in solution.

Fifth, consider the use of a $x^\nabla = 1$ *mole fraction reference concentration* for *solutions*, as opposed to a $b^\circ = 1$ mol·kg^{-1} molal reference concentration in the $bbme_T$ convention, *i.e.* the conversion $bbxe_T \rightarrow bbme_T$. Here again, the conversion only concerns the solution standard state. It can be viewed as a change of the reference mole fraction from x^∇ to $x^\circ = b^\circ/(b^\circ + M_{svt}^{-1})$ where M_{svt} is the molar mass of the solvent, leading to

$$\Delta_c^\circ h_d = \int_{x^\nabla}^{x^\circ} dx\, \partial_x h_d(x, P^\circ, T) \quad (bbxe_T \rightarrow bbme_T) \tag{4.200}$$

and

$$\Delta_c^\circ s_d = \int_{x^\nabla}^{x^\circ} dx\, \partial_x s_d(x, P^\circ, T) \quad (bbxe_T \rightarrow bbme_T) \quad . \tag{4.201}$$

The evaluation can be performed analytically for the solution standard state, assuming ideal-mixture behavior (Section 4.2.3), which implies $\partial_x s_d(x, P, T) = -Rx^{-1}$ and $\partial_b h_d(x, P, T) = 0$, leading to

$$\Delta_c^\circ h_d = 0 \quad (bbxe_T \rightarrow bbme_T) \tag{4.202}$$

and

$$\Delta_c^\circ s_d = R \ln[x^\nabla (M_{svt} b^\circ + 1)/(M_{svt} b^\circ)] \quad (bbxe_T \rightarrow bbme_T) \quad . \tag{4.203}$$

For an alternative solution reference mole fraction $x^\nabla = 1$ and in the case of water as a solvent, $\Delta_c^\circ s_d$ evaluates to 33.54 J·mol^{-1}·K^{-1} irrespective of T (using $M_{svt} = 0.01802$ kg·mol^{-1} from Table 1.1). The corresponding value of $\Delta_c^\ominus \mu_d$ is -10.00 kJ·mol^{-1} at T^-.

The special case of a *concentration-consistent* $fffe_T$ standard-state convention, where f is a measure of pressure (a or b) or concentration (m, M or x), *e.g.* the $aaae_T$ and $bbbe_T$ conventions discussed above, deserves some further comments in terms of their connection with the semi-standard point-to-point solvation parameters introduced in Section 2.2 and discussed in Section 4.2.16. Although the gas-phase and dissolved species have the same effective molar volumes in a $fffe_T$ standard-state convention, so that

$$\Delta_s G^{\triangledown} = \Delta_s G^{\ominus} \quad (fffe_T) \quad , \tag{4.204}$$

corresponding identities do not hold for the derivative thermodynamic properties. In both the standard and density-corrected variants of the solute standard-state definition (Section 4.2.5), $\Delta_s Y^{\triangledown}$ encompasses, in addition to $\Delta_s Y^{\ominus}$, a contribution related to the ideal-gas influence of the species on the molar volume, present in the gas state but absent in the dissolved state. In the standard variant, a further contribution to $\Delta_s Y^{\triangledown}$ is related to the pressure or/and temperature dependence of the solvent density. Considering the density-corrected variant, the correct relationship between $fffe_T$ standard and point-to-point parameters reads

$$\Delta_s Y^{\triangledown *} = \Delta_s Y^{\ominus} + \Delta_s Y_{std}^{\triangledown *} \quad (fffe_T) \quad , \tag{4.205}$$

where $\Delta_s Y_{std}^{\triangledown *}$ is given in analogy with Eqs. 4.178 and 4.183 by 0, $-RT^{-}$, $-R$, $-R$, $-RT^{-}(P^{\nabla})^{-1}$, $-RT^{-}(P^{\nabla})^{-2}$ and $-R(P^{\nabla})^{-1}$ for $Y = G$, H, S, C_P, V, K_T and A_P, respectively. Corresponding correction terms $\Delta_s Y_{std}^{\triangledown}$ appropriate for the standard variant are easily derived in analogy with Eq. 4.184 along with the expressions provided in Table 4.1, and include the additional contribution related to the pressure or/and temperature dependence of the solvent density. Finally, because $\Delta_s Y_{std}^{\triangledown *}$ is identical to the corresponding ideal gas-phase molar variable for $Y = V$, K_T and A_P, one has (see Eq. 4.187)

$$y^{\triangledown *} = \Delta_s Y^{\ominus} \quad (fffe_T), \quad \text{for } Y = V, K_T, A_P \quad , \tag{4.206}$$

implying that the point-to-point solvation volumes, volume-compressibilities and volume-expansivities are identical to the corresponding partial molar parameters of the dissolved ion according to a $fffe_T$ convention.

The sixth case concerns alternative standard-state definitions for the *gas-phase electron* e_g^-. As discussed in Section 4.2.13, the choice of a convention will affect the standard ionization parameters of an atom, as well as the standard formation (but not molar or partial molar) parameters of any gas-phase or solvated ionic species[259,1723,1750].

The treatment of the electron in the $bbme_T$ definition (Section 4.2.4) relies on what is usually referred to as the "electron convention"[294,1723], termed in this book the *warm-electron convention*, and denoted here by e_T. In this convention, the standard state of the electron is the ideal gas-phase electron at temperature T (Section 4.2.13). The alternative convention is usually referred to as the "ion convention"[1723,1730], termed in this book the *cold-electron*

Table 4.3 *Correction terms for (partial) molar variables corresponding to alternative standard-state conventions.* For different alternative conventions, the correction terms (Eq. 4.188) to be applied to the (partial) molar enthalpy ($\Delta_c^\circ h$), entropy ($\Delta_c^\circ s$) and chemical potential ($\Delta_c^\circ \mu$) of a species at a given temperature T so as to obtain a value appropriate for the $bbme_T$ convention used throughout this book are reported. The correction terms to be applied to distinct states are listed separately. Unless otherwise specified (footnotes), absent entries indicate states for which no correction is required, and the numerical values refer to water as a solvent, based on data from Table 1.1 and to a reference temperature T^-, as well as to $P^\nabla = 1$ atm, $c^\nabla = 1$ mol·dm^{-3} and $x^\nabla = 1$.

Convention	State	$\Delta_c^\circ h$ [kJ·mol^{-1}]		$\Delta_c^\circ s$ [J·mol^{-1}·K^{-1}]		$\Delta_c^\circ \mu$ [kJ·mol^{-1}]	Eqs.	Notes
		Eq.	Num.	Eq.	Num.	Num.		
$aame_T$	g	0	0	$-R\ln(P^\circ/P^\nabla)$	0.11	-0.03	4.192,4.193	a,b,c
$aaMe_T$	g	0	0	$-R\ln(P^\circ/P^\nabla)$	0.11	-0.03	4.192,4.193	a,b,c
id.	d	0	0	$-R\ln(b^\circ\rho_{svt}^\ominus/c^\nabla)$	0.02	-0.006	4.198,4.199	c
$bMme_T$	g	0	0	$-R\ln[P^\circ/(RTc^\nabla)]$	26.69	-7.96	4.194,4.195	a
$bmme_T$	g	0	0	$-R\ln[P^\circ/(RTb^\circ\rho_{svt}^\ominus)]$	26.67	-7.95	4.194,4.195	
$bMMe_T$	g	0	0	$-R\ln[P^\circ/(RTc^\nabla)]$	26.69	-7.96	4.194,4.195	a
id.	d	0	0	$-R\ln(b^\circ\rho_{svt}^\ominus/c^\nabla)$	0.02	-0.006	4.198,4.199	
$bbMe_T$	d	0	0	$-R\ln(b^\circ\rho_{svt}^\ominus/c^\nabla)$	0.02	-0.006	4.198,4.199	
$bbxe_T$	d	0	0	$-R\ln\{(M_{svt}b^\circ)/[x^\nabla(M_{svt}b^\circ+1)]\}$	33.54	-10.00	4.202,4.203	b
$bbbe_T$	d	0	0	$-R\ln(RTb^\circ\rho_{svt}^\ominus/P^\circ)$	-26.67	7.95	4.192,4.193	d
$aaae_T$	g	0	0	$-R\ln(P^\circ/P^\nabla)$	0.11	-0.03	4.192,4.193	a,c
id.	d	0	0	$-R\ln(RTb^\circ\rho_{svt}^\ominus/P^\nabla)$	-26.56	7.92	4.198,4.199	c

(a) the reported numerical values for $\Delta_c^\circ h$, $\Delta_c^\circ s$ and $\Delta_c^\circ \mu$ do not depend on the choice of the solvent; (b) the reported numerical values for $\Delta_c^\circ h$ and $\Delta_c^\circ s$ do not depend on the choice of the temperature T^-; (c) corrections to standard properties of the liquid, solid, solvent and solute standard state due to the slight change in reference pressure are in principle also required (Eqs. 4.190 and 4.191), but typically negligible; (d) see also Eq. 2.9 (or 4.178).

convention, and denoted here by e_0. In this convention, the standard state of the electron is the ideal gas-phase electron at temperature $T = 0$ K, rather than T^- (Section 4.2.12). If this change represents the only alteration in the standard-state definition compared to the $bbme_T$ convention, the resulting convention will be noted $bbme_0$. To convert thermodynamic properties of the e_0 to the e_T convention ($bbme_0 \rightarrow bbme_T$), it is necessary to heat up the electron from the absolute zero temperature to the temperature T (Section 4.2.9). The gas-phase single-ion formation enthalpies can thus be converted as

$$\Delta_f H^\circ[I_g^z] = \Delta_f H^\nabla[I_g^z] + z\Delta_h H^\circ[e_g^-] \quad (bbme_0 \rightarrow bbme_T) \quad . \tag{4.207}$$

The e_0 convention is mostly used in the context of gas-phase ion spectroscopy[1730,1751,1752]. As already mentioned in Section 4.2.13, the reason is that spectroscopic experiments concerned with ionization processes typically rely on the absorption or emission of an electron with a measurable kinetic energy. After subtraction of this kinetic energy, one is left with values characterizing a semi-adiabatic process, *i.e.* a process involving ions at finite temperature and electrons at zero temperature. The choice of a convention affects electrochemistry in the context of absolute electrode potentials (Section 4.3.6) *via* the spectroscopic determination of metal work functions (Section 4.3.5), see Point B_{12} in Section 5.14, or *via* the free energies of formation of gas-phase ions (Section 4.2.10), see Point A_{12} in Section 5.14. It also affects single-ion solvation thermodynamics[483] in the context of cluster-based approximations to estimate real solvation parameters, see Point E_{12} in Section 5.14, or intrinsic solvation parameters, see Point E_{14} in Section 5.18, as well as of the air-liquid interfacial potential of a solvent, based on these parameters (Section 5.17). The major drawback of the e_0 convention is that entropies cannot be consistently defined[1723]. This is because the electron standard state is temperature-independent, *i.e.* it always refers to the electron at the absolute zero temperature, so that its standard-state molar entropy (and isobaric heat capacity) is zero at all temperatures.

The issue of the choice of a convention, e_T or e_o, for the standard-state electron should not be confused with another issue concerning the appropriate calculation of molar parameters for the gas-phase electron (and proton). As discussed in Section 4.2.13, this calculation should rely on Fermi-Dirac rather than Boltzmann statistics[1723]. Irrespective of the selected electron convention, thermodynamic parameters derived based on Boltzmann statistics for the ideal electron gas will be slightly erroneous, and must be corrected as described in Section 4.2.13.

In the context of gas-phase ion spectroscopy, the problem of appropriate electron statistics has been recognized for about two decades[1723]. In the context of electrochemistry and ion solvation, the problem of appropriate electron statistics has been raised only recently[278,483,1727,1728], and recent tabulated data, most commonly within the e_T convention, typically incorrectly refers to Boltzmann statistics.

4.2.18 Determination of thermodynamic parameters

To obtain a set of *standard thermodynamic parameters* that is as complete as possible, the following *key approaches* are used in combination: (*i*) determination of the standard (partial) molar volume, volume-compressibility or volume-expansivity of a species in a given phase (densimetry, compressimetry, dilatometry or analytical formula); (*ii*) determination of the standard (partial) molar heat capacity of a species in a given phase (calorimetry or SBS); (*iii*) determination of the standard molar entropy of a pure species in a given phase (TCI or SBS); (*iv*) determination of the standard enthalpy (and heat capacity) of a reaction (RXC or SBS); (*v*) determination of the standard entropy of a reaction involving pure species (TLC); (*vi*) determination of all standard reaction parameters (chemical or electrochemical equilibrium measurements). Unknown thermodynamic parameters may also often be deduced from previously determined ones through the application of the Hess law and the use of thermodynamic cycles (Section 4.2.15). These approaches are briefly sketched below. In the context of derivative thermodynamic parameters involving solutes, it should be kept in mind that standard results may be formulated either in the standard or in the density-corrected variant, the natural variant depending on the type of determination method employed (Section 4.2.5).

The determination of the standard *molar volume* of a pure solid or liquid species is in principle a straightforward task (densimetry), requiring the knowledge of the molar mass and the measurement of the density at P° and T^-. For gases, including the gas-phase electron and proton, the standard molar volume is analytical $v^\ominus = RT^-(P^\circ)^{-1}$ (Sections 4.2.12 and 4.2.13). The determination of the *standard partial molar volume* of a species in a mixture (*e.g.* $v^\ominus$ for a solute in solution) is also relatively easy, requiring measurements of the volume of the mixture as a function of the concentration of the solute at P° and T^-, and extrapolation to infinite dilution. For dissolved ions, only sums over overall neutral ion sets can be determined in this way. Alternatively, one may evaluate the standard reaction volume of a reaction (see below) involving this species along with other species for which standard (partial) molar volumes are known. The evaluation of the *(partial) molar volume-compressibilities and volume-expansivities* of pure species or species within mixtures requires to perform the evaluation of (partial) molar volumes (or reaction volumes) at different pressures (compressimetry) or at different temperatures (dilatometry), respectively. For gases, including the gas-phase electron and proton, the standard molar volume-compressibility and volume-expansivity are analytical, $k_T^\ominus = RT^-(P^\circ)^{-2}$ and $a_P^\ominus = R(P^\circ)^{-1}$ (Sections 4.2.12 and 4.2.13).

The determination of the *standard molar heat capacity* of a pure liquid or solid species is a straightforward task (calorimetry), requiring the measurement of the temperature increase produced in one mole of the substance by the supply of an infinitesimal amount of heat at P° and T^-. For pure gaseous species, however, it is often either more convenient (neutral species) or only possible (charged species) to use SBS (Section 4.2.12; see also Section 4.2.13 for the electron and proton). Finally, the determination of the *standard partial molar heat capacity* of a species in a mixture (*e.g.* solute in solution) is also relatively easy, requiring measurements of the heat capacity of the mixture as a function of the concentration of the solute at constant P° and T^-, and extrapolation to infinite dilution. For dissolved ions, only sums over overall neutral ion sets can be determined. Alternatively, one may evaluate the standard reaction heat capacity of a reaction (see below) involving this species along with other species for which standard (partial) molar heat capacities are known.

The *standard molar entropy* of a pure species can be determined using TCI (Section 4.2.11). For pure gaseous species, however, it is often either more convenient (neutral species) or only possible (charged species) to use SBS (Section 4.2.12; see also Section 4.2.13 for the electron and proton). The *standard partial molar entropy* of a species in a mixture (*e.g.* solute in solution) cannot be determined using these two methods. It must be inferred from the molar entropies of the pure constituents and the entropy of the mixing reaction (see below) monitored as a function of the concentration of the solute at constant P° and T^-, and extrapolated to infinite dilution.

The heat absorbed or released by a spontaneous reaction in a closed container with identical initial and final pressures is related to the corresponding enthalpy change (Eq. 4.43, approximated as $\Delta H \approx Q$ for nearly isobaric conditions; alternatively, the reaction can be performed under isochoric conditions, and an exact expression can then be applied along with compressibility corrections for the reactants and products). This permits the determination of a *standard reaction enthalpy* through calorimetry, after extrapolations or/and corrections to match the appropriate standard-state definition. For the ease of further reference, this procedure will be referred to as *reaction calorimetry* (RXC). The *standard reaction heat capacity* can also be determined through an analysis of the temperature dependence of the reaction enthalpy (Eq. 4.22). Note that formation reactions are special cases of reactions, so that standard formation enthalpies (and heat capacities) are amenable to a similar determination by RXC. Due to the absence of gas-phase ions (Section 2.1), RXC cannot be applied to processes involving such species (*e.g.* ionization). However, in this case, reaction enthalpies can be determined using SBS (Section 4.2.12; see also Section 4.2.13 for the gas-phase proton and electron). Finally, *standard reaction entropies* for reactions involving pure species can be inferred from the molar entropies of the involved species *via* a TLC (Section 4.2.15).

The most favorable situation for the determination of the *standard reaction parameters* associated with a given reaction is when this reaction can be carried out under reversible conditions. This happens in two main situations: (*i*) uncoupled system at chemical equilibrium; (*ii*) coupled system at electrochemical equilibrium (electrochemistry at zero current). When an uncoupled system is at chemical equilibrium, its composition is constrained by the equality between the composition-dependent reaction quotient (Eq. 4.138) and the thermodynamically-determined equilibrium constant (Eq. 4.137). If the composition of the system at equilibrium can be determined, along with estimates for the activity coefficients, the *standard reaction free energy* becomes directly accessible through Eq. 4.137. When an electrochemically-coupled system is at equilibrium, the measured electric potential difference is also constrained to a unique value related to the reaction quotient and equilibrium constant *via* the Nernst equation (Eq. 4.144). Using an independent determination of the equilibrium system composition, as well as estimates for the activity coefficients or suitable extrapolations to infinite dilution, the standard reaction free energy becomes directly accessible through Eq. 4.137. The electrochemistry approach is generally easier, since the solute concentrations can be varied arbitrarily, permitting extrapolations or measurements in a low concentration regime, where the activity coefficients can be neglected or appropriately approximated. In both cases, further analysis of the temperature dependence of the reaction free energy leads to the *standard reaction enthalpy and entropy.*

In principle, the determination of the *standard reaction heat capacity, volume, volume-compressibility and volume-expansivity* is also possible by considering the second derivatives of the reaction free energy with respect to pressure or/and temperature. Note again that formation reactions are special cases of reactions, and therefore often amenable to such types of equilibrium chemical or electrochemical determinations. This approach is sometimes referred to as the *second-law method* for the determination of formation enthalpies[1733,1753] (see Section 5.9 for the related approach called the *third-law* method).

The combined application of the five above types of approaches to obtain standard thermodynamic parameters relevant to ionic processes, as well as the resulting values in the context of alkali and halides with water as a solvent, will be further discussed in Chapter 5.

4.3 Interfacial effects

As discussed in Section 2.2, one of the main factors of complexity in the determination of thermodynamic single-ion solvation properties is related to the presence of *surface effects*, leading to the necessity of distinguishing between *real* and *intrinsic* parameters, the latter ones being more valuable in terms of physical interpretation, but *experimentally elusive* (Section 2.3). The only relevant component of these surface effects or, more generally, of interfacial effects in the context of single-ion solvation at infinite dilution is *electrostatic* in nature, and the goal of the present section is to illustrate how this component can be understood on its own, *i.e.* independently from bulk solvation effects. To this purpose, one first needs to introduce clear definitions for concepts related to the electrostatic influence of phase interfaces in molecular systems containing ions, leading to various forms of *interfacial charge distributions, electric potentials, chemical potentials* and *absolute electrode potentials*.

The different quantities introduced are distinct in three possible respects: (*i*) whether they *include or exclude specific interfacial contributions*, *e.g.* contributions of free and bound charges to the interfacial charge distribution, contributions of surface charge and polarization to the electric potential or the effective chemical potential of charged species; (*ii*) the choice of a *reference point for the electric potential of the standard-state electron* (ideal electron gas), required for the definition of absolute electrode potentials; (*iii*) the consideration of *test charges with or without excluded volume*, *i.e.* probing the potential outside the envelope of atomic or molecular species or at any point within a phase, respectively.

Accordingly, the concepts and definitions introduced include: (*i*) the *free interfacial charge distribution* $\tilde{\rho}_F$ in a system of phases, including the surface charges and polarizations of all phases, but excluding the bound interfacial charge distribution resulting from their (di)electric response; (*ii*) the *total interfacial charge distribution* ρ_T in a system of phases, also including the bound interfacial charge distribution resulting from the bulk (di)electric response of the phase; (*iii*) the *Galvani (or inner) potential* of a conducting phase, either *external* ϕ, corresponding to a test charge with excluded volume, or *internal* $\breve{\phi}$, corresponding to a test charge without excluded volume; (*ii*) the *Volta (or outer) potential* of a conducting phase, either *lateral* ψ, *i.e.* measured at an interface to vacuum (air), or *frontal* $\tilde{\psi}$, *i.e.* measured at an interface with another conducting phase; (*iii*) the *surface (or dipolar) potential* χ of a phase, either *lateral* χ or *frontal* $\tilde{\chi}$ and, for the lateral one, either *external* χ or *internal* $\breve{\chi}$; (*iv*) the *effective chemical potential* of a species in a conducting phase, either *chemical* (intrinsic) μ, *real* $\hat{\mu}$, or *electrochemical* $\tilde{\mu}$, depending on the inclusion or exclusion of specific electric potential contributions; (*v*) the *work function* of a metal electrode, either *spectroscopic* w, *i.e.* associated to the cold-electron convention and thus non-standard according to the definitions used in this book, or *standard* $\omega^{\ominus}$, *i.e.* associated to the warm-electron convention; (*vi*) the standard *absolute potential of an electrode*, either *vacuum* $\tilde{\mathcal{V}}^{\ominus}$, *intrinsic* $\mathcal{V}^{\ominus}$, or *real* $\hat{\mathcal{V}}^{\ominus}$, depending on the reference point selected for the electric potential of the standard-state electron.

4.3.1 Free and bound interfacial charge distributions

A *phase* is a region of a molecular system characterized by homogeneous values for all the local observables corresponding to intensive quantities (Section 4.2.1), except in its surface region (see below). In a *closed system of phases* at equilibrium under *constant pressure and temperature* conditions, *mechanical equilibrium* implies a constant effective pressure and *thermal equilibrium* a constant temperature throughout the entire system, assuming that volume work and heat can be freely exchanged between all phases. The pressure considered is effective in the sense that it includes a contribution due to interfacial-tension effects. *Chemical equilibrium* also implies identical effective chemical potentials in all phases, for any species present in multiple phases and that can be freely exchanged between them. The chemical potential considered is also effective in the sense that it includes effects related to the electrostatic influence of interfaces and will be later referred to as the electrochemical potential (Section 4.3.4). However, besides these constraints, the phases may differ in terms of all other intensive variables as a result of their distinct compositions.

Consider a *system of phases* Ω at equilibrium, such as the one illustrated schematically in Figure 4.3a. The different *phases* occupy finite regions of space that will be assumed here to be of macroscopic extent. Each phase is characterized by a specified homogeneous *composition*, constant in time for a system at equilibrium. The *system of phases* is defined by a given *geometry*, as defined by the extents, shapes, orientations and locations of the different phases, that will also be assumed constant in time for simplicity, *i.e.* possible macroscopic relative motions or shape distortions of the phases are not considered. The system is naturally bounded by a *vacuum (air) phase*, and may encompass parts of phases that extend beyond the system boundary. These phases will be referred to as *electrodes*. As illustrated in Figure 4.3a, the portions of the electrodes that are inside the system may be surrounded by the vacuum (air) phase (left) or be in direct contact with other phases of the system (right). The physical properties of the electrode, both inside and outside the system, are also assumed time-independent. The two-dimensional surface Σ encompassing the contact regions of all pairs of phases in the system, including those involving the bounding vacuum (air) phase, will be termed the *interfacial region*. The portion of this region separating two specific phases P_1 and P_2 will be referred to as the *interface* between P_1 and P_2. As also illustrated in Figure 4.3a, it is convenient to define a complementary system $\overline{\Omega}$ corresponding to the surroundings of Ω, along with an associated interfacial region $\overline{\Sigma}$. It is assumed that $\overline{\Omega}$ solely contains the portions of the electrodes

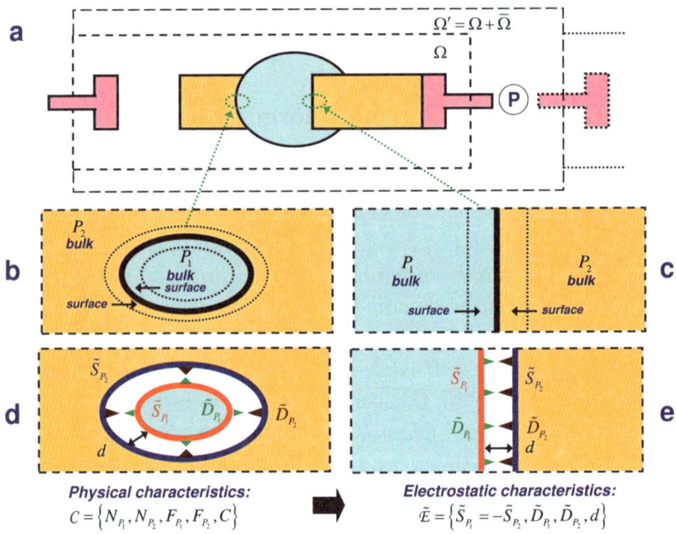

Figure 4.3 *Schematic illustrations regarding the properties of a system of phases at equilibrium.* Panel (a) represents a system of phases Ω (interfacial region Σ) and its exterior $\overline{\Omega}$ (interfacial region $\overline{\Sigma}$), constituting an extended system $\Omega' = \Omega + \overline{\Omega}$ (surface region $\Sigma' = \Sigma + \overline{\Sigma}$). The system encompasses four types of phases (indicated here by white, orange, blue and violet colors), including a bounding vacuum (air) phase (white) and two electrodes (violet). The possible presence of a potentiostat (circled P) between the exposed surface of the two electrodes (the left electrode is drawn again on the right) is also indicated. Panels (b) and (c) represent a system of two phases P_1 and P_2 in contact at a spherical (b) or planar (c) interface with homogeneous physical interface characteristics $\mathcal{C}$ (Eq. 4.209), in the hypothetical situation where the two phases are assumed exempt of any other interface. The physical characteristics include the natures N_{P_1} and N_{P_2} of the phases in contact, the types F_{P_1} and F_{P_2} of the crystallographic faces they expose at the interface (monocrystalline solids only, implying a planar interface), and the interface curvature C. Panels (d) and (e) represent the same systems as in panels (b) and (c), respectively, in terms of electrostatic interface characteristics $\tilde{\mathcal{E}}$ (Eq. 4.210). The phases are assumed to be separated by an infinitesimal vacuum (air) gap of width d, the compensating surface charges $\tilde{S}_{P_1}$ and $\tilde{S}_{P_2}$ (thick red and blue lines) being situated in the phases, immediately at their surface, and the surface polarizations (green and brown triangles) in the vacuum (air) gap, immediately at the surface of the phases.

protruding outside Ω, possibly along with a device capable of maintaining a net non-equilibrium charge separation between their exposed surfaces. This device will be referred to as a *potentiostat*. In some experimental situations, a *potentiometer* is used instead, but since this device performs a potential-difference measurement ideally without perturbing the system, its possible presence need not be considered here. In most practical situations, there will be at most two electrodes consisting of conducting phases of the same nature (typically metal), and at most one potentiostat or potentiometer. The extended system $\Omega' = \Omega + \overline{\Omega}$ will be assumed to be itself surrounded by vacuum (air) and free of external electrostatic influences. The extended interfacial region will be noted $\Sigma' = \Sigma + \overline{\Sigma}$. Finally, the assumption that the system of phases Ω is at equilibrium not only implies the time-invariance of all properties, but also the fulfillment of a *local electroneutrality constraint* (Section 2.1). This constraint implies that only volume elements of truly microscopic sizes within the system, including at interfaces, may contain an excess of positive or negative charges at a given instant. Time-invariance of all properties is also assumed for $\overline{\Omega}$, but not local electroneutrality, which may be violated when a potentiostat is active. According to the definition formulated in Section 2.1, this situation will still be referred to as an equilibrium situation for the system. It is sufficient to capture the design of equilibrium electrochemistry experiments (Sections 4.4.3 and 4.4.4) but excludes the consideration of electrophysics or electrochemistry at non-zero current (*e.g.* electrophoresis, electrolysis), even in the case of a time-independent steady-state system configuration.

Intuitively, one expects the local properties of the interface at any point $\mathbf{x} \in \Sigma$ to depend on the following properties: (*i*) the *nature of the phases* P_1 and P_2 in contact at that point; (*ii*) the *surface characteristics* of P_1 and P_2; (*iii*) the *interface curvature* at that point. For points $\mathbf{x} \in \overline{\Sigma}$, one must add to this list the effect of a possible external potentiostat, maintaining a net (non-equilibrium) charge separation between the exposed surfaces of the electrodes.

In practice, *surface characteristics* are only relevant for *monocrystalline solids*, which may expose distinct types of crystallographic faces at an interface. On the other hand, *interface curvature* is relevant for all condensed phases, including liquids, solids with amorphous or polycrystalline surfaces (assembly of microcrystallites in random orientations), and monocrystalline solids (where the edges and vertices associated with planar faces can be viewed as sharply curved regions). These properties will be termed the *local physical interface characteristics* and collectively denoted by a set $\mathcal{C}(\mathbf{x})$, with

$$\mathcal{C}(\mathbf{x}) = \{N_{P_1}, N_{P_2}, F_{P_1}, F_{P_2}, C\}_{\mathbf{x}} \qquad \mathbf{x} \in \Sigma' \quad , \tag{4.208}$$

where N_{P_1} and N_{P_2} indicate the natures of the phases P_1 and P_2 in contact at $\mathbf{x}$, F_{P_1} and F_{P_2} the type of crystallographic face they expose to the interface at this point (only relevant for monocrystalline solids), and C the corresponding local interface curvature.

To examine the nature of local interfacial effects in the simplest possible context, it is convenient to refer first to the hypothetical situation of a *single interface* between two phases P_1 and P_2 *exempt of any other interfaces*, and that is in addition characterized by *homogeneous local physical interface characteristics*, *i.e.* with

$$\mathcal{C}(\mathbf{x}) = \mathcal{C} = \{N_{P_1}, N_{P_2}, F_{P_1}, F_{P_2}, C\} \qquad \forall \mathbf{x} \in \Sigma \quad . \tag{4.209}$$

This interface may be either a sphere enclosing a finite phase P_1 inside an infinite phase P_2, as depicted in Figure 4.3b, or an infinite plane between two phases P_1 and P_2 of infinite extents, as depicted in Figure 4.3c. If P_1 or/and P_2 is a monocrystalline solid, the interface must be planar (*i.e.* only the second situation is possible), and a single type of crystallographic face must be exposed to the interface by the specific phase. Clearly, the situation considered is entirely specified by the set $\mathcal{C}$ of homogeneous physical interface characteristics, *i.e.* this set encompasses all the information specifying the physics of the problem. This situation is, however, *hypothetical* in the sense that no real system of phases can exactly adopt such a geometry. In the planar case, unless both P_1 and P_2 are vacuum (air), the two phases and the interface must unavoidably terminate with the appearance of other phases (in a non-periodic and Euclidean space). Similarly, in the spherical case and unless the enclosing phase P_2 is vacuum (air), it must also be surrounded by other phases. Nevertheless, this hypothetical situation provides the framework for the discussion of homogeneous single-interface effects, the implications of inhomogeneities in the interface characteristics, of correlations between multiple interfaces, and of a non-equilibrium charge distribution at the exposed surfaces of the electrodes, as enforced by a possible external potentiostat being considered subsequently.

The *bulk region* of a phase represents the region of this phase that is, in terms of microscopic forces, beyond the range of influence of all its interfaces. A species (atom, molecule, ion) in this region is "unaware" of the finite extent of the phase, and subject to an environment that is, in terms of time average, either homogeneous and isotropic (liquids and gases), periodic and symmetric (crystalline solids), or a mixture of both ("semi-ordered" phases, *e.g.* liquid crystals or lipid bilayers). In contrast, the *surface region* of a phase is under the influence of microscopic forces from the bulk on one side, and from another phase across the interface on the other side. For this reason, a species in this region is subject to a heterogeneous and anisotropic (liquids and gases) or non-periodic and asymmetric (crystalline solids) environment. Taking the bulk environment as a reference, the alteration of the local properties of a phase in its surface region, induced by the proximity of an interface to the surface region of another phase, will be referred to as the *surface alteration*. In an arbitrary system of phases, the surface alteration of a phase at a given interface is generally not a local property, *i.e.* it depends on properties of the entire system of phases, including the nature of all phases, the system geometry, and the possible presence of a potentiostat. However, in the hypothetical situation of Figure 4.3b,c, the surface alteration of the two phases in contact is obviously uniquely determined by the nature of the single interface *via* the unique set $\mathcal{C}$ of physical interface characteristics. Other important exceptions to the above general statement will be discussed in Section 4.3.3.

For *liquids or gases*, the positional and, whenever relevant, orientational distributions of the different species in the surface region are not uniform, as they were in the bulk region. The electron clouds and geometries of these species may also be somewhat distorted. In the bulk region, the time-averaged charge density, contributed by the nuclei and electrons of the different species, is homogeneous and isotropic. Furthermore, due to the electroneutrality constraint for macroscopic matter at equilibrium (Section 2.1), it is vanishing. In the surface region, however, this average charge density is generally neither homogeneous and isotropic nor vanishing, *i.e.* it displays variations as a function of the distance to the interface and can possibly integrate to a net charge, provided that this charge is locally compensated by an opposite charge in the surface region of the adjacent phase, on the other side of the interface. This non-uniform charge density, possibly involving a net charge, represents the surface alteration in this case. Considering the hypothetical situation of Figure 4.3b,c, the only property of the interface that can affect the forms and magnitudes of the surface alterations in the two phases in contact, besides the natures N_{P_1} and N_{P_2} of P_1 and P_2, is the *interface curvature* C.

For *crystalline solids*, it is predominantly the electron density which is affected by the presence of the interface, with more limited changes in the positional distributions of the nuclei. In the bulk region, the time-averaged charge density, contributed by the nuclei and electrons within the crystal, is periodic and symmetric. Furthermore, due to the electroneutrality constraint for macroscopic matter at equilibrium (Section 2.1), its integral vanishes over any volume encompassing an integer number of unit cells. In the surface region, however, this charge density is generally no longer exactly periodic and symmetric, and may present a non-vanishing integral over unit cells. In this case, it is the difference between the bulk-like charge density, taken as a reference, and the corresponding local charge density in the surface region that represents the surface alteration. This difference in charge density, which is zero by definition in the bulk region, becomes non-vanishing in the surface region, *i.e.* it displays variations as a function of the distance to the interface and can possibly integrate to a net charge, provided that this charge is locally compensated by an opposite charge in the surface region of the adjacent phase, on the other side of the interface. In the context of a single planar interface (Figure 4.3c) between a crystalline solid and vacuum (air), the surface alteration is often referred to as an *electron spillout*[98,1754], because it predominantly originates from an expansion of the electron density away from the crystal. This expansion is caused by the electron-electron repulsions between atoms of successive layers on the bulk side, which is absent on the vacuum (air) side. The magnitude of the spillout phenomenon is expected to depend on the type of the exposed crystallographic face, *via* the corresponding packing density of the successive atom layers parallel to this face[98]. Considering the hypothetical situation of Figure 4.3b,c, when phase P_1 or/and P_2 is a monocrystalline solid, the interface must be planar (Figure 4.3c) and the only property of the interface that can affect the form and magnitude of the surface alterations of the two phases in contact, besides the natures N_{P_1} and N_{P_2} of P_1 and P_2, is the type F_{P_1} or/and F_{P_2} of *crystallographic face exposed at the interface* by P_1 or/and P_2.

From the point of view of interfacial properties, *amorphous solids* as well as *solids with polycrystalline surfaces* behave similarly to liquids and gases (see above), with the possibility of a curved interface and surface properties affected by the corresponding interface curvature.

The surface alteration of the charge density can be expanded locally in terms of its *surface multipole moments.* The most relevant moments (see further below) are the zeroth-order moment, which is a *surface charge density*, and the first-order moment, which is a normal *surface polarization.* At the molecular level and in the context of liquids or gases, the surface charge density may be viewed as resulting from an excess of positive or negative species in the surface region (conductor only, *e.g.* electrons in a liquid metal, ions in an electrolyte solution, or charged particles in a gas subjected to ionizing radiation), while the surface polarization may be viewed as resulting from inhomogeneous positional and, whenever relevant, orientational distributions of charged and dipolar species, respectively, in this region, along with a smaller contribution related to electronic polarization effects and distortion of the molecular geometries. In the context of crystalline solids, the surface charge density may be viewed as resulting from an excess or lack of surface electrons (conductor only, *e.g.* electrons in a solid metal), while the surface polarization may be viewed as resulting principally from the surface alteration of the electron density (*e.g.* electron spillout at the surface of an isolated crystal), along with a smaller contribution related to altered positional distributions of the nuclei.

Considering the hypothetical situation of Figure 4.3b,c, the fact that the bulk regions of the two phases are electroneutral, while the corresponding surface regions may present non-vanishing surface charge densities, implies that the two phases may formally carry an *overall net charge.* This observation, however, does not violate the electroneutrality constraint for macroscopic matter at equilibrium (Section 2.1). As long as the corresponding surface charges are of equal magnitudes and opposite signs, the system remains locally electroneutral. A similar situation would be encountered, for example, by mentally dividing a crystal through a plane that does not coincide with a symmetry plane. The two subcrystals may be formally highly charged without violating electroneutrality as a whole. However, this observation implies that when two isolated and electroneutral phases are brought in contact, they have to *exchange some charge* across the interface, if opposing surface charge densities are to be established while preserving the electroneutrality of the bulk. As will be seen later (Sections 4.3.2 and 4.3.3), when the interface permits an exchange of charge between the phases (non-polarizable interface), this charge equilibration process takes place directly across the interface, and ensures a unique resulting magnitude of the surface charge densities for two given types of phases, *i.e.* independent of any other property of the system of phases or of the setting of a possible potentiostat. The equilibrium situation at a non-polarizable interface is referred to[1755] as a *Nernstian (Faradaical) equilibrium.* When the interface does not permit such an exchange (polarizable interface), the magnitude of the surface charge densities depends in principle on the history of the system, *i.e.* on the way the phases were assembled. However, since the system is assumed to be at equilibrium, one has to take for granted that the charge equilibration process was permitted *via* an external circuit upon system assembly, also ensuring a unique resulting magnitude of the surface charge density for a given interface in a specific system, *i.e.* now possibly dependent on other phases in contact and on the setting of a possible potentiostat. Since the charge-carrying species of two phases at a polarizable interface are generally distinct (*e.g.* electrons *vs.* ions), the external circuit typically includes a non-polarizable interface at which a reaction (*e.g.* redox) equilibrates the two types of species. The equilibrium situation at a polarizable interface is referred to[1755,1756] as an *electrostatic (non-Faradaical) equilibrium at a given potential.*

These *charge-equilibrium conditions* are implied in most of the *electrochemistry experiments* considered in this book, that involve systems exclusively encompassing non-polarizable interfaces, prominently equilibrium Galvanic cell measurements (Section 4.4.3), or systems also encompassing polarizable interfaces (or air gaps), prominently equilibrium Voltaic cell measurements (Section 4.4.4) and electrocapillary measurements (Section 4.4.5). An interface involving locally compensating surface charge densities is often referred to as an *electrified interface* and the corresponding interfacial charge distribution, referring to surface charge densities only, as an *electric double-layer*[98,1754,1756–1767].

Considering the hypothetical situation of Figure 4.3b,c, the simplicity and symmetry of the systems considered imposes that: (*i*) the compensating surface charges $\tilde{S}_{P_1}$ and $\tilde{S}_{P_2}$ of the two phases P_1 and P_2 are homogeneous over the interface; (*ii*) the corresponding surface polarizations $\tilde{D}_{P_1}$ and $\tilde{D}_{P_2}$ are also homogeneous; (*iii*) these quantities are uniquely determined by the set $\mathcal{C}$ of homogeneous physical interface characteristics. In other words, as illustrated in Figure 4.3d,e, this set translates directly into a corresponding set of *electrostatic interface characteristics* $\tilde{\mathcal{E}}$, with

$$\tilde{\mathcal{E}} = \left\{ \tilde{S}_{P_1}, \tilde{S}_{P_2}, \tilde{D}_{P_1}, \tilde{D}_{P_2}, d \right\} \qquad \forall \mathbf{x} \in \Sigma \quad , \tag{4.210}$$

where the local electroneutrality constraint imposes $\tilde{S}_{P_2} = -\tilde{S}_{P_1}$ and d represents an *effective separation distance* between the compensating surface charges of the electric double-layer. This separation can be interpreted as an *infinitesimal vacuum gap* separating the two phases, the surface charges being situated inside the phases immediately at the start of the gap, and the polarizations inside the gap immediately at the start of the phases. Obviously, in the limit of an infinitesimal gap width, $\tilde{S}_{P_1}$ and $\tilde{S}_{P_2}$ must be increased proportionally to d, so as to preserve the dipole strength of the surface charges at the interface. This suggests that the electric double-layer can actually be interpreted as an *effective interfacial polarization* of magnitude

$$\tilde{D}^{eff}_{P_1,P_2} = d\tilde{S}_{P_1} = -d\tilde{S}_{P_2} \quad , \tag{4.211}$$

as measured in the direction $P_2 \rightarrow P_1$, *i.e.* that $\tilde{D}^{eff}_{P_1,P_2}$ is the physically relevant quantity rather than $\tilde{S}_{P_1}$, $\tilde{S}_{P_2}$ and d. The interfacial charge distribution can then be viewed as exclusively resulting in a total interfacial dipole moment of magnitude

$$\tilde{D}^{tot}_{P_1,P_2} = \tilde{D}_{P_2} - \tilde{D}_{P_1} + \tilde{D}^{eff}_{P_1,P_2} \tag{4.212}$$

as measured in the $P_2 \rightarrow P_1$ direction. However, the contributions $\tilde{D}_{P_1}$ and $\tilde{D}_{P_2}$ *vs.* $\tilde{D}^{eff}_{P_1,P_2}$ are of very distinct physical origins (see above). For this reason, they are kept separate in the following, and the quantities $\tilde{S}_{P_1}$, $\tilde{S}_{P_2}$ and d are retained instead of $\tilde{D}^{eff}_{P_1,P_2}$.

Coming back to the arbitrary system of phases of Figure 4.3a, it is now possible to assign to any point $\mathbf{x}$ of the interfacial region Σ a corresponding set $\tilde{\mathcal{E}}(\mathbf{x})$ of electrostatic characteristics based on the corresponding *local physical characteristics* $\mathcal{C}(\mathbf{x})$ and the *mapping* $\mathcal{C} \rightarrow \tilde{\mathcal{E}}$ defined by the hypothetical situations of Figure 4.3b (mapping to 4.3d) and 4.3c (mapping to 4.3e). This mapping is not uniquely defined on $\overline{\Sigma}$ when a potentiostat is active, because the surface charges on the exposed surfaces of the electrodes can be altered by this device and may violate the local electroneutrality constraint. In other words, the knowledge of the phases locally in contact at $\mathbf{x}$, of the exposed crystallographic faces at this point (for monocrystalline solids), and of the interface curvature at this point is sufficient to assign to the two phases surface charges and polarizations defining the local electrostatic characteristics at the considered point, *i.e.*

$$\tilde{\mathcal{E}}(\mathbf{x}) = \left\{ \tilde{S}_{P_1}, \tilde{S}_{P_2}, \tilde{D}_{P_1}, \tilde{D}_{P_2}, d \right\}_{\mathbf{x}} \qquad \mathbf{x} \in \Sigma' \quad , \tag{4.213}$$

with

$$\tilde{\mathcal{E}}(\mathbf{x}) = \tilde{\mathcal{E}}(\mathcal{C}(\mathbf{x})) \tag{4.214}$$

and

$$\tilde{S}_{P_2}(\mathbf{x}) = -\tilde{S}_{P_1}(\mathbf{x}) \qquad \mathbf{x} \in \Sigma \quad . \tag{4.215}$$

The latter equation is not valid for points of $\overline{\Sigma}$ when a potentiostat is active. In this case, Eq. 4.215 must be generalized to

$$\tilde{\mathcal{E}}(\mathbf{x}) = \tilde{\mathcal{E}}\left(\mathcal{C}(\mathbf{x}), \mathcal{P}\right) \qquad \mathbf{x} \in \overline{\Sigma} \quad , \tag{4.216}$$

where $\mathcal{P}$ denotes the potentiostat setting. In the absence of a potentiostat, Eq. 4.215 is also valid on $\overline{\Sigma}$.

The local and locally-determined electrostatic interface characteristics can be equivalently expressed in the form of a local interfacial charge density $\tilde{\rho}_F(\mathbf{x})$ on Σ', which will be termed the *free interfacial charge distribution*. This distribution is only non-vanishing at interfaces and encompasses a *free surface charge density* component $\tilde{\rho}_S(\mathbf{x})$ and a *free surface polarization* component $\tilde{\rho}_D(\mathbf{x})$, as contributed by all phases, *i.e.*

$$\tilde{\rho}_F(\mathbf{x}) = \tilde{\rho}_S(\mathbf{x}) + \tilde{\rho}_D(\mathbf{x}) \qquad \mathbf{x} \in \Sigma' \quad . \tag{4.217}$$

Note that possible electrodes are also characterized by free surface charge distributions, including at their exposed surfaces $\overline{\Sigma}$. Since it has been assumed that the nature of the phases, including the physical properties of the electrodes, and the geometry of the system of phases are constant in time, $\tilde{\rho}_F$ is a time-independent function.

According to the laws of classical electrostatics[421], only the *first two surface multipole moments* of the surface alteration (see above) need to be considered for a description of the effect of the charge distribution at interfaces on the electric field and potential within the bulk regions of the different phases in contact. This explains why only these two terms have been retained in the definition of the free interfacial charge distribution (Eq. 4.217). As will be discussed in more detail in Section 4.3.2, the *free surface charge density* of a phase causes a *discontinuity in the normal component of the electric field* upon entering this phase, but no discontinuity in the tangential electric field and electric potential, while the *free surface polarization* causes a *discontinuity in the electric potential* upon entering this phase, corresponding to a Dirac delta function in the normal electric field at the interface. Thus, for example, due to the electron spillout phenomenon at the surface of a crystalline solid (surface dominated by a single type of crystallographic face or polycrystalline) in vacuum (air), the potential will present a positive jump upon entering the phase, *i.e.* the surface potential of such a solid is expected to be positive in the air-to-solid direction. Note that for a monocrystalline sample, the magnitude of this jump will depend on a specific choice for the dominant surface crystallographic face of the solid. Higher-order moments such as *e.g.* the surface quadrupole-moment density affect the electric field and potential within the surface region, but not outside. Consequently, these higher-order contributions need not be included in a model that only aims at a correct description of the electrostatic influence of interfacial effects on bulk properties, and not at a description of the interface itself.

The free interfacial charge distribution is only one of the two components of the total interfacial charge distribution. The other component arises from the *(di)electric response of the phases* themselves to the free interfacial charge distribution. This second component $\rho'_B(\mathbf{x})$ will be termed the *bound interfacial charge distribution*. This additional charge density is the one responsible for all the *non-local effects* generally present in a system of phases. More precisely, whereas $\tilde{\rho}_F(\mathbf{x})$ at a given interface point $\mathbf{x} \in \Sigma'$ solely depends on the local physical characteristics $\mathcal{C}(\mathbf{x})$ at $\mathbf{x}$, possibly along with the potentiostat setting $\mathcal{P}$ for $\mathbf{x} \in \overline{\Sigma}$, which entirely determine the corresponding local electrostatic characteristics $\tilde{\mathcal{E}}(\mathbf{x})$ *via* Eqs. 4.215 and 4.216, $\rho'_B(\mathbf{x})$ is influenced at any point $\mathbf{x} \in \Sigma$ by the electrostatic characteristics of the entire interfacial region Σ' of the system of phases. In mathematical terms, one could say that $\tilde{\rho}_F(\mathbf{x})$ is a *local function* of $\mathbf{x}$, namely

$$\tilde{\rho}_F(\mathbf{x}) = \tilde{\rho}_F\left(\mathcal{C}(\mathbf{x})[,\mathcal{P}]_{\mathbf{x}\in\overline{\Sigma}}\right) \qquad \mathbf{x} \in \Sigma' \quad , \tag{4.218}$$

i.e. solely determined by local system properties at $\mathbf{x}$, and that $\rho'_B(\mathbf{x})$ is a *non-local functional* of $\tilde{\rho}_F$, namely

$$\rho'_B(\mathbf{x}) = \rho'_B\left[\tilde{\rho}_F(\mathbf{x})\right] \qquad \mathbf{x} \in \Sigma' \quad , \tag{4.219}$$

i.e. determined by the global knowledge of $\tilde{\rho}_F(\mathbf{x})$ at all points $\mathbf{x} \in \Sigma'$. In particular, the influence of a possible potentiostat will be perceived throughout the entire system of phases *via* the dependence of ρ'_B on the free interfacial charge distribution $\tilde{\rho}_F$, imposed by this device on the exterior interfacial region $\overline{\Sigma}$, *i.e.* on the exposed surfaces of the electrodes. Since it has been assumed that the nature of the phases, including the physical properties of the electrodes, and the geometry of the system of phases are constant in time, $\tilde{\rho}_F$ is a time-independent function, as well as ρ'_B considering that this charge distribution is a functional of $\tilde{\rho}_F$. Note that there is in principle no reason at this point to believe that ρ'_B, which arises from the (di)electric response of the phases to the free interfacial charge distribution $\tilde{\rho}_F$, is itself also an interfacial charge distribution. This, turns out, however, to be the case for all practically relevant situations (see further below).

As a result, only the first two surface multipole moments of ρ'_B are again relevant in a model that only aims at a correct description of interfacial effects on bulk properties (see above). In principle, these are the local *bound surface charge densities* S'_{P_1} and S'_{P_2}, as well as the *bound surface polarizations* D'_{P_1} and D'_{P_2} at an interface between any two phases P_1 and P_2. The corresponding set of electrostatic interface characteristics associated with the bound interfacial charge distribution can thus be written, in analogy with Eq. 4.213 for the free characteristics,

$$\mathcal{E}'(\mathbf{x}) = \left\{S'_{P_1}, S'_{P_2}, D'_{P_1}, D'_{P_2}, d\right\}_{\mathbf{x}} \qquad \mathbf{x} \in \Sigma' \quad , \tag{4.220}$$

and the corresponding bound charge distribution, in analogy with Eq. 4.217 for the free charge distribution, as

$$\rho'_B(\mathbf{x}) = \rho'_S(\mathbf{x}) + \rho'_D(\mathbf{x}) \qquad \mathbf{x} \in \Sigma' \quad . \tag{4.221}$$

In contrast to Eq. 4.215, it cannot be assumed that $S'_{P_2}(\mathbf{x}) = -S'_{P_1}(\mathbf{x})$ at any point $\mathbf{x} \in \Sigma$. This is because, unlike the free interfacial charge distribution within the system, the bound charge distribution is not subject to a local electroneutrality constraint (see further below). Note also that the contributions of $D'_{P_1}(\mathbf{x})$, $D'_{P_2}(\mathbf{x})$ and $\rho'_D(\mathbf{x})$ in Eqs. 4.220 and 4.221 are usually neglected, *i.e.* the bound interfacial charge distribution is generally assumed to only consist of surface charge distributions (see further below).

To understand the properties of both the free and the bound interfacial charge densities, one has to examine the physical properties of the different phases in the system. A schematic classification of different types of phases in terms of their (di)electric properties is provided in Figure 4.4. Here, the response of a phase to an applied electric field *via* the displacement of free charges is referred to as an *electric response*, while the corresponding response *via* the reorientation of molecular dipoles, possibly accompanied by geometric distortions and electronic polarization, is referred to as a *dielectric response* (Section 2.2 and Figure 2.1)

One may distinguish different types of phases in terms of their (di)electric properties. A phase is an *insulant* (Figure 4.4a,b) when it does not contain free charge-carrying particles such as electrons, ions or ionized particles (*e.g.* vacuum, air, aprotic liquid), and is therefore characterized by a vanishing conductivity. In the opposite, a phase will be termed here a *perfect conductor* (Figure 4.4d) when it contains free charge-carrying particles at a sufficient concentration (*e.g.* metal, aqueous electrolyte solution, gas in the presence of ionizing radiation), and is therefore characterized by a high conductivity.

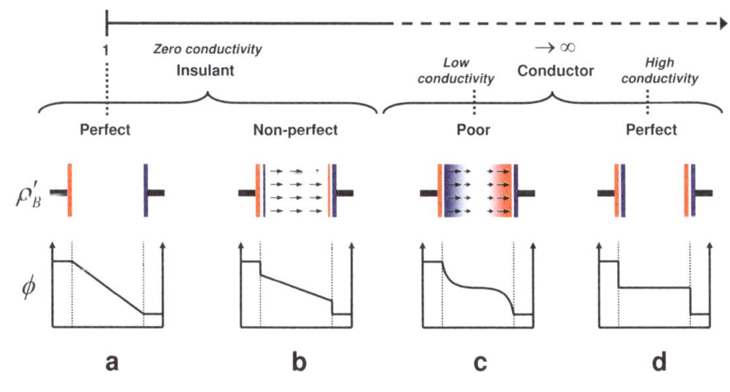

Figure 4.4 *Schematic classification of different types of phases in terms of their (di)electric properties.* The classification (top) is based on the static dielectric permittivity ϵ and on the conductivity properties (presence and concentration of free charge-carrying particles). The situation of a corresponding phase placed between two oppositely charged electrodes is also illustrated in terms of the bound charge distribution ρ'_B (middle) and electric potential ϕ (bottom). For simplicity, it is assumed that there is no free interfacial charge distribution at the interfaces to the electrodes and that these interfaces are polarizable (no charge transfer possible). The bound charge density at the electrode surfaces results from the free surface charge imposed by a potentiostat on their exposed surfaces (Figure 4.3a). The bound charge densities (red and blue) and the surfaces of the phase result from the displacement of free charge-carrying particles (whenever present) and the reorientation of molecular dipoles (arrows, whenever present, possibly along with geometric distortion and electronic polarization effects).

As will be discussed in more detail in Section 4.3.2, according to the laws of classical electrostatics[421], a perfect conductor is characterized by an infinite relative dielectric permittivity value and a vanishing electric field. As a result, the electric potential is constant within the bulk region of a conductor. This is no longer the case for an insulant, which is characterized by a finite permittivity value. As a result, the electric field within an insulant may be non-vanishing and inhomogeneous, in which case the electric potential varies with position. As important special cases of insulants, vacuum and, to a very good approximation, gases (*e.g.* air) in the absence of ionizing radiation, are characterized by a unit permittivity as well as the absence of dipole-carrying particles at a sufficient concentration (in addition to the total absence of charge-carrying particles), and will be termed here *perfect insulants* (Figure 4.4a). The absence of dipole-carrying particles is only exact in the context of non-dipolar molecules, and represents an approximation otherwise, justified by the low density of the gas phase. Note that even in the case of non-dipolar molecules, geometric distortions and electronic polarization could still provide a marginal contribution to the dielectric response. *Non-perfect insulants* (Figure 4.4b) include crystalline or amorphous solids as well as aprotic liquids, characterized by the absence of free charge-carrying particles and a finite permittivity value. Finally, a phase characterized by the presence of free-charge carrying particles at a very low concentration, resulting in a low conductivity, can be termed a *poor conductor* (Figure 4.4c). Although the dielectric permittivity of such a phase is still infinite and the electric potential constant within the bulk, its surface region may become very diffuse, extending deep into the inside of the phase. As will be discussed further below, pure water as well as other protic liquids should be considered in this context as perfect conductors due to autoprotolysis.

The bound charge distribution ρ'_B is the effective charge density resulting from the (di)electric response of the phases under the influence of the free interfacial charge distribution $\tilde{\rho}_F$. Obviously, *perfect insulants* provide *no contribution* to ρ'_B (Figure 4.4a). *Perfect conductors* react to external electrostatic influences by the displacement of their free charge-carrying particles. Since the bulk remains electroneutral and is characterized by a *vanishing electric field*, this redistribution only occurs at the surface of the conductor and corresponds to a *complete electrostatic shielding* of the bulk from interfacial and other external influences (Faraday cage). As a result, ρ'_B should in principle exclusively encompass a surface charge contribution ρ'_S in this case (Figure 4.4d). One might argue, however, that this redistribution of charges has also an indirect influence on the surface polarization properties. For example, in an electrolyte solution, the mobilization of the surface solvent dipoles for the solvation of additional charges is likely to reduce their contribution to the surface polarization. Therefore, the inclusion of a corresponding correction term ρ'_D in Eq. 4.221 might seem justified. This second-order effect is, however, typically neglected and ρ'_D assumed to be zero in this case. As will be seen in Section 4.3.2, this approximation is actually built into Maxwell's equations for electrostatics in continuous media. *Non-perfect insulants* react to external electrostatic influences by the reorientation of their constituting dipoles, along with possible geometric distortion and electronic polarization effects. Since the bulk remains electroneutral, the dipolar charges locally compensating each other, this reorganization also only results in an effective alteration of the charge distribution at the surface of the insulant and leads to a *partial electrostatic shielding* from interfacial and external influences (Figure 4.4b). In contrast to the situation of the perfect conductor, this shielding is only partial and only leads to a *reduction of the electric field* within the bulk, the extent of the reduction being related to the value of the dielectric permittivity of the insulant. As a result, ρ'_B should again in principle exclusively encompass a surface charge contribution ρ'_S, a possible indirect influence of the reorganization on the surface polarization properties, *i.e.* a corresponding term ρ'_D in Eq. 4.221, being here also typically neglected. Finally, *poor conductors* technically behave as perfect conductors in fully screening the bulk from electrostatic interfacial and other external influences. However, in the absence of a sufficient concentration of free charge-carrying particles, the surface charge reorganization, in combination with dipole reorientation and counteracted by thermal motion, may occur within a *surface region that is extremely diffuse*, extending deep into the phase (Figure 4.4c). Obviously, below a certain limiting concentration of free charge-carrying species, depending on the extent of the phase and on the temperature, this surface region formally encompasses the entire phase and the poor conductor ultimately behaves as an insulant.

The above discussion establishes that the bound charge distribution ρ'_B is in most cases a purely interfacial charge distribution, as was the case for the free interfacial charge distribution $\tilde{\rho}_F$. The only ambiguous case is that of the poor conductor, and is of no practical relevance in the context of this book. Furthermore, as a result of the neglect of a surface polarization contribution to the bound charge distribution, Eqs. 4.220 and 4.221 can be simplified to

$$\mathcal{E}'(\mathbf{x}) = \{S'_{P_1}, S'_{P_2}, d\}_{\mathbf{x}} \qquad \mathbf{x} \in \Sigma' \tag{4.222}$$

and

$$\rho'_B(\mathbf{x}) = \rho'_S(\mathbf{x}) \qquad \mathbf{x} \in \Sigma' \quad , \tag{4.223}$$

respectively. Equivalently, one may keep these general equations with the constraints

$$D'_P(\mathbf{x}) = 0 \quad \text{and} \quad D_P(\mathbf{x}) = \tilde{D}_P(\mathbf{x}) \qquad \mathbf{x} \in \Sigma' \tag{4.224}$$

and

$$\rho'_D(\mathbf{x}) = 0 \qquad \mathbf{x} \in \Sigma' \quad , \tag{4.225}$$

respectively.

As a practical and important example concerning the progressive transition from good to poor conductors, and ultimately to near-insulants, upon reducing the concentration of free charge-carrying particles, an estimate for the distance range from an interface up to which the bound charge distribution is expected to be significant in the context of an electrolyte solution can be obtained from the *Gouy-Chapman double-layer model*[1768,1769]. In the latter model, this effective range is related to the *inverse Debye screening length* κ_{DH}, also involved in the subsequently developed Debye-Hückel theory [113] (Section 4.2.6) given by

$$\kappa_{DH} = \left(\frac{2N_A e^2 \rho_{svt} b^\circ I_s}{\epsilon_o \epsilon_{svt} k_B T} \right)^{1/2} , \tag{4.226}$$

where the unitless ionic strength I_s is defined by Eq. 4.116. For example, for a 10^{-7} molal aqueous solution of two monovalent ions under standard conditions at temperature T^-, using values for $\rho^\ominus_{svt}$ and $\epsilon^\ominus_{svt}$ from Table 1.1, one finds $(\kappa^\ominus_{DH})^{-1} = 9.6 \cdot 10^{-7}$ m. This range, on the order of a μm, is very short on the macroscopic scale, indicating that the assumption of a limited extent of the bound charge distribution at the interface region is justified even in the case of pure water, taking into account the autoprotolysis equilibrium. This calculation suggests that *pure water*, as well as other protic liquids, should actually be considered as a *perfect conductor* from the point of view of interfacial properties. As discussed in Section 2.1, the corresponding static permittivity value $\epsilon^\ominus_{wat} = 78.4$ (Table 1.1) is in fact an effective solvent permittivity, excluding the contribution of conduction by the H^+_{aq} and OH^-_{aq} ions. Note that a more accurate description of the electric double-layer may be provided by the *Stern double-layer model*[1770], which combines the Gouy-Chapman (diffuse double-layer) model[1768,1769] with the earlier Quincke-Helmholtz (plate-capacitor) model[1771–1773].

Based on the above considerations, the following statements can be made regarding the *free interfacial charge distribution* $\tilde{\rho}_F$ at specific types of interfaces within the system, *i.e.* for $\mathbf{x} \in \Sigma$:

(A_8) the free charge distribution is always *purely interfacial*;

(B_8) this charge distribution is *locally determined*, *i.e.* uniquely defined by the local properties of the system at the point considered;

(C_8) this charge distribution is characterized by a *vanishing integral* over the interfacial region Σ' of the extended system, which is a consequence of the overall electroneutrality of the considered system;

(D_8) the *free surface charge densities* of two phases at any point of their interface are locally of *identical magnitudes* and *opposite signs*, which is a consequence of the local electroneutrality constraint;

(E_8) a *perfect insulant* presents *neither a free surface charge density nor a free surface polarization* at any of its interfaces, which is a consequence of the absence of free charge- or dipole-carrying particles;

(F$_8$) an *insulant* presents *no free surface charge density* at any of its interfaces, which is a consequence of the absence of free charge-carrying particles;

(G$_8$) *at an interface with an insulant*, a phase presents *no free surface charge density*, which is a consequence of Points C$_8$ and E$_8$.

Note that Points B$_8$, D$_8$ and G$_8$ cannot be generalized to the exposed surfaces of electrodes in the presence of a potentiostat, *i.e.* to $\mathbf{x} \in \overline{\Sigma}$ (see further below). In contrast, Points A$_8$, C$_8$, E$_8$ and F$_8$ are actually more generally valid for the extended system, *i.e.* for $\mathbf{x} \in \Sigma'$. Point C$_8$ excludes the possibility of a net charge for the extended system, which would correspond to a non-equilibrium situation as encountered, for example, in static electricity experiments (Section 2.1). In other words, a possible potentiostat is allowed to generate a non-equilibrium charge separation between the exposed surfaces of the electrodes that violates electroneutrality locally, but not to induce a net charge in the extended system, which would violate electroneutrality globally.

Corresponding statements can also be made regarding the *bound charge distribution* ρ'_B at specific types of interfaces within the extended system, *i.e.* for $\mathbf{x} \in \Sigma'$:

(A$_9$) the bound charge distribution is *purely interfacial* in most *practically relevant situations*, *i.e.* for any system involving phases that are all either perfect conductors or (perfect or non-perfect) insulants, but not poor conductors;

(B$_9$) this charge distribution is *globally determined*, *i.e.* it is a functional of $\tilde{\rho}_F$, determined at the considered point by the value of the latter function over all points of Σ';

(C$_9$) this charge distribution is, to a good approximation, defined by *surface charge densities* only, with no corresponding contribution from surface polarizations (Eqs. 4.222-4.225);

(D$_9$) this charge distribution is characterized by a *vanishing integral* over the interfacial region Σ' of the extended system, which is a consequence of an electric or dielectric response of the phases corresponding to a pure charge reorganization process;

(E$_9$) a *perfect insulant* presents *neither a bound surface charge density nor a bound surface polarization* at any of its interfaces, which is a consequence of the absence of free charge- or dipole-carrying particles;

(F$_9$) the *bound surface charge densities* of two *conductors* at any point of their interface are locally of *identical magnitudes* and *opposite signs*, and the *effective dipole moment* resulting from the total (free plus bound) interfacial charge distribution is *constant over the entire interface*, which is a consequence of the vanishing electric field within the two conductors;

(G_9) the *bound surface charge density* of a *conductor* at an interface with a (perfect or non-perfect) *insulant* may be non-vanishing, and its local value is *proportional to the normal electric field* at the surface of the insulant, which is a consequence of the vanishing electric field within the conductor.

Note that Points B_9-G_9 implicitly assume a system exempt of phases that are poor conductors, a restriction that will be adopted from here on. A number of the above statements regarding the bound interfacial charge distribution deserve additional comments.

Point D_9 will be further refined in Sections 4.3.2 and 4.3.3. As will be seen there, some interfaces permit an exchange of charges between the phases (non-polarizable interfaces) while others do not (polarizable interfaces). The charge reorganization process giving rise to the bound charge distribution can extend across all interfaces of the former, but not of the latter type. As a result, one may state in addition that the bound charge distribution is also characterized by a vanishing integral over any subset of phases connected through a series of interfaces that are all non-polarizable.

Points F_9 and G_9 follow from the constraint of zero field within a conductor. For an interface between two conductors, the contribution of the free interfacial charge distribution can be viewed as resulting in an effective interfacial polarization, corresponding to the free surface polarizations plus the effective polarization associated with the compensating free surface charge densities (Eq. 4.212). If the bound surface charge densities are to result in no discontinuity in the normal electric field, which must vanish on both sides of the interface, the bound surface charges must compensate each other. If they are to result in no corresponding variations in the tangential electric field, which must also vanish on both sides of the interface, the total (free plus bound) effective interfacial polarization, causing a discontinuity of proportional magnitude across the interface, must be constant over the entire interface. For the interface between a conductor and an insulant (Point G_9), there are no free surface charge densities. There may be a free surface polarization at the surface of the conductor, which causes a discontinuity in the electric potential, but not in the electric field across the interface. As a result, if the normal electric field is to be zero within the conductor, the bound surface charge density must be proportional to the magnitude of the normal electric field at the surface of the insulant. The presence of a possibly non-vanishing bound surface charge at an interface between a conductor and an insulant may at first sight appear to represent a violation of the local electroneutrality constraint. This is not the case, however, because the bound charge density is an effective charge density contributed by the "termination" of locally electroneutral charge distributions (*e.g.* chain of dipolar molecules, concentration gradient of free charge-carrying particles), so that it does not need to satisfy a further electroneutrality constraint, in contrast to the free charge density.

As stated above, external interfaces corresponding to the interfacial region $\overline{\Sigma}$ are allowed to violate three of the above statements concerning $\tilde{\rho}_F$ (Points B_8, D_8 and G_8). This is due to the possible presence of a *potentiostat* controlling the free surface charge densities at the exposed surfaces of electrodes. Considering the situation where there are two conducting electrodes separated by a *vacuum (air) gap* in the exterior of the system and under the action of a potentiostat at this gap, one accepts in this case that: (*i*) the free surface charge densities of the two electrodes may be non-vanishing at the gap, even if the corresponding interfaces involve an insulant; (*ii*) these surface charge densities are not determined by local properties of the corresponding interfaces, but controlled by the potentiostat. Note that the surface charge densities are *still compensating*, but on a *non-local basis* due to the presence of the vacuum (air) gap. This clearly represents a non-equilibrium situation as defined in Section 2.1, in the sense that it violates local electroneutrality at the gap. However, the overall electroneutrality of the extended system is still preserved, in contrast to the situation encountered in static electricity experiments. According to the definition formulated in Section 2.1, this situation is still referred to as an *equilibrium situation* for the system, as long as no corresponding violation of the local electroneutrality constraint occurs inside the system itself. The reason is that even if it can last for some time, any macroscopic charge separation within the system will spontaneously evolve towards a more stable state exempt of such a separation. In contrast, the macroscopic charge separation at the level of the external vacuum (air) gap is actively maintained by the potentiostat, and is thus viewed as an external constraint on the state of the system, just like the pressure or temperature.

The *potentiostat* may be imagined as a van de Graaff generator with some feedback mechanism permitting to maintain the magnitude of the compensating surface charges on the electrodes at the external vacuum (air) gap constant in time. In practical applications, the potentiostat will generally be a battery or a more complicated electric device, equipped with some voltage regulation mechanism, and the quantity under control (potentiostat setting referred to as $\mathcal{P}$ above) will be the *Volta potential difference* between the two electrodes, noted $\Delta\mathcal{V}$ in this case. Note that although the role of the potentiostat is to impose non-equilibrium free surface charge densities on the exposed surfaces of the electrodes, the quantity $\Delta\mathcal{V}$ actually characterizes the corresponding total charge densities, including the bound contribution. This situation is typical for *Voltaic cell experiments* (Section 4.4.4), where $\Delta\mathcal{V}$ is referred to as a *compensating potential difference*[1774]. In other types of experiments, the potentiostat will be replaced by a *potentiometer*, measuring $\Delta\mathcal{V}$ without imposing a non-equilibrium free surface charge density at the exposed surfaces of the electrodes. In this case, the entire extended system is at equilibrium and $\Delta\mathcal{V}$ is solely determined by the bound charge density of the electrodes at the vacuum (air) gap. This situation is typical for *Galvanic cell experiments* (Section 4.4.3), where $\Delta\mathcal{V}$ is referred to as an *electromotive force*[1775]. In practical applications, the potentiometer will generally be a device of very high internal resistance where $\Delta\mathcal{V}$ is inferred *via* Ohm's law from the flow of a current assumed to be of a negligible intensity. The exact meaning of a Volta potential difference will be explained in Section 4.3.2. As discussed in Section 4.4.2, when the electrodes or, more usually, their connecting wires are made of the same conducting material (*e.g.* metal, like copper, with polycrystalline surface), $\Delta\mathcal{V}$ is also equal to the corresponding Galvani potential difference.

The combined contributions of the free and bound interfacial charge distributions represents the *total interfacial charge distribution* ρ_T, *i.e.*

$$\rho_T(\mathbf{x}) = \tilde{\rho}_F(\mathbf{x}) + \rho'_B(\mathbf{x}) \qquad \mathbf{x} \in \Sigma' \quad . \tag{4.227}$$

Following from Eqs. 4.217 and 4.223, it can be decomposed as

$$\rho_T(\mathbf{x}) = \rho_S(\mathbf{x}) + \rho_D(\mathbf{x}) \qquad \mathbf{x} \in \Sigma' \quad , \tag{4.228}$$

where

$$\rho_S(\mathbf{x}) = \tilde{\rho}_S(\mathbf{x}) + \rho'_S(\mathbf{x}) \qquad \mathbf{x} \in \Sigma' \tag{4.229}$$

encompasses the locally-determined free surface charge density contribution and the globally-determined bound surface charge density contribution, while

$$\rho_D(\mathbf{x}) = \tilde{\rho}_D(\mathbf{x}) \quad \text{with} \quad \rho'_D(\mathbf{x}) = 0 \qquad \mathbf{x} \in \Sigma' \tag{4.230}$$

only encompasses the locally-determined free surface polarization contribution. For simplicity, the latter distribution will be further referred to as the surface polarization contribution, no distinction between free and total being required in this case. Since the total surface charge densities at the interface between two conductors are locally compensating, just as was the case for the corresponding free and bound surface charge densities, the terms *electrified interface* and *electric double-layer* can also be employed when referring to the total surface charge densities. From here on, these terms will be employed by default as referring to the *total interfacial charge distribution.* Accordingly, because the interface between a conductor and an insulant is generally characterized by a non-vanishing bound surface charge density of the conductor, analogous concepts of *charged interface* and *electric single-layer* could be employed as well, referring again to the total interfacial charge distribution.

The total interfacial charge distribution ρ_T entirely accounts for the electrostatic influence of interfaces within the system, as well as of a possible potentiostat, onto charged species within the bulk of the individual phases. In other words, the electric potential it generates according to Maxwell's equations in free space is the relevant quantity to characterize this influence. Besides the influence of this potential, the interactions determining the chemical potential of charged species are entirely local, *i.e.* characteristic of their immediate environment within the bulk of a phase that can be assumed exempt of interfaces. The characterization of this interface-induced electric potential will be the concern of Section 4.3.2.

4.3.2 Galvani, Volta and surface potentials

According to the discussion of Section 4.3.1, the *electrostatic properties* of a *system of phases at equilibrium* can be described for most practically relevant situations within a simplified representation involving: (*i*) phases that are either perfect insulants (unit permittivity, zero surface charge and polarization, *i.e.* vacuum or, to a very good approximation, air) or perfect conductors (infinite permittivity, possible presence of free and bound surface charge densities and of a surface polarization, *e.g.* metal, ionized gas, electrolyte solution, or even pure water considering autoprotolysis); (*ii*) phases that are all of macroscopic extents; (*iii*) a surface polarization associated with each conducting phase, in the general case inhomogeneous over the surface of the phase; (*iv*) a bound surface charge density (electric single-layer) at the surface of each conducting phase in contact with an insulant phase, in the general case inhomogeneous over the surface of the phase; (*v*) locally compensating total surface charge densities (electric double-layer, locally compensating also in terms of their free and bound components), associated with the two sides of each interface between conducting phases, generally inhomogeneous over the interface; (*vi*) the possible presence of conducting electrodes protruding out of the system, and involving exterior conductor-vacuum interfaces with non-vanishing free surface charge densities imposed by an external potentiostat. The possible existence of non-perfect insulant phases with a finite permittivity or of poorly conducting phases with a diffuse surface region is disregarded here as being of limited relevance in the context of this book. As discussed in Section 4.3.1, the common approximation that the surface polarization of a conducting phase is everywhere independent of the local surface charge density is also adopted (see discussion of Eqs. 4.222-4.225 and Point C_9 in Section 4.3.1). Thus, the surface polarization is described as a free surface polarization and there is no corresponding bound polarization (Eq. 4.230). Finally, the extended system of phases, including the electrodes and potentiostat (whenever present), is assumed to be of finite extent, surrounded by vacuum (air), and overall electroneutral. This simplified representation will be referred to here as the *electrostatic interface model* (EIM).

Assuming a *linear (di)electric response*, the EIM is described by *Maxwell's equations*[651–653] (see also Refs.[1453, 1776, 1777]) of electrodynamics[421] for *continuous media*, considering exclusively the *free interfacial charge distribution* as a source term. In the electrostatic limit (zero magnetic field and electric current), these can be written in the form of *Poisson's equation*[654], already encountered in Section 3.2 (Eq. 3.1), namely

$$\nabla \cdot [\epsilon(\mathbf{x})\nabla\phi(\mathbf{x})] = -\epsilon_o^{-1}\tilde{\rho}_F(\mathbf{x}) \quad , \tag{4.231}$$

where $\mathbf{x}$ is the coordinate vector of a point in space, $\tilde{\rho}_F(\mathbf{x})$ the free interfacial charge distribution at this point, non-vanishing only in the extended interfacial region Σ' (Section 4.3.1), $\epsilon(\mathbf{x})$ the corresponding local static relative dielectric permittivity and $\phi(\mathbf{x})$ the electric potential. To uniquely define ϕ,

Eq. 4.231 requires the introduction of appropriate boundary conditions. Since the extended system of phases is finite and assumed to be surrounded by vacuum, *Dirichlet boundary conditions*[421], setting the potential to zero over a sphere of infinite radius surrounding the system, will be employed, *i.e.*

$$\lim_{|\mathbf{x}|\to\infty} \phi(\mathbf{x}) = 0 \quad , \tag{4.232}$$

where $\mathbf{x} = \mathbf{0}$ corresponds to an arbitrary point within the system. The solutions of Eqs. 4.231 and 4.232 only depend on the choice of this reference point within a trivial translation.

In the EIM, the free charge distribution $\tilde{\rho}_F$ comprises a contribution $\tilde{\rho}_S$ from the free surface charge densities and a contribution $\tilde{\rho}_D$ from the surface polarizations, *i.e.* (Eq. 4.217)

$$\tilde{\rho}_F(\mathbf{x}) = \tilde{\rho}_S(\mathbf{x}) + \tilde{\rho}_D(\mathbf{x}) \quad . \tag{4.233}$$

As discussed in Section 4.3.1, an interface between two phases is considered here as a hypothetical *infinitesimal vacuum gap* (see *e.g.* Figure 4.3d,e). The free surface charge density of a conducting phase is assumed to be located just inside the phase, *i.e.* in the region of infinite permittivity. In contrast, the surface polarization of a phase is assumed to be located just outside the phase, *i.e.* in the vacuum gap of unit permittivity.

Conducting phases are characterized by an infinite permittivity value. In this case, it follows immediately from Eq. 4.231 that the electric field within such a phase must vanish or, equivalently, that the electric potential is constant. The constant potential within a given conducting phase P will be referred to as its *Galvani (or inner) potential*[266,406,455,1754,1778] ϕ_P. A further distinction between internal and external Galvani potentials will be made in Section 4.3.8. At this point, the discussion refers to the external quantity. The Galvani potentials of the different conducting phases in the system depend on the free surface charge densities and polarizations of all the phases in contact, possibly including the non-equilibrium free surface charge density imposed on the exposed surfaces of electrodes by a potentiostat. The aim of the EIM will be the description of this relationship.

Consider an oriented surface element $d^2\boldsymbol{\sigma}$ of a phase at a location $\mathbf{x}$ within the surface of a phase P, where the orientation is defined by the in-to-out direction relative to P. A *surface charge density* of magnitude $S_P(\mathbf{x})$ is defined by the presence of a charge of magnitude $S_P(\mathbf{x})d^2\boldsymbol{\sigma}$ over the surface element. A normal *surface polarization* of magnitude $D_P(\mathbf{x})$ is defined by the presence of a dipole moment of magnitude $D_P(\mathbf{x})d^2\boldsymbol{\sigma}$ over the surface element. Alternatively, one may view the surface polarization as originating from two charge densities $+(dn)^{-1}D_P(\mathbf{x})$ and $-(dn)^{-1}D_P(\mathbf{x})$ on two surface elements parallel to $d^2\boldsymbol{\sigma}$, located slightly outside and slightly inside relative to the surface, respectively, and separated by an infinitesimal distance dn.

In vacuum, the potential generated at a point $\mathbf{x}'$ by a surface element $d^2\boldsymbol{\sigma}$

at $\mathbf{x}$ bearing a *surface charge density* of magnitude $S_P(\mathbf{x})$ is given by

$$d\phi(\mathbf{x}') = (4\pi\epsilon_o)^{-1} d^2\boldsymbol{\sigma}\, |\mathbf{x}' - \mathbf{x}|^{-1}\, S_P(\mathbf{x}) \quad . \tag{4.234}$$

It is easily shown[421] that the electric potential as well as the tangential electric field are continuous across the surface element, but that there is a *discontinuity* of magnitude $\epsilon_o^{-1} S_P(\mathbf{x})$ in the *normal electric field* (in the in-to-out direction) across the surface element, *i.e.* a "kink" in the potential monitored along the line perpendicular to the surface element, corresponding to a potential maximum if $S_P > 0$, or to a minimum otherwise.

The corresponding potential generated by a surface element bearing a *surface polarization* $D_P(\mathbf{x})$ is given by

$$d\phi(\mathbf{x}') = (4\pi\epsilon_o)^{-1} d^2\boldsymbol{\sigma} \cdot \frac{\mathbf{x}' - \mathbf{x}}{|\mathbf{x}' - \mathbf{x}|^3} D_P(\mathbf{x}) = -(4\pi\epsilon_o)^{-1} d\Omega D_P(\mathbf{x}) \quad , \tag{4.235}$$

where $d\Omega$ represents the element of solid angle at which $d^2\boldsymbol{\sigma}$ is seen from $\mathbf{x}'$. It is easily seen that the tangential electric field is continuous but that there is a *discontinuity* of magnitude $\epsilon_o^{-1} D_P(\mathbf{x})$ in the *electric potential* (in the in-to-out direction) across the surface element, corresponding to a Dirac delta function in the normal electric field. Furthermore, in the special case where the entire surface of P is characterized by a *homogeneous surface polarization* D_P, one sees easily by integrating Eq. 4.235 over this closed surface, resulting in a total solid angle of either 4π or 0, that the potential generated by this polarization is *constant* both inside and outside P, with values of $-\epsilon_o^{-1} D_P$ and 0, respectively.

Within the EIM, the Galvani potentials of the different conducting phases can in principle be calculated using Eq. 4.231, with the boundary conditions of Eq. 4.232. The conducting or insulant properties, along with the extents, shapes, orientations and locations of the different phases in the system determine $\epsilon(\mathbf{x})$. The free surface charge densities $\tilde{S}_P(\mathbf{x})$ and polarizations $\tilde{D}_P(\mathbf{x})$ on both sides of the corresponding interfaces, possibly along with the free surface charge densities imposed on the exterior surface of electrodes by a potentiostat, provide the free interfacial charge density $\tilde{\rho}_F(\mathbf{x})$ *via* Eqs. 4.233-4.235.

Alternatively, Eq. 4.231 can be recast into an *equivalent vacuum problem*, provided that one adds an extra effective charge distribution, namely the *bound interfacial charge distribution* ρ'_B (Section 4.3.1), corresponding to the *(di)electric response of the conducting phases* to $\tilde{\rho}_F$. This is easily seen by rewriting Eq. 4.231 as

$$\nabla\epsilon(\mathbf{x}) \cdot \nabla\phi(\mathbf{x}) + \epsilon(\mathbf{x})\nabla^2\phi(\mathbf{x}) = -\epsilon_o^{-1}\tilde{\rho}_F(\mathbf{x}) \quad . \tag{4.236}$$

Dividing by $\epsilon(\mathbf{x})$ and rearranging the terms leads to

$$\nabla^2\phi(\mathbf{x}) = -\epsilon_o^{-1}\rho_T(\mathbf{x}) \quad , \tag{4.237}$$

with

$$\rho_T(\mathbf{x}) = \rho_S(\mathbf{x}) + \rho_D(\mathbf{x}) \tag{4.238}$$

where

$$\rho_D(\mathbf{x}) = \tilde{\rho}_D(\mathbf{x}) \tag{4.239}$$

and

$$\rho_S(\mathbf{x}) = \epsilon_o[\epsilon(\mathbf{x})]^{-1}\nabla\epsilon(\mathbf{x}) \cdot \nabla\phi(\mathbf{x}) \quad . \tag{4.240}$$

In the derivation of Eq. 4.237, it has been considered, as discussed above, that the free surface charge densities are encompassed within the conducting phases (*i.e.* associated with $\epsilon(\mathbf{x}) \to \infty$), while the surface polarizations are encompassed within a hypothetical infinitesimal vacuum gap between the phases (*i.e.* associated with $\epsilon(\mathbf{x}) = 1$). Setting

$$\rho'_S(\mathbf{x}) = \rho_S(\mathbf{x}) - \tilde{\rho}_S(\mathbf{x}) \tag{4.241}$$

and comparing Eqs. 4.238-4.239 and 4.241 with Eqs. 4.228-4.230, one easily identifies ρ_T with the *total interfacial charge distribution* introduced in Section 4.3.1 and ρ'_S with the corresponding bound interfacial component. In other words, recasting Eq. 4.231 into the equivalent vacuum problem of Eq. 4.237 has reformulated the influence of the (di)electric response of the conducting phases from a *permittivity effect* into an *additional source term* for the charge density. Note that the above reformulation already implies the absence of a bound surface polarization contribution (neglect of ρ'_D in Eq. 4.217, see discussion of Eqs. 4.222-4.225 and Point C_9 in Section 4.3.1), because the assumption that the charge density exerts no influence on the permittivity is already a built-in approximation of continuum electrostatics. The tilde is omitted in the symbols ρ_S and ρ_D, as it will be in the corresponding symbols S_P and D_P, so as to underline that these charge densities are now total charge densities, appropriate for the calculation of the electric potential using the Poisson equation in vacuum (Eq. 4.237). Note that owing to Eq. 4.239, the symbols D_P and $\tilde{D}_P$ can be used interchangeably for the surface polarization, and the former one will be retained from here on for simplicity.

At this point, Eqs. 4.231 and 4.237 deserve an important proviso. As will be discussed in Section 4.3.3, some interfaces permit the exchange of charges between the phases, and are referred to as *non-polarizable interfaces*, while others do not, and are referred to as polarizable interfaces. This distinction is only relevant for interfaces between conducting phases, any interface involving an insulant being automatically polarizable. The above equations only apply directly to systems of phases exclusively *presenting polarizable interfaces*. At *non-polarizable interfaces*, the (di)electric charge reorganization process giving rise to the bound charge distribution can extend *across the interface* in its electric component, namely in the form of a net charge transfer between the phases. This possible additional electric response contribution cannot be described as a permittivity effect, and is thus overlooked in Eqs. 4.231 and 4.237. As will be seen in Section 4.3.3, this charge transfer leads to the *dis-*

appearance of the bound surface charge distribution over any non-polarizable interface at equilibrium. The proper reformulations of Eqs. 4.231 and 4.237 taking these effects into account read

$$\nabla \cdot [\epsilon(\mathbf{x})\nabla\phi(\mathbf{x})] = -\epsilon_o^{-1}\left[\tilde{\rho}_F(\mathbf{x}) - \rho_S''(\mathbf{x})\right] \quad , \tag{4.242}$$

and

$$\nabla^2\phi(\mathbf{x}) = -\epsilon_o^{-1}\left[\rho_T(\mathbf{x}) - \rho_S''(\mathbf{x})\right] \quad , \tag{4.243}$$

where

$$\rho_S''(\mathbf{x}) = \begin{cases} \rho_S'(\mathbf{x}) & \mathbf{x} \in \mathrm{non-pol} \\ 0 & \mathbf{x} \in \mathrm{pol} \end{cases} \quad , \tag{4.244}$$

with ρ_S' defined by Eqs. 4.239 and 4.241. Note that Eq. 4.242, with the boundary conditions of Eq. 4.232, must be solved self-consistently for ϕ, considering that ρ_S'' depends itself on ϕ *via* Eqs 4.239, 4.240, 4.241 and 4.244. For simplicity, however, only the situation of Eqs. 4.231 and 4.237 will be discussed in the remainder of this section, and the case of the non-polarizable interface will be considered again in Section 4.3.3.

Due to the presence of the permittivity gradient in Eq. 4.240, the total interfacial charge density ρ_S is immediately seen to represent an effective interfacial charge distribution, corresponding to surface charge densities associated with the surfaces of the conducting phases within the system. Considering that the free charge density is purely interfacial, the same applies to the bound component ρ_S', in agreement with the considerations already made in Section 4.3.1. This total surface charge density can be written for each conducting phase P as (see Eq. 4.240)

$$S_P(\mathbf{x}) = -\epsilon_o \frac{d^2\boldsymbol{\sigma}}{d^2\sigma} \cdot \nabla\phi(\mathbf{x}) \quad , \tag{4.245}$$

i.e. it is directly proportional *via* a factor ϵ_o to the normal electric field (in the in-to-out direction) at the surface of the phase. Owing to Gauss' law[421, 1453], the integral of this charge density over the surface of any conducting phase is also equal to the total charge of this phase as provided by the surface charge component $\tilde{\rho}_S$ to $\tilde{\rho}_F$, which implies that the corresponding integral of the bound component ρ_S' vanishes. However, although the two distributions present the *same integral* over the surface of any conducting phase, ρ_S is *not equal* to $\tilde{\rho}_S$. The transformation from $\tilde{\rho}_S$ to ρ_S corresponds to a redistribution of the surface charge by appearance of the bound charge density within the conductor, so as to enforce the constraint of a vanishing electric field throughout its volume (Faraday cage). The total surface charge distribution of a conducting phase P depends on the geometry and surroundings of the phase, but it depends on the value of $\tilde{\rho}_S$ within P exclusively *via* its integral, *i.e.* any $\tilde{\rho}_S$ with a given integral over P will produce the same ρ_S for a given system of phases. These statements must be adjusted in the presence of *non-polarizable interfaces* (see above, Eqs. 4.243 and 4.244). In this case, the redistribution can occur across interfaces and only the integral of ρ'_S over subsets of conducting phases connected by interfaces of this type vanishes (Sections 4.3.1 and 4.3.3).

In the EIM, there are two possible sources of *discontinuities in the electric potential* within a system of phases. The *first* one originates from the *surface polarizations* of the conducting phases. The discontinuity in the electric potential upon entering a conducting phase P at a given point $\mathbf{x}$, *i.e.* the potential difference measured between a point at an infinitesimal distance outside P at $\mathbf{x}$ and a point anywhere inside P, will be referred to as its *local surface (or dipolar) potential* $\chi_P(\mathbf{x})$. A further distinction between internal and external surface potentials will be made in Section 4.3.8. At this point, the discussion refers to the external quantity. The local surface potential $\chi_P(\mathbf{x})$ is *locally determined*, *i.e.* it exclusively depends on the *local surface polarization* $D_P(\mathbf{x})$ as

$$\chi_P(\mathbf{x}) = -\epsilon_o^{-1} D_P(\mathbf{x}) \quad , \tag{4.246}$$

which follows from Eq. 4.235, considering that the element of solid angle $d\Omega$ changes from -2π to 2π upon crossing $d^2\boldsymbol{\sigma}$.

Complementary to the local surface potential, the potential immediately outside a given conducting phase P at a point $\mathbf{x}$, *i.e.* the potential difference measured between a point at an infinite distance from the system and a point at an infinitesimal distance outside P at $\mathbf{x}$, will be referred to as its *local Volta (or outer) potential* $\psi_P(\mathbf{x})$. Here, no distinction needs to be made between internal and external (Section 4.3.8). Unlike the local surface potential, the local Volta potential is not related to a unique local property of the system at $\mathbf{x}$, but to the total *interfacial charge distribution* (surface charge densities and polarizations) *within the entire extended system of phases*, possibly including the exposed surfaces of electrodes under the influence of a potentiostat, *i.e.* it is only local because its value is measured locally, but it is *not locally determined.*

The *second* possible source of discontinuity in the electric potential concerns interfaces between two conducting phases. Here, the *electric double-layer* surrounding the hypothetical infinitesimal vacuum gap effectively behaves as a *capacitor* with infinitesimal spacing, *i.e.* corresponds to an effective interfacial polarization,

$$D^{\mathit{eff}}_{P_1,P_2} = dS_{P_1} = -dS_{P_2} \quad , \tag{4.247}$$

(see Eq. 4.211; here considering total surface charge densities), that leads to a corresponding discontinuity in the electric potential. The above definitions for the local surface and local Volta potentials can also be applied to such an interface, provided that the infinitesimal distance between the measurement point outside the surface of a phase (within the gap) is still infinitely smaller than the infinitesimal width of the gap itself, *i.e.* taking the limit of surface proximity before the limit of gap closing. In this case, introducing the notation $\Delta\tilde{\psi}_{P_1,P_2}(\mathbf{x})$ for the Volta potential difference at a point $\mathbf{x}$ of the interface, one may write in analogy with Eq. 4.246

$$\Delta\tilde{\psi}_{P_1,P_2}(\mathbf{x}) = \psi_{P_2}(\mathbf{x}) - \psi_{P_1}(\mathbf{x}) = -\epsilon_o^{-1} D^{eff}_{P_1,P_2}(\mathbf{x}) \quad (4.248)$$

for the magnitude of this second type of discontinuity in the $P_1 \rightarrow P_2$ direction.

Considering the above definitions of the Galvani potential, and of the local surface and Volta potentials, one sees immediately that

$$\phi_P = \psi_P(\mathbf{x}) + \chi_P(\mathbf{x}) \quad (4.249)$$

for any point $\mathbf{x}$ at the surface of the phase P. In the special situation where the phase P is characterized by a *homogeneous surface polarization*, the local surface potential becomes independent of $\mathbf{x}$ and may then be simply referred to as *the* surface (or dipolar) potential[266,272,277,406,455,1754,1778] χ_P of the phase. Since ϕ_P is a constant, Eq. 4.249 implies that the same property holds for the local Volta potential, which may then be simply referred to as *the* Volta (or outer) potential[266,406,455,1754,1778] ψ_P of the phase. In this special case, Eq. 4.249 becomes

$$\phi_P = \psi_P + \chi_P \quad . \quad (4.250)$$

However, as will be seen in the following, this situation is in fact *very exceptional*, and the use of the simple terminology solely associating the above potentials with a phase, without reference to a specific measurement point at its surface, is the source of many *confusions* and *misunderstandings*.

The above considerations are illustrated in Figures 4.5-4.7, in the simplified context of *numerical calculations* in *two dimensions* using a finite-difference Poisson solver written by the authors of this book in the C language. Note that the situations considered in these figures only exemplify the case of interfaces precluding any charge transfer between the phases (polarizable interfaces). The situation of interfaces permitting charge transfer between the phases (non-polarizable interfaces) will be considered in Section 4.3.3.

Figure 4.5 illustrates the case of one square conductor exempt of surface polarization and surrounded by vacuum, of two such conductors in contact, except for panels F and G, where two rectangular conductors are considered instead. In this situation, there is no surface potential χ_P, and the Volta and Galvani potentials of a conductor are equal and independent of the choice of a surface point (Eq. 4.250), *i.e.* $\phi_P = \psi_P$.

In panels A and B, a single square conductor is represented with either a negative or a positive free surface charge density $\tilde{\rho}_S$, respectively, on one of its faces. As expected, the electric potential is constant within the conductor, negative in the first case and positive in the second, and progressively decreases in magnitude to zero beyond its surface. The Volta potential ψ_P, independent of the choice of a surface point and equal to the Galvani potential ϕ_P in this situation, is thus negative in the first case and positive in the second one. The total surface charge density ρ_S, generating the same electric potential for the equivalent vacuum problem, is a surface charge density integrating to the same net charge as $\tilde{\rho}_S$. It is inhomogeneous, being larger in magnitude at the corners of the square. Note that the same results for the electric potential ϕ and for the total surface charge density ρ_S would be obtained irrespective of the shape of the free charge density $\tilde{\rho}_S$, *e.g.* any point, volume or surface charge distribution integrating to the same net charge and entirely encompassed within the conductor.

Panels C and D illustrate a similar situation, but where the charged square conductor is in contact with (*i.e.* separated by a tiny vacuum gap from) an uncharged square conductor. It can be seen that the electric potential ϕ is affected by the presence of the second conductor. The Volta potential ψ_P of the charged conductor is decreased in magnitude while that of the un-

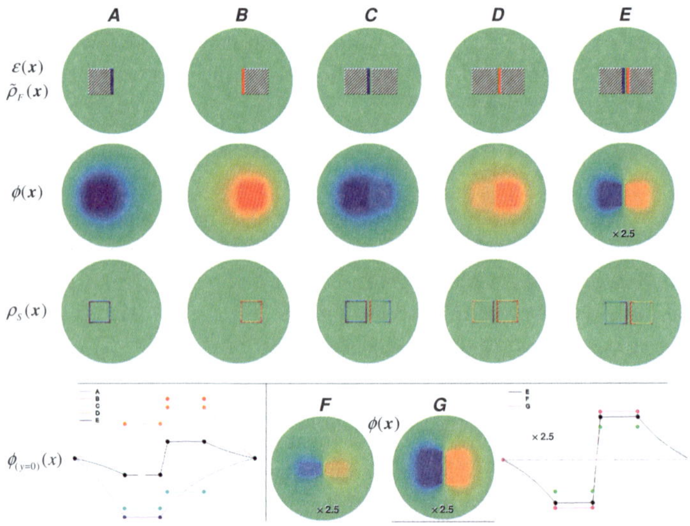

Figure 4.5 *Results of illustrative two-dimensional electrostatics calculations involving square or rectangular conductors surrounded by vacuum.* The quantities displayed are the dielectric permittivity $\epsilon(\mathbf{x})$, the free charge density $\tilde{\rho}_F(\mathbf{x})$, the total charge density $\rho_S(\mathbf{x})$, the electric potential $\phi(\mathbf{x})$, and its projection on a horizontal line passing through the centers of the conductors, noted $\phi_{(y=0)}(x)$. The calculations involve a finite-difference solution of Eq. 4.231 with Dirichlet boundary conditions (analogous to Eq. 4.232), namely $\phi(\mathbf{x}) = 0$ at the surface of the displayed circle. Note that in two dimensions, the Coulomb potential takes the form $V_{CB}(\mathbf{x}) = -(2\pi\epsilon_o)^{-1} q \ln(x/x_o)$ and the potential discontinuity at a dipole layer is $-(2\epsilon_o)^{-1} D_P$. The situations displayed correspond to a negatively-charged square conductor (A), a positively-charged square conductor (B), a negatively-charged square conductor next to an uncharged square conductor (C), a positively-charged square conductor next to an uncharged square conductor (D), two oppositely-charged square conductors in contact (E), two oppositely-charged rectangular conductors in contact, with an interface length increased by 50% compared to panel E (F), and two oppositely-charged rectangular conductors in contact, with an interface length decreased by 50% compared to panel E (G). The results are reported in arbitrary units, but on identical scales unless otherwise indicated. The total charges of the charged conductors are identical in magnitude in all situations displayed, except for panels F and G, where this charge was adjusted proportionally to the interface length.

charged conductor is non-vanishing, negative in the first case and positive in the second one. Thus, even if uncharged, this second conductor has acquired a non-vanishing but overall neutral total surface charge distribution ρ_S (bound contribution exclusively), that plays the role of a Faraday cage by screening the inside of this conductor from the electrostatic influence of the charged conductor.

Panel E illustrates the situation where the two square conductors in contact are oppositely charged. In contrast to the situations of panels A-D, this represents an acceptable equilibrium situation as defined in Section 2.1 and further discussed in Section 4.3.1, considering that local electroneutrality is satisfied in terms of the free interfacial charge distribution at all interfaces. Due to the principle of superposition of electric potentials, the electric potential ϕ is the sum of the corresponding potentials in panels C and D.

Panels A-E show in particular that it is incorrect to define the Volta potential of a phase in a system of phases, as sometimes done in textbooks, as either the potential of the singled out charged phase in vacuum (A, B), or as the potential of the charged phase in the hypothetical situation where the other phases are uncharged (C, D). In the situation depicted in Figure 4.5, both definitions spuriously suggest Volta potentials (and a Volta potential difference) much larger in magnitude than the true values. The only correct definition of ψ_P is the electrostatic potential infinitesimally close to the surface of the phase in the real system of phases (E) which is, in the special case of Figure 4.5 (no surface polarization), equal to the corresponding Galvani potential ϕ_P within the phase.

Another point of interest is illustrated in panels F and G, where the shape of the conductors and the length of the interface are varied at constant free interfacial surface charge densities. It can be seen that the Volta potentials of the two phases, as well as their difference, are affected by the change. This effect, which might seem surprising at first sight, is caused by the dependence of the bound surface charge densities at the interface on the object shapes. Together with the free surface charge densities, they define the overall strength of the electric double-layer and cause the dependence of the Volta potential difference on these shapes *via* Eqs. 4.247 and 4.248. As a result, even in the present case (no surface polarization), the *Volta potential difference* across the interface is *in general not uniquely defined by the nature of the two conducting phases* in contact, but also depends on the system geometry. As will be discussed in Section 4.3.3, this remark does not apply to an interface permitting a charge transfer between the phases (non-polarizable interface). In this case, the charge reorganization process giving rise to the bound charge distribution ρ'_S can extend across the interface. As a result, ρ'_S vanishes at the interface and the Volta potential difference between the two conducting phases in contact *at a non-polarizable interface* is *uniquely determined by their natures*, and independent of the system geometry.

Figure 4.6 illustrates the case of square or rectangular conductors exempt of free surface charge densities, but presenting surface polarizations, and surrounded by vacuum.

Panels H-J consider a square conductor. In panel H, the polarization is homogeneous over the entire surface of the conductor. As discussed previously in the context of Eq. 4.235, the electric potential generated by such a charge distribution is constant both inside and outside the object, with a discontinuity at its surface. As a result, the total surface charge distribution ρ_S vanishes. In this situation, the Volta potential ψ_P is zero over the entire surface of the conductor, and the surface and Galvani potentials of the conductor are equal and independent of the choice of a surface point (Eq. 4.250), *i.e.* $\phi_P = \chi_P$. However, in the general case, the surface polarization of a conductor will not be homogeneous, but *e.g.* depend on the local surface curvature (liquid, or solid with amorphous or polycrystalline surface), on the locally-exposed crystallographic face (monocrystalline solid), or/and on the perturbation of the surface properties at a specific interface (contact with a given adjacent phase), as discussed in Section 4.3.1. Panels I and J illustrate

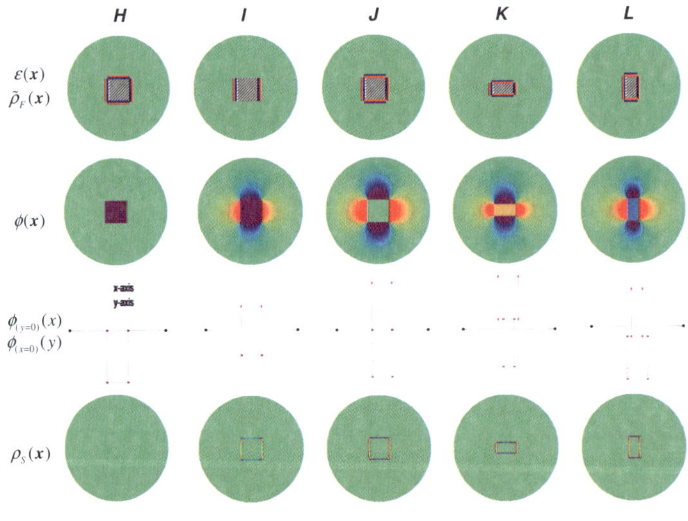

Figure 4.6 *Results of illustrative two-dimensional electrostatics calculations involving square or rectangular conductors surrounded by vacuum.* The quantities displayed are the dielectric permittivity $\epsilon(\mathbf{x})$, the free charge density $\tilde{\rho}_F(\mathbf{x})$, the total surface charge density $\rho_S(\mathbf{x})$, the electric potential $\phi(\mathbf{x})$, and its projection on a horizontal or on a vertical line passing through the center of the conductors, noted $\phi_{(y=0)}(x)$ and $\phi_{(x=0)}(y)$, respectively. The calculations involve a finite-difference solution of Eq. 4.231 with Dirichlet boundary conditions (analogous to Eq. 4.232), namely $\phi(\mathbf{x}) = 0$ at the surface of the displayed circle. Note that in two dimensions, the Coulomb potential takes the form $V_{CB}(\mathbf{x}) = -(2\pi\epsilon_o)^{-1} q \ln(x/x_o)$ and the potential discontinuity at a dipole layer is $-(2\epsilon_o)^{-1} D_P$. The situations displayed correspond to a square conductor with homogeneous surface polarization (H), a square conductor with polarization on the x-faces only (I), a square conductor with opposite surface polarizations on the x- and y-faces (J), and rectangular conductors with opposite surface polarizations on the x- and y-faces (K, L). The results are reported in arbitrary units, but on identical scales. The surface polarizations of the polarized faces are identical in magnitude in all situations displayed.

two extreme situations of this kind. In panel I, only the x-faces of the square present a surface polarization. This induces the appearance of a non-vanishing but overall neutral total surface charge distribution ρ_S (bound contribution exclusively), and thus, of a non-vanishing Volta potential. Furthermore, this potential ψ_P, just like χ_P, depends on the choice of a surface point (Eq. 4.249) in such a way that $\phi_P = \psi_P(\mathbf{x}) + \chi_P(\mathbf{x})$. Along the x-axis, the Volta potential progressively increases towards the conductor and reaches its negative value ϕ_P after a sharp drop caused by the surface polarization. Along the y-axis, the Volta potential progressively decreases towards the conductor to reach the final value ϕ_P without discontinuity. In panel J, the x- and y-faces present opposite polarizations. This inhomogeneity is again compensated by the appearance of a non-vanishing but overall neutral total surface charge distribution ρ_S (bound contribution exclusively). Due to the symmetry of the problem, $\phi_P = 0$ in this case.

The two latter examples illustrate two important points, namely that: (*i*) an *electroneutral conductor may present a non-vanishing Volta potential* even if it is not exposed to any external electrostatic influence (as was the case in panels C and D of Figure 4.5); (*ii*) the *Volta potential depends* in the general case on the *choice of a specific surface point* where it is evaluated. Only when the surface polarization is homogeneous can one say that an isolated electroneutral conductor has a vanishing Volta potential, and that the Volta potential of a conductor in a system of phases is independent of the choice of a surface point. The latter situation corresponds to the one illustrated in Figure 4.5 in the special case where χ_P is homogeneously vanishing and thus $\psi_P = \phi_P$. However, the situation of a conducting phase presenting a homogeneous surface polarization within a system of phases is in practice *extremely unlikely*. Even if the polarization is homogeneous when the phase is isolated in vacuum (air), contact with other phases will almost inevitably lead to alterations of the surface polarization, of distinct magnitudes for the interfaces of the phase with other phases of different types or/and with different interface physical characteristics.

Finally, panels K and L illustrate another somewhat counter-intuitive feature concerning *isolated electroneutral conductors* with a non-homogeneous surface polarization, namely that the Galvani potential may be *non-vanishing and shape-dependent*. In these panels, the square of panel J is contracted into a rectangle along either the y- or the x-axis. In the former case, a smaller fraction of the surface is covered by outwards-directed polarization. As a result, ϕ_P becomes positive. In the latter case, a larger fraction of the surface is covered by outwards-directed polarization and ϕ_P becomes negative. Such a situation occurs *e.g.* in the case of a monocrystal exposing different extents of specific crystallographic faces at its surface, in the case of a non-spherical sample of a liquid (*e.g.* ellipsoidal water droplet) or of a solid with polycrystalline surface properties (*e.g.* cube of metal, considering edges and vertices as sharply curved regions), or in the case of a phase that is exposed to different adjacent phases inducing interface-specific changes in its surface polarization (independently of effects related to surface charge densities).

Finally, Figure 4.7 illustrates the situation of an interface between two conducting phases where both free surface charge density and polarization effects are active. In panel M, the compensating free surface charge densities at the interface are zero. They are then progressively increased from panel N to panel P. For simplicity, it is assumed that both phases, when isolated, present homogeneous and identical surface polarizations. However, for realism, the surface polarizations are altered upon interface formation (here, decreased by a factor five). Panel M shows that even in the absence of free surface charge densities at the interface, both phases present non-vanishing and position-dependent Volta potentials $\psi_P(\mathbf{x})$, see the discussion of Figure 4.6 above. Since the surface polarizations are identical for the two phases and outwards-directed in the illustrated situation, both phases have negative and identical Galvani potentials ϕ_P. Because the surface potentials $\chi_P(\mathbf{x})$ are smaller at the interface for both phases, the corresponding Volta potentials, which compensate for this difference, are negative in the interface region,

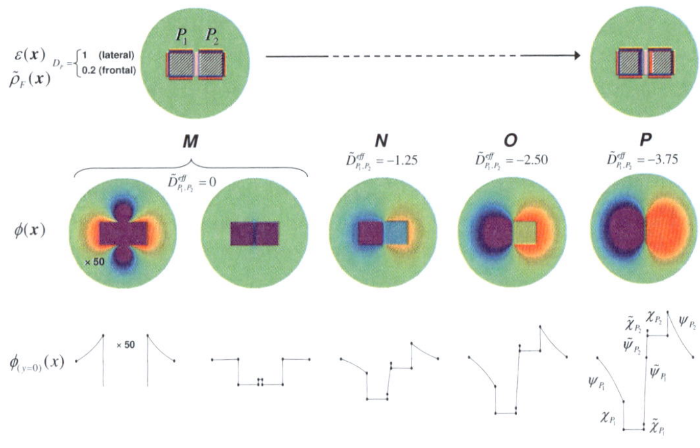

Figure 4.7 *Results of illustrative two-dimensional electrostatics calculations involving square conductors surrounded by vacuum.* The quantities displayed are the dielectric permittivity $\epsilon(\mathbf{x})$, the free charge density $\tilde{\rho}_F(\mathbf{x})$, the electric potential $\phi(\mathbf{x})$, and its projection on a horizontal line passing through the center of the conductors, noted $\phi_{(y=0)}(x)$. The calculations involve a finite-difference solution of Eq. 4.231 with Dirichlet boundary conditions (analogous to Eq. 4.232), namely $\phi(\mathbf{x}) = 0$ at the surface of the displayed circle. Note that in two dimensions, the Coulomb potential takes the form $V_{CB}(\mathbf{x}) = -(2\pi\epsilon_o)^{-1} q \ln(x/x_o)$ and the potential discontinuity at a dipole layer is $-(2\epsilon_o)^{-1} D_P$. The situations displayed correspond to two square conductors in contact. The surface polarizations of the conductors (D_P) are identical for all free faces, but reduced by a factor five at the contact face. The opposite total charges of the two conductors are progressively increased in the series of panels M to P, as indicated by the corresponding effective polarization $\tilde{D}^{eff}_{P_1,P_2}$ corresponding to the free surface charge densities at the interface (Eq. 4.211, where d is the gap width). The results are reported in arbitrary units, but on identical scales unless otherwise indicated. The lateral surface potentials χ_P and Volta potentials ψ_P, as well as the frontal surface potentials $\tilde{\chi}_P$ and Volta potentials $\tilde{\psi}_P$, are also indicated for the two phases P_1 and P_2 in the bottom right drawing.

and positive at the exposed surfaces of the conductors. Upon increasing the magnitude of the free surface charge densities at the interface (negative for the object on the left and positive for the object on the right), the surface potentials are unaffected, while the Volta and Galvani potentials increase in magnitude above their zero-charge values, in proportion to the free surface charge densities at the interface.

The configuration illustrated in this figure represents a realistic approximation to the situation of two conductors at an interface. The key observation is that there are *two sets of surface points* characterized by common values of $\chi_P(\mathbf{x})$ and $\psi_P(\mathbf{x})$ for each of the two phases, the latter also depending on the charge state of the interface, namely the surface points *at the interface*, which will be called *frontal* and the surface points *at the exposed surfaces* of the conductors, which will be called *lateral*. These concepts are further developed in the following paragraphs.

As discussed above, the situation of a phase with homogeneous surface polarization is rather exceptional. As a result, in most practical situations, *surface and Volta potentials* are locally-defined quantities that depend on the *choice of a reference point* at the surface of the phase (Eq. 4.249), and their introduction without clear specification of such a reference point (Eq. 4.250) is *extremely confusing* and, therefore, to be avoided. Following from the discussion of Section 4.3.1, the surface potential $\chi_P(\mathbf{x})$ and thus the Volta potential $\psi_P(\mathbf{x})$ *via* Eq. 4.249 at a given surface point $\mathbf{x}$ of a phase exclusively depends, through the corresponding surface polarization, on the *local physical characteristics* $\mathcal{C}(\mathbf{x})$ of the interface at this point (Eqs. 4.208, 4.213-4.216, 4.224 and 4.246). Thus, a unique definition for χ_P and ψ_P solely involves the choice of a *reference* for the set of interface characteristics $\mathcal{C}$. Two possible references will be introduced here.

The set of *lateral reference interface characteristics* $\mathcal{L}_P$ of any conducting phase P of nature N_P is defined by (see Section 4.3.1 and Eq. 4.208)

$$\mathcal{L}_P = \{N_P, \text{vac}, F_P, -, C\} \quad , \tag{4.251}$$

where "vac" indicates vacuum (air) for the second phase. This corresponds to any point at the surface of a phase P of nature N_P that is in contact with vacuum (air), locally exposes a crystallographic face of a given type F_P (only relevant if P is a monocrystalline solid), and corresponds to a local curvature C (only relevant if P is a liquid, or a solid with amorphous or polycrystalline surface).

Analogously, the set of *frontal reference interface characteristics* $\mathcal{F}_{P_1(P_2)}$ of any conducting phase P_1 of nature N_{P_1} at an interface with any other conducting phase P_2 of nature N_{P_2} is defined by

$$\mathcal{F}_{P_1(P_2)} = \{N_{P_1}, N_{P_2}, F_{P_1}, F_{P_2}, C\} \quad . \tag{4.252}$$

This corresponds to any point at the surface of a phase P_1 of nature N_{P_1} that is at an interface with a phase P_2 of nature N_{P_2}, where P_1 and P_2 locally expose crystallographic faces of given types F_{P_1} and F_{P_2} (only relevant if P_1 or/and P_2 is a monocrystalline solid), respectively, and corresponds to a local curvature C (only relevant if both P_1 and P_2 are either liquids, or solids with amorphous or polycrystalline surfaces). Most commonly, systems of phases do not involve monocrystalline solids and the reference interface characteristics refer *by default* to *amorphous or polycrystalline surfaces* in the context of solid phases. Similarly, curved interfaces are not commonly considered and the reference interface characteristics also refer *by default* to a *planar surface or interface patch*. For simplicity, these defaults will be assumed to hold in the following text whenever no explicit reference is made to exposed crystallographic faces or interface curvature, so that $\mathcal{L}_P$ and $\mathcal{F}_{P_1(P_2)}$ represent uniquely defined sets for a given phase or phase pair. It is important to stress that these reference sets solely depend on phase types, and are independent of the specific system of phases considered.

Given those choices of reference interface characteristics, it is now possible to uniquely define surface and Volta potentials for specific phases and interfaces within a given system. The *lateral surface and Volta potentials* of a given conducting phase P in the system, for which the symbols χ_P and ψ_P will be retained, are those corresponding to surface points of P with characteristics $\mathcal{L}_P$ (Eq. 4.251). The *frontal surface and Volta potentials* of a given conducting phase P_1 at an interface with a given conducting phase P_2 in the system, for which the corresponding symbols $\tilde{\chi}_{P_1(P_2)}$ and $\tilde{\psi}_{P_1(P_2)}$ will be introduced, are those corresponding to surface points of P_1 with characteristics $\mathcal{F}_{P_1(P_2)}$ (Eq. 4.252). When the identity of the interface considered is obvious from the context, the reference to the second phase may be made implicit and the symbols $\tilde{\chi}_P$ and $\tilde{\psi}_P$ used for the frontal quantities of the phase P at this interface. Note that the use of the tilde overscript to distinguish frontal and lateral potentials is not related to its previous use for distinguishing free and total interfacial charge distributions (see above).

Both the frontal and lateral Volta potentials are determined by the *total* interfacial charge distribution. For the frontal and lateral surface potentials, no distinction needs to be made since the two distributions are equivalent in terms of surface polarization. An illustration of the different potentials in the context of two square conductors in contact is presented in Figure 4.7 (bottom right drawing). Obviously, in a given system, all conducting phases need not present lateral interfaces and all pairs of conducting phases need not present frontal interfaces. However, such interfaces will be involved in all practically relevant situations.

Given the above definitions, Eq. 4.250 can now be applied in an *unambiguous* fashion to *lateral potentials*, *i.e.* the equation

$$\phi_P = \psi_P + \chi_P \tag{4.253}$$

is now valid in entire generality for any conducting phase P within an arbitrary system of phases, whereas Eq. 4.250 was initially restricted to phases characterized by a homogeneous surface polarization. An analogous *unam-*

biguous equation holds in terms of *frontal potentials*, namely

$$\phi_{P_1} = \tilde{\psi}_{P_1(P_2)} + \tilde{\chi}_{P_1(P_2)} \quad , \tag{4.254}$$

valid in entire generality for any conducting phase P_1 at an interface with any other conducting phase P_2 within an arbitrary system of phases.

Considering the hypothetical nature of the infinitesimal vacuum gap assumed to be present between conducting phases at an interface, and the inaccessibility of the interfacial region to experimental probes, it should not come as a surprise that the frontal potentials $\tilde{\chi}_{P_1(P_2)}$ and $\tilde{\psi}_{P_1(P_2)}$ are not measurable. Neither are the differences in these quantities across the interface. These differences are nevertheless very useful from a theoretical point of view. Given the definition of the frontal reference interface characteristics (Eq. 4.252), the *frontal surface potential difference*

$$\Delta\tilde{\chi}_{P_1,P_2} = \tilde{\chi}_{P_2(P_1)} - \tilde{\chi}_{P_1(P_2)} \tag{4.255}$$

at an interface between two conducting phases P_1 and P_2 exclusively depends on the natures of the two phases, *via* their surface polarizations at contact. In contrast, in the general case, the corresponding *frontal Volta potential difference*

$$\Delta\tilde{\psi}_{P_1,P_2} = \tilde{\psi}_{P_2(P_1)} - \tilde{\psi}_{P_1(P_2)} \tag{4.256}$$

depends in the general case on the properties of the entire system of phases, *i.e.* on the natures of all phases, on the geometry of the entire system and on the possible influence of a potentiostat, *via* the magnitude of the electric double-layer at the interface. This double-layer encompasses contributions from the free surface charge densities, which only depend on the natures of the two phases P_1 and P_2 in contact, and from the bound surface charge densities, which are in principle affected by the properties of the entire system. If the compensating total, free and bound surface charge densities at the interface are $S_{P_2} = -S_{P_1}$, $\tilde{S}_{P_2} = -\tilde{S}_{P_1}$ and $S'_{P_2} = -S'_{P_1}$, respectively, and the width of the vacuum gap is d, the corresponding frontal Volta potential difference can formally be written

$$\Delta\tilde{\psi}_{P_1,P_2} = -\epsilon_o^{-1} d S_{P_1} = -\epsilon_o^{-1} d \left(\tilde{S}_{P_1} + S'_{P_1}\right) \quad , \tag{4.257}$$

which is the analog of Eqs. 4.247 and 4.248 after proper definition of the frontal Volta potential. However, as will be discussed in Section 4.3.3, $\Delta\tilde{\psi}_{P_1,P_2}$ becomes, just like $\Delta\tilde{\chi}_{P_1,P_2}$, a unique characteristic of the nature of the phases P_1 and P_2 in contact in many practically relevant situations.

The lateral surface potential of a phase is in principle not experimentally accessible. In fact, it belongs to the set of quantities termed experimentally elusive in Section 2.3. Given the definition of the lateral interface characteristics (Eq. 4.251), the lateral surface potential difference

$$\Delta\chi_{P_1,P_2} = \chi_{P_2} - \chi_{P_1} \tag{4.258}$$

between two conducting phases P_1 and P_2 that are not necessarily in contact exclusively depends on the natures of the two phases, *via* their surface polarizations in vacuum (air). The corresponding lateral Volta potential difference is defined as

$$\Delta\psi_{P_1,P_2} = \psi_{P_2} - \psi_{P_1} \quad . \tag{4.259}$$

Although the absolute lateral Volta potential of a phase is not accessible experimentally, the difference between the lateral Volta potentials of two phases in a given system is accessible *via* Voltaic cell experiments (Section 4.4.4).

Defining the Galvani potential difference between two conducting phases P_1 and P_2 in a given system as

$$\Delta\phi_{P_1,P_2} = \phi_{P_2} - \phi_{P_1} \quad , \tag{4.260}$$

the five potential differences are immediately seen to be connected by the equation

$$\Delta\phi_{P_1,P_2} = \Delta\psi_{P_1,P_2} + \Delta\chi_{P_1,P_2} = \Delta\tilde{\psi}_{P_1,P_2} + \Delta\tilde{\chi}_{P_1,P_2} \quad , \tag{4.261}$$

the second equality only applying to a system presenting an interface between P_1 and P_2. The connection provided by Eq. 4.261 is illustrated graphically in Figure 4.8. Following from this connection and the above considerations, one has to conclude that, in the general case, the lateral Volta potential difference $\Delta\psi_{P_1,P_2}$, just like the corresponding frontal difference $\Delta\tilde{\psi}_{P_1,P_2}$, depends on the properties of the entire system of phases (nature of all phases, geometry of the entire system, possible influence of a potentiostat), and not only on the natures of the two phases considered. Here also, however, $\Delta\psi_{P_1,P_2}$ becomes, just like $\Delta\tilde{\chi}_{P_1,P_2}$ and $\Delta\chi_{P_1,P_2}$, and simultaneously with $\Delta\tilde{\psi}_{P_1,P_2}$, a unique characteristic of the nature of the phases P_1 and P_2 in many practical situations, as will be discussed in Section 4.3.3.

The above potentials and potential differences, as defined for each conducting phase or interface between conducting phases, are single-valued intensive properties characteristic of a given system of phases. It is therefore possible to define corresponding standard values, referring to a specific standard-state definition for the content of the different phases (Section 4.2.4). Most relevant are the corresponding standard Galvani and lateral potentials, which will be written $\phi_P^{\ominus}$, $\psi_P^{\ominus}$ and $\chi_P^{\ominus}$ considering the $bbme_T$ convention at temperature T^-. The corresponding frontal potentials can be written $\tilde{\psi}_{P_1(P_2)}^{\ominus}$ and $\tilde{\chi}_{P_1(P_2)}^{\ominus}$, but will be mostly encountered in the form of the corresponding differences $\Delta\psi_{P_1,P_2}^{\ominus}$ and $\Delta\tilde{\chi}_{P_1,P_2}^{\ominus}$. A further distinction will be made subsequently (Section 4.3.8) between two possible alternative definitions of the Galvani and lateral surface potentials, namely internal or external. In principle, a similar distinction should be made for the frontal surface potential. It is, however, omitted in this text in view of the limited practical relevance of this quantity. At this point, the discussion refers exclusively to the external quantities.

In terms of *standard quantities*, the following statements are generally valid: (*i*) $\chi^{\ominus}_{P_1}$, $\chi^{\ominus}_{P_2}$, $\Delta\chi^{\ominus}_{P_1,P_2}$ and $\Delta\tilde{\chi}^{\ominus}_{P_1,P_2}$ depend on the standard-state natures of P_1 and P_2, but not on the properties of a specific system of phases; (*ii*) $\Delta\phi^{\ominus}_{P_1,P_2}$, $\Delta\psi^{\ominus}_{P_1,P_2}$ and $\Delta\tilde{\psi}^{\ominus}_{P_1,P_2}$ depend on the properties of a specific system of phases, *i.e.* on the standard-state nature of all phases, on the geometry of the entire system, and on the possible influence of a potentiostat. As a special

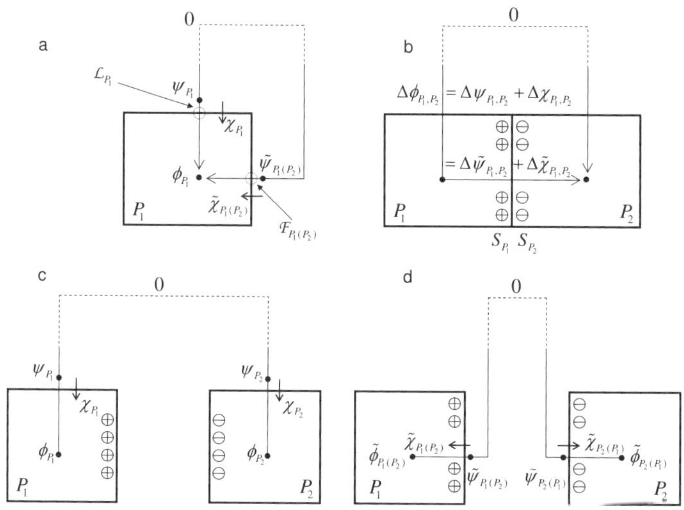

Figure 4.8 *Definitions of the lateral and frontal Volta and surface potentials and potential differences for conducting phases.* (a) The introduction of lateral and frontal reference interface characteristics $\mathcal{L}_{P_1}$ and $\mathcal{F}_{P_1(P_2)}$ (Eqs. 4.251 and 4.252) leads to unique values for the frontal and lateral surface potentials χ_{P_1} and $\tilde{\chi}_{P_1(P_2)}$, as well as for the frontal and lateral surface potentials ψ_{P_1} and $\tilde{\psi}_{P_1(P_2)}$, of a conducting phase P_1 (possibly given an adjacent conducting phase P_2) in a specific system of phases. The values of χ_{P_1} and $\tilde{\chi}_{P_1(P_2)}$ solely depend on the natures of P_1 (and P_2). They are defined by the corresponding reference interface characteristics, including lateral and frontal reference surface polarizations, *via* Eq. 4.246. The values of ψ_{P_1} and $\tilde{\psi}_{P_1(P_2)}$ depend on properties of the entire system of phases. The sums of χ_{P_1} and ψ_{P_1} or $\tilde{\chi}_{P_1(P_2)}$ and $\tilde{\psi}_{P_1(P_2)}$ are both equal to the Galvani potential ϕ_{P_1} of the phase (Eqs. 4.253 and 4.254). (b) The interface between two conducting phases P_1 and P_2 involves an electric double-layer characterized by the total surface charge densities S_{P_1} and $S_{P_2} = -S_{P_1}$. The differences between the Galvani, surface (lateral or frontal) and Volta (lateral or frontal) potentials of the two phases (Eqs. 4.255, 4.256, 4.258, 4.259 and 4.260) are related to each other *via* Eq. 4.261. The values of $\Delta\chi_{P_1,P_2}$ and $\Delta\tilde{\chi}_{P_1,P_2}$ solely depend on the natures of P_1 and P_2. The values of $\Delta\psi_{P_1,P_2}$, $\Delta\tilde{\psi}_{P_1,P_2}$, and thus $\Delta\phi_{P_1,P_2}$, depend in the general case on properties of the entire system of phases (see, however, Section 4.3.3 for important exceptions). The frontal Volta potential difference is related to the strength of the electric double-layer *via* Eq. 4.257. (c,d) Illustration of the decomposition of the lateral quantities (a) and of the frontal quantities (b). The hypothetical infinitesimal vacuum gap separating the phases has been artificially enlarged to reveal the surface points for the potential evaluation.

case of Eqs. 4.253 and 4.261, these quantities obey

$$\phi_P^{\ominus} = \psi_P^{\ominus} + \chi_P^{\ominus} \quad , \tag{4.262}$$

and

$$\Delta\phi_{P_1,P_2}^{\ominus} = \Delta\psi_{P_1,P_2}^{\ominus} + \Delta\chi_{P_1,P_2}^{\ominus} = \Delta\tilde{\psi}_{P_1,P_2}^{\ominus} + \Delta\tilde{\chi}_{P_1,P_2}^{\ominus} \quad . \tag{4.263}$$

The general dependence of $\Delta\phi_{P_1,P_2}^{\ominus}$, $\Delta\psi_{P_1,P_2}^{\ominus}$ and $\Delta\tilde{\psi}_{P_1,P_2}^{\ominus}$ on the specific geometric features of a system of phases may appear to represent a serious obstacle in the characterization of the interfacial properties of conducting phases. However, this discussion will be pursued in Section 4.3.3, where it will be shown that systems of phases can be designed in which this dependence is entirely removed. This special case actually encompasses most of the practically relevant situations encountered in electrochemistry.

4.3.3 Characteristic situations

The discussion of section 4.3.2 was closed on the preliminary conclusion that the differences in Galvani potential $\Delta\phi_{P_1,P_2}$, lateral Volta potential $\Delta\psi_{P_1,P_2}$ and frontal Volta potential $\Delta\tilde{\psi}_{P_1,P_2}$ between pairs of conducting phases P_1 and P_2 (in contact, for the latter difference) within a given system of phases depend on the specific geometric features of the system, *i.e.* do not uniquely characterize the phases P_1 and P_2. In most practically relevant situations, however, systems of phases are investigated in which this dependence is entirely removed. To address this point, six *characteristic situations* will be considered, that are encountered in various combinations in typical experiments, namely: (*i*) the isolated phase with homogeneous surface polarization; (*ii*) the non-polarizable interface at equilibrium; (*iii*) the polarizable interface after charge equilibration; (*iv*) the polarizable interface at zero charge; (*v*) the air gap at zero field; (*vi*) the potentiostat or potentiometer. These situations are discussed in turn below. The first situation is of relevance for *work function measurements* (Sections 4.3.5 and 5.13). The five other situations define the *characteristic elements* or *equilibrium electrochemistry* (Sections 4.4.3-4.4.4).

As seen in Section 4.3.2 (see *e.g.* the discussion of Eq. 4.235 and of Figure 4.6, panel A), an *isolated phase with homogeneous surface polarization* (IHSP), which must be electroneutral at equilibrium, is characterized by a vanishing Volta potential and a surface potential that is constant over its entire surface, and directly proportional to the magnitude of the surface polarization. In principle, due to the influence of the local curvature on the polarization (Section 4.3.1), a rigorously homogeneous surface polarization can only be achieved for a *spherical object*. In practice, however, a *polyhedron with macroscopic faces* also represents a very good approximation to this situation, neglecting the small perturbation induced by the altered polarization at the edges and vertices. In the case of monocrystalline solids, due to the influence of the locally-exposed crystallographic face on the polarization

(Section 4.3.1), an additional constraint must be imposed, namely that the surface of the phase is dominated by a single type of crystallographic face. This can be achieved by considering a *thin monocrystalline lamella*, sliced along a specific crystallographic face. In all the above situations, one may write, for the isolated phase P,

$$\phi_P^{\text{IHSP}} = \chi_P^{\text{IHSP}} \quad \text{and} \quad \psi_P^{\text{IHSP}} = 0 \quad , \tag{4.264}$$

where it is assumed that the lateral reference interface characteristics (Section 4.3.2, Eq. 4.251) are chosen in a compatible way with the (dominant) surface type of the phase considered, *i.e.* with the corresponding surface curvature for a liquid or for a solid with amorphous or polycrystalline surface, and the corresponding exposed crystallographic face type for a monocrystalline solid. The key point is that the quantities involved in Eq. 4.264 are *independent of the shape and extent* of the phase, provided that the condition of homogeneous surface polarization is satisfied. In contrast, in the case of an isolated phase with inhomogeneous surface polarization (Section 4.3.2), the choice of lateral reference interface characteristics is not obvious and, irrespective of this choice, the lateral Volta potential is typically non-vanishing, while both the Volta and Galvani potentials usually depend on the shape and extent of the phase (see *e.g.* Figure 4.6, panels I-L).

A *non-polarizable* interface is an interface permitting the transfer of charges between two conducting phases in contact, *e.g.* by electronic conduction for two metals, by ionic conduction for two solutions, or *via* a redox reaction for a metal-solution interface, the latter possibly involving other phases in contact with the metal. For a *non-polarizable interface at equilibrium* (NPIE), this situation implies an equality constraint for the effective chemical (*i.e.* electrochemical) potential of the species involved in the charge transfer, and one says that the interface is *reversible* with respect to this species. Although the concept of electrochemical potential will only be explicited in Section 4.3.4, the principle of the corresponding equilibrium constraint is easy to understand. The motion across the interface of the charged species for which the interface is reversible is controlled by a driving force that has two components. The first one is purely chemical and arises from differences in the interaction of the species within the bulk of the two phases. The second one is of electrostatic nature and arises from the difference in the energy of the charged species considering the different values of the Galvani potential in the two phases. The motion of the charges across the interface in response to the resultant of these two components has a negative feedback on the driving force itself. Equilibrium is reached when the driving force becomes zero. In this situation, the Galvani potential difference $\Delta\phi_{P_1,P_2}$ is immediately related to the corresponding intrinsic chemical potential difference, and thus exclusively depends on the natures of the phases P_1 and P_2 in contact. This is also immediately seen as a consequence of Eq. 4.275 that will be provided in Section 4.3.4. Based on the discussion of Section 4.3.2 concerning Eq. 4.261, it follows that the other potential differences considered there, namely $\Delta\psi_{P_1,P_2}$, $\Delta\chi_{P_1,P_2}$,

$\Delta\tilde{\psi}_{P_1,P_2}$ and $\Delta\tilde{\chi}_{P_1,P_2}$ present the same characteristics, *i.e.* one may write

$$\Delta\phi^{\text{NPIE}}_{P_1,P_2} = \Delta\psi^{\text{NPIE}}_{P_1,P_2} + \Delta\chi^{\text{NPIE}}_{P_1,P_2} = \Delta\tilde{\psi}^{\text{NPIE}}_{P_1,P_2} + \Delta\tilde{\chi}^{\text{NPIE}}_{P_1,P_2} \quad , \tag{4.265}$$

the key point being that Eq. 4.265 involves quantities that now *solely depend on the natures of the conducting phases in contact*, provided that they share a non-polarizable interface at equilibrium. In fact, it is easily seen that the first equality in this equation also remains valid if P_1 and P_2 are connected *via* other phases by a series of such interfaces. This situation is characteristic of equilibrium *Galvanic cell measurements* (Section 4.4.3) which involve systems exclusively encompassing non-polarizable interfaces at equilibrium. In contrast, in the general case of an arbitrary interface type (Section 4.3.2), both the frontal and the lateral Volta potential differences, as well as the Galvani potential difference, depend on the characteristics of the entire system of phases, *i.e.* on the nature of all phases, on the geometry of the entire system, and on the possible influence of a potentiostat.

The *physical interpretation* of Eq. 4.265 in terms of the concepts developed in Sections 4.3.1 and 4.3.2 is the following. Coming back to the hypothetical situation depicted in Figure 4.3 b-e, where there is no bound interfacial charge distribution ρ'_S, it is first observed that if the Galvani potential difference between the conducting phases is to be independent of the interface physical characteristics $\mathcal{C}$ for a non-polarizable interface at equilibrium, as implied by Eq. 4.249, there must be a correlation between the surface charge density and surface polarization components of the free interfacial charge distribution $\tilde{\rho}_F$. More precisely, the effective dipole moment of $\tilde{\rho}_F$ at the interface must be independent of $\mathcal{C}$. Thus, for example, if the curvature of such a non-polarizable interface is changed, thereby altering the surface polarization contribution to $\tilde{\rho}_F$, a charge transfer across the interface will occur and adjust the corresponding surface charge density contribution so as to preserve this effective dipole moment. Now, consider the case of a real non-polarizable interface between two conductors within a system of phases, with possibly heterogeneous interface characteristics and where the bound interfacial charge distribution ρ'_S is in principle non-vanishing. Since the phases can exchange charges, ρ'_S is not constrained to a vanishing integral over the two phases separately, but only over the two phases jointly. As a result, the contribution of this charge distribution to the electric double-layer at the interface is only constrained by the condition that the effective dipole moment resulting from the total (free plus bound) interfacial charge distribution must be constant over the interface. Considering that this condition is already satisfied by the free component $\tilde{\rho}_F$ (see above), the latter constraint applies directly to ρ'_S, which must therefore be homogeneous over the interface. Since the free charge distribution accounts for the appropriate single-interface equilibrium distribution, any additional bound contribution represents an unnecessary charge-separation process and will be associated with a free-energy increase. It follows that, for a non-polarizable interface at equilibrium, ρ'_S must be homogeneously vanishing, *i.e.* there is *no bound charge density* contribution. In other words, the *electric double-layer* is exclusively defined by the *free surface charge densities*, and the potential difference $\Delta\tilde{\psi}_{P_1,P_2}$ exclusively probes these densities, given a specific choice for the frontal reference interface characteristics (Section 4.3.2, Eq. 4.252).

Considering the discussion of Section 4.3.2, this special property of non-polarizable interfaces at equilibrium may seem surprising at first sight. For example, in the general case, the Galvani potential difference between two conducting phases was found to depend on the phase shapes and interface extent, even in the absence of surface polarization effects (Figure 4.5, panels F and G). Inhomogeneities in the surface polarization were also shown to induce a dependence of the Galvani potential on the phase shape (Figure 4.6, panels J-L), and are nearly automatically present when a phase presents interfaces with distinct other phases. The reason for the disappearance of these dependences in the case of a non-polarizable interface at equilibrium is that the free interfacial charge distribution is alone responsible for the Galvani potential difference at the interface, and adjusts locally by charge transfer across the interface to fulfill a charge-equilibrium condition that solely depends on the natures of the phases in contact.

A *polarizable interface* is an interface that does not permit the transfer of charges between the two conducting phases in contact, because the free charge-carrying particles in the two phases are distinct, *e.g.* electrons in a metal and ions in a solution, in the absence of a redox reaction. Although the distinction between polarizable and non-polarizable interfaces is typically only applied to interfaces between two conductors, one could also say by extension that any interface involving an insulant is always polarizable, considering that no charge transfer is possible to and from an insulant phase, which contains no free charge-carrying particles.

A *polarizable interface after charge equilibration* (PICE) refers to the situation where the two conducting phases separated by the interface have been allowed to exchange charges *via* an external circuit upon system assembly, also ensuring a unique resulting magnitude of the surface charge density for a given interface in a specific system, *i.e.* now possibly dependent on other phases in contact and on the setting of a possible potentiostat. Since the charge-carrying species of two phases at a polarizable interface are generally distinct (*e.g.* electrons *vs.* ions), the external circuit typically includes a non-polarizable interface at which a reaction (*e.g.* redox) equilibrates the two types of species.

A particularly interesting characteristic situation of the latter kind, formally restricted to the case of a non-polarizable interface with homogeneous physical interface characteristics, implying uniform surface polarization of the two conducting phases (interface that is planar or of constant curvature, and involves a single type of exposed crystallographic face if involving monocrystalline solids), is that of a *polarizable interface at zero charge* (PZIC), which corresponds to the absence of any electric double-layer. As will be seen in

Section 4.4.4, this condition can be verified by means of interfacial tension measurements involved in equilibrium *electrocapillary measurements* (Section 4.4.5). An immediate consequence of the absence of electric double-layer is that the frontal Volta potential difference $\Delta\tilde{\psi}_{P_1,P_2}$ vanishes given a choice of frontal reference interface characteristics (Section 4.3.2, Eq. 4.252) compatible with the interface considered. Based on the discussion of Section 4.3.2 concerning Eq. 4.261, it follows that all the other potential differences considered there, namely $\Delta\phi_{P_1,P_2}$, $\Delta\psi_{P_1,P_2}$, $\Delta\chi_{P_1,P_2}$ and $\Delta\tilde{\chi}_{P_1,P_2}$ are dependent on the natures of the phases in contact, *i.e.* one may write

$$\Delta\phi^{\mathrm{PIZC}}_{P_1,P_2} = \Delta\psi^{\mathrm{PIZC}}_{P_1,P_2} + \Delta\chi^{\mathrm{PIZC}}_{P_1,P_2} = \Delta\tilde{\psi}^{\mathrm{PIZC}}_{P_1,P_2} + \Delta\tilde{\chi}^{\mathrm{PIZC}}_{P_1,P_2} \quad , \tag{4.266}$$

the key point being that Eq. 4.266 involves quantities that now *solely depend on the natures of the phases in contact*, provided that they share a polarizable interface at zero charge.

The physical interpretation of Eq. 4.266 in terms of the concepts developed in Sections 4.3.1 and 4.3.2 is the following. The requirement of vanishing electric double-layer implies that the bound charge distribution ρ'_S exactly offsets the free charge distribution $\tilde{\rho}_S$. This is of course only possible if the interface is homogeneous in terms of the surface polarization properties of the two phases, for otherwise, this cancellation would be incompatible with the constraint of vanishing tangential electric field in the two phases.

An *air gap* is a separation between two conducting phases involving a perfect insulant such as vacuum, or, to a good approximation, air or any other gas. A particularly interesting characteristic situation, formally restricted to the case of an air gap involving homogeneous physical interface characteristics for the surfaces of the conducting phases exposed to the air gap, implying uniform surface polarizations of these two conducting phases (surface that is planar or of constant curvature and involves a single type of exposed crystallographic face if involving monocrystalline solids), is that of an *air gap* at *zero field* (AGZF), which corresponds to the absence of any electric field in the gap. As will be seen in Section 4.4.4, this condition can be verified by means of specific approaches involved in equilibrium *Voltaic cell measurements*, *e.g.* the Kelvin probe, ionizing probe and Kenrick methods. An immediate consequence of the absence of electric field in the gap is that the lateral Volta potential difference $\Delta\psi_{P_1,P_2}$ between the two phases surrounding the gap vanishes, given a choice of lateral reference interface characteristics (Section 4.3.2, Eq. 4.251) compatible with the surfaces of these phases. Based on the discussion of Section 4.3.2 concerning Eq. 4.261, it follows that all the other potential differences considered there and relevant for two phases exempt of a direct interface, namely $\Delta\phi_{P_1,P_2}$ and $\Delta\chi_{P_1,P_2}$, depend in this case exclusively on the natures of the phases at the gap, *i.e.* one may write

$$\Delta\phi^{\mathrm{AGZF}}_{P_1,P_2} = \Delta\chi^{\mathrm{AGZF}}_{P_1,P_2} \quad \text{and} \quad \Delta\psi^{\mathrm{AGZF}}_{P_1,P_2} = 0 \quad , \tag{4.267}$$

the key point being that Eq. 4.267 involves quantities that now *solely depend on the natures of the phases at the gap*, provided that they are separated by an air gap at zero field.

The physical interpretation of Eq. 4.267 in terms of the concepts developed in Sections 4.3.1 and 4.3.2 is the following. The requirement of vanishing electric field implies that the bound charge distribution ρ'_S vanishes at the surface of the two conducting phases. This is of course only possible if their surfaces are homogeneous in terms of the surface polarization properties of the two phases, for otherwise, this requirement would be incompatible with the constraint of vanishing tangential electric field in the two phases.

A *potentiostat* (POT), as already introduced in Section 4.3.1, is a device capable of enforcing a non-equilibrium free surface charge separation on the exterior surfaces of two electrodes, *i.e.* conducting phases typically of the same nature, protruding out of the system. From a conceptual point of view, a potentiostat can be modeled as an air gap at non-vanishing and fixed electric field. The integral of this electric field across the gap can be controlled, *i.e.* the corresponding lateral Volta potential difference $\Delta\psi_{M_1,M_2}$ between two electrodes M_1 and M_2 can be constrained to a specified value $\Delta\mathcal{V}$, referred to as a *compensating potential difference*[1774]. In contrast to the potentiostat, the *potentiometer* is a passive device capable of measuring $\Delta\mathcal{V}$ ideally without altering its value. In this case, $\Delta\mathcal{V}$ is referred as an *electromotive force*[1775]. Following similar arguments as those leading to Eq. 4.267 in the context of the air gap at zero field, one may write for an air gap between electrodes M_1 and M_2, that is under potentiostatic control, or for a potentiometric measurement,

$$\Delta\phi^{\text{POT}}_{M_1,M_2} = \Delta\psi^{\text{POT}}_{M_1,M_2} + \Delta\chi^{\text{POT}}_{M_1,M_2} \quad \text{and} \quad \Delta\psi^{\text{POT}}_{M_1,M_2} = \Delta\mathcal{V} \quad . \qquad (4.268)$$

In contrast to Eqs. 4.264-4.267, $\Delta\chi^{\text{POT}}_{M_1,M_2}$ is the only quantity in Eq. 4.268 that *solely depends on the natures of the metal electrodes*. Note that this difference *vanishes* in the *usual situation* where the electrodes are connected to a potentiometer or potentiostat *via* metallic wires made of the same metal, provided that the wires are viewed as M_1 and M_2 and the actual electrodes considered part of the system of phases. When a potentiostat is used, Eq. 4.268 is to be interpreted in the sense that the selected compensating potential difference $\Delta\mathcal{V}$ imposes a particular value of $\Delta\psi_{M_1,M_2}$. When a potentiometer is used, Eq. 4.268 is to be interpreted in the sense that the equilibrium value $\Delta\psi_{M_1,M_2}$ characteristic of the system of phases determines the measured $\Delta\mathcal{V}$.

The characteristic situation of the isolated phase with homogeneous surface polarization (IHSP) is relevant in the context of the measurement of *metal work functions*, *i.e.* of the reversible work required for ejecting an electron from the conduction band of an isolated metal sample and taking it to infinity. In this context, Eq. 4.264 establishes that the quantity measured is independent of the shape and size of the sample, and uniquely characterizes the metal, along with a selected (dominant) surface type. The five other characteristic situations represent *characteristic elements* in the context of *equilibrium electrochemical measurements*, as will be discussed in Section 4.4. In this context, Eqs. 4.265-4.268 establish that the quantity $\Delta\mathcal{V}$ corresponding to a system of phases consisting of an arbitrary sequence of phases separated by non-polarizable interfaces at equilibrium (NPIE), polarizable interfaces at zero charge (PIZC), or air gaps at zero field (AGZF), and connected *via* electrodes (wires) to a potentiostat or potentiometer (POT) is *independent of the shapes and sizes of the phases* in a specific system. Furthermore, a sequence of N elements of this type involves N constraints (Eqs. 4.265-4.267) on corresponding Volta potential differences, from which it follows that $\Delta\mathcal{V}$ is *uniquely determined by a selected sequence* (Eq. 4.268). This implies that any potentiostat setting differing from this value will lead either to a violation of the equilibrium condition at a non-polarizable interface (non-zero electric current), of the zero-charge constraint at a polarizable interface (presence of an electric double-layer), or of the zero-field constraint at an air gap (presence of a lateral Volta potential difference). Note that these considerations do not apply to a system involving a polarizable interface after charge equilibration (PICE, of which PIZC is a special case), where the magnitude of the electric double-layer is unconstrained (and determined by $\Delta\mathcal{V}$).

When the sequence exclusively comprises non-polarizable interfaces, the electromotive force $\Delta\mathcal{V}$ is entirely dictated by the equilibrium condition and a potentiometer can be used. This is equivalent to the use of a potentiostat, the settings of which would be controlled by the constraint of zero current. This is the situation encountered in equilibrium Galvanic cell measurements (Section 4.4.3). When this sequence comprises a polarizable interface or an air gap, along with non-polarizable interfaces, an arbitrary compensating potential difference $\Delta\mathcal{V}$ will usually differ from the value appropriate for the corresponding zero-charge, or zero-field constraint. The potentiostat setting is then adjusted until the required constraint is fulfilled. This is the situation encountered in Voltaic cell and electrocapillary measurements (Sections 4.4.4 and 4.4.5). In other experiments, the inclusion of a polarizable interface after charge equilibration in electrocapillary measurements also permits to investigate the response of the electric double-layer, as probed *e.g.* by interfacial tension measurements, to changes in $\Delta\mathcal{V}$ (Section 4.4.5). Although in principle possible, the inclusion of multiple polarizable interfaces or air gaps is not of any practical relevance.

4.3.4 Chemical potentials

The concept of *chemical potential* has been introduced in Section 4.2.1 (Eq. 4.16) and defined as the *(partial) molar free energy* of a species in a given phase (Section 4.2.2). This quantity, which is only defined within an arbitrary constant, is characteristic of the molecular nature of the species, *i.e.* of the free energy of the isolated molecule (*e.g.* relative to the zero-point energy of its constituting atoms), as well as of the interaction of the species with its environment in the given phase. For charged species, this interaction includes in principle a contribution due to the electrostatic influence of all interfaces within the system of phases encompassing the phase of interest (Section 4.3.1). Since this contribution is physically distinct from the chemical contribution defined by the interaction of the species with its local environment within the bulk of the phase, it is convenient to introduce three alternative definitions for different variants of the chemical potential.

The *intrinsic chemical potential* $\mu_{\Sigma:P}$ of a species Σ in a phase P accounts for the interaction of the species with its local environment within the bulk of the phase, excluding the contribution of interfacial effects. Following the usual nomenclature, this quantity will be simply termed the chemical potential, intrinsic being implied.

In contrast, the *electrochemical potential* $\tilde{\mu}_{\Sigma:P}$ of the species also accounts for the electrostatic influence of interfaces, with reference to a specified system of phases encompassing the phase P of interest. For a neutral species Σ, there is no distinction to be made between electrochemical and chemical potentials. However, if the phase P is conducting and the species Σ^z characterized by an integer charge z, one must write

$$\tilde{\mu}_{\Sigma^z:P} = \mu_{\Sigma^z:P} + zF\phi_P \quad , \tag{4.269}$$

where ϕ_P is the Galvani potential of phase P (Section 4.3.2). Note that the latter quantity depends in the general case on the characteristics of the entire system of phases encompassing P, the same statement obviously holding for $\tilde{\mu}_{\Sigma^z:P}$. The situation of a non-conducting phase P, where $\tilde{\mu}_{\Sigma^z:P}$ generally becomes position-dependent, will not be considered here.

Finally, the *real chemical potential* $\hat{\mu}_{\Sigma:P}$ of a species Σ is introduced with reference to the lateral surface potential of the phase, as defined by the choice of lateral reference interface characteristics (Section 4.3.2, Eq. 4.251). Recall that these characteristics include the specification of a surface curvature for liquids or for solids with amorphous or polycrystalline surfaces, and of an exposed crystallographic face for monocrystalline solids. Most commonly, however, the choice refers by default to a planar and polycrystalline (for solids) surface. For a neutral species, there is again no distinction to be made between real chemical and electrochemical or (intrinsic) chemical potentials. However, if the species Σ^z is characterized by an integer charge z, one defines

$$\hat{\mu}_{\Sigma^z:P} = \mu_{\Sigma^z:P} + zF\chi_P \quad , \tag{4.270}$$

where χ_P is the lateral surface potential of phase P. Note that the latter quantity does not depend on the nature of the system of phases encompassing P, the same statement obviously holding for $\hat{\mu}_{\Sigma^z:P}$. Although this definition is in principle also applicable to non-conducting phases, it will also be assumed here that P is conducting, for consistency with the restriction involved in Eq. 4.269.

Considering Eqs. 4.269 and 4.270, it follows from Eq. 4.253 that

$$\tilde{\mu}_{\Sigma^z:P} = \hat{\mu}_{\Sigma^z:P} + zF\psi_P \quad , \tag{4.271}$$

where ψ_P is the lateral Volta potential of phase P (Section 4.3.2), measured at a point presenting the same lateral reference interface characteristics as those used for defining χ_P. Note that the latter quantity depends in the general case on the nature of the system of phases encompassing P. Considering two conducting phases P_1 and P_2 in a given system of phases, one may define corresponding variants of *chemical potential differences* as

$$\Delta\mu_{\Sigma^z:P_1,P_2} = \mu_{\Sigma^z:P_2} - \mu_{\Sigma^z:P_1} \quad , \tag{4.272}$$

$$\Delta\tilde{\mu}_{\Sigma^z:P_1,P_2} = \tilde{\mu}_{\Sigma^z:P_2} - \tilde{\mu}_{\Sigma^z:P_1} \tag{4.273}$$

and

$$\Delta\hat{\mu}_{\Sigma^z:P_1,P_2} = \hat{\mu}_{\Sigma^z:P_2} - \hat{\mu}_{\Sigma^z:P_1} \quad . \tag{4.274}$$

Based on Eqs. 4.269-4.271, these differences are connected to the corresponding Galvani, *lateral* surface and lateral Volta *potential differences* of Eqs. 4.258-4.260 as

$$\begin{aligned} \Delta\tilde{\mu}_{\Sigma^z:P_1,P_2} &= \Delta\mu_{\Sigma^z:P_1,P_2} + zF\Delta\phi_{P_1,P_2} \\ &= \Delta\hat{\mu}_{\Sigma^z:P_1,P_2} + zF\Delta\psi_{P_1,P_2} \end{aligned} \tag{4.275}$$

and

$$\Delta\hat{\mu}_{\Sigma^z:P_1,P_2} = \Delta\mu_{\Sigma^z:P_1,P_2} + zF\Delta\chi_{P_1,P_2} \quad . \tag{4.276}$$

When the conducting phases P_1 and P_2 are in contact, corresponding connections involving the *frontal* surface and Volta potential *differences* are easily derived using Eq. 4.261, namely

$$\Delta\tilde{\mu}_{\Sigma^z:P_1,P_2} = \Delta\hat{\mu}_{\Sigma^z:P_1,P_2} + zF\Delta\tilde{\psi}_{P_1,P_2} + zF\Delta\Delta\tilde{\chi}_{P_1,P_2} \tag{4.277}$$

and

$$\Delta\hat{\mu}_{\Sigma^z:P_1,P_2} = \Delta\mu_{\Sigma^z:P_1,P_2} + zF\Delta\tilde{\chi}_{P_1,P_2} - zF\Delta\Delta\tilde{\chi}_{P_1,P_2} \quad , \tag{4.278}$$

where

$$\Delta\Delta\tilde{\chi}_{P_1,P_2} = \Delta\tilde{\chi}_{P_1,P_2} - \Delta\chi_{P_1,P_2} \quad . \tag{4.279}$$

In the above equations and in the general case (Section 4.3.2), the differences $\Delta\phi_{P_1,P_2}$, $\Delta\psi_{P_1,P_2}$ and $\Delta\tilde{\psi}_{P_1,P_2}$ depend on the characteristics of the entire system of phases encompassing P_1 and P_2, whereas the quantities $\Delta\chi_{P_1,P_2}$, $\Delta\tilde{\chi}_{P_1,P_2}$ and $\Delta\Delta\tilde{\chi}_{P_1,P_2}$ only depend on the natures of the phases P_1 and P_2 and on the definition of lateral and frontal reference interface characteristics. However, in many important *characteristic situations* (Section 4.3.3), the former quantities lose their dependence on a specific system, to also become *unique properties* of the *type of phase pair considered.* Note, finally, that a further distinction will be made subsequently (Section 4.3.8) between two possible alternative definitions of the Galvani and lateral surface potentials, namely internal or external. This choice obviously also affects the values of the real chemical and electrochemical potentials. But it alters neither their difference nor the value of the (intrinsic) chemical potential. However, it turns out to be of central importance when real parameters are tentatively partitioned into surface and bulk (intrinsic) contributions (Section 2.2). At this point, the discussion refers exclusively to the corresponding external quantities.

The values of the above variants of chemical potentials corresponding to *standard conditions* according to the *bbme*$_T$ convention at temperature T^- for the composition of the different phases will be written $\mu^\ominus_{\Sigma:P}$, $\tilde{\mu}^\ominus_{\Sigma:P}$ and $\hat{\mu}^\ominus_{\Sigma:P}$. They depend on standard values $\phi^\ominus_P$, $\psi^\ominus_P$ and $\chi^\ominus_P$ of the corresponding potentials (Section 4.3.2). In terms of standard quantities, Eqs. 4.269-4.271 become

$$\tilde{\mu}^\ominus_{\Sigma^z:P} = \mu^\ominus_{\Sigma^z:P} + zF\phi^\ominus_P \quad , \tag{4.280}$$

$$\hat{\mu}^\ominus_{\Sigma^z:P} = \mu^\ominus_{\Sigma^z:P} + zF\chi^\ominus_P \quad , \tag{4.281}$$

and

$$\tilde{\mu}^\ominus_{\Sigma^z:P} = \hat{\mu}^\ominus_{\Sigma^z:P} + zF\psi^\ominus_P \quad . \tag{4.282}$$

Based on the above considerations, the standard (intrinsic) chemical potential $\mu^\ominus_{\Sigma^z:P}$ and the standard real chemical potential $\hat{\mu}^\ominus_{\Sigma^z:P}$ solely depend on the standard-state properties of the species Σ^z in phase P, but not on the standard-state properties of the system of phases considered. This is not the case for the standard electrochemical potential $\tilde{\mu}^\ominus_{\Sigma^z:P}$. In the general situation considered in Section 4.3.2, even the difference $\Delta\tilde{\mu}^\ominus_{\Sigma^z:P_1,P_2}$ in the standard electrochemical potentials of a species Σ^z between two conducting phases P_1 and P_2 depends on the detailed geometry of the system of phases, namely the extents, shapes, orientations and locations of the different phases, as well as the possible influence of a potentiostat. However, in the six *characteristic situations* considered in Section 4.3.3, which are the ones relevant in the context of electrochemistry (Section 4.4), this dependence disappears in the differences of the electrochemical potential between adjacent phases. Thus, one may write (Eq. 4.275)

$$\begin{aligned}\Delta\tilde{\mu}^\ominus_{\Sigma^z:P_1,P_2} &= \Delta\mu^\ominus_{\Sigma^z:P_1,P_2} + zF\Delta\phi^\ominus_{P_1,P_2} \\ &= \Delta\hat{\mu}^\ominus_{\Sigma^z:P_1,P_2} + zF\Delta\psi^\ominus_{P_1,P_2} \quad ,\end{aligned} \tag{4.283}$$

where all the quantities involved are *independent of the system of phases considered*, provided that P_1 and P_2 are connected in this system by one of the *characteristic elements* described in Section 4.3.3. This equation can be complemented by (Eq. 4.276)

$$\Delta\hat{\mu}^\ominus_{\Sigma^z:P_1,P_2} = \Delta\mu^\ominus_{\Sigma^z:P_1,P_2} + zF\Delta\chi^\ominus_{P_1,P_2} \quad , \tag{4.284}$$

which does not require reference to any system of phases.

If these differences are solely determined, for adjacent phases, by the natures of the species Σ^z, of the phases P_1 and P_2, and of the characteristic element involved in their connection, it follows that corresponding differences for non-adjacent phases P_1 and P_2 are determined similarly, given the knowledge of the *sequence of phases and characteristic elements* separating them. Such a sequence represents the definition of an *electrochemical cell*, and a consistent notation for the corresponding *cell schemes* will be introduced in Section 4.4.1. As seen in Section 4.3.3, in the case of an *equilibrium* electrochemical cell, the setting of a possible potentiostat (or reading of a potentiometer) is entirely specified by the equilibrium condition along with additional constraints on the properties of the various characteristic elements. As a result, one may say that given the scheme of an electrochemical cell at equilibrium, the standard electrochemical potential $\tilde{\mu}^{\ominus}_{\Sigma^z:P}$ of any species Σ^z in any phase P solely depends on the standard-state properties of Σ^z in P, within an unknown constant (offset in the Galvani potential) that depends on the geometry of the specific system considered. Since chemical potentials have in any case no defined zero point (Section 4.2), the latter offset constant is irrelevant in practice. In summary, in situations relevant for electrochemistry, the *three variants of chemical potentials* $\mu^{\ominus}_{\Sigma^z:P}$, $\hat{\mu}^{\ominus}_{\Sigma^z:P}$ and $\tilde{\mu}^{\ominus}_{\Sigma^z:P}$, all *solely depend on the standard-state properties of species* Σ^z *in phase* P, but the latter one only after specification of the *cell scheme* of an electrochemical cell, *exclusively consisting of the characteristic elements* and with a *potentiostat setting consistent with an equilibrium situation* and with the additional *constraints on these elements* (Section 4.3.3).

The (intrinsic) chemical potential is the most relevant quantity to characterize the *physics of solvation* for charged species (Section 2.3). However, it is also an *experimentally-elusive* quantity, when expressed in the form of an intrinsic single-ion solvation free energy, because it corresponds to a hypothetical (non-experimental) situation where all interfaces have been removed from the system. The *real chemical potential* is a less relevant quantity to characterize the physics of solvation, because it is *contaminated by interfacial effects* in the form of the air-liquid potential of the pure solvent (Section 2.2). It is nevertheless an interesting quantity because in principle *experimentally accessible*, when expressed in the form of a real solvation free energy, and still exclusively characteristic of a given ion in a given solvent without reference to any specific system of phases or electrochemical cell. Finally, the *electrochemical potential* is the least relevant of the three quantities to characterize the physics of solvation, because it is contaminated by interfacial effects and specific to a given system of phases or electrochemical cell. However, it is the one immediately involved in *free-energy changes* and *equilibrium conditions* relevant for *experimental situations* encountered in electrochemistry (Section 4.4).

4.3.5 Standard and spectroscopic work functions

As a particular case of standard real chemical potential (Section 4.3.4), the *standard work function* $\omega_M^\ominus$ of a solid or liquid metal M at temperature T^- is defined here as minus the real chemical potential of the electron in this metal, divided by F. According to Eq. 4.281, one has thus

$$\omega_M^\ominus = -F^{-1}\hat{\mu}_{e^-:M}^\ominus = -F^{-1}\mu_{e^-:M}^\ominus + \chi_M^\ominus \quad , \tag{4.285}$$

where $\chi_M^\ominus$ is the standard lateral surface potential of the metal (Section 4.3.2), a unique quantity for a liquid metal or a solid metal with polycrystalline surface, in both cases referring by default to a planar surface of the isolated metal in vacuum (air). The standard work function introduced here is directly related to the more usual *spectroscopic work function* w_M, that corresponds to the work required for ejecting an electron from the conduction band (Fermi level) of the isolated metal surrounded by vacuum (air), at a given pressure and temperature, and bringing it to infinite distance with zero kinetic energy, divided by F.

The spectroscopic work function characterizes a semi-adiabatic free-energy change (Section 4.2.12) in the sense that the initial state of the process corresponds to an electron "solvated" in the metal at a well-defined pressure and temperature, while the final state of the process corresponds to a free electron exempt of kinetic energy, *i.e.* corresponding to a temperature $T = 0$ K. In other words, the spectroscopic work function at $P°$ and T^- is the analog of the standard work function in the *cold-electron convention*, the latter quantity corresponding to the *warm-electron convention* adopted in this book (Sections 4.2.12 and 4.2.17; see also Ref.[1719] for a discussion of these issues). To obtain a standard work function $\omega_M^\ominus$, the value of w_M measured at $P°$ and T^- must be corrected to match the same pressure and temperature for the standard-state electron. The two quantities are thus connected *via* the *heat-up parameters of the electron* (Sections 4.2.9 and 4.2.13), namely as

$$\omega_M^\ominus = w_M + F^{-1}(\Delta_h H^\ominus[e_g^-] - T^- s^\ominus[e_g^-]) \ , \tag{4.286}$$

where $\Delta_h H^\ominus[e_g^-]$ is the heat-up enthalpy of the gas-phase electron (3.145 kJ·mol^{-1}; Table 4.2) and $s^\ominus[e_g^-]$ the corresponding absolute entropy (22.734 J·mol^{-1}·K^{-1}; Table 4.2), resulting numerically in $\omega_M^\ominus = w_M - 0.038$ V.

Spectroscopic work functions have been determined experimentally for many metals (Section 5.13), and the corresponding standard work functions $\omega_M^\ominus$ are thus readily available. Note that the spectroscopic work functions are usually tabulated in literature sources[98, 221] as $w'_M = e w_M$ in units of eV, rather than as w_M in units of V. Obviously, the corresponding numerical values are identical, but the use of w_M (or $\omega_M^\ominus$) with units of an electric potential is more consistent with the conventions adopted in the present book, *i.e.* leads to simpler equations.

Although seldom stated explicitly in the literature, there is an *impor-*

tant proviso concerning the *appropriate measurement* of work functions. The quantity $\omega_M^\ominus$ derived from the spectroscopic experiment actually corresponds to the electrochemical potential $\tilde{\mu}_{e^-:M}^\ominus$ of the electron in an isolated and electroneutral metal sample, and not to its real chemical potential as implied by Eq. 4.285. Hence, the resulting estimate for $\omega_M^\ominus$ will only be appropriate in experimental situations for which (see Eq. 4.271)

$$\tilde{\mu}_{e^-:M}^\ominus - \hat{\mu}_{e^-:M}^\ominus = -F\psi_M^\ominus = 0 \quad . \tag{4.287}$$

As discussed in Section 4.3.3 (Eq. 4.264), this condition is only exactly fulfilled when the metal sample presents *homogeneous surface polarization properties* and is *exactly spherical*, provided that the lateral reference interface characteristics (Section 4.3.2, Eq. 4.251) are chosen according to the surface properties and curvature of the sample. In practice, considering that the definition of $\omega_M^\ominus$ in Eq. 4.285 typically refers to the lateral surface potential at a planar surface, the error can be minimized by taking a *polyhedral sample* that is sufficiently large for its surface to be considered as entirely *dominated by planar faces*, neglecting the influence of edges and vertices, or a *spherical sample* that is sufficiently large for its surface to be considered as *quasi-planar*. For *monocrystalline samples*, $\omega_M^\ominus$ depends on the choice of a *crystallographic face* for the lateral reference interface characteristics, and the error can be minimized by taking a polyhedral sample with a surface dominated by planar faces of this type, *e.g.* a *thin monocrystalline lamella* sliced along the specific crystallographic face.

It is important to stress that the spectroscopic work function measured on a given polyhedral metal sample *never depends* on the face through which the electron is *ejected*, as sometimes suggested in this context[98, 221, 278, 1779, 1780]. This would imply a violation of the first law of thermodynamics or of Maxwell's equations, which stipulate that the electric field is irrotational, *i.e.* the electron could be ejected from one face and (hypothetically) reinserted through another one with a net change in energy over the cyclic process. However, in situations with inhomogeneous surface polarization, it depends on the *shape and size* of the sample *via* the *distribution of distinct surface types*. This follows from the observations made in Section 4.3.2 concerning the dependence of the Galvani potential of an isolated conductor on its shape and size in these situations (see Figure 4.6). Since the measured work function actually probes the electrochemical potential of the electron in the metal, it is sensitive to the Galvani potential (Eq. 4.269) and presents the same possible dependence. Only when the sample is dominated by one type of surface (see above) will this measured work function become *independent of the sample shape and size*, as well as appropriate for the definition of $\omega_M^\ominus$ in Eq. 4.285, but not automatically equivalent to the reversible work of electron ejection from an arbitrary sample through a surface presenting these properties, unless the sample is (quasi-)homogeneously covered by a surface of this type.

The spectroscopic work functions w_M, from which corresponding standard work functions $\omega_M^\ominus$ can easily be deduced *via* Eq. 4.286, have been determined for many metals[221, 403], most commonly considering *polycrystalline surfaces*. The resulting values can be viewed as averages over microcrystallites in random orientations, the exact nature of this averaging being difficult to characterize in detail and possibly dependent on the experimental determination method[402, 1781]. The spectroscopic and standard work functions of the alkali (as well as some other) metals with polycrystalline surfaces are reported in Table 5.13 and discussed in Section 5.13.

Spectroscopic work functions w_M are positive numbers, because metals do not spontaneously eject electrons and, in view of the small difference (0.038 V), so are standard work functions $\omega_M^{\ominus}$. Considering the definition of Eq. 4.285, two factors positively affect their *magnitude*: (*i*) a *tight binding of the electrons* within the metal (low energy of the conduction band); (*ii*) a *large inwards-directed surface polarization* (more pronounced electron spillout). As mentioned in Section 4.3.1, for monocrystalline solids, the magnitude of the electron spillout phenomenon is expected to depend on the type of crystallographic face *via* the corresponding packing density of the successive atom layers parallel to this face[98]. For example, the spectroscopic work functions of monocrystalline silver associated to the crystallographic faces with Miller indices 100, 110 and 111 are 4.64, 4.52 and 4.74 V, respectively, in line with the corresponding order in packing density[98]. Note that this density is not solely determined by the plane spacing, but also by the in-plane packing density.

The naming of the spectroscopic and standard work functions is introduced by the authors of this book. Most literature sources refer to w_M, simply as the work function (*e.g.* defined in Ref.[221] as "a measure of the minimum energy required to extract an electron from the surface of a solid"). As mentioned earlier, however, this distinction is analogous to a similar distinction commonly made in gas-phase ion thermochemistry, namely between the *ion convention* and the *electron convention* for the thermodynamic formation parameters of gaseous ions[1723, 1752], referred to in this book as the *cold-electron* and *warm-electron conventions*, respectively (Sections 4.2.12 and 4.2.17). Note also that the distinction introduced later concerning internal and external Galvani and surface potentials (Section 4.3.8) does not affect work functions, since these depend on a real chemical potential and thus, simultaneously characterize the bulk "solvation" of the electron in the metal and the reversible work of surface crossing. As mentioned in Section 4.3.8, such a distinction would only become relevant if one attempted to single out the chemical and interfacial components of a work function (Eq. 4.285). Work functions will be further discussed in Section 5.13 (see also Table 5.13 for selected values).

4.3.6 Absolute electrode potentials

Considering a *metal electrode* M (solid, M_s, or liquid, M_ℓ) in contact with a *solution* S at a *non-polarizable interface* in equilibrium, there are three different conventions for defining an *absolute electrode potential*, as illustrated in

Figure 4.9, namely: (*i*) the vacuum absolute electrode potential $\tilde{\mathcal{V}}_{M,S}$; (*ii*) the real absolute electrode potential $\hat{\mathcal{V}}_{M,S}$; (*iii*) the intrinsic absolute electrode potential $\mathcal{V}_{M,S}$. Note that the definition of such an absolute electrode potential appears *a priori* meaningful, considering that the various metal-solution potential differences are uniquely determined by the natures of M and S if the interface is non-polarizable and the system at equilibrium, and independent of the shapes, extents and surroundings of these two components (Section 4.3.3, Eq. 4.265). In all the above potential definitions, the reference state for the electron if defined to be the *ideal electron gas* at temperature T, according to the *bbme*$_T$ standard (Section 4.2.4), including in particular a *warm-electron convention.* The difference arises from the "location" of this electron gas relative to the metal-solution system, *i.e.* to the value of the *reference Galvani potential* associated with this ideal gas. This issue was already mentioned in the discussion of Section 4.2.4 concerning the definition of standard states, but left open until here, because of its irrelevance in the limited context of ion thermodynamics. In electrochemistry, however, this point turns out to be of central importance[1714,1716–1718,1721,1782–1788].

When the metal and solution phases refer to standard conditions at temperature T^-, the above potentials may be called *standard absolute electrode potentials* and noted $\tilde{\mathcal{V}}^{\ominus}_{M,S}$, $\hat{\mathcal{V}}^{\ominus}_{M,S}$ and $\mathcal{V}^{\ominus}_{M,S}$, respectively. These standard absolute electrode potentials should be clearly distinguished from *standard relative electrode potentials*, which represent differences with respect to a common electrode type, *e.g.* the reference hydrogen electrode and will be called in this book *redox potentials* (Sections 4.4.3 and 5.5).

The *standard vacuum electrode potential* $\tilde{\mathcal{V}}^{\ominus}_{M,S}$ will be defined as minus the electrochemical potential of the electron in the metal under standard conditions, divided by F, assuming a zero Galvani potential for the standard-state ideal electron gas. According to Eq. 4.280, one has thus

$$\tilde{\mathcal{V}}^{\ominus}_{M,S} = -F^{-1}\tilde{\mu}^{\ominus}_{e^-:M} = -F^{-1}\mu^{\ominus}_{e^-:M} + \phi^{\ominus}_{M(S)} \quad , \tag{4.288}$$

where $\phi^{\ominus}_{M(S)}$ is the standard Galvani potential of the metal electrode in contact with the solution. Considering Eqs. 4.262 and 4.285, one may further write

$$\tilde{\mathcal{V}}^{\ominus}_{M,S} = \omega^{\ominus}_{M} + \psi^{\ominus}_{M(S)} \quad , \tag{4.289}$$

where $\psi^{\ominus}_{M(S)}$ is the standard lateral Volta potential of the metal electrode in contact with the solution. In this convention, the reference point for the ideal electron gas is situated in vacuum, *infinitely far from any material*[1718,1721], as illustrated in Figure 4.9a. This choice may appear to be the most natural one for the definition of absolute electrode potentials, because it introduces into the standard no reference to any specific type of material, *i.e.* the reference potential for the electron does not depend on the specific metal and solution involved. However, it results in a quantity that is *not unambiguously* defined by the natures of M and S. This is easily seen based on considerations made in Sections 4.3.1-4.3.5 concerning phases with inhomogeneous surface polarization properties, and on the observation that $\tilde{\mathcal{V}}^{\ominus}_{M,S}$ depends on the lateral Volta potential $\psi^{\ominus}_{M(S)}$ of the metal in contact with the solution. Since the metal is partially immersed, its surface potential is not homogeneous, evaluating to $\chi^{\ominus}_{M}$ for the emersed part and $\tilde{\chi}^{\ominus}_{M(S)}$ for the immersed part. As a result, its Galvani potential $\phi^{\ominus}_{M(S)}$ depends on the extent of immersion, and so does the corresponding lateral Volta potential $\psi^{\ominus}_{M(S)}$ as a consequence of Eq. 4.262. It follows that $\tilde{\mathcal{V}}^{\ominus}_{M,S}$ characterizes not only M and S, but also the *extent of immersion of the electrode.* For this reason, this type of absolute electrode potential does not represent a useful quantity.

Interestingly, however, a method has been proposed[1789,1790] to evaluate the absolute potential of an electrode by measuring its "wet" work function $\tilde{\omega}^{\ominus}_{M(S)}$ across a (dry) lateral face immediately after emersion, the thin solution layer remaining attached to the electrode being viewed as a "miniature" solution isolated from external influences (see Section 5.14, Point C_{12}). This effective work function provides in principle a direct estimate for $\tilde{\mathcal{V}}^{\ominus}_{M,S}$. However, taking into account the arguments given above concerning the ill-defined nature of $\tilde{\mathcal{V}}^{\ominus}_{M,S}$, as well as other main shortcomings of this approach[1791,1792], these types of measurements are bound to give results that are extremely sensitive to the detailed experimental setup. Note that, for similar reasons, the Galvani

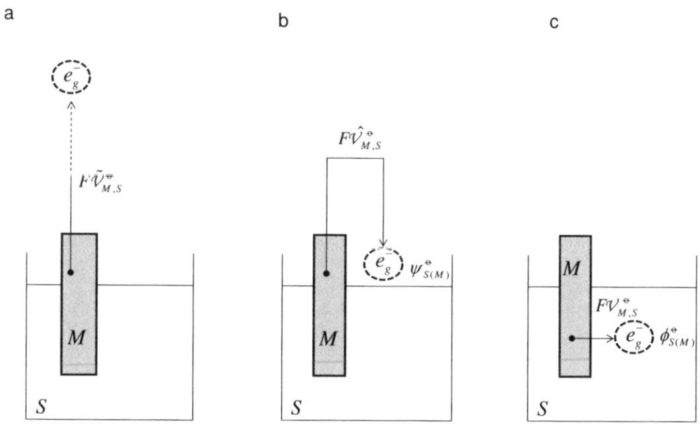

Figure 4.9 *Definition of the standard vacuum, real and intrinsic absolute potentials of an electrode.* A metal electrode M is partially immersed in a solution S, defining a non-polarizable interface at equilibrium, and conditions appropriate for the $bbme_T$ standard-state convention at temperature T^- are selected (Section 4.2.4). The absolute potential of the electrode is defined as minus the electrochemical potential of the electron in the electrode (scaled by F), with reference to an ideal gas of electrons (e_g^-). The reference Galvani potential for this ideal gas of electrons may be either in vacuum at infinite distance, leading to the vacuum electrode potential $\tilde{\mathcal{V}}^{\ominus}_{M,S}$ (a), immediately outside the surface of the solution, leading to the real electrode potential $\hat{\mathcal{V}}^{\ominus}_{M,S}$ (b) or in the solution, leading to the intrinsic electrode potential $\mathcal{V}^{\ominus}_{M,S}$ (c). Note that in the latter case, the situation does not correspond to solvated electrons, but to an ideal gas of electrons at the Galvani potential of the solution. The quantity indicated on the arrows represents the reversible work required for the transfer of one mole of electrons along the arrow under standard conditions.

potential $\phi^{\ominus}_{S(M)}$ of the solution in contact with the metal electrode depends on the details of the experiment, including the extent of electrode immersion, the solvent volume and shape, and the nature of the solvent container. Only the difference $\Delta\phi^{\ominus}_{M,S}$ is well defined. In other words, the ill-defined nature of $\tilde{\mathcal{V}}^{\ominus}_{M,S}$ arises from its dependence on absolute rather than relative potentials.

A similar issue probably affects the evaluation of absolute electrode potentials *via* two other approaches, the first one[482] based on the determination of the "contact" work function of a semi-conductor interfaced with the metal of an electrode at the potential of zero charge (see Section 5.14, Point D_{12}) and the second one[1793–1795] based on the nanocalorimetric determination of the electron capture enthalpies of gas-phase ion-water clusters (see Section 5.14, Point E_{12}). The above considerations also apply in this case. It has been argued[1796], however, that vacuum absolute electrode potentials may still represent acceptable estimates for real ones in practically relevant situations.

The *standard real electrode potential* $\hat{\mathcal{V}}_{M,S}$ will be defined as the standard vacuum electrode potential minus the Volta potential of the solution, *i.e.*

$$\hat{\mathcal{V}}^{\ominus}_{M,S} = \tilde{\mathcal{V}}^{\ominus}_{M,S} - \psi^{\ominus}_{S(M)} = -F^{-1}\mu^{\ominus}_{e^-:M} + \phi^{\ominus}_{M(S)} - \psi^{\ominus}_{S(M)} \quad , \tag{4.290}$$

where $\psi^{\ominus}_{S(M)}$ is the standard lateral Volta potential of the solution in contact with the metal electrode. Considering Eq. 4.289 one may write

$$\hat{\mathcal{V}}^{\ominus}_{M,S} = \omega^{\ominus}_{M} - \Delta\psi^{\ominus}_{M,S} \ , \tag{4.291}$$

where

$$\Delta\psi^{\ominus}_{M,S} = \psi^{\ominus}_{S(M)} - \psi^{\ominus}_{M(S)} \tag{4.292}$$

is the standard lateral Volta potential difference between the metal and the solution (see Eq. 4.259). Using Eq. 4.263, Eq. 4.291 can also be written in terms of frontal quantities at the metal-solution interface, namely

$$\hat{\mathcal{V}}^{\ominus}_{M,S} = \omega^{\ominus}_{M} - \Delta\tilde{\psi}^{\ominus}_{M,S} - \Delta\Delta\tilde{\chi}^{\ominus}_{M,S} \quad , \tag{4.293}$$

with

$$\begin{aligned} \Delta\Delta\tilde{\chi}^{\ominus}_{M,S} &= \Delta\tilde{\chi}^{\ominus}_{M,S} - \Delta\chi^{\ominus}_{M,S} \\ &= \left(\tilde{\chi}^{\ominus}_{S(M)} - \tilde{\chi}^{\ominus}_{M(S)}\right) - \left(\chi^{\ominus}_{S} - \chi^{\ominus}_{M}\right) \\ &= \left(\chi^{\ominus}_{M} - \tilde{\chi}^{\ominus}_{M(S)}\right) - \left(\chi^{\ominus}_{S} - \tilde{\chi}^{\ominus}_{S(M)}\right) \quad , \end{aligned} \tag{4.294}$$

where $\Delta\tilde{\chi}^{\ominus}_{M,S}$ and $\Delta\tilde{\psi}^{\ominus}_{M,S}$ are the standard frontal surface and Volta potential differences at the interface, respectively (see Eqs. 4.255 and 4.256), and $\Delta\chi^{\ominus}_{M,S}$ is the standard lateral surface potential difference corresponding to free surfaces of the metal and solution (Eq. 4.258). The third equality in Eq. 4.294 follows from a mere rearrangement of the terms and underlines that the difference between $\Delta\psi^{\ominus}_{M,S}$ and $\Delta\tilde{\psi}^{\ominus}_{M,S}$ can be viewed as resulting from the perturbation of the surface potentials of M and S corresponding to free surfaces in vacuum (air) in response to their contact at the interface. In this convention, the reference point for the ideal electron gas is situated immediately *at the surface of the solution*[461, 1718, 1721], as illustrated in Figure 4.9b. This choice has the disadvantage of introducing the specification of a given liquid (by default water) into the standard, *via* the definition of the reference potential for the electron, but turns out to be convenient for its relationship with *real single-ion solvation parameters* (Sections 4.3.7 and 4.5.2). Unlike the standard vacuum potential, the standard real potential of an electrode is a well-defined quantity, *i.e.* it only depends on the natures of M and S. This follows from the observation that in Eq. 4.291, $\omega_M^\ominus$ is uniquely defined for a given metal M under standard conditions (Section 4.3.5), including the specification of a surface curvature or exposed crystallographic face type whenever relevant (by default, planar and polycrystalline surface), while $\Delta\psi_{M,S}^\ominus$ is solely determined by the standard-state natures of M and S (Section 4.3.3), given that the interface is non-polarizable and at equilibrium, and including the specification of lateral reference interface characteristics for M (same as those considered for $\omega_M^\ominus$) and S (by default, planar surface). It is also in principle *experimentally accessible*, although with a relatively *limited accuracy* at present, *via* the determination of the standard work function $\omega_M^\ominus$ of the isolated metal using SBS (Sections 4.2.12, 4.3.5 and 5.13) and of the standard metal-solution lateral Volta potential difference $\Delta\psi_{M,S}^\ominus$ using Voltaic cell experiments (Sections 4.4.4 and 5.14).

The *standard intrinsic electrode potential* $\mathcal{V}_{M,S}^\ominus$ will be defined as the standard vacuum electrode potential minus the Galvani potential of the solution, *i.e.*

$$\mathcal{V}_{M,S}^\ominus = \tilde{\mathcal{V}}_{M,S}^\ominus - \phi_{S(M)}^\ominus = -F^{-1}\mu_{e^-:M}^\ominus + \phi_{M(S)}^\ominus - \phi_{S(M)}^\ominus \quad , \tag{4.295}$$

where $\phi_{S(M)}^\ominus$ is the standard Galvani potential of the solution in contact with the metal electrode. Considering Eqs. 4.262 and 4.289, one may write

$$\mathcal{V}_{M,S}^\ominus = \omega_M^\ominus - \Delta\psi_{M,S}^\ominus - \chi_S^\ominus \ , \tag{4.296}$$

where $\chi_S^\ominus$ is the standard lateral surface potential of the solution. Note that this quantity is a property of the solution only, or, rather, of the solvent (see further below). In this convention, the reference point for the ideal electron gas is situated *within the solution*[1718, 1721], as illustrated in Figure 4.9c. It should be stressed, however, that the reference state for the electron corresponds to an ideal gas of electrons *at the Galvani potential of the solution* and *not to solvated electrons*. This choice, just as was the case for the real absolute potential, has the disadvantage of introducing the specification of a given liquid (by default water) into the standard, *via* the definition of the reference potential for the standard-state electron, but turns out to be convenient for its relationship with *intrinsic single-ion solvation parameters* (Sections 4.3.7 and 4.5.3). The standard intrinsic potential of an electrode is also in principle a *well-defined quantity*, *i.e.* it only depends on the natures of M and S. This follows from the considerations made above concerning the unambiguous nature of the real absolute electrode potential and the observation that $\chi_S^\ominus$ is uniquely defined for a given solution S under standard conditions, including the specification of lateral reference interface characteristics (same as those considered for $\Delta\psi_{M,S}^\ominus$; see above).

Comparing Eq. 4.291 with Eq. 4.296, one has

$$\mathcal{V}_{M,S}^\ominus = \hat{\mathcal{V}}_{M,S}^\ominus - \chi_S^\ominus \quad . \tag{4.297}$$

This equation shows that real and intrinsic absolute electrode potentials are related *via* the standard-state surface potential of the solution. Since standard quantities refer to the limit of *infinite dilution*, $\chi_S^\ominus$ is nothing else than the *air-liquid interfacial potential of the pure solvent* $\chi_{svt}^\ominus$ encountered previously (Section 2.2), referring by default to a planar surface at equilibrium. There is thus a close parallel between real and intrinsic quantities in the context of single-ion solvation (Eqs. 2.6 and 2.8) and of absolute electrode potentials (Eq. 4.297). Considering the discussion of Section 2.2, if $\hat{\mathcal{V}}_{M,S}^\ominus$ is in principle experimentally accessible (see above), just as was the case for $\Delta_s\hat{G}_{svt}^\ominus[I_g^z]$ and $\hat{\mathcal{G}}_{H,svt}^\ominus$, it follows that $\mathcal{V}_{M,S}^\ominus$ belongs to the set of *experimentally-elusive quantities* evading direct experimental determination without the help of an extra-thermodynamic assumption (Section 2.3), just as was the case for $\Delta_s G_{svt}^\ominus[I_g^z]$, $\mathcal{G}_{H,svt}^\ominus$, and $\chi_{svt}^\ominus$. To our knowledge, however, no experimental procedure and associated extra-thermodynamic assumption have been formulated for the direct determination of $\mathcal{V}_{M,S}^\ominus$, which would define a corresponding access route to $\chi_{svt}^\ominus$ and $\mathcal{G}_{H,svt}^\ominus$ (Sections 5.17 and 5.18; see, however Ref.[483]).

In summary, the three types of absolute electrode potentials are related among each other and to the work function of the metal and surface potential of the pure solvent by the equations

$$\tilde{\mathcal{V}}_{M,S}^\ominus - \psi_{M(S)}^\ominus = \hat{\mathcal{V}}_{M,S}^\ominus + \Delta\psi_{M,S}^\ominus = \mathcal{V}_{M,S}^\ominus + \Delta\psi_{M,S}^\ominus + \chi_S^\ominus = \omega_M^\ominus \ . \tag{4.298}$$

Note that the names used here for the three variants of absolute electrode potentials (vacuum, real and intrinsic) do not reflect any standard terminology and were introduced by the authors of this book. Unfortunately, many scientific articles concerned with the problem simply refer to *the* absolute potential of an electrode without any further specification[482,1789,1792,1793,1797,1798], sometimes stating in addition the convention employed for the electron standard state[278,483,1719,1727,1728] (ion or electron convention, referred to here as cold- and warm-electron conventions, respectively, based on either Boltzmann or Fermi-Dirac statistics), or/and for the reference point defining the Galvani potential of this state[1714,1716–1718,1721,1782] (in solution or at the solution surface, referred to here as intrinsic and real, respectively). The intrinsic absolute potential has been called a "reduced absolute" single-electrode potential in Refs.[1718,1721].

The *cold-electron* absolute electrode potentials are obtained when $\omega_M^\ominus$ is replaced by w_M at P° and T^- in Eqs. 4.289, 4.291 and 4.296. Cold-electron absolute potentials can thus be converted to *warm-electron* values by addition of -0.038 V (Eq. 4.286). The standard work function was actually introduced in Section 4.3.5 precisely with the purpose of directly generating derived quantities appropriate for the *bbme*$_T$ definition used throughout this book (Section 4.2.4), including a warm-electron convention. Note that in some sources[419], absolute electrode potentials are meant to be reported in the warm-electron convention, but rely on heat-up parameters for the electron inappropriately calculated using *Boltzmann statistics* rather than *Fermi-Dirac statistics* (Section 4.2.13). The correction to be applied to a warm-electron potential relying on Boltzmann statistics, so as to convert it to a warm-electron potential relying on Fermi-Dirac statistics can be deduced from the corresponding heat-up parameters of the electron (Table 4.2) and evaluates to -0.037 V. This value is *coincidentally* almost equal to the cold- to warm-electron conversion using Fermi-Dirac statistics. As a result, a *warm-electron potential incorrectly relying on Boltzmann statistics* is numerically *almost equal* to a *cold-electron potential.* A further distinction will be made subsequently (Section 4.3.8) between two possible alternative definitions of the Galvani and lateral surface potentials, namely internal or external. This choice also affects the values of the intrinsic absolute electrode potential, but not those of the real one, which follows from Eq. 4.298 and the observation that this choice affects $\chi_S^\ominus$ but neither $\omega_M^\ominus$ nor $\Delta\psi_{M,S}^\ominus$. At this point, the discussion of the intrinsic absolute potential refers exclusively to the corresponding external quantity. Absolute electrode potentials will be further discussed in Section 5.14.

4.3.7 Connection to single-ion solvation properties

As previously stated in Section 4.3.3, a *non-polarizable interface* is an interface allowing the transfer of charges between two conducting phases in contact, while a *polarizable interface* does not allow this transfer. In the former case and at equilibrium, the interface obeys an equality constraint for the electrochemical potentials of the species involved in the charge transfer, and one says that this interface is *reversible* relative to these species. As a result, for two phases of a given composition in contact at such an interface, the Galvani potential difference and thus, the magnitude of the electric double-layer at the interface and the corresponding frontal Volta potential difference $\Delta\tilde{\psi}$, are fixed and cannot be varied arbitrarily. There is no such constraint on the electrochemical potentials for a polarizable interface. In this case, the Galvani potential differences can be varied arbitrarily, *e.g.* by means of an external voltage source (potentiostat), and the magnitude of the electric double-layer at the interface and the corresponding frontal Volta potential difference $\Delta\tilde{\psi}$ will respond to the applied potential difference. Non-polarizable interfaces are the ones relevant in the context of absolute electrode potentials (Section 4.3.6) and single-ion thermodynamics, the latter connection being considered in the present section. A brief discussion of polarizable interfaces will be provided later in Section 4.4.5, in the context of electrocapillary and related measurements.

The first important case of non-polarizable interface is the interface between *two metals* M and M'. In this case, the charge transfer is permitted by the *electrons*, as

$$e^- : M \rightleftharpoons e^- : M' \tag{4.299}$$

and the *electrochemical equilibrium condition* under standard conditions reads

$$\tilde{\mu}^{\ominus}_{e^-:M} = \tilde{\mu}^{\ominus}_{e^-:M'} \quad . \tag{4.300}$$

Considering Eqs. 4.282 and 4.285, one may rewrite this condition as

$$\Delta\psi^{\ominus}_{M,M'} = -(\omega^{\ominus}_{M'} - \omega^{\ominus}_{M}) \quad . \tag{4.301}$$

This equation shows that the standard lateral *Volta potential difference between two metals in contact* is uniquely determined by the *standard work functions of the two isolated metals.*

The second important case of non-polarizable interface is the interface between a metal M and a solution S at which an *oxido-reduction (redox) reaction* may take place, possibly including other phases P in contact with both M and S. In this case, the charge transfer is permitted by the *redox reaction.* Writing this reaction in a generic fashion as

$$\Sigma_{rn} + \Sigma^{z}_{rc} : S \rightleftharpoons \Sigma_{on} + \Sigma^{z'}_{oc} : S + (z' - z)e^- : M \quad , \tag{4.302}$$

where Σ_{rn}, Σ^{z}_{rc}, Σ_{on}, $\Sigma^{z'}_{oc}$ denote neutral and charged or reduced and oxidized species, respectively, and it has been assumed that the charged species are all to be found in the solution, while neutral species may correspond to atoms of the metal M, atoms or molecules in the solution S, or atoms or molecules in the separate phase P. The corresponding *electrochemical equilibrium condition* under standard conditions reads

$$\mu^{\ominus}_{\Sigma_{rn}} + \tilde{\mu}^{\ominus}_{\Sigma^{z}_{rc}:S} = \mu^{\ominus}_{\Sigma_{on}} + \tilde{\mu}^{\ominus}_{\Sigma^{z'}_{oc}:S} + (z' - z)\tilde{\mu}_{e^-:M} \quad . \tag{4.303}$$

Considering Eqs. 4.282, 4.285 and 4.291, one may rewrite this condition as

$$\begin{aligned} \hat{\mu}^{\ominus}_{\Sigma^{z'}_{oc}:S} - \hat{\mu}^{\ominus}_{\Sigma^{z}_{rc}:S} + \mu^{\ominus}_{\Sigma_{on}} - \mu^{\ominus}_{\Sigma_{rn}} &= (z' - z)F(\omega^{\ominus}_{M} - \Delta\psi^{\ominus}_{M,S}) \\ &= (z' - z)F\hat{\mathcal{V}}^{\ominus}_{M,S} \quad . \end{aligned} \tag{4.304}$$

This equation is important, since it provides the connection, *via* the measurement of the *standard work function of a metal*, between *real solvation free energies of ions*, *real absolute electrode potentials* and *metal-solution Volta potential differences*. As an illustration for this statement, Eq. 4.304 can be explicitly rewritten in the context of two important reversible electrodes: a *metal-ion electrode* and the *reference hydrogen electrode*, as illustrated in Figure 4.10.

For the metal-ion electrode, and restricting the discussion to a monovalent ion, the oxidation reaction considered reads

$$M : M \rightleftharpoons M^+ : S + e^- : M \quad . \tag{4.305}$$

In this case, the left-hand side of Eq. 4.304 may be rewritten

$$\hat{\mu}^{\ominus}_{M^+:S} - \mu^{\ominus}_{M:M} = \Delta_f G^{\ominus}[M^+_g] + \Delta_s \hat{G}^{\ominus}[M^+_g] \quad , \tag{4.306}$$

where $\Delta_f G^{\ominus}[M^+_g]$ and $\Delta_s \hat{G}^{\ominus}[M^+_g]$ are the formation and real solvation free energies of the gas-phase ion M^+_g (Sections 2.2 and 4.2.10), and one obtains

$$\begin{aligned} \Delta_s \hat{G}^{\ominus}[M^+_g] &= F(\omega^{\ominus}_M - \Delta\psi^{\ominus}_{M,S}) - \Delta_f G^{\ominus}[M^+_g] \\ &= F\hat{\mathcal{V}}^{\ominus}_{M,S} - \Delta_f G^{\ominus}[M^+_g] \quad . \end{aligned} \tag{4.307}$$

This equation provides in particular a simple physical interpretation of the *real absolute potential* of a reversible metal-solution electrode as the *real formation free energy* of the corresponding *dissolved metal ion* (assumed here

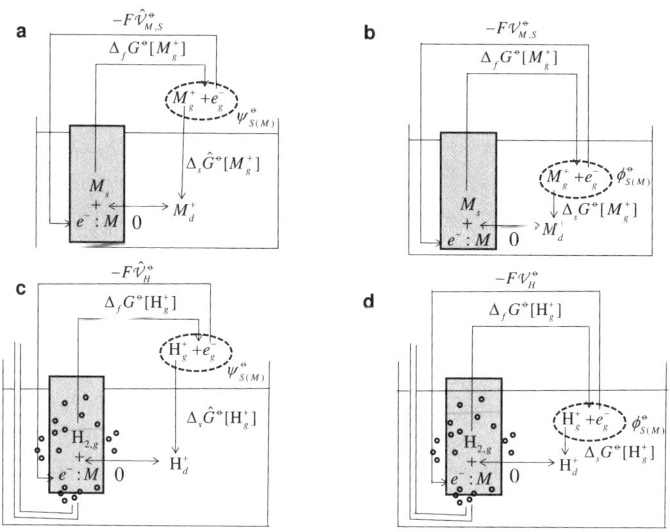

Figure 4.10 *Illustration of the connection between real and intrinsic absolute electrode potentials and corresponding real and intrinsic single-ion solvation free energies.* The top drawings illustrate the thermodynamic cycles relating the real potential of a metal-ion electrode to the real solvation free energy of the ion (a), or the intrinsic potential of the electrode to the intrinsic solvation free energy of the ion (b). The bottom drawings illustrate the similar connection between real (c) and intrinsic (d) quantities for the reference hydrogen electrode.

monovalent), divided by F, *i.e.* namely

$$\hat{\mathcal{V}}^{\ominus}_{M,S} = F^{-1}\left(\Delta_f G^{\ominus}[M_g^+] + \Delta_s \hat{G}^{\ominus}[M_g^+]\right) = F^{-1}\Delta_f \hat{G}[M_d^+] \quad . \tag{4.308}$$

Considering Eqs. 2.6 and 2.8, where $\chi^{\ominus}_S$ is equivalent to $\chi^{\ominus}_{svt}$, as well as Eq. 4.297, a corresponding equation relates the absolute intrinsic electrode potential to the intrinsic solvation free energy of the ion, namely

$$\begin{aligned} \Delta_s G^{\ominus}[M_g^+] &= F(\omega^{\ominus}_M - \Delta\psi^{\ominus}_{M,S} - \chi^{\ominus}_S) - \Delta_f G^{\ominus}[M_g^+] \\ &= F\mathcal{V}^{\ominus}_{M,S} - \Delta_f G^{\ominus}[M_g^+] \; . \end{aligned} \tag{4.309}$$

This equation provides in particular a simple physical interpretation of the *intrinsic absolute potential* of a reversible metal-solution electrode as the *intrinsic formation free energy* of the corresponding *dissolved metal ion* (assumed here monovalent), divided by F, *i.e.* namely

$$\mathcal{V}^{\ominus}_{M,S} = F^{-1}\left(\Delta_f G^{\ominus}[M_g^+] + \Delta_s G^{\ominus}[M_g^+]\right) = F^{-1}\Delta_f G[M_d^+] \quad . \tag{4.310}$$

The *thermodynamic cycles* corresponding to Eqs. 4.307 and 4.309 are illustrated in Figure 4.10a,b. In the real case, a metal atom (M_s) is evaporated and ionized into a gas-phase ion-electron pair ($M_g^+ + e_g^-$) at the lateral Volta potential $\psi^{\ominus}_{S(M)}$ of the solution. The free-energy change $\Delta_f G^{\ominus}[M_g^+]$ associated with this process is independent of the choice of the latter potential, because the pair is overall neutral. The electron is returned to the metal, the free-energy change $-F\hat{\mathcal{V}}^{\ominus}_{M,S}$ associated with this process being related to the real electrode potential. The ion is transferred to the solution, the free-energy change $\Delta_s \hat{G}^{\ominus}[M_g^+]$ associated with this process corresponding to the real single-ion solvation free energy, *i.e.* including the crossing of the air-liquid interface. The solvated ion and the electron in the metal are then recombined into a metal atom with zero free-energy change, owing to the electrochemical equilibrium condition, thereby completing the thermodynamic cycle. Application of the Hess law to this cycle leads to Eq. 4.307. In the intrinsic case, a metal atom (M_s) is evaporated and ionized into a gas-phase ion-electron pair ($M_g^+ + e_g^-$) at the Galvani potential $\phi^{\ominus}_{S(M)}$ of the solution. Recall that neither the ion nor the electron are solvated in this state, which corresponds to ideal gases, but at a reference electric potential $\phi^{\ominus}_{S(M)}$. The free-energy change $\Delta_f G^{\ominus}[M_g^+]$ associated with this process is again independent of the choice of the latter potential, because the pair is overall neutral. The electron is returned to the metal, the free-energy change $-F\mathcal{V}^{\ominus}_{M,S}$ associated with this process being related to the intrinsic electrode potential. The ion is transferred (or rather "grown") into the solution, the free-energy change $\Delta_s G^{\ominus}[M_g^+]$ associated with this process corresponding to the intrinsic single-ion solvation free energy, *i.e.* excluding the crossing of the air-liquid interface. The solvated ion and the electron in the metal are then recombined into a metal atom with zero free-energy change, owing to the electrochemical equilibrium condition, thereby completing the thermodynamic cycle. Application of the Hess law to this cycle leads to Eq. 4.309.

For the reference hydrogen electrode the oxidation reaction considered reads

$$^1/_2\mathrm{H}_2 : G \rightleftharpoons \mathrm{H}^+ : S + e^- : M \quad , \tag{4.311}$$

where G stands for an additional gas phase in contact with the metal. In this case, the left-hand side of Eq. 4.304 may be rewritten

$$\hat{\mu}^{\ominus}_{\mathrm{H}^+:S} - {}^1/_2\mu^{\ominus}_{\mathrm{H}_2,g} = \Delta_f G^{\ominus}[\mathrm{H}^+_g] + \hat{\mathcal{G}}^{\ominus}_H \quad , \tag{4.312}$$

where $\hat{\mathcal{G}}^{\ominus}_H$ is the short notation for $\Delta_s \hat{G}^{\ominus}[\mathrm{H}^+_g]$ introduced in Section 2.2, and one obtains

$$\begin{aligned}\hat{\mathcal{G}}^{\ominus}_H &= F(\omega^{\ominus}_M - \Delta\psi^{\ominus}_{\mathrm{H}(M)}) - \Delta_f G^{\ominus}[\mathrm{H}^+_g] \\ &= F\hat{\mathcal{V}}^{\ominus}_H - \Delta_f G^{\ominus}[\mathrm{H}^+_g] \,,\end{aligned} \tag{4.313}$$

where $\hat{\mathcal{V}}^{\ominus}_H$ is the real absolute potential of the reference hydrogen electrode, and $\Delta\psi^{\ominus}_{\mathrm{H}(M)}$ a notation for $\Delta\psi^{\ominus}_{M,S}$ in the specific context of this electrode, the metal usually being platinum Pt. In analogy with Eq. 4.307, this equation implies in particular

$$\hat{\mathcal{V}}^{\ominus}_H = F^{-1}(\Delta_f G^{\ominus}[\mathrm{H}^+_g] + \hat{\mathcal{G}}^{\ominus}_H) = F^{-1}\Delta_f \hat{G}^{\ominus}[\mathrm{H}^+_d] \quad , \tag{4.314}$$

which relates the *real absolute electrode potential* of the *reference hydrogen electrode* to the *real formation free energy* of the *dissolved proton*, divided by F. Here also, a corresponding equation relates the absolute intrinsic electrode potential $\mathcal{V}^{\ominus}_H$ to the intrinsic solvation free energy of the proton, namely

$$\begin{aligned}\mathcal{G}^{\ominus}_H &= F(\omega^{\ominus}_M - \Delta\psi^{\ominus}_{\mathrm{H}(M)} - \chi^{\ominus}_S) - \Delta_f G^{\ominus}[\mathrm{H}^+_g] \\ &= F\mathcal{V}^{\ominus}_H - \Delta_f G^{\ominus}[\mathrm{H}^+_g] \,,\end{aligned} \tag{4.315}$$

where $\mathcal{G}^{\ominus}_H$ is the short notation for $\Delta_s G^{\ominus}[\mathrm{H}^+_g]$ introduced in Section 2.2. In analogy with Eq. 4.309, this equation implies in particular

$$\mathcal{V}^{\ominus}_H = F^{-1}(\Delta_f G^{\ominus}[\mathrm{H}^+_g] + \mathcal{G}^{\ominus}_H) = F^{-1}\Delta_f G^{\ominus}[\mathrm{H}^+_d] \quad , \tag{4.316}$$

which relates the *intrinsic absolute electrode potential* of the *reference hydrogen electrode* to the *intrinsic formation free energy* of the *dissolved proton*, divided by F.

The thermodynamic cycles corresponding to Eqs. 4.313 and 4.315 are illustrated in Figure 4.10c,d, and closely parallel those described above for the metal-ion electrode. Finally, Eqs. 4.313 and 4.315 can also be combined into an expression connecting the surface potential $\chi^{\ominus}_S$ of the solution and the intrinsic absolute potential $\mathcal{V}_H^{\ominus}$, namely

$$\begin{aligned}\chi_S^{\ominus} &= \hat{\mathcal{V}}_H^{\ominus} - \mathcal{V}_H^{\ominus} \\ &= F^{-1}(\hat{\mathcal{G}}_H^{\ominus} - \mathcal{G}_H^{\ominus}) \\ &= F^{-1}\left(\hat{\mathcal{G}}_H^{\ominus} + \Delta_f G^{\ominus}[\mathrm{H}_g^+]\right) - \mathcal{V}_H^{\ominus} \quad .\end{aligned} \tag{4.317}$$

Keeping in mind that, under standard conditions, $\chi_S^{\ominus}$ is simply the air-liquid interfacial potential of the pure solvent $\chi_{svt}^{\ominus}$, and replacing $\hat{\mathcal{G}}_H^{\ominus}$, $\mathcal{G}_H^{\ominus}$, $\hat{\mathcal{V}}_H^{\ominus}$ and $\mathcal{V}_H^{\ominus}$ by $\hat{\mathcal{G}}_{H,svt}^{\ominus}$, $\mathcal{G}_{H,svt}^{\ominus}$, $\hat{\mathcal{V}}_{H,svt}^{\ominus}$ and $\mathcal{V}_{H,svt}^{\ominus}$, respectively, to make the solvent explicit, Eqs. 4.313-4.317 are the source of Eqs. 2.31-2.34 introduced and discussed in Section 2.2. As mentioned above, these equations permit in particular to formulate a simple and intuitive definition of $\hat{\mathcal{V}}_{H,svt}^{\ominus}$ and $\mathcal{V}_{H,svt}^{\ominus}$, as the real and intrinsic formation free energies, respectively, of the dissolved proton in the given solvent, divided by F.

Corresponding equations connecting the derivative thermodynamic solvation properties of the proton to pressure or/and temperature derivatives of the corresponding absolute potentials are easily derived by application of the thermodynamic-derivative operator $\hat{O}_Y$ (Section 4.2.1) to these three equations. The resulting equations connecting the derivatives of $\hat{\mathcal{G}}_H^{\ominus}$, $\mathcal{G}_H^{\ominus}$ and $\chi_S^{\ominus}$ based on Eq. 4.317 have already been provided in Section 2.2 (Eqs. 2.28 and 2.29). Two additional equations may be of relevance, based on Eqs. 4.313 and 4.315, namely

$$\hat{O}_Y \hat{\mathcal{G}}_H^{\ominus} = F \hat{O}_Y \hat{\mathcal{V}}_H^{\ominus} - \Delta_f Y^{\ominus}[\mathrm{H}_g^+] \tag{4.318}$$

and

$$\hat{O}_Y \mathcal{G}_H^{\ominus} = F \hat{O}_Y \mathcal{V}_H^{\ominus} - \Delta_f Y^{\ominus}[\mathrm{H}_g^+] \quad . \tag{4.319}$$

Note that Eq. 4.313 encompasses an additional and important connection between $\hat{\mathcal{G}}_H^{\ominus}$ or $\hat{\mathcal{V}}_H^{\ominus}$ and the *standard lateral Volta potential difference* $\Delta\psi_{\mathrm{H}(M)}^{\ominus}$ at the *reference hydrogen electrode*, *via* the standard work function $\omega_M^{\ominus}$ of the metal involved, usually platinum Pt, and the formation free energy of the gas-phase proton. This connection, or the analogous connection of Eq. 4.307 for the metal-ion electrode, is exploited in the context of Voltaic cell measurements (Section 4.4.4). It should also be noted that in contrast to the quantities $\hat{\mathcal{V}}_{H,svt}^{\ominus}$, $\mathcal{V}_{H,svt}^{\ominus}$ and $\omega^{\ominus}$ (Sections 5.13 and 5.14), the quantities $\hat{\mathcal{G}}_{H,svt}^{\ominus}$, $\mathcal{G}_{H,svt}^{\ominus}$, $\chi_{svt}^{\ominus}$ and $\Delta\psi_{\mathrm{H}(M)}^{\ominus}$ are independent of any convention regarding the *standard-state electron* (cold- *vs.* warm-electron convention, Boltzmann *vs.* Fermi-Dirac statistics for the ideal electron gas). In addition, the quantities $\mathcal{V}_{H,svt}^{\ominus}$, $\mathcal{G}_{H,svt}^{\ominus}$ and $\chi_{svt}^{\ominus}$ depend on the choice of an *internal or external definition* for the Galvani and lateral surface potential (Section 4.3.8), while the quantities $\hat{\mathcal{V}}_{H,svt}^{\ominus}$, $\hat{\mathcal{G}}_{H,svt}^{\ominus}$, $\omega^{\ominus}$ and $\Delta\psi_{\mathrm{H}(M)}^{\ominus}$ do not. Finally, the quantities $\hat{\mathcal{V}}_{H,svt}^{\ominus}$, $\hat{\mathcal{G}}_{H,svt}^{\ominus}$, $\chi_{svt}^{\ominus}$, $\omega^{\ominus}$ and $\Delta\psi_{\mathrm{H}(M)}^{\ominus}$ depend on the choice of lateral reference interface characteristics (Section 4.3.2) for the metal (by default, planar polycrystalline surface) and the pure solvent (by default, planar surface), while

$\mathcal{V}^{\ominus}_{H,svt}$ and $\mathcal{G}^{\ominus}_{H,svt}$ do not.

Note that an incorrect calculation of the standard gas-phase parameters of the proton using Boltzmann rather than Fermi-Dirac statistics for the proton itself (Section 4.2.13), which may affect some of the above properties *via* $\Delta_f G^{\ominus}[\mathrm{H}_g^+]$ depending on their determination method, has a negligible impact *per se* ($6 \cdot 10^{-4}$ V, corresponding to 0.06 kJ·mol^{-1}, based on the data of Table 4.2), as long as the electron parameters are appropriately calculated using Fermi-Dirac statistics. However, the choice of Boltzmann or Fermi-Dirac for the electron also affects the value of $\Delta_f G^{\ominus}[\mathrm{H}_g^+]$, and this much more significantly (0.038 V, corresponding to 3.7 kJ·mol^{-1}), because the formation of a standard-state proton implies that of a standard-state electron.

4.3.8 External *versus* internal potentials

The working definitions introduced in Section 4.3.1 for the *surface alteration* of the charge density of a phase at an interface, and in Section 4.3.2 for the *Galvani and (lateral or frontal) surface potentials* of this phase are apparently acceptable. However, they actually hide a *fundamental ambiguity*, which is not apparent at the continuum-electrostatics level, but reveals itself when trying to interpret these concepts at the *microscopic level*. The ambiguity arises when mapping the surface alteration of the charge density in the neighborhood of an interface to a surface polarization. In the context of pure liquids and vacuum-liquid (air-liquid) interfaces, it is the source of the discrepancies previously encountered during the calculation using AT simulations of the *intrinsic electric potential* ζ *at a cavity within the liquid* (*via* recalculation of the cavity potential, using M-summation for $\zeta_{AT,M}$ *vs.* P-summation for $\zeta_{AT,P}$; Section 3.3.3), of the *intrinsic bulk electric potential* η *of the pure liquid* (*via* potential averaging, using external M-averaging for $\eta_{AT,M}$ *vs.* internal M-averaging for $\check{\eta}_{AT,M}$ *vs.* external P-averaging for $\eta_{AT,P}$ *vs.* internal P-averaging for $\check{\eta}_{AT,P}$; Section 3.3.4), or of the *air-liquid interfacial potential* χ *of the pure liquid* (*via* surface integration, using M-integration for $\chi_{AT,M}$ *vs.* P-integration for $\chi_{AT,P}$; Section 3.3.5). Note that a similar problem has been encountered more recently in the context of QM computations (Section 6.3.4). This issue is intimately related to a specific electrostatic property of charge distributions, namely their *quadrupole-moment trace*. The meaning of this quantity is most simply illustrated in the context of an isotropic charge distribution[707].

Consider an isotropic charge distribution $\tau(r)$ in vacuum, r indicating the distance from the center of the distribution, that is overall neutral and of finite spatial extent, *i.e.* with

$$4\pi \int_0^\infty dr\, r^2\, \tau(r) = 0 \qquad (4.320)$$

and

$$\tau(r) = 0 \quad \text{for} \quad r \geq r_o \quad . \qquad (4.321)$$

The monopole-, dipole-, octupole- and the off-diagonal elements of the quadrupole-moment tensors characterizing such a distribution all vanish. However, this is not the case for the diagonal elements of the quadrupole-moment tensor. These three elements are identical, while the tensor trace $\mathcal{Q}$ is

$$\mathcal{Q} = 4\pi \int_0^\infty dr\, r^4 \tau(r) \quad . \tag{4.322}$$

The distribution τ will therefore be referred to as an *isotropic quadrupole* (IQ) of trace $\mathcal{Q}$. For Coulombic interactions, the radial electric field E generated by the distribution can be evaluated by the application of Gauss' law[421,1453], leading to

$$E(x) = \epsilon_o^{-1}\, x^{-2} \int_0^x dr\, r^2\, \tau(r) \quad , \tag{4.323}$$

while the corresponding electric potential ϕ, with its zero set in the limit $x \to \infty$, is given by

$$\begin{aligned} \phi(x) &= \int_x^\infty dx'\, E(x') \\ &= \epsilon_o^{-1} \left[x^{-1} \int_0^x dr\, r^2\, \tau(r) + \int_x^\infty dr\, r\, \tau(r) \right] \quad , \end{aligned} \tag{4.324}$$

the latter equality following from integration by parts. An immediate consequence of Eqs. 4.320, 4.321, 4.323 and 4.324 is that

$$\phi(x) = 0 \quad \text{and} \quad E(x) = 0 \qquad \text{for} \qquad x \geq r_o \quad . \tag{4.325}$$

The potential ϕ can be integrated over a spherical volume of radius R around the center of the distribution, leading to

$$\begin{aligned} \Phi(R) &= 4\pi \int_0^R dx\, x^2\, \phi(x) \\ &= 4\pi (6\epsilon_o)^{-1} \left[3R^2 \int_0^R dr\, r^2\, \tau(r) \right. \\ &\quad \left. + 2R^3 \int_R^\infty dr\, r\, \tau(r) - \int_0^R dr\, r^4\, \tau(r) \right] \quad , \end{aligned} \tag{4.326}$$

the last equality following from integration by parts. An immediate consequence of Eqs. 4.320, 4.321, and 4.326 is that

$$\Phi(R) = -(6\epsilon_o)^{-1}\, \mathcal{Q} \quad \text{for} \quad R \geq r_o \quad . \tag{4.327}$$

Thus, the volume integral of the potential generated by an IQ over the entire space, or any finite subvolume encompassing the IQ, is a finite quantity proportional to its trace $\mathcal{Q}$. Note that non-vanishing higher-order moments, *e.g.* the twenty-one non-zero elements of the hexadecapole-moment tensor, do not contribute to this integral. Eqs. 4.325 and 4.327 also imply[707] that the infinitesimal point IQ (PIQ) of unit trace is the source charge density generating a potential spike (Dirac delta function) of magnitude $-(6\epsilon_o)^{-1}$ through the Poisson equation, namely

$$\nabla^2[-(6\epsilon_o)^{-1}\delta(\mathbf{r})] = -\epsilon_o^{-1}\tau_{PIQ}(r) \quad , \tag{4.328}$$

where $\delta(\mathbf{r})$ is the three-dimensional Dirac delta function.

As will be shown below, the *integrated-potential contribution* of Eq. 4.327 is at the origin of the differences encountered between M- and P-scheme results, as well as between results of internal and external averaging for the intrinsic bulk electric potential, in AT simulations[707, 708, 1056, 1429, 1433, 1434, 1436, 1437] (Sections 3.3.3-3.3.5) and, more recently, in QM computations (Sections 6.3.3-6.3.5).

It is important to stress that the above discussion refers to the conventional definition of the quadrupole-moment tensor rather than to its traceless version[421]. For example, the atom Ar and the molecules O_2 and CH_4 in their ground states have vanishing monopole, dipole and traceless quadrupole moments by symmetry. But the traces of their conventional quadrupole moments are non-vanishing, because the radial charge distribution (nuclei and electrons) around the center of symmetry is not uniformly zero. For example, for Ar, $\mathcal{Q}$ will be negative owing to the fact that the positive charge is concentrated at the nucleus while the negative charge is located in the more diffuse electron cloud. Since the quadrupole-moment trace induces no electric field outside the molecule, it is usually removed and the traceless tensor version is considered instead as the only relevant component to characterize intermolecular interactions. Referring to this quantity, one would loosely say that Ar, O_2 and CH_4 have no quadrupole moment. This is not done here, because *the effect of $\mathcal{Q}$ on the potential within the molecule* is precisely the concern of the discussion. Note, finally, that for a species presenting a net charge or dipole moment, the quadrupole moment, including the trace, is origin-dependent.

Now, consider a finite sample of a hypothetical fluid consisting of identical hard spheres at a density ρ_L, confined by a hard wall and surrounded by vacuum, which will be referred to as a liquid L. Each hard sphere is of radius r_o, and carries an IQ charge distribution τ of arbitrary form and trace $\mathcal{Q}$, confined within the sphere (Eq. 4.321). This situation is illustrated in Figure 4.11a,b considering a spherical fluid volume of radius X and a model distribution τ consisting of a negative point charge $-q$ at the center and a positive spherical charge q at distance r_o from the center, in which case one has $\mathcal{Q} = qr_o^2$ (Eq. 4.322). Irrespective of the form chosen for τ within the spheres, Eq. 4.325 implies the absence of any electrostatic interaction between them. In other words, the configurational distribution of this system is entirely dictated by the hard-sphere and hard-wall repulsions, and rigorously insensitive to τ and $\mathcal{Q}$.

Assuming that sphere-packing constraints represent a negligible perturbation, there are no surface effects influencing the spatial distribution of the spheres. Thus, the distribution of the sphere centers is homogeneous throughout the sample, up to a distance r_o from the surface, and zero beyond. Introducing the notations $\tilde{X} = X - r_o$ and $\tilde{\rho}_L = (X/\tilde{X})^3 \rho_L$, the radial charge density ρ_F^- of negative charges, all located at the sphere centers, is thus

$$\rho_F^-(x) = -N_A q \tilde{\rho} M_L^{-1} \times \begin{cases} 1 & \text{if} \quad x \leq \tilde{X} \\ 0 & \text{otherwise} \end{cases} \quad , \tag{4.329}$$

M_L being the molar mass of the hard sphere liquid, used to convert the mass density ρ_L into a number density after amplification by N_A. A simple geometric calculation[708] shows that the corresponding radial charge density ρ_F^+ of positive charges, all located at the sphere surfaces, is

$$\rho_F^+(x) = +N_A q \tilde{\rho}_L M_L^{-1} \times \begin{cases} 1 & \text{if} \quad x \leq \tilde{X} - r_o \\ \frac{1}{2} + \frac{X^2 - r_o^2 - x^2}{4 r_o x} & \text{if} \quad \tilde{X} - r_o < x \leq \tilde{X} + r_o \\ 0 & \text{otherwise} \end{cases} \quad . \tag{4.330}$$

Note that in the common situation where $r_o \ll X$, ρ_F^+ is essentially linear in the range $\tilde{X} - r_o$ to $\tilde{X} + r_o$. These two charge distributions are illustrated

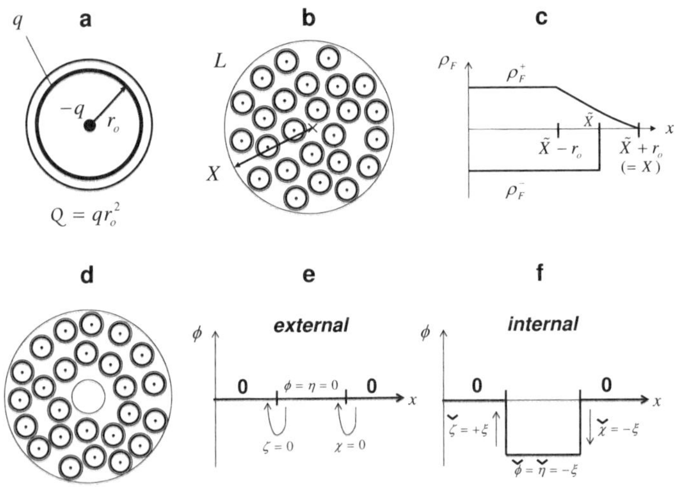

Figure 4.11 *Potentials in a hypothetical fluid of hard spheres bearing isotropic quadrupole charge distributions.* The charge distribution of a hard disk of radius r_o (a) is assumed to consist of a negative point charge $(-q)$ at the center and a positive charge shell $(+q)$ at (drawn as just below) its surface (quadrupole-moment trace $\mathcal{Q} = qr_o^2$). The situations of a sample of hard sphere liquid L (density ρ_L, hard disk molar mass M_L), either pure (b) or containing a hard cavity (d) is illustrated (exclusion potential $\xi_L = (6\epsilon_o)^{-1} \rho M^{-1} \mathcal{Q}$). The charge distributions ρ_F^- and ρ_F^+ of positive and negative charges are displayed for the pure liquid case (c). The relevant potentials in the cavity situation are also displayed according to external (e) or internal (f) definitions.

in Figure 4.11c. In the surface range $\tilde{X} - r_o$ to $\tilde{X} + r_o$, there exists a local imbalance between positive and negative charge densities. The first non-vanishing moment of this surface charge distribution is a surface polarization, which results in a potential jump of magnitude[708]

$$\xi_L = N_A(6\epsilon_o)^{-1} q\tilde{\rho}_L M_L^{-1} r_o^2 \quad , \tag{4.331}$$

in the in-to-out direction. In the common situation where $r_o \ll X$, $\tilde{\rho}_L$ can be identified with ρ_L. Considering that $\mathcal{Q} = qr_o^2$, Eq. 4.331 can be rewritten as

$$\xi = N_A(6\epsilon_o)^{-1} \rho_L M_L^{-1} \mathcal{Q} \quad . \tag{4.332}$$

The quantity ξ_L will be called the *exclusion potential* of the fluid, a quantity that was already encountered in Sections 3.3.3-3.3.5 in the context of AT simulations. At this point, *two distinct perspectives* can be adopted for the definition of the surface alteration, as well as of the Galvani and surface potentials in this simple context. These are illustrated in Figure 4.11e,f considering in addition the presence of a hard-sphere cavity within the solvent, a situation that will be discussed subsequently.

The first perspective (Figure 4.11e) regards *entire hard spheres* as the sources of electric potential and the *free volume of the sample*, outside the spheres, as the region for averaging the electric potential, which is equivalent to assuming that the potential is probed by test charges *with an excluded volume*. It is based on the observation that there are no surface effects on the spatial distribution of the hard spheres and thus, that there is *no surface alteration*. Consequently, the Galvani and surface potentials are both zero. These potentials will be referred to as *external*, and the notations ϕ and χ will be retained for these external Galvani and surface potentials, respectively, in consistency with the discussions of Sections 4.3.1-4.3.7. In summary, in this perspective, one finds

$$\phi_L = \chi_L = 0 \quad . \tag{4.333}$$

The second perspective (Figure 4.11f) regards the *true charge density*, as contributed by ρ_F^- and ρ_F^+ separately, as the source of electric potential and the *entire volume of the sample*, including the inside of the spheres, as the region for averaging the electric potential, which is equivalent to assuming that the potential is probed by test charges *without excluded volume*. It is based on the observation that the charge distribution near the surface is actually inhomogeneous. There is an excess of negative charge density in the distance range from $\tilde{X} - r_o$ to $\tilde{X}$ that is compensated by a neutralizing excess of positive charge density in the distance range from $\tilde{X}$ to $\tilde{X} + r_o$, *i.e.* there is a net outwards-directed surface polarization. Correspondingly, within the bulk, a significant fraction of the total sample volume is located inside the hard spheres, under the electrostatic influence of the negative centers, but not of the positive shells. Thus, the surface and Galvani potentials are expected to be negative. These potentials will be referred to as *internal*, and the notations $\breve{\phi}$ and $\breve{\chi}$ will be used for these internal Galvani and surface potentials, respectively, to distinguish them from the corresponding external values (see above). The values of these potentials are easily deduced from Eq. 4.332 and can be written

$$\breve{\phi}_L = \breve{\chi}_L = -\xi_L \quad . \tag{4.334}$$

Now consider the same fluid sample encompassing a hard cavity, as illustrated in Figure 4.11d. In the external perspective, the potential of the cavity relative to the bulk is $\zeta_L = 0$. In the internal perspective, the cavity is surrounded by an inwards-directed polarization and $\breve{\zeta}_L = +\xi_L$. In both cases, the potential at the cavity center relative to a zero point outside the sample is

$$\chi_L + \zeta_L = \breve{\chi}_L + \breve{\zeta}_L = 0 \quad . \tag{4.335}$$

Thus, if a charge ze is progressively transferred to the cavity so as to obtain a real solvation free energy for an ion I^z according to the unphysical solvation process (Figure 2.2), both the external and internal conventions will agree that this real solvation free energy is zero, also in line with intuitive expectations. However, if the surface contribution is removed so as to lead to an intrinsic value, only the external convention will lead to a value of zero, again in line with intuitive expectations. The internal convention, in contrast, will suggest a value equal to $zF\xi_L$ instead, where $F\xi_L$ is a positive quantity.

This difference is the heart of the problem and can be further illustrated in a somewhat more realistic context[707]. Consider a spherical sample L of pure *liquid argon* (number of atoms N, radius R) at 87 K (density[1322] $\rho_L = 1430$ kg·m^{-3}) surrounded by vacuum, as depicted in Figure 4.12. For simplicity,

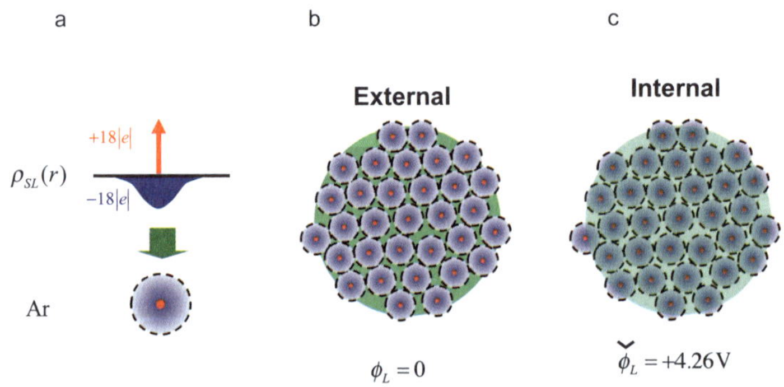

Figure 4.12 *External and internal Galvani potentials within a spherical sample of non-polarizable liquid argon.* Liquid argon (spherical sample L surrounded by vacuum) is modeled here using a Slater-type description of the electronic density and excluding electronic polarizability effects (a). The Galvani potential is calculated by averaging the electric potential generated by all argon atoms (nuclei and electrons) over liquid configurations. In the external convention (b), the potential averaging only includes the region outside the argon atoms (*i.e.* where the electron density is essentially negligible), leading to a value ϕ_L. In the internal convention (c), the potential averaging includes the entire region, leading to a value $\breve{\phi}_L$.

it will be assumed that the charge density in the system for a given configuration of the argon nuclei ($3N$-dimensional vector $\mathbf{r}$, coordinates relative to the sphere center) can be appropriately described as a superposition of point nuclear charge densities and Slater-type electronic charge densities [1799, 1800] (Figure 4.12a), *i.e.*

$$\rho_F(\mathbf{x};\mathbf{r}) = \sum_{n=1}^{N} \rho_{SL}(|\mathbf{x}-\mathbf{r}_n|) \quad , \tag{4.336}$$

with the single-atom spherically-symmetric charge density ρ_{SL} given by

$$\begin{aligned} \rho_{SL}(x) &= \frac{9e}{2\pi} x^{-2}\delta(x) - \frac{2e}{\pi}\alpha_1^3 e^{-2\alpha_1 x} \\ &\quad - \frac{e}{12\pi}\alpha_2^5 x^2 e^{-\alpha_2 x} - \frac{16e}{98415\pi}\alpha_3^7 x^4 e^{-(2/3)\alpha_3 x} \quad , \end{aligned} \tag{4.337}$$

where $\alpha_n = a_o^{-1} Z_n$, a_o being the Bohr radius ($0.52917725 \cdot 10^{-10}$ m) and Z_n the shielded nuclear charge associated to electrons with principal quantum number n ($Z_1 = 17.70$, $Z_2 = 13.85$ and $Z_3 = 6.75$). Note that this simple model does not account for electronic polarization effects, *i.e.* the charge density of individual argon atoms is always rigorously spherically-symmetric, even in the neighborhood of the interface to vacuum (or of a possible dissolved ion). Note that a sphere of radius $x_o = 0.156$ nm centered at the nucleus of an argon atom already encompasses 99% of the integrated electron density (0.198 nm for 99.9%). As a result, such a sphere is essentially electroneutral and due to Gauss' law [421, 1453] under spherical symmetry, the corresponding single-atom spherically-symmetric potential Φ_{SL} essentially vanishes beyond this distance. Given an ensemble of liquid configurations ($\mathbf{r}$-vectors) corresponding to a homogeneous distribution of the N argon atoms within the spherical sample (assuming positional correlations appropriate for bulk liquid argon, even in the surface region), one can calculate the Galvani potential ϕ_L. Obviously, in the absence of net charge, the Volta potential ψ_L is zero so that the surface potential χ_L is equal to the Galvani potential ϕ_L.

As previously discussed in the context of Figure 4.11, the first consistent option (Figure 4.12b) is to consider the Galvani potential as an *external* potential, *i.e.* as probed by a test charge *with an excluded volume.* The potential ϕ_L is then obtained by averaging the potential generated by the argon atoms over all configurations and *over the free region of the sample*, *i.e.* the region that is outside all argon atoms in each configuration, *e.g.* all points beyond a distance x_o from any argon nucleus in this configuration. To give a more formal definition of what is meant by external region, one may write,

$$\phi_L = \lim_{\epsilon \to 0} \left\langle \frac{\int_L d^3\mathbf{x}\, h(\epsilon - \rho(\mathbf{x};\mathbf{r})) \sum_{n=1}^{N} h(R - |\mathbf{r}_n|)\Phi_{SL}(|\mathbf{x}-\mathbf{r}_n|)}{\int_L d^3\mathbf{x}\, h(\epsilon - \rho(\mathbf{x};\mathbf{r}))} \right\rangle_{\mathbf{r}} , \tag{4.338}$$

where h is the Heaviside step function, *i.e.* $h(x) = 0$ for $x \leq 0$ and $h(x) = 1$ for $x > 0$, and $\langle ... \rangle_{\mathbf{r}}$ denotes averaging over all argon configurations. In this equation, ϵ is a *charge density threshold* below which a point is considered to be external, and the limit should be viewed as an *extrapolation* towards zero threshold. This is because, in a quantum system, the electron charge density is nowhere exactly zero, so that if ϵ is taken infinitesimally small, the averaging volume becomes essentially empty. Due to the overall neutrality of the charge density ρ_{SL}, leading to an essentially vanishing single-atom potential ϕ_{SL} beyond the distance x_o, the result of the external averaging defined by Eq. 4.338 is $\phi_L = 0$. Since the Volta potential is zero, one also has $\chi_L = 0$. Based on similar arguments, the internal average bulk potential η_L (potential in the liquid excluding the contribution of a possible surface alteration) and cavity potential ζ_L (potential at a cavity within the sample relative to η_L) are also given by $\eta_L = \zeta_L = 0$. These results are those that would be obtained by M-integration for the surface potential $\chi_{AT,M}$ (Section 3.3.5; Figure 3.7) by M-summation for the (recalculated) potential in a cavity[707] $\zeta_{AT,M}$ (Section 3.3.3; Figure 3.5), or by M-averaging for the external potential $\eta_{AT,M}$ (Section 3.3.4; Figure 3.6). In the case of $\chi_{AT,M}$, this observation is obvious from the fact that the charge density ρ_{SL} is exempt of dipole moment. Thus, the result in terms of external quantities in the present case is

$$\phi_L = \chi_L = \eta_L = \zeta_L = 0 \ , \tag{4.339}$$

and will correspond to the results of *M-summation*, *M-integration* and *M-averaging of the external potential* in AT simulations, *i.e.*

$$\phi_{AT,M} = \chi_{AT,M} = \eta_{AT,M} = \zeta_{AT,M} = 0 \ , \tag{4.340}$$

where $\phi_{AT,M}$ is defined by $\eta_{AT,M} + \chi_{AT,M}$. From here on, the symbols χ and ϕ, as well as the intrinsic chemical potentials (Section 4.3.4), intrinsic electrode potentials (Section 4.3.6) and intrinsic solvation free energies (Section 4.3.7), will be bound to this external definition, with a proviso concerning quantities involving dissolved subatomic particles (see further below).

As previously discussed in the context of Figure 4.11, the second consistent option (Figure 4.12c) is to consider the Galvani potential as an *internal* potential, *i.e.* the potential as probed by a test charge *without excluded volume*. The symbols $\breve{\phi}$, $\breve{\chi}$, $\breve{\eta}$ and $\breve{\zeta}$ will be used for this internal definition, to distinguish them from the corresponding external values (see above). The potential $\breve{\phi}_L$ is obtained by averaging the potential generated by the argon atoms over all configurations and *over the entire sample*, *i.e.* including the points that are inside the argon atoms. To give a more formal definition, one may write

$$\breve{\phi}_L = \lim_{\epsilon \to 0} \left\langle \frac{\int_L d^3\mathbf{x} \sum_{n=1}^{N} H(R - |\mathbf{r}_n|) \Phi_{SL}(|\mathbf{x} - \mathbf{r}_n|)}{(4\pi/3)R^3} \right\rangle_{\mathbf{r}} \ . \tag{4.341}$$

In contrast to the external case, the result of the internal averaging defined by Eq. 4.341 does not vanish. In fact, it is easily shown[707] that $\breve{\phi}_L = -\xi_L$, where ξ_L is the *exclusion potential* of the solvent, and is in the present case given by

$$\begin{aligned} \xi_L &= -N_A \rho_L M_L^{-1} \int_0^\infty dx 4\pi x^2 \Phi_{SL}(x) \\ &= -N_A \epsilon_o^{-1} \rho_L M_L^{-1} e(\alpha_1^{-2} + 40\alpha_2^{-2} + 168\alpha_3^{-2}) \quad , \end{aligned} \tag{4.342}$$

where ρ_L and M_L are the density and molar mass of the liquid. Since the Volta potential is zero, one also has $\chi_L = -\xi_L$. Using the appropriate values of the density, molar mass and α coefficients (see above), one finds $\xi_L = -4.26$ V and thus $\breve{\phi}_L = \xi_L = +4.26$ V for the considered argon system. A positive internal Galvani potential is expected in this case because the averaging region for the potential has been extended to include the interior of the argon atoms, which present an excess of positive charge (nuclei) compared to the negative ones (electrons). Based on similar arguments, the internal average bulk potential $\breve{\eta}_L$ ($\breve{\phi}_L$ excluding the contribution of a possible surface alteration) and the cavity potential ζ_L (potential at a cavity within the sample relative to $\breve{\eta}_L$) are found to be $\breve{\eta}_L = -\xi_L$ and $\breve{\zeta}_L = +\xi_L$. These results are those that would be obtained by P-integration for the surface potential $\chi_{AT,P}$ (Section 3.3.5; Figure 3.7), by P-summation for the (recalculated) potential in a cavity[707] $\zeta_{AT,P}$ (Section 3.3.3; Figure 3.5), or by M-averaging for the internal potential $\breve{\eta}_{AT,M}$ (Section 3.3.4; Figure 3.6). In the case of $\chi_{AT,P}$, this observation results from the fact that the surface charge density at the surface of a sample is imbalanced, *i.e.* the argon electronic clouds protrude towards the outside compared to the nuclear positions, resulting in a more positive internal potential within the phase. Thus, the result in terms of internal quantities in the present case is

$$\breve{\phi}_L = \breve{\chi}_L = \breve{\eta}_L = -\breve{\zeta}_L = -\xi_L \ , \tag{4.343}$$

and will correspond to the results of *P-summation*, *P-integration* and *M-averaging of the internal potential* in AT simulations, *i.e.*

$$\breve{\phi}_{AT,P} = \breve{\chi}_{AT,P} = \breve{\eta}_{AT,M} = -\zeta_{AT,P} = \xi_{AT} \ , \tag{4.344}$$

where $\breve{\phi}_{AT,P}$ is defined by $\breve{\eta}_{AT,M} + \breve{\chi}_{AT,P} - \xi_{AT}$.

The two other approaches suggested in Section 3.3.5, namely P-averaging of the external potential $\eta_{AT,P}$ or of the internal potential $\breve{\eta}_{AT,P}$ will not be discussed here. They are not physically meaningful since argon atoms are indissociable entities, *i.e.* the concept of the potential generated by a fractional argon atom makes little sense. In the present context, the results of P-averaging for the internal potential $\breve{\eta}_{AT,P}$ coincidentally agree with those of M-averaging for the external potential $\eta_{AT,M}$. The results of P-averaging for the external potential $\eta_{AT,P}$ differ from those of M-averaging for the external potential $\eta_{AT,M}$ by ξ. Note that these relationships are not always satisfied in more complex situations (Table 3.5).

In the more general case of an atomic or molecular liquid L, where (i) the charge density in a given configuration of a pure liquid sample is (approximated as) a sum of contributions from individual species (*e.g.* atoms, molecules) and (ii) the orientational distribution of a given species, relative to a selected atomic or molecular center, is uncorrelated with the positional distributions of the surrounding species in the sample, orientationally-averaged with respect to the same center, it is possible to provide a simple expression for the exclusion potential ξ_L of the liquid[707], excluding a possible internal contribution form the nuclei. When these conditions are satisfied, as was the case for the argon liquid (Eq. 4.336 along with the choice of the nucleus as an atomic center), ξ_L, represented here at the QM level by a value ξ_{QM}, is directly related to the quadrupole-moment trace $\mathcal{Q}$ of the charge distribution $\rho_{SP}(\mathbf{x})$ of a single species. More precisely, one has (Eq. 4.332)

$$\xi_{QM} = N_A(6\epsilon_o)^{-1}\rho_L M_L^{-1}\mathcal{Q} \tag{4.345}$$

with

$$\mathcal{Q} = \int_0^\infty d^3\mathbf{x}\, x^2 \rho_{SP}(\mathbf{x}) \ . \tag{4.346}$$

In the argon case with the Slater-type approximation ($\rho_{SP}(\mathbf{x})$ isotropic and given by $\rho_{SL}(x)$ in Eq. 4.337), $\mathcal{Q}$ evaluates to

$$\mathcal{Q} = 4\pi \int_0^\infty dx x^4 \rho_{SL}(x) = -6e(\alpha_1^{-2} + 40\alpha_2^{-2} + 168\alpha_3^{-2}) \ . \tag{4.347}$$

Using the appropriate values of the density, molar mass and α coefficients (see above), one finds $\mathcal{Q} = -0.0655$ $e{\cdot}\text{nm}^2$ and $\xi_L = -4.26$ V, consistent with the results of Eq. 4.342. When the charge density is not partitionable among species, even as an approximation, or when no center can be found for which the orientational distribution of the species is uncorrelated with the positional distributions of the surrounding species, ξ_L must be computed using an appropriate QM computation method along with the use of Eqs. 4.341, 4.343 and 4.345. Note that the value estimated for argon based on a simple Slater-type description should only be taken as indicative. The application of a more accurate QM treatment would probably lead to a significantly different value.

Note also the distinction made here between ξ_L, which refers to the "true" value of the exclusion potential, ξ_{QM}, which refers to a corresponding QM estimate, and ξ_{AT}, which represents the corresponding value calculated for a classical solvent model used in AT simulations. The quantity ξ_L appears at first sight to be a physical observable, although experimentally essentially inaccessible. In principle, one could evaluate a corresponding estimate ξ_{QM} using a QM approach (*e.g.* above Slater-type treatment in the case of liquid argon). Although one could believe that ξ_{AT}, which refers to a given classical solvent model, represents a plausible classical estimate for ξ_{QM}, this is not the case in practice for reasons that will be explained further below. Similarly, if ξ_{QM} is to be viewed as an estimate for the true ξ_L value, it should in principle also include an intramolecular contribution from the averaging of the electric potential inside the nuclei (*i.e.* between quarks), because nuclei may also be characterized by a charge distribution with a non-vanishing quadrupole-moment trace. Such a contribution would obviously be both very difficult (or impossible) to measure or calculate, and appears to be entirely irrelevant for chemical (ionic solvation) problems. In fact, a number of arguments presented below justify that ξ_L is actually itself irrelevant and should not appear as a contribution to "meaningful" estimates for intrinsic single-ion solvation properties, air-liquid interfacial potentials or absolute electrode potentials, be they evaluated based on experimental or theoretical data (in the latter case, in the form of ξ_{QM} or ξ_{AT}). The arguments make use of the word "meaningful" in an intuitive sense, *i.e.* based on intuitive expectations one may have on these quantities in specific limiting situations, the violation of which appears "absurd". As usual, it turns out that an *extra-thermodynamic assumption* is actually lurking behind these intuitive expectations. The nature of this assumption will be discussed in Chapter 7.

The *first argument* concerns the above model calculation for liquid argon. In the external convention, the potential in a cavity within a pure argon sample relative to the bulk, *i.e.* the quantity relevant for the evaluation of an intrinsic solvation free energy, evaluates to zero. In the internal convention, this potential evaluates to -4.26 V. Because the argon atoms present no dipole and are assumed to be non-polarizable, the intrinsic solvation free energy of an ion I^z in the external convention will be zero, while it will be $zF\xi_L = -411z$ kJ·mol^{-1} in the internal convention. These numbers follow from the reversible transfer of the ionic charge from vacuum to a cavity within the fluid (Figure 2.2), excluding the surface contribution. In both cases, the real value, including the surface contribution, is zero. However, the surface contribution to be added is $zF\chi_L = 0$ in the internal convention and $zF\check{\chi}_L = -zF\xi_L = +411z$ kJ·mol^{-1} in the external convention. In other words, the external convention suggests an intuitively reasonable picture where an ion does not interact with a non-polarizable argon solvent. In contrast, the internal convention suggests a rather surprising picture where a monovalent cation would be voraciously swallowed by this solvent, with a solvation free energy of -411 kJ·mol^{-1}, *i.e.* of similar magnitude as that found in water, while a monovalent anion would be aggressively vomited by this solvent ("antisolvation"), with a solvation free energy of $+411$ kJ·mol^{-1}. Both perspectives agree on the vanishing real work of the charge transfer, and thus on the real single-ion solvation free energy. But they differ by its partitioning into bulk and surface contribution, and thus on the intrinsic single-ion solvation free energy and on the surface potential of the liquid. The external convention regards the surface term as a transfer of charge from vacuum to the external potential of the bulk liquid, which is zero in the present situation where isotropically-charged atoms exert no electrostatic influence in their excluded volume, and the bulk term as a transfer of charge from this potential to the potential in the cavity, which is also zero in the present situation. In contrast, the internal convention regards the surface term as a transfer of the charge from vacuum to the internal potential of the bulk liquid, highly positive due to the partially shielded nuclear charge within the atoms, and the bulk term as a transfer of charge from this potential to the potential in the cavity, which is zero in the present case, resulting in a highly negative potential difference. Note that in reality, liquid argon would probably slightly solvate both cations and anions due to polarization effects, neglected in the present discussion. For example, a QM study[1801] evaluates the pairwise binding free energy of Na^+ to Ar in the gas phase to about -3.2 $kJ{\cdot}mol^{-1}$.

A *second argument* arises from the observation that ions are species of atomic sizes, excepting charged subatomic particles like the solvated proton or electron (see further below). Thus, solvated ions are essentially located within the excluded volume of the solvent molecules, and probe the external rather than the internal potential of the liquid. Furthermore, since ions do not have "x-ray view", they can only be sensitive to the detailed internal electronic structure of the solvent molecules *via* its influence on the electric potential they generate outside the molecule (*e.g.* dipole, traceless quadrupole). However, the quadrupole-moment trace of a solvent molecule solely generates a variation of the electric potential within the molecule itself. Taking again the example of argon, the core electrons provide a substantial contribution to the quadrupole-moment trace and thus to the exclusion potential of the atom, although they exert no influence on the electric potential surrounding the atom. This argument can be extended to the internal structure of the nuclei, which may also provide a component to the quadrupole-moment trace of the atom. The internal potentials, because they encompass the exclusion potential, suggest in the opposite that ions are sensitive to the electronic structure of the solvent molecules (or atoms) down to the innermost electrons, and possibly even to the internal structure of the nuclei.

Finally, a *third argument* arises from the observation that solvent models for AT simulations can be constructed that present arbitrary values of the quadrupole-moment trace, and thus of the exclusion potential, without affecting the way they reproduce experimental data concerning the pure liquid, as illustrated in Figure 4.13 in the context of the SPC water model. The original SPC water model is characterized by an exclusion potential ξ_{AT} of +0.82 V (Sections 3.3.3-3.3.5 and Table 3.1). In contrast to the quantum-mechanical argon case, this potential is now positive due to the presence of a negative charge at the molecular center (oxygen atom) along with positive charges closer to the surface (hydrogen atoms). The alternative model I is produced by increasing the oxygen charge by 1000 e and introducing a compensating spherical charge distribution of radius 0.05 nm and magnitude -1000 e centered on this atom. The alternative model II is produced by decreasing the oxygen charge by 1000 e and introducing a compensating spherical charge distribution of radius 0.05 nm and magnitude 1000 e centered on this atom. The corresponding exclusion potentials ξ_{AT} are on the order of -250 V and $+250$ V for the two alternative models. The above alterations will not produce any noticeable difference in the simulated properties of liquid water, because the charge modifications do not affect the electrostatic potential of a water molecule beyond 0.05 nm, while no charge of a neighboring molecule (or ion) can penetrate so deeply within the molecule, owing to van der Waals repulsions. For example, in classical simulations using the SPC water model[1274] at 300 K, the pair (radial) distribution functions typically become negligible (< 0.001) below about 0.24 nm (oxygen-oxygen) and about 0.14 nm (hydrogen-oxygen). The three models are thus equally valid to represent water in the liquid state. However, AT calculations of intrinsic single-ion solvation free energies using the internal convention will suggest that one alternative model (I) solvates cations and antisolvates anions, *e.g.* by about $\mp$ 24000 kJ$\cdot$mol^{-1} for a monovalent ion, the opposite being true for the other alternative model (II). In contrast, calculations of intrinsic single-ion solvation free energies using the external convention will not reveal any difference between the three models. Similarly, AT calculations of the air-liquid interfacial potential using the internal convention will suggest that one alternative model (I) has a highly positive surface potential of about $+250$ V, while the other alternative model (II) has a highly negative surface potential of about -250 V. In contrast, calculations of the air-liquid interfacial potential using the external convention will not reveal any difference between the three models. As discussed previously, the difference between the two conventions vanishes when real single-ion solvation free energies are calculated. However, many AT simulation studies to date use the internal convention and report intrinsic single-ion solvation free energies[708,1056,1423,1429,1430] or surface potentials[430,468–470,833,850,947,973,1442,1443,1445,1447–1449,1451,1454,1456]. Since these quantities depend on a property of the solvent model, the quadrupole-moment trace, that does not influence in any way its ability to reproduce experimental data and can

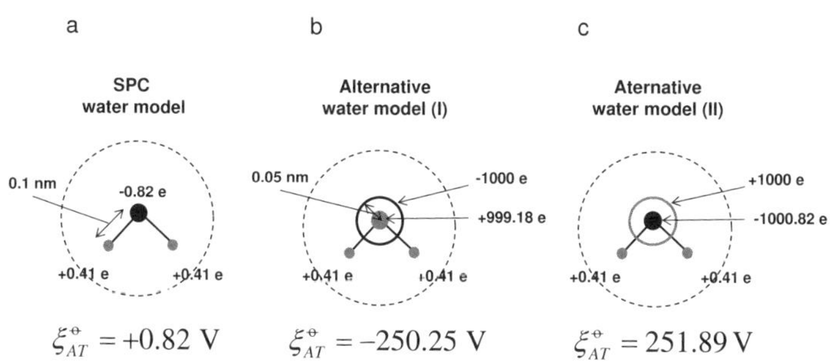

Figure 4.13 *Analogs of the SPC water model with altered charge distributions.* The original SPC water model[1274] (a) is compared with two alternatives presenting a much larger negative (b) or positive (c) value of the quadrupole-moment trace and thus, a much larger negative (b) or positive (c) value of the exclusion potential. Yet, the three models would give an essentially equivalent representation of the properties of liquid water.

be varied arbitrarily, these numbers are plainly meaningless. In contrast, although the validity of the external convention can still be discussed (concept of excluded volume ambiguous at the quantum-mechanical level, need to define a molecular center and envelope), it should at least lead to consistent and reproducible results for any classical solvent model reproducing accurately the experimental properties of the pure liquid.

The three above arguments clearly suggest that only the external convention can be used to define a system of "meaningful" intrinsic single-ion solvation properties and solvent air-liquid interfacial potentials. The exact significance of the word "meaningful" will be explicited in Chapter 7. It should be stressed, however, that the situation may be different if dissolved charged particles of subatomic dimensions are considered, such as the proton and the electron. In this case, even if an external definition for the air-liquid interfacial potential should be maintained for consistency, the corresponding intrinsic solvation properties will be difficult to interpret in a theoretical sense, because they incorrectly assume that these species are excluded from the volume of the solvent molecules. The same consideration applies to the *electron solvated in a metal*. This problem is, however, largely irrelevant in the context of ionic solvation, since the surface potential of a metal is encompassed within the experimentally-accessible work function of the corresponding electrode and never appears explicitly during the determination of ionic solvation properties. Nevertheless, the problem of the appropriate convention to be applied for the electron remains of significance to phenomena such as *electron solvation*[1802–1809] (Section 5.21), electron work functions[402, 403, 1810], electron holography[1811–1814], electron diffraction and x-ray crystallography[1815–1817], where an internal convention (omitting the intranuclear contribution) might still be more relevant than an external one considering that electrons, unlike solvated ions of atomic sizes, do have (some) "x-ray" view.

Note that the distinction between internal and external conventions, a terminology introduced by the authors of the present text (see, however, Ref.[707]) is nowhere to be found in the scientific literature concerned with the area. However, this distinction implicitly underlies the hot debate taking place between advocates of the M- and P-schemes in AT simulations both in the context of ionic solvation[707,708,1056,1081,1187,1424,1429,1432–1439] and surface potential calculations[470,707,833,1439,1445,1469]. The issue has also recently emerged in the context of QM computations (Sections 6.3.3-6.3.5).

Note, in this context, that QM estimates for the internal (P-integration) air-liquid interfacial potential of water suggest[1448–1450] (Table 5.18) a positive value of relatively large magnitude, *i.e.* of[1448,1449] 3.1 or[1450] 3.63 V. The positive sign can be interpreted in a similar way as the corresponding positive sign for the QM argon model of Figure 4.12. On the other hand, QM computational studies[1450, 1571, 1573, 1574, 1818] (Section 6.3.3) suggest negative values of comparable magnitude for the exclusion potential of water ξ_{QM}, namely -3.85 and -3.81 V based on Refs.[1571, 1818], -3.62 V based on Ref.[1450] and -3.7 V based on Ref.[1573, 1574]. Due to the use of pseudo-potentials to represent the nuclei and core electrons, these estimates will differ from QM values explicitly including the contributions of all electrons *via* a term depending on the quadrupole-moment trace of the nucleus plus core electron system.

Note, finally, that the problem of the appropriate convention to be applied also concerns the *solvated proton*. Paradoxically, it appears that the ion that has been selected as an anchoring point for intrinsic solvation thermodynamics (Sections 2.2 and 4.5.1), is also the one for which the physical interpretation of intrinsic parameters is most ambiguous.

4.4 Electrochemistry

From a broad perspective, and in addition to its technological relevance, *electrochemistry*[98,1754,1819] is concerned with the relationship between the potential difference and the current intensity in a system of phases separated by interfaces. However, since this book is exclusively concerned with thermodynamic properties, only the situation of *equilibrium electrochemistry*, *i.e.* measurements in the limit of zero current, will be considered. The following sections discuss in turn: (*i*) the notation employed here to describe a system of phases separated by interfaces (electrochemical cell); (*ii*) the principle of a potential difference measurement; (*iii*) Galvanic cell measurements; (*iv*) Voltaic cell measurements; (*v*) electrocapillary and related measurements.

4.4.1 Notations

An *electrochemical cell* involves a series of N phases (P_1, P_2, ..., P_{N-1}, P_N) separated by interfaces of different types. The *characteristic elements* composing a typical equilibrium electrochemical cell have been described in Section 4.3.3. It has also been seen in Section 4.3.4 that in a cell consisting exclusively of these elements, with the exception of the polarizable interface after charge equilibration (PICE), the various forms of electric (surface, Volta and Galvani) and chemical (intrinsic chemical, real chemical and electrochemical) potentials of the different phases (the latter with respect to species in these phases) are entirely specified, within common arbitrary constants, by the *cell scheme* and the equilibrium condition, along with constraints specific to the characteristic elements. The present section introduces a consistent *notation* for such a scheme as defined by its successive phases and characteristic elements. Note that the scheme itself does not assume equilibrium conditions, but that the above statements only hold under such conditions.

The symbol " $\vdots$ " will be used to denote a *generic interface* of unspecified type. Two different symbols will be used to distinguish specific types of interfaces, namely polarizable or non-polarizable (Section 4.3.3). The symbol " $||$ " will be used to denote a *polarizable interface* between two conducting phases, *i.e.* susceptible to the formation of an electric double-layer, but allowing no charge transfer between the two phases. The symbol " $|$ " will be used to denote a *non-polarizable interface* between two conducting phases, *i.e.* also susceptible to the formation of an electric double-layer, but allowing the transfer of charges between phases, *e.g.* metal-metal electron transfer or metal-solution redox reaction, the latter possibly involving species in other phases in contact (Section 4.3.7). The symbol " $\square$ " will be used to denote an *air gap* between two phases, *i.e.* an insulant space presenting interfaces with the two phases that are not susceptible to the formation of an electric double-layer, but may still present an electric single-layer induced by a bound surface charge density (Section 4.3.1). Finally, when the system cannot be represented as a linear succession of phases, the notation "$P_1 \vdots [P_2]P_3$" will be used to indicate that P_2 and P_3 are simultaneously in contact with P_1 as well as in contact with each other, and the notation "$P_1 \vdots P_2(P_3) \vdots P_4$" to indicate that P_3 is solely in contact with P_2 and neither with P_1 nor with P_4, while P_2 is in contact with P_1, P_3 and P_4 at three distinct interfaces. Different species present in the same phase (*e.g.* multiple solutes in a solution) will be separated by ",". Finally, the symbol " $\sim$ " will be used in the context of *half-cells*, right after the phase that is in common between the given half-cell and another half-cell.

The *phases* themselves will be indicated by the constituting species, as described in Section 4.1. In addition, the notations P (generic notation for a phase), M (conducting solid metal), L (conducting liquid, *e.g.* liquid metal or electrolyte solution), S (conducting solution, *e.g.* electrolyte solution) will also be employed. Primes will be used to distinguish different types of materials. For example, M_1 and M_2 indicate the same metal M in phases 1 and 2, while M_1 and M_2' indicated two different metals in the two phases.

4.4.2 Potential difference measurement

Typically, the measurement device applied to a system of phases separated by interfaces consists of a potentiostat or potentiometer along with a galvanometer connected in series to the terminal phases of the system, *i.e.* the electrodes, by a wire of the same metal, as illustrated in Figure 4.14. If the terminal phases of the system, which will always be assumed to be metals, are of a different nature than the wires, the external circuit also encompasses one or two metal-metal interfaces. The *galvanometer* is a device capable of measuring the current intensity through the circuit. In equilibrium measurements, this intensity is required to vanish. As previously introduced in Sections 4.3.1-4.3.3, the *potentiostat* is a device capable of imposing a tunable potential difference between the two connecting wires, while the *potentiometer* exclusively measures an existing potential difference without ideally altering it. This potential difference will be noted $\Delta\mathcal{V}$, as measured from left to right, *i.e.* from the wire connected to the left terminal phase to the wire connected to the right terminal phase of the cell. Since the wires connected to the potentiometer are made of the same metal, *i.e.* characterized by the same surface potential, there is no distinction to be made between a Galvani ($\Delta\phi$) and a lateral Volta ($\Delta\psi$) potential difference in terms of $\Delta\mathcal{V}$. From a conceptual point of view, the potentiometer and galvanometer at zero current can be thought of as an additional polarizable interface or air gap between the two wires of the identical metal in the exterior of the system, with the special property that the free surface charge density can be tuned to achieve a specified value for $\Delta\mathcal{V} = \Delta\phi = \Delta\psi$. The potentiometer and galvanometer at zero current can be viewed similarly, except that no alteration of the free surface charge density occurs and the device merely records the existing value of $\Delta\mathcal{V} = \Delta\phi = \Delta\psi$.

Consider a cell defined by a system of phases

$$M_1 \vdots P_2 \ldots P_{N-1} \vdots M'_N \quad , \tag{4.348}$$

as depicted in Figure 4.14. If the left terminal phase consists of a metal M and the right terminal phase of a metal M', while the connecting wires consist of a third metal M'', the circuit includes two non-polarizable metal-metal interfaces. At equilibrium, following from Eq. 4.300, one has for the $M - M''_L$ interface, where M''_L denotes the wire on the left of the potentiometer

$$\tilde{\mu}_{e^-:M_1} = \tilde{\mu}_{e^-:M''_L} \quad , \tag{4.349}$$

and at the $M'_N - M''_R$ interface, where M''_R denotes the wire on the right of the potentiometer,

$$\tilde{\mu}_{e^-:M_N} = \tilde{\mu}_{e^-:M''_R} \quad . \tag{4.350}$$

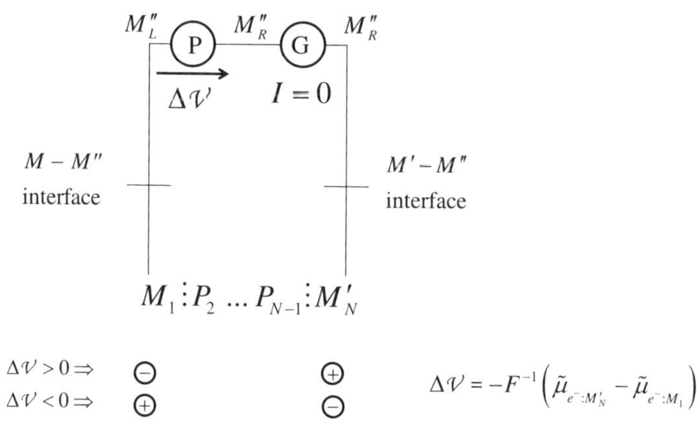

Figure 4.14 *Schematic representation of an electrochemical equilibrium measurement.* A cell (system of N phases) separated by interfaces is connected by its two metallic terminal phases (M_1 and M'_N) to a potentiostat or potentiometer (P) and galvanometer (G) in series through a wire of a metal M''. The potentiostat enforces or the potentiometer indicates a potential difference $\Delta\mathcal{V}$ (from left to right) and the galvanometer a zero electric current ($I = 0$).

Since $\Delta\mathcal{V}$ is the Galvani potential difference between M''_L and M''_R, which are made of the same metal, and thus characterized by the same (intrinsic) chemical potential for the electron, it follows from Eq. 4.275 that

$$\tilde{\mu}_{e^-:M''_R} - \tilde{\mu}_{e^-:M''_L} = -F\Delta\mathcal{V} \quad , \tag{4.351}$$

and one obtains, using Eqs. 4.349 and 4.350,

$$\Delta\mathcal{V} = -F^{-1}(\tilde{\mu}_{e^-:M'_N} - \tilde{\mu}_{e^-:M_1}) \quad . \tag{4.352}$$

Thus, $\Delta\mathcal{V}$ measures the *difference* in the *electrochemical potential of the electron* between the two metallic terminal phases, scaled by $-F$, irrespective of the nature of the connecting wires. In the special case where the two terminal phases are also made of the same metal M, the chemical potential of the electron is identical at both termini. In this case, one may further write

$$\Delta\mathcal{V} = \phi_{M_N} - \phi_{M_1} \quad , \tag{4.353}$$

which implies that in this special situation, $\Delta\mathcal{V}$ measures the *difference* in the *Galvani potential* of the identical terminal phases.

As already outlined in Section 4.3.3, two situations should be distinguished. A *Galvanic cell* (Section 4.4.3) solely encompasses non-polarizable interfaces. In this case, the Galvani potential difference across each interface is dictated by an electrochemical equilibrium condition, and is thus fixed. As a result, the Galvani potential difference between the two terminal phases and thus, the difference in the electrochemical potential of the electron in these phases, is also fixed, and there is only one possible value of $\Delta\mathcal{V}$ satisfying the equilibrium (zero-current) condition. In contrast, a *Voltaic cell* (Section 4.4.4) or an *electrocapillary cell* (Section 4.4.5) encompasses one air gap or polarizable interface, respectively, in addition to non-polarizable interfaces. In this case, any change in $\Delta\mathcal{V}$ will lead to the passing of a small current within the circuit, but this current will decay rapidly due to the accumulation of charges at the air gap or at the polarizable interface, *i.e.* to a compensating change in the magnitude of the two electric single-layers or of the electric double-layer, respectively. In other words, the zero-current condition is here compatible with any $\Delta\mathcal{V}$ value. In the latter types of measurements, however, one usually imposes a further equilibrium (electroneutrality) constraint on a physical property of the air gap (zero electric field in the gap, corresponding to a vanishing lateral Volta potential difference $\Delta\psi$; Section 4.4.4) or of the polarizable interface (maximum interfacial tension resulting from the absence of an electric double-layer, corresponding to a vanishing frontal Volta potential difference $\Delta\tilde{\psi}$). This constraint in turn also imposes a unique value of $\Delta\mathcal{V}$. In the context of electrocapillary measurements, the zero-charge condition is sometimes omitted. In this case, according to the terminology of Section 4.3.3, the polarizable interface can be referred to as a polarizable interface after charge equilibration (PICE), but not at zero charge (PIZC), and the response of the interface to the applied $\Delta\mathcal{V}$ can be used to study the property of the electric double-layer *e.g.* *via* interfacial tension measurements.

In Galvanic cells, the unique $\Delta\mathcal{V}$ value compatible with the equilibrium (zero-current) condition is referred to as the *electromotive force*[1775] of the cell. It can be measured by means of a potentiostat, the setting of which is determined by the constraint of zero current at the galvanometer, or, more simply, by means of a potentiometer. Note that in practice, potentiometers require the circulation of a small current to operate, and it is then assumed that the current intensity is negligible. In Voltaic or electrocapillary cells, the unique $\Delta\mathcal{V}$ value compatible with the equilibrium (zero current plus electroneutrality at the air gap or polarizable interface) is referred to as the *compensating potential difference*[1774] of the cell. It can be measured by means of a potentiostat, the setting of which is determined by the constraint of zero current at the Galvanometer along with that of zero field at the air gap or zero charge at the polarizable interface. A potentiometer cannot be used in this case.

In both situations, when $\Delta\mathcal{V} > 0$, electrons would flow spontaneously through the circuit (outside the system) from the left to the right in the absence of a compensating potential difference at the potentiostat. The left-terminal phase is then the negative ("electron-rich") and the right terminal the positive ("electron-poor") pole. When $\Delta\mathcal{V} < 0$, the situation is reversed. The standard electromotive force of a Galvanic cell or the compensating potential difference of a Voltaic or electrocapillary cell, $\Delta\mathcal{V}^{\ominus}$, corresponds to the value of $\Delta\mathcal{V}$ appropriate for a cell where all species involved are in their standard state (*bbme*$_T$ convention; Section 4.2.4). In general, $\Delta\mathcal{V}^{\ominus}$ is determined by extrapolating measurements at P° and T^{-} for non-standard compositions of the system towards standard-state compositions. As a final note, it should be stressed again that the above considerations refer to *equilibrium situations* only. At finite (and possible oscillatory) currents, phases and interfaces must be considered in terms of their resistance and capacitance (and possibly inductive and capacitive reactance, *i.e.* impedance), and the description becomes much more complex.

4.4.3 Galvanic cell measurements

A *Galvanic cell* measurement corresponds to the measurement of the *electromotive force*[1775] $\Delta\mathcal{V}$ of a cell solely encompassing *non-polarizable interfaces* (Sections 4.3.3 and 4.4.2). In the present context, only equilibrium measurements are considered, in which case the value of $\Delta\mathcal{V}$ is entirely dictated by the cell scheme, along with the equilibrium (zero-current) condition. Most commonly, the cell considered involves two non-polarizable interfaces between pieces of metal (electrodes) and an aqueous solution (electrolyte). These systems are conveniently described in terms of two *half-cells*, each involving one electrode and facing a common solution phase at their right termini. Each half-cell involves an electrochemical equilibrium defined by a specific *redox reaction*. Three simple examples are

$$M_s \mid M^z_{aq} \sim \quad , \tag{4.354}$$

$$M_s \mid I^z_{aq}, I^{z'}_{aq} \sim \quad , \tag{4.355}$$

and

$$M_s \mid [I_{el}] I^z_{aq} \sim \quad . \tag{4.356}$$

In the last two cases, the metal electrode is passive, *i.e.* it does not directly participate in the redox reaction beyond supplying or absorbing electrons.

By convention, the half-cell redox reactions are formulated as *reduction* reactions. The corresponding *electrochemical equilibrium conditions* are easily formulated as detailed in Section 4.3.7. In the first case, where the metal electrode participates directly in the redox reaction, one has (with $z > 0$ for a metal)

$$M^z_{aq} + ze^- \rightleftharpoons M_s \quad , \tag{4.357}$$

corresponding to the electrochemical equilibrium condition

$$\tilde{\mu}_{M^z_{aq}} + z\tilde{\mu}_{e^-:M_s} = \mu_{M_s} \quad . \tag{4.358}$$

In the second case, where the redox reaction only involves species in solution and the electrode plays a passive role of electron supplier in the reduction reaction, one has (for $z > z'$)

$$I^z_{aq} + (z - z')e^- \rightleftharpoons I^{z'}_{aq} \quad , \tag{4.359}$$

or (for $z < z'$)

$$I^{z'}_{aq} + (z' - z)e^- \rightleftharpoons I^z_{aq} \quad , \tag{4.360}$$

corresponding to the electrochemical equilibrium condition

$$\tilde{\mu}_{I^z_{aq}} + (z - z')\tilde{\mu}_{e^-:M_s} = \tilde{\mu}_{I^{z'}_{aq}} \quad . \tag{4.361}$$

In the third case, where the redox reaction involves a third phase and the electrode also plays a passive role of electron supplier in the reduction reaction, one has (for $z > 0$)

$$I^z_{aq} + ze^- \rightleftharpoons I_{el} \quad , \tag{4.362}$$

or (for $z < 0$)

$$I_{el} + (-z)e^- \rightleftharpoons I^z_{aq} \quad , \tag{4.363}$$

corresponding to the electrochemical equilibrium condition

$$\tilde{\mu}_{I^z_{aq}} + z\tilde{\mu}_{e^-:M_s} = \mu_{I_{el}} \quad . \tag{4.364}$$

As a simple example of Galvanic cell combining half-cells of the three above types (third and first type), one may consider the system

$$\mathrm{Pt}_s \mid [\mathrm{H}_{el}]\mathrm{H}^+_{aq}, \mathrm{Zn}^{2+}_{aq} \mid \mathrm{Zn}_s \quad , \tag{4.365}$$

where $\mathrm{H}_{el} = (1/2)\ \mathrm{H}_{2,g}$, as illustrated in Figure 4.15. The reduction reaction at the left interface is

$$\mathrm{H}^+_{aq} + e^- \rightleftharpoons \mathrm{H}_{el} \quad . \tag{4.366}$$

The reduction reaction at the right interface is

$$\mathrm{Zn}^{2+}_{aq} + 2\ e^- \rightleftharpoons \mathrm{Zn}_s \quad . \tag{4.367}$$

The two corresponding electrochemical equilibrium conditions read

$$\tilde{\mu}_{\mathrm{H}^+_{aq}} + \tilde{\mu}_{e^-:\mathrm{Pt}_s} = \mu_{\mathrm{H}_{el}} \tag{4.368}$$

and

$$\tilde{\mu}_{\mathrm{Zn}^{2+}_{aq}} + 2\ \tilde{\mu}_{e^-:\mathrm{Zn}_s} = \mu_{\mathrm{Zn}_s} \quad . \tag{4.369}$$

Combining Eqs. 4.368 and 4.369, the equilibrium electromotive force is given by

$$\begin{aligned} \Delta\mathcal{V} &= -F^{-1}(\tilde{\mu}_{e^-:\mathrm{Zn}_s} - \tilde{\mu}_{e^-:\mathrm{Pt}_s}) \\ &= -(2F)^{-1}(2\ \tilde{\mu}_{\mathrm{H}^+_{aq}} + \mu_{\mathrm{Zn}_s} - 2\ \mu_{\mathrm{H}_{el}} - \tilde{\mu}_{\mathrm{Zn}^{2+}_{aq}}) \\ &= -(2F)^{-1}(2\ \mu_{\mathrm{H}^+_{aq}} + \mu_{\mathrm{Zn}_s} - 2\ \mu_{\mathrm{H}_{el}} - \mu_{\mathrm{Zn}^{2+}_{aq}}) \quad , \end{aligned} \tag{4.370}$$

where the last equality arises from the cancellation of the Galvani potential contributions to the electrochemical potentials of the H^+_{aq} and Zn^{2+}_{aq} ions (Eq.

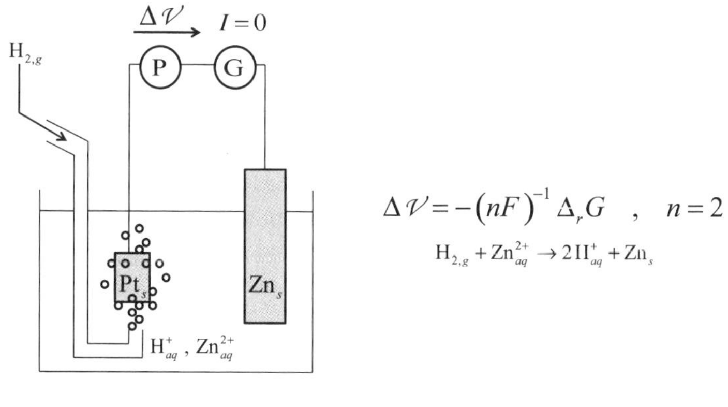

Figure 4.15 *Example of a Galvanic cell.* The cell considered is the one described by the scheme of Eq. 4.365, where $\mathrm{H}_{el} = (1/2)\mathrm{H}_{2,g}$.

4.269). Eq. 4.370 can be rewritten as

$$\Delta_r G = -nF\Delta\mathcal{V} \quad , \tag{4.371}$$

where $\Delta_r G$ is the reaction free energy (Section 4.2.7) of the cell reaction

$$2\ \mathrm{H}_{el} + \mathrm{Zn}^{2+}_{aq} \rightarrow 2\ \mathrm{H}^{+}_{aq} + \mathrm{Zn}_s \quad , \tag{4.372}$$

and $n = 2$ denotes the number of electrons transferred during this reaction. Eq. 4.371 remains valid for arbitrary cells, when applied to the reaction that combines the oxidation of the species in the left half-cell (*i.e.* considering the corresponding electrode as the anode) and the reduction of the species in the right half-cell (*i.e.* considering the corresponding electrode as the cathode). Thus, for instance, a negative $\Delta\mathcal{V}$ for the cell of Figure 4.15 implies a positive $\Delta_r G$ for the reaction of Eq. 4.372 under the pressure, temperature and concentration conditions prevailing in the cell, *i.e.* it indicates that this reaction would spontaneously evolve backward if the current is allowed to flow freely in an external circuit.

Given the introduction of standard states ($bbme_T$ convention; Section 4.2.4), the *standard electromotive force* $\Delta\mathcal{V}^{\ominus}$ of a cell at equilibrium can be defined as the $\Delta\mathcal{V}$ value at P° and T^{-} in the situation where all components of the cell are in their standard states. In this case, it follows from the discussion of Section 4.2.5 (Eqs. 4.144 and 4.145) that $\Delta\mathcal{V}^{\ominus}$ is directly related to the *equilibrium constant* of the corresponding redox reaction as

$$\Delta\mathcal{V}^{\ominus} = -\frac{1}{nF}\Delta_r G^{\ominus} = \frac{RT^{-}}{nF}\ln K_E^{\ominus} \quad , \tag{4.373}$$

so that for a cell with non-standard composition at P° and T^{-},

$$\Delta\mathcal{V} = \Delta\mathcal{V}^{\ominus} - \frac{RT^{-}}{nF}\ln K_R = -\frac{RT^{-}}{nF}\ln\frac{K_R}{K_E^{\ominus}} \quad , \tag{4.374}$$

where K_R is the *reaction quotient* of the redox reaction at the specific system composition. The second equation is the *Nernst equation*[1732], already seen in Section 4.2.7 (Eq. 4.144), that relates the electromotive force of a cell under non-standard conditions to the activities of the species in the cell. It can be used in particular to extrapolate the measured $\Delta\mathcal{V}$ to standard-state conditions for the determination of $\Delta\mathcal{V}^{\ominus}$ (and $\Delta_r G^{\ominus}$). Other applications include in particular the determination of the composition of a system *via* potentiometric or indicator-based measurements (*e.g.* potentiometric or redox titration). Although half-cells of the three above types (Eqs. 4.354-4.356) are common, three types of complications may also frequently occur.

A first type of complication may occur when one of the species formed by a hypothetical oxidation or reduction reaction spontaneously and quantitatively *reacts with other species* present in the half-cell, typically other ions or the solvent, *e.g.* commutation or dismutation, reduction of the aqueous proton to hydrogen gas, formation of complex ion-oxygen species. In this case, the half-cell may not be usable or, otherwise, characterizes the complex redox reaction rather than the simple hypothetical one. An important example of this kind in the context of this book are *alkali metals*, which spontaneously and quantitatively react with water to produce gaseous hydrogen, hydroxide anions and alkali cations. As a result, the hypothetical half-cell $M_s \mid M^+_{aq}$ cannot be used to characterize the electrochemical properties of alkali metals and alkali cations, and indirect methods must be used instead[1820–1824] (Section 5.5).

A second type of complication may arise when one of the species is involved in *chemical equilibria* within the half-cell, *e.g.* protonation of a basic anion, deprotonation of an acidic cation, protonation or deprotonation of water molecules in the solvation shells of a hydrated ion. In this case, the electrochemical properties of the half-cell simultaneously characterize all the involved chemical equilibria. These effects are typically relevant for polyatomic ions (direct protonation or deprotonation) and ions with a high charge density (protonation or deprotonation of neighboring water molecules). Ion-ion binding, possibly solvent-mediated, could also be viewed as a possible complication of this kind, as well as reactions of the previous type that are not quantitative.

Finally, a third type of complication may arise from the possible direct redox reaction between the species within the full cell, *i.e.* not involving the transport of electrons through the external circuit. For example, a hypothetical cell

$$M_s \mid M^z_{aq}, M'^z_{aq} \mid M'_s \quad , \tag{4.375}$$

where M is a stronger reducer than M', will undergo the spontaneous reaction

$$M_s + M'^z_{aq} \rightarrow M^z_{aq} + M'_s \tag{4.376}$$

until equilibrium is reached, in which case the concentrations of the M'^z_{aq} and M^z_{aq} ions are defined by the equilibrium constant of the reaction and $\Delta\mathcal{V} = 0$. In some cases, this does not represent a problem, because the involved reaction is slow (*e.g.* cell of Eq. 4.365 at low H^+_{aq} concentrations). In the opposite case, the problem may be remedied by the introduction of a salt bridge between the half-cells. A *salt bridge* is an electrolyte solution containing "inert" ions, *i.e.* not of the types involved in the redox reactions at the electrodes, and confined by porous walls at its extremities. This wall should be sufficiently porous to allow for electric conduction between the two phases, *i.e.* to represent a non-polarizable interface, but sufficiently tight to prevent a significant extent of mixing between the solution within the salt bridge and the contacting solutions. In this case, the cell of Eq. 4.375 becomes

$$M_s \mid M^z_{aq} \mid S \mid M'^z_{aq} \mid M'_s \quad . \tag{4.377}$$

A salt bridge introduces two additional interfaces in the cell that formally represent non-equilibrium components, namely porous walls between solutions of distinct compositions with a spontaneous tendency to mix, albeit at a very slow rate. Since ions diffuse through the interface at different rates, depending on their mobility, this may create a local *steady-state situation* resulting in an effective electric double-layer. For this reason, accurate electrochemistry measurements must correct for the resulting potential differences, termed *liquid-junction potentials*. Various approaches, involving models for the above steady-state situation, have been proposed to this purpose[1825–1828]. In general, liquid-junction potentials are relatively small when: (*i*) the ionic activities (concentrations) in the salt bridge and solutions in contact are of comparable magnitudes; (*ii*) the ionic mobilities (transferences) of the anionic and cationic species within the salt bridge are comparable (use of an equitransferrent salt, *e.g.* KCl).

The standard electromotive force of a cell can be deduced from potentials attributed to the individual half-cells in two different ways. If *absolute electrode potentials*, either vacuum, real or intrinsic (Section 4.3.6), can be attributed to the two half-cells A and B, one has

$$\Delta\mathcal{V}^{\ominus}_{AB} = \tilde{\mathcal{V}}^{\ominus}_{B} - \tilde{\mathcal{V}}^{\ominus}_{A} = \hat{\mathcal{V}}^{\ominus}_{B} - \hat{\mathcal{V}}^{\ominus}_{A} = \mathcal{V}^{\ominus}_{B} - \mathcal{V}^{\ominus}_{A} \quad . \tag{4.378}$$

However, as seen previously, vacuum electrode potentials are ill-defined and intrinsic electrode potentials are experimentally elusive, while real potentials are affected by relatively large errors. For this reason, it is more convenient to work with a *relative scale*, defined by the standard electromotive forces of cells involving a common reference half-cell. This reference cell is normally chosen to be the *reference hydrogen electrode*,

$$M_s \mid [\mathrm{H}_{el}]\mathrm{H}^+_{aq} \sim \quad , \tag{4.379}$$

where $\mathrm{H}_{el} = (1/2)\ \mathrm{H}_{2,g}$, and the metal M_s is usually taken to be platinum (Pt_s). The standard electromotive force $\Delta\mathcal{V}^{\ominus}$ of a Galvanic cell involving the reference hydrogen electrode as its left half-cell will be called the *redox potential* of the right half-cell and written $\Delta_H\mathcal{V}^{\ominus}$. Obviously, the reference hydrogen electrode itself is characterized by $\Delta_H\mathcal{V}^{\ominus} = 0$ V, which is the anchoring point of the scale of *relative electrode potentials*, also sometimes called the *electrochemical series*. Based on this relative scale, the standard electromotive force of a cell consisting of two half-cells A and B can be deduced as

$$\Delta\mathcal{V}^{\ominus}_{AB} = \Delta_H\mathcal{V}^{\ominus}_{B} - \Delta_H\mathcal{V}^{\ominus}_{A} \quad , \tag{4.380}$$

which represents a practical alternative to Eq. 4.378. For example, for the cell of Figure 4.15 with half-cells A and B defined by Eqs. 4.366 and 4.367, respectively, one has under standard conditions $\Delta_H\mathcal{V}^{\ominus}_{A} = 0$ V (by definition) and[221] $\Delta_H\mathcal{V}^{\ominus}_{B} = -0.76$ V so that $\Delta\mathcal{V}^{\ominus}_{AB} = -0.76$ V. Considering Eq. 4.373, the reaction of Eq. 4.372 is thus spontaneous in the backwards direction under standard conditions. Two other types of half-cells are commonly used as reference electrodes in electrochemistry, namely the saturated calomel and silver chloride electrodes.

The *saturated calomel electrode* corresponds to the half-cell

$$\mathrm{Hg}_\ell \mid \mathrm{Hg}^+_{aq}, \mathrm{K}^+_{aq}, \mathrm{Cl}^-_{aq}(\mathrm{KCl}_s, \mathrm{Hg_2Cl}_{2,s}) \sim \quad . \tag{4.381}$$

The presence of KCl_s in equilibrium indicates that the aqueous solution is saturated in KCl. The presence of $Hg_2Cl_{2,s}$ in equilibrium indicates that the concentration of Hg^+_{aq} is fixed, as dictated by the small solubility product of Hg_2Cl_2 (calomel) and the saturation of the solution in Cl^-_{aq}. The corresponding reduction reaction can formally be written

$$1/2\ \mathrm{Hg_2Cl}_{2,s} + e^- \rightleftharpoons \mathrm{Hg}_\ell + \mathrm{Cl}^-_{aq} \quad . \tag{4.382}$$

The standard electrode potential of the saturated calomel electrode relative to the reference hydrogen electrode is[221] $\Delta_H \mathcal{V}^\ominus = +0.241$ V. A variant of this electrode is the non-saturated calomel electrode, where a 1 M KCl solution (molar, *i.e.* 1 mol·dm^{-3}) is used instead, in which case $\Delta_H \mathcal{V}^\ominus = +0.280$ V. The *silver chloride electrode* corresponds to the half-cell

$$\mathrm{Ag}_s \mid \mathrm{Ag}^+_{aq}, \mathrm{K}^+_{aq}, \mathrm{Cl}^-_{aq}(\mathrm{AgCl}_s) \sim \quad , \tag{4.383}$$

involving a 1 M KCl solution. The presence of $AgCl_s$ in equilibrium indicates that the concentration of Ag^+_{aq} is fixed, as dictated by the small solubility product of AgCl and the standard 1 M concentration of Cl^-_{aq}. The corresponding reduction reaction can formally be written

$$\mathrm{AgCl}_s + e^- \rightleftharpoons \mathrm{Ag}_s + \mathrm{Cl}^-_{aq} \quad . \tag{4.384}$$

The standard electrode potential of the silver chloride electrode relative to the reference hydrogen electrode is[221] $\Delta_H \mathcal{V}^\ominus = +0.222$ V. Standard electrode potentials measured relative to these two alternative reference electrodes are trivially converted to values relative to the reference hydrogen electrode, *i.e.* to standard redox potentials $\Delta_H \mathcal{V}^\ominus$, *via* Eq. 4.380. Note that real molar KCl concentrations of 1 M rather than molal reference activities of 1, are used here, following usual practices in electrochemistry.

Standard redox potentials $\Delta_H \mathcal{V}^\ominus[I_{el}, I^z_{aq}]$ of element-ion couples (I_{el}, I^z_{aq}, with $|z| = 1$) relative to the reference hydrogen electrode can be determined experimentally *via Galvanic cell measurements*. This involves equilibrium electromotive-force measurements at known system compositions, along with suitable extrapolations or corrections for activity effects (Section 4.2.18). Redox potentials in the context of *alkali* and *halide* element-ion couples in water will be discussed in Section 5.5 (see Table 5.5 for numerical values). As mentioned above, for aqueous solutions, the reference electrode is most commonly taken to be the reference hydrogen electrode, leading to a standard but relative scale with the anchoring point $\Delta_H \mathcal{V}^\ominus[\mathrm{H}_{el},\mathrm{H}^+_{aq}] = 0$ V, where H_{el} stands for $(1/2)H_{2,g}$. Note that redox potentials are commonly tabulated[221] using a 1 atm (rather than $P^\circ = 1$ bar) reference pressure and a 1 molar (1 mol·dm^{-3}; rather than $b^\circ = 1$ mol·kg^{-1}) reference concentration, corresponding to a $aaMe_T$ convention (Section 4.2.17) rather than the $bbme_T$ convention employed in this book. The conventions are, however, essentially interchangeable for aqueous solutions in view of the small size of the correction terms (Table 4.3). The standard state of the electron need not be specified here since it is irrelevant on such a relative scale.

For *polyvalent ions*, the redox potential should formally be generalized to $\Delta_H \mathcal{V}^\ominus[I_{aq}^{z-1}, I_{aq}^z]$ for cations ($z > 1$) or $\Delta_H \mathcal{V}^\ominus[I_{aq}^{z+1}, I_{aq}^z]$ for anions ($z < -1$). In many cases, the situation is also further complicated by *protonation equilibria* involving the polyvalent ion itself (*e.g.* protonation of O^{2-} to OH^- in water) or the tight hydration shell of highly charged cations (*e.g.* partial deprotonation of $[Al(H_2O)_6]^{3+}$ in water[808]). Protonation effects are less important in the case of monovalent anions (*e.g.* halides), which are typically very weak bases in aqueous solution, an exception being F^-, with[221] a pK_a of 3.17. In cases where such protonation effects are important, the standard-state convention must be enlarged by a specification concerning the standard pH. However, this more general situation, possibly involving polyvalent ions and coupled protonation equilibria, will not be considered here, and the following discussion exclusively focuses on the case of monovalent ions.

The redox potential $\Delta_H \mathcal{V}^\ominus[I_{el}, I_{aq}^z]$ characterizes the *driving force of the reduction reaction, i.e.* $I_{aq}^+ + e^- \rightarrow I_{el}$ for monovalent cations ($z = 1$) or $I_{el} + e^- \rightarrow I_{aq}^-$ for monovalent anions ($z = -1$), relative to the driving force of the reduction reaction $H_{aq}^+ + e^- \rightarrow H_{el}$. For monovalent cations, a negative redox potential, as encountered *e.g.* for all alkali metals, indicates that I_{el} is a stronger reducer than H_{el} or, equivalently, that I_{aq}^+ is a weaker oxidant than H_{aq}^+. For monovalent anions, a positive redox potential, as encountered *e.g.* for all halide molecules, indicates that I_{el} is a stronger oxidant than H_{aq}^+ or, equivalently, that I_{aq}^- is a weaker reducer than H_{el}. As will be seen in Sections 4.5.1 and 5.5, the redox scale is in this context a disguised form of *conventional scale* for single-ion solvation properties.

The *temperature dependences* of the standard electrode potentials can also be determined. However, it is important to distinguish: (*i*) the so-called *isothermal* temperature derivative $\partial_T \Delta_H \mathcal{V}^\circ[I_{el}, I_{aq}^z]$, which involves the reference hydrogen electrode and a test electrode that are both at the same temperature T upon differentiation (standard value at T^-: $\partial_T \Delta_H \mathcal{V}^\ominus[I_{el}, I_{aq}^z]$); (*ii*) the so-called *thermal* temperature derivative[1829] $\partial_{T'} \Delta_{T,T'} \mathcal{V}^\circ[I_{el}, I_{aq}^z]$, which involves two identical electrodes (half-cells), the reference one at temperature T and the test one at temperature T' upon differentiation with respect to T' only at $T' = T$ (standard value at T^-: $\partial_{T'} \Delta_{T,T'} \mathcal{V}^\ominus[I_{el}, I_{aq}^z]$). Note that $\partial_T \Delta_H \mathcal{V}^\ominus[H_{el}, H_{aq}^+]$, as well as all higher derivatives, is zero by definition, but $\partial_{T'} \Delta_{T,T'} \mathcal{V}^\ominus[H_{el}, H_{aq}^+]$ is not. Isothermal derivatives are useful in view their connection with reaction entropies (first derivative) and heat capacities (second derivative). Thermal derivatives are useful in view their connection with entropy transport[1830–1832].

4.4.4 Voltaic cell measurements

A *Voltaic cell* measurement corresponds to the measurement of the *compensating potential difference*[1774] $\Delta\mathcal{V}$ of a cell comprising an *air gap*, along with non-polarizable interfaces (Sections 4.3.3 and 4.4.2). In the present context, only equilibrium measurements are considered, in which case the value $\Delta\mathcal{V}$ is entirely dictated by the cell scheme along with the equilibrium (zero current, zero field at the air gap) condition. The main purpose of Voltaic cell measurements is to determine the *lateral Volta potential difference* between two phases in contact through a *non-polarizable* interface within the cell. Depending on the method employed to assess the zero-field condition at the air gap, the choice of this method typically being correlated with the nature of the phases defining the non-polarizable interface, one can distinguish three main types of experiments of this kind, as illustrated in Figure 4.16: (*i*) the Kelvin probe method[403,1781], also sometimes called the vibrating condenser or vibrating electrode method, aims at measuring the lateral Volta potential difference between two metals; (*ii*) the ionizing probe method[1781]

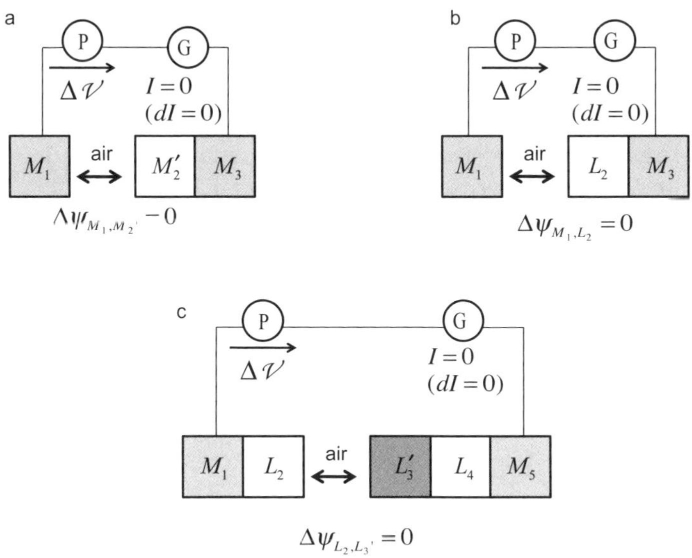

Figure 4.16 *Schematic representation of Voltaic cell measurements.* Illustration of the basic cells for the Kelvin probe (a), ionizing probe (b) and Kenrick (c) methods. In the three cases, the compensating potential difference $\Delta\mathcal{V}$ at the potentiostat is selected in such a way that no current I flows across the galvanometer and that the air gap is at zero field. The latter condition is verified by ensuring a vanishing variation dI of the current upon altering properties of the air gap, *e.g.* the air gap width *via* vibration of the probing electrode (a), the air gap conductivity *via* ionization of the gas (b), the evaluation of surface charges *via* streaming (c).

[,1833,1834] aims at measuring the lateral Volta potential difference between a metal and a conducting liquid (liquid metal or electrolyte solution); (*iii*) the Kenrick method[70,1781,1835] aims at measuring the lateral Volta potential difference between two conducting liquids (liquid metal or electrolyte solution). These methods are described in turn below, assuming standard conditions for simplicity.

The *Kelvin probe method*[403,1781] (Figure 4.16a) corresponds to the cell

$$M_1 \ \square \ M_2' \mid M_3 \quad , \tag{4.385}$$

where M_1 and M_3 are the same metal M (distinct from M_2') and the Volta potential difference is measured for the non-polarizable M_2'-M_3 interface. The phases M_1 and M_2' are separated by an air gap, and the compensating potential difference $\Delta\mathcal{V}^\ominus$ is selected in such a way that the electric field in the air gap is zero, implying that

$$\Delta\psi^\ominus_{M_1,M_2'} = 0 \quad . \tag{4.386}$$

In this case, one may write for the M_1-M_2' interface (Eqs. 4.280 and 4.285)

$$\begin{aligned} \tilde{\mu}^\ominus_{e^-:M_2'} - \tilde{\mu}^\ominus_{e^-:M_1} &= \mu^\ominus_{e^-:M'} - \mu^\ominus_{e^-:M} - F\Delta\chi^\ominus_{M,M'} \\ &= -F(\omega^\ominus_{M'} - \omega^\ominus_M) \quad , \end{aligned} \tag{4.387}$$

where $\Delta\chi^\ominus_{M_3,M_2'}$ has been written $\Delta\chi^\ominus_{M,M'}$ since it only depends on the natures of the two metals M and M', along with the specification of corresponding lateral reference interface characteristics. The equilibrium condition at the M_2'-M_3 interface can be written (Eqs. 4.280 and 4.285)

$$\begin{aligned} \tilde{\mu}^\ominus_{e^-:M_2'} - \tilde{\mu}^\ominus_{e^-:M_3} &= \mu^\ominus_{e^-:M'} - \mu^\ominus_{e^-:M} - F\Delta\chi^\ominus_{M,M'} - F\Delta\psi^\ominus_{M,M'} \\ &= -F(\omega^\ominus_{M'} - \omega^\ominus_M) - F\Delta\psi^\ominus_{M,M'} \\ &= 0 \quad , \end{aligned} \tag{4.388}$$

where $\Delta\psi^\ominus_{M_3,M_2'}$ has been written $\Delta\psi^\ominus_{M,M'}$ since it only depends on the natures of the two metals M and M' in contact, along with the specification of corresponding lateral reference interface characteristics. Therefore, one has (Eqs. 4.352 and 4.353)

$$\begin{aligned} \Delta\mathcal{V}^\ominus &= \phi^\ominus_{M_3} - \phi^\ominus_{M_1} = -F^{-1}(\tilde{\mu}^\ominus_{e^-:M_3} - \tilde{\mu}^\ominus_{e^-:M_1}) \\ &= -\Delta\psi^\ominus_{M,M'} = \omega^\ominus_{M'} - \omega^\ominus_M \quad . \end{aligned} \tag{4.389}$$

Note that only the last equality in this equation relies on the equilibrium condition at the M_2'-M_3 interface (see also Eq. 4.300 in Section 4.3.7). Owing to this last equality, the Kelvin probe method is also sometimes used to evaluate non-spectroscopically the work function of a metal by reference to another metal with spectroscopically known work function. For the same reason, a measurement using the Kelvin probe method can be replaced by the spectroscopic measurement of the work functions of the two metals. The condition of vanishing field in the air gap is verified in this approach by ensuring that *small variations in the width of the air gap* do not induce any electric current in the circuit. In practice, this is achieved by making the probing electrode (M_1) vibrate relative to the sample (M_2'). The compensating potential difference is then adjusted until no alternating current is recorded in the galvanometer. When this condition is fulfilled, the air gap is at zero electric field. Considering this experimental setup, the Kelvin probe method is also sometimes called the *vibrating condenser*[444,1781] or *vibrating electrode*[445] method. Note that this approach has also been applied to metal-liquid interfaces[444,1797].

The *ionizing probe method*[1781,1833,1834] (Figure 4.16b) corresponds to the cell

$$M_1 \,\square\, L_2 \mid M_3 \quad , \tag{4.390}$$

where M_1 and M_3 are the same metal M, L_2 is a conducting liquid (liquid metal or electrolyte solution) and the Volta potential difference is measured for the non-polarizable L_2-M_3 interface. Here also, the phases M_1 and L_2 are separated by an air gap, and the compensating potential difference $\Delta\mathcal{V}^\ominus$ is selected in such a way that the electric field in the air gap is zero, implying

$$\Delta\psi^\ominus_{M_1,L_2} = 0 \quad . \tag{4.391}$$

In this case, analogous considerations show that

$$\Delta\mathcal{V}^\ominus = -\Delta\psi^\ominus_{M,L} \quad , \tag{4.392}$$

where $\Delta\psi^\ominus_{M_3,L_2}$ has been written $\Delta\psi^\ominus_{M,L}$, since it only depends on the natures of the metal M and liquid phase L in contact, along with the specification of corresponding lateral reference interface characteristics. In the case where L is a conducting liquid metal, the L_2-M_3 interface is reversible with respect to the electron, and one may further write in analogy with Eq. 4.389

$$\Delta\mathcal{V}^\ominus = -\Delta\psi^\ominus_{M,L} = \omega^\ominus_L - \omega^\ominus_M \quad . \tag{4.393}$$

In the case where L is a solution containing dissolved ions M^+ of the metal M (assumed monovalent), the L_2-M_3 interface is reversible *via* the redox reaction

$$M^+{:}L + e^-{:}M \rightleftharpoons M{:}M \quad , \tag{4.394}$$

the equilibrium condition at the L_2-M interface reads

$$\begin{aligned} & \tilde{\mu}^\ominus_{M^+:L} + \tilde{\mu}^\ominus_{e^-:M} - \tilde{\mu}^\ominus_{M:M} \\ = \; & \mu^\ominus_{M^+:L} + F\chi^\ominus_L - \mu^\ominus_{M:M} + \mu^\ominus_{e^-:M} - F\chi^\ominus_M + F\Delta\psi^\ominus_{M,L} \\ = \; & \Delta_f G^\ominus[M^+_g] + \Delta_s \hat{G}^\ominus[M^+_g] - \omega^\ominus_M + F\Delta\psi^\ominus_{M,L} = 0 \quad , \end{aligned} \tag{4.395}$$

from which one obtains

$$\Delta\mathcal{V}^{\ominus} = -\Delta\psi^{\ominus}_{M,L} = F^{-1}\left(\Delta_f G^{\ominus}[M^+_g] + \Delta_s\hat{G}^{\ominus}[M^+_g]\right) - \omega^{\ominus}_M$$
$$= \hat{\mathcal{V}}^{\ominus}_{M,L} - \omega^{\ominus}_M \quad , \tag{4.396}$$

the last equality following from Eq. 4.298. This equation was already derived previously (see Eq. 4.307 in Section 4.3.7). In practice, the condition of vanishing field in the air gap is verified in this approach by checking that the *ionization of the air* in the gap through *insertion of radioactive material* does not induce any current in the circuit[98,1781]. Alternatively, the use of a radioactive probing electrode (M_1), *e.g.* radium ^{226}Ra, can serve the same purpose[451].

The *Kenrick method*[70,1781,1835] (Figure 4.16c) corresponds to the circuit

$$M_1 \mid L_2 \,\square\, L'_3 \mid L_4 \mid M_5 \quad , \tag{4.397}$$

where M_1 and M_5 are the same metal M, and L_2 and L_4 are the same liquid L, distinct from L'_3 (liquid metals or electrolyte solutions). Here also, the phases L_2 and L'_3 are separated by an air gap, and the compensating potential

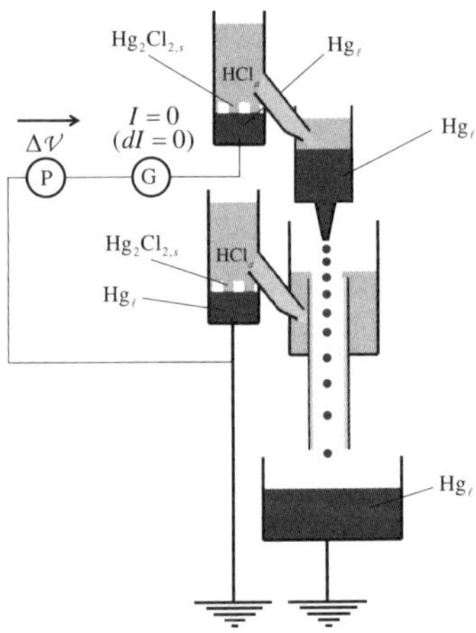

Figure 4.17 *Practical implementation of the Kenrick cell.* The cell illustrated corresponds to the scheme of Eq. 4.402, as an example of the general scheme of Eq. 4.397 and Figure 4.16c. The calomel-type electrode and mercury reservoir at the top correspond to phases M_5, L_4 and L'_3. Besides the connection to the circuit, they are insulated. The calomel-type electrode and aqueous solution reservoir in the middle correspond to phases M_1 and L_2. Besides the connection to the circuit, they are grounded, as well as the mercury collection reservoir at the bottom. When the mercury is streaming, the presence of surface charges at the air gap (non-zero field), which would be progressively evacuated to the ground, is incompatible with the condition of zero electric current[455]. As a result, the compensating potential difference $\Delta\mathcal{V}$ is uniquely defined by the zero-current condition when mercury is streaming or, equivalently, that the current remains zero irrespective whether the mercury is streaming or not.

difference $\Delta\mathcal{V}^{\ominus}$ is selected in such a way that the electric field in the air gap is zero, implying

$$\Delta\psi^{\ominus}_{L_2,L'_3} = 0 \quad . \tag{4.398}$$

In this case, analogous considerations show that

$$\Delta\mathcal{V}^{\ominus} = -\Delta\psi^{\ominus}_{L,L'} \quad , \tag{4.399}$$

where $\Delta\psi^{\ominus}_{L_4,L'_3}$ has been written $\Delta\psi^{\ominus}_{L,L'}$ since it only depends on the natures of the liquid phases L and L' in contact, along with the specification of corresponding lateral reference interface characteristics. In the case where L and L' are conducting liquid metals, the L'_3-L_4 interface is reversible with respect to the electron, and one may further write in analogy with Eqs. 4.389 and 4.393

$$\Delta\mathcal{V}^{\ominus} = -\Delta\psi^{\ominus}_{L,L'} = \omega^{\ominus}_{L'} - \omega^{\ominus}_{L} \quad . \tag{4.400}$$

In the case where L' is a liquid metal and L is a solution containing dissolved ions L'^{+} of this metal (assumed monovalent), the L'_3-L_4 interface is reversible *via* a corresponding redox reaction similar to Eq. 4.394, and one has in analogy with Eq. 4.396

$$\begin{aligned}\Delta\mathcal{V}^{\ominus} &= -\Delta\psi^{\ominus}_{L,L'} = \omega^{\ominus}_{L'} - F^{-1}\left(\Delta_f G[L'^{+}_{g}] + \Delta_s\hat{G}^{\ominus}[L'^{+}_{g}]\right) \\ &= \omega^{\ominus}_{L'} - \hat{\mathcal{V}}^{\ominus}_{L',L} \quad ,\end{aligned} \tag{4.401}$$

the last equality following from Eq. 4.298. The inverse situation where L is a liquid metal and L' a solution containing the corresponding ions would lead to similar results, but is of no practical relevance. Similar considerations apply to the situation where both L and L' are electrolyte solutions. In practice, the condition of vanishing field in the air gap is verified in this approach by ensuring that the surfaces of the two liquids are *exempt of electric double-layers*. This is realized in an experimental setup where the liquid phase L'_3 is *streaming* from an isolated container past the liquid phase L_2 that is itself grounded[70].

A possible practical realization of this cell involving a *dropping mercury electrode* is illustrated in Figure 4.17. Here, the considered system is

$$\mathrm{Hg}_\ell \mid \mathrm{HCl}_{aq}(\mathrm{Hg_2Cl}_{2,s}) \;\square\; \mathrm{Hg}_\ell \mid \mathrm{HCl}_{aq}(\mathrm{Hg_2Cl}_{2,s}) \mid \mathrm{Hg}_\ell \quad , \tag{4.402}$$

where the concentrations of HCl are identical in the two liquid phases. This situation corresponds to Eq. 4.399, with $\Delta\psi^{\ominus}_{L,L'} = \Delta\psi^{\ominus}_{\mathrm{HCl}_{aq},\mathrm{Hg}_\ell}$. The justification for the use of streaming as a means to enforce the zero field condition at the air gap is the following. If there were opposite surface charges on the Hg_ℓ (L'_3) and HCl_{aq} (L_2) phases surrounding the air gap, one would expect the dripping mercury to carry away part of the surface charge of the Hg_ℓ phase to the ground. In response, part of the surface charge of the HCl_{aq} phase, which is grounded, would also be removed. To maintain a constant

$\Delta\mathcal{V}$, the potentiostat would need to compensate for these losses of charge by the circulation of a steady electric current. The absence of such a current therefore implies the absence of surface charges on the two phases and thus, the zero field condition at the air gap.

The interest of Voltaic cell experiments resides in the connection between the measured *lateral Volta potential differences* and either *real absolute electrode potentials* (Section 4.3.6, Eq. 4.291) or *real single-ion solvation free energies* (Section 4.3.7, Eq. 4.307; see also Eqs. 4.396 and 4.401 in this section), as will be further discussed in Sections 5.14 and 5.15.

Both the ionizing probe and the Kenrick method can be used to determine the lateral Volta potential difference $\Delta\psi^{\ominus}_{M,S}$ between a metal M and a solution S, assumed aqueous for simplicity, containing its ions M^+_{aq}, assumed monovalent for simplicity. However, the practical realization of the cell typically imposes constraints on the choice of M. Most commonly, the quantity determined directly is $\Delta\psi^{\ominus}_{\mathrm{Hg}_\ell,\mathrm{Hg}^+_{aq}}$, *i.e.* concerns mercury Hg_ℓ in equilibrium with its monovalent aqueous ion Hg^+_{aq}. However, a quantity $\Delta\psi^{\ominus}_{M',S'}$ for a given pair $M'-S'$ can be directly translated into a $\Delta\psi^{\ominus}_{M,S}$ value for another pair $M-S$ based on an equation derived from Eqs. 4.298, 4.378 and 4.380, namely

$$\Delta\psi^{\ominus}_{M,S} = \Delta\psi^{\ominus}_{M',S'} + \omega^{\ominus}_{M} - \omega^{\ominus}_{M'} + \left(\Delta_H\mathcal{V}^{\ominus}_{M',S'} - \Delta_H\mathcal{V}^{\ominus}_{M,S}\right) \quad , \tag{4.403}$$

where it is assumed that S and S' are solutions involving the same solvent (typically, water). Application of this equation requires knowledge of the difference in the workfunctions of the two metals, obtained *via* spectroscopy (Section 4.3.5) or Voltaic cell measurements (this section, Eqs. 4.389, 4.393 and 4.400), and of the corresponding redox potentials (Section 4.4.3), *i.e.* quantities which are readily available. Of particular interest is the connection between $\Delta\psi^{\ominus}_{\mathrm{Hg}_\ell,\mathrm{Hg}^+_{aq}}$, most commonly the direct target of Voltaic cell measurements (see above), and $\Delta\psi^{\ominus}_{\mathrm{H(Pt)}}$, namely the *lateral Volta potential difference* at the *reference hydrogen electrode* with *platinum* as a metal. In this special case, Eq. 4.403 reads

$$\Delta\psi^{\ominus}_{\mathrm{H(Pt)}} = \Delta\psi^{\ominus}_{\mathrm{Hg}_\ell,\mathrm{Hg}^+_{aq}} + \omega^{\ominus}_{\mathrm{Pt}} - \omega^{\ominus}_{\mathrm{Hg}_\ell} + \Delta_H\mathcal{V}^{\ominus}_{\mathrm{Hg}_\ell,\mathrm{Hg}^+_{aq}} \quad . \tag{4.404}$$

This equation can also be written in terms of the *real absolute potential* of the *reference hydrogen electrode* as (using Eq. 4.298)

$$\hat{\mathcal{V}}^{\ominus}_{H,wat} = -\Delta\psi^{\ominus}_{\mathrm{Hg}_\ell,\mathrm{Hg}^+_{aq}} + \omega^{\ominus}_{\mathrm{Hg}_\ell} - \Delta_H\mathcal{V}^{\ominus}_{\mathrm{Hg}_\ell,\mathrm{Hg}^+_{aq}} \quad . \tag{4.405}$$

Note that the latter quantity, unlike the former one, is independent of the choice of a metal for the reference hydrogen electrode, as expected. The experimental values of $\omega^{\ominus}_{\mathrm{Pt}_s}$ (polycrystalline surface) and $\omega^{\ominus}_{\mathrm{Hg}_\ell}$ (planar surface) are 5.60 and 4.437 V, respectively (Table 5.13), and the experimental value of $\Delta_H\mathcal{V}^{\ominus}_{\mathrm{Hg}_\ell,\mathrm{Hg}^+_{aq}}$ is[221] 0.7973 V (the Hg^+_{aq} cation actually forming Hg_2^{2+} dimers in aqueous solution). Estimates for $\Delta\psi^{\ominus}_{\mathrm{Hg}_\ell,\mathrm{Hg}^+_{aq}}$ are on the order of[411] -0.729

V with an uncertainty on the order of[411] 0.001 V depending on the source (Section 5.14; older estimates suggesting[404] −0.69 V with an uncertainty of 0.03 V), resulting in $\Delta\psi^{\ominus}_{\mathrm{H(Pt)}}$ and $\hat{\mathcal{V}}^{\ominus}_{H,wat}$ on the order of 1.23 and 4.37 V, respectively, with the same uncertainty. More accurate estimates for these quantities will be presented and discussed in Sections 5.14 (Table 5.14) and 5.19 (Table 5.25).

4.4.5 Electrocapillary and related measurements

Electrocapillary and related measurements correspond to the measurement of the *compensating potential difference* $\Delta\mathcal{V}$ of a cell comprising a polarizable interface, along with non-polarizable ones (Sections 4.3.3 and 4.4.2). In the present context, only *zero-charge* measurements are considered, in which the value of $\Delta\mathcal{V}$ is entirely dictated by the cell scheme along with the equilibrium (zero current, zero charge at the polarizable interface) condition. Situations of *non-zero charge* are briefly mentioned at the end of the section.

The principle of an electrocapillary cell is somewhat *analogous to that of a Voltaic cell* (Section 4.4.4). The differences are that: (*i*) the air gap is replaced by a *polarizable interface*; (*ii*) the potentiometer setting is adjusted to match a constraint of *zero charge* (*i.e.* no electric double-layer) at the polarizable interface, instead of a constraint of zero field at the air gap; (*iii*) this constraint translates to a condition of vanishing *frontal* Volta potential difference $\Delta\tilde{\psi}$ at the polarizable interface rather than a constraint of vanishing lateral potential difference $\Delta\psi$ at the air gap. As a consequence of this last difference, electrocapillary measurements, contrary to Voltaic cell experiments, *cannot be used on their own* to determine the *lateral Volta potential difference* at a non-polarizable interface within the cell (see further below). For this reason, they are somewhat less relevant in the context of this book.

Most commonly, the cell considered involves a polarizable interface between a metal (typically liquid, *e.g.* mercury Hg) and an electrolyte solution, and a non-polarizable interface between this solution and another metal. Generally, this second half-cell is taken to be the reference hydrogen electrode (Section 4.4.3, Eq. 4.379; typically involving platinum Pt as a metal). Restricting the discussion to this specific setup, the cell may be written

$$M_1 \,||\, S_2[\mathrm{H}_{el}] \,|\, M_3' \quad , \tag{4.406}$$

where S_2 is a solution that contains H^+_{aq} but not the ions of the metal M, and the metal M' is typically Pt_s. Assuming standard conditions, the compensating potential difference $\Delta\mathcal{V}^{\ominus}$ is selected in such a way that the electric double-layer at the M_1-S_2 interface vanishes, implying that

$$\Delta\tilde{\psi}^{\ominus}_{M_1,S_2} = 0 \quad . \tag{4.407}$$

This specific value of $\Delta\mathcal{V}^{\ominus}$ is referred to as the *potential of zero charge*[98, 1836–1839] of the polarizable metal-solution interface *relative to the reference hydrogen electrode*, and will be further noted $\Delta_H \overline{\mathcal{V}}^{\ominus}_{M,S}$ (the symbols "$E_{\sigma=0}$" or "$U_{\sigma=0}$" are often used in the literature for this quantity[1713,1836]).

Under zero-charge conditions at the M_1-S_2 interface, Eq. 4.407 holds for the frontal Volta potential difference. Using Eq. 4.261, this equation can be translated into a corresponding equation for the lateral potential difference at zero charge, noted here $\Delta\overline{\psi}^{\ominus}_{M_1,S_2}$, namely

$$\Delta\overline{\psi}^{\ominus}_{M_1,S_2} = \Delta\Delta\tilde{\chi}^{\ominus}_{M,S} \tag{4.408}$$

with (see Eq. 4.294 in Section 4.3.6)

$$\begin{aligned} \Delta\Delta\tilde{\chi}^{\ominus}_{M,S} &= \Delta\tilde{\chi}^{\ominus}_{M,S} - \Delta\chi^{\ominus}_{M,S} \\ &= \left(\tilde{\chi}^{\ominus}_{S(M)} - \tilde{\chi}^{\ominus}_{M(S)}\right) - \left(\chi^{\ominus}_{S} - \chi^{\ominus}_{M}\right) \\ &= \left(\chi^{\ominus}_{M} - \tilde{\chi}^{\ominus}_{M(S)}\right) - \left(\chi^{\ominus}_{S} - \tilde{\chi}^{\ominus}_{S(M)}\right) \quad , \end{aligned} \tag{4.409}$$

where $\Delta\tilde{\chi}^{\ominus}_{M,S}$ is the standard frontal surface potential difference at the interface (see Eq. 4.255), and $\Delta\chi^{\ominus}_{M,S}$ is the standard lateral surface potential difference corresponding to free surfaces of the metal and solution (Eq. 4.258). The third equality in Eq. 4.409 follows from a mere rearrangement of the terms, and underlines that the difference between $\Delta\chi^{\ominus}_{M,S}$ and $\Delta\tilde{\chi}^{\ominus}_{M,S}$ can be viewed as resulting from the perturbation of the surface potentials of M and S corresponding to free surfaces in vacuum (air) in response to their contact at the interface.

Following a reasoning similar to that for the ionizing probe method in Section 4.4.4, one obtains, in analogy with Eq. 4.396,

$$\Delta_H \overline{\mathcal{V}}^{\ominus}_{M,S} = \omega^{\ominus}_{M} - \Delta\overline{\psi}^{\ominus}_{M,S} - \hat{\mathcal{V}}^{\ominus}_{H} \quad , \tag{4.410}$$

which, using Eqs. 4.291 and 4.408, can be rewritten as

$$\Delta_H \overline{\mathcal{V}}^{\ominus}_{M,S} = \omega^{\ominus}_{M} - \Delta\Delta\tilde{\chi}^{\ominus}_{M,S} - \omega^{\ominus}_{M'} + \Delta\psi^{\ominus}_{M',S} \quad , \tag{4.411}$$

where the metal M' is typically platinum Pt. Thus, complementing $\Delta_H \overline{\mathcal{V}}^{\ominus}$ with data of Voltaic cell experiments (Section 4.4.4) is useful for understanding the double-layer and surface alteration characteristics of a metal-solution interface. Besides being an aid in the above investigations of interface characteristics on a microscopic level, the potential of zero charge is also an important quantity in the general study of the thermodynamics of polarizable metal-solution interfaces. Based on Lippmann's equation[1836,1840],

$$\left(\frac{\partial\gamma}{\partial\Delta\mathcal{V}}\right)_{P,T,\{\mu_i\}} = -\rho_Q \quad , \tag{4.412}$$

where γ is the interfacial tension, $\Delta\mathcal{V}$ the potential of the electrode of interest with respect to a chosen reference electrode, $\{\mu_i\}$ a chosen set of chemical potentials of compounds in the system, and ρ_Q the charge density on the interface, it follows that for polarizable interfaces, where no charge exchange takes place across the interface, the potential of zero charge coincides with the maximum of interfacial tension. The potential of zero charge can also be used to investigate adsorption effects *via* measurement of interfacial tension as a function of potential and electrolyte concentration.

Finally, it should be mentioned that investigations of the dependence of interfacial tension on the potential ("electrocapillary curves") have a long history, dating back to the 19th century. *Electrocapillary measurements*[98,455,1754,1756,1769,1841–1844], pioneered by Lippmann[1841,1842] (see also Refs.[1845–1852] for subsequent related investigations), rely on the measurement of the interfacial tension at a polarizable liquid-liquid interface (typically between mercury and an electrolyte solution) as a function of the applied potential difference (electrocapillary curve). Assuming that the presence of an electric double-layer at the interface always reduces the interfacial tension, the maximum of the electrocapillary curve (referred to as the potential at zero charge) should correspond to the applied potential difference for which there is no electric double-layer, *i.e.* it should be opposite to the natural potential difference between the two liquids[1853]. In principle, ensuring a maximum in the interfacial tension should offer yet another means to enforce a zero Volta potential difference between two phases in a cell (here, at a polarizable interface rather than an air gap), so as to measure the Volta potential difference between two phases at a given non-polarizable interface. In practice, the so-called "Lippmann capillary electrometer"[1841,1842] seeks to determine the potential that has to be applied to mercury (in contact with an electrolyte solution) to achieve the maximum height of the mercury in the column. However, the constraint refers here to the frontal Volta potential difference $\Delta\tilde{\psi}$ rather than the lateral one $\Delta\psi$, requiring the evaluation (or neglect) of a contribution $\Delta\tilde{\chi} - \Delta\chi$. Furthermore, interfacial tension effects may be altered by specific adsorption effects. In addition, near the maximum of interfacial tension, significant changes in the applied potential cause only small changes of interfacial tension, leading to inaccuracies in the measurements. As a result, evaluations of Volta potential differences *via* electrocapillary measurements are rather imprecise[69,74].

For the sake of completeness, three other ancient methods in the investigation of metal-solution Volta potential differences shall be mentioned.

Dropping electrode methods[69,70,1835,1854,1855] pioneered by Thomson in the study of electric potentials in air (see Ref.[1835]; the authors of this book could not find the original reference), Bichat & Blondlot[1835] and Kenrick[70] (see also Refs.[73,1856,1857] for subsequent related investigations), are the ancestors of the modern Kenrick approach[411,419].

Endosmotic measurements[75,1858–1863], pioneered by Billitzer[1858], rely on the enforcement of zero electric field within an electrolyte solution encompassed between two metal electrodes by variation of the electrolyte concentration and assessment of the absence of motion for fine wires or small colloidal particles in suspension. In the absence of motion, a situation has been realized where the Volta potential difference between the two electrodes is zero. This should offer yet another means to enforce a zero Volta potential difference between two phases in a cell (here, at a solution gap rather than an air gap), so as to measure the Volta potential difference between two phases at a given non-polarizable interface. Here, the constraint correctly refers to a lateral Volta potential (if the two electrodes are made of the same metal, implying identical $\Delta\tilde{\chi}$ values at contact with the electrolyte solution). However, evaluations of Volta potential differences *via* endosmotic measurements have proven sensitive to details of the experimental setup and difficult to reproduce[75, 76, 1860, 1861].

Finally, *capacitance measurements*[74], based on studies of polarization capacity[1864–1868] of electrodes, rely on the measurement of the polarization capacity of a polarizable electrode as a function of the applied potential difference (polarization capacity curves). Assuming that a vanishing potential difference between the (polarizable) metal electrode and the solution and the correspondingly vanishing electric double-layer is accompanied by a maximum in capacity[74], the maximum of the polarization capacity curve should indicate the potential difference opposite to the natural potential difference between the metal and the solution. Such measurements are affected by many sources of error[74] and have not found practical applications.

4.5 Single-ion properties

In this section the three types of *single-ion solvation free energies* and *derivative thermodynamic properties*, including corresponding *partial molar variables* in solution, are discussed, namely: (*i*) *conventional* (relative) parameters; (*ii*) *real* (absolute) parameters; (*iii*) *intrinsic* (absolute) parameters. In a last section, *conversion formulae* are provided to transform conventional parameters to intrinsic ones based on an estimate for the intrinsic proton solvation free energy $\mathcal{G}^{\ominus}_{H,svt}$ (and its derivatives) in the given solvent, to transform conventional parameters to real ones based on an estimate for the corresponding real proton solvation free energy $\hat{\mathcal{G}}^{\ominus}_{H,svt}$ (and its derivatives), or to transform real values to intrinsic ones based on an estimate for the air-liquid interfacial potential $\chi^{\ominus}_{svt}$ of the solvent (and its derivatives). The connection between conventional solvation data and redox potentials $\Delta_H\mathcal{V}^{\ominus}$, intrinsic solvation data and intrinsic absolute electrode potentials $\mathcal{V}^{\ominus}_{M,S}$ (including $\mathcal{G}^{\ominus}_{H,svt}$ *vs.* $\mathcal{V}^{\ominus}_{H,svt}$), as well as real solvation data and real absolute electrode potentials $\hat{\mathcal{V}}^{\ominus}_{M,S}$ (including $\hat{\mathcal{G}}^{\ominus}_{H,svt}$ *vs.* $\hat{\mathcal{V}}^{\ominus}_{H,svt}$) or metal-solution Volta potential differences $\Delta\psi^{\ominus}_{M,S}$, are also summarized in this last section.

The present discussion focuses on the concepts and definitions. Further details concerning the determination methods and corresponding results in the restricted context of alkali and halide hydration will be provided in Chapter 5. Although this discussion is in principle general, it is also somewhat biased towards the investigation of the alkali and halide ion series with water as a solvent. In particular, the description often refers to an arbitrary neutral ion pair IJ, of possibly polyvalent ions, in a given solvent, or more specifically to an alkali-halide pair MX in water. If necessary, the corresponding definitions can easily be generalized to ion sets involving cations and anions of distinct valences, *e.g.* $I^{2z}J_2^{-z}$, and to ternary or higher-order sets, *e.g.* $I^z+J^z+K^{-2z}$ (Section 4.1). In the latter case, the notations M or M^+ (alkali element or ion) and X or X^- (halide element or ion) are used instead of a more general notation including possible polyvalent ions. The special case of sodium chloride (NaCl) is also used as an illustrative numerical example. Finally, although the present discussion mainly focuses on free energy as a thermodynamic parameter, it should be stressed that the distinction between conventional, real and intrinsic values also pertains to all single-ion derivative thermodynamic solvation parameters derived from *e.g.* H, S, C_P, V, K_T and A_P (Section 4.2.1), as well as to all single-ion formation parameters and single-ion partial molar variables in solution (Sections 4.2.2 and 4.2.10).

For the ease of retrieval, the equations discussed in the first three sections are summarized in Section 4.5.4 and Figure 4.18.

4.5.1 Conventional values

As discussed in Section 2.1, using equilibrium experiments performed at usual pressures and temperatures and spectroscopic experiments (SBS) to determine the properties of gas-phase ions, along with appropriate thermodynamic cycles, it is possible to determine sums of solvation parameters for overall neutral sets of ions. Of course, these values are subject to experimental errors, and the results reported by different investigators may evidence non-negligible discrepancies, see *e.g.* Section 5.12 (Tables 5.11 and 5.12) for alkali-halide pair hydration parameters. However, there is no fundamental ambiguity in these numbers in the sense that surface effects, *i.e.* the need to distinguish between intrinsic and real values, do not affect these sums (Section 2.2). This is because the surface term $\Delta_s Y^{\ominus}_{srf}$, that distinguishes the two types of quantities, is linear in the ion charge (Eqs. 2.10 and 2.12). In other words, real and intrinsic solvation free energies refer to a specific partitioning of the sum, but are irrelevant in the sum itself.

In the context of the solvation free energy and considering alkali halides (MX) and water as a solvent, this sum is described by the *salt solvation free energy* $\Delta_s G^{\ominus}[MX_g]$ corresponding to the transfer, expressed on a per mole basis, of one alkali ion and one halide ion from the standard-state gas phase (MX_g, *i.e.* M_g^+ and X_g^-) to the standard-state aqueous phase (MX_{aq}, *i.e.* M_{aq}^+ and X_{aq}^-), the standard states being defined according to the $bbme_T$ convention at temperature T^- (ideal gas and solution, reference pressure P° and molality b°; Section 4.2.4). As a numerical example, an experimental estimate for sodium chloride (NaCl) in water is[221,294] $\Delta_s G^{\ominus}[\mathrm{NaCl}_g] = -728.0\ \mathrm{kJ{\cdot}mol^{-1}}$

(Table 5.11; Set 1G).

It might seem at first sight that salt solvation data should be sufficient to uniquely define single-ion contributions. Considering a set of m cations and n anions, one may determine a total of $m \times n$ salt solvation free energies, to be reproduced as sums over a set of only $m + n$ single-ion solvation free energies. For the alkali-halides, one has $m = 5$ (excluding Fr) and $n = 4$ (excluding At), so that only 9 single-ion parameters must be adjusted to reproduce 20 pair parameters. This could in principle be done using an optimization scheme relying on minimization of the corresponding mean-squared deviation, resulting in a set of equations to be solved by inversion of a 9×9 matrix. Note, however, that the latter optimization would only serve the purpose of minimizing the effect of experimental errors. Given the standard-state definitions of Section 4.2.4, a perfect set of pairwise data should be *exactly additive* in a corresponding set of single-ion parameters, because pair parameters are defined as sums over single-ion properties involving no cross-terms. This perfect set would thus include 20 values connected by 11 constraint equations. Unfortunately, this problem remains underdetermined due to the existence of a 12th constraint, related to the *electroneutrality of all salts* in the pairwise data set. This is easily seen by observing that if an optimal set of single-ion parameters is determined, any set obtained by addition of an arbitrary constant to all monovalent cation values and subtraction of the same constant from all monovalent anion values will produce an equally optimal set, considering that the arbitrary constant cancels out in all pairwise sums. In other words, a perfect set of pairwise data over all the alkali-halide salts actually comprises 8 rather than 9 independent values. The above optimization process for minimizing the experimental errors thus involves the removal of one redundant equation, *e.g.* by setting the single-ion solvation free energy of the Li^+ cation to zero, followed by the inversion of an 8×8 matrix. It results in a set of single-ion parameters defined within an unknown constant, the solvation free energy of Li^+, additive for all cations and subtractive for all anions.

The above reasoning is easily generalized to arbitrary sets of salts, possibly involving polyvalent or/and polyatomic ions, and shows that salt solvation free energies can only be partitioned into single-ion contributions within an unknown constant, entering into the single-ion parameters with a weight equal to their signed integer ionic charges. It is also applicable to all derivative thermodynamic solvation parameters, each corresponding to its own unknown constant. Due to the presence of these unknown constants, single-ion solvation parameters are usually tabulated on *relative scales*, where the corresponding values for a specific ion are set to a given constant by definition. These relative scales of single-ion solvation parameters are called *conventional scales*, and the usual choice for this special ion is the *proton* H^+, *i.e.* the simplest possible ion in terms of internal structure (no electrons at all). Note, however, that in spite of this apparent simplicity, H^+ solvation typically involves complex electronic effects, which are actually far less significant in *e.g.* alkali and halide solvation (Section 4.3.8). There are, however, different possible definitions for the reference proton values, leading to as many distinct conventional scales.

The most common conventional scale[260,262,264,269,274,275,417,819,1750,1869–1874], and the one adopted in this book, will be referred to as the *standard conventional scale.* This scale relies on *standard* parameters according to the $bbme_T$ convention (Section 4.2.4), where the conventional anchoring is performed by setting *all solvation parameters of the proton to zero* at any temperature. Conventional parameters reported according to this scale will be indicated with a "$\bullet$" superscript (instead of "$\circ$" for standard absolute parameters), at a given temperature T, or a "$\bullet\!\!\!-$" superscript (instead of "$\ominus$") at the reference temperature T^-. For all arbitrary thermodynamic parameters Y ($Y = G$, H, S, C_P, V, K_T or A_P), and considering the temperature T^-, one has thus by definition

$$\Delta_s Y^{\bullet\!\!\!-} = 0 \quad . \tag{4.413}$$

As discussed above, the unknown constant in single-ion parameters enters with a weight equal to the signed integer charge of the ion. It follows that *all conventional single-ion solvation parameters of the proton are zero* at any pressure and temperature. For an ion I^z, $\Delta_s Y^{\bullet\!\!\!-}[I_g^z]$ is given by

$$\begin{aligned} \Delta_s Y^{\bullet\!\!\!-}[I_g^z] &= \Delta_s Y^{\ominus}[I_g^z] - z\mathcal{Y}_H^{\ominus} \\ &= \Delta_s \hat{Y}^{\ominus}[I_g^z] - z\hat{\mathcal{Y}}_H^{\ominus} \quad , \end{aligned} \tag{4.414}$$

where $\Delta_s \hat{Y}$ and $\Delta_s Y$ represent absolute solvation parameters, on real (Section 4.5.2) and intrinsic (Section 4.5.3) scales, respectively. In the special case of the solvation free energy, Eq. 4.414 becomes

$$\begin{aligned} \Delta_s G^{\bullet\!\!\!-}[I_g^z] &= \Delta_s G^{\ominus}[I_g^z] - z\mathcal{G}_H^{\ominus} \\ &= \Delta_s \hat{G}^{\ominus}[I_g^z] - z\hat{\mathcal{G}}_H^{\ominus} \quad . \end{aligned} \tag{4.415}$$

Standard conventional solvation parameters can be viewed as thermodynamic reaction parameters associated with the exchange process ($z > 0$)

$$I_g^z + z\mathrm{H}_d^+ \rightarrow I_d^z + z\mathrm{H}_g^+ \quad , \tag{4.416}$$

or ($z < 0$)

$$I_g^z + (-z)\mathrm{H}_g^+ \rightarrow I_d^z + (-z)\mathrm{H}_d^+ \quad . \tag{4.417}$$

Note that the quantity $|z|\,\mathcal{G}_H^{\ominus}$ (or $|z|\,\hat{\mathcal{G}}_H^{\ominus}$) is *added* to the intrinsic (or real) solvation free energy of an anion, but *subtracted* from that of a cation. It is important to stress that although conventional values are relative, they still represent standard values in the sense that they obey the $bbme_T$ convention (ideal gas and solution, reference pressure P° and molality b°; Section 4.2.4). The conventional solvation parameters of monovalent cations are actually independent of the choice of a reference pressure and concentration, because the number of gas-phase and dissolved particles is the same on both sides of the stoechiometric equation (Eq. 4.416) in this case. However, this statement applies neither to polyvalent cations nor to anions. For example, for monovalent anions, the change in the number of gas-phase and dissolved particles in the stoechiometric equation (Eq. 4.417) is twice that for the standard single-ion solvation process. As a result, possible standard-state corrections (Sections 4.2.16 and 4.2.17) affect conventional values twice compared to absolute values.

Single-ion solvation parameters reported in standard conventional form are subject to experimental errors, and the results reported by different investigators may evidence non-negligible discrepancies, see *e.g.* Section 5.16 (Tables 5.16 and 5.17) for alkali and halide conventional hydration parameters. These errors are also affected by the optimization procedure leading from salt solvation parameters to conventional single-ion data, *via* the numbers of salts considered, the experimental uncertainties on the corresponding salt parameters, and the inclusion or exclusion of proton compounds in this set (requiring an additional alignment against other data including proton compounds in the latter case). Similar considerations apply to conventional data for derivative thermodynamic properties, which is also affected by the ambiguity related to standard *vs.* density-corrected variants (Sections 4.2.4 and 4.2.5), except in the special case of monovalent cations. Arguably, the optimization procedure could result in a decrease of the uncertainty compared to the salt data, due to error cancellations during the optimization. This of course only applies if the error in the salt data is not systematic. However, as was the case for salt solvation free energies (see above), there is no fundamental ambiguity in the resulting conventional values in the sense that surface effects, *i.e.* the need to distinguish between intrinsic and real values, do not affect the differences involved in Eq. 4.414, because the surface term $\Delta_s Y_{srf}^{\ominus}$ that distinguishes the two types of quantities is linear in the ion charge (Eqs. 2.10 and 2.12).

As a numerical example, one of the most recent conventional sets of hydration free energies attributes to sodium and chloride (Na^+ and Cl^-) values of[1875, 1876] $\Delta_s G^{\bullet}[Na_g^+] = 680.7$ kJ·mol^{-1} and $\Delta_s G^{\bullet}[Cl_g^-] = -1408.7$ kJ·mol^{-1} (Table 5.16; Set 20g$_1$), with an estimated experimental uncertainty[1875, 1876] of about ± 8 kJ·mol^{-1}. The sum of these two values is -728 kJ·mol^{-1} which is exactly equal to the experimental estimate for the salt value[221, 294] (see above).

If standard conventional values are to be treated on an equal footing with other standard thermodynamic reaction parameters, requiring in particular, that the Hess law and the Gibbs equation are satisfied (Section 4.2.7), Eq. 4.414 imposes the use of consistent conventional definitions for all thermodynamic parameters involving single atoms and single ions. In addition to single-ion solvation parameters, these include atomization and ionization parameters (Section 4.2.14), formation parameters of gas-phase atoms, gas-phase ions and solvated ions (Sections 4.2.10 and 4.2.14), as well as (partial) molar variables of gas-phase atoms, gas-phase ions and solvated ions (Section 4.2.2). For these quantities, the conventional value is defined by subtracting z-times the standard parameter for the proton from the standard parameter for the given species, namely

$$\begin{aligned}\Delta_f Y^{\bullet}[\Sigma^z] &= \Delta_f Y^{\ominus}[\Sigma^z] - z\Delta_f Y^{\ominus}[\mathrm{H}^+] \\ &= \Delta_f \hat{Y}^{\ominus}[\Sigma^z] - z\Delta_f \hat{Y}^{\ominus}[\mathrm{H}^+]\ ,\end{aligned} \tag{4.418}$$

$$\begin{aligned}y^{\bullet}[\Sigma^z] &= y^{\ominus}[\Sigma^z] - zy^{\ominus}[\mathrm{H}^+] \\ &= \hat{y}^{\ominus}[\Sigma^z] - z\hat{y}^{\ominus}[\mathrm{H}^+]\ ,\end{aligned} \tag{4.419}$$

and

$$\begin{aligned}&\Delta_r Y^{\bullet}[\Sigma^z \rightarrow \Sigma^{z'} + (z'-z)e_g^-] \\ =\ &\Delta_r Y^{\ominus}[\Sigma^z \rightarrow \Sigma^{z'} + (z'-z)e_g^-] \\ &-\Delta_r Y^{\ominus}[\mathrm{H} \rightarrow \mathrm{H}^+ + e_g^-] \\ =\ &\Delta_r \hat{Y}^{\ominus}[\Sigma^z \rightarrow \Sigma^{z'} + (z'-z)e_g^-] \\ &-\Delta_r \hat{Y}^{\ominus}[\mathrm{H} \rightarrow \mathrm{H}^+ + e_g^-]\quad ,\end{aligned} \tag{4.420}$$

where Σ^z refers to any neutral (I_{el}, I_g) or ionized species (I_g^z or I_g^z), H^+ being in the same phase (H_{el}, H_g, H_g^+ or H_d^+), and $\Sigma^z \rightarrow \Sigma^{z'} + (z'-z)e_g^-$ to any reaction possibly involving ionized species (Σ^z in the reactants and $\Sigma^{z'}$ in the products), H^+ being in the same phase (H_{el}, H_g, H_g^+ or H_d^+, depending on the reactants and products). This leads to the definitions of the quantities $\Delta_a Y^{\bullet}[I_{el}] = \Delta_a Y^{\ominus}[I_{el}]$, $\Delta_i Y^{\bullet}[I_g]$, $\Delta_s Y^{\bullet}[I_g^z]$, $\Delta_f Y^{\bullet}[I_g] = \Delta_f Y^{\ominus}[I_g]$, $\Delta_f Y^{\bullet}[I_g^z]$, $\Delta_f Y^{\bullet}[I_d^z]$, $y^{\bullet}[I_{el}] = y^{\ominus}[I_{el}]$, $y^{\bullet}[I_g] = y^{\ominus}[I_g]$, $y^{\bullet}[I_g^z]$ and $y^{\bullet}[I_d^z]$. These variables all evaluate to zero for the proton. If such a conventional scale is used consistently, the corresponding symbols in Figures 4.1 and 4.2 can be changed from standard "$\ominus$" to conventional "$\bullet$" and the thermodynamic relationships discussed in Section 4.2 including the Gibbs equation, the Hess law and the thermodynamic derivative relationships (including the distinction between standard and density-corrected variants) remain generally valid with one important exception. The conventional formation parameters of the electron in its standard state are not zero (Sections 4.2.4 and 4.2.10) but, according to Eq. 4.418

$$\Delta_f Y^{\bullet}[e_g^-] = \Delta_f Y[\mathrm{H}_g^+]\quad . \tag{4.421}$$

Note, however, that the conventional scale does not merely result from a different setting for the electron standard state, but from a shifting of the thermodynamic properties of any charged species in any phase by z-times the corresponding proton value.

The standard conventional scale is just one possible convention, and at least two other conventions can be found in the literature. In the first alternative convention[259,262], which will be referred to here as the *gaseous-proton conventional scale*, the conventional values for the proton solvation parameters are set to zero, as in the standard conventional scale. However, the conventional values for the formation parameters of the solvated proton are set to $\Delta_f^{\ominus} Y[\mathrm{H}_g^+]$ rather than zero. In the second alternative convention[259,262,1875], which will be referred to here as the *aqueous-proton conventional scale*, the conventional values for the formation parameters of the solvated proton are set to zero. However, the conventional values for the proton solvation parameters are set to $-\Delta_f Y^{\ominus}[\mathrm{H}_g^+]$ rather than zero. The values of the parameters $\Delta_f Y[\mathrm{H}_g^+]$ can be found in Section 4.2.13 (Table 4.2). The advantage of these two alternative conventional scales over the standard conventional scale is that they do not require adjustments in the parameters concerning gas-phase ions as in Eqs. 4.418-4.420 (formation and ionization parameters, molar variables), *i.e.* the standard and conventional values of these parameters are identical. The drawback is that the Hess law and the Gibbs equation may be violated in some cases when applied to these alternative conventional values, which must therefore be handled with some care[259]. These alternative scales will not be used in the present book.

Finally, there is yet another (slightly disguised) form of conventional scale, namely the *redox scale* of electrode potentials relative to the reference hydrogen electrode (Section 4.4.3). The connection between the redox potentials $\Delta_H \mathcal{V}^{\ominus}[I_{el}, I_d^z]$ for the element-ion couple involving the element I_{el} and the dissolved ion I_d^z and standard conventional solvation free energies is easily deduced from Eqs. 4.378, 4.309 and 4.415, and reads

$$zF\Delta_H \mathcal{V}^{\ominus}[I_{el}, I_d^z] = \Delta_s G^{\bullet}[I_g^z] + \Delta_f G^{\ominus}[I_g^z] - z\Delta_f G^{\ominus}[\mathrm{H}_g^+] \quad . \qquad (4.422)$$

Note also that Eq. 4.422 maps to a corresponding connection between the isothermal temperature derivative $\partial_T \Delta_H \mathcal{V}^{\ominus}[I_{el}, I_d^z]$ and the conventional solvation entropy, *i.e.*

$$-zF\partial_T \Delta_H \mathcal{V}^{\ominus}[I_{el}, I_d^z] = \Delta_s S^{\bullet}[I_d^z] + \Delta_f S^{\ominus}[I_g^z] - z\Delta_f S^{\ominus}[\mathrm{H}_g^+] \quad . \qquad (4.423)$$

Note also that Eq. 4.422 maps to a corresponding connection between the second isothermal temperature derivative $\partial_T^2 \Delta_H \mathcal{V}^{\ominus}[I_{el}, I_d^z]$ and the conventional solvation heat capacity, *i.e.*

$$-zFT^{-}\partial_T^2 \Delta_H \mathcal{V}^{\ominus}[I_{el}, I_d^z] = \Delta_s C_P^{\bullet}[I_d^z] + \Delta_f C_P^{\ominus}[I_g^z] - z\Delta_f C_P^{\ominus}[\mathrm{H}_g^+] \quad . \qquad (4.424)$$

Considering the quantity $zF\Delta_H \mathcal{V}^{\ominus}[I_{el}, I_d^z]$ as a conventional solvation free energy of ion I^z on what could called the *redox conventional scale*, one sees that this scale has the same anchoring point as the standard conventional scale (or the gaseous-proton one), namely a zero solvation free energy for the proton, but a different scaling, involving information on the gas-phase formation free energy. The relationship between this redox and the standard conventional scale reads

$$zF\Delta_H\mathcal{V}^\ominus[I_{el}, I_d^z] = \Delta_s G^\bullet[I_g^z] + \Delta_f G^\bullet[I_g^z] \quad . \tag{4.425}$$

The main advantage of conventional scales over an intrinsic (absolute) one (Section 4.5.3) is that they are unambiguous in the sense that their definition does not rely on the specification of any experimentally-elusive property (Section 2.3). Since the exact value of $\mathcal{G}_H^\ominus$ (or in general $\Delta_s Y^\ominus[H_g^+]$, related to the dependence of $\mathcal{G}_H^\ominus$ on P and T) remains controversial, *i.e.* not known with any accuracy, these scales avoid the propagation of this uncertainty into tabulated single-ion parameters. To some extent, this also represents an advantage over a real (absolute) scale (Section 4.5.2) considering that the quantity $\hat{\mathcal{G}}_H^\ominus$, although not formally elusive, is not known with a very high accuracy at present. Of course, ambiguity may still be present in the form of alternative conventions (see above). In the present book, solely the standard conventional scale will be further considered.

4.5.2 Real values

As described in Section 2.1, using equilibrium experiments performed at usual pressures and temperatures (or equivalent calorimetric procedures) and spectroscopic experiments (SBS) to determine the properties of gas-phase ions, along with appropriate thermodynamic cycles, it is possible to determine single-ion thermodynamic parameters (*e.g.* single-ion solvation free energies), albeit only in the form of real quantities, *i.e.* sums over contributions from bulk and surface effects (Section 2.2). In practice, this determination relies on Voltaic cell experiments (Section 4.4.4) along with the spectroscopic determination of a metal work function (Section 4.3.5) and of gas-phase ion formation parameters (Section 4.2.10). Voltaic cell experiments are relatively difficult to undertake and may be affected by many error sources. Furthermore, these experiments have been performed in a number of experimental groups, and most of them decades ago, see *e.g.* Section 5.15 (Table 5.15) for real solvation parameters of the alkali and halide ions or of the proton. As a result, although there is no fundamental ambiguity in these numbers in the sense that they are not experimentally elusive (Section 2.3), the corresponding experimental estimates are typically affected by relatively large errors.

As a numerical example, one of the most recent sets of real hydration free energies attributes to sodium and chloride (Na^+ and Cl^-) values of[415] $\Delta_s\hat{G}^\ominus[Na_g^+] = -402.4$ kJ·mol^{-1} and $\Delta_s\hat{G}^\ominus[Cl_g^-] = -322.8$ kJ·mol^{-1} (Table 5.15; Set 6ĝ). The sum of these two values is -725.2 kJ·mol^{-1}, very close to the experimental estimate of -728 kJ·mol^{-1} for the salt value (Table 5.11; Set 1G).

The real solvation free energy $\Delta_s\hat{G}^\ominus[I_g^z]$ of an ion I^z at temperature T^- is straightforwardly connected to the corresponding conventional value

$\Delta_s G^{\bullet}[I_g^z]$ (Section 4.5.1) as

$$\Delta_s \hat{G}^{\ominus}[I_g^z] = \Delta_s G^{\bullet}[I_g^z] + z\hat{\mathcal{G}}_H^{\ominus} \quad . \tag{4.426}$$

Recall that the quantity $\hat{\mathcal{G}}_H^{\ominus}$ is directly connected to the absolute real potential $\hat{\mathcal{V}}_H^{\ominus}$ of the reference hydrogen electrode through

$$\hat{\mathcal{G}}_H^{\ominus} = F\hat{\mathcal{V}}_H^{\ominus} - \Delta_f G^{\ominus}[\mathrm{H}_g^+] \quad . \tag{4.427}$$

Equations analogous to Eq. 4.426 apply to all other single-ion real thermodynamic parameters, *i.e.* (see Eqs. 4.414 and 4.418-4.420)

$$\Delta_s \hat{Y}^{\ominus}[I_g^z] = \Delta_s Y^{\bullet}[I_g^z] + z\hat{\mathcal{Y}}_H^{\ominus} \quad , \tag{4.428}$$

$$\Delta_f \hat{Y}^{\ominus}[\Sigma^z] = \Delta_f Y^{\bullet}[\Sigma^z] + z\Delta_f \hat{Y}^{\ominus}[\mathrm{H}^+] \quad , \tag{4.429}$$

$$\hat{y}^{\ominus}[\Sigma^z] = y^{\bullet}[\Sigma^z] + z\hat{\mathcal{y}}_H^{\ominus} \quad , \tag{4.430}$$

and

$$\begin{aligned} \Delta_r \hat{Y}^{\ominus}[\Sigma^z \rightarrow \Sigma^{z'} + (z'-z)e_g^-] &= \Delta_r Y^{\bullet}[\Sigma^z \rightarrow \Sigma^{z'} + (z'-z)e_g^-] \\ &+ \Delta_r \hat{Y}^{\ominus}[\mathrm{H} \rightarrow \mathrm{H}^+ + e_g^-] \quad , \end{aligned} \tag{4.431}$$

where it is recalled that $\hat{\mathcal{Y}}_H^{\ominus}$ and $\hat{\mathcal{y}}_H^{\ominus}$ are short notations for $\Delta_s \hat{Y}^{\ominus}[\mathrm{H}_g^+]$ and $\hat{y}^{\ominus}[\mathrm{H}_d^+]$, respectively (Section 2.3). For derivative thermodynamic properties, these equations are written in the standard variant, but real parameters are also affected by the ambiguity related to standard *vs.* density-corrected variants (Sections 4.2.4 and 4.2.5). Analogous equations also hold in the density-corrected variant.

The main advantage of the real scale over an intrinsic (absolute) one (Section 4.5.3) is that it is unambiguous in the sense that its definition does not rely on the specification of any experimentally-elusive property (Section 2.3). Since the exact value of $\mathcal{G}_H^{\ominus}$ (or in general $\Delta_s Y^{\ominus}[\mathrm{H}_g^+]$, related to the dependence of $\mathcal{G}_H^{\ominus}$ on P and T) remains controversial, *i.e.* not known with any accuracy, this scale avoids the propagation of this uncertainty into tabulated single-ion parameters. Furthermore, in contrast to conventional scales (Section 4.5.1), the real scale is absolute and the corresponding values possess a direct physical significance, although intermingling bulk and surface effects. However, the conventional scale is probably to be preferred for the tabulation of relative single-ion data, since the experimental error on $\hat{\mathcal{G}}_H^{\ominus}$ is comparatively larger than that on conventional values for individual ions.

4.5.3 Intrinsic values

As discussed in Sections 4.5.1 and 4.5.2, both conventional and real single-ion solvation parameters can be measured experimentally. However, conventional parameters are relative, *i.e.* anchored to an arbitrary zero point, so that they do not possess any direct physical meaning, while real parameters account for a mixture of two physically very different effects, *i.e.* bulk solvation and liquid-surface properties, so that their physical significance is limited. The estimation of intrinsic solvation parameters, which exclusively characterize the ion-solvent interactions within the bulk, would thus be highly desirable.

Single-ion intrinsic thermodynamic parameters are the most relevant from a theoretical point of view, because they account for bulk solvation effects without any contamination from surface properties. However, because intrinsic properties are not directly measurable experimentally[257–261,269,271,274,275], all approaches to date to establish intrinsic scales of single-ion solvation parameters have relied on extra-thermodynamic assumptions, *i.e.* on the use of more or less accurate physical models[257–261,269,271,274,275] (Section 2.3). To transform the conventional (or real) scale into an absolute scale, it is necessary to evaluate intrinsic solvation parameters for at least one ion, which in turn fixes the corresponding values for the proton. Although many attempts have been made to determine such a value, so as to fix an absolute scale of single-ion properties, the resulting numbers are affected by considerable uncertainties, see *e.g.* Section 5.18 (Tables 5.19 and 5.20) for estimates of the intrinsic proton solvation parameters. As a numerical example, one of the most recent sets of tentative intrinsic hydration free energies (based on a cluster-pair extra-thermodynamic assumption) attributes to sodium and chloride (Na^+ and Cl^-) values of[1875, 1876] $\Delta_s G^{\ominus}[Na_g^+] = -420.2$ kJ·mol^{-1} and $\Delta_s G^{\ominus}[Cl_g^-] = -307.8$ kJ·mol^{-1} (Table 5.16; Set 20g$_1$ along with $\mathcal{G}_H^{\ominus} = -1100.9$ kJ·mol^{-1} from Table 5.19, same set name). The sum of these two values, which should individually be taken with caution, is exactly equal to the experimental estimate for the ion-pair value of -728 kJ·mol^{-1} (Table 5.11; Set 1G).

The intrinsic solvation free energy $\Delta_s G^{\ominus}[I_g^z]$ of an ion I^z at temperature T^- is straightforwardly connected to the corresponding conventional value (Section 4.5.1) $\Delta_s G^{\bullet}[I_g^z]$ (Eq. 4.415) as

$$\Delta_s G^{\ominus}[I_g^z] = \Delta_s G^{\bullet}[I_g^z] + z\mathcal{G}_H^{\ominus} \quad . \tag{4.432}$$

Recall that the quantity $\mathcal{G}_H^{\ominus}$ is directly connected to the equally elusive intrinsic absolute potential $\mathcal{V}_H^{\ominus}$ of the reference hydrogen electrode through

$$\mathcal{G}_H^{\ominus} = F\mathcal{V}_H^{\ominus} - \Delta_f G^{\ominus}[H_g^+] \quad . \tag{4.433}$$

It is also connected to the experimentally-accessible real solvation free energy $\hat{\mathcal{G}}_H^{\ominus}$ of the proton *via* the equally elusive surface potential of the pure solvent, as

$$\mathcal{G}_H^{\ominus} = \hat{\mathcal{G}}_H^{\ominus} - F\chi_{svt}^{\ominus} \quad . \tag{4.434}$$

Equations analogous to Eq. 4.432 apply to all other single-ion intrinsic thermodynamic parameters, *i.e.* (see Eqs. 4.414 and 4.418-4.420)

$$\Delta_s Y^{\ominus}[I_g^z] = \Delta_s Y^{\bullet}[I_g^z] + z\mathcal{Y}_H^{\ominus} \quad , \tag{4.435}$$

$$\Delta_f Y^{\ominus}[\Sigma^z] = \Delta_f Y^{\bullet}[\Sigma^z] + z\Delta_f Y^{\ominus}[\mathrm{H}^+] \quad , \tag{4.436}$$

$$y^{\ominus}[\Sigma^z] = y^{\bullet}[\Sigma^z] + z y_H^{\ominus} \quad , \tag{4.437}$$

and

$$\begin{aligned} \Delta_r Y^{\ominus}[\Sigma^z \to \Sigma^{z'} + (z'-z)e_g^-] &= \Delta_r Y^{\bullet}[\Sigma^z \to \Sigma^{z'} + (z'-z)e_g^-] \\ &\quad + \Delta_r Y^{\ominus}[\mathrm{H} \to \mathrm{H}^+ + e_g^-] \quad , \end{aligned} \tag{4.438}$$

where it is recalled that $\mathcal{Y}_H^{\ominus}$ and $y_H^{\ominus}$ are short notations for $\Delta_s Y^{\ominus}[\mathrm{H}_g^+]$ and $y^{\ominus}[\mathrm{H}_d^+]$, respectively (Section 2.3). For derivative thermodynamic properties, these equations are written in the standard variant, but intrinsic parameters are also affected by the ambiguity related to standard *vs.* density-corrected variants (Sections 4.2.4 and 4.2.5). Analogous equations also hold in the density-corrected variant.

The main advantage of the intrinsic (absolute) scale in contrast to conventional (Section 4.5.1) and real (Section 4.5.2) scales is that the corresponding values possess a direct physical significance by solely characterizing bulk solvation (rather than intermingling bulk and surface effects as implied by the real scale). However the intrinsic scale relies on the specification of an experimentally-elusive property for each given solvent (Section 2.3) and thus lacks the unambiguous character inherent to the real and conventional scales. Since the exact value of $\mathcal{G}_H^{\ominus}$ (or in general $\Delta_s Y^{\ominus}[\mathrm{H}_g^+]$, related to the dependence of $\mathcal{G}_H^{\ominus}$ on P and T) remains controversial, *i.e.* not known with any accuracy, the intrinsic scale is afflicted by a propagation of this uncertainty into tabulated single-ion parameters. Hence, the conventional scale is definitely to be preferred for the tabulation of relative single-ion data, in view of the extremely large uncertainty (related to the choice of an extra-thermodynamic assumption) on the value of $\mathcal{G}_H^{\ominus}$.

4.5.4 Conversion formulae

A synoptic summary of the connection between the key quantities related to single-ion solvation is presented in Figure 4.18, considering the simple situation of an ion pair IJ consisting of ions I^z and J^{-z} and the free energy as a basic quantity. The generalization to arbitrary overall neutral sets of ions is straightforward, and the generalization to other derivative thermodynamic properties is briefly mentioned at the end of this section. The definition and measurement of the pair solvation free energy $\Delta_s G_{svt}^{\ominus}[IJ_g]$ has been discussed in Sections 2.1, 4.2.14, 4.2.18 and 4.5.1 and will be further detailed, in the context of alkali-halide hydration, in Section 5.12. There is no fundamental ambiguity and limited experimental uncertainty concerning this quantity. As explained in Sections 2.2, 4.2.14 and 4.5.1-4.5.3, this parameter can be partitioned into single-ion values in three different ways. The conventional single-ion solvation free energy $\Delta_s G^{\bullet}_{svt}[I^z_g]$ is a relative value, with the proton as (arbitrary) reference. There is also no fundamental ambiguity and limited experimental uncertainty concerning this quantity. However, the corresponding numerical value is not absolute and provides little direct insight into the nature and magnitude of ion-solvent interactions for the ion I^z, because of its comparative nature with the proton as a reference. The real single-ion solvation free energy $\Delta_s \hat{G}^{\ominus}_{svt}[I^z_g]$ is an absolute value, which however, simultaneously characterizes surface effects and bulk solvation. There is no fundamental ambiguity concerning this quantity, but its measurement is relatively difficult and the resulting estimate is typically affected by a larger uncertainty. Here also, the corresponding numerical value, although formally absolute, provides little direct insight into the nature and magnitude of ion-solvent interactions for the ion I^z, because of the inclusion of surface effects. Finally, the intrinsic single-ion solvation free energy $\Delta_s G^{\ominus}_{svt}[I^z_g]$ is also an absolute value, and characterizes only bulk solvation effects. This quantity is elusive, *i.e.* affected by

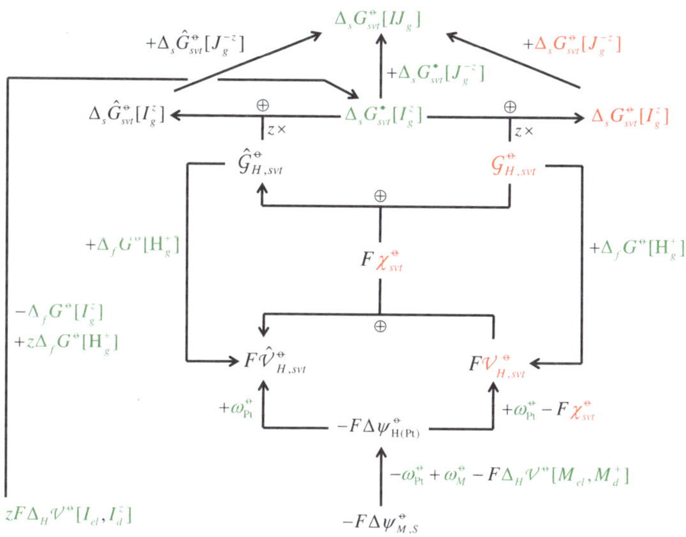

Figure 4.18 *Summary of the relationships connecting key quantities related to single-ion solvation.* Quantities indicated in green are experimentally accessible with a limited uncertainty. Quantities indicated in black (real quantities) are experimentally accessible, albeit with larger uncertainties. Quantities indicated in red (intrinsic quantities) are experimentally inaccessible, and the requirement for an extra-thermodynamic assumption leads to very large uncertainties. The symbols and notations can be found in Table I. The quantities $\Delta\psi^{\ominus}_{\mathrm{H(Pt)}}$ and $\Delta\psi^{\ominus}_{M,S}$ stand for the standard metal-solution Volta potential difference of the reference hydrogen electrode (assuming a platinum electrode Pt) or of an arbitrary electrode (metal M, solution S containing the (monovalent) ions M^+ of the metal), while $\omega^{\ominus}_{\mathrm{Pt}}$ and $\omega^{\ominus}_{M}$ are the standard work functions of the selected metals.

a very large uncertainty, because it depends on some extra-thermodynamic assumption for its evaluation, *i.e.* some intuitive model-based expectation on the physics of single-ion solvation or of surface effects (Section 2.3). On the other hand, it is the only value that provides direct quantitative insight into the nature and magnitude of ion-solvent interactions for ion I^z. The three forms of single-ion solvation free energies can be interconverted *via* Eq. 4.415 based on the knowledge of the real and intrinsic solvation free energies of the proton, $\hat{\mathcal{G}}^{\ominus}_{H,svt}$ and $\mathcal{G}^{\ominus}_{H,svt}$, respectively.

What has been said above concerning the uncertainty and possible ambiguity affecting $\Delta_s\hat{G}^{\ominus}_{svt}[I^z_g]$ and $\Delta_s G^{\ominus}_{svt}[I^z_g]$ also applies to the specific cases of $\hat{\mathcal{G}}^{\ominus}_{H,svt}$ and $\mathcal{G}^{\ominus}_{H,svt}$, respectively. Obviously, $\Delta_s G^{\bullet}_{svt}[\mathrm{H}^+_g]$ is uninteresting since it is set to zero by definition. The real and intrinsic forms of proton solvation free energies can be interconverted *via* Eq. 4.434 based on the knowledge of the surface potential $\chi^{\ominus}_{svt}$ of the pure solvent, a quantity that is obviously also elusive and affected by a very large uncertainty.

The quantity $\hat{\mathcal{G}}^{\ominus}_{H,svt}$ is directly related to the real absolute potential $\hat{\mathcal{V}}^{\ominus}_{H,svt}$ of the reference hydrogen electrode in the given solvent *via* Eq. 4.427 based on the knowledge of the formation free energy $\Delta_f G^{\ominus}[\mathrm{H}^+_g]$ of the gas-phase proton (readily accessible, Section 4.2.10). The quantity $\mathcal{G}^{\ominus}_{H,svt}$ is directly related to the corresponding intrinsic (absolute) potential $\mathcal{V}^{\ominus}_{H,svt}$ *via* Eq. 4.433, also based on $\Delta_f G^{\ominus}[\mathrm{H}^+_g]$.

What has been said above concerning the uncertainty and possible ambiguity affecting $\Delta_s\hat{G}^{\ominus}_{svt}[I^z_g]$ and $\Delta_s G^{\ominus}_{svt}[I^z_g]$, or $\hat{\mathcal{G}}^{\ominus}_{H,svt}$ and $\mathcal{G}^{\ominus}_{H,svt}$, obviously also applies to $\hat{\mathcal{V}}^{\ominus}_{H,svt}$ and $\mathcal{V}^{\ominus}_{H,svt}$, respectively. The two forms of absolute potentials can be interconverted *via* Eq. 4.297, based on the value of $\chi^{\ominus}_{svt}$.

The quantities $\hat{\mathcal{V}}^{\ominus}_{H,svt}$ and $\mathcal{V}^{\ominus}_{H,svt}$ are connected to the absolute metal-solution Volta potential difference $\Delta\psi^{\ominus}_{\mathrm{H(Pt)}}$ of the reference hydrogen electrode *via* Eqs. 4.313 and 4.315, respectively. In both cases, the connection involves the standard work function $\omega^{\ominus}_{\mathrm{Pt}}$ of the platinum (Pt) electrode (readily accessible, Sections 4.3.5 and 5.13). In the latter case, knowledge of $\chi^{\ominus}_{svt}$ is also required. The quantity $\Delta\psi^{\ominus}_{\mathrm{H(Pt)}}$ can in principle be determined using Voltaic cell measurements analogous to those described in Section 4.4.4, but considering the reference hydrogen electrode. In practice, however, these measurements target the quantity $\Delta\psi^{\ominus}_{M,S}$ relevant for other types of electrodes. The value $\Delta\psi^{\ominus}_{\mathrm{H(Pt)}}$ can then be inferred *via* Eq. 4.403 based on the knowledge of $\omega^{\ominus}_{M}$, $\omega^{\ominus}_{\mathrm{Pt}}$ and $\Delta_H\mathcal{V}^{\ominus}_{M}$ (the latter also readily accessible, Sections 4.3.5 and 5.5).

Note, finally, the presence of a direct connection between the redox potential $\Delta_H\mathcal{V}^{\ominus}[I_{el}, I^z_d]$ relative to the reference hydrogen electrode (Section 4.4.3) and the conventional single-ion solvation free energy $\Delta_s G^{\bullet}_{svt}[I^z_g]$, according to Eq. 4.422. Ultimately, the relatively high experimental uncertainties in the series of real quantities ($\hat{\mathcal{V}}^{\ominus}_{H,svt}$, $\hat{\mathcal{G}}^{\ominus}_{H,svt}$, $\Delta_s\hat{G}^{\ominus}_{svt}[I^z_g]$) originates from the relative complexity of measuring $\Delta\psi^{\ominus}_{M,S}$ values accurately in Voltaic cell experiments (and related approaches). Similarly, the elusive nature in the series of intrinsic quantities ($\mathcal{V}^{\ominus}_{H,svt}$, $\mathcal{G}^{\ominus}_{H,svt}$, $\Delta_s G^{\ominus}_{svt}[I^z_g]$) can be viewed as resulting from the fact that they all encompass the elusive quantity $\chi^{\ominus}_{svt}$. Finally, the other quantities involved in Figure 4.18 ($\omega^{\ominus}_M$, $\omega^{\ominus}_{\mathrm{Pt}}$, $\Delta_f G^{\ominus}[M^+_g]$, $\Delta_f G^{\ominus}[\mathrm{H}^+_g]$, $\Delta_s G^{\bullet}_{svt}[I^z_g]$, $\Delta_H \mathcal{V}^{\ominus}[I_{el}, I^z_d]$ and $\Delta_s G^{\ominus}_{svt}[IJ_g]$) are not elusive and can be determined accurately by independent methods (*i.e.* not relying on Voltaic cell experiments).

Figure 4.18 is easily generalized to derivative thermodynamic quantities by application of Eq. 4.185 (standard variant) or Eq. 4.186 (density-corrected variant). This also results in the appearance of potential and work function derivatives of the form $\hat{O}_Y \Delta_H \mathcal{V}^{\ominus}$, $\hat{O}_Y \chi^{\ominus}_{svt}$, $\hat{O}_Y \hat{\mathcal{V}}^{\ominus}_{H,svt}$, $\hat{O}_Y \mathcal{V}^{\ominus}_{H,svt}$, $\hat{O}_Y \omega^{\ominus}_M$, $\hat{O}_Y \omega^{\ominus}_{\mathrm{Pt}}$, $\hat{O}_Y \Delta\psi^{\ominus}_{\mathrm{H(Pt)}}$ and $\hat{O}_Y \Delta\psi^{\ominus}_{M,S}$. Note that the distinction between standard *vs.* density-corrected variants in derivative thermodynamic properties (Section 4.2.4) pertains to conventional real and intrinsic values, except for conventional values of monovalent cations. This scheme is also easily further generalized to formation parameters, by subtraction of the corresponding diagram for elements in their standard state, or to partial molar variables in solution, by addition of the corresponding diagram for gas-phase ions.

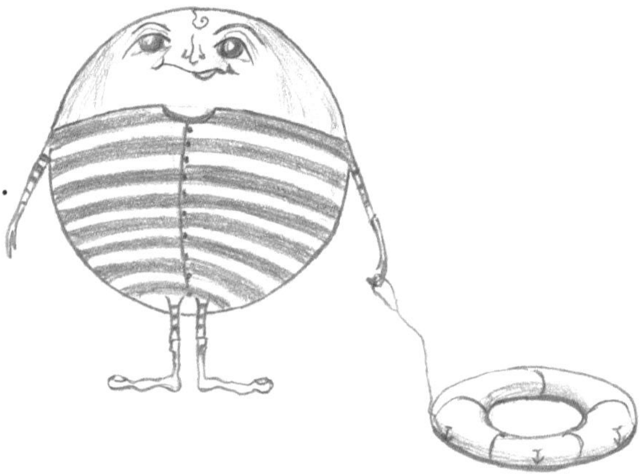

After a careful study of all the required concepts and definitions, a single ion feels the urge to bring its knowledge into practice and to go for a swim in Chapter 5. Considering that it has not yet experienced real solvation up to this point in the book, it carries along a lifebelt for safety.

Chapter 5
Experimental determination

In this chapter, the approaches employed for the *experimental determination* of thermodynamic (and some structural) parameters related to ionic solvation are reviewed, and the results of investigations employing these methods are discussed in the restricted context of *alkali and halide hydration.* Closely related properties formally pertaining to the pure solvent (water), including the absolute potential of the reference hydrogen electrode in a solvent, the air-liquid interfacial potential of this solvent, and the absolute solvation parameters of the proton in this solvent, are also considered in view of their central role in single-ion solvation thermodynamics.

The key problems and relevant concepts have been presented in Chapters 2 and 4, respectively. Although the description of the experimental methods is in principle general, it is still somewhat biased towards the investigation of the alkali and halide ion series with water as a solvent. In particular, most of the discussion explicitly refers to the case of a neutral ion pair I^zJ^{-z}, *i.e.* a binary combination of isovalent ions or, more specifically, to an alkali-halide ion pair M^+X , rather than to an arbitrary neutral set of ions, which would require a more general and cumbersome notation (Section 4.1). The reported results, on the other hand, exclusively concern alkali and halide hydration, along with relevant pure-solvent properties (see above). For experimental data concerning the hydration thermodynamics of *other ions*, possibly polyvalent or/and polyatomic, the reader is referred to *e.g.* Refs.[273,275,1877]. For experimental data concerning the solvation thermodynamics of alkali, halide and other ions in *non-aqueous solvents*, the reader is referred to *e.g.* Refs.[265,267,285,413,473,478,816,1878–1891]. Finally, information concerning the *transfer properties* of ions across different solvents can be found in *e.g.* Refs.[274,1887,1892–2019]. See also Table 2.1 for estimates of the air-liquid interfacial potential of various non-aqueous solvents.

The present description of experimental methods and results proceeds

RSC Theoretical and Computational Chemistry Series No. 3
Single-Ion Solvation
Experimental and Theoretical Approaches to Elusive Thermodynamic Quantities
By Philippe Hünenberger and Maria Reif

Published by the Royal Society of Chemistry, www.rsc.org

roughly incrementally, from the most directly accessible quantities to the more elusive ones, namely: (*i*) molar thermodynamic parameters of the elements; (*ii*) structural and molar thermodynamic parameters of the salts (ionic crystals); (*iii*) gas-phase equilibrium ion-pair distances; (*iv*) effective ionic radii; (*v*) relative electrode (redox) potentials (anchored based on the reference hydrogen electrode); (*vi*) thermodynamic parameters of salt formation; (*vii*) thermodynamic parameters of dissolved salt formation; (*viii*) thermodynamic parameters of salt dissolution; (*ix*) thermodynamic parameters of atomization; (*x*) thermodynamic parameters of ionization; (*xi*) thermodynamic parameters of reticulation; (*xii*) thermodynamic parameters of salt solvation; (*xiii*) metal work functions; (*xiv*) real absolute potential of the reference hydrogen electrode; (*xv*) real single-ion solvation parameters; (*xvi*) conventional single-ion solvation parameters; (*xvii*) air-liquid interfacial potential of the pure solvent; (*xviii*) intrinsic proton solvation parameters. Note that although *ionic radii* are actually very ambiguous quantities, they are introduced relatively early in the text, in view of their relevance for the discussion of many other properties.

Three additional sections at the rear of the chapter provide: (*i*) a summary of *recommended data* concerning alkali and halide hydration, along with relevant pure-solvent properties; (*ii*) a discussion of the *suggested intrinsic single-ion solvation parameters* for the alkali and halide ions, in connection with the *Hofmeister series*; (*iii*) a brief discussion of the properties of the *solvated electron*.

5.1 Molar thermodynamic parameters of the elements

The *standard molar thermodynamic parameters of the elements*, including the molar entropies $s^{\ominus}[I_{el}]$, heat capacities $c_P^{\ominus}[I_{el}]$, volumes $v^{\ominus}[I_{el}]$, volume-compressibilities $k_T^{\ominus}[I_{el}]$ and volume-expansivities $a_P^{\ominus}[I_{el}]$, are accessible experimentally (Section 4.2.18). Recall that the corresponding values refer to *one mole of the element (atom)*, irrespective of the compound (*e.g.* diatomic molecule) in which this element is found in its standard state (Section 4.2.4). The determination of these parameters involves calorimetric ($c_P^{\ominus}$ and $s^{\ominus}$ for solids and liquids, the latter property *via* TCI), spectroscopic ($c_P^{\ominus}$ and $s^{\ominus}$ for gases, *via* SBS), densimetric ($v^{\ominus}$ for solids and liquids), and compressimetric ($k_T^{\ominus}$ for solids and liquids) or dilatometric ($a_P^{\ominus}$ for solids and liquids) measurements. For gases, which are ideal in their standard state, one has $v^{\ominus} = (n_a P^{\circ})^{-1} R T^{-}$, $k_T^{\ominus} = (P^{\circ})^{-1} v^{\ominus}$ and $a_P^{\ominus} = (T^{-})^{-1} v^{\ominus}$, where n_a is the number of atoms of the element in the molecule of the standard-state compound. The *standard formation parameters of the elements*, $\Delta_f Y^{\ominus}[I_{el}]$ are *zero by definition* (Section 4.2.10). The same applies to the ideal electron gas at temperature T^{-}, within the warm-electron convention employed in this book (Sections 4.2.12, 4.2.13 and 4.2.17), which can be viewed as an additional kind of "element" (Section 4.2.4). The values of the above parameters have been determined[221, 2020, 2021, 2023, 2024] for all alkali (M_{el}) and halide (X_{el})

Table 5.1 *Standard molar thermodynamic parameters of the alkali and halide elements as well as hydrogen.* The reported data includes the molar entropy $s^\ominus$, the molar heat capacity $c_P^\ominus$, the molar volume $v^\ominus$, the molar volume-compressibility $k_T^\ominus$, the molar volume-expansivity $a_P^\ominus$, the compressibility $\kappa_T^\ominus$, the expansivity (thermal expansion coefficient) $\alpha_P^\ominus$, and the molar (atomic) mass M. The notation I_{el} refers to an alkali (M_{el}) or halide (X_{el}) element, or to hydrogen (H_{el}). The standard states are according to the $bbme_T$ convention (Section 4.2.4; reference pressure $P^\circ = 1$ bar and reference temperature $T^- = 298.15$ K). Note that the standard state for the element I_{el} refers to one mole of the element (atom), irrespective of the compound (*e.g.* diatomic molecule) in which this element is found in its standard state (Section 4.2.4). To make this clear, the entries corresponding to diatomic molecules are reported in the form of values appropriate for one mole of the diatomic molecule multiplied by a prefactor one-half (except for the molar mass, which is reported directly on a per-atom basis).

Property	Li_s	Na_s	K_s	Rb_s	Cs_s	Ref.	Notes
$s^\ominus[I_{el}]$ [$J{\cdot}mol^{-1}{\cdot}K^{-1}$]	29.12	51.30	64.68	76.78	85.23	221	
$c_P^\ominus[I_{el}]$ [$J{\cdot}mol^{-1}{\cdot}K^{-1}$]	24.8	28.2	29.6	31.1	32.2	221	
$v^\ominus[I_{el}]$ [10^{-6} $m^3{\cdot}mol^{-1}$]	13.00	23.70	43.93	55.86	70.96	(221)	a
$k_T^\ominus[I_{el}]$ [10^{-10} $m^3{\cdot}mol^{-1}{\cdot}bar^{-1}$]	1.10	3.41	13.6	17.4	34.3		b
$a_P^\ominus[I_{el}]$ [10^{-8} $m^3{\cdot}mol^{-1}{\cdot}K^{-1}$]	0.179	0.505	1.10	1.51	2.06		c
$\kappa_T^\ominus[I_{el}]$ [10^{-5} bar^{-1}]	0.848	1.44	3.09	3.12	4.83	2020	d
$\alpha_P^\ominus[I_{el}]$ [10^{-6} K^{-1}]	138	213	250	270	291	221,2021	e, f
M [10^{-3} $kg{\cdot}mol^{-1}$]	6.941	22.989768	39.0983	85.4678	132.90543	2022	j
Property	**$1/2H_{2,g}$**	**$1/2F_{2,g}$**	**$1/2Cl_{2,g}$**	**$1/2Br_{2,\ell}$**	**$1/2I_{2,s}$ (rhombic)**	**Ref.**	**Notes**
$s^\ominus[I_{el}]$ [$J{\cdot}mol^{-1}{\cdot}K^{-1}$]	$1/2{\cdot}130.680$	$1/2{\cdot}202.791$	$1/2{\cdot}223.081$	$1/2{\cdot}152.21$	$1/2{\cdot}116.14$	221	
$c_P^\ominus[I_{el}]$ [$J{\cdot}mol^{-1}{\cdot}K^{-1}$]	$1/2{\cdot}28.8$	$1/2{\cdot}31.3$	$1/2{\cdot}33.9$	$1/2{\cdot}75.7$	$1/2{\cdot}54.4$	221	
$v^\ominus[I_{el}]$ [10^{-3} $m^3{\cdot}mol^{-1}$]	$1/2{\cdot}24.790$	$1/2{\cdot}24.790$	$1/2{\cdot}24.790$	$1/2{\cdot}0.05150$	$1/2{\cdot}0.05145$	(221)	a
$k_T^\ominus[I_{el}]$ [10^{-2} $m^3{\cdot}mol^{-1}{\cdot}bar^{-1}$]	$1/2{\cdot}2.48$	$1/2{\cdot}2.48$	$1/2{\cdot}2.48$	$1/2{\cdot}2.72 \cdot 10^{-7}$	$1/2{\cdot}6.68 \cdot 10^{-8}$		b
$a_P^\ominus[I_{el}]$ [10^{-5} $m^3{\cdot}mol^{-1}{\cdot}K^{-1}$]	$1/2{\cdot}8.31$	$1/2{\cdot}8.31$	$1/2{\cdot}8.31$	$1/2{\cdot}5.76 \cdot 10^{-3}$	$1/2{\cdot}12.86 \cdot 10^{-4}$		c
$\kappa_T^\ominus[I_{el}]$ [bar^{-1}]	1.00	1.00	1.00	$53 \cdot 10^{-6}$	$13 \cdot 10^{-6}$	(2023)	g
$\alpha_P^\ominus[I_{el}]$ [10^{-3} K^{-1}]	3.35	3.35	3.35	1.12^i	0.25	(2023,2024)	h
M [10^{-3} $kg{\cdot}mol^{-1}$]	1.00794	18.9984032	35.4527	79.904	126.90447	2022	j

(a) calculated as $v^\ominus = RT^-(2P^\circ)^{-1}$ for the diatomic gaseous species or as $v^\ominus = M/\rho^\ominus$ for the liquid and solid species, where $\rho^\ominus$ is the density of the standard-state compound at P° and T^- (taken from the indicated reference); (b) calculated as $k_T^\ominus = v^\ominus \kappa_T^\ominus$; (c) calculated as $a_P^\ominus = v^\ominus \alpha_P^\ominus$; (d) values quoted in the indicated reference as corresponding to a low-pressure limit; (e) calculated from the linear isobaric thermal expansion coefficient reported in the indicated reference (multiplication by a factor three); (f) all values are from Ref.[221] except for Rb_s from Ref.[2021]; (g) calculated as $\kappa_T^\ominus = (P^\circ)^{-1}$, except for $Br_{2,\ell}$ and $I_{2,s}$ (taken from the indicated reference); (h) calculated as $\alpha_P^\ominus = (T^-)^{-1}$, except for $Br_{2,\ell}$ and $I_{2,s}$ (taken from Ref.[2023] and Ref.[2024], respectively); (i) value reported for a temperature of 20°C; (j) the values are reported for the elements (atoms, averaged over the natural isotopic distribution) and the same values are used in this book for the ions, resulting in negligible errors considering the small molar mass of the electron ($5.48580 \cdot 10^{-7}$ $kg{\cdot}mol^{-1}$; Table 4.2).

elements. These standard molar thermodynamic parameters, as well as those of hydrogen (H_{el}), are reported in Table 5.1.

For the monoatomic solids (alkali), $c_P^{\ominus}$ is close to $3R = 24.9$ J·mol^{-1}·K^{-1}, the value expected for classical three-dimensional harmonic oscillations of the atoms about lattice positions according to the law of Dulong and Petit[2025] (see Ref.[2026] for historical details). The value for the only diatomic solid (iodine) is also close to $3R$, suggesting that the intramolecular vibration is thermally active and essentially harmonic in $I_{2,s}$ at room temperature. For the diatomic gases (hydrogen, fluorine and chlorine), $c_P^{\ominus}$ can be compared to $(7/4)R = 14.6$ J·mol^{-1}·K^{-1}, the value expected on a per-atom basis for a classical linear and rigid three-dimensional rotor[2027], taking into account the pressure-volume contribution R. Positive deviations for F_{el} and Cl_{el} arise from the partial thermal activation of the intramolecular vibration at room temperature, with a limiting value of $(9/4)R = 18.7$ J·mol^{-1}·K^{-1} for full activation of a harmonic vibration. For the only liquid (bromine), $c_P^{\ominus}$ is significantly larger than for all other elements, due to strong and anharmonic intermolecular interactions.

Expectedly, the molar volume $v^{\ominus}$ is much lower for the solids and liquid compared to the gases, the opposite trend being observed for the molar entropy $s^{\ominus}$ (except for hydrogen). The molar volume-compressibilities and volume-expansivities of the gases are also about 4-8 orders of magnitude larger than those of the solids and liquid. Recall that the molar volume-compressibility $k_T^{\ominus}$ and the molar volume-expansivity $a_P^{\ominus}$ are related to the corresponding compressibility $\kappa_T^{\ominus}$ and expansivity (thermal expansion coefficient) $\alpha_P^{\ominus}$, the two latter quantities being more commonly reported for pure substances, as $k_T^{\ominus} = v^{\ominus}\kappa_T^{\ominus}$ and $a_P^{\ominus} = v^{\ominus}\alpha_P^{\ominus}$ (Sections 4.2.1 and 4.2.2). The corresponding standard values $\kappa_T^{\ominus}[I_{el}]$ and $\alpha_P^{\ominus}[I_{el}]$ are also reported in Table 5.1 for completeness, as well as the molar masses $M[I_{el}]$ of the elements (atoms, averaged over the natural isotopic distribution) for the ease of further reference. Note that the same values are used in this book for the ions, resulting in negligible errors considering the small molar mass of the electron ($5.48580 \cdot 10^{-7}$ kg·mol^{-1}; Table 4.2).

5.2 Structural and molar thermodynamic parameters of the salts

Relevant *standard parameters* characterizing the *structure and thermodynamics of a salt* (ionic crystal) include: (*i*) the lattice type (at P° and T^{-}); (*ii*) the lattice parameters (at P° and T^{-}), *e.g.* the unit-cell edge length $L^{\ominus}[IJ_s]$ for cubic lattices; (*iii*) the molar heat capacity $c_P^{\ominus}[IJ_s]$; (*iv*) the molar volume $v^{\ominus}[IJ_s]$; (*v*) the molar volume-compressibility $k_T^{\ominus}[IJ_s]$; (*vi*) the molar volume-expansivity $a_P^{\ominus}[IJ_s]$; (*vii*) the melting temperature $T_m^{\circ}[IJ_s]$; (*viii*) the melting enthalpy $\Delta_m H^{\circ}[IJ_s]$ at $T_m^{\circ}[IJ_s]$. The lattice type, along with the values of $L^{\ominus}$ for a cubic lattice and of $v^{\ominus}$, related to $L^{\ominus}$ *via* the number of atoms per unit cell (see Eq. 5.1 below), can be obtained from crystallographic x-ray diffraction experiments[68]. Note that $v^{\ominus}$ can also be accessed through straightforward densimetric measurements. The values of all other parameters are measured using various calorimetric ($c_P^{\ominus}$, T_m°, $\Delta_m H^{\circ}$), compressimetric ($k_T^{\ominus}$), dilatometric ($a_P^{\ominus}$), piston-cylinder ($k_T^{\ominus}$), acoustic (adiabatic compressibility) and x-ray diffraction ($a_P^{\ominus}$, $k_T^{\ominus}$) experiments[68, 221, 2028]. The standard formation parameters and molar entropies of the salts are of course also relevant, but will be discussed in Section 5.6. The values of the above structural and molar thermodynamic parameters have been determined[68, 221, 2029–2032] for all alkali-halide salts (MX_s). These standard parameters are reported in Table 5.2.

All alkali-halide salts crystallize in a *cubic crystal lattice*, namely face-centered cubic (fcc) or body-centered cubic (bcc). In the fcc case, the positions of the M^+ and X^- ions both define intertwined fcc lattices. Taking M^+ as the central ion, the fcc unit cell encompasses contributions from 1 M^+ ion at the cube center (entirely within the cell), 6 X^- ions at the centers of the cube faces (shared between 2 cells), 12 M^+ ions at the midpoints of the cube edges (shared between 4 cells), and 8 X^- ions at the cube corners (shared between 8 cells). As a result, the number n_L of ion pairs MX per unit cell and the ratio f_L of the closest anion-cation distance to the lattice parameter are $n_L = 4$ and $f_L = 1/2$ for the fcc lattice. The corresponding space group, point group and coordination number are Fm3m, m3m (O_h) and 6. In the bcc case, the positions of the M^+ and X^- ions define together a bcc lattice. Taking M^+ as the central ion, the bcc unit cell encompasses contributions from 1 M^+ ion at the cube center (entirely within the cell) and 8 X^- ions at the cube corners (shared between 8 cells). As a result, the values of n_L and f_L are $n_L = 1$ and $f_L = 3^{1/2}/2$ for the bcc lattice. The corresponding space group, point group and coordination number are Pm3m, m3m (O_h) and 8. For cubic unit cells, the relationship between molar volume and unit-cell edge length reads

$$v^{\ominus} = N_A n_L^{-1} (L^{\ominus})^3 \quad . \tag{5.1}$$

The $c_P^{\ominus}$ values are close to $6R = 49.9$ J·mol^{-1}·K^{-1}, the value expected for classical three-dimensional harmonic oscillations of the two ions about their lattice positions according to a law analogous to that of Dulong and Petit[2025] (Section 5.1). Expectedly, the values of $f_L L^{\ominus}$ (closest anion-cation distance) systematically increase with the effective sizes (atomic numbers) of the involved ions, *i.e.* along the series $Li^+ \rightarrow Cs^+$ for salts involving a common anion and $F^- \rightarrow I^-$ for salts involving a common cation. The same trend holds for $v^{\ominus}$, except for three inversions concerning the chloride, bromide and iodide salts of rubidium and cesium. In this case, $v^{\ominus}$ decreases from Rb^+ to Cs^+ in spite of the increase in effective cation size, due to the lattice change from fcc to bcc. A similar trend, although not entirely systematic, is observed in terms of the molar volume-compressibilities $k_T^{\ominus}$ and volume-expansivities $a_P^{\ominus}$. Recall that the molar volume-compressibility $k_T^{\ominus}$ and the molar volume-expansivity $a_P^{\ominus}$ are related to the corresponding compressibility $\kappa_T^{\ominus}$ and expansivity (thermal expansion coefficient) $\alpha_P^{\ominus}$, the two latter quan-

Table 5.2 *Standard structural and molar thermodynamic parameters of the alkali-halide salts.* The reported data includes the (cubic) crystal lattice type (fcc: face-centered cubic; bcc: body-centered cubic) and unit-cell edge length $L^\ominus$, the molar heat capacity $c_P^\ominus$, the molar volume $v^\ominus$, the molar volume-compressibility $k_T^\ominus$, the molar volume-expansivity $a_P^\ominus$, the compressibility $\kappa_T^\ominus$, the expansivity (thermal expansion coefficient) $\alpha_P^\ominus$, the melting temperature T_m° and the melting enthalpy $\Delta_m H^\circ$ at T_m°. The notation MX_s refers to an alkali-halide salt. Standard formation parameters and molar entropies of the alkali-halide salts can also be found in Table 5.6. The standard states are according to the $bbme_T$ convention (Section 4.2.4; reference pressure $P^\circ = 1$ bar and reference temperature $T^- = 298.15$ K).

Salt MX_s	Lattice type	$L^\ominus[MX_s]$ [nm]	$c_P^\ominus[MX_s]$ [J·mol^{-1}·K^{-1}]	$v^\ominus[MX_s]$ [10^{-6} m^3·mol^{-1}]	$k_T^\ominus[MX_s]$ [10^{-10} m^3·mol^{-1}·bar^{-1}]	$a_P^\ominus[MX_s]$ [10^{-8} m^3·mol^{-1}·K^{-1}]	$\kappa_T^\ominus[MX_s]$ [10^{-6} bar^{-1}]	$\alpha_P^\ominus[MX_s]$ [10^{-4} K^{-1}]	$T_m^\circ[MX_s]$ [K]	$\Delta_m H^\circ[MX_s]$ [kJ·mol^{-1}]
LiF	fcc	0.402620	41.6	9.83	0.153	0.0973	1.56 (1.62)	0.99^b	1118	26.4
LiCl	fcc	0.513988	48.0	20.44	0.734	0.270	3.59 (3.71)	1.32^b	878	19.9
LiBr	fcc	0.5501	48.1	25.06	1.14	0.368	4.55 (4.67)	1.47^b	823	17.7
LiI	fcc	0.6012	51.0	32.71	2.12	0.579	6.47 (6.69)	1.77	722	14.6
NaF	fcc	0.46344^a	46.9	14.99	0.340	0.148	2.27 (2.30)	0.99^b	1266	33.6
NaCl	fcc	0.564009	50.5	27.01	1.19	0.316	4.41 (4.53)	1.17^b	1074	28.0
NaBr	fcc	0.597229	51.4	32.07	1.72	0.433	5.37 (5.53)	1.35^b	1020	26.1
NaI	fcc	0.64728	52.1	40.83	2.89	0.551	7.07 (7.30)	1.35^c	934	23.6
KF	fcc	0.5344	49.0	22.98	0.788	0.221	3.43 (3.51)	0.96^b	1131	28.2
KCl	fcc	0.629294	51.3	37.52	2.26	0.394	6.02 (6.20)	1.05^b	1043	26.5
KBr	fcc	0.65982	52.3	43.25	3.06	0.480	7.08 (7.32)	1.11^b	1007	25.5
KI	fcc	0.706555	52.9	53.10	4.77	0.621	8.99 (9.33)	1.17^b	954	24.0
RbF	fcc	0.56516	50.7	27.18	1.08	0.277	3.98 (4.12)	1.02	1068	25.7
RbCl	fcc	0.65810^b	52.4	42.91	2.91	0.515	6.77 (7.00)	1.20^b	991	23.7
RbBr	fcc	0.6889	52.8	49.22	3.97	0.546	8.06 (8.34)	1.11^b	966	23.3
RbI	fcc	0.73466^b	53.2	59.70	5.99	0.698	10.03 (10.37)	1.17^b	920	22.0
CsF	fcc	0.6014	51.1	32.75	1.46	0.334	4.45 (-)	1.02^c	955	21.7
CsCl	bcc	0.41230	52.5	42.21	2.57	0.595	6.08 (6.19)	1.41^b	918	20.3
CsBr	bcc	0.42953	52.9	47.72	3.24	0.659	6.79 (6.99)	1.38^b	909	23.6
CsI	bcc	0.45679^a	52.8	57.40	4.86	0.792	8.46 (8.67)	1.38^b	899	23.6
Ref.	68	68	221, 2029, 2030				68	68	68	68, 2031, 2032
Notes			d	e	f	g	h	i	j	k

(a) measured at 26°C; (b) measured at 27°C; (c) measured at 10°C; (d) values from Ref.[221], except LiBr from Ref.[2029] and RbF from Ref.[2030] (the value for RbF is difficult to measure accurately in view of the high hygroscopicity of this salt; the estimate of 50.7 J·mol^{-1}·K^{-1} reported in Ref.[2030] was retained here rather than the estimate of 43.8 J·mol^{-1}·K^{-1} reported in Ref.[2029] upon consideration of the trends within the alkali-halide series); (e) per mole ion pair MX, calculated as $v^\ominus = N_A n_L^{-1}(L^\ominus)^3$ where the number n_L of pairs per unit cell is 4 for fcc and 1 for bcc; (f) calculated as $k_T^\ominus = v^\ominus \kappa_T^\ominus$; (g) calculated as $a_P^\ominus = v^\ominus \alpha_P^\ominus$; (h) calculated based on equations and data from the indicated reference, as $\kappa_T^\ominus = (1 + T^- c_{GR}^\ominus \alpha_P^\ominus)/c_{AB}^\ominus$, where $c_{GR}^\ominus$ is the Grüneisen parameter (from thermal expansion measurements or, between parentheses, from elastic properties measurements) and $c_{AB}^\ominus$ the adiabatic bulk modulus; (i) calculated from the linear isobaric thermal expansion coefficients reported in the indicated reference (multiplication by a factor three; source data from temperature-dependent x-ray diffractometry and high-precision differential dilatometry measurements); (j) converted from original data reported in the indicated reference in degrees Celsius by adding 273 K (rather than 273.15 K); (k) values from Ref.[2032], except LiF and NaF from Ref.[68], and KF from Ref.[2031].

Table 5.3 *Equilibrium distances of the alkali-halide gas-phase ion-pair complexes.* The reported value is the equilibrium distance d_g referring to a hypothetical isolated and vibrationless pair at $T = 0$ K. The notation $[M^+ \cdot X^-]_g$ refers to an alkali-halide gas-phase ion-pair complex. The values are taken from the data compilation in Ref.[298].

	$d_g[[M^+ \cdot X^-]_g]$ [nm]				
	Li^+	Na^+	K^+	Rb^+	Cs^+
F^-	0.15639	0.19259	0.21715	0.22703	0.23454
Cl^-	0.20207	0.23608	0.26667	0.27867	0.29063
Br^-	0.21704	0.25020	0.28208	0.29447	0.30723
I^-	0.23919	0.27115	0.30478	0.31769	0.33152

tities being more commonly reported for pure substances, as $k_T^\ominus = v^\ominus \kappa_T^\ominus$ and $a_P^\ominus = v^\ominus \alpha_P^\ominus$ (Sections 4.2.1 and 4.2.2). The corresponding standard values $\kappa_T^\ominus[MX_s]$ and $\alpha_P^\ominus[MX_s]$ are also reported in Table 5.2 for completeness. Finally, the melting temperatures T_m° and enthalpies ΔH_m° are correlated, and tend to decrease when the effective sizes of the involved ions increase, as expected considering the dependence of the Coulombic interactions within the crystal, approximating the ions as point charges, on the inverse of the closest anion-cation distance. Standard formation parameters and molar entropies of the alkali-halide salts are also reported in Table 5.6, and will be discussed in Section 5.6.

5.3 Gas-phase equilibrium ion-pair distances

The interaction between oppositely-charged ions in the gas phase can be probed experimentally using molecular-beam spectroscopy measurements[2033–2036], along with a quantum-mechanical data analysis involving the fitting to an analytical potential-energy function[985,2037]. Obviously, these experiments represent non-equilibrium measurements since free gaseous ions cannot exist at equilibrium under usual conditions of temperature and pressure (Section 2.1). However, this approach permits an evaluation of the *equilibrium distance of the ion-pair complex* $d_g[[I^z \cdot J^{-z}]_g]$, referring to a hypothetical *isolated and vibrationless pair* at $T = 0$ K.

Gas-phase equilibrium ion-pair distances d_g have been determined[298] for all alkali-halide gas-phase ion-pair complexes ($[M^+ \cdot X^-]_g$). The corresponding values are reported in Table 5.3. Expectedly, d_g increases systematically with the effective sizes (atomic numbers) of the involved ions, *i.e.* along the series $Li^+ \rightarrow Cs^+$ for pairs involving a common anion and $F^- \rightarrow I^-$ for pairs involving a common cation.

5.4 Effective ionic radii

The *effective radius* $R_I[I^z]$ of an ion I^z is *not an experimental observable.* Quantum-mechanically, and for a monoatomic ion, it can possibly be defined as the radius of a sphere centered at the nucleus and encompassing a given fraction f_E of the total electronic charge $-(Z-z)$ of the ion, where Z is the atomic number. This definition can be applied to isolated ions, but also in principle to gas-phase ion-pair complexes (Section 5.3) or to ions in condensed phases (see *e.g.* the minimum-overlap division method[2038] for ionic crystals under Point B_{10} below), even though the electronic charge density itself cannot be unambiguously partitioned into contributions from the individual ions. Obviously, however, the choice of the *electronic charge fraction* f_E introduced in the definition of such a *quantum-mechanical radius* represents a *factor of arbitrariness*, *i.e.* the use of different values will lead to as many distinct and equally acceptable sets of effective ionic radii. This fundamental ambiguity persists when *empirical definitions* are adopted instead (see further below).

In spite of its ill-defined nature, the concept of effective ionic radius turns out to be very useful in practice, *e.g.* for the qualitative rationalization of the lattice types adopted by ionic crystals[66], for the estimation of ionic solvation free energies *via* CE calculations (Section 3.2), and for the evaluation of possible corrections to the latter estimates (Section 6.1.1) or to corresponding estimates based on AT simulations (Section 6.2.1). In any case, ionic radii should probably be considered as *fitting parameters* related to a *given model*, *e.g.* hard-sphere packing model of ionic crystals or Born solvation model, rather than as intrinsic properties of the individual ions bearing a physical meaning of their own. For this reason, a set of radii should always be quoted along with the physical model it refers to.

In general terms, one may distinguish three types of ionic radii, depending on the *environment* of the ion: (*i*) *in-crystal radii*, appropriate for ions within ionic crystals; (*ii*) *in-solution radii*, appropriate for solvated ions; (*iii*) *gas-phase radii*, appropriate for isolated ions or, possibly, ions within isolated ion pairs (see Point N_{10} below). The radii of the two former types should in principle be somewhat smaller than those of the latter type, because the transfer of an ion from the gas phase to a condensed phase is expected to result in a size decrease of its electron cloud caused by the denser environment and by the attractive ion-environment interactions, *e.g.* anion-cation attraction in a crystal or ion-solvent attraction in solution, a phenomenon previously referred to as *ion compression* (Section 3.2). This effect is, however, not unambiguously evidenced by the available sets of radii for the alkali and halide ions (see further below). Effective ionic radii are also expected to depend on *pressure and temperature*, and the notations $R_I^{\ominus}[I_s^z]$, $R_I^{\ominus}[I_d^z]$ and $R_I^{\ominus}[I_g^z]$ will be used to denote in-crystal, in-solution and gas-phase radii appropriate for P° and T^{-}. In-crystal radii are generally derived under the constraint that the radius of an ion should be the same for all salts of this ion (see, however, Refs.[2039, 2040]). In contrast, the in-solution radius of an ion is meant to be appropriate only for a given solvent, typically water, leading to a corresponding in-water radius $R_I^{\ominus}[I_{aq}^z]$.

In spite of the above remarks concerning the ambiguous nature of the concept of effective ionic radius, the physical relevance of in-crystal radii may still be *partially justified* based on the experimental observation of an *additivity rule* concerning the lattice parameters of ionic salts (Section 5.2). In the restricted context of the alkali-halide salts, this observation may be expressed in the form of the law

$$\begin{aligned} & f_{L,MX_s'} L^{\ominus}[MX_s'] - f_{L,MX_s} L^{\ominus}[MX_s] \\ = \; & f_{L,M'X_s'} L^{\ominus}[M'X_s'] - f_{L,M'X_s} L^{\ominus}[M'X_s] \end{aligned} \quad \forall M, M', X, X' \;, \tag{5.2}$$

where f_L is 1/2 for a fcc lattice or $3^{1/2}/2$ for a bcc lattice. In plain words, the substitution of an anion X^- by a different anion X'^- leads to the same change in the closest anion-cation distance within the crystal, irrespective of the cation. A similar result holds for cation substitution, which follows from a simple rearrangement of Eq. 5.2. This law is obeyed up to a remarkable level of precision for the alkali-halide salts (see further below), suggesting that some intrinsic additivity principle indeed holds in this context. A similar law also holds for gas-phase equilibrium ion-pair distances (Section 5.3), although the level of precision is slightly lower in this case (see further below).

Most of the available sets of in-crystal radii are designed in such a way that the sum of the cation and anion radii for a given salt are essentially equal to the experimental closest anion-cation distance in the corresponding crystal, as derived for the given lattice type from the crystallographic unit-cell parameter $L^{\ominus}$ (Section 5.2). This constraint can of course only be satisfied within a given tolerance, because a set of m cations and n anions involves $m \times n$ lattice parameters, corresponding to all combined salts, to be reproduced using only $m + n$ adjustable radii. For alkali and halide ions, however, the minimal error is very low due to the nearly exact satisfaction of Eq. 5.2. Yet, this problem remains underdetermined, because two sets of radii differing by a common offset constant, additive for the cations and subtractive for the anions, are indistinguishable in terms of their ability to reproduce the lattice parameters of the combined salts, as an immediate consequence of Eq. 5.2. Note the strong parallel between this situation and that encountered in the partitioning of salt solvation parameters (Section 4.5.1).

In-crystal radii obeying the above *additivity constraint*, at least to a good level of precision, are referred to as *hard-sphere radii*, because cations and anions within a salt crystal can be viewed in this case as hard-spheres in contact. However, some in-crystal sets are also derived based on different principles[1869, 2041–2044] and without enforcement of this additivity constraint. These are referred to as *soft-sphere radii*, because cations and anions within a salt crystal are viewed in this case as soft-spheres that can partially interpenetrate. The ionic radius is then a parameter entering into the definition of some effective ion-ion interaction potential. The distinction between hard-sphere and soft-sphere radii should also in principle apply to in-solution radii.

For example, in-solution radii used in CE calculations (Section 3.2) define hard cavities, while ion-solvent van der Waals interaction parameters used in AT simulations (Section 3.3.2) can be viewed as defining soft radii. However, the classification remains somewhat artificial in the case of in-solution radii, due to the absence of an obvious expected additivity criterion. Finally, the distinction between hard-sphere and soft-sphere radii does not apply to gas-phase radii, except for those constrained to optimally reproduce gas-phase equilibrium ion-pair distances[2044] (see Point N_{10} below).

Many different sets of effective ionic radii have been derived following various approaches, which can be broadly classified as follows:

(A_{10}) *In-crystal radii*, derived using an *empirical (physically-motivated) criterion* for *partitioning the experimental closest anion-cation distances in ionic crystals* (Section 5.2) into contributions from the cations and the anions. Typical assumptions used to define this splitting include: (*i*) an assumption of hard-sphere contact for the anions (I^- in LiI_s may be selected[2045], assuming that the very small Li^+ cations have a negligible influence on the packing of the much larger I^- anions) or for the cation-anion pair (Li^+ and F^- may be selected[2046], considering that these low-polarizability ions present a predominantly ionic bonding character) in a reference salt; (*ii*) a procedure called hard-sphere or soft-sphere division[2043, 2044], which relies (exclusively for hard-sphere division, in first line for soft-sphere division) on treating the partitioning for a set of n cations and m anions as an optimization problem for $m + n$ adjustable radii against $m \times n$ lattice parameters (for alkali and halide ions, this approach is hazardous in view of the high level of precision with which Eq. 5.2 is obeyed, *i.e.* the determinant of the matrix to be inverted is nearly zero, and the unknown offset constant that would appear in the radii if this equation was exactly satisfied is solely determined by the corresponding very small violation); (*iii*) an expected correlation between the ionic radius and some (experimental or/and theoretical) observable (*e.g.* partial molar refractivity of ions in solution[2047, 2048], effective nuclear charge[66,2046,2047,2049], Thomas-Fermi[2050,2051] self-energy[2039], partial molar magnetic susceptibility of ions in crystals[2052], interatomic distance in the cation-forming metal[2043, 2053], crystal and gas-phase ion spectroscopic properties[2054], or limiting equivalent conductivity in solution[2055]); (*iv*) the use of previously derived radii as anchoring points for a new partitioning approach (*e.g.* F^- from Ref.[2048] for Ref.[2056], radii from Ref.[2047, 2056] for Ref.[2057], K^+ and Cl^- from Refs.[2047, 2056] for Ref.[2049], radii from Ref.[2047] and F^- from Ref.[2048] for Ref.[2046], radii from Ref.[2056] for Ref.[398], F^- from Ref.[2048] and O^{2-} from Ref.[1332] for Refs.[2058–2060], or radii from Ref.[2056] for Ref.[778]). To our knowledge, only four sets of effective ionic radii proposed in this category rely on a soft-sphere approach. Three of them are based on expressing experimental interionic distances as a (5/3)-power mean of the cation and anion radii (Ref.[2043] and two sets of Ref.[2044]). The fourth one[1869] relies on the (at present obsolete) concept of cubic atoms[1869,2045] ("Würfelatome" in German).

(B_{10}) *In-crystal radii*, derived based on the *partitioning of electron density maps from x-ray diffraction experiments on ionic crystals*[822,2038,2061–2063]. Typically, the effective ionic radius is defined in this case as the distance between the nucleus of an ion and the minimum-density point along the line connecting this ion to an oppositely-charged nearest-neighbor ion (minimum-density division). An alternative approach[2038] involves the attribution of (part of) the electron density in the crystal to the cations and anions based on contacting spherical volumes centered at the corresponding nuclei and the optimization of the sphere radii, *i.e.* of the contact distance along the line connecting oppositely-charged nearest-neighbor ions, for an attribution of the electron density matching the ratio of the net ionic charges (minimum-overlap division). To our knowledge, all sets of ionic radii proposed in this category rely on a hard-sphere approach.

(C_{10}) *In-crystal radii*, derived based on the *fitting of experimental data using a (physically-motivated) equation* that depends on the set of ionic radii[2040, 2064–2066]. In this approach, all radii are optimized simultaneously to reproduce some experimental data set based on the connecting equation (*e.g.* Kapustinskii equation connecting so-called thermochemical ionic radii[398] given in Refs.[221, 2056] to the reticulation enthalpy of a salt crystal, equation connecting radii adapted from Refs.[2047,2056] to the atomic numbers of the ions[2064], Born-like equation plus van der Waals term connecting the lattice parameters and permittivity of a salt crystal to its reticulation enthalpy[2065], Phillips-van Vechten theory[2067–2069] connecting the lattice parameters of a crystal and quantum-mechanical ionic properties to its refraction index[2040], model connecting the elastic properties of a cation-forming metal and quantum-mechanical atomic properties to its compressibility[2066]; see also Section 5.11, Point D_{11}, for the Kapustinskii equation). To our knowledge, all sets of ionic radii proposed in this category rely on a hard-sphere approach.

(D_{10}) *In-crystal radii*, derived based on the *optimization of a radius-dependent interionic repulsion potential*, leading at equilibrium and in balance with electrostatic interactions to the *experimental lattice parameters*[1332,2041]. This repulsive potential has been modeled using the Born-Landé potential[395, 396] augmented by a term for Pauli repulsion in Ref.[2041, 2042] or using the Huggins-Mayer potential[180,987] in Ref.[1332], in both cases reformulated so as to depend explicitly on the ionic radii. The derivation of ionic radii is then formulated as an optimization problem, namely that of determining a set of values for which predicted equilibrium crystal structures, given lattice electrostatics and the selected repulsive potential, reproduce the experimental lattice parameters with the highest accuracy (see also Section 5.11, Points B_{11} and C_{11}). This approach should in principle lead to soft-sphere ionic radii. However, for one of the two sets of ionic radii proposed in this category[1332], the radius definition and corresponding values are further adjusted so as to lead to hard-sphere radii.

(E_{10}) *In-crystal radii*, derived from the *in-water sets L, M and H* described under Point J_{10} (Ref.[1289], defined by R^{is}_{min}) *via addition of constant offsets* of −0.229, −0.234 and −0.243 nm, respectively. The offset constants were optimized against the experimental unit-cell parameters $L^{\ominus}$ for the alkali-halide salts. These sets represent hard-sphere radii and are discussed in Section 6.2.3.

(F_{10}) *In-crystal radii*, derived from the *in-water sets L, M and H* described under Point K_{10} (Ref.[1289], defined by the first-peak positions of radial distribution functions in water as obtained from AT simulations) *via addition of constant offsets* of −0.133, −0.136 and −0.139 nm, respectively. The offset constants were optimized against the experimental unit-cell parameters $L^{\ominus}$ for the alkali-halide salts. These sets represent hard-sphere radii and are discussed in Section 6.2.3.

(G_{10}) *In-solution radii* (water), derived based on the *Stokes law*[2070,2071] and the *Einstein-Sutherland equation*[2072,2073], along with a *size correction*[2074]. The Stokes law connects the friction coefficient, *i.e.* the ratio of the frictional force exerted by an ideal fluid (stick boundary conditions) on a macroscopic sphere divided by the sphere velocity, to the radius of the sphere and the viscosity of the solvent. The Einstein-Sutherland equation in turn connects this friction coefficient to the self-diffusion constant of the sphere. When combined into the so-called Stokes-Einstein-Sutherland equation, these relations connect the self-diffusion constant of the sphere to the viscosity of the solvent *via* the sphere radius. Although the Stokes law is not strictly applicable to spheres of microscopic sizes, the resulting radii may tentatively be corrected by assuming that[2074]: (*i*) the corrected Stokes radius of large and weakly hydrated tetraalkylammonium ions (for which in-crystal and uncorrected Stokes radii are found to be essentially identical) is equal to their uncorrected values; (*ii*) the minimal corrected Stokes radius (corresponding to the limit of a vanishing uncorrected value) is about equal to the diameter of a water molecule (taken in Ref.[2074] to be 0.269 nm); (*iii*) the corrected values for arbitrary ions can be linearly interpolated between these two extremes. This correction scheme is rather *ad hoc*, but compensates for the fact that the uncorrected Stokes radii of small ions tend to be underestimated. Note, however, that the corrected Stokes radius still characterizes the effective size of the solvated ion, possibly including tightly bound solvent molecules, rather than the effective size of the ion itself within the solvent.

(H_{10}) *In-solution radii* (water), derived by application of the *Born equation* (Section 3.2) based on experimental conventional single-ion solvation free energies (Section 5.16) along with an estimate for the intrinsic solvation free energy $\mathcal{G}_H^{\ominus}$ of the proton (Section 5.18). This approach[259,819] is certainly questionable in view of: (*i*) the very large uncertainty range in the available estimates for the experimentally-elusive quantity $\mathcal{G}_H^{\ominus}$ (Section 2.3); (*ii*) the approximate nature of the Born model for evaluating charging free energies (Section 3.2); (*iii*) the usual neglect of the cavitation contribution and standard-state correction term (Section 3.2) upon comparing the theoretical charging free energies with the estimated experimental single-ion solvation free energies.

(I_{10}) *In-solution radii* (water), derived based on the *averaging* (arithmetic mean) of a *selected set of in-crystal radii* with a corresponding set of *first-peak positions in the ion-water (oxygen atom) radial distribution functions around the hydrated ions*[778,779,781]. This approach has been applied[778] on the basis of the Goldschmidt in-crystal radii[2056] along with radial distribution functions evaluated from AT simulations. The use of the corresponding experimentally-determined functions has also been considered. It was initially motivated by the excellent correlation observed between charging free energies evaluated by application of the Born equation (Section 3.2) using these radii, and experimental conventional single-ion solvation free energies (Section 5.16). Although some theoretical justification has been given in terms of a two-sphere Born model[778,779,781], this approach remains somewhat *ad hoc*. In particular, it should be noted that: (*i*) it depends on the choice of one specific set of in-crystal radii; (*ii*) it also depends on the choice of a solvent model and set of ion-solvent van der Waals interaction parameters (Section 3.3.2) in the AT simulation used to evaluate the radial distribution functions (the latter parameters already define some effective in-solution radii, and many different sets are also available, see Tables 3.3 and 3.4); (*iii*) the cavitation contribution and standard-state correction terms (Section 3.2) are usually neglected upon comparing theoretical charging free energies with the estimated experimental single-ion solvation free energies.

(J_{10}) *In-solution radii* (water), defined by the *minimum* R_{min}^{is} *of the ion-solvent Lennard-Jones curve* (Section 3.3.1), considering ion-solvent Lennard-Jones interaction parameter sets fitted to reproduce *estimated experimental intrinsic single-ion solvation free energies* in *AT simulations*[1289]. The target intrinsic single-ion solvation free energies are obtained here using conventional data from Refs.[1875,1876] (or Ref.[1750] for Cs^+) along with three plausible values of −1100, −1075 or −1050 $kJ \cdot mol^{-1}$ for the intrinsic solvation free energy $\mathcal{G}_H^{\ominus}$ of the proton, thereby defining three different parameter sets noted L, M and H, respectively (Tables 3.3 and 3.4). The simulations are carried out using the SPC water model[1274] and include corrections for methodological errors (Section 6.2.1), as well as the cavitation contribution and the standard-state correction term (Section 3.3.3). Although these radii formally represent in-solution radii appropriate for water (assuming a specific $\mathcal{G}_H^{\ominus}$ value), they can in principle easily be converted to in-crystal radii or in-solution radii for other solvents by application of combination rules (Section 3.3.2). These sets are discussed in more detail in Sections 6.2.2 and 6.2.3.

(K_{10}) In-solution radii (water), defined by the *first-peak positions in the ion-water (oxygen atom) radial distribution functions around hydrated ions* obtained from *AT simulations using the same sets L, M and H* of Point J_{10} (based on Ref. [1289]). In contrast to the previous sets (defined by R^{is}_{min}), these radii formally encompass specific information on the hydration structure and cannot be easily converted to in-crystal radii or in-solution radii for other solvents. These sets are discussed in Section 6.2.3.

(L_{10}) *Gas-phase radii*, derived *from in-crystal radii* by *empirical (physically-motivated) adjustments or extrapolations*[2053,2075,2076]. Examples of such adjustments or extrapolations include: (*i*) no adjustment for cations[2053] (assumed to be unaffected by the change of environment between crystal and gas-phase due to their limited polarizability); (*ii*) small adjustment for cations[2075] based on estimated electronic polarizabilities (relying on fitting to the thermodynamic parameters of salt formation); (*iii*) adjustments for all ions[2076] according to the model of overlapping atoms[2077,2078] (relying on correlations connecting interatomic distances in molecules or crystals with estimated covalent and repulsive bonding contributions); (*iv*) extrapolation for anions[2053] (relying on expressing the in-crystal radius of a common anion in a series of salts as a function of the in-crystal radius of the corresponding cation, and extrapolating this expression to infinite cation sizes, the assumption being that the influence of the cation on the in-crystal radius of the anion decreases with the cation size due to a reduction in the intensity of the cation-generated electric field).

(M_{10}) *Gas-phase radii*, derived from *quantum-mechanical calculations* and analysis of the electron density, electric potential or Kohn-Sham potential[660,661,2079–2082]. The various approaches are the following: (*i*) scaling of the van der Waals radius of the isoelectronic noble gas[660] as determined from real-gas equilibrium and transport properties[660,2083] under consideration of the effective atomic number (atomic number minus screening parameter[1799]); (*ii*) location of the minimum of the radial electric potential[2080] (radius of an overall neutral sphere centered at the nucleus and encompassing the free ion), based on relativistic self-consistent field quantum-mechanical calculations[2084]; (*iii*) location of the maximum of the radial charge density distribution function for the outermost orbital, either[2081] according to Slater-type quantum-mechanical calculations[1799,2085] or[2079] according to Dirac-Slater quantum-mechanical calculations[2086]; (*iv*) location of the classical turning point and virial turning point in the Kohn-Sham potential[2082], based on Kohn-Sham density functional theory calculations[1569]; (*v*) identification[661] of the electrostatic self-energy of an isolated ion with its free energy of ionization[407].

(N_{10}) *Gas-phase radii*, derived using an *empirical (physically-motivated) criterion* for *partitioning the observed gas-phase equilibrium ion-pair distances* d_g (Section 5.3) into contributions from the cations and the anions[1295, 2044, 2087], using either a hard-sphere[2044] or a soft-sphere[1295, 2087] approach. Note that such a set accounts for an effective ionic radius within an isolated cation-anion complex at $T = 0$ K, and will not be strictly comparable to the previous gas-phase sets, derived by consideration of isolated ions.

A collection of *63 available sets* of effective in-crystal, in-water and gas-phase ionic radii $R_I^{\ominus}$ for the alkali and halide ions, labeled 1r-63r for the ease of further reference, is provided in Table 5.4. The corresponding values are also illustrated graphically in Figure 5.1. This extensive compilation spans a range of 91 years (1920 to 2011) of scientific literature. It results from a very thorough scanning of this literature, and is probably close to exhaustive (at least up to year 2009).

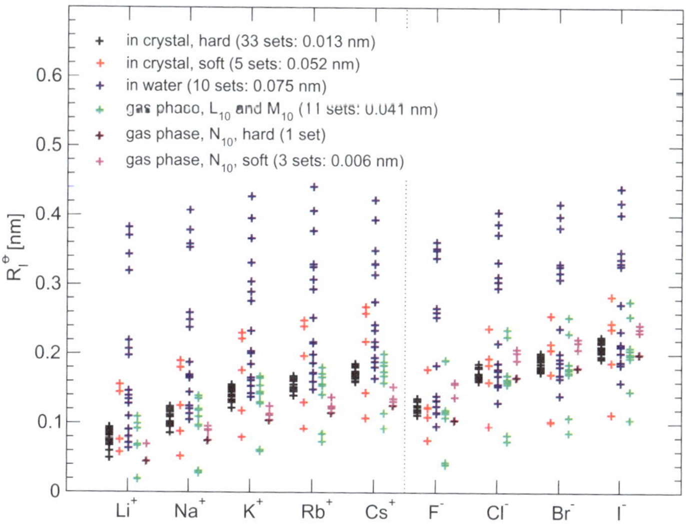

Figure 5.1 *Available sets of effective ionic radii for the alkali and halide ions.* This figure is a graphical illustration of the ionic radii belonging to the 63 sets reported numerically in Table 5.4. The numbers indicated between parentheses in the legend refer to the number of sets in the category and to the root-mean-square deviation of the ionic radii available in these different sets for a given ion, averaged over the nine ions.

Table 5.4 *Available sets of effective ionic radii for the alkali and halide ions.* The reported values are estimates $R_I^{\ominus}$ for the effective radius of an ion I^z in the indicated environment (in-crystal: I_s^z; in-water: I_{aq}^z; gas-phase: I_g^z), referring to standard conditions (reference pressure $P^\circ = 1$ bar and reference temperature $T^- = 298.15$ K). The notation I^z refers to an alkali (M^+) or halide (X^-) ion. The methods refer to the literal enumeration of Section 5.4. The sets are listed in chronological order of publication for each type of method. The entries δ and σ correspond to the comparison between predicted (sums of cation and anion radius from the set) and experimental nearest-neighbor cation-anion distances. The latter values are given by the experimental crystal lattice parameters $L^{\ominus}$ (Table 5.2) amplified by f_L, where f_L is 1/2 for a fcc lattice or $3^{1/2}/2$ for a bcc lattice. The parameter δ represents the corresponding average difference (predicted minus experimental) and the parameter σ the corresponding root-mean-square difference, over the set of all possible salt combinations for which radii are available in the set (at most 20). The numbers between parentheses in these two columns (gas-phase radii only) refer to a similar comparison with experimental gas-phase equilibrium ion-pair distances d_g (Table 5.3). See also Figure 5.1 for a graphical illustration of the data presented in this table.

		$R_I^{\ominus}[I^z]$ [nm]												
Set	Method	Li^+	Na^+	K^+	Rb^+	Cs^+	F^-	Cl^-	Br^-	I^-	δ [nm]	σ [nm]	Ref.	Notes
		$R_I^{\ominus}[I_s^z]$ (in crystal)												
1r	A_{10}	-	0.11	0.145	0.165	0.16	0.12	0.165	0.18	0.205	-0.009	0.015	2045	a
2r	A_{10}	-	0.0517	0.0794	0.0914	0.107	0.075	0.0953	0.1021	0.1122	-0.143	0.145	1869	b
3r	A_{10}	-	0.101	0.130	0.150	0.175	0.133	0.172	0.192	0.219	-0.004	0.006	2048	a
4r	A_{10}	0.078	0.098	0.133	0.149	0.165	0.133	0.181	0.196	0.220	-0.002	0.005	2056	a
5r	A_{10}	0.060	0.095	0.133	0.148	0.169	0.136	0.181	0.195	0.216	-0.006	0.010	66, 2047	a
6r	A_{10}	0.069	0.096	0.133	0.148	0.167	0.134	0.181	0.195	0.218	-0.005	0.007	2057	a
7r	A_{10}	0.068	0.098	0.133	0.148	0.167	0.133	0.181	0.196	0.219	-0.004	0.006	2049	a
8r	A_{10}	-	0.106-0.112	0.129-0.133	0.144-0.150	0.158-0.163	-	0.175-0.197	0.191-0.212	0.210-0.231	0.001	0.008	2039	a,c

(continues on the next page)

Table 5.4 *(continued)*

		$R_I^{\ominus}[I^z]$ [nm]												
Set	Method	Li^+	Na^+	K^+	Rb^+	Cs^+	F^-	Cl^-	Br^-	I^-	δ [nm]	σ [nm]	Ref.	Notes
9r	A_{10}	0.068	0.097	0.133	0.147	0.167	-	-	-	-	-	-	2046	a
10r	A_{10}	0.073	0.100	0.132	0.146	0.171	0.132	0.182	0.198	0.224	-0.001	0.003	2052	a
11r	A_{10}	0.092	0.118	0.145	0.156	0.168	0.112	0.164	0.179	0.202	-0.009	0.012	2053	a
12r	A_{10}	0.076	0.102	0.138	0.152	0.167	0.133	0.181	0.196	0.220	0.000	0.005	2058, 2059, 2060	a,d
13r	A_{10}	0.090	0.116	0.152	0.166	0.181	0.119	0.167	0.182	0.206	0.000	0.005	2058, 2059, 2060	a,e
14r	A_{10}	0.1548	0.1889	0.2298	0.2480	0.2676	0.1085	0.1835	0.2053	0.2363	0.092	0.093	2043	b
15r	A_{10}	-	0.12085	0.15137	0.16319	-	-	0.16115	0.17426	0.19442	-0.005	0.006	2054	a
16r	A_{10}	0.0756	0.0998	0.1327	0.1481	0.1655	0.1327	0.1805	0.1970	0.2232	-0.001	0.004	2044	a
17r	A_{10}	0.0752	0.1244	0.1754	0.1961	0.2175	0.1771	0.2367	0.2551	0.2826	0.086	0.088	2044	b,f
18r	A_{10}	0.1439	0.1795	0.2215	0.2394	0.2583	0.1214	0.1930	0.2139	0.2444	0.093	0.094	2044	b,g
19r	A_{10}	0.082	0.112	0.148	0.163	0.182	0.119	0.169	0.186	0.211	-0.001	0.004	2055	a
20r	B_{10}	-	0.118	-	-	-	-	0.163	-	-	-0.001	0.001	2061	a
21r	B_{10}	0.093	0.117	0.149	0.164	0.183	0.116	0.164	0.180	0.204	-0.002	0.004	2062	a
22r	B_{10}	0.094	0.117	0.149	0.163	0.186	0.116	0.164	0.180	0.205	-0.001	0.003	2063	a
23r	B_{10}	0.070	0.098	0.134	-	-	0.135	0.181	0.198	-	-0.001	0.004	2038	a,h
24r	B_{10}	0.085	0.112	0.145	-	-	0.126	0.167	0.181	-	-0.001	0.006	2038	a,i
25r	B_{10}	0.071	0.097	0.141	0.150	0.173	0.124	0.180	0.198	0.225	-0.001	0.006	822	a,j
26r	C_{10}	0.050	0.086	0.122	0.140	0.161	-	-	-	-	-	-	2064	a
27r	C_{10}	0.078	0.098	0.133	0.149	0.165	0.126	0.168	0.190	0.211	-0.011	0.012	221, 2056	a,k

(continues on the next page)

Table 5.4 *(continued)*

Set	Method	Li^+	Na^+	K^+	Rb^+	Cs^+	F^-	Cl^-	Br^-	I^-	δ [nm]	σ [nm]	Ref.	Notes
$R_I^{\ominus}[I^z]$ [nm]														
28r	C_{10}	-	-	-	-	-	0.116	0.161	0.182	0.209	-	-	2065	a
29r	C_{10}	0.078-	0.108-	0.131-	0.140-	-	0.123-	0.162-	0.169-	0.185-	-0.001	0.012	2040	a,c
		0.098	0.132	0.166	0.182		0.143	0.166	0.183	0.202				
30r	C_{10}	0.086	0.118	0.1555	0.165	0.1735	-	-	-	-	-	-	2066	a
31r	D_{10}	0.0574	0.0873	0.1173	0.1294	0.1434	0.1225	0.1589	0.1702	0.1867	-0.043	0.044	2041, 2042	b
32r	D_{10}	0.094	0.124	0.154	0.168	0.183	0.116	0.162	0.176	0.197	-0.002	0.007	1332	a
33r	E_{10}(L)	0.090	0.124	0.137	0.148	0.164	0.132	0.175	0.187	0.210	0.000	0.013	1289	a,l
34r	E_{10}(M)	0.109	0.144	0.161	0.172	0.188	0.115	0.153	0.164	0.184	0.000	0.013	1289	a,l
35r	E_{10}(H)	0.128	0.164	0.185	0.199	0.218	0.096	0.130	0.140	0.159	0.000	0.013	1289	a,l
36r	F_{10}(L)	0.064	0.105	0.143	0.160	0.181	0.133	0.178	0.193	0.214	0.000	0.009	1289	a,l
37r	F_{10}(M)	0.071	0.113	0.153	0.171	0.194	0.124	0.168	0.181	0.201	0.000	0.012	1289	a,l
38r	F_{10}(H)	0.080	0.120	0.165	0.186	0.210	0.114	0.156	0.169	0.188	0.000	0.014	1289	a,l
$R_I^{\ominus}[I_{aq}^z]$ (in water)														
39r	G_{10}	0.382	0.358	0.331	0.329	0.329	0.352	0.332	0.330	0.331	0.373	0.379	2074	
40r	H_{10}	0.134	0.167	0.203	0.217	0.242	0.183	0.215	0.237	0.272	0.110	0.111	259	m
41r	H_{10}	0.146	0.187	0.233	0.252	0.275	0.139	0.186	0.200	0.233	0.099	0.100	819	n
42r	I_{10}	0.139	0.169	0.200	0.215	0.234	-	-	-	-	-	-	778	
43r	J_{10}(L)	0.319	0.353	0.366	0.377	0.393	0.361	0.404	0.416	0.439	0.457	0.458	1289	l

(continues on the next page)

Table 5.4 *(continued)*

Set	Method	Li^+	Na^+	K^+	Rb^+	Cs^+	F^-	Cl^-	Br^-	I^-	δ [nm]	σ [nm]	Ref.	Notes
							$R_I^{\ominus}[I^z]$ [nm]							
44r	J_{10}(M)	0.343	0.378	0.395	0.406	0.422	0.349	0.387	0.398	0.418	0.468	0.468	1289	l
45r	J_{10}(H)	0.370	0.407	0.427	0.441	0.460	0.338	0.372	0.382	0.401	0.485	0.485	1289	l
46r	K_{10}(L)	0.197	0.238	0.276	0.293	0.314	0.266	0.311	0.326	0.347	0.267	0.267	1289	l
47r	K_{10}(M)	0.207	0.249	0.289	0.307	0.330	0.260	0.304	0.317	0.337	0.272	0.272	1289	l
48r	K_{10}(H)	0.219	0.259	0.304	0.325	0.349	0.253	0.295	0.308	0.327	0.278	0.278	1289	l
							$R_I^{\ominus}[I_g^z]$ (gas phase)							
49r	L_{10}	0.092	0.118	0.145	0.156	0.168	0.19	0.235	0.253	0.276	0.065 (0.118)	0.066 (0.119)	2053	
50r	L_{10}	0.067	0.097	0.130	-	-	-	-	-	-	-	-	2075	
51r	L_{10}	0.069	0.095	0.128	0.141	0.158	0.119	0.169	0.185	0.209	-0.020 (0.033)	0.021 (0.033)	2076	
52r	M_{10}	-	0.1352	0.1671	0.1801	0.1997	0.1909	0.2252	0.2298	0.2548	0.074 (0.127)	0.075 (0.127)	660	
53r	M_{10}	0.0189	0.0278	0.0592	0.0734	0.0921	0.0400	0.0741	0.0863	0.105	-0.179 (-0.125)	0.179 (0.126)	2079	o
54r	M_{10}	-	0.1391	0.1645	0.1709	0.1832	-	-	-	-	-	-	661	
55r	M_{10}	-	-	-	-	-	0.108	0.163	0.179	0.205	-	-	2080	
56r	M_{10}	0.0196	0.0309	0.0615	0.0847	0.1144	0.0436	0.0828	0.1080	0.1460	-0.152 (-0.099)	0.152 (0.100)	2081	
57r	M_{10}	0.099	0.109	0.142	0.154	0.174	0.114	0.156	0.173	0.198	-0.013 (0.040)	0.016 (0.041)	2082	p

(continues on the next page)

Table 5.4 *(continued)*

		$R_I^{\ominus}[I^z]$ [nm]												
Set	Method	Li^+	Na^+	K^+	Rb^+	Cs^+	F^-	Cl^-	Br^-	I^-	δ [nm]	σ [nm]	Ref.	Notes
58r	M_{10}	0.099	0.109	0.142	0.154	0.174	0.115	0.159	0.171	0.194	-0.014 (0.039)	0.017 (0.041)	2082	q
59r	M_{10}	0.109	0.120	0.154	0.166	0.188	0.116	0.161	0.177	0.200	0.002 (0.055)	0.008 (0.056)	2082	r
60r	N_{10}	0.069	0.095	0.124	0.137	0.152	0.137	0.190	0.206	0.231	-0.003 (0.051)	0.007 (0.051)	2087	b,s,t
61r	N_{10}	0.069	0.089	0.111	0.121	0.131	0.158	0.206	0.221	0.242	0.002 (0.055)	0.013 (0.056)	1295	b,s,u
62r	N_{10}	0.069	0.090	0.113	0.124	0.135	0.156	0.201	0.216	0.236	-0.001 (0.053)	0.012 (0.053)	1295	b,s,v
63r	N_{10}	0.04461	0.07474	0.10397	0.11468	0.12537	0.10397	0.16585	0.17971	0.19908	-0.054 (0.000)	0.055 (0.006)	2044	a,w

(a) hard-sphere radii; (b) soft-sphere radii; (c) value ranges refer to distinct radii attributed to the same ion in different alkali-halide salts (δ and σ are calculated using the range midpoint); (d) labeled "effective ionic radii" in the indicated references; (e) labeled "crystal radii" in the indicated references; (f) labeled "set A" in the indicated reference; (g) labeled "set B" in the indicated reference; (h) based on minimum-overlap division; (i) based on minimum-density division; (j) labeled "diffraction data" the indicated reference; (k) so-called thermochemical radii (alkali ions from Ref.[2056], halide ions from Ref.[221]; based on the approach of Ref.[398]); (l) based on conventional values from Refs.[1875, 1876] (except Cs^+, from Ref.[1750]) and $\mathcal{G}_H^{\ominus} = -1100$, -1075 or -1050 kJ·mol^{-1}, respectively, for sets L, M and H (these sets are discussed in Sections 6.2.2 or/and 6.2.3); (m) based on conventional values from Ref.[294] and $\mathcal{G}_H^{\ominus} = -1082.0$ kJ·mol^{-1}; (n) based on conventional values from Ref.[221] and $\mathcal{G}_H^{\ominus} = -1052.0$ kJ·mol^{-1}; (o) halide radii are as reported in Ref.[1295] (average of the two sets obtained by Ref.[2079] from the $p^{1/2}$ and $p^{3/2}$ orbitals); (p) using the BZ model[2088] for the local spin density (LSD) correction to the exchange potential; (q) using the LB model[2089] for the LSD correction; (r) using the γSIC model[2090] for the LSD correction; (s) see the quoted reference for the source data concerning gas-phase equilibrium ion-pair distances; (t) based on the Rittner potential[2091] for Pauli repulsion and a $R_I^{\ominus}[Li_g^+]$ value from quantum-chemical SCF calculations; (u) based on the T-Rittner potential[2092] for Pauli repulsion and the $R_I^{\ominus}[Li_g^+]$ value from Ref.[2087]; (v) based on the DD potential[2092] for Pauli repulsion and the $R_I^{\ominus}[Li_g^+]$ value from Ref.[2087]; (w) based on gas-phase equilibrium ion-pair distances from Ref.[298] (Table 5.3).

Note that most of the references used as data sources for Table 5.4, as well as Refs.[2093–2098], also report radii estimates for *other ions* than alkali and halides.

All the reported sets present the expected trend of an increase in $R_I^{\ominus}$ when the atomic number of the ion increases, *i.e.* along the series $Li^+ \rightarrow Cs^+$ and $F^- \rightarrow I^-$. They also all respect the expected trend of a larger $R_I^{\ominus}$ for anions compared to cations when considering ions with the same electronic structure, *i.e.* Na^+ *vs.* F^-, K^+ *vs.* Cl^-, Rb^+ *vs.* Br^- and Cs^+ *vs.* I^-, *i.e.* a cation is smaller than an anion at identical electronic structure due to its excess, as opposed to deficit of positive nuclear charge compared to the negative electronic charge (increased nucleus-electron attraction, decreased electron-electron repulsion).

The compatibility of the different sets with the crystal lattice parameters $L^{\ominus}$ reported in Table 5.2 is indicated in Table 5.4 by the quantities δ and σ. These refer to the comparison between predicted (sum of the cation and anion radius from the set) and experimental nearest-neighbor cation-anion distances. The latter values are given by the experimental lattice parameters $L^{\ominus}$ (Table 5.2) amplified by f_L, where $f_L = (1/2)L^{\ominus}$ and $f_L = (3^{1/2}/2)L^{\ominus}$ for fcc and bcc lattices, respectively. The parameter δ represents the corresponding average difference (predicted minus experimental) and the parameter σ the corresponding root-mean-square difference, over the set of all possible salt combinations for which radii are available in the set (at most 20). The alternative δ and σ values reported between parentheses (gas-phase radii only) result from a similar comparison with the experimental gas-phase equilibrium ion-pair distances d_g (Table 5.3).

The quantity σ referring to crystallographic parameters has a lower bound of 0.003 nm for a complete set of 9 ionic radii, as determined by the level of precision with which Eq. 5.2 is satisfied based on the $L^{\ominus}$ values of Table 5.2. This minimal deviation is very low, indicating that Eq. 5.2 is satisfied nearly exactly in the context of the alkali-halide salts. The corresponding lower bound of σ referring to the gas-phase equilibrium ion-pair distances d_g of Table 5.3 is 0.006 nm, which is only slightly higher, suggesting that an additivity principle analogous to Eq. 5.2 also holds with a nearly equal precision in this case. Note, however, that d_g values, unlike $L^{\ominus}$ values, correspond to experiment-based data involving a significant extent of theoretical post-processing. In contrast to the root-mean-square difference σ, the average difference δ can be brought arbitrarily close to zero by inclusion of a common additive constant to the radii of a given set. Choosing this offset in such a way that δ vanishes will lead to a derived set presenting the smallest possible σ value achievable based on the given original set. A few sets in Table 5.4 (Sets 33r-38r with respect to $L^{\ominus}$ and Set 63r with respect to d_g) were explicitly refined here using this principle.

Most sets of hard-sphere in-crystal radii in Table 5.4 are, as expected, associated with low δ and σ values referring to the $L^{\ominus}$ parameters. Similarly, the single set of hard-sphere gas-phase radii (Set 63r of type N_{10}) is associated with a δ value of zero and the smallest possible σ value referring to the d_g parameters, because this set was designed for such an optimal partitioning. As evident from Figure 5.1, the spread between alternative sets of hard-sphere in-crystal radii is relatively limited, due to the enforcement of the additivity constraint. The spread is somewhat more important between alternative sets of soft-sphere in-crystal radii, due to the variability in the physical meaning of the effective ionic radius in the different models considered for its definition. The discussion of hard- and soft-sphere gas-phase radii of type N_{10} is not meaningful here (only 4 sets in total). Finally, isolated-ion gas-phase radii of types L_{10} and M_{10} and, to an even larger extent, in-water radii, are affected by a considerable variability across alternative sets, due to the absence of any additivity constraint (even loosely enforced in the context of soft-sphere radii) and to a strong dependence on the model considered for the radius definition.

As a final note, a few in-crystal sets of Table 5.4 are historically particularly "famous" and usually referred to by the names of their authors. These are the Goldschmidt (Set 4r; 1926), Pauling (Set 5r; 1927), Goldschmidt-Ahrens (cations from Set 9r, 1952; anions from Set 4r, 1926), Pauling-Ahrens (cations from Set 9r, 1952; anions from Set 5r, 1927), Gourary-Adrian (Set 22r; 1960), and Shannon-Prewitt (Set 12r or Set 13r; 1976) radii, to which one may add the thermochemical radii[221,2056] (Set 27r; 1926-2008) used in the Kapustinskii approach[398].

5.5 Relative electrode (redox) potentials

The *standard relative electrode (redox) potentials* $\Delta_H \mathcal{V}^\ominus[I_{el}, I_d^z]$ of element-ion couples (I_{el}, I_d^z, with $|z| = 1$) relative to the reference hydrogen electrode in a given solvent can be measured *via* Galvanic cell measurements (Section 4.4.3). Also experimentally accessible are the corresponding standard *isothermal temperature derivatives*, $\partial_T \Delta_H \mathcal{V}^\ominus[I_{el}, I_d^z]$, and *thermal temperature derivatives* $\partial_{T'}^\ominus \Delta_{T,T'} \mathcal{V}^\ominus[I_{el}, I_d^z]$. In the context of aqueous systems, standard redox potentials $\Delta_H \mathcal{V}[I_{el}, I_{aq}]$ are commonly tabulated[221] using a 1 atm (1.013 bar, rather than $P^\circ = 1$ bar) reference pressure and a 1 molar (1 mol·dm^{-3}, rather than $b^\circ = 1$ mol·kg^{-1}) reference concentration. This corresponds to the *aaMe$_T$* convention, rather than the *bbme$_T$* convention employed in this book (Section 4.2.4). The conventions are, however, essentially interchangeable for aqueous solutions in view of the small sizes of the correction terms (Section 4.2.17). The measurement of standard redox potentials for the *alkali* element-ion couples is possible, but poses a specific *technical problem.* Due to their negative redox potentials, alkali metals react with water, *i.e.* they spontaneously reduce H_{aq}^+ to H_{el}. For this reason, they cannot be used as electrodes in an aqueous environment. To overcome this problem, the measurement is performed in two steps, relying on the use of *alkali amalgams* that are non-reactive in water[1820–1824]: (*i*) measurement of the electrode potential of the amalgam relative to the pure alkali metal in a solvent where the alkali metal is non-reactive and a salt of its cation is soluble (*e.g.* liquid amine

Table 5.5 *Standard relative electrode (redox) potentials and temperature derivatives for the alkali and halide element-ion couples as well as the hydrogen-proton couple in water.* The reported data includes the electrode potential $\Delta_H \mathcal{V}^\ominus$ relative to the reference hydrogen electrode, the isothermal first temperature derivative $\partial_T \Delta_H \mathcal{V}^\ominus$ of this potential, the isothermal second temperature derivative $\partial_T^2 \Delta_H \mathcal{V}^\ominus$ of this potential, and the thermal temperature derivative $\partial_{T'} \Delta_{T,T'} \mathcal{V}^\ominus$ of the potential measured between two identical electrodes (half-cells), the reference one at temperature T^- and the test one at temperature T' upon differentiation with respect to T' only at $T' = T^-$. The notation I_{el}, I_{aq}^z refers to an alkali (M_{el}, M_{aq}^z) or halide (X_{el}, X_{aq}^z) element-ion couple in water, or to the hydrogen-proton (H_{el}, H_{aq}^+) couple. The standard states are according to the $bbme_T$ convention (Section 4.2.4; reference pressure $P^\circ = 1$ bar, reference molality $b^\circ = 1$ mol·kg^{-1} and reference temperature $T^- = 298.15$ K). Note, however, that the reported data is actually quoted for the $aaMe_T$ convention (Section 4.2.17; reference pressure 1 atm and reference molarity 1 mol·dm^{-3}). It was not corrected in view of the small sizes of the correction terms.

I_{el}, I_{aq}^z	Reduction reaction	$\Delta_H \mathcal{V}^\ominus[I_{el}, I_{aq}^z]$ [V]	$\partial_T \Delta_H \mathcal{V}^\ominus[I_{el}, I_{aq}^z]$ [10^{-3} V·K^{-1}]	$\partial_T^2 \Delta_H \mathcal{V}^\ominus[I_{el}, I_{aq}^z]$ [10^{-6} V·K^{-2}]	$\partial_{T'} \Delta_{T,T'} \mathcal{V}^\ominus[I_{el}, I_{aq}^z]$ [10^{-3} V·K^{-1}]
$1/2 H_{2,g}, H_{aq}^+$	$H_{aq}^+ + e^- \rightleftharpoons 1/2\ H_{2,g}$	0^a	0^a	0^a	0.871
Li_s, Li_{aq}^+	$Li_{aq}^+ + e^- \rightleftharpoons Li_s$	-3.0401	-0.534	-	0.337
Na_s, Na_{aq}^+	$Na_{aq}^+ + e^- \rightleftharpoons Na_s$	-2.71	-0.772	-	0.099
K_s, K_{aq}^+	$K_{aq}^+ + e^- \rightleftharpoons K_s$	-2.931	-1.080	-	-0.209
Rb_s, Rb_{aq}^+	$Rb_{aq}^+ + e^- \rightleftharpoons Rb_s$	-2.98	-1.245	-	-0.374
Cs_s, Cs_{aq}^+	$Cs_{aq}^+ + e^- \rightleftharpoons Cs_s$	-3.026	-1.197	-	-0.326
$1/2 F_{2,g}, F_{aq}^-$	$1/2 F_{2,g} + e^- \rightleftharpoons F_{aq}^-$	2.866	-1.830	-5.339	-0.959
$1/2 Cl_{2,g}, Cl_{aq}^-$	$1/2 Cl_{2,g} + e^- \rightleftharpoons Cl_{aq}^-$	1.35827	-1.260	-5.454	-0.389
$1/2 Br_{2,\ell}, Br_{aq}^-$	$1/2 Br_{2,\ell} + e^- \rightleftharpoons Br_{aq}^-$	1.066	-0.629	-6.210	0.242
$1/2 I_{2,s}, I_{aq}^-$	$1/2 I_{2,s} + e^- \rightleftharpoons I_{aq}^-$	0.5355	-0.148	-5.965	0.723
Ref.		221	1829	2099	1829
Notes		b	b	b	b

(a) by definition; (b) standard values in the $bbme_T$ convention are approximated here by standard values in the $aaMe_T$ convention.

solvent and iodide salt); (*ii*) measurement of the electrode potential of the amalgam relative to the reference hydrogen electrode in an aqueous solution containing a dissolved salt of the alkali cation.

All the above quantities, *i.e.* redox potentials, isothermal and thermal first derivatives, as well as isothermal second derivatives for the anions, have been determined[221,1829,2099] for the alkali (M_{el}, M_{aq}^+) and halide (X_{el}, X_{aq}^-) element-ion couples in water. These standard potentials and temperature derivatives for the alkali and halide element-ion couples as well as the hydrogen-proton couple in water are reported in Table 5.5.

The values of $\Delta_H \mathcal{V}^\ominus$ are negative for all alkali metal-ion couples, indicating that M_{el} is a stronger reducer than H_{el} or, equivalently, that M_{aq}^+ is a weaker oxidant than H_{aq}^+. The reducing power of M_{el} increases very slightly and monotonically along the series from Na_{el} to Cs_{el}. However, Li_{el} is about as strong a reducer as Cs_{el}. The values of $\Delta_H \mathcal{V}^\ominus$ are positive for all halide element-ion couples, indicating that X_{el} is a stronger oxidant than H_{aq}^+ or, equivalently, that X_{aq}^- is a weaker reducer than H_{el}. The oxidizing power of X_{el} decreases markedly and monotonically along the series from F_{el} to I_{el}.

The isothermal first temperature derivatives $\partial_T \Delta_H \mathcal{V}^\ominus$ are negative for all ions considered, indicating that a small temperature increase above T^- enhances the reducing power of M_{el} relative to H_{el} and diminishes the oxidizing power of X_{el} relative to H^+_{aq}.

As discussed in Section 4.5.1, the redox potential scale, recast in the form $zF\Delta_H\mathcal{V}^\ominus[I_{el}, I^z_d]$, is nothing but a form of conventional single-ion solvation free-energy scale, that was called there the *redox conventional scale*. This scale can be related to the standard conventional scale employed in this book *via* the free energies of formation of I^z_g and H^+_g, *i.e.* the sums of the corresponding atomization and ionization free energies, as (Eq. 4.422)

$$zF\Delta_H\mathcal{V}^\ominus[I_{el}, I^z_d] = \Delta_s G^\bullet[I^z_g] + \Delta_f G^\ominus[I^z_g] - z\Delta_f G^\ominus[\mathrm{H}^+_g] \quad , \tag{5.3}$$

with

$$\Delta_f G^\ominus[I^z_g] = \Delta_a G^\ominus[I_{el}] + \Delta_i G^\ominus[I_g] \quad . \tag{5.4}$$

Similarly, the isothermal first temperature derivative of the redox potential can be related to the corresponding conventional solvation entropy as (Eq. 4.423)

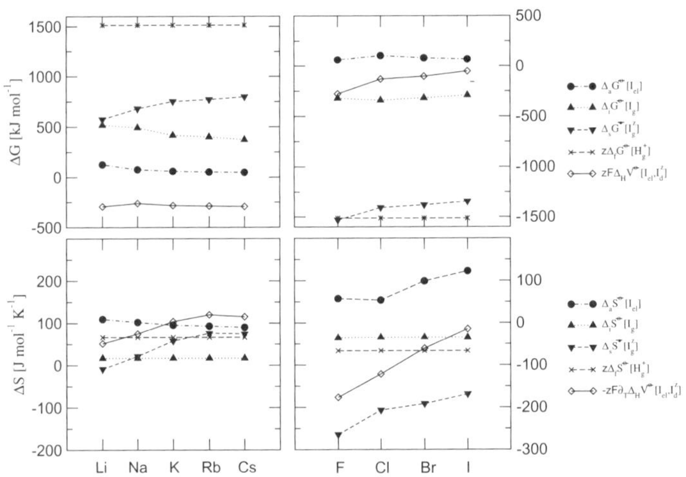

Figure 5.2 *Standard electrode (redox) potentials and first isothermal temperature derivatives for the alkali and halide element-ion couples in water, along with contributions of more fundamental thermodynamic parameters to these quantities.* The partitioning is performed according to Eqs. 5.3-5.6 (see also Eqs. 4.422 and 4.423). The numerical values for the redox potential $\Delta_H\mathcal{V}^\ominus[I_{el}, I^z_{aq}]$ and its first isothermal temperature derivative $\partial_T\Delta_H\mathcal{V}^\ominus[I_{el}, I^z_{aq}]$ are taken from Table 5.5. The numerical values for the atomization and ionization free energies ($\Delta_a G^\ominus[I_{el}]$ and $\Delta_i G^\ominus[I_g]$) and entropies ($\Delta_a S^\ominus[I_{el}]$ and $\Delta_i S^\ominus[I_g]$) are taken from Table 5.9. The numerical values for the gas-phase proton formation free energy ($\Delta_f G^\ominus[\mathrm{H}^+_g]$) and entropy ($\Delta_f S^\ominus[\mathrm{H}^+_g]$) are also taken from Table 5.9 (see also Table 4.2). The conventional solvation free energies ($\Delta_s G^\bullet[I^z_g]$) and entropies ($\Delta_s S^\bullet[I^z_g]$) are deduced by solving Eqs. 5.3-5.6 for these parameters based on the other quantities. These values, along with corresponding $\Delta_s H^\bullet[I^z_g]$ values obtained by application of the Gibbs equation, are also reported numerically in Table 5.16 (Sets 28g$_0$, 28h$_0$ and 28s̄$_0$).

$$-zF\partial_T \Delta_H \mathcal{V}^{\ominus}[I_{el}, I_d^z] = \Delta_s S^{\bullet}[I_g^z] + \Delta_f S^{\ominus}[I_g^z] - z\Delta_f S^{\ominus}[\mathrm{H}_g^+] \quad , \tag{5.5}$$

with

$$\Delta_f S^{\ominus}[I_g^z] = \Delta_a S^{\ominus}[I_{el}] + \Delta_i S^{\ominus}[I_g] \quad . \tag{5.6}$$

These relationships are illustrated in Figure 5.2 in the context of the alkali and halide series with water as a solvent. The connection provided by Eqs. 5.3-5.6 is interesting, because it permits the interpretation of trends observed in redox potentials on the basis of *more fundamental thermodynamic properties*, related to atomization, ionization and solvation. For example, it may be used to rationalize the above trends in $\Delta_H \mathcal{V}^{\ominus}$ and $\partial_T \Delta_H \mathcal{V}^{\ominus}$ concerning the alkali and halide ion series in water.

For both ion series, the sign and magnitude of the redox potential $\Delta_H \mathcal{V}^{\ominus}$ is seen to be dominated by the terms $\Delta_i G^{\ominus}[I_g]$, $\Delta_s G^{\bullet}[I_g^z]$ and $-z\Delta_f G^{\ominus}[\mathrm{H}_g^+]$, the term $\Delta_a G^{\ominus}[I_{el}]$ being comparatively much smaller. For the alkali, these terms are on the order of +450, +700, −1500 and +50 kJ·mol^{-1}, respectively, resulting in $zF\Delta_H \mathcal{V}^{\ominus}$ on the order of −300 kJ·mol^{-1}. For the halides, these terms are on the order of −300, −1400, +1500 and +50 kJ·mol^{-1}, respectively, resulting in $zF\Delta_H \mathcal{V}^{\ominus}$ on the order of −150 kJ·mol^{-1}. The difference in $F\Delta_H \mathcal{V}^{\ominus}$ between alkali and halides (halide minus alkali) can then be partitioned into four corresponding contributions on the order of −150, +700, 0 and −100 kJ·mol^{-1}, respectively, for a total difference of about +450 kJ·mol^{-1}. In other words, the observation that alkali M_{el} are strong reducers and halides X_{el} are strong oxidants follows mainly from solvation effects (the oxidation of M_{el} produces a hydrated cation M_{aq}^+, while the reduction of X_{aq}^- removes a hydrated anion to form X_{el}), partly counteracted by differences in atomization *vs.* "deatomization" and ionization *vs.* "deionization" free energies. This of course does not explain why alkali M_{el} are poor oxidants and halides X_{el} poor reducers. The reason is that although a similar solvation term would favor the reduction of M_{el} to M_{aq}^- over that of X_{aq}^+ to X_{el}, this effect is counteracted here by a strong opposite ionization term disfavoring the formation of the gas-phase ions M_g^- and X_g^+ which, unlike M_g^+ and X_g^-, are not closed-shell ions.

Further consideration of Figure 5.2 shows that the very slight decrease of $\Delta_H \mathcal{V}^{\ominus}$ along the alkali series from Na to Cs results from the partial compensation of a significant decrease in $\Delta_i G^{\ominus}$ (larger cations are more easily formed in the gas phase by electron abstraction) and a significant increase in $\Delta_s G^{\bullet}$ (larger cations are less efficiently hydrated), $\Delta_a G^{\ominus}$ being nearly constant (slight decrease). The Li anomaly results from a somewhat lower $\Delta_s G^{\bullet}$ value relative to that extrapolated from the series of larger ions, along with corresponding slightly positive and negative deviations for $\Delta_a G^{\ominus}$ and $\Delta_i G^{\ominus}$, respectively. The marked decrease of $\Delta_H \mathcal{V}^{\ominus}$ along the halide series from F to I is due to a significant increase in both $\Delta_i G^{\ominus}$ (larger anions are less easily formed in the gas phase by electron addition, F_g^- representing an exception to this trend) and $\Delta_s G^{\bullet}$ (larger anions are less efficiently hydrated, the hydration of F^-_{aq} being particularly favorable), $\Delta_a G^{\ominus}$ displaying more limited variations.

For both ion series and in contrast to the case of the redox potentials, the sign and magnitude of the first isothermal temperature derivative $\partial_T \Delta_H \mathcal{V}^{\ominus}$ is seen to be dominated by the terms $\Delta_a S^{\ominus}[I_{el}]$, $\Delta_s S^{\bullet}[I^z_g]$ and $-z\Delta_f S[\mathrm{H}^+_g]$, the term $\Delta_i S^{\ominus}[I_g]$ being here comparatively smaller, and essentially constant within the alkali and halide series. Along the alkali series, $\Delta_a S^{\ominus}$ decreases with the ion size while $\Delta_s S^{\bullet}$ increases more markedly, resulting in the progressive decrease of $\partial_T \Delta_H \mathcal{V}^{\ominus}$ (except for Cs). Along the halide series, both $\Delta_a S^{\ominus}$ (except for F) and $\Delta_s S^{\bullet}$ markedly increase with the ionic size, resulting in the progressive increase of $\partial_T \Delta_H \mathcal{V}^{\ominus}$.

An equation similar to Eqs. 5.3 and 5.5 also connects the isothermal second temperature derivative $\partial^2_T \Delta_H \mathcal{V}^{\ominus}$ of the redox potential to the conventional solvation heat capacity of the ion (Eq. 4.424). Written in terms of the conventional partial molar heat capacity of the solvated ion, the corresponding equation reads

$$-zFT^{-}\partial^2_T \Delta_H \mathcal{V}^{\ominus}[I_{el}, I^z_d] = c_P^{\bullet}[I^z_d] - c_P^{\ominus}[I_{el}] + zc_P^{\ominus}[\mathrm{H}_{el}] \quad . \tag{5.7}$$

Note that Eq. 5.5 may also be rewritten in the form of an equation for the partial molar entropy, as

$$-zF\partial_T \Delta_H \mathcal{V}^{\ominus}[I_{el}, I^z_d] = s^{\bullet}[I^z_d] - s^{\ominus}[I_{el}] + zs^{\ominus}[\mathrm{H}_{el}] \quad , \tag{5.8}$$

The conventional parameters $\Delta_s G^{\bullet}[I^z_g]$, $\Delta_s H^{\bullet}[I^z_g]$, $\Delta_s S^{\bullet}[I^z_g]$, $s^{\bullet}[I^z_{aq}]$ and $c_P^{\bullet}[I^z_{aq}]$ (the latter for anions only) deduced from the electrochemical data provided in Table 5.5 for alkali and halide element-ion couples in water, along with the Gibbs equation and using additional data from Tables 5.1 and 5.9, are reported later in Tables 5.16 (Sets 28g$_0$, 28h$_0$ and 28$\bar{\mathrm{s}}_0$) and 5.17 (Sets 28s$_6$ and 28c$_0$).

The pressure dependence of the redox potential may also be of interest, considering its connection with the conventional partial molar volume (first derivative), volume-compressibility (second derivative) and volume-expansivity (mixed pressure-temperature derivative) of a solvated ion. However, although pressure derivatives of redox potentials have been reported for some systems [2100], these quantities have, to our knowledge, never been determined for the alkali and halide element-ion couples in water.

5.6 Thermodynamic parameters of salt formation

The *standard formation parameters and molar entropies of the salts*, namely $\Delta_f Y^{\ominus}[IJ_s]$ and $s^{\ominus}[IJ_s]$, respectively, are accessible experimentally (Section 4.2.18). The former parameters that are most relevant include the formation free energies $\Delta_f G^{\ominus}[IJ_s]$, enthalpies $\Delta_f H^{\ominus}[IJ_s]$ and entropies $\Delta_f S^{\ominus}[IJ_s]$. The corresponding standard molar heat capacities, volumes, volume-compressibilities and volume-expansivities are also relevant, but have been discussed previously in Section 5.2.

The determination of $s^\ominus$ involves calorimetric measurements (TCI). The parameter $\Delta_f Y^\ominus$ characterizes the reaction of the elements to form the salt, *e.g.* the reaction of an alkali metal with a halide gas, liquid or solid. These reactions can seldom be performed under equilibrium conditions, which would give access to all formation parameters simultaneously, including derivative parameters *via* pressure and temperature analysis. Therefore, RXC is typically used to determine $\Delta_f H^\ominus$. The corresponding formation entropy $\Delta_f S^\ominus$ is then calculated as the difference between the molar entropy of the salt and those of the constituting elements (Section 5.1), *i.e.* by means of a TLC, while the formation free energy $\Delta_f G^\ominus$ is calculated by application of the Gibbs equation. If required, the formation heat capacity, volume, volume-compressibility or volume-expansivity can also be evaluated as a difference between the corresponding molar property of the salt (Section 5.2) and those of the constituting elements (Section 5.1). Alternatively, these parameters could be obtained from a further analysis of the pressure and temperature dependence of $\Delta_f G^\ominus$. In special cases where the formation parameters of a salt cannot be determined directly, they can still in principle be deduced from the formation parameters of the dissolved salt (Section 5.7) and the dissolution parameters of the salt (Section 5.8) by application of the FDR cycle. On the other hand, the application of the BFH cycle to deduce salt formation parameters, although in principle possible, is of no practical relevance since this cycle involves reaction parameters that are more difficult to access experimentally (ionization parameters $\Delta_i^\ominus Y[I_g]$ and $\Delta_i^\ominus Y[J_g]$, Section 5.10) or experimentally inaccessible (reticulation parameters $\Delta_l Y^\ominus[IJ_g]$, Section 5.11).

The values of the formation parameters and molar entropies have been determined[221] for all alkali-halide salts (MX_s) except RbF_s, for which only $\Delta_f H^\ominus$ is available to our knowledge. These standard thermodynamic parameters are reported in Table 5.6. Estimates of $s^\ominus$, $\Delta_f S^\ominus$ and $\Delta_f G^\ominus$ for RbF_s are also included, based on the BFH cycle using a non-experimental (interpolated) estimate for $\Delta_l S^\ominus$, as described in Section 5.11. Note that, by construction, the reported data satisfies the Gibbs equation ($\Delta_f G^\ominus[MX_s] = \Delta_f H^\ominus[MX_s] - T^- \Delta_f S^\ominus[MX_s]$) and the Hess law in terms of entropies ($\Delta_f S^\ominus[MX_s] = s^\ominus[MX_s] - s^\ominus[M_{el}] - s^\ominus[X_{el}]$; also using data from Table 5.1).

The $\Delta_f G^\ominus$ and $\Delta_f H^\ominus$ values are all very large and negative, indicating that the reaction of an alkali (solid) with a halide (gas, liquid or solid) is associated with both a strong driving force and an important release of heat. The $\Delta_f S^\ominus$ values are all negative. Except for the bromide and iodide salts, this predominantly results from the entropy reduction accompanying the association of a solid (alkali) with a gas (halide) to form a solid (salt). For the bromides, the reaction combines a solid with a liquid, and the entropy decrease is accordingly smaller. For the iodides, the reaction combines two solids and the

Table 5.6 *Standard thermodynamic parameters of salt formation and corresponding molar entropies for the alkali-halide salts.* The reported data includes the formation free energy $\Delta_f G^\ominus$, the formation enthalpy $\Delta_f H^\ominus$, the formation entropy $\Delta_f S^\ominus$, and the molar entropy $s^\ominus$. The notation MX_s refers to an alkali-halide salt. Standard molar heat capacities, volumes, volume-compressibilities and volume-expansivities of the alkali-halide salts can also be found in Table 5.2. If required, the formation heat capacity, volume, volume-compressibility or volume-expansivity can be obtained as a difference between the corresponding molar property of the salt (Table 5.2) and those of the constituting elements (Table 5.1). The standard states are according to the $bbme_T$ convention (Section 4.2.4; reference pressure $P^\circ = 1$ bar and reference temperature $T^- = 298.15$ K).

Salt MX_s	$\Delta_f G^\ominus[MX_s]$ [kJ·mol^{-1}]	$\Delta_f H^\ominus[MX_s]$ [kJ·mol^{-1}]	$\Delta_f S^\ominus[MX_s]$ [J·mol^{-1}·K^{-1}]	$s^\ominus[MX_s]$ [J·mol^{-1}·K^{-1}]
LiF	-587.7	-616.0	-94.9	35.6 [35.7]
LiCl	-384.4	-408.6	-81.2	59.5 [59.3]
LiBr	-342.0	-351.2	-30.8	74.4 [74.3]
LiI	-270.3	-270.4	-0.3	86.9 [86.8]
NaF	-546.3	-576.6	-101.6	51.1 [51.1]
NaCl	-384.1	-411.2	-90.9	71.9 [72.1]
NaBr	-349.0	-361.1	-40.6	86.8 [86.8]
NaI	-286.1	-287.8	-5.7	103.7 [98.5]
KF	-537.8	-567.3	-99.0	67.1 [66.6]
KCl	-408.5	-436.5	-93.9	82.3 [82.6]
KBr	-380.7	-393.8	-44.0	96.8 [95.9]
KI	-324.9	-327.9	-10.0	112.7 [106.3]
RbF	(-529.0)	-557.7	(-96.3)	(81.9) [-]
RbCl	-407.8	-435.4	-92.6	95.7 [95.9]
RbBr	-381.8	-394.6	-42.9	110.0 [110.0]
RbI	-328.9	-333.8	-16.4	118.4 [118.4]
CsF	-525.5	-553.5	-93.9	92.7 [92.8]
CsCl	-414.5	-443.0	-95.6	101.2 [101.2]
CsBr	-391.4	-405.8	-48.3	113.0 [113.1]
CsI	-340.6	-346.6	-20.1	123.2 [123.1]
Ref.	221	221		
Notes	a,b		a,c	a,d,e

(a) the value for RbF_s, indicated between parentheses, is based on the BFH cycle using a non-experimental (interpolated) estimate for $\Delta_l S^\ominus[RbF_g]$, see footnote (b) in Table 5.10, *i.e.* this value is calculated based on $\Delta_f S^\ominus[RbF_s] = \Delta_f S^\ominus[Rb_g^+] + \Delta_f S^\ominus[F_g^-] + \Delta_l S^\ominus[RbF_g]$ using data from Table 5.10 for the latter quantity and from Table 5.9 for the two former quantities; (b) calculated using the Gibbs equation ($\Delta_f G^\ominus[MX_s] = \Delta_f H^\ominus[MX_s] - T^- \Delta_f S^\ominus[MX_s]$) for RbF_s; (c) calculated using the Gibbs equation ($\Delta_f S^\ominus[MX_s] = (T^-)^{-1}(\Delta_f H^\ominus[MX_s] - \Delta_f G^\ominus[MX_s])$), except for RbF_s; (d) calculated from the salt formation entropy and the molar entropies of the constituting elements ($s^\ominus[MX_s] = s^\ominus[M_{el}] + s^\ominus[X_{el}] + \Delta_f S^\ominus[MX_s]$; using values from Table 5.1 for the two former quantities); (e) the molar entropies quoted in Ref.[221], indicated between square brackets for comparison, may slightly differ from the reported (recalculated) values (*i.e.* the recalculated values correct for a slight numerical inconsistency in the data set of Ref.[221] for $s^\ominus[M_{el}]$, $s^\ominus[X_{el}]$, $s^\ominus[MX_s]$, $\Delta_f G^\ominus[MX_s]$ and $\Delta_f H^\ominus[MX_s]$).

entropy decrease is accordingly even smaller. The $s^{\ominus}$ values systematically increase with the effective sizes (atomic numbers) of the involved ions, *i.e.* along the series $Li^+ \rightarrow Cs^+$ for salts involving a common anion and $F^- \rightarrow I^-$ for salts involving a common cation. The latter trend is probably related to the weakening of the Coulombic interactions within the crystal (Section 5.2), the magnitude of which decreases with increasing nearest-neighbor cation-anion distance, approximating the ions as point charges. This enables more pronounced oscillations of the ions about their lattice positions, resulting in the observed entropy increase.

5.7 Thermodynamic parameters of dissolved salt formation

The *standard formation parameters and partial molar variables of the dissolved salts*, namely $\Delta_f Y^{\ominus}[IJ_d]$ and $y^{\ominus}[IJ_d]$, respectively, where IJ_d represents the dissociated species I_d^z and J_d^{-z} in solution, are accessible experimentally (Section 4.2.18). The former parameters that are most relevant include the formation free energies $\Delta_f G^{\ominus}[IJ_d]$, enthalpies $\Delta_f H^{\ominus}[IJ_d]$ and entropies $\Delta_f S^{\ominus}[IJ_d]$. The latter variables include the partial molar entropies $s^{\ominus}[IJ_d]$, heat capacities $c_P^{\ominus}[IJ_d]$, volumes $v^{\ominus}[IJ_d]$, volume-compressibilities $k_T^{\ominus}[IJ_d]$ and volume-expansivities $a_P^{\ominus}[IJ_d]$.

The parameter $\Delta_f Y^{\ominus}$ characterizes the reaction of the elements to form the dissolved salt, *e.g.* the reaction upon mixing an alkali metal with a halide gas, liquid or solid in water. In practice, however, it would be difficult to avoid the side reaction of the alkali metal with the solvent itself. These reactions can formally be performed under equilibrium conditions using Galvanic cell measurements (Section 4.4.3), giving access to all formation parameters simultaneously, including derivative parameters *via* pressure and temperature analysis. In practice, however, the measurements concerning alkali metals must be performed indirectly, using alkali amalgams that are non-reactive in water (Section 5.5). The formation free energy $\Delta_f G^{\ominus}$ is directly related to the *difference in redox potentials* $\Delta_H \mathcal{V}^{\ominus}$ between the involved element-ion couples. In the specific case of alkali and halide ions with water as a solvent, this connection reads (Eqs. 4.373 and 4.380)

$$\Delta_f G^{\ominus}[MX_{aq}] = -F\{\Delta_H \mathcal{V}^{\ominus}[X_{el}, X_{aq}^-] - \Delta_H \mathcal{V}^{\ominus}[M_{el}, M_{aq}^+]\} \ . \tag{5.9}$$

The corresponding formation entropies $\Delta_f S^{\ominus}$ are thus related to the isothermal first temperature derivatives $\partial_T \Delta_H \mathcal{V}^{\ominus}$ of these redox potentials, as

$$\Delta_f S^{\ominus}[MX_{aq}] = F\{\partial_T \Delta_H \mathcal{V}^{\ominus}[X_{el}, X_{aq}^-] - \partial_T \Delta_H \mathcal{V}^{\ominus}[M_{el}, M_{aq}^+]\} \ . \tag{5.10}$$

The corresponding formation enthalpies $\Delta_f H^{\ominus}$ may then be calculated by application of the Gibbs equation.

Finally, the partial molar entropy $s^{\ominus}$ of the dissolved salt may be calculated as the sum of its formation entropy and the molar entropies of the constituting elements (Section 5.1). Note that $s^{\ominus}$ cannot be obtained directly *via* TCI, which is only applicable to pure substances (Section 4.2.11). If required, the formation heat capacity, volume, volume-compressibility or volume-expansivity can also be evaluated as a difference between the corresponding partial molar property of the dissolved salt (see below) and the corresponding molar properties of the constituting elements (Section 5.1). In special cases where the formation parameters of a dissolved salt cannot be determined experimentally, they can still in principle be deduced from the formation parameters (Section 5.6) and the dissolution parameters (Section 5.8) of the salt by application of the FDR cycle. On the other hand, the application of the ISR cycle to deduce dissolved salt formation parameters, although in principle possible, is of now practical relevance since this cycle involves reaction parameters that are more difficult to access experimentally (ionization parameters $\Delta_i Y[I_g]$ and $\Delta_i Y[J_g]$, Section 5.10) or experimentally inaccessible (salt solvation parameters $\Delta_s Y[IJ_g]$, Section 5.12).

In principle, an approach similar to that of Eqs. 5.9 and 5.10 could be employed to evaluate the partial molar heat capacity $c_P^{\ominus}$, volume $v^{\ominus}$, volume-compressibility $k_T^{\ominus}$ and volume-expansivity $a_P^{\ominus}$ of the dissolved salt *via* further analysis of the pressure and temperature dependence of the redox potential. However, as discussed in Section 5.5, only the first temperature derivatives of $\Delta_H \mathcal{V}^{\ominus}$ have been determined experimentally for all alkali and halides in water, the second temperature derivatives being available for the halides only, and the pressure derivatives unavailable. Fortunately, all the above quantities can in principle be accessed by alternative routes.

The standard partial molar heat capacity $c_P^{\ominus}$ of a dissolved salt can be evaluated *via* determination of the heat capacity of salt solutions at various concentrations (calorimetry[2101,2102] or flow microcalorimetry[2103–2105]) and extrapolation to infinite dilution, typically by fitting to a Debye-Hückel approximation[98, 113, 2101]. The raw quantity monitored in these experiments is the so-called apparent molar heat capacity of the solute, defined as

$$c_P^{app} = n^{-1} \left(C_P - n_{svt} c_{P,svt} \right) \quad , \tag{5.11}$$

where C_P is the heat capacity of a solution containing n moles of the solute and n_{svt} moles of the solvent, and $C_{P,svt}$ the molar heat capacity of the pure solvent. The apparent molar heat capacity becomes equal to the partial molar one in the limit of infinite dilution.

The standard partial molar volume $v^{\ominus}$ of a dissolved salt can be evaluated *via* determination of the volume of salt solutions at various concentrations (densimetry[2105–2117], *e.g.* using a magnetic float densimeter[2118]) and extrapolation to infinite dilution, typically again by fitting to a Debye-Hückel approximation[98, 113, 2119, 2120]. Here again, the raw quantity monitored in these experiments is the so-called apparent molar volume of the solute, defined as

$$v^{app} = n^{-1} \left(V - n_{svt} v_{svt} \right) \quad , \tag{5.12}$$

where V is the volume of a solution containing n moles of the solute and n_{svt} moles of the solvent, and v_{svt} the molar volume of the pure solvent. The apparent molar volume becomes equal to the partial molar one in the limit of infinite dilution. Alternatively, these quantities can be inferred from the results of ultrasonic wave propagation velocity measurements on electrolyte solutions[2121–2123].

The standard partial molar volume-compressibility $k_T^\ominus$ of a dissolved salt can be evaluated *via* analysis of the pressure dependence of $v^\ominus$, the latter evaluated by either of the two above methods (compressimetry[2108,2123–2128]). Finally, the standard partial molar volume-expansivity $a_P^\ominus$ of a dissolved salt can be evaluated *via* analysis of the temperature dependence of $v^\ominus$, the latter evaluated by either of the two above methods (dilatometry[2123,2127,2129]).

As discussed in Section 4.2.5, partial molar quantities derived from an analysis of the pressure or/and temperature dependence of the dissolved salt formation (or solvation, Section 5.12) free energy, *e.g.* *via* Eq. 5.10 and analogous expressions for other derivatives, correspond to the standard-variant of the solute standard-state definition. The same applies to the $c_P^\ominus$ values determined using the above calorimetric methods. However, the standard values derived by extrapolation of electrolyte solution properties to infinite dilution, *e.g.* $v^\ominus$, $k_T^\ominus$, and $a_P^\ominus$ as described above, correspond to the density-corrected variant. The same would apply to $c_P^\ominus$ values evaluated by monitoring the concentration dependence of the heat of mixing, rather than of the solution heat capacity. Note also that the use of partial molar volume-compressibilities $k_T^\ominus$ and expansivities $a_P^\ominus$ is certainly to be preferred here over that of compressibilities $\kappa_T^\ominus$ and expansivities (thermal expansion coefficients) $\alpha_P^\ominus$, the quantities most commonly reported for pure substances, due to the ambiguity in the choice of a scaling volume in this case (Section 4.2.2, Eqs. 4.50-4.52). For consistency with this choice, $k_T^\ominus$ and $a_P^\ominus$ are reported and discussed as primary quantities throughout this chapter, including for pure substances, rather than $\kappa_T^\ominus$ and $\alpha_P^\ominus$.

The values of the formation parameters and partial molar variables have been determined[221] for all dissolved alkali-halide salts in water (MX_{aq}). These standard thermodynamic parameters are reported in Table 5.7. Note that, by construction, the reported data satisfies the Gibbs equation ($\Delta_f G^\ominus[MX_{aq}] = \Delta_f H^\ominus[MX_{aq}] - T^- \Delta_f S^\ominus[MX_{aq}]$) and the Hess law in terms of entropies ($\Delta_f S^\ominus[MX_{aq}] = s^\ominus[MX_{aq}] - s^\ominus[M_{el}] - s^\ominus[X_{el}]$; also using data from Table 5.1). The values reported in this table represent an experimental data set considered by the authors of this book as the most trustworthy to date, because it is based on the most recent available measurements, namely from Ref.[221] for $\Delta_f G^\ominus$ and $\Delta_f H^\ominus$, as well as the derived parameters $\Delta_f S^\ominus$ and $s^\ominus$, Ref.[2104] for $c_P^\ominus$, Refs.[2108,2122] for $v^\ominus$, Refs.[2126,2127] for $k_T^\ominus$ and Refs.[2123,2127,2129] for $a_P^\ominus$, along with gas-phase data from Refs.[294,1723,1750] (Table 5.9), and consistent with the data reported in the other tables of this chapter.

The recommended sets for $\Delta_f G^\ominus$, $\Delta_f H^\ominus$, $\Delta_f S^\ominus$ and $s^\ominus$ are adapted directly from Ref.[221], after correction of minor internal inconsistencies. The construction of the recommended sets for $c_P^\ominus$, $v^\ominus$, $k_T^\ominus$ and $a_P^\ominus$, as well as the consideration of alternative experimental data sets for all the above quantities, will be presented later in the context of the salt solvation parameters (Section 5.12). This ordering of the discussion was selected considering that the solvation contribution to the formation parameters is the one affected by the largest experimental uncertainty. No attempt was made to distinguish between experimental data appropriate for the standard or density-corrected variants of the solute standard-state definition (Section 4.2.5) on a case-to-case basis. The experimental values as reported in the original references are simply assumed to pertain to the standard variant for $\Delta_f H^\ominus$, $\Delta_f S^\ominus$, $s^\ominus$ and $c_P^\ominus$ and to the density-corrected variant for $v^\ominus$, $k_T^\ominus$ and $a_P^\ominus$, which represents the most likely situation (see above), but may not necessarily be the case in reality. According to the above assumption, the three latter standard parameters are thus indicated as $v^{\ominus *}$, $k_T^{\ominus *}$ and $a_P^{\ominus *}$ in the tables. Considering the differences between the two variants in the context of water (Table 4.1; values to be multiplied by two in the context of binary salts) and the magnitude of the values reported in Table 5.7, the errors possibly incurred by this ambiguity are limited but non-negligible. Taking $NaCl_{aq}$ as an example, the differences between the two variants amount to about 0.09, 2.7, 1.1, 19.9, 13.7, 11.5 and 2.6% for $\Delta_f H^\ominus$, $\Delta_f S^\ominus$, $s^\ominus$, $c_P^\ominus$, $v^{\ominus *}$, $k_T^{\ominus *}$ and $a_P^{\ominus *}$, respectively, $\Delta_f G^\ominus$ being identical in the two variants.

The $\Delta_f G^\ominus$ and $\Delta_f H^\ominus$ values are all very large and negative, indicating that the reaction of an alkali (solid) with a halide (gas, liquid or solid) in the presence of water to form the salt solution is associated with both a strong driving force and an important release of heat. The corresponding $\Delta_f S^\ominus$ values are negative for all fluorides and chlorides, and positive for all bromides and iodides (except LiBr). As was the case for the salt formation entropies (Section 5.6), these differences can be predominantly ascribed to the standard-state phase of the halide, namely gas for F_{el} and Cl_{el}, liquid for Br_{el} and solid for I_{el}. The $s^\ominus$ values are all positive (except for LiF) and increase systematically with the effective sizes (atomic numbers) of the involved ions, *i.e.* along the series $Li^+ \rightarrow Cs^+$ for salts involving a common anion and $F^- \rightarrow I^-$ for salts involving a common cation. The observation of positive partial molar entropies might appear surprising at first sight, because it seems to imply that dissolved ions promote a destructuring of their solvent environment. This intuitive reasoning is, however, incorrect.

The standard partial molar entropy $s^\ominus$ of a dissolved salt cannot be directly interpreted in terms of its solvent structuring or destructuring propensity, because it encompasses an ideal-gas contribution that depends on the different effective molar volumes accessible to the ions in the gas phase and dissolved states, as defined by the standard-state reference pressure P°, temperature T^- and molality b°, along with the solvent density $\rho_{svt}^\ominus$ at P° and T^-. Furthermore, it also includes a contribution related to the ideal-gas influence of the species on the molar volume, present in the gas-phase state but absent in the dissolved state. Finally, in the standard variant, a further contribution is related to the pressure or/and temperature dependence of the solvent density. This issue has already been encountered in Section 4.2.17, where it was observed that even in a standard-state convention equating the effective molar volumes accessible to species in the gas-phase and dissolved states, thereby eliminating the first of the three above influences, derivative thermodynamic solvation properties were still not identical to corresponding point-to-point solvation properties (Eq. 4.205).

In the present context, the relevant quantity to characterize exclusively the structuring effect of the ion onto the solvent is the semi-standard point-to-point solvation entropy $\Delta_s S^{\ominus\!\!\!\!\ominus}$ (Sections 2.2 and 4.2.16), which can be derived from $s^{\ominus}$ for an ion pair MX_{aq} as

$$\Delta_s S^{\ominus\!\!\!\!\ominus}[MX_g] = s^{\ominus}[MX_{aq}] - s^{\ominus}[MX_g] - 2\Delta_s S^{\ominus}_{std} \quad , \tag{5.13}$$

where the gas-phase absolute molar entropies are given in Table 5.9, and one has (Eqs. 4.183 and 4.184)

$$\Delta_s S^{\ominus}_{std} = -(T^-)^{-1}\Delta_s G^{\ominus}_{std} - R - RT^- \partial_T \ln \rho^{\ominus}_{svt} \quad . \tag{5.14}$$

Using numerical data for water given in Table 1.1, $\Delta_s S^{\ominus}_{std}$ evaluates to -34.34 J·mol^{-1}·K^{-1}. Similar considerations apply to the heat capacity, where the discussion should be based on the semi-standard point-to-point solvation heat capacity $\Delta C_P^{\ominus\!\!\!\!\ominus}$ of the salt. In this case, one has

$$\Delta_s C_P^{\ominus\!\!\!\!\ominus}[MX_g] = c_P^{\ominus}[MX_{aq}] - c_P^{\ominus}[MX_g] - 2\Delta_s C^{\ominus}_{P,std} \quad , \tag{5.15}$$

where, assuming the absence of thermally-accessible electronically-excited states, the gas-phase molar heat capacities are analytical and given by $c_P^{\ominus} = (5/2)R$, and (Eqs. 4.183 and 4.184)

$$\Delta_s C^{\ominus}_{P,std} = -R - RT^-[2\partial_T \ln \rho^{\ominus}_{svt} + T^- \partial_T^2 \ln \rho^{\ominus}_{svt}] \quad . \tag{5.16}$$

Using numerical data for water given in Table 1.1, $\Delta_s C^{\ominus}_{P,std}$ evaluates to 0.08 J·mol^{-1}·K^{-1}. Note that the very small value of this correction term is coincidental, and specific to the properties of water. On the other hand, the discussion of partial molar volumes, volume-compressibilities and volume-expansivities in the dissolved state can be performed directly on the basis of the corresponding standard quantities considering that these quantities are exactly equal to the corresponding point-to-point solvation properties in the density-corrected variant (Eq. 4.187)

$$\Delta_s Y^{\ominus\!\!\!\!\ominus}[MX_g] = y^{\ominus *}[MX_d] \quad . \tag{5.17}$$

The values $\Delta_s S^{\ominus\!\!\!\!\ominus}$ and $\Delta_s C_P^{\ominus\!\!\!\!\ominus}$ are also reported in Table 5.7.

The quantity $\Delta_s S^{\ominus\!\!\!\!\ominus}$ is negative for all alkali-halide salts, showing that the dissolved ion pairs indeed promote a structuring of their solvent environment. It decreases systematically in magnitude with the effective sizes (atomic num-

Table 5.7 *Standard thermodynamic parameters of dissolved salt formation and corresponding partial molar variables for the alkali-halide salts in water.* The reported data includes the formation free energy $\Delta_f G^\ominus$, the formation enthalpy $\Delta_f H^\ominus$, the formation entropy $\Delta_f S^\ominus$, the partial molar entropy $s^\ominus$, the partial molar heat capacity $c_P^\ominus$, the partial molar volume $v^\ominus$, the partial molar volume-compressibility $k_T^\ominus$ and the partial molar volume-expansivity $a_P^\ominus$, as well as the semi-standard point-to-point solvation entropy and heat capacity, $\Delta_s S^\oplus$ and $\Delta_s C_P^\oplus$, respectively (Sections 2.2 and 4.2.16). The notation MX_{aq} refers to an alkali-halide salt dissolved in water (where MX_{aq} represents the dissociated species M_{aq}^+ and X_{aq}^- in aqueous solution). If required, the formation heat capacity, volume, volume-compressibility or volume-expansivity can be obtained as a difference between the corresponding partial molar property of the dissolved salt and the corresponding molar properties of the constituting elements (Table 5.1). The reported values represent recommended data sets. These sets are also listed as Sets 1G, 1H, 1$\overline{\text{S}}$, 1S, 1C, 1V, 1K and 1A in Tables 5.11 and 5.12, along with alternative data sets. The standard states are according to the $bbme_T$ convention (Section 4.2.4; reference pressure $P^\circ = 1$ bar, reference molality $b^\circ = 1$ mol·kg^{-1} and reference temperature $T^- = 298.15$ K). The experimental values as reported in the original references are assumed here to pertain to the standard ($\Delta_f H^\ominus$, $\Delta_f S^\ominus$, $s^\ominus$, $c_P^\ominus$) or the density-corrected ($v^{\ominus *}$, $k_T^{\ominus *}$, $a_P^{\ominus *}$) variant (Section 4.2.5), which represents the most likely situation, but may not necessarily be the case in reality. The formation free energy ($\Delta_f G^\ominus$) is identical in the two variants.

Salt MX_{aq}	$\Delta_f G^\ominus[MX_{aq}]$ [kJ·mol^{-1}]	$\Delta_f H^\ominus[MX_{aq}]$ [kJ·mol^{-1}]	$\Delta_f S^\ominus[MX_{aq}]$ [J·mol^{-1}·K^{-1}]	$s^\ominus[MX_{aq}]$ [J·mol^{-1}·K^{-1}]
LiF	-571.9 (-569.9)	-611.1 (-607.1) [-611.3]	-131.5 (-125.0) [-132.1]	-1.0 (5.5) [-1.6]
LiCl	-424.6 (-424.4)	-445.6 (-445.3) [-445.6]	-70.4 (-70.0) [-70.4]	70.3 (70.6) [70.3]
LiBr	-397.3 (-396.2)	-400.0 (-398.9) [-400.0]	-9.1 (-9.2) [-9.1]	96.1 (96.1) [96.1]
LiI	-344.8 (-345.0)	-333.7 (-333.9) [-333.7]	37.2 (37.2) [37.2]	124.4 (124.4) [124.4]
NaF	-540.7 (-538.0)	-572.8 (-568.4) [-575.7]	-107.7 (-102.1) [-117.4]	45.0 (50.6) [35.3]
NaCl	-393.1 (-392.5)	-407.3 (-406.6) [-407.3]	-47.6 (-47.1) [-47.6]	115.2 (115.8) [115.2]
NaBr	-365.8 (-364.3)	-361.7 (-360.2) [-361.7]	13.8 (13.8) [13.8]	141.2 (141.2) [141.2]
NaI	-313.5 (-313.1)	-295.4 (-295.2) [-295.3]	60.7 (60.2) [61.0]	170.1 (169.6) [170.4]
KF	-562.1 (-559.3)	-585.0 (-580.9) [-585.0]	-76.8 (-72.4) [-76.8]	89.3 (93.7) [89.3]
KCl	-414.5 (-413.9)	-419.5 (-419.0) [-419.3]	-16.8 (-17.4) [-16.1]	159.4 (158.9) [160.1]
KBr	-387.2 (-385.7)	-373.9 (-372.7) [-373.9]	44.6 (43.5) [44.6]	185.4 (184.3) [185.4]
KI	-334.9 (-334.5)	-307.6 (-307.7) [-307.6]	91.6 (89.9) [91.6]	214.3 (212.7) [214.3]
RbF	-562.8 (-564.1)	-583.8 (-580.9) [-583.8]	-70.4 (-56.4) [-70.4]	107.8 (121.7) [107.8]
RbCl	-415.2 (-418.6)	-418.3 (-419.0) [-418.1]	-10.4 (-1.4) [-9.7]	177.9 (186.9) [178.6]
RbBr	-387.9 (-390.4)	-372.7 (-372.7) [-372.7]	51.0 (59.4) [51.0]	203.9 (212.3) [203.9]
RbI	-335.6 (-339.2)	-306.4 (-307.6) [-308.7]	97.9 (105.8) [90.2]	232.8 (240.7) [225.1]
CsF	-570.8 (-568.5)	-590.0 (-586.7) [-590.4]	-64.4 (-61.1) [-65.7]	122.2 (125.6) [120.9]
CsCl	-423.2 (-423.0)	-425.4 (-424.8) [-425.2]	-7.4 (-6.1) [-6.7]	189.4 (190.7) [190.1]
CsBr	-396.0 (-394.8)	-379.8 (-378.5) [-379.8]	54.2 (54.8) [54.3]	215.5 (216.1) [215.6]
CsI	-343.6 (-343.6)	-313.5 (-313.5) [-313.3]	101.0 (101.2) [101.6]	244.3 (244.5) [244.9]
Ref.	221	221		
Notes	a,b	a,c,d	a,e,f	a,g,h

(continues on the next page)

Table 5.7 *(continued)*

Salt	$c_P^{\ominus}[MX_{aq}]$	$v^{\ominus *}[MX_{aq}]$	$k_T^{\ominus *}[MX_{aq}]$	$a_P^{\ominus *}[MX_{aq}]$	$\Delta_s S^{\ominus}[MX_{aq}]$	$\Delta_s C_P^{\ominus}[MX_{aq}]$
MX_{aq}	[J·mol^{-1}·K^{-1}]	[10^{-6} m^3·mol^{-1}]	[10^{-10} m^3·mol^{-1}·bar^{-1}]	[10^{-8} m^3·mol^{-1}·K^{-1}]	[J·mol^{-1}·K^{-1}]	[J·mol^{-1}·K^{-1}]
LiF	(-54.2)	(-2.5)	(-64.7)	(2.2)	-210.9	-96.0
LiCl	-65.2	17.0	-41.0	2.5	-147.7	-107.0
LiBr	-69.1	24.0	-33.4	(3.6)	-131.7	-110.9
LiI	(-59.6)	35.5	-22.0	(6.3)	-109.2	-101.4
NaF	-74.9	-3.5	-71.6	6.7	-179.8	-116.7
NaCl	-84.4	16.4	-49.6	7.7	-117.4	-126.2
NaBr	-88.1	23.5	-39.7	8.2	-101.6	-129.9
NaI	-76.5	35.1	-27.7	10.1	-78.4	-118.3
KF	-104.0	6.6	(-64.2)	(6.6)	-142.2	-145.8
KCl	-114.2	26.5	-41.3	6.3	-79.9	-156.0
KBr	-118.3	33.7	-33.0	8.0	-64.0	-160.1
KI	-108.7	45.4	-20.4	11.5	-40.9	-150.5
RbF	-124.9	12.8	(-61.2)	(5.9)	-133.4	-166.7
RbCl	-135.9	31.9	-38.3	6.2	-71.1	-177.7
RbBr	-139.0	38.7	(-29.7)	(7.3)	-55.2	-180.8
RbI	-131.5	50.3	(-17.7)	(10.0)	-32.1	-173.3
CsF	-136.3	20.1	(-57.8)	(5.5)	-124.5	-178.1
CsCl	-148.9	39.2	-34.9	5.8	-65.1	-190.7
CsBr	-155.3	46.1	(-26.3)	(6.9)	-49.2	-197.1
CsI	-145.0	57.7	(-14.3)	(9.6)	-26.1	-186.8
Ref.	2104	2108, 2122	2126, 2127	2129, 2123, 2127		
Notes	i,j	i,k	i,l	i,m	n	n

(a) the formation parameters and partial molar entropies calculated from the standard redox potentials and their isothermal first temperature derivatives (Table 5.5) according to Eqs. 5.9 and 5.10, along with the Gibbs equation, are reported between parentheses (the difference represents a slight numerical inconsistency in the data set of Ref.[221,1829] for $\Delta_H \mathcal{V}^{\ominus}$, $\partial_T \Delta_H \mathcal{V}^{\ominus}$, $\Delta_f G^{\ominus}[MX_{aq}]$ and $\Delta_f H^{\ominus}[MX_{aq}]$); (b) corresponding to Set 1G in Table 5.11; (c) values calculated from the salt formation and salt dissolution enthalpies ($\Delta_f H^{\ominus}[MX_{aq}] = \Delta_f H^{\ominus}[MX_s] + \Delta_d H^{\ominus}[MX_s]$; using values from Tables 5.6 for the former quantity and directly from Ref.[221] for the latter quantity, also reported in square brackets in Table 5.8) are indicated between square brackets for comparison and may slightly differ from the reported values (*i.e.* there is a slight numerical inconsistency in the data set of Ref.[221] for $\Delta_f H^{\ominus}[MX_s]$, $\Delta_f H^{\ominus}[MX_{aq}]$ and $\Delta_d H^{\ominus}[MX_s]$); (d) corresponding to Set 1H in Table 5.11; (e) calculated using the Gibbs equation ($\Delta_f S^{\ominus}[MX_{aq}] = (T^{-})^{-1}(\Delta_f H^{\ominus}[MX_{aq}] - \Delta_f G^{\ominus}[MX_{aq}])$); (f) corresponding to Set 1S in Table 5.11; (g) calculated from the dissolved salt formation entropy and the molar entropies of the constituting elements ($s^{\ominus}[MX_{aq}] = s^{\ominus}[M_{el}] + s^{\ominus}[X_{el}] + \Delta_f S^{\ominus}[MX_{aq}]$; using values from Table 5.1 for the two former quantities); (h) Set 1S in Table 5.12; (i) the derivation of the set based on the indicated references is detailed in Section 5.12 (values between parentheses correspond to non-experimental estimates inferred *via* optimal partitioning); (j) Set 1C in Table 5.12; (k) Set 1V in Table 5.12; (l) Set 1K in Table 5.12; (m) Set 1A in Table 5.12; (n) the quantities $\Delta_s S^{\ominus}[MX_{aq}]$ and $\Delta_s C_P^{\ominus}[MX_{aq}]$ are calculated using Eqs. 5.13 and 5.15, respectively.

bers) of the involved ions, *i.e.* along the series $Li^+ \rightarrow Cs^+$ for salts involving a common anion and $F^- \rightarrow I^-$ for salts involving a common cation. These observations suggest that the salts of the smallest alkali and halide ions are powerful "structure makers", *i.e.* they promote a large decrease in the entropy of the surrounding solvent, while the salts of the largest alkali and halide ions are less powerful "structure makers" but in no case "structure breakers", *i.e.* none of them promotes an increase in the entropy of the surrounding solvent (see Section 5.20 for further discussion of this topic).

The quantity $\Delta_s C_P^{\boxminus}$ is negative for all salts considered and nearly systematically increases in magnitude upon increasing the sizes of the ions, the transition from bromides to iodides representing a notable exception to this trend. The negative sign is compatible with a local stiffening of the water structure in the neighborhood of the ions. The ion-size dependence and the iodide anomaly are more difficult to rationalize. On the one hand, smaller ions exert a more intense electric field on the surrounding solvent, which is expected to enhance the stiffening effect. On the other hand, they are also surrounded by fewer solvent molecules in their first hydration shells, which is expected to reduce it.

The partial molar volumes $v^{\ominus *}$ (equal to $\Delta_s V^{\boxminus}$) are positive for all salts except LiF and NaF and nearly systematically increase upon increasing the sizes of the ions, the transition from lithium to sodium salts representing a notable exception. The generally positive sign is compatible with the expected excluded-volume effect upon inserting a new particle in a dense environment. The negative sign for LiF and NaF, as well as the ion-size dependence, are easy to understand. Smaller ions are characterized by smaller excluded volumes, but they also exert a more intense electric field on the surrounding solvent, which is expected to decrease the partial molar volume, possibly towards negative values, due to electrostriction (Section 3.2, Point E_3). However, the lithium anomaly is more difficult to rationalize.

The partial molar volume-compressibilities $k_T^{\ominus *}$ (equal to $\Delta_s K_T^{\boxminus}$) are negative for all salts considered and tendentially decrease in magnitude upon increasing the sizes of the ions, the transition from lithium to sodium salts representing again a notable exception. The negative sign is compatible with a local stiffening of the water structure in the neighborhood of the ions. However, the ion-size dependence and the lithium anomaly are more difficult to rationalize. In particular, if the negative values of $\Delta_s C_P^{\boxminus}$ and $k_T^{\ominus *}$ are both predominantly ascribed to a stiffening of the water structure, the corresponding trends along the series are generally opposite to each other, suggesting the influence of other factors.

Finally, the partial molar volume-expansivities $a_P^{\ominus *}$ (equal to $\Delta_s A_P^{\boxminus}$) are positive for all salts, tendentially increase upon increasing the size of the anion, and tendentially decrease upon increasing the size of the cation, except for the transition from lithium to sodium salts. The positive sign suggests that the perturbed water structure in the neighborhood of the ions renders it more prone to thermal expansion. The ion-size dependence is again more difficult to rationalize. A more detailed discussion of the trends observed in the partial molar variables along the alkali and halide series in water, including a tentative partitioning into single-ion contributions, will be provided in Section 5.20. Note, however, that the analysis is performed there on the basis of single-ion values corresponding to exactly-partitionable salt (pairwise) data, which slightly differs from the values presented in Table 5.7 and discussed in this section.

5.8 Thermodynamic parameters of salt dissolution

The *standard thermodynamic parameters of salt dissolution* $\Delta_d Y^\ominus[IJ_s]$ are accessible experimentally (Section 4.2.18). These parameters include the dissolution free energies $\Delta_d G^\ominus[IJ_s]$, enthalpies $\Delta_d H^\ominus[IJ_s]$ and entropies $\Delta_d S^\ominus[IJ_s]$. They characterize the dissolution of a salt in a solvent to form the dissolved salt, *e.g.* by plain mixing.

The dissolution reaction can be performed under equilibrium conditions, giving in principle access to all dissolution parameters simultaneously, including derivative parameters *via* temperature and pressure analysis. The relevant equilibrium parameter is the *saturation concentration* of the salt in the given solvent. If the salt is only poorly soluble, the corresponding activity coefficients of the solvated ions can be set to one or accurately evaluated using the Debye-Hückel approximation (Section 4.2.6), and the value of the equilibrium constant, called in this context the *solubility product*, is immediately accessible. Unfortunately, many salts, including the alkali-halide salts, are well soluble in water so that activity effects cannot be neglected or adequately approximated at the saturation concentration. Therefore, RXC is typically used instead to determine $\Delta_d H^\ominus$. The corresponding dissolution entropy $\Delta_d S^\ominus$ is then calculated as the difference between the partial molar entropy of the dissolved salt (Section 5.7) and the molar entropy of the salt (Section 5.6), *i.e.* by means of a TLC, and the dissolution free energy $\Delta_f G^\ominus$ is obtained by application of the Gibbs equation. If required, the dissolution heat capacity, volume, volume-compressibility or volume-expansivity can also be evaluated as a difference between the corresponding partial molar property of the dissolved salt (Section 5.7) and the corresponding molar property of the salt (Section 5.2). Alternatively, these parameters could be obtained from a further analysis of the pressure and temperature dependence of $\Delta_d G^\ominus$. In special cases where the dissolution parameters of a salt cannot be determined directly, they can still in principle be deduced from the formation parameters of the salt (Section 5.6) and the formation parameters of the dissolved salt (Section 5.7) by application of the FDR cycle. In the normal case, however, all these parameters are accessible independently, and the FDR cycle can be used to check their mutual consistency.

The values of the above parameters have been determined[221] for the dissolution of all alkali-halide salts (MX_s) in water except RbF_s, for which only $\Delta_d H^\ominus$ is available to our knowledge. These standard thermodynamic parameters are reported in Table 5.8. Estimates of $\Delta_d S^{\ominus}$ and $\Delta_d G^{\ominus}$ for RbF_s are also included, based on the BFH and FDR cycles using a non-experimental (interpolated) estimate for $\Delta_l S^{\ominus}$, see Section 5.11. Note that, by construction, the reported data satisfies the Gibbs equation ($\Delta_d G^{\ominus}[MX_s] = \Delta_d H^{\ominus}[MX_s] - T^{-}\Delta_d S^{\ominus}[MX_s]$), the Hess law in terms of entropies ($\Delta_d S^{\ominus}[MX_s] = s^{\ominus}[MX_{aq}] - s^{\ominus}[MX_s]$; also using data from Tables 5.6 and 5.7), and the closure of the FDR cycle in terms of free energies, enthalpies and entropies ($\Delta_f Y^{\ominus}[MX_s] + \Delta_d Y^{\ominus}[MX_s] - \Delta_f Y^{\ominus}[MX_{aq}] = 0$; also using data from Tables 5.6 and 5.7). As was the case for the values reported in Table 5.7 (see discussion in Section 5.7), the data reported in Table 5.8 represents an experimental data set considered by the authors of this book as the most trustworthy. The recommended sets for $\Delta_d G^{\ominus}$, $\Delta_d H^{\ominus}$ and $\Delta_d S^{\ominus}$ are adapted directly from Ref.[221], after correction of minor internal inconsistencies. The consideration of alternative experimental data sets will be presented later in the context of the salt solvation parameters (Section 5.12). As was the case for Table 5.7, no attempt was made here to distinguish between experimental data appropriate for the standard or density-corrected variants of the solute standard-state definition (Section 4.2.5) on a case-to-case basis. The experimental values as reported in the original reference are simply assumed to pertain to the standard variant for $\Delta_d H^{\ominus}$ and $\Delta_d S^{\ominus}$.

The trends in dissolution parameters along the alkali-halide salt series are illustrated graphically in Figure 5.3. The $\Delta_d G^{\ominus}$ values are all negative except for LiF and NaF, indicating that the dissolution of nearly all alkali-halide salts in water is associated with a favorable driving force under standard-state conditions, corresponding to the formation of an ideal solution at molality b°.

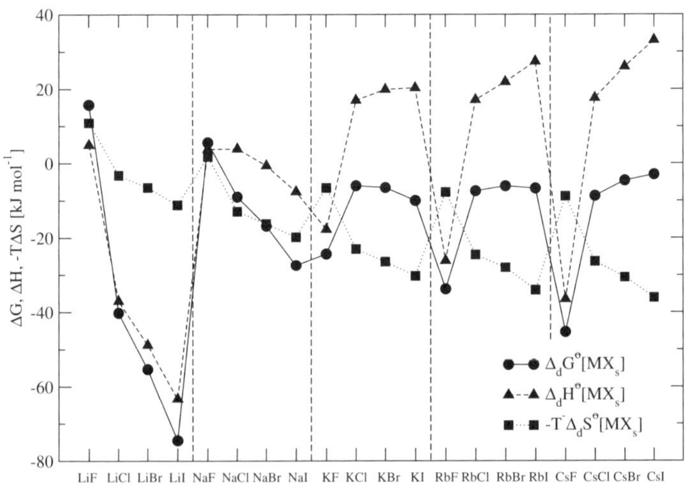

Figure 5.3 *Standard thermodynamic parameters of dissolution for the alkali-halide salts in water.* The numerical values for the dissolution free energy $\Delta_d G^{\ominus}[MX_s]$, enthalpy $\Delta_d H^{\ominus}[MX_s]$ and entropy $\Delta_d S^{\ominus}[MX_s]$ are taken from Table 5.8.

Table 5.8 *Standard thermodynamic parameters of dissolution for the alkali-halide salts in water.* The reported data includes the dissolution free energy $\Delta_d G^{\ominus}$, enthalpy $\Delta_d H^{\ominus}$ and entropy $\Delta_d S^{\ominus}$, as well as the saturation concentration D at P° and T^{-}. The notation MX_s refers to an alkali-halide salt. If required, the dissolution heat capacity, volume, volume-compressibility or volume-expansivity can be obtained as a difference between the corresponding partial molar property of the dissolved salt (Table 5.7) and the corresponding molar property of the salt (Table 5.2). The standard states are according to the $bbme_T$ convention (Section 4.2.4; reference pressure $P^{\circ} = 1$ bar, reference molality $b^{\circ} = 1$ mol·kg^{-1} and reference temperature $T^{-} = 298.15$ K). The experimental values as reported in the original references are assumed here to pertain to the standard variant (Section 4.2.5), which represents the most likely situation but may not necessarily be the case in reality.

Salt MX_s	$\Delta_d G^{\ominus}[MX_s]$ [kJ·mol^{-1}]	$\Delta_d H^{\ominus}[MX_s]$ [kJ·mol^{-1}]	$\Delta_d S^{\ominus}[MX_s]$ [J·mol^{-1}·K^{-1}]	$D[MX_s]$ [mol·kg^{-1}]
LiF	15.8	4.9 [4.73]	-36.6	0.05
LiCl	-40.2	-37.0 [-37.03]	10.8	10.2
LiBr	-55.3	-48.8 [-48.83]	21.7	18.8
LiI	-74.5	-63.3 [-63.30]	37.5	12.3
NaF	5.6	3.8 [0.91]	-6.1	0.98
NaCl	-9.0	3.9 [3.88]	43.3	6.1
NaBr	-16.8	-0.6 [-0.60]	54.4	9.1
NaI	-27.4	-7.6 [-7.53]	66.4	12.2
KF	-24.3	-17.7 [-17.73]	22.2	16.5
KCl	-6.0	17.0 [17.22]	77.1	4.6
KBr	-6.5	19.9 [19.87]	88.6	5.5
KI	-10.0	20.3 [20.33]	101.6	8.7
RbF	(-33.8)	26.1 [26.11]	(25.9)	12.6
RbCl	-7.4	17.1 [17.28]	82.2	7.5
RbBr	-6.1	21.9 [21.88]	93.9	6.5
RbI	-6.7	27.4 [25.10]	114.3	7.4
CsF	-45.3	-36.5 [-36.86]	29.5	24.0
CsCl	-8.7	17.6 [17.78]	88.2	10.9
CsBr	-4.6	26.0 [25.98]	102.5	5.8
CsI	-3.0	33.1 [33.35]	121.1	2.7
Ref.				2130
Notes	a,b	c,d	b,d	e

(a) calculated using the Gibbs equation ($\Delta_d G^{\ominus}[MX_s] = \Delta_d H^{\ominus}[MX_s] - T^{-}\Delta_d S^{\ominus}[MX_s]$); (b) the value for RbF_s, indicated between parentheses, is based *via* the FDR cycle on $\Delta_f S^{\ominus}[RbF_s]$, see footnote (a) in Table 5.6, itself based *via* the BFH cycle on a non-experimental (interpolated) estimate for $\Delta_l S^{\ominus}[RbF_g]$, see footnote (b) in Table 5.10, *i.e.* this value is calculated as $\Delta_d S^{\ominus}[RbF_s] = \Delta_f S^{\ominus}[RbF_{aq}] - \Delta_l S^{\ominus}[RbF_g] - \Delta_f S^{\ominus}[Rb_g^+] - \Delta_f S^{\ominus}[F_g^-]$, using data from Table 5.7 for the first quantity, from Table 5.10 for the second quantity and from Table 5.9 for the two last quantities; (c) the dissolution enthalpies quoted in Ref. [221], indicated between square brackets for comparison, may slightly differ from the reported values (*i.e.* there is a slight numerical inconsistency in the data set of Ref. [221] for $\Delta_f H^{\ominus}[MX_s]$, $\Delta_f H^{\ominus}[MX_{aq}]$ and $\Delta_d H^{\ominus}[MX_s]$); (d) calculated from the dissolved salt and salt formation parameters ($\Delta_d Y^{\ominus}[MX_s] = \Delta_f Y^{\ominus}[MX_{aq}] - \Delta_f Y^{\ominus}[MX_s]$; using values from Tables 5.6 and 5.7 for the two quantities); (e) solubility in moles of salt per kg water at P° and T^{-}.

This is in qualitative agreement with the observation that the corresponding saturation concentrations $D[MX_s]$ at P° and T^-, also reported in Table 5.8, are all well above b°, except for LiF and NaF. The $\Delta_d H^\ominus$ values are negative for LiCl, LiBr, LiI, NaBr, NaI, KF, RbF and CsF, indicating that the dissolution process under standard-state conditions is exothermic. This process is endothermic for all other salts and thus entropy-driven, except for LiF and NaF, where the entropy change is also negative. The $\Delta_d S^\ominus$ values are all positive except for LiF and NaF, which accounts for the disorder increase upon releasing the ions from their fixed lattice positions and turning them into freely diffusing species in solution, within an effective molar volume appropriate for an ideal solution at molality b°.

5.9 Thermodynamic parameters of atomization

The *standard thermodynamic parameters of atomization* $\Delta_a Y^\ominus[I_{el}]$, which are identical to the formation parameters $\Delta_f Y^\ominus[I_g]$ of the gas-phase atoms, as well as the *standard molar variables of the gas-phase atoms* $y^\ominus[I_g]$, are accessible experimentally (Section 4.2.18). The former parameters that are most relevant include the atomization free energies $\Delta_a G^\ominus[I_{el}]$, $\Delta_a H^\ominus[I_{el}]$ and entropies $\Delta_a S^\ominus[I_{el}]$. The latter variables include the molar entropies $s^\ominus[I_g]$, heat capacities $c_P^\ominus[I_g]$, volumes $v^\ominus[I_g]$, volume-compressibilities $k_T^\ominus[I_g]$ and volume-expansivities $a_T^\ominus[I_g]$.

The gas-phase molar thermodynamic parameters of isolated atoms can be determined using SBS (Section 4.2.12). In the absence of thermally-accessible electronically-excited states, the molar entropy $s^\ominus$ is an analytical function of the molar mass (Eq. 4.157 with $q_{in}^- = 2J_0 + 1$), that also depends on the total angular-momentum quantum number J_0 of the atom in its electronic ground state. This number is non-zero for the alkali and halide atoms, namely[2131] 1/2 for the alkali and 3/2 for the halides. Under the same assumption, the molar heat capacity $c_P^\ominus$, which does not depend on the molar mass and angular-momentum degeneracy, evaluates to $(5/2)R$ (Eq. 4.161), corresponding to a contribution $(3/2)R$ for the isochoric heat capacity associated with the three translational degrees of freedom of the atom, and a contribution R for the difference between the isochoric and isobaric heat capacities. An accurate evaluation of $s^\ominus$ and $c_P^\ominus$ in the presence of thermally-accessible electronically-excited states involves spectroscopic measurements of the electronic energy levels of the atom and application of statistical mechanics (Eqs. 4.157 and 4.160). Irrespective of the electronic properties of the atom, the molar volume $v^\ominus$ is given by $RT^-(P^\circ)^{-1}$ (Eq. 4.164), the molar volume-compressibility by $k_T^\ominus = RT^-(P^\circ)^{-2}$ (Eq. 4.165), corresponding to a compressibility $\kappa_T^\ominus = (P^\circ)^{-1}$, and the molar volume-expansivity by $a_P = R(P^\circ)^{-1}$ (Eq. 4.166), corresponding to an expansivity (thermal expansion coefficient) $\alpha_P^\ominus = (T^-)^{-1}$. The values of the above parameters are available[221] for all alkali (M_g) and halide (X_g) gas-phase atoms, based on a SBS treatment including consideration of electronically-excited states and angular-momentum degeneracy. Note that the consideration of excited states is required for the obtension of accurate results in the case of the F and Cl atoms[221], since these possess low-energy excited states[2131–2137]. The determination of thermodynamic parameters of atomization (gas-phase atom formation) is more complicated.

For the *alkali metals*, atomization is related to *sublimation*, with the difference that it produces ideal gas-phase atoms at the standard pressure P° rather than at the equilibrium vapor pressure P_{vap} of the metal. However, because the vapor pressure of metals is typically extremely low even at relatively high temperatures, the sublimation parameters must be determined indirectly. In practice, the *vaporization rate* of the metal is measured at different elevated temperatures over a certain range. This kinetic parameter can be related to the corresponding thermodynamic vapor pressure at a given temperature[2138,2139], *e.g.* *via* the Hertz-Knudsen-Langmuir equation[2140–2142] or a modified version thereof[2139]. The temperature dependence of the calculated vapor pressure may then be used to evaluate the sublimation parameters in either of two ways. In the so-called *second-law method*[1733,1753], the vapor pressure curve is fitted to the equation $R \ln P_{vap}(T) = -T^{-1}\Delta H + \Delta S$, where ΔH and ΔS are sublimation parameters in the form of mean values over the considered temperature interval. A more accurate expression involving the first-order temperature dependence of ΔH and ΔS can also be used instead. In the so-called *third-law method*[1733,1753], the same equation is used, but the entropy change ΔS is calculated at each temperature from the molar entropy of the ideal gas at pressure P_{vap} (SBS) and the molar entropy of the metal (TCI). As a result, ΔH and ΔS are obtained separately at each temperature, without fitting over a given interval. In both cases, the sublimation parameters are then extrapolated to the reference temperature T^{-}, *e.g.* using free-energy functions[1733,1753]. After correction for the different gas-phase pressures (equilibrium vapor pressure P_{vap} for sublimation *vs.* P° for atomization), the atomization parameters are obtained.

For the *halogen diatomic molecules*, atomization involves *bond dissociation*, possibly preceded by *vaporization* (Br_2) or *sublimation* (I_2). The parameters for the latter processes are evaluated from straightforward equilibrium measurements (vapor pressure at T^{-}). The parameters for the former bond-dissociation processes are obtained from SBS, *via* spectroscopic measurement of the electronic, vibrational and rotational energy levels of the molecule and application of statistical mechanics.

If required, the atomization (gas-phase atom formation) heat capacity, volume, volume-compressibility or volume-expansivity can also be evaluated as a difference between the corresponding molar property of the gas-phase atom (see above) and the corresponding molar property of the element (Section 5.1). Note that the application of the BFH or ISR cycles to deduce atomization parameters, although in principle possible, is of no practical relevance since these cycles involve reaction parameters that are more difficult to access experimentally (ionization parameters $\Delta_i Y^{\ominus}[I_g]$ and $\Delta_i Y^{\ominus}[J_g]$ for both cycles, Section 5.10) or experimentally inaccessible (reticulation parameters $\Delta_l Y^{\ominus}[IJ_g]$ for the BFH cycle, Section 5.11, and salt solvation parameters $\Delta_s Y^{\ominus}[IJ_g]$ for the ISR cycle, Section 5.12).

The values of the atomization (gas-phase atom formation) parameters and corresponding gas-phase atom molar variables have been determined[221] for all alkali (M_{el}) and halide (X_{el}) elements. These standard thermodynamic parameters, as well as those of hydrogen, are reported in Table 5.9. Note that, by construction, the reported data satisfies the Gibbs equation ($\Delta_f G^{\ominus}[I_g] = \Delta_f H^{\ominus}[I_g] - T^{-}\Delta_f S^{\ominus}[I_g]$) and the Hess law in terms of entropies ($\Delta_f S^{\ominus}[I_g] = s^{\ominus}[I_g] - s^{\ominus}[I_{el}]$; also using data from Table 5.1). Also recall that the symbols $\Delta_a Y^{\ominus}[I_{el}]$ and $\Delta_f Y^{\ominus}[I_g]$ represent the same quantities and can be used interchangeably. The $\Delta_f G^{\ominus}$ and $\Delta_f H^{\ominus}$ values are all very large and positive, indicating that the atomization process is a non-spontaneous and endothermic process, due to the disruption of interatomic interactions within a metal (alkali), intramolecular interactions within a diatomic molecule (halides) or/and intermolecular interactions within a liquid (bromine) or solid (iodine). The $\Delta_f S^{\ominus}$ values are all positive and large, due to the increase in disorder upon releasing individual gas-phase atoms from condensed phase or/and diatomic molecules. The $s^{\ominus}$ values are all positive and increase with the atomic number, as expected from Eq. 4.157. The $c_P^{\ominus}$ values are nearly identical for all atoms except F and Cl, due to the presence of thermally-accessible electronically-excited states. The $v^{\ominus}$, $k_T^{\ominus}$ and $a_P^{\ominus}$ values are all positive and exactly identical for all atoms, because they account for ideal-gas properties.

5.10 Thermodynamic parameters of ionization

The *standard thermodynamic parameters of ionization* $\Delta_i Y^{\ominus}[I_g]$, as well as the *standard molar variables of the gas-phase ions* $y^{\ominus}[I_g^z]$, are accessible experimentally (Section 4.2.18). The former parameters that are most relevant include the ionization free energies $\Delta_i G^{\ominus}[I_g]$, enthalpies $\Delta_i H^{\ominus}[I_g]$ and entropies $\Delta_i S^{\ominus}[I_g]$. The latter variables include the molar entropies $s^{\ominus}[I_g^z]$, heat capacities $c_P^{\ominus}[I_g^z]$, volumes $v^{\ominus}[I_g^z]$, volume-compressibilities $k_T^{\ominus}[I_g^z]$ and volume-expansivities $a_P^{\ominus}[I_g^z]$. The *standard formation parameters of the gas-phase ions* $\Delta_f Y^{\ominus}[I_g^z]$ are also of relevance, and immediately related to the atomization (gas-phase atom formation) parameters (Table 5.9) and ionization parameters as $\Delta_f Y^{\ominus}[I_g^z] = \Delta_a Y^{\ominus}[I_{el}] + \Delta_i Y^{\ominus}[I_g]$.

The gas-phase molar thermodynamic parameters of isolated ions can be determined using SBS (Section 4.2.12), including a special treatment for the electron and proton (Section 4.2.13). For monoatomic ions and in the absence of thermally-accessible electronically-excited states, $s^{\ominus}$ and $c_P^{\ominus}$ can be calculated analytically, according to Eq. 4.157 with $q_{in}^{-} = 2J_0 + 1$ and Eq. 4.161, respectively. Note that even under this assumption and neglecting the very small mass difference, the entropy of a monoatomic gas-phase ion may differ significantly from that of the corresponding gas-phase atom, because this

Table 5.9 *Standard thermodynamic parameters of gas-phase atom or ion formation and corresponding molar parameters for the alkali and halide atoms or ions, as well as the hydrogen atom and the proton.* The reported data includes the formation free energy $\Delta_f G^\ominus$, the formation enthalpy $\Delta_f H^\ominus$, the formation entropy $\Delta_f S^\ominus$, the heat-up enthalpy $\Delta_h H^\ominus$, the absolute molar entropy $s^\ominus$, the molar heat capacity $c_P^\ominus$, the molar volume $v^\ominus$, the molar volume-compressibility $k_T^\ominus$, the molar volume-expansivity $a_P^\ominus$, the compressibility $\kappa_T^\ominus$, the expansivity (thermal expansion coefficient) $\alpha_P^\ominus$, and the static electric dipole polarizability volume $\tilde{\alpha}_{el}$ for both gas-phase atoms and gas-phase ions, as well as estimates for the effective electron numbers n_{el} of the gas-phase ions (values corresponding to the isoelectronic noble gases, *i.e.* He→Li^+, Ne→Na^+/F^-, Ar→K^+/Cl^-, Kr→Rb^+/Br^-, and Xe→Cs^+/I^-). The (first) ionization potentials E_I and electron affinities E_A of the gas-phase atoms are also reported, as well as the corresponding ionization free energy $\Delta_i G^\ominus$, enthalpy $\Delta_i H^\ominus$ and entropy $\Delta_i S^\ominus$. The notation I_g refers to an alkali (M_g), halide (X_g) or hydrogen (H_g) gas-phase atom. The notation I_g^z refers an alkali (M_g^+), halide (X_g^-) or hydrogen (H_g^+; proton) gas-phase ion. The atomization parameters ($\Delta_a Y^\ominus[I_{el}]$; where the notation I_{el} refers to an alkali or halide element, or to hydrogen) are identical to the gas-phase atom formation parameters ($\Delta_f Y^\ominus[I_g]$). If required, the gas-phase atom formation (atomization) heat capacity, volume, volume-compressibility or volume-expansivity can be obtained as a difference between the corresponding molar property of the gas-phase atom and that of the corresponding element (Table 5.1). Similarly, the analogous gas-phase ion formation parameter can be obtained as a difference between the corresponding molar property of the gas-phase ion and that of the constituting element (Table 5.1) and electron (Table 4.2), and the analogous ionization parameter as a difference between the corresponding molar property of the gas-phase ion and that of the constituting gas-phase atom and electron (Table 4.2). The standard states are according to the $bbme_T$ convention (Section 4.2.4; reference pressure $P^\circ = 1$ bar and reference temperature $T^- = 298.15$ K; standard state for the free electron according to the warm-electron convention: ideal-gas at temperature T^-, according to Fermi-Dirac statistics; Fermi-Dirac statistics is also applied for the gas-phase standard state of the proton).

Property	H	Li	Na	K	Rb	Cs	F	Cl	Br	I	Ref.	Notes
$\Delta_f G^\ominus[I_g]$ [kJ·mol^{-1}]	203.28	126.60	76.96	60.48	53.08	49.56	62.28	105.31	82.38	70.17		a
$\Delta_f H^\ominus[I_g]$ [kJ·mol^{-1}]	217.998	159.3	107.5	89.0	80.9	76.5	79.38	121.301	111.87	106.76	221	
$\Delta_f S^\ominus[I_g]$ [J·mol^{-1}·K^{-1}]	49.377	109.662	102.418	95.661	93.314	90.371	57.3555	53.6495	98.913	122.717		b
$\Delta_h H^\ominus[I_g]$ [kJ·mol^{-1}]	6.197	6.197	6.197	6.197	6.197	6.197	6.519	6.272	6.197	6.197	294	
$s^\ominus[I_g]$ [J·mol^{-1}·K^{-1}]	114.717	138.782	153.718	160.341	170.094	175.601	158.751	165.190	175.018	180.787	221	
$c_P^\ominus[I_g]$ [J·mol^{-1}·K^{-1}]	20.8	20.8	20.8	20.8	20.8	20.8	22.7	21.8	20.8	20.8	221	
$v^\ominus[I_g]$ [10^{-3} m^3·mol^{-1}]	24.8	24.8	24.8	24.8	24.8	24.8	24.8	24.8	24.8	24.8		c
$k_T^\ominus[I_g]$ [10^{-2} m^3·mol^{-1}·bar^{-1}]	2.48	2.48	2.48	2.48	2.48	2.48	2.48	2.48	2.48	2.48		d
$a_P^\ominus[I_g]$ [10^{-5} m^3·mol^{-1}·K^{-1}]	8.31	8.31	8.31	8.31	8.31	8.31	8.31	8.31	8.31	8.31		e
$\kappa_T^\ominus[I_g]$ [bar^{-1}]	1.00	1.00	1.00	1.00	1.00	1.00	1.00	1.00	1.00	1.00		f
$\alpha_P^\ominus[I_g]$ [10^{-3} K^{-1}]	3.35	3.35	3.35	3.35	3.35	3.35	3.35	3.35	3.35	3.35		g
$\tilde{\alpha}_{el}[I_g]$ [10^{-30} m^3]	0.666793	24.33	24.11	43.4	47.3	59.42	0.557	2.18	3.05	5.35	221	h,i
$E_I[I_g]$ [eV]	13.598443	5.391719	5.139076	4.3406633	4.177128	3.893905	-	-	-	-	221	
$E_A[I_g]$ [eV]	-	-	-	-	-	-	3.4011895	3.612724	3.363588	3.059037	221	
$\Delta_i G^\ominus[I_g]$ [kJ·mol^{-1}]	1310.0	518.3	493.7	417.1	401.0	373.3	-321.0	-341.1	-317.7	-290.5		c

(continues on the next page)

Table 5.9 *(continued)*

Property	H^+	Li^+	Na^+	K^+	Rb^+	Cs^+	F^-	Cl^-	Br^-	I^-	Ref.	Notes
$\Delta_i H^\ominus[I_g]$	1315.1	523.4	498.8	422.2	406.1	378.4	-331.7	-351.4	-327.9	-300.7		j
	(1315.2)	(523.4)	(499.0)	(422.0)	(406.2)	(378.8)	(-331.6)	(-351.8)	(-327.7)	(-298.3)		k
[kJ·mol^{-1}]												
$\Delta_i S^\ominus[I_g]$	17.0	17.0	17.0	17.0	17.0	17.0	-35.9	-34.6	-34.3	-34.3		j
[J·mol^{-1}·K^{-1}]												
$\Delta_f G^\ominus[I_g^z]$	1513.32	644.98	570.71	477.63	454.17	422.91	-258.73	-235.77	-235.29	-220.32		l
[kJ·mol^{-1}]												
$\Delta_f H^\ominus[I_g^z]$	1533.10	682.73	606.31	511.21	487.05	454.91	-252.34	-230.08	-216.02	-193.95	294	m
[kJ·mol^{-1}]												
$\Delta_f S^\ominus[I_g^z]$	66.34	126.63	119.38	112.63	110.28	107.34	21.45	19.09	64.65	88.46		n
[J·mol^{-1}·K^{-1}]												
$\Delta_h H^\ominus[I_g^z]$	6.199	6.197	6.197	6.197	6.197	6.197	6.197	6.197	6.197	6.197	294	o
[kJ·mol^{-1}]												
$s^\ominus[I_g^z]$	108.95	133.02	147.95	154.58	164.33	169.84	145.58	153.36	163.49	169.26	1750	
[J·mol^{-1}·K^{-1}]												
$c_P^\ominus[I_g^z]$	20.8	20.8	20.8	20.8	20.8	20.8	20.8	20.8	20.8	20.8		p
[J·mol^{-1}·K^{-1}]												
$v^\ominus[I_g^z]$	24.8	24.8	24.8	24.8	24.8	24.8	24.8	24.8	24.8	24.8		c
[10^{-3} m^3·mol^{-1}]												
$k_T^\ominus[I_g^z]$	2.48	2.48	2.48	2.48	2.48	2.48	2.48	2.48	2.48	2.48		d
[10^{-2} m^3·mol^{-1}·bar^{-1}]												
$a_P^\ominus[I_g^z]$	8.31	8.31	8.31	8.31	8.31	8.31	8.31	8.31	8.31	8.31		e
[10^{-5} m^3·mol^{-1}·K^{-1}]												
$\kappa_T^\ominus[I_g^z]$	1.00	1.00	1.00	1.00	1.00	1.00	1.00	1.00	1.00	1.00		f
[bar^{-1}]												
$\alpha_P^\ominus[I_g^z]$	3.35	3.35	3.35	3.35	3.35	3.35	3.35	3.35	3.35	3.35		g
[10^{-3} K^{-1}]												
$\tilde{\alpha}_{el}[I_g^z]$	-	0.029	0.181	0.840	1.42	2.44	1.05	3.69	4.81	7.16	727	i
[10^{-30} m^3]												
$n_{el}[I_g^z]$	-	1.42	3.81	5.40	6.31	7.25	3.81	5.40	6.31	7.25	1321	i,q
		(1.92)	(3.31)	(5.36)	(6.78)	(8.46)	(3.31)	(5.36)	(5.36)	(8.46)	1289	i,r

(a) calculated using the Gibbs equation ($\Delta_f G^\ominus[I_g] = \Delta_f H^\ominus[I_g] - T^- \Delta_f S^\ominus[I_g]$); (b) calculated from the molar entropy of the gas-phase atom and that of the corresponding element ($\Delta_f S^\ominus[I_g] = s^\ominus[I_g] - s^\ominus[I_{el}]$, using the value from Table 5.1 for the latter quantity); (c) calculated as $v^\ominus = RT^-(P^\circ)^{-1}$; (d) calculated as $k_T^\ominus = RT^-(P^\circ)^{-2}$; (e) calculated as $a_P^\ominus = R(P^\circ)^{-1}$; (f) calculated as $\kappa_T^\ominus = (P^\circ)^{-1}$; (g) calculated as $\alpha_P^\ominus = (T^-)^{-1}$; (h) value for I_g based on the refraction index (an alternative calculated value of $4.7 \cdot 10^{-30}$ m^3 is also reported in the indicated reference); (i) see Sections 3.3.1 (Eq. 3.15) for the introduction of these quantities, Sections 3.3.2 and 6.2.2 for their use in force-field calibration (Eq. 3.53), and Section 6.1.1 for their use in corrections to the BO+ST model (Eq. 6.4); (j) calculated as the difference between gas-phase ion formation and gas-phase atom formation (atomization) parameters ($\Delta_i Y[I_g] = \Delta_f Y[I_g^z] - \Delta_f Y[I_g]$); (k) the alternative values for $\Delta_i H^\ominus[I_g]$ reported between parentheses are recalculated based on the raw E_I and E_A values, *i.e.* as $\Delta_i H^\ominus[I_g] = E_I[I_g] - \Delta_h H^\ominus[I_g] + \Delta_h H^\ominus[I_g^+] + \Delta_h H^\ominus[e^-]$ for cations (Eq. 5.18) and $\Delta_i H^\ominus[I_g] = -E_A[I_g] - \Delta_h H^\ominus[I_g] + \Delta_h H^\ominus[I_g^-] - \Delta_h H^\ominus[e^-]$ for anions (Eq. 5.19), where $\Delta_h H^\ominus[e^-]$ is taken from Table 4.2; (l) calculated using the Gibbs equation ($\Delta_f G^\ominus[I_g^z] = \Delta_f H^\ominus[I_g^z] - T^- \Delta_f S^\ominus[I_g^z]$); (m) the original data from the indicated reference was corrected from Boltzmann statistics to proper Fermi-Dirac statistics in the treatment of the electron standard state ($\Delta_f H^\ominus[I_g^z] = \Delta_f H^{(\ominus)}[I_g^z] - z\Delta_h H^{(\ominus)}[e_g^-] + z\Delta_h H^\ominus[e_g^-]$, using values from Table 4.2 for the two latter quantities), except for H_g^+ (value directly taken from Table 4.2); (n) calculated from the molar entropy of the gas-phase ion and that of the constituting gas-phase atom and electron ($\Delta_f S^\ominus[I_g^z] = s^\ominus[I_g^z] - s^\ominus[I_g] + zs^\ominus[e_g^-]$, using the value from Table 4.2 for the latter quantity); (o) all values are from the indicated reference, except $\Delta_h H^\ominus[H_g^+]$, which is taken from Table 4.2, as well as $\Delta_h H^\ominus[Cl_g^-]$ and $\Delta_h H^\ominus[I_g^-]$, set equal to $(5/2)RT^-$ assuming the absence of thermally-accessible electronically-excited states in the gas-phase ions; (p) calculated as $c_P^\ominus = (5/2)R$, assuming the absence of thermally-accessible electronically-excited states in the gas-phase ions; (q) values based on experimental estimates for the isoelectronic noble-gas dispersion coefficients and polarizabilities; (r) the values between parentheses are based on quantum-mechanical estimates for the isoelectronic noble-gas dispersion coefficients[1323] and polarizabilities from Ref.[1322].

quantity depends on the total angular-momentum quantum number J_0 of the species in its electronic ground state. This number is non-zero for the alkali and halide atoms (Section 5.9), but zero for the closed-shell alkali and halide ions. In contrast, the absence of thermally-accessible electronically-excited states is sufficient to guarantee that the heat capacity of a monoatomic gas-phase ion is identical to that of the corresponding gas-phase atom, namely $(5/2)R$. An accurate evaluation of $s^\ominus$ and $c_P^\ominus$ in the presence of thermally-accessible electronically-excited states involves spectroscopic measurements of the electronic energy levels of the ion and application of statistical mechanics (Eqs. 4.157 and 4.160). Note that these considerations do not entirely apply to the gas-phase proton and electron, for which Fermi-Dirac rather than Boltzmann statistics must be applied (Section 4.2.13). Irrespective of the electronic properties of the ion (and including the electron and proton), the quantities $v^\ominus$, $k_T^\ominus$ and $a_P^\ominus$ are analytical, evaluating to $RT^-(P^\circ)^{-1}$, $RT^-(P^\circ)^{-2}$ and $R(P^\circ)^{-1}$, respectively (Eqs. 4.164-4.166), and exactly identical to the corresponding parameters for the gas-phase atoms.

The values of the above parameters are available[1750] for all alkali (M_g^+) and halide (X_g^-) gas-phase ions, based on a SBS treatment omitting consideration of electronically-excited states. Assuming the absence of thermally-accessible electronically-excited states is reasonable for the alkali cations, which possess no low-energy excited states[2131]. However, the quality of this approximation[1750] in the context of the halide anions (in particular F^- and Cl^-, for which the corresponding atoms possess low-energy excited states[221]; Section 5.9) cannot be assessed since there is, to our knowledge, no available data concerning the electronic spectra of halide anions[2131,2133–2135,2143]. However, considering that these anions, unlike the corresponding atoms, possess a closed-shell electronic structure, the assumption nevertheless appears reasonable in this case.

The determination of the thermodynamic parameters of ionization relies on spectroscopic data. Using optical or (photo-)electron spectroscopy experiments, it is possible to measure the *ionization potentials*[392] of cation-forming species, with an accuracy[2144] on the order of 1 meV, *i.e.* about 0.1 kJ·mol^{-1} (estimate based on reported uncertainties for the first ionization potentials of alkali atoms). Using photo-electron or threshold spectroscopy (*e.g.* photo-ionization or laser photo-detachment) experiments, it is possible to measure the *electron affinities*[393] of anion-forming species, with an accuracy[221] on the order of 10^{-5} eV, *i.e.* about 10^{-3} kJ·mol^{-1} (estimated based on reported uncertainties for the electron affinities of halide atoms). In the specific context of alkali and halides, only the first ionization potential E_I and the single electron affinity E_A are relevant. As already mentioned in Sections 4.2.12 and 4.2.14, the first ionization potential is defined as the adiabatic energy change associated with the process $I_g \rightarrow I_g^+ + e_g^-$. The electron affinity is defined as the adiabatic energy change associated with the process $I_g^- \rightarrow I_g + e_g^-$. Here, adiabatic means that these quantities refer to ideal motionless species, *i.e.* E_I and E_A represent purely energetic quantities corresponding to a temperature

$T = 0$ K. Note that in the context of *thermodynamic parameters*, the word *ionization* will be used here to refer to the *conversion of an atom into an ion* either by addition or by abstraction of an electron, whereas in the context of spectroscopy, the term ionization potential is reserved to the formation of a cation by abstraction of an electron. Conversely, the spectroscopic term electron affinity refers to the formation of a neutral atom by abstraction of an electron from the anion, *i.e.* to a process occurring in the opposite direction compared to the thermodynamic ionization process.

To obtain standard ionization enthalpies, the values of E_I and $-E_A$ must be corrected to match a finite temperature T. For the transformation of a gas-phase atom I_g into a monovalent gas-phase cation I_g^+, one has

$$\begin{aligned} \Delta_i H^\ominus[I_g] &= E_I[I_g] - \Delta_h H^\ominus[I_g] + \Delta_h H^\ominus[I_g^+] + \Delta_h H^\ominus[e_g^-] \\ &\approx E_I[I_g] + \Delta_h H^\ominus[e_g^-] \,, \end{aligned} \tag{5.18}$$

where E_I is the ionization potential, expressed on a per mole basis (values are usually tabulated in terms of energy per atom, *e.g.* units of eV), and $\Delta_h H^\ominus$ is the heat-up enthalpy of a species (Section 4.2.9). Similarly, for the transformation of a gas-phase atom I_g into a monovalent gas-phase anion I_g^-, one has

$$\begin{aligned} \Delta_i H^\ominus[I_g] &= -E_A[I_g] - \Delta_h H^\ominus[I_g] + \Delta_h H^\ominus[I_g^-] - \Delta_h H^\ominus[e_g^-] \\ &\approx -E_A[I_g] - \Delta_h H^\ominus[e_g^-] \,, \end{aligned} \tag{5.19}$$

where E_A is the electron affinity expressed on a per mole basis (values are usually tabulated in terms of energy per atom, *e.g.* units of eV). The calculation of $\Delta_h H^\ominus$ has been discussed in Sections 4.2.12 (gas-phase atom and ion) and 4.2.13 (gas-phase electron and proton; see Table 4.2). The approximation in Eqs. 5.18 and 5.19 applies to the situation where neither the atom nor the ion possess thermally-accessible electronically-excited states at room temperature, which is reasonable for alkali and halide atoms and ions, except for the F and Cl atoms (Section 5.9). Note that these quantities do not include any volume work correction, *i.e.* the term RT^- usually occurring in energy to enthalpy conversions for ideal gases, because this term is already encompassed within the heat-up enthalpy of the electron.

The corresponding ionization entropies are obtained from the absolute entropies of the participating species, namely (for both cation and anion)

$$\Delta_i S^\ominus[I_g] = -s^\ominus[I_g] + s^\ominus[I_g^z] + z s^\ominus[e_g^-] \,. \tag{5.20}$$

The ionization free energy $\Delta_i G^\ominus$ is then calculated by application of the Gibbs equation. If required, the ionization heat capacity, volume, volume-compressibility or volume-expansivity can also be evaluated as a difference between the corresponding molar property of the gas-phase ion (see above) and that of the gas-phase atom (Section 5.9), taking also into account the corresponding contribution of the absorbed ($z < 0$) or produced ($z > 0$)

electrons (Section 4.2.13), in an analogous way to Eq. 5.20.

The ionization parameters are related to the formation parameters of the gas-phase atom (atomization parameters) and gas-phase ion *via* the Hess law, namely

$$\Delta_i Y^{\ominus}[I_g] = \Delta_f Y^{\ominus}[I_g^z] - \Delta_f Y^{\ominus}[I_g] \quad , \tag{5.21}$$

and in particular

$$\Delta_i S^{\ominus}[I_g] = \Delta_f S^{\ominus}[I_g^z] - \Delta_f S^{\ominus}[I_g] \quad . \tag{5.22}$$

In contrast to Eq. 5.20, as well as corresponding equations for other molar quantities, there is no contribution related to the electron here, since its corresponding formation parameters are zero by definition in the electron standard state (Section 4.2.13).

In special cases where the ionization parameters of a gas-phase atom cannot be determined directly, they can still in principle be deduced from thermodynamic cycles[392]. For this, one considers an atom pair consisting of the specific atom and a second atom, capable of forming an ion of opposite charge and for which the ionization parameters are known. The ionization parameters of the gas-phase atom pair are then deduced from the atomization (gas-phase atom formation) parameters of the atom pair (Section 5.9), the formation parameters of the salt (Section 5.6), and the reticulation parameters of the gas-phase ion pair (Section 5.11) by application of the BFH cycle. Since the ionization parameters of the second atom are known, those of the atom of interest are obtained by mere subtraction. This approach is, however, moderately accurate since the reticulation parameters of the gas-phase ion pair are not experimentally accessible and must be calculated using theoretical models (Section 5.11). On the other hand, the application of the ISR cycle to deduce ionization parameters, although in principle possible, is of no practical relevance since this cycle involves reaction parameters that are inaccessible experimentally and difficult to calculate theoretically with a reasonable accuracy (salt solvation parameters $\Delta_s Y^{\ominus}[IJ_g]$, Section 5.12). The evaluation of experimentally-inaccessible ionization enthalpies (in particular, those related to the electron affinities of the halogens[79,81]) based on theoretical estimates for the reticulation enthalpies, in view of their subsequent use to evaluate single-ion solvation enthalpies *via* the ISR cycle, is actually one of the main reasons for the historical importance of the BFH cycle[79–81] (1919). The other main point of relevance of the work of Born on the theoretical calculation of reticulation enthalpies[395,396,2145,2146] (1918-1919) was to permit the evaluation of salt solvation enthalpies based on experimentally-accessible salt dissolution enthalpies, as done in the work of Fajans[82,83,1869] (1919-1921).

The values of the ionization parameters, and corresponding gas-phase ion formation parameters and molar variables, have been determined[294,1723,2144] for all alkali (M_g) and halide (X_g) gas-phase atoms. These standard thermodynamic parameters, as well as those of the proton, are reported in Table 5.9. Note that, by construction, the reported data satisfies the Gibbs equation in terms of formation ($\Delta_f G^\ominus[I_g^z] = \Delta_f H^\ominus[I_g^z] - T^- \Delta_f S^\ominus[I_g^z]$) and ionization ($\Delta_i G^\ominus[I_g] = \Delta_i H^\ominus[I_g] - T^- \Delta_i S^\ominus[I_g]$) parameters, as well as the Hess law in terms of entropies ($\Delta_f S^\ominus[I_g^z] = s^\ominus[I_g^z] - s^\ominus[I_{el}] + zs^\ominus[e_g^-]$; also using data from Tables 4.2 and 5.1) and ionization parameters ($\Delta_i Y^\ominus[I_g] = \Delta_f Y^\ominus[I_g^z] - \Delta_f Y^\ominus[I_g]$). The reported ionization enthalpies $\Delta_i H^\ominus[I_g]$ are also essentially (although not exactly) consistent with the reported ionization potentials E_I and electron affinities E_A *via* Eqs. 5.18 and 5.19 (also using data from Table 4.2). The values recalculated based on E_I and E_A data from Ref.[221] and $\Delta_h H$ data from Refs.[294, 1723] are also provided in Table 5.9 for comparison (between parentheses), and deviate from the $\Delta_i H^\ominus$ values calculated from the atom[221] and gas-phase ion[294] formation parameters by 0.4 kJ·mol^{-1} at most (except for iodide: 2.4 kJ·mol^{-1}).

The $\Delta_i G^\ominus$ and $\Delta_i H^\ominus$ values are large and positive for the gaseous alkali atoms, indicating that the ionization process is non-spontaneous and endothermic, as expected for a charge separation process involving the conversion of a neutral species (atom) into two charged species (ion and electron). The values are large and negative for the gaseous halide atoms, indicating that the ionization process is spontaneous and exothermic, as expected for a process involving the transfer of a non-interacting electron into a neutralizing electron cloud surrounding a positively charged nucleus, *i.e.* the interaction of the additional electron with the nucleus is only partially offset by the repulsion originating from the other electrons. For both alkali and halides, these values nearly systematically decrease in magnitude with increasing atomic number, reflecting the higher principal quantum number associated with the abstracted (alkali) or added (halides) electron. A notable exception to this trend is the F atom, the ionization of which is less favorable than that of the Cl atom. This results from a particularly important electron-electron repulsion in the tight valence shell of the F$^-$ anion.

The fact that the halide ions are characterized by negative $\Delta_i G^\ominus$ values implies by no means that X_g^- is the most stable species of these elements in the gas phase, because the ideal gas of non-interacting electrons is an abstraction. Due to the electroneutrality constraint, these electrons must be removed from other atoms (*e.g.* alkali, thereby forming M_g^+), rendering the coupled ionization process unfavorable in common situations. For the alkali-halide pairs, the sums of $\Delta_i G^\ominus[M_g]$ and $\Delta_i G^\ominus[X_g]$ values are all positive, indicating that the neutral gas-phase atom pair $M_g + X_g$ is systematically more stable than the corresponding gas-phase ion pair MX_g (*i.e.* $M_g^+ + X_g^-$), by 32.3 (Cs_g+Cl_g) to 227.9 (Li_g+I_g) kJ·mol^{-1}. However, these values remain relatively low, because the coupled ionization process of the two atoms converts two open-shell species into two closed-shell species. As a result, the presence of strong interionic interactions within a crystal or ion-solvent interactions within a solution can easily shift the above equilibrium in the direction of the ion pair.

The $\Delta_i S$ values are positive and identical for all gaseous alkali atoms, and exclusively reflect the entropy of the released electron and the decrease in the angular-momentum degeneracy $2J_0 + 1$ from 2 to 1. The values are negative and nearly identical for all gaseous halide atoms, and reflect the entropy of the absorbed electron and the decrease in angular-momentum degeneracy from 4 to 1, along with a small contribution from electronically-excited states for F and Cl (Section 5.9).

The $s^\ominus$ values are all positive and increase with the atomic number, as expected from Eq. 4.157. The $c_P^\ominus$ values are positive and identical for all ions. They are analytical, assuming the absence of thermally-accessible electronically-excited states in the gas-phase ions (see above), and also equal to those of the corresponding gas-phase atoms except for F and Cl, due to the presence of thermally-accessible electronically-excited states in these gas-phase atoms (Section 5.9). The $v^\ominus$, $k_T^\ominus$ and $a_P^\ominus$ values are all positive, exactly identical for all ions, and also equal to those of the corresponding gas-phase atoms (Section 5.9), because they account for ideal-gas properties.

5.11 Thermodynamic parameters of reticulation

The *standard thermodynamic parameters of reticulation (lattice formation)* $\Delta_l Y^\ominus[IJ_g]$ are not directly accessible experimentally, due to the absence of free gaseous ions at equilibrium (Section 2.1). The most relevant of these parameters are the reticulation free energies $\Delta_l G^\ominus[IJ_g]$, enthalpies $\Delta_l H^\ominus[IJ_g]$, and entropies $\Delta_l S^\ominus[IJ_g]$. The formation parameters $\Delta_f Y^\ominus[IJ_g]$ and molar variables $y^\ominus[IJ_s]$ are also relevant, but have been discussed previously in Sections 5.2 and 5.6. The parameters $\Delta_l Y^\ominus$ characterize the aggregation of the gaseous ions into a periodic crystal lattice. Although a direct measurement is not possible, the corresponding enthalpies can be either calculated using *theoretical models* or *inferred from the BFH cycle*. The corresponding entropies can be accessed by means of a TLC or, if possible, also be estimated using theoretical models. These approaches are discussed in turn below.

The *theoretical determination* of reticulation parameters involves the calculation of the reticulation energy E_l, the application of a correction to obtain the corresponding reticulation enthalpy $\Delta_l H^\ominus$ and, possibly, the calculation of the reticulation entropy $\Delta_l S^\ominus$. The approaches used for the theoretical evaluation of the *lattice energy* can be broadly classified as follows:

(A_{11}) The *hard-sphere model* assumes that the ions in the crystal are rigid spheres in contact, with the net ion charge located at a single point in their center. In this case, the interaction energy within the crystal is purely electrostatic in nature and can be evaluated through lattice (Madelung [394]) summation of the Coulomb potential. For a neutral lattice of the salt $I_{n_I} J_{n_J}$ made of ions I^{z_I} and J^{z_J} with $n_I z_I + n_J z_J = 0$, the resulting expression for the lattice energy E_l, expressed on a per mole basis, is of the form

$$E_l = -\frac{N_A C_M e^2 z_I z_J}{4\pi\epsilon_o R_{IJ}} , \qquad (5.23)$$

Table 5.10 *Standard thermodynamic parameters of reticulation for the alkali-halide gas-phase ion pairs.* The reported data includes the reticulation free energy $\Delta_l G^{\ominus}$, enthalpy $\Delta_l H^{\ominus}$, approximate enthalpy $\Delta_l H^{(\ominus)}$ (estimated using the Kapustinskii equation) and entropy $\Delta_l S^{\ominus}$. The notation MX_g refers to an alkali-halide gas-phase ion pair (sum over properties of the species M_g^+ and X_g^- in the gas phase). If required, the reticulation heat capacity, volume, volume-compressibility or volume-expansivity can be obtained as a difference between the corresponding molar property of the salt (Table 5.2) and that of the constituting gas-phase ions (Table 5.9). The standard states are according to the *bbme*$_T$ convention (Section 4.2.4; reference pressure $P^\circ = 1$ bar and reference temperature $T^- = 298.15$ K).

Salt MX_g	$\Delta_l G^{\ominus}[MX_g]$ [kJ·mol^{-1}]	$\Delta_l H^{\ominus}[MX_g]$ [kJ·mol^{-1}]	$\Delta_l H^{(\ominus)}[MX_g]$ [kJ·mol^{-1}]	$\Delta_l S^{\ominus}[MX_g]$ [J·mol^{-1}·K^{-1}]
LiF	-973.9	-1046.4 [-1046.5]	(-986.4)	-243.0
LiCl	-793.6	-861.3 [-861.5]	(-846.1)	-226.9
LiBr	-751.7	-817.9 [-817.5]	(-786.9)	-222.1
LiI	-695.0	-759.2 [-761.5]	(-737.4)	-215.4
NaF	-858.3	-930.6 [-927.5]	(-914.5)	-242.4
NaCl	-719.0	-787.4 [-787.5]	(-791.9)	-229.4
NaBr	-684.4	-751.4 [-751.5]	(-739.6)	-224.6
NaI	-636.5	-700.2 [-702.5]	(-695.6)	-213.5
KF	-756.7	-826.2 [-826.5]	(-810.1)	-233.1
KCl	-650.3	-717.6 [-717.5]	(-711.7)	-225.6
KBr	-623.0	-689.0 [-688.5]	(-668.9)	-221.3
KI	-582.2	-645.2 [-647.5]	(-632.5)	-211.1
RbF	(-724.4)	-792.4 [-792.5]	(-769.7)	(-228.0)
RbCl	-626.2	-692.4 [-692.5]	(-680.1)	-222.0
RbBr	-600.7	-665.6 [-665.5]	(-640.9)	-217.8
RbI	-562.8	-626.9 [-629.5]	(-607.3)	-215.1
CsF	-689.7	-756.1 [-756.5]	(-733.0)	-222.7
CsCl	-601.6	-667.8 [-667.5]	(-651.1)	-222.0
CsBr	-579.0	-644.7 [-644.5]	(-615.0)	-220.3
CsI	-543.2	-607.6 [-610.5]	(-584.0)	-215.9
Notes	a, b	c, d	e	b, f

(a) calculated using the Gibbs equation ($\Delta_l G^{\ominus}[MX_g] = \Delta_l H^{\ominus}[MX_g] - T^- \Delta_l S^{\ominus}[MX_g]$); (b) the value of $\Delta_l S^{\ominus}[\text{RbF}_g]$, reported between parentheses, is not experimental, but derived by the authors of this book *via* an empirical correlation between $\Delta_l S^{\ominus}[MX_g]$ and T_m° for the eleven sodium, potassium and rubidium salts with known $\Delta_l S^{\ominus}[MX_g]$ (correlation coefficient -0.97, data from Table 5.2 for T_m°, estimated uncertainty of about 1 J·mol^{-1}·K^{-1}); (c) calculated from the enthalpies of formation of the salt and of the gas-phase ions ($\Delta_l H^{\ominus}[MX_g] = -\Delta_f H^{\ominus}[M_g^z] - \Delta_f H^{\ominus}[X_g^{-z}] + \Delta_f H^{\ominus}[MX_s]$, using values from Table 5.9 for the two former quantities and values from Table 5.6 for the latter quantity); (d) the reticulation enthalpies derived from Ref.[221] (calculated from the reported reticulation energies $E_l[MX_g]$ by addition of RT^-, indicated between square brackets for comparison) may slightly differ from the reported (recalculated) values (*i.e.* the recalculated values correct for a slight numerical inconsistency between the data set of Ref.[221] for $\Delta_f H^{\ominus}[MX_s]$ and $E_l[MX_g]$, and the data set of Ref.[294] for $\Delta_f H^{\ominus}[M_g^z]$ and $\Delta_f H^{\ominus}[X_g^{-z}]$); (e) approximate values calculated using the Kapustinskii equation (Eq. 5.26, using radii of Set 27r in Table 5.4) and increased by RT^-; (f) calculated from the molar entropies of the salt and of the gas-phase ions ($\Delta_l S^{\ominus}[MX_g] = -s^{\ominus}[M_g^z] - s^{\ominus}[X_g^{-z}] + s^{\ominus}[MX_s]$, using values from Table 5.9 for the two former quantities and values from Table 5.6 for the latter quantity).

where R_{IJ} is the experimental interionic distance in the crystal (Section 5.2) and C_M is the Madelung constant associated with the given lattice type[179,221,2147] ($C_M = 1.74756$ and 1.76267, respectively, for the fcc and bcc lattices characteristic of alkali-halide salts).

(B_{11}) The *Born-Landé model*[395,396] introduces Pauli repulsion in addition to electrostatic interactions (soft-sphere model). The repulsive term is assumed to be of the form $C_{BL} r^{-n_{BL}}$ (Born-Landé potential), where appropriate values of n_{BL} and C_{BL} can be inferred from experimental crystal compressibilities. However, rather than using the resulting value of C_{BL}, this parameter is readjusted to ensure that the experimental interionic distance R_{IJ} in the crystal corresponds to a minimum-energy configuration under the action of both electrostatic and repulsion interactions. The resulting expression for the lattice energy E_l is of the form

$$E_l = -\frac{N_A C_M e^2 z_I z_J (1 - n_{BL}^{-1})}{4\pi\epsilon_o R_{IJ}} . \tag{5.24}$$

See also Section 5.4 (Point D_{10}) for effective ionic radii derived using this approach.

(C_{11}) The *Born-Mayer model*[397] also introduces Pauli repulsion in addition to electrostatic interactions (soft-sphere model). The repulsive term is assumed here to be of the form $C_{BM} \exp[-R_{BM}^{-1} r]$ (Born-Mayer potential), which represents a more adequate description from the quantum-mechanical point of view. Appropriate values of R_{BM} and C_{BM} can also be inferred from experimental crystal compressibilities. However, these parameters are again readjusted to ensure that the experimental interionic distance R_{IJ} in the crystal corresponds to a minimum-energy configuration under the action of both electrostatic and repulsion interactions. The resulting expression for the lattice energy E_l is of the form

$$E_l = -\frac{N_A C_M e^2 z_I z_J (R_{IJ} - R_{BM})}{4\pi\epsilon_o R_{IJ}^2} . \tag{5.25}$$

See also Section 5.4 (Point D_{10}) for effective ionic radii derived using this approach.

(D_{11}) The *Kapustinskii equation*[398] is an empirical equation inspired from the Born-Mayer model[397], but trying to avoid the input of the Madelung constant, *i.e.* of any prior knowledge concerning the crystallographic lattice type. This equation is

$$E_l = -\frac{C_K n_K z_I z_J}{R_I + R_J}\left(1 - \frac{R_K}{R_I + R_J}\right) , \tag{5.26}$$

where n_K is the number of ions in the formula (two for a binary salt of isovalent ions), R_I and R_J are adjusted ionic radii (so-called thermochemical radii[221,398]; see *e.g.* Table 5.4, Set 27r, for values in the context of alkali and halide ions), while C_K and R_K are empirical parameters (C_K = 121.4 kJ·nm·mol^{-1}, R_K = 0.0345 nm). See also Section 5.4 (Point C_{10}) for a discussion of thermochemical radii.

The calculated reticulation energies E_l represent estimates for the adiabatic energy change associated with the process $I_g^z + J_g^{-z} \rightarrow IJ_s$. Here, adiabatic means that they refer to ideal motionless species, *i.e.* these parameters represent purely energetic quantities corresponding to a temperature $T = 0$ K. To obtain standard reticulation enthalpies, the value of E_l must be corrected[2148] to match the finite temperature T^-, as

$$\begin{aligned} \Delta_l H^{\ominus}[IJ_g] &= E_l[IJ_g] - \Delta_h H^{\ominus}[I_g^z] \\ &\quad -\Delta_h H^{\ominus}[J_g^{-z}] + \Delta_h H^{\ominus}[IJ_s] \\ &\approx E_l[IJ_g] + RT^- \quad , \end{aligned} \tag{5.27}$$

where the heat-up enthalpy of a solid is defined in analogy with that of a gas (Section 4.2.9) as

$$\Delta_h H^{\ominus} = \int_0^{T^-} dT' c_P^{\circ}(T') \quad , \tag{5.28}$$

$c_P^{\circ}(T')$ being the molar isobaric heat capacity of the solid at pressure P° and temperature T'.

The approximation in Eq. 5.27 is justified for binary salts of monoatomic ions when the Debye temperature of the salt is much lower than T^- and when the gas-phase ions have no thermally-accessible electronically-excited states. With the first assumption, the heat capacity of the salt (binary salt of monoatomic ions) can be approximated as $6R$ at any temperature, the isobaric heat capacity being assumed identical to the isochoric one, and integration from the absolute zero temperature leads to $\Delta_h H^{\ominus}[IJ_s] \approx 6RT^-$. With the second assumption, the heat capacity of each gas-phase ion (monoatomic ion) can be approximated as $(5/2)R$ at any temperature, corresponding to the isochoric heat capacity plus R, and integration from the absolute zero temperature leads to $\Delta_h H^{\ominus}[I_g^z] \approx \Delta_h H^{\ominus}[J_g^{-z}] \approx (5/2)RT^-$. With these approximations, the correction term in Eq. 5.27 evaluates to RT^-. Note that Ref.[2148] defines the reticulation reaction as forming the gas-phase ions from the solid, leading to a correction term $-RT^-$ instead. If required, the approximation in Eq. 5.27 can be bypassed by explicit integration of the temperature-dependent heat capacity of the ions *via* SBS (Eq. 4.147) and of the salt *via* calorimetry or/and approximate analytical models (Einstein[1735,1736] or Debye[1734] model) from the absolute zero temperature. Note that the latter approach may be preferable in the context of alkali-halide salts, some of which being characterized by a relatively elevated Debye temperature[2148], in particular LiF, NaF, KF, LiCl and NaCl, with Debye temperatures of 722, 491, 334, 422 and 321 K, respectively[2149–2152].

The reticulation entropy is related to the absolute entropies of the participating species *via* a TLC, as

$$\Delta_l S^{\ominus}[IJ_g] = -s^{\ominus}[I_g^z] - s^{\ominus}[J_g^{-z}] + s^{\ominus}[IJ_s] \ . \tag{5.29}$$

The absolute molar entropies of the gas-phase ions are available *via* SBS (Section 5.10) and the lattice entropies $s^{\ominus}[IJ_s]$ entering in this equation can also in principle be calculated on the basis of statistical mechanics, *e.g.* from (calculated or experimental) vibrational frequencies using (quasi-)harmonic models[196,399–401]. However, it is generally much more accurate to use the experimental entropies based on TCI (see further below).

Calculating the reticulation enthalpies (and possibly entropies) using a theoretical model does not represent a very accurate procedure, in particular because it neglects[394] or only approximately includes[395–397] the Pauli repulsion contribution to the lattice energy, and entirely overlooks contributions from electronic polarization and dispersion interactions. In practice, it is by far preferable to deduce the reticulation enthalpy from the atomization (gas-phase atom formation) parameters of the atom pair (Section 5.9), the formation parameters of the salt (Section 5.6), and the ionization parameters of the gas-phase atom pair (Section 5.10) by application of the BFH cycle. Similarly, it is much more accurate to evaluate the reticulation entropy *via* a TLC (Eq. 5.29) using the experimental molar entropy of the salt based on TCI (Section 5.9). The reticulation free energy $\Delta_l G^{\ominus}$ is then calculated by application of the Gibbs equation. The alternative approach *via* calculation (see above) is only used in practice when the parameters for one other leg of the BFH cycle cannot be determined, typically the ionization parameters of one of the species, or the molar entropy of the salt. Note that even if the reticulation enthalpy is deduced using the BFH cycle, this cycle can still be used to check the internal consistency of the corresponding entropies, when these are accessible independently for all participating species *via* TCI or SBS. If required, the reticulation heat capacity, volume, volume-compressibility or volume-expansivity can also be evaluated as a difference between the corresponding molar property of the salt (Section 5.2) and that of the gas-phase ions (Section 5.10).

The values of the reticulation parameters have been determined[221] for nearly all alkali-halide gas-phase ion pairs (MX_g) by application of the BFH cycle based on quantities previously described (Sections 5.6, 5.9 and 5.10). These standard thermodynamic parameters are reported in Table 5.10 except for RbF_s, for which only $\Delta_l H^{\ominus}$ is available to our knowledge. For this salt, a non-experimental estimate of $\Delta_l S^{\ominus}$ was inferred by the authors of this book from the observation of an approximate linear relationship between $\Delta S_l^{\ominus}$ and the melting temperature T_m° of the salt (Table 5.2) for the eleven sodium, potassium and rubidium halides with known $\Delta_l S^{\ominus}$ (correlation coefficient of -0.97). Interpolation to the T_m° value of RbF_s led to the value reported between parentheses in the table, with an estimated uncertainty of about 1 J·mol^{-1}·K^{-1}. This estimate was used to deduce corresponding values for the reticulation free energy, as well as for the salt formation and dissolution enthalpies and free energies reported in Tables 5.6 and 5.8. Note that, by construction, the data reported in Table 5.10 satisfies the Gibbs equation ($\Delta_l G^\ominus[MX_g] = \Delta_l H^\ominus[MX_g] - T^- \Delta_l S^\ominus[MX_g]$), the Hess law in terms of entropies ($\Delta_l S^\ominus[MX_g] = s^\ominus[MX_s] - s^\ominus[M_g^+] - s^\ominus[X_g^-]$; also using data from Tables 5.6 and 5.9), and the closure of the BFH cycle in terms of free energies, enthalpies and entropies ($\Delta_a Y^\ominus[M_{el}] + \Delta_a Y^\ominus[X_{el}] + \Delta_i Y^\ominus[M_g] + \Delta_i Y^\ominus[X_g] + \Delta_l Y^\ominus[MX_g] - \Delta_f Y^\ominus[MX_s] = 0$; also using data from Tables 5.6 and 5.9).

The $\Delta_l G^\ominus$ and $\Delta_l H^\ominus$ values are very large and negative, indicating that the reticulation process is associated with both a strong driving force and an important release of heat, as expected for a process associating gas-phase species of opposite charges into a compact crystal lattice. They systematically decrease in magnitude with increasing effective ion sizes (atomic numbers), *i.e.* along the series $Li^+ \rightarrow Cs^+$ for salts involving a common anion and $F^- \rightarrow I^-$ for salts involving a common cation. This trend is related to the corresponding increase in the nearest-neighbor cation-anion distance within the crystal (Section 5.2), leading to reduced Coulombic interactions. The $\Delta_l S^\ominus$ values are all negative, resulting from the decrease in disorder upon combining two gas-phase species at reference pressure P° into a solid salt at the same reference pressure. They nearly systematically decrease in magnitude with increasing ion sizes. The latter general trend is probably related to the weakening of the Coulombic nearest-neighbor interactions within the crystal (Section 5.2), the magnitude of which decreases with the cation-anion distance, approximating the ions as point charges. This enables more pronounced oscillations of the ions about their lattice positions. Compared to the experimental values $\Delta_l H^\ominus$ determined *via* the BFH cycle, the theoretical estimates $\Delta_l H^{(\ominus)}$ obtained *via* the Kapustinskii equation, also reported in the table for comparison, are seen to be of reasonable magnitude but of limited accuracy, with errors on the order of 5-60 kJ·mol^{-1}.

5.12 Thermodynamic parameters of salt solvation

The *standard parameters of salt solvation* $\Delta_s Y^\ominus[IJ_g]$, where IJ_g represents a gas-phase ion pair, corresponding to sums over corresponding parameters of the species I_g^z and J_g^{-z}, are not directly accessible experimentally due to the absence of free gaseous ions at equilibrium (Section 2.1). The most relevant of these parameters are the salt solvation free energies $\Delta_s G^\ominus[IJ_g]$, enthalpies $\Delta_s H^\ominus[IJ_g]$ and entropies $\Delta_s S^\ominus[IJ_g]$. The parameters $\Delta_s Y^\ominus$ characterize the dissolution of free gaseous ions into a given solvent. Since a direct measurement is not possible, these parameters can only be *inferred from the ISR cycle* (see further below). The formation parameters $\Delta_f Y^\ominus[IJ_d]$ and partial molar variables $y^\ominus[IJ_d]$ of the dissolved salts are also relevant, but have been discussed previously in Section 5.7. In contrast to the solvation parameters, the formation free energies $\Delta_f G^\ominus[IJ_d]$, enthalpies $\Delta_f H^\ominus[IJ_d]$ and entropies $\Delta_f S^\ominus[IJ_d]$, as well as the partial molar entropies $s^\ominus[IJ_d]$, heat capacities $c_P^\ominus[IJ_d]$, volumes $v^\ominus[IJ_d]$, volume-compressibilities $k_T^\ominus[IJ_d]$ and volume-expansivities $a_P^{\ominus}[IJ_d]$ of dissolved salts, where IJ_d represents the dissociated species I_d^z and J_d^{-z} in solution, are accessible experimentally. As noted in Section 5.7, however, $\Delta_f S^{\ominus}[IJ_d]$ and $s^{\ominus}[IJ_d]$ cannot be obtained directly *via* TCI, which is only applicable to pure substances. As a result, application of the ISR cycle is the only way to evaluate $\Delta_s S^{\ominus}$, and no cross-checking by means of a TLC is possible in this case. A slightly different approach to access salt solvation parameters relies on the summation of corresponding real single-ion solvation parameters determined based on Voltaic cell measurements (Section 5.15). However, considering Eqs. 4.307 (along with a similar equation for the anion X_g^-) and 4.378, one sees easily that this approach is also a slightly disguised version of the ISR cycle, where the solvation parameters of the salt are deduced from an indirect Galvanic-cell potential difference measurement, and the formation parameters of the gas-phase ions are still required as input. On the other hand, as discussed in detail in Chapters 3 and 6, the use of theoretical models to evaluate thermodynamic solvation parameters, although promising for the future, is not reliable enough at present to allow for quantitative determinations.

Upon application of the ISR cycle, the salt solvation parameters can be deduced from the formation parameters of the dissolved salt (Section 5.7), the atomization (gas-phase atom formation) parameters of the atom pair (Section 5.9), and the ionization parameters of the gas-phase atom pair (Section 5.10) as

$$\begin{aligned}\Delta_s Y^{\ominus}[IJ_g] &= -(\Delta_a Y^{\ominus}[I_{el}] + \Delta_i Y^{\ominus}[I_g]) \\ &\quad -(\Delta_a Y^{\ominus}[J_{el}] + \Delta_i Y^{\ominus}[J_g]) + \Delta_f Y^{\ominus}[IJ_d] \quad . \end{aligned} \tag{5.30}$$

For the solvation entropy, this expression is more usually formulated in terms of (partial) molar entropies, as

$$\Delta_s S^{\ominus}[IJ_g] = -s^{\ominus}[I_g^z] - s^{\ominus}[J_g^{-z}] + s^{\ominus}[IJ_d] \quad . \tag{5.31}$$

Obviously, the solvation free energy $\Delta_s G^{\ominus}$ is also related to the corresponding enthalpies and entropies *via* the Gibbs equation. If required, the salt solvation heat capacity, volume, volume-compressibility or volume-expansivity can also be evaluated as a difference between the corresponding partial molar property of the dissolved salt (Section 5.7) and the corresponding molar properties of the gas-phase ions (Section 5.10).

The values of the salt solvation parameters have been determined for all alkali-halide gas-phase ion pairs (MX_g) in water by application of the ISR cycle, based on quantities previously discussed (Sections 5.7, 5.9 and 5.10). The corresponding values of the dissolved salt partial molar variables have also been determined for all alkali-halide dissolved ion pairs in water (MX_{aq}), using approaches described in Section 5.7.

The solvation parameters $\Delta_s G^{\ominus}[MX_g]$, $\Delta_s H^{\ominus}[MX_g]$ and $\Delta_s S^{\ominus}[MX_g]$ derived as described above for the alkali-halide pairs in water based on the data reported earlier in this chapter define only one data set for these quantities, considered by the authors of this book as the most trustworthy to date. However, *many alternative experimental sets* have been proposed based on the same approach, but relying on different raw data. Similar considerations apply to the dissolved salt partial molar parameters for the alkali-halide pairs in water, namely the partial molar entropy $s^{\ominus}[MX_{aq}]$, heat capacity $c_P^{\ominus}[MX_{aq}]$, volume $v^{\ominus}[MX_{aq}]$, volume-compressibility $k_T^{\ominus}[MX_{aq}]$ and volume-expansivity $a_P^{\ominus}[MX_{aq}]$. Here also, a single data set for these quantities, considered by the authors of this book as the most trustworthy to date, was presented and discussed in Section 5.7. However, many alternative experimental sets have been proposed.

In the literature concerning ionic solvation (hydration), salt solvation parameters and dissolved salt partial molar variables can be constructed from data presented in at least five alternative forms: (*i*) directly as salt solvation parameters for a given set of salts; (*ii*) as conventional single-ion solvation parameters (Sections 4.5.1 and 5.16) for a given set of ions; (*iii*) as real single-ion solvation parameters (Sections 4.5.2 and 5.15) for a given set of ions; (*iv*) as tentative intrinsic single-ion solvation parameters (Sections 4.5.3, 5.18 and 5.19) for a given set of ions; (*v*) as redox potentials and derivatives for electrodes (half-cells) involving the reduction of a given set of ions (equivalent to conventional data, see Sections 4.5.1 and 5.5). As a result, there are numerous independent sets of salt solvation parameters available in a direct or direct form, a number of which present high similarities because they are derived from the same raw experimental data.

Standard thermodynamic parameters of salt solvation for the alkali-halide gas-phase ion pairs in water corresponding to various experimental data sets are reported in Table 5.11. Similarly, standard partial molar variables of the aqueous alkali-halide ion pairs corresponding to various experimental data sets are reported in Table 5.12. For the ease of further reference, these sets are numbered from 1 to 44 based on the literature source, along with the letter G, H, $\overline{\text{S}}$, S, C, V, K or A to indicate the type of property, namely, $\Delta_s G^{\ominus}$, $\Delta_s H^{\ominus}$, $\Delta_s S^{\ominus}$, $s^{\ominus}$, $c_P^{\ominus}$, $v^{\ominus}$, $k_T^{\ominus}$ or $a_P^{\ominus}$, respectively. Only sources *explicitly reporting pairwise values* are listed in these tables, although it is possible to construct many additional pairwise sets (see above) based on the conventional data provided later in Tables 5.16 and 5.17, which themselves include conventional values derived from real data provided in Table 5.15, and from redox potentials and derivatives provided in Table 5.5.

The recommended pairwise sets deduced directly from the entries of Tables 5.7 and 5.9 are labeled 1G, 1H and 1$\overline{\text{S}}$ in Table 5.11. The recommended sets discussed in Section 5.7 and reported in Table 5.7 are labeled 1S, 1C, 1V, 1K and 1A in Table 5.12. Note that, by construction, the recommended values of Sets 1G, 1H, 1$\overline{\text{S}}$ and 1S satisfy the Gibbs equation ($\Delta_s G^{\ominus}[MX_g] = \Delta_s H^{\ominus}[MX_g] - T^{-}\Delta_s S^{\ominus}[MX_g]$), the Hess law in terms of entropies ($\Delta_s S^{\ominus}[MX_g] = s^{\ominus}[MX_{aq}] - s^{\ominus}[M_g^+] - s^{\ominus}[X_g^-]$; also using data from Table 5.9), and the closure of the ISR cycle in terms of free energies, enthalpies and entropies ($\Delta_a Y^{\ominus}[M_{el}] + \Delta_a Y^{\ominus}[X_{el}] + \Delta_i Y^{\ominus}[M_g] + \Delta_i Y^{\ominus}[X_g] + \Delta_s Y^{\ominus}[MX_g] - \Delta_f Y^{\ominus}[MX_{aq}] = 0$; also using data from Tables 5.7 and 5.9).

Table 5.11 *Standard thermodynamic parameters of solvation for the alkali-halide gas-phase ion pairs in water.* The reported data includes the solvation free energy $\Delta_s G^{\ominus}$, enthalpy $\Delta_s H^{\ominus}$ and entropy $\Delta_s S^{\ominus}$. The notation MX_g refers to an alkali-halide gas-phase ion pair (sum over properties of the species M_g^+ and X_g^- in the gas phase). The different sets correspond to values quoted in different literature sources, listed in chronological order of publication for $\Delta_s G^{\ominus}$, $\Delta_s H^{\ominus}$ and $\Delta_s S^{\ominus}$ separately, except for Sets 1G, 1H and 1S (listed first), which represent recommended data and are calculated using the ISR cycle based on the values reported in Tables 5.7 and 5.9. Only sources explicitly reporting pairwise values are listed, although it is possible to construct many additional pairwise sets based on single-ion conventional solvation parameters $\Delta_s Y^{\bullet}[I_g^z]$ from Table 5.16. On the other hand, single-ion conventional solvation parameters deduced from the pairwise sets given in the present table are reported in Table 5.16. The entry σ refers to the ability of the pairwise set to be partitioned into single-ion contributions within an unknown constant. This parameter represents the minimal achievable root-mean-square difference, reached for an optimal partitioning, between the sum of two single-ion values and the corresponding pairwise value, over all possible combinations for which pairwise data is available in the set (at most 20). Non-experimental data inferred by optimal partitioning is indicated between parentheses. The σ values pertain to the complemented sets, values for the original sets also being indicated between parentheses whenever relevant. The lines NUM, AVE, RMS, ERR and REC report for each pair the number of entries, average value, root-mean-square deviation, estimated error ($NUM^{-1/2}\cdot RMS$) and recommended value (Sets 1, after minimal adjustments to generate exactly partitionable sets) for the preceding block. If required, the solvation heat capacity, volume, volume-compressibility or volume-expansivity can be obtained as a difference between the corresponding partial molar property of the dissolved salt (Tables 5.7 and 5.12) and the corresponding molar property of the constituting gas-phase ions (Table 5.9). The standard states are according to the *bbme*$_T$ convention (Section 4.2.4; reference pressure $P^{\circ} = 1$ bar, reference molality $b^{\circ} = 1$ mol·kg^{-1} and reference temperature $T^{-} = 298.15$ K). The experimental values reported in the original references are assumed here to pertain to the standard variant for $\Delta_s H^{\ominus}$ and $\Delta_s S^{\ominus}$ (Section 4.2.5), which represents the most likely situation, but may not necessarily be the case in reality. The solvation free energy $\Delta_s G^{\ominus}$ is identical in the two variants.

Set	Li^+				Na^+				K^+				Rb^+				Cs^+				σ	Ref.	Notes
	F^-	Cl^-	Br^-	I^-	F^-	Cl^-	Br^-	I^-	F^-	Cl^-	Br^-	I^-	F^-	Cl^-	Br^-	I^-	F^-	Cl^-	Br^-	I^-			
$\Delta_s G^{\ominus}[MX_g]$ [kJ·mol^{-1}]																							
1G	-958.1	-833.8	-807.0	-769.5	-852.7	-728.0	-701.2	-663.9	-781.0	-656.4	-629.5	-592.2	-758.2	-633.6	-606.8	-569.5	-735.0	-610.3	-583.6	-546.2	0.05		a
2G	-943.1	-815.5	-785.3	-747.3	-831.8	-709.6	-683.2	-649.3	-763.6	-641.4	-614.6	-582.0	-733.4	-618.0	-590.8	-559.8	-715.0	-581.6	-562.3	-530.9	3.19	814	b,c

(continues on the next page)

Table 5.11 *(continued)*

Set	Li$^+$				Na$^+$				K$^+$				Rb$^+$				Cs$^+$				σ	Ref.	Notes
	F$^-$	Cl$^-$	Br$^-$	I$^-$	F$^-$	Cl$^-$	Br$^-$	I$^-$	F$^-$	Cl$^-$	Br$^-$	I$^-$	F$^-$	Cl$^-$	Br$^-$	I$^-$	F$^-$	Cl$^-$	Br$^-$	I$^-$			
3G	-948.0	-829.2	-803.3	-767.7	-841.3	-722.5	-696.6	-661.0	-768.1	-649.3	-623.4	-587.8	-750.5	-631.7	-605.8	-570.6	-733.8	-615.0	-589.5	-553.9	0.10	1873	d
4G	-936.4	-814.2	-785.3	-763.2	-837.6	-715.5	-687.0	-664.8	-765.3	-643.5	-614.6	-592.9	-739.7	-617.6	-588.7	-566.5	-725.5	-604.6	-575.7	-553.5	0.22	2153	
NUM	4	4	4	4	4	4	4	4	4	4	4	4	4	4	4	4	4	4	4	4			
AVE	-946.4	-823.2	-795.2	-761.9	-840.9	-718.9	-692.0	-659.8	-769.5	-647.6	-620.5	-588.7	-745.5	-625.2	-598.0	-566.6	-727.3	-602.9	-577.8	-546.1			
RMS	7.9	8.5	10.0	8.8	7.6	7.0	7.2	6.2	6.8	5.8	6.3	4.3	9.6	7.5	8.3	4.2	8.0	12.8	10.2	9.3			
ERR	4.0	4.3	5.0	4.4	3.8	3.5	3.6	3.1	3.4	2.9	3.2	2.2	4.8	3.7	4.2	2.1	4.0	6.4	5.1	4.7			
REC	-958.3	-833.7	-806.9	-769.5	-852.6	-728.1	-701.3	-663.9	-781.0	-656.4	-629.6	-592.2	-758.2	-633.6	-606.8	-569.4	-735.0	-610.4	-583.6	-546.2			
$\Delta_s H^\ominus[MX_g]$ [kJ·mol^{-1}]																							
1H	-1041.5	-898.2	-866.7	-822.5	-926.8	-783.5	-752.0	-707.8	-843.9	-700.6	-669.1	-624.9	-818.5	-675.3	-643.7	-599.5	-792.6	-650.2	-618.7	-574.5	0.16		a
5H	-	-782.4	-744.8	-702.9	-	-755.2	-715.5	-665.3	-	-665.3	-627.6	-581.6	-	-627.6	(-589.2)	(-543.2)	-	-631.8	(-593.4)	(-547.4)	1.44 (1.68)	83	
6H	-994.1	-861.1	-823.8	-778.6	-882.4	-750.6	-719.2	-679.9	-801.7	-663.2	-633.9	-597.9	-773.6	-641.8	-603.3	-568.2	-750.6	-605.4	-572.0	-536.0	2.98	2154	
7H	-1012.5	-880.7	-849.4	-805.4	-901.7	-771.9	-744.8	-702.9	-824.2	-686.2	-661.1	-623.4	-799.1	-665.3	-633.9	-596.2	-771.9	-627.6	-598.3	-564.8	2.60	2155	

(continues on the next page)

Table 5.11 *(continued)*

Set	Li⁺				Na⁺				K⁺				Rb⁺				Cs⁺				σ	Ref.	Notes
	F^-	Cl^-	Br^-	I^-	F^-	Cl^-	Br^-	I^-	F^-	Cl^-	Br^-	I^-	F^-	Cl^-	Br^-	I^-	F^-	Cl^-	Br^-	I^-			
8H	-1034.7	-899.6	-868.2	-826.8	-918.8	-783.2	-751.9	-710.9	-835.1	-699.6	-667.8	-626.8	-808.8	-672.8	-641.4	-600.4	-791.2	-655.2	-623.8	-582.8	0.15	1872	
3H	-1035.5	-898.7	-868.2	-825.1	-922.6	-783.2	-753.1	-710.0	-839.3	-700.0	-669.4	-626.3	-829.3	-670.7	-639.7	-599.1	-778.2	-641.4	-610.0	-568.2	3.51	1873	
9H	(-1016.1)	-878.9	-848.8	-805.5	(-908.0)	-773.5	-739.5	-695.8	-827.3	-689.5	-655.4	-611.9	-802.2	-663.9	-630.4	-587.9	-759.5	-631.6	-597.4	-553.7	1.67 (1.76)	1882	
4H	-1019.2	-878.2	-844.8	-815.9	-912.1	-771.1	-737.6	-708.4	-828.4	-687.9	-654.4	-625.5	-799.1	-658.1	-624.7	-595.8	-784.1	-644.3	-610.9	-582.0	0.23	2153	
10H	-1049.9	-899.1	-868.1	-827.8	-934.4	-783.6	-752.6	-712.3	-850.7	-699.9	-668.9	-628.6	-826.1	-675.3	-644.3	-604.0	-800.9	-650.1	-619.1	-578.8	0.00	1874	
NUM	8	9	9	9	8	9	9	9	8	9	9	9	8	9	9	9	8	9	9	9			
AVE	-1025.5	-875.2	-842.5	-801.2	-913.3	-772.9	-740.7	-699.2	-831.3	-688.0	-656.4	-616.3	-807.1	-661.2	-627.9	-588.3	-778.6	-637.5	-604.8	-565.4			
RMS	17.0	35.1	37.4	37.7	15.3	11.8	13.6	15.4	14.0	13.8	14.8	15.4	16.8	15.5	18.2	18.8	16.1	14.5	15.4	15.5			
ERR	6.0	11.7	12.5	12.6	5.4	3.9	4.5	5.1	4.9	4.6	4.9	5.1	5.9	5.2	6.1	6.3	5.7	4.8	5.1	5.2			
REC	-1041.4	-898.3	-866.8	-822.5	-926.6	-783.6	-752.0	-707.8	-843.7	-700.7	-669.1	-624.9	-818.4	-675.3	-643.8	-599.5	-793.1	-650.0	-618.5	-574.3			
$\Delta_s S^{\ominus}[MX_g]$ [J·mol⁻¹·K⁻¹]																							
$1\overline{S}$	-279.6	-216.1	-200.4	-177.9	-248.5	-186.1	-170.2	-147.1	-210.9	-148.5	-132.7	-109.5	-202.1	-139.8	-123.9	-100.8	-193.2	-133.8	-117.8	-94.8	0.57		a

(continues on the next page)

Table 5.11 *(continued)*

Set	Li^+				Na^+				K^+				Rb^+				Cs^+				σ	Ref.	Notes
	F^-	Cl^-	Br^-	I^-	F^-	Cl^-	Br^-	I^-	F^-	Cl^-	Br^-	I^-	F^-	Cl^-	Br^-	I^-	F^-	Cl^-	Br^-	I^-			
$11\overline{S}$	(-273.9)	-216.8	-201.4	-178.7	-242.8	-185.7	-170.3	-147.6	(-207.1)	-150.1	-134.7	-112.0	(-195.1)	-138.1	(-122.7)	(-100.0)	(-191.9)	-134.8	-119.4	-96.7	0.00 (0.00)	1881	e
$3\overline{S}$	-293.5	-233.2	-217.7	-192.5	-272.4	-203.7	-189.7	-164.4	-238.8	-170.0	-154.6	-129.3	-264.0	-130.7	-113.9	-95.6	-149.0	-88.6	-69.0	-47.9	11.79	1873	f
$4\overline{S}$	-277.8	-215.5	-199.6	-176.6	-248.1	-185.4	-169.9	-146.9	-211.3	-148.5	-133.1	-109.6	-199.2	-136.8	-120.9	-97.9	-195.8	-133.5	-117.6	-94.6	0.10	2153	
NUM	4	4	4	4	4	4	4	4	4	4	4	4	4	4	4	4	4	4	4	4			
AVE	-281.2	-220.4	-204.8	-181.4	-253.0	-190.2	-175.0	-151.5	-217.0	-154.3	-138.7	-115.1	-215.1	-136.3	-120.3	-98.6	-182.5	-122.7	-105.9	-83.5			
RMS	7.4	7.4	7.5	6.4	11.5	7.8	8.5	7.5	12.7	9.1	9.2	8.3	28.4	3.4	3.9	2.0	19.4	19.7	21.4	20.6			
ERR	3.7	3.7	3.7	3.2	5.7	3.9	4.2	3.7	6.3	4.5	4.6	4.1	14.2	1.7	1.9	1.0	9.7	9.8	10.7	10.3			
REC	-278.7	-216.7	-200.8	-177.8	-248.2	-186.2	-170.3	-147.3	-210.6	-148.6	-132.7	-109.7	-201.8	-139.8	-124.0	-101.0	-195.1	-133.1	-117.2	-94.2			

(a) calculated using the ISR cycle from the atomization (gas-phase atom formation), ionization and dissolved salt formation parameters reported in Tables 5.7 and 5.9 ($\Delta_s Y^{\ominus}[MX_g] = \Delta_f Y^{\ominus}[MX_{aq}] - \Delta_a Y^{\ominus}[M_{el}] - \Delta_a Y^{\ominus}[X_{el}] - \Delta_i Y^{\ominus}[M_g] - \Delta_i Y^{\ominus}[X_g]$); (b) including standard-state conversion from 1 molar to 1 bar gas-phase reference pressure (Eqs. 4.194 and 4.195); (c) including standard-state conversion from 1 mol·dm^{-3} (molar) to 1 mol·kg^{-1} (molal) reference concentration in solution (Eqs. 4.198 and 4.199); (d) including standard-state conversion from 1 atm to 1 bar gas-phase reference pressure (Eqs. 4.192 and 4.193); (e) calculated from the dissolved salt partial molar entropies reported in the indicated reference and listed in Table 5.12 (S-set with the same number) using the gas-phase ion molar entropies from Table 5.9 ($\Delta_s S^{\ominus}[MX_g] = s^{\ominus}[MX_{aq}] - s^{\ominus}[M_g^+] - s^{\ominus}[X_g^-]$); (f) calculated from the salt solvation free energy, enthalpy or/and entropy reported in the indicated reference (G-, H- or/and $\overline{S}$-sets with the same number) using the Gibbs equation ($\Delta_s G^{\ominus}[MX_g] = \Delta_s H^{\ominus}[MX_g] - T^- \Delta_s S^{\ominus}[MX_g]$).

Table 5.12 *Standard partial molar variables for the dissolved alkali-halide salts in water.* The reported data includes the partial molar entropy $s^\ominus$, heat capacity $c_P^\ominus$, volume $v^\ominus$, volume-compressibility $k_T^\ominus$, and volume-expansivity $a_P^\ominus$. The notation MX_{aq} refers to an alkali-halide dissolved ion pair (sum over properties of the dissociated species M_{aq}^+ and X_{aq}^- in the solution phase). The different sets correspond to values quoted in different literature sources, listed in chronological order of publication for $c_P^\ominus$, $v^\ominus$, $k_T^\ominus$ and $a_P^\ominus$ separately, except for Sets 1S, 1C, 1V, 1K and 1A (listed first), which represent recommended data. Only sources explicitly reporting pairwise values are listed, although it is possible to construct many additional pairwise sets based on single-ion conventional parameters from Table 5.17. On the other hand, single-ion conventional solvation parameters deduced from the pairwise sets given in the present table are reported in Table 5.17. The entry σ refers to the ability of the pairwise set to be partitioned into single-ion contributions within an unknown constant. This parameter represents the minimal achievable root-mean-square difference, reached for an optimal partitioning, between the sum of two single-ion values and the corresponding pairwise value, over all possible combinations for which pairwise data is available in the set (at most 20). Non-experimental data inferred by optimal partitioning is indicated between parentheses. The σ values pertain to the complemented set, values for the original sets also being indicated between parentheses, whenever relevant. The lines NUM, AVE, RMS, ERR and REC report for each pair the number of entries, average value, root-mean-square deviation, estimated error ($\text{NUM}^{-1/2}\cdot\text{RMS}$) and recommended value (Sets 1, after minimal adjustments to generate exactly partitionable sets) for the preceding block. The standard states are according to the $bbme_T$ convention (Section 4.2.4; reference pressure $P^\circ = 1$ bar, reference molality $b^\circ = 1$ mol·kg^{-1} and reference temperature $T^- = 298.15$ K). The experimental values reported in the original references are assumed here to pertain to the standard ($s^\ominus$, $c_P^\ominus$), or the density-corrected ($v^{\ominus *}$, $k_T^{\ominus *}$ and $a_P^{\ominus *}$) variant (Section 4.2.5), which represents the most likely situation, but may not necessarily be the case in reality.

Set	Li$^+$				Na$^+$				K$^+$				Rb$^+$				Cs$^+$				σ	Ref.	Notes
	F$^-$	Cl$^-$	Br$^-$	I$^-$	F$^-$	Cl$^-$	Br$^-$	I$^-$	F$^-$	Cl$^-$	Br$^-$	I$^-$	F$^-$	Cl$^-$	Br$^-$	I$^-$	F$^-$	Cl$^-$	Br$^-$	I$^-$			
$s^\ominus[MX_{aq}]$ [J·mol^{-1}·K^{-1}]																							
1S	-1.0	70.3	96.1	124.4	45.0	115.2	141.2	170.1	89.3	159.4	185.4	214.3	107.8	177.9	203.9	232.8	122.2	189.4	215.5	244.3	0.57		a
11S	(4.7)	69.6	95.1	123.6	50.8	115.6	141.1	169.6	(93.0)	157.9	183.4	211.8	(114.8)	179.6	(205.2)	(233.6)	(123.6)	188.4	213.9	242.4	0.00 (0.00)	1881	b
3S	-14.9	53.2	78.8	109.8	21.1	97.6	121.8	152.8	61.4	137.9	163.5	194.5	45.9	187.0	213.9	238.0	166.5	234.6	264.4	291.2	11.79	1873	a
4S	0.8	70.9	96.9	125.7	45.4	116.0	141.6	170.4	88.9	159.4	185.0	214.2	110.8	180.9	206.9	235.7	119.6	189.7	215.8	244.5	0.10	2153	a

(continues on the next page)

Table 5.12 *(continued)*

Set	Li$^+$				Na$^+$				K$^+$				Rb$^+$				Cs$^+$				σ	Ref.	Notes
	F$^-$	Cl$^-$	Br$^-$	I$^-$	F$^-$	Cl$^-$	Br$^-$	I$^-$	F$^-$	Cl$^-$	Br$^-$	I$^-$	F$^-$	Cl$^-$	Br$^-$	I$^-$	F$^-$	Cl$^-$	Br$^-$	I$^-$			
NUM	4	4	4	4	4	4	4	4	4	4	4	4	4	4	4	4	4	4	4	4			
AVE	-2.6	66.0	91.7	120.9	40.6	111.1	136.4	165.7	83.1	153.7	179.3	208.7	94.8	181.3	207.5	235.0	133.0	200.5	227.4	255.6			
RMS	7.4	7.4	7.5	6.4	11.5	7.8	8.5	7.5	12.7	9.1	9.2	8.3	28.4	3.4	3.9	2.0	19.4	19.7	21.4	20.6			
ERR	3.7	3.7	3.7	3.2	5.7	3.9	4.2	3.7	6.3	4.5	4.6	4.1	14.2	1.7	1.9	1.0	9.7	9.8	10.7	10.3			
REC	-0.1	69.7	95.7	124.5	45.4	115.1	141.1	169.9	89.6	159.4	185.3	214.1	108.1	177.9	203.8	232.6	120.3	190.1	216.1	244.9			
$c_P^{\ominus}[MX_{aq}]$ [J·mol^{-1}·K^{-1}]																							
1C	(-54.2)	-65.2	-69.1	(-59.6)	-74.9	-84.4	-88.1	-76.5	-104.0	-114.2	-118.3	-108.7	-124.9	-135.9	-139.0	-131.5	-136.3	-148.9	-155.3	-145.0	0.99 (1.04)		d
12C	-	-	-	-	-	-97.5	-102.5	-103.8	-	-119.2	-124.3	-125.9	-	-	-	-	-	-	-	-	0.10	2101	e
13C	-	-65.4	-	-	-	-	-	-	-	-	-	-	-	-	-	-	-	-	-	-	0.00	2102	e
14C	-	-65.4	-	-	-	-97.5	-	-	-	-119.2	-	-	-	-	-	-	-	-	-	-	0.00	2124	f
15C	(-62.8)	-75.3	-79.5	-83.7	(-83.7)	-96.2	-100.4	-104.6	-104.6	-117.2	-121.3	-125.5	(-150.6)	(-163.2)	-167.4	(-171.5)	(-192.5)	(-205.0)	-209.2	(-213.4)	0.00 (0.00)	2156	f,g
16C	-	-65.4	-	-	-	-99.6	-	-	-	-121.3	-	-	-	-	-	-	-	-	-	-	0.00	2123	f
17C	(-54.8)	-62.5	-65.8	(-55.5)	-73.2	-81.4	-84.5	-72.2	-102.6	-109.8	-113.2	-102.9	-123.5	-130.6	-133.1	-124.8	-135.3	-142.8	-148.5	-137.5	0.68 (0.72)	2103	h
18C	-	-	-	-	-73.2	-81.4	-84.2	-73.5	(-101.8)	-108.9	-113.2	-102.9	(-123.0)	-130.6	(-134.2)	(-123.7)	(-135.2)	-142.8	(-146.3)	(-135.8)	0.25 (0.33)	2127	f

(continues on the next page)

Table 5.12 *(continued)*

Set	Li$^+$				Na$^+$				K$^+$				Rb$^+$				Cs$^+$				σ	Ref.	Notes
	F$^-$	Cl$^-$	Br$^-$	I$^-$	F$^-$	Cl$^-$	Br$^-$	I$^-$	F$^-$	Cl$^-$	Br$^-$	I$^-$	F$^-$	Cl$^-$	Br$^-$	I$^-$	F$^-$	Cl$^-$	Br$^-$	I$^-$			
19C	-	-	-	-	-	-81.4	-	-	-	-	-	-	-	-	-	-	-	-	-	-	0.00	2116	h
20C	(-54.2)	-65.2	-69.1	(-59.6)	-74.9	-84.4	-88.1	-76.5	-104.0	-114.2	-118.3	-108.7	-124.9	-135.9	-139.0	-131.5	-136.3	-148.9	-155.3	-145.0	0.99 (1.04)	2104	h
21C	-	-	-	-	-	-	-88.5	-	-	-	-	-	-	-	-	-	-	-	-	-	0.00	2105	h
NUM	4	7	4	4	5	9	7	6	5	8	6	6	5	5	5	5	5	5	5	5			
AVE	-56.5	-66.3	-70.9	-64.6	-76.0	-89.3	-90.9	-84.5	-103 4	-115.5	-118.1	-112.4	-129.4	-139.2	-142.5	-136.6	-147.1	-157.7	-162.9	-155.3			
RMS	3.6	3.8	5.2	11.2	3.9	7.6	6.9	14.0	1.0	4.2	4.0	9.7	10.7	12.2	12.7	17.8	22.7	23.8	23.4	29.3			
ERR	1.8	1.4	2.6	5.6	1.8	2.5	2.6	5.7	0.5	1.5	1.6	4.0	4.8	5.5	5.7	7.9	10.1	10.7	10.5	13.1			
REC	-54.2	-65.0	-69.3	-59.6	-73.1	-84.0	-88.2	-78.5	-103.5	-114.3	-118.6	-108.9	-125.0	-135.8	-140.1	-130.4	-138.5	-149.4	-153.6	-143.9			
$v^{\ominus *}[MX_{aq}]$ $[10^{-6}\ m^3{\cdot}mol^{-1}]$																							
1V	(-2.5)	17.0	24.0	35.5	-3.5	16.4	23.5	35.1	6.6	26.5	33.7	45.4	12.8	31.9	38.7	50.3	20.1	39.2	46.1	57.7	0.21 (0.21)		i
22V	-	17.0	24.1	35.5	-	16.4	23.5	35.1	-	26.5	33.7	45.4	-	31.9	38.7	50.3	-	39.1	46.2	57.7	0.07	2157	f
44V	-	16.9	23.7	35.1	-	16.8	23.6	35.0	-	27.0	33.9	45.2	-	32.2	39.0	50.4	-	39.4	46.3	57.6	0.00	2120	j
14V	-	17.1	-	-	-	16.3	-	-	-	26.4	-	-	-	-	-	-	-	-	-	-	0.00	2124	f
23Va	-	-	-	-	-	-	23.5	-	-	-	-	-	-	-	-	-	-	-	-	-	0.00	2106	j

(continues on the next page)

Table 5.12 *(continued)*

Set	Li^+				Na^+				K^+				Rb^+				Cs^+				σ	Ref.	Notes
	F^-	Cl^-	Br^-	I^-	F^-	Cl^-	Br^-	I^-	F^-	Cl^-	Br^-	I^-	F^-	Cl^-	Br^-	I^-	F^-	Cl^-	Br^-	I^-			
23Vb	-	-	-	-	-	-	-	-	-	26.8	-	-	-	-	-	-	-	-	-	-	0.00	2106	j
24V	-	-	-	-	-	16.4	23.4	-	-	-	-	-	-	-	-	-	-	-	-	-	0.00	2107	j
25V	(-3.1)	17.1	24.0	35.6	(-3.6)	16.6	23.5	35.1	6.6	26.8	33.7	45.3	(11.6)	31.8	38.7	50.3	(19.0)	39.2	46.1	57.7	0.00 (0.00)	2108	f
26V	-	-	-	-	-	(16.4)	-	35.4	-	26.5	-	45.4	-	-	-	-	-	-	-	-	0.00 (0.00)	2109	j
27V	-	-	-	-	-	16.5	23.5	35.0	-	26.4	33.6	45.2	-	31.7	38.7	50.3	-	39.1	46.2	57.8	0.05	409	k
28V	-	-	-	-	-	16.4	23.4	-	-	26.4	33.4	-	-	-	-	-	-	39.3	45.8	-	0.12	2110	l
29V	-	-	-	-	-1.2	-	-	35.2	-	-	-	-	-	-	-	-	-	-	-	-	0.00	2111	j
30V	-	17.0	-	-	-	16.6	-	-	-	26.9	-	-	-	31.9	-	-	-	39.1	-	-	0.00	2112	j
31V	-	17.0	(24.1)	(35.8)	-	16.4	23.5	35.1	-	26.5	33.7	45.4	-	31.9	(39.0)	(50.7)	-	39.2	(46.4)	(58.0)	0.03 (0.03)	2121	m
32V	-	-	-	-	-	-	23.5	-	-	-	33.7	-	-	-	-	-	-	-	-	-	0.00	2117	j
33V	-	-	-	-	-2.3	-	-	-	-	-	-	-	-	-	-	-	-	-	-	-	0.00	2113	j
34V	(-2.7)	17.0	(24.1)	35.5	-3.5	16.4	23.5	35.1	6.6	26.5	33.7	45.4	12.8	31.9	(39.4)	(51.0)	20.1	39.2	(46.7)	(58.3)	0.15 (0.17)	2122	m
35V	-	-	-	-	-	16.6	(23.6)	(35.0)	-	26.8	33.8	45.2	-	-	-	-	-	-	-	-	0.00 (0.00)	2114	j
36V	(-1.9)	17.0	(23.8)	(35.4)	-2.3	16.6	(23.5)	(35.0)	(8.0)	26.9	33.7	45.3	(13.0)	31.9	(38.8)	(50.4)	(20.2)	39.1	(46.0)	(57.6)	0.00 (0.00)	2129	f
16V	-	17.0	-	-	-	16.6	-	-	-	26.5	-	-	-	-	-	-	-	-	-	-	0.00	2123	n
37V	-	-	-	-	-	16.6	-	-	-	-	-	-	-	-	-	-	-	-	-	-	0.00	2115, 2158	j

(continues on the next page)

Table 5.12 *(continued)*

Set	Li⁺				Na⁺				K⁺				Rb⁺				Cs⁺				σ	Ref.	Notes
	F^-	Cl^-	Br^-	I^-	F^-	Cl^-	Br^-	I^-	F^-	Cl^-	Br^-	I^-	F^-	Cl^-	Br^-	I^-	F^-	Cl^-	Br^-	I^-			
19V	-	-	-	-	-	16.6	-	-	-	-	-	-	-	-	-	-	-	-	-	-	0.00	2116	j
38V	-	17.1	(23.9)	35.3	-	16.6	23.5	35.0	-	26.8	(33.7)	45.2	-	-	-	-	-	39.2	(46.1)	57.6	0.04 (0.04)	2159	o
39V	-	17.1	(24.0)	(35.4)	-	16.6	23.5	35.0	-	26.8	33.8	45.2	-	-	-	-	-	-	-	-	0.01 (0.01)	2160	f
40V	-	-	-	-	-	16.7	-	-	-	-	-	-	-	-	-	-	-	-	-	-	0.00	2161	f
21V	-	-	-	-	-	-	23.6	-	-	-	-	-	-	-	-	-	-	-	-	-	0.00	2105	j
NUM	4	12	9	9	6	20	16	13	4	17	13	12	4	9	8	8	4	11	10	9			
AVE	-2.6	17.0	24.0	35.5	-2.7	16.5	23.5	35.1	6.9	26.6	33.7	45.3	12.6	31.9	38.9	50.5	19.9	39.2	46.2	57.8			
RMS	0.4	0.1	0.1	0.2	0.9	0.1	0.1	0.1	0.6	0.2	0.1	0.1	0.6	0.1	0.2	0.2	0.5	0.1	0.2	0.2			
ERR	0.2	0.0	0.0	0.1	0.4	0.0	0.0	0.0	0.3	0.0	0.0	0.0	0.3	0.0	0.1	0.1	0.2	0.0	0.1	0.1			
REC	-2.5	17.0	24.0	35.6	-3.2	16.4	23.4	35.0	7.0	26.5	33.5	45.1	12.4	31.9	38.9	50.5	19.7	39.3	46.3	57.9			
$k_T^{\ominus *}[MX_{aq}]$ $[10^{-10}\ m^3 \cdot mol^{-1} \cdot bar^{-1}]$																							
1K	(-64.7)	-41.0	-33.4	-22.0	-71.6	-49.6	-39.7	-27.7	(-64.2)	-41.3	-33.0	-20.4	(-61.2)	-38.3	(-29.7)	(-17.7)	(-57.8)	-34.9	(-26.3)	(-14.3)	0.36 (0.47)		p
14K	-	-36.4	-	-	-	-44.2	-	-	-	-38.3	-	-	-	-	-	-	-	-	-	-	0.00	2124	f
25K	-	-41.5	(-32.8)	-17.0	-	-50.0	-42.0	(-25.3)	-	-44.5	-34.0	-18.0	-	-	-	-	-	(-50.2)	(-41.2)	-25.0	0.41 (0.51)	2108	q

(continues on the next page)

Table 5.12 *(continued)*

Set	Li⁺				Na⁺				K⁺				Rb⁺				Cs⁺				σ	Ref.	Notes
	F^-	Cl^-	Br^-	I^-	F^-	Cl^-	Br^-	I^-	F^-	Cl^-	Br^-	I^-	F^-	Cl^-	Br^-	I^-	F^-	Cl^-	Br^-	I^-			
41K	-	-	-	-	-	-46.6	-	-	-	-40.7	-	-	-	-	-	-	-	-	-	-	0.00	2125	r
42K	-	-41.0	-33.4	-22.0	-	-55.1	-40.7	-28.6	-	-47.0	-37.5	-23.8	-	-	-	-	-	-	-	-	1.44	2126	r
16K	-	-38.7	-	-	-	-46.7	-	-	-	-39.3	-	-	-	-	-	-	-	-	-	-	0.00	2123	r
18K	-	-	-	-	-71.6	-49.6	-39.7	-27.7	(-64.2)	-41.3	-33.0	-20.4	(-60.7)	-38.3	(-29.2)	(-16.9)	(-57.3)	-34.9	(-25.8)	(-13.5)	0.20 (0.27)	2127	r
43K	-	-	-	-	-	-48.6	-	-	-	-	-	-	-	-	-	-	-	-	-	-	0.00	2128	s
NUM	1	5	3	3	2	8	4	4	2	7	4	4	2	2	2	2	2	3	3	3			
AVE	-64.7	-39.7	-33.2	-20.3	-71.6	-48.8	-40.5	-27.3	-64.2	-41.8	-34.4	-20.6	-61.0	-38.3	-29.4	-17.3	-57.5	-40.0	-31.1	-17.6			
RMS	0.0	1.9	0.3	2.4	0.0	3.0	0.9	1.2	0.0	2.8	1.8	2.1	0.2	0.0	0.2	0.4	0.2	7.2	7.1	5.2			
ERR	0.0	0.9	0.2	1.4	0.0	1.1	0.5	0.6	0.0	1.1	0.9	1.0	0.2	0.0	0.2	0.3	0.2	4.2	4.1	3.0			
REC	-64.7	-41.9	-33.3	-21.3	-71.6	-48.7	-40.1	-28.1	-64.2	-41.3	-32.7	-20.7	-61.2	-38.3	-29.7	-17.7	-57.8	-34.9	-26.3	-14.3			
$a_P^{\ominus *}[MX_{aq}]$ $[10^{-8}\ m^3 \cdot mol^{-1} \cdot K^{-1}]$																							
1A	(2.2)	2.5	(3.6)	(6.3)	6.7	7.7	8.2	10.1	(6.6)	6.3	8.0	11.5	(5.9)	6.2	(7.3)	(10.0)	(5.5)	5.8	(6.9)	(9.6)	0.31 (0.44)		t
36A	-	2.5	(3.3)	(6.7)	-	9.3	(10.1)	(13.5)	-	8.5	9.3	12.7	-	7.0	(7.8)	(11.2)	-	7.0	(7.8)	(11.2)	0.00 (0.00)	2129	f
16A	-	2.5	-	-	-	9.3	-	-	-	8.5	-	-	-	-	-	-	-	-	-	-	0.00	2123	f

(continues on the next page)

Table 5.12 *(continued)*

Set	Li$^+$				Na$^+$				K$^+$				Rb$^+$				Cs$^+$				σ	Ref.	Notes
	F$^-$	Cl$^-$	Br$^-$	I$^-$	F$^-$	Cl$^-$	Br$^-$	I$^-$	F$^-$	Cl$^-$	Br$^-$	I$^-$	F$^-$	Cl$^-$	Br$^-$	I$^-$	F$^-$	Cl$^-$	Br$^-$	I$^-$			
18A	-	-	-	-	6.7	7.7	8.2	10.1	(6.6)	6.3	8.0	11.5	(5.9)	6.2	(7.3)	(10.0)	(5.5)	5.8	(6.9)	(9.6)	0.35 (0.47)	2127	f
NUM	1	3	2	2	2	4	3	3	2	4	3	3	2	3	3	3	2	3	3	3			
AVE	2.2	2.5	3.5	6.5	6.7	8.5	8.8	11.2	6.6	7.4	8.4	11.9	5.9	6.5	7.5	10.4	5.5	6.2	7.2	10.1			
RMS	0.0	0.0	0.1	0.2	0.0	0.8	0.9	1.6	0.0	1.1	0.6	0.6	0.0	0.4	0.2	0.6	0.0	0.6	0.4	0.8			
ERR	0.0	0.0	0.1	0.1	0.0	0.4	0.5	0.9	0.0	0.6	0.4	0.3	0.0	0.2	0.1	0.3	0.0	0.3	0.2	0.4			
REC	2.2	2.5	3.6	6.3	6.7	7.0	8.1	10.8	6.6	7.0	8.1	10.8	5.9	6.2	7.3	10.0	5.5	5.8	6.9	9.6			

(a) calculated from the solvation entropies reported in the indicated reference (or in this book for Set 1S) and listed in Table 5.11 ($\overline{\text{S}}$-set with the same number) using gas-phase ion molar entropies from Table 5.9 ($s^{\ominus}[MX_{aq}] = s^{\ominus}[M_g^+] + s^{\ominus}[X_g^-] + \Delta_s S^{\ominus}[MX_g]$); (b) including standard-state conversion from a mole fraction of unity to 1 mol·kg^{-1} (molal) reference concentration in solution (Eqs. 4.202 and 4.203); (d) based on data from Ref.[2104] as described in the text; (e) based on calorimetry measurements; (f) see quoted reference for various data sources; (g) the values for RbBr and CsBr are apparently taken from Ref.[2162] (it is unclear whether the values for the Li salts and for KF are also from the same source); (h) based on flow microcalorimetry measurements; (i) based on data from Refs.[2108,2122] as described in the text; (j) based on densimetric measurements; (k) apparent molar volume data and nature of the measurement unclear in the quoted source; (l) from various sources of densimetric data indicated in the quoted reference (the value for NaCl corresponds to an average over a pair of values reported therein based on different sources of raw data); (m) data pertains to 22°C and is based on ionic vibration potentials (ultrasound wave propagation velocity) measurements; (n) based on ultrasonic velocity measurements; (o) based on the conventional data compilation of Ref.[2163]; (p) based on Refs.[2126,2127] as described in the text; (q) based on densimetric measurements at different pressures (the values for LiCl, NaCl and KCl correspond to averages over pairs of values reported therein based on different sources of raw data); (r) based on adiabatic compressibilities obtained from sound velocity measurements; (s) based on piezometric measurements (the value for NaCl corresponds to an average over a pair of values reported therein based on different extrapolation methods to infinite dilution); (t) based on Refs.[2123,2127,2129] as described in the text.

As was the case for Tables 5.7 and 5.8, no attempt was made here to distinguish between experimental data appropriate for the standard or density-corrected variants of the solute standard-state definition (Section 4.2.5) on a case-to-case basis. The experimental values as reported in the original references are simply assumed to pertain to the standard variant for $\Delta_s H^{\ominus}$, $\Delta_s S^{\ominus}$, $s^{\ominus}$ and $c_P^{\ominus}$ and to the density-corrected variant for $v^{\ominus}$, $k_T^{\ominus}$ and $a_P^{\ominus}$, which represents the most likely situation (see above), but may not necessarily be the case in reality. According to the above assumption, the three latter standard parameters are thus indicated as $v^{\ominus *}$, $k_T^{\ominus *}$ and $a_P^{\ominus *}$ in the tables. Considering the differences between the two variants in the context of water (Table 4.1; values to be multiplied by two for binary salts) and the magnitude of the values reported in Tables 5.11 and 5.12 the errors possibly incurred by this ambiguity are limited but non-negligible. Taking $NaCl_g$ ($\Delta_s Y^{\ominus}$, $Y = H$, S; recommended sets labeled "REC" in Table 5.11) and $NaCl_{aq}$ ($y^{\ominus}$, $y = s$, c_P, v, k_T, a_P; recommended sets labeled "REC" in Table 5.12) as an example, the differences between the two variants amount to about 0.05, 0.7, 1.1, 20.0, 13.7, 11.7 and 2.8% for $\Delta_s H^{\ominus}$, $\Delta_s S^{\ominus}$, $s^{\ominus}$, $c_P^{\ominus}$, $v^{\ominus *}$, $k_T^{\ominus *}$ and $a_P^{\ominus *}$, respectively, $\Delta_s G^{\ominus}$ being identical in the two variants.

In both tables, the entry σ refers to the ability of the pairwise set to be partitioned into single-ion contributions within an unknown constant (Section 4.5.1). This parameter corresponds to the minimal achievable root-mean-square difference, reached for an optimal partitioning, between the sum of two single-ion values and the corresponding pairwise value, over the set of all possible combinations for which pairwise data is available in the set (at most 20). Given the definitions of the gas-phase and dissolved standard states (Section 4.2.4), an ideal pairwise set should be exactly partitionable. The quantity σ therefore provides a lower bound on the magnitude of the experimental errors affecting a particular set. The entries indicated between parentheses in the tables are not experimental, but deduced from the other values in the set given this optimal partitioning. The σ value refers to these complemented sets, the corresponding σ value for the original non-complemented set also being indicated between parentheses.

Another estimate of the error is provided by the violations of the Gibbs equation ($\Delta_s G^{\ominus}[MX_g] - \Delta_s H^{\ominus}[MX_g] + T^{-}\Delta_s S^{\ominus}[MX_g] = 0$, for sources reporting simultaneously the three quantities) and of the predicted ion-pair gas-phase entropies ($s^{\ominus}[M_g^+] + s^{\ominus}[X_g^-] - s^{\ominus}[MX_{aq}] + \Delta_s S^{\ominus}[MX_g] = 0$, for sources reporting simultaneously the two latter quantities and based on the recommended values of Table 5.9 for the two former quantities). In the present case, no significant violations of either types are found, because the two above relationships are actually used to construct missing sets whenever possible. Only for Set 4G, 4H and 4$\overline{\text{S}}$ were the quantities explicitly reported in the original literature source, and are found to satisfy the Gibbs equation within at most 0.5 kJ·mol^{-1} for all ion pairs.

In the two tables, the sets labeled 1G, 1H, 1$\overline{\text{S}}$, 1S, 1C, 1V, 1K and 1A represent experimental data sets that are considered by the authors of this book as the most trustworthy to date, because they are: (*i*) based on the most recent available measurements, namely from Ref.[221] for $\Delta_s G^\ominus$ and $\Delta_s H^\ominus$, as well as the derived parameters $\Delta_s S^\ominus$ and $s^\ominus$, Ref.[2104] for $c_P^\ominus$, Refs.[2108,2122] for $v^{\ominus *}$, Refs.[2126,2127] for $k_T^{\ominus *}$ and Refs.[2123,2127,2129] for $a_P^{\ominus *}$, along with gas-phase data taken from Refs.[294,1723,1750] (Table 5.9); (*ii*) consistent with the data reported in the other tables of this chapter; (*iii*) complete over the alkali-halide series. Note, however, that these sets are not yet entirely satisfactory because they are not exactly partitionable into single-ion contributions, *i.e.* they are characterized by non-zero values for σ. The ultimate recommended pairwise sets proposed in this book also encompass a slight correction to remedy this problem (see records "REC" discussed further below).

As mentioned above, sets 1G, 1H, $1\overline{S}$ and 1S are derived from the data reported in Tables 5.7 and 5.9 by application of the ISR cycle. They are thus essentially based on Refs.[221, 294, 1723, 1750]. These sets are partitionable into single-ion contributions within remarkably small errors σ of 0.05 kJ·mol^{-1}, 0.16 kJ·mol^{-1}, 0.57 J·mol^{-1}·K^{-1} and 0.57 J·mol^{-1}·K^{-1}.

Sets 1C, 1V, 1K and 1A are those that were discussed in Section 5.7 and included in Table 5.7. Set 1C is based on Ref.[2104] (Set 20C, 1976). In the absence of reliable experimental data concerning $c_P^\ominus$ for LiF and LiI (the value of Set 15C for LiI based on Ref.[2156] was considered inaccurate and not retained), this set is complemented by corresponding non-experimental values, inferred *via* optimal partitioning of the available values. Set 1C is partitionable within an error σ of 0.99 J·mol^{-1}·K^{-1}. Set 1V is based on Ref.[2122] (Set 34V, 1967), complemented by Ref.[2108] (Set 25V, 1941) for LiBr, RbBr, RbI, CsBr and CsI. In the absence of any experimental data concerning $v^{\ominus *}$ for LiF, this composite set is further complemented by a non-experimental value inferred *via* optimal partitioning of the available values. Set 1V is partitionable within an error σ of $0.21 \cdot 10^{-6}$ m^3·mol^{-1}. Set 1K is based on Ref.[2127] (Set 18K, 1974), complemented by Ref.[2126] (Set 42K, 1966) for LiCl, LiBr and LiI. In the absence of reliable experimental data concerning $k_T^{\ominus *}$ for LiF, KF, RbF, RbBr, RbI, CsF, CsBr and CsI (the value of Set 25K for CsI based on Ref.[2108] was considered inaccurate and not retained), this composite set is further complemented by non-experimental values inferred *via* optimal partitioning of the available values. Set 1K is partitionable within an error σ of $0.36 \cdot 10^{-10}$ m^3·mol^{-1}·bar^{-1}. Set 1A is based on Ref.[2127] (Set 18A, 1974), complemented by Ref.[2129] or Ref.[2123] (Sets 36A and 16A, 1968 and 1969, respectively) for LiCl. In the absence of any experimental data concerning $a_P^{\ominus *}$ for LiF, LiBr, LiI, KF, RbF, RbBr, RbI, CsF, CsBr and CsI, this composite set is further complemented by non-experimental values inferred *via* optimal partitioning of the available values. Set 1A is partitionable within an error σ of $0.31 \cdot 10^{-8}$ m^3·mol^{-1}·K^{-1}.

Finally, in both tables, average values ("AVE" in the tables) over the number of available sets for each salt ("NUM") are indicated, along with corresponding root-mean-square deviations ("RMS"), error estimates ("ERR") and recommended values ("REC"). The error estimates were evaluated by dividing the root-mean-square deviation by the square root of the number of available sets for the corresponding salts. The recommended values correspond to the above sets 1G, 1H, 1$\overline{\text{S}}$, 1S, 1C, 1V, 1K and 1A, after minimal adjustments to make them exactly partitionable. This adjustment relies on the definition of conventional single-ion values according to an optimal partitioning and alignment procedure described in Section 5.16, followed by resummation into salt values. In other words, the sets labeled "REC" are very close to the above recommended sets, exactly partitionable and exactly compatible with the recommended single-ion conventional values that will be provided later in Tables 5.16 and 5.17 as sets 1g$_3$, 1h$_3$, 1$\bar{\text{s}}_3$, 1s$_3$, 1c$_3$, 1v$_3$, 1k$_3$ and 1a$_3$ (also listed in these tables as "REC"). They are also compatible with the final compilation of recommended data provided in Section 5.19.

The $\Delta_s G^\ominus$ and $\Delta_s H^\ominus$ values are very large and negative, indicating that the hydration process is associated with both a strong driving force and an important release of heat, as expected for a process involving the formation of favorable electrostatic interactions between ions and solvent dipoles, along with a less important cost associated with the perturbation of the solvent-solvent interactions. They systematically decrease in magnitude with increasing effective ion sizes (atomic numbers), *i.e.* along the series $Li^+ \rightarrow Cs^+$ for salts involving a common anion and $F^- \rightarrow I^-$ for salts involving a common cation. This trend is related to the corresponding increase in the effective closest ion-solvent distance in solution, leading to reduced charge-dipole interactions, and can be very simply rationalized on the basis of the BO+ST model (Section 3.2). The $\Delta_s S^\ominus$ values are all negative, primarily due to the reduction of the effective volume accessible to the ions upon transfer from the gas phase at reference pressure P° to the solution at reference molality b°. Another entropy contribution arises from the perturbation of the solvent structure, as mentioned in Section 5.7 and further discussed in Section 5.20. The $\Delta_s S^\ominus$ values systematically decrease in magnitude with increasing ion sizes, a trend which arises predominantly from the above contribution related to the perturbation of the solvent structure.

The $\Delta_s G^\ominus$, $\Delta_s H^\ominus$ and $\Delta_s S^\ominus$ values corresponding to distinct experimental data sets are essentially consistent, although significant variations of up to 33 kJ·mol^{-1} (CsCl), 125 kJ·mol^{-1} (LiI) and 69 J·mol^{-1}·K^{-1} (RbF), respectively, can be observed. The corresponding estimated errors are in ranges of about $2-6$ kJ·mol^{-1}, $4-13$ kJ·mol^{-1} and $1-14$ J·mol^{-1}·K^{-1}, respectively, for the different salts. Finally, the differences between the recommended set ("REC") and the average of all sets ("AVE") is at most 1.7 times as large as the corresponding root-mean-square difference ("RMS") for the three properties and for all salts.

The $s^\ominus$, $c_P^\ominus$, $v^{\ominus *}$, $k_T^{\ominus *}$ and $a_P^{\ominus *}$ values have been discussed previously in terms of their signs, magnitudes and trends (Section 5.7). Here, the available data sets are typically sparser. The values corresponding to distinct data sets are also essentially consistent, although significant variations of up to 69 J·mol^{-1}·K^{-1} (RbF), 78 J·mol^{-1}·K^{-1} (CsI), 2·10^{-6} m^3·mol^{-1} (NaF),

$15 \cdot 10^{-10}$ m^3·mol^{-1}·bar^{-1} (CsBr), and $3 \cdot 10^{-8}$ m^3·mol^{-1}·K^{-1} (NaI), respectively, can be observed. The corresponding estimated errors are in ranges of about $1 - 14$ J·mol^{-1}·K^{-1}, $1 - 13$ J·mol^{-1}·K^{-1}, $0 - 1 \cdot 10^{-6}$ m^3·mol^{-1}, $0 - 4 \cdot 10^{-10}$ m^3·mol^{-1}·bar^{-1}, and $0 - 1 \cdot 10^{-8}$ m^3·mol^{-1}·K^{-1}, respectively, for the different salts. Finally, the differences between the recommended set ("REC") and the average of all sets ("AVE") is at most 2.0 times as large as the corresponding root-mean-square difference ("RMS") for the three properties and for all salts.

5.13 Metal work functions

The *standard work function* $\omega^{\ominus}[M]$ of a (solid or liquid) metal ($M = M_s$ or M_ℓ) is accessible experimentally. It can be derived from the corresponding spectroscopic work function $w[M]$, as detailed in Section 4.3.5. The two quantities are essentially equivalent, the former one being appropriate for the warm-electron convention used in this book, and the latter one for the alternative cold-electron convention (Sections 4.2.4 and 4.2.17). Differences in work functions between two metals may also be accessed *via* Voltaic cell experiments (Section 4.4.4).

The *spectroscopic work function* may be determined using photo-electron, thermionic or field-emission spectroscopy[402,403] on metal samples with highly clean surfaces. As discussed in Sections 4.2.12 and 4.3.5, this quantity represents the *semi-adiabatic free-energy change* associated with the *ejection of an electron* from the conduction band of the metal through the metal surface, divided by F. It is semi-adiabatic in the sense that the initial state of the process corresponds to an electron "solvated" in the metal at a well-defined pressure and temperature, while the final state of the process corresponds to a free electron exempt of kinetic energy, *i.e.* corresponding to a temperature $T = 0$ K (see Ref.[1719] for a discussion of these issues). To obtain a standard work function, the value $w[M]$, measured for the metal at a pressure P° and a temperature T^-, must be corrected so that the final state of the process involves an ideal electron gas at the same pressure and temperature (Eq. 4.286).

As detailed in Section 4.3.5, the appropriate measurement of a work function should also involve a (solid or liquid) metal sample that is *dominated by (quasi-)planar surface patches* (*e.g.* polyhedron with large faces or droplet with large radius) and, for monocrystalline solids, *by a single type of crystallographic face* (*e.g.* thin lamella with the two largest opposite surfaces corresponding to a given crystallographic face type). Otherwise, the measured work function will depend on the sample shape *via* the curvature or/and face type dependences of the local surface potential, resulting in a non-vanishing and shape-dependent Galvani potential (Figure 4.6).

In the case of *monocrystalline solid metals*, the work function therefore depends in the general case on the shape of the sample *via* the proportion of crystallographic faces of different types exposed at its surface. This depen-

Table 5.13 *Spectroscopic and standard work functions for the alkali (and some other) metals.* The reported data includes the spectroscopic work function w (at P° and T^-) and the standard work function $\omega^\ominus$ of the metal. These parameters refer to metal samples dominated by (quasi-)planar surface patches and, for solids, presenting polycrystalline surfaces. The notation M refers to a solid (M_s) or liquid (M_ℓ) metal. The quantity $F\omega^\ominus$ is also reported. The standard states are according to the $bbme_T$ convention (Section 4.2.4; reference pressure $P^\circ = 1$ bar and reference temperature $T^- = 298.15$ K; standard state for the free electron, applying to $\omega^\ominus$ only, according to the warm-electron convention corresponding to an ideal electron gas at temperature T^- with properties calculated according to Fermi-Dirac statistics).

M	$w[M]$ [V]	$\omega^\ominus[M]$ [V]	$F\omega^\ominus[M]$ [kJ·mol^{-1}]
Li_s	2.93	2.89	279.1
Na_s	2.36	2.32	224.1
K_s	2.29	2.25	217.3
Rb_s	2.26	2.22	214.4
Cs_s	1.95	1.91	184.5
Cr_s	4.5	4.46	430.6
Mn_s	4.1	4.06	392.0
Fe_s	4.5	4.46	430.6
Co_s	5.0	4.96	478.8
Ni_s	5.15	5.11	493.3
Pt_s	5.64	5.60	540.5
Cu_s	4.65	4.61	445.0
Ag_s	4.0	3.96	382.3
Zn_s	3.63	3.59	346.6
Hg_ℓ	4.475	4.437	428.1
Refs.	221,2164		
Notes	a	b	

(a) values from Ref.[221], except Fe_s, Ni_s, Cu_s and Ag_s from Ref.[2164]; (b) calculated according to Eq. 4.286 using values from Table 4.2.

dence is often incorrectly interpreted as a dependence on the crystallographic face of the metal through which the electron is ejected[98, 221, 278, 1779, 1780, 1786]. However, for a given sample, the work function must be the same irrespective of the choice of this face, since otherwise the electron could be ejected through one face by supplying a certain amount of work and hypothetically reinjected through another one with recovery of a different amount of work, resulting in a violation of the first law of thermodynamics. The effect observed experimentally is not a dependence on the ejection face, but on the dominant exposed crystallographic face of the sample considered. This sample is typically a thin metal lamella where the largest surfaces are of a specific crystallographic face type, one of these surfaces also being selected as the ejection surface for practical reasons. The work function measured for such a lamella solely depends on the nature of the metal and on type of the dominant exposed crystallographic face, and can be loosely referred to as the work function of the metal relative to this face type.

In the case of *solid metals with polycrystalline surfaces*, the surface properties are averaged over an assembly of microcrystallites with random orientations, and therefore no longer depend on a specified crystallographic face type. On the other hand, the sample surfaces are not necessarily planar. This situation is similar to that of *liquid metals*. In both cases, the measured work function does not depend on the sample shape, provided that the sample is dominated by surface patches of identical curvature. When this condition is satisfied, the work function solely depends on the nature of the metal and on the dominant surface curvature, and can be referred to as the work function of the metal at this curvature. As an illustrative example, the curvature (tube diameter) dependence of the work function of carbon nanotubes has recently been investigated by density functional calculations[2165,2166]. In practice, most commonly, a situation of zero curvature is considered, involving a sample that is a large droplet or a polyhedron with large faces (see above), and the mention of the curvature is omitted. As a result, the work function no longer depends on the sample shape. An interesting question in the context of metals with polycrystalline surfaces is whether the corresponding work function represents an unbiased average over the monocrystal work functions associated with all possible crystallographic faces, or whether it is biased *e.g.* by non-homogeneous exposure of face types at the level of the microcrystallites or by overweighting of low workfunction paths upon electron ejection[1781].

The spectroscopic work functions, from which corresponding standard work functions are easily derived, have been determined[221] for all alkali metals (M_s) in the context of polycrystalline samples. Spectroscopic and standard work functions for these (and some other[221,2164]) metals are reported in Table 5.13. The reader is referred to *e.g.* Ref.[221] for examples of spectroscopic work functions measured on monocrystalline samples with reference to distinct crystallographic face types (considering the metals Ag, Al, Au, Cu, Fe, Ir, Mo, Nb, Ni, Pt, Ta, U and W).

Note that spectroscopic work functions are usually tabulated in literature sources[98,221] in the form of $w'_M = ew_M$ in units of eV, rather than w_M in units of V. Obviously, the corresponding numerical values are identical, but the use of w_M with the unit of an electric potential is more consistent with the conventions adopted in the present book, *i.e.* leads to simpler equations. Similarly, $\omega^\ominus$ is reported here in units of V and the quantity $F\omega^\ominus$, also provided in Table 5.13 for reference, has units of kJ·mol^{-1}. Uncertainties in measured work functions are on the order of 0.02 V (value estimated[1718] for Hg_ℓ), *i.e.* about 2 kJ·mol^{-1} in terms of $F\omega^\ominus$.

As mentioned previously, w implicitly refers to a cold-electron convention. As such, this purely spectroscopic quantity does not depend on any standard-state convention, except for the fact that it is usually measured on metal samples at P° and T^-. In contrast, $\omega^\ominus$ refers to a warm-electron convention and depends strictly on the choice of P° and T^- as determining the standard-state properties of the electron *via* Eq. 4.286. In addition, these properties must have been appropriately calculated using Fermi-Dirac rather than Boltzmann statistics (Section 4.2.13). Although the standard work function has been introduced by the authors of this book, *i.e.* is not an established quantity in the current literature, it is implicitly relied upon in the determination of many other properties, including *e.g.* real absolute electrode potentials (Section 5.14) and real single-ion solvation free energies (Section 5.15). A similar issue also exists in the context of properties derived from gas-phase cluster formation thermodynamics (*e.g.* Point E_{12} in Section 5.14 and Point E_{14} in Section 5.18). As a result, errors in the electron statistics employed may propagate into these derived properties and must be corrected when analyzing the corresponding literature results, which is (as much as possible) done in the tables of this book.

5.14 Real absolute potential of the hydrogen electrode

The *standard absolute real potential of the reference hydrogen electrode* $\hat{\mathcal{V}}_H^\ominus$ is in principle experimentally accessible, although its determination is *relatively difficult*. First, it should be observed that the measurement of the absolute real potential $\hat{\mathcal{V}}_{M,S}^\ominus$ of any reversible metal electrode (M) in contact with a solution (S) containing its cations (assumed monovalent for simplicity) is sufficient to determine $\hat{\mathcal{V}}_H^\ominus$, since the electrode potentials relative to the reference hydrogen electrode, *i.e.* the redox potentials $\Delta_H\mathcal{V}^\ominus$, are known accurately (Section 5.5) and are equal to the corresponding differences in absolute potentials (Section 4.4.2, Eq. 4.378). Second, it has been seen that $\hat{\mathcal{V}}_{M,S}^\ominus$ is equal to the difference between the standard work function $\omega_M^\ominus$ of the metal and the standard lateral Volta potential difference $\Delta\psi_{M,S}^\ominus$ between the metal electrode and the solution (Section 4.3.6, Eq. 4.291). Since standard metal work functions are known accurately (Section 5.13), the measurement of a metal-solution lateral Volta potential difference $\Delta\psi_{M,S}^\ominus$ is also sufficient to determine $\hat{\mathcal{V}}_H^\ominus$. Third, it has also been shown that the absolute real potential $\hat{\mathcal{V}}_{M,S}^\ominus$ is equal to the sum of the formation free energy $\Delta_f G^\ominus[M_g^+]$ of the gas-phase ion and of its real solvation free energy $\Delta_s\hat{G}^\ominus[M_g^+]$, *i.e.* to the real formation free energy $\Delta_f\hat{G}^\ominus[M_d^+]$ of the dissolved ion, divided by F (Section 4.3.7, Eq. 4.307). Since the gas-phase ion formation free energies are known accurately (Section 5.10), there is a direct connection between $\hat{\mathcal{V}}_{M,S}^\ominus$ and $\Delta_s\hat{G}^\ominus[M_g^+]$ and, in particular, between $\hat{\mathcal{V}}_H^\ominus$ and $\hat{\mathcal{G}}_H^\ominus$ (Section 4.3.7, Eq. 4.313). Finally, the sum of the real solvation free energies $\Delta_s\hat{G}^\ominus[M_g^+]$ and $\Delta_s\hat{G}^\ominus[X_g^-]$ of the metal cation and of a given anion (assumed monovalent for simplicity) is equal to the solvation free energy $\Delta_s G^\ominus[MX_g]$ of the corresponding binary salt (Section 2.2, Eq. 2.23), which is also an unambiguous quantity (Section 5.12).

Considering the four above relationships, there is no clear-cut borderline between the methods aiming at the measurement of any of the above quantities $\hat{\mathcal{V}}_H^\ominus$, $\hat{\mathcal{V}}_{M,S}^\ominus$, $\Delta\psi_{M,S}^\ominus$, $\Delta_s\hat{G}^\ominus[M_g^+]$, $\Delta_s\hat{G}^\ominus[X_g^-]$ and $\hat{\mathcal{G}}_H^\ominus$. The connecting equations are repeated here for convenience, and read

$$\hat{\mathcal{G}}_H^\ominus = F\hat{\mathcal{V}}_H^\ominus - \Delta_f G^\ominus[\mathrm{H}_g^+] \quad , \tag{5.32}$$

$$\Delta_s\hat{G}^\ominus[M_g^+] = F\hat{\mathcal{V}}_{M,S}^\ominus - \Delta_f G^\ominus[M_g^+] \quad , \tag{5.33}$$

$$\hat{\mathcal{V}}_H^\ominus = \hat{\mathcal{V}}_{M,S}^\ominus - \Delta_H\mathcal{V}^\ominus[M_{el}, M_g^+] \quad , \tag{5.34}$$

$$\hat{\mathcal{V}}_{M,S}^\ominus = \omega_M^\ominus - \Delta\psi_{M,S}^\ominus \tag{5.35}$$

and

$$\Delta_s G^\ominus[MX_g] = \Delta_s\hat{G}^\ominus[M_g^+] + \Delta_s\hat{G}^\ominus[X_g^-] \quad . \tag{5.36}$$

Based on these considerations, the methods to determine the real absolute electrode potential $\hat{\mathcal{V}}_H^\ominus$ of the reference hydrogen electrode can be broadly classified as follows:

(A_{12}) Determination of the *real solvation free energy of the proton* and combination with the free energy of formation of the gas-phase proton *via* Eq. 5.32 (as done in Ref.[419] and Ref.[1718] based on original data from Ref.[411] and Ref.[445], respectively, and in this book based on original data from Refs.[264,407–409,411,415] as reported in Table 5.15), or determination of the *real solvation free energy of another ion* and combination with the free energy of formation of this gaseous ion and the redox potential of the corresponding element-ion couple *via* Eqs. 5.33 and 5.34 (as done in Ref.[2167] using Ag^+ as reference ion, based on original data from Ref.[467]). Note that this approach does not formally represent a determination method *per se*, because the required real solvation parameters must still previously have been evaluated using one of the other methods (Points B_{12}-E_{12} below).

(B_{12}) Combination of a *metal work function* with the *lateral Volta potential difference* corresponding to a reversible electrode of this metal in contact with a solution containing the corresponding ions, and correction using the redox potential of the corresponding element-ion couple, *via* Eqs. 5.34 and 5.35. The Voltaic cell measurements involved in the determination of such a metal-solution lateral Volta potential difference have been described in Section 4.4.4. The specific types of measurements that have been applied in this context include (Section 4.4.4, Figure 4.16) the Kenrick method[70,1781,1835] (as done in Refs.[411,445]) or the differential application of the Kelvin probe method[403,1781] separately to the metal and the solution (as done in Ref.[1797]). In the case of Ref.[2168] and Ref.[1718], the metal-solution lateral Volta potential difference measured in Ref.[411] and Ref.[445], respectively, for the calomel electrode at the potential of the reference hydrogen electrode and for the ideally-polarizable mercury-water electrode at the potential of zero charge with respect to the reference hydrogen electrode, respectively, were employed. This approach implicitly corrects for the redox potential of the metal-electrode considered (Eq. 5.34), directly leading to an estimate for $\hat{\mathcal{V}}_H^{\ominus}$ rather than $\hat{\mathcal{V}}_{M,S}^{\ominus}$. In the case of Ref.[1797], the metal-solution lateral Volta potential difference for the calomel electrode was determined in own measurements, and a subsequent correction for the redox potential of the calomel electrode leads to an estimate for $\hat{\mathcal{V}}_H^{\ominus}$. Note that Ref.[411] and Ref.[445] did not directly report $\hat{\mathcal{V}}_H^{\ominus}$ estimates, which were only derived subsequently in Ref.[2168] (as well as Ref.[419], Point A_{12}) and Ref.[1718], respectively.

(C_{12}) Determination of the *"wet" work function* of an electrode at the potential of the reference hydrogen electrode, *immediately after emersion* from an aqueous solution[1789–1792] (In_2O_3 electrode in NaF solution for Ref.[1789] or Au electrode in $NaClO_4$ solution for Ref.[1790]). The basic principle of this determination is that a comparison between the usual work function of a metal, corresponding to a dry and electroneutral metal sample in vacuum or air, and the analogous "wet" work function of an electrically-insulated and electroneutral metal-solution sample (measured across a dry surface of the metal) provides in principle a direct access to the metal-solution lateral Volta potential difference. In practice, such an isolated metal-solution sample is generated by slow emersion (about[1792] 2 $mm{\cdot}s^{-1}$) of the electrode under potentiostatic control[1780], *i.e.* at constant potential difference relative to the reference hydrogen electrode (enforced using a control voltage source) into an insulating medium (nitrogen at atmospheric pressure in Ref.[1789], ultra-high vacuum in Ref.[1790]). After full emersion, the "wet" work function is determined, *e.g.* directly *via* photo-electron spectroscopy[1790] or indirectly *via* a Kelvin probe approach[1789], providing access to the absolute potential of the hydrogen electrode. Although quite ingenious, this approach is affected by numerous sources of errors[1792], most of which are related to the fact that the properties of the thin solution layer at the surface of the emersed electrode may differ from those of a solution of macroscopic extent. In principle, alterations may affect: (*i*) the surface potential of the metal at contact with the solution; (*ii*) the surface potential of the solution at contact with the metal; (*iii*) the electric double-layer (ions *vs.* electrons) at the metal-solution interface; (*iv*) the surface potential of the solvent at contact with the insulating medium. The largest sources of error appear to originate from water loss at the solution surface[1791] and from the fact that the interface between solution and insulating medium is perturbed by the influence of the nearby metal[1792], resulting in a preferential orientation of the water dipoles with their oxygen atoms pointing towards the electrode, rather than towards the insulating phase for an unperturbed air-water interface. Correction terms for these effects have been proposed on the basis of additional experiments, namely in Ref.[1792] for the effect of solution surface perturbation in the experiments of Ref.[1789] and in Ref.[1791] for the effect of water loss in the experiments of Ref.[1790]. As discussed in Section 4.3.6, another issue affecting these measurements concerns the nature of the absolute electrode potential they probe. Based on arguments presented there, these experiments probably represent an attempt to evaluate the vacuum absolute potential $\tilde{\mathcal{V}}_H^{\ominus}$ rather than the real absolute potential $\hat{\mathcal{V}}_H^{\ominus}$, the former quantity being ill-defined in view of its possible sensitivity to the extents and shapes of the phases in contact. Assuming a weak influence of this source of error[1796], the results of these determinations are nevertheless retained here as acceptable estimates for $\hat{\mathcal{V}}_H^{\ominus}$.

(D_{12}) Determination of the *"contact" work function* of a semi-conductor (indium phosphide with p-type conductivity p-InP) interfaced with the metal (Pt) of a reference hydrogen electrode at the potential of zero charge[482]. The basic principle of this determination is to transpose the problem of evaluating the surface potential at an air-liquid interface into that of evaluating the interfacial potential at a solid-solid interface. In this approach, the "contact" work function of p-InP interfaced with the metal is measured and corrected to account for the density of holes in the valence band of the semi-conductor. This corrected work function is combined with an estimate for the solid-solid interfacial potential at the junction to the hydrogen-saturated Pt electrode (flat-band potential of the Schottky barrier), and with the value of the potential of zero charge of the reference hydrogen electrode, providing access to the absolute potential of the hydrogen electrode. Note that because the reference hydrogen electrode is reversible, its potential of zero charge is not directly accessible and must be inferred by extrapolation according to a procedure described in Refs.[2169–2171]. As pointed out in Ref.[1796], and similarly to the situation encountered for the methods of Point C_{12}, this study probably also represents an attempt to evaluate a vacuum absolute potential $\tilde{\mathcal{V}}_H^{\ominus}$ rather than the real absolute potential $\hat{\mathcal{V}}_H^{\ominus}$, the former quantity being ill-defined in view of its possible sensitivity to the extents and shapes of the phases in contact. Assuming a weak influence of this source of error[1796], the result of this determination is nevertheless retained here as an acceptable estimate for $\hat{\mathcal{V}}_H^{\ominus}$.

(E_{12}) Measurement of the *electron-capture enthalpy of gas-phase ion-water clusters*[1793–1795] *via* nanocalorimetry[1808], and combination with theoretical estimates (Born model) for the intrinsic solvation enthalpy of the clusters in the bulk[1793,1794] to obtain the reduction enthalpy in solution, or direct determination of the reduction enthalpy in solution *via* extrapolation of the cluster results to infinite cluster size[1795]. The combination of the reduction enthalpies evaluated in this way with experimental estimates for the entropy changes of the corresponding redox reactions, derived from isothermal temperature derivatives of the associated redox potentials, leads to estimates for the corresponding (absolute) half-cell potentials in solution. These can then be combined with the (relative) half-cell redox potentials to obtain an estimate for the absolute potential of the reference hydrogen electrode. In a first approach [1793], nanocalorimetry is applied to determine the enthalpy change of the gas-phase reduction reaction

$$[M(\mathrm{NH_3})_6(\mathrm{H_2O})_{55}]_g^{3+} + e_g^- \rightarrow [M(\mathrm{NH_3})_6(\mathrm{H_2O})_{55-n}]_g^{2+} + n\mathrm{H_2O} \quad (5.37)$$

considering the trivalent ions of the metals M = Ru, Co, Os, Cr or Ir, or

$$[M(\mathrm{H_2O})_{32}]_g^{2+} + e_g^- \rightarrow [M(\mathrm{H_2O})_{32-n}]_g^{+} + n\mathrm{H_2O} \quad (5.38)$$

considering the divalent ion of the metal M = Cu. The solvation enthalpies of the reactant and product clusters are then estimated using the Born model for the enthalpy (charging term in Eq. 3.9), resulting in estimates for the enthalpy change of the corresponding half-cell reaction

$$M_{aq}^{3+} + e_g^- \rightarrow M_{aq}^{2+} \quad (5.39)$$

for M = Ru, Co, Os, Cr or Ir, or

$$M_{aq}^{2+} + e_g^- \rightarrow M_{aq}^{+} \quad (5.40)$$

for M = Cu. As a last step, the associated entropy changes are evaluated based on the isothermal temperature coefficients of the corresponding redox reactions, and conversion to non-isothermal temperature coefficients using an estimate for the absolute temperature coefficient of the reference hydrogen electrode[1829]. Comparison of the free-energy change of the reaction, divided by F, with the redox potential of the corresponding electrode provides the required estimate for $\hat{\mathcal{V}}_H^\ominus$. In a second approach [1794], a slightly different gas-phase reduction reaction is analyzed, namely

$$[M(\mathrm{H_2O})_{24}]_g^{2+} + e_g^- \rightarrow [M(\mathrm{OH})(\mathrm{H_2O})_{23}]_g^{+} + \mathrm{H}_g \quad , \quad (5.41)$$

considering the divalent ions of the metals M = Mn, Fe, Co, Ni, Cu, Zn, Mg, Ca or Sr. In this case, the calculation involves three other quantities, namely the enthalpy change associated with the reaction converting $[M(\mathrm{H_2O})]_{aq}^{2+}$ to $[M(\mathrm{OH})]_{aq}^{+}$ and H_g obtained by application of the Born model to the involved gas-phase clusters, the enthalpy of the first hydrolysis reaction of M^{2+} in solution, *i.e.* $M_{aq}^{2+} \rightarrow [M(\mathrm{OH})]_{aq}^{+} + \mathrm{H}_{aq}^{+}$, and the formation enthalpy of the gas-phase hydrogen atom. It leads again, after combination with entropy estimates[1829], to a value of $\hat{\mathcal{V}}_H^\ominus$. Finally, in a third approach[1795], the use of the Born model is avoided by estimating absolute in-solution redox potentials *via* extrapolation of the measured gas-phase cluster values to infinite cluster size. Here, the determination relies on $[\mathrm{Eu}(\mathrm{H_2O})_n]^{3+}$ clusters, with n ranging from 55 to 140. The reduction enthalpies of the gas-phase reactions

$$[\mathrm{Eu}(\mathrm{H_2O})_n^{3+}]_g + e_g^- \rightarrow [\mathrm{Eu}(\mathrm{H_2O})_n^{2+}]_g \tag{5.42}$$

are fitted against $n^{-1/3}$. The intercept of the resulting (approximately linear) function gives the absolute reduction enthalpy of the in-solution reaction. After combination with an entropy estimate [1829] and with the redox potential of the Eu^{3+}/Eu^{2+} couple, a value of $\hat{\mathcal{V}}_H^{\ominus}$ is again obtained. The accuracy of the $\hat{\mathcal{V}}_H^{\ominus}$ estimates resulting from the three above (closely related) approaches may be questioned in four main respects. First, the Born model is used in Refs. [1793,1794] to evaluate the intrinsic solvation enthalpies of ion-water clusters. As discussed in Section 3.2, this model is moderately accurate, especially in the context of thermodynamic derivative properties (such as the enthalpy). In addition, the cavitation contribution, potentially non-negligible for nanoclusters, is not included and the results may be very sensitive to the ambiguous choice of a cluster radius. Arguably, this issue is eliminated in the extrapolation approach of Ref. [1795]. Second, the application of the Born model, which overlooks interfacial potential contributions at the cavity-solvent and solvent-air interfaces, or of an extrapolation procedure, will only lead to a real solvation enthalpy provided that the clusters under investigation are sufficiently large to have reached the macroscopic-system regime in terms of surface potential, an assumption explicitly stated in Ref. [1794]. Third, this approach raises an important issue concerning the gas-phase standard state of the electron. To define a real absolute potential, this state should refer to an ideal gas of electrons at the Volta potential of the solution (Section 4.3.6 and Figure 4.9). However, in the present approach, this reference state is at infinity. As a result, one may argue, as in Points C_{12} and D_{12} above, that the vacuum absolute potential $\tilde{\mathcal{V}}_H^{\ominus}$ is evaluated instead of the real one $\hat{\mathcal{V}}_H^{\ominus}$. Finally, another ambiguity concerns the question whether a cold- or a warm-electron convention (Section 4.2.17) is applied in these determinations, and in the latter case, whether the standard-state properties of the ideal electron gas are appropriately calculated based on Fermi-Dirac statistics (Section 4.2.13).

A collection of 20 available estimates for the real absolute potential $\hat{\mathcal{V}}_H^{\ominus}$ of the reference hydrogen electrode in water, labeled $1\hat{\mathrm{V}}$-$20\hat{\mathrm{V}}$ for the ease of further reference, is presented in Table 5.14. Note, however, that some of the reported values may actually rather pertain to the corresponding vacuum absolute potential $\tilde{\mathcal{V}}_H^{\ominus}$ (see Points C_{12}, D_{12} and E_{12}). These estimates include values reported explicitly in the original literature sources after application of various corrections or/and conversions (conversion to the warm-electron convention, correction from Boltzmann to Fermi-Dirac statistics for the ideal electron gas, use of up-to-date gas-phase data, other corrections suggested in Refs.[1791,1792]), along with values deduced by the authors of this book from the real proton solvation free energies reported subsequently in Table 5.15.

Conversely, real proton solvation free energies deduced from the values given in the present table (including corrections) are also reported in Table 5.15. The reported $\hat{\mathcal{V}}_H^\ominus$ values refer to the warm-electron convention adopted in the present book, using Fermi-Dirac statistics for the ideal electron gas. The corresponding values in the cold-electron convention are also reported for comparison. The difference between the two quantities (cold minus warm) is $-F^{-1}(\Delta_h H^\ominus[e_g^-] - T^- s^\ominus[e_g^-]) = 0.038$ V (see Eq. 4.286, based on data from Table 4.2). Note that for the estimates of type E_{12} ($18\hat{V}$-$20\hat{V}$), the convention used in the quoted literature source is not entirely obvious, and the most reasonable assumption was made. For a number of estimates ($5\hat{V}$, $9\hat{V}$, $14\hat{V}$, $15\hat{V}$) the values in this table differ from those originally reported, due to application of additional corrections (see above). The data presented in Table 5.14 is illustrated graphically in Figure 5.4.

The average of the 20 estimates for $\hat{\mathcal{V}}_H^\ominus$ is 4.41 V, with a standard deviation of 0.15 V and an estimated error on the mean of $\pm$0.03 V, assuming that all measurements are independent. Considering the above discussion of potential shortcomings in the various determination methods, the authors of the present book tend to trust preferentially estimates based on approaches A_{12}, B_{12}, C_{12} (after correction according to Refs.[1791,1792]) and D_{12}. The average of this restricted set of estimates is 4.42 V. The value recommended by IUPAC[1718] is that of Ref.[1718], namely 4.44 V (Sets $8\hat{V}$ and $11\hat{V}$) according to the cold-electron convention, which corresponds to 4.40 V in the warm-

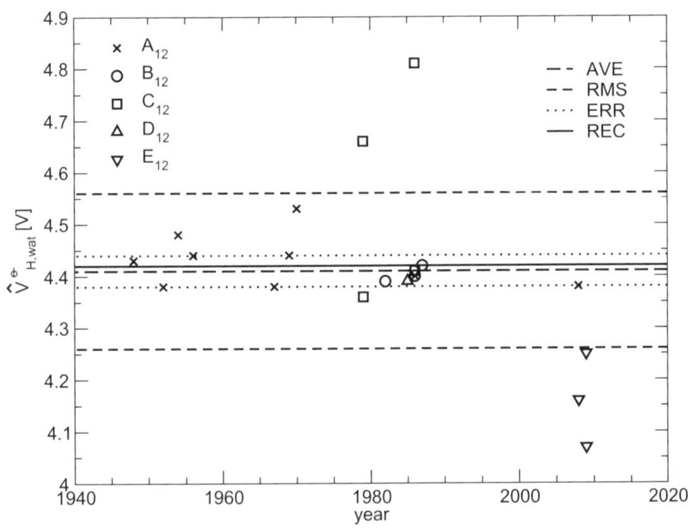

Figure 5.4 *Available estimates for the real absolute potential $\hat{\mathcal{V}}_H^\ominus$ of the reference hydrogen electrode in water.* This graph illustrates the data reported in Table 5.14, the estimates being sorted by publication year and by methods.

Table 5.14 *Real absolute potential of the reference hydrogen electrode in water.* The reported data represents estimates for the real absolute potential $\hat{\mathcal{V}}_H^{\ominus}$ of the reference hydrogen electrode in water. The methods refer to the literal enumeration of Section 5.14. The values rely on the use of the standard work function $\omega^{\ominus}$ as appropriate for the warm-electron standard-state convention adopted in this book. The values in parentheses rely on the use of a spectroscopic work function w (at P° and T^{-}) instead, *i.e.* correspond to a cold-electron convention. The values in boldface are the ones reported in the quoted reference (possibly modified by application of corrections, see footnotes), while the values in italics were recalculated from the latter. The different sets correspond to values quoted in different literature sources, listed in chronological order of publication for each type of method. Sources explicitly reporting absolute potentials for the reference hydrogen electrode in water are listed, along with additional $\hat{\mathcal{V}}_H^{\ominus}$ values calculated from real proton solvation free-energy estimates $\hat{\mathcal{G}}_H^{\ominus}$ from Table 5.15. Conversely, real proton solvation free energies deduced from the explicitly-reported $\hat{\mathcal{V}}_H^{\ominus}$ values given in the present table are also to be found in Table 5.15. The lines NUM, AVE, RMS, ERR and REC report the number of entries, average value, root-mean-square deviation, estimated error (NUM$^{-1/2}$·RMS) and recommended value. The standard states are according to the $bbme_T$ convention (Section 4.2.4; reference pressure $P^{\circ} = 1$ bar and reference temperature $T^{-} = 298.15$ K; standard state for the free electron according to the warm-electron convention corresponding to an ideal electron gas at temperature T^{-}, with properties calculated according to Fermi-Dirac statistics; absolute potential definitions: Section 4.3.6).

Set	Method	Potential [V]	Source	Ref.	Notes	Year
		$\hat{\mathcal{V}}_H^{\ominus}$				
1V̂	A_{12}	4.43 (4.47)	1ĝ	407	a	1948
2V̂	A_{12}	4.38 (4.42)	2ĝ	408	a	1952
3V̂	A_{12}	4.48 (4.52)	3ĝ	409	a	1954
4V̂	A_{12}	4.44 (4.48)	4ĝ	411	a	1956
5V̂	A_{12}	*4.38* (**4.42**)	-	2167	b,c	1967
6V̂	A_{12}	4.44 (4.48)	5ĝ	264	a	1969
7V̂	A_{12}	4.53 (4.57)	6ĝ	415	a	1970
8V̂	A_{12}	*4.40* (**4.44**)	-	1718	d	1986
9V̂	A_{12}	**4.38** (*4.42*)	-	419	e,f	2008
10V̂	B_{12}	*4.39* (**4.43**)	-	2168	g	1982
11V̂	B_{12}	*4.40* (**4.44**)	-	1718	h	1986
12V̂	B_{12}	*4.42* (**4.46**)	-	1797	i	1987
13V̂	C_{12}	*4.66* (**4.7**)	-	1789	j,k	1979
14V̂	C_{12}	*4.36* (**4.4**)	-	1789	k,l	1979
15V̂	C_{12}	*4.81* (**4.85**)	-	1790	k,m	1986
16V̂	C_{12}	*4.41* (**4.45**)	-	1790	k,n	1986
17V̂	D_{12}	*4.39* (**4.43**)	-	482	k	1985
18V̂	E_{12}	*4.16* (**4.2**)	-	1793	k,o	2008

(continues on the next page)

Table 5.14 *(continued)*

Set	Method	Potential [V]	Source	Ref.	Notes	Year
		$\hat{\mathcal{V}}_H^{\ominus}$				
19$\hat{\text{V}}$	E_{12}	*4.25* (**4.29**)	-	1794	k,o	2009
20$\hat{\text{V}}$	E_{12}	*4.07* (**4.11**)	-	1795	k,o	2009
NUM		20				
AVE		4.41				
RMS		0.15				
ERR		0.03				
REC		4.42				

(a) derived from the $\hat{\mathcal{G}}_H^{\ominus}$ estimate of the indicated source set in Table 5.15 using Eq. 5.32 along with data from Table 4.2; (b) based on the real solvation free energy of the silver ion Ag_g^+ of Ref.[467] (itself relying on method of type B_{12}); (c) corrected by using an up-to-date value for the first ionization potential of the gaseous silver atom[221], $E_I[Ag_g] = 7.576$ eV (the cold-electron value thereby changes from 4.48 V in the indicated reference to 4.42 V); (d) based on the real solvation free energy of the proton H_g^+ reported by Ref.[445] (itself relying on method of type B_{12}); (e) based on the real solvation free energy of the proton H_g^+ deduced in the indicated reference as an average of real proton solvation free energy values inferred from real electrode potentials of the alkali metals in water by Ref.[411] (relying on method of type B_{12}); (f) corrected from Boltzmann to Fermi-Dirac statistics for the ideal electron gas (the warm-electron value thereby changes from 4.42 V in the indicated reference to 4.38 V); (g) based on the Hg-water lateral Volta potential difference at the potential of the reference hydrogen electrode using data from Ref.[411], based on Kenrick cell experiments with $Hg_\ell \;\square\; KCl_{aq}(HgCl_s) \mid Hg_\ell$ using various molalities of the KCl_{aq} solution; (h) based on the Hg-water lateral Volta potential difference at zero charge using data from Ref.[445], based on Kenrick cell experiments with $Hg_\ell \;\square\; HCl_{aq}(AgCl_s) \mid Ag_s$ and $Hg_\ell \;\square\; HCl_{aq} \mid Pt_s(H_{2,g})$ cells, using various molalities of the HCl_{aq} solution, combined with the potential of zero charge of the Hg-water electrode with respect to the reference hydrogen electrode; (i) based on the reference calomel electrode; (j) not considering the alteration of the solution air-liquid interfacial potential at the emersed electrode; (k) indicated value probably represents an estimate for $\tilde{\mathcal{V}}_H^{\ominus}$ rather than $\hat{\mathcal{V}}_H^{\ominus}$; (l) considering the alteration of the solution air-liquid interfacial potential at the emersed electrode and correcting for it according to Ref.[1792] (correction of -0.3 V); (m) not considering water loss at the emersed electrode; (n) considering water loss at the emersed electrode and correcting for it according to Ref.[1791] (correction of -0.4 V); (o) the value reported in the original article was assumed to pertain to the cold-electron convention.

electron convention with Fermi-Dirac statistics. Also considering the subsequent discussion of real proton solvation, of the surface potential of water and of intrinsic proton solvation (Sections 5.15, 5.17 and 5.18), the authors of the present book recommend a value $\hat{\mathcal{V}}_H^{\ominus} = 4.42$ V for the real absolute potential of the reference hydrogen electrode in water, with an estimated error of ± 0.03 V in the present standard-state convention, *i.e.* involving a standardized work function appropriate for the warm-electron convention and Fermi-Dirac statistics for the ideal electron gas.

There is, to our knowledge, no literature source explicitly reporting estimates for the pressure or temperature derivatives of $\hat{\mathcal{V}}_H^{\ominus}$. However, based on the single available estimate for the real proton hydration entropy $\hat{\mathcal{S}}_H^{\ominus}$ (Section

5.15, Set 6ŝ in Table 5.15), it is possible to evaluate $\partial_T \hat{\mathcal{V}}_H^\ominus$ according to

$$-F\partial_T \hat{\mathcal{V}}_H^\ominus = \hat{\mathcal{S}}_H^\ominus + \Delta_f S[\mathrm{H}_g^+] \quad , \tag{5.43}$$

which is the negative derivative of Eq. 5.33 with respect to temperature. Using the corresponding numerical value $\hat{\mathcal{S}}_H^\ominus = -90.1$ J·mol^{-1}·K^{-1} with an estimated error of ± 2 J·mol^{-1}·K^{-1} along with data from Table 5.9, one finds $\partial_T \hat{\mathcal{V}}_H^\ominus = 0.25$ mV·K^{-1} with an estimated error of ± 0.02 mV·K^{-1}. As discussed in Section 5.15, this estimate pertains to the standard variant of the solute standard-state definition (Section 4.2.5). The above values and error estimates for $\hat{\mathcal{V}}_H^\ominus$ and $\partial_T \hat{\mathcal{V}}_H^\ominus$ are those that will be retained within the final compilation of recommended data in Section 5.19.

Finally, recently an estimate has been explicitly reported in the literature concerning the intrinsic absolute potential $\mathcal{V}_H^\ominus$ of the reference hydrogen electrode in water [483]. This estimate of 4.275 V is calculated based on a value of -1000.9 kJ·mol^{-1} for the intrinsic proton hydration free energy $\mathcal{G}_H^\ominus$, itself derived from the estimate of -1104.5 kJ·mol^{-1} suggested by Refs. [1875, 1876] (cluster-pair extra-thermodynamic assumption, set 20g$_1$ in Table 5.19) after application of a correction from Boltzmann to Fermi-Dirac statistics for the ideal electron gas. Just as $\hat{\mathcal{V}}_H^\ominus$ and $\hat{\mathcal{G}}_H^\ominus$, the quantities $\mathcal{V}_H^\ominus$ and $\mathcal{G}_H^\ominus$ are trivially connected by parameters that are experimentally easily accessible. The quantities $\hat{\mathcal{V}}_H^\ominus$ and $\hat{\mathcal{G}}_H^\ominus$ are themselves related to $\mathcal{V}_H^\ominus$ and $\mathcal{G}_H^\ominus$, respectively, *via* the air-liquid interfacial potential $\chi_{wat}^\ominus$ of pure water. However, while $\hat{\mathcal{V}}_H^\ominus$ and $\hat{\mathcal{G}}_H^\ominus$ are in principle experimentally accessible, albeit difficult to determine accurately in practice (see above), $\mathcal{V}_H^\ominus$, $\mathcal{G}_H^\ominus$ and $\chi_{wat}^\ominus$ are experimentally elusive (Section 2.3), their determination requiring the application of an extra-thermodynamic assumption. For this reason, the present book does not contain any table explicitly reporting $\mathcal{V}_H^\ominus$ values. Instead, all estimates available to our knowledge for $\mathcal{G}_H^\ominus$ are gathered in Table 5.19 and discussed in Section 5.18, including those derived from explicit $\chi_{wat}^\ominus$ estimates which are reported in Table 5.18. If desired, these 58 estimates could be trivially converted to as many $\mathcal{V}_H^\ominus$ and $\chi_{wat}^\ominus$ estimates *via* Eqs. 2.27 and 2.31.

5.15 Real single-ion solvation parameters

The *standard real single-ion solvation parameters* $\Delta_s \hat{Y}^\ominus[I_g^z]$ are in principle accessible experimentally, although their determination is relatively difficult. The most relevant of these parameters are the real solvation free energies $\Delta_s \hat{G}^\ominus[I_g^z]$, enthalpies $\Delta_s \hat{H}^\ominus[I_g^z]$ and entropies $\Delta_s \hat{S}^\ominus[I_g^z]$. Of particular importance are the corresponding parameters for the proton $\Delta_s \hat{Y}^\ominus[\mathrm{H}_g^+]$, which are also written here $\hat{\mathcal{Y}}_H^\ominus$, *i.e.* $\hat{\mathcal{G}}_H^\ominus$, $\hat{\mathcal{H}}_H^\ominus$ and $\hat{\mathcal{S}}_H^\ominus$ for $Y = G$, H and S. The parameters $\Delta_s \hat{Y}^\ominus[I_g^z]$ characterize the solvation of a free gaseous ion in a given solvent, including the contribution from the crossing of the corresponding air-liquid interface (Section 2.2). As discussed in Section 5.14, there is no clear-cut boundary between the methods aiming at the measurement of the real absolute potential $\hat{\mathcal{V}}_H^\ominus$ of the reference hydrogen electrode, of the real absolute potential $\hat{\mathcal{V}}^{\ominus}_{M,S}$ of any reversible metal-solution electrode, of the lateral Volta potential difference $\Delta\psi^{\ominus}_{M,S}$ of such an electrode, of the real solvation free energy $\hat{\mathcal{G}}^{\ominus}_{H}$ of the proton, or of the real solvation free energy $\Delta_s\hat{G}^{\ominus}[I^z_g]$ of any other ion, since the quantities required for the interconversion of these properties *via* Eqs. 5.32-5.36 are available experimentally with a sufficient accuracy. Therefore, the possible approaches for the determination of $\Delta_s\hat{G}^{\ominus}$ are essentially the same as those listed in Section 5.14 (Points A_{12}-E_{12}). In practice, the most common determination method relies on Voltaic cell measurements (Point B_{12}). The corresponding evaluation of $\Delta_s\hat{S}^{\ominus}$ relies on a measurement of the temperature derivative of $\Delta_s\hat{G}^{\ominus}$, $\Delta_s\hat{H}^{\ominus}$ being subsequently calculated using the Gibbs equation. The quantity $\Delta_s\hat{S}^{\ominus}$ also gives access to the real partial molar entropy $\hat{s}^{\ominus}[I^z_d]$ of the dissolved ion, by addition of the gas-phase ion entropy $s^{\ominus}[I^z_g]$. To our knowledge, however, real single-ion solvation free energies have never been investigated in terms of their pressure or higher-order temperature derivatives. As a result, there are no estimates available for the real partial molar heat capacities $\hat{c}^{\ominus}_P[I^z_d]$, volumes $\hat{v}^{\ominus}[I^z_d]$, volume-compressibilities $\hat{k}^{\ominus}_T[I^z_d]$ and volume-expansivities $\hat{a}^{\ominus}_P[I^z_d]$ of dissolved ions.

Real single-ion solvation parameters have been determined for all alkali (M^+_g) and halide (X^-_g) gas-phase ions with water as a solvent. A number of *independent determinations* have led to as many distinct sets. These standard parameters, as well as those of the proton, are reported in Table 5.15. For the ease of further reference, these sets are numbered from 1 to 22 based on the literature source, along with the letter ĝ, ĥ or ŝ to indicate the type of property, namely $\Delta_s\hat{G}^{\ominus}$, $\Delta_s\hat{H}^{\ominus}$ or $\Delta_s\hat{S}^{\ominus}$, respectively. Only sources *explicitly reporting real single-ion values* are listed, as well as $\hat{\mathcal{G}}^{\ominus}_{H}$ values deduced from the real absolute potential $\hat{\mathcal{V}}^{\ominus}_{H}$ of the reference hydrogen electrode in water (Table 5.14, *via* Eq. 5.32, using data from Table 5.9), although it is possible to construct many additional real sets based on single-ion conventional parameters $\Delta_s Y^{\bullet}[I^z_g]$ provided later in Table 5.16, along with a given choice of real proton solvation parameters $\hat{\mathcal{Y}}^{\ominus}_{H}$ from the present table. On the other hand, single-ion conventional parameters deduced from the real sets given in this table are also reported in Table 5.16. Because they are derived directly from the temperature derivative of the free energy $\Delta_s\hat{G}^{\ominus}$, the derivative properties $\Delta_s\hat{H}^{\ominus}$ and $\Delta_s\hat{S}^{\ominus}$ unambiguously pertain to the standard variant of the solute standard-state definition (Section 4.2.5). This is to be expected considering that any experimental number encompassing information on a derivative of $\chi^{\ominus}_{svt}$ will automatically encompass the influence of the corresponding derivative of $\rho^{\ominus}_{svt}$, unless it is explicitly corrected to remove this dependence.

For two sets (1ĝ, 3ĝ), the quantity $\hat{\mathcal{G}}_H$ was not provided in the original literature source and had to be reconstructed based on the data reported for other ions, by optimal alignment against recommended conventional data as described in Section 5.16. Sets 7ĝ-20ĝ are derived directly from sets 5V̂ and 8V̂-20V̂ in Table 5.14 by application of Eq. 5.32 using data from Table 5.9. The corresponding determination methods are discussed in Section 5.14. Sets 1ĝ-6ĝ, 6ĥ, 21ĥ-22ĥ and 6ŝ involve determination methods of the type described under Point B_{12} in Section 5.14, including analysis of the temperature dependence for the enthalpy sets and the single entropy set.

For the above sets, the reference electrodes considered are the reference hydrogen electrode (as done in Refs.[82,83]), the calomel electrode (as done in Ref.[2172] and Refs.[264,404,405,407–409,411,417]; the results of Ref.[408] are based on the original data from Refs.[404,405]; the results of Ref.[409] also include consideration of the data from Refs.[404,405,408], Ref.[417] provides a correction to the original data of Ref.[411]) and the silver chloride electrode (as done in Refs.[415,2174]; the results of Ref.[415] rely on original data from Ref.[2174]). The types of Voltaic cell experiments employed are the ionizing probe method[1781,1833,1834] (as done in Refs.[404,405]) and the Kenrick method[70,1781,1835] (as done in Ref.[411] and in Ref.[2174]). The determination of Ref.[2172] relies on an average over previous enthalpy estimates from Refs.[71,73,74,2173] in combination with an approximation for the entropy component. The determination of Refs.[81,82] relies on assuming zero enthalpy change for the reaction $^1/_2H_{2,g} \rightarrow H^+_{aq}$, based on consideration of electrocapillary and dropping electrode studies of Ostwald[71].

No experimental source was found that explicitly reported free energies, enthalpies and entropies simultaneously. Therefore, the consistency of the sets cannot be checked against the Gibbs equation. In contrast, one set ($6\hat{g}$) was derived from two other sets explicitly reported by the same source ($6\hat{h}$, $6\hat{s}$) by application of the Gibbs equation. The data presented in Table 5.15 concerning $\hat{\mathcal{G}}^{\ominus}_{H}$ is illustrated graphically in Figure 5.5. This graph is opti-

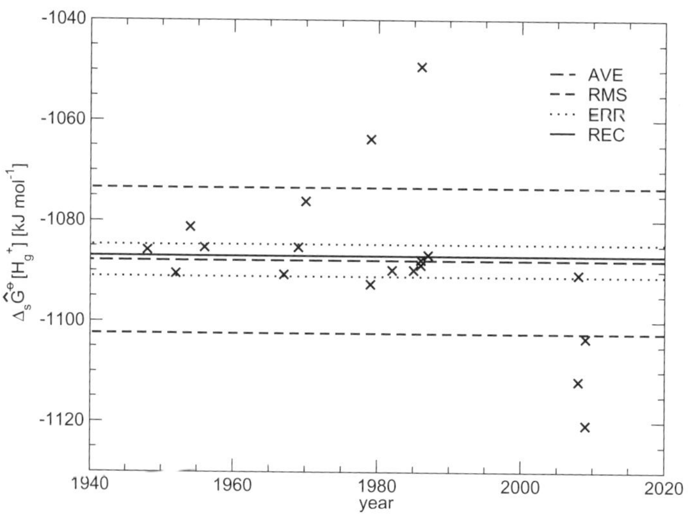

Figure 5.5 *Available estimates for the real single-ion solvation free energy $\hat{\mathcal{G}}^{\ominus}_{H}$ of the proton in water.* This graph illustrates the data reported in Table 5.15 concerning $\hat{\mathcal{G}}^{\ominus}_{H}$, the estimates being sorted by publication year. Note that the representative points of Sets $12\hat{g}$ and $13\hat{g}$ (1986, -1088.8 kJ·mol^{-1}) are superimposed. Corresponding graphs for $\hat{\mathcal{H}}^{\ominus}_{H}$ and $\hat{\mathcal{S}}^{\ominus}_{H}$ are not shown in view of the sparseness of the data (see numerical values in Table 5.15).

Table 5.15 *Standard real single-ion solvation parameters for alkali and halide gas-phase ions as well as the proton in water.* The reported data includes the real solvation free energy $\Delta_s\hat{G}^\ominus$, enthalpy $\Delta_s\hat{H}^\ominus$ and entropy $\Delta_s\hat{S}^\ominus$. The notation I_g^z refers to an alkali (M_g^+) or halide (X_g^-) gas-phase ion, or to the proton (H_g^+). The different sets correspond to values quoted in different literature sources listed in chronological order of publication for $\Delta_s\hat{G}^\ominus$, $\Delta_s\hat{H}^\ominus$ and $\Delta_s\hat{S}^\ominus$ separately (chronological order is also separate for Sets 1ĝ-6ĝ and 7ĝ-20ĝ). Only sources explicitly reporting real single-ion values are listed, as well as $\Delta_s\hat{G}^\ominus[H_g^+]$ values deduced from the absolute real potential $\hat{\mathcal{V}}_H^\ominus$ of the reference hydrogen electrode in water from Table 5.14, although it is possible to construct many additional real sets based on single-ion conventional parameters $\Delta_s Y^\bullet[I_g^z]$ from Table 5.16 along with a given choice of real proton solvation parameters $\Delta_s\hat{Y}^\ominus[H_g^+]$ from the present table. On the other hand, single-ion conventional parameters deduced from the real sets given in the present table are reported in Table 5.16. Non-experimental proton values inferred by alignment against consensus conventional data (Section 5.16) are indicated between parentheses. The lines NUM, AVE, RMS, ERR and REC report for each ion the number of entries, average value, root-mean-square deviation, estimated error ($\text{NUM}^{-1/2}\cdot\text{RMS}$) and recommended value (see explanations in Section 5.15) for the preceding block. The standard states are according to the $bbme_T$ convention (Section 4.2.4; reference pressure $P^\circ = 1$ bar, reference molality $b^\circ = 1$ mol·kg^{-1} and reference temperature $T^- = 298.15$ K), the reported data pertaining to the standard variant (Section 4.2.5).

Sets	H^+	Li^+	Na^+	K^+	Rb^+	Cs^+	F^-	Cl^-	Br^-	I^-	Source	Ref.	Notes	Year
					$\Delta_s\hat{G}^\ominus[I_g^z]$ [kJ·mol^{-1}]									
1ĝ	-1085.8	-511.8	-411.4	-338.1	-322.7	-297.1	-437.1	-313.7	-286.9	-253.4	-	407	a,b,c,d,e	1948
2ĝ	(-1090.5)	-510.9	-393.7	-360.2	-314.2	-	-	-	-	-	-	404,405,408	b,f,g	1952
3ĝ	(-1081.2)	-510.9	-393.7	-322.6	-301.7	-276.6	-448.1	-314.2	-297.5	-264.0	-	409	b,d,e,f,h	1954
4ĝ	-1085.3	-506.2	-406.2	-332.6	-311.3	-279.0	-419.2	-300.4	-276.1	-243.9	-	411,417	a,i	1956
5ĝ	-1085.3	-506.2	-406.2	-332.6	-311.3	-277.8	-438.9	-321.7	-307.9	-261.5	-	264	a,d,e,j,k	1969
6ĝ	-1076.1	-502.2	-402.4	-328.1	-	-	-446.8	-322.8	-296.6	-262.3	-	415	l	1970
7ĝ	-1090.7	-	-	-	-	-	-	-	-	-	5V̂	2167	m	1967
8ĝ	-1063.7	-	-	-	-	-	-	-	-	-	13V̂	1789	m	1979
9ĝ	-1092.6	-	-	-	-	-	-	-	-	-	14V̂	1789	m	1979
10ĝ	-1089.8	-	-	-	-	-	-	-	-	-	10V̂	2168	m	1982
11ĝ	-1089.8	-	-	-	-	-	-	-	-	-	17V̂	482	m	1985
12ĝ	-1088.8	-	-	-	-	-	-	-	-	-	11V̂	1718	m	1986
13ĝ	-1088.8	-	-	-	-	-	-	-	-	-	8V̂	1718	m	1986
14ĝ	-1049.2	-	-	-	-	-	-	-	-	-	15V̂	1790	m	1986
15ĝ	-1087.8	-	-	-	-	-	-	-	-	-	16V̂	1790	m	1986
16ĝ	-1086.8	-	-	-	-	-	-	-	-	-	12V̂	1797	m	1987
17ĝ	-1090.7	-	-	-	-	-	-	-	-	-	9V̂	419	m	2008
18ĝ	-1111.9	-	-	-	-	-	-	-	-	-	18V̂	1793	m	2008
19ĝ	-1103.3	-	-	-	-	-	-	-	-	-	19V̂	1794	m	2009
20ĝ	-1120.6	-	-	-	-	-	-	-	-	-	20V̂	1795	m	2009
NUM	20	6	6	6	5	4	5	5	5	5				
AVE	-1087.9	-508.0	-402.3	-335.7	-312.2	-282.6	-438.0	-314.6	-293.0	-257.0				
RMS	14.5	3.4	6.6	11.9	6.7	8.4	10.3	8.0	10.7	7.5				
ERR	3.2	1.4	2.7	4.9	3.0	4.2	4.6	3.6	4.8	3.4				
REC	-1087.0	-	-	-	-	-	-	-	-	-				
					$\Delta_s\hat{H}^\ominus[I_g^z]$ [kJ·mol^{-1}]									
21ĥ	-1514.6	-878.6	-849.4	-761.5	-723.8	-728.0	-	-96.2	-133.9	-179.9	-	82,83	n	1919
22ĥ	-1012.5	-497.9	-401.7	-322.2	-301.2	-	-	-	-	-	-	2172	o	1926
6ĥ	-1103.0	-532.0	-423.0	-338.0	-	-	-499.0	-358.0	-327.0	-291.0	-	415	p	1970
NUM	3	3	3	3	2	1	1	2	2	2				
AVE	-1210.0	-636.2	-558.0	-473.9	-512.5	-728.0	-499.0	-227.1	-230.4	-235.5				
RMS	218.5	172.0	206.2	203.5	211.3	-	-	130.9	96.6	55.5				
ERR	126.2	99.3	119.0	117.5	149.4	-	-	92.5	68.3	39.3				
REC	-1113.9	-	-	-	-	-	-	-	-	-				

(continues on the next page)

Table 5.15 *(continued)*

Sets	H^+	Li^+	Na^+	K^+	Rb^+	Cs^+	F^-	Cl^-	Br^-	I^-	Source	Ref.	Notes	Year
					$\Delta_s\hat{S}^\ominus[I_g^z]$ [J·mol^{-1}·K^{-1}]									
6ŝ	-90.1	-100.1	-69.1	-33.1	-	-	-175.1	-118.1	-102.1	-96.1	-	415	a,q	1970
NUM	1	1	1	1	0	0	1	1	1	1				
AVE	-90.1	-100.1	-69.1	-33.1	-	-	-175.1	-118.1	-102.1	-96.1				
RMS	-	-	-	-	-	-	-	-	-	-				
ERR	-	-	-	-	-	-	-	-	-	-				
REC	-90.1	-	-	-	-	-	-	-	-	-				

(a) including standard-state conversion from 1 atm to 1 bar gas-phase reference pressure (Eqs. 4.192 and 4.193); (b) including standard-state conversion from 1 mol·dm^{-3} (molar) to mol·kg^{-1} (molal) reference concentration in the aqueous phase; (c) this determination also relied on data from Refs.[404,405], and was corrected for a calculation error (in Eq. 13 of the indicated reference, -103.92 kcal·mol^{-1} should be replaced by -101.70 kcal·mol^{-1}); (d) gas-phase reference concentration or pressure not stated explicitly in the quoted reference and assumed by the authors of this book (see possible additional footnotes on standard-state corrections); (e) solution reference concentration not stated explicitly in the quoted reference and assumed by the authors of this book (see possible additional footnotes on standard-state corrections); (f) including standard-state conversion from 1 mol·dm^{-3} (molar) to 1 bar for the gas-phase reference concentration or pressure (Eqs. 4.194 and 4.195); (g) the reported values are corrected by Ref.[408] based on original data from Refs.[404,405]; (h) this determination also relied on data from Refs.[404,405,408] (standard-state convention not indicated in the quoted reference and assumed identical to that of Ref.[408]); (i) data from Ref.[411] was corrected here for the use of inaccurate data according to Ref.[417] (see also Ref.[262]); (j) the data from the quoted reference, which is based on data from Ref.[411], was corrected here for the use of inaccurate data in this reference according to Ref.[417]; (k) standard states not stated explicitly in the quoted reference and assumed to be the same as in the reference providing the proton data employed[411]; (l) calculated using the Gibbs equation ($\Delta_s\hat{G}^\ominus[I_g^z] = \Delta_s\hat{H}^\ominus[I_g^z] - T^- \Delta_s\hat{S}^\ominus[I_g^z]$); (m) derived from the $\hat{\mathcal{V}}^\ominus$ estimate of the indicated source set in Table 5.14 using Eq. 5.32; (n) based on an assumption of zero enthalpy change for the reaction $1/2 H_{2,g} \rightarrow H^+_{aq}$ (this assumption is based on the electrocapillary and dropping electrode studies of Ostwald[71], and is highly questionable); (o) this determination also relied on data from Refs.[71,73,74,2173] in combination with an approximation for the entropy component; (p) obtained by the indicated reference *via* application of the Gibbs equation, using real solvation free energies of unspecified source; (q) based on data from Ref.[2174].

cally very similar to that of Figure 5.4 concerning $\hat{\mathcal{V}}_H^\ominus$, considering the simple connection between the two quantities according to Eq. 5.32. Corresponding illustrations concerning $\hat{\mathcal{H}}_H^\ominus$ and $\hat{\mathcal{S}}_H^\ominus$ are omitted in view of the sparseness of the data.

The average of the 20 estimates for $\hat{\mathcal{G}}_H^\ominus$ is -1087.9 kJ·mol^{-1}, with a standard deviation of 14.5 kJ·mol^{-1} and an estimated error on the mean of 3.2 kJ·mol^{-1}, assuming that all the measurements are independent. The 3 estimates for $\hat{\mathcal{H}}_H^\ominus$ differ considerably. However, the most recent value (Set 6ĥ, 1970) of -1103 kJ·mol^{-1} is probably more reliable than the two earlier and relatively old ones (Sets 21ĥ and 22ĥ, 1919 and 1926, respectively; based on a highly questionable assumption for the former one). For the corresponding entropy $\hat{\mathcal{S}}_H^\ominus$, only one data set (Set 6ŝ, 1970) is available to date, suggesting a value of -90.1 J·mol^{-1}·K^{-1} with an error estimated in the original source[415] to be about 2 J·mol^{-1}·K^{-1}.

Considering the discussion of potential shortcomings in the various determination methods, as well as of the relationship with the real absolute potential of the hydrogen electrode, the surface potential of water and the intrinsic proton solvation properties (Sections 5.14, 5.17 and 5.18), the authors of the present book recommend values $\hat{\mathcal{G}}_H^\ominus = -1087.0 \pm 3.2$ kJ·mol^{-1}, $\hat{\mathcal{H}}_H^\ominus = -1113.9 \pm 3.8$ kJ·mol^{-1} and $\hat{\mathcal{S}}_H^\ominus = -90.1 \pm 2.0$ J·mol^{-1}·K^{-1} for the real solvation parameters of the proton in water, along with $\hat{s}_H^\ominus = 18.8 \pm 2.0$ J·mol^{-1}·K^{-1} for the real partial molar entropy of the aqueous proton. These values satisfy the Gibbs equation. The error on $\hat{\mathcal{H}}_H^\ominus$ is constructed considering worst-case errors on $\hat{\mathcal{G}}_H^\ominus$ and $\hat{\mathcal{S}}_H^\ominus$. The $\hat{\mathcal{G}}_H^\ominus$ value and error are also compatible with the $\hat{\mathcal{V}}_H^\ominus$ value of 4.42 ± 0.03 V recommended in Section 5.14. Similarly, the $\hat{\mathcal{S}}_H^\ominus$ value and error are compatible with the corresponding $\partial_T \hat{\mathcal{V}}_H^\ominus$ value of 0.25 ± 0.02 mV·K^{-1} recommended in this section. Finally, the difference $\hat{\mathcal{S}}_H^\ominus - \hat{s}_H^\ominus$ is also equal to the entropy of the gas-phase proton given in Table 5.9. The above values and error estimates for $\hat{\mathcal{G}}_H^\ominus$, $\hat{\mathcal{H}}_H^\ominus$ and $\hat{\mathcal{S}}_H^\ominus$ are those that will be retained within the final compilation of recommended data in Section 5.19. The real solvation parameters of the other ions will not be discussed here, since they basically result from a combination of real proton parameters and conventional values, discussed in Section 5.16.

5.16 Conventional single-ion solvation parameters

The *standard conventional single-ion solvation parameters* $\Delta_s Y^{\bullet}[I_g^z]$, namely the free energy $\Delta_s G^{\bullet}[I_g^z]$, enthalpy $\Delta_s H^{\bullet}[I_g^z]$ and entropy $\Delta_s S^{\bullet}[I_g^z]$, are accessible experimentally, based on the partitioning of salt solvation parameters (Section 5.12) into single-ion contributions with the anchoring point $\Delta_s Y^{\bullet}[H_g^+] = 0$ (Section 4.5.1). This implies the use of thermodynamic information considering a data set involving compounds of anions and hydrogen. Conventional values can also be directly deduced from measurements of redox potentials against the reference hydrogen electrode (Section 5.5) or from corresponding real values (Section 5.15), the latter provided that the real proton solvation parameter $\hat{\mathcal{G}}_H^\ominus$ has been determined. Similarly, *standard conventional single-ion partial molar variables in solution* $y^{\bullet}[I_d^z]$, namely the entropy $s^{\bullet}[I_d^z]$, heat capacity $c_P^{\bullet}[I_d^z]$, volume $v^{\bullet}[I_d^z]$, volume-compressibility $k_T^{\bullet}[I_d^z]$ and volume-expansivity $a_P^{\bullet}[I_d^z]$, are accessible experimentally, based on the partitioning of dissolved salt partial molar variables (Sections 5.7 and 5.12) into single-ion contributions with the anchoring $y^{\bullet}[H_d^+] = 0$ (Section 4.5.1). Alternatively, pressure or/and temperature derivatives of redox potentials or real single-ion solvation free energies can in principle also be used. In the specific case of alkali-halide hydration, only first temperature derivatives are available in both cases (Sections 5.5 and 5.15), giving access to conventional partial molar entropies, along with second temperature derivatives for the redox potentials and for the halide anions only, giving access to conventional partial molar heat capacities.

Conventional solvation parameters and partial molar variables in solution have been determined for all alkali (M_g^+) and halide (X_g^-) gas-phase ions with water as a solvent and for all alkali (M^+_{aq}) and halide (X^-_{aq}) aqueous ions, respectively. It should be stressed, however, that *numerous independent determinations* have led to as many distinct sets, a number of which present high similarities because they are derived from the same raw experimental data. Similarly, standard conventional single-ion solvation parameters for alkali and halide gas-phase ions in water are reported in Table 5.16. Standard conventional single-ion partial molar variables for the alkali and halide ions dissolved in water are reported in Table 5.17. For the ease of further reference, these sets are numbered from 1 to 221 based on the literature source, along with the letter g, h, $\bar{s}$, s, c, k or a to indicate the type of property, namely $\Delta_s G^\bullet$, $\Delta_s H^\bullet$, $\Delta_s S^\bullet$, $s^\bullet$, $c_P^\bullet$, $v^\bullet$, $k_T^\bullet$ and $a_P^\bullet$, respectively, and a subscript in the range $0 - 12$ characterizing the determination procedure (see further below). Note that the source numbering is joint with that of Tables 5.18, 5.19 and 5.20 presented subsequently, so that Tables 5.16 and 5.17 only encompass data from 144 sources (out of 221), for a total of 317 sets for either of the eight above parameters. In analogy with the choices made in Section 5.12, the data concerning $\Delta_s G^\bullet$, $\Delta_s H^\bullet$, $\Delta_s S^\bullet$ and $c_P^\bullet$ is assumed to pertain to the standard variant of the solute standard-state definition (Section 4.2.5), while the data concerning $v^\bullet$, $k_T^\bullet$ and $a_P^\bullet$ is assumed to pertain to the density-corrected variant, and therefore noted $v^{\bullet *}$, $k_T^{\bullet *}$ and $a_P^{\bullet *}$. Note that in terms of conventional values, there is no distinction required between the two variants for monovalent cations, the difference in terms of intrinsic values canceling out between the ion and the proton, while the difference is doubled for the anions.

The data reported in these two tables represents an extensive, probably close to exhaustive, compilation of *all conventional parameters* reported in the literature *in a direct or indirect form* over the past *nearly hundred years*. Note, however, that this data may also be partly redundant, considering that different literature sources have relied on often partially overlapping raw calorimetric or/and electrochemical data to construct the reported conventional sets.

In these two tables, the sets that are reported in the original literature source in an explicitly conventional form were simply copied. The corresponding set names are marked with a "0" subscript. The sets that are reported in the original literature source in the form of real or (tentative) intrinsic values, along with a corresponding proton estimate, were converted to conventional data by mere subtraction (cations) or addition (anions) of the proton value. The resulting sets are marked with a "1" (derived from intrinsic) or a "2" (derived from real) subscript. The corresponding proton values are reported in Table 5.15 (real) or Tables 5.19 and 5.20 (intrinsic).

The conventional sets of the three above types were then used to define consensus conventional values in the form of averages ("ave" in the tables) over the number of available estimates for each ion ("num"), along with corresponding root-mean-square deviations ("rms"). Sets 1G, 1H, 1$\bar{S}$, 1C, 1K and 1A (Tables 5.11 and 5.12) described in Section 5.12 were then converted to corresponding conventional Sets $1g_3$, $1h_3$, $1\bar{s}_3$, $1c_3$, $1v_3$, $1k_3$ and $1a_3$ by optimal partitioning, followed by alignment against the above consensus conventional data. For the entropy, the corresponding conventional Set $1s_3$ derived from the pairwise Set 1S was obtained directly from $1\bar{s}_3$ using gas-phase single-ion data from Table 5.9, rather than resorting to alignment against a corresponding consensus. The alignments were performed by minimizing the sum over all ions of the square difference to the consensus value divided by the mean-square deviation of the consensus data.

The sets that are reported in the original literature source in a pairwise form (Tables 5.11 and 5.12), as well as those reported in the form of real or (tentative) intrinsic values, but without a corresponding proton estimate, were converted to conventional data by optimal partitioning (pairwise sets only) and alignment against the above recommended Sets $1g_3$, $1h_3$, $1\bar{s}_3$, $1s_3$, $1c_3$, $1v_3$, $1k_3$ and $1a_3$. Here, the alignments were performed by minimizing the sum over all ions of the square difference to the recommended value. The resulting sets are marked with a "3" (derived from pairwise), a "4" (derived from intrinsic) or a "5" (derived from real) subscript. The corresponding inferred proton values (real or intrinsic sets only) are then also reported in Table 5.15 (real) or in Tables 5.19 and 5.20 (intrinsic), between parentheses.

Finally, since conventional solvation entropies $\Delta_s S^{\bullet}$ and partial molar entropies in solution $s^{\bullet}$ are directly connected through the gas-phase entropies of the ion and of the proton (Table 5.9), the above procedure was applied first for $\Delta_s S^{\bullet}$ considering the original literature sources explicitly reporting information on the solvation entropies. The resulting 41 sets were then converted to corresponding conventional partial molar values and added to the data sets for $s^{\bullet}$, along with those derived from original literature sources explicitly reporting information on the conventional partial molar entropies. These converted sets are labeled with a subscript that is increased by 6 relative to that of the source $\Delta_s S^{\bullet}$ set. As a result, the data reported for $s^{\bullet}$ is complete, while that reported for $\Delta_s S^{\bullet}$ is only partial.

In both tables, final average conventional values ("AVE" in the tables) over the number of available sets for each ion ("NUM") are indicated, along with corresponding root-mean-square deviations ("RMS"), error estimates ("ERR" in the tables) and recommended values ("REC" in the tables), corresponding to the above sets $1g_3$, $1h_3$, $1\bar{s}_3$, $1s_3$, $1c_3$, $1v_3$, $1k_3$ and $1a_3$. The error estimates were evaluated by dividing the root-mean-square deviation by the square root of f^{-1} times the number of available sets for the corresponding ion, assuming a redundancy factor $f = 5$, *i.e.* an estimated ratio of one raw data source for five proposed conventional data sets. Since some of the data sets correspond to rather old determinations with presumably large errors, corresponding corrected entries ("NUMC", "AVC", "RMC" and "ERC") report the statistics after removal of outliers. These were produced by iteratively removing, for each ion, the entry with the largest deviation from the current average if this deviation exceeded three times the current root-mean-square deviation. These outliers are indicated between square brackets in the table.

Another estimate of the error is provided by the violations of the Gibbs equation ($\Delta_s G^{\minuso}[I_g^z] - \Delta_s H^{\minuso}[I_g^z] + T^- \Delta_s S^{\minuso}[I_g^z] = 0$, for sources reporting simultaneously the three quantities) and of the predicted gas-phase entropies of the ions ($s^{\ominus}[I_g^z] - z s^{\ominus}[H_g^+] - s^{\minuso}[I_{aq}^z] + \Delta_s S^{\minuso}[I_g^z] = 0$, for sources reporting simultaneously the two latter quantities and based on the recommended values of Table 5.9 for the two former quantities). These violations are also reported in the tables in the form of a maximum value over the set (ϵ_g^{max} and ϵ_e^{max} entries, the value between parentheses being evaluated after removal of the outliers).

The $\Delta_s G^{\minuso}$, $\Delta_s H^{\minuso}$ and $\Delta_s S^{\minuso}$ values corresponding to distinct experimental data sets are essentially consistent, although significant variations (excluding outliers) of up to 171 kJ·mol^{-1} (F^-), 107 kJ·mol^{-1} (F^-) and 25 J·mol^{-1}·K^{-1} (I^-), respectively, can be observed. The corresponding estimated errors are in ranges of about $1 - 10$ kJ·mol^{-1}, $0 - 12$ kJ·mol^{-1} and $0 - 2$ J·mol^{-1}·K^{-1}, respectively, for the different ions. Finally, the differences between the recommended set ("REC") and the average of all sets ("AVC") is at most 2.0 times as large as the corresponding root-mean-square difference ("RMC"), considering the three properties and all ions.

The $s^{\minuso}$, $c_P^{\minuso}$, $v^{\minuso *}$, $k_T^{\minuso *}$ and $a_P^{\minuso *}$ values corresponding to distinct data sets are also essentially consistent, although significant variations (excluding outliers) of up to 14 J·mol^{-1}·K^{-1} (Li^+), 35 J·mol^{-1}·K^{-1} (Rb^+), 3·10^{-6} m^3·mol^{-1} (F^-), 37·10^{-10} m^3·mol^{-1}·bar^{-1} (Cs^+), and 6·10^{-8} m^3·mol^{-1}·K^{-1} (K^+), respectively, can be observed. The corresponding estimated errors are in ranges of about $0-1$ J·mol^{-1}·K^{-1}, $2-6$ J·mol^{-1}·K^{-1}, $0-1\cdot10^{-6}$ m^3·mol^{-1}, $0 - 17 \cdot 10^{-10}$ m^3·mol^{-1}·bar^{-1}, and $1 - 2 \cdot 10^{-8}$ m^3·mol^{-1}·K^{-1}, respectively, for the different ions. Finally, the differences between the recommended set ("REC") and the average of all sets ("AVC") is at most 3.2 times (2.0 times, if the Cs^+ ion is excluded from the $s^{\minuso}$ statistics) as large as the corresponding root-mean-square difference ("RMC"), considering the five properties and all ions.

The single-ion conventional hydration parameters and partial molar variables in aqueous solution recommended by the authors of this book are those listed in the "REC" entries of the two tables, associated error estimates being provided in the "ERC" entries. These values are those retained within the final compilation of recommended data in Section 5.19. Since conventional data relies on the use of an arbitrary relative scale, the magnitudes of these quantities are not immediately relevant to characterize the physics of the solvation process. Since the corresponding trends are unaffected by addition of an estimate for the corresponding intrinsic proton parameters (Section 5.18), single-ion solvation properties will be discussed later in Section 5.20, in the context of suggested intrinsic single-ion solvation parameters.

Table 5.16 *Standard conventional single-ion solvation parameters for the alkali and halide gas-phase ions in water.* The reported data includes the conventional solvation free energy $\Delta_s G^{\bullet}$, enthalpy $\Delta_s H^{\bullet}$ and entropy $\Delta_s S^{\bullet}$. The notation I_g^z refers to an alkali (M_g^+) or halide (X_g^-) gas-phase ion. The different sets correspond to values derived from different literature sources, listed in chronological order of publication separately for each block of the table. The lines NUM, AVE, RMS, ERR and REC report for each ion the number of entries, average value, root-mean-square deviation, estimated error ($f^{-1/2}$NUM$^{-1/2}$RMS, with a redundancy factor $f = 5$) and recommended values (Sets 1) for all preceding entries corresponding to a given parameter. The lines num, ave and rms report the corresponding quantities for the preceding block. The lines NUC, AVC, RMC and ERC are similar to NUM, AVE, RMS and ERR, but exclude outliers (between square brackets in the table). The entries ϵ_g^{max} and ϵ_e^{max} report maximum violations of the Gibbs equation ($\Delta_s G^{\bullet}[I_g^z] - \Delta_s H^{\bullet}[I_g^z] + T^- \Delta_s S^{\bullet}[I_g^z]$) and of the predicted gas-phase entropies ($s^{\ominus}[I_g^z] - zs^{\ominus}[H_g^+] - s^{\bullet}[I_{aq}^z] + \Delta_s S^{\bullet}[I_g^z]$, using data from Tables 5.9 and 5.17), whenever relevant. The column "Source" provides the source set of a derived set (from Table 5.11 or Table 5.15). The data concerning $\Delta_s S^{\bullet}$ only includes source references explicitly reporting conventional solvation entropies, excluding values deduced from partial molar entropies of dissolved ions (see footnote (z) for exceptions). A complete list including the latter is given in Table 5.17. The standard states are according to the $bbme_T$ convention (Section 4.2.4; reference pressure $P^{\circ} = 1$ bar, reference molality $b^{\circ} = 1$ mol·kg^{-1} and reference temperature $T^- = 298.15$ K). The experimental values reported in the original references are assumed here to pertain to the standard variant (Section 4.2.5), which represents the most likely situation, but may not necessarily be the case in reality. The conventional solvation free energy $\Delta_s G^{\bullet}$ is identical in the two variants.

Set	Li^+	Na^+	K^+	Rb^+	Cs^+	F^-	Cl^-	Br^-	I^-	ϵ_g^{max}	ϵ_e^{max}	Source	Ref.	Notes	Year
					$\Delta_s G^{\bullet}$ [kJ·mol^{-1}]										
$2g_2$	574.0	674.5	747.7	763.2	788.7	-1522.9	-1399.5	-1372.7	-1339.2	-	-	1ġ	407	a,b,c,d,e	1948
$3g_0$	579.0	678.9	752.8	774.0	806.3	-1453.9	-1397.4	-1363.9	-1331.7	4.38(4.38)	-	-	1871/ 2175	d,f	1954
$4g_2$	579.1	679.1	752.7	774.0	806.3	-1504.5	-1385.7	-1361.4	-1329.2	-	-	4ġ	411,417	a,d	1956
$5g_1$	577.4	677.8	753.1	774.0	807.5	-1440.1	-1385.7	-1348.1	[-1314.6]	0.00(0.00)	0.00(0.00)	-	754	c,g,h	1957
$6g_0$	578.9	678.4	752.8	774.0	806.1	-1453.4	-1397.3	-1368.5	-1331.2	0.03(0.03)	-	-	259/ 1744,2175 - 2178	d,f,i,j	1962
$6g_1$	578.9	678.4	752.8	774.0	806.1	-1448.5	-1392.4	-1363.6	-1326.3	0.10(0.10)	-	-	259	g,h	1962
$7g_1$	-	669.0	[743.1]	764.4	796.6	-1443.8	-1387.8	-1358.9	-1321.7	-	-	-	660/ 259	d,g,j,k	1964
$8g_0$	579.1	679.1	752.7	774.0	806.3	-1524.2	-1407.0	-1393.2	-1346.8	4.37(4.37)	-	-	262/ 2175	d,f	1965
$9g_1$	579.1	679.1	752.7	774.0	807.5	-1524.2	-1407.0	-1393.2	-1346.8	4.37(4.37)	-	-	264	b,c,d,g	1969
$9g_2$	579.1	679.1	752.7	774.0	807.5	-1524.2	-1407.0	-1393.2	-1346.8	-	-	5ġ	264	a,b,c,d	1969
$10g_0$	574.0	680.3	753.5	771.1	787.8	-1521.7	-1402.8	-1377.3	-1341.8	0.00(0.00)	-	-	1873	d,j,l	1970
$10g_1$	574.0	680.3	753.5	771.1	787.8	-1521.7	-1402.8	-1377.3	-1341.8	0.00(0.00)	-	-	1873	d,g	1970
$11g_2$	574.0	673.7	748.0	-	-	-1522.9	-1398.9	-1372.7	-1338.5	0.00(0.00)	-	6ġ	415	a,m	1970
$12g_1$	573.2	678.6	751.9	769.4	792.9	-1524.2	-1407.0	-1393.2	-1346.8	4.37(4.37)	-	-	269/ 262,1744	d,e,g,j	1973
$13g_0$	581.6	679.9	752.3	778.2	792.5	-1518.0	-1395.8	-1367.3	-1345.2	0.37(0.37)	-	-	2153	f,j	1976
$13g_1$	581.6	680.3	752.7	778.6	792.5	-1518.0	-1395.8	-1366.9	-1345.6	0.70(0.70)	-	-	2153	g	1976
$14g_1$	565.4	680.2	753.5	-	-	-1522.5	-1399.9	-1377.8	-	-	-	-	444	d,g	1977
$15g_1$	578.5	678.4	752.0	773.3	805.5	-1522.7	-1405.6	-1391.8	-1345.4	3.24(3.24)	-	-	274/ 262	d,g,j	1978
$16g_0$	579.1	679.1	752.7	774.0	806.3	-1524.2	-1407.0	-1393.2	-1346.8	0.00(0.00)	-	-	417/ 262	d,j	1981
$16g_1$	579.1	679.1	752.7	774.0	806.3	-1524.2	-1407.0	-1393.2	-1346.8	0.00(0.00)	-	-	417	d,g	1981
$17g_0$	574.5	680.4	753.4	775.1	798.2	-1540.8	-1408.5	-1382.3	-1347.4	0.00(0.00)	0.00(0.00)	-	1874/ 2179,2180	m	1982
$17g_1$	574.5	680.3	753.3	775.1	798.2	-1540.6	-1408.4	-1382.2	-1347.4	0.00(0.00)	0.39(0.39)	-	1874	d,g	1982
$18g_0$	575.2	680.7	752.5	775.1	798.4	-1527.6	-1403.4	-1376.7	-1339.3	0.00(0.00)	316.09(17.77)	-	275/ 294	d	1985
$18g_1$	575.0	681.0	752.0	775.0	798.0	-1527.9	-1402.9	-1376.9	-1338.9	0.00(0.00)	19.11(18.11)	-	275	d,f,g	1985
$19g_1$	588.0	668.0	759.0	772.0	803.0	-1523.0	-1411.0	-1386.0	-1355.0	-	-	-	822	b,c,g	1992
$20g_1$	575.1	680.7	752.5	775.2	-	-1533.3	-1408.7	-1381.9	-1344.5	0.00(0.00)	-	-	1875,1876	b,g,n	1998
$21g_0$	575.1	680.8	752.6	775.2	798.4	-1533.6	-1408.5	-1382.2	-1347.1	0.08(0.08)	-	-	1750/ 294	f	1999
$22g_0$	575.6	680.6	753.6	775.2	798.7	-1536.7	-1408.6	-1382.2	-1347.3	0.00(0.00)	-	-	819/ 2181	m	2000
$22g_1$	576.0	681.0	754.0	776.0	799.0	-1537.1	-1409.1	-1382.1	-1347.1	0.76(0.76)	-	-	819	e,g,o	2000
$23g_1$	[552.0]	[635.7]	756.0	786.5	[837.1]	-1611.0	[-1632.7]	[-1610.7]	[-1582.1]	-	-	-	668	b,g,p	2003
$24g_1$	[631.8]	663.3	755.3	780.6	818.2	-1486.6	[-1512.4]	[-1491.6]	[-1465.3]	-	-	-	668	b,g,p	2003
$25h_1$	563.1	[641.4]	759.2	[791.3]	[835.0]	-1593.0	[-1615.3]	[-1593.0]	[-1564.2]	-	-	-	668	b,g,p	2003
$26g_1$	[637.9]	667.3	755.1	776.9	817.6	-1498.2	[-1523.9]	[-1503.4]	[-1477.1]	-	-	-	668	b,g,p	2003
$27g_1$	575.3	680.7	752.7	775.3	798.3	-1533.4	-1408.3	-1382.4	-1347.2	-	-	-	1747/ 1750,1875	e,g,j,o,n	2006
$28g_0$	575.0	681.1	752.9	771.6	798.4	-1531.1	-1408.6	-1380.9	-1344.7	0.09(0.09)	-	-		q	2011
num	34	35	35	33	32	35	35	35	34						
ave	577.4	675.1	752.8	774.5	800.4	-1517.1	-1407.8	-1386.4	-1353.5						
rms	15.2	10.1	2.7	4.9	11.2	35.3	57.0	58.4	61.5						
$1g_3$	574.9	680.5	752.2	775.0	798.2	-1533.2	-1408.6	-1381.8	-1344.4	0.08(0.08)	0.08(0.08)	1G		r	2011
$29g_4$	[218.4]	[498.7]	[695.3]	[745.6]	799.9	[-1696.2]	[-1553.9]	[-1537.2]	[-1503.7]	-	-	-	1778	h,s	1930
$30g_4$	[625.5]	692.8	[789.3]	[824.6]	[833.2]	-1475.4	[-1355.3]	[-1337.0]	[-1315.9]	2.04(2.04)	-	-	644	m,s	1938
$31g_4$	583.1	687.3	755.1	780.2	808.2	-1523.3	-1399.0	-1373.1	-1339.6	0.85(0.85)	1.97(1.67)	-	814	e,o,s	1939
$31g_3$	584.3	688.6	756.7	781.6	809.6	-1521.5	-1397.3	-1371.4	-1338.0	-	-	2G	814	e,o,r	1939
$32g_4$	585.2	685.6	754.6	777.6	809.0	-1516.9	-1399.8	-1376.8	-1343.3	1.27(1.27)	-	-	2155	e,o,s	1940
$33g_4$	[603.1]	686.0	750.0	770.1	794.3	-1506.8	-1405.6	-1381.8	-1351.2	-	-	-	748	h,s	1950
$34g_5$	579.6	696.8	[730.3]	776.3	-	-	-	-	-	-	-	2ĝ	404,405,408	e,o,t	1952
$35g_4$	568.2	685.4	760.7	781.6	806.7	-	-1385.0	-1372.4	-1355.7	-	-	-	409	h,s	1954
$35g_5$	570.3	687.4	758.6	779.5	804.6	-1529.3	-1395.4	-1378.6	-1345.2	-	-	3ĝ	409	b,c,e,o,t	1954
$36g_4$	589.2	679.1	751.9	781.6	806.7	-1527.2	-1401.7	-1372.4	-1338.9	1.80(1.80)	-	-	821	d,s,u	1955
$37g_4$	[605.9]	681.2	[731.4]	-	-	-1527.2	-1405.9	-1380.8	-1343.1	-	-	-	821	h,s	1955
$38g_4$	583.0	688.9	757.1	780.1	816.9	-1513.5	-1398.5	-1370.9	-1339.9	-	-	-	2182	b,c,e,o,s	1963

(continues on the next page)

Table 5.16 *(continued)*

Set	Li^+	Na^+	K^+	Rb^+	Cs^+	F^-	Cl^-	Br^-	I^-	ϵ_g^{max}	ϵ_e^{max}	Source	Ref.	Notes	Year
					$\Delta_s G^\bullet$ [kJ·mol^{-1}]										
39g_4	-	684.5	759.4	782.4	813.4	-1456.4	-1404.1	-1400.8	[-1372.8]	-	-	-	660	d,s	1964
10g_3	578.1	684.8	758.0	775.4	792.1	-1526.0	-1407.2	-1381.4	-1345.9	0.07(0.07)	0.07(0.04)	3G	1873	d,r	1970
13g_3	586.0	684.5	756.7	782.6	795.9	-1522.1	-1400.2	-1371.4	-1349.3	0.12(0.12)	0.06(0.06)	4G	2153	r	1976
40g_4	-	683.2	757.7	778.6	810.8	-	-1401.8	-1373.0	-1335.7	-	-	-	662	d,p,s	1978
41g_4	573.2	679.0	752.0	773.8	796.9	-1539.4	-1407.2	-1381.0	-1346.2	1.15(1.15)	-	-	2183	c,d,s	1984
42g_4	578.7	678.7	762.7	782.7	815.7	-1479.8	-1403.8	-1385.8	-1360.8	-	-	-	2184	h,s	1998
21g_4	[614.2]	694.2	[764.2]	781.2	809.2	-1477.3	-1391.3	-1372.3	-1345.3	0.07(0.07)	-	-	1750	h,s	1999
43g_4	571.0	-	-	-	-	-	-1412.4	-	-	-	-	-	1571	h,s	2009
NUM	53	55	55	52	50	52	55	54	53						
AVE	574.7	675.9	752.8	776.5	804.6	-1519.0	-1416.3	-1393.0	-1359.3						
RMS	51.8	26.0	10.7	9.3	10.8	40.5	50.8	52.5	54.2						
ERR	15.9	7.8	3.2	2.9	3.4	12.6	15.3	16.0	16.6						
REC	574.9	680.5	752.2	775.0	798.2	-1533.2	-1408.6	-1381.8	-1344.4						
NUC	45	52	49	49	47	51	49	48	45						
AVC	577.2	680.7	754.1	775.8	802.6	-1515.5	-1402.0	-1378.1	-1343.3						
RMC	5.2	6.0	3.0	4.4	7.7	32.3	6.9	10.2	7.2						
ERC	1.7	1.9	0.9	1.4	2.5	10.1	2.2	3.3	2.4						
					$\Delta_s H^\bullet$ [kJ·mol^{-1}]										
44h_2	[636.0]	[665.3]	[753.1]	790.8	[786.6]	-	[-1610.8]	[-1648.5]	[-1694.5]	-	-	$21\hat{h}$	82,83	a	1919
45h_1	-	[644.3]	[711.3]	[736.4]	[753.1]	-1594.1	[-1422.6]	[-1384.9]	[-1338.9]	-	-	-	1869	g,l	1921
46h_1	[481.2]	[548.1]	[615.0]	[640.2]	-	-	-	-	-	-	-	-	2172	g,h	1926
46h_2	[514.6]	[610.9]	[690.4]	[711.3]	-	-	-	-	-	-	-	$22\hat{h}$	2172	a	1926
47h_1	585.8	[677.8]	[761.5]	790.8	820.1	-1560.6	[-1426.7]	-1393.3	[-1351.4]	-	-	-	643	g	1933
48h_0	-	-	-	-	-	-1571.5	-1446.8	-1412.5	-1373.2	-	-	-	1870	j,l,v	1940
49h_0	-	-	-	-	-	-1584.5	-1461.5	-1430.5	-1391.2	-	-	-	1870	j,l,w	1940
3h_0	576.0	685.3	769.8	794.5	827.6	-1532.6	-1459.4	-1425.5	-1382.0	4.38(4.38)	-	-	1871/ 2175	f	1954
5h_1	574.5	684.1	770.1	794.6	829.0	-1518.9	-1447.5	-1405.2	-1365.0	0.00(0.00)	0.00(0.00)	-	754	c,g,m	1957
6h_0	576.0	685.3	769.7	794.4	827.5	-1532.5	-1459.3	-1425.4	-1381.9	0.03(0.03)	-	-	259/ 1744,2175 - 2178	m	1962
6h_1	576.1	685.3	769.8	794.3	827.4	-1527.8	-1454.6	-1420.7	-1377.2	0.10(0.10)	-	-	259	g	1962
50h_0	570.4	684.8	768.8	789.9	814.2	-1596.2	-1453.9	-1426.7	-1385.7	-	-	-	260	j,x	1963
8h_0	576.1	685.3	769.9	795.0	827.6	-1602.9	-1469.0	-1454.8	-1397.0	4.37(4.37)	-	-	262/ 2175	f	1965
51h_0	569.9	686.2	770.3	796.6	814.2	-1605.4	-1469.4	-1438.0	-1397.0	-	-	-	1872	j,x	1968
51h_1	569.9	686.2	770.3	796.6	814.2	-1605.4	-1469.4	-1438.0	-1397.0	-	-	-	1872	g	1968
52h_1	569.9	686.2	769.9	796.6	814.2	-1605.8	-1469.0	-1437.6	-1397.0	-	-	-	1872	g	1968
53h_1	569.9	686.2	770.7	797.1	814.6	-1605.4	-1469.8	-1438.5	-1397.0	-	-	-	1872	g	1968
9h_1	576.1	685.3	769.9	795.0	827.6	-1602.9	-1469.0	-1454.8	-1397.0	4.37(4.37)	-	-	264	b,c,g,j,y	1969
10h_0	572.0	686.6	770.3	799.6	829.3	-1609.2	-1465.7	-1447.2	-1393.7	0.00(0.00)	-	-	1873	j,l	1970
10h_1	572.0	686.6	770.3	800.0	829.3	-1609.2	-1465.7	-1447.2	-1393.7	0.00(0.00)	-	-	1873	g	1970
11h_2	571.0	680.0	[765.0]	-	-	-1602.0	-1461.0	-1430.0	-1394.0	0.00(0.00)	-	$6\hat{h}$	415	a	1970
12h_1	570.3	684.9	768.6	789.9	814.2	-1602.9	-1469.0	-1454.8	-1397.0	4.37(4.37)	-	-	269/ 262,1744	g,j	1973
13h_0	579.1	686.2	769.4	799.1	814.2	-1598.3	-1457.3	-1423.8	-1394.5	0.37(0.37)	-	-	2153	f,j	1976
13h_1	579.1	686.2	769.9	799.1	814.2	-1598.3	-1457.3	-1423.8	-1394.5	0.70(0.70)	-	-	2153	g	1976
15h_1	575.6	684.7	769.1	794.2	826.8	-1601.4	-1467.6	-1453.4	-1395.7	3.24(3.24)	-	-	274/ 262	g,j	1978
16h_0	576.1	685.3	769.9	795.0	827.6	-1602.9	-1469.0	-1454.8	-1397.0	0.00(0.00)	-	-	417/ 262	j	1981
16h_1	576.1	685.3	769.9	795.0	827.6	-1602.9	-1469.0	-1454.8	-1397.0	0.00(0.00)	-	-	417	g	1981
17h_0	570.7	686.2	769.9	794.5	819.7	-1620.6	-1469.8	-1438.8	-1398.5	0.00(0.00)	0.00(0.00)	-	1874/ 2179,2180	f,j	1982
17h_1	570.7	686.2	769.9	794.5	819.7	-1620.5	-1469.8	-1438.8	-1398.5	0.00(0.00)	0.39(0.39)	-	1874	g	1982
41h_1	570.7	686.2	769.9	794.5	819.7	-1620.6	-1469.8	-1438.8	-1398.5	-	-	-	2183	c,g	1984
18h_0	571.9	687.1	769.5	794.9	820.0	-1513.4	-1470.3	-1438.7	-1394.4	0.00(0.00)	316.09(17.77)	-	275/ 294		1985
18h_1	572.0	687.0	770.0	795.0	820.0	-1613.0	-1470.0	-1439.0	-1394.0	0.00(0.00)	19.11(18.11)	-	275	f,g	1985
54h_1	572.0	687.0	769.0	795.0	820.0	-1613.0	-1470.0	-1439.0	-1394.0	-	-		2105 - 2187	g	1987
55h_1	576.4	686.0	771.2	794.6		1600.4	-1408.8	-1435.2	-1398.4	-	-	-	164	g	1994
20h_1	571.2	686.6	769.9	795.1	-	-1614.0	-1470.0	-1439.0	-1397.8	0.00(0.00)	-	-	1875,1876	b,g,n	1998
21h_0	571.9	686.7	769.6	794.9	820.0	-1613.6	-1469.8	-1438.8	-1396.9	0.08(0.08)	-	-	1750/ 294	j	1999
22h_0	572.1	686.4	770.1	795.0	819.9	-1616.7	-1469.9	-1438.8	-1398.5	0.00(0.00)	-	-	819/ 2181	j	2000
22h_1	572.0	687.0	770.0	795.0	820.0	-1617.0	-1470.0	-1439.0	-1399.0	0.76(0.76)	-	-	819	g	2000
28h_0	572.4	687.5	770.2	794.3	820.7	-1609.9	1470.3	-1438.0	-1395.1	0.09(0.09)	-	-		q	2011
num	36	37	37	36	32	36	37	37	37						
ave	574.4	681.4	769.7	795.4	822.8	-1593.5	-1470.0	-1445.0	-1404.8						
rms	21.1	25.8	29.1	29.9	14.2	30.0	26.5	38.1	51.1						
1h_3	571.8	686.6	769.5	794.8	820.1	-1613.2	-1470.1	-1438.6	-1394.4	0.08(0.08)	0.08(0.08)	1H		r	2011
44h_3	[657.1]	688.5	[775.6]	[813.8]	809.6	-	[-1441.4]	-1403.0	[-1357.0]	-	-	5H	83	r	1919
56h_4	580.1	693.0	[776.7]	793.5	826.9	-	-1471.3	-1425.2	-1379.2	-	-	-	1778	s	1930
57h_3	593.2	[699.6]	[783.4]	[810.9]	841.6	-1586.2	-1450.2	-1416.2	-1377.8	-	-	6H	2154	r	1932
58h_4	[619.8]	682.6	[783.0]	803.9	837.4	-1565.8	-1452.8	-1436.1	-1377.5	-	-	-	643	h,s	1933
30h_4	[619.5]	[694.9]	[799.5]	[837.1]	[870.6]	-1556.8	[-1422.9]	-1393.6	-1364.3	2.04(2.04)	-	-	644	s	1938
31h_4	582.0	693.3	771.9	799.5	829.7	-1602.0	-1460.2	-1429.7	-1390.7	0.85(0.85)	1.97(1.67)	-	814	s	1939
32h_4	584.0	690.7	772.3	797.4	830.8	-1596.5	-1460.5	-1433.3	-1393.6	1.27(1.27)	-	-	2155	s	1940
32h_3	584.0	690.7	772.3	797.4	830.4	-1596.9	-1461.3	-1432.4	-1393.5	-	-	7H	2155	r	1940
59h_4	588.4	[697.2]	[780.9]	806.0	839.4	-1546.5	-1471.2	-1437.7	-1391.7	-	-	-	2188	s	1953
36h_4	586.3	685.9	769.6	801.0	830.3	-1604.3	-1462.1	-1428.6	-1390.9	1.80(1.80)	-	-	821	s	1955
60h_4	[899.7]	[909.9]	[925.0]	[931.0]	[936.4]	[-1256.0]	[-1239.1]	[-1234.3]	[-1227.7]	-	-	-	717	h,s	1957
51h_3	571.5	687.6	771.5	797.9	815.5	-1606.5	-1470.9	-1439.4	-1398.3	-	-	8H	1872	r	1968
10h_3	571.0	685.7	769.1	793.2	828.1	-1610.5	-1468.3	-1437.6	-1395.2	0.07(0.07)	0.07(0.04)	3H	1873	r	1970
61h_3	582.5	690.6	[773.8]	798.7	834.3	-1598.6	-1463.4	-1430.3	-1386.9	-	-	9H	1882	r	1971
13h_3	583.2	690.4	[773.7]	803.3	817.4	-1602.2	-1461.5	-1428.1	-1399.1	0.12(0.12)	0.06(0.06)	4H	2153	r	1976
17h_3	569.6	685.1	768.8	793.4	818.6	-1619.5	-1468.7	-1437.7	-1397.4	-	-	10H	1874	r	1982
21h_4	588.2	684.9	767.4	787.1	819.3	-1572.2	-1471.6	-1449.6	-1418.8	0.07(0.07)	-	-	1750	m,s	1999
NUM	54	55	55	54	50	52	55	55	55						
AVE	583.1	686.0	768.5	794.1	824.3	-1586.3	-1460.8	-1433.1	-1392.3						
RMS	49.2	37.6	32.9	32.2	21.8	53.8	37.9	42.3	48.4						
ERR	15.0	11.3	9.9	9.8	6.9	16.7	11.4	12.8	14.6						
REC	571.8	686.6	769.5	794.8	820.1	-1613.2	-1470.1	-1438.6	-1394.4						
NUC	47	46	40	47	46	51	49	52	50						
AVC	575.8	686.6	769.9	795.8	823.3	-1592.8	-1465.0	-1433.7	-1392.2						
RMC	6.0	2.3	0.9	3.5	7.4	27.7	6.7	14.0	9.4						
ERC	2.0	0.8	0.3	1.1	2.4	8.7	2.1	4.3	3.0						

(continues on the next page)

Table 5.16 *(continued)*

Set	Li^+	Na^+	K^+	Rb^+	Cs^+	F^-	Cl^-	Br^-	I^-	ϵ_g^{max}	ϵ_e^{max}	Source	Ref.	Notes	Year
					$\Delta_s S^\bullet$ [kJ·mol^{-1}]										
$62\bar{s}_1$	[-4.2]	19.7	55.6	64.9	72.4	-265.8	-205.9	-189.6	-172.0	-	-	-	2189	d,g,p	1945
$3\bar{s}_0$	-10.0	21.3	56.9	69.0	72.0	-264.2	-207.3	-191.8	-168.9	4.38(4.38)	-	-	1871/ 2175	d,f	1954
$5\bar{s}_1$	-9.8	21.2	56.9	68.9	72.2	-264.2	-207.1	-191.7	-169.0	0.00(0.00)	0.00(0.00)	-	754	c,g,z	1957
$6\bar{s}_0$	-9.9	23.1	56.8	68.6	71.5	-265.5	-207.7	-191.0	-170.1	0.03(0.03)	-	-	259/ 1744,2175 - 2178	d,i	1962
$6\bar{s}_1$	-9.8	23.1	56.9	68.7	71.6	-266.0	-208.2	-191.5	-170.6	0.10(0.10)	-	-	259	f,g,α	1962
$8\bar{s}_0$	-10.0	21.3	56.9	69.0	72.0	-264.2	-207.3	-191.8	-168.8	4.37(4.37)	-	-	262/ 2175	d,f	1965
$9\bar{s}_1$	-10.0	21.3	56.9	69.0	72.0	-264.2	-207.3	-191.8	-168.8	4.37(4.37)	-	-	264	b,c,d,g,j,y	1969
$10\bar{s}_0$	-7.0	21.1	56.1	[95.4]	[138.9]	[-293.5]	-210.7	[-234.6]	-174.2	0.00(0.00)	-	-	1873	m	1970
$10\bar{s}_1$	-7.0	21.0	56.1	[96.8]	[138.9]	[-293.5]	-210.7	[-234.6]	-174.2	0.00(0.00)	-	-	1873	g,m	1970
$11\bar{s}_2$	-10.0	21.0	57.0	-	-	-265.2	-208.2	-192.2	-186.2	0.00(0.00)	-	$6\tilde{s}$	415	a,d	1970
$12\bar{s}_1$	-10.0	21.3	56.9	69.0	72.0	-264.2	-207.3	-191.8	-168.8	4.37(4.37)	-	-	269/ 262,1744	d,e,g,j	1973
$13\bar{s}_0$	-9.6	20.1	56.9	69.0	72.0	-268.2	-205.4	-190.0	-166.5	0.37(0.37)	-	-	2153	f,j	1976
$13\bar{s}_1$	-10.0	20.1	56.9	68.2	72.0	-268.2	-205.4	-189.5	-166.5	0.70(0.70)	-	-	2153	g	1976
$15\bar{s}_1$	-10.0	21.3	56.8	68.9	71.9	-268.0	-211.1	-195.6	-172.6	3.24(3.24)	-	-	274/ 262	d,g,j	1978
$63\bar{s}_1$	-	23.0	56.9	68.6	71.5	-	-207.6	-190.9	-170.0	-	-	-	663	d,g,p	1978
$16\bar{s}_0$	-9.8	21.1	57.5	70.2	71.6	-264.0	-207.9	-206.5	-168.6	0.00(0.00)	-	-	417/ 262	m	1981
$16\bar{s}_1$	-9.8	21.1	57.5	70.2	71.6	-264.0	-207.9	-206.5	-168.6	0.00(0.00)	-	-	417	g,m	1981
$17\bar{s}_0$	-12.8	19.4	55.4	65.1	72.0	-267.7	-205.6	-189.6	-171.5	0.00(0.00)	0.00(0.00)	-	1874/ 2179,2180	f,z	1982
$17\bar{s}_1$	-12.7	19.8	55.7	65.1	72.1	-267.9	-205.8	-189.7	-171.3	0.00(0.00)	0.39(0.39)	-	1874	g,m	1982
$18\bar{s}_0$	-11.1	21.5	57.0	66.4	72.4	[47.8]	[-224.2]	-207.8	-184.7	0.00(0.00)	316.09(17.77)	-	275/ 294	m	1985
$18\bar{s}_1$	-10.1	20.1	[60.4]	67.1	73.8	[-285.3]	[-224.9]	-208.2	-184.7	0.00(0.00)	19.11(18.11)	-	275	g,m	1985
$64\bar{s}_1$	-9.6	23.0	56.9	68.6	71.5	-	-207.4	-190.7	-169.7	-	-	-	2190	b,g,p	1995
$20\bar{s}_1$	-13.1	19.8	58.4	66.7	-	-270.7	-205.6	-191.5	-178.8	0.00(0.00)	-	-	1875,1876	b,g,m	1998
$21\bar{s}_0$	-10.7	20.0	56.9	66.1	72.2	-268.3	-205.8	-190.1	-166.9	0.08(0.08)	-	-	1750/ 294	f	1999
$22\bar{s}_0$	-11.9	19.4	55.5	66.3	71.2	-268.2	-205.5	-189.7	-171.6	0.00(0.00)	-	-	819/ 2181	e,f,o,β	2000
$22\bar{s}_1$	-12.0	19.3	55.4	66.2	71.1	-268.1	-205.4	-189.6	-171.5	0.76(0.76)	-	-	819	e,g,o	2000
$28\bar{s}_0$	-8.8	21.4	57.9	[76.2]	[74.5]	-264.4	-207.0	-191.7	-169.1	0.09(0.09)	-	-		q	2011
num	26	27	27	26	25	25	27	27	27						
ave	-10.0	21.0	56.9	70.1	75.1	-256.2	-208.5	-196.5	-172.0						
rms	1.8	1.1	1.0	7.8	18.2	62.7	4.8	12.2	5.3						
$1\bar{s}_3$	-10.3	20.2	57.8	66.6	73.3	-268.4	-206.4	-190.5	-167.5	0.08(0.08)	0.08(0.08)	$1\bar{S}$		r	2011
$65\bar{s}_4$	-	[35.5]	[53.9]	[56.4]	-	-	[-200.5]	-188.4	-174.6	-	-	-	733	d,s,u	1929
$66\bar{s}_4$	-	[33.4]	[43.8]	[45.9]	-	-	-203.7	-192.4	-189.9	-	-	-	733	h,s	1929
$67\bar{s}_4$	-	21.5	56.3	-	-	-	-205.0	-189.9	-169.9	-	-	-	1748	d,s,u	1936
$68\bar{s}_4$	-	23.9	[33.5]	-	-	-	[-198.8]	-195.5	-190.9	-	-	-	1748	h,s	1936
$30\bar{s}_4$	-13.3	[13.4]	[40.6]	[48.6]	-	[-279.3]	[-233.3]	-196.5	-168.8	2.04(2.04)	-	-	644	d,e,s	1938
$31\bar{s}_4$	[-2.9]	18.0	55.6	64.0	72.4	-263.8	-205.2	-192.7	-171.7	0.85(0.85)	1.97(1.67)	-	814	e,o,s	1939
$32\bar{s}_4$	[-1.4]	19.1	55.1	64.3	71.9	-264.1	-205.5	-189.6	-172.1	1.27(1.27)	-	-	2155	e,o,s	1940
$36\bar{s}_4$	-9.8	21.6	57.2	69.3	73.1	-262.4	-206.7	-191.3	-168.7	1.80(1.80)	-	-	821	d,s,u	1955
$69\bar{s}_3$	-9.9	21.2	56.8	68.9	72.1	-264.0	-206.9	-191.5	-168.8	-	0.08(0.08)	$11\bar{S}$	1881	r,z	1968
$10\bar{s}_3$	[-23.7]	[3.0]	[37.4]	[59.5]	[121.9]	[-283.2]	-204.9	-188.6	-165.6	0.07(0.07)	0.07(0.04)	$3\bar{S}$	1873	m,r	1970
$13\bar{s}_3$	-9.6	20.2	57.2	69.1	72.4	-268.3	-205.8	-190.1	-167.0	0.12(0.12)	0.06(0.06)	$4\bar{S}$	2153	r	1976
$41\bar{s}_4$	-12.1	20.3	56.3	65.7	72.8	-268.5	-206.5	-190.4	-171.9	1.15(1.15)	-	-	2183	c,d,s	1984
$21\bar{s}_4$	[-87.1]	[-31.1]	[10.9]	[19.9]	[33.9]	[-318.5]	[-269.5]	[-259.5]	[-246.5]	0.07(0.07)	-	-	1750	h,s	1999
NUM	36	41	41	38	34	35	41	41	41						
AVE	-12.2	19.7	53.8	66.5	76.4	-261.6	-209.5	-196.5	-174.1						
RMS	13.1	9.2	8.8	11.5	19.0	54.3	11.2	14.3	13.0						
ERR	4.9	3.2	3.1	4.2	7.3	20.5	3.9	5.0	4.5						
REC	-10.3	20.2	57.8	66.6	73.3	-268.4	-206.4	-190.5	-167.5						
NUC	31	36	34	30	29	28	35	38	40						
AVC	-10.3	20.9	56.7	67.6	72.1	-266.1	-206.8	-192.8	-172.3						
RMC	1.5	1.3	0.8	1.8	0.6	2.1	1.6	5.2	6.3						
ERC	0.6	0.5	0.3	0.7	0.2	0.9	0.6	1.9	2.2						

(continues on the next page)

Table 5.16 *(continued)*

(a) calculated from a set of real single-ion solvation parameters (Table 5.15) by subtracting (cations) or adding (anions) the corresponding $\hat{\mathcal{Y}}_H^{\ominus}$ parameter (value provided in the same reference; Table 5.15); (b) gas-phase reference concentration or pressure not stated explicitly in the quoted reference and assumed by the authors of this book (see possible additional footnotes on standard-state corrections); (c) solution reference concentration not stated explicitly in the quoted reference and assumed by the authors of this book (see possible additional footnotes on standard-state corrections); (d) including standard-state conversion from 1 atm to 1 bar gas-phase reference pressure (Eqs. 4.192 and 4.193); (e) including standard-state conversion from 1 mol·dm^{-3} (molar) to 1 mol·kg^{-1} (molal) reference concentration in solution (Eqs. 4.198 and 4.199); (f) based on thermodynamic tables (secondary reference) for formation parameters of gaseous ions and conventional formation parameters of aqueous ions; (g) calculated from a set of intrinsic single-ion solvation parameters by subtracting (cations) or adding (anions) the corresponding $\mathcal{Y}_H^{\ominus}$ parameter (value provided in the same reference; Table 5.19); (h) original value from a theoretical model and corrected by the authors of this book to the standard-variant (addition of 7.95 kJ·mol^{-1} for free energies, -2.29 kJ·mol^{-1} for enthalpies, and -34.34 J·mol^{-1}·K^{-1} for entropies); (i) converted from values on an alternative conventional scale where the conventional values for the proton solvation parameters are set to $-\Delta_f Y^{\ominus}[\mathrm{H}_g^+]$ (Section 4.5.1); (j) primary reference quotes value based on secondary reference (see primary reference for the original data sources and both references for indications concerning the determination method, if available); (k) the proton value is reported as the free energy of formation of the aqueous proton and the authors of this book converted this value to a solvation free energy using $\Delta_f G^{\ominus}[\mathrm{H}_g^+]$ from Table 5.9; (l) based on salt dissolution parameters, along with $\mathrm{H}X_g$ dissolution parameters, and reticulation parameters, or, for entropies, absolute entropies of the MX_s compound (or $\mathrm{H}X_g$); (m) calculated using the Gibbs equation ($\Delta_s G^{\bullet} = \Delta_s H^{\bullet} - T^{-}\Delta_s S^{\bullet}$) based on the corresponding sets as reported in this table; (n) the source data was corrected for the improper use of Boltzmann statistics rather than Fermi-Dirac statistics for the thermodynamic formation parameters of the gaseous proton by adding $-(\Delta_f Y^{\ominus}[\mathrm{H}_g^+] - \Delta_f Y^{(\ominus)}[\mathrm{H}_g^+])$, where the former corresponds to Fermi-Dirac and the latter to Boltzmann statistics (3.64 kJ·mol^{-1}, 3.10 kJ·mol^{-1} and -1.82 J·mol^{-1}·K^{-1} for $Y = G, H, S$, respectively, using data from Table 4.2); (o) including standard-state conversion from 1 mol·dm^{-3} (molar) to 1 bar for the gas-phase reference concentration or pressure (Eqs. 4.194 and 4.195); (p) including standard-state conversion from a mole fraction of unity to 1 mol·kg^{-1} (molal) reference concentration in solution (Eqs. 4.202 and 4.203); (q) calculated from redox potentials and derivatives of Table 5.5 by application of Eqs. 5.3-5.6; (r) calculated by optimal partitioning of a pairwise set (Table 5.11), followed by alignment against the consensus (for Sets 1) or recommended (otherwise) conventional data, as described in the text; (s) calculated by alignment of a set of intrinsic single-ion solvation parameters against recommended conventional data, as described in the text (the corresponding proton value can be found in Table 5.19 between parentheses); (t) calculated by alignment of a set of real single-ion solvation parameters (Table 5.15) against the recommended conventional data, as described in the text (the corresponding proton value can be found in Table 5.15 between parentheses); (u) including standard-state conversion from 0.041 mol·dm^{-3} (molar) to 1 mol·kg^{-1} (molal) reference concentration in solution (Eqs. 4.198 and 4.199); (v) partitioning also relied on considering electronegativities from Ref.[2154]; (w) partitioning also relied on considering electronegativities from Ref.[2191]; (x) based on thermodynamic tables (secondary reference) for conventional parameters, along with reticulation and dissolution parameters to get pair solvation parameters for MX_g, followed by combination of the conventional cation solvation parameters and the pair solvation parameters to get conventional solvation parameters for anions; (y) standard states not stated explicitly in the quoted reference and assumed to be the same as in the reference providing source proton data[411]; (z) calculated from the conventional single-ion partial molar entropies reported in the original source using the gas-phase ion molar entropies of Table 5.9; (α) before fitting the data, the authors of the quoted reference remove a "non-electrostatic contribution" of -26.57 J·mol^{-1}·K^{-1}, -2.38 kJ·mol^{-1}, and 5.54 kJ·mol^{-1} from the solvation entropies, enthalpies and free energies, respectively, in order to consider only the charging contribution in the fitting procedure, and the authors of this book correct the resulting values by adding -26.67 J·mol^{-1}·K^{-1}, 0 kJ·mol^{-1}, and 7.95 kJ·mol^{-1} (corresponding to Eqs. 4.192 and 4.193), neglecting the difference; (β) converted from values on an alternative conventional scale where the conventional values for the proton solvation parameters are set to $y^{\ominus}[\mathrm{H}_g^+]$ (Section 4.5.1), possibly at a pressure differing from P°.

Table 5.17 *Standard conventional single-ion partial molar variables for the alkali and halide ions dissolved in water.* The reported data includes the conventional partial molar entropy $s^{\bullet}$, heat capacity $c_P^{\bullet}$, volume $v^{\bullet}$, volume-compressibility $k_T^{\bullet}$ and volume-expansivity $a_P^{\bullet}$. The notation I_d^z refers to an alkali (M_d^+) or halide (X_d^-) ion in solution. The different sets correspond to values derived from different literature sources, listed in chronological order of publication separately for each block of the table. The lines NUM, AVE, RMS, ERR and REC report for each ion the number of entries, average value, root-mean-square deviation, estimated error ($f^{-1/2}$NUM$^{-1/2}$RMS, with a redundancy factor $f = 5$) and recommended values (Sets 1) for all preceding entries corresponding to a given parameter. The lines num, ave and rms report the corresponding quantities for the preceding block. The lines NUC, ABC, RMC and ERC are similar to NUM, AVE, RMS and ERR, but exclude outliers (between square brackets in the table). The entries ϵ_g^{max} and ϵ_e^{max} report maximum violations of the Gibbs equation ($\Delta_s G^{\bullet}[I_g^z] - \Delta_s H^{\bullet}[I_g^z] + T^{-}\Delta_s S^{\bullet}[I_g^z]$, using data from Table 5.16) and of the predicted gas-phase entropies ($s^{\ominus}[I_g^z] - zs^{\ominus}[H_g^+] - s^{\bullet}[I_{aq}^z] + \Delta_s S^{\bullet}[I_g^z]$, using data from Tables 5.9 and 5.16), whenever relevant. The column "Source" provides the source set of a derived set (from Table 5.12). The data concerning $s^{\bullet}$ includes source references explicitly reporting conventional single-ion partial molar entropies as well as values derived from conventional solvation entropies (Table 5.16), listed in the first block. The standard states are according to the $bbme_T$ convention (Section 4.2.4; reference pressure $P^{\circ} = 1$ bar, reference molality $b^{\circ} = 1$ mol·kg^{-1} and reference temperature $T^{-} = 298.15$ K). The experimental values reported in the original references are assumed to pertain to the standard variant for $s^{\bullet}$ and $c_P^{\bullet}$ and to the density-corrected variant for $v^{\bullet *}$, $k_T^{\bullet *}$ and $a_P^{\bullet *}$ (Section 4.2.5), which represents the most likely situation, but may not necessarily be the case in reality.

Set	Li^+	Na^+	K^+	Rb^+	Cs^+	F^-	Cl^-	Br^-	I^-	ϵ_g^{max}	ϵ_e^{max}	Source	Ref.	Notes	Year
				$s^{\bullet}[I_d^z]$ [J·mol^{-1}·K^{-1}]											
$28s_6$	15.3	60.4	103.5	[131.6]	[135.4]	-9.9	55.3	80.7	109.1	-	-	$28\bar{s}_0$		a,b	2011
$65s_{10}$	-	[74.5]	[99.5]	[111.8]	-	-	[61.8]	84.1	103.6	-	-	$65\bar{s}_4$	733	a,c,d	1929
$66s_{10}$	-	[72.4]	[89.5]	[101.3]	-	-	58.6	80.0	[88.3]	-	-	$66\bar{s}_4$	733	a,e	1929
$67s_{10}$	-	60.5	101.9	-	-	-	57.3	82.5	108.4	-	-	$67\bar{s}_4$	1748	a,c,d	1936
$68s_{10}$	-	62.9	[79.1]	-	-	-	[63.5]	77.0	[87.3]	-	-	$68\bar{s}_4$	1748	a,e	1936
$30s_{10}$	10.7	[52.4]	[86.3]	[104.0]	-	[-24.8]	[29.0]	[76.0]	109.4	-	-	$30\bar{s}_4$	644	a,c,f	1938
$31s_{10}$	21.1	57.0	101.3	119.4	133.3	-9.3	57.1	79.8	106.5	-	-	$31\bar{s}_4$	814	a,f,g	1939
$32s_{10}$	[22.7]	58.1	100.8	119.7	132.7	-9.6	56.8	82.8	106.2	-	-	$32\bar{s}_4$	2155	a,f,g	1940
$62s_7$	19.9	58.7	101.3	120.2	133.3	-11.2	56.4	82.8	106.2	-	-	$62\bar{s}_1$	2189	a,c,h	1945
$3s_6$	14.0	60.3	102.5	124.4	132.9	-9.7	55.0	80.6	109.3	-	-	$3\bar{s}_0$	1871/ 2175	a,c	1954
$36s_{10}$	14.3	60.6	102.8	124.7	134.0	-7.9	55.6	81.2	109.5	-	-	$36\bar{s}_4$	821	a,c,d	1955
$5s_7$	14.2	60.2	102.5	124.3	133.1	-9.6	55.2	80.8	109.2	-	-	$5\bar{s}_1$	754	a,i,j	1957
$6s_6$	14.2	62.1	102.5	124.0	132.4	-10.9	54.6	81.4	108.1	-	-	$6\bar{s}_0$	259/ 1744,2175 - 2178	a,c,k	1962
$6s_7$	14.2	62.1	102.5	124.0	132.5	-11.4	54.1	80.9	107.6	-	-	$6\bar{s}_1$	259	a,l	1962
$8s_6$	14.0	60.3	102.5	124.4	132.9	-9.7	55.0	80.6	109.4	-	-	$8\bar{s}_0$	262/ 2175	a,c	1965
$9s_7$	14.0	60.3	102.5	124.4	132.9	-9.7	55.0	80.6	109.4	-	-	$9\bar{s}_1$	264	a,c,i,m	1969
$10s_6$	17.1	60.0	101.8	[150.8]	[199.8]	[-39.0]	[51.6]	[37.9]	104.0	-	-	$10\bar{s}_0$	1873	a,n	1970
$10s_7$	17.1	60.0	101.8	[152.2]	[199.8]	[-39.0]	[51.6]	[37.9]	104.0	-	-	$10\bar{s}_1$	1873	a,n	1970
$11s_8$	14.1	60.0	102.6	-	-	-10.7	54.1	80.2	[92.0]	-	-	$11\bar{s}_2$	415	a,c	1970
$12s_7$	14.0	60.3	102.5	124.4	132.9	-9.7	55.0	80.6	109.4	-	-	$12\bar{s}_1$	269/ 262,1744	a,c,f	1973
$13s_6$	14.4	59.1	102.5	124.4	132.9	-13.7	56.9	82.5	111.7	-	-	$13\bar{s}_0$	2153	a	1976
$13s_7$	14.0	59.1	102.5	123.6	132.9	-13.7	56.9	82.9	111.7	-	-	$13\bar{s}_1$	2153	a	1976
$15s_7$	14.1	60.3	102.4	124.3	132.8	-13.5	[51.2]	76.8	105.6	-	-	$15\bar{s}_1$	274/ 262	a,c	1978
$63s_7$	-	62.0	102.5	124.0	132.4	-	54.7	81.6	108.3	-	-	$63\bar{s}_1$	663	a,c,h	1978
$16s_6$	14.2	60.0	103.2	125.5	132.5	-9.5	54.4	[66.0]	109.6	-	-	$16\bar{s}_0$	417/ 262	a,n	1981
$16s_7$	14.2	60.0	103.2	125.5	132.5	-9.5	54.4	[66.0]	109.6	-	-	$16\bar{s}_1$	417	a,n	1981
$17s_6$	11.3	58.4	101.0	120.5	132.8	-13.2	56.7	82.8	106.7	-	-	$17\bar{s}_0$	1874/ 2179,2180	a,j	1982
$17s_7$	11.3	58.8	101.3	120.4	133.0	-13.3	56.5	82.7	107.0	-	-	$17\bar{s}_1$	1874	a,n	1982
$41s_{10}$	12.0	59.3	102.0	121.1	133.7	-13.9	55.8	82.1	106.3	-	-	$41\bar{s}_4$	2183	a,c,i	1984
$18s_6$	13.0	60.5	102.6	121.8	133.3	[302.3]	[38.1]	[64.6]	[93.5]	-	-	$18\bar{s}_0$	275/ 294	a,n	1985
$18s_7$	14.0	59.1	[106.0]	122.5	[134.7]	[-30.8]	[37.4]	[64.3]	[93.5]	-	-	$18\bar{s}_1$	275	a,n	1985
$64s_7$	14.4	62.0	102.5	124.0	132.4	-	54.9	81.8	108.5	-	-	$64\bar{s}_1$	2190	a,h,m	1995
$20s_7$	11.0	58.8	104.0	122.1	-	-16.1	56.7	80.9	[99.4]	-	-	$20\bar{s}_1$	1875,1876	a,m,n	1998
$21s_6$	13.4	59.0	102.5	121.5	133.1	-13.8	56.5	82.3	111.3	-	-	$21\bar{s}_0$	1750/ 294	a	1999
$21s_{10}$	[-63.1]	[7.9]	[56.5]	[75.2]	[94.7]	[-64.0]	[-7.2]	[12.9]	[31.7]	-	-	$21\bar{s}_4$	1750	a,e	1999
$22s_6$	12.2	58.4	101.1	121.7	132.1	-13.7	56.8	82.7	106.6	-	-	$22\bar{s}_0$	819/ 2181	a,f,g,o	2000
$22s_7$	12.1	58.3	101.0	121.6	132.0	-13.6	56.9	82.8	106.7	-	-	$22\bar{s}_1$	819	a,f,g	2000
$70s_0$	[4.2]	62.8	[92.0]	[104.6]	-	[-20.9]	[64.9]	85.8	113.0	-	-	-	2192		1926
$71s_0$	[-8.4]	[54.4]	[97.5]	[96.2]	-	[-25.1]	[61.5]	86.2	110.9	-	-	-	2193		1930
$72s_0$	7.5	61.5	102.9	[95.4]	-	[-20.9]	56.5	81.2	107.5	-	-	-	2194		1934
$73s_0$	19.7	58.6	101.3	120.1	133.1	-	56.5	82.4	105.9	-	-	-	2195		1938
$74s_0$	14.2	60.2	102.5	124.3	133.1	-9.6	55.1	80.5	109.4	-	-	-	820	i	1951
$75s_0$	14.2	60.2	102.5	124.3	133.1	-9.6	55.2	80.7	109.4	-	-	-	1744		1952
$5s_1$	14.2	60.2	102.5	124.3	133.1	-9.6	55.2	80.8	109.2	0.00(0.00)	0.00(0.00)	-	754	i,p	1957
$76s_1$	14.2	60.2	102.5	120.1	133.1	-9.6	55.2	80.8	109.2	-	-	-	1749	p	1965
$69s_1$	14.2	60.2	102.5	124.3	133.1	-9.5	55.4	80.9	109.3	-	-	-	1881,2196	h,p	1968
$77s_1$	14.2	60.2	102.9	127.2	[136.4]	-	55.4	80.9	109.3	-	-	-	2197	h,p	1969

(continues on the next page)

Table 5.17 *(continued)*

Set	Li^+	Na^+	K^+	Rb^+	Cs^+	F^-	Cl^-	Br^-	I^-	ϵ_g^{max}	ϵ_e^{max}	Source	Ref.	Notes	Year
					$s^\bullet[I_d^z]$ [$J\cdot mol^{-1}\cdot K^{-1}$]										
$17s_0$	11.3	58.4	101.0	120.5	132.8	-13.2	56.7	82.8	106.7	0.00(0.00)	0.00(0.00)	-	1874 / 2179,2180		1982
$17s_1$	11.3	58.4	101.0	120.5	132.8	-13.2	56.7	82.8	106.7	0.00(0.00)	0.39(0.39)	-	1874	p	1982
$18s_0$	13.4	59.0	102.5	121.5	133.1	-13.8	56.5	82.4	111.3	0.00(0.00)	316.09(17.77)	-	275 / 294		1985
$18s_1$	13.4	59.0	102.5	121.5	133.1	-13.8	56.5	82.4	111.3	0.00(0.00)	19.11(18.11)	-	275	p	1985
$1s_3$	13.8	59.2	103.4	121.9	134.2	-13.8	55.9	81.9	110.7	0.08(0.08)	0.08(0.08)	1S		q	2011
$78s_4$	-	-	-	-	-	-	55.9	-	-	-	-	-	2198	r	1928
$31s_4$	19.2	56.9	102.9	119.6	132.2	-10.8	56.1	81.2	106.3	0.85(0.85)	1.97(1.67)	-	814	f,r	1939
$79s_4$	-	-	-	-	-	-	55.9	-	-	-	-	-	2199	r	1942
$80s_4$	-	-	-	-	-	-	55.9	-	-	-	-	-	2200	r	1950
$69s_3$	14.2	60.2	102.5	124.2	133.0	-9.5	55.4	80.9	109.4	-	0.08(0.08)	11S	1881	h,q	1968
$10s_3$	[0.4]	[42.0]	[83.0]	[114.9]	[182.8]	[-28.6]	57.4	83.8	112.6	0.07(0.07)	0.07(0.04)	3S	1873	j,q	1970
$13s_3$	14.5	59.2	102.8	124.5	133.3	-13.8	56.5	82.4	111.2	0.12(0.12)	0.06(0.06)	4S	2153	j,q	1976
NUM	51	56	56	53	46	48	59	56	56						
AVE	11.8	58.8	100.0	120.7	136.2	-8.9	54.0	77.6	105.4						
RMS	11.6	7.9	7.7	11.2	16.4	46.5	9.6	12.6	11.4						
ERR	3.6	2.4	2.3	3.5	5.4	15.0	2.8	3.8	3.4						
REC	13.8	59.2	103.4	121.9	134.2	-13.8	55.9	81.9	110.7						
NUC	46	50	47	42	39	38	48	48	49						
AVC	14.1	59.8	102.3	122.8	132.9	-11.5	55.9	81.7	108.5						
RMC	2.4	1.3	0.7	2.0	0.4	2.1	1.0	1.7	2.2						
ERC	0.8	0.4	0.2	0.7	0.2	0.7	0.3	0.5	0.7						
					$c_P^\bullet[I_d^z]$ [$J\cdot mol^{-1}\cdot K^{-1}$]										
$81c_0$	56.8	22.6	0.8	-	-	-120.1	-122.2	-124.3	-127.2	-	-	-	261	s	1964
$81c_1$	56.9	22.6	0.8	-	-	-120.1	-122.2	-124.3	-127.2	-	-	-	261	p	1964
$82c_1$	54.4	37.7	12.6	8.4	-8.4	-121.3	-129.7	-138.1	-142.3	-	-	-	2196	p	1964
$83c_0$	60.6	41.9	12.9	-8.1	-21.6	-115.2	-123.1	-126.1	-115.4	-	-	-	2103	t	1974
$84c_0$	62.0	43.0	13.0	-8.5	-23.0	-117.0	-127.0	-131.0	-121.5	-	-	-	2104	t	1976
$85c_1$	-	-	-	-	-	-	-123.0	-	-	-	-	-	2201	p	1977
$15c_1$	-	-	-	-	-	-	-123.0	-	-	3.24(3.24)	-	-	274	p	1978
$86c_1$	62.0	43.0	13.0	-9.0	-23.0	-116.0	-127.0	-131.0	-121.0	-	-	-	2202	p	1986
$87c_1$	-	-	-	-	-	-	-	-131.0	-	-	-	-	2105	p	2004
$28c_0$	-	-	-	-	-	-123.5	-125.6	-126.3	-130.0	0.09(0.09)	-	-		b	2011
num	6	6	6	4	4	7	9	8	7						
ave	58.8	35.1	8.9	-4.3	-19.0	-119.0	-124.8	-129.0	126.1						
rms	2.9	9.0	5.7	7.3	6.2	2.8	2.5	4.4	7.9						
$1c_3$	61.0	42.0	11.7	-9.8	-23.4	-115.2	-126.0	-130.3	-120.6	0.08(0.08)	0.08(0.08)	1C		q	2011
$88c_3$	-	29.7	7.8	-	-	-	-127.2	-132.2	-133.6	-	-	12C	2101	q	1929
$89c_3$	60.8	-	-	-	-	-	-126.2	-	-	-	-	13C	2102	q	1932
$90c_3$	65.3	33.2	11.5	-	-	-	-130.7	-	-	-	-	14C	2124	q	1933
$91c_3$	61.4	40.5	19.6	-26.5	[-68.3]	-124.2	-136.7	-140.9	-145.1	-	-	15C	2156	q	1948
$92c_3$	66.4	32.2	10.4	-	-	-	-131.8	-	-	-	-	16C	2123	q	1969
$83c_3$	61.1	43.0	13.7	-7.2	-20.2	-116.0	-123.5	-127.1	-116.7	-	-	17C	2103	q	1974
$93c_3$	-	42.6	13.9	-7.2	-19.4	-115.8	-123.4	-127.0	-116.5	-	-	18C	2127	q	1974
$94c_3$	-	43.3	-	-	-	-	-124.7	-	-	-	-	19C	2116	q	1975
$84c_3$	61.0	42.0	11.7	-9.8	-23.4	-115.2	-126.0	-130.3	-120.6	-	-	20C	2104	q	1976
$95c_4$	-	-	-	-	-22.1	-	-	-128.9	-	-	-	-	2203	r	1980
$96c_4$	61.8	42.8	12.8	-8.7	-23.2	-114.8	-124.8	-128.8	-119.3	-	-	-	2204	r	1982
$97c_4$	-	-	-	-	-	-	-	-130.3	-	-	-	-	2204	r	1982
$98c_4$	-	-	-	-	-	-	-	-130.3	-	-	-		2204	r	1982
$99c_4$		-	-	-	-	-	-	-130.3	-	-	-	-	2204	r	1982
$87c_3$	-	41.9	-	-	-	-	-	-130.4	-	-	-	21C	2105	q	2004
NUM	14	17	15	10	11	13	20	20	14						
AVE	60.8	37.9	11.1	-8.6	-25.1	-118.0	-126.2	-129.9	-125.5						
RMS	3.0	6.9	4.7	7.8	14.3	3.2	3.6	3.9	9.0						
ERR	1.8	3.8	2.7	5.5	9.6	2.0	1.8	2.0	5.4						
REC	61.0	42.0	11.7	-9.8	-23.4	-115.2	-126.0	-130.3	-120.6						
NUC	14	17	15	10	10	13	20	20	14						
AVC	60.8	37.9	11.1	-8.6	-20.8	-118.0	-126.2	-129.9	-125.5						
RMC	3.0	6.9	4.7	7.8	4.3	3.2	3.6	3.9	9.0						
ERC	1.8	3.8	2.7	5.5	3.1	2.0	1.8	2.0	5.4						

(continues on the next page)

Table 5.17 *(continued)*

Set	Li^+	Na^+	K^+	Rb^+	Cs^+	F^-	Cl^-	Br^-	I^-	ϵ_g^{max}	ϵ_e^{max}	Source	Ref.	Notes	Year
				$v^{\bullet*}[I_d^z]$ $[10^{-6}\ m^3{\cdot}mol^{-1}]$											
$47v_1$	[0.0]	-0.5	[16.0]	[23.5]	[35.0]	-2.0	[29.0]	[39.5]	[58.0]	-	-	-	643	p	1933
$100v_0$	-1.0	-1.5	8.7	13.7	21.1	-2.1	18.1	25.0	36.6	-	-	-	2108	u	1941
$101v_1$	-1.1	-1.9	[8.2]	13.5	20.9	-1.6	18.2	[25.3]	36.9	-	-	-	2205	p	1942
$102v_1$	-1.0	-1.5	8.7	13.7	21.1	-2.1	18.1	25.0	36.6	-	-	-	2206	p,v	1956
$103v_1$	-	-0.9	8.7	-	21.6	-	17.7	[24.2]	-	-	-	-	2110	p	1956
$104v_1$	-1.0	-1.5	8.7	13.6	21.1	-2.1	18.1	25.0	36.6	-	-	-	2207	p	1957
$105v_0$	-0.7	-1.2	9.0	14.0	21.4	-2.4	17.8	24.7	36.3	-	-	-	736	w	1961
$106v_1$	-0.7	-1.2	9.0	14.0	21.4	-2.4	17.8	24.7	36.3	-	-	-	736,2208	p,x	1961
$107v_1$	-1.2	-1.8	8.6	13.8	20.9	-2.0	18.2	[25.3]	36.7	-	-	-	737,2209	p	1964
$81v_0$	-0.9	-1.5	8.8	13.8	21.2	-2.1	18.1	25.0	36.6	-	-	-	261	y	1964
$81v_1$	-0.9	-1.5	8.8	13.8	21.2	-2.1	18.1	25.0	36.6	-	-	-	261	p,y	1964
$108v_1$	[-5.8]	-2.0	8.8	14.4	20.9	-2.1	18.3	24.8	36.0	-	-	-	2122	p	1967
$109v_1$	-	-	-	-	-	-	-	24.8	-	-	-	-	2210	p	1968
$110v_0$	-0.9	-1.2	9.0	14.1	21.3	-1.2	17.8	24.7	36.2	-	-	-	2163	z	1971
$85v_1$	-	-	-	-	-	-	18.1	-	-	-	-	-	2201	p	1977
$111v_1$	-0.9	-1.2	9.0	14.1	21.3	-1.2	17.8	24.7	36.2	-	-	-	274	p,α	1978
$112v_1$	-0.9	-1.2	9.0	14.1	21.3	-1.2	17.8	24.7	36.2	-	-	-	274	p,α	1978
$113v_1$	-0.9	-1.2	8.9	14.0	21.4	-1.2	17.8	24.8	36.2	-	-	-	2211	p,β	2008
num	15	16	16	15	16	15	17	17	15						
ave	-1.2	-1.4	9.2	14.5	22.1	-1.9	18.6	25.7	37.9						
rms	1.3	0.4	1.8	2.4	3.3	0.4	2.6	3.5	5.4						
$1v_3$	-0.7	-1.4	8.8	14.2	21.5	-1.8	17.7	24.7	36.3	0.08(0.08)	0.08(0.08)	1V		q	2011
$114v_3$	-0.7	-1.2	9.0	14.0	21.4	-	17.7	24.8	36.3	-	-	22V	2157	q	1931
$163v_3$	-1.0	-1.2	9.1	14.2	21.5	-	18.0	24.8	36.1	-	-	44V	2120	q	1931
$90v_3$	-0.6	-1.4	8.7	-	-	-	17.7	-	-	-	-	14V	2124	q	1933
$115v_{3a}$	-	-1.3	-	-	-	-	-	24.8	-	-	-	23Va	2106	q	1934
$115v_{3b}$	-	-	9.0	-	-	-	17.9	-	-	-	-	23Va	2106	q	1934
$100v_3$	-0.8	-1.3	8.9	13.9	21.3	-2.3	17.9	24.8	36.4	-	-	25V	2108	q	1941
$116v_3$	-	-1.3	-	-	-	-	17.7	24.7	-	-	-	24V	2107	q	1941
$117v_3$	-	-1.2	8.9	-	-	-	17.6	-	36.6	-	-	26V	2109	q	1952
$35v_4$	-	-1.2	8.9	14.0	21.3	-	17.5	24.8	36.3	-	-	-	409	q	1954
$35v_3$	-	-1.2	8.8	14.0	21.5	-	17.6	24.7	36.3	-	-	27V	409	q	1954
$103v_3$	-	-1.3	8.7	-	21.4	-	17.7	24.6	-	-	-	28V	2110	q	1956
$118v_4$	-0.8	-1.3	8.9	13.9	21.3	-2.3	17.9	24.8	36.4	-	-	-	710	r,γ	1957
$105v_4$	[-6.8]	[1.0]	[10.1]	14.7	[22.5]	-0.2	[16.5]	[23.6]	35.8	-	-	-	736	r	1961
$119v_4$	-0.8	-	-	-	-	-	17.6	-	-	-	-	-	2212	r	1965
$120v_4$	-0.7	-	-	-	-	-	-	-	-	-	-	-	2212	r	1965
$121v_4$	-	-	-	-	-	-	17.6	24.9	36.2	-	-	-	2213	r	1966
$122v_3$	-	-0.6	-	-	-	-0.6	-	-	35.8	-	-	29V	2111	q	1966
$123v_3$	-0.8	-1.2	9.1	14.1	21.3	-	17.8	-	-	-	-	30V	2112	q	1966
$124v_3$	-0.7	-1.3	8.9	14.2	21.5	-	17.7	24.8	36.5	-	-	31V	2121	q	1966
$108v_3$	-0.8	-1.4	8.7	14.5	21.8	-1.9	17.7	24.9	36.5	-	-	34V	2122	q	1967
$125v_3$	-	-1.3	8.9	-	-	-	-	24.8	-	-	-	32V	2117	q	1967
$126v_3$	-	-0.9	-	-	-	-1.4	-	-	-	-	-	33V	2113	q	1967
$127v_4$	-	-	-	-	-	-	17.7	-	-	-	-	-	2214	r	1968
$128v_3$	-	-1.3	8.9	-	-	-	17.9	24.8	36.3	-	-	35V	2114	q	1968
$129v_3$	-0.8	-1.1	9.1	14.2	21.4	-1.2	17.8	24.6	36.2	-	-	36V	2129	q	1968
$92v_3$	-0.8	-1.2	8.7	-	-	-	17.8	-	-	-	-	16V	2123	q	1969
$130v_3$	-	-1.2	-	-	-	-	17.8	-	-	-	-	37V	2115,2158	q	1971
$131v_4$	-0.9	-0.9	9.1	-	[22.1]	-	[16.9]	[25.9]	36.9	-	-	-	2159	r	1975
$131v_3$	-0.8	-1.2	9.0	-	21.4	-	17.9	24.7	36.2	-	-	38V	2159	q	1975
$94v_3$	-	-1.2	-	-	-	-	17.8	-	-	-	-	19V	2116	q	1975
$132v_3$	-0.8	-1.3	9.0	-	-	-	17.9	24.8	36.3	-	-	39V	2160	q	1978
$133v_4$	-	-	-	-	-	-	-	24.7	-	-	-	-	2215	r	1980
$134v_4$	-0.9	-1.2	9.0	-	[22.3]	-1.2	17.8	24.7	36.2	-	-	-	2204	r	1982
$135v_3$	-	-1.2	-	-	-	-	17.9	-	-	-	-	40V	2161	q	1989
$136v_4$	[-3.2]	[-4.4]	[4.3]	[7.6]	[14.3]	-3.1	[12.0]	[17.3]	[26.6]	-	-	-	457,879	r	1994
$137v_4$	[-8.9]	[4.0]	[18.6]	-	[29.9]	[2.1]	[21.6]	[28.6]	[40.2]	-	-	-	1710	r	2002
$87v_3$	-	-1.2	-	-	-	-	-	24.8	-	-	-	21V	2105	q	2004
NUM	35	48	41	28	34	26	48	42	36						
AVE	-1.4	-1.2	9.2	14.1	21.8	-1.6	18.0	25.1	36.8						
RMS	1.8	1.0	2.0	2.2	3.0	1.0	1.9	2.6	4.0						
ERR	0.7	0.3	0.7	0.9	1.1	0.4	0.6	0.9	1.5						
REC	-0.7	-1.4	8.8	14.2	21.5	-1.8	17.7	24.7	36.3						
NUC	30	45	36	26	28	25	43	34	33						
AVC	-0.9	-1.3	8.9	14.0	21.3	-1.7	17.8	24.8	36.4						
RMC	0.1	0.3	0.1	0.3	0.2	0.6	0.2	0.1	0.2						
ERC	0.1	0.1	0.0	0.1	0.1	0.3	0.1	0.0	0.1						

(continues on the next page)

Table 5.17 *(continued)*

Set	Li^+	Na^+	K^+	Rb^+	Cs^+	F^-	Cl^-	Br^-	I^-	ϵ_g^{max}	ϵ_e^{max}	Source	Ref.	Notes	Year
			$k_T^{\bullet *}[I_d^z]$ $[10^{-10}\ m^3 \cdot mol^{-1} \cdot bar^{-1}]$												
$100k_0$	-34.0	-42.0	-37.0	-	-27.0	-	-8.0	2.0	18.0	-	-	-	2108	δ	1941
$81k_1$	-34.0	-42.0	-37.0	-	-27.0	-	-8.0	2.0	18.0	-	-	-	261	p,ϵ	1964
num	2	2	2	0	2	0	2	2	2						
ave	-34.0	-42.0	-37.0	-	-27.0	-	-8.0	2.0	18.0						
rms	0.0	0.0	0.0	-	0.0	-	0.0	0.0	0.0						
$1k_3$	-35.6	-42.4	-35.0	-32.0	-28.6	-29.2	-6.3	2.3	14.3	0.08(0.08)	0.08(0.08)	1K		q	2011
$90k_3$	-33.4	-41.2	-35.3	-	-	-	-3.0	-	-	-	-	14K	2124	q	1933
$100k_3$	-32.1	-40.7	-33.8	-	-40.4	-	-9.8	-0.7	15.4	-	-	25K	2108	q	1941
$138k_3$	-	-41.3	-35.3	-	-	-	-5.4	-	-	-	-	41K	2125	q	1957
$139k_3$	-34.4	-43.7	-38.4	-	-	-	-8.9	1.6	14.0	-	-	42K	2126	q	1966
$127k_4$	-	-	-	-	-	-	-6.3	-	-	-	-	-	2214	r	1968
$92k_3$	-34.2	-42.2	-34.8	-	-	-	-4.5	-	-	-	-	16K	2123	q	1969
$140k_4$	-	-	-	-	-	-	-6.3	-	-	-	-	-	2216	r	1970
$141k_4$	-	-39.2	-	-	-	-	-3.0	-	-	-	-	-	2128	r	1974
$141k_3$	-	-42.4	-	-	-	-	-6.2	-	-	-	-	43K	2128	q	1974
$93k_3$	-	-42.5	-35.1	-31.7	-28.3	-29.1	-6.6	2.5	14.8	-	-	18K	2127	q	1974
$137k_4$	-47.9	[-28.9]	-13.2	-	-3.9	-1.0	[11.7]	10.7	7.9	-	-	-	1710	r	2002
NUM	8	12	10	2	6	3	14	7	7						
AVE	-35.7	-40.7	-33.5	-31.8	-25.9	-19.8	-5.0	2.9	14.6						
RMS	4.7	3.7	6.9	0.2	10.9	13.2	5.0	3.3	3.2						
ERR	3.7	2.4	4.9	0.3	9.9	17.1	3.0	2.8	2.7						
REC	-35.6	-42.4	-35.0	-32.0	-28.6	-29.2	-6.3	2.3	14.3						
NUC	8	11	10	2	6	3	13	7	7						
AVC	-35.7	-41.8	-33.5	-31.8	-25.9	-19.8	-6.3	2.9	14.6						
RMC	4.7	1.1	6.9	0.2	10.9	13.2	2.0	3.3	3.2						
ERC	3.7	0.8	4.9	0.3	9.9	17.1	1.2	2.8	2.7						
			$a_P^{\bullet *}[I_d^z]$ $[10^{-8}\ m^3 \cdot mol^{-1} \cdot K^{-1}]$												
$142a_0$	1.0	10.0	7.0	7.0	7.0	-	0.0	1.0	4.0	-	-	-	261		1964
$81a_0$	-0.9	5.9	5.1	-	-	-	3.4	4.2	7.6	-	-	-	261	ζ	1964
$81a_1$	1.0	10.0	7.0	7.0	7.0	-	0.0	1.0	4.0	-		-	261	p,η	1964
$143a_1$	-0.2	4.4	3.5	2.8	2.4	2.3	3.4	3.8	5.7	-	-	-	2210,2217	p,θ	1968
num	4	4	4	3	3	1	4	4	4						
ave	0.2	7.6	5.7	5.6	5.5	2.3	1.7	2.5	5.3						
rms	0.8	2.5	1.5	2.0	2.2	0.0	1.7	1.5	1.5						
$1a_3$	0.9	5.4	5.4	4.6	4.2	1.3	1.6	2.7	5.4	0.08(0.08)	0.08(0.08)	1A		q	2011
$129a_3$	0.2	7.0	6.2	4.7	4.7	-	2.3	3.1	6.5	-	-	36A	2129	q	1968
$92a_3$	-0.1	6.7	5.9	-	-	-	2.6	-	-	-	-	16A	2123	q	1969
$93a_3$	-	5.4	5.4	4.6	4.2	1.3	1.6	2.7	5.4	-	-	18A	2127	q	1974
NUM	7	8	8	6	6	3	8	7	7						
AVE	0.3	6.9	5.7	5.1	4.9	1.6	1.9	2.7	5.5						
RMS	0.7	2.0	1.1	1.5	1.6	0.5	1.2	1.2	1.2						
ERR	0.6	1.6	0.8	1.4	1.5	0.6	1.0	1.0	1.0						
REC	0.9	5.4	5.4	4.6	4.2	1.3	1.6	2.7	5.4						
NUC	7	8	8	6	6	3	8	7	7						
AVC	0.3	6.9	5.7	5.1	4.9	1.6	1.9	2.7	5.5						
RMC	0.7	2.0	1.1	1.5	1.6	0.5	1.2	1.2	1.2						
ERC	0.6	1.6	0.8	1.4	1.5	0.6	1.0	1.0	1.0						

(continues on the next page)

Table 5.17 *(continued)*

(a) original reference provides entropies of solvation rather than partial molar entropies of aqueous ions; (b) calculated from the isothermal temperature derivatives of the redox potentials of Table 5.5 by application of Eqs. 5.7 and 5.8; (c) including standard-state conversion from 1 atm to 1 bar gas-phase reference pressure (Eqs. 4.192 and 4.193); (d) including standard-state conversion from 0.041 mol·dm^{-3} (molar) to 1 mol·kg^{-1} (molal) reference concentration in solution (Eqs. 4.198 and 4.199); (e) corrected to correspond to the standard-variant; (f) including standard-state conversion from 1 mol·dm^{-3} (molar) to 1 mol·kg^{-1} (molal) reference concentration in solution (Eqs. 4.198 and 4.199); (g) including standard-state conversion from 1 mol·dm^{-3} (molar) to 1 bar for the gas-phase reference concentration or pressure (Eqs. 4.194 and 4.195); (h) including standard-state conversion from a mole fraction of unity to 1 mol·kg^{-1} (molal) reference concentration in solution (Eqs. 4.202 and 4.203); (i) solution reference concentration not stated explicitly in the quoted reference and assumed by the authors of this book (see possible additional footnotes on standard-state corrections); (j) calculated from the conventional single-ion partial molar entropies reported in the original source using the gas-phase ion molar entropies of Table 5.9; (k) converted from values on an alternative conventional scale where the conventional values for the proton solvation parameters are set to $-\Delta_f Y^\ominus[\mathrm{H}_g^+]$ (Section 4.5.1); (l) before fitting the data, the authors of the quoted reference remove a "non-electrostatic contribution" of -26.57 J·mol^{-1}·K^{-1}, -2.38 kJ·mol^{-1}, and 5.54 kJ·mol^{-1} from the solvation entropies, enthalpies and free energies, respectively, in order to consider only the charging contribution in the fitting procedure, and the authors of this book correct the resulting values by adding -26.67 J·mol^{-1}·K^{-1}, 0 kJ·mol^{-1}, and 7.95 kJ·mol^{-1} (corresponding to Eqs. 4.192 and 4.193), neglecting the difference; (m) gas-phase reference concentration or pressure not stated explicitly in the quoted reference and assumed by the authors of this book (see possible additional footnotes on standard-state corrections); (n) the source conventional solvation entropy was calculated using the Gibbs equation ($\Delta_s S^\bullet = (-T^-)^{-1}(\Delta_s G^\bullet - \Delta_s H^\bullet)$) based on the corresponding sets reported in Table 5.16; (o) converted from values on an alternative conventional scale where the conventional values for the proton solvation parameters are set to $y^\ominus[\mathrm{H}_g^+]$ (Section 4.5.1), possibly at a pressure differing from P°; (p) calculated from the reported intrinsic single-ion partial molar parameters by subtracting (cations) or adding (anions) the corresponding $y_H^\ominus$ parameter (value provided in the same reference; Table 5.20); (q) calculated by optimal partitioning of a pairwise set (Table 5.12), followed by alignment against the consensus (for Sets 1) or recommended (otherwise) conventional data, as described in the text; (r) calculated by alignment of an intrinsic set against consensus conventional data, as described in the text (the corresponding proton value can be found in Table 5.20 between parentheses); (s) data compilation based on Ref.[1826]; (t) based on flow microcalorimetry experiments; (u) based on density measurements; (v) conventional data of Ref.[2108]; (w) see indicated reference for data sources; (x) conventional data of Ref.[2108], increased by 0.3 cm^3·mol^{-1} for cations and decreased by 0.3 cm^3·mol^{-1} for anions, corresponding to a more accurate value of the partial molar volume of HCl_{aq}; (y) data compilation based on Refs.[1826, 2108]; (z) data compilation based on Ref.[2218]; (α) conventional data of Ref.[2163]; (β) using conventional hydration volume data[2211]; (γ) conventional data of Refs.[1826, 2205, 2206]; (δ) based on density measurements at different pressures; (ϵ) conventional data of Ref.[2108]; (ζ) data of Ref.[1826], complemented by Ref.[261]; (η) conventional data of Ref.[1826]; (θ) salt data from Ref.[2217].

5.17 Air-liquid interfacial potential of the pure solvent

The *air-liquid interfacial potential* $\chi_{svt}^\ominus$ of a pure solvent is one of the three quantities, besides the intrinsic solvation free energy $\mathcal{G}_{H,svt}^\ominus$ of the proton in this solvent and the intrinsic absolute potential $\mathcal{V}_{H,svt}^\ominus$ of the reference hydrogen electrode in this solvent, that *eludes a purely experimental determination* (Section 2.3). Note that the two latter quantities are often simply noted $\mathcal{G}_H^\ominus$ and $\mathcal{V}_H^\ominus$ for simplicity, when the identity of the solvent is obvious from the context. These three quantities are related to each other and to corresponding real quantities *via* (Section 4.5.4 and Figure 4.18)

$$\mathcal{G}_{H,svt}^\ominus = \hat{\mathcal{G}}_{H,svt}^\ominus - F\chi_{svt}^\ominus \tag{5.44}$$

and

$$\mathcal{V}^{\ominus}_{H,svt} = \hat{\mathcal{V}}^{\ominus}_{H,svt} - \chi^{\ominus}_{svt} \quad , \tag{5.45}$$

with

$$F\mathcal{V}^{\ominus}_{H,svt} = \mathcal{G}^{\ominus}_{H,svt} + \Delta_f G^{\ominus}[\mathrm{H}^+_g] \tag{5.46}$$

and

$$F\hat{\mathcal{V}}^{\ominus}_{H,svt} = \hat{\mathcal{G}}^{\ominus}_{H,svt} + \Delta_f G^{\ominus}[\mathrm{H}^+_g] \quad , \tag{5.47}$$

the two last equations implying that $\hat{\mathcal{V}}^{\ominus}_{H,svt}$ (or $\mathcal{V}^{\ominus}_{H,svt}$) represents the real (or intrinsic) formation free energy of the dissolved proton in the given solvent, divided by F. The determination of $\chi^{\ominus}_{svt}$ is only possible if experimental data is interpreted using some *extra-thermodynamic assumption*, *i.e.* a reasonable but formally unprovable hypothesis concerning the expected outcome of an experiment in some limiting situation, or an interpretation of the experimental data in terms of a given theoretical model.

The present discussion focuses principally on the case of *water* as a solvent, and the generic subscript "*svt*" will sometimes be replaced here by "*wat*" to underline this restriction. A more general discussion of the surface potential for arbitrary solvents was provided in Section 2.2, including available data for non-aqueous solvents in Table 2.1. The calculation of $\chi^{\ominus}_{wat}$ from the real hydration free energy $\hat{\mathcal{G}}^{\ominus}_{H,wat}$ of the proton (Section 5.15) *via* estimation of the corresponding intrinsic value $\mathcal{G}^{\ominus}_{H,wat}$ (Eq. 5.44) will not be considered here, except when discussing literature sources explicitly reporting $\chi^{\ominus}_{svt}$ values based on this approach (Point A_{13} below). Instead, the evaluation of $\mathcal{G}^{\ominus}_{H,wat}$, including estimates derived from $\hat{\mathcal{G}}^{\ominus}_{H,wat}$ and $\chi^{\ominus}_{svt}$, will be discussed in Section 5.18. In principle, the calculation of $\chi^{\ominus}_{wat}$ from the real absolute potential $\hat{\mathcal{V}}^{\ominus}_{H,wat}$ of the reference hydrogen electrode in water (Section 5.14) *via* estimation of the corresponding intrinsic value $\mathcal{V}^{\ominus}_{H,wat}$ (Eq. 5.45) would also be possible. However, no direct tentative determination of the latter quantity has been reported to date (see, however, Ref.[483] discussed in Section 5.14 for an indirect estimation relying on a corrected $\mathcal{G}^{\ominus}_{H,wat}$ estimate from Refs.[1875,1876]). On the other hand, since the air-liquid interfacial potential of a liquid is an experimentally-elusive quantity, the calculation of $\chi^{\ominus}_{wat}$ using theoretical models as a form of extra-thermodynamic assumption is included here, but with a limited amount of details. The discussion of these approaches is the main concern of Chapters 3 and 6 (Sections 3.3.5, 6.2.4 and 6.3.4).

The methods attempting the determination of $\chi^{\ominus}_{svt}$ for a given solvent, prominently water, and the involved extra-thermodynamic assumptions can be broadly classified as follows[135,266,272,277,455,1762]:

(A_{13}) Comparison between the experimental *real single-ion solvation free energy of an ion* (Section 5.15) and *an estimate for the corresponding intrinsic value*[272,277,406–410,412–414,416,418–420,2219] (Eqs. 2.6 and 2.8). Since the associated conventional value is typically available experimentally (Section 5.16), formulating an estimate for the intrinsic solvation free energy of any ion is equivalent to fixing the value of the intrinsic solvation free energy $\mathcal{G}^{\ominus}_{H,svt}$ of the proton in the solvent considered. The extrathermodynamic assumption involved in this approach is then the one applied in the evaluation of $\mathcal{G}^{\ominus}_{H,svt}$ (Section 5.18). Due to the considerable uncertainty in estimated $\mathcal{G}^{\ominus}_{H,svt}$ values, this approach is notoriously inaccurate[272,277,454,465].

(B_{13}) Measurement of the *temperature derivative* $\partial_T \chi^{\ominus}_{svt}$ of $\chi^{\circ}_{svt}(T)$ at T^- and extrapolation to the critical temperature[272,415,446,448,465,2174]. Although the absolute value of the air-liquid interfacial potential $\chi^{\circ}_{svt}(T)$ of a pure solvent at a given temperature T cannot be accessed directly, it has been suggested that its temperature derivative $\partial_T \chi^{\ominus}_{svt}$ at T^- can be measured experimentally. The question whether this derivative is actually accessible or also elusive remains, however, relatively intricate, and will be addressed in Chapter 7. In the context of water, the basic experimental setup of this approach is that of a Voltaic cell measurement (Section 4.4.4), more specifically a Kenrick cell (Figures 4.16 and 4.17), where the temperature of the solution at the air gap can be controlled and varied with the constraint that the electric field remains zero within the gap at any temperature[448,2174]. For example, in Ref.[448], the cell setup is

$$\begin{aligned} & M_1(T^-) \mid L_2(T^-) \mid L_3'(T^-) \\ & \square\, L_4'(T) \mid L_5'(T^-) \mid L_6(T^-) \mid M_7(T^-) \quad , \end{aligned} \tag{5.48}$$

or, more specifically,

$$\begin{aligned} & \mathrm{Hg}_\ell \mid \mathrm{KCl}_{aq}(\mathrm{Hg_2Cl}_{2,s}) \mid \mathrm{KCl}_{aq} \\ & \square\, \mathrm{KCl}_{aq} \mid \mathrm{KCl}_{aq} \mid \mathrm{KCl}_{aq}(\mathrm{Hg_2Cl}_{2,s}) \mid \mathrm{Hg}_\ell \quad , \end{aligned} \tag{5.49}$$

while in Ref.[2174] the setup is

$$M_1(T^-) \mid M_2'(T^-) \mid L_3(T^-) \,\square\, L_4'(T) \mid M_5(T) \mid M_6(T) \quad , \tag{5.50}$$

or, more specifically,

$$\mathrm{Cu}_s \mid \mathrm{Ag}_s \mid \mathrm{KCl}_{aq}(\mathrm{AgCl}_s) \,\square\, \mathrm{KCl}_{aq}(\mathrm{AgCl}_s) \mid \mathrm{Ag}_s \mid \mathrm{Cu}_s \quad . \tag{5.51}$$

The first approach (Eqs. 5.48 and 5.49) has the disadvantage of presenting a thermal gradient at the L_4'-L_5' junction, which may induce a thermoelectric effect, as well as the appearance of liquid-junction potentials at the L_2-L_3' and L_5'-L_6 junctions. These two sources of errors may affect the accuracy of the results[272]. In the second approach (Eqs. 5.50 and 5.51), the thermal gradient occurs between the two copper wires connected to the potentiostat, and the corresponding thermoelectric effect can be assumed to be negligible[2174]. In both cases, the temperature derivative $\partial_T \Delta\mathcal{V}^{\ominus}$ of the compensating potential difference of such a cell is then used to evaluate $\partial_T \chi^{\ominus}_{wat}$ by extrapolation to low electrolyte concentration (KCl_{aq} in the above cells) or neglect of concentration effects. In the first approach (Eqs. 5.48 and 5.49), the temperature derivatives of the two quantities are simply equal. However, in the second approach (Eqs. 5.50 and 5.51), $\partial_T \Delta \mathcal{V}^\ominus$ involves, in addition to $\partial_T \chi^\ominus_{wat}$, a contribution from the entropy change of the reaction involved in the half-cell at the right of the air gap, namely

$$\mathrm{Ag}_s + \mathrm{Cl}^-_{aq} \rightarrow \mathrm{AgCl}_s + e^-_{\mathrm{Ag}} \quad . \tag{5.52}$$

This entropy change may be evaluated based on conventional data and on an estimate for the intrinsic partial molar entropy $s^\ominus_{H,wat}$ of the aqueous proton (set to -23.0 J·mol^{-1}·K^{-1} in Ref.[2174] or -22.2 J·mol^{-1}·K^{-1} in Ref.[272]). Because this second approach relies on an estimated intrinsic value $s^\ominus_{H,wat}$ to evaluate $\partial_T \chi^\ominus_{wat}$, it involves additional (hidden) extra-thermodynamic assumptions. However, a simple consistency check against the corresponding real value $\hat{s}^\ominus_{H,wat}$ *via*

$$\hat{s}^\ominus_{H,svt} = s^\ominus_{H,svt} - F \partial_T \chi^\ominus_{svt} \quad , \tag{5.53}$$

using $s^\ominus_{H,wat} = -23.0$ J·mol^{-1}·K^{-1} (as in Ref. [2174]), $\hat{s}_{H,wat} = 18.9$ J·mol^{-1}·K^{-1} (as deduced from $\hat{\mathcal{S}}^\ominus_{H,wat} = -90.1$ J·mol^{-1}·K^{-1} from Ref.[415] in Table 5.15 and $s^\ominus[\mathrm{H}^+_g] = 108.95$ J·mol^{-1}·K^{-1} from Table 5.9) with $\partial_T \chi^\ominus_{wat} = -0.4$ mV·K^{-1} (deduced in the experiment of Ref.[2174]) gives very good agreement (error of 3.2 J·mol^{-1}·K^{-1} in Eq. 5.53). Subsequent studies or/and reviews[272,465] have proposed tentative corrections for various error sources in the above determinations (*e.g.* Ref.[465] based on Ref.[2174] and Ref.[272] based on Refs.[448,2174]). Note also that the value of $\hat{s}^\ominus_{H,wat}$ based on Ref.[415] (1970) was not available at the time of the measurements of Ref.[2174] (1965). In other words, if Ref.[2174] in 1965 had taken $\hat{s}^\ominus_{H,wat}$ based on Ref.[415] (which was not yet available in 1965), and used his $s^\ominus_{H,wat}$ estimate of -23 J·mol^{-1}·K^{-1}, he could have applied Eq. 5.53 to get $\partial_T \chi^\ominus_{svt} = -0.43$ V·K^{-1} without doing any measurements. This very same approach has actually been used by others[415], namely to evaluate $\partial_T \chi^\ominus_{svt}$ from a comparison between the experimental real single-ion solvation entropy of an ion (Section 5.15) and an estimate for the corresponding intrinsic value (temperature derivatives of Eqs. 2.6 and 2.8), *i.e.* using an equation analogous to Eq. 5.53 for the selected ion, namely

$$\Delta_s \hat{S}^\ominus[I^z_g] = \Delta_s S^\ominus[I^z_g] - zF \partial_T \chi^\ominus_{svt} \quad , \tag{5.54}$$

or in the particular case of the proton,

$$\hat{\mathcal{S}}^\ominus_{H,svt} = \mathcal{S}^\ominus_{H,svt} - F \partial_T \chi^\ominus_{svt} \quad . \tag{5.55}$$

This approach is analogous to A_{13}, here in terms of temperature derivatives, and involves the extra-thermodynamic assumption applied in the evaluation of the intrinsic single-ion solvation entropy of the chosen ion

(Section 5.18), in addition to the one specific to the present approach (see below). Once $\partial_T \chi^{\ominus}_{svt}$ has been evaluated, this derivative is used to extrapolate the air-liquid interfacial potential $\chi^{\circ}_{svt}(T)$ from T^{-} (at pressure P°) to the critical temperature of the solvent ($T_c = 647.14$ K, *i.e.* 373.99°C, for water; Table 1.1), where it is assumed that the interfacial potential vanishes. Typical functional forms employed for this extrapolation are[448,465] $\partial_T \chi^{\circ}_{svt}(T) = a - bT^{-2}$ or[272] $\partial_T \chi^{\circ}_{svt}(T) = -bT^{-2}$. The extra-thermodynamic assumption involved in this extrapolation is the expectation that surface effects at an air-liquid interface should vanish at the critical point, where liquid and gas phases are characterized by indistinguishable properties. However, this approach involves other assumptions in practice: (*i*) the electrochemical determination of $\partial_T \chi^{\ominus}_{svt}$ (Refs. [272, 446, 448, 465, 2174]) assumes that the interfacial-potential derivative measured for a dilute electrolyte solution is identical to that of the pure solvent (see, however Refs.[446,465]; Ref.[465] suggests a correction for finite-concentration effects based on the data of Ref.[2174] and on the analysis of Ref.[446]); (*ii*) the cells used for these measurements involve thermal gradients and the corresponding thermoelectric effects and, possibly liquid-junction potentials, are neglected (this gradient can occur at the connecting wires[272,446,2174], where it is thought to be negligible, or at a liquid junction[448], where it might represent a non-negligible error source[272]); (*iii*) the determination of $\partial_T \chi^{\ominus}_{svt}$ by comparison of real and intrinsic solvation entropies[415] relies on other extra-thermodynamic assumptions specific to the evaluation of the intrinsic single-ion solvation entropies (Section 5.18); (*iv*) the extrapolation of the interfacial potential from T^{-} to T_c relies on the fitting of $\partial^{\circ}_T \chi_{svt}(T)$ to an arbitrary (*ad hoc*) function[465]; (*v*) the extrapolation is performed at constant pressure P°, although the critical point generally occurs at a different pressure P_c ($P_c = 220.6$ bar for water; Table 1.1), and would be more meaningful if it was performed along the boiling curve of the liquid, *i.e.* involving a simultaneous variation and extrapolation of the pressure and temperature (in fact, the extrapolation at P° leads for water to a state point at T_c where the liquid is a thermodynamically-unstable phase and the concept of an air-liquid interface makes little physical sense).

(C_{13}) Measurement of the *dependence* of $\chi^{\ominus}_{stn}(c_e)$ *on the concentration* c_e of *dissolved electrolytes* in a solution (*stn*) of the solvent and *extrapolation to zero concentration*[441–447, 465] (see Refs.[441, 442] for early work in this area and Ref.[465] for an overview). Although the absolute value of the air-solution interfacial potential $\chi^{\ominus}_{stn}(c_e)$ of an electrolyte solution at concentration c_e cannot be accessed directly, it has been suggested that its concentration dependence can be measured experimentally within an unknown constant. The question whether the derivative $\partial_{c_e} \chi^{\ominus}_{stn}$ is actually accessible or elusive remains, however, relatively intricate, and will be addressed in Chapter 7. In the context of water, the basic experimental setup of this approach is again that of a Voltaic cell measurement

(Section 4.4.4), more specifically either a Kenrick cell in Refs.[443,445–447], or the Kelvin probe method in Ref.[444] (see Figures 4.16 and 4.17). In the approach of Ref.[443], the cell setup is

$$M_1 \mid L_2 \,\square\, L_3'(c_e) \mid L_4 \mid M_5 \quad , \tag{5.56}$$

or, more specifically,

$$\begin{aligned} &\mathrm{Hg}_\ell \mid \mathrm{NaCl}_{aq}(\mathrm{Hg_2Cl}_{2,s}) \\ &\square\, \mathrm{NaCl}_{aq}(c_e) \mid \mathrm{NaCl}_{aq}(\mathrm{Hg_2Cl}_{2,s}) \mid \mathrm{Hg}_\ell \quad . \end{aligned} \tag{5.57}$$

In the approach of Refs.[445–447], the setup is

$$M_1 \,\square\, L_2(c_e) \mid M_3 \quad , \tag{5.58}$$

or, more specifically, in the case of Ref.[445],

$$\mathrm{Hg}_\ell \,\square\, \mathrm{HCl}_{aq}(c_e)(\mathrm{AgCl}_s) \mid \mathrm{Ag}_s \quad , \tag{5.59}$$

where the air gap is actually filled with nitrogen gas. Finally, in the approach of Ref.[444], the setup is

$$M_1 \,\square\, L_2(c_e) \mid M_3 \quad , \tag{5.60}$$

or, more specifically (as tentatively inferred from the original article by the authors of this book),

$$\mathrm{Ag}_s \,\square\, \mathrm{KCl}_{aq}(c_e)(\mathrm{AgCl}_s) \mid \mathrm{Ag}_s \quad . \tag{5.61}$$

In all cases, the concentration dependence $\partial_{c_e} \Delta\mathcal{V}^\ominus$ of the compensating potential difference of such a cell is then used to evaluate $\partial_{c_e}\chi^\ominus_{wat}$. By considering solutions at different concentrations c_e, it is possible to establish the dependence of $\Delta\chi^\ominus_{stn}(c_e) = \chi^\ominus_{stn}(c_e) - \chi^\ominus_{wat}$ on c_e, *i.e.* with the anchoring point $\Delta\chi^\ominus_{stn}(0) = 0$. The extra-thermodynamic assumption involved in this approach occurs in the form of a model predicting the concentration-dependent distribution of ions at the air-solution interface, and thus $\Delta\chi^\ominus_{stn}(c_e)$, with $\chi^\ominus_{svt}$ entering as a parameter. The concentration dependence of $\Delta\mathcal{V}^\ominus$ in such a cell can only arise from the interfacial potential at the non-polarizable interface between the solution and the metal electrode (Eqs. 5.58 or 5.60) or another solution (Eq. 5.56). The basic model relies on assuming that the interfacial-potential derivative measured at this junction is identical to that which would be observed at an air-solution interface, in which case one might write $\partial_{c_e}\chi^\ominus_{stn} = \partial_{c_e}\Delta\mathcal{V}^\ominus$. The model involved in Ref.[443] involves a correction for the diffusion potential (presence of a liquid-liquid junction) between two electrolytes at different concentrations in this experimental setup and an extrapolation to zero concentration using an arbitrary curve (about quadratic). The model involved in Refs.[444–447] is due to Ref.[425], and involves a representation of the electric potential in the vicinity of the electrode based on the linearized Poisson-Boltzmann equation (similar in spirit to the Gouy-Chapman model mentioned in Section 4.3.1), the approximation of a concentration-independent Debye screening length, and a linear extrapolation to zero concentration (of chloride ions). However, this approach involves other assumptions in practice: (*i*) the electrochemical determination of $\Delta\chi^{\ominus}_{stn}(c_e)$ in Refs.[444–447] assumes that the interfacial-potential derivative measured at a metal-solution interface is identical to that which would be observed at an air-solution interface; (*ii*) the correction of Ref.[443] for the diffusion potential is approximate; (*iii*) in both cases, the determination requires an extrapolation to zero concentration, but it has been suggested that sharp variations may in fact exist in the very low concentration regime[2220,2221], because the dimension of the surface region is infinitesimal compared to that of the bulk region, so that even very small bulk concentrations of an electrolyte may still be associated with significant surface concentrations of the corresponding ions.

(D$_{13}$) Measurement of the *dependence* of $\chi^{\ominus}_{stn}(c_o)$ *on the concentration* c_o of *tensioactive organic molecules* in a solution (stn) of the solvent and *extrapolation to zero surface fraction of organic molecules*[448–454] (see Ref.[448] for early work in this area and Ref.[453] for an overview). Although the absolute value of $\chi^{\ominus}_{stn}(c_o)$ cannot be measured, its variation upon altering the organic solute concentration can be monitored. In the context of water, the basic experimental setup of this approach is that of a Voltaic cell measurement (Section 4.4.4), more specifically the ionizing probe method (Figure 4.16). In the approach of Ref.[451] (see also Ref.[452]), the cell setup is

$$M_1 \,\square\, L_2 \mid M_3 \quad , \tag{5.62}$$

or, more specifically (as tentatively inferred from the original article by the authors of this book),

$$^{226}\mathrm{Ra} \,\square\, \mathrm{NaCl}_{aq}(\mathrm{AgCl}_s) \mid \mathrm{Ag}_s \quad , \tag{5.63}$$

where the use of a ^{226}Ra radioactive radium electrode ensures the ionization of the gas in the air gap. The concentration dependence $\partial_{c_o}\Delta\mathcal{V}^{\ominus}$ of the corresponding compensating potential difference is then used to evaluate $\partial_{c_o}\chi^{\ominus}_{svt}$. The extra-thermodynamic assumption involved in this approach relies on some expected relationship between the concentration of organic molecules and the surface potential of the solution *via* the surface fraction of organic molecules. In Ref.[451], benzene is used as a solute, the surface fraction of benzene molecules at a given benzene concentration is estimated using the Gibbs adsorption equation[456], and the change in the surface potential caused by benzene adsorption is interpreted as a scaling of the pure-water surface potential by the relative surface fraction left to the water molecules *via* the Helmholtz equation[455]. It is thus assumed that benzene molecules do not contribute to the surface potential. In Ref.[452], terphenyl esters are used as solutes. The extrapolation method is similar to that of Ref.[451], except that a contribution of the terphenyl esters to the surface potential of the solution (proportional to their surface fraction) is also included, considering molecular dipole moment estimates based on idealized geometries and approximate bond dipole moments. These two methods rely, among other approximations, on the assumption that the surface potential associated with a surface mixture of water with an organic species is additive in the contributions of the two species weighted by their surface fractions, *i.e.* on the Helmholtz equation. Obviously, this neglects the presence of mutual correlations between the orientational (and possibly spatial) preferences of the two species. A few other sources[449, 450, 453, 454] also report estimates based on related approaches. In Refs.[449, 450], dependences of the surface potential and of the surface tension on c_o are correlated, and an estimate for $\chi^{\ominus}_{wat}$ is derived from an assumption of zero surface potential at zero surface tension. The latter assumption is rather difficult to justify. Although the absence of surface tension for a given concentration c_o indicates that the chemical potential differences between the two species (water and organic solute) are identical in the bulk and surface regions, this by no means guarantees the absence of preferential positional and orientational ordering in the latter region. In particular, the extremely high value[449,450] of +1.0 V obtained for water using this approach has been shown to be unlikely[2222]. In Ref.[453], a value is suggested based on an overview of previous data involving experiments of the present type. In Ref.[454], an experimental estimate is obtained for the difference in the air-liquid interfacial potentials of water and ethanol based on measurements with Kenrick cells, and combined with an estimate[2223] for the air-liquid interfacial potential of ethanol, itself based on the so-called immersion method[2223].

(E_{13}) Measurement of the *dependence* of $\chi^{\ominus}_{svt,neq}(t)$, corresponding to a non-equilibrium value at P° and T^{-}, *on the age* t of an interface initially prepared in a disordered state[2224] and *extrapolation to zero time* $t = 0$. In this approach, water jet experiments are used to produce water droplets with off-equilibrium surface properties, and the variation $\chi^{\ominus}_{svt,neq}(t)$ of the air-liquid interfacial potential with time is monitored. The results are then extrapolated to $t = 0$. The extra-thermodynamic assumption involved in this approach is the expectation that the surface of the droplets is entirely disordered at the onset of the experiments ($t = 0$) and that the corresponding surface potential should vanish. In these experiments, which are performed in the context of water, the age of the droplets is determined from the jet velocity profile in order to ensure the absence of a dead time. Upon aging, the surface potential recovers its equilibrium value within times on the order of 3 ms. The relaxation of $\chi^{\ominus}_{wat,neq}(t)$ is fitted to an exponential curve (relaxation time of 0.44 ms in Ref.[2224]) and the assumption $\chi^{\ominus}_{wat,neq}(0) = 0$ leads to the required estimate for $\chi^{\ominus}_{wat}$. Of course, the assumption of initial disorder may be questioned, *i.e.* it is unclear to which extent the surface order can be destroyed by a plain mechanical manipulation. The accuracy of this approximation may actually depend significantly on the detailed setup of the experiment. Recall also that even in the absence of preferential orientation of the surface molecules, electronic polarization effects (and, possibly, geometric distortions), which have much faster relaxation times, may also contribute to the surface potential (see *e.g.* the non-polar liquids in Table 2.1). In earlier work, a similar assumption of surface disorder has also been used in the context of water adsorbed at the surface of a porous body[2225]. In this approach, water droplets fall from a water reservoir inside a copper-dropper into a copper tube (in some experiments, a glass dropper with a zinc rod and a zinc tube are used instead). The copper (zinc) tube is itself coated with wet filter paper. A potential difference is applied between the copper dropper and the tube. In the limit of zero electric current through the external circuit, the compensating potential difference is assumed to be equal to $\chi^{\ominus}_{wat}$. The justification of this procedure relies on the assumption that there is no surface potential at the surface of water soaked into filter paper, due to an assumed randomization of the orientation of surface water molecules. Here also, the assumption of surface disorder as well as the basic setup of the experiment have been subsequently questioned[277,454]. In the opinion of the authors of this book, even the theoretical justification of this approach is unclear, considering that dipolar molecules cannot carry charges across an electric circuit. The resulting current is thus probably rather due to an excess of H^+ or OH^- ions, or to charged impurities.

(F_{13}) Measurement of the *potential of zero charge* at a mercury-solvent interface and application of a *correction* for the effect of mercury, rather than air, at the interface[461, 1713, 1762, 2226] (see also Ref.[2226] for early work in this area). In the context of water, the basic experimental setup is that of the determination of a potential of zero charge (Section 4.4.5), where the solution in contact with the metal (mercury) contains an organic solute (ethylene glycol) at a certain concentration c_o. The concentration dependence of the potential of zero charge along with an estimate for the intrinsic hydration free energy $\mathcal{G}^{\ominus}_{H,wat}$ of the proton are then used to evaluate $\chi^{\ominus}_{wat}$. More specifically, the analysis of the dependence of the lateral Volta potential difference $\Delta\overline{\psi}^{\ominus}_{M,S}$ at zero charge on c_o is interpreted *via* Eqs. 4.408 and 4.409, namely

$$\Delta\overline{\psi}^{\ominus}_{M,S} = \Delta\tilde{\chi}^{\ominus}_{M,S} - \Delta\chi^{\ominus}_{M,S} = (\tilde{\chi}^{\ominus}_{S(M)} - \chi^{\ominus}_{S}) - (\tilde{\chi}^{\ominus}_{M(S)} - \chi^{\ominus}_{M}) \quad , \quad (5.64)$$

as an alteration of $\tilde{\chi}^{\ominus}_{S}$ by surface adsorption of the organic molecules, under the assumption that $\tilde{\chi}_M^{\ominus}$ remains unaffected. The use of an approach similar to that presented under Point D_{13} (Gibbs adsorption equation[456], Helmholtz equation[455], assumption of negligible contribution of the ethylene glycol molecules to $\tilde{\chi}_S^{\ominus}$) leads to an estimate for the value of $\tilde{\chi}_S^{\ominus}$ in the absence of ethylene glycol. Eq. 5.64 is then sufficient to evaluate $\chi_S^{\ominus}$, *i.e.* $\chi_{wat}^{\ominus}$ in the present case, based on $\Delta\overline{\psi}_{M,S}^{\ominus}$ measured in the absence of ethylene glycol, provided that an estimate for $\tilde{\chi}_M^{\ominus} - \chi_M^{\ominus}$ can be formulated. This difference is evaluated as[1713] (Eqs. 4.291, 4.296, 4.409 and 4.411)

$$\tilde{\chi}_M - \chi_M = \Delta_H \overline{\mathcal{V}}_{M,S}^{\ominus} - \omega_M^{\ominus} + \tilde{\chi}_S + \mathcal{V}_{H,wat}^{\ominus} \quad , \tag{5.65}$$

using an estimate for $\mathcal{V}_{H,wat}^{\ominus}$ corresponding to the value for $\mathcal{G}_{H,wat}^{\ominus}$ recommended by Ref.[413] (-1100.4 kJ·mol^{-1}, based on a range of available estimates for $\chi_{wat}^{\ominus}$, the constraint of having a positive value for $\chi_{wat}^{\ominus}$ considering its negative temperature derivative, and the value of $\hat{\mathcal{G}}_{H,wat}^{\ominus}$ from Ref. [411]). The method relies on the assumption that the surface potential associated with a surface mixture of water with an organic species is additive in the contributions of the two species weighted by their surface fractions, *i.e.* on the Helmholtz equation. This neglects the presence of mutual correlations between the orientational (and possibly spatial) preferences of the two species. In addition, the possible influence of the ethylene glycol on $\tilde{\chi}_M^{\ominus}$ is neglected. More importantly, the method requires an estimate for $\mathcal{V}_{H,wat}^{\ominus}$ (or, equivalently, $\mathcal{G}_{H,wat}^{\ominus}$) as input, similarly to the approaches of Point A_{13}, and adding the extra-thermodynamic assumption inherent to this specific determination. The authors[1713] argue, however, that the present approach is advantageous because the independent evaluation of $\tilde{\chi}_S^{\ominus}$ and $\tilde{\chi}_S^{\ominus} - \chi_M^{\ominus}$ provides a way to assess the relative magnitudes of these contributions to $\chi_S^{\ominus}$.

(G_{13}) Use of *theoretical models*, approximate analytical models[1446,1454,1457,2227], AT simulations[430,465,470,707,833,850,947,973,1440–1446,1451,1473], classical density-functional theory[1447] (DFT) or QM computations[1448–1450]. The AT simulation approaches to evaluate $\chi_{svt}^{\ominus}$ are described in Section 3.3.5, and results briefly discussed in Section 6.2.4. The corresponding QM computation approaches are described and discussed in Section 6.3.4. Two major issues concerning these types of calculations are: (*i*) the ambiguity in the calculated potential, namely internal *vs.* external (Section 4.3.8), apparent in both AT and QM evaluations (Sections 3.3.5 and 6.3.4), and resulting in a controversy regarding the appropriate way to evaluate $\chi_{svt}^{\ominus}$ based on the sampled solvent configurations (P- *vs.* M-schemes; Sections 3.3.3-3.3.5); (*ii*) the dependence of the results on the sample geometry and on the nature of the confinement potential (Section 3.3.5); (*iii*) the difficulty of reaching appropriate system sizes and simulation timescales in QM computations (Section 3.4); (*iv*) the use of implicit-polarization models in AT simulations, likely to result in overestimated surface potential values (Section 6.2.4). The simple analytical models represent early attempts at modeling the surface structure of liquids and evaluating the associated surface potential. In a first type of approach[1446,1454], a surface water molecule is represented as a sphere bearing a point dipole and a point quadrupole, embedded in a continuous medium of heterogeneous dielectric permittivity, linearly varying from the bulk value to one across the surface. This results in an approximate expression for the (free) energy as a function of molecular orientation. In a second type of approach[2227], a set of surface water molecules is represented as Lennard-Jones spheres in contact, with geometric packing parameters inferred from x-ray scattering data, bearing a point dipole and a point quadrupole, along with a mean-field representation of the bulk solvent. This results in an approximate expression for the free energy as a function of the distribution of molecular orientations. In a third type of approach[1457], a heuristic equation is developed directly for the free energy of a polarized surface, that attempts to encompass the interaction between constituting point dipoles plus quadrupoles along with an entropy penalty upon alignment and, possibly, contributions from adsorbed ions. In a fourth of approach[1447], classical DFT is used to formulate an equation for the free energy of a polarized surface accounting for intermolecular bulk-like interactions (bulk equation of state), intermolecular hydrogen bonding, a gradient correction for the effect of the anisotropic surface environment and local anisotropy in the preferential orientations of the water molecules. The parameters entering into this description are calibrated based on the results of AT simulations (from Ref. [1446], using the SPC/E water model[1275]). In the four cases, an estimate for the average orientation of the surface molecules is obtained and the corresponding field is integrated, leading to the required estimate for $\chi^{\ominus}_{svt}$. The above analytical models involve numerous assumptions, *e.g.* choice of approximate molecular dipole and quadrupole moments and neglect of electronic polarization (all approaches), continuum-electrostatics or mean-field approximation at the microscopic level (for Refs. [1446, 1454, 2227]), assumed surface geometry and Lennard-Jones interactions (for Ref.[2227]), heuristic or empirically-parameterized free-energy expression (for Refs. [1447, 1457]).

Note that approaches similar to that of Point D_{13} have also been used to evaluate *differences in air-liquid interfacial potentials* between water and other water-miscible solvents[413,454,473], by measurement of the dependence of $\chi^{\ominus}_{mix}(x)$ for the solvent mixture on the mole fraction x of the cosolvent in the mixture and extrapolation to $x = 0$ or $x = 1$. The results can be used to assess the internal consistency between $\chi^{\ominus}_{svt}$ values estimated for different solvents *via* other approaches (*e.g.* A_{13}-G_{13} above), or to convert an estimate of $\chi^{\ominus}_{svt}$ for a given reference solvent into a corresponding estimate for another solvent (as done for water in Ref.[454] using ethanol as a reference solvent, based on a corresponding estimate from Ref.[2223]; see Point D_{13}). The question whether such differences are experimentally elusive is again an intricate issue, which will be returned to in Chapter 7. A collection of 65 available estimates for the standard air-liquid interfacial potential $\chi_{wat}^{\ominus}$ of water or/and its temperature derivative $\partial_T \chi_{wat}^{\ominus}$, labeled 1X-65X for the ease of further reference, is presented in Table 5.18. Only sources explicitly reporting a value of $\chi_{wat}^{\ominus}$ or/and $\partial_T \chi_{wat}^{\ominus}$ are listed, although it is possible to construct many additional $\chi_{wat}^{\ominus}$ (or $\partial_T \chi_{wat}^{\ominus}$) estimates of type A_{13} (B_{13}) from other intrinsic proton hydration free-energy estimates $\mathcal{G}_{H,wat}^{\ominus}$ listed subsequently in Table 5.19 (or intrinsic aqueous proton entropies $s_{H,wat}^{\ominus}$ listed subsequently in Table 5.20) using real data from Table 5.15, *via* Eqs. 5.44 or 5.53. On the other hand, intrinsic proton hydration free energies (or intrinsic aqueous proton entropies) deduced from the values given in the present table are reported in Table 5.19 (or Table 5.20).

The values corresponding to theoretical approaches (Point G_{13}) are reported in the form of external surface potentials (Section 4.3.8), corresponding to the M-scheme (Sections 3.3.3-3.3.5). Whenever possible, values reported in the original reference in the form of internal potentials (P-scheme) were converted to external values *via* Eq. 3.86. In 7 cases, this conversion could not be performed in the absence of an obvious estimate for the exclusion potential $\xi_{AT}^{\ominus}$ corresponding to the solvent representation employed in the model (*e.g.* polarizable water model or quantum-mechanical representation). Inter-

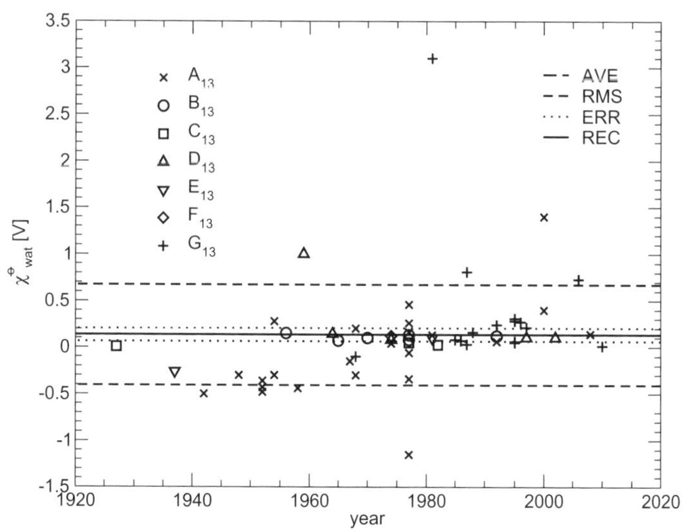

Figure 5.6 *Available estimates for the standard air-liquid interfacial potential $\chi_{wat}^{\ominus}$ of water.* This graph illustrates the data reported in Table 5.18 concerning $\chi_{wat}^{\ominus}$, the estimates being sorted by publication year and by methods. A corresponding graph for $\partial_T \chi_{svt}^{\ominus}$ is not shown in view of the sparseness of the data (see numerical values in Table 5.18).

nal values, from the original source or *via* conversion from the external value, whenever possible, are also reported between parentheses. No corresponding distinction was made concerning the temperature derivatives, under the assumption that $\xi^{\ominus}_{AT}$ is (nearly) temperature-independent. The data presented in Table 5.18 concerning $\chi^{\ominus}_{wat}$ is illustrated graphically in Figure 5.6. A corresponding illustration concerning $\partial_T \chi^{\ominus}_{wat}$ is omitted in view of the sparseness of the data.

The average of the 58 external estimates for $\chi^{\ominus}_{wat}$ is 0.13 V, with a standard deviation of 0.54 V and an estimated error on the mean of ± 0.07 V, assuming that all measurements are independent. The average of the 10 corresponding estimates for $\partial_T \chi^{\ominus}_{wat}$ is -0.60 mV·K^{-1}, with a standard deviation of 0.30 mV·K^{-1} and an estimated error on the mean of 0.09 mV·K^{-1}. Considering the discussion of potential shortcomings in the various determination methods, the authors of the present book tend to trust preferentially estimates based on approaches B_{13}-F_{13}, except Sets 35X and 40X, which have been seriously questioned[277,454,2222]. The 15 corresponding $\chi^{\ominus}_{svt}$ estimates present an average of 0.09 V with a standard deviation of 0.04 V and an estimated error on the mean of 0.01 V. For the 7 corresponding $\partial_T \chi^{\ominus}_{svt}$ the average is -0.48 mV·K^{-1} with a standard deviation of 0.23 mV·K^{-1} and an estimated error on the mean of 0.06 mV·K^{-1}. Also considering the relationship with the absolute potential of the reference hydrogen electrode, and with real and intrinsic proton solvation properties (Sections 5.14, 5.15 and 5.18), the authors of the present book recommend the values $\chi^{\ominus}_{wat} = 0.13 \pm 0.10$ V and $\partial_T \chi^{\ominus}_{wat} = -0.42 \pm 0.10$ mV·K^{-1} for its temperature derivative. In particular, the indicated errors are approximately compatible with worst-case errors on $\hat{\mathcal{G}}^{\ominus}_{H,wat}$, $\mathcal{G}^{\ominus}_{H,wat}$, $\hat{\mathcal{S}}^{\ominus}_{H,wat}$ and $\mathcal{S}^{\ominus}_{H,wat}$ (Sections 5.15 and 5.18; 0.15 V for $\chi^{\ominus}_{wat}$ and 0.07 mV·K^{-1} for $\partial_T \chi^{\ominus}_{wat}$, respectively). These values and error estimates for $\chi^{\ominus}_{wat}$ are those that will be retained within the final compilation of recommended data in Section 5.19. Note that these quantities are independent of any standard-state definition (Section 4.2.4), including the choice of a variant for the solutes (Section 4.2.5) and of a temperature for the ideal-gas electron (Section 4.2.13), except for the specification of P° and T^{-} and the condition that $\chi^{\ominus}_{wat}$ and $\partial_T \chi^{\ominus}_{wat}$ pertain to a planar air-water interface at equilibrium.

Although there is a *considerable uncertainty* on the actual value of $\chi^{\ominus}_{wat}$, so that even the sign of this quantity is arguably not known with absolute certainty, the results presented in Table 5.18 and Figure 5.6, along with the above considerations concerning the relative reliabilities of different experimental approaches, clearly support a positive value close to 0.1 V. Note in this respect that the negative sign observed for $\partial_T \chi^{\ominus}_{wat}$ in the 10 available independent determinations (7 experimental and 3 theoretical) of this quantity (Table 5.18) provides strong evidence, if not a definitive proof, for a positive value of $\chi^{\ominus}_{wat}$, as suggested in Ref.[413]. The reason is that due to enhanced thermal motion, a temperature increase is very likely to decrease the influence of the (predominantly dipole-orientational) effects at the origin of the interfacial potential, and thus to decrease the magnitude of $\chi^{\ominus}_{wat}$ itself. However, if $\chi_{wat}^{\ominus}$ was negative, a negative value of $\partial_T \chi_{wat}^{\ominus}$ would imply a temperature-induced enhancement of these effects which, although not formally impossible over a certain temperature interval, remains nevertheless improbable.

As previously discussed in Sections 2.2 and 4.3.8, one may distinguish *four main mechanisms* altering the properties of the interface region relative to the bulk in the case of a dipolar protic solvent, and resulting into corresponding contributions to the air-liquid interfacial potential: (*i*) preferential orientation of the solvent dipoles; (*ii*) anisotropy of electronic polarization effects; (*iii*) distortion of the molecular geometry; (*iv*) preferential adsorption or exclusion of ions issued from autoprotolysis.

The contribution of *preferential dipole orientation* is most likely the dominant one in the case of water. The results of a number of independent AT simulations of (spherical or planar) water-vapor interfaces[465,470,707,833,850,1440–1443,1445,1446,1451,1473] indicate that: (*i*) on the vapor side of the interface, water molecules preferentially project one hydrogen atom towards the vapor phase; (*ii*) on the liquid side of the interface, water molecules preferentially orient with their symmetry axis close to the plane of the interface, but with both hydrogen atoms pointing slightly towards the liquid phase; (*iii*) due to the higher density on the latter side of the interface, there is an excess surface dipole moment pointing towards the liquid phase. Thus, broadly speaking, water molecules preferentially orient the positive ends of their dipoles towards the bulk liquid. These observations are in line with a positive value of $\chi_{wat}^{\ominus}$, as indeed found for all classical solvent models considered in Table 5.18 (Sets 46X-57X and 63X) when applying the M-scheme to obtain an external potential. Note, however, that AT simulations relying on implicit-polarization models are likely to overestimate $\chi_{wat}^{\ominus}$, because their atomic partial charges are artificially enhanced to account for the mean effect of electronic polarization in the bulk (Sections 3.3.2 and 6.2.4). It is also interesting to observe that the behavior of water is atypical for a dipolar solvent (protic or aprotic), considering that the $\chi_{wat}^{\ominus}$ values are probably negative for most other solvents (Table 2.1). For non-aqueous solvents, the typically negative value of $\chi_{svt}^{\ominus}$ suggests a preferential orientation of the positive end of the solvent dipoles away from the liquid phase.

The contribution to $\chi_{svt}^{\ominus}$, and in particular $\chi_{wat}^{\ominus}$, originating from the *anisotropy of electronic polarization effects* and from possible *distortions of the molecular geometry* at the interface are probably much less significant than preferential dipole orientation effects, as can be inferred from the very small $\chi_{svt}^{\ominus}$ values suggested for non-dipolar solvents (Table 2.1).

Finally, the nature and magnitude of a possible contribution arising from the *preferential adsorption or exclusion of ions issued from autoprotolysis*, *i.e.* H_3O^+ and OH^- in the case of water, is difficult to evaluate in the present state of knowledge. Classical AT simulations cannot be used to investigate the autoprotolysis equilibrium of water, and hence are not able to capture these effects in the context of pure water. However, some classical and quantum-mechanical (Car-Parrinello) simulations of aqueous OH^- and H_3O^+ ions have suggested preferential adsorption of the latter ion at air-liquid interfaces[2232–

Table 5.18 *Estimates for the standard air-liquid interfacial potential and its temperature derivative for water.* The reported data includes estimates for the air-liquid interfacial potential $\chi^{\ominus}_{wat}$ of pure water (planar surface at equilibrium) and its temperature derivative $\partial_T \chi^{\ominus}_{wat}$ under standard conditions (reference pressure $P^{\circ} = 1$ bar and reference temperature $T^{-} = 298.15$ K). Corresponding estimates for non-aqueous solvents can also be found in Table 2.1. The methods and involved extra-thermodynamic assumptions refer to the literal enumeration of Section 5.17. The different sets correspond to values quoted in different literature sources, listed in chronological order of publication for the different methods separately. Only sources explicitly reporting values for the air-liquid interfacial potential or/and its temperature derivative are listed, although it is possible to construct many additional values of type A_{13} for $\chi^{\ominus}_{wat}$ based on the comparison of real proton solvation free-energy estimates $\hat{\mathcal{G}}^{\ominus}_{H,wat}$ in Table 5.15 and intrinsic proton solvation free-energy estimates $\mathcal{G}^{\ominus}_{H,wat}$ in Table 5.19, or of type B_{13} for $\partial_T \chi^{\ominus}_{wat}$ based on $\hat{s}^{\ominus}_{H,wat}$ (*via* $\hat{\mathcal{S}}^{\ominus}_{H,wat}$ in Table 5.15 and $s^{\ominus}[H^+_g]$ in Table 5.9) and $s^{\ominus}_{H,wat}$ estimates in Table 5.20. On the other hand, intrinsic proton solvation parameters deduced from the values given in the present table using recommended estimates for the corresponding real values (Section 5.15) are reported in Tables 5.19 and 5.20. The values calculated using AT simulations (method G_{13}) correspond to the external surface potential $\chi^{\ominus}_{wat}$ calculated based on the M-scheme ($\chi^{\ominus}_{AT,M}$; Sections 3.3.5 and 4.3.8). The corresponding estimates for the internal surface potential $\check{\chi}^{\ominus}_{wat}$ calculated based on the P-scheme ($\chi^{\ominus}_{AT,P}$) are reported between parentheses (note that these values depend strongly on the solvent representation employed). The value in boldface ($\chi^{\ominus}_{AT,M}$ or $\chi^{\ominus}_{AT,P}$) is the one reported in the quoted reference, while a value in italics was calculated from the latter using Eq. 3.86 based on the exclusion potential $\xi^{\ominus}_{AT}$ appropriate for the water model employed at the experimental pure water density (Table 3.1). This interconversion could not always be performed, an analytical expression for $\xi^{\ominus}_{AT}$ only being available for non-polarizable water models with a single Lennard-Jones interaction site. The lines NUM, AVE, RMS, ERR and REC report the number of entries, average value, root-mean-square deviation, estimated error ($\text{NUM}^{-1/2}\cdot$RMS) and recommended value.

Set	Method	$\chi^{\ominus}_{wat}$ [V]	$\partial_T \chi^{\ominus}_{wat}$ [10^{-3} V·K^{-1}]	Ref.	Notes	Year
1X	A_{13}	-0.5	-	406 / 404; 2155	a	1942
2X	A_{13}	-0.3	-	407 / 407; 814	a	1948
3X	A_{13}	-0.36	-	408 / 404; 814	a,b,c	1952
4X	A_{13}	-0.48	-	408 / 404; 2155	a,b	1952
5X	A_{13}	-0.43	-	408 / 404; 814	a,b	1952
6X	A_{13}	0.28	-	409 / 408,409; 814, 730, 731	a,c,d	1954
7X	A_{13}	-0.3±0.1	-	410		1954
8X	A_{13}	-0.44	-	412 / 404; 814	a	1958
9X	A_{13}	-0.15±0.05	-	413 / 411; 2228, 2229, 2230, 261	a,e	1967
10X	A_{13}	0.2	-	414 / 411; 258,260	a,f	1968
11X	A_{13}	-0.3	-	2219 / 407,411; 2219	a,g	1968
12X	A_{13}	0.04	-	416 / 411; 416	a,j	1974
13X	A_{13}	0.46	-	272 / 411; 644	a,h,i	1977
14X	A_{13}	0.46	-	272 / 411; 643	a,h,i	1977
15X	A_{13}	0.26	-	272 / 411; 660	a,h,i	1977
16X	A_{13}	-0.06	-	272 / 411; 261	a,f,h	1977
17X	A_{13}	-0.34	-	272 / 411; 814	a,h,i	1977
18X	A_{13}	-1.15	-	272 / 411; 1873	a,h,i	1977
19X	A_{13}	0.16	-	272 / 411; 409	a,h,i	1977
20X	A_{13}	0.13	-	417 / 411; 417	a,g	1981
21X	A_{13}	0.06	-	418 / 418;–	a,j	1992
22X	A_{13}	0.4	-	277 /–; 643	a,k	2000
23X	A_{13}	1.4	-	277 /–; 754	a,k	2000
24X	A_{13}	0.14	-	419 / 411; 1875,1876	a,g	2008
25X	A_{13}	0.14	-	420/-,-	j,l	2008
26X	B_{13}	0.15±0.05	-0.27	448		1956
27X	B_{13}	0.07	-0.4	2174		1965
28X	B_{13}	0.1	-0.39	415		1970
29X	B_{13}	0.08	-0.27	272	n	1977

(continues on the next page)

Table 5.18 *(continued)*

Set	Method	$\chi^{\ominus}_{wat}$ [V]	$\partial_T \chi^{\ominus}_{wat}$ [10^{-3} V·K^{-1}]	Ref.	Notes	Year
30X	B_{13}	0.13	-0.43	272	m	1977
31X	B_{13}	0.12±50	-0.57	465	m	1992
32X	C_{13}	0.0055	-	443		1927
33X	C_{13}	0.05	-	444		1977
34X	C_{13}	0.025±0.01	-1.0±0.5	445, 446, 447		1982
35X	D_{13}	1.0	-	449,450	o	1959
36X	D_{13}	0.142	-	451	p	1964
37X	D_{13}	0.8	-	452	q	1974
38X	D_{13}	0.1	-	453	r	1997
39X	D_{13}	0.1	-	454	r,s	2002
40X	E_{13}	-0.26	-	2225		1937
41X	E_{13}	0.1	-	2224		1981
42X	F_{13}	0.12	-	461,1713		1974
43X	G_{13}	- (**0.029**)	-	1454	t	1967
44X	G_{13}	**-0.1** (-)	-	1457	t	1968
45X	G_{13}	**3.1** (-)	-	2227	t	1981
46X	G_{13}	**0.08** (*-0.53*)	-	1440 /ROW 1281	u	1985
47X	G_{13}	**0.07** (*-0.89*)	-	1441 /TIPS2 1277	u	1986
48X	G_{13}	*0.81±0.05* (**-0.13±0.05**)	-	1442,1443,1445 /TIP4P 861	u,v	1987
49X	G_{13}	*0.03* (**-0.14**)	-0.8	1451 /ST2 1282	u,w	1987
50X	G_{13}	**0.16** (*-0.6*)	-0.65	1444 /CC 1283	u,x	1988
51X	G_{13}	**0.24** (*-0.59*)	-	465 /SPC 1274	u	1992
52X	G_{13}	*0.31* (**-0.63**)	-	470 /TIP4P 861	u	1995
53X	G_{13}	*0.05* (**-0.12**)	-	470,833 /ST2 1282	u	1995
54X	G_{13}	**0.28** (*-0.66*)	-	1473 /TIP4P 861	u	1995
55X	G_{13}	**0.06** (*-0.11*)	-	1473 /ST2 1282	u	1995
56X	G_{13}	*0.27* (**-0.5**)	-	850 /TIP3P 861	u	1996
57X	G_{13}	**0.21** (*-0.64*)	-	1446 /SPC/E 1275	u	1997
58X	G_{13}	- (**-0.01**)	-	1446	t,y	1997
59X	G_{13}	- (**-0.5±0.01**)	-	430 /RPOL 920,1335	u	2002
60X	G_{13}	- (**-0.54**)	-	947 /SWM4-DP 947	u	2003
61X	G_{13}	- (**-0.758**)	-	1447 /classical DFT	u	2004
62X	G_{13}	- (**-0.48**)	-1.2	973 /RPOL 920,1335	u,z	2006
63X	G_{13}	**0.73** (*-0.1*)	-	707 /SPC 1274	u	2006
64X	G_{13}	- (**3.1**)	-	1448,1449 /CPMD	u	2008
65X	G_{13}	**0.012** (**3.63**)	-	1450 /CPMD		2010
NUM		58	10			
AVE		0.13	-0.60			
RMS		0.54	0.30			
ERR		0.07	0.09			
REC		0.13	-0.42			

(a) the listed secondary references indicate the sources of real (listed first) and intrinsic (listed second) data; (b) real values from Ref.[404] corrected as described in Ref.[408]; (c) intrinsic values based on quoted secondary reference with further adjustments by primary reference; (d) standard-state convention not indicated in Ref.[409] and assumed identical to that of Ref.[408]; (e) also using data from Refs.[257, 260]; (f) considering the ion H^+; (g) real values based on quoted secondary reference with further adjustments by primary reference; (h) real values from[411] corrected as described in Ref.[417]; (i) considering the ion K^+; (j) intrinsic values involving theory; (k) real values from unspecified source; (l) real values obtained from experimental data as reported in Ref.[420]; (m) based on the temperature derivative from Ref.[2174]; (n) based on the temperature derivative from Ref.[448]; (o) using various organic solutes; (p) using benzene as organic solute; (q) using terphenyl esters as organic solutes; (r) using ethanol; (s) also based on data from Refs.[2223, 2231]; (t) analytical model (point dipole and quadrupole); (u) the acronym and the secondary reference indicate the method (classical DFT: classical statistical-mechanical density functional theory; CPMD: Car-Parrinello molecular dynamics) or water model (AT simulations) employed along with the reference for this model; (v) at 325 K; (w) data at 314 K, temperature derivative by finite difference between 289 and 314 K; (x) temperature derivative by finite difference between 250 and 350 K; (y) based the model of Ref.[1454] (with adjusted quadrupole moment); (z) temperature derivative by finite difference between 298 and 323 K.

[2234]. But there are also quantum-mechanical (Car-Parrinello) simulations supporting in the opposite a slight preferential affinity of OH^- for the surface[2235]. Some experimental spectroscopic studies, involving second-harmonic generation spectroscopy on liquid water, indicated an excess of H_3O^+ ions at the air-water interface[2236, 2237]. In contrast, experimental studies discussing the surface affinity of OH^- ions did not yet reach an unambiguous conclusion[2237]. On the one hand, studies at the microscopic level, involving vibrational sum-frequency generation and second-harmonic generation spectroscopy on liquid water, as well as infra-red spectroscopy on ice nanoparticles, concluded against a preferential affinity of OH^- ions for the surface[2236–2240]. On the other hand, studies at the macroscopic level, including electrophoretic and titration measurements on oil droplets and air bubbles in water, as well as protonation experiments on aqueous microjets, supported a preferential affinity of OH^- ions for the surface[2237, 2241–2245].

The apparent contradiction between simulation studies, most of which support a preferential surface affinity of H_3O^+, and macroscopic experiments, most of which support in the opposite a preferential surface affinity of OH^-, has been suggested[2244] to result from the microscopic system sizes considered in the simulations and the failure to reproduce the relatively large dielectric decrement of OH^- in simulation studies. Such a reduction in the relative dielectric permittivity of water in the vicinity of OH^- ions, *i.e.* a reduction in dipole-moment fluctuations, induces a force that attracts the ions towards regions of low relative dielectric permittivity, *i.e.* the air-liquid interfacial region. The apparent contradiction between spectroscopic and macroscopic experimental studies has also been suggested[2244] to follow from the fact that the OH^- ions might not be situated in the outermost molecular layers of the air-liquid interface, focused upon by spectroscopic studies, but not by the other experimental approaches, which probe a wider region up to a depth of about 2-3 nm. In summary, given the experimental observation of a negative charge on the water surface in the macroscopic measurements, it is likely[2244] that in pure water at neutral pH, although both H_3O^+ and OH^- are adsorbed at the air-liquid interface, the OH^- ion evidences a preferential affinity compared to the H_3O^+ ion. Accepting this suggestion, the contribution of ions issued from autoprotolysis to the air-liquid interfacial potential of pure water $\chi_{wat}^{\ominus}$ would be positive, just as that of the preferential dipole orientation.

5.18 Intrinsic proton solvation parameters

The *intrinsic solvation free energy of the proton in a given solvent* $\mathcal{G}_{H,svt}^{\ominus}$ is one of the three quantities, besides the air-liquid interfacial potential $\chi_{svt}^{\ominus}$ of the pure solvent and the intrinsic absolute potential $\mathcal{V}_{H,svt}^{\ominus}$ of the reference hydrogen electrode in this solvent, that *eludes a purely experimental determination* (Section 2.3). The relationships between these three quantities and with corresponding real quantities have been presented previously (Section 4.5.4 and Figure 4.18) and are summarized in Section 5.17 (Eqs. 5.44-5.47).

The determination of $\mathcal{G}^{\ominus}_{H,svt}$ is only possible if experimental data is interpreted using some *extra-thermodynamic assumption*, *i.e.* a reasonable but formally unprovable hypothesis concerning the expected outcome of an experiment in some limiting situation, or an interpretation of the experimental data in terms of a given theoretical model.

The present discussion focuses principally on the case of *water* as a solvent, and the generic notation "*svt*" will sometimes be replaced by "*wat*" to underline this restriction. The calculation of $\mathcal{G}^{\ominus}_{H,wat}$ from the corresponding real value $\hat{\mathcal{G}}^{\ominus}_{H,wat}$ (Section 5.15), including estimates for the latter quantity derived from the real absolute potential $\hat{\mathcal{V}}^{\ominus}_{H,wat}$ of the reference hydrogen electrode in water (Section 5.14), *via* estimation of the air-liquid interfacial potential $\chi^{\ominus}_{wat}$ of pure water (Section 5.17) will also be considered here (Eq. 5.44). In principle, the calculation of $\mathcal{G}^{\ominus}_{H,wat}$ based on an estimate for the intrinsic absolute potential $\mathcal{V}^{\ominus}_{H,wat}$ of the reference hydrogen electrode in water would also be possible (Eq. 5.45). However, no direct tentative determination of the latter quantity has been reported to date (see, however, Ref.[483] discussed in Section 5.14 for an indirect estimation relying on a corrected $\mathcal{G}^{\ominus}_{H,wat}$ estimate from Refs.[1875,1876]). On the other hand, because the intrinsic proton solvation free energy is an experimentally-elusive quantity, the calculation of $\mathcal{G}^{\ominus}_{H,wat}$ using theoretical models as a form of extra-thermodynamic assumption will be partially included. More precisely, a distinction is made here between theoretical approaches involving a dominant experimental component (model used as fitting device) and approaches aiming at a more fundamental prediction (some QM computations and many AT simulations). The former ones are considered here, and the corresponding data included in the tables, while the discussion of the latter ones is deferred to Chapter 6, and the corresponding data omitted from the tables. A few exceptions are made concerning purely theoretical estimates from approaches considered by the authors of this book as promising[1289,1571,1572]. Note that this distinction is somewhat arbitrary and reflects a personal choice of the authors of this book rather than a clear-cut boundary.

The present discussion is also concerned with the evaluation of corresponding *intrinsic derivative thermodynamic solvation properties* of the gas-phase proton in water, namely the solvation enthalpy $\mathcal{H}^{\ominus}_{H,wat}$ and entropy $\mathcal{S}^{\ominus}_{H,wat}$, as well as of the corresponding *intrinsic partial molar variables* of the hydrated proton, namely the partial molar entropy $s^{\ominus}_{H,wat}$, heat capacity $c^{\ominus}_{P,H,wat}$, volume $v^{\ominus}_{H,wat}$, volume-compressibility $k^{\ominus}_{T,H,wat}$ and volume-expansivity $a^{\ominus}_{P,H,wat}$. The question whether these derivative quantities are also *experimentally elusive* is, however, relatively intricate. As discussed in Section 4.2.5, in cases where the determination method delivers data appropriate for the standard variant of the solute standard-state definition, the difference to the corresponding real quantity involves pressure or/and temperature derivatives of the air-liquid interfacial potential of the pure solvent. The question then turns into that of the possibly elusive nature of these derivatives, and will be addressed in Chapter 7. In cases where this determination method delivers data appropriate for the density-corrected variant, the contribution of the surface potential derivatives is excluded, along with that of the solvent-density derivatives. As a result, these parameters are intrinsic but in principle not experimentally elusive. As mentioned in Section 4.2.5, this only remains a formal consideration. In practice, due to the electroneutrality constraint, no currently-available experimental technique permits the measurement of $c_P^{\ominus *}$, $v^{\ominus *}$, $k_T^{\ominus *}$ and $a_P^{\ominus *}$ (*e.g.* *via* heats of mixing or corresponding volume changes, and pressure or temperature derivatives thereof) for single ions in solution. The first situation will typically arise for $\mathcal{G}_{H,svt}^{\ominus}$, which is elusive, although no distinction is required here between standard and density-corrected variants, as well as for $\mathcal{S}_{H,svt}^{\ominus}$, $\mathcal{H}_{H,svt}^{\ominus}$, $\mathscr{s}_{H,svt}^{\ominus}$ and $\mathscr{c}_{P,H,svt}^{\ominus}$, when evaluated *via* temperature analysis of $\mathcal{G}_{H,svt}^{\ominus}$. The second situation will typically arise for $\mathscr{v}_{H,svt}^{\ominus}$, $\mathscr{k}_{T,H,svt}^{\ominus}$ and $\mathscr{a}_{P,H,svt}^{\ominus}$, which are therefore noted $\mathscr{v}_{H,svt}^{\ominus *}$, $\mathscr{k}_{T,H,svt}^{\ominus *}$ and $\mathscr{a}_{P,H,svt}^{\ominus *}$, since these quantities are typically determined by extrapolation of solution volumes to the infinite dilution regime and analysis of the corresponding pressure or/and temperature dependence. It may also arise for $\mathscr{c}_{P,H,svt}^{\ominus}$, then noted $\mathscr{c}_{P,H,svt}^{\ominus *}$, in the special case of a determination involving the temperature derivative of the mixing enthalpy extrapolated to the infinite dilution regime.

The methods attempting the determination of $\mathcal{G}_{H,svt}^{\ominus}$, $\mathcal{H}_{H,svt}^{\ominus}$ and $\mathcal{S}_{H,svt}^{\ominus}$ for a given solvent (prominently water), and the involved extra-thermodynamic assumptions, can be broadly classified as follows[258,260,264,267,274,275]:

(A_{14}) Assumption of *equal intrinsic solvation parameters* (or of a specific ratio) *for a selected anion-cation pair*[643,1778,1870,2182,2185–2188,2197]. For this specific pair, considering an equal partitioning and restricting the discussion to the case of monovalent ions M^+ and X^-, one has

$$\Delta_s Y^{\ominus}[M_g^+] + \Delta_s Y^{\ominus}[X_g^-] = \Delta_s Y^{\bullet}[M_g^+] + \Delta_s Y^{\bullet}[X_g^-] \qquad (5.66)$$

and

$$\Delta_s Y^{\ominus}[M_g^+] - \Delta_s Y^{\ominus}[X_g^-] = 0 \quad , \qquad (5.67)$$

and therefore

$$\Delta_s Y^{\ominus}[M_g^+] = \Delta_s Y^{\ominus}[X_g^-] = \frac{1}{2}(\Delta_s Y^{\bullet}[M_g^+] + \Delta_s Y^{\bullet}[X_g^-]) \quad , \qquad (5.68)$$

from which one infers

$$\mathcal{Y}_{H,svt}^{\ominus} = \frac{1}{2}(\Delta_s Y^{\bullet}[X_g^-] - \Delta_s Y^{\bullet}[M_g^+]) \; . \qquad (5.69)$$

In the context of hydration, possible choices of pairs include *e.g.* Cs^+/I^- (Ref. [1778], equal partitioning, justified by the comparison of solvation enthalpies from the Born-Bjerrum equation[874], based on radii averaged from Refs. [2047,2056] with experimental pair solvation enthalpy data and discussion of asymmetric hydration effects for Cs^+ and I^-; note that

I^- is noticeably larger than Cs^+ in this radius set), K^+/F^- (Ref.[643], non-equal partitioning with ratio 17.5:18 for K^+ and F^-, justified by the solvation enthalpy ratio predicted using own molecular-mechanics calculations with the radii of Ref.[2056]), K^+/F^- (Ref.[1870], equal partitioning, justified by similar solvation enthalpies predicted in Ref.[644] using a molecular-mechanics model, probably based on the radii of Ref.[2056]), K^+/Br^- (Ref.[2188], non-equal partitioning with ratio 0.63:0.37 for K^+ and Br^-, justified by the solvation enthalpy ratio predicted in Ref.[644] using a molecular-mechanics model[644], probably based on the radii of Ref.[2056]), Rb^+/Cl^- (Ref.[2182], equal partitioning, justified by the similar radii in the set of Ref.[2063]), K^+/Cl^- (Ref.[2197], equal partitioning, justification not provided) and $[Ph_4As]^+/[Ph_4B]^-$ (Ref.[2185–2187], equal partitioning, justified by the fact that these are large hydrophobic ions of similar sizes). The last assumption[2185–2187] is commonly referred to as the *tetraphenylarsonium-tetraphenylborate or TATB assumption* and sometimes also applied to $[Ph_4P]^+/[Ph_4B]^-$. Although probably most reasonable in the context of large nearly-spherical monovalent ions with similar sizes and shapes, and apolar periphery[2186,2187], the extra-thermodynamic assumption of equal partitioning is probably inaccurate in water even in this case, due to *anisotropic solvation effects*[786,787,789,791–793,796–798,803]. For this solvent, the anisotropic nature of the molecular charge distribution relative to the van der Waals envelope induces an intrinsically more favorable solvation of anions compared to cations, as evidenced by experiment[785–789,1988] and by AT simulations[94,427–429,780,792,795,797,798,800,803]. There is also a considerable amount of literature concerning partitioning methods of this type (typically the TATB assumption) in the context of free energies of transfer of ions across solvents (see *e.g.* Refs.[427,429,789,1884,1897,1902,1932,1939,1951,1961,1980,2011,2014,2019,2187,2246,2247] for some key references).

(B_{14}) *Fitting of the conventional solvation parameters* increased by unknown constants, namely the corresponding intrinsic proton solvation parameters, weighted by the integer ion charge for a series of ions *using a physical model* relating intrinsic solvation parameters to specific properties of the ion (*e.g.* radius) and solvent (*e.g.* permittivity)[408,409,643,644,660,662,668,717,733,748,754,814,821,822,1750,1778,1871,2184,2188,2248]. Based on this fitting to a theoretical intrinsic estimate $\Delta_s Y^{\ominus}_{TH}[I^z_g]$, as

$$\Delta_s Y^{\bullet}[I^z_g] \approx \Delta_s Y^{\ominus}_{TH}[I^z_g] + zC \quad , \tag{5.70}$$

one infers

$$\mathcal{Y}^{\ominus}_{H,svt}[H^+_g] = -C \quad . \tag{5.71}$$

The physical models may be based on continuum electrostatics (Born equation)[84,2172], on empirical modifications of the Born equation[259–261,754,755,814,821,822,1750,1872,1873,2182,2184,2249,2250] (Section 6.1.1), on the mean-spherical approximation (MSA; Section 6.1.2) model[822,1750,2184,2249,2251,2252], on molecular-mechanics (plus continuum-electrostatics) calculations[643–646,712,713,717,2253,2254], or on AT simulations[1289]. Note that QM approaches could also in principle be used for such a correlation against experimental conventional data. These approaches are discussed in Sections 6.3.1 (quasi-chemical theory), 6.3.2 (hybrid quantum-mechanical molecular-mechanical simulations) and 6.3.3 (Car-Parrinello molecular dynamics simulations). Examples of theoretical approaches that have been applied in practice in the context of hydration involve (the indication "own fitting" refers to a fitting against recommended conventional data performed by the authors of this book as described in Section 5.16): the Born model[84] (Ref.[1778], based on radii that are arithmetic means of those of Refs[2047,2056]; own fitting), molecular-mechanics calculations (Ref.[643], using a model by Ref.[643], probably based on radii of Ref.[2056]; own fitting), molecular-mechanics calculations (Ref.[644], using a model by Ref.[644], based on radii of Ref.[2056]; own fitting), Born model[84] (Ref.[814], based on radii of Ref.[2047] modified according to Ref.[814] by an additional contribution of 0.01 nm for anions and 0.085 nm for cations; own fitting), Born model[84] (Ref.[748], using a model with modifications by Ref.[748], based on the same radii as Ref.[814]; own fitting), Born model[84] (Ref.[408], based on radii of Ref.[2047], modified according to Ref.[408] by an additional contribution of 0.025 nm for anions and 0.085 nm for cations; own fitting), molecular-mechanics calculations (Ref.[2188], using a model by Ref.[644], based on radii from Ref.[2255]; own fitting), modified Born model[730,731] (Ref.[409], using a model with modifications by Ref.[409], based on radii by Ref.[409]; own fitting), Born model[84] (Ref.[821], based on the same radii as Ref.[814]; own fitting), Born model[84] (Ref.[717], using a model with modifications by Ref.[717], based on radii of Ref.[2056]; own fitting), Born model[84] (Ref.[754], using a model with modifications by Ref.[754], based on radii of Ref.[2056] multiplied by a factor of 1.25; fitting against conventional data of Ref.[1871] after removal of a "non-electrostatic" contribution), Born model[84] (Ref.[660], using a model with modifications by Ref.[660], based on gas-phase radii of Ref.[660] and in-solution radii approximated by the in-crystal-radii of Ref.[66]; own fitting), Born model[84] (Ref.[662], using a model with modifications by Ref.[662], based on radii of Ref.[2256]; own fitting), MSA model[822,1750,2184,2249,2251,2252] (Ref.[822], based on radii of Ref.[822]; fitting against conventional data from Ref.[2257]), MSA model[822,1750,2184,2249,2251,2252] (Ref.[2184], based on radii of Refs.[2058–2060]; own fitting), MSA model[822,1750,2184,2249,2251,2252] (Ref.[1750], based on radii of Refs.[2058–2060]; own fitting) and Born model[84] (Ref.[668], using a model with own modifications, based on radii from either Refs.[66,2047], either omitting or including additional terms for ion-dipole, ion-induced dipole and ion-quadrupole interactions; fitting against standard electrode potentials from Ref.[1744]), and Car-Parrinello free-energy calculations of Refs.[1571,1572] (Section 6.3.3; own fitting). Some studies also rely on an indirect correlation against tentative intrinsic data from previous work (*e.g.* Ref.[1871] against the data of Ref.[814]). Obviously, the accuracy of the intrinsic proton solvation parameters evaluated based on these theoretical approaches is limited by the ability of the selected model to capture the physics of the ionic solvation process. Shortcomings of these models, including the approximate nature of the parameters they rely on, have been detailed in Chapter 3 (Sections 3.2, 3.3.6 and 3.4). Another common source of error arises from a possible discrepancy in the standard-state conventions associated with the theoretical model and the conventional data employed. In general, the raw results of a theoretical calculation correspond to the point-to-point transfer involved in the hypothetical solvation process discussed in Section 2.2. In contrast, experimental conventional data is usually tabulated based on some specified standard-state convention, as detailed in Sections 4.2.4 and 4.2.17. If no care is taken to adjust the raw theoretical data to the latter convention by application of an appropriate standard-state correction (Section 4.2.16), or if the applied correction is inconsistent with the standard-state of the conventional data, the fitting approach of Eqs. 5.70 and 5.71 becomes formally incorrect. In general, the resulting error on the calculated proton value is not a simple offset equal to the omitted standard-state correction, because in terms of conventional values, standard-state corrections cancel out for monovalent cations, while they add up for monovalent anions (Section 4.5.1). Thus, standard-state inconsistencies affect the entire fitting procedure and result in rather unpredictable errors on the final estimate for $\mathcal{Y}^{\ominus}_{H,wat}$.

(C_{14}) *Fitting of the salt (pairwise) solvation parameters* for a series of cations with a common anion (or for a series of anions with a common cation) using a *physical model*, and *extrapolation to infinite cation (or anion) size*, where the pairwise solvation parameter can be identified with the intrinsic solvation parameter of the anion (or cation) only[259,1872,1874,1882,2155]. Restricting the discussion to the case of monovalent ions, if the approach relies on a common cation M^+, the fitting equation is given by

$$\Delta_s Y^{\bullet}[M_g^+] + \Delta_s Y^{\bullet}[X_g^-] \approx f_{M^+}(R_I[X^-]) \quad , \tag{5.72}$$

where $R_I[X^-]$ represents the effective ionic radius of the anion X^-, and one takes the limit

$$\Delta_s Y^{\ominus}[M_g^+] = \lim_{R\to\infty} f_{M^+}(R) = C \quad , \tag{5.73}$$

from which one infers

$$\mathcal{Y}^{\ominus}_{H,svt} = C - \Delta_s Y^{\bullet}[M_g^+] \quad . \tag{5.74}$$

The physical models used here to represent $f(R)$ are similar to those mentioned under Point B_{14}, but it is assumed in addition that the intrinsic solvation parameters of an ion vanish in the limit of infinite size. Obviously, this statement cannot be entirely correct if the cavitation contribution is included, so that in most cases, only charging contributions[259, 1872, 1874] rather than total solvation parameters[1882] are fitted. Examples of approaches of this kind that have been applied in practice in the context of hydration involve fitting to the inverse ionic radius with subsequent adjustments (Ref.[2155], fitting the total contribution to the pairwise solvation enthalpy, considering both anion extrapolation at common cation and cation extrapolation at common anion, based on radii which are averages of the radii of Refs.[66,2047] and Ref.[2056] and using own pairwise data; see also discussion in Ref.[258]), fitting to the inverse ionic radius plus a constant (Ref.[1872], fitting the charging contribution to the pairwise solvation enthalpy, considering cation extrapolation at common anion, based on radii of Ref.[2062] and using own conventional data), fitting to the inverse ionic radius plus a constant (Ref.[1872], fitting the charging contribution to the pairwise solvation enthalpy, considering anion extrapolation at common cation, based on radii of Ref.[2062] and using own conventional data), fitting to the inverse ionic radius plus a constant (Ref.[1882], fitting the total contribution to the pairwise solvation enthalpy, considering anion extrapolation at common cation, based on cation radii of Ref.[2046] and anion radii of Ref.[2056] and using own pairwise data), fitting to the inverse ionic radius plus a constant (Ref.[1874], fitting the charging contribution to the pairwise solvation enthalpy, considering both anion extrapolation at common cation and cation extrapolation at common anion, based on radii of Ref.[2062] and using own conventional data) and fitting to the inverse ionic radius plus a constant (Ref.[259], fitting the charging contribution to the conventional solvation free energy, see Ref.[2155], the common ion here being the proton, based on radii of Ref.[66] and using own conventional data). The main shortcomings and sources of error of this approach include the extra-thermodynamic assumption of vanishing intrinsic solvation parameters at infinite ion size (which overlooks the dispersion and, possibly, the cavitation contribution to solvation), the approximate nature of the theoretical model employed for the fitting (see Point B_{14}), the use of a very limited range of ion sizes for the extrapolation to infinite size (see also Point D_{14}), and possible standard-state inconsistencies between model results and experimental data (see Point B_{14}).

(D_{14}) *Fitting of the conventional solvation parameters* for series of cations and anions, respectively, using a *physical model*, and estimation of the *difference* (anions minus cations) *extrapolated to infinite ion sizes*, the difference being related to the intrinsic proton solvation parameters in this limit[260, 1872, 1873, 1882, 2258]. Restricting the discussion to the case of monovalent ions, if the separate fitting equations are given by

$$\begin{aligned}\Delta_s Y^{\bullet}[M_g^+] &\approx f_{M^+}(R_I[M^+]) \\ \Delta_s Y^{\bullet}[X_g^-] &\approx f_{X^-}(R_I[X^-]) \quad ,\end{aligned} \tag{5.75}$$

where $R_I[M^+]$ and $R_I[X^-]$ represent the effective ionic radii of the cation M^+ and anion X^-, respectively, one infers

$$\mathcal{Y}^{\ominus}_{H,svt} = \frac{1}{2} \lim_{R\to\infty} [f_{X^-}(R) - f_{M^+}(R)] \ . \tag{5.76}$$

The physical models used here to represent $f_{M^+}(R)$ and $f_{X^-}(R)$ are similar to those mentioned under Point B_{14}, but it is assumed in addition that anions and cations of identical sizes are characterized by identical intrinsic solvation parameters in the limit of large sizes (see Point A_{14} and Eq. 5.69 above). Examples of approaches of this kind that have been applied in the context of hydration involve fitting to the inverse third power of the adjusted ionic radius[260, 717] plus a constant (Ref.[260], considering a function $(R + R_W)^{-3} + C$ following the reasoning of Ref.[717], where R_W is an effective radius of the water molecule taken to be $R_W = 0.138$ nm, based on cation radii of Ref.[2046] and anion radii of Ref.[66, 2047] and using own conventional data), fitting to the inverse third power of the adjusted ionic radius[260, 717] plus a constant (Ref.[1872], considering the same function as Ref.[260], based on radii of Ref.[2062] and using own conventional data), fitting to the inverse second and third powers of the adjusted ionic radius[1873] plus constant (Ref.[1873], considering a function $(R + R_W)^{-2} + (R + R_W)^{-3} + C$, where R_W is an effective radius of the water molecule taken to be $R_W = 0.138$ nm, based on radii from Ref.[2063] and using own conventional data), fitting to the inverse third power of the adjusted ionic radius[260, 717] plus a constant (Ref.[1882], based on cation radii from Ref.[2046] and anion radii from Ref.[2056] and using own conventional data) and fitting to the inverse third power the adjusted ionic radius[260, 717] plus a constant (Ref.[2258], based on cation radii from Ref.[2046] and anion radii from Refs.[66, 2047] and using own conventional data). The main shortcomings and sources of error of this approach include the extra-thermodynamic assumption of identical cation and anion intrinsic solvation parameters in the limit of infinite ion size (which overlooks possible asymmetric solvation effects, see Point A_{14}), the approximate nature of the theoretical model employed for the fitting (see Point B_{14}), the use of a very limited range of ion sizes for the extrapolation to infinite size (see also Point C_{14}), and possible standard-state inconsistencies between model results and experimental data (see Point A_{14}).

(E_{14}) *Ion-water cluster methods* using comparisons of *gas-phase ion-water cluster data* with *bulk solvation data* to determine intrinsic single-ion solvation parameters[164,170,417,1747,1875,1876,2259]. These approaches involve a measurement [160, 161, 2260] of the stepwise thermodynamic parameters $\Delta Y_{n-1,n}[I_g^z]$ associated to the gas-phase reactions

$$[I(\mathrm{H_2O})_{n-1}]^z + \mathrm{H_2O} \rightarrow [I(\mathrm{H_2O})_n]^z \quad , \tag{5.77}$$

and, possibly, of the stepwise thermodynamic parameters $\Delta Y_{n-1,n}[\mathrm{H_2O}]$ associated to the corresponding reaction for neutral water clusters, namely

$$(\mathrm{H_2O})_{n-1} + \mathrm{H_2O} \rightarrow (\mathrm{H_2O})_n \quad . \tag{5.78}$$

In the early work of Ref. [417], the stepwise parameters $\Delta Y_{n-1,n}[I_g^z]$ were measured for the ions Na^+ to Cs^+, NH_4^+, Cl^- and Br^- in water, with $Y = G$ or H and $n = 1 - 5$. It was assumed that in the limit $n \rightarrow \infty$, monovalent cations and anions have same stepwise solvation parameters, and that this infinite-cluster limit was reached at $n = 5$. Considering the case of monovalent ions M^+ and X^-, one has thus

$$\begin{aligned} &\Delta_s Y^\ominus[M_g^+] - \Delta_s Y^\ominus[X_g^-] \\ \approx\ &\sum_{n=1}^{5} \left(\Delta Y_{n-1,n}[M_g^+] - \Delta Y_{n-1,n}[X_g^-]\right) = C \quad . \end{aligned} \tag{5.79}$$

Comparison of Eq. 5.79 with the equivalence of intrinsic and conventional solvation parameters at the level of pair data, namely,

$$\Delta_s Y^\ominus[M_g^+] + \Delta_s Y^\ominus[X_g^-] = \Delta_s Y^\bullet[M_g^+] + \Delta_s Y^\bullet[X_g^-] \quad , \tag{5.80}$$

one obtains

$$\Delta_s Y^\ominus[M_g^+] = \frac{1}{2}\left(\Delta_s Y^\bullet[M_g^+] + \Delta_s Y^\bullet[X_g^-] + C\right) \tag{5.81}$$

or

$$\Delta_s Y^\ominus[X_g^-] = \frac{1}{2}\left(\Delta_s Y^\bullet[M_g^+] + \Delta_s Y^\bullet[X_g^-] - C\right) \quad , \tag{5.82}$$

from which one infers

$$\mathcal{Y}_{H,svt}^\ominus = \frac{1}{2}\left(\Delta_s Y^\bullet[X_g^-] - \Delta_s Y^\bullet[M_g^+] + C\right) \quad . \tag{5.83}$$

In other words, the information on the difference between the solvation free energies of M^+ and X^- in the gas-phase clusters, assumed applicable to the bulk already for $n = 5$, is used to partition the pairwise solvation free energy of the salt into intrinsic single-ion contributions. However, the method was rightfully criticized [164, 1875] considering that the stepwise parameters for cations and anions are actually not identical at the considered maximal cluster size, but also in view of the high uncertainties in the cluster data involving large halide anions, which might actually not be located at the cluster centers. More elaborate approaches[164, 170, 1747, 1875, 1876, 2259] rely on cluster solvation parameters $\Delta Y_n[I_g^z]$, associated with the solvation of the ion into a water cluster of n water molecules, *i.e.* with the gas-phase reaction

$$I_g^z + [(\mathrm{H_2O})_n]_g \rightarrow [I(\mathrm{H_2O})_n]_g^z \quad , \tag{5.84}$$

and are related to the previous stepwise parameters *via*

$$\Delta Y_n[I_g^z] = \sum_{m=1}^{n} \Delta Y_{m-1,m}[I_g^z] - \sum_{m=1}^{n} \Delta Y_{m-1,m}[\mathrm{H_2O}] \quad . \tag{5.85}$$

More specifically, these approaches rely on the consideration of cluster solvation parameter differences between anions and cations. Restricting the discussion to the case of monovalent ions M^+ and X^-, the relevant quantity is thus

$$\Delta\Delta Y_n[M^+, X^-] = \Delta Y_n[M_g^+] - \Delta Y_n[X_g^-] \quad . \tag{5.86}$$

Three variants of the method are briefly described in the following. In Ref.[164], it is assumed that the differences $\Delta\Delta Y_n[M^+, X^-]$ in the limit $n \rightarrow \infty$ are representative for the corresponding bulk solvation differences. These limiting differences are determined using an extrapolation against $n^{-1/3}$ for the ions OH^-, Li^+ to Rb^+ and F^- to I^- in water, with $Y = H$ and $n = 1 - 9$. A least-squares fit of the obtained limiting cluster solvation differences to the bulk differences (omitting the large halide ions) gives the intrinsic absolute solvation parameters of the proton according to

$$\Delta_s Y^\ominus[M_g^+] - \Delta_s Y^\ominus[X_g^-] \approx \lim_{n\rightarrow\infty} \Delta\Delta Y_n[M^+, X^-] = C \quad , \tag{5.87}$$

along with Eqs. 5.80-5.83. Besides the consideration of larger clusters and the possible error reduction following from the application of a least-squares fit procedure over multiple ions, the key advantage of this method over the previous one is the application of an extrapolation procedure to $n \rightarrow \infty$, replacing the questionable assumption of convergence at the maximal cluster size. Note that the reformulation in terms of cluster solvation reactions (Eq. 5.84), involving the addition of an ion to a preexisting water cluster, rather than in terms of a sum of stepwise water addition reactions (Eq. 5.77), involving the combination of an ion with a number of isolated gas-phase water molecules to form the ion-water cluster, is an essential ingredient of this extrapolation procedure, because it leads to the approximate $n^{-1/3}$ dependence. In Refs.[1875, 1876], a so-called *cluster-pair approximation* to the solvation parameter $\mathcal{Y}_{H,svt}^\ominus$, noted here $\tilde{\mathcal{Y}}_{H,svt}$, is introduced in analogy with Eqs. 5.83 and

5.87, namely as

$$\tilde{\mathcal{Y}}_{H,svt}(M^+, X^-, n) = \frac{1}{2}\Big(\Delta_s Y^{\bullet}[X_g^-] - \Delta_s Y^{\bullet}[M_g^+] + \Delta\Delta Y_n[M^+, X^-]\Big) \quad . \tag{5.88}$$

However, instead of assuming as done previously (Eq. 5.87) that

$$\mathcal{Y}^{\ominus}_{H,svt} = \lim_{n\to\infty} \tilde{\mathcal{Y}}_{H,svt}(M^+, X^-, n) \tag{5.89}$$

for any cation-anion pair, one considers instead the existence of a hypothetical pair $\tilde{M}^+$ and $\tilde{X}^-$ characterized by

$$\lim_{n\to\infty} \Delta\Delta Y_n(\tilde{M}^+, \tilde{X}^-, n) = 0 \quad . \tag{5.90}$$

For such a pair, Eq. 5.88 implies

$$\lim_{n\to\infty} \tilde{\mathcal{Y}}_{H,svt}(\tilde{M}^+, \tilde{X}^-, n) = \frac{1}{2}\Big(\Delta_s Y^{\bullet}[\tilde{X}_g^-] - \Delta_s Y^{\bullet}[\tilde{M}_g^+]\Big) \quad . \tag{5.91}$$

Therefore, on a graph representing $\tilde{\mathcal{Y}}_{H,svt}(M^+, X^-, n)$ as a function of $(1/2)(\Delta_s Y^{\bullet}[X_g^-] - \Delta_s Y^{\bullet}[M_g^+])$ for a number of monovalent ion pairs, individual (fitted) curves corresponding to different n values should all cross at a single point in the limit of large n. According to the assumption of Eq. 5.89, now only applied to the hypothetical pair $\tilde{M}^+$ and $\tilde{X}^-$, the ordinate of this point provides the required estimate for $\mathcal{Y}^{\ominus}_{H,svt}$. In Refs.[170,1875,1876], this approach was applied to the alkali and halide ions except Cs^+, as well as the OH^- ion, considering water as a solvent and with $Y = G$ or H and $n = 1 - 5$ in Ref.[170] or $n = 1 - 6$ in Refs.[1875,1876]. In practice, the curves in the above graph are very close to straight lines and already share a common intersection point for $n = 0, 1$ and 2. In Refs.[1747,2259], a variant of this approach is introduced, the so-called *cluster-pair correlation approximation*. Here, Eq. 5.88 is rewritten as

$$\tilde{\mathcal{Y}}'_{H,svt}(\tilde{M}^+, \tilde{X}^-, n) = \frac{1}{2}\Big(\Delta_s \tilde{Y}_n^{\bullet}[\tilde{X}^-] - \Delta_s \tilde{Y}_n^{\bullet}[\tilde{M}^+]\Big) \quad , \tag{5.92}$$

where $\Delta_s \tilde{Y}_n^{\bullet}[I^z]$ represents the conventional solvation parameter of the cluster-ion $[I(H_2O)_n]_g^z$ into the bulk, *i.e.* corresponding to the process

$$[I(H_2O)_n]_g^z \rightarrow I_{aq}^z + nH_2O_{aq} \quad . \tag{5.93}$$

It is easily verified that Eqs. 5.88 and 5.92 are in principle equivalent. In practice, conventional solvation parameters of the cluster-ions are obtained by combining the conventional solvation parameters of the ion with the solvation parameters of the solvent (in itself) and the stepwise cluster parameters of the ion, as

$$\Delta_s \tilde{Y}_n^{\bullet}[I^z] = \Delta_s \tilde{Y}^{\bullet}[I^z] + n\Delta_s Y^{\ominus}[\mathrm{H_2O}_g] - \sum_{m=1}^{n} \Delta Y_{m-1,m}[I_g^z] \quad . \quad (5.94)$$

Here also, on a graph representing $\tilde{\mathcal{Y}}'_{H,svt}(M^+, X^-, n)$ as a function of $(1/2)(\Delta_s \tilde{Y}_n^{\bullet}[X^-] - \Delta_s \tilde{Y}_n^{\bullet}[M^+])$ for a number of ion pairs, the required estimate for $\mathcal{Y}_{H,svt}^{\ominus}$ is found as the crossing point of all (fitted) curves in the limit of large n. In Ref.[2259], this approach was applied to the alkali and halide ions except Cs^+, considering water as a solvent and with $Y = G$ or H, and $n = 1 - 6$. In Ref.[1747], the same approach was applied to the alkali and halide ions, as well as many other ions, including polyatomic ions such as *e.g.* alkyloxonium or alkylammonium ions, considering water as a solvent and with $Y = G$ and $n = 1 - 6$. Here also, the curves in the above graph are very close to straight lines and already share a common intersection point for $n = 0, 1$ and 2. The cluster-pair approximation[170, 1875, 1876] and the cluster-pair correlation[1747, 2259] methods are often quoted as not relying on any extra-thermodynamic assumption. Considering the entire content of the present book, this is of course very unlikely. In fact, the assumption resides in the acceptance that the microscopic clusters considered have not reached their macroscopic-system value in terms of surface potential, and that the corresponding contribution to the cluster solvation parameters can be neglected. This assumption is opposite to that made in the context of cluster-based methods to evaluate the absolute potential of the reference hydrogen electrode (Refs.[1793–1795], see Point E_{12} in Section 5.14), where significantly larger clusters are considered. If the limit of Eq. 5.90 was really taken towards n values representative of the macroscopic regime, these approaches should deliver an estimate for the real rather than the intrinsic solvation parameters of the proton. In other words, the graphs considered for the evaluation of $\mathcal{Y}_{H,svt}^{\ominus}$ only present apparent convergence to a single point for small n values, and the ordinate of this point should actually progressively shift from $\mathcal{Y}_{H,svt}^{\ominus}$ to $\hat{\mathcal{Y}}_{H,svt}^{\ominus}$ upon considering higher n values. Nevertheless, the convergence to a single point observed for distinct small n values provides convincing evidence that the extrapolation procedure is licit within a limited range of cluster sizes, and likely to provide a reasonable estimate for the intrinsic solvation parameters of the proton. Still, as will be discussed further below, the resulting entropy estimates (*e.g.* $\mathcal{S}_{H,wat}^{\ominus} = -154.8$ J·mol^{-1}·K^{-1} for Refs.[1875,1876]) present significant discrepancies with established estimates based on thermocell measurements (*e.g.* the recommendation of Ref.[274], $\mathcal{S}_{H,wat}^{\ominus} = -131.15$ J·mol^{-1}·K^{-1}), which might in part be due to the issue discussed above, *i.e.* even if the surface potential is indeed negligible in the context of the small clusters considered, the same might not be true for its temperature derivative. Finally, as pointed out recently by Ref.[483], some cluster-pair studies inappropriately rely on Boltzmann rather than Fermi-Dirac statistics for the ideal electron gas in the determination of the formation parameters of gaseous ions. The corresponding correction shifts the resulting free-energy estimate (*e.g.* $\mathcal{G}_H^\ominus = -1104.5$ kJ·mol^{-1} for Refs.[1875,1876]) by about +3.64 kJ·mol^{-1} (*e.g.* to −1100.9 kJ·mol^{-1} in this case).

(F_{14}) Many *other ingenious approaches* and associated extra-thermodynamic assumptions have been devised for the estimation of the intrinsic proton solvation parameters $\mathcal{Y}_{H,svt}^\ominus$. These approaches are too diverse to be described in detail here, and only a few examples are given below. The reader is referred to the original references for more information. In the context of hydration, examples include the approaches of Ref.[1869] (anchoring of conventional data based on the assumption that salts have a minimum solubility if the hydration enthalpies of the constituting ions are equal, and that the solubility increases with increasing difference in hydration enthalpies), of Ref.[2183] (partitioning of pair data using a least-squares fitting procedure relying on Fajans' principle of minimum solubility at matching heats of hydration for the alkali-halides[1869]), of Ref.[2172] (use of the Born equation[84] with an estimate for the radius of the proton $R_I[H_{aq}^+] = 0.073$ nm, based on crystallographic data concerning ice), of Ref.[819] (value for $\mathcal{S}_{H,wat}^\ominus$ chosen based on consideration of experimental data for the thermodynamics of water dissociation, value for $\mathcal{H}_{H,wat}^\ominus$ deduced based on an empirical relationship between the entropy and the formation enthalpy of the aqueous proton as formulated by Ref.[2261]), of Refs.[416,2219] (based on experimental/statistical-mechanical data, and an estimate for the difference in hydration free energies between the neutral ferrocene and the ferricinium cation obtained from the Born equation), and of Ref.[1289] (obtained by the authors of this book using the atomistic-consistency assumption, see Section 6.2.3).

(G_{14}) Combination of an experimental value for the *real solvation free energy* $\hat{\mathcal{G}}_{H,svt}^\ominus$ or *entropy* $\hat{\mathcal{S}}_{H,svt}^\ominus$ of the proton with an estimate for the *air-liquid interfacial potential* $\chi_{svt}^\ominus$ of the pure solvent or its temperature derivative $\partial_T \chi_{svt}^\ominus$ according to Eqs. 5.44 or 5.55. Recommended values for $\hat{\mathcal{G}}_{H,wat}^\ominus$ and $\hat{\mathcal{S}}_{H,wat}^\ominus$ have been provided in Section 5.15, based on a careful analysis of the available experimental data. The extra-thermodynamic assumption involved in this approach is then the one applied in the evaluation of $\chi_{wat}^\ominus$ or $\partial_T \chi_{wat}^\ominus$ as described in Section 5.17. Due to the considerable uncertainty in the estimated values for these parameters, this approach is moderately reliable.

The methods attempting the determination of $s_{H,svt}^\ominus$ for a given solvent (prominently water) and the involved extra-thermodynamic assumptions can be broadly classified as follows:

(A_{15}) Assumption of *equal partial molar entropies for a selected cation-anion pair*, *e.g.* in the case of water, K^+ and Cl^- (Ref.[2197]) or H^+ and

OH^-(Ref.[819]), or for a *selected redox pair*, *e.g.* dipyridyl or di-*o*-phenantroyl complexes of Fe(II) and Fe(III) or Ru(II) and Ru(III), respectively (Ref.[2262]). This approach is the analog, in terms of partial molar entropies, of method A_{14} for solvation parameters, and the description and comments provided there also apply here.

(B_{15}) *Correlation of conventional partial molar entropies* with some (experimental, estimated or theoretical) *quantity assumed monotonically related to the corresponding intrinsic partial molar entropies*, *e.g.* the single-ion viscosity B-coefficient entering into the Jones-Dole equation[2263] (Ref.[257, 2190]; the viscosity B-coefficient must be estimated, which requires an additivity[2264] and an extra-thermodynamic[111] assumption), the intrinsic single-ion solvation enthalpy (Ref.[2153]; the enthalpy must be estimated, which requires an extra-thermodynamic assumption), the intrinsic single-ion partial molar heat capacity (Ref.[1881, 2196]; the heat capacity must be estimated, which requires an extra-thermodynamic assumption), or the electrostatic contribution to the intrinsic single-ion solvation entropy estimated using a modified Born equation (Ref.[814]; theoretical). This approach is the analog, in terms of partial molar entropies, of method B_{14} for solvation parameters, and the description and comments provided there also apply here.

(C_{15}) *Fitting of pairwise partial molar entropies at common cation* (or anion) against the *inverse of the effective ionic radius* of the cation (or anion) and *extrapolation to infinite ion size*[259, 261, 1748]. This approach is the analog, in terms of partial molar entropies, of method C_{14} for solvation parameters and the description and comments presented there also apply here.

(D_{15}) Based on *thermocell measurements*[2193,2198,2200,2265–2268] (see Refs.[2198,2269,2270] for the principle of these measurements and Ref.[274] for a review). A thermocell is a Galvanic cell (Section 4.4.3) where the two half-cells are identical but maintained at different temperatures. Such cells have already been encountered in the context of the determination of the thermal temperature derivatives of redox potentials (Sections 4.4.3 and 5.5). At first sight, one would expect the derivative of the electromotive force with respect to the temperature T of the second half-cell, the first one being held at temperature T^-, to be immediately related to the entropy of the half-cell reaction, from which one could easily deduce $s^{\ominus}_{H,svt}$ using appropriate entropy data concerning the metal of the electrode, the corresponding gas-phase cation and the gas-phase electron. The situation is, however, not so simple due to the presence of a liquid junction in the cell that is subject to a temperature gradient, and to the presence of a corresponding temperature gradient between the wires connecting the potentiometer to the cell. At the level of the junction, the temperature gradient induces the appearance of a spurious entropy contribution related to the transfer of ions between the two half-cells. This contribution can be evaluated and removed according to a procedure described in Refs.[2198, 2269, 2270]. At the level of both the junction and the wires, this gradient induces additional spurious contributions related to the appearance of a corresponding concentration or potential gradient, respectively, *via* the Soret effect. This contribution is typically neglected at the level of the wires, while it can be minimized at the level of the liquid junction by adjustment of the experimental setup[2198]. As noted previously[274], all thermocell methods of this type rely at some point on the application of some model or extra-thermodynamic assumption. Nevertheless, these approaches probably still represent the most reliable methods to date for the estimation of $s^{\ominus}_{H,svt}$.

(E_{15}) *Data compilations* or *updated estimates* based on a consideration of previous studies[258, 264, 269, 274, 275, 663, 733, 754, 821, 1749, 1874, 2155, 2189, 2193].

(F_{15}) *Theoretical approaches, e.g.* modified Born equations[644,733,1748] or mean-spherical-approximation (MSA) model[822,2184,2246,2251,2252], or approaches[416,2219] involving *differences between the partial molar entropies of solvated neutral and ionic compounds* estimated using the Born equation. In the context of water, examples of this type include the estimates of Ref.[733] (modified Born equation, accounting for electrostriction based on Ref.[732]), of Ref.[1748] (modified Born equation, in analogy with Ref.[2172]), of Ref.[644] (sum of a charging term from the Born equation and a term accounting for water motional restriction and ionic translational entropy), of Ref.[1750] (MSA model[822, 2184, 2246, 2251, 2252]), or of Refs.[416, 2219] (based on an estimate for the difference between the solvation entropies of neutral ferrocene and of the ferricinium cation obtained using the Born equation).

(G_{15}) *Miscellaneous approaches, e.g.* Ref.[2183] (partitioning of pair data using a least-squares fit procedure relying on Fajans' principle of minimum solubility at matching heats of hydration for the alkali-halides[1869]).

(H_{15}) Estimates derived from $\mathcal{S}^{\ominus}_{H,svt}$, as determined by methods of type A_{14}-G_{14}, by combination with gas-phase ion entropies (Table 5.9) according to

$$s^{\ominus}_{H,svt} = \mathcal{S}^{\ominus}_{H,svt} + s^{\ominus}[H^+_g] \quad . \tag{5.95}$$

The methods attempting the determination of $c^{\ominus}_{P,H,svt}$ for a given solvent (prominently water) and the involved extra-thermodynamic assumptions can be broadly classified as follows:

(A_{16}) Assumption of *equal partial molar heat capacities for a selected cation-anion pair*[261,2105,2202,2204] *e.g.*, in the case of water, K^+ and F^- with equal partitioning in Ref.[261] or Ph_4P^+ and Ph_4B^- with equal partitioning in Refs.[2105, 2202, 2204]. This approach is the analog, in terms of partial molar heat capacities, of method A_{14} for solvation parameters, and the description and comments provided there also apply here (see also Point A_{15} for partial molar entropies).

(B_{16}) Assumption of *equal partial molar heat capacities for a given ion and a homologous neutral compound*[274, 2201, 2204] *e.g.*, in the case of water, $(HOCH_2)_4C$ and $(HOCH_2)_4P^+$ with equal partitioning in Ref.[2201] or Ref.[274], in the latter case including a correction due to the electrostatic influence of the charged species on the solvent, as derived from the Born equation, or Ph_4C and Ph_4P^+ with equal partitioning in Ref.[2204].

(C_{16}) *Fitting of pairwise partial molar heat capacities at common anion* and *extrapolation to zero molecular weight of the cation*[2203, 2204] *e.g.*, in the case of water, using R_4NBr salts, where R is an alkyl group, in Refs.[2203, 2204]. This approach is the analog, in terms of partial molar heat capacities, of method C_{14} for solvation parameters, and the description and comments provided there also apply here (see also Point C_{15} for partial molar entropies). It has been criticized by Refs.[274, 2204].

(D_{16}) Based on *temperature derivatives of thermocell potentials*, as in Ref.[2196]. See Point D_{15} above.

(E_{16}) *Theoretical approaches e.g.*, in the case of water, second temperature derivative of the Born equation used to formulate an estimate for the difference in partial molar heat capacities of aqueous ferrocene and the aqueous ferricinium cation in Refs.[416, 2219].

The methods attempting the determination of $v^{\ominus}_{H,svt}$ for a given solvent (prominently water) and the involved extra-thermodynamic assumptions can be broadly classified as follows:

(A_{17}) Assumption of *equal partial molar volumes for a selected cation-anion pair* (or of a specific ratio) *e.g.*, in the case of water, division of the data for CsCl in proportion to estimated ionic volumes of Cs^+ and Cl^- in the CsCl crystal for Refs.[643, 2110], equal partitioning for Cl^- and NH_4^+ in Ref.[2205], asymmetric partitioning for Ph_4As^+ and Ph_4B^- based on Stokes or van der Waals radii ratios in Ref.[1302] , asymmetric partitioning for Ph_4As^+ and Ph_4B^- based on a value for the volume difference of As and B in tetrahedral covalent species in Ref.[1302], asymmetric partitioning for Ph_4As^+ and Ph_4B^- based on van der Waals volume ratios in Ref.[2159], partitioning for Ph_4P^+ and Ph_4B^- in equal ratios[2105], asymmetric partitioning for Ph_4P^+ and Ph_4B^- based on van der Waals volume ratios in Ref.[2105], and asymmetric partitioning for Ph_4P^+ and Ph_4B^- using corrections for the effect of charge and data on corresponding neutral model compounds in Ref.[2204]. This approach is the analog, in terms of partial molar volumes, of method A_{14} for solvation parameters, and the description and comments provided there also apply here (see also Points A_{15} and A_{16} for partial molar entropies and heat capacities, respectively).

(B_{17}) *Theoretical approaches e.g.*, in the case of water, using intrinsic ionic volumes and electrostriction corrections based on the Drude-Nernst equation[728] in Refs.[736,879], electrostriction corrections according to Ref.[730] with own modifications in Ref.[409], electrostriction corrections according to Ref.[735] in Ref.[737], or electrostriction corrections based on own equations[2214], or using[1710] expressions for the partial molar ionic volumes derived in the context of Kirkwood-Buff theory[2271] and applied to polyatomic molecules with the help of integral equation (RISM) theory[2272] in Refs.[2273,2274].

(C_{17}) *Fitting of pairwise partial molar volumes* at common cation (or anion) reduced by the excluded (hard-sphere) volume of the anion (or cation), using a *physical model* for this excluded volume and the electrostrictive volume reduction, and *extrapolation to infinite anion (or cation) size*, where the pairwise partial molar volume reduced by the excluded volume of the anion (or cation) can be identified with the partial molar volume of the anion (or cation) only[2212]. The physical models are *e.g.* based on geometric considerations of dead space around spheres for the intrinsic volumes[2212] and on electrostriction volumes based on the pressure expression of Ref.[732]. This approach is the analog, in terms of partial molar volumes, of method C_{14} for solvation parameters, and the description and comments provided there also apply here (see also Points C_{15} and C_{16} for partial molar entropies and heat capacities, respectively).

(D_{17}) *Theoretical approaches* applied to *a chosen subset of ions* with characteristics that render them *particularly suitable for a theoretical treatment*, and obtension of the partial molar volumes of the other ions from experimental pairwise data *e.g.*, in the case of water, approximation of the partial molar volume of an anion by its excluded volume based on the assumption that anion solvation is not affected by electrostriction effects[2207], or approximation of the partial molar volume of large tetraalkylammonium cations as a scaled version of their van der Waals volumes (common proportionality constant), which allows the determination of the partial molar volume of the proton as the intercept in a graph representing $v^{\ominus}[\mathrm{RNH_4}X_{aq}] - v^{\ominus}[\mathrm{H}X_{aq}]$ against van der Waals volumes of the cations[2275].

(E_{17}) Based on the measurement of *ultrasonic vibration potentials*[2121,2276], as in Ref.[2122]. See Section 5.7 for a discussion of these approaches.

(F_{17}) *Fitting of the conventional partial molar volumes* by an empirical function of the effective ionic radius, and *addition of a constant which makes the anion and cation values fall onto the same line e.g.*, in the case of water, a function of R_I^3, using conventional values for partial molar volumes in Refs.[736,2206], or a function of R_I using conventional values for hydration volumes in Ref.[2211] (*i.e.* the difference between conventional partial molar volumes and excluded ionic volumes). This approach is the analog, in terms of partial molar volumes, of method B_{14} for solvation parameters, and the description and comments provided there also apply here (see also Point B_{15} for partial molar entropies).

(G_{17}) *Fitting of the conventional partial molar volumes* to a model equation for the corresponding intrinsic partial molar volumes (for cations and anions separately) *e.g.*, in the case of water, using the pressure derivative of the Born equation and a cavitation contribution in Ref.[261], or using an own equation parameterized as a function of ionic radii against pairwise data from Ref.[2277], with subsequent use of these parameters in another own equation expressing single-ion partial molar ionic volumes in Ref.[2278]. This approach is also the analog, in terms of partial molar volumes, of method B_{14} for solvation parameters, and the description and comments provided there also apply here (see also Point B_{15} for partial molar entropies).

(H_{17}) *Fitting of pairwise partial molar volumes at common anion* and *extrapolation to zero molecular weight or zero ionic radius of the cation e.g.*, in the case of water, using R_4NI salts in Ref.[263], where R is an alkyl group, R_4NX salts in Ref.[2213], where X is a halide anion, R_4NBr salts in Refs.[2210,2215], MCl salts in Ref.[2212], where M is an alkali ion. This approach is the analog, in terms of partial molar volumes, of method C_{14} for solvation parameters, and the description and comments provided there also apply here (see also Points C_{15} and C_{16} for partial molar entropies and heat capacities, respectively).

(I_{17}) Assumption of *equal partial molar volumes for a given ion and a homologous neutral compound e.g.*, in the case of water, $(HOCH_2)_4C$ and $(HOCH_2)_4P^+$, as done in Ref.[2201]. This approach is the analog, in terms of partial molar volumes, of method B_{16} for partial molar heat capacities, and the description and comments provided there also apply here.

(J_{17}) *Data compilations* or *updated estimates* based on a consideration of previous studies[274,710].

The methods attempting the determination of $\kappa^{\ominus}_{T,H,svt}$ for a given solvent (prominently water) and the involved extra-thermodynamic assumptions can be broadly classified as follows:

(A_{18}) *Fitting of the conventional partial molar volume-compressibilities* to a model equation for the corresponding intrinsic partial molar volume-compressibilities *e.g.*, in the case of water, the Born model, including an estimate for the non-electrostatic contribution, as done in Ref.[261]. This approach is the analog, in terms of partial molar volume-compressibilities, of method B_{14} for solvation parameters, and the description and comments provided there also apply here (see also Points B_{15} and G_{17} for partial molar entropies and volumes, respectively).

(B_{18}) *Fitting of pairwise partial molar volume-compressibilities at common anion* and *extrapolation to zero molecular weight or zero ionic radius of the cation e.g.*, in the case of water, using R_4N^+ halides, where R is an alkyl group, as done in Ref.[2216]. This approach is the analog, in terms of partial molar volume-compressibilities, of method C_{14} for solvation parameters, and the description and comments provided there also apply here (see also Points C_{15}, C_{16} and C_{17} for partial molar entropies, heat capacities and volumes, respectively).

(C_{18}) *Theoretical approaches, e.g.* integral-equation theory[1710], or empirical models[2214] relating water compressibility and field strength.

(D_{18}) Assumption of *a certain ratio of partial molar volume-compressibilities for a selected cation-anion pair* and *partitioning according to this ratio e.g.* in the case of water, division of pair data according to a certain ratio of ionic radii[2128]. This approach is the analog, in terms of partial molar volume-compressibilities, of method A_{14} for solvation parameters, and the description and comments provided there also apply here (see also Points A_{15}, A_{16} and A_{17} for partial molar entropies, heat capacities and volumes, respectively).

The methods attempting the determination of $\boldsymbol{a}^{\ominus}_{P,H,svt}$ for a given solvent (prominently water) and the involved extra-thermodynamic assumptions can be broadly classified as follows:

(A_{19}) *Fitting of pairwise partial molar volume-expansivities at common anion* and *extrapolation to zero molecular weight or zero ionic radius of the cation e.g.*, in the case of water, using R_4N^+ chlorides, where R is an alkyl group[2210, 2217]. This approach is the analog, in terms of partial molar volume-expansivities, of method C_{14} for solvation parameters, and the description and comments provided there also apply here (see also Points C_{15}, C_{16}, C_{17} and B_{18} for partial molar entropies, heat capacities, volumes and volume-compressibilities, respectively).

(B_{19}) *Fitting of the conventional partial molar volume-expansivities* to a model equation for the corresponding intrinsic absolute partial molar volume-expansivities *e.g.*, in the case of water, the Born model, including an estimate for the non-electrostatic contribution[261]. This approach is the analog, in terms of partial molar volume-expansivities, of method B_{14} for solvation parameters, and the description and comments provided there also apply here (see also Points B_{15}, G_{17} and A_{18} for partial molar entropies, volumes and volume-compressibilities, respectively).

Estimates for the *standard intrinsic hydration parameters* $\mathcal{G}^{\ominus}_{H,wat}$, $\mathcal{H}^{\ominus}_{H,wat}$, and $\mathcal{S}^{\ominus}_{H,wat}$ of the *proton* are reported in Table 5.19. Estimates for the *standard intrinsic partial molar parameters* $s^{\ominus}_{H,wat}$, $v^{\ominus}_{H,wat}$, $c^{\ominus}_{P,H,wat}$, $k^{\ominus}_{T,H,wat}$ and $a^{\ominus}_{P,H,wat}$, of the *aqueous proton* are reported in Table 5.20. For the ease of further reference, these sets are numbered from 1 to 221 based on the literature source, along with the letter g, h, $\bar{\text{s}}$, s, c, k or a to indicate the type of property, namely $\mathcal{G}^{\ominus}_{H,wat}$, $\mathcal{H}^{\ominus}_{H,wat}$, $\mathcal{S}^{\ominus}_{H,wat}$, $s^{\ominus}_{H,wat}$, $c^{\ominus}_{P,H,wat}$, $v^{\ominus}_{H,wat}$, $k^{\ominus}_{T,H,wat}$ and $a^{\ominus}_{P,H,wat}$, respectively, and a subscript in the range $0-12$ characterizing the determination procedure (see further below). Note that the source numbering is joint with that of Tables 5.16, 5.17 and 5.18, so that Tables 5.19 and 5.20 only encompass data from 179 sources (out of 221), for a total of 284 estimates for either of the eight above parameters. The data concerning $\mathcal{G}^{\ominus}_{H,wat}$, $\mathcal{H}^{\ominus}_{H,wat}$, $\mathcal{S}^{\ominus}_{H,wat}$, $s^{\ominus}_{H,wat}$ and $c^{\ominus}_{P,H,wat}$ is assumed to pertain to the standard variant of the solute standard-state definition (Section 4.2.5), while the data concerning $v^{\ominus}_{H,wat}$, $k^{\ominus}_{T,H,wat}$ and $a^{\ominus}_{P,H,wat}$ is assumed to pertain to the density-corrected variant, and therefore noted $v^{\ominus *}_{H,wat}$, $k^{\ominus *}_{T,H,wat}$ and $a^{\ominus *}_{P,H,wat}$. The possible errors incurred by this ambiguity will be discussed in Section 5.19 in the context of recommended values for the above quantities. Note that the estimates of type B_{14} could only be corrected to the standard-state convention adopted in this book when a distinct convention was applied consistently between model estimates and source conventional data.

The data reported in these two tables represents an extensive, probably close to exhaustive, compilation of all intrinsic proton parameters that can be deduced from the literature over the past nearly hundred years. Note, however, that about one half of the individual entries reported in these tables correspond to values that were not directly reported in original literature sources, but deduced by the authors of this book from related data. The corresponding deduced entries are indicated between parentheses or square brackets.

In these two tables, the estimates that are reported in the original literature source explicitly were just copied. These sets are marked with a "1" subscript. The estimates that are reported in the original literature source in the form of tentative intrinsic values for a set of ions, without a corresponding proton estimate, were aligned onto recommended conventional data as described in Section 5.16. The corresponding inferred proton values are reported between parentheses in the tables, and the associated conventional data can be found in Tables 5.16 and 5.17. These sets are marked with a "4" subscript. In Table 5.19, values deduced from recommended real proton solvation free energies and entropies (Section 5.14; $\hat{\mathcal{G}}^{\ominus}_{H,wat} = -1087.0$ kJ·mol^{-1}, $\hat{\mathcal{S}}^{\ominus}_{H,wat} = -90.1$ J·mol^{-1}·K^{-1}) and estimates for the air-liquid interfacial potential $\chi^{\ominus}_{wat}$ of water or its temperature derivative $\partial_T \chi^{\ominus}_{wat}$ in Table 5.18 (method G_{14} above) are also reported between parentheses. These sets are marked with a "12" subscript. In Table 5.20, values derived from the solvation entropies reported in Table 5.19 (method H_{15} above) are also added between parentheses and the associated conventional data can be found in Table 5.17.

These sets are marked with a subscript that is increased by 6 relative to that of the source set. As a result, the data reported for $s^{\ominus}_{H,wat}$ in Table 5.20 is complete, while that reported for $\mathcal{S}^{\ominus}_{H,wat}$ in Table 5.19 is only partial.

Another estimate of the error is provided by the violations of the Gibbs equation ($\mathcal{G}^{\ominus}_{H,wat} - \mathcal{H}^{\ominus}_{H,wat} + T^{-}\mathcal{S}^{\ominus}_{H,wat} = 0$ for sources reporting simultaneously the three quantities) and of the predicted gas-phase entropy of the proton ($s^{\ominus}_{H,wat} - \mathcal{S}^{\ominus}_{H,wat} - s^{\ominus}[H^+_g] = 0$, for sources reporting simultaneously the two former quantities and based on the recommended value of Table 5.9 for the latter quantity). These violations are also reported in the tables (ϵ^{max}_g and ϵ^{max}_e entries) whenever relevant. The data presented in Tables 5.19 and 5.20 is illustrated graphically in Figure 5.7.

The averages of the 98, 40 and 30 estimates for $\mathcal{G}^{\ominus}_{H,wat}$, $\mathcal{H}^{\ominus}_{H,wat}$ and $\mathcal{S}^{\ominus}_{H,wat}$ reported in Table 5.19 are -1098.2 kJ·mol^{-1}, -1116.1 kJ·mol^{-1} and -131.4 J·mol^{-1}·K^{-1}, respectively, with standard deviations of 50.0 kJ·mol^{-1}, 49.7 kJ·mol^{-1} and 12.4 J·mol^{-1}·K^{-1}, respectively, and estimated errors on the mean of ± 5.1 kJ·mol^{-1}, ± 7.9 kJ·mol^{-1} and ± 12.4 J·mol^{-1}·K^{-1}, respectively, assuming that all measurements are independent. Note, however, that the spreads in the suggested values are extremely large, spanning corresponding ranges with widths of about 410 kJ·mol^{-1}, 309 kJ·mol^{-1} and 48 J·mol^{-1}·K^{-1}, respectively. Considering the above discussion of potential shortcomings of the different approaches, as well as the relationship with absolute electrode potential, real proton and surface potential data (Sections 5.14, 5.15 and 5.17), the authors of the present book recommend the values $\mathcal{G}^{\ominus}_{H,wat} = -1100 \pm 5$ kJ·mol^{-1}, $\mathcal{H}^{\ominus}_{H,wat} = -1139 \pm 8$ kJ·mol^{-1} and $\mathcal{S}^{\ominus}_{H,wat} = -131 \pm 2$ J·mol^{-1}·K^{-1} for the standard proton hydration parameters, the latter two in the standard variant of the solute standard-state definition. These values are those retained for the final compilation of recommended data in Section 5.19. Note that the suggested error on $\mathcal{S}^{\ominus}_{H,wat}$ is actually derived from the analysis of the more complete data on the partial molar entropy $s^{\ominus}_{H,wat}$ of the aqueous proton (see below).

Although the convictions as to which value is the most appropriate for anchoring the intrinsic scale of single-ion solvation free energies relative to the conventional scale have differed widely throughout the last century, *two estimates* have been particularly *popular* in the recent years. Many studies[1464, 1726, 1747, 2248, 2279, 2280] consider that the best estimate for $\mathcal{G}^{\ominus}_{H,wat}$ to date is that of Tissandier *et al.*[1875, 1876] (-1104.5 kJ·mol^{-1}), derived from conventional values and experiments on ion-water clusters in vacuum (cluster-pair extra-thermodynamic assumption, Point E_{14}), generally without correction from Boltzmann to Fermi-Dirac statistics for the standard-state electron[483, 1723] (corrected value of -1100.9 kJ·mol^{-1}). This value is significantly more negative than the other commonly employed[877, 1327, 1329, 1331] value of Marcus[275] (-1056 kJ·mol^{-1}), derived from the Gibbs equation with $\mathcal{H}^{\ominus}_{H,wat} = -1094 \pm 6$ kJ·mol^{-1} (a value that Marcus[275] attributes to the approach of Halliwell & Nyburg[260], corresponding to Point C_{14}; note, however, that the

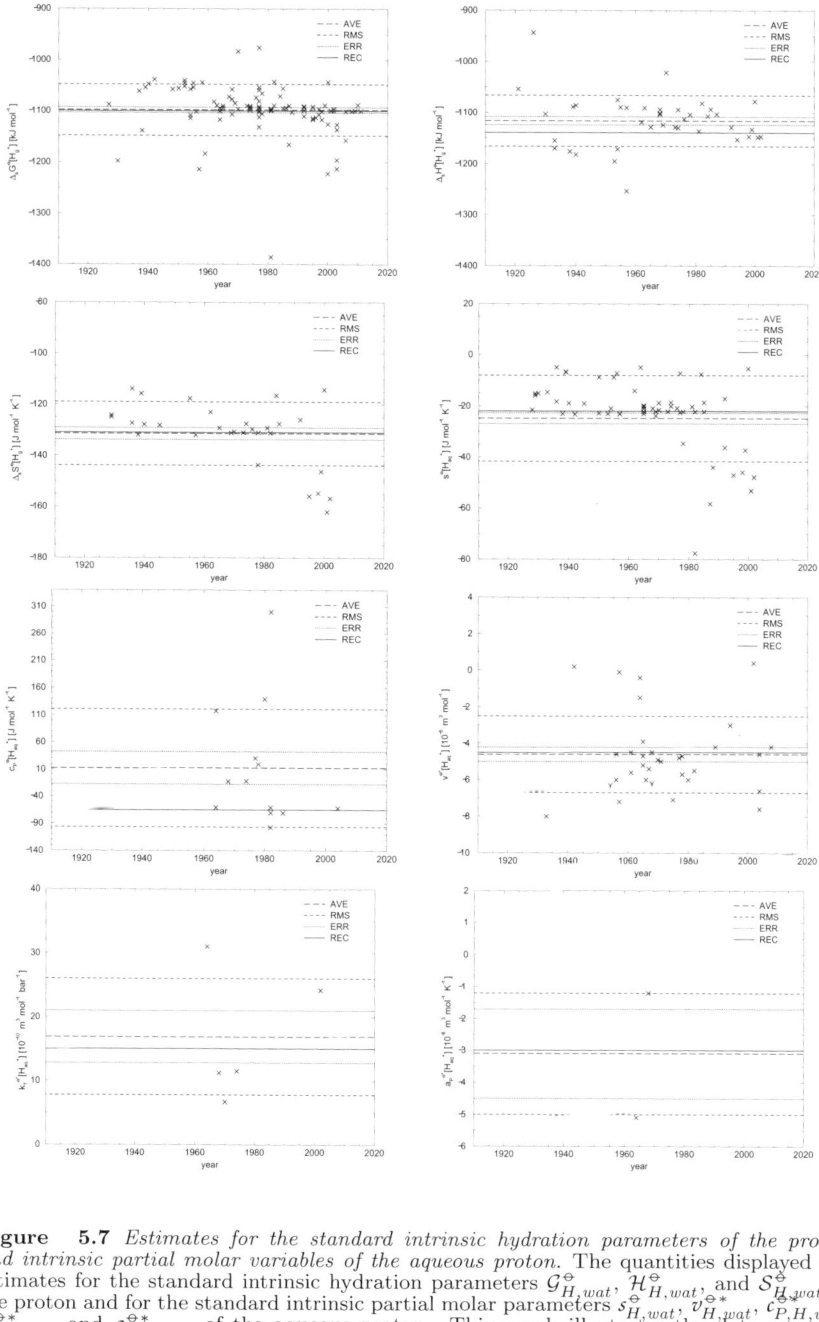

Figure 5.7 *Estimates for the standard intrinsic hydration parameters of the proton and intrinsic partial molar variables of the aqueous proton.* The quantities displayed are estimates for the standard intrinsic hydration parameters $\mathcal{G}^{\ominus}_{H,wat}$, $\mathcal{H}^{\ominus}_{H,wat}$, and $\mathcal{S}^{\ominus}_{H,wat}$ of the proton and for the standard intrinsic partial molar parameters $s^{\ominus}_{H,wat}$, $v^{\ominus*}_{H,wat}$, $c^{\ominus*}_{P,H,wat}$, $\kappa^{\ominus*}_{T,H,wat}$ and $a^{\ominus*}_{P,H,wat}$ of the aqueous proton. This graph illustrates the data reported in Tables 5.19 concerning the former three quantities and 5.20 concerning the latter five quantities, the estimates being sorted by publication year.

Table 5.19 *Available estimates for the intrinsic solvation parameters of the proton in water.* The reported data represents estimates for the proton hydration free energy $\mathcal{G}^{\ominus}_{H,wat}$, enthalpy $\mathcal{H}^{\ominus}_{H,wat}$ and entropy $\mathcal{S}^{\ominus}_{H,wat}$. The methods and involved extra-thermodynamic assumptions refer to the literal enumeration of Section 5.18. The different sets correspond to values quoted in different literature sources, listed in chronological order of publication for the different methods separately. Sources explicitly reporting intrinsic proton solvation parameters are listed, as well as values deduced by optimal alignment of reported intrinsic sets omitting a proton value against conventional data (Table 5.16 and Section 5.16), and values deduced from the recommended real proton hydration parameters (Table 5.15 and Section 5.15), along with estimates for the air-liquid interfacial potential of water (Table 5.18 and Section 5.17). Values deduced in this way are indicated between parentheses. Solvation entropy estimates deduced from additional estimates for the partial molar entropy of the aqueous proton in Table 5.20 are not included. The lines NUM, AVE, RMS, ERR and REC report the number of entries, average value, root-mean-square deviation, estimated error (NUM$^{-1/2}$·RMS) and recommended value for the preceding block. The entries ϵ_g^{max} and ϵ_e^{max} report maximum violations of the Gibbs equation ($\mathcal{G}^{\ominus}_{H,wat} - \mathcal{H}^{\ominus}_{H,wat} + T^{-}\mathcal{S}^{\ominus}_{H,wat}$) and of the predicted gas-phase entropies ($s^{\ominus}_{H,wat} - \mathcal{S}^{\ominus}_{H,wat} - s^{\ominus}[\mathrm{H}^{+}_{g}]$, using data from Table 5.9), whenever relevant. The values reported between square brackets were calculated based on the Gibbs equation or the gas-phase entropy of the proton. The set numbering corresponds to that used in Tables 5.16 and 5.17. The column "Source" provides the source set of a set derived from Table 5.18. The standard states are according to the $bbme_T$ convention (Section 4.2.4; reference pressure $P^{\circ} = 1$ bar, reference molality $b^{\circ} = 1$ mol·kg^{-1} and reference temperature $T^{-} = 298.15$ K). The experimental values reported in the original references are assumed here to pertain to the standard variant (Section 4.2.5) for $\mathcal{H}^{\ominus}_{H,wat}$ and $\mathcal{S}^{\ominus}_{H,wat}$, which represents the most likely situation, but may not necessarily be the case in reality. The solvation free energy $\mathcal{G}^{\ominus}_{H,wat}$ is identical in the two variants.

Set	Method	$\mathcal{G}^{\ominus}_{H,wat}$ [kJ·mol^{-1}]	$\mathcal{H}^{\ominus}_{H,wat}$ [kJ·mol^{-1}]	$\mathcal{S}^{\ominus}_{H,wat}$ [J·mol^{-1}·K^{-1}]	ϵ_g^{max}	ϵ_e^{max}	Source	Ref.	Notes	Year
$56h_4$	A_{14}	-	(-1103.1)	-	-	-	-	1778	a	1930
$47h_1$	A_{14}	-	-1154.8	-	-	-	-	643	b	1933
$48h_1$	A_{14}	-	-1182.0	-	-	-	-	1870	b	1940
$59h_4$	A_{14}	-	(-1195.1)	-	-	-	-	2188	a	1953
$54h_1$	A_{14}	-	-1103.0	-	-	-	-	2185 - 2187	b	1987
$148h_1$	B_{14}	-	-1171.5	-	-	-	-	1871	b,c	1954
$5g_1,h_1,\bar{s}_1$	B_{14}	-1213.8	[-1253.1]	[-132.0]	0.06	0.05	-	754	b,d,e,f,g,h	1957
$19g_1,h_1,\bar{s}_1$	B_{14}	-1091.0	[-1128.6]	-126.0	0.03	0.05	-	822	b,e,f,i	1992
$50h_1$	C_{14}	-	-1090.8	-	-	-	-	260	b	1963
$51h_1$	C_{14}	-	-1103.3	-	-	-	-	1872	b	1968
$10g_1,h_1,\bar{s}_1$	C_{14}	-983.2	-1022.2	[-130.6]	0.06	-	-	1873	b,f,j	1970
$6g_1,h_1,\bar{s}_1$	D_{14}	-1082.0	-1118.6	-123.0	0.07	-	-	259	b,d,k	1962
$52h_1$	D_{14}	-	-1094.1	-	-	-	-	1872	b,m	1968
$53h_1$	D_{14}	-	-1101.2	-	-	-	-	1872	b,n	1968
$17g_1,h_1,\bar{s}_1$	D_{14}	-1043.3	-1082.4	[-131.2]	0.02	0.05	-	1874	b,f,j,n	1982
$16g_1,h_1,\bar{s}_1$	E_{14}	-1097.9	-1136.4	[-129.2]	0.02	-	-	417	b,f,j	1981
$55h_1$	E_{14}	-	-1152.5	-	-	-	-	164	b	1994
$20g_1,h_1,\bar{s}_1$	E_{14}	-1100.9	-1147.0	[-154.8]	0.05	-	-	1875,1876	b,f,i,o	1998
$145g_1,h_1,\bar{s}_1$	E_{14}	-1099.6	-1147.9	[-162.1]	0.03	0.05	-	170	b,e,f,i,o	2001
$146g_1,h_1,\bar{s}_1$	E_{14}	-1100.1	-1146.8	[-156.8]	0.05	0.05	-	2259	b,e,f,i,o	2002
$27g_1$	E_{14}	-1100.9	-	-	-	-	-	1747	b,o,p,q	2006
$45h_1$	F_{14}	-	-1054.4	-	-	-	-	1869	b,r	1921
$46h_1$	F_{14}	-	-943.7	-	-	-	-	2172	b,d,s	1926
$65\bar{s}_4$	F_{14}	-	-	(-124.9)	-	-	-	733	a,j,t,u	1929
$66\bar{s}_4$	F_{14}	-	-	(-124.2)	-	-	-	733	a,d	1929
$29g_4$	F_{14}	(-1197.8)	-	-	-	-	-	1778	a,d	1930
$58h_4$	F_{14}	-	(-1170.2)	-	-	-	-	643	a,d	1933
$67\bar{s}_4$	F_{14}	-	-	(-127.3)	-	-	-	1748	a,j,t,u	1936
$68\bar{s}_4$	F_{14}	-	-	(-113.9)	-	-	-	1748	a,d,u	1936
$30g_4,h_4,\bar{s}_4$	F_{14}	[-1138.6]	(-1176.0)	(-131.9)	1.93	-	-	644	a,f,j,q,u	1938
$31g_4,h_4,\bar{s}_4$	F_{14}	(-1054.7)	(-1089.1)	(-115.8)	0.13	0.25	-	814	a,p,q,u	1939
$32g_4,h_4,\bar{s}_4$	F_{14}	(-1047.9)	(-1086.1)	(-127.8)	0.10	-	-	2155	a,p,q,u	1940
$62\bar{s}_1$	F_{14}	-	-	-128.1	-	-	-	2189	b,j,u,v	1945
$33g_4$	F_{14}	(-1056.2)	-	-	-	-	-	748	a,d	1950
$3h_1$	F_{14}	-	-1075.3	-	-	-	-	1871	b	1954
$35g_4$	F_{14}	(-1108.4)	-	-	-	-	-	409	a,d	1954
$36g_4,h_4,\bar{s}_4$	F_{14}	(-1054.0)	(-1089.7)	(-117.7)	0.61	-	-	821	a,j,t,u	1955
$37g_4$	F_{14}	(-1045.6)	-	-	-	-	-	821	a,d	1955
$60h_4$	F_{14}	-	(-1090.5)	-	-	-	-	717	a,d	1957
$38g_4$	F_{14}	(-1089.3)	-	-	-	-	-	2182	a,e,i,p,q	1963
$39g_4$	F_{14}	(-1117.1)	-	-	-	-	-	660	a,j	1964
$7g_1$	F_{14}	-1094.9	-	-	-	-	-	660	b,j,w	1964
$8g_1,h_1,\bar{s}_1$	F_{14}	-1089.9	-1128.4	[-129.2]	0.02	0.65	-	262	b,j,f	1965
$9g_1,h_1,\bar{s}_1$	F_{14}	-1085.3	-1124.2	-131.1	0.19	-	-	264	b,e,i,j,u	1969
$12g_1,h_1,\bar{s}_1$	F_{14}	-1089.9	-1128.8	-131.0	0.16	-	-	269	b,j,q,u	1973
$144g_1,h_1,\bar{s}_1$	F_{14}	-1092.0	-1130.0	[-127.5]	0.01	1.45	-	416	b,e,f,i	1974

(continues on the next page)

Table 5.19 *(continued)*

Set	Method	$\mathcal{G}^{\ominus}_{H,wat}$ [kJ·mol^{-1}]	$\mathcal{H}^{\ominus}_{H,wat}$ [kJ·mol^{-1}]	$\mathcal{S}^{\ominus}_{H,wat}$ [J·mol^{-1}·K^{-1}]	ϵ_g^{max}	ϵ_e^{max}	Source	Ref.	Notes	Year
$13g_1,h_1,\bar{s}_1$	F_{14}	-1074.0	-1112.9	-129.7	0.23	-	-	2153	b,u	1976
$14g_1$	F_{14}	-1059.4	-	-	-	-	-	444	b,j	1977
$15g_1,h_1,\bar{s}_1$	F_{14}	-1066.0	-1104.0	-131.1	1.09	-	-	274	b,j,u	1978
$40g_4$	F_{14}	(-1105.7)	-	-	-	-	-	662	a,j,v	1978
$63\bar{s}_1$	F_{14}	-	-	-143.6	-	-	-	663	b,j,u,v	1978
$41g_4,h_1,\bar{s}_4$	F_{14}	(-1071.2)	-1107.0	(-116.6)	1.04	-	-	2183	a,b,e,j	1984
$18g_1,h_1,\bar{s}_1$	F_{14}	-1056.0	-1094.0	[-127.6]	0.04	3.55	-	275	b,f,j	1985
$64\bar{s}_1$	F_{14}	-	-	-156.0	-	-	-	2190	b,i,u,v	1995
$42g_4$	F_{14}	(-1116.8)	-	-	-	-	-	2184	a,d	1998
$21g_4,h_4,\bar{s}_4$	F_{14}	(-1089.2)	[-1132.8]	(-146.2)	0.01	-	-	1750	a,d,f,u	1999
$22g_1,h_1,\bar{s}_1$	F_{14}	-1044.0	-1078.0	-114.3	0.08	-	-	819	b,p,q,u	2000
$23g_1$	F_{14}	-1213.0	-	-	-	-	-	668	b,i,v,x,y	2003
$24g_1$	F_{14}	-1126.5	-	-	-	-	-	668	b,i,v,z,α	2003
$25h_1$	F_{14}	-1195.8	-	-	-	-	-	668	b,i,v,y,α	2003
$26g_1$	F_{14}	-1136.5	-	-	-	-	-	668	b,i,v,x,z	2003
$43g_4$	F_{14}	(-1099.8)	-	-	-	-	-	1571	a,d	2009
$147g_1$	F_{14}	-1100.0	-	-	-	-	-	1289	b	2011
$164x_{12}$	G_{14}	(-1038.8)	-	-	-	-	1X	406	β	1942
$165x_{12}$	G_{14}	(-1058.0)	-	-	-	-	2X	407	β	1948
$166x_{12}$	G_{14}	(-1052.3)	-	-	-	-	3X	408	β	1952
$167x_{12}$	G_{14}	(-1040.7)	-	-	-	-	4X	408	β	1952
$168x_{12}$	G_{14}	(-1045.5)	-	-	-	-	5X	408	β	1952
$169x_{12}$	G_{14}	(-1114.0)	-	-	-	-	6X	409	β	1954
$170x_{12}$	G_{14}	(-1058.0)	-	-	-	-	7X	410	β	1954
$171x_{12}$	G_{14}	(-1044.5)	-	-	-	-	8X	412	β	1958
$172x_{12}$	G_{14}	(-1072.5)	-	-	-	-	9X	413	β	1967
$173x_{12}$	G_{14}	(-1106.3)	-	-	-	-	10X	414	β	1968
$174x_{12}$	G_{14}	(-1058.0)	-	-	-	-	11X	2219	β	1968
$175x_{12}$	G_{14}	(-1090.9)	-	-	-	-	12X	416	β	1974
$176x_{12}$	G_{14}	(-1131.4)	-	-	-	-	13X	272	β	1977
$177x_{12}$	G_{14}	(-1131.4)	-	-	-	-	14X	272	β	1977
$178x_{12}$	G_{14}	(-1112.1)	-	-	-	-	15X	272	β	1977
$179x_{12}$	G_{14}	(-1081.2)	-	-	-	-	16X	272	β	1977
$180x_{12}$	G_{14}	(-1054.2)	-	-	-	-	17X	272	β	1977
$181x_{12}$	G_{14}	(-976.0)	-	-	-	-	18X	272	β	1977
$182x_{12}$	G_{14}	(-1102.4)	-	-	-	-	19X	272	β	1977
$183x_{12}$	G_{14}	(-1099.5)	-	-	-	-	20X	417	β	1981
$184x_{12}$	G_{14}	(-1092.8)	-	-	-	-	21X	418	β	1992
$185x_{12}$	G_{14}	(-1125.6)	-	-	-	-	22X	277	β	2000
$186x_{12}$	G_{14}	(-1222.1)	-	-	-	-	23X	277	β	2000
$187x_{12}$	G_{14}	(-1100.5)	-	-	-	-	24X	419	β	2008
$188x_{12}$	G_{14}	(-1100.5)	-	-	-	-	25X	420	β	2008
$189x_{12}$	G_{14}	(-1101.5)	[-1136.13]	(-116.15)	-	-	26X	448	β	1956
$190x_{12}$	G_{14}	(-1093.8)	[-1132.19]	(-128.75)	-	-	27X	2174	β	1965
$191x_{12}$	G_{14}	(-1096.7)	[-1134.79]	(-127.75)	-	-	28X	415	β	1970
$192x_{12}$	G_{14}	(-1094.7)	[-1129.33]	(-116.15)	-		29X	272	β	1977
$193x_{12}$	G_{14}	(-1099.5)	[-1138.72]	(-131.55)	-	-	30X	272	β	1977
$194x_{12}$	G_{14}	(-1098.6)	[-1141.88]	(-145.15)	-	-	31X	465	β	1992
$195x_{12}$	G_{14}	(-1087.5)	-	-	-	-	32X	443	β	1927
$196x_{12}$	G_{14}	(-1091.8)	-	-	-		33X	444	β	1977
$197x_{12}$	G_{14}	(-1089.4)	[-1145.02]	(-186.55)	-	-	34X	445	β	1982
$198x_{12}$	G_{14}	(-1183.5)	-	-	-	-	35X	449	β	1959
$199x_{12}$	G_{14}	(-1100.7)	-	-	-	-	36X	451	β	1964
$200x_{12}$	G_{14}	(-1094.7)	-	-	-	-	37X	452	β	1974
$201x_{12}$	G_{14}	(-1096.7)	-	-	-	-	38X	453	β	1997
$202x_{12}$	G_{14}	(-1096.7)	-	-	-	-	39X	454	β	2002
$203x_{12}$	G_{14}	(-1061.9)	-	-	-	-	40X	2225	β	1937
$204x_{12}$	G_{14}	(-1096.7)	-	-	-	-	41X	2224	β	1981
$205x_{12}$	G_{14}	(-1098.6)	-	-	-	-	42X	1713	β	1974
$206x_{12}$	G_{14}	(-1077.3)	-	-	-	-	44X	1457	β	1968
$207x_{12}$	G_{14}	(-1386.1)	-	-	-	-	45X	2227	β	1981
$208x_{12}$	G_{14}	(-1094.7)	-	-	-	-	46X	1440	β	1985
$209x_{12}$	G_{14}	(-1093.8)	-	-	-	-	47X	1441	β	1986
$210x_{12}$	G_{14}	(-1165.[illegible])		-	-	-	48X	1442	β	1987
$211x_{12}$	G_{14}	(-1089.9)	[-1139.77]	(-167.25)	-	-	49X	1451	β	1987
$212x_{12}$	G_{14}	(-1102.4)	[-1147.97]	(-152.85)	-	-	50X	1444	β	1988
$213x_{12}$	G_{14}	(-1110.2)	-	-	-	-	51X	465	β	1992
$214x_{12}$	G_{14}	(-1116.9)	-	-	-	-	52X	470	β	1995
$215x_{12}$	G_{14}	(-1091.8)	-	-	-	-	53X	470	β	1995
$216x_{12}$	G_{14}	(-1114.0)	-	-	-	-	54X	1473	β	1995
$217x_{12}$	G_{14}	(-1092.8)	-	-	-	-	55X	1473	β	1995
$218x_{12}$	G_{14}	(-1113.0)	-	-	-	-	56X	850	β	1996
$219x_{12}$	G_{14}	(-1107.3)	-	-	-	-	57X	1446	β	1997

(continues on the next page)

Table 5.19 *(continued)*

Set	Method	$\mathcal{G}^{\ominus}_{H,wat}$ [kJ·mol^{-1}]	$\mathcal{H}^{\ominus}_{H,wat}$ [kJ·mol^{-1}]	$\mathcal{S}^{\ominus}_{H,wat}$ [J·mol^{-1}·K^{-1}]	ϵ_g^{max}	ϵ_e^{max}	Source	Ref.	Notes	Year
220x$_{12}$	G$_{14}$	-	-	(-205.85)	-	-	62X	973	β	2006
221x$_{12}$	G$_{14}$	(-1157.4)	-	-	-	-	63X	707	β	2006
NUM		98	40	30						
AVE		-1098.2	-1116.1	-131.4						
RMS		50.0	49.7	12.4						
ERR		5.1	7.9	2.3						
REC		-1100.0	-1139.0	-131.0						

(a) the indicated reference reports intrinsic values, but not for the proton (the proton value was obtained from optimal alignment); (b) the indicated reference reports an intrinsic value; (c) based on comparison of own conventional data with the absolute data of Ref.[2281]; (d) original value from a theoretical model and corrected by the authors of this book to the standard-variant (addition of 7.95 kJ·mol^{-1} for free energies, -2.29 kJ·mol^{-1} for enthalpies, and -34.34 J·mol^{-1}·K^{-1} for entropies); (e) solution reference concentration not stated explicitly in the quoted reference and assumed by the authors of this book (see possible additional footnote on standard-state corrections); (f) the value in square brackets is calculated using the Gibbs equation ($\mathcal{G}_H^{\ominus} = \mathcal{H}_H^{\ominus} - T^{-}\mathcal{S}_H^{\ominus}$); (g) $\mathcal{S}_H^{\ominus}$ is calculated from $s_H^{\ominus}$ given in the quoted reference ($\mathcal{S}_H^{\ominus} = s_H^{\ominus} - s^{\ominus}[\mathrm{H}_g^+]$, using the value from Table 5.9 for the latter quantity); (h) before fitting the data, the authors of the quoted reference remove a "non-electrostatic contribution" of (presumably) 8.37 kJ·mol^{-1}, in order to consider only the charging contribution in the fitting procedure, and the authors of this book correct the resulting value with 7.95 kJ·mol^{-1} (corresponding to Eqs. 4.192 and 4.193), neglecting the difference; (i) gas-phase reference concentration or pressure not stated explicitly in the quoted reference and assumed by the authors of this book (see possible additional footnote on standard-state corrections); (j) including standard-state conversion from 1 atm to 1 bar gas-phase reference pressure (Eqs. 4.192 and 4.193); (k) before fitting the data, the authors of the quoted reference remove a "non-electrostatic contribution" of -26.57 J·mol^{-1}·K^{-1}, -2.38 kJ·mol^{-1}, and 5.54 kJ·mol^{-1} from the solvation entropies, enthalpies and free energies, respectively, in order to consider only the charging contribution in the fitting procedure, and the authors of this book correct the resulting values by adding -26.67 J·mol^{-1}·K^{-1}, 0 kJ·mol^{-1}, and 7.95 kJ·mol^{-1} (corresponding to Eqs. 4.192 and 4.193), neglecting the difference; (l) based on extrapolation of the charging contribution to the salt (pairwise) solvation enthalpy to infinite cation size for a common anion; (m) based on extrapolation of the charging contribution to the salt (pairwise) solvation enthalpy to infinite anion size for a common cation; (n) since the fitting is performed using the electrostatic contribution to enthalpies of solvation (the aqueous entropy of the proton is taken from Ref.[274]), no correction was applied; (o) original value was corrected for the improper use of Boltzmann statistics for the thermodynamic properties of the ideal-gas electron by adding $-(\Delta_f Y^{\ominus}[\mathrm{H}_g^+] - \Delta_f Y^{(\ominus)}[\mathrm{H}_g^+])$, where the former corresponds to Fermi-Dirac and the latter to Boltzmann statistics (3.64 kJ·mol^{-1}, 3.10 kJ·mol^{-1} and -1.82 J·mol^{-1}·K^{-1} for $Y = G, H, S$, respectively, using data from Table 4.2); (p) including standard-state conversion from 1 mol·dm^{-3} (molar) to 1 bar for the gas-phase reference concentration or pressure (Eqs. 4.194 and 4.195); (q) including standard-state conversion from 1 mol·dm^{-3} (molar) to 1 mol·kg^{-1} (molal) reference concentration in solution (Eqs. 4.198 and 4.199); (r) calculated from the value for $\Delta_s G^{\ominus}[\mathrm{K}^+]$ which is used for anchoring the conventional scale in the indicated reference; (s) the value is actually a free energy, but is compared to enthalpies in this study; (t) including standard-state conversion from 0.041 mol·dm^{-3} (molar) to 1 mol·kg^{-1} (molal) reference concentration in solution (Eqs. 4.198 and 4.199); (u) see also Table 5.20 for a classification of the entropy determination method; (v) including standard-state conversion from a mole fraction of unity to 1 mol·kg^{-1} (molal) reference concentration in solution (Eqs. 4.202 and 4.203); (w) based on consideration of the literature, a value of 418.4 kJ·mol^{-1} was chosen by the quoted reference for the free energy of formation of the aqueous proton and the authors of this book converted this value to a solvation free energy using $\Delta_f G^{\ominus}[\mathrm{H}_g^+]$ of Table 5.9; (x) calculation using radii from Ref.[66]; (y) calculation including additional terms for ion-dipole, ion-induced dipole and ion-quadrupole interactions; (z) calculation omitting additional terms for ion-dipole, ion-induced dipole and ion-quadrupole interactions; (α) calculation using radii from Refs.[2058–2060]; (β) calculated from the surface potential and its temperature derivative (if available) reported in the indicated reference (Table 5.18) according to Eq. 5.44 for $\mathcal{G}_H^{\ominus}$ or Eq. 5.55 for $\mathcal{S}_H^{\ominus}$, using $\hat{\mathcal{G}}_H^{\ominus} = -1087.0$ kJ·mol^{-1} and $\hat{\mathcal{S}}_H^{\ominus} = -90.1$ J·mol^{-1}·K^{-1}.

Table 5.20 *Estimates for the standard intrinsic partial molar variables of the aqueous proton.* The reported data represents estimates for the partial molar entropy $s^{\ominus}_{H,wat}$, heat capacity $c^{\ominus}_{P,H,wat}$, volume $v^{\ominus}_{H,wat}$, volume-compressibility $\kappa^{\ominus}_{T,H,wat}$ and volume-expansivity $a^{\ominus}_{P,H,wat}$ of the aqueous proton. The methods and involved extra-thermodynamic assumptions refer to the literal enumerations of Section 5.18. The different sets correspond to values quoted in different literature sources, listed in chronological order of publication for the different methods separately. Sources explicitly reporting partial molar variables of the aqueous proton are listed, as well as values deduced by optimal alignment of reported intrinsic sets omitting a proton value against conventional data (Table 5.17 and Section 5.16), values deduced from the recommended real proton hydration parameters (for entropies only; Table 5.15 and Section 5.15), along with estimates for the temperature derivative of the air-liquid interfacial potential of water (Table 5.18 and Section 5.17). Deduced values are indicated between parentheses. Note that this table also contains estimates for the partial molar entropy deduced from proton solvation entropy estimates of Table 5.19. The lines NUM, AVE, RMS, ERR and REC report the number of entries, average value, root-mean-square deviation, estimated error ($\mathrm{NUM}^{-1/2}\cdot\mathrm{RMS}$) and recommended value for the preceding block. The set numbering corresponds to that used in Tables 5.16 and 5.17. The column "Source" provides the source set of a set derived from Tables 5.18 or 5.19. The standard states are according to the $bbme_T$ convention (Section 4.2.4; reference pressure $P^{\circ} = 1$ bar, reference molality $b^{\circ} = 1\ \mathrm{mol\cdot kg^{-1}}$ and reference temperature $T^{-} = 298.15$ K). The experimental values reported in the original references are assumed to pertain to the standard variant for $s^{\ominus}_{H,wat}$ and $c^{\ominus}_{P,H,wat}$, and to the density-corrected variant for $v^{\ominus *}_{H,wat}$, $\kappa^{\ominus *}_{T,H,wat}$ and $a^{\ominus *}_{P,H,wat}$ (Section 4.2.5), which represents the most likely situation, but may not necessarily be the case in reality.

Set	Method	$s^{\ominus}_{H,wat}$ [$\mathrm{J\cdot mol^{-1}\cdot K^{-1}}$]	Source	Ref.	Notes	Year
$155s_1$	A_{15}	(-23.0)	-	2262	a	1965
$77s_1$	A_{15}	(-23.8)	-	2197	a,b	1969
$22s_7$	A_{15}	(-5.3)	$22\bar{s}_1$	819	c,d,e	2000
$31s_{10}$	B_{15}	(-6.8)	$31\bar{s}_4$	814	d,e,f	1939
$152s_1$	B_{15}	(-23.0)	-	257	a	1953
$69s_1$	B_{15}	(-20.9)	-	1881,2196	a,b	1968
$13s_7$	B_{15}	(-20.8)	$13\bar{s}_1$	2153	c,g	1976
$64s_7$	B_{15}	(-47.0)	$64\bar{s}_1$	2190	b,c,h	1995
$78s_4$	D_{15}	(-21.5)	-	2198	i,j	1928
$150s_1$	D_{15}	(-14.6)	-	2265	a	1933
$79s_4$	D_{15}	(-23.2)	-	2199	i	1942
$151s_1$	D_{15}	(-8.8)	-	2267	a	1950
$80s_4$	D_{15}	(-22.8)	-	2200	i,k	1950
$154s_1$	D_{15}	(-22.2)	-	2268	a	1965
$65s_{10}$	E_{15}	(-15.9)	$65\bar{s}_4$	733	f,l,m,n	1929
$71s_1$	E_{15}	(-15.1)	-	2193	a,o	1930
$67s_{10}$	E_{15}	(-18.3)	$67\bar{s}_4$	1748	f,l,m,p	1936
$32s_{10}$	E_{15}	(-18.9)	$32\bar{s}_4$	2155	d,e,f,q	1940
$62s_7$	E_{15}	(-19.1)	$62\bar{s}_1$	2189	b,c,l,r	1945
$153s_1$	E_{15}	(-20.9)	-	258	a,s	1954
$36s_{10}$	E_{15}	(-8.8)	$36\bar{s}_4$	821	f,l,m,t	1955
$5s_1$	E_{15}	(-23.0)	-	754	a,u,v	1957
$76s_1$	E_{15}	(-23.0)	-	1749	a,v	1965
$8s_1$	E_{15}	(-20.9)	-	262	a,w	1965
$9s_7$	E_{15}	(-22.1)	$9\bar{s}_1$	264	c,l,h,u,x	1969
$12s_7$	E_{15}	(-22.1)	$12\bar{s}_1$	269	c,e,l,y	1973
$15s_7$	E_{15}	(-22.1)	$15\bar{s}_1$	274	c,l,z	1978
$63s_7$	E_{15}	(-34.6)	$63\bar{s}_1$	663	b,c,l,α	1978
$17s_1$	E_{15}	(-22.2)	-	1874	a,β	1982
$18s_1$	E_{15}	(-22.2)	-	275	a,β	1985
$68s_{10}$	F_{15}	(-4.9)	$68\bar{s}_4$	1748	f,γ	1936
$30s_{10}$	F_{15}	(-23.0)	$30\bar{s}_4$	644	e,f,l	1938
$144s_1$	F_{15}	(-20.0)	-	416	a,h,u,δ	1974
$21s_{10}$	F_{15}	(-37.2)	$21\bar{s}_4$	1750	f,γ	1999

(continues on the next page)

Table 5.20 *(continued)*

Set	Method	$s^{\ominus}_{H,wat}$ [J·mol^{-1}·K^{-1}]	Source	Ref.	Notes	Year
$66s_{10}$		(-15.2)	$66\bar{s}_4$	733	f,γ	1929
$31s_4$		(-6.6)	-	814	e,i	1939
$189x_{12}$		-7.2	26X	448	ϵ	1956
$5s_7$		(-23.1)	$5\bar{s}_1$	754	c,u,ζ	1957
$6s_7$		(-14.1)	$6\bar{s}_1$	259	c,η	1962
$81s_1$		(-4.9)	-	261	a	1964
$190x_{12}$		-19.8	27X	2174	ϵ	1965
$8s_7$		(-20.2)	$8\bar{s}_1$	262	c,θ	1965
$10s_7$		(-21.6)	$10\bar{s}_1$	1873	c,θ	1970
$191x_{12}$		-18.8	28X	415	ϵ	1970
$144s_7$		(-18.6)	$144\bar{s}_1$	416	c,h,u,θ	1974
$192x_{12}$		-7.2	29X	272	ϵ	1977
$193x_{12}$		-22.6	30X	272	ϵ	1977
$16s_7$		(-20.2)	$16\bar{s}_1$	417	c,θ	1981
$17s_7$		(-22.2)	$17\bar{s}_1$	1874	c,θ	1982
$197x_{12}$		-77.6	34X	445 - 447	ϵ	1982
$41s_{10}$		(-7.6)	$41\bar{s}_4$	2183	f,l,u	1984
$18s_7$		(-18.6)	$18\bar{s}_1$	275	c,θ	1985
$211x_{12}$		-58.3	49X	1451	ϵ	1987
$212x_{12}$		-43.9	50X	1444	ϵ	1988
$19s_1$		(-17.1)	-	822	a,h,u,ζ	1992
$194x_{12}$		-36.2	31X	465	ϵ	1992
$20s_7$		(-45.9)	$20\bar{s}_1$	1875,1876	c,h,θ	1998
$145s_1$		(-53.2)	-	170	a,h,u,ζ	2001
$145s_7$		(-53.1)	$145\bar{s}_1$	170	c,h,u,θ	2001
$146s_1$		(-47.8)	-	2259	a,h,u,ζ	2002
$146s_7$		(-47.9)	$146\bar{s}_1$	2259	c,h,u,θ	2002
$220x_{12}$		-96.9	62X	973	ϵ	2006
NUM		63				
AVE		-24.8				
RMS		16.8				
ERR		2.1				
REC		-22.0				

Set	Method	$c^{\ominus}_{P,H,wat}$ [J·mol^{-1}·K^{-1}]	Source	Ref.	Notes	Year
$81c_1$	A_{16}	-60.7	-	261	a,ι	1964
$96c_4$	A_{16}	(-60.8)	-	2204	i	1982
$99c_4$	A_{16}	(-70.3)	-	2204	i	1982
$86c_1$	A_{16}	-71.0	-	2202	a	1986
$87c_1$	A_{16}	-62.0	-	2105	a	2004
$85c_1$	B_{16}	30.5	-	2201	a	1977
$15c_1$	B_{16}	19.0	-	274	a	1978
$98c_4$	B_{16}	(-97.3)	-	2204	i	1982
$95c_4$	C_{16}	(139.1)	-	2203	i	1980
$97c_4$	C_{16}	(299.7)	-	2204	i	1982
$82c_1$	D_{16}	117.2	-	2196	a,κ	1964
$156c_1$	E_{16}	-12.6	-	2219	a	1968
$144c_1$	E_{16}	-12.0	-	416	a,δ	1974
NUM		13				
AVE		12.2				
RMS		108.6				
ERR		30.1				
REC		-65.0				

(continues on the next page)

Table 5.20 *(continued)*

Set	Method	$v^{\ominus *}_{H,wat}$ [10^{-6}m^3·mol^{-1}]	Source	Ref.	Notes	Year
$47v_1$	A_{17}	(-8.0)	-	643	a,λ	1933
$101v_1$	A_{17}	(0.2)	-	2205	a	1942
$103v_1$	A_{17}	(-4.6)	-	2110	a	1956
$130v_1$	A_{17}	(-5.0)	-	2115	a,μ	1971
$131v_4$	A_{17}	(-7.1)	-	2159	i	1975
$134v_4$	A_{17}	(-5.5)	-	2204	i	1982
$160v_1$	A_{17}	(-7.6)	-	2105	a,ν	2004
$161v_1$	A_{17}	(-4.6)	-	2105	a,ξ	2004
$162v_1$	A_{17}	(-6.6)	-	2105	a,o	2004
$35v_4$	B_{17}	(-6.3)	-	409	i	1954
$105v_4$	B_{17}	(-5.6)	-	736	i	1961
$107v_1$	B_{17}	(-0.4)	-	737,2209	a	1964
$120v_4$	B_{17}	(-3.9)	-	2212	i	1965
$127v_4$	B_{17}	(-4.5)	-	2214	i	1968
$136v_4$	B_{17}	(-3.0)	-	457,879	i	1994
$137v_4$	B_{17}	(0.4)	-	1710	i	2002
$104v_1$	D_{17}	(-7.2)	-	2207	a	1957
$158v_1$	D_{17}	(-4.9)	-	2275	a	1970
$108v_1$	E_{17}	(-5.4)	-	2122	a	1967
$102v_1$	F_{17}	(-6.0)	-	2206	a	1956
$106v_1$	F_{17}	(-4.5)	-	736,2208	a	1961
$113v_1$	F_{17}	(-4.2)	-	2211	a	2008
$81v_1$	G_{17}	(-1.5)	-	261	a	1964
$159v_1$	G_{17}	(-4.2)	-	2278	a	1989
$85v_1$	H_{17}	(-4.8)	-	2201	a	1977
$119v_4$	I_{17}	(-4.7)	-	2212	i	1965
$157v_1$	I_{17}	(-5.2)	-	263	a	1965
$121v_4$	I_{17}	(-6.0)	-	2213	i	1966
$109v_1$	I_{17}	(-6.2)	-	2210	a	1968
$133v_4$	I_{17}	(-6.0)	-	2215	i	1980
$118v_4$	J_{17}	(-0.1)	-	710	i,π	1957
$111v_1$	J_{17}	(-5.7)	-	274	a,ρ	1978
$112v_1$	J_{17}	(-4.7)	-	274	a,σ	1978
NUM		33				
AVE		-4.6				
RMS		2.1				
ERR		0.4				
REC		-4.5				

Set	Method	$\kappa^{\ominus *}_{T,H,wat}$ [10^{-10}m^3·mol^{-1}·bar^{-1}]	Source	Ref.	Notes	Year
$81k_1$	A_{18}	(31.0)	-	261	a,τ	1964
$140k_4$	B_{18}	(6.7)	-	2216	i,υ	1970
$127k_4$	C_{18}	(11.2)	-	2214	i	1968
$137k_4$	C_{18}	(24.2)	-	1710	i,ϕ	2002
$141k_4$	D_{18}	(11.5)	-	2128	i,χ	1974
NUM		5				
AVE		16.9				
RMS		9.1				
ERR		4.1				
REC		15.0				

Set	Method	$a^{\ominus *}_{P,H,wat}$ [10^{-8}m^3 mol^{-1}·K^{-1}]	Source	Ref.	Notes	Year
$143a_1$	A_{19}	(-1.2)	-	2210,2217	a,ψ	1968
$81a_1$	B_{19}	(-5.1)	-	261	a,ω	1964
NUM		2				
AVE		-3.1				
RMS		1.9				
ERR		1.4				
REC		-3.0				

(continues on the next page)

Table 5.20 *(continued)*

(a) the indicated reference reports an intrinsic value; (b) including standard-state conversion from a mole fraction of unity to 1 mol·kg^{-1} (molal) reference concentration in solution (Eqs. 4.202 and 4.203); (c) $s_H^{\ominus}$ is calculated from $\mathcal{S}_H^{\ominus}$ given in the quoted reference ($s_H^{\ominus} = \mathcal{S}_H^{\ominus} + s^{\ominus}[\mathrm{H}_g^+]$, using the value from Table 5.9 for the latter quantity); (d) including standard-state conversion from 1 mol·dm^{-3} (molar) to 1 bar for the gas-phase reference concentration or pressure (Eqs. 4.194 and 4.195); (e) including standard-state conversion from 1 mol·dm^{-3} (molar) to 1 mol·kg^{-1} (molal) reference concentration in solution (Eqs. 4.198 and 4.199); (f) the indicated reference reports intrinsic values for $\Delta_s S^{\ominus}[I_g^z]$, but not for the proton (the proton value was obtained from optimal alignment); (g) it appears that the original article has a typo in the value for F^- (wrong sign, corrected in this table); (h) gas-phase reference concentration or pressure not stated explicitly in the quoted reference and assumed by the authors of this book (see possible additional footnotes on standard-state corrections); (i) the indicated reference reports intrinsic values, but not for the proton (the proton value was obtained from optimal alignment); (j) at 15°C; (k) at 30°C; (l) including standard-state conversion from 1 atm to 1 bar gas-phase reference pressure (Eqs. 4.192 and 4.193); (m) including standard-state conversion from 0.041 mol·dm^{-3} (molar) to 1 mol·kg^{-1} (molal) reference concentration in solution (Eqs. 4.198 and 4.199); (n) based on data from Ref.[2198] corresponding to a temperature of 15°C at 0.01 mol·dm^{-3} (molar) concentration, which is converted in the indicated reference to 25°C and a 1 mol·dm^{-3} concentration, using an estimate of $c_{P,H}^{\ominus} = -50.2$ J·mol^{-1}·K^{-1} (the latter value is referenced as "Based on the work of Randall and Rossini, unpublished"); (o) based on a $s^{\ominus}[\mathrm{Cl}_{aq}^-]$ value from the studies of Refs.[733, 2198]; (p) based on data from Refs.[2198,2266]; (q) based on data from Refs.[644,2195]; (r) based on data from Ref.[2266]; (s) based on the $s^{\ominus}[\mathrm{Cl}_{aq}^-]$ value from the studies of Ref.[2198]; (t) based on data from Ref.[2267]; (u) solution reference concentration not stated explicitly in the quoted reference and assumed by the authors of this book (see possible additional footnotes on standard-state corrections); (v) based on data from Ref.[257]; (w) based on consideration of Refs.[257, 258, 2206]; (x) based on consideration of Refs.[257, 1749, 2196, 2198, 2200, 2265, 2268]; (y) based on consideration of Refs.[262, 1744]; (z) based on consideration of Refs.[257, 259, 1749, 2196, 2198, 2200, 2265, 2268]; (α) based on data from Ref.[259]; (β) based on consideration of Ref.[274]; (γ) original value from a theoretical model and corrected by the authors of this book to the standard-variant (addition of 7.95 kJ·mol^{-1} for free energies, -2.29 kJ·mol^{-1} for enthalpies, and -34.34 J·mol^{-1}·K^{-1} for entropies); (δ) it is confusing that Ref.[416] uses a Δ symbol to denote the partial molar entropy and heat capacity of the aqueous proton; (ϵ) calculated from the temperature derivative of the surface potential reported in the indicated reference (Table 5.18) according to Eq. 5.55, using $\hat{S}_H^{\ominus} = -90.1$ J·mol^{-1}·K^{-1} and the gas-phase entropy of the proton reported in Table 5.9; (ζ) calculated from the conventional single-ion partial molar entropies reported in the original source using the gas-phase ion molar entropies of Table 5.9; (η) before fitting the data, the authors of the quoted reference remove a "non-electrostatic contribution" of -26.57 J·mol^{-1}·K^{-1}, -2.38 kJ·mol^{-1}, and 5.54 kJ·mol^{-1} from the solvation entropies, enthalpies and free energies, respectively, in order to consider only the charging contribution in the fitting procedure, and the authors of this book correct the resulting values by adding -26.67 J·mol^{-1}·K^{-1}, 0 kJ·mol^{-1}, and 7.95 kJ·mol^{-1} (corresponding to Eqs. 4.192 and 4.193), neglecting the difference; (θ) the $\mathcal{S}_H^{\ominus}$ value from which $s_H^{\ominus}$ is obtained was calculated using the Gibbs equation ($\mathcal{S}_H^{\ominus} = -(T^-)^{-1}(\mathcal{G}_H^{\ominus} - \mathcal{H}_H^{\ominus})$); ($\iota$) the values for K_{aq}^+ and F_{aq}^- are not equal, because the indicated reference, although saying that they are partitioned equally, assigns -14.3 to the former and -14.2 cal·mol^{-1}·K^{-1} to the former and latter ion, respectively; (κ) according to the original study[2196], these values are "subject to large errors"; (λ) based on apparent volumes in dilute rather than infinitely dilute solution; (μ) average of three different splitting schemes for $\mathrm{Ph_4As^+}$ and $\mathrm{Ph_4B^-}$; (ν) equal partitioning for $\mathrm{Ph_4P^+}$ and $\mathrm{Ph_4B^-}$; (ξ) asymmetric partitioning of $\mathrm{Ph_4P^+}$ and $\mathrm{Ph_4B^-}$ according to the ratio of van der Waals volumes; (o) asymmetric partitioning of $\mathrm{Ph_4P^+}$ and $\mathrm{Ph_4B^-}$ to correct for the effect of charge; (π) original data is based on conventional values of Refs.[1826, 2205, 2206] and an absolute value for Cl_{aq}^- of Ref.[2205]; (ρ) based on consideration of Refs.[263, 2121, 2122, 2213]; (σ) based on the mean of absolute proton values reviewed in Ref.[2163]; (τ) conventional data of Ref.[2108]; (υ) the indicated reference reports adiabatic compressibilities and the value given here is reported by Ref.[2128] for isothermal conditions; (ϕ) the original data is given in units of 10^{-9} Pa^{-1} and was here converted to the indicated units based on ionic partial molar volumes given in Ref.[1710]; (χ) division of NaCl data (for a temperature of 30°C; radius set not specified); (ψ) source data specified in Ref.[2217]; (ω) conventional data of Ref.[1826].

reference quoted for this value of $\mathcal{H}^{\ominus}_{H,wat}$ seems to inappropriately correspond to Glueckauf[2214]) and $\mathcal{S}^{\ominus}_{H,wat} = -131 \pm 1.4$ J·mol^{-1}·K^{-1} (based on the partial molar entropy of -22.2 J·mol^{-1}·K^{-1} for the aqueous proton suggested by Conway[274], itself derived as an average over thermocell measurements[2198] from different sources). Most other commonly quoted experimental estimates are comprised between those of Tissandier *et al.*[1875,1876] and Marcus[275].

The choice of the *recommended value* for $\mathcal{G}^{\ominus}_{H,wat}$ in the present book follows from four main considerations: (*i*) it is close to the average of the 98 independent estimates in Table 5.19; (*ii*) it is the value consistently obtained by application of cluster approaches (Point E_{14}, six estimates[164, 170, 417, 1747, 1875, 1876, 2259], after application of an appropriate correction for Boltzmann to Fermi-Dirac statistics for the ideal electron gas[483]); (*iii*) it is the value supported by preliminary application of the atomistic-consistency assumption (ACA) based on AT simulations[1289] (Section 6.2.3); (*iv*) this value is compatible with the recommended values of -1087 kJ·mol^{-1} and 0.13 V for $\hat{\mathcal{G}}^{\ominus}_{H,wat}$ and $\chi^{\ominus}_{wat}$ suggested in Sections 5.15 and 5.17. The motivations for the choice of the recommended value for $\mathcal{S}^{\ominus}_{H,wat}$ will be provided further below, considering the more complete data concerning $s^{\ominus}_{H,wat}$, related to the latter quantity by Eq. 5.95. Finally, the recommended value for $\mathcal{H}^{\ominus}_{H,wat}$ is deduced by application of the Gibbs equation.

The averages of the 63, 13, 33, 5 and 2 estimates for $s^{\ominus}_{H,wat}$, $c^{\ominus}_{P,H,wat}$, $v^{\ominus*}_{H,wat}$, $k^{\ominus*}_{T,H,wat}$ and $a^{\ominus*}_{P,H,wat}$ reported in Table 5.20 are -24.8 J·mol^{-1}·K^{-1}, 12.2 J·mol^{-1}·K^{-1}, $-4.6 \cdot 10^{-6}$ m^3·mol^{-1}, $16.9 \cdot 10^{-10}$ m^3·mol^{-1}·bar^{-1} and $-3.1 \cdot 10^{-8}$ m^3·mol^{-1}·K^{-1}, respectively, with standard deviations of 16.8 J·mol^{-1}·K^{-1}, 108.6 J·mol^{-1}·K^{-1}, $2.1 \cdot 10^{-6}$ m^3·mol^{-1}, $9.1 \cdot 10^{-10}$ m^3·mol^{-1}·bar^{-1} and $1.9 \cdot 10^{-8}$ m^3·mol^{-1}·K^{-1}, respectively, and estimated errors on the mean of 2.1 J·mol^{-1}·K^{-1}, 30.1 J·mol^{-1}·K^{-1}, $0.4 \cdot 10^{-6}$ m^3·mol^{-1}, $4.1 \cdot 10^{-10}$ m^3·mol^{-1}·bar^{-1} and $1.4 \cdot 10^{-8}$ m^3·mol^{-1}·K^{-1}, respectively, assuming that all measurements are independent. Note, however, that the spreads in the suggested values are extremely large, spanning corresponding ranges with widths of about 92 J·mol^{-1}·K^{-1}, 397 J·mol^{-1}·K^{-1}, $8 \cdot 10^{-6}$ m^3·mol^{-1}, $24 \cdot 10^{-10}$ m^3·mol^{-1}·bar^{-1} and $4 \cdot 10^{-8}$ m^3·mol^{-1}·K^{-1}, respectively. Considering the above discussion of potential shortcomings of the different approaches, the authors of this book recommend the values $s^{\ominus}_{H,wat} = -22 \pm 2$ J·mol^{-1}·K^{-1}, $c^{\ominus}_{P,H,wat} = -65$ J·mol^{-1}·K^{-1} (tentative value only), $v^{\ominus*}_{H,wat} = -4.5 \pm 0.5 \cdot 10^{-6}$ m^3·mol^{-1}, $k^{\ominus*}_{T,H,wat} = 15 \pm 5 \cdot 10^{-10}$ m^3·mol^{-1}·bar^{-1} and $a^{\ominus*}_{P,H,wat} = -3.0 \pm 1.5 \cdot 10^{-8}$ m^3·mol^{-1}·K^{-1} for the standard partial molar variables of the aqueous proton, the former two in the standard variant of the solute standard-state definition and the three latter in the density-corrected variant. These values are those retained for the final compilation of recommended data in Section 5.19. The choice of the recommended value for $s^{\ominus}_{H,wat}$ follows from three main considerations: (*i*) it is close to the average over the 63 independent measurements in Table 5.20; (*ii*) it is the value nearly consistently obtained by application of thermocell experiments (Point D_{15}, four estimates[2198–2200, 2268] considered by the authors of this book to be the most reliable); (*iii*) this value is compatible with the recommended values of -90.1 J·mol^{-1}·K^{-1} and -0.42 mV·K^{-1} for $\hat{s}^{\ominus}_{H,wat}$ and $\partial_T \chi^{\ominus}_{wat}$ suggested in Sections 5.15 and 5.17. Note, however, that the three most recent cluster studies (Point E_{14}) led to estimates that are noticeably larger in magnitude for $\mathcal{S}^{\ominus}_{H,wat}$ and $s^{\ominus}_{H,wat}$, namely[170, 1875, 1876, 2259] about -160 and -50 J·mol^{-1}·K^{-1}, respectively. The recommended value for $v^{\ominus *}_{H,wat}$ appears to be about consensual over the 33 estimates in Table 5.20. The data is sparser concerning $k^{\ominus *}_{T,H,wat}$ and $a^{\ominus *}_{P,H,wat}$, and the above recommended values should be taken with some caution. In terms of $c^{\ominus}_{P,H,wat}$, the 13 suggested values in Table 5.20 span an extremely wide range of positive and negative values. The selected estimate is only tentative and corresponds to a value that is close to consensual for six of these estimates[261, 2105, 2202, 2204].

5.19 Recommended data

The goal of the present section is to provide a compilation of *recommended data* for thermodynamic properties characterizing *alkali and halide hydration* in the infinitely dilute regime, along with *proton hydration*, *pure solvent* and *electrochemical* properties also relevant in this context.

As evident from the content of the previous sections and tables in this chapter, two major difficulties in the area of single-ion thermodynamics and electrochemistry are: (*i*) the existence of many *alternative standard-state conventions*; (*ii*) the existence of many *alternative types of parameters* characterizing the same physical properties from a slightly different perspective. These two features render the comparative analysis of the literature in the area relatively tedious, and are the source of numerous mistakes.

Such a comparative analysis was nevertheless performed in the preceding sections taking into account the two above features as carefully as possible. The present summary attempts to restore simplicity to the problem by reporting data: (*i*) using a single *consistent standard-state convention*, namely the $bbme_T$ convention (Section 4.2.4) including the *standard conventional convention* (Section 4.5.1), and along with specified *solute standard-state variants* (Section 4.2.5) for the different quantities; (*ii*) based on a *non-redundant set of parameters*. The second constraint implies that a minimal set of independent parameters is selected, from which all alternative types of parameters can be derived. A few exceptions to these rules have been accepted for important derived quantities, but the corresponding redundant parameters are explicitly indicated as such in the tables.

Recommended values for the *reticulation parameters* and *salt partial molar variables* of the *alkali-halides* are reported in Table 5.21. These parameters are discussed in Sections 5.11 and 5.12, respectively. The values must be reported in a pairwise form, because they are not partitionable into single-ion contributions (Section 4.2.14).

Recommended values for the *atomization* (gas-phase atom formation), *ionization* and *conventional hydration parameters*, as well as the *element*, *gas-*

phase atom, *gas-phase ion* and *conventional aqueous ion partial molar variables* for the *alkali* and *halides*, along with *hydrogen*, are reported in Table 5.22. The *standard work functions* relevant for the corresponding electrodes (alkali metals or platinum for halides and hydrogen) are also reported. These parameters are discussed in Sections 5.1, 5.9, 5.10, 5.13 and 5.16. Although conventional hydration parameters are not elusive quantities, their determination is still affected by relatively large errors. As discussed in Section 5.16, the recommended data corresponds to sets considered by the authors of this book as the most trustworthy to date, and error estimates derived from the statistical analysis of literature data are also provided in the table. Note that derived pairwise values may slightly differ from those reported in Tables 5.7, 5.11 and 5.12, considering that they were slightly adjusted for exact partitioning into single-ion contributions.

Recommended values for the *real hydration parameters of the proton* and *real partial molar variables of the aqueous proton* are reported in Table 5.23. These parameters are discussed in Section 5.15. In principle, only the values of $\hat{\mathcal{G}}^{\ominus}_{H,wat}$ and $\hat{s}^{\ominus}_{H,wat}$ are non-redundant. Derived values of $\hat{\mathcal{H}}^{\ominus}_{H,wat}$ and $\hat{\mathcal{S}}^{\ominus}_{H,wat}$, as well as of the *real absolute potential of the reference hydrogen electrode* $\hat{\mathcal{V}}^{\ominus}_{H}$ and its first temperature derivative $\partial_T\hat{\mathcal{V}}^{\ominus}_{H}$ (Section 5.14) and of the *lateral metal-solution Volta potential difference at this electrode* $\Delta\psi^{\ominus}_{\mathrm{H(Pt)}}$ (assuming platinum as a metal) and its first temperature derivative $\partial_T\Delta\psi^{\ominus}_{\mathrm{H(Pt)}}$ (Section 4.4.4) are nevertheless included in view of their relevance. There are, to our knowledge, no available estimates for the other real partial molar variables $\hat{c}^{\ominus}_{P,H,wat}$, $\hat{v}^{\ominus}_{H,wat}$, $\hat{k}^{\ominus}_{T,H,wat}$ and $\hat{a}^{\ominus}_{P,H,wat}$. The choice of a recommended data set and the procedure to estimate errors based on the statistical analysis of literature data are detailed in Section 5.15. Note that even if $\hat{\mathcal{G}}^{\ominus}_{H,wat}$, $\hat{\mathcal{H}}^{\ominus}_{H,wat}$ and $\hat{\mathcal{S}}^{\ominus}_{H,wat}$ contain one redundant quantity, the corresponding errors are evaluated independently. The errors possibly incurred by the ambiguity between standard *vs.* density-corrected solute standard-state variants for $\hat{\mathcal{S}}^{\ominus}_{H,wat}$ and $\hat{s}^{\ominus}_{H,wat}$ are of about 0.7 and 3.4 %, respectively. The real proton hydration free energy $\hat{\mathcal{G}}^{\ominus}_{H,wat}$ is identical in the two variants. Although they are relatively difficult to determine accurately, these parameters are not experimentally elusive.

Recommended values for the *intrinsic hydration parameters of the proton* and *intrinsic partial molar variables of the aqueous proton* are reported in Table 5.24. These parameters are discussed in Section 5.18. In principle, only the values of $\mathcal{G}^{\ominus}_{H,wat}$, $s^{\ominus}_{H,wat}$, $c^{\ominus}_{P,H,wat}$, $v^{\ominus}_{H,wat}$, $k^{\ominus}_{T,H,wat}$ and $a^{\ominus}_{P,H,wat}$ are non-redundant. Derived values of $\mathcal{H}^{\ominus}_{H,wat}$ and $\mathcal{S}^{\ominus}_{H,wat}$, as well as of the *intrinsic absolute potential of the reference hydrogen electrode* $\mathcal{V}^{\ominus}_{H}$ and its first temperature derivative $\partial_T\mathcal{V}^{\ominus}_{H}$ (Section 4.3.6) and of the *air-liquid interfacial potential of pure water* $\chi^{\ominus}_{wat}$ and its first temperature derivative $\partial_T\chi^{\ominus}_{wat}$ (Section 5.17) are nevertheless included in view of their relevance. In contrast to conventional and real parameters, these parameters are *experimentally elusive* and this recommended set should thus be taken with some caution (see below). The choice of a recommended data set and the procedure to estimate errors

Table 5.21 *Recommended data for the reticulation parameters and salt partial molar variables of the alkali-halides.* The reported data includes the reticulation enthalpy $\Delta_l H^\ominus$ of the alkali-halide gas-phase ions (MX_g), as well as the molar entropy $s^\ominus$, the molar heat capacity $c_P^\ominus$, the molar volume $v^\ominus$, the molar volume-compressibility $k_T^\ominus$ and the molar volume-expansivity $a_P^\ominus$ of the alkali-halide salts (MX_s). The standard states are according to the $bbme_T$ convention (Section 4.2.4; reference pressure $P^\circ = 1$ bar, reference molality $b^\circ = 1$ mol·kg^{-1} and reference temperature $T^- = 298.15$ K).

Property	Li$^+$				Na$^+$				K$^+$			
	F$^-$	Cl$^-$	Br$^-$	I$^-$	F$^-$	Cl$^-$	Br$^-$	I$^-$	F$^-$	Cl$^-$	Br$^-$	I$^-$
$\Delta_l H^\ominus[MX_g]$ [kJ·mol^{-1}]	-1046.4	-861.3	-817.9	-759.2	-930.6	-787.4	-751.4	-700.2	-826.2	-717.6	-689.0	-645.2
$s^\ominus[MX_s]$ [J·mol^{-1}·K^{-1}]	35.6	59.5	74.4	86.9	51.1	71.9	86.8	103.7	67.1	82.3	96.8	112.7
$c_P^\ominus[MX_s]$ [J·mol^{-1}·K^{-1}]	41.6	48.0	48.1	51.0	46.9	50.5	51.4	52.1	49.0	51.3	52.3	52.9
$v^\ominus[MX_s]$ [10^{-6} m^3·mol^{-1}]	9.83	20.44	25.06	32.71	14.99	27.01	32.07	40.83	22.98	37.52	43.25	53.10
$k_T^\ominus[MX_s]$ [10^{-10} m^3·mol^{-1}·bar^{-1}]	0.153	0.734	1.14	2.12	0.340	1.19	1.72	2.89	0.788	2.26	3.06	4.77
$a_P^\ominus[MX_s]$ [10^{-8} m^3·mol^{-1}·K^{-1}]	0.0973	0.270	0.368	0.579	0.148	0.316	0.433	0.551	0.221	0.394	0.480	0.621

Property	Rb$^+$				Cs$^+$				Source	Notes
	F$^-$	Cl$^-$	Br$^-$	I$^-$	F$^-$	Cl$^-$	Br$^-$	I$^-$	(Table)	
$\Delta_l H^\ominus[MX_s]$ [kJ·mol^{-1}]	-792.4	-692.4	-665.6	-626.9	-756.1	-667.8	-644.7	-607.6	5.6	-
$s^\ominus[MX_s]$ [J·mol^{-1}·K^{-1}]	(81.9)	95.7	110.0	118.4	92.7	101.2	113.0	123.2	5.6	a
$c_P^\ominus[MX_s]$ [J·mol^{-1}·K^{-1}]	50.7	52.4	52.8	53.2	51.1	52.5	52.9	52.8	5.2	-
$v^\ominus[MX_s]$ [10^{-6} m^3·mol^{-1}]	27.18	42.91	49.22	59.70	32.75	42.21	47.72	57.40	5.2	-
$k_T^\ominus[MX_s]$ [10^{-10} m^3·mol^{-1}·bar^{-1}]	1.08	2.91	3.97	5.99	1.46	2.57	3.24	4.86	5.2	-
$a_P^\ominus[MX_s]$ [10^{-8} m^3·mol^{-1}·K^{-1}]	0.277	0.515	0.546	0.698	0.334	0.595	0.659	0.792	5.2	-

(a) the value for RbF is based on a non-experimental (interpolated) estimate for $\Delta_l S^\ominus[\text{RbF}_g]$, see footnote (b) in Table 5.10.

based on the statistical analysis of literature data are detailed in Section 5.18. The errors possibly incurred by the ambiguity between standard *vs.* density-corrected solute standard-state variants for $\mathcal{H}^\ominus_{H,wat}$, $\mathcal{S}^\ominus_{H,wat}$, $s^\ominus_{H,wat}$, $c^\ominus_{P,H,wat}$, $v^\ominus_{H,wat}$, $k^\ominus_{T,H,wat}$ and $a^\ominus_{P,H,wat}$ are 0.02, 0.5, 2.9, 12.9, 24.9, 19.0 and 3.3%, respectively. The intrinsic proton hydration free energy $\mathcal{G}^\ominus_{H,wat}$ is identical in the two variants.

Even if the recommended values reported in Table 5.24 are suggested on the basis of a careful analysis of the literature, they nevertheless remain bound to a specific *extra-thermodynamic assumption*. In the opinion of the authors of this book, the nature of this assumption ultimately resides in a specific choice for the partitioning of electric potentials between bulk and surface contributions. As will be discussed in Chapter 7, the most "reasonable" partitioning is the one implicitly built into the Born model for CE calculations (Section 3.2) or the M-scheme for AT simulations (Sections 3.3.4 and 4.3.8), as well as in most extra-thermodynamic assumptions used in the interpretation of specific experiments (Sections 5.17 and 5.18). This assumption is the most intuitive and can actually be viewed as a *definition*.

Finally, recommended values for the *redox potentials* of the alkali-halide element-ion couples, along with their pressure and temperature derivatives, as well as for the *real absolute electrode potentials* and *metal-solution Volta*

Table 5.22 *Recommended data for the standard thermodynamic parameters of alkali, halides and hydrogen.* The reported data includes the atomization enthalpy $\Delta_a H^\ominus$, the molar entropy $s^\ominus$, the molar heat capacity $c_P^\ominus$, the molar volume $v^\ominus$, the molar volume-compressibility $k_T^\ominus$ and the molar volume-expansivity $a_P^\ominus$ of the alkali and halide or hydrogen elements (I_{el}), the ionization enthalpy $\Delta_i H^\ominus$, the molar entropy $s^\ominus$, the molar heat capacity $c_P^\ominus$, the molar volume $v^\ominus$, the molar volume-compressibility $k_T^\ominus$ and the molar volume-expansivity $a_P^\ominus$ of the alkali and halide or hydrogen gas-phase atoms (I_g), the molar entropy $s^\ominus$, the molar heat capacity $c_P^\ominus$, the molar volume $v^\ominus$, the molar volume-compressibility $k_T^\ominus$ and the molar volume-expansivity $a_P^\ominus$ of the alkali and halide gas-phase ions or of the gas-phase proton (I_g^z), the conventional hydration free energy $\Delta_s G^\bullet$, the conventional hydration enthalpy $\Delta_s H^\bullet$ and conventional hydration entropy $\Delta_s S^\bullet$ of the alkali and halide gas-phase ions or of the proton (I_g^z), the conventional partial molar entropy $s^\bullet$, the conventional partial molar heat capacity $c_P^\bullet$, the conventional partial molar volume $v^\bullet$, the conventional partial molar volume-compressibility $k_T^\bullet$ and the conventional partial molar volume-expansivity $a_P^\bullet$ of the aqueous alkali and halide ions or of the aqueous proton (I_{aq}^z), as well as the standard work function $\omega^\ominus$ of the alkali metals (M_s) and of platinum Pt_s (values between parentheses). The entries ϵ provide error estimates (Section 5.16; Tables 5.16 and 5.17) on the conventional solvation parameters and partial molar variables in aqueous solution reported in the preceding lines. The standard states are according to the $bbme_T$ convention (Section 4.2.4; reference pressure $P^\circ = 1$ bar, reference molality $b^\circ = 1$ mol·kg^{-1} and reference temperature $T^- = 298.15$ K; standard state for the free electron according to the warm-electron convention: ideal-gas at temperature T^-, according to Fermi-Dirac statistics; Fermi-Dirac statistics is also applied for the gas-phase standard state of the proton; the work functions refer to polycrystalline surfaces). The solvation parameters and partial molar variables in solution are assumed here to pertain to the standard ($\Delta_s G^\bullet$, $\Delta_s H^\bullet$, $\Delta_s S^\bullet$, $s^\bullet$, $c_P^\bullet$) or the density-corrected ($v^{\bullet *}$, $k_T^{\bullet *}$, $a_P^{\bullet *}$) variant (Section 4.2.5), which represents the most likely situation, but may not necessarily be the case in reality. The conventional solvation free energy ($\Delta_s G^\bullet$) is identical in the two variants.

Property	H	Li	Na	K	Rb	Cs	F	Cl	Br	I	Source (Table)	Notes
$s^\ominus[I_{el}]$ [J·mol^{-1}·K^{-1}]	65.34	29.12	51.30	64.68	76.78	85.23	101.40	111.54	76.11	58.07	5.1	-
$c_P^\ominus[I_{el}]$ [J·mol^{-1}·K^{-1}]	14.4	24.8	28.2	29.6	31.1	32.2	15.7	16.9	37.9	27.2	5.1	-
$v^\ominus[I_{el}]$ [10^{-3} m^3·mol^{-1}]	12.395	0.01300	0.02370	0.04393	0.05586	0.07096	12.395	12.395	0.02575	0.02573	5.1	-
$k_T^\ominus[I_{el}]$ [10^{-10} m^3·mol^{-1}·bar^{-1}]	$1.24 \cdot 10^8$	1.10	3.41	13.6	17.4	34.3	$1.24 \cdot 10^8$	$1.24 \cdot 10^8$	13.6	3.34	5.1	-
$a_P^\ominus[I_{el}]$ [10^{-8} m^3·mol^{-1}·K^{-1}]	4150	0.179	0.505	1.10	1.51	2.06	4150	4150	2.88	0.643	5.1	-
$\Delta_a H^\ominus[I_{el}]$ [kJ·mol^{-1}]	217.998	159.3	107.5	89.0	80.9	76.5	79.38	121.301	111.87	106.76	5.9	-
$s^\ominus[I_g]$ [J·mol^{-1}·K^{-1}]	114.717	138.782	153.718	160.341	170.094	175.601	158.751	165.190	175.018	180.787	5.9	-
$c_P^\ominus[I_g]$ [J·mol^{-1}·K^{-1}]	20.8	20.8	20.8	20.8	20.8	20.8	22.7	21.8	20.8	20.8	5.9	-
$v^\ominus[I_g]$ [10^{-3} m^3·mol^{-1}]	24.8	24.8	24.8	24.8	24.8	24.8	24.8	24.8	24.8	24.8	5.9	-
$k_T^\ominus[I_g]$ [10^{-2} m^3·mol^{-1}·bar^{-1}]	2.48	2.48	2.48	2.48	2.48	2.48	2.48	2.48	2.48	2.48	5.9	-
$a_P^\ominus[I_g]$ [10^{-5} m^3·mol^{-1}·K^{-1}]	8.31	8.31	8.31	8.31	8.31	8.31	8.31	8.31	8.31	8.31	5.0	
$\Delta_i H^\ominus[I_g]$ [kJ·mol^{-1}]	1315.1	523.4	498.8	422.2	406.1	378.4	-331.7	-351.4	-327.9	-300.7	5.9	-
$s^\ominus[I_g^z]$ [J·mol^{-1}·K^{-1}]	108.95	133.02	147.95	154.58	164.33	169.84	145.58	153.36	163.49	169.26	5.9	-
$c_P^\ominus[I_g^z]$ [J·mol^{-1}·K^{-1}]	20.8	20.8	20.8	20.8	20.8	20.8	20.8	20.8	20.8	20.8	5.9	-
$v^\ominus[I_g^z]$ [10^{-3} m^3·mol^{-1}]	24.8	24.8	24.8	24.8	24.8	24.8	24.8	24.8	24.8	24.8	5.9	-
$k_T^\ominus[I_g^z]$ [10^{-2} m^3·mol^{-1}·bar^{-1}]	2.48	2.48	2.48	2.48	2.48	2.48	2.48	2.48	2.48	2.48	5.9	-
$a_P^\ominus[I_g^z]$ [10^{-5} m^3·mol^{-1}·K^{-1}]	8.31	8.31	8.31	8.31	8.31	8.31	8.31	8.31	8.31	8.31	5.9	-
$\Delta_s G^\bullet[I_g^z]$ [kJ·mol^{-1}]	0.0	574.9	680.5	752.2	775.0	798.2	-1533.2	-1408.6	-1381.8	-1344.4	5.16	a,b
$\epsilon(\Delta_s G^\bullet)$	-	1.7	1.9	0.9	1.4	2.5	10.1	2.2	3.3	2.4	5.16	a
$\Delta_s H^\bullet[I_g^z]$ [kJ·mol^{-1}]	0.0	571.8	686.6	769.5	794.8	820.1	-1613.2	-1470.1	-1438.6	-1394.4	5.16	a
$\epsilon(\Delta_s H^\bullet)$	-	2.0	0.8	0.3	1.1	2.4	8.7	2.1	4.3	3.0	5.16	a
$\Delta_s S^\bullet[I_g^z]$ [J·mol^{-1}·K^{-1}]	0.0	-10.3	20.2	57.8	66.6	73.3	-268.4	-206.4	-190.5	-167.5	5.16	a,b
$\epsilon(\Delta_s S^\bullet)$	-	0.6	0.5	0.3	0.7	0.2	0.9	0.6	1.9	2.2	5.16	a
$s^\bullet[I_{aq}^z]$ [J·mol^{-1}·K^{-1}]	0.0	13.8	59.2	103.4	121.9	134.2	-13.8	55.9	81.9	110.7	5.17	c
$\epsilon(s^\bullet)$	-	0.8	0.4	0.2	0.7	0.2	0.7	0.3	0.5	0.7	5.17	c
$c_P^\bullet[I_{aq}^z]$ [J·mol^{-1}·K^{-1}]	0.0	61.0	42.0	11.7	-9.8	-23.4	-115.2	-126.0	-130.3	-120.6	5.17	c
$\epsilon(c_P^\bullet)$	-	1.8	3.8	2.7	5.5	3.1	2.0	1.8	2.0	5.4	5.17	c
$v^{\bullet *}[I_{aq}^z]$ [10^{-6} m^3·mol^{-1}]	0.0	-0.7	-1.4	8.8	14.2	21.5	-1.8	17.7	24.7	36.3	5.17	c
$\epsilon(v^{\bullet *})$	-	0.1	0.1	0.0	0.1	0.1	0.3	0.1	0.0	0.1	5.17	c
$k_T^{\bullet *}[I_{aq}^z]$ [10^{-10} m^3·mol^{-1}·bar^{-1}]	0.0	-35.6	-42.4	-35.0	-32.0	-28.6	-29.2	-6.3	2.3	14.3	5.17	c
$\epsilon(k_T^{\bullet *})$	-	3.7	0.8	4.9	0.3	9.9	17.1	1.2	2.8	2.7	5.17	c
$a_P^{\bullet *}[I_{aq}^z]$ [10^{-8} m^3·mol^{-1}·K^{-1}]	0.0	0.9	5.4	5.4	4.6	4.2	1.3	1.6	2.7	5.4	5.17	c
$\epsilon(a_P^{\bullet *})$	-	0.6	1.6	0.8	1.4	1.5	0.6	1.0	1.0	1.0	5.17	c
$F\omega^\ominus[M_s]$ [kJ·mol^{-1}]	(540.50)	279.10	224.10	217.30	214.40	184.50	(540.50)	(540.50)	(540.50)	(540.50)	5.13	d

(a) lines REC (recommended data) and ERC (estimated error ϵ) in Table 5.16; (b) redundant data; (c) lines REC (recommended data) and ERC (estimated error ϵ) in Table 5.17; (d) value refers to alkali metals or to platinum (Pt_s) for halides and hydrogen (between parentheses).

Table 5.23 *Recommended data for the real hydration parameters of the proton and the real partial molar variables of the aqueous proton.* The reported data includes the real hydration free energy $\hat{\mathcal{G}}^{\ominus}_{H,wat}$, the real hydration enthalpy $\hat{\mathcal{H}}^{\ominus}_{H,wat}$ and the real hydration entropy $\hat{\mathcal{S}}^{\ominus}_{H,wat}$ of the gas-phase proton, the real partial molar entropy $\hat{s}^{\ominus}_{H,wat}$, the real partial molar heat capacity $\hat{c}^{\ominus}_{P,H,wat}$, the real partial molar volume $\hat{v}^{\ominus}_{H,wat}$, the real partial molar volume-compressibility $\hat{k}^{\ominus}_{T,H,wat}$ and the real partial molar volume-expansivity $\hat{a}^{\ominus}_{P,H,wat}$ of the aqueous proton, the real absolute electrode potential of the reference hydrogen electrode $\hat{\mathcal{V}}^{\ominus}_{H,wat}$ and its temperature derivative $\partial_T \hat{\mathcal{V}}^{\ominus}_{H,wat}$, as well as the metal-solution Volta potential difference of the reference hydrogen electrode $\Delta\psi^{\ominus}_{\mathrm{H(Pt)}}$, assuming platinum Pt as metal. The standard states are according to the $bbme_T$ convention (Section 4.2.4; reference pressure $P^{\circ} = 1$ bar, reference molality $b^{\circ} = 1$ mol·kg^{-1} and reference temperature $T^{-} = 298.15$ K; standard state for the free electron according to the warm-electron convention: ideal-gas at temperature T^{-}, according to Fermi-Dirac statistics; Fermi-Dirac statistics is also applied for the gas-phase standard state of the proton). The solvation parameters and partial molar variables in solution are assumed here to pertain to the standard ($\hat{\mathcal{G}}^{\ominus}_{H,wat}$, $\hat{\mathcal{H}}^{\ominus}_{H,wat}$, $\hat{\mathcal{S}}^{\ominus}_{H,wat}$, $\hat{s}^{\ominus}_{H,wat}$, $\hat{c}^{\ominus}_{P,H,wat}$) or the density-corrected ($\hat{v}^{\ominus *}_{H,wat}$, $\hat{k}^{\ominus *}_{T,H,wat}$, $\hat{a}^{\ominus *}_{P,H,wat}$) variant (Section 4.2.5), which represents the most likely situation, but may not necessarily be the case in reality. The real solvation free energy $\hat{\mathcal{G}}^{\ominus}_{H,wat}$ is identical in the two variants.

Property		Source (Section)	Notes
$\hat{\mathcal{G}}^{\ominus}_{H,wat}$ [kJ·mol^{-1}]	-1087.0±3.2	5.15	-
$\hat{\mathcal{S}}^{\ominus}_{H,wat}$ [J·mol^{-1}·K^{-1}]	-90.1±2.0	5.15	a
$\hat{\mathcal{H}}^{\ominus}_{H,wat}$ [kJ·mol^{-1}]	-1113.9±3.8	5.15	a
$\hat{s}^{\ominus}_{H,wat}$ [J·mol^{-1}·K^{-1}]	18.8±2.0	5.15	-
$\hat{c}^{\ominus}_{P,H,wat}$ [J·mol^{-1}·K^{-1}]	-	-	-
$\hat{v}^{\ominus}_{H,wat}$ [10^{-6} m^3·mol^{-1}]	-	-	-
$\hat{k}^{\ominus}_{T,H,wat}$ [10^{-10} m^3·mol^{-1}·bar^{-1}]	-	-	-
$\hat{a}^{\ominus}_{P,H,wat}$ [10^{-8} m^3·mol^{-1}·K^{-1}]	-	-	-
$\hat{\mathcal{V}}^{\ominus}_{H,wat}$ [V]	4.42±0.03	5.14	a
$\partial_T \hat{\mathcal{V}}^{\ominus}_{H,wat}$ [10^{-3} V·K^{-1}]	0.25±0.02	-	a
$\Delta\psi^{\ominus}_{\mathrm{H(Pt)}}$ [V]	1.18	-	a,b

(a) redundant data; (b) calculated from $\hat{\mathcal{V}}^{\ominus}_{H,wat}$ (this table) and $\omega^{\ominus}_{\mathrm{Pt}}$ (Table 5.22), using Eq. 4.291.

Table 5.24 *Recommended data for the intrinsic hydration parameters of the proton and the intrinsic partial molar variables of the aqueous proton.* The reported data includes the intrinsic hydration free energy $\mathcal{G}^{\ominus}_{H,wat}$, the intrinsic hydration enthalpy $\mathcal{H}^{\ominus}_{H,wat}$ and the intrinsic hydration entropy $\mathcal{S}^{\ominus}_{H,wat}$ of the gas-phase proton, the intrinsic partial molar entropy $s^{\ominus}_{H,wat}$, the intrinsic partial molar heat capacity $c^{\ominus}_{P,H,wat}$, the intrinsic partial molar volume $v^{\ominus}_{H,wat}$, the intrinsic partial molar volume-compressibility $k^{\ominus}_{T,H,wat}$ and the intrinsic partial molar volume-expansivity $a^{\ominus}_{P,H,wat}$ of the aqueous proton, the intrinsic absolute electrode potential of the reference hydrogen electrode $\mathcal{V}^{\ominus}_{H,wat}$ and its temperature derivative $\partial_T \mathcal{V}^{\ominus}_{H,wat}$, as well as the air-liquid interfacial potential jump of pure water $\chi^{\ominus}_{wat}$ and its temperature derivative $\partial_T \chi^{\ominus}_{wat}$. The standard states are according to the $bbme_T$ convention (Section 4.2.4; reference pressure $P^{\circ} = 1$ bar, reference molality $b^{\circ} = 1$ mol·kg^{-1} and reference temperature $T^{-} = 298.15$ K; standard state for the free electron according to the warm-electron convention: ideal-gas at temperature T^{-}, according to Fermi-Dirac statistics; Fermi-Dirac statistics is also applied for the gas-phase standard state of the proton). The solvation parameters and partial molar variables in solution are assumed here to pertain to the standard ($\mathcal{G}^{\ominus}_{H,wat}$, $\mathcal{H}^{\ominus}_{H,wat}$, $\mathcal{S}^{\ominus}_{H,wat}$, $s^{\ominus}_{H,wat}$, $c^{\ominus}_{P,H,wat}$) or the density-corrected ($v^{\ominus*}_{H,wat}$, $k^{\ominus*}_{T,H,wat}$, $a^{\ominus*}_{P,H,wat}$) variant (Section 4.2.5), which represents the most likely situation, but may not necessarily be the case in reality. The intrinsic solvation free energy $\mathcal{G}^{\ominus}_{H,wat}$ is identical in the two variants. Note that the data in this table corresponds to experimentally-elusive properties and has therefore to be used with great caution.

Property		Source (Section)	Notes
$\mathcal{G}^{\ominus}_{H,wat}$ [kJ·mol^{-1}]	-1100±5	5.18	-
$\mathcal{S}^{\ominus}_{H,wat}$ [J·mol^{-1}·K^{-1}]	-131±2	5.18	a
$\mathcal{H}^{\ominus}_{H,wat}$ [kJ·mol^{-1}]	-1113±8	5.18	a
$s^{\ominus}_{H,wat}$ [J·mol^{-1}·K^{-1}]	-22±2	5.18	-
$c^{\ominus}_{P,H,wat}$ [J·mol^{-1}·K^{-1}]	(-65)	5.18	b
$v^{\ominus}_{H,wat}$ [10^{-6} m^3·mol^{-1}]	-4.5±5	5.18	-
$k^{\ominus}_{T,H,wat}$ [10^{-10} m^3·mol^{-1}·bar^{-1}]	15±5	5.18	-
$a^{\ominus}_{P,H,wat}$ [10^{-8} m^3·mol^{-1}·K^{-1}]	-3.0±1.5	5.18	-
$\mathcal{V}^{\ominus}_{H,wat}$ [V]	4.28±0.13	-	a
$\partial_T \mathcal{V}^{\ominus}_{H,wat}$ [10^{-3} V·K^{-1}]	0.67±0.01	-	a
$\chi^{\ominus}_{wat}$ [V]	0.13±0.10	5.17	a
$\partial_T \chi^{\ominus}_{wat}$ [10^{-3} V·K^{-1}]	-0.42±0.20	5.17	a

(a) redundant data; (b) tentative value.

Table 5.25 *Recommended data for standard redox potentials and their first and second pressure and temperature derivatives for the alkali and halide half-cell elements considering water as a solvent.* The reported data includes the redox potential $\Delta_H \mathcal{V}^\ominus$ (relative to the reference hydrogen electrode), its first and second pressure derivatives $\partial_P \Delta_H \mathcal{V}^\ominus$ and $\partial_P^2 \Delta_H \mathcal{V}^\ominus$, respectively, its first and second temperature derivatives $\partial_T \Delta_H \mathcal{V}^\ominus$ and $\partial_T^2 \Delta_H \mathcal{V}^\ominus$, respectively, its cross-second derivative with respect to temperature and pressure $\partial_{P,T}^2 \Delta_H \mathcal{V}^\ominus$, as well as the metal-solution lateral Volta potential difference $\Delta\psi_{M,S}^\ominus$ and the real absolute electrode potential $\hat{\mathcal{V}}_{M,S}^\ominus$ of the alkali and halide half-cell elements. The standard states are corresponding to the $bbme_T$ convention (Section 4.2.4; reference pressure $P^\circ = 1$ bar, reference molality $b^\circ = 1$ mol·kg^{-1} and reference temperature $T^- = 298.15$ K; standard state for the free electron according to the warm-electron convention: ideal-gas at temperature T^-, according to Fermi-Dirac statistics; Fermi-Dirac statistics is also applied for the gas-phase standard state of the proton).

Property	Li	Na	K	Rb	Cs	F	Cl	Br	I	Source (Table)	Notes
$\Delta_H \mathcal{V}^\ominus$ [V]	-3.04	-2.72	-2.94	-2.95	-3.03	2.89	1.36	1.08	0.53	-	a,b
$\Delta_H \mathcal{V}^\ominus$ [V]	-3.04	-2.71	-2.93	-2.98	-3.03	2.87	1.36	1.07	0.54	-	c
$\partial_T \Delta_H \mathcal{V}^\ominus$ [10^{-3} V·K^{-1}]	-0.52	-0.76	-1.08	-1.15	-1.18	-1.87	-1.25	-0.62	-0.13	-	a,b
$\partial_T \Delta_H \mathcal{V}^\ominus$ [10^{-3} V·K^{-1}]	-0.53	-0.77	-1.08	-1.25	-1.20	-1.83	-1.26	-0.63	-0.15	-	c
$\partial_P \Delta_H \mathcal{V}^\ominus$ [10^{-3} V·bar^{-1}]	12.06	11.41	21.94	27.50	35.10	27.55	7.34	-12.74	-24.76	-	a,b
$\partial_T^2 \Delta_H \mathcal{V}^\ominus$ [10^{-6} V·K^{-2}]	-1.76	-0.98	0.12	0.92	1.43	-5.05	-5.47	-6.35	-5.64	-	a,b
$\partial_T^2 \Delta_H \mathcal{V}^\ominus$ [10^{-6} V·K^{-2}]	-	-	-	-	-	-5.34	-5.45	-6.21	-5.97	-	c
$\partial_P^2 \Delta_H \mathcal{V}^\ominus$ [10^{-2} V·bar^{-2}]	-1.29	-1.29	-1.29	-1.29	-1.29	-2.57	-2.57	-1.29	-1.29	-	a,b
$\partial_{P,T}^2 \Delta_H \mathcal{V}^\ominus$ [10^{-6} V·K^{-1}·bar^{-1}]	42.99	42.96	42.90	42.86	42.80	47.31	47.31	43.01	43.01	-	a,b
$\Delta\psi_{M,S}^\ominus$ [V]	1.52	0.62	0.77	0.75	0.52	(-1.70)	(-0.17)	(0.11)	(0.65)	-	a,d
$\hat{\mathcal{V}}_{M,S}^\ominus$ [V]	1.38	1.70	1.48	1.47	1.39	(7.31)	(5.78)	(5.49)	(4.95)	-	a,d

(a) redundant data; (b) calculated from the recommended conventional data reported in Table 5.22 based on Eqs. 5.3, 5.5 and 5.7, as well as analogous equations for the other derivatives; (c) experimental data from Refs.[221,1829,2099] as reported in Table 5.5; (d) value refers to alkali metals or to platinum (Pt_s) for halides (between parentheses).

potential differences relevant for the corresponding electrodes (alkali metals, or platinum for halides and hydrogen) are reported in Table 5.25. These parameters are discussed in Sections 5.5 and 5.14. This data is redundant with the one presented in the previous tables. It is nevertheless included in view of its practical relevance and considering that only the first temperature derivatives (and second for the halides only) have been determined experimentally (Section 5.5). These experimental values are also reported in the tables for comparison.

5.20 Suggested intrinsic single-ion solvation parameters

In Section 5.19, recommended data for conventional single-ion hydration parameters and partial molar variables of aqueous ions have been provided for the alkali and halides (Table 5.22), along with estimates for the intrinsic proton hydration parameters and partial molar variables of the aqueous proton (Table 5.24). It is therefore possible to derive suggested values for the *intrinsic hydration parameters and partial molar variables of aqueous ions* for the *alkali and halides.* The corresponding values are reported in Table 5.26.

As already mentioned in Section 5.7, these parameters do not permit a discussion of the physics of solvation in the simplest possible context. The reason is that they encompass an ideal-gas contribution that depends on the different effective molar volumes accessible to the ion in the gas-phase and dissolved states, as defined by the standard-state reference pressure P°, temperature T^- and molality b°, along with the solvent density $\rho_{svt}^\ominus$ at P° and T^-. Furthermore, they also include a contribution related to the ideal-gas influence of the species on the molar volume, present in the gas state but absent in the dissolved state. Finally, in the standard variant, a further contribution is related to the pressure or/and temperature dependence of the solvent density. This issue has already been encountered in Section 4.2.17, where it was observed that even in a standard-state convention equating the effective molar volumes accessible to species in the gas and dissolved states, thereby eliminating the first of the three above influences, derivative thermodynamic solvation properties were still not identical to corresponding point-to-point solvation properties (Eq. 4.205).

In the present context, the relevant quantities to characterize exclusively the physics of the solvation process and the influence of the ion on the solvent are the semi-standard *point-to-point solvation parameters* $\Delta_s Y^{\boxminus}$ (Sections 2.2 and 4.2.16). These parameters are given by

$$\Delta_s Y^{\boxminus}[I_g^z] = \Delta_s Y^\ominus[I_g^z] - \Delta_s Y_{std}^\ominus \tag{5.96}$$

for an intrinsic solvation parameter $\Delta_s Y^\ominus[I_g^z]$ pertaining to the standard variant,

$$\Delta_s Y^{\boxminus}[I_g^z] = y^\ominus[I_d^z] - y^\ominus[I_g^z] - \Delta_s Y_{std}^\ominus \tag{5.97}$$

for an intrinsic partial molar variable $y^\ominus[I_d^z]$ pertaining to the standard variant, or

$$\Delta_s Y^{\boxminus}[I_g^z] = y^{\ominus *}[I_d^z] - y^\ominus[I_g^z] - \Delta_s Y_{std}^{\ominus *} \tag{5.98}$$

for an intrinsic partial molar variable $y^{\ominus *}[I_d^z]$ pertaining to the density-corrected variant, where $\Delta_s Y_{std}^\ominus$ and $\Delta_s Y_{std}^{\ominus *}$ are standard-state corrections terms appropriate for the standard and density-corrected solute variants, respectively, and are defined by Eqs. 4.178, 4.183 and 4.184 (along with the expressions provided in Table 4.1).

In the present case, Eq. 5.96 applies to the solvation free energy, enthalpy and entropy, Eq. 5.97 to the partial molar entropy and heat capacity, and Eq. 5.98 to the partial molar volume, volume-compressibility and volume-expansivity. As discussed in Section 4.2.16 (Eq. 4.187), in the latter case, Eq.

Table 5.26 *Recommended data for the single-ion intrinsic hydration parameters of the gas-phase proton and alkali and halide ions, and the single-ion intrinsic partial molar variables of the aqueous proton and alkali and halide ions.* The reported data includes the intrinsic hydration free energy $\Delta_s G^\ominus$, the intrinsic hydration enthalpy $\Delta_s H^\ominus$ and the intrinsic hydration entropy $\Delta_s S^\ominus$ of the alkali and halide and hydrogen gas-phase ions (I_g^z), the intrinsic partial molar entropy $s^\ominus$, the intrinsic partial molar heat capacity $c_P^\ominus$, the intrinsic partial molar volume $v^\ominus$, the intrinsic partial molar volume-compressibility $k_T^\ominus$ and the intrinsic partial molar volume-expansivity $a_P^\ominus$ of the aqueous alkali and halide and hydrogen ions (I_{aq}^z), the intrinsic absolute electrode potential of the reference hydrogen electrode $\mathcal{V}_{H,wat}^\ominus$ and its temperature derivative $\partial_T \mathcal{V}_{H,wat}^\ominus$, as well as the semi-standard point-to-point solvation parameters $\Delta_s G^\boxminus$, $\Delta_s H^\boxminus$, $\Delta_s S^\boxminus$ and $\Delta_s C_P^\boxminus$ of the gas-phase alkali and halide and hydrogen ions (I_g^z). The standard states are according to the *bbme*$_T$ convention (Section 4.2.4; reference pressure $P^\circ = 1$ bar, reference molality $b^\circ = 1$ mol·kg^{-1} and reference temperature $T^- = 298.15$ K). The solvation parameters and partial molar variables in solution are assumed here to pertain to the standard ($\Delta_s G^\ominus$, $\Delta_s H^\ominus$, $\Delta_s S^\ominus$, $s^\ominus$, $c_P^\ominus$) or the density-corrected ($v^{\ominus *}$, $k_T^{\ominus *}$, $a_P^{\ominus *}$) variant (Section 4.2.5), which represents the most likely situation, but may not necessarily be the case in reality. The intrinsic solvation free energy $\Delta_s G^\ominus$ is identical in the two variants. Note that the three latter density-corrected standard partial molar variables in solution are identical to the corresponding semi-standard point-to-point solvation parameters ($\Delta_s V^\boxminus$, $\Delta_s K_T^\boxminus$, $\Delta_s A_P^\boxminus$). Note, finally, that the data in this table characterizes elusive properties and has hence to be used with great caution.

Property	H	Li	Na	K	Rb	Cs	F	Cl	Br	I
$\Delta_s G^\ominus[I_g^z]$ [kJ·mol^{-1}]	-1100.00	-525.11	-419.46	-347.78	-325.02	-301.79	-433.18	-308.60	-281.80	-244.41
$\Delta_s H^\ominus[I_g^z]$ [kJ·mol^{-1}]	-1139.04	-567.21	-452.49	-369.59	-344.23	-318.96	-474.15	-331.09	-299.55	-255.32
$\Delta_s S^\ominus[I_g^z]$ [J·mol^{-1}·K^{-1}]	-130.95	-141.24	-110.74	-73.15	-64.40	-57.66	-137.43	-75.43	-59.58	-36.59
$s^\ominus[I_{aq}^z]$ [J·mol^{-1}·K^{-1}]	-22.00	-8.22	37.21	81.43	99.93	112.19	8.15	77.93	103.91	132.67
$c_P^\ominus[I_{aq}^z]$ [J·mol^{-1}·K^{-1}]	-65	-4.00	-22.97	-53.30	-74.82	-88.37	-50.16	-61.03	-65.27	-55.56
$v^{\ominus *}[I_{aq}^z]$ (= $\Delta_s V^\boxminus[I_g^z]$) [$10^{-6}$ m^3·mol^{-1}]	-4.50	-5.25	-5.86	4.32	9.69	17.04	2.70	22.21	29.21	40.81
$k_T^{\ominus *}[I_{aq}^z]$ (= $\Delta_s K_T^\boxminus[I_g^z]$) [$10^{-10}$ m^3·mol^{-1}·bar^{-1}]	15.00	-20.58	-27.45	-20.01	-17.01	-13.61	-44.15	-21.29	-12.69	-0.69
$a_P^{\ominus *}[I_{aq}^z]$ (= $\Delta_s A_P^\boxminus[I_g^z]$) [$10^{-8}$ m^3·mol^{-1}·K^{-1}]	-3.00	-2.12	2.42	2.35	1.58	1.18	4.28	4.62	5.72	8.42
$\Delta_s G^\boxminus[I_g^z]$ [kJ·mol^{-1}]	-1107.95	-533.06	-427.41	-355.73	-332.97	-309.74	-441.13	-316.55	-289.75	-252.36
$\Delta_s H^\boxminus[I_g^z]$ [kJ·mol^{-1}]	-1136.76	-564.92	-450.20	-367.30	-341.94	-316.67	-471.86	-328.80	-297.26	-253.03
$\Delta_s S^\boxminus[I_g^z]$ [J·mol^{-1}·K^{-1}]	-96.61	-106.90	-76.40	-38.81	-30.06	-23.32	-103.09	-41.09	-25.24	-2.25
$\Delta_s C_P^\boxminus[I_g^z]$ [J·mol^{-1}·K^{-1}]	-85.87	-24.87	-43.84	-74.16	-95.69	-109.24	-71.03	-81.90	-86.14	-76.43

5.98 simplifies to

$$\Delta_s Y^\boxminus[I_g^z] = y^{\ominus *}[I_d^z] \quad , \tag{5.99}$$

i.e. for these three variables, the partial molar variables of the dissolved ion in the density-corrected variant are equal to point-to-point solvation parameters.

The point-to-point hydration parameters $\Delta_s G^\boxminus$, $\Delta_s H^\boxminus$, $\Delta_s S^\boxminus$ and $\Delta_s C_P^\boxminus$ are also reported in Table 5.26. The parameters $\Delta_s V^\boxminus$, $\Delta_s K_T^\boxminus$ and $\Delta_s A_P^\boxminus$ are equal to $v^{\ominus *}$, $k_T^{\ominus *}$ and $a_P^{\ominus *}$ (Eq. 5.99), and are thus not repeated in the table. The values of all these quantities are illustrated graphically in Figure 5.8.

The $\Delta_s G^\boxminus$ and $\Delta_s H^\boxminus$ values are very large and negative for all alkali and halide ions, indicating that the hydration process is associated with both a strong driving force and an important release of heat, as expected for a process involving the formation of favorable electrostatic interactions between

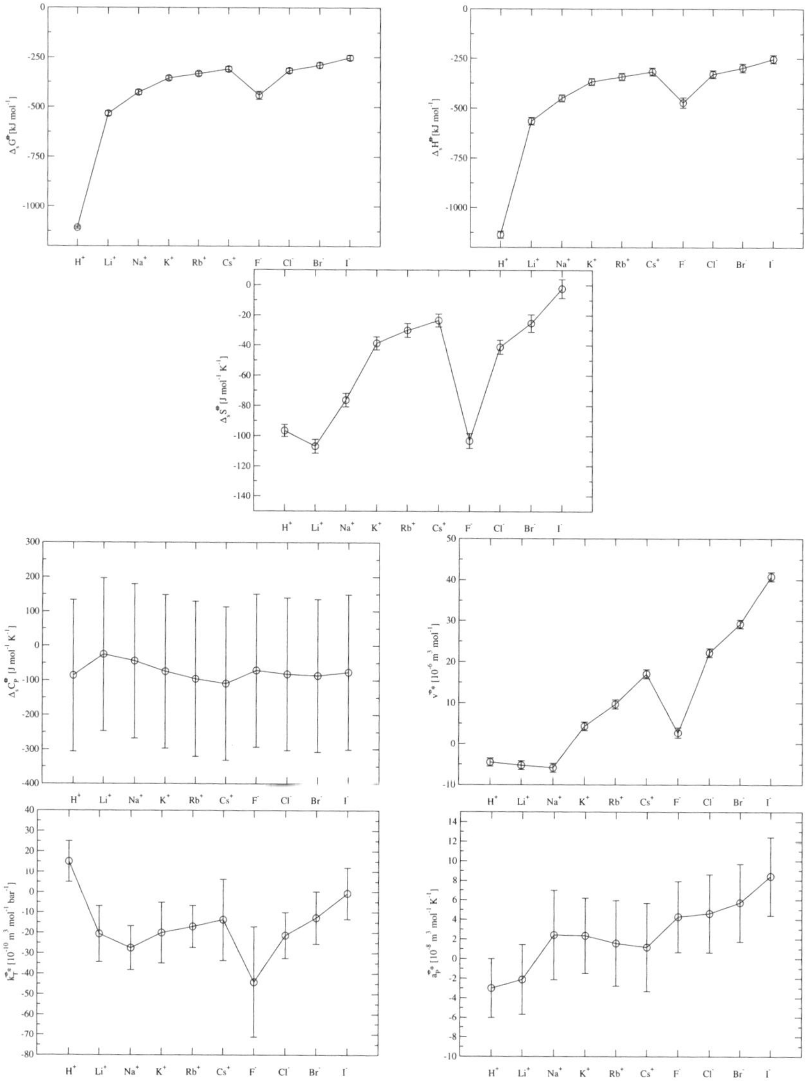

Figure 5.8 *Recommended data for the single-ion intrinsic hydration parameters of the gas-phase proton and alkali and halide ions, and the single-ion intrinsic partial molar variables of the aqueous proton and alkali and halide ions.* The graphs display the thermodynamic solvation parameters $\Delta_s G^{\ominus}[I_g^z]$, $\Delta_s H^{\ominus}[I_g^z]$, $\Delta_s S^{\ominus}[I_g^z]$, $\Delta_s C_P^{\ominus}[I_g^z]$, and the partial molar variables $v^{\ominus *}[I_{aq}^z]$ $(= \Delta_s V^{\ominus}[I_g^z])$, $k_T^{\ominus *}[I_{aq}^z]$ $(= \Delta_s K_T^{\ominus}[I_g^z])$ and $a_P^{\ominus *}[I_{aq}^z]$ $(= \Delta_s A_P^{\ominus}[I_g^z])$ for the proton and the alkali and halide ions. The displayed proton values and error bars correspond to Table 5.24. The displayed alkali and halide data corresponds to Table 5.26, the indicated errors being the sum of the rms error on the corresponding conventional data (noted "ERC" in Tables 5.16 and 5.17) and the error on the corresponding proton data.

ions and solvent dipoles, along with a less important cost associated with the perturbation of solvent-solvent interactions. They systematically decrease in magnitude with increasing effective ion size, *i.e.* along the series $Li^+ \rightarrow Cs^+$ for the alkali and $F^- \rightarrow I^-$ for the halides. This trend is related to the corresponding increase in the effective closest ion-solvent distance in solution, leading to reduced charge-dipole interactions and can be very simply rationalized on the basis of the BO+ST model (Section 3.2), considering the trend in effective ionic radii observed in all sets reported in Section 5.4 (Table 5.4), *i.e.* ion size increases with the number of electron shells.

The $\Delta_s G^{\ominus}$ and $\Delta_s H^{\ominus}$ values are also systematically larger in magnitude for cations compared to anions considering ions with the same electronic structure, *i.e.* Na^+ *vs.* F^-, K^+ *vs.* Cl^-, Rb^+ *vs.* Br^- and Cs^+ *vs.* I^-, a notable exception being Na^+ *vs.* F^-. This general trend is also easily rationalized in terms of the BO+ST model considering corresponding trends in effective ionic radii, *i.e.* a cation is smaller than an anion at identical electronic structure due to its excess, as opposed to deficit, of positive nuclear charge compared to the negative electronic charge (increased nucleus-electron attraction, decreased electron-electron repulsion). However, asymmetric solvation effects in water, which intrinsically favor the solvation of anions over that of cations, partly counteract this trend, and are the cause for the Na^+ *vs.* F^- inversion. Due to its very small size, the F^- anion is extremely well solvated in water[1590], ion-solvent interactions presenting in this special case a partially covalent character. Finally, the proton is characterized by the most negative $\Delta_s G^{\ominus}$ and $\Delta_s H^{\ominus}$ values of all ions considered here. This is not surprising in view of the fact that this species is an electron-free (naked) ion, and can thus directly interact with the electron clouds of the water molecules, presenting a largely covalent solvation mode.

The $\Delta_s S^{\ominus}$ values are negative for all alkali and halide ions, showing that the corresponding dissolved ions all promote a structuring of their solvent environment. They decrease systematically in magnitude with increasing effective ion size, *i.e.* along the series $Li^+ \rightarrow Cs^+$ for the alkali and $F^- \rightarrow I^-$ for the halides. These observations suggest that the smallest alkali and halide ions are powerful "structure makers", *i.e.* they promote a large decrease in the entropy of the surrounding solvent, while the largest ions are less powerful "structure makers" but in no case "structure breakers", *i.e.* none of them promotes an increase in the entropy of the surrounding solvent.

Just as was the case for $\Delta_s G^{\ominus}$ and $\Delta_s H^{\ominus}$, the $\Delta_s S^{\ominus}$ values are also systematically larger in magnitude for cations compared to anions considering ions with the same electronic structure, *i.e.* Na^+ *vs.* F^-, K^+ *vs.* Cl^-, Rb^+ *vs.* Br^- and Cs^+ *vs.* I^-. Here also, the F^- anion sticks out as being characterized by a negative value of $\Delta_s S^{\ominus}$ that is extremely large in magnitude. The three ions with the most negative point-to-point solvation entropies (close to -100 $J{\cdot}mol^{-1}{\cdot}K^{-1}$) are, in this order, Li^+, F^- and H^+. At least in the two latter cases, the important structuring effect is certainly related to the partially covalent nature of the ion-solvent interactions. Interestingly, however, the point-to-point solvation free energies and enthalpies of Li^+ and F^- are considerably less negative than those of H^+, suggesting that the usual enthalpy-entropy correlations are not valid in the case of these three ions.

The discussion of the $\Delta_s C_P^{\ominus}$ values in terms of signs and relative values for cations *vs.* anions is somewhat hazardous, considering that the value reported in Table 5.24 for the partial molar heat capacity $c_{P,H,wat}^{\ominus}$ of the aqueous proton is only tentative. The trends along the alkali and halide ion series considered separately are, however, independent of this value. Along both series, the $\Delta_s C_P^{\ominus}$ values systematically decrease (in an absolute sense) with increasing effective ion size, *i.e.* along the series $Li^+ \rightarrow Cs^+$ for the alkali and $F^- \rightarrow I^-$ for the halides, the transition from Br^- to I^- representing an exception (noted previously in Section 5.7). Accepting the tentative estimate of -65 J·mol^{-1}·K^{-1} for $c_{P,H,wat}^{\ominus}$, the $\Delta_s C_P^{\ominus}$ values are negative for all ions considered. The negative sign is compatible with a local stiffening of the water structure in the neighborhood of the ions. The ion-size dependence and the iodide anomaly are more difficult to rationalize. On the one hand, smaller ions exert a more intense electric field on the surrounding solvent, which is expected to enhance the stiffening effect. On the other hand, they are also surrounded by fewer solvent molecules in their first hydration shells, which is expected to reduce it. The observed trends suggest that the latter effect is dominant.

The partial molar volumes $v^{\ominus *}$ (equal to $\Delta_s V^{\ominus}$) are positive for all ions considered except Li^+, Na^+, as well as H^+, and nearly systematically increase with increasing effective ion size, the transition from Li^+ to Na^+ representing a notable exception. The generally positive sign is compatible with the expected excluded-volume effect upon inserting a new particle in a dense environment. The negative sign for Li^+, Na^+ and H^+, as well as the ion size dependence, are easy to understand. Smaller ions are characterized by smaller excluded volumes, but they also exert a more intense electric field on the surrounding solvent, which is expected to decrease the partial molar volume, possibly towards negative values, due to electrostriction (Section 3.2, Point E3). However, the lithium anomaly (previously noted in Section 5.7) is more difficult to rationalize.

The partial molar volume-compressibilities $k_T^{\ominus *}$ (equal to $\Delta_s K_T^{\ominus}$) are negative for all ions considered, except H^+, and tendentially decrease in magnitude upon increasing the sizes of the ions, the transition from Li^+ to Na^+ representing again an exception (previously noted in Section 5.7). The generally negative sign is compatible with a local stiffening of the water structure in the neighborhood of the ions. However, the ion-size dependence and the lithium anomaly are more difficult to rationalize. In particular, if the negative values of $\Delta_s C_P^{\ominus}$ and $k_T^{\ominus *}$ are both predominantly ascribed to a stiffening of the water structure, the corresponding trends along the series are generally opposite to each other, suggesting the influence of other factors.

Finally, the partial molar volume-expansivities $a_P^{\ominus *}$ (equal to $\Delta_s A_P^{\ominus}$) are positive for all ions considered, except Li^+ and H^+, tendentially increase upon increasing the size of the anion, and tendentially decrease upon increasing the size of the cation, except for the transition from Li^+ to Na^+. The generally positive sign suggests that the perturbed water structure in the neighborhood of the ions renders it more prone to thermal expansion. The ion-size dependence is again more difficult to rationalize.

A popular qualitative framework for the comparative assessment of ionic properties in terms of hydration and mutual affinity in aqueous solution is provided by the so-called *Hofmeister series*. Recent reviews on the topic can be found in Refs.[87, 99, 140, 149, 280, 282, 2282–2286]. The Hofmeister series[1751, 2287–2292] ranks ions in an empirical fashion according to their *influence on certain phenomena* in aqueous solution or, equivalently, according to a common *underlying principle* explaining these phenomena. Whereas the former can usually be defined unambiguously based on experimental observations, the latter is still a matter of debate. The most prominent phenomena mirroring Hofmeister effects involve *e.g.* protein and colloidal solubility[87, 99, 140, 280, 2282, 2283, 2285], interfacial tension[87, 99, 2283, 2285, 2293], interfacial potential[87, 140, 2283, 2293] or counter-ion binding specificity in polyelectrolyte systems[2286]. Since these processes are omnipresent in daily life[87] (*e.g.* cryoprotectants, diseases involving the precipitation of biomolecules or salt crystals, protein purification and crystallization, or anesthetics), many studies have attempted to clarify the molecular mechanisms underlying Hofmeister effects.

The orginal work of Franz Hofmeister and his co-workers is collected in a series of seven articles[2287–2292, 2294], which are reviewed by Ref.[2284]. The latter reference also provides an English translation of two of the most important articles[2287, 2288], which investigate the *ability of various salts* to promote the precipitation of polyelectrolytes such as proteins[2287,2288] and colloids[2288]. Their conclusions can be summarized by the observation that this ability correlates with the water ordering capacity of the salt. For instance, the carbonate anion lies on one end of the Hofmeister series, because it decreases the solubility of proteins, *i.e.* induces protein precipitation. In the opposite, the thiocyanate anion lies on the other end of the Hofmeister series, because it increases the solubility of proteins, *i.e.* counteracts precipitation. Note, in this context, that Hofmeister effects typically affect anions more strongly than cations.

These differences are commonly assumed correlated with the *water-structuring propensity* of the different ions. This propensity was often formulated[2285] in terms of the phenomenological notion of *kosmotropic* and *chaotropic* behavior[2264]. Kosmotropic compounds are defined as water *structure makers*, while chaotropic compounds are defined as water *structure breakers*. They are also sometimes referred to as *strongly hydrated* and *weakly hydrated* compounds, respectively. In the context of ions, one would intuitively think that small or/and highly charged ions are kosmotropic (structure maker, strongly hydrated), whereas large or/and weakly charged ions are chaotropic (structure breakers, weakly hydrated). The relationship with the Hofmeister series is then interpreted as a tendency for ions with *matching water structuring propensities* (either both kosmotropic or both chaotropic) to associate more tightly, leading to lower solubility and precipitation.

Besides Hofmeister effects on polyelectrolyte precipitation, the above qualitative principles have been used to tentatively rationalize the relative solubilities of alkali-halide salts [1869] (principle of lowest solubility at matching hydration enthalpy by Fajans, in 1921), the relative binding propensities of counter-ions to protein carboxylate groups[280] (highest at comparable hydration free energies), the influence of different ions in protein crystallization[2283], or[282] the low solubilities (based on the similar magnitudes of hydration free energies of these ions) of calcium carbonate (egg shell), calcium oxalate (kidney stones) or calcium hydroxyphosphate (bones, teeth). A tentative classification of the alkali and halide ions as kosmotropes or chaotropes, *i.e.* an abbreviated Hofmeister series, was also defined[282], suggesting the Li^+, Na^+ and F^- ions as kosmotropic, and the K^+, Rb^+, Cs^+, Cl^-, Br^- and I^- ions as chaotropic. Note that an alternative classification[109] based on the hydrogen-bonding extent (rather than strength or dynamics), expressed in terms of the change in the standard chemical potential of the ion upon transfer from H_2O to D_2O using the tetraphenylarsonium tetraphenylborate assumption, suggested the Li^+ ion as kosmotropic, the Na^+ and F^- ions as borderline cases, and the K^+, Rb^+, Cs^+, Cl^-, Br^- and I^- ions as chaotropic.

Although interesting from a theoretical point of view, these classifications are difficult to recast into a quantitative or even rigorous physical framework. One reason is that they rely on a rather "fuzzy" terminology. While the Hofmeister series is originally a measure of *ionic binding affinities*, the terms kosmotropic and chaotropic refer to *solvation structure*, the expressions structure maker and structure breaker may be viewed as referring to *solvation entropies*, and the expressions strong hydration and weak hydration to a *solvation enthalpy* or *free energy*.

As evident from Table 5.26 and Figure 5.8, in the context of the alkali and halide ions in water, these four types of properties do not need to be correlated, and the ambiguity of the classification has been stressed previously[247,248] (see, however, Ref.[109]). For example[248], the Na^+ cation can, in comparison to the K^+ cation, be called a structure maker considering the narrower orientational distributions and tighter attachment of water molecules in its first hydration shell, as opposed to the K^+ ion, which shows broader orientational distributions and looser attachment. However, the Na^+ cation can, in comparison to the Cl^- ion, be called a structure breaker considering its more perturbing effect on water structure beyond the first hydration shell, arising from its relative inability to fit into the hydrogen bonding network of water, as opposed to the Cl^- ion, which accommodates smoothly in the hydrogen bond network of water and does not lead to severe disruptions of water structure beyond the first hydration shell. Considering the data of Table 5.26, one may also observe that the H^+, Li^+ and F^- ions appear to be strong "structure makers" in terms of entropy, but that the Li^+ and F^- ions are more "weakly hydrated" than H^+ in terms of enthalpy and free energy. Note also that the negative signs of the solvation entropies and enthalpies suggest that all alkali and halide ions should be described as "structure makers" and "strongly hydrated", in contrast to the scales of Ref.[282] or Ref.[109].

Experimental findings on the structure and dynamics of the hydrogen-bond network of water molecules in salt solutions investigated by ultrafast infra-red spectroscopy[243,2295–2298] have also suggested that an alteration of bulk water structure can actually not be the basis of the ion-induced effects implied by the Hofmeister series, because such an impact of the investigated ionic solutes on bulk water structure appears to be absent. Similarly, experiments on the bulk water dielectric relaxation times in salt solutions investigated by dielectric spectroscopy[2299] have shown no influence of added salts on the water dynamics. Instead, it has been suggested that Hofmeister effects can only be explained in terms of ion-specific interactions with *e.g.* a solvated biomolecule or colloidal particle, *i.e.* of phenomena occurring in the first hydration shell rather in the bulk[2285]. See, however, the infra-red study of Ref.[249], which suggests in the opposite that salts do influence the hydrogen bonding network of water, and that this influence correlates well with expectations from the Hofmeister series.

5.21 The solvated electron

The *hydrated electron* e^-_{aq} was discovered in 1962, in a spectroscopic study of pure (deaerated) water that revealed an ephemeral absorption band peaking near 700 nm after irradiation of the sample with a pulse of 1.8 MeV electrons[2300]. As a strong reductant (standard redox potential −2.87 V)[1803], it is a very reactive and thus transient species in water, and rapidly reacts with trace impurities in the solvent (rate constants on the order of $10^9 - 10^{10}$ $dm^3{\cdot}mol^{-1}{\cdot}s^{-1}$)[2301].

The thermodynamic properties of the hydrated electron have recently been investigated *via* nanocalorimetric studies in the gas phase[1808] using water clusters containing trivalent metals ions and 42 – 160 water molecules[1809]. Extrapolation to the bulk situation suggests[1809] a hydration enthalpy of −1.3 eV (-125.4 $kJ{\cdot}mol^{-1}$). A QM computational study using continuum models for the solvation of electron-water clusters (1 – 12 water molecules) suggested a hydration free energy of the electron of −1.52 eV (-146.7 $kJ{\cdot}mol^{-1}$)[1807]. Based on estimates for the formation free energy of the aqueous proton of Ref.[261], another study[1802] found the hydration free energy of the electron to be -164.8 $kJ{\cdot}mol^{-1}$.

The surprisingly low magnitude of the hydration free energy and enthalpy compared to *e.g.* $\mathcal{G}^{\ominus}_{H,wat} = -1100$ $kJ{\cdot}mol^{-1}$ for the aqueous proton (Section 5.19) probably arises from the smeared-out character of the negative charge. Whereas a former *ab initio* study of Ref.[1804] suggested that the hydrated electron resides in a short-lived cavity, with six water molecules in its first solvation shell, this view has been challenged by hybrid quantum-mechanical molecular-mechanical (QM/MM) molecular dynamics simulations. These simulations showed that the hydrated electron is associated in the opposite with a region of increased water density[2302] (radius of about 0.6 nm, encompassing about 37 water molecules), *i.e.* this species is not adequately described by a point-particle inhabiting a cavity.

Many other computational studies of the hydrated electron have been performed with Car-Parrinello molecular dynamics (CPMD) simulations[1804] or hybrid quantum-mechanical molecular-mechanical (QM/MM) molecular dynamics simulations[2302–2307]. Note that a pseudo-potential commonly applied to describe the electron-water interactions[2303] has recently been shown to be erroneous[2306, 2307], affecting the conclusions of a number of theoretical investigations concerning the hydrated electron. Using a corrected pseudo-potential[2306], the hydrated electron is found to inhabit an approximately spherical region with a radius of gyration of about 0.26 nm (in contrast to the previous estimate of 0.21 nm; the experimental value is[1805] 0.25 nm) and the calculated absorption spectrum peaks at approximately 1.5 eV (in contrast to the previous estimate of 2.4 eV; the experimental value is[2308] 1.7 eV). Structural properties (radial distribution functions) turn out to also be significantly different when using the corrected pseudo-potential[2306].

Experimental studies concerning hydrated electrons near the planar air-water interface have shown that the electron is fully hydrated, *i.e.* it is not located at the Gibbs dividing surface (as it is observed in gas-phase clusters and in ice on metal surfaces), but just below[2309].

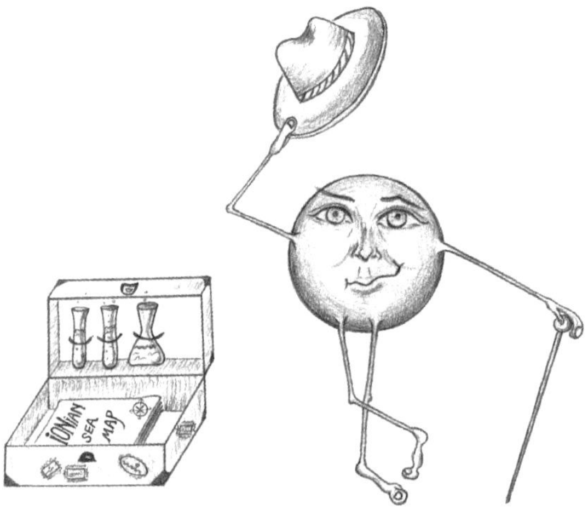

A single ion simply raises its hat in acknowledgment for the amazing amount of human energy and time that has been dedicated to the experimental investigation of ionic properties. Owing to this massive effort, physical chemists are nowadays equipped with a remarkably detailed map of the "ionian sea".

CHAPTER 6
THEORETICAL DETERMINATION

In this chapter, the approaches employed for the *theoretical determination* of thermodynamic parameters related to ionic solvation are reviewed, and the results of these investigations are discussed, mainly in the restricted context of alkali and halide hydration. The relevant features of the basic physical models employed to this purpose have been summarized in Chapter 3, and the present chapter focuses principally on *improvements, refinements and corrections* of these basic models. This discussion of theoretical methods and results follows the outline of Chapter 3, namely: (*i*) *continuum-electrostatics (CE) calculations* and related semi-atomistic approaches; (*ii*) *classical atomistic (AT) simulations*, including correction schemes and improvements of the force-field description; (*iii*) *quantum-mechanical (QM) computations* (calculations or simulations), including quasi-chemical theory, hybrid approaches and Car-Parrinello molecular dynamics. Although not discussed in detail here, it should be kept in mind that the theoretical evaluation of properties to be compared with experimental data always involves *standard-state definition issues*. These issues have been discussed in Sections 2.2, 4.2.4, 4.2.5, 4.2.16 and 4.2.17. For all thermodynamic parameters, the key issue resides in the distinction between semi-standard point-to-point parameters $\Delta_s Y^{\oplus}$ at $P^{\ominus}$ and T^{-}, which typically represent the natural outcome of a calculation, and standard parameters $\Delta_s Y^{\ominus}$, which represent the usual format of tabulated experimental data (Sections 2.2 and 4.2.16). In the context of derivative thermodynamic solvation parameters, the distinction between standard and density-corrected variants of the solute standard-state[1412] is also of particular relevance (Sections 4.2.5 and 4.2.16). In practice, ambiguities concerning standard-state definitions, namely the existence of different conventions and the incomplete specification of the convention applied in a given study, often represent a significant factor of "uncertainty" when considering the results of theoretical calculations.

RSC Theoretical and Computational Chemistry Series No. 3
Single-Ion Solvation
Experimental and Theoretical Approaches to Elusive Thermodynamic Quantities
By Philippe Hünenberger and Maria Reif

Published by the Royal Society of Chemistry, www.rsc.org

6.1 Continuum-electrostatics calculations

Due to the *numerous shortcomings* discussed in Section 3.2, *CE calculations* generally fail to provide *quantitative results* concerning ionic solvation properties, mainly because they overlook the microscopic granularity of the solvent. There exist two main approaches attempting to improve the basic CE description: (*i*) inclusion of *correction terms* to the Born plus surface tension (BO+ST) model; (*ii*) design of *semi-atomistic approaches*, including these solvent granularity effects (microscopic structure) in an implicit fashion. These approaches are discussed in turn in the following sections.

6.1.1 Correction terms

The basic model for the calculation of intrinsic single-ion solvation free energies using CE is the *BO+ST model* (Section 3.2 and Eq. 3.7). However, it is well known that this model *overlooks many important aspects of solvation* (Points A_3-I_3 in Section 3.2), and that the incurred errors prevent the calculated values from reaching quantitative agreement with experiment, even considering the very large range of uncertainty in the experimental values to compare with. There is, however, a considerable amount of literature suggesting empirical (physically-motivated or *ad hoc*) approaches to remedy one or the other of these problems, by addition of corresponding *correction terms* to the BO+ST estimate $\Delta_s G^{\ominus}_{BO+ST}[I^z_g]$ for the standard intrinsic solvation free energy of an ion I^z_g. These suggested correction terms will be written $\Delta_s G^{cor}_{\mathrm{A}_3}$ to $\Delta_s G^{cor}_{\mathrm{I}_3}$ in the following discussion, based on the corresponding neglected effects A_3 to I_3 in Section 3.2. However, for reasons detailed further below, none of these corrections, taken individually or combined together, is able to turn the BO+ST model into a refined model that is at the same time *quantitative* and *predictive*. Thus, these terms should probably rather be viewed as more or less reliable estimates for the magnitude of the error incurred by a specific neglected effect.

The methods suggested to correct the BO+ST estimate can be broadly classified in the same sequence (Points A_3-I_3) as used in Section 3.2:

(A_3) Correction for the neglect of the *solvent polarization at the ion-solvent interface*[707]: the electric potential at the center of an uncharged cavity within a finite liquid sample is affected by the solvent polarization at both the air-liquid and liquid-cavity interfaces. In the case of water, these effects have been investigated[707] using AT simulations with the SPC water model[1274], resulting in an empirical equation relating the curvature radius R of an air-water interface (with $R < 0$ for a concave interface, *e.g.* cavity, and $R > 0$ for a convex interface, *e.g.* droplet) to the potential difference $\tilde{\chi}^{\ominus}_{svt}(R)$ across this interface (in the air to liquid direction) at 300 K, namely[707]

$$\tilde{\chi}^{\ominus}_{svt}(R) = \chi^{\ominus}_{svt} + \begin{cases} \tilde{\chi}^{\ominus}_{svt,+} R^{-1} & \text{(convex, } R > 0) \\ \tilde{\chi}^{\ominus}_{svt,-} \, |R|^{-1} & \text{(concave, } R < 0) \end{cases} , \tag{6.1}$$

with $\tilde{\chi}^{\ominus}_{svt}(\pm\infty) = \chi^{\ominus}_{svt} = 0.73$ V (planar interface), $\tilde{\chi}^{\ominus}_{svt,+} = -1.11$ V·nm (convex interface, *e.g.* droplet surface; fit in the range $R = 2$ nm $\rightarrow \infty$) and $\tilde{\chi}^{\ominus}_{svt,-} = -0.11$ V·nm (concave interface, *e.g.* cavity surface; fit in the range $|R| = 0.18$ nm $\rightarrow \infty$), see also the corresponding discussion in Section 6.2.1. This approximate equation, which is not to be applied outside its calibration range (*i.e.* obviously invalid for -0.18 nm $< R <$ 2 nm), suggests that for the SPC water model[1274]: (*i*) the air-water interfacial potential is positive, *i.e.* the surface water molecules are preferentially oriented with the positive end of their molecular dipole (hydrogen atoms) pointing towards the bulk liquid; (*ii*) the potential is maximal for a planar interface and decreases in magnitude with the interface curvature for both convex and concave interfaces. The above expression can be used to formulate a correction term for the neglect of the solvent polarization, as

$$\Delta_s G^{cor}_{\mathrm{A}_3} = -zF(C^{cor}_{\mathrm{A}_3} + \tilde{C}^{cor}_{\mathrm{A}_3} R_I^{-1}) , \tag{6.2}$$

with the coefficients $C^{cor}_{\mathrm{A}_3} = \chi^{\ominus}_{svt}$ and $\tilde{C}^{cor}_{\mathrm{A}_3} = \tilde{\chi}^{\ominus}_{svt,-}$, where it has been assumed that the corresponding interfacial potential is constant during the charging process of the ion. Note that the correction only accounts for the polarization at the ion-solvent interface and not at the air-solvent interface, which is the appropriate choice if the corrected model is still to provide an estimate for the intrinsic solvation free energy of the ion.

(B_3) Correction for the neglect of the *ion-compression work*[660]: a simple estimate for the ion-compression work can be formulated by assuming that the ionic charge density is compressed homogeneously when the ion is transferred from the gas-phase to a condensed phase, the compression factor being determined by the ratio $R_I[I^z_d]/R_I[I^z_g]$ of the in-solution radius $R_I[I^z_d]$, possibly approximated by a corresponding in-crystal radius, to the corresponding gas-phase radius $R_I[I^z_g]$. This assumption must be combined with an approximation concerning the charge density within the ion, given its radius R_I. If the ion is taken to be a conductor with its entire charge ze located at the surface, the self-energy of the charge distribution is $(8\pi\epsilon_o)^{-1}z^2e^2R_I^{-1}$ and the ion-compression work may be evaluated as[660]

$$\Delta_s G^{cor}_{\mathrm{B}_3} = N_A \frac{z^2e^2}{4\pi\epsilon_o} C^{cor}_{\mathrm{B}_3} \left(\frac{1}{R_I[I^z_d]} - \frac{1}{R_I[I^z_g]} \right) , \tag{6.3}$$

where $C_{\mathrm{B}_3}^{cor} = 1/2$. Alternatively, one might consider the ion as homogeneously charged in its volume. A simple electrostatic calculation shows that in this case, the self-energy becomes $3(20\pi\epsilon_o)^{-1}z^2e^2R_I^{-1}$, and Eq. 6.3 should be applied with $C_{\mathrm{B}_3}^{cor} = 3/5$, which is numerically close to 1/2. An even more refined model would consider the ion as a nuclear point charge of magnitude Ze surrounded by a homogeneous electron cloud of charge $(-Z+z)e$. In this case, the sum of the electronic self-energy and nucleus-electron interaction, both calculated classically, becomes $-3(40\pi\epsilon_o)^{-1}(3Z^2+Zz+2z^2)e^2R_I^{-1}$, and Eq. 6.3 should be applied with $C_{\mathrm{B}_3}^{cor} = -(3/10)(3Z^2-Zz-2z^2)$. Unfortunately, this expression suggests that the ion-compression work is negative for all realistic values of Z and z. This arises because the charge density within an ion is governed by the laws of quantum mechanics (classically, the electrons should progressively lose their energy in the form of electromagnetic radiation and collapse onto the nucleus). For this reason, the application of an equation like Eq. 6.3 for evaluating the ion-compression work is very hazardous. In addition, the large variations observed in sets of ionic radii appropriate for gas-phase and condensed-phase environments (Section 5.4 and Table 5.4) suggest that even assuming that the physics of the correction is reasonable, no accurate value can be obtained in practice. In fact, the suggested compression effect is not even obvious when considering the various available sets of gas-phase, in-crystal and in-solution radii. However, an accurate estimation of $\Delta_s G_{\mathrm{B}_3}^{cor}$ could in principle be achieved based on QM calculations.

(C_3) Correction for the neglect of the *ion polarizability*[715,716]: to our knowledge, no correction terms have been suggested for this effect. Note that, as discussed in Section 3.2, increasing the value of the dielectric permittivity within the ion above one in the BO+ST model is not a viable approach to evaluate this contribution.

(D_3) Correction for the neglect of the *ion-solvent dispersion interactions*[712,713,716–718]: in principle, the leading term of the ion-solvent dispersion interactions could be estimated using the Slater-Kirkwood formula (Section 3.3.1, Eq. 3.15) based on estimates for the polarizability volumes of the ion and of the solvent molecule in solution, $\tilde{\alpha}_I$ and $\tilde{\alpha}_{svt}$, respectively, and of the corresponding effective electron numbers, $n_{el,I}$ and $n_{el,svt}$, respectively, leading, after r^{-6}-weighted spherical averaging, to

$$\Delta_s G_{\mathrm{D}_3}^{cor} = -\frac{8\pi}{3} C_{SK} \frac{N_A \tilde{\alpha}_I \tilde{\alpha}_{svt} n_{el,I}^{1/2} n_{el,svt}^{1/2}}{(n_{el,I}\tilde{\alpha}_I)^{1/2} + (n_{el,svt}\tilde{\alpha}_{svt})^{1/2}} M_{svt}^{-1} \rho_{svt} R_I^{-3} \quad , \tag{6.4}$$

where M_{svt} is the molar mass of the solvent, ρ_{svt} its density, and $C_{SK} = 23.97$ kJ·mol^{-1}·nm$^{3/2}$. For the alkali and halide ions, experimental values[727] of $\tilde{\alpha}_I$ and estimated values[1321] of $n_{el,I}$ can be found in Table 5.9. For water, corresponding values are $\tilde{\alpha}_{svt} = 1.45 \cdot 10^{-30}$ m^3

based on Ref.[1322] and $n_{el,svt} = 4.08$ based on Ref.[1321].

(E$_3$) Correction for the neglect of *electrostriction*: electrostriction is not accounted for in the BO+ST model, because this effect is related to the local field-induced compression of the solvent molecules in the neighborhood of the ion, while CE assumes a homogeneous dipole density. Tentative evaluations of the contribution $\Delta_s G^{cor}_{\mathrm{E}_3}$ of electrostriction to the solvation free energy of an ion have been reported[730,731,742]. A simple way to introduce electrostriction into the BO+ST model, and thereby estimate the contribution $\Delta_s G^{cor}_{\mathrm{E}_3}$ by comparison with the unmodified BO+ST equation, relies on considering the thermodynamics of water in an applied electric field according to Ref.[735]. Neglecting the dependence of the relative dielectric permittivity on the applied field strength, *i.e.* assuming the absence of dielectric saturation (see Point F$_3$ below), the effective permittivity of the solvent around an ion at a given temperature is directly related to its local density. The latter quantity is in turn determined by the local pressure increase resulting from the electric field due to the ion *via* the compressibility of water. Thus, given an appropriate function describing the dependence of the density on the strength of an applied electric field, and considering the distance dependence of the electric field due to the ion, an electrostriction correction to the Born equation may be obtained[742]. A corresponding first-order correction term, *i.e.* which assumes that the relative dielectric permittivity of water is linear in the density, is

$$\Delta_s G^{cor}_{\mathrm{E}_3} = N_A \frac{\partial_P \epsilon_{svt}}{160\pi\epsilon^4_{svt} R^5_I} \left[\frac{(ze)^2}{4\pi\epsilon_o}\right]^2 \left(\frac{\partial_P \epsilon_{svt}}{\kappa_T} + \epsilon_{svt} - 1\right) \quad . \tag{6.5}$$

Estimates for the quantities $\epsilon^{\ominus}_{svt}$, $\partial_P \epsilon^{\ominus}_{svt}$ and $\kappa^{\ominus}_T$ appropriate for water can be found in Table 1.1. These calculations suggest that the term $\Delta_s G^{cor}_{\mathrm{E}_3}$ slightly favors solvation for both cations and anions, due to an increase in the effective solvent permittivity *via* the density increase in the neighborhood of the ion. However, its magnitude is essentially negligible for monovalent ions in water, because electrostriction effects are intrinsically weak and only active in the immediate vicinity of the ion.

(F$_3$) Correction for the neglect of *dielectric saturation*: dielectric saturation is not accounted for in the BO+ST model, because this effect is intimately related to the discrete nature of the solvent molecules, inducing a non-linear dielectric response at high fields, while CE assumes linearity at all field strengths (linear-response approximation). There is no upper bound to the polarization of a volume element within a dielectric continuum, which always grows linearly with the local field. In contrast, there is an upper bound to the polarization that can be produced by microscopic solvent dipoles at a given effective density, namely, the polarization corresponding to a complete alignment of all dipoles. If electronic polarization (and, possibly, molecular geometric distortion) effects are active, the polarization can grow beyond this limit, but likewise, this increase is linear for low fields only and also becomes saturated beyond a certain field strength. Dielectric saturation is particularly significant in the case of small or/and highly-charged ions, which give rise to a strong electric field in their vicinity. A simple way to introduce dielectric saturation into the BO+ST model, and thereby estimate the contribution $\Delta_s G^{cor}_{\mathrm{F}_3}$ by comparison with the unmodified BO+ST equation, relies on altering the representation of the solvent dielectric permittivity in the model. Such approaches include: (*i*) the use of an effective (reduced) solvent permittivity value in the original BO+ST equation; (*ii*) the use of a multiple-shell description involving concentric volumes with different permittivities[660, 662–665, 675, 760, 761, 768, 770]; (*iii*) the introduction of an explicit (radial) dependence of the solvent permittivity on the distance from the ion[661,666,717,730,731,744,748,752,754,757–759,762,764–766,769]. For example, a simple approach of the third kind relies on expanding the local dielectric permittivity as a series in the squared local electric field strength[753]. This leads to a correction of the form[717]

$$\Delta_s G^{cor}_{\mathrm{F}_3} = N_A^2 \left[\frac{(ze)^2}{4\pi\epsilon_o} \right]^2 \frac{C^{cor}_{\mathrm{F}_3}}{20\epsilon^4_{svt} R_I^5} \quad , \tag{6.6}$$

where[717, 753] $C^{cor}_{\mathrm{F}_3} = 5.0 \cdot 10^{-3}\ \mathrm{J}^{-1}\cdot\mathrm{nm}^3 \cdot \mathrm{mol}$ is the coefficient of the squared field strength in the expansion, considering water as a solvent. This quantity is related to the static relative dielectric permittivity of the solvent, its number density, and the molecular principal polarizabilities of the solvent molecule together with the components of the dipole moment along the axes of the polarizability ellipsoid. The quantity $\epsilon^{\ominus}_{svt}$ for water can be found in Table 1.1. These calculations suggest that the term $\Delta_s G^{cor}_{\mathrm{F}_3}$ decreases the magnitude of the solvation free energy for both cations and anions, due to a decrease in the effective solvent permittivity *via* saturation of the solvent dielectric response in the neighborhood of the ion. This term is most significant for small ions with high charges. Although dielectric saturation is only important within the first hydration shell of the ion, it is non-negligible even for monovalent ions in water. In particular, dielectric saturation effects, reducing the magnitude of the solvation free energy, typically overweight electrostriction effects (see Point E_3 above), which have the opposite influence. Note that the field dependence of the relative dielectric permittivity has also been investigated in AT simulations[743, 767, 771, 773, 2310].

(G_3) Correction for the neglect of *solvation structure* and of *specific ion-solvent interactions*[260,325,643,644,712–714,717,718,758,759,775–781]: Solvation structure and specific solute-solvent interactions are entirely disregarded in the BO+ST model, which models the solvent as a homogeneous distribution of point dipoles at a given density and neglects all the microscopic details of the solvation structure (continuous-medium approximation), as well as the possibility of specific (other than purely charge-dipole) ion-solvent interactions. Because this approximation is so fundamentally anchored in the CE treatment, relaxing it requires a change to an explicit AT or QM representation of the solvent, or at least a semi-atomistic description (Section 6.1.2). An alternative approach relies on the *ad hoc* adjustment of the effective ionic radii, as an attempt to capture the corresponding mean effect in an empirical fashion (Point H_3 below).

(H_3) Correction for the ambiguity in the *choice of the ionic radius*[646,661,714,716,746,748,754,759,765,770,772,776,778–781,790,791,813–819]: as discussed in Section 5.4, the effective ionic (in-solution) radius R_I, which characterizes the effective extent of the ionic cavity in the solution environment, is not an experimental observable, *i.e.* it does not possess an unambiguous physical definition. For this reason, a common approach relies on treating the ionic radii as adjustable parameters of the model. This is typically done by introducing one or more correction factors to the crystallographic radii, *e.g.* a common additive correction for the cations and the anions separately[408,748,776,814,816,818,820–822], or a modification of the radii using information from ion-solvent radial distribution functions in solution[778,779,781]. For example, in the *Latimer approach*[814], the Pauling radii[2047] are offset by +0.01 nm for monovalent anions and +0.085 nm for monovalent cations, accounting in an effective way for asymmetric solvation effects in water. The optimization of the correction terms follows from comparison of the BO+ST prediction with experimental data. However, this choice has the disadvantages that: (*i*) the model loses predictive ability to rather become a fitting tool; (*ii*) the correction terms encompass in an effective way all the sources of errors discussed in the present enumeration, so that the physical significance of these terms becomes unclear; (*iii*) the correction terms also encompass experimental errors on the set of intrinsic single-ion solvation free energies employed in the fitting, which can be extremely large (Section 2.3). In view of the inability of the BO+ST model to capture accurately the temperature dependence of single-ion solvation free energies (*e.g.* enthalpy-entropy partitioning[733,754,758,776,777,791,814,819,872,874,2311], Section 3.2), the definition of temperature-dependent ionic radii has also been explored[791,819,875,876]. Alternative semi-atomistic approaches are also mentioned in Section 6.1.2.

(I_3) Correction for the ambiguity in the choice of the *microscopic surface tension coefficient*[823–840,840–843]: besides the (usually small) pressure-volume contribution, the surface tension model for the calculation of the cavitation free energy assumes a linear dependence of this free energy on the surface area of the ion-sized cavity (Section 3.2, Eq. 3.6). This relationship is based on thermodynamic considerations for macroscopic interfaces and may not be accurate for microscopic ones, where *e.g.* an additional volume term accounting for missing solvent-solvent interactions might have to be considered[844]. Even accepting the linear dependence on the surface area, the microscopic (air-liquid) surface tension coefficient $\tilde{\gamma}_{svt}^{\ominus}$ need not be immediately related to the corresponding macroscopic coefficient $\gamma_{svt}^{\ominus}$, although the latter value can be used as a first estimate in the absence of a better guess. An alternative approach involves the estimation of the work of cavitation by application of the *Tolman equation*[2312,2313]. This method relies on the application at the microscopic level[2314,2314–2318] of the macroscopic theory of surface tension at curved interfaces (see also Refs.[827,829,838,840–842,2319–2326] for refined approaches). Note that this formalism rests on a parameter called the Tolman length, which cannot be determined experimentally (see, however, Ref.[845]) and is the subject of intense debate[2319–2327]. Alternative semi-atomistic approaches are also mentioned in Section 6.1.2.

Various forms for the term $\Delta_s G_{chg}$, alternative to the simplest form of Eq. 3.4, have been suggested in the literature. Their purpose was to approximately account for the above effects, and in particular electrostriction, dielectric saturation and specific solute-solvent interactions, or/and to empirically improve the agreement with experiment, based on a given set of crystallographic radii and estimated intrinsic single-ion solvation free energies. Illustrative results for the different correction terms are presented in Table 6.1. Note that, for reasons detailed below (highly approximate nature and non-additivity of the correction terms) the values presented therein should not be taken very seriously and the resulting corrected estimate has very little physical meaning.

Although interesting from a conceptual point of view, none of the resulting approaches can be regarded as entirely satisfactory. One reason is that individual studies typically suggest a correction for *one single effect*, leaving out the others. Another reason is that the derivation of physically-motivated correction terms for the different neglected effects relies on *additional assumptions and approximations*, that are themselves also subject to large errors. Similarly, the suggestion of *ad hoc* modifications introduces *additional fitting parameters* optimized for agreement with experimental data, so that it is no longer obvious whether the improvement really results from a more accurate physical description of the problem. Finally, it should be stressed that the various sources of error in the BO+ST model are not independent but *largely interrelated*, *i.e.* it is difficult to draw an entirely clear-cut boundary between them. For example: (*i*) ion-solvent interface polarization (Point A_3) is the consequence of a specific solvation structure (Point G_3); (*ii*) ion-compression (Point B_3) and electrostriction (Point E_3) both result from a density increase caused by large ion-solvent attractive forces; (*iii*) ion polarizability (Point C_3), dispersion interactions (Point D_3) and specific interactions (Point G_3) are all active components of ion-solvent interactions at the microscopic level; (*iv*) electrostriction (Point E_3), dielectric saturation (Point F_3) and solvation

Table 6.1 *Continuum-electrostatics estimates for the hydration free energies of the alkali and halide ions, along with suggested correction terms for various neglected effects.* The standard hydration free energy $\Delta_s G^{\ominus}_{BO+ST}$ is calculated according to the Born plus surface tension (BO+ST) model (Eq. 3.7). Various estimates for suggested correction terms $\Delta_s G^{cor}_{\mathrm{A}_3}$-$\Delta_s G^{cor}_{\mathrm{I}_3}$, the subscript referring to the literal enumeration A_3-I_3 of Section 3.3.6, are also provided. Inclusion of all these corrections leads to a tentative corrected estimate $\Delta_s G^{cor\ominus}_{BO+ST}$. Note that in view of the highly approximate nature and non-additivity of the different correction terms, this quantity is not very meaningful. Unless otherwise specified, the calculations are based on the radius Set 33r in Table 5.4 and on the density $\rho^{\ominus}_{wat}$, static relative dielectric permittivity $\epsilon^{\ominus}_{wat}$ and macroscopic surface tension coefficient $\gamma^{\ominus}_{wat}$ reported for water in Table 1.1. The standard states are according to the $bbme_T$ convention (Section 4.2.4; reference pressure $P^{\circ} = 1$ bar and reference temperature $T^{-} = 298.15$ K).

Ion	$\Delta_s G^{\ominus}_{BO+ST}$	$\Delta_s G^{cor}_{\mathrm{A}_3}$	$\Delta_s G^{cor}_{\mathrm{B}_3}$	$\Delta_s G^{cor}_{\mathrm{D}_3}$	$\Delta_s G^{cor}_{\mathrm{E}_3}$	$\Delta_s G^{cor}_{\mathrm{F}_3}$	$\Delta_s G^{cor}_{\mathrm{H}_3}$	$\Delta_s G^{cor}_{\mathrm{I}_3}$	$\Delta_s G^{cor\ominus}_{BO+ST}$
	[kJ·mol^{-1}]								
Li^+	-749.6 (-762.0)	47.5	92.6	-11.1	1.7	21.7	370.1	-2.3	-229.4
Na^+	-536.7 (-553.1)	15.2	55.7	-35.2	0.3	4.4	224.9	-4.3	-275.8
K^+	-482.4 (-500.6)	7.0	109.6	-103.1	0.2	2.7	191.7	-5.3	-279.6
Rb^+	-443.5 (-463.4)	1.3	100.4	-125.7	0.1	1.8	169.0	-6.1	-302.6
Cs^+	-395.6 (-418.2)	-5.7	90.9	-139.0	0.1	1.1	142.8	-7.5	-313.1
F^-	-502.1 (-519.6)	10.0	194.8	-124.6	0.3	3.2	36.6	-4.9	-386.7
Cl^-	-367.3 (-391.9)	-9.8	106.2	-143.8	0.1	0.8	21.2	-8.6	-401.2
Br^-	-339.7 (-366.7)	-13.7	83.0	-144.2	0.0	0.6	18.6	-9.8	-405.1
I^-	-294.6 (-326.6)	-19.9	69.8	-133.9	0.0	0.3	14.8	-12.3	-375.7
Eq.	3.7	6.2	6.3	6.4	6.5	6.6	-	-	
Notes	a	b	c	d	e	f	g	h	i

(a) the number indicated between parentheses is the charging free-energy contribution (Eq. 3.4) to $\Delta_s G^{\ominus}_{BO+ST}$; (b) using $C^{cor}_{\mathrm{A}_3} = 0.73$ V and $\tilde{C}^{cor}_{\mathrm{A}_3} = -0.11$ V·nm; (c) using $C^{cor}_{\mathrm{B}_3} = 3/5$ and gas-phase radii of set 52r, except for Li^+, where a gas-phase radius of 0.1 nm was selected; (d) using $\tilde{\alpha}_{el,wat} = 1.45 \cdot 10^{-30}$ m^3 and $n_{el,wat} = 4.08$ for water, along with $n_{el,I}$ as given by Ref.[1321] (Table 5.9), gas-phase ion polarizabilities $\tilde{\alpha}_{el,I}$ as given by Ref.[727] (Table 5.9), as well as M_{wat} and ρ°_{wat} from Table 1.1; (e) using $\partial_T \epsilon^{\ominus}_{wat}$ and $\kappa^{\ominus}_T$ from Table 1.1; (f) using $C^{cor}_{\mathrm{F}_3} = 5.0 \cdot 10^{-3}$ J^{-1}·nm^3 · mol for water[717, 753], and $\epsilon^{\ominus}_{wat}$ from Table 1.1; (g) the reported values are the difference $\Delta_s G^{BO}_{chg}(R'_I) - \Delta_s G^{BO}_{chg}(R_I)$ (the quantities $\Delta_s G^{BO}_{chg}$ being evaluated according to Eq. 3.4), where R_I is taken from set 33r (Table 5.4) and R'_I is an altered effective ionic radius given by $R'_I = R_I + C$, with $C = 0.085$ nm for cations and $C = 0.01$ nm for anions, as suggested by Ref.[814]; (h) using a tentative estimate for the microscopic surface tension coefficient of $\tilde{\gamma}^{\ominus}_{wat} = 35 \cdot 10^{-3}$ J·m^{-2} based on positronium-induced nanobubble formation[845] (Section 3.2); (i) sum of the values of columns 2-9.

structure (Point G_3) simultaneously affect the solvent dielectric response. For this reason, the errors incurred in the BO+ST model by all these neglected features are certainly *not additive*, and a patched model that would combine the best possible correction terms taken individually would certainly involve double counting and not be very meaningful. Furthermore, in the absence of unambiguous values for the ionic radii (Point H_3) and the microscopic surface tension coefficient of the solvent (Point I_3), as well as for the experimental intrinsic single-ion solvation free energies to compare with, even the resultant of all these effects cannot be characterized accurately.

In summary, there is no well-established improved version of the BO+ST model. The *original BO+ST model* is *simple and analytical*, and captures the dominant component of solvation, namely the electrostatic interaction of the ion charge with its polarized solvent environment. It represents a great *qualitative tool* for the basic discussion of ionic solvation, because it enables a correct first-order reasoning concerning the effect of ionic charge and radius on the solvation free energy, and provides a solid reference point for the discussion of more complex secondary effects. It also permits the formulation of long-range corrections to approaches where short-range effects are represented at a higher-level of resolution (AT or QM, see Sections 6.2 and 6.3). However, although a considerable amount of work has been dedicated in the past to the tentative definition of a *corrected variant* achieving simultaneously *quantitative accuracy* and *predictive power*, this probably represents a *hopeless endeavor*. The reason is that the main flaw of the Born model is the *neglect of the solvent granularity* (microscopic structure) and of the details of ion-solvent interactions. The inclusion of these features appears to be simply incompatible with the very principle of the CE approach. Some tentative methods to include implicit reference to the solvent granularity are nevertheless briefly mentioned in the next section.

6.1.2 Semi-atomistic approaches

In this section, a few approaches are briefly mentioned that attempt to partially include effects related to the microscopic *granularity of the solvent* into the CE approach, while retaining an *implicit description* of the solvent degrees of freedom. Such approaches are termed here *semi-atomistic.*

A simple semi-atomistic approach is *scaled-particle theory* (SPT) [862–868], which introduces a semi-atomistic representation through the description of the solvent as a fluid of hard-spheres characterized by a given molecular radius (see also Refs.[869–871] for refined approaches). This theory does not include electrostatic interactions and can be applied to the evaluation of the cavitation term (Point I_3 in Section 6.1.1), as an alternative to the ST model (Section 3.2, Eq. 3.6) or the Tolman equation. In practice, SPT will provide an estimate for the microscopic surface tension coefficient of the model, taking into account its discrete nature. Microscopic surface tension has also been directly investigated by computer simulations [831–837], density functional theory [838–840], or in the framework of integral equation [841] or gradient theory [840,842,843]. See also Point I_3 in Section 3.3.1 and references therein for additional considerations on this issue.

A semi-atomistic scheme including electrostatic effects is the *integral-equation approach*, relying for example on the *mean-spherical approximation*[98,822,2249,2251] (MSA) or on site-site interaction potentials taken from *classical force fields*[655,791,1166,1270]. These methods represent alternatives to the Born equation for calculating single-ion solvation free energies. The MSA relies on the introduction of mean spatial and orientational correlations between the molecules within the solvent continuum, based on a hard-sphere description of the solvent molecules, the ion also being described as a rigid non-polarizable sphere. Although physically motivated, the model leads to the introduction of additional parameters that cannot be unambiguously determined experimentally, and are thus fitted. In this sense, they are comparable to the *ad hoc* radius increments of the Latimer approach (Point H_3 in Section 6.1.1). For example, in the MSA model presented in Refs.[822,1750,2251], the resulting equation for the charging free energy involves an additive solvent-dependent correction term δ_S to the ionic radius. This parameter is given by $\lambda_S^{-1} R_S$, where R_S is an estimate for the effective radius of the solvent molecule (*e.g.*[2328] 0.142 nm) and λ_S is the Wertheim polarization parameter[2329], a quantity related to the relative dielectric permittivity of the solvent. The Born model is recovered in the limit $R_S \rightarrow 0$ or $\lambda_S \rightarrow \infty$. In the case of water, δ_S is estimated to 0.054 nm in this model. A similar MSA approach considers the extent to which solvent-solvent dipole interactions reduce ion-solvent dipole interactions[1750,2249]. This involves a scaling factor $(1 - f_{dd})$ applied to the charging free energy including the δ_S correction, which can be calibrated[1750] by fitting against experimental data. For instance[1750], considering experimental solvation free energy data for the alkali and alkali earth ions as well as the halide ions and the sulfide ion, fitted values $f_{dd} = 0$ and $\delta_S = 0.049$ nm for the cations and $f_{dd} = 0.26$ and $\delta_S = 0$ nm for the anions were obtained, to be used in combination with the Shannon-Prewitt radii[2058]. This difference effectively introduces *asymmetric solvation effects* in water into the model, the original BO+ST model being symmetric in terms of cation *vs.* anion solvation. It is concluded that cations have a more disruptive effect on water structure compared to anions, and that repulsive solvent-solvent dipole interactions act to significantly reduce ion-solvent interactions in the case of anion solvation, but not in the case of cation solvation.

6.2 Classical atomistic simulations

As summarized in Chapter 3 (Sections 3.3.1-3.3.6), the raw ionic solvation free energies calculated using AT simulations are affected by six *major sources of error* (Points A_4-F_4 in Section 3.3.6). As already stated there, errors due to finite sampling (Point F_4) are no longer an issue given the computing resources available nowadays, at least for the calculation of solvation free energies, the calculations of solvation entropies requiring significantly longer simulations[877,878,1405,1406]. The five remaining sources of error will be discussed in three groups, namely: (*i*) Points A_4-C_4, which will be referred to as *methodology-dependent errors*, a minor component of Point D_4 being also included here, related to the inaccurate permittivity of the solvent model employed; (*ii*) Point E_4, which will be referred to as the *ion-size ambiguity error*; (*iii*) Point D_4, which will be referred to as the *force-field approximation error*. Tentative approaches to remedy these errors are: (*i*) inclusion of *correction terms* to remove the methodology-dependence in raw single-ion solvation free energies and derivative thermodynamic properties calculated by AT simulations; (*ii*) *consistent calibration* of ion-solvent van der Waals interaction parameters against intrinsic single-ion solvation free-energy estimates, assuming a specific value for the proton; (*iii*) application of the *atomistic-consistency assumption* as an attempt to relieve the ion-size ambiguity problem and to derive a consistent value for the proton; (*iv*) use of *improved force-field descriptions* as an attempt to refine the AT approach by inclusion of *e.g.* explicit electronic polarization or more accurate functional forms and combination rules for van der Waals interactions. These four approaches are addressed in turn in Sections 6.2.1-6.2.4. The discussion is principally focused on the direct calculation of single-ion solvation properties (Section 3.3.3). The alternative route to access intrinsic properties *via* calculation of the air-liquid interfacial potential of the pure solvent (Section 3.3.5) is only briefly mentioned in Sections 6.2.1 and 6.2.4.

6.2.1 Correction terms

Until recently, AT simulations did not permit a reliable calculation of single-ion solvation free energies, for a given solvent model and set of ion-solvent van der Waals interaction parameters, due to the approximate treatment of electrostatic interactions and boundary conditions in these simulations. Over the last few years, however, Kastenholz, Reif and Hünenberger[707,1081,1289,1412] have shown that the raw simulation results (ionic charging free energies evaluated *e.g. via* thermodynamic integration[628]), which are extremely sensitive to the simulation methodology employed, can be corrected *ex post* so that *methodological independence* is achieved. The corresponding additive corrections, summarized in Figure 6.1, may be evaluated using CE, or empirical equations fitted against the results of such calculations[1412], and analytical models, and must account for a number of errors sources namely, following the same sequence (Points A_4-D_4) as used in Section 3.3.6[1081]:

(A_4) The *deviation of the solvent polarization* around the ion relative to the polarization in an *ideal Coulombic (CB) system*[331,333,334,792,1080,1081,1166,1167,1169,1170,1192,1194,1199,1430,1468] and the *incomplete or/and inexact interaction* of the ion with the polarized solvent[1081,1171,1194,1199,1468], a consequence of possible approximations made in the representation of electrostatic interactions during the simulation, *e.g.* use of non-Coulombic interactions involving cutoff truncation[490] (CT), possibly along with a shifting[792,1019,1071,1166,1173,1182,1197,1208,1221,1222,1225–1232], switching[1019,1166,1173,1182,1197,1221–1224] or reaction-field[1013,1050,1082,1185,1213–1215] modification. This *type-A correction* is evaluated for the given approximate electrostatic scheme used in the simulation, but in the idealized context of a macroscopic system. For CT schemes, it can be subdivided into a type-A_1 correction, acting beyond the cutoff sphere of the ion, and a type-A_2 correction, acting within this sphere[1081]. There is no such correction for lattice-sum (LS) schemes[330,1072,1112–1114], which are Coulombic in the limit of infinite system (box) sizes[329–331,1079–1081].

(B$_4$) The *deviation of the solvent polarization* around the ion relative to the polarization in an *ideal macroscopic system*[329–331,333,334,1056,1057,1079–1081,1423,1430,1431,1468], a consequence of the use of a finite (microscopic and possibly periodic) system during the simulation, *e.g.* considering a finite droplet simulated under fixed boundary conditions[565,1042–1044,1046,1048,1049,1052,1053,1058–1061,1063–1065] (FBC) or a computational box simulated under periodic boundary conditions[490,564] (PBC), see Section 3.3.1. This *type-B correction* is evaluated for the given approximate electrostatic scheme used in the simulation, and is typically much larger for FBC/CB and PBC/LS schemes compared to PBC/CT schemes[1081].

(C$_4$) The *deviation of the solvent-generated electric potential* at the ion site as calculated from the simulated trajectory relative to the *"correct" electric potential*, a consequence of the possible application of an inappropriate summation scheme (along with a possibly non-Coulombic potential function) for the contributions of individual solvent atomic

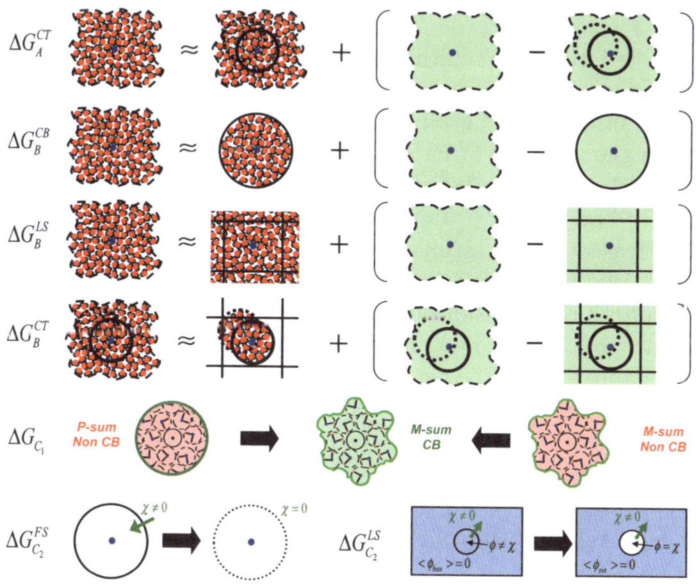

Figure 6.1 *Correction terms to raw ionic solvation free energies calculated using atomistic simulations.* This figure illustrates schematically the physical nature of correction terms to be applied to raw ionic solvation free energies calculated using AT simulations, namely types-A (including A$_1$ and A$_2$, only the sum being illustrated), B (three variants shown, corresponding to FBC/CB, PBC/LS and PBC/CT situations) and C (including C$_1$ and C$_2$, both illustrated) corrections. The type-D correction (not illustrated) accounts for a (usually very small) error in the permittivity of the solvent model employed. The basic principle of the correction scheme is to approximate the perturbation induced by a specific error source in the AT simulation (ion surrounded by explicit water molecules in the illustration) by the corresponding perturbation within a CE representation (ion surrounded by a homogeneous dielectric medium shown in green in the illustration).

charges to this potential (*i.e.* P-summation, summing over individual charges *vs.* M-summation, summing over charges within individual solvent molecules[707, 708, 1056, 1187, 1424, 1429, 1432–1434, 1436, 1438, 1469], see Section 3.3.3), as well as of the possible presence of a constant offset in this potential[707] (*e.g.* due to an interfacial potential at the surface of a FBC/CB system or to the constraint of vanishing average potential in PBC/LS schemes). This *type-C correction* is evaluated for the given approximate electrostatic scheme and choice of boundary conditions used in the simulation, and can be subdivided into a type-C_1 correction for improper potential summation and a type-C_2 correction for the potential offset[1081].

(D_4) The possibly *inaccurate dielectric permittivity of the solvent model employed.* The incurred error is typically very small for highly polar solvents (*e.g.* water), and was neglected in Ref.[1081]. Note that this correction only accounts for a marginal contribution to the overall force-field approximation error. The leading contribution to this error will be discussed in Section 6.2.4.

These suggested correction terms will be written as $\Delta_s G_A^{cor}$ to $\Delta_s G_D^{cor}$ in the following based on the corresponding error sources A_4 to D_4 listed above. Note that the subscript "4" is omitted in these symbols for simplicity and consistency with the notation of Refs.[707,1289,1412]. The term $\Delta_s G_A^{cor}$ includes two contributions $\Delta_s G_{A_1}^{cor}$ and $\Delta_s G_{A_2}^{cor}$ and the term $\Delta_s G_C^{cor}$ includes two contributions $\Delta_s G_{C_1}^{cor}$ and $\Delta_s G_{C_2}^{cor}$. In addition, the corrected charging free energies must be increased by an estimate for the work of cavitation (independent simulation involving the growth of a neutral ion-sized cavity) and adjusted to match the appropriate definition of standard states (differing for gas phase and solution), so as to lead to standard single-ion solvation free energies[1081] (Sections 2.2 and 3.3.3). The correction scheme is designed to produce intrinsic (rather than real) solvation free energies[407,409,411], *i.e.* accounting for bulk solvation effects only and excluding the reversible work of crossing a polarized air-liquid interface[430, 454, 463–465, 468–470, 833, 850, 947, 973, 1170, 1176, 1439–1451, 1454–1457, 1469, 1473]. It is also meant to be applied to raw solvation free energies exempt of any contribution from a possible self-term V_{slf} in the potential-energy function (Section 3.3.1).

The above sources of errors and thus, the requirement for correction terms, as well as the distinction between intrinsic and real values, affect not only the calculated solvation free energies, but also all *derivative thermodynamic parameters for single-ion solvation* and the corresponding *single-ion partial molar variables in solution* (*e.g.* enthalpy, entropy, heat capacity, volume, volume-compressibility and volume-expansivity). This is easily seen by observing that all these variables can be expressed as functions of the solvation free energy and its first and second derivatives with respect to temperature or pressure, along with gas-phase single-ion parameters when considering the partial molar variables in solution (Sections 2.2 and 4.2.1). Corresponding correction terms for these derivative quantities have also been derived recently[1412].

In the PBC/LS case, it is further assumed that the potential-energy function excludes the LS self-energy term[330,792,1082,1084,1085] (Eq. 3.27, *i.e.* $V_{slf} = 0$ in Eq. 3.16) and that tinfoil boundary conditions[330] are used in the simulations (Section 3.3). In the PBC/CT case, only two specific schemes are considered: straight cutoff (SC) truncation of Coulombic interactions[490] and truncation with a Barker-Watts (BW) reaction-field correction[1013,1082,1185,1213–1215] (pairwise interaction functions given by Eqs. 3.34 and 3.35, respectively; the notation BW is synonymous to the notation RF in Chapter 3). In both cases, the truncation can be applied on the basis of interatomic distances[707, 1013, 1081, 1216, 1220] (atom-based cutoff, Eq. 3.23; SC→CA, BW→BA) or on the basis of intermolecular distances[490, 492] (molecule-based cutoff, Eq. 3.24; SC→CM, BW→BM). When a molecule-based cutoff is applied, the ion is also considered as a "molecule" and the truncation relies on the definition of a specific molecular center (M) for the solvent molecule, which is assumed here to coincide with its single van der Waals interaction site. In the PBC/CT case, it is also assumed that the potential-energy function excludes any self-energy term[1082, 1083] (Eq. 3.37 for BW, with $\epsilon_{RF} = 1$ for SC, *i.e.* $V_{slf} = 0$ in Eq. 3.16).

The solvent-dependent quantities involved in the evaluation of the correction terms may sometimes differ significantly, depending on whether they are calculated by simulation for a given solvent model or measured experimentally for the real solvent. To avoid possible ambiguities, the former quantities will be marked with a prime, and the latter left unprimed (*e.g.* ρ'_S *vs.* ρ_S for the mass density, ϵ'_S *vs.* ϵ_S for the static relative dielectric permittivity, or χ'_S *vs.* χ_S for the air-liquid interfacial potential). Note that a "S" rather than "*svt*" subscript is used in this section for compactness of notation and consistency with Refs.[1289, 1412].

The *type-A correction* (Figure 6.1) accounts for the error in the solvent polarization around the ion and the incomplete or/and inexact interaction of the ion with the polarized solvent, arising from the use of an effective (non-Coulombic) electrostatic interaction scheme during the simulation, in the idealized situation of a non-periodic system of infinite extent. The resulting perturbation cannot be evaluated directly in the context of AT simulations, because these simulations always involve finite systems, that are also too small to permit a meaningful extrapolation to infinite system sizes. However, it can be evaluated in the context of CE, by comparing the outcome of a Born charging free energy calculation (Coulombic interactions, infinite non-periodic system), which is analytical, to an analogous calculation involving CT (rather than Coulombic) interactions. The latter model has been formulated by Baker *et al.*[1194,1199], and results in a one-dimensional integral equation (see Eq. 6.12 below) taking advantage of the spherical symmetry of the problem. This equation can no longer be solved analytically, and must be processed numerically, *e.g.* by discretization and matrix inversion. Alternatively, an analytical (fitted) equation has also been suggested[1412]. The difference between the two charging free energies is assumed to provide a good estimate for the corresponding perturbation in the AT simulation, and added to the raw charging free energy as a correction term $\Delta_s G_A^{CT}$. This term can be further subdivided into an analytical term $\Delta_s G_{A_1}^{CT}$, correcting for the omission of the exactly Coulombic interaction between the ion and the polarized solvent beyond its cutoff sphere, and a term $\Delta_s G_{A_2}^{CT}$, correcting for incorrect solvent polarization and a non-Coulombic potential function within this sphere.

Obviously, there is no type-A correction in the FBC/CB case, which is Coulombic, *i.e.*

$$\Delta_s G_A^{CB} = 0 \quad . \tag{6.7}$$

Although the PBC/LS scheme is formally non-Coulombic for any finite box size, the LS interaction between two charges separated by a finite distance becomes exactly Coulombic in the limit of an infinite box size[329–331, 1079–1081]. For this reason, there is also no type-A correction in this case, *i.e.*

$$\Delta_s G_A^{LS} = 0 \quad . \tag{6.8}$$

In the PBC/CT case, all ion-solvent interactions beyond the cutoff sphere of the ion are ignored, irrespective of the exact nature of the CT scheme employed during the simulation. Thus, a first type-A correction term[1081, 1171, 1194, 1199, 1468] corresponds to the inclusion of these neglected interactions as calculated from CE in a non-periodic system (Born model[84]), namely

$$\Delta_s G_{A_1}^{CT} = -(8\pi\epsilon_o)^{-1} N_A \, q_I^2 \, (1 - {\epsilon'_S}^{-1}) \, R_C^{-1} \quad . \tag{6.9}$$

A second type-A correction term must be applied to compensate for the error in the polarization within the cutoff sphere of the ion caused by the approximate treatment of ion-solvent and solvent-solvent electrostatic interactions[331, 333, 334, 792, 1080, 1081, 1166, 1167, 1169, 1170, 1192, 1194, 1199, 1430, 1468], as well as the calculation of the resulting potential at the ion site using an effective (possibly non-Coulombic) potential function[1081, 1171, 1194, 1199]. The way to compute such a correction using CE in the context of both Coulombic and non-Coulombic truncated electrostatic interactions for a non-periodic system of infinite extent has been presented elsewhere[1194, 1199], and only the result is summarized below. The effective interaction energy between two charges q_i and q_j at a distance r_{ij} in a given CT scheme can be written (Eqs. 3.16, 3.30, 3.34 and 3.35)

$$V_{CT}(q_i, q_j, r_{ij}) = (4\pi\epsilon_o)^{-1} \, q_i q_j \, h(R_C - r_{ij}) \sum_{k=-1}^{N_\alpha} \alpha_k R_C^{-k-1} r_{ij}^k \quad , \tag{6.10}$$

with

$$\alpha_{-1} = 1 \;\; \text{and} \;\; \alpha_0 = -\left(1 + \sum_{k=1}^{N_\alpha} \alpha_k\right) \quad , \tag{6.11}$$

where h is the Heaviside step function, *i.e.* $h(x) = 1$ for $x > 0$, $h(x) = 0$ otherwise. Eqs. 6.10 and 6.11 assume: (*i*) a non-periodic system (otherwise, minimum-image distances should be considered); (*ii*) an atom-based truncation (otherwise, molecule-based distances may be involved in the Heaviside function); (*iii*) a limiting Coulombic behavior at short distances; (*iv*) the continuity of the effective interaction function (but not necessarily of its derivative) at R_C. The dimensionless coefficients α_k entering into Eq. 6.10 are (see Eqs. 3.34 and 3.35): $\alpha_0 = -1$ with $N_\alpha = 0$ (SC) or $\alpha_0 = -3\epsilon_{BW}(2\epsilon_{BW}+1)^{-1}$, $\alpha_1 = 0$ and $\alpha_2 = (\epsilon_{BW} - 1)(2\epsilon_{BW} + 1)^{-1}$ with $N_\alpha = 2$ (BW), where it is recalled that BW is used as a synonym for RF in this section. In the BW case, ϵ_{BW} represents the permittivity assigned to the medium outside the cutoff sphere of each particle, ideally set self-consistently[2330] equal to that of the solvent model ϵ'_S. Within the CE approximation, the radial polarization $P(r)$ around the ion (for $r > R_I$) is found to be a solution of the integral equation

$$u(r) = v(r) + (1 - \epsilon_S'^{-1}) \int_{\max\{R_I, |r-R_C|\}}^{R_C+r} dr' K(r, r') u(r') \quad , \tag{6.12}$$

with $u(r) = q_I^{-1}(1 - \epsilon_S'^{-1})^{-1} P(r)$, where the inhomogeneous term and kernel are given by

$$v(r) = -(4\pi)^{-1} h(R_C - r) \sum_{k=-1}^{N_\alpha} k\alpha_k R_C^{-k-1} r^{k-1} \tag{6.13}$$

and

$$\begin{aligned} K(r, r') \quad = \quad & \sum_{k=-1}^{N_\alpha} \frac{\alpha_k}{8(k+2)R_C^{k+1} r^2} \Big[2(k+2) R_C^k (r^2 + r'^2) \\ & + k(k+2) R_C^{k-2} (r^2 - r'^2)^2 \\ & - 4(k+1)(r^2 + krr' + r'^2) \left| r - r' \right|^k - k^2 R_C^{k+2} \Big] \quad . \end{aligned} \tag{6.14}$$

The second type-A correction is then obtained according to a procedure described by Baker *et al.*[1194, 1199], by solving Eq. 6.12 numerically, *e.g. via* discretization and matrix inversion, for the specific effective interaction function used in the simulation, calculating the corresponding approximate contribution of the solvent within the cutoff sphere to the charging free energy, and comparing this result to the ideal value for Coulombic interactions in a non-periodic system (Born model[84]), leading to

$$\begin{aligned} \Delta_s G_{A_2}^{CT} \quad = \quad & -(8\pi\epsilon_o)^{-1} N_A q_I^2 (1 - \epsilon_S'^{-1}) \\ & \times \left[R_I^{-1} - R_C^{-1} + 4\pi \sum_{k=-1}^{N_\alpha} k\alpha_k R_C^{-k-1} \int_{R_I}^{R_C} dr\, r^{k+1} u(r) \right] \quad . \end{aligned} \tag{6.15}$$

Table 6.2 *Fitting coefficients for the empirical evaluation of specific correction terms to raw single-ion solvation free energies calculated by atomistic simulations.* These unitless coefficients enter into the evaluation of $\Delta_s G_{A_2}^{CT}$ and $\Delta_s G_B^{CT}$ *via* Eqs. 6.16 and 6.17 or Eqs. 6.22 and 6.23-6.26, respectively, in the context of cutoff-based (CT) electrostatic interactions with straight cutoff (SC) truncation or with the Barker-Watts (BW) reaction-field scheme.

$\Delta_s G_{A_2}^{SC}$ (Eq. 6.16)		$\Delta_s G_{A_2}^{BW}$ (Eq. 6.17)		$\Delta_s G_B^{SC}$ (Eqs. 6.22, 6.23 and 6.24)		$\Delta_s G_B^{BW}$ (Eqs. 6.22, 6.25 and 6.26)	
a_1	3.336	b_1	5.100	$c_{\mu,1}$	-4.937	$d_{\mu,1}$	-2.376
a_2	2.581	b_2	1.447	$c_{\mu,2}$	-5.441	$d_{\mu,2}$	-8.275
a_3	3.968	b_3	2.390	$c_{\nu,1}$	2.793	$d_{\nu,1}$	0.472
a_4	7.774	b_4	3.442	$c_{\nu,2}$	3.578	$d_{\nu,2}$	4.706
a_5	1.633	b_5	1.688				
a_6	2.021	b_6	4.445				
		b_7	3.764				
		b_8	2.376				
		b_9	1.245				
		b_{10}	1.308				
		b_{11}	4.454				

Note that due to the form of Eq. 6.12, the third term within brackets in Eq. 6.15 depends on ϵ_S', so that $\Delta_s G_{A_2}^{CT}$ is not exactly proportional to $(1 - {\epsilon_S'}^{-1})$. An alternative empirical (fitted) expression[1412] for $\Delta_s G_{A_2}^{CT}$ is given by

$$\Delta_s G_{A_2}^{SC} = (8\pi\epsilon_o)^{-1} N_A q_I^2 R_C^{-1} 10^{-1}\, a_1 \Big[1 - a_2 {\epsilon_S'}^{-1} + 10\, a_3 \left(1 + a_4 \epsilon_S' + 10\, a_5\, {\epsilon_S'}^2\right)^{-1}\Big] \times \left[1 - a_6 \left(R_C^{-1} R_I\right)^2\right] \tag{6.16}$$

and

$$\Delta_s G_{A_2}^{BW} = (8\pi\epsilon_o)^{-1} N_A q_I^2 R_C^{-1} \Big\{ - 10^{-1} b_1 \Big[1 - b_2 {\epsilon_S'}^{-1} - b_3 {\epsilon_S'}^{-2} + b_4 \left({\epsilon_S'}^2 \left(1 + 10^{-1}\, b_5\, \epsilon_S' + 10^{-2}\, b_6\, {\epsilon_S'}^2\right)\right)^{-1}\Big] + b_7 \left(R_C^{-1} R_I\right)^3 \left(1 - b_8\, {\epsilon_S'}^{-1} + b_9 {\epsilon_S'}^{-2} + 10^{-1}\, b_{10} \epsilon_S'\right) \times \left(\epsilon_S' + b_{11}\right)^{-1} \Big\} \quad . \tag{6.17}$$

The unitless fitting coefficients a_i, $i = 1...6$ and b_j, $j = 1...11$ are reported in Table 6.2. The overall correction for type-A errors in CT simulations is then

$$\Delta_s G_A^{CT} = \Delta_s G_{A_1}^{CT} + \Delta_s G_{A_2}^{CT} \quad . \tag{6.18}$$

The type-A correction term, non-vanishing for PBC/CT schemes only, is quadratic in the ionic charge q_I and generally depends on: (*i*) the specific CT electrostatic scheme used in the simulation and the corresponding cutoff distance R_C (plus the reaction-field permittivity ϵ_{BW} for the BW scheme); (*ii*) the effective ionic radius R_I; (*iii*) the model solvent permittivity ϵ'_S. This correction does not depend on the type of solvent model (*e.g.* non-polarizable or polarizable, rigid or flexible, single or multiple van der Waals interaction sites) and on the way the cutoff truncation is applied in the simulation (atom- or molecule-based). The generalization of the type-A correction to other CT schemes is in principle straightforward, provided that these have the form of Eqs. 6.10 and 6.11 ($\Delta_s G_{A_1}^{CT}$ is unchanged; see *e.g.* Ref.[1081] for $\Delta_s G_{A_2}^{CT}$ in the so-called generalized reaction-field[792,1227,1228]). The generalization to rigid polyatomic ions is in principle also possible, but requires the use of a CE approach that does not rely on spherical averaging[333], and is thus computationally more expensive.

The *type-B correction* (Figure 6.1) accounts for the error in the solvent polarization arising from the finite (microscopic) size of the simulated system and its possible periodicity, given the effective electrostatic scheme employed during the simulation. Here also, the resulting perturbation cannot be evaluated directly in the context of an AT simulation. However, it can be evaluated in the context of CE. In the FBC/CB and PBC/LS cases, the effective electrostatic interaction scheme is Coulombic or becomes Coulombic in the limit of infinite system sizes. As a result, one has to compare the outcome of a Born charging free energy calculation (Coulombic interactions, infinite non-periodic system), which is analytical, to an analogous calculation involving CB or LS interactions, respectively, in the context of a finite droplet or box size, respectively. In both cases, the latter model can also be formulated analytically (FBC/CB: finite-system variant of the Born model[84]; PBC/LS: analytical equation formulated by Hünenberger & McCammon in Ref.[329]). In the PBC/CT case, the effective electrostatic interaction is non-Coulombic, and one has to compare the outcome of a charging free energy calculation with CT electrostatics within an infinite non-periodic system to an analogous calculation involving CT electrostatics in the context of a finite box size. The former quantity can in principle be calculated using the Baker *et al.* algorithm[1194, 1199] mentioned previously. However, to benefit from error cancellation, it is actually preferable to perform the two above calculations using a common algorithm capable of handling CT interactions within both non-periodic or periodic boundary conditions. The equations of the latter models have been formulated by Bergdorf *et al.*[333] in the form of a three-dimensional field equation (see also Peter *et al.*[334] for an alternative Fourier transform algorithm in the periodic case). This equation cannot be solved analytically, and must be processed numerically, *e.g.* by discretization and a finite-element approach. Alternatively, an analytical (fitted) equation has also been suggested[1412]. For the three schemes considered, the difference between the two charging free energies (non-periodic *vs.* periodic) is assumed to provide a good estimate for the corresponding perturbation in the AT simulation, and added to the raw charging free energy as a correction term $\Delta_s G_B$ (*i.e.* $\Delta_s G_B^{CB}$, $\Delta_s G_B^{LS}$ or $\Delta_s G_B^{CT}$). Note that these corrections have been formulated based on an interaction function excluding the self-energy term V_{slf} in the LS case[792, 1081, 1423, 1430, 1431, 1437], where its inclusion is actually common in standard simulation programs, as well as in the CT case[1082, 1083], where it is less common (Section 3.3.1). In both cases, this implies that $\Delta_s G_B^{LS}$ does not encompass the contribution $-V_{slf}$ corresponding to the subtraction of the non-vanishing charging free energy of the ion "in vacuum" (Section 3.3.3) which is assumed to have already been removed from the raw charging free energy. The choice of the excluded-site term V_{exc} is also assumed to be consistent with that of the electrostatic scheme (Eq. 3.26 for LS and Eq. 3.31 for CT), while that of the surface term V_{srf} (vacuum, adjusted or tinfoil boundary conditions; the latter being used by default in all subsequent numerical examples) is left unspecified, but not expected to significantly affect the results (Section 3.3.1).

In the FBC/CB case (spherical droplet), the lack of polarization beyond the surface of the system (vacuum) may be corrected by the inclusion of these neglected interactions as calculated from CE (Born model[84]), namely

$$\Delta_s G_B^{CB} = -(8\pi\epsilon_o)^{-1} N_A q_I^2 (1 - \epsilon_S'^{-1}) S^{-1} \quad . \tag{6.19}$$

In the PBC/LS case (cubic box), the comparison between the Born solvation free energy (Coulombic, non-periodic system) and the solvation free energy of a spherical ion surrounded by a periodic dielectric continuum leads to an analytical correction[329]

$$\begin{aligned}\Delta_s G_B^{LS} &= (8\pi\epsilon_o)^{-1} N_A q_I^2 (1 - \epsilon_S'^{-1}) L^{-1} \\ &\times \left[\alpha_{LS} + \frac{4\pi}{3} \left(\frac{R_I}{L} \right)^2 - \frac{16\pi^2}{45} \left(\frac{R_I}{L} \right)^5 \right] \quad ,\end{aligned} \tag{6.20}$$

where $\alpha_{LS} \approx -2.837297$ is the LS self-term constant[330,792,1084,1085,2331,2332] (see Eq. 3.28).

In the PBC/CT case, periodicity effects are generally significantly reduced compared to LS, but not removed[333,1468]. The corresponding type-B correction depends on the effective functional form employed for the CT electrostatic interactions during the simulation. This term can be calculated according to the procedure described by Bergdorf *et al.*[333], by comparing the charging free energy of the ion evaluated numerically (*e.g.* *via* three-dimensional discretization and finite-element calculation), using CE with the given CT scheme, under either non-periodic (NPBC) or periodic (PBC) boundary conditions, *i.e.*

$$\Delta_s G_B^{CT} = \Delta_s G^{(CT,NPBC)} - \Delta_s G^{(CT,PBC)} \quad . \tag{6.21}$$

For a given CT scheme, the parameters of such a calculation are R_I, L, R_C and ϵ_S' (along with ϵ_{BW} for the BW scheme). A detailed numerical analysis[333] suggests that the type-B correction term can be related to the type-A correction term (Eq. 6.18) through the empirical relationship

$$\Delta_s G_B^{CT} = -\frac{R_I}{R_C} \exp\left[\mu^{CT}(\epsilon_S')\frac{L}{2R_C} + \nu^{CT}(\epsilon_S')\right] \times \left[(8\pi\epsilon_o)^{-1} N_A q_I^2 (1 - {\epsilon_S'}^{-1}) R_I^{-1} + \Delta_s G_A^{CT}\right] \quad , \quad (6.22)$$

where μ^{CT} and ν^{CT} are functions of ϵ_S' specific for a given CT scheme. This equation shows that the magnitude of $\Delta_s G_B^{CT}$ increases with the ratio of the cutoff distance to the half box edge. An alternative empirical (fitted) expression[1412] for $\Delta_s G_B^{CT}$ is given by Eq. 6.22, along with

$$\mu^{SC}(\epsilon_S') = c_{\mu,1} + c_{\mu,2}{\epsilon_S'}^{-1} \quad , \quad (6.23)$$

$$\nu^{SC}(\epsilon_S') = c_{\nu,1} + c_{\nu,2}{\epsilon_S'}^{-1} \quad , \quad (6.24)$$

$$\mu^{BW}(\epsilon_S') = d_{\mu,1} + d_{\mu,2}{\epsilon_S'}^{-1} \quad (6.25)$$

and

$$\nu^{BW}(\epsilon_S') = d_{\nu,1} + d_{\nu,2}{\epsilon_S'}^{-1} \quad . \quad (6.26)$$

The unitless fitting coefficients $c_{\mu,i}$, $c_{\nu,i}$, $d_{\mu,i}$, $d_{\nu,i}$, $i = 1, 2$ are reported in Table 6.2.

The type-B correction term, typically large for FBC/CB and PBC/LS, and smaller for PBC/CT, is quadratic in the ionic charge q_I, and generally depends on: (*i*) the specific boundary conditions and approximate electrostatic scheme used in the simulation (including the reaction-field permittivity ϵ_{BW} for the PBC/BW scheme); (*ii*) the cutoff distance R_C (PBC/CT); (*iii*) the effective ionic radius R_I (PBC/LS and PBC/CT); (*iv*) the droplet radius S (FBC/CB); (*v*) the box edge length L (PBC/LS and PBC/CT); (*vi*) the model solvent permittivity ϵ_S' (all schemes). This correction does not depend on the type of solvent model (*e.g.* non-polarizable or polarizable, rigid or flexible, single or multiple van der Waals interaction sites) and, for CT schemes, on the way the cutoff truncation is applied in the simulation (atom- or molecule-based). The generalization of the type-B correction to non-spherical droplets (FBC/CB), to non-cubic boxes (PBC/LS), to other CT schemes (PBC/CT; see *e.g.* Ref.[1081] for the so-called generalized reaction-field[792,1227,1228]) or to rigid polyatomic ions is in principle possible based on various numerical CE approaches (either non-periodic[329,333,682,686,695] or periodic[329,333,334,1079,2333]).

The *type-C correction* (Figure 6.1) accounts for the error in the potential at the ion site caused by its possible evaluation using an improper summation method (along with a possibly non-Coulombic potential function), and due to the possible presence of an explicit air-liquid interface in the simulated system (FBC/CB) or of a constraint of vanishing average potential over the computational box (PBC/LS). These effects have been discussed in detail previously (Sections 2.2, 3.3.3-3.3.5 and 4.3.8) and only the main results are summarized below. Note that the type-C correction term is introduced here so as to lead to intrinsic solvation free energies, excluding the surface contribution involved in the crossing of a polarized air-liquid interface.

It is also convenient to recall the definition of the quantity ξ'_S (previously noted ξ_{AT}) called the exclusion potential of the solvent model[707] (at a given density ρ'_S), defined as (see Eq. 4.332)

$$\xi'_S = (6\epsilon_o)^{-1} \, N_A \, M_S^{-1} \, \rho'_S \, \overline{\mathcal{Q}}'_S \quad , \tag{6.27}$$

where M_S is the molar mass of the solvent, and $\overline{\mathcal{Q}}'_S = \mathcal{Q}'_S$ the quadrupole-moment trace of the solvent model[708, 1429] relative to its molecular center M, *i.e.*

$$\overline{\mathcal{Q}}'_S = \sum_i^{N_s} q_i r_i^2 \quad , \tag{6.28}$$

where $\{q_i \, | i = 1..N_s\}$ and $\{r_i \, | i = 1..N_s\}$ are the partial charges and distances from M associated to the N_s point charges within the solvent molecule.

The type-C_1 correction term accounts for the error committed in the evaluation of the electric potential at the ion site based on the sampled configurations. The proper summation scheme would involve considering the potential generated by entire solvent molecules (M-scheme) over a certain region of the bulk surrounding the ion (in the limit of a sufficiently large region) according to a Coulombic potential form. However, as discussed in Section 3.3.3, the potential is seldom calculated in this way in practice (except for the FBC/CB scheme and, except for an offset constant $-R_C^{-1}$ in Eq. 3.34, the PBC/CT/SC schemes), and many schemes implicitly involve P-summation or/and summation based on a non-Coulombic potential form. Because these effects are largely independent of the ionic charge state[708], they are most conveniently evaluated in the context of an uncharged ion-sized cavity[707], the resulting potential discrepancies being amplified by the full ionic charge q_I to obtain a free-energy correction. Because these effects are also largely independent of the temperature[707], they are also most conveniently evaluated in the context of solvent configurations corresponding to an infinite rotational temperature, *i.e.* exempt of intermolecular orientational correlations between the solvent molecules, a situation called the *orientational disorder limit*[707] (ODL). In the general case, the potential at the cavity site can be evaluated numerically for a given electrostatic interaction (and possibly truncation) scheme *via* simulations in the ODL ensemble, realized in practice by turning off the atomic partial charges of the solvent molecules during the sampling. In the special (but common for water) case where the solvent model involves a single van der Waals interaction site, and molecule-based truncation, if applied, relies on this specific site for the truncation, the potential at the cavity center can be evaluated analytically. In both cases, comparison to the ideal situation of M-truncation based on exactly Coulombic interactions, which leads to a vanishing potential in the cavity, leads to the appropriate correction term $\Delta_s G_{C_1}$, *i.e.* $\Delta_s G_{C_1}^{LS}$ and $\Delta_s G_{C_1}^{CT}$, the latter differing between the cases of atomic and molecular truncation. The exact form of the correction term can be deduced from the quotations in Table I of Ref.[707].

In the FBC/CB case, which is Coulombic and involves summation over all charges in the finite system, obviously grouped in terms of intact molecules, one has

$$\Delta_s G_{C_1}^{CB} = 0 \quad . \tag{6.29}$$

In the PBC/LS case (cubic box), one has

$$\Delta_s G_{C_1}^{LS} = -N_A q_I \left(1 - \frac{4\pi R_I^3}{3L^3}\right) \xi_S' \quad , \tag{6.30}$$

where the quantity between parentheses is typically very close to unity, because $R_I \ll L$ in all practically relevant situations.

In the PBC/CT case, the correction term depends on the specific electrostatic and truncation scheme employed. For truncated Coulombic interactions with a molecule-based cutoff (CM), where the summation involves intact molecules, one has

$$\Delta_s G_{C1}^{CM} = 0 \quad . \tag{6.31}$$

Coulombic interactions with an atom-based cutoff (CA) cannot be used to generate configurations in explicit-solvent simulations involving dipolar solvent molecules, due to strong cutoff artifacts[1013]. However, if this scheme is used for the analysis of configurations generated with another scheme, one has

$$\Delta_s G_{C_1}^{CA} = -N_A q_I \xi_S' \quad . \tag{6.32}$$

For Barker-Watts reaction-field corrected interactions with a molecule-based cutoff (BM), one has

$$\Delta_s G_{C_1}^{BM} = -N_A q_I \frac{2(\epsilon_{BW} - 1)}{2\epsilon_{BW} + 1} \left(1 - \frac{R_I^3}{R_C^3}\right) \xi_S' \quad . \tag{6.33}$$

When atom-based truncation is applied instead (BA), one has

$$\Delta_s G_{C_1}^{BA} = \Delta_s G_{C_1}^{BM} - N_A q_I \frac{3}{2\epsilon_{BW} + 1} \xi_S' \quad . \tag{6.34}$$

The type-C$_2$ correction term accounts for the possible presence of a constant offset in the potential at the cavity center. Such an offset is present in the FBC/CB case, where it is caused by the air-liquid interfacial potential at the surface of the finite system (droplet). This potential offset is an artifact when intrinsic solvation free energies are sought after, and should be removed by the correction scheme. Such an offset is also present in the PBC/LS case, where it is caused by the constraint of vanishing average potential over the system (computational box), an implicit boundary condition in the LS approach, resulting from the use of a Fourier series and the omission of the zero term in the series. In this case, the potential offset is due to the interfacial potential at the ion-solvent interface (cavity), which causes the average potential within the bulk solvent to slightly differ from zero, so as to satisfy the constraint of a vanishing average potential over the entire computational box. This potential offset is an artifact, and should be removed by the correction scheme. In practice, the evaluation of these two correction terms requires a determination of the function representing the air-liquid interfacial potential of the solvent at a curved (convex or concave) interface (see Point A_3 in Section 6.1.1). This function can be determined using AT simulations of solvent droplets or solvent systems containing cavities, *e.g.* by potential averaging (Section 3.3.5), which permits the determination of the correction term $\Delta_s G_{C_2}$, *i.e.* $\Delta_s G_{C_2}^{CB}$ and $\Delta_s G_{C_2}^{LS}$, the latter being very small in practical situations, due to the small size of the ion relative to the computational box.

In the FBC/CB case, the type-C_2 correction is required to compensate for the potential jump at the air-liquid interface in the simulation, which is an artifact when intrinsic solvation free energies are sought after. The magnitude of this finite-temperature effect was estimated for the SPC water model[1274] at 300 K from a series of FBC/CB simulations involving water confined within planar slabs (thickness 2-4 nm) and within droplets of varying sizes (radii 2-3 nm)[707]. The resulting correction can (for a spherical droplet) be cast into the empirical form (see Eq. 6.1)

$$\Delta_s G_{C_2}^{CB} = -N_A q_I \left(\chi'_S + \tilde{\chi}'_{S+} S^{-1} \right) \quad , \tag{6.35}$$

where χ'_S (previously noted χ_{svt}) is the interfacial potential at a planar air-liquid interface, measured in the air-to-liquid direction, and $\tilde{\chi}'_{S+}$ (previously noted $\tilde{\chi}_{svt,+}$) a factor characterizing the approximately linear dependence of this interfacial potential on the inverse curvature radius for a convex interface. Obviously, the approximation of Eq. 6.35 is only valid for "large-enough" droplets. The values of χ'_S and $\tilde{\chi}'_{S+}$ calculated[707] for the SPC water model[1274] at 300 K are $\chi'_S = 0.73$ V and $\tilde{\chi}'_{S+} = -1.11$ V·nm.

In principle, there should be no such correction in the PBC case, because an infinite periodic system is exempt of any interface to an external medium. This observation applies in the PBC/CT case and one has

$$\Delta_s G_{C_2}^{CT} = 0 \quad . \tag{6.36}$$

However, in the PBC/LS case, a type-C_2 correction is still required to account for the fact that the boundary condition on the LS electric potential is a vanishing average value over the entire computational box, rather than on the solvent-occupied region of this box. This term thus depends on the potential jump at the ion-solvent interface. The magnitude of this finite-temperature effect was estimated for the SPC water model[1274] at 300 K from a series of PBC simulations involving cavities (radius $0.2-0.8$ nm) within cubic solvent boxes (edge 4 nm)[707]. The resulting correction (for a cubic box) can be cast into the empirical form

$$\Delta_s G_{C_2}^{LS} = -N_A q_I \frac{4\pi R_I^3}{3L^3} \left(\chi'_S + \tilde{\chi}'_{S-} R_I^{-1} \right) \quad , \tag{6.37}$$

where $\tilde{\chi}'_{S-}$ (previously noted $\tilde{\chi}_{svt,-}$) is a factor characterizing the (approximately linear) dependence of the air-liquid interfacial potential, measured in the air-to-liquid direction, on the inverse curvature radius for a concave interface. Obviously, the approximation of Eq. 6.37 is only valid for "large-enough" cavities. The value of $\tilde{\chi}'_{S-}$ calculated[707] for the SPC water model[1274] at 300 K is $\tilde{\chi}'_{S-} = -0.11$ V·nm. Due to its dependence on $(R_I/L)^3$, $\Delta_s G_{C_2}^{LS}$ is typically very small, because $R_I \ll L$ in all practically relevant situations. An alternative numerical procedure to evaluate $\Delta_s G_{C_2}^{LS}$ relies on considering the raw solvent-generated electrostatic potential Φ_o^{raw} at the center of an uncharged ion-sized cavity, the computation of which is anyway required for that of the raw charging free energy. The air-liquid interfacial potential χ_{ion} at the ion surface is equal to the value of Φ_o corrected for types-B and C_1 errors, *i.e.*

$$\chi_{ion} = \Phi_o^{raw} + \frac{2}{N_A q_I} \Delta_s G_B^{LS} + \frac{1}{N_A q_I} \Delta_s G_{C_1}^{LS} \quad , \tag{6.38}$$

so that one obtains

$$\Delta_s G_{C_2}^{LS} = \frac{4\pi R_I^3}{3L^3} \left(N_A q_I \Phi_o^{raw} + 2\Delta_s G_B^{LS} + \Delta_s G_{C_1}^{LS} \right) \quad . \tag{6.39}$$

The overall correction for type-C errors is then given by

$$\Delta_s G_C = \Delta_s G_{C_1} + \Delta_s G_{C_2} \quad . \tag{6.40}$$

The type-C correction term, non-vanishing except for the PBC/CM scheme, is linear in the ionic charge q_I, and generally depends on: (*i*) the specific boundary conditions and approximate electrostatic scheme used in the simulation (including the reaction-field permittivity ϵ_{BW} for the PBC/BW scheme); (*ii*) the atom- or molecule-based truncation procedure and the cutoff distance R_C (PBC/CT); (*iii*) the effective ionic radius R_I (except FBC/CB and PBC/CA); (*iv*) the droplet radius S (FBC/CB); (*v*) the box edge length L (PBC/LS); (*vi*) the solvent model employed (all schemes; *via* the density ρ'_S, the quadrupole-moment trace $\mathcal{Q}'_S$ and the surface properties χ'_S, $\tilde{\chi}'_{S+}$ and $\tilde{\chi}'_{S-}$). The generalization of the type-C correction to other CT schemes (PBC/CT) is in principle straightforward, provided that these schemes have the form of Eqs. 6.10 and 6.11 ($\Delta_s G_{C_2}^{CT}$ still vanishes; see *e.g.* Refs.[707,1081] for the so-called generalized reaction-field[792,1227,1228]). Molecule-based truncation schemes relying on an alternative molecular center (M′) can also be handled[707,1081]. The generalization to non-spherical droplets (FBC/CB) or to non-cubic boxes (PBC/LS) would also be possible given appropriate modifications of Eqs. 6.30, 6.35, 6.37 and 6.39 (for PBC/CT, there is no dependence on the box shape). On the other hand, the consideration of (possibly polarizable and flexible) solvent molecules involving multiple van der Waals interaction sites is more complicated. In this case, Eq. 6.28 can no longer be used, and the molecular quadrupole-moment trace $\mathcal{Q}'_S$ must be replaced by a corresponding effective quantity $\overline{\mathcal{Q}}'_S$. This parameter may still in principle be evaluated numerically *via* Eq. 6.27 based on the determination of the exclusion potential ξ'_S using simulations in the ODL, as discussed in Refs.[803,1081]. The generalization of the type-C correction to rigid polyatomic ions involves a significantly higher level of complexity and has not been investigated up to now.

Finally, the *type-D correction* (not shown in Figure 6.1) accounts for a possible inaccuracy in the relative dielectric permittivity of the solvent model compared to its experimental value ϵ_S. It is identical for all schemes and easily evaluated by comparison of the results of the Born equation[84] applied with the two permittivities, namely

$$\Delta_s G_D = (8\pi\epsilon_o)^{-1} N_A q_I^2 \left(\epsilon_S^{-1} - {\epsilon'_S}^{-1}\right) R_I^{-1} \quad . \tag{6.41}$$

The inclusion of such a correction term (which was neglected in Ref.[1081]) is consistent with the formulation of the types-A and B correction terms based on the model solvent permittivity ϵ'_S rather than the corresponding experimental value ϵ_S. Note that this correction term, usually very small, is only a minor component of force-field errors (Section 6.2.4).

To convert the (corrected) charging free energy $\Delta_s G_{chg}$ into a *standard intrinsic solvation free energy* $\Delta_s G^{\ominus}$, the simulations must be carried out at $P^{\circ} = 1$ bar and $T^{-} = 298.15$ K, the correction terms being evaluated using parameters appropriate for these same conditions, and a cavitation contribution and standard-state correction term must be added (Sections 2.2, 3.3.3 and 4.2.16), resulting in

$$\Delta_s G^{\ominus} = \Delta_s G_{chg}^{raw} + \Delta_s G_{cor} + \Delta_s G_{cav} + \Delta_s G_{std}^{\ominus} \quad . \tag{6.42}$$

The term $\Delta_s G_{cav}$ accounts for the reversible work associated with the formation of the neutral ion-sized cavity in the solvent and can be computed by thermodynamic integration[628] over the derivative of the Hamiltonian with respect to a scaling parameter λ applied to the ion-solvent van der Waals interactions (at zero ionic charge), as described in Section 3.3.3. Since the cavitation term involves short-range interactions (at the ion-solvent interface), it is relatively insensitive to the chosen simulation methodology and does not require the application of correction terms. The term $\Delta_s G_{std}^{\ominus}$ is the free energy of "isothermal ideal-gas compression" corresponding to the standard-state conversion from $P^{\circ} = 1$ bar for the gaseous ion to $b^{\circ} = 1$ mol·kg^{-1} (molality) for the solvated ion (Eq. 2.9), and has been discussed in Sections 2.2, 3.3.3 and 4.2.16.

In the context of the boundary conditions, approximate electrostatic schemes and types of solvent models considered here, $\Delta_s G_{cor}$ is given by

$$\begin{aligned}
\Delta_s G_{cor} &= \left[\Delta_s G_{A_1}\left(|q_I|, R_C, \epsilon'_S\right)\right]^{\mathrm{CT}} \\
&+ \left[\Delta_s G_{A_2}\left(|q_I|, R_I, R_C, \epsilon'_S[, \epsilon_{BW}]^{\mathrm{BW}}\right)\right]^{\mathrm{CT}} \\
&+ \Delta_s G_B\left(|q_I|\, [, S]^{\mathrm{CB}}[, R_I, L]^{\mathrm{LS,CT}}[, R_C]^{\mathrm{CT}}, \epsilon'_S[, \epsilon_{BW}]^{\mathrm{BW}}\right) \\
&+ \left[\Delta_s G_{C_1}\left(q_I, R_I, L, N_S, \overline{\mathcal{Q}}'_S[, R_C, \epsilon_{BW}]^{\mathrm{BW}}\right)\right]^{\mathrm{LS,BW,CA}} \\
&+ \left[\Delta_s G_{C_2}\left(q_I, \chi'_S[, S, \tilde{\chi}'_{S+}]^{\mathrm{CB}}[, R_I, L, \tilde{\chi}'_{S-}]^{\mathrm{LS}}\right)\right]^{\mathrm{CB,LS}} \\
&+ \Delta_s G_D\left(|q_I|, R_I, \epsilon'_S, \epsilon_S\right) \quad ,
\end{aligned} \tag{6.43}$$

where N_S is the number of solvent molecules in the computational box. The square brackets are used here to indicate terms or dependences that are only present in the specified scheme. The different terms are explicited in Eqs. 6.9, 6.15, 6.19-6.21, 6.30, 6.32-6.35, 6.37 and 6.41, Eqs. 6.15 and 6.21 referring to the original (numerical) scheme for $\Delta_s G_{A_2}^{CT}$ and $\Delta_s G_B^{CT}$ and Eqs. 6.16, 6.17, 6.22 and 6.23-6.26 to the empirical (fitted) scheme[1412], the latter being entirely analytical. Note that: (*i*) $\overline{\mathcal{Q}}'_S$ is equal to $\mathcal{Q}'_S$ as defined by Eq. 6.28 in the context of a non-polarizable rigid solvent model with a single van der Waals interaction site; (*ii*) the dependence of $\Delta_s G_{C_1}$ on ρ'_S (*via* Eq. 6.27) has been replaced by a dependence on R_I, L and N_S (see below); (*iii*) the use of Eq. 6.22 instead of Eq. 6.21 for $\Delta_s G_B^{CT}$ induces an additional dependence of this term on $\Delta_s G_A^{CT}$; (*iv*) the use of Eq. 6.39 instead of Eq. 6.37 for $\Delta_s G_{C_2}^{LS}$ removes the explicit dependence on χ'_S and $\tilde{\chi}'_{S-}$, but renders the function dependent on other parameters. The connection between ρ'_S, R_I, L and N_S is given by the equation

$$\rho'_S = \frac{N_A^{-1} M_S N_S}{L^3 - \frac{4\pi}{3} R_I^3} \quad , \tag{6.44}$$

which can be approximated as

$$\rho'_S \approx \frac{N_A^{-1} M_S (N_S + 1)}{L^3} \quad , \tag{6.45}$$

where it is assumed that the volume of the ion is about equal to that of one solvent molecule. Note that $\Delta_s G_{C_1}$ does not depend on M_S, which cancels out between Eqs. 6.27 and 6.44 (or 6.45). In constant-volume simulations, L and ρ'_S are constant, with the usual choice $\rho'_S = \rho_S$, and independent of the charge state of the ion. In constant-pressure simulations, the box-edge length L may fluctuate and vary with the charge state of the ion. In this case, L will refer to the average value of this parameter in the simulation at zero ionic charge. In the absence of electrostriction effects (uncharged cavity), ρ'_S in Eq. 6.44 will be nearly identical to the equilibrium density of the pure solvent model, which may, for some water models, significantly differ from the experimental value ρ_S. Expressions analogous to Eq. 6.43 have been developed recently[1412] for derivative thermodynamic quantities, namely the solvation entropy, heat capacity, volume, volume-compressibility and volume-expansivity.

As an illustration of the ability of this correction scheme to generate results *nearly exempt of any methodological dependence*, single-ion hydration free energies calculated based on AT simulations (Na^+ parameters of Straatsma & Berendsen[1171] with the SPC water model[1274]) considering 5 different electrostatic schemes (CB, LS, BA, BM and CM) and 48 combinations of the associated parameters (droplet radius S for CB, box size L for LS, BA, BM and CM, cutoff distance R_C for BA, BM and CM) are reported in Table 6.3. The quantities listed are the raw charging free energies $\Delta_s G_{chg}^{raw}$, the cavitation free energy $\Delta_s G_{cav}$, and the different correction terms (types-A_1, A_2, B, C_1, C_2 and D, standard-state correction). The simulations were performed at constant volume (corresponding to a density $\rho'_S = 1000$ kg·m^{-3}) and at a temperature of 298.15 K, using a protocol as detailed in Refs.[1081,1412].

It is immediately obvious that the *raw charging free energies* $\Delta_s G_{chg}^{raw}$ are affected by *extremely large variations* across the different electrostatic schemes and parameter combinations. Considering only the systems with $N_S \geq 512$, the corresponding average value is -356.3 kJ·mol^{-1} with a standard deviation as large as 63.8 kJ·mol^{-1} (maximum difference 176.0 kJ·mol^{-1}). This quantity is thus extremely sensitive to the simulation methodology (choice of boundary conditions and treatment of electrostatic interactions) and parameters (box size, cutoff distance). However, this dependence is successfully removed by the different correction terms, leading to *remarkably consistent results* for the *corrected hydration free energies* $\Delta_s G^{\ominus}$. Considering again the systems with $N_S \geq 512$, and $\Delta_s G_B^{CT}$ estimates, the corresponding average value is -464.8 kJ·mol^{-1} with a standard deviation as low as 2.0 kJ·mol^{-1} (maximum difference 8.2 kJ·mol^{-1}). In Ref.[1081], more extensive investigation evidenced a corresponding reduction of the standard deviation from 82.8 kJ·mol^{-1} to 11.6 kJ·mol^{-1} upon application of the same correction scheme, considering 72 combinations of boundary conditions, electrostatic schemes and corresponding parameters.

Remembering the observations made in Section 3.2 concerning the poor quantitative accuracy of the Born model (*e.g.* neglect of electrostriction, dielectric saturation and specific ion-solvent interactions), it is licit to question the *adequacy of the CE approximation* for the evaluation of the correction terms $\Delta_s G_A$ and $\Delta_s G_B$ (see above). Yet, the correction scheme is remarkably successful at reducing the dramatic spread in raw charging free energies issued from AT simulations (Table 6.3). There are two main reasons for this success: (i) the correction terms derived from CE benefit from a large *cancellation of errors*, because they rely on a difference between two such calculations involving different electrostatic schemes or boundary conditions, as shown pictorially in Figure 6.1; (ii) this difference concerns mainly *long-range effects*, for which the CE approximation is most accurate, the inaccurate short-range component essentially canceling out between the two calculations (with the

Table 6.3 *Standard intrinsic hydration free energy of the sodium cation evaluated by atomistic simulations with application of correction terms.* The simulations are performed using the Na^+ ion-solvent Lennard-Jones interaction parameters of Straatsma & Berendsen[1171] and the SPC water model[1274], and at constant volume and temperature (NVT ensemble, T^-=298.15 K, density $\rho'_S = 1000$ kg·m^{-3}). The boundary conditions and electrostatic schemes considered are FBC/CB (droplet of radius S containing N_S water molecules), PBC/LS (cubic box of edge L containing N_S water molecules) and PBC/CT (cubic box of edge L containing N_S water molecules, cutoff distance R_C; $\epsilon_{BW} = \epsilon'_S$ for the BA and BM schemes). The correction terms are calculated according to Eqs. 6.43 and 6.45, using the parameters[1289] $R_I = 0.16$ nm, $\epsilon_S = 78.4$, $\epsilon'_S = 66.6$, $\mathcal{Q}'_S = 0.0082$ e·nm^2, $\chi'_S = 0.73$ V, $\tilde{\chi}'_{S+} = -1.11$ V·nm and $\tilde{\chi}'_{S-} = -0.11$ V·nm. The correction terms $\Delta_s G_{A_2}^{CT}$ and $\Delta_s G_B^{CT}$ were calculated using the empirical (fitted) approach[1412] according to Eqs. 6.16 and 6.17 (CT schemes only). See Refs.[1081,1412] for more details.

Scheme	N_S	S or L [nm]	R_C [nm]	$\Delta_s G_{chg}^{raw}$	$\Delta_s G_{A_1}$	$\Delta_s G_{A_2}$	$\Delta_s G_B$	$\Delta_s G_{C_1}$ [kJ·mol^{-1}]	$\Delta_s G_{C_2}$	$\Delta_s G_D$	$\Delta_s G_{cav}$	$\Delta_s G_{std}^{\ominus}$	$\Delta_s G^{\ominus}$
CB	818	1.80	-	-428.66	-	-	-38.01	-	-10.86	-0.78	9.68	7.95	-460.68
CB	3790	3.00	-	-422.56	-	-	-22.81	-	-34.63	-0.78	9.68	7.95	-463.16
CB	8918	4.00	-	-423.22	-	-	-17.11	-	-43.55	-0.78	9.68	7.95	-467.03
CB	17454	5.00	-	-421.37	-	-	-13.68	-	-48.90	-0.78	9.68	7.95	-467.11
LS	4	0.53	-	-45.84	-	-	-292.76	-61.79	-3.56	-0.78	9.68	7.95	-387.11
LS	8	0.65	-	-118.73	-	-	-257.95	-70.01	-1.93	-0.78	9.68	7.95	-431.77
LS	16	0.79	-	-164.76	-	-	-222.81	-74.32	-1.07	-0.78	9.68	7.95	-446.12
LS	32	0.99	-	-209.85	-	-	-184.37	-76.99	-0.55	-0.78	9.68	7.95	-454.90
LS	64	1.25	-	-248.99	-	-	-149.46	-78.37	-0.27	-0.78	9.68	7.95	-460.25
LS	128	1.57	-	-279.79	-	-	-120.70	-79.05	-0.14	-0.78	9.68	7.95	-462.83
LS	256	1.97	-	-303.76	-	-	-97.05	-79.39	-0.07	-0.78	9.68	7.95	-463.42
LS	512	2.49	-	-324.80	-	-	-77.23	-79.57	-0.03	-0.78	9.68	7.95	-464.78
LS	1024	3.13	-	-340.49	-	-	-61.65	-79.65	-0.02	-0.78	9.68	7.95	-464.97
LS	2048	3.94	-	-353.47	-	-	-49.09	-79.70	-0.01	-0.78	9.68	7.95	-465.42
BA	512	2.49	0.8	-271.22	-85.53	-42.60	-1.88	-78.52	0.00	-0.78	9.68	7.95	-462.91
BA	512	2.49	1.0	-294.41	-68.42	-34.36	-3.66	-79.12	0.00	-0.78	9.68	7.95	-463.12
BA	1024	3.13	0.8	-272.46	-85.53	-42.60	-0.69	-78.52	0.00	-0.78	9.68	7.95	-462.96
BA	1024	3.13	1.0	-298.77	-68.42	-34.36	-1.65	-79.12	0.00	-0.78	9.68	7.95	-465.47
BA	1024	3.13	1.2	-313.23	-57.02	-28.73	-2.82	-79.38	0.00	-0.78	9.68	7.95	-464.33
BA	2048	3.94	0.8	-273.49	-85.53	-42.60	-0.19	-78.52	0.00	-0.78	9.68	7.95	-463.50
BA	2048	3.94	1.0	-299.06	-68.42	-34.36	-0.60	-79.12	0.00	-0.78	9.68	7.95	-464.71
BA	2048	3.94	1.2	-314.96	-57.02	-28.73	-1.21	-79.38	0.00	-0.78	9.68	7.95	-464.46
BA	2048	3.94	1.4	-326.92	-48.87	24.67	-1.96	-79.51	0.00	-0.78	9.68	7.95	-465.09
BA	2048	3.94	1.6	-334.94	-42.77	-21.61	-2.75	-79.59	0.00	-0.78	9.68	7.95	-464.80
BM	512	2.49	0.8	-275.87	-85.53	-42.60	-1.88	-76.74	0.00	-0.78	9.68	7.95	-465.78
BM	512	2.49	1.0	-298.57	-68.42	-34.36	-3.66	-77.33	0.00	-0.78	9.68	7.95	-465.50
BM	512	2.49	1.2	-311.03	-57.02	-28.73	-5.49	-77.60	0.00	-0.78	9.68	7.95	-463.02
BM	1024	3.13	0.8	-278.38	-85.53	-42.60	-0.69	-76.74	0.00	-0.78	9.68	7.95	-467.10
BM	1024	3.13	1.0	-300.79	-68.42	-34.36	-1.65	-77.33	0.00	-0.78	9.68	7.95	-465.70
BM	1024	3.13	1.2	-315.61	-57.02	-28.73	-2.82	-77.60	0.00	-0.78	9.68	7.95	-464.93
BM	1024	3.13	1.4	-326.41	-48.87	-24.67	-4.03	-77.73	0.00	-0.78	9.68	7.95	-464.87
BM	2048	3.94	0.8	-277.92	-85.53	-42.60	-0.19	-76.74	0.00	-0.78	9.68	7.95	-466.14
BM	2048	3.94	1.0	-301.72	-68.42	-34.36	-0.60	-77.33	0.00	-0.78	9.68	7.95	-465.59
BM	2048	3.94	1.2	-318.42	-57.02	-28.73	-1.21	-77.60	0.00	-0.78	9.68	7.95	-466.13
BM	2048	3.94	1.4	-329.10	-48.87	-24.67	-1.96	-77.73	0.00	-0.78	9.68	7.95	-465.48
BM	2048	3.94	1.6	-337.66	-42.77	-21.61	-2.75	-77.80	0.00	-0.78	9.68	7.95	-465.74
CM	512	2.49	0.8	-423.65	-85.53	24.34	-0.49	-	-	-0.78	9.68	7.95	468.48
CM	512	2.49	1.0	-431.26	-68.42	20.49	-1.96	-	-	-0.78	9.68	7.95	-464.31
CM	512	2.49	1.2	-434.06	-57.02	17.53	-4.76	-	-	-0.78	9.68	7.95	-461.47
CM	1024	3.13	0.8	-424.47	-85.53	24.34	-0.07	-	-	-0.78	9.68	7.95	-468.88
CM	1024	3.13	1.0	-435.70	-68.42	20.49	-0.39	-	-	-0.78	9.68	7.95	-467.19
CM	1024	3.13	1.2	-440.34	-57.02	17.53	-1.25	-	-	-0.78	9.68	7.95	-464.23
CM	1024	3.13	1.4	-442.33	-48.87	15.26	-2.78	-	-	-0.78	9.68	7.95	-461.88
CM	2048	3.94	0.8	-424.32	-85.53	24.34	-0.01	-	-	-0.78	9.68	7.95	-468.67
CM	2048	3.94	1.0	-434.19	-68.42	20.49	-0.05	-	-	-0.78	9.68	7.95	-465.33
CM	2048	3.94	1.2	-438.78	-57.02	17.53	-0.23	-	-	-0.78	9.68	7.95	-461.65
CM	2048	3.94	1.4	-446.09	-48.87	15.26	-0.65	-	-	-0.78	9.68	7.95	-463.51
CM	2048	3.94	1.6	-447.19	-42.77	13.49	-1.40	-	-	-0.78	9.68	7.95	-461.02

possible exception of the type-A_2 correction in the PBC/SC scheme when using small R_C distances). For the same reason, the correction scheme is expected to be relatively insensitive to the exact choice of R_I, ϵ_S and ϵ'_S, within reasonable bounds. In other words, CE is not applied in a single calculation to evaluate an absolute solvation free energy, but in two closely related calculations to evaluate a difference in free energies (Figure 6.1). The difference predominantly occurs in the long-distance range, the short-range polarization being essentially correct in the atomistic description (with the possible exception of the PBC/CT/SC scheme, see Figure 3.8), where the CE approach tends to be more accurate, as opposed to the short-distance range, where the shortcomings of this approach are more pronounced. As a result, the differential CE evaluation involved in the correction terms is substantially more accurate than the two involved calculations taken separately (in addition to a possible cancellation of discretization errors in the case of numerical evaluations). The contributions of electrostriction, dielectric saturation and specific ion-solvent interactions (*e.g.* asymmetric hydration of equisized anions and cations in water[803]), which are short-range effects overlooked in the Born model, are actually present in the calculated $\Delta_s G^\ominus$ *via* the AT contributions $\Delta_s G_{chg}^{raw}$ and $\Delta_s G_{cav}$. The same considerations may be applied to the analogous correction scheme used in the calculation of derivative thermodynamic properties[1412], which may remain accurate in spite of the known deficiency of the Born model to account (even qualitatively) for ionic solvation entropies (Section 3.2).

When using empirical (fitted) expressions for the correction terms $\Delta_s G_A^{cor}$ and $\Delta_s G_B^{cor}$ (CT schemes only), the application of the correction scheme to raw solvation free energies calculated by AT simulation is actually extremely simple. In this case, the entire scheme is analytical and encompassed within Eqs. 6.9, 6.19, 6.30, 6.32-6.35, 6.37, 6.16, 6.17, 6.22 and 6.23-6.26 (along with the coefficients given in Table 6.2), and can be implemented in a small program or script in a very straightforward fashion. Furthermore, when using approximate (finite-difference) estimates for the correction terms (along with the above empirical expressions), the application of the correction scheme to calculated derivative properties (solvation entropies, heat capacities, volumes, volume-compressibilities and volume-expansivities) is also relatively easy[1412]. Finally, the numerical assessment of this scheme clearly shows that the *application of correction terms* is an *absolute necessity* if AT simulations are to provide any meaningful result concerning single-ion solvation properties.

Irrespective of the force-field representation considered, the comparison between simulated and experimental single-ion solvation free energies (and, possibly, of corresponding derivative quantities such as solvation enthalpies, entropies, heat capacities, volumes, volume-compressibilities and volume-expansivities) should represent a very important test for the accuracy of a given parameter set. However, agreement or discrepancies in this context have been be extremely deceptive in the past, because: (*i*) in the absence of a full understanding of the artifacts associated with approximate electrostatics, microscopic-system boundary conditions, improper potential summation and spurious potential offsets in the simulations, the calculated numbers were typically affected by very large methodology-dependent errors; (*ii*) this comparison did not always properly account for the cavitation free energy and involve a consistent choice of standard states between simulation and experiment.

The discussion of the present section shows that these issues can nowadays be solved in a satisfactory fashion. The possibilities opened by this achievement will be considered in Sections 6.2.2-6.2.4. Finally, considering the alternative route to access single-ion intrinsic solvation free energies *via* evaluation of the air-liquid interfacial potential of the pure solvent (Section 3.3.5), the methodological errors encompass the improper calculation of the interfacial potential (analogous to Point C_4 above) and the influence of the confinement potential (Point G_5 in Section 3.3.6). As seen in Section 4.3.8, the former issue can be eliminated by the use of M-integration rather than P-integration. The latter issue should be investigated by analyzing the sensitivity of the results on the nature of the confinement potential. Besides these two points, there are no serious methodological issues in these calculations. However, the influence of force-field representation errors, and in particular the use of a mean-field treatment for electronic polarization effects, represent another serious problem, which will be mentioned in Section 6.2.4.

6.2.2 Consistent calibration of ion-solvent interactions

As described in Section 6.2.1, the raw results of AT simulations in terms of single-ion solvation free energies (and derivative thermodynamic properties) can be corrected for errors related to approximate electrostatics, microscopic system boundary conditions, improper potential summation and inaccurate solvent-model permittivity. Using this correction scheme, the strong methodological dependence in the calculated raw solvation free energies can be essentially eliminated[1081, 1412]. The resulting standard intrinsic solvation properties are then essentially *methodology-independent* and exclusively characteristic of the *underlying molecular model* (force-field). They can thus be validly *compared to experimental data* for the purpose of *force-field validation or refinement.*

As discussed in Section 3.3.2, in the context of monoatomic ions, the force field is typically defined by *four choices*: (*i*) the explicit (*e.g.* inducible dipoles[915, 916], charges on spring[722, 914, 917, 947–949, 955, 976, 977, 980] or fluctuating charges[919, 926, 935, 956, 975]) or implicit (mean-field approach by enhancement of the solvent charges[861, 1274–1276]) *representation of electronic polarization effects*[940,943,959]; (*ii*) the functional form used to represent the *non-electrostatic (van der Waals) component of the interactions*; (*iii*) the chosen *solvent model*; (*iv*) the *combination rules* used (if any) to define ion-ion van der Waals interaction parameters based on ion-solvent and solvent-solvent ones[719, 726, 992, 1297, 1299, 1314, 1315, 1318, 1319]. The last choice is irrelevant when computing single-ion solvation properties, but becomes crucial when extending the application range to ionic solutions[130, 246, 346, 368, 1168, 1228, 1328, 1330, 2334–2344], liquids[24, 199–206, 2345], melts[209, 978, 1168, 1174, 2346–2362] or crystals[246, 1286, 1331, 2341, 2362–2366]. Given the four above choices, the calculated (corrected) ionic solvation parameters are exclusively determined by the set of *ion-solvent van der Waals interaction parameters* employed in the simulation, which control the *effective size of the solvated ion* and thus, the magnitude of its *electrostatic solvation* (similarly to the effective ionic radius R_I in CE), as well as the magnitude of the *ion-solvent dispersion interactions.*

For the sake of simplicity, the present discussion is restricted to the hydration properties of monoatomic ions and to: (i) a force field with implicit (mean-field) polarization; (ii) van der Waals interactions described by a Lennard-Jones function[490, 881–883], involving interaction sites at the ion center and at the water oxygen atoms; (iii) the SPC[1274] and SPC/E[1275] water models; (iv) the application of a geometric-mean combination rule[1297]. Given this limited scope, there is a one-to-one correspondence between the *Lennard-Jones interaction coefficients* C_6^{ii} and C_{12}^{ii} of an ion (Section 3.3.2) and the *calculated (corrected) single-ion solvation properties* of this ion. The fundamental remaining sources of error are thus those related to the force-field approximation (Point D_4 in Section 3.5) and to the ion-size ambiguity (Point E_4 in Section 3.3.6). The former type of error will be discussed in Section 6.2.4, while the latter one is discussed here.

The *ion-size ambiguity* resides in the fact that even under the assumption that the above force-field choices lead to an exact description of solvation, any AT calculation still relies on specified values for C_6^{ii} and C_{12}^{ii}, two quantities which are *a priori* not known and potentially elusive, just as the ionic radius R_I in CE calculations.

The most obvious way to calibrate the parameters C_6^{ii} and C_{12}^{ii} for an ion is to optimize them by comparison between simulated (corrected) intrinsic single-ion solvation free energies and corresponding experimental estimates. However, in the absence of a precise experimental value for the standard intrinsic hydration free energy $\mathcal{G}^{\ominus}_{H,wat}$ of the proton in a given solvent, which anchors these experimental conventional (relative) scale of values for all cationic and anionic species[170, 707, 1081, 1747, 1750, 1875, 1888, 2259], the experimental numbers are also affected by a very large uncertainty (Sections 2.3 and 5.18). In other words, the optimization of the ion-solvent Lennard-Jones interaction parameters based on the comparison between methodology-independent simulation results for a given force-field representation and experimentally-accessible conventional values amounts to *transferring the elusiveness* of $\mathcal{G}^{\ominus}_{H,wat}$ onto C_6^{ii} and C_{12}^{ii}.

This observation was explicitly acknowledged in a recent study[1289], the goal of which was to optimize new sets of Lennard-Jones interaction parameters for the alkali and halide ions leading to agreement between the calculated (corrected) single-ion hydration free energies and corresponding experimental conventional values. Since the value of the intrinsic proton hydration free energy $\mathcal{G}^{\ominus}_{H,wat}$ is experimentally elusive, this optimization was not performed using a single (arbitrary) estimate for this quantity, but using *different plausible values*. More specifically, the comparison of calculated (corrected) hydration free energies with experimental data (along with the consideration of ionic polarizabilities) was used to calibrate six new sets of ion-water Lennard-Jones interaction parameters for the alkali (Li^+, Na^+, K^+, Rb^+, Cs^+) and halide (F^-, Cl^-, Br^-, I^-) ions along with either the SPC or the SPC/E water models, the resulting C_6^{is} and C_{12}^{is} coefficients being connected to corresponding C_6^{ii} and C_{12}^{ii} parameters by application of a geometric-mean combination rule[1297,1300,1301].

The experimental data set was defined by conventional single-ion hydration free energies of Tissandier *et al.*[1875,1876] (all ions except Cs^+) and Fawcett[1750] (Cs^+), along with three plausible choices for $\mathcal{G}^{\ominus}_{H,wat}$, namely -1100, -1075 or -1050 kJ·mol^{-1}, resulting in three sets L (low), M (medium) and H (high), respectively, for the SPC water model and three sets L_E, M_E and H_E, respectively, for the SPC/E water model. The Lennard-Jones potential[490,881–883] accounts for (London) dispersion[720] (C_6 coefficient) and (Pauli) exchange-repulsion[880] (C_{12} coefficient) effects. As discussed in Section 3.3.2, although the choice of a r^{-12} dependence for the latter contribution is largely arbitrary, C_{12} thus being essentially an empirical parameter, a r^{-6} dependence for the leading term of dispersion effects is physically motivated[720,722]. Therefore, the choice was made to exclusively optimize the C_{12}^{is} coefficients, while the C_6^{ii} coefficients were selected by application of the Slater-Kirkwood formula (Eq. 3.15) using quantum-mechanical (*ab initio*) data for gas-phase cation polarizabilities[2367] and noble gas homoatomic C_6 coefficients[1323], along with experimental data for in-solution anion polarizabilities[2368] and isoelectronic noble gas polarizabilities[1322]. The C_6^{ii} and C_{12}^{is} coefficients could be converted to C_6^{is} and C_{12}^{ii} coefficients by application of a geometric-mean combination rule[1297] (Eq. 3.47) involving the corresponding C_6^{ss} and C_{12}^{ss} coefficients of the water-water (oxygen-oxygen) potential for the SPC and SPC/E water models, these parameters being identical for the two models. The optimization of the C_{12}^{is} coefficient relied on simulations using the LS scheme, the NPT ensemble, $N_S = 1024$ water molecules, and correction terms evaluated with $R_{mean} = (R_{rdf} + R_G)/2$, where R_{rdf} is the position of the first peak in the calculated ion-water (oxygen) radial distribution (Table 5.4, Sets 46r-48r) and R_G the Goldschmidt radius[2056] (Table 5.4, Set 4r) as an effective ionic radius. The C_{12}^{is} coefficient was adjusted for each ion so as to reach agreement between simulated and experimental solvation free energies given one of the two water molecules and one of the three selected values of $\mathcal{G}^{\ominus}_{H,wat}$. This optimization procedure was facilitated by the observation that the (corrected) ionic charging free energy ($\Delta_s G_{chg}^{raw} + \Delta_s G_{cor}$) at constant C_6^{is} is approximately proportional to $(C_{12}^{is})^{-1/6}$. This relationship follows from the Born model[84] (the corrected charging free energy is inversely proportional to the effective ionic radius, within the limit of validity of CE) and from the form of the Lennard-Jones curve (assuming ionic radii linearly related to R_{min}^{is} and not-

Table 6.4 *Conventional and intrinsic hydration free energies of the alkali and halide ions calculated from atomistic simulations based on different ion-water Lennard-Jones interaction parameter sets or derived from experiment.* The quantities reported are estimated standard conventional ($\Delta_s G^{\bullet}$) and standard intrinsic ($\Delta_s G^{\ominus}$) hydration free energies. The simulation results correspond to free-energy calculations at $P^{\circ} = 1$ bar and $T^{-} = 298.15$ K, using optimized Lennard-Jones interaction parameter sets L, H and M together with the SPC water model[1274], or L_E, M_E and H_E together with the SPC/E water model[1275] (Tables 3.3 and 3.4, sets Re). The conventional values were derived using the $\mathcal{G}^{\ominus}_{H,wat}$ values for which each set was optimized. The experimental data ("exp.") is taken from Tissandier *et al.*[1875,1876] (all ions except Cs^+) and Fawcett[1750] (Cs^+). The recommended data from Sections 5.19 and 5.20 ("rec.") is also reported for comparison (Tables 5.22 and 5.26). See Ref.[1289] for more details.

	Li^+	Na^+	K^+	Rb^+	Cs^+	F^-	Cl^-	Br^-	I^-
					$\Delta_s G^{\bullet}$ [kJ·mol^{-1}]				
L	573.6 ± 0.8	679.7 ± 0.8	751.1 ± 0.9	775.0 ± 1.0	798.0 ± 1.0	-1534.9 ± 1.1	-1410.2 ± 1.1	-1382.7 ± 1.2	-1345.4 ± 1.4
M	573.9 ± 0.9	679.4 ± 0.9	752.0 ± 0.9	774.4 ± 1.0	797.6 ± 1.0	-1533.5 ± 1.1	-1409.7 ± 1.1	-1383.0 ± 1.1	-1345.6 ± 1.2
H	574.9 ± 0.9	679.7 ± 0.9	751.2 ± 1.0	774.6 ± 1.0	798.3 ± 1.1	-1534.7 ± 1.2	-1409.7 ± 1.2	-1383.0 ± 1.2	-1345.4 ± 1.2
L_E	574.9 ± 0.9	680.1 ± 1.0	752.1 ± 1.1	774.8 ± 1.1	798.2 ± 1.1	-1533.5 ± 1.4	-1409.6 ± 1.4	-1382.7 ± 1.4	-1344.9 ± 1.5
M_E	574.4 ± 1.0	679.7 ± 1.0	752.7 ± 1.0	774.7 ± 1.2	798.6 ± 1.2	-1533.3 ± 1.1	-1408.4 ± 1.3	-1381.7 ± 1.2	-1344.3 ± 1.6
H_E	575.0 ± 1.1	680.3 ± 1.0	751.6 ± 1.1	774.7 ± 1.3	798.1 ± 1.4	-1534.4 ± 1.2	-1408.8 ± 1.4	-1382.6 ± 1.3	-1345.2 ± 1.3
exp.	575.1[a]	680.7[a]	752.5[a]	775.2[a]	798.4[b]	-1533.3[a]	-1408.7[a]	-1381.9[a]	-1344.5[a]
rec.	574.9	680.5	752.2	775.0	798.2	-1533.2	-1408.6	-1381.8	-1344.4
					$\Delta_s G^{\ominus}$ [kJ·mol^{-1}]				
L	-526.4 ± 0.8	-420.3 ± 0.8	-348.9 ± 0.9	-325.0 ± 0.9	-302.0 ± 1.0	-434.9 ± 1.1	-310.2 ± 1.1	-282.7 ± 1.2	-245.4 ± 1.4
M	-501.1 ± 0.9	-395.6 ± 0.9	-323.0 ± 0.9	-300.6 ± 1.0	-277.4 ± 1.0	-458.5 ± 1.1	-334.7 ± 1.1	-308.0 ± 1.1	-270.6 ± 1.2
H	-475.1 ± 0.9	-370.3 ± 0.9	-298.8 ± 1.0	-275.4 ± 1.0	-251.7 ± 1.1	-484.7 ± 1.1	-359.7 ± 1.2	-333.0 ± 1.2	-295.4 ± 1.2
L_E	-525.1 ± 0.9	-419.9 ± 1.0	-347.9 ± 1.1	-325.2 ± 1.1	-301.8 ± 1.1	-433.5 ± 1.4	-309.6 ± 1.4	-282.7 ± 1.4	-244.9 ± 1.5
M_E	-500.6 ± 1.0	-395.3 ± 1.0	-322.3 ± 1.0	-300.3 ± 1.2	-276.4 ± 1.2	-458.3 ± 1.1	-333.4 ± 1.3	-306.7 ± 1.2	-269.3 ± 1.6
H_E	-475.0 ± 1.0	-369.7 ± 1.0	-298.4 ± 1.1	-275.3 ± 1.3	-251.9 ± 1.4	-484.4 ± 1.2	-358.8 ± 1.4	-332.6 ± 1.3	-295.2 ± 1.3
rec.	-525.11	-419.46	-347.78	-325.02	-301.79	-433.18	-308.60	-281.80	-244.41

(a) Refs.[1875,1876]; (b) Ref.[1750].

ing that R^{is}_{min} is proportional to $(C^{is}_{12})^{1/6}$ at constant C^{is}_{6}). The six resulting reoptimized sets of Lennard-Jones interaction coefficients (Table 3.2) are reported in Tables 3.3 and 3.4 (sets Re), while corresponding effective ionic radii are reported in Table 5.4 (Sets 43r-45r; see also 33r-35r for derived in-crystal radii).

The standard intrinsic hydration free energies of the alkali and halide ions calculated using the six reoptimized parameter sets, along with the associated water model, are reported in Table 6.4. Using the $\mathcal{G}^{\ominus}_{H,wat}$ value for which each set was optimized, this data can be converted to conventional values and compared with experimental results. This comparison is also performed in Table 6.4. The resulting conventional values are essentially equal to the corresponding experimental values by Tissandier *et al.*[1875,1876] (all ions except Cs^+) and Fawcett [1750] (Cs^+), considering that the root-mean-square deviation from the experimental data, 0.8 kJ·mol^{-1}, is smaller than the maximum statistical error of 1.6 kJ·mol^{-1}. This is of course not surprising considering that the achievement of such an agreement was the target of the optimization.

However, the corresponding intrinsic values are bound to a specific choice of $\mathcal{G}^{\ominus}_{H,wat}$ and, therefore, differ for sets L and L$_E$ *vs.* M and M$_E$ *vs.* H and H$_E$. A less negative value of $\mathcal{G}^{\ominus}_{H}$ leads to more weakly hydrated cations and more strongly hydrated anions. These differences are also reflected in the corresponding effective radii (Tables 3.3 and 3.4), as illustrated graphically in Figure 6.2 for sets L, M and H in terms of the effective radius R^{is}_{min} corresponding to the minimum of the ion-water Lennard-Jones curve. Here, a less negative value of $\mathcal{G}^{\ominus}_{H,wat}$ leads to larger cations and smaller anions.

The three values of $\mathcal{G}^{\ominus}_{H}$ selected here for the optimization procedure (-1100, -1075 and -1050 kJ·mol^{-1}) span a reasonable range in terms of experimental estimates, but are nevertheless *arbitrary*. Depending on personal convictions or on the possible future availability of a more definitive experimental estimate, it may be necessary to reoptimize a new set for a different $\mathcal{G}^{\ominus}_{H,wat}$ value. In order to avoid a major reparameterization effort, a simple way to design such a new set would be to *interpolate* between the C^{is}_{12} values derived here for sets L and M or M and H, as appropriate, using the approximate linear relationship between $\Delta_s G^{\ominus} - \Delta_s G^{\ominus}_{std}$ and $(C^{is}_{12})^{-1/6}$ at constant C^{is}_{6} (see above). To investigate the accuracy of such an approach, it was applied to predict the $\Delta_s G^{\ominus}$ values of sets M and M$_E$ (based on the C^{ii}_{12} parameters of Tables 3.3 and 3.4) using the $\Delta_s G^{\ominus}$ values of sets L and H, or L$_E$ and

Figure 6.2 *Illustration of the ion sizes corresponding to the three optimized ion-water Lennard-Jones interaction parameter sets.* The drawings refer to the ion-water Lennard-Jones interaction parameters of sets L, M and H, optimized for $\mathcal{G}^{\ominus}_{H,wat} = -1100$, -1075 and -1050 kJ·mol^{-1}, respectively, together with the SPC water model[1274]. The effective ionic radius R^{is}_{min} (minimum of the Lennard-Jones curve for the ion-water interactions) and the calculated standard intrinsic hydration free energy $\Delta_s G^{\ominus}$ of the ion (Table 6.4) are indicated.

H_E, respectively. All predicted $\Delta_s G^\ominus$ values were found[1289] to be within RT (2.5 kJ·mol^{-1} at T^-) of the computed $\Delta_s G^\ominus$ counterparts. Note that this interpolation, if justified for $\Delta_s G_{chg}^{raw} + \Delta_s G_{cor}$, involves an identical but less justified scaling for $\Delta_s G_{cav}$. However, the consideration of a separate scaling scheme for $\Delta_s G_{cav}$, *e.g.* linear in $(C_{12}^{is})^{-1/3}$ or $(C_{12}^{is})^{-1/2}$, renders the interpolation more complicated without noticeably altering the results[1289]. The remarkable accuracy of the approximate relationship between the ion-water Lennard-Jones C_{12}^{is} parameter and the $\Delta_s G^\ominus$ value for a fixed C_6^{is} parameter suggests that, once a definitive value for $\mathcal{G}_{H,wat}^\ominus$ is known (if ever), a simple calculation can immediately yield appropriate Lennard-Jones interaction parameters for non-polarizable alkali and halide ions in SPC or SPC/E water. Note that in this case, the interpolation would involve sets L and M or M and H, *i.e.* a smaller interpolation interval as that used above, so that the resulting error should be even more limited.

As a final note, it should be kept in mind that the calibration of ion-solvent Lennard-Jones interaction parameters based on corrected solvation free energies may not necessarily be the most appropriate strategy if these parameters are to be used in simulations that are *still biased* by these effects, at the level of the interatomic forces. For example, considering that the correction terms typically provide an overall negative contribution to the solvation free energy, reaching agreement with experiment in terms of corrected values may lead to ions that are in effect *too weakly hydrated* given the approximate electrostatic scheme and boundary conditions actually used in the simulation. Assuming that the corresponding artifacts introduced in direct ion-ion interactions are of weaker magnitudes, this may for example result in too strong mean interionic interactions when simulating an electrolyte solution at finite concentration.

6.2.3 The atomistic-consistency assumption

As described in Sections 6.2.1 and 6.2.2, the results of AT simulations in terms of intrinsic single-ion solvation properties can be corrected for methodological dependence, and the resulting one-to-one correspondence between ion-solvent or, equivalently (given combination rules), ion-ion van der Waals (typically Lennard-Jones) interaction parameters exploited to calibrate the latter parameters against experimental estimates for intrinsic single-ion solvation free energies. The latter calibration depends, however, on a specific choice for the experimentally-elusive intrinsic proton hydration free energy $\mathcal{G}_{H,wat}^\ominus$.

A priori, it might seem a strange idea to generate *multiple parameter sets* (see, however, Ref.[877]) or, by interpolation, a continuous spectrum thereof, indexed by the associated $\mathcal{G}_{H,wat}^\ominus$ value, rather than selecting a single value of this parameter considered as trustworthy. However, consideration of the literature in the domain over the last century suggests that such a trustworthy value is not yet available (Section 5.18). Therefore, a possible long-term goal is rather to seek for the *validation of a specific set* or of an interpolated variant thereof, and thus of the associated $\mathcal{G}^{\ominus}_{H,wat}$ value, *in terms of other properties* depending on the corresponding ionic parameters. This approach can be viewed as equivalent to an extra-thermodynamic assumption, which will be termed here the *atomistic-consistency assumption*[1289] (ACA), and can be stated in the following way:

It is possible to construct classical atomistic models capable of accounting consistently for all experimental (thermodynamic, dynamic, structural) data concerning ionic properties for which such a classical atomistic modeling approach is adequate; provided that the experimental data set is large enough and that the agreement between simulation results and experiment is precise enough, all these models will be compatible with a single value of the absolute intrinsic hydration free energy of the proton $\mathcal{G}^{\ominus}_{H,wat}$.

As illustrated in Figure 6.3, the experimental properties to be considered may include: (*i*) properties of *hydrated ions at infinite dilution* (*e.g.* solvation free energy or enthalpy, partial molar entropy, volume, heat capacity, volume-compressibility or volume-expansivity, hydration structure, water exchange dynamics); (*ii*) properties of *aqueous electrolyte solutions* at finite concentrations (density, partial molar volume, heat capacity, volume-compressibility or volume-expansivity, dissolution free energy or enthalpy, solvent-vapor pressure, osmotic pressure, activity coefficients, dielectric permittivity, surface tension, transport coefficients); (*iii*) properties of *ions in non-aqueous solvents* (transfer free energy from aqueous solution, *i.e.* partition coefficient, or transfer enthalpy and entropy); (*iv*) properties of *ionic crystals* (crystal structure, lattice parameters, lattice energy, volume-compressibility, volume-expansivity, heat capacity, melting temperature and enthalpy); (*v*) properties of *ionic melts* (density, partial molar volume, heat capacity, volume-compressibility or volume-expansivity, transport coefficients); (*vi*) properties of *gas-phase ion-water clusters* (formation or stepwise-binding free energy, enthalpy or entropy, geometry); (*vii*) properties of *ions at interfaces*, *e.g.* solid-liquid, liquid-liquid or air-liquid (density profiles, preferential adsorption, ion exchange, ion or solvent longitudinal *vs.* transverse diffusion, dipole orientational relaxation times at the interface, interfacial tension, interfacial potential).

The *adequacy of atomistic modeling* to represent a specific property is an important condition within the ACA. It is likely that a number of the properties listed above are actually inaccessible to such a classical approach, *e.g.* because they involve specific electronic or other quantum effects. Furthermore, even among the ones that are in principle accessible, a number may also turn out to require a more accurate atomistic representation compared to the one applied in the present study, *e.g.* explicit inclusion of electronic polarizability, more accurate representation of van der Waals interactions, more complex solvent models, improved combination rules. Nevertheless, it is probable that relatively simple atomistic modeling may properly account for *a sufficient set* of experimental properties for the ACA to be applicable, and ultimately lead to the suggestion of a unique $\mathcal{G}^{\ominus}_{H,wat}$ value. Still, it should be kept in mind that the ACA is truly an extra-thermodynamic assumption in two respects.

First, it *encompasses a model*, namely that of a classical atom-based representation with force-field based interactions. Although significantly more complex and probably more accurate than the extra-thermodynamic assumptions encountered previously (*e.g.* in Sections 5.17 and 5.18), it nevertheless only represents an idealization of reality. Of particular relevance is the fact that a M-scheme has been retained to partition bulk and surface contributions to the electric potential (Section 4.3.8), a point that will be returned to in Chapter 7. Second, the ACA is a *postulate*, which may be correct or incorrect. For example, a comparable postulate suggesting that salt lattice parameters unambiguously determine ionic radii or that salt solvation parameters unambiguously determine intrinsic single-ion solvation parameters, provided that a sufficient number of salts are considered, would be plainly incorrect, in view of the exact (lattice parameters) or nearly exact (standard salt solvation parameters) additivity constraint on these properties (Sections 5.4 and 5.7, respectively).

As a first step towards the application of the ACA, an assessment of the optimized parameters of Section 6.2.2 based on experimental quantities other than the hydration free energies was performed[1289], namely in terms of the first peak position in the ion-oxygen (water) radial distribution function, of the salt partial molar volumes in water and of crystal properties. The results provided some preliminary evidence for the appropriateness of set L. How-

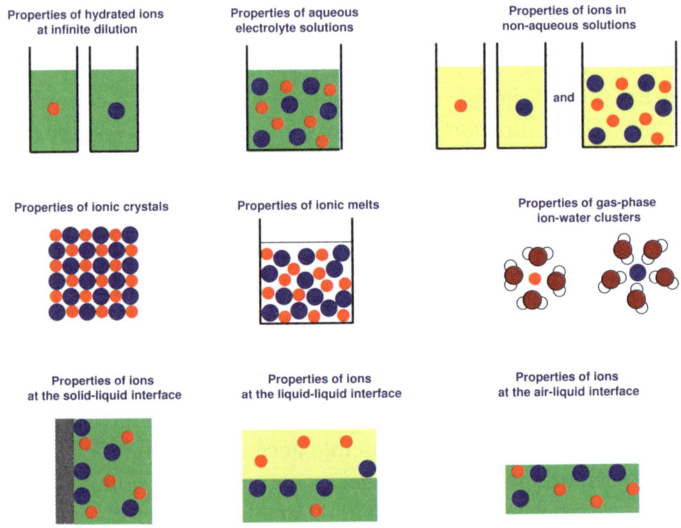

Figure 6.3 *Properties that could in principle be used for the (in)validation of a specific ion-water (and ion-ion) Lennard-Jones interaction parameter set.* It is assumed that the set has been optimized against intrinsic single-ion hydration free energies, assuming a given value for $\mathcal{G}^{\ominus}_{H,wat}$. The comparison against other properties aims at (in)validating this specific $\mathcal{G}^{\ominus}_{H,wat}$ value. This approach can be viewed as a form of extra-thermodynamic hypothesis, termed here the atomistic-consistency assumption (ACA).

ever, it also appeared that agreement with experiment was systematically poorer for the smallest as opposed to the largest ions and that the properties of crystals involving the Li^+ cation were significantly affected by the choice of a specific combination rule. Since small ions are characterized by a high surrounding ionic field and thus, the strongest interaction with surrounding species, this observation may hint towards some inherent flaw in the Lennard-Jones description of the repulsive component of van der Waals interactions, and possibly also of the geometric-mean combination rule. Given this proviso, the three above comparisons provide some preliminary evidence supporting a $\mathcal{G}^{\ominus}_{H,wat}$ value of about -1100 kJ·mol^{-1} (set L) for the intrinsic hydration free energy of the proton (reported as entry 147g$_1$ in Table 5.19), which is essentially the one suggested by Tissandier *et al.*[1875,1876] (-1104.5 kJ·mol^{-1} according to Boltzmann-statistics for the standard-state electron, and -1100.9 kJ·mol^{-1} after correction to Fermi-Dirac statistics[483,1723], reported as entry 20g$_1$ in Table 5.19), and identical to the value recommended by the authors of this book in Sections 5.18 and 5.19 (-1100 kJ·mol^{-1}).

The Lennard-Jones interaction parameters derived from the optimization procedure of Section 6.2.2 can also be mapped to corresponding estimates for the in-solution (in-water) radii of the alkali and halide ions. These sets of radii are presented in Section 5.4 (see Points J_{10} and K_{10} therein), either in the form of the minimum R^{is}_{min} of the corresponding ion-solvent Lennard-Jones curve (Point J_{10}, Sets 43r-45r in Table 5.4), or in the form of first-peak positions in the ion-water (oxygen atom) radial distribution functions (Point K_{10}, Sets 46r-48r in Table 5.4). They can also be converted to tentative in-crystal radii by addition of offset constants, calibrated for optimal partitioning of the salt crystal lattice parameters. These sets of radii are also presented in Section 5.4 (see Points E_{10} and F_{10} therein), considering the two alternative forms of in-water radii (see above and Point E_{10}, Sets 33r-35r or Point F_{10}, Sets 36r-38r in Table 5.4, respectively). The interesting aspect of the latter radii is that they perform remarkably well in terms of the root-mean-square deviations σ between predicted and experimental lattice parameters, with values of at most 0.014 nm. Furthermore, in terms of the radii set of type E_{10}, the value of σ is also noticeably lower for set L (0.009 nm) compared to sets M (0.012 nm) and H (0.014 nm), suggesting again that the former set is more reasonable within the ACA. Note that this approach involving the evaluation of in-crystal radii based on estimated single-ion solvation data is actually opposite to the usual CE approach of evaluating solvation properties based on estimated in-crystal radii.

6.2.4 Improved force-field description

The discussion of Section 6.2.3 concerning the atomistic-consistency assumption (ACA) summarizes what is considered by the authors of this book as the best possible approach to ultimately access intrinsic single-ion solvation properties *via* AT simulation. However, on the long term, *improvements* at the level of the *force-field representation* itself (Point D_4 in Section 3.5) may be desirable, as discussed at the head of Section 6.2. This follows, for example, from three illustrative considerations. First, the optimization procedure followed in Section 6.2.2 to derive unique ion-solvent van der Waals interaction parameters from estimated intrinsic single-ion solvation free energies leads to parameters compensating for the mean effect of the error caused by the approximate representation of the potential-energy function in the simulations, *i.e.* they are *effective* parameters. As a result, they may fail to fulfill the ACA due to force-field representation errors (Point D_4), thereby failing to resolve the ion-size ambiguity error (Point E_4). Second, the alternative approach to intrinsic single-ion solvation free energies relies on the evaluation of the air-liquid interfacial potential of the pure solvent (Section 3.3.5). Errors in the force-field representation at the level of the solvent will affect the reproduction of this quantity, without preventing the derivation of *consistent* ion-solvent *van der Waals interaction parameters* according to the procedure of Section 6.2.2. Thus, the inclusion of this property within the ACA characterizes force-field representation errors (Point D_4) and not ion-size ambiguity errors (Point E_4). Third, the effect of force-field representation errors is revealed by the influence of the force-field definition on *many calculated properties*, *e.g.* by a dependence of the calculated air-liquid interfacial potential (*e.g.* Refs.[470,1473]; Table 5.18), or of the calculated single-ion solvation properties[1289] (even after correction for methodology-dependent errors) on the chosen water model, as well as by a dependence of the simulated crystal lattice properties on the chosen combination rules[1289]. However, before undertaking major adjustments in the force-field functional form, which may increase its complexity and computational cost, and require important reparameterization efforts, it is reasonable to first assess if and where such adjustments are required. For instance, there is little point in using an explicitly polarizable force-field to calculate single-ion solvation free energies, as long as the other (much larger) sources of error have not been eliminated[946], including the methodology-dependent errors discussed in Section 6.2.1.

Given this proviso, in the present context, the most relevant force-field improvement options are probably the explicit representation of *electronic polarization effects* and the refinement of the *representation of van der Waals interactions* in terms of both functional form and combination rules. These two points are briefly discussed below.

Most currently available force fields rely on a *mean-field representation* of *electronic polarization* effects, *via artificial enhancement* of the *fixed atomic partial charges* of the solvent model. In other words, and in the specific context of water, these charges are selected to reproduce the effective dipole moment of the water molecule in a bulk (pure liquid) environment, *i.e.* about 2.9 D (essentially identical to the corresponding value for water molecules in the solid phase), which is significantly larger than the corresponding value of 1.8546 D for the gas-phase water molecule (Table 1.1). This difference

Table 6.5 *Effective dipole moment of water molecules in the first hydration shell of ions, estimated based on quantum-mechanical computations.* The reported values are all based on the results of Car-Parrinello molecular dynamics simulations, *via* an analysis of the instantaneous electron density (*e.g.* localized Wannier function analysis[2369]) and averaging of the estimated effective dipole moments over time.

Ion	μ_d [D]	Ref.	Ion	μ [D]	Ref.	Ion	μ [D]	Ref.
Li^+	2.96	1596	Be^{2+}	3.1	1577	Al^{3+}	4.14	1588
Li^+	≈ 2.95	1599	Mg^{2+}	3.3	1582	Y^{3+}	3.6	1592
Na^+	2.81	1596	Mg^{2+}	≈ 3.2	1599			
Na^+	≈ 2.9	1599	Ca^{2+}	3.35	1608	F^-	3.07	1590
K^+	2.77	1596	Ca^{2+}	3.4	1604	Cl^-	3.14	1587
K^+	≈ 2.9	1599	Ca^{2+}	3.4...3.5	1585	Br^-	2.9	1584
K^+	2.78	1604	Ca^{2+}	≈ 3.1	1599	I^-	3	1591

arises from the fact that the local environment of a water molecule in the bulk is instantaneously heterogeneous. The electric field generated by the water molecule induces preferential orientations (as well as electronic polarization and slight geometric distortions) of its neighbor molecules, resulting in a non-zero reaction field at its location. This reaction field triggers in turn the reorientation of the molecule itself, and induces its electronic polarization (as well as slight geometric distortions). Due to the preferential interaction geometries with favorable dipole-dipole interactions (*e.g.* water hydrogen-bonding network), the reaction field exerted on a water molecule by its surroundings has (on average) a leading component that is aligned with its dipole moment. As a result, the mean effect of electronic polarization is to *enhance the effective molecular dipole moment.* The required artificial charge enhancement in mean-field polarization models of water is evident in all water models presented in Table 3.1, the dipole moments of which range from 2.12 (CC water model[1283]) to 2.35 D (ST2, TIP3P and SPC/E water models[861, 1275, 1282]), except the ROW water model (1.84 D; originally developed for the study of water in the gas phase and ice, and including an approximate term to account for induced multipoles in the latter case). In the context of ionic solvation calculations, the bulk value of about 2.9 D is expected to represent a *lower bound* for the effective dipole moment of water molecules in the *first solvation shells* of an ion. Since these water molecules are exposed to the ionic field, which can be extremely large (especially for small or/and highly charged ions), their polarization response is expected to be more important than in the bulk, as well as anisotropic, *i.e.* biased along the direction of the ion. These considerations are illustrated by the results of a number of QM calculations summarized in Table 6.5. These results suggest that mono- and divalent ions do not dramatically alter the magnitude of the effective dipole moment of water molecules in their first hydration shell. The monovalent cations considered even appear to slightly decrease this magnitude in comparison to bulk water, whereas the monovalent anions appear to slightly increase it. However, divalent and trivalent cations lead to a pronounced increase in the magnitude of the effective first-shell water dipole moment in comparison to bulk water, by about 10 and 30%, respectively.

In the context of air-liquid interfacial potential calculations, the bulk value of about 2.9 D is expected to represent an *upper bound* for the effective dipole moment of a *surface water molecule*. Since these water molecules are exposed to the bulk liquid only on one side, the other side representing an interface to a non-polar medium (vacuum or air), their polarization response is expected to be less important than in the bulk, as well as anisotropic, *i.e.* biased along the direction of the bulk liquid. As a result, their effective dipole moment will be closer to the gas-phase value of 1.85 D.

These issues can be remedied by an *explicit representation* of *electronic polarizability*[480, 940, 943, 959, 2370] (solvent, and possibly also ion) into the force field. An explicit inclusion of polarizability in a simulation can be achieved by means of *e.g.* inducible dipoles[915,916], charges on spring[722,914,917,947–949,955,976,977,980] or fluctuating charges[919, 926, 935, 956, 975]. In contrast to the implicit mean-field approach by charge enhancement[861, 1274–1276], an explicit inclusion of polarizability enhances the transferability of the model, *i.e.* its ability to properly account for the distinct electronic properties of species (*e.g.* effective dipole moment) in different environments (*e.g.* gas phase, condensed phase, non-polar environment, mixtures, vicinity of charged species, or interfaces) and at different state points (*e.g.* pressure, temperature). However, this explicit treatment also poses a number of methodological problems. First, because polarization is a many-body term, the polarization of a given molecule depending simultaneously on that of all other molecules in the system, its contribution must be evaluated *self-consistently* (*i.e.* iteratively) at each step of the simulation. Irrespective of the approach employed, this induces a significant increase in the computational cost, roughly by a factor $3-5$ for pure water systems of the sizes typically considered nowadays. Second, polarization is a quantum-mechanical effect and its approximate description by means of semi-classical approaches is only appropriate above a certain distance between the interacting species. At short range, a so-called *polarization catastrophe*[2371] may result from the unlimited mutual enhancement of the electronic polarization of two species in close proximity, requiring the need for damping schemes, *e.g.* Thole-type damping[2372], introducing a distance-dependent damping factor for short-range dipole-dipole interactions, or use of a sublinear dependence of the induced dipoles on the electric field for large field strengths[959, 980, 2373]. Alternatively, the same damping can be effectively achieved by artificially increasing Lennard-Jones repulsion between the species[949]. Third, the design of polarizable force fields requires a more *intricate parameterization procedure*, including the selection of a specific methodological approach, along with a given damping scheme, the specification of polarizable sites and with the requirement to calibrate the additional parameters controlling polarizability against a number of experimental observables that is still limited.

As a result, although an explicit treatment of electronic polarization is certainly required on the long term to improve transferability, the design of polarizable force fields is by no means trivial and the resulting models remain largely empirical, involving in particular strongly correlated methodological assumptions and effective force-field parameters. In the context of solvation, it should also be kept in mind that apparent agreement or disagreement between simulated and experimental single-ion solvation properties, which are often used in the calibration or validation of polarizable force fields for ionic systems, may be extremely deceptive in the absence of a thorough characterization of the physical nature of the experimental data (Chapter 5) and of the systematic application of corrections for methodology dependence (Section 6.2.1).

Yet, resulting from the relevance of the polarizability issue in the present context, many force fields employing polarizable ions, to be used with a polarizable solvent, have been developed[373,905,920,921,924,928,929,931,937,961,969,970,1336], and numerous studies involving polarizable ions in polarizable solvents[141,240,242,354,359,366,439,440,798,799,908–910,918,921,923,927,932–934,938,939,941,942,944,946,951–954,958,960,962–967,971–974,978,979,1467,2374–2378], non-polarizable ions in polarizable solvents[922,925,945,957,2379], or comparing the explicit and implicit inclusion of polarizability[141,240,354,359,366,439,440,799,908,918,925,932,937–939,941,942,944–946,952–954,966,972,979,1424,1439,1467,1470,2376,2380,2381] have been reported to date. It has been shown in particular that explicit polarizability is important *e.g.* in the study of surface properties (*e.g.* water orientation and hydrogen bonding properties, air-liquid interfacial potential, preferential affinity or exclusion of ions relative to the surface; see Refs.[152,1612,2382] for a recent controversy concerning the latter point) in the context of pure water[2383] or salt solutions[439,440,973,2383], of ion-water clusters[141,240,942,953,966], of ionic diffusion[945], of asymmetric solvation effects[799], of the specific geometric features of the first hydration shell around solvated ions[937,944,972,2378] and of ion channels[2384,2385].

In the present context, the replacement of implicit by explicit polarizability at the level of the solvent is likely to have a significant impact on: (*i*) the charging free energy of an ion (Section 3.3.3), where an enhancement is expected for small or/and highly charged ions due to the increase in the effective dipole moment of the first-shell solvent molecules relative to the bulk; (*ii*) the representation of the air-liquid interfacial potential at the surface of an uncharged cavity (Point A_3 in Section 3.2), where a reduction is expected due to the decrease of the effective dipole moment of the interfacial solvent molecules relative to the bulk; (*iii*) the representation of the air-liquid interfacial potential at the surface of a liquid sample (Section 3.3.5), where a reduction is expected due to the decrease of the effective dipole moment of the surface molecules relative to the bulk. Concerning the latter point, estimates of the air-liquid interfacial potential of water estimated using common implicit-polarization models (based on M-integration; Sections 3.3.5 and 4.3.8), *e.g.* χ_{wat} = 0.27, 0.21 or 0.73 V, for the TIP3P, SPC/E, SPC water models, respectively (Table 5.18) are likely to represent upper bounds to the true value of this potential, the value recommended by the authors of this book being +0.13 V (Section 5.17). In rough terms, one would expect the χ_{wat} value calculated by a classical implicit-polarization model to be overestimated by about 50%, *i.e.* the extent of enhancement of the effective molecular dipole moment when comparing bulk and gas-phase environments (2.9 *vs.* 1.85 D). This is of course a very crude estimate, considering that: (*i*) most water models only have a dipole enhancement of about 10-30% relative to the gas-phase dipole (see *e.g.* values in Table 3.1); (*ii*) the true effective surface dipole moment is probably still larger than the gas-phase value; (*ii*) a too high dipole moment in classical water models may also lead to overstructuring of the surface. At the level of the ion, explicit polarizability is expected to have a more limited impact, *via* alteration of specific ion-solvent interactions in instantaneous configurations that are not spherically symmetric (see Points C_3 and G_3 in Section 3.2).

A second issue of particular importance in the study of ionic properties *via* AT simulations when extending the application range to ionic solutions[130,246,346,368,1168,1228,1328,1330,2334–2344], liquids[24,199–206,2345], melts[209,978,1168,1174,2346–2362] or crystals[246,1286,1331,2341,2362–2366] concerns the representation of *van der Waals interactions*, including the choice of a *functional form* and *combination rules*. Common functional forms for the representation of van der Waals interactions have been mentioned in Section 3.3.1 and include the Lennard-Jones (12-6) function[881–883], the 9-6 function[981–984], the Morse function[985], the Kihara function[986], the Born-Mayer-Huggins function[180,987,988], the Buckingham exp-6 function[989,990], the exp-6-8 function[991] and the buffered-14-7 function[992]. Common choices of combination rules have been mentioned[992,1297,1299–1303,1303–1305,1305–1319] in Section 3.3.2.

Note that many force fields partly bypass the use of mathematical combination rules. This typically happens in the following situations: (*i*) when combination rules are defined by a table instead of a mathematical formula; (*ii*) when van der Waals interaction parameters are provided explicitly for a subset of atom type pairs[2386]; (*iii*) when special or scaled interaction parameters are introduced for third covalent neighbors[1297]; (*iv*) when special interaction parameters override those defined by the generic combination rule for specific atom type pairs[2387]. However, to our knowledge, no existing force field entirely bypasses the use of combination rules, which would imply the specification of individual van der Waals interaction parameters for all pairs of atoms within the simulated system.

Among the different possible choices concerning the representation of van der Waals interactions, the Lennard-Jones function[881–883] and the geometric-mean[1297,1300,1301] or Lorentz-Berthelot[1303,1305] combination rules are the ones most typically encountered in current AT simulation programs. However, experimental data concerning *real-gas properties* of noble gases suggests that the Lennard-Jones function is inaccurate in this context[992], and that the simple arithmetic or geometric combination rules cannot accurately reproduce the connection between homo- and heteroatomic interaction potentials[992,

[1319]. Based on these observations, many alternative functional forms[180, 981–992] and combination rules[992, 1299–1319] that more accurately reproduce real-gas experimental data have been proposed. However, the validity of these considerations in the context of condensed-phase systems is *limited* by the fact that in this case, the relevant interactions involve sites that in general represent atoms within molecules rather than isolated closed-shell atoms, and may in addition carry partial charges. These alternative (generally more complex) functional forms and combination rules may be computationally more expensive and often involve additional parameters, which must be calibrated against a number of experimental observables that is still limited. Furthermore, the choice of a van der Waals functional form, of combination rules and of interaction parameters are strongly *correlated* with each other, as well as with other methodological choices made in the simulation, in particular concerning the treatment of electrostatic interactions, and the choice of atomic partial charges, and the possible inclusion of a long-range (tail) correction to truncated van der Waals interactions[490, 856, 1258, 1261–1269]. The selected effective van der Waals interaction parameters are adjusted by calibration against experimental data, so as to partly correct for errors related to all the above methodological choices. It follows that "improving" the *description of van der Waals interactions* in terms of functional form or/and combination rules will generally *deteriorate* the accuracy of a force field, unless it is accompanied by a *consistent reoptimization* of all force-field parameters.

Possible *shortcomings* of simple van der Waals functional forms in studies of condensed-phase systems involving ions have been reported. For example, the optimization of ion-solvent Lennard-Jones interaction parameters against single-ion hydration free energies and entropies simultaneously was found impossible for heavy alkali metal cations[877] (see, however, Ref. [878]). It was also observed[1328] that Kirkwood-Buff integrals of electrolyte solutions could only be reproduced using Lennard-Jones interactions when abandoning the geometric-mean combination rule for ion-water interactions. In the case of crystal properties (lattice energies and parameters), a strong dependence of the simulation results on the selected combination rule was found, suggesting that the importance of this choice should not be overlooked[1289].

Although the impact of the van der Waals functional form and combination rules in AT simulations of systems involving ion-ion interactions is probably as important as that of electronic polarization (see above), the investigation of this influence is even more complicated, due to the strong correlation between choices in the representation of these interactions and all other methodological choices and parameters used in the simulation.

For completeness, one may mention a third possible extension of simple force fields, namely the use of flexible water models[884–900]. This choice is probably not so crucial, considering that intramolecular vibrations of the water molecule are only weakly activated at room temperature and that geometric distortion is a minor component of the dielectric response in the case of this solvent (Section 2.1).

In the opinion of the authors, the next major progresses in the field of AT simulations concerning single-ion thermodynamics and liquid surface potential properties will come from the use of explicitly polarizable force fields with more adequate van der Waals functional forms and combination rules. Along with the systematic application of correction terms for methodology dependence (Section 6.2.1) or the consideration of electrostatic schemes and system sizes where the corresponding errors can be neglected (far future), such improvements should permit the inclusion of a sufficient number of accurately calculated observables for the ACA (Section 6.2.3) to become reliably applicable in practice.

6.3 Quantum-mechanical computations

As summarized in Section 3.4, due to their high inherent computational cost, QM computations (calculations or simulations) must resort in practice to a number of *time-saving approximations* (Points A_6-E_6), leading to *serious shortcomings* (Points A_7-D_7) in the context of problems related to ionic solvation.

For the calculation of single-ion solvation free energies, the main types of approaches employed are: (*i*) *quasi-chemical theory*; (*ii*) *hybrid quantum-classical simulations* or more precisely, quantum-mechanical molecular-mechanical simulations (QM/MM); (*iii*) *Car-Parrinello molecular dynamics simulations* (CPMD). The latter type of approach has also been employed for the evaluation of the *air-liquid interfacial potential* of a liquid, as well as of *deprotonation and redox* parameters (see also Refs. [1673, 2388–2390] for QM/MM calculations of the latter parameters). These five types of calculations are described in turn in the following sections.

6.3.1 Quasi-chemical theory

In *quasi-chemical theory*[2391–2393], also referred to as the *supermolecule* or *hybrid-cluster* approach[1729], a charged cluster consisting of the ion and the tightly-bound solvent molecules of its first solvation shell is regarded as a quasi-chemical entity, *i.e.* it is viewed as a single molecular ion, which will be referred to here as a *cluster-ion*. The solvation free energy of the ion itself is then evaluated according to one of the two thermodynamic cycles[1729] presented in Figure 6.5.

The first cycle is termed the *monomer cycle*[2391–2394]. It involves the formation of the cluster-ion from the ion and n individual solvent molecules in the gas-phase ($\Delta_s G_{MG}[IS^z_{n,g}]$), the solvation of this cluster into the bulk solvent ($\Delta_s G[IS^z_{n,g}]$), the dissociation of the solvated cluster-ion into a solvated ion and n individual solvent molecules ($-\Delta_s G_{MS}[IS^z_{n,d}]$), and the back-transfer the n solvent molecules to the gas-phase ($-n\Delta_s G[S_g]$). The sum of these contributions is equal to the solvation free energy of the ion ($\Delta_s G[I^z_g]$).

The second cycle is termed the *cluster cycle*[1484, 1729, 2279, 2395–2398]. It involves the dissolution of the ion into a cluster of n solvent molecules in the gas-phase ($\Delta_s G_{CG}[IS^z_{n,g}]$), the solvation of the resulting cluster-ion into the bulk solvent ($\Delta_s G[IS^z_{n,g}]$), the dissociation of the solvated cluster-ion into a solvated ion and a cluster of n solvent molecules ($-\Delta_s G_{CS}[IS^z_{n,sol}]$), and the back-transfer of the solvent cluster to the gas-phase ($-\Delta_s G[S_{n,g}]$). The sum of these contributions is also equal to the solvation free energy of the ion ($\Delta_s G[I^z_g]$).

The contributions $\Delta_s G_{MS}[IS^z_{n,d}]$ and $\Delta_s G_{CS}[IS^z_{n,d}]$ are exclusively related to the "tagging" of specific solvent molecules within a reference molar volume v° of the solution and thus, purely entropic, namely

$$\Delta_s G_{MS}[IS^z_{n,d}] = nRT \ln[v^\circ M^{-1}_{svt} \rho_{svt}] \tag{6.46}$$

and

$$\Delta_s G_{CS}[IS^z_{n,d}] = RT \ln[n^{-1} v^\circ M^{-1}_{svt} \rho_{svt}] \quad , \tag{6.47}$$

where M_{svt} and ρ_{svt} are the molar mass and density of the solvent, respectively. Note that the dependence of these quantities on the arbitrary volume v° will be removed in the final $\Delta_s G[I^z_g]$ value, by cancellation with a corresponding dependence on v° in $n\Delta_s G[S_g]$ (monomer cycle) or $\Delta_s G[S_{n,g}]$ (cluster cycle). The free-energy changes involved in the gas-phase legs of the two cycles ($\Delta_s G_{MG}[IS^z_{n,g}]$ or $\Delta_s G_{CG}[IS^z_{n,g}]$) are evaluated using quantum mechanics[1484,1729,2279,2391–2393,2395–2398]. This involves a calculation of the binding energy, based on the minimum-energy cluster structure, along with an

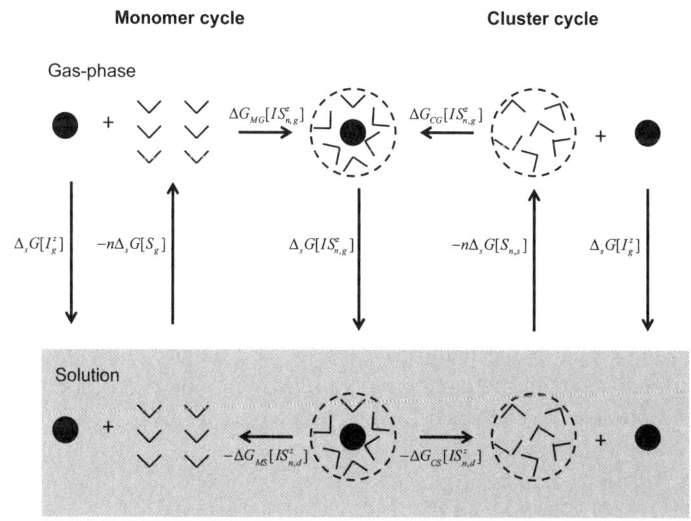

Figure 6.5 *Thermodynamic cycles involved in the quantum-mechanical calculation of single-ion solvation free energies using quasi-chemical theory.* The two cycles represent the monomer (left) and the cluster (right) cycles involved in this approach. The solvent is schematized by water molecules.

evaluation of the corresponding entropic contribution, typically based on a harmonic analysis of the vibrational frequencies within the cluster. The solvation free energy of the cluster-ion ($\Delta_s G[IS^z_{n,g}]$) is evaluated using either CE calculations[1484, 1729, 2279, 2391–2393, 2395–2398] or AT simulations[2391–2393]. The atomic partial charges required for this calculation are typically obtained from an electrostatic potential fit based on the QM results for the cluster-ion. Finally, the solvation free energy of the solvent molecule ($\Delta_s G[S_g]$; monomer cycle) and of the solvent cluster ($\Delta_s G[S_{n,g}]$, minimum-energy cluster structure based on quantum mechanics; cluster cycle) is calculated in a similar way.

This type of approach has been applied within the framework of density functional theory (DFT) to the $[K(H_2O)_4]^+$ (Ref.[2392]), $[H(H_2O)_2]^+$ (Refs.[2391,2393]), $[Li(H_2O)_4]^+$ (Ref.[2391]), $[Na(H_2O)_4]^+$ (Ref.[2391]) and $[OH(H_2O)_3]^-$ (Ref.[2391]) cluster-ions. The major advantage of the quasi-chemical approach is that it restricts the expensive QM component of the calculation to the interactions where this level of description is expected to be most relevant, *i.e.* to the interaction of the ion with its first solvation shell. However, this approach also involves a number of important shortcomings, concerning[1729,2391–2393]: (*i*) the dependence of the results on the number of water molecules n included in the cluster-ion (in principle, the final result of the calculation should not depend on this number); (*ii*) the (partial) inclusion or exclusion of the air-liquid interfacial potential contribution into the calculated solvation free energy[2391] (*i.e.* it is somewhat unclear whether an intrinsic or a real value is obtained); (*iii*) the compatibility of the results with a specified standard-state definition[1729]; (*iv*) the ability of the QM method and of the harmonic vibrational analysis to accurately capture the free energy of cluster-ion formation (considering in particular the problems summarized in Section 3.4); (*v*) the ability of the CE or AT method to accurately capture the solvation free energy of the cluster-ion (considering in particular the problems summarized in Sections 3.2 and 3.3.6, respectively).

6.3.2 Hybrid quantum-classical simulations

In the so-called *quantum-mechanical molecular-mechanical approach* (QM/MM)[649,650], the QM treatment is restricted to a small region of interest, typically the *ion alone* or including its *first or first few solvation shells*, the remainder of the system being treated classically, following a force-field based AT simulation scheme, denoted MM in this context. When the QM system includes first-shell solvent molecules, these can either be restrained to the QM region, or be allowed to freely diffuse between the QM and MM regions.

The QM/MM scheme can be applied within a molecular dynamics (MD) simulation scheme or a Monte Carlo (MC) sampling scheme. For example, within MD, the nuclear positions are propagated according to Newton's equations of motion using a conventional MD integrator. The dynamics of the nuclei in the QM subsystem relies on the Born-Oppenheimer approximation[487]

,[647]. The QM electronic energy and energy gradient relative to the nuclear positions are computed for a given nuclear configuration based on the selected QM method, taking into account the electrostatic influence of the MM subsystem. The resulting forces, along with contributions from internuclear repulsion and from interaction with the MM subsystem, are used to propagate the nuclear QM positions. Similarly, the MM particles are propagated according to forces originating from both the MM and the QM subsystems. The evaluation of the coupling contributions, *i.e.* the forces induced by the MM subsystem in the QM subsystem and conversely, can be quite complicated when covalent bonds are shared between the two subsystems[649]. Fortunately, this is not the case in the present context, where solvent molecules can be treated as integral entities, belonging to either of the two subsystems, but never split across them.

Like quasi-chemical theory (Section 6.3.1), the QM/MM approach has the advantage of restricting the expensive QM component of the calculation to the region of the system where this level of description is expected to be the most relevant, *i.e.* to the interaction of the ion with its first or first few solvation shells. However, the major difficulty in this approach is the coupling between the two subsystems. This treatment becomes in particular quite complicated when the QM subsystem includes solvent molecules and when these molecules are allowed to diffuse freely to and from the MM subsystem[103], even as integral entities (no split covalent bonds). Among a number of issues encountered in this case, one may mention that of the rigidity of the solvent molecules, which is generally enforced by constraints in the MM but not in the QM subsystem. Finally, because QM/MM simulations remain expensive, especially when a large number of water molecules are included in the QM subsystem and when relatively long simulation timescales are required, both conditions being relevant in the context of the evaluation of single-ion solvation properties, only relatively "cheap" QM methods can be employed in practice at present.

Numerous QM/MM studies of the *structure and dynamics* of ionic solvation[236,239,244,250,1478,1613–1671] have been reported. These calculations typically include the ion and at least its first solvation shell in the QM region. Recent years have also witnessed important progresses in QM/MM *free-energy calculation* methodologies [1373, 1673, 2388, 2389, 2399, 2400]. Fractional-electron approaches [1673] are now available, which permit to handle unphysical species with non-integer charges at the QM level (this approach has actually been used earlier in the framework of CPMD free-energy calculations[602,1672]; Section 6.3.3). The advantage of the fractional-electron method in the context of free-energy calculations is that the alchemical Hamiltonian of Eq. 3.55 appropriate for the charging process needs no longer be defined as a superposition (*e.g.* linear) of the charged and uncharged states, thereby requiring two electronic structure calculations per simulation step. In the context of thermodynamic integration[628] (TI) and provided that the derivative of the electronic energy with respect to the fractional charge can be calculated, a fractional-electron calculation only requires a single electronic structure calculation per step, resulting in a doubling of the computational efficiency. This approach has been applied to the computation of redox potentials[1673] (Section 6.3.5).

In spite of these promising methodological achievements, the authors of this book are only aware of two applications of QM/MM methods to the direct calculation of single-ion solvation free energies[1614, 1618] (the calculation of redox potentials and deprotonation constants can, however, be viewed as an indirect approach to the same problem; Section 6.3.5). In these two studies, the QM subsystem is defined by the ion alone and a two-state (linear) Hamiltonian coupling scheme is employed. In Ref.[1614], the hydration free energy of the chloride ion was calculated using AM1/MNDO QM models and a MC sampling procedure (cubic box containing 216 TIP3P water molecules, 0.9 nm straight cutoff for both ion-solvent and solvent-solvent interactions, $1.5 \cdot 10^6$ configurations for each of 10 λ-points). In Ref.[1618] the hydration free energies of a number of molecular ions were calculated using AM1, MNDO or PM3 QM models and a MD simulation procedure (cubic boxes containing $800 - 1000$ TIP3P water molecules, 0.9 nm straight cutoff for solvent-solvent interactions, 1.2 nm straight cutoff for ion-solvent interactions, slow growth forward and backward over 60 ps).

Although interesting from a historical perspective, the results of these studies are probably rather inaccurate due to the low level of the QM theory, approximate MM electrostatics, short sampling times and small system sizes. In addition, a QM treatment of the ion alone is not likely to bring a significant improvement over a classical treatment, considering that the major source of error in the AT description at this level resides in the approximate description of specific ion-solvent interactions and first-shell solvation properties.

6.3.3 Car-Parrinello molecular dynamics simulations

Car-Parrinello molecular dynamics[648, 1533] (CPMD) relies on equations of motion that combine the advantages of Born-Oppenheimer molecular dynamics, involving a classical propagation of the nuclei using a relatively large timestep, with those of Ehrenfest molecular dynamics, involving a simultaneous propagation of the electronic wave function requiring a single electronic structure calculation at each timestep. In this way, a full electronic structure optimization is only required for the initial configuration[487]. Although implemented in a density-functional theory (DFT) framework, CPMD is also sometimes called *ab initio* molecular dynamics. In contrast to the quasi-chemical theory and QM/MM approaches (Sections 6.3.1 and 6.3.2), the CPMD method attempts to treat the entire system at the QM level, thereby eliminating all issues related to the coupling between a QM and an AT (or CE) description.

Although extremely ingenious, the CPMD scheme nevertheless presents a number of shortcomings, namely: (*i*) the requirement of a still relatively small timestep (about $0.1 - 0.2$ fs *vs.* $1 - 2$ fs for classical AT simulations), due to the explicit propagation of electronic degrees of freedom; (*ii*) the limitation on system sizes (typically at most about 100 solvent molecules) and simulation times (typically a few tens of picoseconds), resulting from a relatively high computational cost (electronic structure propagation at each timestep, small timestep size); (*iii*) the requirement of balancing accuracy with efficiency *via* the magnitude of the fictitious masses and temperature of the electronic degrees of freedom, given the chosen simulation timestep (see above); (*iv*) the shortcomings of the DFT description itself (choice of exchange-correlation functional, approximate nature of exchange interactions, unsystematic error component) and of the pseudo-potentials (previously referred to as effective core potentials in Section 3.3.6) possibly employed to describe nuclei plus core electrons.

With respect to the third point, the use of smaller masses or/and of a higher temperature enhances the decoupling of nuclear and electronic motions, which leads to a dynamics staying closer to the Born-Oppenheimer surface, but requires smaller integration timesteps. In the opposite, larger masses or/and a lower temperature decrease the electronic velocities, which leads to a dynamics that may leave the Born-Oppenheimer surface, *i.e.* the energy corresponding to the given electron density is no longer a solution of the electronic Schrödinger equation for the given nuclear positions, but enables larger integration timesteps.

Numerous CPMD studies of the *structure and dynamics* of ionic solvation[141,808,810,811,1467,1575–1604,1604–1612] have been reported, see *e.g.* Table 6.5 for results concerning effective solvent dipole moment magnitudes in the first hydration shells of ions. Recent years have also witnessed important progresses in CPMD *free-energy calculation* methodologies[602,1571,1672]. Alchemical λ-path techniques[602,1672] are now available, which permit to change pseudo-potentials as well as the number of electrons. These approaches are relatively straightforward (and similar to classical thermodynamic-integration calculations) if the pseudo-potential replaces all the core electrons of the ion. However, in cases where the growth of an ion in the solvent is not only associated with a change in the ion pseudo-potential, but, in addition, with a change in the number of electrons, more complicated methods are required[602,1672].

Recently, hydration free energies of the Li^+, Cl^- and Ag^+ ions have been calculated using a thermodynamic integration (TI) scheme within the CPMD framework [1571,1572] (cubic computational boxes containing 32 or 64 water molecules, two or six λ-points, sampling of at least 30 – 40 ps per λ-point). Because the pseudo-potential of the Li^+ ion replaces all the core electrons (*i.e.* there are no explicit 1*s* electrons in these simulations), the TI procedure merely involves a charge scaling, as in the classical case. In contrast, the pseudo-potential of the Cl^- ion does not replace all the core electrons (*i.e.* there are still eight explicit 3*s* and 3*p* core electrons in these simulations), which requires a more complex procedure. First, a classical Cl^- ion is solvated in the QM water system. Second, a one-step free-energy perturbation is performed in order to evaluate the difference in solvation free energies between the classical and the quantum-mechanical Cl^- ions. This approach is justified under the assumption that the classical and quantum-mechanical Cl^- ions present similar hydration structures.

Although these calculations offer promising perspectives, the post-processing of the raw simulation results into corresponding estimates for intrinsic single-ion solvation free energies constitutes a serious factor of inaccuracy in these studies[1571,1572,2401], the issues being analogous to those encountered in AT simulations, and discussed in Section 6.2.1. First, for systems of the sizes considered, and due to the enforcement of strict periodic boundary conditions, the inclusion of correction terms for finite-size effects are absolutely essential (type-B correction term in Section 6.2.1). For example, Ref.[1571] includes approximate finite-size corrections according to Ref.[792]. Second, the computed hydration free energies have to be corrected for improper potential summation, *i.e.* for a spurious offset in the electric potential, considering that the LS-like scheme implicitly included in these calculations enforces a vanishing average electric potential over the computational box rather than a vanishing value at infinity (type-C_1 correction term in Section 6.2.1). In both the AT and QM approaches, the magnitude of the corresponding potential correction is sizeable. In the case of AT simulations employing rigid solvent models with a single van der Waals interaction site, this correction is straightforward to evaluate. However, it is not so trivial in the case of QM simulations. The authors of this book are aware of two approaches for the estimation of this correction term in CPMD simulations, one applied in the context of hydration free energies[1571,1572], and one applied in the context of redox free energies[1573].

In the first approach[1571,1572], a quantity $q_I \xi_{QM}$, ξ_{QM} being the quantum-mechanical analog of the classical exclusion potential ξ_{AT} discussed in Section 4.3.8, is subtracted from the simulation results. In the study under consideration[1571,1572], an estimate for ξ_{QM} is taken from Refs.[1571,1818], namely -3.85 V for a water density of 1000 kg·m^{-3} or -3.81 V for a water density of 990 kg·m^{-3}. Note that Refs.[1571,1572] do not explicitly subtract $q_I \xi_{QM}$, but rather define a quantity $\phi_q = -\xi_{QM}$, called there the "quadrupolar contribution to the surface potential", and add $q_I \phi_q$ to the simulation results, which is obviously equivalent. The evaluation of ξ_{QM} was also performed using CPMD simulations of pure water and involves a local Wannier function analysis[1450] of the simulated configurations to dissect the dipolar and quadrupolar contributions[1571,1818].

In the second approach[1573], the CPMD formalism is employed to evaluate the intrinsic hydration free energy of the proton $\mathcal{G}^{\ominus}_{H,wat}$, and the resulting (finite-size and standard-state corrected) value is compared to an experimental estimate for the same quantity, namely[1574] the one of Refs.[1875,1876]. The difference is identified with the correction term and amounts to 3.7 eV, suggesting a quantum-mechanical exclusion potential $\xi_{QM} = -3.7$ V.

Note, however, that the overall correction scheme employed in the hydration free energy calculations of Refs.[1571,1572] involves a number of other terms (finite-size, standard-state, cavitation and additional corrections), the nature of some of which is somewhat confusing. In particular, a correction based on the quadrupole-moment trace of the SPC/E water model is added prior to comparison with the experimental estimate for $\mathcal{G}^{\ominus}_{H,wat}$ of Ref.[457] and a correction based on the air-liquid interfacial potential of the SPC/E water model (P-integration) added prior to comparison with the corresponding estimate from Refs.[1875,1876], the latter being inappropriately considered as real (Section 5.18). Both correction terms involve a contribution of the quadrupole-moment trace $\mathcal{Q}_{AT}$ of the SPC/E water model which is, as discussed in Section 4.3.8 (see *e.g.* Figure 4.13), an essentially arbitrary quantity, in particular entirely unrelated to $\mathcal{Q}_{QM}$. This correction scheme is therefore rather questionable.

Taking the two approaches into consideration, and considering in addition the air-liquid interfacial potential evaluated in the CPMD simulation study of Ref.[1450] (Section 6.3.4), one is left with four estimates for ξ_{QM}, namely -3.85 and -3.81 V based on Refs.[1571,1818] (corresponding to water densities of 1000 kg·m^{-3} and 990 kg·m^{-3}, respectively), -3.62 V based on Ref.[1450] (corresponding to a water density of about 920 kg·m^{-3}) and -3.7 V based on Ref.[1573,1574] (as evaluated from a comparison between calculated and tentative experimental data for $\mathcal{G}^{\ominus}_{H,wat}$).

6.3.4 Calculation of the air-liquid interfacial potential

Car-Parrinello molecular dynamics (CPMD) simulations (Section 6.3.3) have also been used to estimate the *air-liquid interfacial potential* of water at the QM level[1448–1450]. These calculations involved the integration of the normal component of the electric field across the interface, the field itself being evaluated by integration of the charge density (nuclei and electrons) in the successive configurations of a CPMD trajectory. This procedure has been referred to in Section 3.3.5 as P-integration, and is expected to lead to an estimate for the internal surface potential $\check{\chi}_{wat}$ rather than the external potential χ_{wat} (Section 4.3.8). Note, however, that due to the use of pseudo-potentials to represent the nuclei and core electrons, this estimate will differ from a QM value explicitly including the contributions of all electrons *via* a term depending on the quadrupole-moment trace of the nucleus plus core electron system (see also Section 4.3.8 for the discussion of a possible contribution of the nuclei themselves suggesting a rather ill-defined nature of $\check{\chi}_{wat}$ at the QM level).

In a first study[1448,1449], a planar slab (confinement potential not specified) containing 216 water molecules was simulated for 7 ps, and a value of 3.1 V was obtained for the surface potential (pseudo-potential representation for the oxygen atoms). Although the simulation timescale may appear very short, analogous AT simulations using the SPC/E water model suggested that a 5 ps timescale is sufficient to reach an satisfactory level of convergence in this observable. In a second study[1450], 100 water configurations were extracted from an AT simulation of 1 ns duration considering a slab of 128 SPC/E water molecules (confinement potential not specified). These configurations were analyzed using density-functional theory (DFT) and the air-liquid interfacial potential evaluated as described above, resulting in a comparable estimate of 3.63 V (again using a pseudo-potential representation for the oxygen atoms). Note that the two studies involve systems at a slightly erroneous density.

In view of the extremely high value calculated for the surface potential, the authors of the second study [1450] considered the possibility of calculating the surface potential by integration of the surface polarization, *i.e.* the procedure referred to in Section 3.3.5 as M-integration, leading to an estimate for the external surface potential χ_{wat} (Section 4.3.8). This was undertaken based on a local Wannier function analysis [1450] of the simulated configurations, resulting in an estimate of +0.012 V for the external surface potential. In other words, by comparison to the internal estimate of 3.63 V, the exclusion potential of the QM solvent evaluated to $\xi_{QM} = -3.62$ V (based on a pseudo-potential representation for the oxygen atoms, and corresponding to a density of about 920 kg·m^{-3}). The simulations of Refs. [1571,1818] mentioned in Section 6.3.3 suggested comparable values of −3.85 and −3.81 V for water densities of 1000 and 990 kg·m^{-3}, respectively, and those of Refs. [1573,1574] an estimate of −3.7 V. The fact that this quantity is negative, whereas it was positive for classical solvent models (Section 4.3.8) is not surprising, considering that a classical solvent model has an excess of negative charge at its center (negative partial charge of the oxygen atom), whereas the QM charge density of a QM water molecule has an excess of positive charge at its center (nuclei or pseudo-nuclei). The positive estimate of about[1450] +0.01 V for χ_{wat} is also not entirely unreasonable considering the discussion of Section 5.17, and the estimate of about +0.13 V recommended by the authors of this book.

6.3.5 Calculation of deprotonation and redox parameters

Car-Parrinello molecular dynamics (CPMD) simulations (Section 6.3.3) have also been used to estimate *deprotonation constants* of organic and inorganic acids[1574,2402,2403], as well as absolute *redox potentials* of aqueous metal complexes with *e.g.* water[1573,2404,2405], bipyridine[1573] or tyrosinate[2406] ligands, or of aqueous molecules such as benzoquinone[2403]. These calculations rely on a thermodynamic integration (TI) scheme and involve (for the most recent studies) systems containing about[1573,2402] 30 or up to 60 water molecules[2402].

A prototypical *deprotonation reaction* in water (example of a neutral monoprotic acid) reads

$$I\mathrm{H}_{aq} \rightarrow I^-_{aq} + \mathrm{H}^+_{aq} \quad . \tag{6.48}$$

The corresponding standard free-energy change can be written

$$\begin{aligned} \Delta_r G^{\ominus}_{dep} &= \Delta_f G^{\ominus}[I^-_g] + \Delta_f G^{\ominus}[\mathrm{H}^+_g] - \Delta_f G^{\ominus}[I\mathrm{H}^+_g] \\ &\quad -\Delta_s G^{\ominus}[I\mathrm{H}^+_g] + \Delta_s G^{\bullet}[I^-_g] \quad . \end{aligned} \tag{6.49}$$

The first three quantities on the right-hand side are generally unproblematic and can be accessed by experiment (Sections 5.9 and 5.10) or *via* gas-phase QM calculations. The fourth quantity corresponds to the solvation of a neutral species. It is also generally unproblematic and can usually be accessed by experiment or a QM calculation. As a result, the calculation of $\Delta_s G^\ominus_{dep}$ amounts in principle to the evaluation of the conventional solvation free energy $\Delta_s G^\bullet[I_g^-]$.

A prototypical *redox reaction* (example of a neutral diatomic oxidant in the gas-phase being reduced to a monoanionic species), formulated relative to the redox reaction at the reference hydrogen electrode, reads

$$^1/_2 I_{2,g} + {}^1/_2 \mathrm{H}_{2,g} \to I_{aq}^- + \mathrm{H}_{aq}^+ \quad . \tag{6.50}$$

The corresponding absolute redox potential of the $(1/2)I_{2,g} - I_{aq}^-$ couple characterizes the half-cell (reduction) reaction

$$^1/_2 I_{2,g} + e_M^- \to I_{aq}^- \quad . \tag{6.51}$$

The corresponding standard free-energy changes can be written

$$\Delta_r G^\ominus_{rel} = \Delta_f G^\ominus[I_g^z] + \Delta_f G^\ominus[\mathrm{H}_g^+] + \Delta_s G^\bullet[I_g^z] \tag{6.52}$$

and

$$\Delta_r G^\ominus_{abs} = \Delta_f G^\ominus[I_g^z] + \Delta_s G^\ominus[I_g^z] \tag{6.53}$$

for the reactions given in Eqs. 6.50 and 6.51, respectively. Here again, the quantities $\Delta_f G^\ominus[I_g^z]$ and $\Delta_f G^\ominus[\mathrm{H}_g^+]$ are unproblematic and the calculations of $\Delta_r G^\ominus_{rel}$ and $\Delta_r G^\ominus_{abs}$ amount in principle to the evaluation of the conventional and intrinsic solvation free energies $\Delta_s G^\bullet[I_g^z]$ and $\Delta_s G^\ominus[I_g^z]$, respectively.

Absolute half-cell redox potentials can thus be compared to experimental estimates for intrinsic single-ion solvation free energies, but this requires an estimate for the intrinsic solvation free energy of the proton. These absolute potentials can also be directly compared to experimental estimates for the corresponding potentials, but this requires an estimate for the intrinsic absolute electrode potential of the reference hydrogen electrode. For instance, Ref.[1573] compares the absolute half-cell potential for $\mathrm{Ru(bpy)}_3^{2+/3+}$ with experimental data based on a value of 4.36 V for the reference hydrogen electrode. The calculated value for the absolute electrode potential is 5.75 V. The experimental value is $4.36 + 1.24 = 5.6$ V, where 1.24 V is the experimental relative electrode potential.

In contrast, calculated deprotonation constants or relative redox potentials (with respect to the proton or with respect to other couples) can be reliably compared with experimental data (*e.g.* pK_a values or relative electrode potentials), considering that conventional single-ion solvation free energies $\Delta_s G^\bullet$ are not elusive (Section 4.5.1). For instance, Ref.[2404] compares relative electrode potentials *e.g.* for the $\mathrm{Cu}^+/\mathrm{Cu}^{2+}$ and $\mathrm{Ag}^+/\mathrm{Ag}^{2+}$ couples with experimental data. For the redox reaction involving copper oxidation and silver reduction, the calculated value is -1.7 V, to be compared with the experimental value of -1.83 V. As an example of pK_a calculation, Ref.[2402]

evaluates pK_a values for a series of organic and inorganic molecules, resulting in *e.g.* a calculated value of −6.7 for HCl, to be compared with the experimental value of about −7, and a calculated value of −5.0 for H_3O^+, to be compared with the experimental value of about −1.74. Note that the latter pK_a corresponds to the reaction

$$I\mathrm{H}_{aq} + \mathrm{H_2O}_{svt} \rightarrow I^-_{aq} + \mathrm{H_3O}^+_{aq} \quad , \tag{6.54}$$

where, in the example above, both half reactions (deprotonation of the acid, protonation of water) are evaluated with the QM approach.

In view of the small system sizes considered, the calculated raw free energies have to be corrected for finite-size effects according to Ref.[329] (analogous to the finite-size correction of Ref.[1081, 1412]; Section 6.2.1). Alternatively, extrapolation procedures[1573] (solvent reorganization free energy or oxidation free energy as a function of the inverse or inverse square root of the box edge), based on classical simulations in different box sizes, have also been employed to estimate this correction. In addition, the computed redox free energies have to be corrected for a spurious offset in the electric potential, considering that the LS-like scheme implicitly included in these calculations enforces a vanishing average electric potential over the computational box rather than a vanishing value at infinity. Both in AT and QM approaches, the magnitude of the corresponding potential correction is sizeable. In the case of classical simulations employing rigid solvent models with a single van der Waals interaction site, this correction is straightforward to evaluate. However, as discussed in Section 6.3.3, it is not so trivial in the case of QM simulations.

Note that redox potentials and acidity constants are also frequently computed using QM/MM methods[1673, 2388–2390]. Since QM/MM methods permit the treatment of much larger systems in comparison to CPMD, such studies often focus on redox reactions in biomolecules[1673, 2388, 2389] or bio-organic molecules[2390]. However, aqueous aquo-metal complexes have also been considered[1673, 2390].

After finishing this chapter, a single ion formulates its theory of "complication relativity", which states that even in the apparently simple context of single-ion solvation, the theory is "relatively complicated". Equipped with this knowledge, it is nevertheless going to pursue theoretical investigations because, when carried out with awareness of their shortcomings, they provide great insight into the molecular mechanism associated with the solvation process.

Chapter 7
Conclusion

Arrived at the term of this long journey through the mysteries of single-ion solvation, the authors would like to provide a few *concluding remarks*, as well as a *perspective* on still unsettled issues and on progresses expected to result from on-going research in the area.

The problem of the experimental or theoretical determination of *intrinsic single-ion solvation free energies*, as well as of corresponding *derivative thermodynamic solvation parameters*, has represented a challenge to the physical chemistry community for more than one century.

From the *experimental point of view*, the challenging aspect of the problem really results from the combination of the two words *intrinsic* and *single-ion*. The former word implies that the quantities that are sought after should exclusively characterize bulk solvation effects, without any contamination by surface properties. In contrast, the determination of corresponding *real* single-ion solvation parameters encompassing the two types of effects simultaneously, although not easy in practice, does not represent a fundamental challenge, and is possible by means of appropriate spectroscopic and electrochemical measurements. The latter word implies that these quantities concern individual ions. In contrast, the determination of corresponding *sums* over overall neutral sets of ions (salts) does not represent a fundamental challenge, and is possible by application of thermodynamic cycles relying on an appropriate combination of calorimetric, electrochemical and spectroscopic measurements. This simplification results from the fact that bulk solvation is approximately quadratic in the ion charge, while surface effects are rigorously linear in this charge. As a result, they cancel out in the sums, for which a distinction between real and intrinsic quantities is no longer required.

Reformulated in slightly different terms, the experimental problem amounts to determining *conventional* (relative, *e.g.* taking the proton as a reference ion) or *real* (*i.e.* absolute, but including a contribution from surface prop-

RSC Theoretical and Computational Chemistry Series No. 3
Single-Ion Solvation
Experimental and Theoretical Approaches to Elusive Thermodynamic Quantities
By Philippe Hünenberger and Maria Reif

Published by the Royal Society of Chemistry, www.rsc.org

erties) single-ion solvation free energies, which are accessible experimentally, along with either of three other quantities, namely: the *air-liquid interfacial potential* of the pure solvent $\chi^{\ominus}_{svt}$, the *intrinsic absolute potential of the reference hydrogen electrode* in this solvent $\mathcal{V}^{\ominus}_{H,svt}$, or the *intrinsic solvation free energy of the proton* in this solvent $\mathcal{G}^{\ominus}_{H,svt}$. Due to the constraints imposed by the *electroneutrality* of macroscopic matter at equilibrium and by the unavoidable presence of *polarization effects* at interfaces, neither of these three intimately related quantities can be accessed experimentally without the help of some *extra-thermodynamic assumption, i.e.* an apparently reasonable but not strictly provable postulate or model concerning the physics of the bulk solvation process, the physics of surface properties, or the expected outcome of an experiment in some hypothetical limiting situation. Due to this dependence on an unprovable assumption, the three above quantities have been termed here *experimentally elusive.* Unfortunately, considering water at atmospheric pressure and room temperature, the use of different experimental approaches along with distinct (apparently equally reasonable) assumptions leads to a very large uncertainty range on the order of $0.5-1.0$ V (potentials) or $50-100$ kJ·mol^{-1} (free energy) in the estimated values for these quantities. The issue of elusiveness also appears to extend to all derivative thermodynamic solvation parameters (see further below), which are connected to the pressure or/and temperature derivatives of the three above elusive quantities.

One might wonder whether elusiveness really results from *a fundamental impossibility* of determining these quantities on the sole basis of experimental measurements, or rather from a very high level of *technical difficulty* that has prevented the achievement of an accurate determination of this kind until now. At first sight, the unavoidable involvement of the *hypothetical physical solvation process* (Figure 2.2), which intermingles bulk and surface effects, within any experimental measurement concerning single-ion solvation seems to imply that a purely experimental determination of elusive quantities is fundamentally impossible. However, one could still tentatively envision the future invention of new procedures *bypassing this restriction.* For example, one could imagine an experiment probing the transfer of a single ion from air into a liquid sample under conditions where the surface potential of the liquid is made to vanish (see Point E_{13} in Section 5.17 for a related principle). One could as well imagine an experimental technique probing the average charge (nuclei and electron) density at the surface of a liquid sample with such a precision and reaching-depth that the surface potential could be directly calculated based on the results of this measurement.

A key question is of course whether such experiments will ever become feasible in practice. However, an even more relevant question is to which extent these experiments, assuming they become feasible, will still have to rely on unprovable assumptions. For example, in the first hypothetical experiment, one still has to ensure that the conditions have been realized for the surface potential of the liquid to vanish, and to show that this realization did not induce a strong perturbation of other system properties. In fact, experiments on the formation of gas-phase ion-solvent clusters (*e.g.* cluster-pair solvation and correlation methods, see Point E_{14} in Section 5.18) are probably close to realizing these conditions in the context of microscopic ion-water clusters. But there is no way to prove that the surface potential of the clusters really vanishes, assuming that the concept of surface potential remains meaningful when considering microscopic systems, and to examine the progressive transition of this surface potential to the macroscopic regime, from the assumed zero value for microscopic clusters to the bulk value for macroscopic samples. As a result, the postulate of zero surface potential for microscopic clusters and the associated extrapolation procedure to the bulk regime must really be viewed as extra-thermodynamic assumptions. Similarly, in the second hypothetical experiment, one still has to ensure that the connection between charge density and surface potential is correct, and to show that the measurement has not perturbed the surface properties. The former point is particularly crucial, considering the ambiguous nature of the concept of surface potential at the microscopic level (*e.g.* internal *vs.* external definitions; Section 4.3.8). In fact, the connection between surface charge density and surface potential also involves a form of extra-thermodynamic assumption, which can actually be viewed as an arbitrary definition (see further below).

Based on this type of questioning, the authors of this book tend to think that the experimentally-elusive nature of the quantities $\chi^{\ominus}_{svt}$, $\mathcal{V}^{\ominus}_{H,svt}$ and $\mathcal{G}^{\ominus}_{H,svt}$ is not a mere consequence of a high level of technical difficulty involved in their measurement, although this difficulty certainly contributes to the challenge, but rather from a more fundamental problem: *bulk and surface effects* are simply *not unambiguously partitionable.* Although it is in the human nature to believe that the most recent or/and sophisticated determination of an elusive quantity is exact and definitive, *i.e.* void of any unprovable postulate, careful analysis always reveals the involvement of some extra-thermodynamic assumption. However, if one accepts that the non-partitionability of bulk and surface effects is a *fundamental restriction* imposed by *the laws of physics*, rather than a consequence of our inability to perform the required experimental measurement, the concept of extra-thermodynamic assumption takes a different dimension, namely that of an *arbitrary definition* for this partitioning. This interpretation will be developed further below, after the consideration of theoretical approaches.

From the *theoretical point of view*, three approaches have been employed to evaluate intrinsic single-ion solvation free energies : *continuum-electrostatics calculations*, classical *atomistic simulations* and *quantum-mechanical computations* (calculations and simulations). The main features of these approaches can be summarized as follows.

Continuum-electrostatics calculations, based on the Born model and modified versions thereof, provide excellent *qualitative predictions* regarding intrinsic single-ion solvation free energies, within the limits set by a comparison against elusive experimental values, and permit an indispensable *first-order reasoning* concerning ionic solvation properties. However, they do not account for the *granularity of matter*, *i.e.* the microscopic structure of the solvent and the specific details of ion-solvent interactions (including in particular asymmetric solvation effects differentiating cations from anions at identical sizes), and depend on the *ill-defined concept* of *effective ionic radius*. Therefore, they fail, even after application of various corrections, to provide quantitative results, and even qualitative results in terms of derivative thermodynamic solvation properties.

Classical atomistic simulations should in principle be *more accurate*, because they account for the *granularity of matter*, owing to an explicit representation of the solvent molecules at atomistic resolution. These methods have been applied both to the direct calculation of intrinsic single-ion solvation free energies and to the evaluation of the air-liquid interfacial potential of pure solvents. In practice, however, *methodological issues* regarding the choice of boundary conditions, the approximate treatment of electrostatic interactions and the evaluation of electrostatic potentials from the sampled configurations have prevented the obtension of consistent results in both cases up to recently. Although these problems are *essentially solved* nowadays, at least in the opinion of the authors (see the correction scheme of Section 6.2.1), the outcome of these calculations is still affected by three major sources of error: the mean-field treatment of *electronic polarizability* in most calculations, the approximate representation of *van der Waals interactions* (functional form and combination rules), and the dependence of the results on the *ion-solvent van der Waals interaction parameters*, the latter representing the atomistic analog of an effective ionic radius.

Quantum-mechanical computations have also been applied to both the direct calculation of single-ion solvation free energies and the evaluation of the air-liquid interfacial potential of pure solvents. They bear the *promise* of *definitely settling the problem*, *i.e.* of permitting the calculation of $\chi^{\ominus}_{svt}$, $\mathcal{V}^{\ominus}_{H,svt}$ and $\mathcal{G}^{\ominus}_{H,svt}$ in an assumption-free fashion. This is because they are based on the *most accurate physical representation* of nature available at present and do not rely on the specification of ambiguous parameters related to *effective ion size*, such as the effective ionic radius of continuum-electrostatics calculations or the ion-solvent van der Waals interaction parameters of classical atomistic simulations. The question whether quantum-mechanical approaches are, even in principle, able to keep this promise will be addressed further below. In any case, in practice, the exact first-principles calculation of single-ion solvation free energies or air-liquid interfacial potentials is still far beyond reach. The *high computational cost* of quantum-mechanical calculations induces *severe restrictions* concerning the system size, configurational sampling, basis-set size and treatment of electron correlation, resulting in potentially very large errors that are often difficult to characterize and quantify. It may be reduced by interfacing a quantum-mechanical treatment of the ion and its first solvation shells with a classical atomistic or/and continuum-electrostatics approach for the surrounding bulk water. However, even in this case, the computation of single-ion solvation free energies represents a considerable challenge, and the necessary approximations may still for a while compromise the gain in accuracy expected from a more fundamental treatment of the problem.

Summarized in broad terms, it appears that continuum electrostatics has taught us (already in the last century) what it could, and that quantum mechanics has to tell us (probably still in this century) what it can. In the mean time, *great progresses* have been accomplished in the context of *atomistic simulations* over the last three decades. These progresses include: (*i*) the improvement of free-energy calculation methods, *e.g.* thermodynamic integration and variants thereof replacing less accurate schemes such as slow-growth for calculating charging free energies, and soft-core scaling for the calculation of cavitation free energies (Section 3.3.3); (*ii*) the increased sampling enabled by faster computers, reducing sampling errors to a negligible magnitude in the context of the calculation of single-ion solvation free energies and air-liquid interfacial potentials (Section 3.5); (*iii*) the improved understanding of finite-size effects, surface effects and potential summation artifacts (Sections 3.3.3-3.3.5 and 4.3.8), permitting the formulation of an appropriate correction scheme for the raw simulation results (Section 6.2.1); (*iv*) the improved understanding of the fundamental experimental, theoretical and conceptual problems related to single-ion solvation thermodynamics (this book).

As a result, atomistic simulations can at present be applied to calculate converged and corrected single-ion solvation free energies *exempt of any significant methodological dependence* on the approximate electrostatic scheme, boundary conditions and potential summation method applied in the simulation (Section 6.2.1). At a given thermodynamic state point, these corrected single-ion solvation free energies are then uniquely and reproducibly determined by the underlying molecular model (force field), *i.e.* by the choice of a *solvent model* and of a set of *ion-solvent van der Waals interaction parameters* (Section 6.2.2). Classical atomistic simulations have thereby reached the same level as continuum-electrostatics calculations in terms of *consistency*, *i.e.* an unambiguous connection between the solvation free energy of an ion and parameters determining its effective size. However, in contrast to continuum-electrostatics calculations, atomistic simulations account for the microscopic structure of the solvent. Furthermore, whereas the effective ion-size parameter of continuum-electrostatics calculations (effective ionic radius) can solely be connected (at best) with crystallographic parameters of the salts, the effective ion-size parameters of atomistic simulations (ion-solvent van der Waals interaction coefficients) can be connected to a wealth of other properties. For this reason, there is a chance that observables that are experimentally unambiguous and can be validly compared with the outcome of atomistic simulations, *i.e.* for which such a type of classical modeling is appropriate, contain enough information to permit the definitive validation of a single model (or, more likely, of a number of alternative models) compatible with unique values for $\chi^{\ominus}_{svt}$, $\mathcal{V}^{\ominus}_{H,svt}$ and $\mathcal{G}^{\ominus}_{H,svt}$. This assumption has been termed here the *atomistic-consistency assumption* (ACA; Section 6.2.3). The application of the ACA in the context of atomistic simulations bears a similar type of *promise* as that apparently made by quantum-mechanical computations (see above), namely that of definitely *settling the problem*, *i.e.* of permitting the calculation of the three above experimentally-elusive quantities in an assumption-free fashion. Whether both types of approaches are, even in principle, able to keep this promise is the concern of the following paragraphs.

As mentioned above in the context of experimental approaches, the point of view of the authors of this book is that the elusive nature of $\chi^{\ominus}_{svt}$, $\mathcal{V}^{\ominus}_{H,svt}$ and $\mathcal{G}^{\ominus}_{H,svt}$ reflects a fundamental problem, namely that *bulk and surface effects* are simply *not unambiguously partitionable.* Adopting this line of thinking, one is bound to conclude that even an exact first-principles quantum-mechanical computation or a fully consistent application of the ACA in the context of atomistic simulations *cannot deliver assumption-free estimates* for these quantities, *i.e.* there must still be some extra-thermodynamic assumption involved in the interpretation of the corresponding results. This assumption resides in fact precisely in an *arbitrary decision* made concerning the partitioning of bulk and surface effects, as most obvious from the discussion concerning the air-liquid interfacial potential of the pure solvent (Section 4.3.8).

At the classical level, one has to decide whether the electric potential in the liquid phase is the one felt by a test charge *without or with excluded volume*, or, equivalently, whether *atoms or molecules* are viewed as the source entities for this potential. Similarly, at the quantum-mechanical level, one has to decide whether the potential is calculated by averaging over the *entire space* or solely over *regions of low electron density* or, equivalently, whether the *charge density itself* or a charge density *partitioned among neutral solvent molecules* is viewed as the source for this potential. The former choices lead to an *internal* definition and the latter choices to an *external* definition for the surface potential. The external definition is problematic because it seems to rely on other choices concerning the definition of a *test charge size* or *solvent molecular envelope* (classical case), and of a *charge density threshold* or *partitioning scheme* (quantum-mechanical case). However, given "reasonable" choices in this context (see further below), it leads to results that are "intuitive" and in principle compatible when comparing the classical and quantum-mechanical situations, as well as both of these with the results of continuum-electrostatics calculations. The internal definition is also problematic because it leads to a surface potential encompassing information on the *internal structure of the solvent molecule*, that is either *entirely inaccurate* (classical case) or *essentially inaccessible* (quantum-mechanical case, considering the contributions of nuclei, quarks and possible smaller particles), and in any case expected to be entirely irrelevant in terms of its influence on ionic solvation. It leads to results that are "counter-intuitive" and inconsistent when comparing the classical and quantum-mechanical situations, as well as both of these with the results of continuum-electrostatics calculations. Note that there is in principle a continuous spectrum of possible decisions concerning this partitioning, of which the internal and external ones are just the extreme cases. The fundamental reason for this ambiguity is that the laws of physics do not distinguish between bulk and surface regions of a phase. This distinction is actually only *in the eye of the beholder*, and if a borderline is to be artificially drawn between the two regions, its exact location must be specified as well by the beholder. Finally, it is worth observing in this context that the *continuum-electrostatics* description implicitly sustains an *external* definition. This is because Maxwell's equations of electrostatics in continuous media exclude a possible average potential contribution, *via* the quadrupole-moment trace of the constituting species, from the elementary dipoles defining a dielectric medium.

Accepting that the partitioning of bulk and surface effects rests on an arbitrary decision one is bound to conclude that an *extra-thermodynamic assumption* is actually nothing else but a *definition*. The merit of theoretical approaches in this context is not to permit the assumption-free evaluation of experimentally-elusive quantities, which they cannot do, but rather to permit the formulation of this definition in the most fundamental possible context. This formulation may thus be referred to as an *ultimate extra-thermodynamic assumption*. In continuum-electrostatics calculations, this ultimate assumption is implicitly built into the model *via* Maxwell's equations of electrostatics in continuous media, and corresponds to an external potential definition (see above). In classical atomistic simulations, it must be made explicitly in the form of the acceptance of a given summation scheme, *e.g.* the M-scheme also corresponding to an external potential definition. Finally, in quantum-mechanical computations, it must also be made explicitly in the form of an electron density threshold for averaging, *e.g.* extrapolation to infinitesimal density (Eq. 4.338) also corresponding to an external potential definition.

Given the acceptance of such a definition, intrinsic single-ion solvation free energies, intrinsic absolute electrode potentials and air-liquid interfacial potentials are viewed as *theoretical objects* that can in principle be calculated, but remain bound to the adopted definition, *i.e.* to a certain *representation of reality*. They can also be measured experimentally given an experiment-related extra-thermodynamic assumption that is compatible with the selected definition. In principle, there is no restriction concerning the choice of such a definition. However, some definitions will lead to quantities amenable to *intuitive reasoning*, while others will not. In the present situation, our intuitive reasoning is so much rooted in the Born continuum-electrostatics solvation model that the external potential definition appears to be the only meaningful choice. It is also the one compatible with the spirit of most of the experiment-related extra-thermodynamic assumptions employed to date, and the recommended data provided in Sections 5.19 and 5.20 concerning intrinsic quantities is to be viewed as the one compatible with this choice. To make this definition explicit, one could state it in the following way (Section 4.3.8 and Figure 4.11):

The partitioning of bulk and surface effects must be done in such a way that an idealized solvent consisting of identical hard-sphere molecules bearing arbitrary isotropic and electroneutral charge distributions confined to their vol-

umes presents a vanishing surface potential and a vanishing contribution to the solvation free energy of a hard-sphere ion.

The definition is of course partial in the sense that it only provides a limiting case. In practical situations, it will be necessary to design an effective partitioning scheme satisfying this limit. The detailed specification of this scheme, in the form of *e.g.* a solvent molecular envelope and test-charge size in atomistic simulations or an electron-density threshold in quantum-mechanical computations, will then complete the definition.

The relegation of intrinsic solvation parameters, intrinsic absolute electrode potentials and air-liquid interfacial potentials to the rank of *definition-bound properties* rather than *physical observables* is of course unpleasant. However, it should be recalled that a large part of our chemical reasoning is based on such properties. For example, molecular orbitals, natural bonding orbitals, effective atomic charges, electronegativities, and crystallographic radii are extremely useful definition-based chemical concepts. But maybe the most obvious example is provided by Lewis structures, relying on the intuitive assignment of electrons to localized bonding and non-bonding orbitals, and without which any reasoning in chemistry would hardly be possible. In addition, in contrast to the above definition-bound concepts, intrinsic single-ion solvation parameters are not definition-bound in terms of their sums over neutral ion sets, which are not elusive. In other words, if not directly measurable themselves, their sums are unambiguously defined in the framework of thermodynamics.

Closing this discussion concerning the ultimate nature of extra-thermodynamic assumptions, the authors would like to come back to an issue that was raised in Section 2.3 and left essentially open throughout the book, because of its relatively intricate nature. As stated above, the quantities $\chi^{\ominus}_{svt}$, $\mathcal{V}^{\ominus}_{H,svt}$ and $\mathcal{G}^{\ominus}_{H,svt}$ are *experimentally elusive*. Because they are interconnected *via* experimentally-accessible parameters, knowledge of *one of them* automatically *determines the others*. The elusiveness therefore affects a single parameter, say $\mathcal{G}^{\ominus}_{H,svt}$. However, this parameter is relevant for *one type of property*, namely the free energy, and *one specific solvent*, typically water. The questions arise then whether: (*i*) the *derivatives* of $\mathcal{G}^{\ominus}_{H,svt}$ with respect to *pressure or/and temperature*, *i.e.* the intrinsic thermodynamic solvation properties of the proton, are also elusive; (*ii*) the *difference* of $\mathcal{G}^{\ominus}_{H,svt}$ between two *different solvents* are also elusive. Elements of answers to these two questions are provided in the following paragraphs.

Considering the first question, the unavoidable mixing of bulk and surface effects within any experimental measurement concerning single-ion solvation, which applies to both the quantities and their derivatives, seems to imply a positive answer, *i.e.* that thermodynamic derivatives of elusive quantities are also elusive. However, a number of experimental approaches have been claimed to measure the derivative of the surface potential of a pure liquid with respect to temperature (see Point B_{13} in Section 5.17). Although the use of these derivatives to estimate an absolute surface potential always clearly involves some extra-thermodynamic assumption, it remains unclear whether the determination of the derivatives themselves also requires such an assumption. If not, intrinsic derivative solvation properties should actually not be considered as elusive. Based on their current knowledge, the authors of this book tend to think that derivative thermodynamic properties are also elusive quantities. A simple argument supporting this opinion in the context of temperature derivatives is that if $-\partial_T \chi^{\ominus}_{svt}$ and $-T^2\partial_T(T^{-1}\chi^{\ominus}_{svt})$ were both experimentally accessible, then $\Delta_s S^{\ominus}_{srf}$ and $\Delta_s H^{\ominus}_{srf}$ would also be measurable for a given ion charge z (Eq. 2.12). Consequently, $\Delta_s G^{\ominus}_{srf}$ would be known *via* the Gibbs equation and thus, $\chi^{\ominus}_{svt}$ would also be accessible (Eq. 2.6). This suggests that if $\chi^{\ominus}_{svt}$ is experimentally elusive, at least its temperature derivatives must also be elusive.

Considering the second question, the observation that the transfer of an ion between the bulk of two solvents unavoidably involves the crossing of two air-liquid interfaces (or, possibly, of a single liquid-liquid interface for immiscible solvents) seems to imply a positive answer, *i.e.* that differences of elusive quantities between two solvents are also elusive. However, a number of experimental methods have been claimed to measure the difference in air-liquid interfacial potentials between two solvents (see Point D_{13} in Section 5.17 and comment following the corresponding literal enumeration), where the involvement of an extra-thermodynamic assumption is not obvious. If no assumption was involved, the interfacial potential difference between two solvents would be experimentally-measurable, and the elusive quantities would only be those of a single reference solvent (*e.g.* water). In the opposite case, each solvent would be characterized by its own elusive quantities. Based on their current knowledge, the authors of this book tend to think that differences of elusive quantities between two solvents are also elusive.

Considering the content of the present book, the field of single-ion thermodynamics may appear rather complex. Part of this complexity is inherent to the problem and thus unavoidable. However, just as in any other scientific area, unnecessary additional complexity is caused by a number of field-specific *plagues*, *i.e.* continuous sources of complications, confusions, ambiguities and mistakes. In the opinion of the authors, the three plagues of single-ion thermodynamics are: (*i*) the *incomplete or ambiguous definitions* commonly adopted for some *central concepts*; (*ii*) the existence of numerous *redundant parameters* encompassing the *same physical information* in a different format; (*iii*) the *plethora* of alternative *standard-state definitions*.

Clarifying *concepts and definitions* was the main purpose of Chapter 4, which should provide a solid framework of definitions, relationships and notations for future investigations in the field.

By providing connections between physically-related quantities (see *e.g.* Figures 4.1, 4.2 and 4.18), this chapter also outlines the structure of a *minimal* (non-redundant) *set of parameters* to characterize the properties of a specific set of ions, with the consideration of a given solvent (here, water) in the context of solvation properties. Such a minimal set is provided in Chapter

5 (Section 5.19) in the context of recommended data for the alkali and halide ions with reference to water as a solvent. The analysis of the literature in the field would be greatly simplified by the adoption of a standard minimal set for the tabulation of these quantities. To give a simple example, the intrinsic (or real) solvation free energy of an ion in a given solvent is conveniently partitioned into an intrinsic (or real) proton solvation parameter, common to all ions for this solvent, but elusive (or difficult to measure accurately), and a conventional value, specific to this ion. However, in the literature, the conventional value may be encoded in the form of a redox potential, of a real value, of a tentative intrinsic value, or of a set of pairwise values (with other ions). In the three latter cases, the data may further be reported in the form of solvation, of dissolution, or of formation parameters in solution. Similarly, the real proton value may be encoded in the form of an absolute electrode potential or of a metal-solution Volta potential difference, and the intrinsic proton value in the form of an air-liquid interfacial potential or of an absolute electrode potential. In terms of solvation physics, the information content of all these combinations is exactly the same, but this connection is extremely cumbersome to establish on a case-to-case basis, as was done *e.g.* in Chapter 5 (Sections 5.12, 5.14, 5.15, 5.16, 5.17 and 5.18) of this book.

Finally, the plethora of *alternative standard-state definitions* is the third major plague of ion thermodynamics. Along the pages of this book, the existence of alternative definitions has been encountered at six levels: (*i*) choice of a reference pressure and of a reference solution concentration (Sections 4.2.4 and 4.2.17); (*ii*) choice of a standard or density-corrected solute standard-state variant (Sections 4.2.5 and 4.2.16); (*iii*) choice of a warm- or cold-electron convention for the standard-state ideal electron gas (Sections 4.2.4, 4.2.13, 4.2.17 and 4.3.5); (*iv*) choice of Boltzmann or Fermi-Dirac statistics for calculating the properties of the standard-state electron and proton (Section 4.2.13); (*v*) choice of a reference electric potential for the ideal electron gas in the definition of absolute electrode potentials (Section 4.3.6); (*vi*) choice of a specific anchoring point for the conventional scale of single-ion solvation parameters (Section 4.5.1).

The numerous possible combinations of these six choices, and their incomplete or ambiguous specification in many literature articles, represents a continuous source of mistakes in the field. Clearly, the analysis of the literature would also be greatly simplified by the adoption of a unique standard-state convention. The data compilation performed in Chapter 5 (Sections 5.12, 5.14, 5.15, 5.16, 5.17 and 5.18) involved many different types of standard-state conversions. It resulted in recommended data reported in a uniform convention, namely the $bbme_T$ standard (Section 4.2.4), including a warm-electron convention with Fermi-Dirac statistics for the standard-state electron and proton. This convention is the one recommended by the authors of this book for future investigations.

Two additional points are worth mentioning in this respect. If the $bbme_T$ standard convention is recommended for the *tabulation* of standard thermodynamic properties, it should be recalled that the physics of the solvation process is most directly characterized by the *semi-standard point-to-point solvation parameters* (Sections 2.2, 4.2.16, 5.7 and 5.20), which are exempt of standard-state bias, except for the specification of a reference pressure and temperature for the solvent properties. Second, the present book has revealed the presence of an ambiguity in the standard-state definition for (charged as well as neutral) solute species, namely the existence of *standard* and *density-corrected* variants (Section 4.2.5). The natural variant in which a standard quantity is determined is related to the type of measurement performed. However, this issue is not documented in thermodynamics, and the variant is at present never specified in the literature along with the report of standard quantities.

In spite (or maybe because of) its complexity, behind a superficial appearance of simplicity, the field of single-ion thermodynamics remains a thrilling area of research that certainly still also has a long future ahead. As a compact yet extensive account of this field, it is the hope of the authors that this book has provided the reader with a useful and solid basis for further investigations. For there is one truth that escapes any possible elusiveness: ions have always been, are, and will for ever remain fascinating chemical species.

A single ion has made it to the end of this book. Having first naively enjoyed the simple pleasure of lonely solvation and then suffered the difficult climb to the summit of its complexity, it can now relax and enjoy the vision of a lake ahead. It is ready for the next swim, which will for sure feel a bit different.

REFERENCES

[1] Albarède, F. The geochemical behaviour of selected elements. In: *Geochemistry. An Introduction.* Cambridge Univ. Press, Cambridge, UK, pp 191-206 (2003)

[2] Klein, C. & Dutrow, B. Crystal chemistry and systematic descriptions of native elements, sulfides, and sulfosalts. In: *Manual of mineral science.* John Wiley & Sons, Inc., New York, USA, pp 331-367 (2007)

[3] Klein, C. & Dutrow, B. Crystal chemistry and systematic descriptions of oxides, hydroxides, and halides. In: *Manual of mineral science.* John Wiley & Sons, Inc., New York, USA, pp 368-398 (2007)

[4] Klein, C. & Dutrow, B. Crystal chemistry and systematic descriptions of carbonates, nitrates, borates, sulfates, chromates, tungstates, molybdates, phosphates, arsenates, and vanadates. In: *Manual of mineral science.* John Wiley & Sons, Inc., New York, USA, pp 399-433 (2007)

[5] Kohnlein, W. A model of the terrestrial ionosphere in the altitude interval 50-4000 km. 1. Atomic ions (H^+, He^+, N^+, O^+). *Earth Moon Planets* **45**, 53-100 (1989)

[6] Kohnlein, W. A model of the terrestrial ionosphere in the altitude interval 50-4000 km. 2. Molecular ions (N_2^+, NO^+, O_2^+) and electron density. *Earth Moon Planets* **47**, 109-163 (1989)

[7] Smith, D. & Spanel, P. Ions in the terrestrial atmosphere and in interstellar clouds. *Mass Spectrom. Rev.* **14**, 255-278 (1995)

[8] Harrison, R.G. & Tammet, H. Ions in the terrestrial atmosphere and other solar system atmospheres. *Space Sci. Rev.* **137**, 107-118 (2008)

[9] Smith, D. The ion chemistry of interstellar clouds. *Chem. Rev.* **92**, 1473-1485 (1992)

[10] Bochsler, P. Minor ions in the solar wind. *Astron. Astrophys. Rev.* **14**, 1-40 (2007)

[11] Petrie, S. & Bohme, D.K. Ions in space. *Mass Spectrom. Rev.* **26**, 258-280 (2007)

[12] Takahashi, T. Water electrolysis. In: *Solar-hydrogen energy systems.* Ohta, T., Ed. Pergamon Press, New York, USA, pp 35-58 (1979)

[13] Sun, P., Laforge, F.O. & Mirkin, M.V. Scanning electrochemical microscopy in the 21st century. *Phys. Chem. Chem. Phys.* **9**, 802-823 (2007)

[14] Moussallem, I., Jorissen, J., Kunz, U., Pinnow, S. & Turek, T. Chlor-alkali electrolysis with oxygen depolarized cathodes: History, present status and future prospects. *J. Appl. Electrochem.* **38**, 1177-1194 (2008)

[15] Beck, F. & Ruetschi, P. Rechargeable batteries with aqueous electrolytes. *Electrochim. Acta* **45**, 2467-2482 (2000)

[16] Brodd, R.J., Bullock, K.R., Leising, R.A., Middaugh, R.L., Miller, J.R. & Takeuchi, E. Batteries, 1977 to 2002. *J. Electrochem. Soc.* **151**, K1-K11 (2004)

[17] Patil, A., Patil, V., Shin, D.W., Choi, J.W., Paik, D.S. & Yoon, S.J. Issues and challenges facing rechargeable thin film lithium batteries. *Materials Res. Bull.* **43**, 1913-1942 (2008)

[18] Wei, B.Q., D'Arcy-Gall, J., Ajayan, P.M. & Ramanath, G. Tailoring structure and electrical properties of carbon nanotubes using kilo-electron-volt ions. *Appl. Phys. Lett.* **83**, 3581-3583 (2003)

[19] Braga, D., Maini, L., Polito, M. & Grepioni, F. Hydrogen bonding interactions between ions: A powerful tool in molecular crystal engineering. In: *Supramolecular assembly via hydrogen bonds II. Structure and bonding. Volume 111.* Springer-Verlag, Berlin, Germany, pp 1-32 (2004)

[20] Krasheninnikov, A.V. & Banhart, F. Engineering of nanostructured carbon materials with electron or ion beams. *Nature Mat.* **6**, 723-733 (2007)

[21] Sneen, R.A. Substitution at saturated carbon-atom. 17. Organic ion pairs as intermediates in nucleophilic substitution and elimination-reactions. *Acc. Chem. Res.* **6**, 46-53 (1973)

[22] El Abedin, S.Z. & Endres, F. Ionic liquids: The link to high-temperature molten salts? *Acc. Chem. Res.* **40**, 1106-1113 (2007)

[23] Hu, Z.H. & Margulis, C.J. Room-temperature ionic liquids: Slow dynamics, viscosity,

and the red edge effect. *Acc. Chem. Res.* **40**, 1097-1105 (2007)

[24] Hardacre, C., Holbrey, J.D., Nieuwenhuyzen, M. & Youngs, T.G.A. Structure and solvation in ionic liquids. *Acc. Chem. Res.* **40**, 1146-1155 (2007)

[25] Joglekar, H.G., Rahman, I. & Kulkarni, B.D. The path ahead for ionic liquids. *Chem. Engineer. Tech.* **30**, 819-828 (2007)

[26] Plechkova, N.V. & Seddon, K.R. Applications of ionic liquids in the chemical industry. *Chem. Soc. Rev.* **37**, 123-150 (2008)

[27] Weingärtner, H. Understanding ionic liquids at the molecular level: Facts, problems, and controversies. *Angew. Chem. Int. Ed.* **47**, 654-670 (2008)

[28] Pregel, M.J., Dunn, E.J., Nagelkerke, R., Thatcher, G.R.J. & Buncel, E. Alkali-metal ion catalysis and inhibition in nucleophilic displacement reactions of phosphorus-, sulfur- and carbon-based esters. *Chem. Soc. Rev.* **24**, 449-455 (1995)

[29] Bohme, D.K. & Schwarz, H. Gas-phase catalysis by atomic and cluster metal ions: The ultimate single-site catalysts. *Angew. Chem. Int. Ed.* **44**, 2336-2354 (2005)

[30] Gelbard, G. Organic synthesis by catalysis with ion-exchange resins. *Industr. Engineer. Chem. Res.* **44**, 8468-8498 (2005)

[31] Nahar, S. & Tajmirriahi, H.A. Do metal ions alter the protein secondary structure of a light-harvesting complex of thylakoid membranes? *J. Inorg. Biochem.* **58**, 223-234 (1995)

[32] Bregadze, V.G. Metal ion interactions with DNA: Considerations on structure, stability, and effects from metal ion binding. In: *Metal ions in biological systems. Volume 32.* Marcel Dekker, New York, USA, pp 419-451 (1996)

[33] Jiang, M., Shen, T., Xu, H.B. & Liu, C.L. The influences of metal ions on protein folding, recognition, self-assembly and biological functions. *Prog. Chem.* **14**, 263-272 (2002)

[34] Mason, P.E., Neilson, G.W., Dempsey, C.E., Barnes, A.C. & Cruickshank, J.M. The hydration structure of guanidinium and thiocyanate ions: Implications for protein stability in aqueous solution. *Proc. Natl. Acad. Sci. USA* **100**, 4557-4561 (2003)

[35] Andrushchenko, V., Tsankov, D. & Wieser, H. Vibrational circular dichroism spectroscopy and the effects of metal ions on DNA structure. *J. Mol. Struct.* **661**, 541-560 (2003)

[36] Banci, L., Bertini, I., Cramaro, F., del Conte, R. & Viezzoli, M.S. Solution structure of Apo-Cu,Zn superoxide dismutase: Role of metal ions in protein folding. *Biochemistry* **42**, 9543-9553 (2003)

[37] Sissi, C., Marangon, E., Chemello, A., Noble, C.G., Maxwell, A. & Palumbo, M. The effects of metal ions on the structure and stability of the DNA gyrase B protein. *J. Mol. Biol.* **353**, 1152-1160 (2005)

[38] Woodson, S.A. Metal ions and RNA folding: A highly charged topic with a dynamic future. *Curr. Opin. Chem. Biol.* **9**, 104-109 (2005)

[39] Bushmarina, N.A., Blanchet, C.E., Vernier, G. & Forge, V. Cofactor effects on the protein folding reaction: Acceleration of alpha-lactalbumin refolding by metal ions. *Prot. Sci.* **15**, 659-671 (2006)

[40] Sharma, S.K., Goloubinoff, P. & Christen, P. Heavy metal ions are potent inhibitors of protein folding. *Biochem. Biophys. Res. Comm.* **372**, 341-345 (2008)

[41] Draper, D.E. RNA folding: Thermodynamic and molecular descriptions of the roles of ions. *Biophys. J.* **95**, 5489-5495 (2008)

[42] Siuzdak, G., Ichikawa, Y., Caulfield, T.J., Munoz, B., Wong, C.H. & Nicolaou, K.C. Evidence of Ca^{2+}-dependent carbohydrate association through ion spray mass spectrometry. *J. Am. Chem. Soc.* **115**, 2877-2881 (1993)

[43] Fried, M.G. & Stickle, D.F. Ion-exchange reactions of proteins during DNA binding. *Eur. J. Biochem.* **218**, 469-475 (1993)

[44] Arnott, S., Bian, W., Chandrasekaran, R. & Mains, B.R. Lessons for today and tomorrow from yesterday - the structure of alginic acid. *Fibre Diffr. Rev.* **9**, 44-51 (2000)

[45] Saecker, R.M. & Record Jr., M.T. Protein surface salt bridges and paths for DNA wrapping. *Curr. Opin. Struct. Biol.* **12**, 311-319 (2002)

[46] Tan, Z.J. & Chen, S.J. Ion-mediated nucleic acid helix-helix interactions. *Biophys. J.* **91**, 518-536 (2006)

[47] Williams, P.A. Molecular interactions of plant and algal polysaccharides. *Struct. Chem.* **20**, 299-308 (2009)

[48] Endo, M. Calcium ion as a second messenger with special reference to excitation-contraction coupling. *J. Pharm. Sci.* **100**, 519-524 (2006)

[49] Orlov, S.N. & Hamet, P. Intracellular monovalent ions as second messengers. *J. Memb. Biol.* **210**, 161-172 (2006)

[50] Strater, N., Lipscomb, W.N., Klabunde, T. & Krebs, B. Two-metal ion catalysis in enzymatic acyl- and phosphoryl-transfer reactions. *Angew. Chem. Int. Ed.* **35**, 2024-2055 (1996)

[51] Yang, W., Lee, J.Y & Nowotny, M. Making and breaking nucleic acids: Two-Mg^{2+}-ion catalysis and substrate specificity. *Mol. Cell* **22**, 4-13 (2006)

[52] Sigel, R.K.O. & Pyle, A.M. Alternative roles for metal ions in enzyme catalysis and the

implications for ribozyme chemistry. *Chem. Rev.* **107**, 97-113 (2007)
[53] Scott, W.G. RNA structure, metal ions, and catalysis. *Curr. Opin. Chem. Biol.* **3**, 705-709 (1999)
[54] Hanna, R. & Doudna, J.A. Metal ions in ribozyme folding and catalysis. *Curr. Opin. Chem. Biol.* **4**, 166-170 (2000)
[55] Stahley, M.R. & Strobel, S.A. RNA splicing: Group I intron crystal structures reveal the basis of splice site selection and metal ion catalysis. *Curr. Opin. Struct. Biol.* **16**, 319-326 (2006)
[56] Guyton, A.C. & Hall, J.E. Textbook of medicinal physiology. Edition 11. Elsevier, Oxford, UK (2005)
[57] Mann, S. Biomineralization. Principles and concepts in bioinorganic materials chemistry. Edition 1. Oxford Univ. Press, Oxford, UK (2001)
[58] Atwood, C.S., Huang, X.D., Moir, R.D., Tanzi, R.E. & Bush, A.I. Role of free radicals and metal ions in the pathogenesis of Alzheimer's disease. In: *Metal ions in biological systems. Volume 36.* Marcel Dekker, New York, USA, pp 309-364 (1999)
[59] Lehmann, S. Metal ions and prion diseases. *Curr. Opin. Struct. Biol.* **6**, 187-192 (2002)
[60] Martin, S.F. T-lymphocyte-mediated immune responses to chemical haptens and metal ions: Implications for allergic and autoimmune disease. *Int. Arch. Allergy Immunol.* **134**, 186-198 (2004)
[61] Greco, C. & Wolden, S. Current status of radiotherapy with proton and light ion beams. *Cancer* **109**, 1227-1238 (2007)
[62] Schulz-Ertner, D. & Tsujii, H. Particle radiation therapy using proton and heavier ion beams. *J. Clin. Oncol.* **25**, 953-964 (2007)
[63] Thompson, K.H. & Orvig, C. Boon and bane of metal ions in medicine. *Science* **300**, 936-939 (2003)
[64] Sigel, A. & Sigel, H. Metal ions in biological systems: Metal complexes in tumor diagnosis and as anticancer agents. Volume 42. CRC Press, Boca Raton, Florida, USA (2004)
[65] Wells, A.F. The structures of crystals. *Solid State Phys.: Adv. Res. App.* **7**, 425-503 (1958)
[66] Pauling, L. The nature of the chemical bond. Edition 3. Cornell University Press, Ithaca NY, USA (1960)
[67] Tosi, M.P. Cohesion of ionic solids in the Born model. *Solid State Phys.: Adv. Res. App.* **16**, 1-120 (1964)
[68] Sirdeshmukh, D.B., Sirdeshmukh, L. & Subhadra, K.G. Alkali halides. A handbook of physical properties. Springer-Verlag: Berlin, Heidelberg, New York (2001)
[69] Ostwald, W. Studien zur Kontaktelektrizität. *Z. Phys. Chem. (Stöchiometrie u. Verwandtschaftslehre)* **1**, 583-610 (1887)
[70] Kenrick, F.B. Die Potentialsprünge zwischen Gasen und Flüssigkeiten. *Z. Phys. Chem. (Stöchiometrie u. Verwandtschaftslehre)* **19**, 625-656 (1896)
[71] Ostwald, W. Über die absoluten Potentiale der Metalle nebst Bemerkungen über Normalelektroden. *Z. Phys. Chem. (Stöchiometrie u. Verwandtschaftslehre)* **35**, 333-339 (1900)
[72] Wilsmore, N.T.M. & Ostwald, W. Über Elektrodenpotentiale und absolute Potentiale. *Z. Phys. Chem. (Stöchiometrie u. Verwandtschaftslehre)* **36**, 91-98 (1901)
[73] Billitzer, J. Zur Theorie der kapillarelektrischen Erscheinungen. I. Versuche mit Tropfelektroden. *Z. Phys. Chem. (Stöchiometrie u. Verwandtschaftslehre)* **48**, 513-548 (1904)
[74] Billitzer, J. Zur Theorie der kapillarelektrischen Erscheinungen. *Z. Phys. Chem. (Stöchiometrie u. Verwandtschaftslehre)* **51**, 167-192 (1905)
[75] Goodwin, H.M. On Billitzer's method for determining absolute potential differences. *Phys. Rev.* **21**, 129-146 (1905)
[76] Palmaer, W. Über das absolute Potential der Kalomelelektrode. *Z. Phys. Chem. (Stöchiometrie u. Verwandtschaftslehre)* **59**, 129-191 (1907)
[77] Ewell, A.W. Electrostatic measurement of electrode potentials. *Phys. Rev.* **6**, 271-282 (1915)
[78] MacInnes, D.A. The activities of the ions of strong electrolytes. *J. Am. Chem. Soc.* **41**, 1086-1092 (1919)
[79] Born, M. Die Elektronenaffinität der Halogenatome. *Ber. dtsch. physik. Ges.* **21**, 679-685 (1919)
[80] Haber, F. Betrachtungen zur Theorie der Wärmetönung. *Ber. dtsch. physik. Ges.* **21**, 750-768 (1919)
[81] Fajans, K. Die Elektronenaffinität der Halogenatome und die Ionisierungsenergie der Halogenwasserstoffe. *Ber. dtsch. physik. Ges.* **21**, 714-722 (1919)
[82] Fajans, K. Bemerkungen zu meiner Arbeit: Über Hydrationswärmen gasförmiger Atomionen. *Ber. dtsch. physik. Ges.* **21**, 709-713 (1919)
[83] Fajans, K. Über Hydrationswärmen gasförmiger Atomionen. *Ber. dtsch. physik. Ges.* **21**, 549-558 (1919)

[84] Born, M. Volumen und Hydrationswärme der Ionen. *Z. Phys.* **1**, 45-48 (1920)
[85] Conway, B.E. Electrolyte solutions: Solvation and structural aspects. *Ann. Rev. Phys. Chem.* **17**, 481-528 (1966)
[86] Enderby, J.E. & Neilson, G.W. The structure of electrolyte solutions. *Rep. Prog. Phys.* **44**, 593-653 (1981)
[87] Collins, K.D. & Washabaugh, M.W. The Hofmeister effect and the behavior of water at interfaces. *Quart. Rev. Biophys.* **18**, 323-422 (1985)
[88] Ohtaki, H. & Radnai, T. Structure and dynamics of hydrated ions. *Chem. Rev.* **93**, 1157-1204 (1993)
[89] Sacco, A. Structure and dynamics of electrolyte solutions. A NMR relaxation approach. *Chem. Soc. Rev.* **23**, 129-136 (1994)
[90] Enderby, J.E. Ion solvation via neutron scattering. *Chem. Soc. Rev.* **24**, 159-168 (1995)
[91] Lisy, J.M. Spectroscopy and structure of solvated alkali-metal ions. *Int. Rev. Phys. Chem.* **16**, 267-289 (1997)
[92] Helm, L. & Merbach, A.E. Water exchange on metal ions: experiments and simulations. *Coord. Chem. Rev.* **187**, 151-181 (1999)
[93] Ohtaki, H. Ionic solvation in aqueous and nonaqueous solutions. *Monatshefte f. Chemie* **132**, 1237-1268 (2001)
[94] Rasaiah, J.C. & Lynden-Bell, R.M. Computer simulation studies of the structure and dynamics of ions and non-polar solutes in water. *Phil. Trans. R. Soc. Lond. A* **359**, 1545-1574 (2001)
[95] Krienke, H., Fischer, R. & Barthel, J. Ion solvation in nonaqueous solvents on the Born-Oppenheimer level. *J. Mol. Liq.* **98-99**, 329-354 (2002)
[96] Erras-Hanauer, H., Clark, T. & van Eldik, R. Molecular orbital and DFT studies on water exchange mechanisms of metal ions. *Coord. Chem. Rev.* **238**, 233-253 (2003)
[97] Vinogradov, E.V., Smirnov, P.R. & Trostin, V.N. Structure of hydrated complexes formed by metal ions of Groups I-III of the Periodic Table in aqueous electrolyte solutions under ambient conditions. *Russ. Chem. Bull.* **52**, 1253-1271 (2003)
[98] Fawcett, W.R. Liquids, solutions and interfaces. Oxford Univ. Press, Oxford, UK (2004)
[99] Kunz, W., Lo Nostro, P. & Ninham, B.W. The present state of affairs with Hofmeister effects. *Curr. Opin. Colloid. Interface Sci.* **9**, 1-18 (2004)
[100] Holovko, M., Druchok, M. & Bryk, T. Computer modelling of hydration structure of highly charged ions and cationic hydrolysis effects. *Curr. Opin. Coll. Int. Sci.* **9**, 64-66 (2004)
[101] Meot-Ner, M. The ionic hydrogen bond. *Chem. Rev.* **105**, 213-284 (2005)
[102] Rode, B.M., Schwenk, C.F., Hofer, T.S. & Randolf, B.R. Coordination and ligand exchange dynamics of solvated metal ions. *Coord. Chem. Rev.* **249**, 2993-3006 (2005)
[103] Rode, B.M. & Hofer, T.S. How to access structure and dynamics of solutions: The capabilities of computational methods. *Pure Appl. Chem.* **78**, 525-539 (2006)
[104] Soper, A.K. & Weckström, K. Ion solvation and water structure in potassium halide aqueous solutions. *Biophys. Chem.* **124**, 180-191 (2006)
[105] Ansell, S., Barnes, A.C., Mason, P.E., Neilson, G.W. & Ramos, S. X-ray and neutron scattering studies of the hydration structure of alkali ions in concentrated aqueous solutions. *Biophys. Chem.* **124**, 171-179 (2006)
[106] Petersen, P.B. & Saykally, R.J. On the nature of ions at the liquid water surface. *Annu. Rev. Phys. Chem.* **57**, 333-364 (2006)
[107] Burnham, C.J., Petersen, M.K., Day, T.J.F., Iyengar, S.S. & Voth, G.A. The properties of ion-water clusters. II. Solvation structures of Na^+, Cl^- and H^+ clusters as a function of temperature. *J. Chem. Phys.* **124**, 024327/1-024327/9 (2006)
[108] McLain, S.E., Imberti, S., Soper, A.K., Botti, A., Bruni, F. & Ricci, M.A. Structure of 2 molar NaOH in aqueous solution from neutron diffraction and empirical potential structure refinement. *Phys. Rev. B* **74**, 094201/1-094201/8 (2006)
[109] Marcus, Y. Effects of ions on the structure of water: Structure making and breaking. *Chem. Rev.* **109**, 1346-1370 (2009)
[110] Maroncelli, M. The dynamics of solvation in polar liquids. *J. Mol. Liq.* **57**, 1-37 (1993)
[111] Jenkins, H.D.B. & Marcus, Y. Viscosity B-coefficients of ions in solution. *Chem. Rev.* **95**, 2695-2724 (1995)
[112] Dufreche, J.F., Bernard, O., Durand-Vidal, S. & Turq, P. Analytical theories of transport in concentrated electrolyte solutions from the MSA. *J. Phys. Chem. B* **109**, 9873-9884 (2005)
[113] Debye, P., Zur Theorie der Elektrolyte. I. Gefrierpunktserniedrigung und verwandte Erscheinungen. *Phys. Z.* **24**, 185-206 (1923)
[114] The theory of concentrated, aqueous solutions of strong electrolytes. *Phys. Z.* **26**, 93-147 (1925)
[115] Scatchard, G. & Epstein, L.F. The calculation of the thermodynamic properties and the association of electrolyte solutions. *Chem. Rev.* **30**, 211-226 (1942)
[116] Scatchard, G. Osmotic coefficients and activity coefficients in mixed electrolyte solutions. *J. Am. Chem. Soc.* **83**, 2636-2642 (1961)

[117] Scatchard, G. Solutions of electrolytes. *Ann. Rev. Phys. Chem.* **14**, 161-176 (1963)
[118] McCall, D.W. & Douglas, D.C. The effect of ions on the self-diffusion of water. I. Concentration dependence. *J. Phys. Chem.* **69**, 2001-2011 (1965)
[119] Bell, G.M. & Rangecroft, P.D. Theory of surface tension for a 2-1 electrolyte solution. *Trans. Faraday Soc.* **67**, 649-659 (1971)
[120] Pitzer, K.S. Electrolyte theory. Improvements since Debye and Hückel. *Acc. Chem. Res.* **10**, 371-377 (1977)
[121] Pitzer, K.S. Characteristics of very concentrated aqueous solutions. *Phys. Chem. Earth* **13-14**, 249-272 (1981)
[122] Parkhurst, D.L. Ion-association models and mean activity coefficients of various salts. *ACS Symposium Series* **416**, 30-43 (1990)
[123] Wolery, T.J. & Jackson, K.J. Activity coefficients in aqueous salt solutions: Hydration theory equations. *ACS Symposium Series* **416**, 16-29 (1990)
[124] Covington, A.K. & Pethybridge, A.D. Electrolyte solutions. *Ann. Rep. Prog. Chem. A* **74**, 5-21 (1977)
[125] Friedman, H.L. Electrolyte solutions at equilibrium. *Ann. Rev. Phys. Chem.* **32**, 179-204 (1981)
[126] Bates, R.G., Dickson, A.G., Gratzl, M., Hrabeczypall, A., Lindner, E. & Pungor, E. Determination of mean activity coefficients with ion-selective electrodes. *Anal. Chem.* **55**, 1275-1280 (1983)
[127] Kuznetsova, E.M. Basic directions in the theory of activity of strong electrolyte solutions. *Russ. J. Phys. Chem.* **76**, 866-880 (2002)
[128] Boström, M., Kunz, W. & Ninham, B.W. Hofmeister effects in surface tension of aqueous electrolyte solution. *Langmuir* **21**, 2619-2623 (2005)
[129] Marcus, Y. & Hefter, G. Ion pairing. *Chem. Rev.* **106**, 4585-4621 (2006)
[130] Du, H., Rasaiah, J.C. & Miller, J.D. Structural and dynamic properties of concentrated alkali halide solutions: A molecular dynamics simulation study. *J. Phys. Chem. B* **111**, 209-217 (2007)
[131] Buchner, R. What can be learnt from dielectric relaxation spectroscopy about ion solvation and association? *Pure Appl. Chem.* **80**, 1239-1252 (2008)
[132] Buchner, R. & Hefter, G. Interactions and dynamics in electrolyte solutions by dielectric spectroscopy. *Phys. Chem. Chem. Phys.* **11**, 8984-8999 (2009)
[133] Zhang, C., Raugei, S., Eisenberg, B. & Carloni, P. Molecular dynamics in physiological solutions: Force fields, alkali metal ions, and ionic strength. *J. Chem. Theory Comput.* **6**, 2167-2175 (2010)
[134] Kalyuzhnyi, Y.V., Vlachy, V. & Dill, K.A. Aqueous alkali halide solutions: can osmotic coefficients be explained on the basis of the ionic sizes alone? *Phys. Chem. Chem. Phys.* **12**, 6260-6266 (2010)
[135] Conway, B.E. The state of water and hydrated ions at interfaces. *Adv. Colloid. Interface Sci.* **8**, 91-212 (1977)
[136] Sposito, G. Molecular models of ion adsorption on mineral surfaces. *Rev. Mineralogy* **23**, 261-279 (1990)
[137] Bunton, C.A., Nome, F., Quina, F.H. & Romsted, L.S. Ion binding and reactivity at charged aqueous interfaces. *Acc. Chem. Res.* **24**, 357-364 (1991)
[138] Conway, B.E. Individual solvated ion properties and specificity of ion adsorption effects in processes at electrodes. *Chem. Soc. Rev.* **21**, 253-261 (1992)
[139] Benjamin, I. Molecular structure and dynamics at liquid-liquid interfaces. *Ann. Rev. Phys. Chem.* **48**, 407-451 (1997)
[140] Cacace, M.G., Landau, E.M. & Ramsden, J.J. The Hofmeister series: salt and solvent effects on interfacial phenomena. *Quart. Rev. Biophys.* **30**, 241-277 (1997)
[141] Jungwirth, P. & Tobias, D.J. Ions at the air/water interface. *J. Phys. Chem. B* **106**, 6361-6373 (2002)
[142] Wennerström, H. Ion binding to interfaces. *Curr. Opin. Colloid Interface Sci.* **9**, 163-164 (2004)
[143] Bradl, H.B. Adsorption of heavy metal ions on soils and soil constituents. *J. Colloid Interface Sci.* **277**, 1-18 (2004)
[144] Netz, R.R. Water and ions at interfaces. *Curr. Opin. Colloid Interface Sci.* **9**, 192-197 (2004)
[145] Taylor, C.D. & Neurock, M. Theoretical insights into the structure and reactivity of the aqueous/metal interface. *Curr. Opin. Solid State & Mater. Sci.* **9**, 49-65 (2005)
[146] Barbero, C.A. Ion exchange at the electrode/electrolyte interface studied by probe beam deflection techniques. *Phys. Chem. Chem. Phys.* **7**, 1885-1899 (2005)
[147] Gabovich, A.M., Reznikov, Y.A. & Voitenko, A.I. Excess nonspecific Coulomb ion adsorption at the metal electrode/electrolyte solution interface: Role of the surface layer. *Phys. Rev. E* **73**, 021606/1-021606/12 (2006)
[148] Taylor, C.D. & Neurock, M. Theoretical insights into the structure and reactivity of the aqueous/metal interface. *Curr. Opin. Solid State Mater. Sci.* **9**, 49-65 (2006)
[149] Jungwirth, P. & Tobias, D.J. Specific effects at the air/water interface. *Chem. Rev.* **106**,

1259-1281 (2006)

[150] Jungwirth, P. & Winter, B. Ions at aqueous interfaces: From water surface to hydrated proteins. *Ann. Rev. Phys. Chem.* **59**, 343-366 (2008)

[151] McCarty, L.S. & Whitesides, G.M. Electrostatic charging due to separation of ions at interfaces: Contact electrification of ionic electrets. *Angew. Chem. Int. Ed.* **47**, 2188-2207 (2008)

[152] Wick, C.D. Electrostatic dampening dampens the anion propensity for the air-water interface. *J. Chem. Phys.* **131**, 084715/1-084715/6 (2009)

[153] Noah-Vanhoucke, J. & Geissler, P.L. On the fluctuations that drive small ions toward, and away from, interfaces between polar liquids and their vapors. *Proc. Natl. Acad. Sci. USA* **106**, 15125-15130 (2009)

[154] Levin, Y. Polarizable ions at interfaces. *Phys. Rev. Lett.* **102**, 147803/1-147803/4 (2009)

[155] Bauer, B.A. & Patel, S. Molecular dynamics simulations of nonpolarizable inorganic salt solution interfaces: NaCl, NaBr, and NaI in transferable intermolecular potential 4-point with charge dependent polarizability (TIP4P-QDP) water. *J. Chem. Phys.* **132**, 024713/1-024713/13 (2010)

[156] Shamay, E.S. & Richmond, G.L. Ionic disruption of the liquid-liquid interface. *J. Phys. Chem. C* **114**, 12590-12597 (2010)

[157] Tamashiro, M.N. & Constantino, M.A. Ions at the water-vapor interface. *J. Phys. Chem. B* **114**, 3583-3591 (2010)

[158] Callahan, K.M., Casillas-Ituarte, N.N., Xu, M., Roeselová, M., Allen, H.C. & Tobias, D.J. Effect of magnesium cation on the interfacial properties of aqueous salt solutions. *J. Phys. Chem. A* **114**, 8359-8368 (2010)

[159] Arshadi, M., Yamdagni, R. & Kebarle, P. Hydration of the halide negative ions in the gas phase. II. Comparison of hydration energies for the alkali positive and halide negative ions. *J. Phys. Chem.* **74**, 1475-1482 (1970)

[160] Keesee, R.G. & Castelman Jr., A.W. Gas-phase studies of hydration complexes of Cl^- and I^- and comparison to electrostatic calculations in the gas phase. *Chem. Phys. Lett.* **74**, 139-142 (1980)

[161] Keesee, R.G., Lee, N. & Castelman Jr., A.W. Properties of clusters in the gas phase: V. Complexes of neutral molecules onto negative ions. *J. Chem. Phys.* **73**, 2195-2202 (1980)

[162] Castleman, A.W. & Keesee, R.G. Ionic clusters. *Chem. Rev.* **86**, 589-618 (1986)

[163] Castleman, A.W. & Keesee, R.G. Gas-phase clusters: Spanning the states of matter. *Science* **241**, 36-42 (1988)

[164] Coe, J.V. Connecting cluster ions and bulk aqueous solvation. A new determination of bulk single ion solvation enthalpies. *Chem. Phys. Lett.* **229**, 161-168 (1994)

[165] Castleman, A.W. & Bowen, K.H. Clusters: Structure, energetics, and dynamics of intermediate states of matter. *J. Phys. Chem.* **100**, 12911-12944 (1996)

[166] Grunwald, E. & Steel, C. A thermodynamic analysis of the first solvation shells of alkali and halide ions in liquid water and in the gas phase. *Int. Rev. Phys. Chem.* **15**, 273-281 (1996)

[167] Duncan, M.A. Spectroscopy of metal ion complexes: Gas-phase models for solvation. *Ann. Rev. Phys. Chem.* **48**, 69-93 (1997)

[168] Takashima, K. & Riveros, J.M. Gas-phase solvated negative ions. *Mass Spectrom. Rev.* **17**, 409-430 (1998)

[169] Laidig, K.E., Speers, P. & Streitwieser, A. Complexation of Li^+, Na^+, and K^+ by water and ammonia. *Coord. Chem. Rev.* **197**, 125-139 (2000)

[170] Coe, J.V. Fundamental properties of bulk water from cluster ion data. *Int. Rev. Phys. Chem.* **20**, 33-58 (2001)

[171] Bondybey, V.E. & Beyer, M.K. How many molecules make a solution? *Int. Rev. Phys. Chem.* **21**, 277-306 (2002)

[172] Robertson, W.H. & Johnson, M.A. Molecular aspects of halide ion hydration: The cluster approach. *Annu. Rev. Phys. Chem.* **54**, 173-213 (2003)

[173] Chang, H.C., Wu, C.C. & Kuo, J.L. Recent advances in understanding the structures of medium-sized protonated water clusters. *Int. Rev. Phys. Chem.* **24**, 553-578 (2005)

[174] Walker, N.R., Walters, R.S. & Duncan, M.A. Frontiers in the infrared spectroscopy of gas phase metal ion complexes. *New J. Chem.* **29**, 1495-1503 (2005)

[175] Rais, J. & Okada, T. Three empirical correlations connecting gaseous cluster energies and solvation energies of alkali metal and halide ions. *Anal. Sci.* **22**, 533-538 (2006)

[176] Rais, J. & Okada, T. Erratum to "Three empirical correlations connecting gaseous cluster energies and solvation energies of alkali metal and halide ions" [Anal. Sci. 22, 533-538 (2006)]. *Anal. Sci.* **22**, 919-919 (2006)

[177] Beyer, M.K. Hydrated metal ions in the gas phase. *Mass Spectrom. Rev.* **26**, 517-541 (2007)

[178] Bush, M.F., Saykally, R.J. & Williams, E.R. Reactivity and infrared spectroscopy of gaseous hydrated trivalent metal ions. *J. Am. Chem. Soc.* **130**, 9122-9128 (2008)

[179] Sherman, J. Crystal energies of ionic compounds and thermochemical applications.

Chem. Rev. **11**, 93-170 (1932)
[180] Huggins, M.L. Lattice energies, equilibrium distances, compressibilities and characteristic frequencies of alkali halide crystals. *J. Chem. Phys.* **5**, 143-148 (1937)
[181] Wagner, C. The electrochemistry of ionic crystals. *J. Electrochem. Soc.* **99**, 346C-354C (1952)
[182] Delauney, J. The theory of specific heats and lattice vibrations. *Solid State Phys.: Adv. Res. App.* **2**, 219-303 (1956)
[183] Huntington, H.B. The elastic constants of crystals. *Solid State Phys.: Adv. Res. App.* **7**, 213-351 (1958)
[184] Smith, C.S. Macroscopic symmetry and properties of crystals. *Solid State Phys.: Adv. Res. App.* **6**, 175-249 (1958)
[185] Leibfried, G. & Ludwig, W. Theory of anharmonic effects in crystals. *Solid State Phys.: Adv. Res. App.* **12**, 275-444 (1961)
[186] Gilman, J.J. Mechanical behavior of ionic crystals. *Uspekhi Fiz. Nauk* **80**, 455-503 (1963)
[187] Tosi, M.P. & Fumi, F.G. Ionic sizes and Born repulsive parameters in the NaCl-Type alkali halides - II. *J. Phys. Chem. Solids* **25**, 45-52 (1964)
[188] Basu, A.N., Roy, D. & Sengupta, S. Polarizable models for ionic-crystals and effective many-body interaction. *Phys. Stat. Solidi A* **23**, 11-32 (1974)
[189] Dixon, M. & Murthy, C.S.N. Ion pair potentials for alkali halides and some applications to liquids and defect studies. *Phys. Chem. Liq.* **12**, 83-107 (1982)
[190] Martin, T.P. Alkali halide clusters and micro-crystals. *Physics Reports* **95**, 167-199 (1983)
[191] Shanker, J. & Agrawal, G.G. Van der Waals potentials in ionic crystals. *Phys. Stat. Solidi B* **123**, 11-26 (1984)
[192] Pyper, N.C. Relativistic ab initio calculations of the properties of ionic solids. *Philosoph. Trans. Roy. Soc. London A* **320**, 107-158 (1986)
[193] Shanker, J. & Bhende, W.N. Higher-order elastic constants and thermoelastic properties of ionic solids. *Phys. Stat. Solidi B* **136**, 11-30 (1986)
[194] Harding, J.H. Computer simulation of defects in ionic solids. *Rep. Prog. Phys.* **53**, 1403-1466 (1990)
[195] Ohtaki, H. Molecular aspects on the dissolution and nucleation of ionic-crystals in water. *Adv. Inorg. Chem.* **39**, 401-437 (1992)
[196] Rickman, J.M. & le Sar, R. Free-energy calculations in materials research. *Annu. Rev. Mater. Res.* **32**, 195-217 (2002)
[197] Allan, N.L., Barrera, G.D., Purton, J.A., Sims, C.E. & Taylor, M.B. Ionic solids at elevated temperatures and/or high pressures: Lattice dynamics, molecular dynamics, Monte Carlo and ab initio studies. *Phys. Chem. Chem. Phys.* **2**, 1099-1111 (2000)
[198] de Oliveira, C.A.F., Hamelberg, D. & McCammon, J.A. Estimating kinetic rates from accelerated molecular dynamics simulations: Alanine dipeptide in explicit solvent as a case study. *J. Chem. Phys.* **127**, 175105/1-175105/8 (2007)
[199] Lynden-Bell, R.M., del Popolo, M.G., Youngs, T.G.A., Kohanoff, J., Hanke, C.G., Harper, J.B. & Pinilla, C.C. Simulations of ionic liquids, solutions, and surfaces. *Acc. Chem. Res.* **40**, 1138-1145 (2007)
[200] Maginn, E.J. Atomistic simulation of the thermodynamic and transport properties of ionic liquids. *Acc. Chem. Res.* **40**, 1200-1207 (2007)
[201] Wang, Y., Jiang, W., Yan, T. & Voth, G.A. Understanding ionic liquids through atomistic and coarse-grained molecular dynamics simulations. *Acc. Chem. Res.* **40**, 1193-1199 (2007)
[202] Bhargava, B.L., Balasubramanian, S. & Klein, M.L. Modelling room temperature ionic liquids. *Chem. Comm.* **2008**, 3339-3351 (2008)
[203] Shim, Y. & Kim, H.J. Dielectric relaxation, ion conductivity, solvent rotation, and solvation dynamics in a room-temperature ionic liquid. *J. Phys. Chem. B* **112**, 11028-11038 (2008)
[204] Borodin, O. Polarizable force field development and molecular dynamics simulations of ionic liquids. *J. Phys. Chem. B* **113**, 11463-11478 (2009)
[205] Maginn, E.J. Molecular simulation of ionic liquids: Current status and future opportunities. *J. Phys.: Condens. Matter* **21**, 373101/1-373101/17 (2009)
[206] Sambasivarao, S.V. & Acevedo, O. Development of OPLS-AA force field parameters for 68 unique ionic liquids. *J. Chem. Theory Comput.* **5**, 1038-1050 (2009)
[207] Janz, G.J., Solomons, C. & Gardner, H.J. Physical properties and constitution of molten salts: Electrical conductance, transport, and cryoscopy. *Chem. Rev.* **58**, 461-508 (1958)
[208] Inman, D. & White, S.H. Molten salts. *Ann. Rep. Prog. Chem.* **62**, 106-130 (1965)
[209] Sangster, M.J.L. & Dixon, M. Interionic potentials in alkali-halides and their use in simulations of molten salts. *Adv. Phys.* **25**, 247-342 (1976)
[210] Gillan, M.J. Theories of thermodynamic properties of pure molten salts. *Phys. Chem. Liq.* **8**, 121-141 (1978)
[211] Hensel, F. Electrical properties of unusual ionic melts. *Adv. Phys.* **28**, 555-594 (1979)
[212] March, N.H. Melting of ionic crystals, defect energies and fast ion conduction. *Phys.*

Chem. Liq. **10**, 1-9 (1980)
[213] Todheide, K. The influence of density and temperature on the properties of pure molten salts. *Angew. Chem. Int. Ed.* **19**, 606-619 (1980)
[214] Rovere, M. & Tosi, M.P. Structure and dynamics of molten salts. *Rep. Prog. Phys.* **49**, 1001-1081 (1986)
[215] McGreevy, R.L. Experimental studies of the structure and dynamics of molten alkali and alkaline earth halides. *Solid State Phys.: Adv. Res. App.* **40**, 247-325 (1987)
[216] Shanker, J. & Kumar, M. Studies on melting of alkali halides. *Phys. Stat. Solidi B* **158**, 11-49 (1990)
[217] Adya, A.K., Takagi, R. & Gaune-Escard, M. Unravelling the internal complexities of molten salts. *Z. Naturforsch. A* **53**, 1037-1048 (1998)
[218] Tosi, M.P. Melting and liquid structure of polyvalent metal-halides. *Z. Phys. Chem.* **184**, 121-138 (1994)
[219] Guillot, B. & Guissani, Y. Towards a theory of coexistence and criticality in real molten salts. *Mol. Phys.* **87**, 37-86 (1996)
[220] Kirillov, S.A. Interactions and picosecond dynamics in molten salts: A review with comparison to molecular liquids. *J. Mol. Liq.* **76**, 35-95 (1998)
[221] Lide, D.R. CRC Handbook of Chemistry and Physics. Edition 88 (Internet version). CRC Press, Boca Raton, Florida (2008)
[222] Haverlock, T.J., Mirzadeh, S. & Moyer, B.A. Selectivity of calix[4]arene-bis(benzocrown-6) in the complexation and transport of francium ion. *J. Am. Chem. Soc.* **125**, 1126-1127 (2003)
[223] Kritskaya, E.B., Burylev, B.P., Moisov, L.P. & Kritskii, V.E. Thermodynamic properties of binary melts of manganese(II) bromide with lithium, cesium, and francium bromides. *Russ. J. Phys. Chem.* **79**, 299-301 (2005)
[224] Champion, J., Alliot, C., Renault, E., Mokili, B.M., Chérel, M., Galland, N. & Montavon, G. Astatine standard redox potentials and speciation in acidic medium. *J. Phys. Chem. A* **114**, 576-582 (2010)
[225] Dye, J.L., Andrews, C.W. & Ceraso, J.M. Nuclear magnetic resonance studies of alkali metal anions. *J. Phys. Chem.* **79**, 3076-3079 (1975)
[226] Dreyer, I., Dreyer, R. & Chalkin, V.A. Cations of astatine in aqueous solutions. Preparation and properties. *Radiochem. Radioanal. Lett.* **36**, 389-398 (1978)
[227] Dreyer, I., Dreyer, R., Chalkin, V.A. & Milanov, M. Halide-complexes of stable astatine cations in aqueous solutions. *Radiochem. Radioanal. Lett.* **40**, 145-153 (1979)
[228] Dye, J.L. Compounds of alkali metal anions. *Angew. Chem. Int. Ed.* **18**, 587-598 (1979)
[229] Milanov, M., Doberenz, V., Khalkin, V.A. & Marinov, A. Chemical properties of positive singly-charged astatine ion in aqueous solution. *J. Radioanal. Nucl. Chem.* **83**, 291-299 (1984)
[230] Dreyer, R., Dreyer, I., Fischer, S., Hartmann, H. & Rosch, F. Synthesis and characterization of cationic astatine compounds with sulfur-containing ligands stable in aqueous solutions. *J. Radioanal. Nucl. Chem. Lett.* **96**, 333-342 (1985)
[231] Dye, J.L. Physical and chemical properties of alkalides and electrides. *Ann. Rev. Phys. Chem.* **38**, 271-301 (1987)
[232] Mezei, M. & Beveridge, D.L. Monte Carlo studies of the structure of dilute aqueous solutions of Li^+, Na^+, K^+, F^-, and Cl^-. *J. Chem. Phys.* **74**, 6902-6910 (1981)
[233] Abraham, M.H., Liszi, J. & Papp, E. Calculations on ionic solvation. Part 6. Structure-making and structure-breaking effects of alkali halide ions from electrostatic entropies of solvation. Correlation with viscosity B-coefficients, nuclear magnetic resonance B'-coefficients and partial molal volumes. *J. Chem. Soc. Faraday Trans. 1* **78**, 197-211 (1982)
[234] Impey, R.W., Madden, P.A. & McDonald, I.R. Hydration and mobility of ions in solution. *J. Phys. Chem.* **87**, 5071-5083 (1983)
[235] Lee, S.H. & Rasaiah, J.C. Molecular dynamics simulation of ion mobility: 2. Alkali metal and halide ions using the SPC/E model for water at 25°C. *J. Phys. Chem.* **100**, 1420-1425 (1996)
[236] Tongraar, A., Liedl, K.R. & Rode, B.M. Born-Oppenheimer ab initio QM/MM dynamics simulations of Na^+ and K^+ in water: From structure making to structure breaking effects. *J. Phys. Chem. A* **102**, 10340-10347 (1998)
[237] Wasse, J.C., Hayama, S., Skipper, N.T., Benmore, C.J. & Soper, A.K. The structure of saturated lithium- and potassium-ammonia solutions as studied by using neutron diffraction. *J. Chem. Phys.* **112**, 7147-7151 (2000)
[238] Hribar, B., Southall, N.T., Vlachi, V. & Dill, K.A. How ions affect the structure of water. *J. Am. Chem. Soc.* **124**, 12302-12311 (2002)
[239] Schwenk, C.F., Hofer, T.S. & Rode, B.M. "Structure breaking" effect of hydrated Cs^+. *J. Phys. Chem. A* **108**, 1509-1514 (2004)
[240] Vrbka, L., Mucha, M., Minofar, B., Jungwirth, P., Brown, E.C. & Tobias, D.J. Propensity of soft ions for the air/water interface. *Curr. Opin. Coll. Int. Sci.* **9**, 67-73 (2004)
[241] Nickolov, Z.S. & Miller, J.D. Water structure in aqueous solutions of alkali halide salts:

FTIR spectroscopy of the OD stretching band. *J. Colloid. Interface Sci.* **287**, 572-580 (2005)
[242] Tůma, L., Jenícek, D. & Jungwirth, P. Propensity of heavier halides for the water/vapor interface revisited using the Amoeba force field. *Chem. Phys. Lett.* **411**, 70-74 (2005)
[243] Bakker, H.J., Kropman, M.F. & Omta, A.W. Effect of ions on the structure and dynamics of liquid water. *J. Phys.: Condens. Matter* **17**, S3215-S3224 (2005)
[244] Hofer, T.S., Randolf, B.R. & Rode, B.M. Structure-breaking effects of solvated Rb(I) in dilute aqueous solution - An *ab initio* QM/MM MD approach. *J. Comput. Chem.* **26**, 949-956 (2005)
[245] Varma, S. & Rempe, S.B. Coordination numbers of alkali metal ions in aqueous solutions. *Biophys. Chem.* **124**, 192-199 (2006)
[246] Du, H. & Miller, J.D. Interfacial water structure and surface charge of selected alkali chloride salt crystals in saturated solutions: A molecular dynamics modeling study. *J. Phys. Chem. B* **111**, 10013-10022 (2007)
[247] Mancinelli, R., Botti, A., Bruni, F., Ricci, M.A. & Soper, A.K. Perturbation of water structure due to monovalent ions in solution. *Phys. Chem. Chem. Phys.* **9**, 2959-2967 (2007)
[248] Mancinelli, R., Botti, A., Bruni, F., Ricci, M.A. & Soper, A.K. Hydration of sodium, potassium, and chloride ions in solution and the concept of structure maker/breaker. *J. Phys. Chem. B* **111**, 13570-13577 (2007)
[249] Nucci, N.V. & Vanderkooi, J.M. Effects of salts of the Hofmeister series on the hydrogen bond network of water. *J. Mol. Liq.* **143**, 160-170 (2008)
[250] Azam, S.S., Hofer, T.S., Randolf, B.R. & Rode, B.M. Hydration of sodium(I) and potassium(I) revisited: A comparative QM/MM and QMCF MD simulation study of weakly hydrated ions. *J. Phys. Chem. A* **113**, 1827-1834 (2009)
[251] Harned, H.S. The diffusion coefficients of the alkali metal chlorides and potassium and silver nitrates in dilute aqueous solutions at 25°C. *Proc. Natl. Acad. Sci. USA* **40**, 551-556 (1954)
[252] Irish, D.E. & Brooker, M.H. Interactions and residence times in aqueous alkali metal nitrite solutions from Raman spectral measurements. *Trans. Faraday Soc.* **67**, 1916-1922 (1971)
[253] Wei, Y.-Z., Chiang, P. & Sridhar, S. Ion size effects on the dynamic and static dielectric properties of aqueous alkali solutions. *J. Chem. Phys.* **96**, 4569-4573 (1992)
[254] Lu, R. & Leaist, D.G. Mutual diffusion in solutions of alkali metal halides: Aqueous LiF, NaF and KF at 25°C. *J. Chem. Soc. Faraday Trans.* **94**, 111-114 (1998)
[255] Amo, Y., Annaka, M. & Tominaga, Y. Classification of alkali halide aqueous solutions by Kubo number. *J. Mol. Liq.* **100/2**, 143-151 (2002)
[256] Horvat, J., Bešter-Rogač, M., Klofutar, C. & Rudan-Tasic, D. Viscosity of aqueous solutions of lithium, sodium, potassium, rubidium and caesium cyclohexylsulfamates from 293.15 K to 323.15 K. *J. Solut. Chem.* **37**, 1329-1342 (2008)
[257] Gurney, R.W. Ionic processes in solution. McGraw-Hill, New York, USA (1953)
[258] Conway, B.E. & Bockris, J.O'M. Ionic solvation. *Mod. Aspects Electrochem.* **1**, 47-102 (1954)
[259] Noyes, R.M. Thermodynamics of ion hydration as a measure of effective dielectric properties of water. *J. Am. Chem. Soc.* **84**, 513-522 (1962)
[260] Halliwell, H.F. & Nyburg, S.C. Enthalpy of hydration of the proton. *Trans. Farad. Soc.* **59**, 1126-1140 (1963)
[261] Noyes, R.M. Assignment of individual ionic contributions to properties of aqueous ions. *J. Am. Chem. Soc.* **86**, 971-979 (1964)
[262] Rosseinsky, D.R. Electrode potentials and hydration energies. Theories and correlations. *Chem. Rev.* **65**, 467-490 (1965)
[263] Conway, B.E., Verall, R.E. & Desnoyers, J.E. Specificity in ionic hydration and the evaluation of individual ionic properties. *Z. Phys. Chem. (Leipzig)* **230**, 157-178 (1965)
[264] Desnoyers, J.E. & Jolicoeur, C. Hydration effects and the thermodynamic properties of ions. *Mod. Aspects Electrochem.* **5**, 1-89 (1969)
[265] Popovych, O. Estimation of medium effects for single ions in non-aqueous solvents. *Crit. Rev. Anal. Chem.* **1**, 73-117 (1970)
[266] Llopis, J. Surface potential at liquid interfaces. *Mod. Aspects Electrochem.* **6**, 91-158 (1971)
[267] Padova, J.I. Ionic solvation in nonaqueous and mixed solvents. *Mod. Asp. Electrochem.* **7**, 1-82 (1972)
[268] Bockris, J.O'M. & Reddy, A.K.N. Modern electrochemistry. Volume 1. Plenum Publishing Corporation, New York, USA (1973)
[269] Friedman, H.L. & Krishnan, C.V. Thermodynamics of ion hydration. In: *Water: A comprehensive treatise. Volume 3.* Franks, F., Ed. Plenum Press, New York, USA, pp 1-118 (1973)
[270] Kebarle, P. Gas-phase ion equilibria and ion solvation. *Mod. Asp. Electrochem.* **9**, 1-46 (1974)

[271] Desnoyers, J.E. Ionic solute hydration. *Phys. Chem. Liq.* **7**, 63-106 (1977)

[272] Randles, J.E.B. Structure at the free surface of water and aqueous electrolyte solutions. *Phys. Chem. Liq.* **7**, 107-179 (1977)

[273] Burgess, M.A. Metal Ions in Solution. Ellis Horwood, Chichester, UK (1978)

[274] Conway, B.E. The evaluation and use of properties of individual ions in solution. *J. Solut. Chem.* **7**, 721-770 (1978)

[275] Marcus, Y. Ion hydration. In: *Ion solvation.* John Wiley & Sons, New York, USA, pp 87-129 (1985)

[276] Conway, B.E. & Ayranci, E. Effective ionic radii and hydration volumes for evaluation of solution properties and ionic adsorption. *J. Solut. Chem.* **28**, 163-192 (1999)

[277] Paluch, M. Electrical properties of free surface of water and aqueous solutions. *Adv. Colloid. Interface Sci.* **84**, 27-45 (2000)

[278] Llano, J. & Eriksson, L.A. First principles electrochemistry: Electrons and protons reacting as independent ions. *J. Chem. Phys.* **117**, 10193-10206 (2002)

[279] Dill, K.A., Truskett, T.M., Vlachy, V. & Hribar-Lee, B. Modeling water, the hydrophobic effect, and ion solvation. *Ann. Rev. Biophys. Biomol. Struc.* **34**, 173-199 (2005)

[280] Collins, K.D. Ion hydration: Implications for cellular function, polyelectrolytes, and protein crystallization. *Biophys. Chem.* **119**, 271-281 (2006)

[281] Bronsted Nielsen, S. & Andersen, L.H. Properties of microsolvated ions: From the microenvironment of chromophore and alkali metal ions in proteins to negative ions in water clusters. *Biophys. Chem.* **124**, 229-237 (2006)

[282] Collins, K.D., Neilson, G.W. & Enderby, J.E. Ions in water: Characterizing the forces that control chemical processes and biological structure. *Biophys. Chem.* **128**, 96-104 (2007)

[283] Ben-Amotz, D. & Underwood, R. Unraveling water's entropic mysteries: A unified view of nonpolar, polar, and ionic hydration. *Acc. Chem. Res.* **41**, 957-967 (2008)

[284] Ferse, A. Zur Problematik individueller thermodynamischer Aktivitätskoeffizienten einzelner Ionensorten in Elektrolytlösungen hoher Konzentration. *Z. anorg. allg. Chem.* **634**, 797-815 (2008)

[285] Marcus, Y. Tetraalkylammonium ions in aqueous and non-aqueous solutions. *J. Solut. Chem.* **37**, 1071-1098 (2008)

[286] Floriano, W.B. & Nascimento, M.A.C. Dielectric constant and density of water as a function of pressure at constant temperature. *Braz. J. Phys.* **34**, 38-41 (2004)

[287] Kell, G.S. Density, thermal expansivity, and compressibility of liquid water from 0° to 150°C: Correlations and tables for atmospheric pressure and saturation reviewed and expressed on 1968 temperature scale. *J. Chem. Eng. Data* **20**, 97-105 (1975)

[288] Vavruch, I. On the evaluation of the surface tension-pressure coefficient for pure liquids. *J. Colloid. Interface Sci.* **169**, 249-250 (1995)

[289] Cooper, J.R. IAPWS release on surface tension of ordinary water substance. *Issued by the International Association for the Properties of Water and Steam* **1994**, 1-4 (1994)

[290] Fernández, D.P., Goodwin, A.R.H., Lemmon, E.W., Levelt Sengers, J.M.H. & Williams, R.C. A formulation for the static permittivity of water and steam at temperatures from 238 K to 873 K at pressures up to 1200 MPa, including derivatives and Debye-Hückel coefficients. *J. Phys. Chem. Ref. Data* **26**, 1125-1166 (1997)

[291] Badyal, Y.S., Saboungi, M.-L., Price, D.L., Shastri, S.D., Haeffner, D.R. & Soper, A.K. Electron distribution in water. *J. Chem. Phys.* **112**, 9206-9208 (2006)

[292] Coulson, C.A. & Eisenberg, D. Interactions of H_2O molecules in ice. I. The dipole moment of an H_2O molecule in ice. *Proc. Roy. Soc. London A* **291**, 445-453 (1966)

[293] Haar, L., Gallagher, J.S. & Kell, G.S. NBS/NRC steam tables. Hemisphere Publishing, New York, USA (1984)

[294] Wagman, D.D., Evans, W.H., Parker, V.B., Schumm, R.H., Halow, I., Bailey, S.M., Churney, K.L. & Nuttall, R.L. The NBS tables of chemical thermodynamic properties. Selected values for inorganic and C_1 and C_2 organic substances in SI units. *J. Phys. Chem. Ref. Data 11 Suppl.* **2**, 1-392 (1982)

[295] Kell, G.S. Precise representation of volume properties of water at one atmosphere. *J. Chem. Eng. Data* **12**, 66-69 (1967)

[296] Ellison, W.J., Lamkaouchi, K. & Moreau, J.-M. Water: A dielectric reference. *J. Mol. Liq.* **68**, 171-279 (1996)

[297] Coulomb, C.A. Sur l'électricité et le magnétisme, premier mémoire, construction et usage d'une balance électrique, fondée sur la propriété qu'ont les fils de métal, d'avoir une force de réaction de torsion proportionnelle à l'angle de torsion. *Mémoires de l'Académie Royale des Sciences* **1785**, 569-577 (1788)

[298] Huber, K.P. & Herzberg, G. Molecular spectra and molecular structure. Constants of diatomic molecules. Volume 4. Van Nostrand Reinhold, New York, USA (1979)

[299] McCaffrey, P.D., Mawhorter, R.J., Turner, A.R., Brain, P.T. & Rankin, D.W.H. Accurate equilibrium structures obtained from gas-phase electron diffraction data: sodium chloride. *J. Phys. Chem. A* **111**, 6103-6114 (2007)

[300] Feller, D., Glendening, E.D., Woon, D.E. & Feyereisen, M.W. An extended basis set *ab*

initio study of alkali metal cation-water clusters. *J. Chem. Phys.* **103**, 3526-3542 (1995)

[301] Kim, J., Lee, H.M., Suh, S.B., Majumdar, D. & Kim, K.S. Comparative *ab initio* study of the structures, energetics and spectra of $X^- \cdot (H_2O)_{n=1-4}$ [X=F,Cl,Br,I] clusters. *J. Chem. Phys.* **113**, 5259-5272 (2000)

[302] Thompson, A. & Taylor, B.N. Guide for the use of the international system of units (SI). *NIST special publication* **811**, 1-76 (2008)

[303] Glasstone, S. & Dolan, P. The effects of nuclear weapons. Edition 3. US Department of Defense/US Energy Research and Development Administration, Washington D.C., USA (1977)

[304] Debye, P. Polar molecules. The chemical catalog company Inc., New York, USA (1929)

[305] Hill, N., Vaughan, W.E., Price, A.H. & Davies, M.D. Dielectric properties and molecular behaviour. Van Nostrand Co., Ltd., London, UK (1969)

[306] Hasted, J.B. Liquid water: Dielectric properties. In: *Water, a comprehensive treatise. Volume 1.* Franks, F., Ed. Plenum, New York, USA, pp 255-309 (1973)

[307] Hasted, J.B. Aqueous dielectrics. Chapman and Hall, London, UK (1973)

[308] Böttcher, C.J.F., van Belle, O.C., Bordewijk, P. & Rip, A. Theory of electric polarization. Elsevier, Amsterdam, Netherlands (1973)

[309] Fröhlich, H. Theory of dielectrics: Dielectric constant and dielectric loss. Edition 2. Oxford Univ. Press, Oxford, UK (1987)

[310] Scaife, B.K.P. Principles of dielectrics. Oxford University Press, Oxford, UK (1989)

[311] van Hippel, A.R. Dielectrics and waves. Wiley, New York, USA (1954)

[312] Oehme, F. Dielektrische Messmethoden. Verlag Chemie, Weinheim, Germany (1961)

[313] Kaatze, U. Perspectives in dielectric measurement techniques for liquids. *Meas. Sci. Technol.* **19**, 112001/1-112001/4 (2008)

[314] Grove, T.T., Masters, M.F. & Miers, M.F. Determining dielectric constants using a parallel plate capacitor. *Am. J. Phys.* **73**, 52-56 (2005)

[315] Barthel, J. & Buchner, R. Dielectric properties of nonaqueous electrolyte solutions. *Pure Appl. Chem.* **58**, 1077-1090 (1986)

[316] Barthel, J., Buchner, R., Bachhuber, K., Hetzenauer, H., Kleebauer, M. & Ortmaier, H. Molecular processes in electrolyte solutions at microwave frequencies. *Pure Appl. Chem.* **62**, 2287-2296 (1990)

[317] Barthel, J. & Buchner, R. High frequency permittivity and its use in the investigation of solution properties . *Pure Appl. Chem.* **63**, 1473-1482 (1991)

[318] Barthel, J. & Buchner, R. Dielectric permittivity and relaxation of electrolyte solutions and their solvents. *Chem. Soc. Rev.* **21**, 263-270 (1992)

[319] Barthel, J., Hetzenauer, H. & Buchner, R. Dielectric relaxation of aqueous electrolyte solutions. I. Solvent relaxation of 1:2, 2:1 and 2:2 electrolyte solutions. *Ber. Bunsenges. Phys. Chem.* **96**, 988-997 (1992)

[320] Barthel, J., Buchner, R., Eberspächer, P.-N., Münsterer, M., Stauber, J. & Wurm, B. Dielectric relaxation spectroscopy of electrolyte solutions. Recent developments and prospects. *J. Mol. Liq.* **78**, 83-109 (1998)

[321] Buchner, R., Hefter, G.T. & May, P.M. Dielectric relaxation of aqueous NaCl solutions. *J. Phys. Chem. A* **103**, 1-9 (1999)

[322] Buchner, R. & Barthel, J. Dielectric relaxation in solutions. *Annu. Rep. Prog. Chem. Sect. C* **97**, 349-382 (2001)

[323] Barthel, J., Buchner, R. & Münsterer, M. Electrolyte data collection. Part 2: Dielectric properties of water and aqueous electrolyte solutions. Dechema, Frankfurt a. M., Germany (1995)

[324] Barthel, J., Hetzenauer, H. & Buchner, R. Dielectric relaxation of aqueous electrolyte solutions. II. Ion-pair relaxation of 1:2, 2:1 and 2:2 electrolyte solutions. *Ber. Bunsenges. Phys. Chem.* **96**, 1424-1432 (1992)

[325] Chan, D.Y.C., Mitchell, D.J. & Ninham, B.W. A model of solvent structure around ions. *J. Chem. Phys.* **70**, 2946-2957 (1979)

[326] Rashin, A.A. Electrostatics of ion-ion interactions in solution. *J. Phys. Chem.* **93**, 4664-4669 (1989)

[327] Pratt, L.R., Hummer, G. & Garciá, A.E. Ion pair potentials-of mean-force in water. *Biophys. Chem.* **51**, 147-165 (1994)

[328] Resat, H. Correcting for solvent-solvent electrostatic cutoffs considerably improves the ion-pair potential of mean force. *J. Chem. Phys.* **110**, 6887-6889 (1999)

[329] Hünenberger, P.H. & McCammon, J.A. Ewald artifacts in computer simulations of ionic solvation and ion-ion interaction: a continuum electrostatics study. *J. Chem. Phys.* **110**, 1856-1872 (1999)

[330] Hünenberger, P.H. Lattice-sum methods for computing electrostatic interactions in molecular simulations. In: *Simulation and theory of electrostatic interactions in solution: Computational chemistry, biophysics, and aqueous solution.* Hummer, G. & Pratt, L.R., Eds. American Institute of Physics, New York, USA, pp 17-83 (1999)

[331] Boresch, S. & Steinhauser, O. The dielectric self-consistent field method. I. Highways, byways, and illustrative results. *J. Chem. Phys.* **115**, 10780-10792 (2001)

[332] Allen, R. & Hansen, J.-P. Density functional approach to the effective interaction between charges within dielectric cavities. *J. Phys. Condens. Matter* **14**, 11981-11987 (2002)

[333] Bergdorf, M., Peter, C. & Hünenberger, P.H. Influence of cutoff truncation and artificial periodicity of electrostatic interactions in molecular simulations of solvated ions: A continuum electrostatics study. *J. Chem. Phys.* **119**, 9129-9144 (2003)

[334] Peter, C., van Gunsteren, W.F. & Hünenberger, P.H. A fast-Fourier-transform method to solve continuum-electrostatics problems with truncated electrostatic interactions: Algorithm and application to ionic solvation and ion-ion interaction. *J. Chem. Phys.* **119**, 12205-12223 (2003)

[335] Kastenholz, M. & Hünenberger, P.H. Development of a lattice-sum method emulating non-periodic boundary conditions for the treatment of electrostatic interactions in molecular simulations. A continuum electrostatics study. *J. Chem. Phys.* **124**, 124108/1-124108/12 (2006)

[336] Linse, P. Electrostatics in the presence of spherical dielectric discontinuities. *J. Chem. Phys.* **128**, 214505- (2008)

[337] Berkowitz, M., Omar, A.K., McCammon, J.A. & Rossky, P. Sodium chloride ion pair interaction in water: Computer simulation. *Chem. Phys. Lett.* **106**, 577-580 (1984)

[338] van Eerden, J., Briels, W.J., Harkema, S. & Feil, D. Potential of mean force by thermodynamic integration: Molecular-dynamics simulation of decomplexation. *Chem. Phys. Lett.* **164**, 370-376 (1989)

[339] Dang, L.X. & Pettitt, B.M. A theoretical study of like ion pairs in solution. *J. Phys. Chem.* **94**, 4303-4308 (1990)

[340] Boudon, S., Wipff, G. & Maigret, B. Monte Carlo simulations on the like-charged guanidinium-guanidinium ion pair in water. *J. Phys. Chem.* **94**, 6056-6061 (1990)

[341] Dang, L.X., Rice, J.E. & Kollman, P.A. The effect of water models on the interaction of the sodium-chloride ion pair in water: Molecular dynamics simulations. *J. Chem. Phys.* **93**, 7528-7529 (1990)

[342] Guàrdia, E., Rey, R. & Padró, J.A. Na^+-Na^+ and Cl^--Cl^- ion pairs in water: Mean force potentials by constrained molecular dynamics. *J. Chem. Phys.* **95**, 2823-2831 (1991)

[343] Dang, L.X., Pettitt, B.M. & Rossky, P.J. On the correlation between like ion pairs in water. *J. Chem. Phys.* **96**, 4046-4047 (1992)

[344] Dang, L.X. Fluoride-fluoride association in water from molecular dynamics simulations. *Chem. Phys. Lett.* **200**, 21-25 (1992)

[345] Rey, R. & Guàrdia, E. Dynamical aspects of the Na^+-Cl^- ion pair association in water. *J. Phys. Chem.* **96**, 4712-4718 (1992)

[346] Hummer, G., Soumpasis, D.M. & Neumann, M. Computer simulations do not support Cl-Cl pairing in aqueous NaCl solution. *Mol. Phys.* **81**, 1155-1163 (1994)

[347] Dang, L.X. Free energies for association of Cs^+ to 18-crown-6 in water. A molecular dynamics study including counter ions. *Chem. Phys. Lett.* **227**, 211-214 (1994)

[348] Payne, V.A., Forsyth, M., Ratner, M.A., Shriver, D.F. & de Leeuw, S.W. Highly concentrated salt solutions: Molecular dynamics simulations of structure and transport. *J. Chem. Phys.* **100**, 5201-5210 (1994)

[349] Smith, D.E. & Dang, L.X. Interionic potentials of mean force for $SrCl_2$ in polarizable water. A computer simulation study. *Chem. Phys. Lett.* **230**, 209-214 (1994)

[350] Figueirido, F., del Buono, G.S. & Levy, R.M. On finite-size effects in computer simulations using the Ewald potential. *J. Chem. Phys.* **103**, 6133-6142 (1995)

[351] Dang, L.X. Mechanism and thermodynamics of ion selectivity in aqueous solutions of 18-crown-6 ether: A molecular-dynamics study. *J. Am. Chem. Soc.* **117**, 6954-6960 (1995)

[352] Dang, L.X. & Kollman, P.A. Free energy of association of the K^+:18-crown-6 complex in water: A new molecular dynamics study. *J. Phys. Chem.* **99**, 55-58 (1995)

[353] Friedman, R.A. & Mezei, M. The potentials of mean force of sodium chloride and sodium dimethylphosphate in water: An application of adaptive umbrella sampling. *J. Chem. Phys.* **102**, 419-426 (1995)

[354] Laria, D. & Fernández-Prini, R. Molecular dynamics study of water clusters containing ion pairs: From contact to dissociation. *J. Chem. Phys.* **102**, 7664-7673 (1995)

[355] Resat, H., Mezei, M. & McCammon, J.A. Use of the grand canonical ensemble in potential of mean force calculations. *J. Phys. Chem.* **100**, 1426-1433 (1996)

[356] Soetens, J.-C., Millot, C., Chipot, C., Jansen, G., Ángyán, J.G. & Maigret, B. Effect of polarizability on the potential of mean force of two cations. The guanidinium-guanidinium ion pair in water. *J. Phys. Chem. B* **101**, 10910-10917 (1997)

[357] Rozanska, X. & Chipot, C. Modeling ion-ion interaction in proteins: A molecular dynamics free energy calculation of the guanidinium-acetate association. *J. Chem. Phys.* **112**, 9691-9694 (2000)

[358] Martorana, V., la Fata, L., Bulone, D. & San Biagio, P.L. Potential of mean force between two ions in a sucrose rich aqueous solution. *Chem. Phys. Lett.* **329**, 221-227

(2000)
[359] Peslherbe, G.H., Ladanyi, B.M. & Hynes, J.T. Free energetics of NaI contact and solvent-separated ion pairs in water clusters. *J. Phys. Chem. A* **104**, 4533-4548 (2000)
[360] Masunov, A. & Lazaridis, T. Potentials of mean force between ionizable amino acid side chains in water. *J. Am. Chem. Soc.* **125**, 1722-1730 (2003)
[361] Shinto, H., Morisada, S., Miyahara, M. & Higashitani, K. A reexamination of mean force potentials for the methane pair and the constituent ion pairs of NaCl in water. *J. Chem. Engin. Jpn.* **36**, 57-65 (2003)
[362] Liu, W.B., Wood, R.H. & Doren, D.J. Hydration free energy and potential of mean force for a model of the sodium chloride ion pair in supercritical water with *ab initio* solute-solvent interactions. *J. Chem. Phys.* **118**, 2837-2844 (2003)
[363] Maksimiak, K., Rodziewicz-Motowidlo, S., Czaplewski, C., Liwo, A. & Scheraga, H.A. Molecular simulation study of the potentials of mean force for the interactions between models of like-charged and between charged and nonpolar amino acid side chains in water. *J. Phys. Chem. B* **107**, 13496-13504 (2003)
[364] Hassan, S.A. Intermolecular potentials of mean force of amino acid side chain interactions in aqueous medium. *J. Phys. Chem. B* **108**, 19501-19509 (2004)
[365] Shinto, H., Morisada, S. & Higashitani, K. A reexamination of mean force potentials for the constituent ion pairs of tetramethylammonium chloride in water. *J. Chem. Engin. Jpn.* **37**, 1345-1356 (2004)
[366] Zidi, Z.S. Solvation of sodium-chloride ion pair in water cluster at atmospheric conditions: Grand canonical ensemble Monte Carlo simulation. *J. Chem. Phys.* **123**, 064309/1-064309/13 (2005)
[367] Ghoufi, A. & Malfreyt, P. Calculations of the potential of mean force from molecular dynamics simulations using different methodologies: An application to the determination of the binding thermodynamic properties of an ion pair. *Mol. Phys.* **104**, 3787-3799 (2006)
[368] Hess, B., Holm, C. & van der Vegt, N. Osmotic coefficients of atomistic NaCl(aq) force fields. *J. Chem. Phys.* **124**, 164509/1-164509/8 (2006)
[369] Keasler, S.J., Nellas, R.B. & Chen, B. Water mediated attraction between repulsive ions: A cluster-based simulation approach. *J. Chem. Phys.* **125**, 144520/1-144520/5 (2006)
[370] Ghoufi, A., Archirel, P., Morel, J.-P., Morel-Desrosiers, N., Boutin, A., Methodology for the calculation of the potential of mean force for a cation-π complex in water. *Chem. Phys. Chem.* **8**, 1648-1656 (2007)
[371] Jagoda-Cwiklik, B., Vácha, R., Lund, M., Srebro, M. & Jungwirth, P. Ion pairing as a possible clue for discriminating between sodium and potassium in biological and other complex environments. *J. Phys. Chem. B* **111**, 14077-14079 (2007)
[372] Makowski, M., Liwo, A., Maksimiak, K., Makowska, J. & Scheraga, H.A. Simple physics-based analytical formulas for the potentials of mean force for the interaction of amino acid side chains in water. 2. Tests with simple spherical systems. *J. Phys. Chem. B* **111**, 2917-2924 (2007)
[373] Bruneval, F., Donadio, D. & Parrinello, M. Molecular dynamics study of the solvation of calcium carbonate in water. *J. Phys. Chem. B* **111**, 12219-12227 (2007)
[374] Hassan, S.A. Computer simulation of ion cluster speciation in concentrated aqueous solutions at ambient conditions. *J. Phys. Chem. B* **112**, 10573-10584 (2008)
[375] Khavrutskii, I.V., Dzubiella, J. & McCammon, J.A. Computing accurate potentials of mean force in electrolyte solutions with the generalized gradient-augmented harmonic Fourier beads method. *J. Chem. Phys.* **128**, 044106/1-044106/13 (2008)
[376] Baumketner, A. Removing systematic errors in potentials of mean force computed in molecular simulations of solvated ions using reaction field-based electrostatics. *J. Chem. Phys.* **130**, 104106/1-104106/10 (2009)
[377] Timko, J., Bucher, D. & Kuyucak, S. Dissociation of NaCl in water from ab initio molecular dynamics simulations. *J. Chem. Phys.* **132**, 114510/1-114510/8 (2010)
[378] Bredig, M.A., Bronstein, H.R. & Smith Jr., W.T. Miscibility of liquid metals with salts. II. The potassium-potassium fluoride and cesium-cesium halide systems. *J. Am. Chem. Soc.* **77**, 1454-1458 (1955)
[379] Bredig, M.A., Johnson, J.W. & Smith Jr., W.T. Miscibility of liquid metals with salts. I. The sodium-sodium halide systems. *J. Am. Chem. Soc.* **77**, 307-312 (1955)
[380] Johnson, J.W. & Bredig, M.A. Miscibility of metals with salts in the molten state. III. The potassium-potassium halide systems. *J. Phys. Chem.* **62**, 604-607 (1958)
[381] Bredig, M.A. & Johnson, J.W. Miscibility of metals with salts. V. The rubidium-rubidium halide systems. *J. Phys. Chem.* **64**, 1899-1900 (1960)
[382] Bredig, M.A. & Bronstein, H.R. Miscibility of liquid metals with salts. IV. The sodium-sodium halide systems at high temperatures. *J. Phys. Chem.* **64**, 64-67 (1960)
[383] Gellings, P.J., Huiskamp, G.B. & van den Broek, E.G. A model for solutions of non-metals in liquid alkali metals. *J. Chem. Soc. Faraday Trans.* **II**, 531-536 (1972)
[384] Gellings, P.J., van der Scheer, A. & Caspers, W.J. Extension of a model for solutions of non-metals in liquid alkali metals: Calculation of the enthalpy of solvation. *J. Chem. Soc. Faraday Trans.* **II**, 531-536 (1974)

[385] Yokokawa, H. & Kleppa, O.J. Thermodynamics of liquid cesium-cesium halide mixtures at high temperatures. *J. Chem. Phys.* **76**, 5574-5587 (1982)

[386] Pietzko, S. & Schmutzler, R.W. Determination of the thermodynamic activity of potassium in K/KCl melts. *Z. Phys. Chem. - Int. J. Res. Phys. Chem. Chem. Phys.* **193**, 123-138 (1996)

[387] Gurnett, D.A. & Bhattacharjee, A. Introduction to plasma physics: With space and laboratory applications. Cambridge University Press, Cambridge, UK (2005)

[388] Kassabji, F. & Fauchais, P. Les générateurs à plasma. *Revue Phys. Appl.* **16**, 549-577 (1981)

[389] Graf, S., Altwegg, K., Balsiger, H., Fiethe, B. & Fischer, J. First pressure measurements on-board the ESA Rosetta spacecraft. *Geophys. Res. Abstracts* **8**, 02149/1-02149/1 (2006)

[390] van Atta, C.M. & Hablanian, M. Vacuums and vacuum technology. In: *Encyclopedia of Physics*. Lerner, R.G. & Trigg, G.L., Eds. VCH Publisher, New York, USA, pp 1330-1334 (1991)

[391] Chang, J.-S., Kelly, A.J. & Crowley, J.M. Handbook of electrostatic processes. Marcel Dekker, New York (1995)

[392] Rosenstock, H.M. The measurement of ionization and appearance potentials. *Int. J. Mass Spectrom. Ion Phys.* **20**, 139-190 (1976)

[393] Rienstra-Kiracofe, J.C., Tschumper, G.S., Schaefer, H.F., Nandi, S. & Ellison, G.B. Atomic and molecular electron affinities: Photoelectron experiments and theoretical computations. *Chem. Rev.* **102**, 231-282 (2002)

[394] Madelung, E. The electric field in systems of regularly arranged point charges. *Phys. Z.* **19**, 524-533 (1918)

[395] Born, M. & Landé, A. Über die Berechnung der Kompressibilität regulärer Kristalle aus der Gittertheorie. *Verh. dtsch. physik. Ges.* **20**, 210-216 (1918)

[396] Born, M. Eine thermochemische Anwendung der Gittertheorie. *Verh. dtsch. physik. Ges.* **21**, 13-24 (1919)

[397] Born, M. & Mayer, J. Zur Gittertheorie der Ionenkristalle. *Z. Phys.* **75**, 1-18 (1932)

[398] Kapustinskii, A.F. Lattice energy of ionic crystals. *Quart. Rev.* **10**, 283-294 (1956)

[399] Maradudin, A. Theory of lattice dynamics in the harmonic approximation. Edition 2. Academic Press Inc., New York, USA (1971)

[400] Wallace, D.C. Thermodynamics of crystals. Wiley, New York, USA (1972)

[401] Ashcroft, N.W. & Mermin, N.D. Solid state physics. Holt, Rinehart & Winston, New York, USA (1976)

[402] Herring, C. & Nichols, M.H. Thermionic emission. *Rev. Mod. Phys.* **21**, 185-270 (1949)

[403] Riviere, J.C. Work function: Measurements and results. In: *Solid state surface science. Volume 1*. Green, M., Ed. Decker, New York, USA, pp 179-289 (1969)

[404] Klein, O., Die Normal-Voltapotentiale $\Delta\psi_o$ der wichtigsten elektrochemischen Zweiphasensysteme, insbesondere der Elektroden: Metall/Metallsalzlösung. *Z. Elektrochem.* **43**, 570-584 (1937)

[405] Klein, O., Normalvoltapotentiale $\Delta\psi_o$ elektrochemischer Zweiphasensysteme. Zugleich ein Beitrag zur zeitlichen Oberflächenstrukturveränderung hochkonzentrierter Elektrolytlösungen. *Z. Elektrochem.* **44**, 562-568 (1938)

[406] Verwey, E.J.W. The electrical potential drop at the free surface of water, and other phase boundary potentials. *Rec. Trav. Chim.* **61**, 564-572 (1942)

[407] Hush, N.S. The free energies of hydration of gaseous ions. *Austral. J. Sci. Res.* **1**, 480-493 (1948)

[408] Strehlov, H. Zum Problem des Einzelelektrodenpotentials. *Z. Elektrochem.* **56**, 119-129 (1952)

[409] Passoth, G. Über die Hydrationsenergien und die scheinbaren Molvolumen einwertiger Ionen. *Z. Phys. Chem. (Leipzig)* **203**, 275-291 (1954)

[410] Mishchenko, K.P. & Kvyat, E.I. Solvatatsiya ionov v rastvorakh elektrolitov. 3. Skachok potentsiala na granitse vodnyi rastvor-gazovaya faza. *Zh. Fiz. Khim.* **28**, 1451-1457 (1954)

[411] Randles, J.E.B. The real hydration energies of ions. *Trans. Farad. Soc.* **52**, 1573-1581 (1956)

[412] de Bethune, A.J. Recalculation of the Latimer, Pitzer and Slansky absolute electrode potential - A discussion of its operational significance. *J. Chem. Phys.* **29**, 616-625 (1958)

[413] Case, B. & Parsons, R. The real free energies of solvation of ions in some non-aqueous and mixed solvents. *Trans. Faraday Soc.* **63**, 1224-1239 (1967)

[414] Bockris, J.O'M. & Argade, S.D. Work function of metals and the potential at which they have zero charge in contact with solutions. *J. Chem. Phys.* **11**, 5133-5134 (1968)

[415] Schiffrin, D.J. Real standard entropy of ions in water. *Trans. Faraday Soc.* **66**, 2464-2468 (1970)

[416] Nedermeijer-Denessen, H.J.M. & de Ligny, C.L. A revised calculation of the standard chemical free enthalpy and the standard enthalpy of hydration of the hydrogen ion and

of the surface potential of water at 25°C. *J. Electroanal. Chem.* **57**, 265-266 (1974)
[417] Klots, C.E. Solubility of protons in water. *J. Phys. Chem.* **85**, 3585-3588 (1981)
[418] Krishtalik, L.I., Alpatova, N.M. & Ovsyannikova, E.V. Determination of the surface potentials of solvents. *J. Electroanal. Chem.* **329**, 1-9 (1992)
[419] Fawcett, W.R. The ionic work function and its role in estimating absolute electrode potentials. *Langmuir* **24**, 9768-9875 (2008)
[420] Krishtalik, L.I. The surface potential of solvent and the intraphase pre-existing potential. *Russ. J. Electrochem.* **44**, 43-49 (2008)
[421] Jackson, J.D. Classical electrodynamics. Edition 3. John Wiley & Sons, New York, USA (1999)
[422] Randles, J.E.B. The interface between aqueous electrolyte solutions and the gas phase. *Adv. Electrochem. Eng.* **3**, 1-30 (1963)
[423] Harrison, J.A., Randles, J.E.B. & Schiffrin, D.J. Ionic hydration and the thermodynamic cationic surface excess at the mercury-aqueous electrolyte interface. *J. Electroanal. Chem.* **25**, 197-207 (1970)
[424] Conway, B.E. Ion hydration near air/water interfaces and the structure of liquid surfaces. *J. Electroanal. Chem.* **65**, 491-504 (1975)
[425] Madden, W.G., Gomer, R. & Mandell, M.J. The effect of electrolyte on dipole layers at liquid-air interfaces. *J. Phys. Chem.* **81**, 2652-2657 (1977)
[426] Schurhammer, R. & Wipff, G. About the TATB hypothesis: solvation of the $As\Phi_4^+$ and $B\Phi_4^-$ ions and their tetrahedral and spherical analogues in aqueous/nonaqueous solvents and at water-chloroform interfaces. *New. J. Chem.* **23**, 381-391 (1999)
[427] Schurhammer, R. & Wipff, G. About the TATB assumption: effect of charge reversal on transfer of large spherical ions from aqueous to non-aqueous solvents and on their interfacial behaviour. *J. Mol. Struct. (Theochem)* **500**, 139-155 (2000)
[428] Schurhammer, R., Engler, E. & Wipff, G. Hydrophobic ions in TIP5P water and at water-chloroform interfaces: The effect of sign inversion investigated by MD and FEP simulations. *J. Phys. Chem. B* **105**, 10700-10708 (2001)
[429] Schurhammer, R. & Wipff, G. Corrigendum to "About the TATB assumption: effect of charge reversal on transfer of large spherical ions from aqueous to non-aqueous solvents and on their interfacial behaviour." [J. Mol. Struct. (Theochem) 500, 139-155 (2000)] *J. Mol. Struct. (Theochem)* **536**, 289-289 (2001)
[430] Dang, L.X. & Chang, T.-M. Molecular mechanism of ion binding to the liquid/vapor interface of water. *J. Phys. Chem. B* **106**, 235-238 (2002)
[431] Paul, S. & Chandra, A. Dynamics of water molecules at liquid-vapour interfaces of aqueous ionic solutions: effects of ion concentration. *Chem. Phys. Lett.* **373**, 87-93 (2003)
[432] Manciu, M. & Ruckenstein, E. On the interactions of ions with the air/water interface. *Langmuir* **21**, 11312-11319 (2005)
[433] Chang, T.-M. & Dang, L.X. Recent advances in molecular simulations of ion solvation at liquid interfaces. *Chem. Rev.* **106**, 1305-1322 (2006)
[434] Hrobárik, T., Vrbka, L. & Jungwirth, P. Selected biologically relevant ions at the air/water interface: A comparative molecular dynamics study. *Biophys. Chem.* **124**, 238-242 (2006)
[435] Ishiyama, T. & Morita, A. Molecular dynamics study of gas-liquid aqueous sodium halide interfaces. I. Flexible and polarizable molecular modeling and interfacial properties. *J. Phys. Chem.* **111**, 721-737 (2007)
[436] Thomas, J.L., Roeselov, M., Dang, L.X. & Tobias, D.J. Molecular dynamics simulations of the solution-air interface of aqueous sodium nitrate. *J. Phys. Chem. A* **111**, 3091-3098 (2007)
[437] Eggimann, B.L. & Siepmann, J.I. Size effects on the solvation of anions at the aqueous liquid-vapor interface. *J. Phys. Chem. C* **112**, 210-218 (2008)
[438] Brown, M.A., D'Auria, R., Kuo, I.-F.W., Krisch, M.J., Starr, D.E., Bluhm, H., Tobias, D.J. & Hemminger, J.C. Ion spatial distributions at the liquid-vapor interface of aqueous potassium fluoride solutions. *Phys. Chem. Chem. Phys.* **10**, 4778-4784 (2008)
[439] Warren, G.L. & Patel, S. Electrostatic properties of aqueous salt solution interfaces: A comparison of polarizable and nonpolarizable ion models. *J. Phys. Chem. B* **112**, 11679-11693 (2008)
[440] Warren, G.L. & Patel, S. Comparison of the solvation structure of polarizable and nonpolarizable ions in bulk water and near the aqueous liquid-vapor interface. *J. Phys. Chem. C* **113**, 7455-7467 (2008)
[441] Frumkin, A.N. Phasengrenzkräfte und Adsorbtion an der Trennungsfläche Luft | Lösung anorganischer Elektrolyte. *Z. Phys. Chem. (Stöchiometrie u. Verwandtschaftslehre)* **109**, 34-48 (1924)
[442] Randles, J.E.B. Ionic hydration and the surface potential of aqueous electrolytes. *Discuss. Faraday Soc.* **24**, 194-199 (1957)
[443] Buhl, A. Über die Potentialdifferenz in der Doppelschicht an der Oberfläche einfacher Elektrolyte und des reinen Wassers. *Ann. Phys.* **389**, 211-244 (1927)

[444] Gomer, R. & Tryson, G. An experimental determination of absolute half-cell emf's and single ion free energies of solvation. *J. Chem. Phys.* **66**, 4413-4424 (1977)

[445] Farrell, J.R. & McTigue, P. Precise compensating potential difference measurements with a voltaic cell. The surface potential of water. *J. Electroanal. Chem.* **139**, 37-56 (1982)

[446] Farrell, J.R. & McTigue, P. Temperature dependence of the compensating potential difference of a voltaic cell. *J. Electroanal. Chem.* **163**, 129-136 (1984)

[447] Borazio, A., Farrell, J.R. & McTigue, P. Charge distribution at the gas-water interface. The surface potential of water. *J. Electroanal. Chem.* **193**, 103-112 (1985)

[448] Frumkin, A.N., Iofa, Z.A. & Gerovich, M.A. K voprosu o raznosti potentsialov na granitse voda gaz. *Zh. Fiz. Khim.* **10**, 1455-1468 (1956)

[449] Kamieński, B. The nature of the electric potential at the free surface of aqueous solutions. *Electrochim. Acta* **1**, 272-277 (1959)

[450] Kamieński, B. Zależność między napięciem powierzchniowym i elektrycznym na powierzchni swobodnej roztworów (Surface tension and electric potential on the free surface of solutions). *Wiad. Chem.* **14**, 619-648 (1960)

[451] Blank, M. & Ottewil, R.H. The adsorption of aromatic vapors on water surfaces. *J. Phys. Chem.* **68**, 2206-2211 (1964)

[452] Demchak, R.J. & Fort Jr., T. Surface dipole moment of close-packed un-ionized monolayers at the air-water interface. *J. Colloid. Interface Sci.* **46**, 191-202 (1974)

[453] Koczorowski, Z. Surface potentials of liquid solutions and solvents. *Bull. Pol. Acad. Sci. Chem.* **45**, 225-242 (1997)

[454] Parfenyuk, V.I. Surface potential at the gas-aqueous solution interface. *Colloid J.* **64**, 588-595 (2002)

[455] Parsons, R. Equilibrium properties of electrified interphases. *Mod. Aspects Electrochem.* **1**, 103-179 (1954)

[456] Gibbs, J.W. On the equilibrium of heterogeneous substances. *Trans. Connecticut Acad.* **3**, 343-524 (1878)

[457] Marcus, Y. A simple empirical model describing the thermodynamics of hydration of ions of widely varying charges, sizes, and shapes. *Biophys. Chem.* **51**, 111-127 (1994)

[458] Ben-Naim, A. Molecular theory of solutions. Oxford University Press, Oxford, UK (2006)

[459] Damaskin, B.B. & Petrii, O.A. Vvedenie v elektrokhimicheskuyu kinetiku (Electrochemical kinetics: An introduction). Vysshaya Shkola, Moskow, UdSSR (1983)

[460] Koczorowski, Z. & Zagórska, I. Temperature coefficient of the surface potential of methanol. *Rocz. Chém.* **44**, 911-913 (1970)

[461] Trasatti, S. Interfacial behaviour of non-aqueous solvents. *Electrochim. Acta* **32**, 843-850 (1987)

[462] Barraclough, C.G., Borazio, A., McTigue, P.T. & Verity, B. The real standard potential of the aqueous hydrogen scale of the silver/silver chloride electrode in methanol: the surface potential of methanol. *J. Electroanal. Chem.* **243**, 353-365 (1988)

[463] Matsumoto, M. & Kataoka, Y. Molecular orientation near liquid-vapor interface of methanol: Simulational study. *J. Chem. Phys.* **90**, 2398-2407 (1989)

[464] Matsumoto, M. & Kataoka, Y. Erratum to "Molecular orientation near liquid-vapor interface of methanol: Simulational study" [J. Chem. Phys. 90, 2398-2407 (1989)]. *J. Chem. Phys.* **95**, 7782-7782 (1991)

[465] Barraclough, C.G., McTigue, P.T. & Ng, Y.L. Surface potentials of water, methanol and water+methanol mixtures. *J. Electroanal. Chem.* **329**, 9-24 (1992)

[466] Parfenyuk, V.I. Surface potential at the gas-nonaqueous solution interface. *Colloid J.* **66**, 466-469 (2004)

[467] Case, B., Hush, N.S., Parsons, R. & Peover, M.E. The real solvation energies of hydrocarbon ions in acetonitrile and the surface potential of acetonitrile. *J. Electroanal. Chem.* **10**, 360-370 (1965)

[468] Harder, E. & Roux, B. On the origin of the electrostatic potential difference at a liquid-vacuum interface. *J. Chem. Phys.* **129**, 234706/1-234706/9 (2008)

[469] Patel, S. & Brooks III, C.L. Revisiting the hexane-water interface via molecular dynamics simulations using nonadditive alkane-water potentials. *J. Chem. Phys.* **124**, 204706/1-204706/14 (2006)

[470] Brodskaya, E.N. & Zakharov, V.V. Computer simulation study of the surface polarization of pure polar liquids. *J. Chem. Phys.* **102**, 4595-4599 (1995)

[471] Wohlfahrt, C. Pure liquids: Data. In: *Landolt-Börnstein. Numerical data and functional relationships in science and technology. Volume 6.* Madelung, O., Ed. Springer, Berlin, Germany, pp 5-228 (1991)

[472] Jorgensen, W.L. Optimized intermolecular potential functions for liquid alcohols. *J. Phys. Chem.* **90**, 1276-1284 (1986)

[473] Parsons, R. & Rubin, B.T. The medium effect for single ionic species. *J. Chem. Soc. Faraday Trans. I* **70**, 1636-1648 (1974)

[474] Rabalais, J.W., Werme, L.O., Bergmark, T., Karlsson, L., Hussain, M. & Siegbahn, K. Electron spectroscopy of open-shell systems: Spectra of $Ni(C_5H_5)_2$, $Fe(C_5H_5)_2$,

$Mn(C_5H_5)_2$, and $Cr(C_5H_5)_2$. *J. Chem. Phys.* **57**, 1185-1192 (1972)
[475] Frumkin, A.N. Potentsialy Nulevogo Zaryada. Nauka, Moscow, Russia (1979)
[476] Fomenko, V.S. Emissionnye svoistva materialov. Naukova Dumka, Kiev, Ukraine (1970)
[477] Damaskin, B.B. & Kaganovich, R.I. The Volta potential at an electrode/solution interface at the zero-charge potential. *Elektrokhimiya* **13**, 248-250 (1977)
[478] Izmailov, N.A. Energy of solvation and persolvation (lg γ_0) of individual ions in non-aqueous solutions. *Dokl. Akad. Nauk. SSSR* **149**, 1364-1367 (1963)
[479] Zagoruchenko, V.A. & Zhuravlev, A.M. Thermophysical properties of gaseous and liquid methane. Israel Program for Scientific Translations, Jerusalem, Israel (1970)
[480] Patel, S. & Brooks III, C.L. Fluctuating charge force fields: recent developments and applications from small molecules to macromolecular biological systems. *Mol. Simul.* **32**, 231-249 (2006)
[481] Klauda, J.B., Brooks, B.R., MacKerrell, A.D., Venable, R.M. & Pastor, R.W. An ab initio study on the torsional surface of alkanes and its effect on molecular simulations of alkanes and a DPPC bilayer. *J. Phys. Chem. B* **109**, 5300-5311 (2005)
[482] Reiss, H. & Heller, A. The absolute potential of the standard hydrogen electrode: A new estimate. *J. Phys. Chem.* **89**, 4207-4213 (1985)
[483] Isse, A.A. & Gennaro, A. Absolute potential of the standard hydrogen electrode and the problem of interconversion of potentials in different solvents. *J. Phys. Chem. B* **114**, 7894-7899 (2010)
[484] Csányi, G., Abaret, T., Moras, G., Payne, M.C. & de Vita, A. Multiscale hybrid simulation methods for material systems. *J. Phys.: Condens. Matter* **17**, R691-R703 (2005)
[485] Wang, C.Y. & Zhang, X. Multiscale modeling and related hybrid approaches. *Curr. Opin. Solid State & Mat. Sci.* **10**, 2-14 (2006)
[486] Sharma, S., Ding, F. & Dokholyan, N.V. Multiscale modeling of nucleosome dynamics. *Biophys. J.* **92**, 1457-1470 (2007)
[487] Berendsen, H.J.C. Simulating the physical world. Cambridge University Press, Cambridge, UK (2007)
[488] Praprotnik, M., delle Site, L. & Kremer, K. Multiscale simulation of soft matter: From scale bridging to adaptive resolution. *Ann. Rev. Phys. Chem.* **59**, 545-571 (2008)
[489] Pandit, S.A. & Scott, H.L. Multiscale simulations of heterogeneous model membranes. *Biochim. Biophys. Acta Biomembranes* **1788**, 136-148 (2009)
[490] Allen, M.P. & Tildesley, D.J. Computer simulation of liquids. Oxford University Press, New York, USA (1987)
[491] van Gunsteren W.F., Weiner, P.K., Wilkinson, Computer simulation of biomolecular systems: Theoretical and experimental applications. Volume 1. Escom Science Publ., Leiden, The Netherlands (1989)
[492] van Gunsteren, W.F. & Berendsen, H.J.C. Computer simulation of molecular dynamics: Methodology, applications and perspectives in chemistry. *Angew. Chem. Int. Ed.* **29**, 992-1023 (1990)
[493] Lipkowitz, K.B. & Boyd, D.B. Reviews in computational chemistry. Volume 1. Wiley, New York, USA (1990)
[494] Tomasi, J. & Persico, M. Molecular interactions in solution: An overview of methods based on continuous distributions of solvent. *Chem. Rev.* **94**, 2027-2094 (1994)
[495] Smith, P.E. & Pettitt, B.M. Modeling solvent in biomolecular systems. *J. Phys. Chem.* **98**, 9700-9711 (1994)
[496] Hünenberger, P.H. & van Gunsteren, W.F. Empirical classical interaction functions for molecular simulations. In: *Computer simulation of biomolecular systems, theoretical and experimental applications. Volume 3.* van Gunsteren, W.F., Weiner, P.K. & Wilkinson, A.J., Eds. Kluwer/Escom Science Publishers, Dordrecht, The Netherlands., pp 3-82. (1997)
[497] Roux, B. & Simonson, T. Implicit solvent models. *Biophys. Chem.* **78**, 1-20 (1999)
[498] Baschnagel, J., Binder, K., Doruker, P., Gusev, A.A., Hahn, O., Kremer, K., Mattice, W.L., Müller-Plathe, F., Murat, M., Paul, W., Santos, S., Suter, U.W. & Tries, W. Bridging the gap between atomistic and coarse-grained models of polymers: status and perspectives. *Adv. Polym. Sci.* **152**, 41-156 (2000)
[499] Orozco, M. & Luque, F.J. Theoretical methods for the description of the solvent effect in biomolecular systems. *Chem. Rev.* **100**, 4187-4225 (2000)
[500] Leach, A. Molecular modelling: Principles and applications. Edition 2. Prentice Hall, New York, USA (2001)
[501] Cremer, D., Kraka, E. & He, Y. Exact geometries from quantum chemical calculations. *J. Mol. Struct.* **567**, 275-293 (2001)
[502] Friesner, R.A. & Dunietz, B.D. Large-scale ab initio quantum chemical calculations on biological systems. *Acc. Chem. Res.* **34**, 351-358 (2001)
[503] Karplus, M. & McCammon, J.A. Molecular dynamics simulations of biomolecules. *Nature Struct. Biol.* **9**, 646-652 (2002)
[504] Frenkel, D. & Smit, B. Understanding molecular simulation. Edition 2. Academic Press, San Diego, USA (2002)

[505] Schlick, T. Molecular modeling and simulation: An interdisciplinary guide. Springer, New York, USA (2002)

[506] Briels, W.J. & Akkermans, R.L.C. Representation of coarse-grained potentials for polymer simulations. *Mol. Simul.* **28**, 145-152 (2002)

[507] Klopper, W. & Noga, J. Accurate quantum-chemical prediction of enthalpies of formation of small molecules in the gas phase. *Chem. Phys. Chem.* **4**, 32-48 (2003)

[508] Rapaport, D.C. The art of molecular dynamics simulation. Edition 2. Cambridge Univ. Press, Cambridge, UK (2004)

[509] Cramer, C.J. Essentials of computational chemistry: Theories and models. Edition 2. John Wiley & Sons, Chichester, UK (2004)

[510] Nielsen, S.O., Lopez, C.F., Srinivas, G. & Klein, M.L. Coarse grain models and the computer simulation of soft materials. *J. Phys.: Condens. Matter* **16**, R481-R512 (2004)

[511] Friesner, R.A. Ab initio quantum chemistry: Methodology and applications. *Proc. Natl. Acad. Sci. USA* **102**, 6648-6653 (2005)

[512] Tozzini, V. Coarse-grained models for proteins. *Curr. Opin. Struct. Biol.* **15**, 144-150 (2005)

[513] Baker, N.A. Improving implicit solvent simulations: a Poisson-centric view. *Curr. Opin. Struct. Biol.* **15**, 137-143 (2005)

[514] Tomasi, J., Mennucci, B. & Cammi, R. Quantum mechanical continuum solvation models. *Chem. Rev.* **105**, 2999-3094 (2005)

[515] van Gunsteren, W.F., Bakowies, D., Baron, R., Chandrasekhar, I, Christen, M., Daura, X., Gee, P., Geerke, D.P., Glättli, A., Hünenberger, P.H., Kastenholz, M.A., Oostenbrink, C., Schenk, M., Trzesniak, D., van der Vegt, N.F.A. & Yu, H.B. Biomolecular modelling: goals, problems, perspectives. *Angew. Chem. Int. Ed.* **45**, 4064-4092 (2006)

[516] Cavalli, A., Carloni, P. & Recanatini, M. Target-related applications of first principles quantum chemical methods in drug design. *Chem. Rev.* **106**, 3497-3519 (2006)

[517] Peters, M.B., Raha, K. & Merz, K.M. Quantum mechanics in structure-based drug design. *Curr. Opin. Drug Discovery & Develop.* **9**, 370-379 (2006)

[518] Müller, M., Katsov, K. & Schick, M. Biological and synthetic membranes: What can be learned from a coarse-grained description? *Phys. Rep.* **434**, 113-176 (2006)

[519] Bond, P.J., Holyoake, J., Ivetac, A., Khalid, S. & Sansom, M.S.P. Coarse-grained molecular dynamics simulations of membrane proteins and peptides. *J. Struct. Biol.* **157**, 593-605 (2007)

[520] Scheraga, H.A., Khalili, M. & Liwo, A. Protein-folding dynamics: Overview of molecular simulation techniques. *Annu. Rev. Phys. Chem.* **58**, 57-83 (2007)

[521] Bartlett, R.J. & Musial, M. Coupled-cluster theory in quantum chemistry. *Rev. Mod. Phys.* **79**, 291-352 (2007)

[522] Marrink, S.J., Risselada, H.J., Yefimov, S., Tieleman, D.P. & de Vries, A.H. The MARTINI force field: Coarse grained model for biomolecular simulations *J. Phys. Chem. B* **111**, 7812-7824 (2007)

[523] Jensen, F. Introduction to computational chemistry. Edition 2. John Wiley & Sons, Chichester, UK (2007)

[524] Echenique, P. & Alonso, J.L. A mathematical and computational review of Hartree-Fock SCF methods in quantum chemistry. *Mol. Phys.* **105**, 3057-3098 (2007)

[525] Raha, K., Peters, M.B., Wang, B., Yu, N., Wollacott, A.M., Westerhoff, L.M. & Merz, K.M. The role of quantum mechanics in structure-based drug design. *Drug Discovery Today* **12**, 725-731 (2007)

[526] Wenjian, L. New advances in relativistic quantum chemistry. *Prog. Chem.* **19**, 833-851 (2007)

[527] Helgaker, T., Klopper, W. & Tew, D.P. Quantitative quantum chemistry. *Mol. Phys.* **106**, 2107-2143 (2008)

[528] Hu, H. & Yang, W.T. Free energies of chemical reactions in solution and in enzymes with ab initio quantum mechanics/molecular mechanics methods. *Ann. Rev. Phys. Chem.* **59**, 573-601 (2008)

[529] Fabian, W.M.F. Accurate thermochemistry from quantum chemical calculations? *Monatshefte f. Chemie* **139**, 309-318 (2008)

[530] Voth, G.A. Coarse-graining of condensed phase and biomolecular systems. Taylor & Francis, Inc., New York, USA (2008)

[531] Clementi, C. Coarse-grained models of protein folding: Toy models or predictive tools? *Curr. Opin. Struct. Biol.* **18**, 10-15 (2008)

[532] van Gunsteren, W.F., Huber, T. & Torda, A.E. Biomolecular modelling: overview of types of methods to search and sample conformational space. *AIP Conf. Proc.* **330**, 253-268 (1995)

[533] Berne, B.J. & Straub, J.E. Novel methods of sampling phase space in the simulation of biological systems. *Curr. Opin. Struct. Biol.* **7**, 181-189 (1997)

[534] Voter, A.F., Monatlenti, F. & Germann, T.C. Extending the time scale in atomistic simulations of materials. *Annu. Rev. Mater. Res.* **32**, 321-46 (2002)

[535] Tai, K. Conformational sampling for the impatient. *Biophys. Chem.* **107**, 213-220 (2004)

[536] van der Vaart, A. Simulation of conformational transitions. *Theor. Chem. Acc.* **116**, 183-193 (2006)
[537] Cancés, E., Legoll, F. & Stoltz, G. Theoretical and numerical comparison of some sampling methods for molecular dynamics. *ESAIM: M2AN* **41**, 351-389 (2007)
[538] Schettino, V., Chelli, R., Marsili, S., Barducci, A., Faralli, C., Pagliai, M., Procacci, P. & Cardini, G. Problems in molecular dynamics of condensed phases. *Theor. Chem. Acc.* **39**, 265-273 (2007)
[539] Gao, Y.Q., Yang, L., Fan, Y. & Shao, Q. Thermodynamics and kinetics simulations of multi-time-scale processes for complex systems. *Int. Rev. Phys. Chem.* **27**, 201-227 (2008)
[540] Christen, M. & van Gunsteren, W.F. On searching in, sampling of, and dynamically moving through conformational space of biomolecular systems: A review. *J. Comput. Chem.* **29**, 157-166 (2008)
[541] Hansen, H.S. & Hünenberger, P.H. Using the local elevation method to construct optimized umbrella sampling potentials: Calculation of the relative free energies and interconversion barriers of glucopyranose ring conformers in water. *J. Comput. Chem.* **31**, 1-23 (2010)
[542] Hünenberger, P.H. Thermostat algorithms for molecular dynamics simulations. *Adv. Polym. Sci.* **173**, 105-149 (2005)
[543] Micha, D.A. & Burghardt, I. (Eds.) Quantum dynamics of complex molecular systems. Volume 83. Springer series in chemical physics, Springer, Berlin, Germany (2007)
[544] Marx, D. & Hutter, J. Ab initio molecular dynamics: Basic theory and advanced methods. Cambridge University Press, Cambridge, UK (2009)
[545] Wesseling, P. Principles of computational fluid dynamics. Edition 1. Springer, Berlin, Germany (2000)
[546] Valleau, J.P. & Whittington, S.G. A guide to Monte Carlo for statistical mechanics: 2. Byways. In: *Modern theoretical chemistry. Volume 5.* Berne, B.J., Ed. Plenum Press, New York, USA, pp 137-168 (1977)
[547] Swendsen, R.H., Wang, J.S. & Ferrenberg, A.M. New Monte Carlo methods for improved efficiency of computer simulations in statistical mechanics. *Topics Appl. Phys.* **71**, 75-91 (1992)
[548] Frenkel, D. Monte Carlo simulations: A primer. In: *Computer simulation of biomolecular systems, theoretical and experimental applications. Volume 2.* van Gunsteren, W.F., Weiner, P.K. & Wilkinson, A.J., Eds. ESCOM Science Publishers, B.V., Leiden, The Netherlands., pp 37-66 (1993)
[549] Binder, K., Baumgartner, A., Burkitt, A.N., Ceperley, D., Ferrenberg, A.M., Heermann, D.W., Herrmann, H.J., Landau, D.P., von der Linden, W., de Raedt, H., Schmidt, K.E., Selke, W., Stauffer, D. & Young, A.P. Recent developments in the Monte Carlo simulation of condensed matter. *Topics Appl. Phys.* **71**, 385-410 (1995)
[550] Peslherbe, G.H., Wang, H.B. & Hase, W.L. Monte Carlo sampling for classical trajectory simulations. *Adv. Chem. Phys.* **105**, 171-201 (1999)
[551] Siepmann, J.I. An introduction to the Monte Carlo method for particle simulations. *Adv. Chem. Phys.* **105**, 1-12 (1999)
[552] Newman, M.E.J. & Barkema, G.T. Monte Carlo methods in statistical physics. Edition 2. Oxford University Press, Oxford, UK (2001)
[553] van Gunsteren, W.F. & Berendsen, H.J.C. Algorithms for macromolecular dynamics and constraint dynamics. *Mol. Phys.* **34**, 1311-1327 (1977)
[554] Ryckaert, J.-P., Ciccotti, G. & Berendsen, H.J.C. Numerical integration of the Cartesian equations of motion of a system with constraints: Molecular dynamics of n-alkanes. *J. Comput. Phys.* **23**, 327-341 (1977)
[555] Andersen, H.C. Rattle : A "velocity" version of the SHAKE algorithm for molecular dynamics calculations. *J. Comput. Phys.* **52**, 24-34 (1983)
[556] Miyamoto, S. & Kollman, P.A. SETTLE : An analytical version of the SHAKE and RATTLE algorithm for rigid water models. *J. Comput. Chem.* **13**, 952-962 (1992)
[557] Hess, B., Bekker, H., Berendsen, H.J.C. & Fraaije, J.G.E.M. LINCS : A linear constraint solver for molecular simulations. *J. Comput. Chem.* **18**, 1463-1472 (1997)
[558] Kräutler, V., van Gunsteren, W.F. & Hünenberger, P.H. A fast SHAKE algorithm to solve distance constraint equations for small molecules in molecular dynamics simulations. *J. Comput. Chem.* **22**, 501-508 (2001)
[559] Christen, M & van Gunsteren, W.F. An approximate but fast method to impose flexible distance constraints in molecular dynamics simulations. *J. Chem. Phys.* **122**, 144106/1-144106/12 (2005)
[560] Tobias, D.J. & Brooks III, C.L. Molecular dynamics with internal coordinate constraints. *J. Chem. Phys.* **89**, 5115-5127 (1988)
[561] Schlitter, J., Engels, M., Krüger, P., Jacoby, E. & Wollmer, A. Targeted molecular dynamics simulation of conformational change. Application to the T-R transition in insulin. *Mol. Simul.* **10**, 291-308 (1993)
[562] Heinz, T.N., van Gunsteren, W.F. & Hünenberger, P.H. Comparison of four methods to compute the dielectric permittivity of liquids from molecular dynamics simulations. *J.*

Chem. Phys. **115**, 1125-1136 (2001)
[563] Kastenholz, M.A., Schwartz, T.U. & Hünenberger, P.H. The transition between the B and Z conformations of DNA investigated by targeted molecular dynamics simulations with explicit solvation. *Biophys. J.* **91**, 2976-2990 (2006)
[564] Bekker, H. Unification of box shapes in molecular simulations. *J. Comput. Chem.* **18**, 1930-1942 (1997)
[565] Petraglio, G., Ceccarelli, M. & Parrinello, M. Nonperiodic boundary conditions for solvated systems. *J. Chem. Phys.* **123**, 044103/1-044103/7 (2005)
[566] Riihimäki, E.-S., Martínez, J.M. & Kloo, L. An evaluation of non-periodic boundary condition models in molecular dynamics simulations using prion octapeptides as probes. *J. Mol. Struct. (Theochem)* **760**, 91-98 (2006)
[567] Hünenberger, P.H. Calculation of the group-based pressure in molecular simulations. I. A general formulation including Ewald and particle-particle–particle-mesh electrostatics. *J. Chem. Phys.* **116**, 6880-6897 (2002)
[568] Woodcock, L.V. Isothermal molecular dynamics calculations for liquid salts. *Chem. Phys. Lett.* **10**, 257-261 (1971)
[569] Hoover, W.G., Ladd, A.J.C. & Moran, B. High-strain-rate plastic flow studied via nonequilibrium molecular dynamics. *Phys. Rev. Lett.* **48**, 1818-1820 (1982)
[570] Evans, D.J. Computer "experiment" for nonlinear thermodynamics of Couette flow. *J. Chem. Phys.* **78**, 3297-3302 (1983)
[571] Berendsen, H.J.C., Postma, J.P.M., van Gunsteren, W.F., di Nola, A. & Haak, J.R. Molecular dynamics with coupling to an external bath. *J. Chem. Phys.* **81**, 3684-3690 (1984)
[572] Nosé, S. A molecular dynamics method for simulations in the canonical ensemble. *Mol. Phys.* **52**, 255-268 (1984)
[573] Nosé, S. A unified formulation of the constant temperature molecular dynamics methods. *J. Chem. Phys.* **81**, 511-519 (1984)
[574] Hoover, W.G. Canonical dynamics: Equilibrium phase-space distributions. *Phys. Rev. A* **31**, 1695-1697 (1985)
[575] Martyna, G.J., Klein, M.L. & Tuckerman, M. Nosé-Hoover chains: The canonical ensemble via continuous dynamics. *J. Chem. Phys.* **97**, 2635-2643 (1992)
[576] Laird, B.B. & Leimkuhler, B.J. Generalized dynamical thermostating technique. *Phys. Rev. E* **68**, 016704/1-016704/6 (2003)
[577] Creutz, M. Microcanonical Monte Carlo simulations. *Phys. Rev. Lett.* **50**, 1411-1414 (1983)
[578] Ray, J.R. Microcanonical ensemble Monte Carlo method. *Phys. Rev. A* **44**, 4061-4064 (1991)
[579] Lustig, R. Microcanonical Monte Carlo simulation of thermodynamic properties. *J. Chem. Phys.* **109**, 8816-8828 (1998)
[580] Andersen, H.C. Molecular dynamics simulations at constant pressure and/or temperature. *J. Chem. Phys.* **72**, 2384-2393 (1980)
[581] Parrinello, M. & Rahman, A. Crystal structure and pair potentials: A molecular-dynamics study. *Phys. Rev. Lett.* **45**, 1196-1199 (1980)
[582] Parrinello, M. & Rahman, A. Strain fluctuations and elastic constants. *J. Phys. Chem.* **76**, 2662-2666 (1982)
[583] Grünwald, M. & Dellago, C. Ideal gas pressure bath: a method for applying hydrostatic pressure in the computer simulation of nanoparticles. *Mol. Phys.* **104**, 3709-3715 (2006)
[584] Grünwald, M., Dellago, C. & Geissler, P.L. An efficient transition path sampling algorithm for nanoparticles under pressure. *J. Chem. Phys.* **127**, 154718/1-154718/10 (2007)
[585] McDonald, I.R. Monte Carlo calculations for one- and two-component fluids in the isothermal-isobaric ensemble. *Chem. Phys. Lett.* **3**, 241-243 (1969)
[586] Wood, W.W. NPT-ensemble Monte Carlo calculations for hard-disk fluid. *J. Chem. Phys.* **52**, 729-741 (1970)
[587] King, H.F. Isobaric versus canonical ensemble formalism for Monte Carlo studies of liquids. *J. Chem. Phys.* **57**, 1837-1841 (1972)
[588] McDonald, I.R. NPT-ensemble Monte Carlo calculations for binary liquid mixtures. *Mol. Phys.* **23**, 41-58 (1972)
[589] Owicki, J.C. & Scheraga, H.A. Monte Carlo calculations in isothermal-isobaric ensemble. 1. Liquid water. *J. Am. Chem. Soc.* **99**, 7403-7412 (1977)
[590] Escobedo, F.A. & Depablo, J.J. A new method for generating volume changes in isobaric-isothermal Monte Carlo simulations of flexible molecules. *Macromol. Theory Simul.* **4**, 691-707 (1995)
[591] Brennan, J.K. & Madden, W.G. Efficient volume changes in constant-pressure Monte Carlo simulations. *Mol. Simul.* **20**, 139-157 (1998)
[592] Chesnut, D.A. & Salsburg, Z.W. Monte Carlo procedure for statistical mechanical calculations in a grand canonical ensemble of lattice systems. *J. Chem. Phys.* **38**, 2861-2870 (1963)

[593] Çağin, T. & Pettitt, B.M. Molecular dynamics with a variable number of molecules. *Mol. Phys.* **72**, 169-175 (1991)
[594] Ji, J., Çağin, T. & Pettitt, B.M. Dynamic simulations of water at constant chemical potential. *J. Chem. Phys.* **96**, 1333-1342 (1992)
[595] Beutler, T.C. & van Gunsteren, W.F. Molecular dynamics simulations with first order coupling to a bath of constant chemical potential. *Mol. Simul.* **14**, 21-34 (1994)
[596] Lo, C. & Palmer, B. Alternative Hamiltonian for molecular dynamics simulations in the grand canonical ensemble. *J. Chem. Phys.* **102**, 925-931 (1995)
[597] Lynch, G.C. & Pettitt, B.M. Grand canonical ensemble molecular dynamics simulations: Reformulation of extended system dynamics approaches. *J. Chem. Phys.* **107**, 8594-8610 (1997)
[598] Marrone, T.J., Resat, H., Hodge, N., Chang, C.-H. & McCammon, J.A. Solvation studies of DMP323 and A76928 bound to HIV protease: Analysis of water sites using grand canonical Monte Carlo simulations. *Protein Science* **7**, 573-579 (1998)
[599] Börjesson, U. & Hünenberger, P.H. Explicit-solvent molecular simulation at constant pH: Methodology and application to small amines. *J. Chem. Phys.* **114**, 9706-9719 (2001)
[600] Bürgi, R., Kollman, P.A. & van Gunsteren, W.F. Simulating proteins at constant pH: An approach combining molecular dynamics and Monte Carlo simulation. *Proteins* **47**, 469-480 (2002)
[601] Adams, D.J. Grand canonical ensemble Monte Carlo for a Lennard-Jones fluid. *Mol. Phys.* **29**, 307-311 (1975)
[602] von Lilienfeld, O.A. & Tuckerman, M.E. Molecular grand-canonical ensemble density functional theory and exploration of chemical space. *J. Chem. Phys.* **125**, 154104/1-154104/10 (2006)
[603] Deng, Y. & Roux, B. Computation of binding free energy with molecular dynamics and grand canonical MC simulations. *J. Chem. Phys.* **128**, 115103/1-115103/8 (2008)
[604] van Gunsteren, W.F., Nanzer, A.P. & Torda, A.E. Molecular simulation methods for generating ensembles or trajectories consistent with experimental data. In: *Monte Carlo and molecular dynamics of condensed matter systems, Proceedings of the Euroconference, 3-28 July 1995, Como, Italy. Volume 49.* Binder, K. & Ciccotti, G., Eds. SIF, Bologna, Italy., pp 777-788 (1996)
[605] van Gunsteren, W.F., Dolenc, J. & Mark, A.E. Molecular simulation as an aid to experimentalists. *Curr. Opin. Struct. Biol.* **18**, 149-153 (2008)
[606] de Vlieg, J., Boelens, R., Scheek, R.M., Kaptein, R. & van Gunsteren, W.F. Restrained molecular dynamics procedure for protein tertiary structure determination from NMR data: A Lac repressor headpiece structure based on information on J-coupling and from presence and absence of NOE's. *Isr. J. Chem.* **27**, 181-188 (1986)
[607] Gros, P., van Gunsteren, W.F. & Hol, W.G.J. Inclusion of thermal motion in crystallographic structures by restrained molecular dynamics. *Science* **249**, 1149-1152 (1990)
[608] Gros, P. & van Gunsteren, W.F. Crystallographic refinement and structure-factor time-averaging by molecular dynamics in the absence of a physical force field. *Mol. Simul.* **10**, 377-395 (1993)
[609] van Gunsteren, W.F., Brunne, R.M., Gros, P., van Schaik, R.C., Schiffer, C.A. & Torda, A.E. Accounting for molecular mobility in structure determination based on nuclear magnetic resonance spectroscopic and X-ray diffraction data. In: *Methods in enzymology: nuclear magnetic resonance. Volume 239.* James, T.L. & Oppenheimer, N.J., Eds. Academic Press, New York, USA, pp 619-654 (1994)
[610] Schiffer, C.A., Gros, P. & van Gunsteren, W.F. Time-averaging crystallographic refinement: Possibilities and limitations using α-cyclodextrin as a test system. *Acta Crystallogr. D* **51**, 85-92 (1995)
[611] van Gunsteren, W.F., Kaptein, R. & Zuiderweg, E.R.P. Use of molecular dynamics computer simulations when determining protein structure by 2D NMR. In: *Proceedings NATO/CECAM workshop on nucleic acid conformation and dynamics.* Olson, W.K., Ed. Orsay, CECAM, France., pp 79-92 (1984)
[612] Zuiderweg, E.R.P., Scheek, R.M., Boelens, R., van Gunsteren, W.F. & Kaptein, R. Determination of protein structures from nuclear magnetic resonance data using a restrained molecular dynamics approach: The lac repressor DNA binding domain. *Biochimie* **67**, 707-715 (1985)
[613] Kaptein, R., Zuiderweg, E.R.P., Scheek, R.M., Boelens, R. & van Gunsteren, W.F. A protein structure from nuclear magnetic resonance data: lac repressor headpiece. *J. Mol. Biol.* **182**, 179-182 (1985)
[614] Torda, A.E., Scheek, R.M. & van Gunsteren, W.F. Time-dependent distance restraints in molecular dynamics simulations. *Chem. Phys. Lett.* **157**, 289-294 (1989)
[615] Torda, A.E., Scheek, R.M. & van Gunsteren, W.F. Time-averaged nuclear Overhauser effect distance restraints applied to tendamistat. *J. Mol. Biol.* **214**, 223-235 (1990)
[616] Torda, A.E., Brunne, R.M., Huber, T., Kessler, H. & van Gunsteren, W.F. Structure refinement using time-averaged J-coupling constant restraints. *J. Biomol. NMR* **3**, 55-66 (1993)
[617] Keller, B., Christen, M., Oostenbrink, C. & van Gunsteren, W.F. On using oscillating

time-dependent restraints in MD simulation. *J. Biomol. NMR.* **34**, 1-14 (2007)
[618] Christen, M., Keller, B. & van Gunsteren, W.F. Biomolecular structure refinement based on adaptive restraints using local-elevation simulation. *J. Biomol. NMR* **39**, 265-273 (2007)
[619] Beveridge, D.L. & DiCapua, F.M. Free energy via molecular simulation: Applications to chemical and biomolecular systems. *Annu. Rev. Biophys. Biophys. Chem.* **18**, 431-492 (1989)
[620] King, P.M. Free energy via molecular simulation : A primer. In: *Computer simulation of biomolecular systems, theoretical and experimental applications. Volume 2.* van Gunsteren, W.F., Weiner, P.K. & Wilkinson, A.J., Eds. ESCOM Science Publishers, B.V., Leiden, The Netherlands., pp 267-314 (1993)
[621] Kollman, P. Free energy calculations: Applications to chemical and biochemical phenomena. *Chem. Rev.* **93**, 2395-2417 (1993)
[622] van Gunsteren, W.F., Beutler, T.C., Fraternali, F., King, P.M., Mark, A.E. & Smith, P.E. Computation of free energy in practice : Choice of approximations and accuracy limiting factors. In: *Computer simulation of biomolecular systems, theoretical and experimental applications. Volume 2.* van Gunsteren, W.F., Weiner, P.K. & Wilkinson, A.J., Eds. ESCOM Science Publishers, B.V., Leiden, The Netherlands., pp 315-367 (1993)
[623] Straatsma, T.P. Free energy by molecular simulation. In: *Reviews in computational chemistry. Volume 9.* Lipkowitz, K.B. & Boyd, D.B., Eds. VCH Publishers Inc., New York, USA, pp 81-127 (1996)
[624] Mark, A.E., Xu, Y., Liu, H. & van Gunsteren, W.F. Rapid non-empirical approaches for estimating relative binding free energies. *Acta Biochim. Pol.* **42**, 525-536 (1995)
[625] Liu, H., Mark, A.E. & van Gunsteren, W.F. Estimating the relative free energy of different molecular states with respect to a single reference state. *J. Phys. Chem.* **100**, 9485-9494 (1996)
[626] Schäfer, H., van Gunsteren, W.F. & Mark, A.E. Estimating relative free energies from a single ensemble: Hydration free energies. *J. Comput. Chem.* **20**, 1604-1617 (1999)
[627] Pitera, J.W. & van Gunsteren, W.F. One-step perturbation methods for solvation free energies of polar solutes. *J. Phys. Chem.* **105**, 11264-11274 (2001)
[628] Kirkwood, J.G. Statistical mechanics of fluid mixtures. *J. Chem. Phys.* **3**, 300-313 (1935)
[629] Zwanzig, R.W. High-temperature equation of state by a perturbation method. I. Nonpolar gases. *J. Chem. Phys.* **22**, 1420-1426 (1954)
[630] Kong, X. & Brooks III, C.L. λ-dynamics: A new approach to free energy calculations. *J. Chem. Phys.* **105**, 2414-2423 (1996)
[631] Guo, Z., Brooks III, C.L. & Kong, X. Efficient and flexible algorithm for free energy calculations using the λ-dynamics approach. *J. Phys. Chem. B* **102**, 2032-2036 (1998)
[632] Christ, C.D. & van Gunsteren, W.F. Enveloping distribution sampling: A method to calculate free energy differences from a single simulation. *J. Chem. Phys.* **126**, 184110/1-184110/10 (2007)
[633] Christ, C.D. & van Gunsteren, W.F. Multiple free energies from a single simulation: Extending enveloping distribution sampling to nonoverlapping phase-space distributions. *J. Chem. Phys.* **128**, 174112/1-174112/12 (2008)
[634] Christ, C.D. & van Gunsteren, W.F. Simple, efficient, and reliable computation of multiple free energy differences from a single simulation: a reference Hamiltonian parameter update scheme for enveloping distribution sampling (EDS). *J. Chem. Theory Comput.* **5**, 276-286 (2009)
[635] Christ, C.D. & van Gunsteren, W.F. Comparison of three enveloping distribution sampling Hamiltonians for the estimation of multiple free energy differences from a single simulation. *J. Comput. Chem.* **30**, 1664-1679 (2009)
[636] Fukunishi, H., Watanabe, O. & Takada, S. On the Hamiltonian replica exchange method for efficient sampling of biomolecular systems: Application to protein structure prediction. *J. Chem. Phys.* **116**, 9058-9067 (2002)
[637] Piana, S. Atomistic simulation of the DNA helix-coil transition. *J. Phys. Chem. A* **111**, 12349-12354 (2007)
[638] Piana, S. & Laio, A. A bias-exchange approach to protein folding. *J. Phys. Chem. B* **111**, 4553-4559 (2007)
[639] Kannan, S. & Zacharias, M. Enhanced sampling of peptide and protein conformations using replica exchange simulations with a peptide backbone biasing-potential. *Proteins: Struct. Funct. Bioinf.* **66**, 697-706 (2007)
[640] Min, D., Li, H., Li, G., Bitetti-Putzer, R. & Yang, W. Synergistic approach to improve alchemical free energy calculation in rugged energy surface. *J. Chem. Phys.* **126**, 144109/1-144109/12 (2007)
[641] Babin, V., Roland, C. & Sagui, C. Adaptively biased molecular dynamics for free energy calculations. *J. Chem. Phys.* **128**, 134101/1-134101/7 (2008)
[642] Hritz, J. & Oostenbrink, C. Hamiltonian replica exchange molecular dynamics using soft-core interactions. *J. Chem. Phys.* **128**, 144121/1-144121/10 (2008)
[643] Bernal, J.D. & Fowler, R.H. A theory of water and ionic solution, with particular refer-

ence to hydrogen and hydroxyl ions. *J. Chem. Phys.* **1**, 515-547 (1933)
[644] Eley, D.D. & Evans, M.G. Heats and entropy changes accompanying the solution of ions in water. *Trans. Farad. Soc.* **38**, 1093-1112 (1938)
[645] Verwey, E.J.W. The charge distribution in the water molecule and the calculation of the intermolecular forces. *Rec. Trav. Chim. Pays-Bas* **60**, 887-896 (1941)
[646] Verwey, E.J.W. The interaction of ion and solvent in aqueous solutions of electrolytes. *Rec. Trav. Chim.* **61**, 127-142 (1942)
[647] Born, M. & Oppenheimer, R. Zur Quantentheorie der Molekeln. *Ann. Phys.* **389**, 457-484 (1927)
[648] Marx, D. & Hutter, J. Ab initio molecular dynamics: Theory and implementation. In: *Modern methods and algorithms of quantum chemistry. Volume 3.* Grotendorst, J., Ed. John von Neumann Institute for Computing, Jülich, Germany; NIC Series, pp 329-477 (2000)
[649] Sherwood, P. Hybrid quantum mechanics/molecular mechanics approaches. In: *Modern methods and algorithms of quantum chemistry. Volume 3.* Grotendorst, J., Ed. John von Neumann Institute for Computing, Jülich, Germany; NIC Series, pp 285-305 (2000)
[650] Warshel, A. & Karplus, M. Calculation of ground and excited state potential surfaces of conjugated molecules. I. Formulation and parametrization. *J. Am. Chem. Soc.* **94**, 5612-5625 (1972)
[651] Maxwell, J.C. A dynamical theory of the electromagnetic field. *Phil. Trans. Roy. Soc. London* **155**, 459-512 (1865)
[652] Maxwell, J.C. A treatise on electricity and magnetism. Volume 1. Edition 3. Clarendon Press, Oxford, UK (reprinted by Dover, New York, USA in 1954) (1891)
[653] Maxwell, J.C. A treatise on electricity and magnetism. Volume 2. Edition 3. Clarendon Press, Oxford, UK (reprinted by Dover, New York, USA in 1954) (1891)
[654] Poisson, S.D. Remarques sur une équation qui se présente dans la théorie des attractions des sphéroides. *Nouveau bulletin des sciences par la Société Philomatique de Paris* **3**, 388-392 (1813)
[655] Yu, H. & Karplus, M. A thermodynamic analysis of solvation. *J. Chem. Phys.* **89**, 2366-2379 (1988)
[656] Guillot, B. & Guissani, Y. A computer simulation study of the temperature dependence of the hydrophobic hydration. *J. Chem. Phys.* **99**, 8075-8094 (1993)
[657] Gallicchio, E., Kubo, M.M. & Levy, R.M. Enthalpy-entropy and cavity decomposition of alkane hydration free energies: Numerical results and implications for theories of hydrophobic solvation. *J. Phys. Chem. B* **104**, 6271-6285 (2000)
[658] van der Vegt, N.F.A. & van Gunsteren, W.F. Entropic contributions in cosolvent binding to hydrophobic solutes in water. *J. Phys. Chem. B* **108**, 1056-1064 (2004)
[659] van der Vegt, N.F.A., Trzesniak, D., Kasumaj, B. & van Gunsteren, W.F. Energy-entropy compensation in the transfer of nonpolar solutes from water to co-solvent/water mixtures. *Chem. Phys. Chem.* **5**, 144-147 (2004)
[660] Stokes, R.H. The van der Waals radii of gaseous ions of the noble gas structure in relation to hydration energies. *J. Am. Chem. Soc.* **86**, 979-982 (1964)
[661] Millen, W.A. & Watts, D.W. Theoretical calculations of thermodynamic functions of solvation of ions. *J. Am. Chem. Soc.* **89**, 6051-6056 (1967)
[662] Abraham, M.H. & Liszi, J. Calculations on ionic solvation. Part 1. Free energies of solvation of gaseous univalent ions using a one-layer continuum model. *J. Chem. Soc. Faraday Trans.* **74**, 1604-1614 (1978)
[663] Abraham, M.H. & Liszi, J. Calculations on ionic solvation. Part 2. Entropies of solvation of gaseous univalent ions using a one-layer continuum model. *J. Chem. Soc. Faraday Trans.* **74**, 2858-2867 (1978)
[664] Abraham, M.H. & Liszi, J. Calculations on ionic solvation. Part 4. Further calculations in solvation of gaseous univalent ions using one-layer and two-layer continuum models. *J. Chem. Soc. Faraday Trans.* **76**, 1219-1231 (1980)
[665] Abraham, M.H., Matteoli, E. & Liszi, J. Calculation of the thermodynamics of solvation of gaseous univalent ions in water from 273 to 573 K. *J. Chem. Soc. Faraday Trans. 1* **79**, 2781-2800 (1983)
[666] Ehrenson, S. Continuum radial dielectric functions for ion and dipole solution systems. *J. Comp. Chem.* **10**, 77-93 (1989)
[667] Lahiri, S.C. Calculation of Gibbs energies of hydration of monovalent ions: Examination of Born equation and its reevaluation. *Z. Phys. Chem.* **214**, 27-44 (2000)
[668] Lahiri, S.C. Determination of Gibbs energies of solvation of monovalent ions in water, methanol and ethanol and re-evaluation of the interaction energies. *Z. Phys. Chem.* **217**, 13-33 (2003)
[669] Uhlig, H.H. The solubilities of gases and surface tension. *J. Phys. Chem.* **41**, 1215-1225 (1937)
[670] Kirkwood, J.G. Theory of solutions of molecules containing widely separated charges with special application to zwitterions. *J. Chem. Phys.* **2**, 351-361 (1934)
[671] Onsager, L. Electric moments of molecules in liquids. *J. Am. Chem. Soc.* **58**, 1486-1493 (1936)

[672] Harrison, S.W., Nolte, H.J. & Beveridge, D.L. Free energy of a charge distribution in a spheroidal cavity in a polarizable dielectric continuum. *J. Phys. Chem.* **80**, 2580-2585 (1976)

[673] Gersten, J.I. & Sapse, A.M. Electrostatic effect's influence on some amines' basicity. *J. Phys. Chem.* **85**, 3407-3409 (1981)

[674] Gersten, J.I. & Sapse, A.M. Solvent-effect investigations through the use of an extended Born equation. *J. Comp. Chem.* **6**, 481-485 (1985)

[675] Gomez-Jeria, J.S. & Morales-Lagos, D. Free energy of a charge distribution in a spheroidal cavity surrounded by concentric dielectric continua. *J. Phys. Chem.* **94**, 3790-3795 (1990)

[676] Abe, T. A further modification of the Born equation. *Bull. Chem. Soc. Jpn.* **64**, 3035-3038 (1991)

[677] Ehrenson, S. Transmission of substituent effects. Generalization of the ellipsoidal cavity field effect model. *J. Am. Chem. Soc.* **98**, 7510-7514 (1976)

[678] Dillet, V., Rinaldi, D. & Rivail, J.-L. Liquid-state quantum chemistry: An improved cavity model. *J. Phys. Chem.* **98**, 5034-5039 (1994)

[679] Buckingham, A.D. The application of Onsager's theory to spheroidal molecules. *Trans. Faraday Soc.* **49**, 881-886 (1953)

[680] Gersten, J.I. & Sapse, A.M. Generalization of the Born equation to nonspherical solvent cavities. *J. Am. Chem. Soc.* **107**, 3786-3788 (1985)

[681] Onsager, L. Electrostatic interaction of molecules. *J. Phys. Chem.* **43**, 189-196 (1939)

[682] Warwicker, J. & Watson, H.C. Calculation of the electric potential in the active site cleft due to α-helix dipoles. *J. Mol. Biol.* **157**, 671-679 (1982)

[683] Gilson, M.K. & Honig, B. Calculation of the total electrostatic energy of a macromolecular system: solvation energies, binding energies, and conformational analysis. *Proteins: Struct. Funct. Genet.* **4**, 7-18 (1988)

[684] Bashford, D. & Karplus, M. pKa's of ionizable groups in proteins: Atomic detail from a continuum electrostatic model. *Biochemistry* **29**, 10219-10225 (1990)

[685] Bashford, D. & Karplus, M. Multiple-site titration curve of proteins: An analysis of exact and approximate methods for their calculation. *J. Phys. Chem.* **95**, 9556-9561 (1991)

[686] Davis, M.E., Madura, J.D., Luty, B.A. & McCammon, J.A. Electrostatics and diffusion of molecules in solution: Simulations with the University of Houston Brownian dynamics program. *Comput. Phys. Commun.* **62**, 187-197 (1991)

[687] Sharp, K.A. Incorporating solvent and ion screening into molecular dynamics using the finite-difference Poisson-Boltzmann method. *J. Comput. Chem.* **12**, 454-468 (1991)

[688] Zauhar, R.J. The incorporation of hydration forces determined by continuum electrostatics into molecular mechanics simulations. *J. Comput. Chem.* **12**, 575-583 (1991)

[689] Gilson, M.K., Davis, M.E., Luty, B.A. & McCammon, J.A. Computation of electrostatic forces on solvated molecules using the Poisson-Boltzmann equation. *J. Phys. Chem.* **97**, 3591-3600 (1993)

[690] Yang, A.-S. & Honig, B. On the pH dependence of protein stability. *J. Mol. Biol.* **231**, 459-474 (1993)

[691] Yang, A.-S., Gunner, M.R., Sampogna, R., Sharp, K. & Honig, B. On the calculation of pKa's in proteins. *Proteins: Struct. Funct. Genet.* **15**, 252-265 (1993)

[692] Sitkoff, D., Sharp, K.A. & Honig, B. Accurate calculation of hydration free energies using macroscopic solvent models. *J. Phys. Chem.* **98**, 1978-1988 (1994)

[693] Madura, J.D., Davis, M.E., Gilson, M.K., Wade, R.C., Luty, B.A. & McCammon, J.A. Biological applications of electrostatic calculations and Brownian dynamics simulations. In: *Reviews in computational chemistry. Volume 4.* Lipkowitz, K.B. & Boyd, D.B., Eds. VCH Publishers Inc., New York, USA, pp 229-267 (1994)

[694] Antosiewicz, J., McCammon, J.A. & Gilson, M.K. Prediction of pH-dependent properties of proteins. *J. Mol. Biol.* **238**, 415-436 (1994)

[695] Madura, J.D., Briggs, J.M., Wade, R.C., Davis, M.E., Luty, B.A., Ilin, A., Antosiewicz, J., Gilson, M.K., Bagheri, B., Scott, L.R. & McCammon, J.A. Electrostatics and diffusion of molecules in solution: simulations with the University of Houston Brownian Dynamics program. *Comput. Phys. Commun.* **91**, 57-95 (1995)

[696] Honig, B. & Nicholls, A. Classical electrostatics in biology and chemistry. *Science* **268**, 1144-1149 (1995)

[697] Gilson, M.K., McCammon, J.A. & Madura, J.D. Molecular dynamics simulation with a continuum electrostatic model of the solvent. *J. Comput. Chem.* **16**, 1081-1095 (1995)

[698] Simonson, T. Macromolecular electrostatics: Continuum models and their growing pains. *Curr. Opin. Struct. Biol.* **11**, 243-252 (2001)

[699] Wang, J., Tan, C.H., Tan, Y.H., Lu, Q. & Luo, R. Poisson-Boltzmann solvents in molecular dynamics simulations. *Comm. Comp. Phys.* **3**, 1010-1031 (2008)

[700] Lu, B.Z., Zhou, Y.C., Holst, M.J. & McCammon, J.A. Recent progress in numerical methods for the Poisson-Boltzmann equation in biophysical applications. *Comm. Comp. Phys.* **3**, 973-1009 (2008)

[701] Tjong, H. & Zhou, H.X. On the dielectric boundary in Poisson-Boltzmann calculations.

J. Chem. Theory Comput. **4**, 507-514 (2008)
[702] Grochowski, P. & Trylska, J. Continuum molecular electrostatics, salt effects, and counterion binding. A review of the Poisson-Boltzmann theory and its modifications. *Biopolymers* **89**, 93-113 (2008)
[703] Kolafa, J., Moučka, F. & Nezbeda, I. Handling electrostatic interactions in molecular simulations: A systematic study. *Collect. Czech. Chem. Commun.* **73**, 481-506 (2008)
[704] Still, W.C., Tempczyk, A., Hawley, R.C. & Hendrickson, T. Semianalytical treatment of solvation for molecular mechanics and dynamics. *J. Am. Chem. Soc.* **112**, 6127-6129 (1990)
[705] Bashford, D. & Case, D.A. Generalized Born models of macromolecular solvation effects. *Annu. Rev. Phys. Chem.* **51**, 129-152 (2000)
[706] Chen, J.H., Brooks, C.L. & Khandogin, J. Recent advances in implicit solvent-based methods for biomolecular simulations. *Curr. Opin. Struct. Biol.* **18**, 140-148 (2008)
[707] Kastenholz, M.A. & Hünenberger, P.H. Computation of methodology-independent ionic solvation free energies from molecular simulations: I. The electrostatic potential in molecular liquids. *J. Chem. Phys.* **124**, 124106/1-124106/27 (2006)
[708] Ashbaugh, H.S. & Wood, R.H. Effects of long-range electrostatic potential truncation on the free energy of ionic hydration. *J. Chem. Phys.* **106**, 8135-8139 (1997)
[709] Fajans, K. Deformation von Ionen und Molekeln auf Grund refraktometrischer Daten. *Z. Elektrochem.* **34**, 503-520 (1928)
[710] Hepler, L.G. Partial molal volumes of aqueous ions. *J. Phys. Chem.* **61**, 1426-1429 (1957)
[711] Stokes, R.H. Crystal lattice energies and the electrostatic self-energies of gaseous ions. *J. Am. Chem. Soc.* **86**, 982-986 (1964)
[712] Goldman, S. & Bates, R.G. Calculation of thermodynamic functions for ionic hydration. *J. Am. Chem. Soc.* **94**, 1476-1784 (1972)
[713] Goldman, S. & Morss, L.R. Semiempirical calculations on free energy and enthalpy of hydration for trivalent lanthanides and actinides. *Can. J. Chem. - Rev. Can. Chim.* **53**, 2695-2700 (1975)
[714] Conway, B.E. Factors limiting applications of the historically significant Born equation: A critical review. In: *Modern Aspects of Electrochemistry. Volume 35.* R.E. White, B.E. Conway, Ed. Kluwer Academic/Plenum Publishers, New York, USA, pp 295-323 (2002)
[715] Qureshi, P.M. & Varshney, R.K. A very simple empirical modification of the Born equation. *J. Chem. Edu.* **66**, 641-643 (1989)
[716] Boström, M. & Ninham, B.W. Contributions from dispersion and Born self-free energies to the solvation energies of salt solutions. *J. Phys. Chem.* **108**, 12593-12595 (2004)
[717] Buckingham, A.D. A theory of ion-solvent interaction. *Disc. Faraday Soc.* **24**, 151-157 (1957)
[718] Muirhead, J.S. & Laidler, K.J. Discontinuous model for hydration of monatomic ions. *Trans. Faraday Soc.* **63**, 944-952 (1967)
[719] London, F. Über einige Eigenschaften und Anwendungen der Molekularkräfte. *Z. Phys. Chem. B* **11**, 222-251 (1930)
[720] Eisenchitz, R. & London, F. Über das Verhältnis der van der Waalsschen Kräfte zu den homöopolaren Bindungskräften. *Z. Phys.* **60**, 491-477 (1930)
[721] London, F. The general theory of molecular forces. *Trans. Faraday Soc.* **33**, 8-26 (1937)
[722] Drude, P. The theory of optics. Longmans and Green, New York, USA (1901)
[723] Rigby, M., Smith, E.B., Wakeham, W.A. & Maitland, G.C. The forces between molecules. Clarendon Press, Oxford, UK (1986)
[724] Stone, A.J. The theory of intermolecular forces. Oxford Univ. Press, Oxford, UK (2000)
[725] Unsöld, A. Beiträge zur Quantenmechanik der Atome. *Ann. Phys.* **387**, 355-393 (1927)
[726] Slater, J.C. & Kirkwood, J.G. The van der Waals forces in gases. *Phys. Rev.* **37**, 682-697 (1931)
[727] Pauling, L. The theoretical prediction of the physical properties of many-electron atoms and ions. Mole refraction, diamagnetic susceptibility, and extension in space. *Proc. Roy. Soc. London A* **114**, 181-211 (1927)
[728] Drude, P. & Nernst, W. Über Elektrostriktion durch freie Ionen. *Z. Phys. Chem.* **15**, 79-85 (1894)
[729] Gyemant, A. Die Hydration der Ionen. *Z. Phys.* **30**, 240-252 (1924)
[730] Webb, T.J. The free energy of hydration of ions and the electrostriction of the solvent. *J. Am. Chem. Soc.* **48**, 2589-2603 (1926)
[731] Webb, T.J. On the free energy of hydration of ions. *Proc. Natl. Acad. Sci. USA* **8**, 524-529 (1926)
[732] Zwicky, F. Zur Theorie der spezifischen Wärme von Lösungen. *Phys. Z.* **27**, 271-287 (1926)
[733] Latimer, W.M. & Kasper, C. The theoretical evaluation of the entropies of aqueous ions. *J. Am. Chem. Soc.* **51**, 2293-2299 (1929)
[734] Haggis, G.H., Hasted, J.B. & Buchanan, T.J. The dielectric properties of water in solu-

tions. *J. Chem. Phys.* **20**, 1452-1465 (1952)
[735] Frank, H.S. Thermodynamics of a fluid substance in the electrostatic field. *J. Chem. Phys.* **23**, 2023-2032 (1955)
[736] Mukerjee, P. On ion-solvent interactions. Part I. Partial molal volumes of ions in aqueous solution. *J. Phys. Chem.* **65**, 740-744 (1961)
[737] Padova, J. Ion-solvent interaction. II. Partial molar volume and electrostriction: a thermodynamic approach. *J. Chem. Phys.* **39**, 1552-1557 (1963)
[738] Benson, S.W. & Copeland, C.S. Partial molar volumes of ions. *J. Phys. Chem.* **67**, 1194-1197 (1963)
[739] Desnoyers, J.E., Verrall, R.E. & Conway, B.E. Electrostriction in aqueous solutions of electrolytes. *J. Chem. Phys.* **43**, 243-250 (1965)
[740] Mukerjee, P. Ionic partial molal volumes and electrostrictions in aqueous solution. *J. Phys. Chem.* **70**, 2708-2708 (1966)
[741] Padova, J.I. Ion solvent interaction. VIII. Thermodynamic properties of ions in methanol solution. *J. Chem. Phys.* **56**, 1606-1610 (1972)
[742] Wood, R.H., Quint, J.R. & Grolier, J.-P.E. Thermodynamics of a charged hard sphere in a compressible dielectric fluid. A modification of the Born equation to include the compressibility of the solvent. *J. Phys. Chem.* **85**, 3944-3949 (1981)
[743] Jayaryam, B., Fine, R., Sharp, K. & Honig, B. Free energy calculations of ion hydration: An analysis of the Born model in terms of microscopic simulations. *J. Phys. Chem.* **93**, 4320-4327 (1989)
[744] Hyun, J.-K. & Ichiye, T. Nonlinear response in ionic solvation: A theoretical investigation. *J. Chem. Phys.* **109**, 1074-1083 (1998)
[745] Marcus, Y. & Hefter, G. On the pressure and electric field dependencies of the relative permittivity of liquids. *J. Solut. Chem.* **28**, 575-592 (1999)
[746] Danielewicz-Ferchmin, I. & Ferchmin, A.R. Ion hydration and large electrocaloric effect. *J. Solut. Chem.* **31**, 81-96 (2002)
[747] Hasted, J.B., Ritson, D.M. & Collie, C.H. Dielectric properties of aqueous ionic solutions. *J. Chem. Phys.* **16**, 1-21 (1948)
[748] Grahame, D.C. Effects of dielectric saturation upon the diffuse double layer and the free energy of hydration of ions. *J. Chem. Phys.* **18**, 903-909 (1950)
[749] Booth, F. The dielectric constant of water and the saturation effect. *J. Chem. Phys.* **19**, 391-394 (1951)
[750] Booth, F. Erratum to "The dielectric constant of water and the saturation effect" [J. Chem. Phys. 19, 391-394 (1951)]. *J. Chem. Phys.* **19**, 1327-1328 (1951)
[751] Booth, F. Erratum to "The dielectric constant of water and the saturation effect" [J. Chem. Phys. 19, 391-394 (1951)]. *J. Chem. Phys.* **19**, 1615-1615 (1951)
[752] Grahame, D.C. Diffuse double layer theory for electrolytes of unsymmetrical valence types. *J. Chem. Phys.* **21**, 1054-1060 (1953)
[753] Buckingham, A.D. Theory of the dielectric constant at high field strengths. *J. Chem. Phys.* **25**, 428-434 (1956)
[754] Laidler, K.J. & Pegis, C. The influence of dielectric saturation on the thermodynamic properties of aqueous ions. *Proc. R. Soc.* **248**, 80-92 (1957)
[755] Azzam, A.M. Theoretical studies on solvation. Part II. New theory for evaluation of ionic solvation number for divalent ions at 25°C. *Can. J. Chem.* **38**, 993-1002 (1960)
[756] Rosseinsky, D.R. Some factors affecting ion-pair formation in water. *J. Chem. Soc.* **(Mar.)**, 785-791 (1962)
[757] Conway, B.E., Desnoyers, J.E. & Smith, A.C. On the hydration of simple ions and polyions. *Philos. Trans. R. Soc. London Series A* **256**, 389-437 (1964)
[758] Glueckauf, E. Heats and entropies of ions in aqueous solution. *Trans. Faraday Soc.* **60**, 572-577 (1964)
[759] Laidler, K.J. & Muirhead-Gould, J.S. Continuous dielectric model for hydration of monatomic ions. *Trans. Farad. Soc.* **63**, 953-957 (1967)
[760] Beveridge, D.L. & Schnuelle, G.W. Free energy of a charge distribution in concentric dielectric continua. *J. Phys. Chem.* **79**, 2562-2566 (1975)
[761] Abraham, M.H., Liszi, J. & Mészáros, L. Calculations on ionic solvation. III. The electrostatic free energy of solvation of ions, using a multi-layered continuum model. *J. Chem. Phys.* **70**, 2491-2496 (1979)
[762] Stiles, P.J. Contributions from dielectric inhomogeneity to the free energy of ionic solvation. *Aust. J. Chem.* **33**, 1389-1391 (1980)
[763] Klugman, I.Y. Influence of electric saturation on the dielectric permittivity of electrolytes. *Soviet Electrochemistry* **17**, 602-605 (1981)
[764] Abe, T. A modification of the Born equation. *J. Phys. Chem.* **90**, 713-715 (1986)
[765] Bucher, M. & Porter, T.L. Analysis of the Born model for hydration of ions. *J. Phys. Chem.* **90**, 3406-3411 (1986)
[766] Ehrenson, S. Boundary continuity and analytical potentials in continuum solvent models. Implications for the Born model. *J. Phys. Chem.* **91**, 1868-1873 (1987)
[767] Alper, H.E. Field strength dependence of dielectric saturation in liquid water. *J. Phys.*

Chem. **94**, 8401-8403 (1990)

[768] Kumar, A. A modified Born equation for solvation energy of ions. *J. Phys. Chem. Soc. Jpn.* **61**, 4247-4250 (1992)

[769] Bontha, J.R. & Pintauro, P.N. Prediction of ion solvation free-energies in a polarizable dielectric continuum. *J. Phys. Chem.* **96**, 7778-7782 (1992)

[770] Kim, H.-S. & Chung, J.-J. Free energy of ion hydration. *Bull. Korean Chem. Soc.* **14**, 220-225 (1993)

[771] Hyun, J.-K., Babu, C.S. & Ichiye, T. Apparent local dielectric response around ions in water: A method for its determination and its applications. *J. Phys. Chem.* **99**, 5187-5195 (1995)

[772] Sandberg, L. & Edholm, O. Nonlinear response effects in continuum models of the hydration of ions. *J. Chem. Phys.* **116**, 2936-2944 (2002)

[773] Gavryushow, S. & Linse, P. Polarization deficiency and excess free energy of ion hydration in electric fields. *J. Phys. Chem. B* **107**, 7135-7142 (2003)

[774] Gong, H., Hocky, G. & Freed, K.F. Influence of nonlinear electrostatics on transfer energies between liquid phases: Charge burial is far less expensive than Born model. *Proc. Natl. Acad. Sci. USA* **105**, 11146-11151 (2008)

[775] Frank, H.S. & Wen, W.-Y. Ion-solvent interaction. Structural aspects of ion-solvent interaction in aqueous solutions: A suggested picture of water structure. *Discuss. Farad. Soc.* **24**, 133-140 (1957)

[776] Hirata, F., Redfern, P. & Levy, R.M. Viewing the Born model for ion hydration through a microscope. *Int. J. Quant. Chem.* **15**, 179-190 (1988)

[777] Kumar, A. Surface effects in the Born solvation model. *Bull. Chem. Soc. Jpn.* **67**, 3150-3152 (1994)

[778] Babu, C.S. & Lim, C. Theory of ionic hydration: Insights from molecular dynamics simulations and experiment. *J. Phys. Chem. B* **103**, 7958-7968 (1999)

[779] Babu, C.S. & Lim, C. A new interpretation of the effective Born radius from simulations and experiment. *Chem. Phys. Lett.* **310**, 225-228 (1999)

[780] Ashbaugh, H.S. Convergence of molecular and macroscopic continuum descriptions of ion hydration. *J. Phys. Chem. B* **104**, 7235-7238 (2000)

[781] Babu, C.S. & Lim, C. Incorporating nonlinear solvent response in continuum dielectric models using a two-sphere description of the Born radius. *J. Phys. Chem. A* **105**, 5030-5036 (2001)

[782] Hinton, J.F. & Amis, E.S. Solvation numbers of ions. *Chem. Rev.* **71**, 627-674 (1971)

[783] Leyendekkers, J.V. Structure of aqueous electrolyte solutions. Thermodynamic internal pressure. *J. Chem. Soc. Faraday Trans. 1* **79**, 1109-1121 (1983)

[784] Chalikian, T.V. Structural thermodynamics of hydration. *J. Phys. Chem. B* **105**, 12566-12578 (2001)

[785] Jolicoeur, C., The, N.D. & Cabana, A. Near infrared spectra of water in aqueous solutions of organic salts. A solvation study of Bu_4NBr, Φ_4AsCl, and $NaB\Phi_4$. *Can. J. Chem.* **49**, 2008-2013 (1971)

[786] Swain, C.G., Swain, M.S., Powell, A.L. & Alunni, S. Solvent effects on chemical reactivity. Evaluation of anion- and cation-solvation components. *J. Am. Chem. Soc.* **105**, 502-513 (1983)

[787] Stangret, J. & Gampe, T. Ionic hydration behavior derived from infrared spectra in HDO. *J. Phys. Chem. A* **106**, 5393-5402 (2002)

[788] Stangret, J. & Kamieńska-Piotrowicz, E. Effect of tetraphenylphosphonium and tetraphenylborate ions on the water structure in aqueous solutions; FTIR studies of HDO spectra. *J. Chem. Soc. Farad. Trans* **93**, 3463-3466 (1997)

[789] Coetzee, J.F. & Sharpe, W.R. Solute-solvent interactions. VI. Specific interactions of tetraphenylarsonium, tetraphenylphosphonium, and tetraphenylborate ions with water and other solvents. *J. Phys. Chem.* **75**, 3141-3146 (1971)

[790] Rashin, A.A. & Honig, B. Reevaluation of the Born model of ion hydration. *J. Phys. Chem.* **89**, 5588-5593 (1985)

[791] Roux, B., Yu, H.-A. & Karplus, M. Molecular basis for the Born model of ion solvation. *J. Phys. Chem.* **94**, 4683-4688 (1990)

[792] Hummer, G., Pratt, L.R. & Garcia, A.E. Free energy of ionic hydration. *J. Phys. Chem.* **100**, 1206-1215 (1996)

[793] Lynden-Bell, R.M. & Rasaiah, J.C. From hydrophobic to hydrophilic behaviour: A simulation study of solvation entropy and free energy of simple solutes. *J. Chem. Phys.* **107**, 1981-1991 (1997)

[794] Koneshan, S., Rasaiah, J.C., Lynden-Bell, R.M. & Lee, S.H. Solvent structure, dynamics, and ion mobility in aqueous solutions at 25°C. *J. Phys. Chem. B* **102**, 4193-4204 (1998)

[795] Schurhammer, R. & Wipff, G. Are the hydrophobic $As\Phi_4^+$ and $B\Phi_4^-$ ions equally solvated? A theoretical investigation in aqueous and nonaqueous solutions using different charge distributions. *J. Phys. Chem. A* **104**, 11159-11168 (2000)

[796] Lynden-Bell, R.M., Rasaiah, J.C. & Noworyta, J.P. Using simulation to study solvation in water. *Pure Appl. Chem.* **73**, 1721-1731 (2001)

[797] Rajamani, S., Ghosh, T. & Garde, S. Size dependent ion hydration, its asymmetry, and convergence to macroscopic behavior. *J. Chem. Phys.* **120**, 4457-4466 (2004)

[798] Grossfield, A. Dependence of ion hydration on the sign of the ion's charge. *J. Chem. Phys.* **122**, 024506/1-024506/10 (2005)

[799] Öhrn, A. & Karlström, G. Many-body polarization, a cause of asymmetric solvation of ions and quadrupoles. *J. Chem. Theory Comput.* **3**, 1993-2001 (2007)

[800] Gruziel, M., Rudnicki, W.R. & Lesyng, B. Hydration free energy of a model Lennard-Jones solute particle: microscopic Monte Carlo simulation studies, and interpretation based on mesoscopic models. *J. Chem. Phys.* **128**, 064503/1-064503/13 (2008)

[801] Ashbaugh, H.S. & Asthagiri, D. Single ion hydration free energies: A consistent comparison between experiment and classical molecular simulation. *J. Chem. Phys.* **129**, 204501/1-204501/6 (2008)

[802] Mobley, D.L., Barber II, A.E., Fennell, C.J. & Dill, K.A. Charge asymmetries in hydration of polar solutes. *J. Phys. Chem. B* **112**, 2405-2414 (2008)

[803] Reif, M.M. & Hünenberger, P.H. Computation of methodology-independent single-ion solvation properties from molecular simulations. Asymmetric solvation effects. *Manuscript in preparation* (2011)

[804] Thompson, W.H. & Hynes, J.T. Frequency shifts in the hydrogen-bonded OH stretch in halide-water clusters. The importance of charge transfer. *J. Am. Chem. Soc.* **112**, 6278-6286 (2000)

[805] Wu, D.-Y., Duan, S., Liu, X.-M., Xu, Y.-C., Jiang, Y.-X., Ren, B., Xu, X., Lin, S.H. & Tian, Z.-Q. Theoretical study of binding interactions and vibrational Raman spectra of water in hydrogen-bonded anionic complexes: $(H_2O)_n^-$ (n=2 and 3), $H_2O\cdots X^-$ (X=F, Cl, Br, and I), and $H_2O\cdots M$ (M=Cu, Ag, and Au). *J. Phys. Chem. A* **112**, 1313-1321 (2008)

[806] Hawkes, S.J. All positive ions give acid solutions in water. *J. Chem. Educ.* **73**, 516-517 (1996)

[807] Bernasconi, L., Baerends, E.J. & Sprik, M. Long-range solvent effects on the orbital interaction mechanism of water acidity enhancement in metal ion solutions: A comparative study of the electronic structure of aqueous Mg and Zn dications. *J. Phys. Chem. B* **110**, 11444-11453 (2006)

[808] Swaddle, T.W., Rosenqvist, J., Yu, P., Bylaska, E., Phillips, B.L. & Casey, W.H. Kinetic evidence for five-coordination in $AlOH(aq)^{2+}$ ion. *Science* **308**, 1450-1453 (2005)

[809] Robertson, W.H., Johnson, M.A., Myashakin, E.M. & Jordan, K.D. Isolating the charge-transfer component of the anionic H-bond via spin suppression of the intracluster proton transfer reaction in the $NO^-\cdot H_2O$ entrance channel complex. *J. Phys. Chem. A* **106**, 10010-10014 (2002)

[810] Simon, C. & Klein, M.L. Ab initio molecular dynamics simulation of a water-hydrogen fluoride equimolar mixture. *Chem. Phys. Chem.* **6**, 148-153 (2005)

[811] Laasonen, K., Larrucea, J. & Sillap, A. Ab initio molecular dynamics study of a mixture of HF(aq) and HCl(aq). *J. Phys. Chem. B* **110**, 12699-12706 (2006)

[812] Blandamer, M.J. & Fox, M.F. Theory and applications of charge-transfer-to-solvent spectra. *Chem. Rev.* **70**, 59-93 (1970)

[813] Voet, A. Ionic radii and heat of hydration. *Trans. Faraday Soc.* **32**, 1301-1303 (1936)

[814] Latimer, W.M., Pitzer, K.S. & Slansky, C.M. The free energy of hydration of gaseous ions, and the absolute potential of the normal calomel electrode. *J. Chem. Phys.* **7**, 108-111 (1939)

[815] Canady, W.J. Comparison of effective ionic radii in solution. *Can. J. Chem.* **35**, 1073-1075 (1957)

[816] Senozan, N.M. Effective ionic radii and the enthalpies of solvation in liquid ammonia. *J. Inorg. Nucl. Chem.* **35**, 727-736 (1973)

[817] Qureshi, P.M. & Kamoonpuri, S.I.M. Ion solvation. The ionic radii problem. *J. Chem. Edu.* **68**, 109-109 (1991)

[818] Hyun, J.-K. & Ichiye, T. Understanding the Born radius via computer simulations and theory. *J. Phys. Chem. B* **101**, 3596-3604 (1997)

[819] Schmid, R., Miah, A.M. & Sapunov, V.N. A new table for the thermodynamic quantities of ionic hydration: Values and some applications (enthalpy-entropy compensation and Born radii). *Phys. Chem. Chem. Phys.* **2**, 97-102 (2000)

[820] Powell, R.E. & Latimer, W.M. The entropy of aqueous solutes. *J. Chem. Phys.* **19**, 1139-1141 (1951)

[821] Latimer, W.M. Single ion free energies and entropies of aqueous ions. *J. Chem. Phys.* **23**, 90-92 (1955)

[822] Fawcett, W.R. & Blum, L. Application of the mean spherical approximation to the estimation of single ion thermodynamic quantities of solvation for monoatomic monovalent ions in aqueous solutions. *J. Electroanal. Chem.* **328**, 333-340 (1992)

[823] Choi, D.S., Jhon, M.S. & Eyring, H. Curvature dependence of the surface tension and the theory of solubility. *J. Chem. Phys.* **53**, 2608-2614 (1956)

[824] Sinanoğlu, O. Microscopic surface tension down to molecular dimensions and microther-

modynamic surface areas of molecules or clusters. *J. Chem. Phys.* **75**, 463-468 (1981)
[825] Moura-Ramos, J.J., Dionísio, M.S., Gonçalves, R.M.C. & Diogo, H.P. A further view on the calculation of the enthalpy of cavity formation in liquids. The influence of the cavity size and shape. *Can. J. Chem.* **66**, 2894-2902 (1988)
[826] Gonçalves, R.M.C., Simões, A.M.N. & Moura-Ramos, J.J. The cavity models and the curvature dependence of the surface tension. Spherical cavities in anisotropic solvents. *J. Solut. Chem.* **22**, 507-517 (1993)
[827] Schmelzer, J.W.P., Gutzow, I. & Schmelzer Jr., J. Curvature-dependent surface tension and nucleation theory. *J. Colloid. Interface Sci.* **178**, 657-665 (1996)
[828] Marchuk, O.S. & Sysoev, V.M. Influence of the drop size on the coefficient of surface tension in a wide range of pressures and temperatures along the coexistence curve. *J. Mol. Liq.* **105**, 121-125 (2003)
[829] Kashchiev, D. Determining the curvature dependence of surface tension. *J. Chem. Phys.* **118**, 9081-9083 (2003)
[830] Baidakov, V.G., Boltachev, G.S. & Chernykh, G.G. Curvature corrections to surface tension. *Phys. Rev. E* **70**, 011603/1-011603/7 (2004)
[831] Thompson, S.M., Gubbins, K.E., Walton, J.P.R.B., Chantry, R.A.R. & Rowlinson, J.S. A molecular dynamics study of liquid drops. *J. Chem. Phys.* **81**, 530-542 (1984)
[832] Nijmeijer, M.J.P., Bruin, C., van Woerkom, A.B., Bakker, A.F. & van Leeuwen, I.M.J. Molecular dynamics of the surface tension of a drop. *J. Chem. Phys.* **96**, 565-576 (1992)
[833] Brodskaya, E.N., Eriksson, J.C., Laaksonen, A. & Rusanov, A.I. Local structure and work of formation of water clusters studied by molecular dynamics simulations. *J. Colloid. Interface Sci.* **180**, 86-97 (1996)
[834] Zakharov, V.V., Brodskaya, E.N. & Laaksonen, A. Surface tension of water droplets: A molecular dynamics study of model and size dependencies. *J. Chem. Phys.* **107**, 10675-10683 (1997)
[835] Yasuoka, K. & Matsumoto, M. Molecular dynamics of homogeneous nucleation in the vapor phase. II. Water. *J. Chem. Phys.* **109**, 8463-8470 (1998)
[836] Moody, M.P. & Attard, P. Curvature dependent surface tension from a simulation of a cavity in a Lennard-Jones liquid close to coexistence. *J. Chem. Phys.* **115**, 8967-8977 (2001)
[837] Moody, M.P. & Attard, P. Curvature-dependent surface tension of a growing droplet. *Phys. Rev. Lett.* **91**, 056104/1-056104/4 (2003)
[838] Bryk, P., Roth, R., Mecke, K.R. & Dietrich, S. Hard-sphere fluids in contact with curved substrates. *Phys. Rev. E* **68**, 031602/1-031602/8 (2003)
[839] Stewart, M.C. & Evans, R. Wetting and drying at a curved substrate: Long-ranged forces. *Phys. Rev. E* **71**, 011602/1-011602/14 (2005)
[840] Blokhuis, E.M. & Kuipers, J. On the determination of the structure and tension of the interface between a fluid and a curved hard wall. *J. Chem. Phys.* **126**, 054702/1-054702/10 (2007)
[841] Bresme, F. Integral equation study of the surface tension of colloidal-fluid spherical interfaces. *J. Phys. Chem. B* **106**, 7852-7859 (2002)
[842] Falls, A.H., Scriven, L.E. & Davis, H.T. Structure and stress in spherical microstructures. *J. Chem. Phys.* **75**, 3986-4002 (1981)
[843] Guermeur, R., Biquard, F. & Jacolin, C. Density profiles and surface tension of spherical interfaces. Numerical results for nitrogen drops and bubbles. *J. Chem. Phys.* **82**, 2040-2051 (1985)
[844] Pitera, J.W. & van Gunsteren, W.F. The importance of solute-solvent van der Waals interactions with interior atoms of biopolymers. *J. Am. Chem. Soc.* **123**, 3163-3164 (2001)
[845] Byakov, V.M. & Stepanov, S.V. Microscopic surface tension of liquids with curved free boundary studied by positron annihilation. *Radiation Phys. Chem.* **58**, 687-692 (2000)
[846] Sharp, K.A., Nicholls, A., Fine, R.F. & Honig, B. Reconciling the magnitude of the microscopic and macroscopic hydrophobic effects. *Science* **252**, 106-109 (1991)
[847] Matsumoto, M., Takaoka, Y. & Kataoka, Y. Liquid-vapor interface of water-methanol mixture. I. Computer simulation. *J. Chem. Phys.* **98**, 1464-1472 (1993)
[848] Alejandre, J., Tildesley, D.J. & Chapela, G.A. Molecular dynamics simulations of the orthobaric densities and surface tension of water. *J. Chem. Phys.* **102**, 4574-4583 (1995)
[849] Taylor, R.S., Dang, L.X. & Garrett, B.C. Molecular dynamics simulations of the liquid/vapor interface of SPC/E water. *J. Phys. Chem.* **100**, 11720-11725 (1996)
[850] Feller, S.E., Pastor, R.W., Rojnuckarin, A., Bogusz, S. & Brooks Jr., B.R. Effect of electrostatic force truncation on interfacial and transport properties of water. *J. Phys. Chem.* **100**, 17011-17020 (1996)
[851] Huang, D.M., Geissler, P.L. & Chandler, D. Scaling of hydrophobic solvation free energies. *J. Phys. Chem. B* **105**, 6704-6709 (2001)
[852] Rivera, J.L., Predota, M., Chialvo, A.A. & Cummings, P.T. Vapor-liquid equilibrium simulations of the SCPDP model of water. *Chem. Phys. Lett.* **357**, 189-194 (2002)
[853] Wynveen, A. & Bresme, F. Interactions of polarizable media in water: A molecular

dynamics approach. *J. Chem. Phys.* **124**, 104502/1-104502/8 (2006)
[854] Rivera, J.L., Starr, F.W., Paricaud, P. & Cummings, P.T. Polarizable contributions to the surface tension of liquid water. *J. Chem. Phys.* **125**, 094712/1-094712/8 (2006)
[855] Chacón, E., Tarazona, P. & Alejandre, J. The intrinsic structure of the water surface. *J. Chem. Phys.* **125**, 014709/1-014709/10 (2006)
[856] Ismail, A.E., Grest, G.S. & Stevens, M.J. Capillary waves at the liquid-vapor interface and the surface tension of water. *J. Chem. Phys.* **125**, 014702/1-014702/10 (2006)
[857] Garrett, B.C., Schenter G.K., Morita, Molecular simulations of the transport of molecules across the liquid/vapor interface of water. *Chem. Rev.* **106**, 1355-1374 (2006)
[858] Lü, Y.L. & Wei, B. Second inflection point of water surface tension. *Appl. Phys. Lett.* **89**, 164106/1-164106/3 (2006)
[859] Vega, C. & de Miguel, E. Surface tension of the most popular models of water by using the test-area simulation method. *J. Chem. Phys.* **126**, 154707/1-154707/10 (2007)
[860] Mountain, R.D. An internally consistent method for the molecular dynamics simulation of the surface tension: Application to some TIP4P-type models of water. *J. Phys. Chem. B* **113**, 482-486 (2009)
[861] Jorgensen, W.L., Chandrasekhar, J., Madura, J.D., Impey, R.W. & Klein, M.L. Comparison of simple potential functions for simulating liquid water. *J. Chem. Phys.* **79**, 926-935 (1983)
[862] Reiss, H., Frisch, H.L. & Lebowitz, J.L. Statistical mechanics of rigid spheres. *J. Chem. Phys.* **31**, 369-380 (1959)
[863] Reiss, H., Frisch, H.L., Helfand, E. & Lebowitz, J.L. Aspects of the statistical thermodynamics of real fluids. *J. Chem. Phys.* **32**, 119-124 (1960)
[864] Helfand, E., Reiss, H., Frisch, H.L. & Lebowitz, J.L. Scaled particle theory of fluids. *J. Chem. Phys.* **33**, 1379-1385 (1960)
[865] Pierotti, R.A. Solubility of gases in liquids. *J. Phys. Chem.* **67**, 1840-1845 (1963)
[866] Pierotti, R.A. Aqueous solutions of nonpolar gases. *J. Phys. Chem.* **69**, 281-288 (1965)
[867] Stillinger, F.H. Structure in aqueous solutions of nonpolar solutes from the standpoint of scaled-particle theory. *J. Solut. Chem.* **2**, 141-158 (1973)
[868] Pierotti, R.A. A scaled particle theory of aqueous and nonaqueous solutions. *Chem. Rev.* **76**, 717-726 (1976)
[869] Huang, D.M. & Chandler, D. Cavity formation and the drying transition in the Lennard-Jones fluid. *Phys. Rev. E* **61**, 1501-1506 (2000)
[870] Ashbaugh, H.S. & Paulaitis, M.E. Effect of solute size and solute-water attractive interactions on hydration water structure around hydrophobic solutes. *J. Am. Chem. Soc.* **123**, 10721-10728 (2001)
[871] Ashbaugh, H.S. & Pratt, L.R. Scaled particle theory and the length scales of hydrophobicity. *Rev. Mod. Phys.* **78**, 159-178 (2006)
[872] Laidler, K.J. The entropies of ions in aqueous solution. I. Dependence on charge and radius. *Can. J. Chem.* **34**, 1107-1113 (1956)
[873] Askew, F.A., Bullock, E., Smith, H.T., Tinkler, R.K., Gatty, O. & Wolfenden, J.H. The thermochemistry of solutions. Part II. Heats of solution of electrolytes in non-aqueous solvents. *J. Chem. Soc.* **(2)**, 1368-1376 (1934)
[874] Bjerrum, N. & Larsson, E. Studien über Ionenverteilungskoeffizienten. I. *Z. Phys. Chem. (Stöchiometrie u. Verwandtschaftslehre)* **127**, 358-384 (1927)
[875] Yu, H.-A., Roux, B. & Karplus, M. Solvation thermodynamics: An approach from analytic temperature derivatives. *J. Chem. Phys.* **92**, 5020-5033 (1990)
[876] Thomas, A.S. & Elcock, A.H. Molecular simulations suggest protein salt bridges are uniquely suited to life at high temperatures. *J. Am. Chem. Soc.* **126**, 2208-2214 (2004)
[877] Horinek, D., Mamatkulov, S.I. & Netz, R.R. Rational design of ion force fields based on thermodynamic solvation properties. *J. Chem. Phys.* **130**, 124507/1-124507/21 (2009)
[878] Carlsson, J. & Åqvist, J. Absolute hydration entropies of alkali metal ions from molecular dynamics simulations. *J. Phys. Chem. B* **113**, 10255-10260 (2009)
[879] Marcus, Y. Thermodynamics of solvation of ions. 6. The standard partial molar volumes of aqueous ions at 298.15 K. *J. Chem. Soc. Faraday Trans.* **89**, 713-718 (1993)
[880] Pauli, W. Über den Zusammenhang des Abschlusses der Elektronengruppen im Atom mit der Komplexstruktur der Spektren. *Z. Physik* **31**, 765-783 (1925)
[881] Mie, G. Zur kinetischen Theorie der einatomigen Körper. *Ann. Phys.* **316**, 657-697 (1903)
[882] Jones, J.E. On the determination of molecular fields. - II. From the equation of state of a gas. *Proc. R. Soc. London Ser. A* **106**, 463-477 (1924)
[883] Jones, J.E. On the determination of molecular fields. - I. From the variation of the viscosity of a gas with temperature. *Proc. R. Soc. London Ser. A* **106**, 441-462 (1924)
[884] Bopp, P., Jancso, G. & Heinzinger, K. An improved potential for non-rigid water-molecules in the liquid-phase. *Chem. Phys. Lett.* **98**, 129-133 (1983)
[885] Toukan, K. & Rahman, A. Molecular-dynamics study of atomic motions in water. *Phys. Rev. B* **31**, 2643-2648 (1985)
[886] Dang, L.X. & Pettitt, B.M. Simple intramolecular model potentials for water. *J. Phys.*

Chem. **91**, 3349-3354 (1987)
[887] Wallqvist, A. & Teleman, O. Properties of flexible water models. *Mol. Phys.* **74**, 515-533 (1991)
[888] Zhu, S.B., Yao, S., Zhu, J.B., Singh, S. & Robinson, G.W. A flexible polarizable simple point-charge water model. *J. Phys. Chem.* **95**, 6211-6217 (1991)
[889] Zhu, S.B., Singh, S. & Robinson, G.W. A new flexible polarizable water model. *J. Chem. Phys.* **95**, 2791-2799 (1991)
[890] Ferguson, D.M. Parameterization and evaluation of a flexible water model. *J. Comp. Chem.* **16**, 501-511 (1995)
[891] Tironi, I.G., Brunne, R.M. & van Gunsteren, W.F. On the relative merits of flexible versus rigid models for use in computer simulations of molecular liquids. *Chem. Phys. Lett.* **250**, 19-24 (1996)
[892] Lewitt, M., Hirshberg, M., Sharon, R., Laidig, K.E. & Daggett, V. Calibration and testing of a water model for simulation of the molecular dynamics of proteins and nucleic acids in solution. *J. Phys. Chem. B* **101**, 5051-5061 (1997)
[893] Schmitt, U.W. & Voth, G.A. The computer simulation of proton transport in water. *J. Chem. Phys.* **111**, 9361-9381 (1999)
[894] Hess, B., Saint-Martin, H. & Berendsen, H.J.C. Flexible constraints: An adiabatic treatment of quantum degrees of freedom, with application to the flexible and polarizable mobile charge densities in harmonic oscillators model for water. *J. Chem. Phys.* **116**, 9602-9610 (2002)
[895] Burnham, C.J. & Xantheas, S.S. Development of transferable interaction models for water. IV. A flexible, all-atom polarizable potential (TTM2-F) based on geometry dependent charges derived from an ab initio monomer dipole moment surface. *J. Chem. Phys.* **116**, 5115-5124 (2002)
[896] Lawrence, C.P. & Skinner, J.L. Flexible TIP4P model for molecular dynamics simulation of liquid water. *Chem. Phys. Lett.* **372**, 842-847 (2003)
[897] Jeon, J., Lefohn, A.E. & Voth, G.A. An improved polarflex water model. *J. Chem. Phys.* **118**, 7504-7518 (2003)
[898] Amira, S., Spangberg, D. & Hermansson, K. Derivation and evaluation of a flexible SPC model for liquid water. *Chem. Phys.* **303**, 327-334 (2004)
[899] Wu, Y.J., Tepper, H.L. & Voth, G.A. Flexible simple point-charge water model with improved liquid-state properties. *J. Chem. Phys.* **124**, 024503/1-024503/12 (2006)
[900] Zhang, X.B., Liu, Q.L. & Zhu, A.M. An improved fully flexible fixed-point charges model for water from ambient to supercritical condition. *Fluid Phase Equil.* **262**, 210-216 (2007)
[901] Axilrod, B.M. & Teller, E. Interaction of the van der Waals type between three atoms. *J. Chem. Phys.* **11**, 299-300 (1943)
[902] Fitts, D.D. Statistical mechanics - a study of intermolecular forces. *Ann. Rev. Phys. Chem.* **17**, 59-82 (1966)
[903] Barker, J.A., Henderson, D. & Smith, W.R. Pair and triplet interactions in argon. *Mol. Phys.* **17**, 579-592 (1969)
[904] Barker, J.A., Fisher, R.A. & Watts, R.O. Liquid argon - Monte Carlo and molecular dynamics calculations. *Mol. Phys.* **21**, 657-673 (1971)
[905] Lybrand, T.P. & Kollman, P.A. Water-water and water-ion potential functions including terms for many body effects. *J. Chem. Phys.* **83**, 2923-2933 (1985)
[906] Attard, P. Simulation results for a fluid with the Axilrod-Teller triple dipole potential. *Phys. Rev. A* **45**, 5649-5653 (1992)
[907] Elrod, M.J. & Saykally, R.J. Many-body effects in intermolecular forces. *Chem. Rev.* **94**, 1975-1997 (1994)
[908] Su, Y., Caldwell, J.W. & Kollman, P.A. Molecular dynamics and free energy perturbation study of spherand complexation with metal ions employing additive and nonadditive force fields. *J. Phys. Chem.* **99**, 10081-10085 (1995)
[909] Sremaniak, L.S., Perera, L. & Berkowitz, M.L. Thermally induced structural changes in $F^-(H_2O)_{11}$ and $Cl^-(H_2O)_{11}$ clusters: Molecular dynamics computer simulations. *J. Phys. Chem.* **100**, 1350-1356 (1996)
[910] Cieplak, P. & Kollman, P. Monte Carlo simulation of aqueous solutions of Li^+ and Na^+ using many-body potentials. Coordination numbers, ion solvation enthalpies, and the relative free energy of solvation. *J. Chem. Phys.* **92**, 6761-6767 (1990)
[911] Spångberg, D. & Hermansson, K. Effective three-body potentials for Li+(aq) and Mg2+(aq). *J. Chem. Phys.* **119**, 7263-7281 (2003)
[912] Loeffler, H.H. Many-body effects on structure and dynamics of aqueous ionic solutions. *J. Comp. Chem.* **24**, 1232-1239 (2003)
[913] Moghaddam, M.S., Shimizu, S. & Chan, H.S. Temperature dependence of three-body hydrophobic interactions: Potential of mean force, enthalpy, entropy, heat capacity, and nonadditivity. *J. Am. Chem. Soc.* **127**, 303-316 (2005)
[914] Born, M. Dynamic theory of crystal lattices. Oxford University Press, Oxford, UK (1954)
[915] Warshel, A. & Levitt, M. Theoretical studies of enzymic reactions - dielectric, electro-

static and steric stabilization of carbonium-ion in reaction of lysozyme. *J. Mol. Biol.* **103**, 227-249 (1976)

[916] van Belle, D., Couplet, I., Prevost, M. & Wodak, S.J. Calculations of electrostatic properties in proteins - Analysis of contributions from induced protein dipoles. *J. Mol. Biol.* **198**, 721-735 (1987)

[917] Straatsma, T.P. & McCammon, J.A. Molecular dynamics simulations with interaction potentials including polarization development of a noniterative method and application to water. *Mol. Simul.* **5**, 181-192 (1990)

[918] Karim, O.A. Simulation of an anion in water: effect of ion polarizability. *Chem. Phys. Lett.* **184**, 560-565 (1991)

[919] Rappe, A.K. & Goddard, W.A. Charge equilibration for molecular-dynamics simulations. *J. Phys. Chem.* **95**, 3358-3363 (1991)

[920] Dang, L.X., Rice, J.E., Caldwell, J. & Kollman, P.A. Ion solvation in polarizable water: Molecular dynamics simulations. *J. Am. Chem. Soc.* **113**, 2481-2486 (1991)

[921] Dang, L.X. Development of nonadditive intermolecular potentials using molecular dynamics: Solvation of Li^+ and F^- ions in polarizable water. *J. Chem. Phys.* **96**, 6970-6977 (1992)

[922] Zhu, S.-B. & Robinson, G.W. Molecular-dynamics computer simulation of an aqueous NaCl solution: Structure. *J. Chem. Phys.* **97**, 4336-4348 (1992)

[923] Dang, L.X. The nonadditive intermolecular potential for water revised. *J. Chem. Phys.* **97**, 2659-2660 (1992)

[924] Dang, L.X. & Garrett, B.C. Photoelectron spectra of the hydrated iodine anion from molecular-dynamics simulations. *J. Chem. Phys.* **99**, 2972-2977 (1993)

[925] Jorgensen, W.L. & Severance, D.L. Limited effects of polarization for $Cl^-(H_2O)_n$ and $Na^+(H_2O)_n$ clusters. *J. Chem. Phys.* **99**, 4233-4235 (1993)

[926] Rick, S.W., Stuart, S.J. & Berne, B.J. Dynamical fluctuating charge force field: Application to liquid water. *J. Chem. Phys.* **101**, 6141-6156 (1994)

[927] Perera, L. & Berkowitz, M.L. Structures of $Cl^-(H_2O)_n$ and $F^-(H_2O)_n$ ($n = 2, 3, ..., 15$) clusters. Molecular dynamics computer simulations. *J. Chem. Phys.* **100**, 3085-3093 (1994)

[928] Smith, D.E. & Dang, L.X. Computer simulation of NaCl association in polarizable water. *J. Chem. Phys.* **100**, 3757-3766 (1994)

[929] Sremaniak, L.S., Perera, L. & Berkowitz, M.L. Enthalpies of formation and stabilization energies of $Br^-(H_2O)_n$ (n=1,2,...,15) clusters. Comparisons between molecular dynamics computer simulations and experiment. *Chem. Phys. Lett.* **218**, 377-382 (1994)

[930] Roberts, J.E., Woodman, B.L. & Schnitker, J. The reaction field method in molecular dynamics simulations of point polarizable water models. *Mol. Phys.* **88**, 1089-1108 (1996)

[931] Xantheas, S.S. & Dang, L.X. Critical study of fluoride-water interactions. *J. Phys. Chem.* **100**, 3989-3995 (1996)

[932] Stuart, S.J. & Berne, B.J. Effects of polarizability on the hydration of the chloride ion. *J. Phys. Chem.* **100**, 11934-11943 (1996)

[933] Chang, T.M. & Dang, L.X. Ion solvation in polarizable chloroform: A molecular dynamics study. *J. Phys. Chem. B* **101**, 10518-10526 (1997)

[934] Carignano, M.A., Karlström, G & Linse, P. Polarizable ions in polarizable water: A molecular dynamics study. *J. Phys. Chem. B* **101**, 1142-1147 (1997)

[935] Banks, J.L., Kaminski, G.A., Zhou, R.H., Mainz, D.T., Berne, B.J. & Friesner, R.A. Parameterizing a polarizable force field from ab initio data. I. The fluctuating point charge model. *J. Chem. Phys.* **110**, 741-754 (1999)

[936] Bret, C., Field, M.J. & Hemmingsen, L. A chemical potential equalization model for treating polarization in molecular mechanical force fields. *Mol. Phys.* **98**, 751-763 (2000)

[937] Baaden, M., Berny, F., Madic, C. & Wipff, G. M^{3+} lanthanide cation solvation by acetonitrile: The role of cation size, counterions, and polarization effects investigated by molecular dynamics and quantum mechanical simulations. *J. Phys. Chem. A* **104**, 7659-7671 (2000)

[938] Martínez, J.M., Hernández-Cobos, J., Saint-Martin, H., Pappalardo, R.P., Ortega-Blake, I. & Marcos, E.S. Coupling a polarizable water model to the hydrated ion-water interaction potential: A test on the Cr^{3+} hydration. *J. Chem. Phys.* **112**, 2339-2347 (2000)

[939] Peslherbe, G.H., Ladanyi, B.M. & Hynes, J.T. Structure of NaI ion pairs in water clusters. *Chem. Phys.* **258**, 201-224 (2000)

[940] Halgren, T.A. & Damm, W. Polarizable force fields. *Curr. Opin. Struct. Biol.* **11**, 236-242 (2001)

[941] Koneshan, S., Rasaiah, J.C. & Dang, L.X. Computer simulation studies of aqueous solutions at ambient and supercritical conditions using effective pair potential and polarizable potential models for water. *J. Chem. Phys.* **114**, 7544-7555 (2001)

[942] Koch, D.M. & Peslherbe, G.H. On the transition from surface to interior solvation in iodide-water clusters. *Chem. Phys. Lett.* **359**, 381-389 (2002)

[943] Rick, S.W. & Stuart, S.J. Potentials and algorithms for incorporating polarizability in

computer simulations. *Rev. Comp. Chem.* **18**, 89-146 (2002)
[944] Fischer R., Richardi J. & Fries P.H., Krienke H. The solvation of ions in acetonitrile and acetone. II. Monte Carlo simulations using polarizable solvent models. *J. Chem. Phys.* **117**, 8467-8478 (2002)
[945] Kubo, M., Levy, R.M., Rossky, P.J., Matubayasi, N. & Nakahara, M. Chloride ion hydration and diffusion in supercritical water using a polarizable water model. *J. Phys. Chem. B* **106**, 3979-3986 (2002)
[946] Grossfield, A., Ren, P. & Ponder, J.W. Ion solvation thermodynamics from simulation with a polarizable force field. *J. Am. Chem. Soc.* **125**, 15671-15682 (2003)
[947] Lamoureux, G., MacKerrel, A.D. & Roux, B. A simple polarizable model of water based on classical Drude oscillators. *J. Chem. Phys.* **119**, 5185-5197 (2003)
[948] Lamoureux, G. & Roux, B. Modeling induced polarization with classical Drude oscillators: Theory and molecular dynamics simulation algorithm. *J. Chem. Phys.* **119**, 3025-3039 (2003)
[949] Yu, H., Hansson, T. & van Gunsteren, W.F. Development of a simple, self-consistent polarizable model for liquid water. *J. Chem. Phys.* **118**, 221-234 (2003)
[950] Ayala, R., Martinez, J.M. & Pappalardo, R.R. On the halide hydration study: Development of first-principles halide ion-water interaction potential based on a polarizable model. *J. Chem. Phys.* **119**, 9538-9548 (2003)
[951] Carrillo-Tripp, M., Saint-Martin, H. & Ortega-Blake, I. A comparative study of the hydration of Na^+ and K^+ with refined polarizable model potentials. *J. Chem. Phys.* **118**, 7062-7073 (2003)
[952] Lee, S.H. Molecular dynamics simulation of limiting conductance for Li^+ ion in supercritical water using polarizable models. *Mol. Simul.* **29**, 211-221 (2003)
[953] Yoo, S., Lei, Y.A. & Zeng, X.C. Effect of polarizability of halide anions on the ionic solvation in water clusters. *J. Chem. Phys.* **119**, 6083-6091 (2003)
[954] Lee, S.H. Molecular dynamics simulation studies of the limiting conductances of $CaCl_2$ using extended simple point charge and revised polarizable models. *Mol. Simul.* **30**, 669-678 (2004)
[955] Yu, H. & van Gunsteren, W.F. Charge-on-spring polarizable water models revisited: From water clusters to liquid water to ice. *J. Chem. Phys.* **121**, 9549-9564 (2004)
[956] Patel, S. & Brooks, C.L. CHARMM fluctuating charge force field for proteins: I. Parameterization and application to bulk organic liquid simulations. *J. Comp. Chem.* **25**, 1-15 (2004)
[957] Spångberg, D. & Hermansson, K. Many-body potentials for aqueous Li^+, Na^+, Mg^{2+}, and Al^{3+}: Comparison of effective three-body potentials and polarizable models. *J. Chem. Phys.* **120**, 4829-4843 (2004)
[958] Ahn-Ercan, G., Krienke, H. & Kunz, W. Role of polarizability in molecular interactions in ion solvation. *Curr. Opin. Coll. Int. Sci.* **9**, 92-96 (2004)
[959] Yu, H. & van Gunsteren, W.F. Accounting for polarization in molecular simulation. *Comput. Phys. Commun.* **172**, 69-85 (2005)
[960] Archontis, G., Leontidis, E. & Andreou, G. Attraction of iodide ions by the free water surface, revealed by simulations with a polarizable force field based on Drude oscillators. *J. Phys. Chem. B* **109**, 17957-17966 (2005)
[961] Yang, Z.Z. & Li, X. Ion solvation in water from molecular dynamics simulation with the ABEEM/MM force field. *J. Phys. Chem. A* **109**, 3517-3520 (2005)
[962] Li, X. & Yang, Z.Z. Study of lithium cation in water clusters: Based on atom-bond electronegativity equalization method fused into molecular mechanics. *J. Phys. Chem. A* **109**, 4102-4111 (2005)
[963] Li, X. & Yang, Z.Z. Hydration of Li^+-ion in atom-bond electronegativity equalization method-7P water: A molecular dynamics simulation study. *J. Chem. Phys.* **122**, 084514/1-084514/15 (2005)
[964] Hagberg, D., Brdarski, S. & Karlström, G. On the solvation of ions in small water droplets. *J. Phys. Chem. B* **109**, 4111-4117 (2005)
[965] Hagberg, D., Karlström. G., Roos, B.O., Gagliardi, The coordination of uranyl in water: A combined quantum chemical and molecular simulation study. *J. Am. Chem. Soc.* **127**, 14250-14256 (2005)
[966] Herce, D.H., Perera, L., Darden, T. & Sagui, C. Surface solvation for an ion in a water cluster. *J. Chem. Phys.* **122**, 024513/1-024513/10 (2005)
[967] Shevkunov, S.V., Lukyanov, S.I., Leyssale, J.-M. & Millot, C. Computer simulation of Cl^- hydration in anion-water clusters. *Chem. Phys.* **310**, 97-107 (2005)
[968] Ma, Y. & Garofalini, S.H. Iterative fluctuation charge model: A new variable charge molecular dynamics method. *J. Chem. Phys.* **124**, 234102/1-234102/7 (2006)
[969] Jiao, D., King, C., Grossfield, A., Darden, T.A. & Ren, P. Simulation of Ca^{2+} and Mg^{2+} solvation using polarizable atomic multipole potential. *J. Phys. Chem. B* **110**, 18553-18559 (2006)
[970] Lamoureux, G. & Roux, B. Absolute hydration free energy scale for alkali and halide ions established from simulations with a polarizable force field. *J. Phys. Chem. B* **110**,

3308-3322 (2006)

[971] Piquemal, J.P., Perera, L., Cisneros, G.A., Ren, P.Y., Pedersen, L.G. & Darden, T.A. Towards accurate solvation dynamics of divalent cations in water using the polarizable amoeba force field: From energetics to structure. *J. Chem. Phys.* **125**, 054511/1-054511/7 (2006)

[972] San-Román, M., Carrillo-Tripp, M., Saint-Martin, H., Hernández-Cobos, J. & Ortega-Blake, I. A theoretical study of the hydration of Li^+ by Monte Carlo simulations with refined ab initio based model potentials. *Theor. Chem. Acc.* **115**, 177-189 (2006)

[973] Wick, C.D., Dang, L.X. & Jungwirth, P. Simulated surface potentials at the vapor-water interface for the KCl aqueous electrolyte solution. *J. Chem. Phys.* **125**, 024706/1-024706/4 (2006)

[974] Wick, C.D. & Dang, L.X. Molecular mechanism of transporting a polarizable iodide anion across the water-CCl_4 liquid/liquid interface. *J. Chem. Phys.* **126**, 134702/1-134702/4 (2007)

[975] Chen, J.H. & Martinez, T.J. QTPIE: Charge transfer with polarization current equalization. A fluctuating charge model with correct asymptotics. *Chem. Phys. Lett.* **438**, 315-320 (2007)

[976] Geerke, D.P. & van Gunsteren, W.F. On the calculation of atomic forces in classical simulation using the charge-on-spring method to explicitly treat electronic polarization. *J. Chem. Theory Comput.* **3**, 2128-2137 (2007)

[977] Geerke, D.P. & van Gunsteren, W.F. Calculation of the free energy of polarization: Quantifying the effect of explicitly treating electronic polarization on the transferability of force-field parameters. *J. Phys. Chem. B* **111**, 6425-6436 (2007)

[978] Alcaraz, O., Bitrián, V. & Trullàs, J. Molecular dynamics study of polarizable point dipole models for molten sodium iodide. *J. Chem. Phys.* **127**, 154508/1-154508/10 (2007)

[979] Laage, D. & Hynes, J.T. On the residence time for water in a solute hydration shell: Application to aqueous halide solutions. *J. Phys. Chem. B* **112**, 7697-7701 (2008)

[980] Kunz, A.-P. E. & van Gunsteren, W.F. Development of a non-linear classical polarization model for liquid water and aqueous solutions: COS/D. *J. Phys. Chem. A* **113**, 11570-11579 (2009)

[981] Hagler, A.T. & Ewig, C.S. On the use of quantum energy surfaces in the derivation of molecular force-fields. *Comp. Phys. Comm.* **84**, 131-155 (1994)

[982] Hwang, M.-J., Stockfisch, T.P. & Hagler, A.T. Derivation of class-II force-fields. 2. Derivation and characterization of a class-II force-field, CFF93, for the alkyl functional-group and alkane molecules. *J. Am. Chem. Soc.* **116**, 2515-2525 (1994)

[983] Maple, J.R., Hwang, M.-J., Stockfisch, T.P., Dinur, U., Waldman, M., Ewig, C.S. & Hagler, A.T. Derivation of class-II force-fields. 1. Methodology and quantum force-field for the alkyl functional-group and alkane molecules. *J. Comp. Chem.* **15**, 162-182 (1994)

[984] Maple, J.R., Hwang, M.-J., Stockfisch, T.P. & Hagler, A.T. Derivation of class-II force-fields. 3. Characterization of a quantum force-field for alkanes. *Isr. J. Chem.* **34**, 195-231 (1994)

[985] Morse, P.M. Diatomic molecules according to the wave mechanics. II. Vibrational levels. *Phys. Rev.* **34**, 57-64 (1929)

[986] Kihara, T. Virial coefficients and models of molecules in gases. *Rev. Mod. Phys.* **25**, 831-843 (1953)

[987] Huggins, M.L. & Mayer, J.E. Interatomic distances in crystals of the alkali halides. *J. Chem. Phys.* **1**, 643-646 (1933)

[988] Huggins, M.L. Erratum to "Lattice energies, equilibrium distances, compressibilities and characteristic frequencies of alkali halide crystals" [J. Chem. Phys. 5, 143-148 (1937)]. *J. Chem. Phys.* **15**, 212-212 (1947)

[989] Slater, J.C. The normal state of helium. *Phys. Rev.* **32**, 349-360 (1928)

[990] Buckingham, R.A. The classical equation of state of gaseous helium, neon and argon. *Proc. Roy. Soc. Lond. A* **168**, 264-283 (1938)

[991] Buckingham, R.A. & Corner, J. Tables of second virial and low-pressure Joule-Thomson coefficients for intermolecular potentials with exponential repulsion. *Proc. Roy. Soc. Lond. A* **189**, 118-129 (1947)

[992] Halgren, T.A. Representation of van der Waals (vdW) interactions in molecular mechanics force-fields: Potential form, combination rules, and vdW parameters. *J. Am. Chem. Soc.* **114**, 7827-7843 (1992)

[993] Lennard-Jones, J.E. The equation of state of gases and critical phenomena. *Physica* **4**, 941-956 (1937)

[994] Schneider, T. & Stoll, E. Molecular-dynamics study of a three-dimensional one-component model for distortive phase transitions. *Phys. Rev. B* **17**, 1302-1322 (1978)

[995] van Gunsteren, W.F., Berendsen, H.J.C. & Rullmann, J.A.C. Stochastic dynamics for molecules with constraints. Brownian dynamics of n-alkanes. *Mol. Phys.* **44**, 69-95 (1981)

[996] Yun-yu, S., Lu, W. & van Gunsteren, W.F. On the approximation of solvent effects on the conformation and dynamics of cyclosporin A by stochastic dynamics simulation

techniques. *Mol. Simul.* **1**, 369-383 (1988)
[997] Herschbach, D.R., Johnston, H.S. & Rapp, D. Molecular partition functions in terms of local properties. *J. Chem. Phys.* **31**, 1652-1661 (1959)
[998] Gō, N. & Scheraga, H.A. Analysis of the contribution of internal vibrations to the statistical weights of equilibrium conformations of macromolecules. *J. Chem. Phys.* **51**, 4751-4767 (1969)
[999] Fixman, M. Classical statistical mechanics of constraints: A theorem and application to polymers. *Proc. Natl. Acad. Sci. USA* **71**, 3050-3053 (1974)
[1000] Gō, N. & Scheraga, H.A. On the use of classical statistical mechanics in the treatment of polymer chain conformation. *Macromolecules* **9**, 535-542 (1976)
[1001] Gottlieb, M. & Bird, R.B. A molecular dynamics calculation to confirm the incorrectness of the random walk distribution for describing the Kramers freely jointed bead-rod chain. *J. Chem. Phys.* **65**, 2467-2468 (1976)
[1002] Helfand, E. Flexible *vs.* rigid constraints in statistical mechanics. *J. Chem. Phys.* **71**, 5000-5007 (1979)
[1003] Edholm, O. & Berendsen, H.J.C. Entropy estimation from simulations of non-diffusive systems. *Mol. Phys.* **51**, 1011-1028 (1984)
[1004] Carter, E.A., Ciccotti, G., Hynes, J.T. & Kapral, R. Constrained reaction coordinate dynamics for the simulation of rare events. *Chem. Phys. Lett.* **156**, 472-477 (1989)
[1005] Boresch, S. & Karplus, M. The Jacobian factor in free energy simulations. *J. Chem. Phys.* **105**, 5145-5154 (1996)
[1006] den Otter, W.K. & Briels, W.J. The calculation of free-energy differences by constrained molecular-dynamics simulations. *J. Chem. Phys.* **109**, 4139-4146 (1998)
[1007] den Otter, W.K. & Briels, W.J. Free energy from molecular dynamics with multiple constraints. *Mol. Phys.* **98**, 773-781 (2000)
[1008] Straatsma, T.P., Zacharias, M. & McCammon, J.A. Holonomic constraint contributions to free energy differences from thermodynamic integration molecular dynamics simulations. *Chem. Phys. Lett.* **196**, 297-302 (1992)
[1009] Wang, L. & Hermans, J. Change of bond lengths in free-energy simulations: Algorithmic improvements, but when is it necessary. *J. Chem. Phys.* **100**, 9129-9139 (1994)
[1010] Sprik, M. & Ciccotti, G. Free energy from constrained molecular dynamics. *J. Chem. Phys.* **109**, 7737-7744 (1998)
[1011] Darve, E. & Pohorille, A. Calculating free energies using average force. *J. Chem. Phys.* **115**, 9169-9183 (2001)
[1012] Teleman, O. An efficient way to conserve the total energy in molecular dynamics simulations; boundary effects on energy conservation and dynamic properties. *Mol. Simul.* **1**, 345-355 (1988)
[1013] Hünenberger, P.H. & van Gunsteren, W.F. Alternative schemes for the inclusion of a reaction-field correction into molecular dynamics simulations: Influence on the simulated energetic, structural, and dielectric properties of liquid water. *J. Chem. Phys.* **108**, 6117-6134 (1998)
[1014] Berendsen, H.J.C. Treatment of long-range forces in molecular dynamics. In: *Molecular dynamics and protein structure (Proceedings Workshop 13-18 May 1984, at University of North Carolina).* Hermans, J., Ed. Polycrystal Book Service, P.O. Box 27, Western Springs, Illinois 60558, USA, pp 18-22 (1985)
[1015] Berendsen, H.J.C. & Hol, W.G.J. Long-range electrostatic forces. *Europhysics News* **17**, 8-10 (1986)
[1016] Harvey, S.C. Treatment of electrostatic effects in macromolecular modeling. *Proteins: Struct. Funct. Genet.* **5**, 78-92 (1989)
[1017] Davis, M.E. & McCammon, J.A. Electrostatics in biomolecular structure and dynamics. *Chem. Rev.* **90**, 509-521 (1990)
[1018] Warshel, A. & Åqvist, J. Electrostatic energy and macromolecular function. *Annu. Rev. Biophys. Biophys. Chem.* **20**, 267-298 (1991)
[1019] Berendsen, H.J.C. Electrostatic interactions. In: *Computer simulation of biomolecular systems, theoretical and experimental applications. Volume 2.* van Gunsteren, W.F., Weiner, P.K. & Wilkinson, A.J., Eds. ESCOM Science Publishers, B.V., Leiden, The Netherlands., pp 161-181 (1993)
[1020] Smith, P.E. & van Gunsteren, W.F. Methods for the evaluation of long range electrostatic forces in computer simulations of molecular systems. In: *Computer simulation of biomolecular systems, theoretical and experimental applications. Volume 2.* van Gunsteren, W.F., Weiner, P.K. & Wilkinson, A.J., Eds. ESCOM Science Publishers, B.V., Leiden, The Netherlands., pp 182-212 (1993)
[1021] Levy, R.M. & Gallicchio, E. Computer simulations with explicit solvent: Recent progress in the thermodynamic decomposition of free energies, and in modeling electrostatic effects. *Annu. Rev. Phys. Chem.* **49**, 531-567 (1998)
[1022] Warshel, A. & Papazyan, A. Electrostatic effects in macromolecules: Fundamental concepts and practical modeling. *Curr. Opin. Struct. Biol.* **8**, 211-217 (1998)
[1023] Cheatham III, T.E. & Brooks, B.R. Recent advances in molecular dynamics simulation towards the realistic representation of biomolecules in solution. *Theor. Chem. Acc.* **99**,

279-288 (1998)
[1024] Sagui, C. & Darden, T.A. Molecular dynamics simulations of biomolecules: Long-range electrostatic effects. *Annu. Rev. Biophys. Biomol. Struct.* **28**, 155-179 (1999)
[1025] Darden, T., Perera, L., Li, L. & Pedersen, L. New tricks for modelers from the crystallography toolkit: The particle mesh Ewald algorithm and its use in nucleic acid simulations. *Structure* **7**, R55-R60 (1999)
[1026] Hünenberger, P.H., Börjesson, U. & Lins, R.D. Electrostatic interactions in biomolecular systems. *Chimia* **55**, 861-866 (2001)
[1027] Gibbon, P. & Sutmann, G. Long-range interactions in many-particle simulation. In: *Quantum simulations of complex many-body systems: From theory to algorithms, lecture notes, NIC series. Volume 10.* Grotendorst, J., Marx, D. & Muramatsu, A., Eds. John von Neumann institute for Computing, Jülich, Germany, pp 467-506 (2002)
[1028] Kastenholz, M. & Hünenberger, P.H. Influence of artificial periodicity and ionic strength in molecular dynamics simulations of charged biomolecules employing lattice-sum methods. *J. Phys. Chem. B* **108**, 774-788 (2004)
[1029] Koehl, P. Electrostatics calculations: latest methodological advances. *Curr. Opin. Struct. Biol.* **16**, 142-151 (2006)
[1030] Warshel, A., Sharma, P.K., Kato, M. & Parson, W.W. Modeling electrostatic effects in proteins. *Biochim. Biophys. Acta* **1764**, 1647-1676 (2006)
[1031] Reif, M.M., Kräutler, V., Kastenholz, M.A., Daura, X. & Hünenberger, P.H. Explicit-solvent molecular dynamics simulations of a reversibly-folding β-heptapeptide in methanol: Influence of the treatment of long-range electrostatic interactions. *J. Phys. Chem. B* **113**, 3112-3128 (2009)
[1032] Moore, G.E. Cramming more components onto integrated circuits. *Electronics* **38**, 114-117 (1965)
[1033] Gilson, M.K. Theory of electrostatic interactions in macromolecules. *Curr. Opin. Struct. Biol.* **5**, 216-223 (1995)
[1034] Brooks III, C.L. Methodological advances in molecular dynamics simulations of biological systems. *Curr. Opin. Struct. Biol.* **5**, 211-215 (1995)
[1035] Hansson, T., Oostenbrink, C. & van Gunsteren, W.F. Molecular dynamics simulations. *Curr. Opin. Struct. Biol.* **12**, 190-196 (2002)
[1036] Norberg, J. & Nilsson, L. Advances in biomolecular simulations: Methodology and recent applications. *Quart. Rev. Biophys.* **36**, 257-306 (2003)
[1037] Adcock, S.A. & McCammon, J.A. Molecular dynamics: Survey of methods for simulating the activity of proteins. *Chem. Rev.* **106**, 1589-1615 (2006)
[1038] Friedman, H.L. Image approximation to reaction field. *Mol. Phys.* **29**, 1533-1543 (1975)
[1039] Berkowitz, M. & McCammon, J.A. Molecular dynamics with stochastic boundary conditions. *Chem. Phys. Lett.* **90**, 215-217 (1982)
[1040] Brooks III, C.L. & Karplus, M. Deformable stochastic boundaries in molecular dynamics. *J. Chem. Phys.* **79**, 6312-6325 (1983)
[1041] Berkowitz, M., Morgan, J.D. & McCammon, J.A. Generalized Langevin dynamics simulations with arbitrary time-dependent memory kernels. *J. Chem. Phys.* **78**, 3256-3261 (1983)
[1042] Brünger, A., Brooks III, C.L. & Karplus, M. Stochastic boundary conditions for molecular dynamics simulations of ST2 water. *Chem. Phys. Lett.* **105**, 495-500 (1984)
[1043] Belch, A.C. & Berkowitz, M. Molecular-dynamics simulations of TIPS2 water restricted by a spherical hydrophobic boundary. *Chem. Phys. Lett.* **113**, 278-282 (1985)
[1044] Warshel, A. & King, G. Polarization constraints in molecular dynamics simulations of aqueous solutions : The surface constraint all atom solvent (SCAAS) model. *Chem. Phys. Lett.* **121**, 124-129 (1985)
[1045] Rullmann, J.A.C. & van Duijinen, P.T. Analysis of discrete and continuum dielectric models - Application to the calculation of protonation energies in solution. *Mol. Phys.* **61**, 293-311 (1987)
[1046] King, G. & Warshel, A. A surface constrained all-atom solvent model for effective simulations. *J. Chem. Phys.* **91**, 3647-3661 (1989)
[1047] Åqvist, J. Ion-water interaction potentials derived from free energy perturbation simulations. *J. Phys. Chem.* **94**, 8021-8024 (1990)
[1048] Juffer, A.H. & Berendsen, H.J.C. Dynamic surface boundary conditions. A simple boundary model for molecular dynamics simulations. *Mol. Phys.* **79**, 623-644 (1993)
[1049] Wallqvist, A. On the implementation of Friedman boundary-conditions in liquid water simulations. *Mol. Simul.* **10**, 13-17 (1993)
[1050] Alper, H. & Levy, R.M. Dielectric and thermodynamic response of a generalized reaction field model for liquid state simulations. *J. Chem. Phys.* **99**, 9847-9852 (1993)
[1051] Beglov, D. & Roux, B. Finite representation of an infinite bulk system : Solvent boundary potential from computer simulations. *J. Chem. Phys.* **100**, 9050-9063 (1994)
[1052] Wang, L. & Hermans, J. Reaction field molecular dynamics simulations with Friedman's image charge method. *J. Phys. Chem.* **99**, 12001-12007 (1995)
[1053] Essex, J.W. & Jorgensen, W.L. An empirical boundary potential for water droplet sim-

ulations. *J. Comput. Chem.* **16**, 951-972 (1995)
[1054] Montoro, J.C.G. & Abascal, J.L.F. A modulated bulk as fuzzy boundary for the simulation of long-ranged inhomogeneous systems. *Mol. Simul.* **14**, 313-329 (1995)
[1055] Beglov, D. & Roux, B. Dominant solvation effects from the primary shell of hydration: Approximation for molecular dynamics simulations. *Biopolymers* **35**, 171-178 (1995)
[1056] Darden, T., Pearlman, D. & Pedersen, L.G. Ionic charging free energies : Spherical versus periodic boundary conditions. *J. Chem. Phys.* **109**, 10921-10935 (1998)
[1057] Sham, Y.Y. & Warshel, A. The surface constraint all atom model provides size independent results in calculations of hydration free energies. *J. Chem. Phys.* **109**, 7940-7944 (1998)
[1058] Lounnas, V., Lüdemann, S.K., Towards molecular dynamics simulation of large proteins with a hydration shell at constant pressure. *Biophys. Chem.* **78**, 157-182 (1999)
[1059] Kimura, S.R., Brower, R.C., Zhang, C. & Sugimori, M. Surface of active polarons: A semiexplicit solvation method for biomolecular dynamics. *J. Chem. Phys.* **112**, 7723-7734 (2000)
[1060] Sankararamakrishnan, R., Konvicka, K., Mehler, E.L. & Weinstein, H. Solvation in simulated annealing and high-temperature molecular dynamics of proteins: A restrained water droplet model. *Int. J. Quant. Chem.* **77**, 174-186 (2000)
[1061] Im, W., Bernèche, S. & Roux, B. Generalized solvent boundary potential for computer simulations. *J. Chem. Phys.* **114**, 2924-2937 (2001)
[1062] Banavali, N.K., Im, W. & Roux, B. Electrostatic free energy calculations using the generalized solvent boundary potential method. *J. Chem. Phys.* **117**, 7381-7388 (2002)
[1063] Li, Y., Krilov, G. & Berne, B.J. Elastic bag model for molecular dynamics simulations of solvated systems: Application to liquid argon. *J. Phys. Chem. B* **109**, 463-470 (2005)
[1064] Brancato, G., di Nola, A., Barone, V. & Amadei, A. A mean field approach for molecular simulations of fluid systems. *J. Chem. Phys.* **122**, 154109/1-154109/9 (2005)
[1065] Brancato, G., Rega, N. & Barone, V. Reliable molecular simulations of solute-solvent systems with a minimum number of solvent shells. *J. Chem. Phys.* **124**, 21405/1-21405/9 (2006)
[1066] Wang, J., Deng, Y. & Roux, B. Absolute binding free energy calculations using molecular dynamics simulations with restraining potentials. *Biophys. J.* **91**, 2798-2814 (2006)
[1067] Li, Y., Krilov, G. & Berne, B.J. Elastic bag model for molecular dynamics simulations of solvated systems: Application to liquid water and solvated peptides. *J. Phys. Chem. B.* **110**, 13256-13263 (2006)
[1068] Alder, B.J. & Wainwright, T.E. Studies in molecular dynamics. I. General method. *J. Chem. Phys.* **31**, 459-466 (1959)
[1069] Brooks III, C.L., Brünger, A.T. & Karplus, M. Active site dynamics in protein molecules: A stochastic boundary molecular-dynamics approach. *Biopolymers* **24**, 843-865 (1985)
[1070] Brünger, A.T., Brooks III, C.L. & Karplus, M. Active site dynamics of ribonuclease. *Proc. Natl. Acad. Sci. USA* **82**, 8458-8462 (1985)
[1071] Kim, K.S. On effective methods to treat solvent effects in macromolecular mechanics and simulations. *Chem. Phys. Lett.* **156**, 261-268 (1989)
[1072] Ewald, P.P. Die Berechnung optischer und elektrostatischer Gitterpotentiale. *Ann. Phys.* **369**, 253-287 (1921)
[1073] de Leeuw, S.W., Perram, J.W. & Smith, E.R. Simulation of electrostatic systems in periodic boundary conditions. I. Lattice sums and dielectric constants. *Proc. R. Soc. Lond. A* **373**, 27-56 (1980)
[1074] de Leeuw, S.W., Perram, J.W. & Smith, E.R. Simulation of electrostatic systems in periodic boundary conditions. II. Equivalence of boundary conditions. *Proc. R. Soc. Lond. A.* **373**, 57-66 (1980)
[1075] Luty, B.A., Davis, M.E., Tironi, I.G. & van Gunsteren, W.F. A comparison between particle-particle, particle-mesh and Ewald methods for calculating electrostatic interactions in periodic molecular systems. *Mol. Simul.* **14**, 11-20 (1994)
[1076] Darden, T.A., Toukmaji, A. & Pedersen, L.G. Long-range electrostatic effects in biomolecular simulation. *J. Chim. Phys.* **94**, 1346-1364 (1997)
[1077] Deserno, M. & Holm, C. How to mesh up Ewald sums. I. A theoretical and numerical comparison of various particle mesh routines. *J. Chem. Phys.* **109**, 7678-7693 (1998)
[1078] Deserno, M. & Holm, C. How to mesh up Ewald sums. II. An accurate error estimate for the particle-particle-particle-mesh algorithm. *J. Chem. Phys.* **109**, 7694-7701 (1998)
[1079] Hünenberger, P.H. & McCammon, J.A. Effect of artificial periodicity in simulations of biomolecules under Ewald boundary conditions: A continuum electrostatics study. *Biophys. Chem.* **78**, 69-88 (1999)
[1080] Boresch, S. & Steinhauser, O. The dielectric self-consistent field method. II. Application to the study of finite range effects. *J. Chem. Phys.* **115**, 10793-10807 (2001)
[1081] Kastenholz, M.A. & Hünenberger, P.H. Computation of methodology-independent ionic solvation free energies from molecular simulations: II. The hydration free energy of the sodium cation. *J. Chem. Phys.* **124**, 224501/1-224501/20 (2006)
[1082] Heinz, T.N. & Hünenberger, P.H. Combining the lattice-sum and reaction-field ap-

proaches for evaluating long-range electrostatic interactions in molecular simulations. *J. Chem. Phys.* **123**, 034107/1-034107/19 (2005)

[1083] Christen, M., Hünenberger, P.H., Bakowies, D., Baron, R., Bürgi, R., Geerke, D.P., Heinz, T.N., Kastenholz, M.A., Kräutler, V., Oostenbrink, C., Peter, C., Trzesniak, D. & van Gunsteren, W.F. The GROMOS software for biomolecular simulation: GROMOS05. *J. Comput. Chem.* **26**, 1719-1751 (2005)

[1084] Redlack, A. & Grindlay, J. The electrostatic potential in a finite ionic crystal. *Can. J. Phys.* **50**, 2815-2825 (1972)

[1085] Nijboer, B.R.A. & Ruijgrok, T.W. On the energy per particle in three- and two-dimensional Wigner lattices. *J. Stat. Phys.* **53**, 361-382 (1988)

[1086] Roberts, J.E. & Schnitker, J. Boundary conditions in simulations of aqueous ionic solutions: A systematic study. *J. Phys. Chem.* **99**, 1322-1331 (1995)

[1087] Boresch, S. & Steinhauser, O. Presumed versus real artifacts of the Ewald summation technique: The importance of the dielectric boundary condition. *Ber. Bunsenges. Phys. Chem.* **101**, 1019-1029 (1997)

[1088] Boresch, S. & Steinhauser, O. Rationalizing the effects of modified electrostatic interactions in computer simulations : The dielectric self-consistent field method. *J. Chem. Phys.* **111**, 8271-8274 (1999)

[1089] Caillol, J.-M. Comments on the numerical simulations of electrolytes in periodic boundary conditions. *J. Chem. Phys.* **101**, 6080-6090 (1994)

[1090] Euwema, R.N., Whilhite, D.L. & Surratt, G.T. General crystalline Hartree-Fock formalism - Diamond results. *Phys. Rev. B* **7**, 818-831. (1973)

[1091] Euwema, R.N. & Surratt, G.T. The absolute positions of calculated energy bands. *J. Phys. Chem. Solids* **36**, 67-71 (1975)

[1092] Redlack, A. & Grindlay, J. Coulombic potential lattice sums. *J. Phys. Chem. Solids* **36**, 73-82 (1975)

[1093] Stuart, S.N. Depolarization correction for Coulomb lattice sums. *J. Comput. Phys.* **29**, 127-132 (1978)

[1094] Deem, M.W., Newsam, J.M. & Sinha, S.K. The h=0 term in Coulomb sums by the Ewald transformation. *J. Phys. Chem.* **94**, 8356-8359 (1990)

[1095] Roberts, J.E. & Schnitker, J. How the unit cell surface charge distribution affects the energetics of ion-solvent interactions in simulations. *J. Chem. Phys.* **101**, 5024-5031 (1994)

[1096] Fraser, L.M., Foulkes, W.M.C., Rajagopal, G., Needs, R.J., Kenny, S.D. & Williamson, A.J. Finite-size effects and Coulomb interactions in quantum Monte Carlo calculations for homogeneous systems with periodic boundary conditions. *Phys. Rev. B* **53**, 1814-1832 (1996)

[1097] Pillardy, J., Wawak, R.J., Arnautova, Y.A., Czaplewski, C. & Scheraga, H.A. Crystal structure prediction by global optimization as a tool for evaluating potentials: Role of the dipole moment correction term in successful predictions. *J. Am. Chem. Soc.* **122**, 907-921 (2000)

[1098] Herce, H.D. & Garcia, A.E. The electrostatic surface term: (I) Periodic systems. *J. Chem. Phys.* **126**, 124106/1-124106/13 (2007)

[1099] de Wette, F.W. Comments on the electrostatic energy of a Wigner solid. *Phys. Rev. B* **21**, 3751-3753 (1980)

[1100] Neumann, M. Dipole moment fluctuation formulas in computer simulations of polar systems. *Mol. Phys.* **50**, 841-858 (1983)

[1101] Evjen, H.M. On the stability of certain heteropolar crystals. *Phys. Rev.* **39**, 675-687 (1932)

[1102] Gurney, I.D.C. Lattice sums: The validity of Evjen's method. *Phys. Rev.* **90**, 317-318 (1952)

[1103] Krishnan, K.S. & Roy, S.K. Evjen's method of evaluating the Madelung constant. *Phys. Rev.* **87**, 581-582 (1952)

[1104] Dahl, J.P. Correction and extension of Evjen's method for evaluating crystal potentials by means of lattice sums. *J. Phys. Chem. Solids* **26**, 33-40 (1965)

[1105] Lekner, J. Summation of dipolar fields in simulated liquid-vapour interfaces. *Physica A* **157**, 826-838 (1989)

[1106] Lekner, J. Summation of Coulomb fields in computer-simulated disordered systems. *Physica A* **176**, 485-498 (1991)

[1107] Sperb, R. Extension and simple proof of Lekner's summation formula for Coulomb forces. *Mol. Simul.* **13**, 189-193 (1994)

[1108] Gronbech-Jensen, N. Lekner summation of long range interactions in periodic systems. *Int. J. Mod. Phys. C* **8**, 1287-1297 (1997)

[1109] Mazars, M. Lekner summations. *J. Chem. Phys.* **115**, 2955-2965 (2001)

[1110] English, N.J. & Macelroy, J.M.D. Atomistic simulations of liquid water using Lekner electrostatics. *Mol. Phys.* **100**, 3753-3769 (2002)

[1111] Mazars, M. Lekner summations and Ewald summations for quasi-two-dimensional systems. *Mol. Phys.* **103**, 1241-1260 (2005)

[1112] Hockney, R.W. & Eastwood, J.W. Computer simulation using particles. McGraw-Hill, New York, USA (1981)
[1113] Darden, T., York, D. & Pedersen, L. Particle mesh Ewald: An Nlog(N) method for Ewald sums in large systems. *J. Chem. Phys.* **98**, 10089-10092 (1993)
[1114] Essmann, U., Perera, L., Berkowitz, M.L., Darden, T., Lee, H. & Pedersen, L.G. A smooth particle mesh Ewald method. *J. Chem. Phys.* **103**, 8577-8593 (1995)
[1115] Greengard, L. & Rokhlin, V. A fast algorithm for particle simulations. *J. Comput. Phys.* **73**, 325-348 (1987)
[1116] Schmidt, K.E. & Lee, M.A. Implementing the fast multipole method in three dimensions. *J. Stat. Phys.* **63**, 1223-1235 (1991)
[1117] Ding, H.-Q., Karasawa, N. & Goddard III, W.A. The reduced cell multipole method for Coulomb interactions in periodic systems with million-atom unit cells. *Chem. Phys. Lett.* **196**, 6-10 (1992)
[1118] Shimada, J., Kaneko, H. & Takada, T. Efficient calculations of Coulombic interactions in biomolecular simulations with periodic boundary conditions. *J. Comput. Chem.* **14**, 867-878 (1993)
[1119] Shimada, J., Kaneko, H. & Takada, T. Performance of fast multipole methods for calculating electrostatic interactions in biomacromolecular simulations. *J. Comput. Chem.* **15**, 28-43 (1994)
[1120] Pollock, E.L. & Glosli, J. Comments on P3M, FMM, and the Ewald method for large periodic Coulombic systems. *Comput. Phys. Commun.* **95**, 93-110 (1996)
[1121] Lambert, C.G., Darden, T.A. & Board Jr., J.A. A multipole-based algorithm for efficient calculation of forces and potentials in macroscopic periodic assemblies of particles. *J. Comput. Phys.* **126**, 274-285 (1996)
[1122] Challacombe, M., White, C. & Head-Gordon, M. Periodic boundary conditions and fast multipole methods. *J. Chem. Phys.* **107**, 10131-10140 (1997)
[1123] Figueirido, F., Levy, R.M., Zhou, R. & Berne, B.J. Large scale simulation of macromolecules in solution: Combining the periodic fast multipole method with multiple time step integrators. *J. Phys. Chem.* **106**, 9835-9849 (1997)
[1124] Schmidt, K.E. & Lee, M.A. Multipole Ewald sums for the fast multipole method. *J. Stat. Phys.* **89**, 411-424 (1997)
[1125] Kudin, K.N. & Scuseria, G.E. A fast multipole method for periodic systems with arbitrary unit cell geometries. *Chem. Phys. Lett.* **283**, 61-68 (1998)
[1126] Duan, Z.-H. & Krasny, R. An Ewald summation based multipole method. *J. Chem. Phys.* **113**, 3492-3495 (2000)
[1127] Kudin, K.N. & Scuseria, G.E. Analytic stress tensor with the periodic fast multipole method. *Phys. Rev. B* **61**, 5141-5146 (2000)
[1128] Zheng, J., Balasundaram, R., Gehrke, S.H., Heffelfinger, G.S., Goddard III, W.A. & Jiang, S. Cell multipole method for molecular simulations in bulk and confined systems. *J. Chem. Phys.* **118**, 5347-5355 (2003)
[1129] Ogata, S., Campbell, T.J., Kalia, R.K., Nakano, A., Vashishta, P. & Vemparala, S. Scalable and portable implementation of the fast multipole method on parallel computers. *Comput. Phys. Comm.* **153**, 445-461 (2003)
[1130] Kudin, K.N. & Scuseria, G.E. Revisiting infinite lattice sums with the periodic fast multipole method. *J. Chem. Phys.* **121**, 2886-2890 (2004)
[1131] Merrick, M.P., Iyer, K.A. & Beck, T.L. Multigrid method for electrostatic computations in numerical density functional theory. *J. Phys. Chem.* **99**, 12478-12482 (1995)
[1132] Sagui, C. & Darden, T. Multigrid methods for classical molecular dynamics simulations of biomolecules. *J. Chem. Phys.* **114**, 6578-6591 (2001)
[1133] Izaguirre, J.A., Hampton, S.S. & Matthey, T. Parallel multigrid summation for the N-body problem. *J. Parallel. Distrib. Comput.* **65**, 949-962 (2005)
[1134] York, D.M. & Yang, W. The fast Fourier Poisson method for calculating Ewald sums. *J. Chem. Phys.* **101**, 3298-3300 (1994)
[1135] Genovese, L., Deutsch, T. & Godecker, T. Efficient and accurate three-dimensional Poisson solver for surface problems. *J. Chem. Phys.* **127**, 054704/1-054704/6 (2007)
[1136] Maggs, A.C. Dynamics of a local algorithm for simulating Coulomb interactions. *J. Chem. Phys.* **117**, 1975-1981 (2002)
[1137] Maggs, A.C. & Rosetto, V. Local simulation algorithms for Coulomb interactions. *Phys. Rev. Lett.* **88**, 196402/1-196402/4 (2002)
[1138] Rottler, J., A continuum, O(N) Monte Carlo algorithm for charged particles. *J. Chem. Phys.* **120**, 3119-3129 (2004)
[1139] Rottler, J. & Maggs, A.C. Local molecular dynamics with Coulombic interactions. *Phys. Rev. Lett.* **93**, 170201/1-170201/4 (2004)
[1140] Maggs, A.C., Auxiliary field simulation and Coulomb's law. *Comp. Phys. Comm.* **169**, 160-165 (2005)
[1141] Barojas, J., Levesque, D. & Quentrec, B. Simulation of diatomic homonuclear liquids. *Phys. Rev. A* **7**, 1092-1105 (1973)
[1142] Hockney, R.W., Goel, S.P. & Eastwood, J.W. A 10000 particle molecular dynamics

model with long range forces. *Chem. Phys. Lett.* **21**, 589-591 (1973)

[1143] Quentrec, B. & Brot, C. New method for searching for neighbors in molecular dynamics computations. *J. Comput. Phys.* **13**, 430-432 (1973)

[1144] Schofield, P. Computer simulation of the liquid state. *Comput. Phys. Commun.* **5**, 17-23 (1973)

[1145] Hockney, R.W., Goel, S.P. & Eastwood, J.W. Quiet high-resolution computer models of a plasma. *J. Comput. Phys.* **14**, 148-158 (1974)

[1146] Finney, J.L. Long-range forces in molecular dynamics calculations on water. *J. Comput. Chem.* **28**, 92-102 (1978)

[1147] Eastwood, J.W., Hockney, R.W. & Lawrence, D.N. P3MDP - The three-dimensional periodic particle-particle/particle-mesh program. *Comput. Phys. Commun.* **19**, 215-261 (1980)

[1148] van Gunsteren, W.F., Berendsen, H.J.C., Colonna, F., Perahia, D., Hollenberg, J.P. & Lellouch, D. On searching neighbors in computer simulations of macromolecular systems. *J. Comput. Chem.* **5**, 272-279 (1984)

[1149] Boris, J. A vectorized "near neighbors" algorithm of order N using a monotonic logical grid. *J. Comput. Phys.* **66**, 1-20 (1986)

[1150] Yip, V. & Elber, R. Calculation of a list of neighbors in molecular dynamics simulations. *J. Comput. Chem.* **10**, 921-927 (1989)

[1151] Arnold, A. & Mauser, N. An efficient method for bookkeeping next neighbours in molecular dynamics simulations. *Comput. Phys. Commun.* **59**, 267-275 (1990)

[1152] Chialvo, A.A. & Debenedetti, P.G. On the use of the Verlet neighbor list in molecular dynamics. *Comput. Phys. Commun.* **60**, 215-224 (1990)

[1153] Chialvo, A.A. & Debenedetti, P.G. On the performance of an automated Verlet neighbor list algorithm. *Comput. Phys. Commun.* **64**, 15-18 (1991)

[1154] Chialvo, A.A. & Debenedetti, P.G. An automated Verlet neighbor list algorithm with a multiple time-step approach for the simulation of large systems. *Comput. Phys. Commun.* **70**, 467-477 (1992)

[1155] Everaers, R. & Kremet, K. A fast grid search algorithm for molecular dynamics simulations with short-range interactions. *Comput. Phys. Commun.* **81**, 19-55 (1994)

[1156] Forester, T. & Smith, W. On multiple time-step algorithms and the Ewald sum. *Mol. Simul.* **13**, 195-204 (1994)

[1157] Pütz, M. & Kolb, A. Optimization techniques for parallel molecular dynamics using domain decomposition. *Comput. Phys. Commun.* **113**, 145-167 (1998)

[1158] Morales, J.J. & Nuevo, M.J. A technique for improving the linked-cell method. *Comput. Phys. Commun.* **60**, 195-199 (1990)

[1159] Morales, J.J. & Nuevo, M.J. Comparison of linked-cell and neighbourhood tables on a range of computers. *Comput. Phys. Commun.* **69**, 223-228 (1992)

[1160] Mattson, W. & Rice, B.M. Near-neighbour calculations using a modified cell-linked list method. *Comput. Phys. Commun.* **119**, 135-148 (1999)

[1161] Lindahl, E., Hess, B. & van der Spoel, D. GROMACS 3.0: A package for molecular simulation and trajectory analysis. *J. Mol. Model.* **7**, 306-317 (2001)

[1162] Petrella, R.J., Andricioeaei, I., Brooks, B.R. & Karplus, M. An improved method for nonbonded list generation: Rapid determination of near-neighbor pairs. *J. Comput. Chem.* **24**, 222-231 (2003)

[1163] Heinz, T.N. & Hünenberger, P.H. A fast pairlist-construction algorithm for molecular simulations under periodic boundary conditions. *J. Comput. Chem.* **25**, 1474-1486 (2004)

[1164] Kräutler, V. & Hünenberger, P.H. A multiple-timestep algorithm compatible with a large number of distance classes and an arbitrary distance dependence of the timestep size for the fast evaluation of non-bonded interactions in molecular simulations. *J. Comput. Chem.* **124**, 1163-1176 (2006)

[1165] Adams, D.J. Computer simulation of ionic systems: The distorting effects of the boundary conditions. *Chem. Phys. Lett.* **62**, 329-332 (1979)

[1166] Brooks III, C.L., Pettitt, B.M. & Karplus, M. Structural and energetic effect of truncating the long-ranged interactions in ionic and polar fluids. *J. Chem. Phys.* **83**, 5897-5908 (1985)

[1167] Brooks III, C.L. Thermodynamics of ionic solvation: Monte Carlo simulations of aqueous chloride and bromide ions. *J. Phys. Chem.* **90**, 6680-6684 (1986)

[1168] Linse, P. & Andersen, H.C. Truncation of Coulombic interactions in computer simulations of liquids. *J. Chem. Phys.* **85**, 3027-3041 (1986)

[1169] Brooks III, C.L. The influence of long-range force truncation on the thermodynamics of aqueous ionic solutions. *J. Chem. Phys.* **86**, 5156-5162 (1987)

[1170] Madura, J.D. & Pettitt, B.M. Effect of truncating long-range interactions in aqueous ionic solution simulations. *Chem. Phys. Lett.* **150**, 105-108 (1988)

[1171] Straatsma, T.P. & Berendsen, H.J.C. Free energy of ionic hydration: Analysis of a thermodynamic integration technique to evaluate free energy differences by molecular dynamics simulations. *J. Chem. Phys.* **89**, 5876-5886 (1988)

[1172] Kuwajima, S. & Warshel, A. The extended Ewald method: A general treatment of long-range electrostatic interactions in microscopic simulations. *J. Chem. Phys.* **89**, 3751-3759 (1988)

[1173] Loncharich, R.J. & Brooks, B.R. The effects of truncating long-range forces on protein dynamics. *Proteins: Struct. Funct. Genet.* **6**, 32-45 (1989)

[1174] Sloth, P. & Sørensen, T.S. Monte Carlo calculations of chemical potentials in ionic fluids by application of Widom's formula: Correction for finite-system effects. *Chem. Phys. Lett.* **173**, 51-56 (1990)

[1175] Beglov, D.B. & Lipanov, A.A. Charge grouping approaches to calculation of electrostatic forces in molecular dynamics of macromolecules. *J. Biomol. Struct. Dyn.* **9**, 205-214 (1991)

[1176] Bader, J.S. & Chandler, D. Computer simulation study of the mean force between ferrous and ferric ions in water. *J. Phys. Chem.* **96**, 6423-6427 (1992)

[1177] Lee, F.S. & Warshel, A. A local reaction field method for fast evaluation of long-range electrostatic interactions in molecular simulations. *J. Chem. Phys.* **97**, 3100-3107 (1992)

[1178] Schreiber, H. & Steinhauser, O. Molecular dynamics studies of solvated polypeptides: Why the cut-off scheme does not work. *Chem. Phys.* **168**, 75-89 (1992)

[1179] Schreiber, H. & Steinhauser, O. Cutoff size does strongly influence molecular dynamics results on solvated polypeptides. *Biochemistry* **31**, 5856-5860 (1992)

[1180] Schreiber, H. & Steinhauser, O. Taming cut-off induced artifacts in molecular dynamics studies of solvated polypeptides. *J. Mol. Biol.* **228**, 909-923 (1992)

[1181] York, D.M., Darden, T.A. & Pedersen, L.G. The effect of long-range electrostatic interactions in simulations of macromolecular crystals: A comparison of the Ewald and truncated list methods. *J. Chem. Phys.* **99**, 8345-8348 (1993)

[1182] Steinbach, P.J. & Brooks, B.R. New spherical-cutoff methods for long-range forces in macromolecular simulation. *J. Comput. Chem.* **15**, 667-683 (1994)

[1183] Niedermeier, C. & Tavan, P. A structure adapted multipole method for electrostatic interactions in protein dynamics. *J. Chem. Phys.* **101**, 734-748 (1994)

[1184] Perera, L., Essmann, U. & Berkowitz, M.L. Effect of the treatment of long-range forces on the dynamics of ions in aqueous solutions. *J. Chem. Phys.* **102**, 450-456 (1994)

[1185] Tironi, I.G., Sperb, R., Smith, P.E. & van Gunsteren, W.F. A generalized reaction field method for molecular dynamics simulations. *J. Chem. Phys.* **102**, 5451-5459 (1995)

[1186] Auffinger, P. & Beveridge, D.L. A simple test for evaluating the truncation effects in simulations of systems involving charged groups. *Chem. Phys. Lett.* **234**, 413-415 (1995)

[1187] Gabdoulline, R.R. & Zheng, C. Effect of the cutoff center on the mean potential and pair distribution functions in liquid water. *J. Comput. Chem.* **16**, 1428-1433 (1995)

[1188] Cheatham III, T.E., Miller, J.L., Fox, T., Darden, T.A. & Kollman, P.A. Molecular dynamics simulations of solvated biomolecular systems: the particle mesh Ewald method leads to stable trajectories of DNA, RNA, and proteins. *J. Am. Chem. Soc.* **117**, 4193-4194 (1995)

[1189] Louise-May, S., Auffinger, P. & Westhof, E. Calculations of nucleic acid conformations. *Curr. Opin. Struct. Biol.* **6**, 289-298 (1996)

[1190] del Buono, G.S., Figueirido, F.E. & Levy, R.M. Dielectric response of solvent surrounding an ion pair: Ewald potential versus spherical truncation. *Chem. Phys. Lett.* **263**, 521-529 (1996)

[1191] Kalko, S.G., Sese, G. & Padro, J.A. On the effect of truncating the electrostatic interactions: Free energies of ion hydration. *J. Chem. Phys.* **104**, 9578-9585 (1996)

[1192] Resat, H. & McCammon, J.A. Free energy simulations: Correcting for electrostatic cutoffs by use of the Poisson equation. *J. Chem. Phys.* **104**, 7645-7651 (1996)

[1193] Resat, H. & McCammon, J.A. Correcting for electrostatic cutoffs in free energy simulations: Towards consistency between simulations with different cutoffs. *J. Chem. Phys.* **108**, 9617-9623 (1998)

[1194] Baker, N.A., Hünenberger, P.H. & McCammon, J.A. Polarization around an ion in a dielectric continuum with truncated electrostatic interactions. *J. Chem. Phys.* **110**, 10679-10692 (1999)

[1195] Zuegg, J. & Gready, J.E. Molecular dynamics simulations of human prion protein: Importance of correct treatment of electrostatic interactions. *Biochemistry* **38**, 13862-13876 (1999)

[1196] de Souza, O.N. & Ornstein, R.L. MD simulations of a protein-protein dimer PME electrostatics far superior to standard cutoff model. *J. Biomol. Struct. Dyn.* **16**, 1205-1218 (1999)

[1197] Norberg, J. & Nilsson, L. On the truncation of long-range electrostatic interactions in DNA. *Biophys. J.* **79**, 1537-1553 (2000)

[1198] Higo, J., Kono, H., Nakamura, H. & Sarai, A. Solvent density and long-range dipole field around a DNA-binding protein studied by molecular dynamics. *Proteins: Struct. Funct. Genet.* **40**, 193-206 (2000)

[1199] Baker, N.A., Hünenberger, P.H. & McCammon, J.A. Erratum : "Polarization around an ion in a dielectric continuum with truncated electrostatic interactions" [J. Chem. Phys. 110, 10679-10692 (1999)]. *J. Chem. Phys.* **113**, 2510-2511 (2000)

[1200] Walser, R., Hünenberger, P.H. & van Gunsteren, W.F. Comparison of different schemes to treat long-range electrostatic interactions in molecular dynamics simulations of a protein crystal. *Proteins: Struct. Funct. Genet.* **44**, 509-519 (2001)

[1201] Ledauphin, V. & Vergoten, G. New nonbonded interactions calculation strategy for rectangular systems. I. Preliminary molecular dynamics study of solvated Na^+ ion. *Biopolymers* **57**, 373-382 (2001)

[1202] Tieleman, D.P., Hess, B. & Sansom, M.S.P. Analysis and evaluation of channel models: Simulation of alamethicin. *Biophys. J.* **83**, 2393-2407 (2002)

[1203] Anézo, C., de Vries, A.H., Höltje, H.-D., Tieleman, D.P. & Marrink, S.-J. Methodological issues in lipid bilayer simulations. *J. Phys. Chem. B* **107**, 9424-9433 (2003)

[1204] Patra, M., Karttunen, M., Hyvönen, M.T., Falck, E., Lindqvist, P. & Vattulainen, I. Molecular dynamics simulations of lipid bilayers: Major artifacts due to truncating electrostatic interactions. *Biophys. J.* **84**, 3636-3645 (2003)

[1205] Patra, M., Karttunen, M., Hyvönen, M.T., Falck, E. & Vattulainen, I. Lipid bilayers driven to a wrong lane in molecular dynamics simulations by subtle changes in long-range electrostatic interactions. *J. Phys. Chem. B* **108**, 4485-4494 (2004)

[1206] Monticelli, L. & Colombo, G. The influence of simulation conditions in molecular dynamics investigations of model β-sheet peptides. *Theor. Chem. Acc.* **112**, 145-157 (2004)

[1207] Baumketner, A. & Shea, J.-E. The influence of different treatments of electrostatic interactions on the thermodynamics of folding of peptides. *J. Phys. Chem. B* **109**, 21322-21328 (2005)

[1208] Beck, D.A.C., Armen, R.S. & Daggett, V. Cutoff size need not strongly influence molecular dynamics results for solvated polypeptides. *Biochemistry* **44**, 609-616 (2005)

[1209] Lins, R.D. & Röthlisberger, U. Influence of long-range electrostatic treatments on the folding of the N-terminal H4 histone tail peptide. *J. Chem. Theory Comput.* **2**, 246-250 (2006)

[1210] Monticelli, L., Simões, C., Belvisi, L. & Colombo, G. Assessing the influence of electrostatic schemes on molecular dynamics simulations of secondary structure forming peptides. *J. Phys. Condensed Mat.* **18**, 329-345 (2006)

[1211] Fennell, C.J. & Gezelter, J.D. Is the Ewald summation still necessary? Pairwise alternatives to the accepted standard for long-range electrostatics. *J. Chem. Phys.* **124**, 234104/1-234104/12 (2006)

[1212] Cordomí, A., Edholm, O. & Perez, J. Effect of different treatments of long-range interactions and sampling conditions in molecular dynamic simulations of rhodopsin embedded in a dipalmitoyl phosphatidylcholine bilayer. *J. Comp. Chem.* **28**, 1017-1030 (2007)

[1213] Barker, J.A. & Watts, R.O. Monte Carlo studies of the dielectric properties of water-like models. *Mol. Phys.* **26**, 789-792 (1973)

[1214] Barker, J.A. Reaction field, screening, and long-range interactions in simulations of ionic and dipolar systems. *Mol. Phys.* **83**, 1057-1064 (1994)

[1215] Mathias, G., Egwolf, B., Nonella, M. & Tavan, P. A fast multipole method combined with a reaction field for long-range electrostatics in molecular dynamics simulations: The effects of truncation on the properties of water. *J. Chem. Phys.* **118**, 10847-10860 (2003)

[1216] Hummer, G., Soumpasis, D.M. & Neumann, M. Pair correlation in an NaCl-SPC water model. Simulations versus extended RISM computations. *Mol. Phys.* **77**, 769-785 (1992)

[1217] Walser, R., Hünenberger, P.H. & van Gunsteren, W.F. Molecular dynamics simulations of a double unit cell in a protein. *Proteins: Struct. Funct. Genet.* **48**, 327-340 (2002)

[1218] Nina, M. & Simonson, T. Molecular dynamics of the $tRNA^{Ala}$ acceptor stem: Comparison between continuum reaction field and particle-mesh Ewald electrostatic treatment. *J. Phys. Chem. B* **106**, 3696-3705 (2002)

[1219] Kräutler, V. & Hünenberger, P.H. Explicit-solvent molecular dynamics simulations of a DNA tetradecanucleotide duplex: Lattice-sum versus reaction-field electrostatics. *Mol. Simul.* **34**, 491-499 (2008)

[1220] Müller-Plathe, F. YASP - A molecular simulation package. *Comput. Phys. Commun.* **78**, 77-94 (1993)

[1221] Brooks, B.R., Bruccoleri, R.E., Olafson, B.D., States, D.J., Swaminathan, S. & Karplus, M. CHARMM: A program for macromolecular energy, minimization and dynamics calculations. *J. Comput. Chem.* **4**, 187-217 (1983)

[1222] Stote, R.H., States, D.J. & Karplus, M. On the treatment of electrostatic interactions in biomolecular simulations. *J. Chim. Physique* **88**, 2419-2433 (1991)

[1223] Ding, H.-Q., Karasawa, N. & Goddard III, W.A. Optimal spline cutoffs for Coulomb and van der Waals interactions. *Chem. Phys. Lett.* **193**, 197-201 (1992)

[1224] Guenot, J. & Kollman, P.A. Conformational and energetic effects of truncating nonbonded interactions in an aqueous protein dynamics simulation. *J. Comput. Chem.* **14**, 295-311 (1993)

[1225] Adams, D.J. & Dubey, G.S. Taming the Ewald sum in computer simulation of charged systems. *J. Comput. Phys.* **72**, 156-176 (1987)

[1226] Prevost, M., van Belle, D., Lippens, G. & Wodak, S. Computer simulations of liquid water: treatment of long-range interactions. *Mol. Phys.* **71**, 587-603 (1990)

[1227] Hummer, G. & Soumpasis, D.M. Computation of the water density distribution at the ice-water interface using the potentials-of-mean-force expansion. *Phys. Rev. E* **49**, 591-596 (1994)

[1228] Hummer, G. & Soumpasis, D.M. Computer simulation of aqueous Na-Cl electrolytes. *J. Phys.: Condens. Matter* **6**, A141-A144 (1994)

[1229] Yakub, E. & Ronchi, C. An efficient method for computation of long-ranged Coulomb forces in computer simulation of ionic fluids. *J. Chem. Phys.* **119**, 11556-11560 (2003)

[1230] Wu, X. & Brooks, B.R. Isotropic periodic sum: A method for the calculation of long-range interactions. *J. Chem. Phys.* **122**, 044107/1-044107/18 (2005)

[1231] Yakub, E. & Ronchi, C. A new method for computation of long ranged Coulomb forces in computer simulation of disordered systems. *J. Low Temp. Phys.* **139**, 633-643 (2005)

[1232] Yakub, E. Effective computer simulation of strongly coupled Coulomb fluids. *J. Phys. A: Math. Gen.* **39**, 4643-4649 (2006)

[1233] Wolf, D., Keblinski, S.R., Phillpot, S.R. & Eggebrecht, J. Exact method for the simulation of Coulombic systems by spherically truncated, pairwise r^{-1} summation. *J. Chem. Phys.* **110**, 8254-8282 (1999)

[1234] Zahn, D., Schilling, B. & Kast, S.M. Enhancement of the Wolf damped Coulomb potential: Static, dynamic and dielectric properties of liquid water from molecular simulations. *J. Phys. Chem. B* **106**, 10725-10732 (2002)

[1235] Ma, Y. & Garofalini, S.H. Modified Wolf electrostatic summation: Incorporating an empirical charge overlap. *Mol. Simul.* **31**, 739-748 (2005)

[1236] Tasaki, K., McDonald, S. & Brady, J.W. Observations concerning the treatment of long-range interactions in molecular dynamics simulations. *J. Comp. Chem.* **14**, 278-284 (1993)

[1237] Garemyr, R. & Elofsson, A. Study of the electrostatic treatment in molecular dynamics simulations. *Proteins: Struct. Funct. Genet.* **37**, 417-428 (1999)

[1238] Brunsteiner, M. & Boresch, S. Influence of the treatment of electrostatic interactions on the results of free energy calculations of dipolar systems. *J. Chem. Phys.* **112**, 6953-6955 (2000)

[1239] Mark, P. & Nilsson, L. Structure and dynamics of liquid water with different long-range interaction truncation and temperature control methods in molecular dynamics simulations. *J. Comput. Chem.* **23**, 1211-1219 (2002)

[1240] Luty, B.A. & van Gunsteren, W.F. Calculating electrostatic interactions using particle-particle-particle-mesh method with non-periodic long-range interactions. *J. Phys. Chem.* **100**, 2581-2587 (1996)

[1241] Kratky, K.W. New boundary conditions for computer experiments of thermodynamic systems. *J. Comput. Phys.* **37**, 205-217 (1980)

[1242] Kratky, K.W. & Schreiner, W. Computational techniques for spherical boundary conditions. *J. Comput. Phys.* **47**, 313-320 (1982)

[1243] Schreiner, W. & Kratky, K.W. Finiteness effects in computer simulations of fluids with spherical boundary conditions. *Mol. Phys.* **50**, 435-452 (1983)

[1244] Caillol, J.M. & Levesque, D. Numerical simulations of homogeneous and inhomogeneous ionic systems : An efficient alternative to the Ewald method. *J. Chem. Phys.* **94**, 597-607 (1991)

[1245] Caillol, J.M. Structural, thermodynamic, and electrical properties of polar fluids and ionic solutions on a hypersphere: Theoretical aspects. *J. Chem. Phys.* **96**, 1455-1476 (1992)

[1246] Caillol, J.M. & Levesque, D. Structural, thermodynamic, and electrical properties of polar fluids and ionic solutions on a hypersphere: Results of simulations. *J. Chem. Phys.* **96**, 1477-1483 (1992)

[1247] Caillol, J.M. A new potential for the numerical simulation of electrolyte solutions on a hypersphere. *J. Chem. Phys.* **99**, 8953-8963 (1993)

[1248] Caillol, J.M. Numerical simulations of Coulomb systems: A comparison between hyperspherical and periodic boundary conditions. *J. Chem. Phys.* **111**, 6528-6537 (1999)

[1249] Hanassab, S. & Vandernoot, T.J. Spherical boundary conditions: A finite and system size independent geometry for simulations of electrolytic liquids. *Mol. Simul.* **2**, 527-533 (2003)

[1250] Hanassab, S. & Vandernoot, T.J. Monte Carlo simulations of primitive models (PM) electrolytes in non-Euclidean geometries. *Mol. Simul.* **30**, 301-311 (2004)

[1251] Råsmark, P.J., Ekholm, T. & Elvingson, C. Computer simulations of polymer chain structure and dynamics on a hypersphere in four-space. *J. Chem. Phys.* **122**, 184110/1-184110/8 (2005)

[1252] Williams, D.E. Accelerated convergence of crystal-lattice potential sums. *Acta Crystallogr. A* **27**, 452-455 (1971)

[1253] Karasawa, N. & Goddard III, W.A. Acceleration of convergence for lattice sums. *J. Phys. Chem.* **93**, 7320-7327 (1989)

[1254] Monkenbusch, M. A set of routines for efficient and accurate computation of lattice sums of $1/r^n$-potentials. *Comput. Phys. Commun.* **67**, 343-355 (1991)

[1255] Ding, H.-Q., Karasawa, N. & Goddard III, W.A. Atomic level simulation on a million particles: The cell multipole method for Coulomb and London nonbond interactions. *J. Chem. Phys.* **97**, 4309-4315 (1992)
[1256] Chen, Z.-M., Çağin, T. & Goddard III, W.A. Fast Ewald sums for general van der Waals potentials. *J. Comput. Chem.* **18**, 1365-1370 (1997)
[1257] Ko, G.H. & Fink, W.H. Rapidly converging lattice sums for nonelectrostatic interactions. *J. Comput. Chem.* **23**, 477-483 (2002)
[1258] Ou-Yang, W.-Z., Lu, Z.-Y., Shi, T.-F., Sun, Z.-Y. & An, L.-J. A molecular dynamics simulation study of the dependence of Lennard-Jones gas-liquid phase diagram on the long-range part of the interactions. *J. Chem. Phys.* **123**, 234502/1-234502/8 (2005)
[1259] Lopez-Lemus, J., Robero-Bastida, M., Darden, T.A. & Alejandre, J. Liquid-vapour equilibrium of n-alkanes using interface simulations. *Mol. Phys.* **104**, 2413-2421 (2006)
[1260] Shi, B., Sinha, S. & Dir, V.K. Molecular dynamics simulation of the density and surface tension by particle-particle particle-mesh method. *J. Chem. Phys.* **124**, 204715/1-204715/7 (2006)
[1261] In't Veld, P.J., Ismail, A.E. & Grest, G.S. Application of Ewald summation to long-range dispersion forces. *J. Chem. Phys.* **127**, 144711/1-144711/8 (2007)
[1262] Chapela, G.A., Saville, G., Thompson, S.M. & Rowlinson, J.S. Computer simulation of a gas-liquid surface. *J. Chem. Soc. Farad. Trans.* **73**, 1133-1144 (1977)
[1263] Blokhuis, E.M., Bedeaux, D., Holcomb, C.D. & Zollweg, J.A. Tail corrections to the surface-tension of a Lennard-Jones liquid-vapor interface. *Mol. Phys.* **85**, 665-669 (1995)
[1264] Lagüe, P., Pastor, R.W. & Brooks, B.R. Pressure-based long-range correction for Lennard-Jones interactions in molecular dynamics simulations: Application to alkanes at interfaces. *J. Phys. Chem. B* **108**, 363-368 (2004)
[1265] Janeček, J. Long range correction in inhomogeneous simulations. *J. Phys. Chem. B* **110**, 6264-6269 (2006)
[1266] Janeček, J. Interfacial properties of cyclic hydrocarbons: A Monte Carlo study. *J. Phys. Chem. B* **110**, 6916-6923 (2006)
[1267] Shen, V.K., Mountain, R.D. & Errington, J.R. Comparative study of the effect of tail corrections on surface tension determined by molecular simulation. *J. Phys. Chem. B* **111**, 6198-6207 (2007)
[1268] Shirts, M.R., Mobley, D.L., Chodera, J.D. & Pande, V.S. Accurate and efficient corrections for missing dispersion interactions in molecular simulations. *J. Phys. Chem. B* **111**, 13052-13063 (2007)
[1269] Ghoufi, A., Goujon, F., Lachet, V. & Malfreyt, P. Expressions for local contributions to the surface tension from the virial route. *Phys. Rev. E* **77**, 031601/1-031601/13 (2008)
[1270] Pettitt, B.M. & Rossky, P.J. Integral-equation predictions of liquid-state structure for waterlike intermolecular potentials. *J. Chem. Phys.* **77**, 1451-1457 (1982)
[1271] Neria, E., Fischer, S. & Karplus, M. Simulation of activation free energies in molecular systems. *J. Chem. Phys.* **105**, 1902-1921 (1996)
[1272] Glättli, A., Daura, X. & van Gunsteren, W.F. Derivation of an improved simple point charge model for liquid water: SPC/A and SPC/L. *J. Chem. Phys.* **116**, 9811-9828 (2002)
[1273] Glättli, A., Daura, X. & van Gunsteren, W.F. A novel approach for designing simple point charge models for liquid water with three interaction sites. *J. Comput. Chem.* **24**, 1087-1096 (2003)
[1274] Berendsen, H.J.C., Postma, J.P.M., van Gunsteren, W.F. & Hermans, J. Interaction models for water in relation to protein hydration. In: *Intermolecular Forces.* Pullman, B., Ed. Reidel, Dordrecht, The Netherlands., pp 331-342 (1981)
[1275] Berendsen, H.J.C., Grigera, J.R. & Straatsma, T.P. The missing term in effective pair potentials. *J. Phys. Chem.* **91**, 6269-6271 (1987)
[1276] Jorgensen, W.L. Transferable intermolecular potential functions for water, alcohols, and ethers. Application to liquid water. *J. Am. Chem. Soc.* **103**, 335-340 (1981)
[1277] Jorgensen, W.L. Quantum and statistical mechanical studies of liquids. 24. Revised TIPS for simulations of liquid water and aqueous solutions. *J. Chem. Phys.* **77**, 4156-4163 (1982)
[1278] Horn, H.W., Swope, W.C., Pitera, J.W., Madura, J.D., Dick, T.J., Hura, G.L. & Head-Gordon, T. Development of an improved four-site water model for biomolecular simulations: TIP4PEW. *J. Chem. Phys.* **120**, 9665-9678 (2004)
[1279] Mahoney, M.W. & Jorgensen, W.L. A five-site model for liquid water and the reproduction of the density anomaly by rigid, nonpolarizable potential functions. *J. Chem. Phys.* **112**, 8910-8922 (2000)
[1280] Rick, S.W. A reoptimization of the five-site water potential (TIP5P) for use with Ewald sums. *J. Chem. Phys.* **120**, 6085-6093 (2004)
[1281] Rowlinson, J.S. The lattice energy of ice and the second virial coefficient of water vapour. *Trans. Faraday Soc.* **47**, 120-129 (1951)
[1282] Stillinger, F.H. & Rahman, A. Improved simulation of liquid water by molecular dynamics. *J. Chem. Phys.* **60**, 1545-1557 (1974)
[1283] Carravetta, V. & Clementi, E. Water-water interaction potential - An approximation

of the electron correlation contribution by a functional of the SCF density matrix. *J. Chem. Phys.* **81**, 2646-2651 (1984)

[1284] Jorgensen, W.L. & Jenson, C. Temperature dependence of TIP3P, SPC, and TIP4P water from NPT Monte Carlo simulations: Seeking temperatures of maximum density. *J. Comput. Chem.* **19**, 1179-1186 (1998)

[1285] Barker, J.A. & Watts, R.O. Structure of water; A Monte Carlo calculation. *Chem. Phys. Lett.* **3**, 144-145 (1969)

[1286] Peng, Z.W., Ewig, C.S., Hwang, M.-J., Waldman, M. & Hagler, A.T. Derivation of class II force fields. 4. van der Waals parameters of alkali metal cations and halide anions. *J. Phys. Chem. A* **101**, 7243-7252 (1997)

[1287] Engelsen, S.B., Fabricius, J. & Rasmussen, K. The consistent force-field. 1. Methods and strategies for optimization of empirical potential-energy functions. *Acta Chem. Scand.* **48**, 548-552 (1994)

[1288] Engelsen, S.B., Fabricius, J. & Rasmussen, K. The consistent force-field. 2. An optimized set of potential-energy functions for the alkanes. *Acta Chem. Scand.* **48**, 553-565 (1994)

[1289] Reif, M.M. & Hünenberger, P.H. Computation of methodology-independent single-ion solvation properties from molecular simulations. IV. Optimized Lennard-Jones parameter sets for the alkali and halide ions in water. *J. Chem. Phys., in press* (2011)

[1290] Srivastava, B.N. & Srivastava, K.P. Combination rules for potential parameters of unlike molecules on exp-6 model. *J. Chem. Phys.* **24**, 1275-1276 (1956)

[1291] Saxena, S.C. & Mathur, B.P. Central and molecular potentials, combination rules and properties of gases and gas mixtures. *Chem. Phys. Lett.* **1**, 224-226 (1967)

[1292] Calvin, D.W. & Reed, T.M. III Mixture rules for the Mie (n,6) intermolecular pair potential and the Dymond-Alder pair potential. *J. Chem. Phys.* **54**, 3733-3738 (1971)

[1293] Hogervorst, W. Transport and equilibrium properties of simple gases and forces between like and unlike atoms. *Physica* **51**, 77-89 (1971)

[1294] Kong, C.L. & Chakrabarty, M.R. Combining rules for intermolecular potential parameters. 3. Application to exp-6 potential. *J. Phys. Chem.* **77**, 2668-2670 (1973)

[1295] Brumer, P. Combination rules and correlations in repulsive potential parameters for alkali-halide diatomics. *Phys. Rev. A* **10**, 1-8 (1974)

[1296] Wiese, H. & Brickmann, J. Lennard-Jones (12,6) potentials for the non-ionic interaction of alkali cations and halide anions - a test of different combination rules. *Ber. Bunsen Ges. Phys. Chem. Chem. Phys.* **93**, 1464-1467 (1989)

[1297] van Gunsteren, W.F., Billeter, S.R., Eising, A.A., Hünenberger, P.H., Krüger, P., Mark, A.E., Scott, W.R.P. & Tironi, I.G. Biomolecular simulation: The GROMOS96 manual and user guide. Verlag der Fachvereine, Zürich, Switzerland (1996)

[1298] Zarkova, L., Hohm, U. & Damyanova, M. Comparison of Lorentz-Berthelot and Tang-Toennies mixing rules using an isotropic temperature-dependent potential applied to the thermophysical properties of binary gas mixtures of CH_4, CF_4, SF_6, and $C(CH_3)_4$ with Ar, Kr, and Xe. *Int. J. Thermophys.* **25**, 1775-1798 (2004)

[1299] Al-Matar, A.K. & Rockstraw, D.A. A generating equation for mixing rules and two new mixing rules for interatomic potential energy parameters. *J. Comp. Chem.* **25**, 660-668 (2004)

[1300] Good, R.J. & Hope, C.J. New combining rule for intermolecular distances in intermolecular potential functions. *J. Chem. Phys.* **53**, 540-543 (1970)

[1301] Good, R.J. & Hope, C.J. Test of combining rules for intermolecular distances - potential function constants from second virial coefficients. *J. Chem. Phys.* **55**, 111-116 (1971)

[1302] Mirskaya, K.V. Combining rules for interatomic potential functions of Buckingham form. *Tetrahedron* **29**, 679-682 (1973)

[1303] Lorentz, H.A. Über die Anwendung des Satzes vom Virial in der kinetischen Theorie der Gase. *Ann. Phys.* **248**, 127-136 (1881)

[1304] Fuchs, R.R., Mccourt, F.R.W., Thakkar, A.J. & Grein, F. Two new anisotropic potential-energy surfaces for N_2-He. The use of Hartree-Fock SCF calculations and a combining rule for anisotropic long-range dispersion coefficients. *J. Phys. Chem.* **88**, 2036-2045 (1984)

[1305] Berthelot, D. Sur le mélange des gaz. *Comptes rendus de l'académie des sciences Paris* **126**, 1703-1706 (1889)

[1306] Tada, Y., Hiraoka, S., Uemura, T. & Harada, M. Law of corresponding states of uniunivalent molten-salt mixtures. 1. Mixing rule of pair potential parameters. *Indust. Engineer. Chem. Res.* **29**, 1509-1516 (1990)

[1307] Carlton, T.S. & Gittins, C.M. Internuclear distance and repulsive potential between noble-gas atoms - combining rules based on repulsive force. *J. Chem. Phys.* **94**, 2614-2617 (1991)

[1308] Dedkov, G.V. Effective rules for combination of interatomic interaction potentials and forces. *Tech. Phys. Lett.* **23**, 645-648 (1997)

[1309] Srivastava, B.N. & Madan, M.P. Thermal diffusion of gas mixtures and forces between unlike molecules. *Proc. Phys. Soc. Lond. A* **66**, 278-287 (1953)

[1310] Fender, B.E.F. & Halsey, G.D. Second virial coefficients of argon, krypton, and argon-krypton mixtures at low temperatures. *J. Chem. Phys.* **36**, 1881-1888 (1962)

[1311] Kong, C.L. Combining rules for intermolecular potential parameters. II. Rules for the Lennard-Jones (12-6) potential and the Morse potential. *J. Chem. Phys.* **59**, 2464-2467 (1973)

[1312] Kong, C.L. Combining rules for intermolecular potential parameters. I. Rules for the Dymond-Alder potential. *J. Chem. Phys.* **59**, 1953-1958 (1973)

[1313] Chrisman, D.C. & Leach, J.W. Intermolecular parameters and combining rules for square-well potential. *Indust. Engineer. Chem. Fund.* **12**, 423-431 (1973)

[1314] Peña, M.D., Pando, C. & Renuncio, J.A.R. Combination rules for intermolecular potential parameters. I. Rules based on approximations for the long-range dispersion energy. *J. Chem. Phys.* **76**, 325-332 (1982)

[1315] Peña, M.D., Pando, C. & Renuncio, J.A.R. Combination rules for intermolecular potential parameters. II. Rules based on approximations for the long-range dispersion energy and an atomic distortion model for the repulsive interactions. *J. Chem. Phys.* **76**, 333-339 (1982)

[1316] Tang, K.T. & Toennies, J.P. New combining rules for well parameters and shapes of the van der Waals potential of mixed rare-gas systems. *Z. Physik D* **1**, 91-101 (1986)

[1317] Bzowski, J., Mason, E.A. & Kestin, J. On combination rules for molecular van der Waals potential-well parameters. *Int. J. Thermophys.* **9**, 131-143 (1988)

[1318] Ihm, G., Cole, M.W., Toigo, F. & Klein, J.R. Charge-overlap model of physical interactions and a combining rule for unlike systems. *Phys. Rev. A* **42**, 5244-5252 (1990)

[1319] Waldman, M. & Hagler, A.T. New combining rules for rare gas van der Waals parameters. *J. Comp. Chem.* **14**, 1077-1084 (1993)

[1320] Sandler, S.I. & Wheatley, J.K. Intermolecular potential parameter combining rules for Lennard-Lones 6-12 potential. *Chem. Phys. Lett.* **10**, 375-378 (1971)

[1321] Buckingham, A.D., Fowler, P.W. & Hutson, J.M. Theoretical studies of van der Waals molecules and intermolecular forces. *Chem. Rev.* **88**, 963-988 (1988)

[1322] Lide, D.R. CRC Handbook of Chemistry and Physics. Edition 83. CRC Press, Boca Raton, Florida (2002)

[1323] Becke, A.D. & Johnson, E.R. Exchange-hole dipole moment and the dispersion interaction: High-order dispersion coefficients. *J. Chem. Phys.* **124**, 014104/1-014104/6 (2006)

[1324] Chandrasekhar, J., Spellmeyer, D.C. & Jorgensen, W.L. Energy component analysis for dilute aqueous solutions of Li^+, Na^+, F^- and Cl^- ions. *J. Am. Chem. Soc.* **106**, 903-910 (1984)

[1325] Pettitt, B.M. & Rossky, P.J. Alkali halides in water: Ion-solvent correlations and ion-ion potentials of mean force at infinite dilution. *J. Chem. Phys.* **84**, 5836-5844 (1986)

[1326] Roux, B. Valence selectivity of the gramicidin channel: A molecular dynamics free energy perturbation study. *Biophys. J.* **71**, 3177-3185 (1996)

[1327] Borodin, O., Bell, R.L., Bedrov, D. & Smith, G.D. Polarizable and nonpolarizable potentials for K^+ cation in water. *Chem. Phys. Lett.* **336**, 292-302 (2001)

[1328] Weerasinghe, S. & Smith, P.E. A Kirkwood-Buff derived force field for sodium chloride in water. *J. Chem. Phys.* **119**, 11342-11349 (2003)

[1329] Jensen, K.P. & Jorgensen, W.L. Halide, ammonium and alkali metal ion parameters for modeling aqueous solutions. *J. Chem. Theory Comput.* **2**, 1499-1509 (2006)

[1330] Alejandre, J. & Hansen, J.-P. Ions in water: From ion clustering to crystal nucleation. *Phys. Rev. E* **76**, 061505/1-061505/5 (2007)

[1331] Joung, I.S. & Cheatham III, T.E. Determination of alkali and halide monovalent ion parameters for use in explicitly solvated biomolecular simulations. *J. Phys. Chem. B* **112**, 9020-9041 (2008)

[1332] Fumi, F.G. & Tosi, M.P. Ionic sizes and Born repulsive parameters in the NaCl-Type alkali halides - I. *J. Phys. Chem. Solids* **25**, 31-43 (1964)

[1333] van Gunsteren, W.F. GROMOS Reference Manual. BIOMOS B.V., Groningen, The Netherlands. (1987)

[1334] Lybrand, T.P., Ghosh, I. & McCammon, J.A. Hydration of chloride and bromide anions: Determination of relative free-energy by computer simulation. *J. Am. Chem. Soc.* **107**, 7793-7794 (1985)

[1335] Caldwell, J., Dang, L.X. & Kollman, P. Implementation of nonadditive intermolecular potentials by use of molecular dynamics: Development of a water-water potential and water-ion cluster interactions. *J. Am. Chem. Soc.* **112**, 9144-9147 (1990)

[1336] Yu, H., Whitfield, T.W., Harder, E., Lamoureux, G., Vorobyov, I., Anisimov, V.M., MacKerell Jr., A.D. & Roux, B. Simulating monovalent and divalent ions in aqueous solution using a Drude polarizable force field. *J. Chem. Theory Comput.* **6**, 774-786 (2010)

[1337] Reynolds, C.A., King, P.M. & Richards, W.G. Free energy calculations in molecular biophysics. *Mol. Phys.* **76**, 251-275 (1992)

[1338] Straatsma, T.P. & McCammon, J.A. Computational alchemy. *Annu. Rev. Phys. Chem.* **43**, 407-435 (1992)

[1339] Roux, B. The calculation of the potential of mean force using computer simulations. *Comput. Phys. Commun.* **91**, 275-282 (1994)

[1340] van Gunsteren, W.F. & Mark, A.E. Validation of molecular dynamics simulation *J. Chem. Phys.* **108**, 6109-6116 (1998)

[1341] Mark, A.E. Free energy perturbation calculations. In: *Encyclopaedia of computational chemistry.* von Ragué Schleyer, P., Ed. Wiley, New York, USA, pp 1070-1083 (1998)

[1342] Chipot, C. & Pearlman, D.A. Free energy calculations. The long and winding gilded road. *Mol. Simul.* **28**, 1-12 (2002)

[1343] Rodinger, T. & Pomès, R. Enhancing the accuracy, the efficiency and the scope of free energy simulations. *Curr. Opin. Struct. Biol.* **15**, 164-170 (2005)

[1344] Shen, T. & Hamelberg, D. A statistical analysis of the precision of reweighting-based simulations. *J. Chem. Phys.* **129**, 034103/1-034103/9 (2008)

[1345] Christ, C.D., Mark, A.E. & van Gunsteren, W.F. Basic ingredients of free energy calculations: A review. *J. Comput. Chem.* **31**, 1569-1582 (2010)

[1346] Hansen, H.S. & Hünenberger, P.H. Ball-and-stick local elevation umbrella sampling: molecular simulations involving enhanced sampling within conformational or alchemical subspaces of low internal dimensionalities, minimal irrelevant volume and problem-adapted geometries. *J. Chem. Theory Comput.* **6**, 2622-2646 (2010)

[1347] Allinger, N.L. Molecular mechanics. The MM3 force field for hydrocarbons. 1. *J. Am. Chem. Soc.* **111**, 8551-8566 (1989)

[1348] Allinger, N.L. Molecular mechanics. The MM3 force field for hydrocarbons. 2. Vibrational frequencies and thermodynamics. *J. Am. Chem. Soc.* **111**, 8566-8575 (1989)

[1349] Allinger, N.L. Molecular mechanics. The MM3 force field for hydrocarbons. 3. The van der Waals' potentials and crystal data for aliphatic and aromatic hydrocarbons. *J. Am. Chem. Soc.* **111**, 8576-8582 (1989)

[1350] Allinger, N.L., Yuh, Y.H. & Lii, J.-H. Molecular mechanics. The MM3 force field for hydrocarbons. 1. *J. Am. Chem. Soc.* **111**, 8551-8566 (1989)

[1351] Lii, J.-H. & Allinger, N.L. The MM3 force field for hydrocarbons. *J. Am. Chem. Soc.* **111**, 8551-8588 (1989)

[1352] Lii, J.-H. & Allinger, N.L. Molecular mechanics. The MM3 force field for hydrocarbons. 3. The van der Waals' potentials and crystal data for aliphatic and aromatic hydrocarbons. *J. Am. Chem. Soc.* **111**, 8576-8582 (1989)

[1353] Rappe, A.K., Casewit, C.J., Colwell, K.S., Goddard, W.A. & Skiff, W.M. UFF, a full periodic-table force-field for molecular mechanics and molecular-dynamics simulations. *J. Am. Chem. Soc.* **114**, 10024-10035 (1992)

[1354] Halgren, T.A. Merck molecular force field. 1. Basis, form, scope, parameterization, and performance of MMFF94. *J. Comp. Chem.* **17**, 490-519 (1996)

[1355] Halgren, T.A. Merck molecular force field. 2. MMFF94 van der waals and electrostatic parameters for intermolecular interactions. *J. Comp. Chem.* **17**, 520-552 (1996)

[1356] Halgren, T.A. Merck molecular force field. 3. Molecular geometries and vibrational frequencies for MMFF94. *J. Comp. Chem.* **17**, 553-586 (1996)

[1357] Halgren, T.A. & Nachbar, R.B. Merck molecular force field. 4. Conformational energies and geometries for MMFF94. *J. Comp. Chem.* **17**, 587-615 (1996)

[1358] Halgren, T.A. Merck molecular force field. 5. Extension of MMFF94 using experimental data, additional computational data, and empirical rules. *J. Comp. Chem.* **17**, 616-641 (1996)

[1359] Gaedt, K. & Holtje, H.D. Consistent valence force-field parameterization of bond lengths and angles with quantum chemical ab initio methods applied to some heterocyclic dopamine d-3-receptor agonists. *J. Comp. Chem.* **19**, 935-946 (1998)

[1360] Ritschl, F., Fait, M., Fiedler, K., Kohler, J.E.H., Kubias, B. & Meisel, M. An extension of the consistent valence force field (CVFF) with the aim to simulate the structures of vanadium phosphorus oxides and the adsorption of n-butane and of 1-butene on their crystal planes. *Z. Anorg. Allg. Chem.* **628**, 1385-1396 (2002)

[1361] Beutler, T.C., Mark, A.E., van Schaik, R., Gerber, P.R. & van Gunsteren, W.F. Avoiding singularities and numerical instabilities in free energy calculations based on molecular simulations. *Chem. Phys. Lett.* **222**, 529-539 (1994)

[1362] Zacharias, M., Straatsma, T.P. & McCammon, J.A. Separation-shifted scaling, a new scaling method for Lennard-Jones interactions in thermodynamic integration. *J. Chem. Phys.* **100**, 9025-9031 (1994)

[1363] Huber, T., Torda, A.E. & van Gunsteren, W.F. Structure optimization combining soft-core interaction functions, the diffusion equation method, and molecular dynamics. *J. Phys. Chem. A* **101**, 5926-5930 (1997)

[1364] Shao, C.-S., Byrd, R., Eskow, E. & Schnabel, R.B. Global optimization for molecular clusters using a new smoothing approach. *J. Global Optim.* **16**, 167-196 (2000)

[1365] Pitera, J.W. & van Gunsteren, W.F. A Comparison of non-bonded scaling approaches for free energy calculations. *Mol. Simul.* **28**, 45-65 (2002)

[1366] Christen, M., Kunz, A.-P.E. & van Gunsteren, W.F. Sampling of rare events using hidden restraints. *J. Phys. Chem. B* **110**, 8488-8498 (2006)

[1367] Christen, M., Kunz, A.-P.E. & van Gunsteren, W.F. Erratum to: "Sampling of rare events using hidden restraints" [J. Phys. Chem. B 110, 8488-8498 (2006)] *J. Phys. Chem. B* **112**, 11446-11446 (2008)

[1368] Straatsma, T.P. & McCammon, J.A. Multiconfiguration thermodynamic integration. *J. Chem. Phys.* **95**, 1175-1188 (1991)

[1369] Chipot, C., Kollman, P.A. & Pearlman, D.A. Alternative approaches to potential of mean force calculations: free energy perturbation versus thermodynamic integration. Case study of some representative nonpolar interactions. *J. Comput. Chem.* **17**, 1112-1131 (1996)

[1370] Ciccotti, G. & Ferrario, M. Rare events by constrained molecular dynamics. *J. Mol. Liq.* **89**, 1-18 (2000)

[1371] Ciccotti, G. & Ferrario, M. Blue moon approach to rare events. *Mol. Simul.* **30**, 787-793 (2004)

[1372] Ciccotti, G., Kapral, R. & Vanden-Eijnden, E. Blue moon sampling, vectorial reaction coordinates, and unbiased constrained dynamics. *ChemPhysChem* **6**, 1809-1814 (2005)

[1373] Kästner, J., Senn, H.M., Thiel, S., Otte, N. & Thiel, W. QM/MM free-energy perturbation compared to thermodynamic integration and umbrella sampling: Application to an enzymatic reaction. *J. Chem. Theory Comput.* **2**, 452-461 (2006)

[1374] Postma, J.P.M, Berendsen, H.J.C. & Haak, J.R. Thermodynamics of cavity formation in water. A molecular dynamics study. *Faraday Symp. Chem. Soc.* **17**, 55-67 (1982)

[1375] Straatsma, T.P., Berendsen, H.J.C. & Postma, J.P.M. Free energy of hydrophobic hydration: A molecular dynamics study of noble gases in water. *J. Chem. Phys.* **85**, 6720-6727 (1986)

[1376] Mitchell, M.J. & McCammon, J.A. Free energy difference calculations by thermodynamic integration: Difficulties in obtaining a precise value. *J. Comput. Chem.* **12**, 271-275 (1991)

[1377] Hermans, J. Simple analysis of noise and hysteresis in (slow-growth) free energy simulations. *J. Phys. Chem.* **95**, 9029-9032 (1991)

[1378] Wood, R.H. Estimation of errors in free energy calculations due to the lag between the Hamiltonian and the system configuration. *J. Phys. Chem.* **95**, 4838-4842 (1991)

[1379] Hu, H., Yun, R.H. & Hermans, J. Reversibility of free energy simulations: Slow growth may have a unique advantage (with a note on use of Ewald summation). *Mol. Simul.* **28**, 67-80 (2002)

[1380] Jarzynski, C. Nonequilibrium equality for free energy differences. *Phys. Rev. Lett.* **78**, 2690-2693 (1997)

[1381] Hummer, G. Fast-growth thermodynamic integration: Results for sodium ion hydration. *Mol. Simul.* **28**, 81-90 (2002)

[1382] Kosztin, I., Barz, B. & Janosi, L. Calculating potentials of mean force and diffusion coefficients from nonequilibrium processes without Jarzynski's equality. *J. Chem. Phys.* **124**, 064106/1-064106/11 (2006)

[1383] Piccinini, E., Ceccarelli, M., Affinito, F., Brunetti, R. & Jacoboni, C. Biased molecular simulations for free-energy mapping: A comparison on the KcsA channel as a test case. *J. Chem. Theory Comput.* **4**, 173-183 (2008)

[1384] Tobias, D.J. & Brooks III, C.L. Calculation of free energy surfaces using the methods of thermodynamic perturbation theory. *Chem. Phys. Lett.* **142**, 472-476 (1987)

[1385] Brooks III, C.L. Thermodynamic calculations on biological molecules. *Int. J. Quantum Chem.: Quantum Biology Symposium* **15**, 221-234 (1988)

[1386] Tobias, D.J., Brooks III, C.L. & Fleischman, S.H. Conformational flexibility in free energy simulations. *Chem. Phys. Lett.* **156**, 256-260 (1989)

[1387] Widom, B. Some topics in the theory of fluids. *J. Chem. Phys.* **39**, 2808-2812 (1963)

[1388] Kubo, R. Generalized cumulant expansion method. *J. Phys. Soc. Jpn.* **17**, 1100-1120 (1962)

[1389] Smith, P.E. & van Gunsteren, W.F. Predictions of free energy differences from a single simulation of the initial state. *J. Chem. Phys.* **100**, 577-585 (1994)

[1390] Hummer, G. & Szabo, A. Calculation of free-energy differences from computer simulations of initial and final states. *J. Chem. Phys.* **105**, 2004-2010 (1996)

[1391] Tidor, B. Simulated annealing on free energy surfaces by a combined molecular dynamics and Monte Carlo approach. *J. Phys. Chem.* **97**, 1069-1073 (1993)

[1392] Liu, Z. & Berne, B.J. Method for accelerating chain folding and mixing. *J. Chem. Phys.* **99**, 6071-6077 (1993)

[1393] Guo, Z. & Brooks III, C.L. Rapid screening of binding affinities: Applications of the λ-dynamics method to a trypsin-inhibitor system. *J. Am. Chem. Soc.* **120**, 1920-1921 (1998)

[1394] Leitgeb, M., Schröder, C. & Boresch, S. Alchemical free energy calculations and multiple conformational substates. *J. Chem. Phys.* **122**, 084109/1-084109/15 (2005)

[1395] Abrams, J.B., Rosso, L. & Tuckerman, M.E. Efficient and precise solvation free energies via alchemical adiabatic molecular dynamics. *J. Chem. Phys.* **125**, 074115/1-074115/12 (2006)

[1396] Han, K.-K. A new Monte Carlo method for estimating free energy and chemical potential. *Phys. Lett. A* **165**, 28-32 (1992)

[1397] Smith, G.R. & Bruce, A.D. A study of the multi-canonical Monte Carlo method *J. Phys.*

A: Math. Gen. **28**, 6623-6643 (1995)
[1398] Escobeda, F.A. & de Pablo, J.J. Monte Carlo simulation of the chemical potential of polymers in an expanded ensemble. *J. Chem. Phys.* **103**, 2703-2710 (1995)
[1399] Son, H.S., Kim, S.-Y., Lee, J. & Han, K.-K. Application of the multiensemble sampling to the equilibrium folding of proteins. *Bioinformatics* **22**, 1832-1837 (2006)
[1400] Okazaki, S., Nakanishi, K. & Touhara, H. Monte Carlo studies of the hydrophobic hydration in dilute aqueous solutions of nonpolar solutes. *J. Chem. Phys.* **71**, 2421-2429 (1979)
[1401] Jorgensen, W.L. & Madura, J.D. Solvation and conformation of methanol in water. *J. Am. Chem. Soc.* **105**, 1407-1413 (1983)
[1402] Jorgensen, W.L., Gao, J. & Ravimohan, C. Monte Carlo simulations of alkanes in water: Hydration numbers and the hydrophobic effect. *J. Phys. Chem.* **89**, 3470-3473 (1985)
[1403] Smith, D.E. & Haymet, A.D.J. Free energy, entropy, and internal energy of hydrophobic interactions: Computer simulations. *J. Chem. Phys.* **98**, 6445-6454 (1993)
[1404] Lu, N., Kofke, D.A. & Woolf, T.B. Staging is more important than perturbation method for computation of enthalpy and entropy changes in complex systems. *J. Phys. Chem. B* **107**, 5598-5611 (2003)
[1405] Peter, C., Oostenbrink, C., van Dorp, A. & van Gunsteren, W.F. Estimating entropies from molecular dynamics simulations. *J. Chem. Phys.* **120**, 2652-2661 (2004)
[1406] Reif, M.M. & Hünenberger, P.H. Computation of methodology-independent single-ion solvation properties from molecular simulations. V. Calculation of solvation entropies. *Manuscript in preparation* (2011)
[1407] Guillot, B., Guissani, Y. & Bratos, S. A computer-simulation study of hydrophobic hydration of rare gases and of methane. I. Thermodynamic and structural properties. *J. Chem. Phys.* **95**, 3643-3648 (1991)
[1408] Smith, D.E., Zhang, L. & Haymet, A.D.J. Entropy of association of methane in water: A new molecular dynamics computer simulation. *J. Chem. Phys.* **98**, 6445-6454 (1992)
[1409] Wallqvist, A. & Berne, B.J. Computer simulation of hydrophobic hydration forces on stacked plates at short range. *J. Phys. Chem.* **99**, 2893-2899 (1995)
[1410] Tsunekawa, N., Miyagawa, H., Kitamura, K. & Hiwatari, Y. A study of water-water interactions in hydrophobic association by a molecular dynamics simulation with an optimized umbrella sampling method. *J. Chem. Phys.* **116**, 6725-6730 (2002)
[1411] Fleischman, S.H. & Brooks III, C.L. Thermodynamics of aqueous solvation: Solution properties of alcohols and alkanes. *J. Chem. Phys.* **87**, 3029-3037 (1987)
[1412] Reif, M.M. & Hünenberger, P.H. Computation of methodology-independent single-ion solvation properties from molecular simulations. III. Correction terms for the solvation free energies, enthalpies, entropies, heat capacities, volumes, compressibilities and expansivities of solvated ions. *J. Chem. Phys., in press* (2011)
[1413] Hermans, J. & Shankar, S. The free energy of xenon binding to myoglobin from molecular dynamics simulation. *Isr. J. Chem.* **27**, 225-227 (1986)
[1414] Roux, B., Nina, M., Pomès, R. & Smith, J.C. Thermodynamic stability of water molecules in the Bacteriorhodopsin proton channel: A molecular dynamics free energy perturbation study. *Biophys. J.* **71**, 670-681 (1996)
[1415] Gilson, M.K., Given, J.A., Bush, B.L. & McCammon, J.A. The statistical thermodynamic basis for computation of binding. *Biophys. J.* **72**, 1047-1069 (1997)
[1416] Hermans, J. & Wang, L. Inclusion of loss of translational and rotational freedom in theoretical estimates of free energies of binding. Application to a complex of benzene and mutant T4 Lysozyme. *J. Am. Chem. Soc.* **119**, 2707-2714 (1997)
[1417] Boresch, S., Tettinger, F. & Leitgeb, M. Absolute binding free energies: A quantitative approach for their calculation. *J. Phys. Chem. B* **107**, 9535-9551 (2003)
[1418] Swanson, J.M.J., Henchman, R.H. & McCammon, J.A. Revisiting free energy calculations: A theoretical connection to MM/PBSA and direct calculation of the association free energy. *Biophys. J.* **86**, 67-74 (2004)
[1419] Deng, Y. & Roux, B. Calculation of standard binding free energies: Aromatic molecules in the T4 lysozyme L99A Mutant. *J. Chem. Theory Comput.* **2**, 1255-1273 (2006)
[1420] Simonson, T. Free energy of particle insertion. An exact analysis of the origin singularity for simple liquids. *Mol. Phys.* **80**, 441-447 (1993)
[1421] Pillardy, J. & Piela, L. Molecular dynamics on deformed potential energy hypersurfaces. *J. Phys. Chem.* **99**, 11805-11812 (1995)
[1422] Garde, S., Hummer, G. & Paulaitis, M.E. Free energy of hydration of a molecular ionic solute: Tetramethylammonium ion. *J. Chem. Phys.* **108**, 1552-1561 (1998)
[1423] Sakane, S., Ashbaugh, H.S. & Wood, R.H. Continuum corrections to the polarization and thermodynamic properties of Ewald sum simulations for ions and ion pairs at infinite dilution. *J. Phys. Chem. B* **102**, 5673-5682 (1998)
[1424] Herce, D.H., Darden, T. & Sagui, C. Calculation of ionic charging free energies in simulation systems with atomic charges, dipoles, and quadrupoles. *J. Chem. Phys.* **119**, 7621-7632 (2003)
[1425] Almlöf, M., Carlsson, J. & Åqvist, J. Improving the accuracy of the linear interaction energy method for solvation free energies. *J. Chem. Theory Comput.* **3**, 2162-2175 (2007)

[1426] Oostenbrink, C. Efficient free energy calculations on small molecule host-guest systems. A combined linear interaction energy/one-step perturbation approach. *J. Comput. Chem.* **30**, 212-221 (2008)
[1427] Warshel, A. Dynamics of reactions in polar solvents. Semiclassical trajectory studies of electron-transfer and proton-transfer reactions. *J. Phys. Chem.* **86**, 2218-2224 (1982)
[1428] Warshel, A., Sussman, F. & King, G. Free energy of charges in solvated proteins: Microscopic calculations using a reversible charging process. *Biochemistry* **25**, 8368-8372 (1986)
[1429] Hummer, G., Pratt, L.R., Garcia, A.E., Berne, B.J. & Rick, S.W. Electrostatic potentials and free energies of solvation of polar and charged molecules. *J. Phys. Chem. B* **101**, 3017-3020 (1997)
[1430] Hummer, G., Pratt, L.R. & Garcia, A.E. Ion sizes and finite-size corrections for ionic-solvation free energies. *J. Chem. Phys.* **107**, 9275-9277 (1997)
[1431] Figueirido, F., del Buono, G.S. & Levy, R.M. On finite-size corrections to the free energy of ionic hydration. *J. Phys. Chem. B* **101**, 5622-5623 (1997)
[1432] Åqvist, J. & Hansson, T. Analysis of electrostatic potential truncation schemes in simulations of polar solvents. *J. Phys. Chem. B* **102**, 3837-3840 (1998)
[1433] Ashbaugh, H.S. & Wood, R.H. Reply to comment on "Electrostatic potentials and free energies of solvation of polar and charged molecules". *J. Phys. Chem. B* **102**, 3844-3845 (1998)
[1434] Vorobjev, Y.N. & Hermans, J. A critical analysis of methods of calculation of a potential in simulated polar liquids : Strong arguments in favor of "Molecule-based" summation and of vacuum boundary conditions in Ewald summations. *J. Phys. Chem. B* **103**, 10234-10242 (1999)
[1435] Yang, P.-K. & Lim, C. Nonconvergence of the solute potential in an infinite solvent and its implications in continuum models. *J. Phys. Chem. B* **106**, 12093-12096 (2002)
[1436] Babu, C.S., Yang, P.-K. & Lim, C. On the charge and molecule based summations of solvent electrostatic potentials and the validity of electrostatic linear response in water. *J. Biol. Phys.* **28**, 95-113 (2002)
[1437] Hummer, G., Pratt, L.R. & Garcia, A.E. Molecular theories and simulation of ions and polar molecules in water. *J. Phys. Chem. A* **102**, 7885-7895 (1998)
[1438] Hummer, G., Pratt, L.R., Garcia, A.E., Garde, S., Berne, B.J. & Rick, S.W. Reply to comments on "Electrostatic potentials and free energies of solvation of polar and charged molecules". *J. Phys. Chem. B* **102**, 3841-3843 (1998)
[1439] Warren, G.L. & Patel, S. Hydration free energies of monovalent ions in transferable intermolecular potential four point fluctuating charge water: An assessment of simulation methodology and force field performance and transferability. *J. Chem. Phys.* **127**, 064509/1-064509/19 (2007)
[1440] Christou, N.I., Whitehouse, J.S., Nicholson, D. & Parsonage, N.G. Studies of high density water films by computer simulation. *Mol. Phys.* **55**, 397-410 (1985)
[1441] Aloisi, G., Guidelli, R., Jackson, R.A., Clark, S.M. & Barnes, P. The structure of water at a neutral interface. *J. Electroanal. Chem* **206**, 131-137 (1986)
[1442] Wilson, M.A., Pohorille, A. & Pratt, L.R. Molecular dynamics of the water liquid-vapor interface. *J. Phys. Chem.* **91**, 4873-4878 (1987)
[1443] Wilson, M.A., Pohorille, A. & Pratt, L.R. Surface potential of the water liquid-vapor interface. *J. Chem. Phys.* **88**, 3281-3285 (1988)
[1444] Matsumoto, M. & Kataoka, Y. Study on liquid-vapor interface of water. I. Simulation results of thermodynamic properties and orientational structure. *J. Chem. Phys.* **88**, 3233-3245 (1988)
[1445] Wilson, M.A., Pohorille, A. & Pratt, L.R. Comment on "Study on the liquid-vapor interface of water. I. Simulation results of thermodynamic properties and orientational structure". *J. Chem. Phys.* **90**, 5211-5213 (1989)
[1446] Sokhan, V.P. & Tildesley, D.J. The free surface of water: molecular orientation, surface potential and nonlinear susceptibility. *Mol. Phys.* **92**, 625-640 (1997)
[1447] Jaqaman, K., Tuncay, K. & Ortoleva, P.J. Classical density functional theory of orientational order at interfaces: Application to water. *J. Chem. Phys.* **120**, 926-938 (2004)
[1448] Kathmann, S.M., I-Feng, W.K. & Mundy, C.J. Electronic effects on the surface potential at the vapor-liquid interface of water. *J. Am. Chem. Soc.* **130**, 16556-16561 (2008)
[1449] Kathmann, S.M., I-Feng, W.K. & Mundy, C.J. Erratum to "Electronic effects on the surface potential at the vapor-liquid interface of water" [J. Am. Chem. Soc. 130, 16556-16561 (2009)]. *J. Am. Chem. Soc.* **131**, 17522-17522 (2009)
[1450] Leung, K. Surface potential at the air-water interface computed using density functional theory. *J. Phys. Chem. Lett.* **1**, 496-499 (2010)
[1451] Brodskaya, E.N., The molecular dynamics simulation of water clusters. *Mol. Phys.* **62**, 251-265 (1987)
[1452] Kalcher, I., Horinek, D., Netz, R.R. & Dzubiella, J. Ion-specific correlations in bulk and at biointerfaces. *J. Phys.: Condens. Matter* **21**, 424108/1-424108/10 (2009)
[1453] Gauss, C.F. Theoria attractionis corporum sphaeroidicorum ellipticorum homogeneorum. Methodo novo tractata. *Commentationes societatis regiae scientiarum Gottin-*

gensis recentiores II **5**, 1-24 (1813)
[1454] Stillinger Jr., F.H. & Ben-Naim, A. Liquid-vapor interface potential for water. *J. Chem. Phys.* **47**, 4431-4437 (1967)
[1455] Pohorille, A. & Wilson, M.A. Viewpoint 9 - Molecular structure of aqueous interfaces. *J. Mol. Struct. (Theochem)* **284**, 271-298 (1993)
[1456] Ashbaugh, H.S. Influence of potential truncation on anisotropic systems. *Mol. Phys.* **97**, 433-437 (1999)
[1457] Fletcher, N.H. Surface structure of water and ice. II. A revised model. *Philos. Mag.* **18**, 1287-1300 (1968)
[1458] Beglov, D. & Roux, B. Finite representation of an infinite bulk system: Solvent boundary potential for computer simulations. *J. Chem. Phys.* **100**, 9050-9063 (1994)
[1459] Åqvist, J. Comment on "Transferability of ion models" [J. Phys. Chem. 97, 6524-6529 (1993)]. *J. Phys. Chem.* **98**, 8253-8255 (1994)
[1460] Brodskaya, E.N., Molecular dynamics computation of the work of ion solvation: comparison of two models for water. *Mol. Phys.* **101**, 1495-1500 (2003)
[1461] Lukyanov, S.I., Zidi, Z.S. & Shevkunov, S.V. Monte Carlo bicanonical ensemble simulation for sodium cation hydration free energy in liquid water. *Fluid Phase Equilibria* **233**, 34-46 (2005)
[1462] Marrone, T.J. & Merz Jr., K.M. Transferability of ion models. *J. Phys. Chem.* **97**, 6524-6529 (1993)
[1463] Chipot, C., Millot, C., Maigret, B. & Kollman, P.A. Molecular dynamics free energy perturbation calculations: Influence of nonbonded parameters on the free energy of charged and neutral species. *J. Phys. Chem.* **98**, 11362-11372 (1994)
[1464] Pliego, J.R. & Riveros, J.M. On the calculation of the absolute solvation free energy of ionic species: Application of the extrapolation method to the hydroxide ion in aqueous solution. *J. Phys. Chem. B* **104**, 5155-5160 (2000)
[1465] Patra, M. & Karttunen, M. Systematic comparison of force fields for microscopic simulations of NaCl in aqueous solutions: Diffusion, free energy of hydration, and structural properties. *J. Comp. Chem.* **25**, 678-689 (2004)
[1466] de Araujo, A.S., Sonoda, M.T., Piro, O.E. & Castellano, E.E. Development of new Cd^{2+} and Pb^{2+} Lennard-Jones parameters for liquid simulations. *J. Phys. Chem. B* **111**, 2219-2224 (2007)
[1467] Whitfield, T.W., Varma, S., Harder, E., Lamoureux, G., Rempe, S.B. & Roux, B. Theoretical study of aqueous solvation of K^+ comparing ab initio, polarizable, and fixed-charge models. *J. Chem. Theory Comput.* **3**, 2068-2082 (2007)
[1468] Wood, R.H. Continuum electrostatics in a computational universe with finite cutoff radii and periodic boundary conditions: Correction to computed free energies of ionic solvation. *J. Chem. Phys.* **103**, 6177-6187 (1995)
[1469] Gabdoulline, R.R., Zheng, C. & Vanderkooi, G. The mean electrostatic potential difference between liquid water and vacuum by MD simulation. *J. Mol. Liq.* **71**, 1-10 (1997)
[1470] Lukyanov, S.I., Zidi, Z.S. & Shevkunov, S.V. Ion water cluster free energy computer simulation using some of most popular ion-water and water-water pair interaction potentials. *Chem. Phys.* **332**, 188-202 (2007)
[1471] Lyubartsev, A.P., Forrisdahl, O.K. & Laaksonen, A. Solvation free energies of methane and alkali halide ion pairs: An expanded ensemble molecular dynamics simulation study. *J. Chem. Phys.* **108**, 227-233 (1998)
[1472] Lukyanov, S.I., Zidi, Z.S. & Shevkunov, S.V. Study of Aqvist's interaction model in Na^+-water clusters: free energy and structure. *J. Mol. Struct. (Theochem)* **623**, 221-236 (2003)
[1473] Kuzmin, V.L., Calculation of the surface potential for charge models of water. *Colloid J.* **6**, 776-780 (1995)
[1474] Stern, H.A. & Berne, B.J. Quantum effects in liquid water: Path-integral simulations of a flexible and polarizable ab initio model. *J. Chem. Phys.* **115**, 7622-7628 (2001)
[1475] Egelstaff, P.A. Structural quantum effects in hydrogeneous liquids and glasses: Part I. Review of early methods and experiments. *Phys. Chem. Liq.* **40**, 203-219 (2002)
[1476] Truhlar, D.G., Gao, J.L., Alhambra, C., Garcia-Viloca, M., Corchado, J., Sanchez, M.L. & Villa, J. The incorporation of quantum effects in enzyme kinetics modeling. *Acc. Chem. Res.* **35**, 341-349 (2002)
[1477] Schofield, J. Quantum effects in ab initio calculations of rate constants for chemical reactions occurring in the condensed phase. *Theor. Chem. Acc.* **116**, 18-30 (2006)
[1478] Tongraar, A. & Rode, B.M. The role of second shell quantum effects on the preferential solvation of Li^+ in aqueous ammonia: An extended *ab initio* QM/MM MD simulation with enlarged QM region. *Chem. Phys. Lett.* **466**, 61-64 (2008)
[1479] Geerke, D.P., Luber, S., Marti, K.H. & van Gunsteren, W.F. On the direct calculation of the free energy of quantization for molecular systems in the condensed phase. *J. Comp. Chem.* **30**, 514-523 (2008)
[1480] Scheiner, S. Calculation of isotope effects from first principles. *Biochem. Biophys. Acta - Bioenergetics* **1458**, 28-42 (2000)

[1481] Marti, S., Moliner, V., Tunon, M. & Williams, I.H. Computing kinetic isotope effects for chorismate mutase with high accuracy. A new DFT/MM strategy. *J. Phys. Chem. B* **109**, 3707-3710 (2005)
[1482] Hakem, I.F., Boussaid, A., Benchouk-Taleb, H. & Bockstaller, M.R. Temperature, pressure, and isotope effects on the structure and properties of liquid water: A lattice approach. *J. Chem. Phys.* **127**, 224106/1-224106/10 (2007)
[1483] Pabis, A., Paluch, P., Szala, J. & Paneth, P. A DFT study of the kinetic isotope effects on the competing S(N)2 and E2 reactions between hypochlorite anion and ethyl chloride. *J. Chem. Theory Comput.* **5**, 33-36 (2009)
[1484] Zhan, C.-G. & Dixon, D.A. Absolute hydration free energy of the proton from first-principles electronic structure calculations. *J. Phys. Chem. A* **105**, 11534-11540 (2001)
[1485] Gray, H.B. & Winkler, J.R. Electron tunneling through proteins. *Quart. Rev. Biophys.* **36**, 341-372 (2003)
[1486] Liang, Z.X. & Klinman, J.P. Structural bases of hydrogen tunneling in enzymes: Progress and puzzles. *Curr. Opin. Struct. Biol.* **14**, 648-655 (2004)
[1487] Hammes-Schiffer, S. Hydrogen tunneling and protein motion in enzyme reactions. *Acc. Chem. Res.* **39**, 93-100 (2006)
[1488] Mavri, J., Liu, H.B., Olsson, M.H.M. & Warshel, A. Simulation of tunneling in enzyme catalysis by combining a biased propagation approach and the quantum classical path method: Application to lipoxygenase. *J. Phys. Chem. B* **112**, 5950-5954 (2008)
[1489] Jortner, J. & Rosenblit, M. Ultracold large finite systems. *Adv. Chem. Phys.* **132**, 247-343 (2006)
[1490] Barranco, M., Guardiola, R., Hernández, S., Mayol, R., Navarro, J. & Pi, M. Helium nanodroplets: an overview. *J. Low Temp. Phys.* **142**, 1-81 (2006)
[1491] Balibar, S. Supersolidity and superfluidity. *Contemp. Phys.* **48**, 31-39 (2007)
[1492] Prokof'ev, N. What makes a crystal supersolid? *Adv. Phys.* **56**, 381-402 (2007)
[1493] Giorgini, S., Pitaevskii, L.P. & Stringari, S. Theory of ultracold atomic Fermi gases. *Rev. Mod. Phys.* **80**, 1215-1274 (2008)
[1494] Carr, L.D., DeMille, D., Krems, R. & Ye, J. Cold and ultracold molecules: science, technology and applications. *New J. Phys.* **11**, 055049/1-055049/87 (2009)
[1495] Balibar, S. The enigma of supersolidity. *Nature* **464**, 176-182 (2010)
[1496] Huang, K. Statistical mechanics. Edition 2. Wiley, New York, USA (1987)
[1497] Wormer, P.E.S. & van der Avoird, A. Intermolecular potentials, internal motions, and spectra of van der Waals and hydrogen-bonded complexes. *Chem. Rev.* **100**, 4109-4143 (2000)
[1498] Muller-Dethlefs, K. & Hobza, P. Noncovalent interactions: A challenge for experiment and theory. *Chem. Rev.* **100**, 143-167 (2000)
[1499] Keutsch, F.N., Cruzan, J.D. & Saykally, R.J. The water trimer. *Chem. Rev.* **103**, 2533-2577 (2003)
[1500] le Fèvre, R.J.W. Polarization and polarizability in chemistry. *Rev. Pure Appl. Chem.* **20**, 67-79 (1970)
[1501] Berendsen, H.J.C. & van Gunsteren, W.F. Practical algorithms for dynamic simulations. In: *Molecular-dynamics simulation of statistical-mechanical systems, proceedings of the international school of physics "Enrico Fermi", course 97.* Ciccotti, G. & Hoover, W.G., Eds. North-Holland, Amsterdam, The Netherlands., pp 43-65 (1986)
[1502] Nazmutdinov, R.R. Quantum-chemical description of charge transfer processes at the Metal/Solution interface: Yesterday, today, and tomorrow. *Russ. J. Electrochem.* **38**, 111-122 (2002)
[1503] Cramer, T., Steinbrecher, T., Labahn, A. & Koslowski, T. Static and dynamic aspects of DNA charge transfer: A theoretical perspective. *Phys. Chem. Chem. Phys.* **7**, 4039-4050 (2005)
[1504] Vlcek, A. & Zalis, S. Modeling of charge-transfer transitions and excited states in d(6) transition metal complexes by DFT techniques. *Coor. Chem. Rev.* **251**, 258-287 (2007)
[1505] Guillaumont, D. & Daniel, C. A quantum chemical investigation of the metal-to-ligand charge transfer photochemistry. *Coord. Chem. Rev.* **177**, 181-199 (1998)
[1506] Ben-Nun, M., Quenneville, J. & Martinez, T.J. Ab initio multiple spawning: Photochemistry from first principles quantum molecular dynamics. *J. Phys. Chem. A* **104**, 5161-5175 (2000)
[1507] Serrano-Andres, L. & Merchán, M. Photostability and photoreactivity in biomolecules: Quantum chemistry of nucleic acid base monomers and dimers. In: *Radiation induced molecular phenomena in nucleic acids. Volume 5.* Shukla, M. & Leszczynski, J., Eds. Springer, Netherlands, pp 435-472 (2008)
[1508] Zamaraev, K.I. & Khairutdinov, R.F. Photoinduced electron-tunneling reactions in chemistry and biology. *Topics Curr. Chem.* **163**, 1-94 (1992)
[1509] Marian, C.M. & Gilka, N. Performance of the density functional theory/multireference configuration interaction method on electronic excitation of extended pi-systems. *J. Chem. Theory Comput.* **4**, 1501-1515 (2008)
[1510] Fabian, J. Electronic excitation of sulfur-organic compounds. Performance of time-dependent density functional theory. *Theor. Chem. Acc.* **106**, 199-217 (2001)

[1511] Siegman, A.E. Lasers. Edition 1. University Science Books, Sausalito, California, USA (1986)

[1512] Han, J.D., Kaledin, L.A., Goncharov, V., Komissarov, A.V. & Heaven, M.C. Accurate ionization potentials for UO and UO_2: A rigorous test of relativistic quantum chemistry calculations. *J. Am. Chem. Soc.* **125**, 7176-7177 (2003)

[1513] Roca-Sanjuan, D., Rubio, M., Merchan, M. & Serrano-Andres, L. Ab initio determination of the ionization potentials of DNA and RNA nucleobases. *J. Chem. Phys.* **125**, 084302/1-084302/7 (2006)

[1514] Lau, K.C. & Ng, C.Y. Accurate ab initio predictions of ionization energies and heats of formation for the 2-propyl, phenyl, and benzyl radicals. *J. Chem. Phys.* **124**, 044323/1-044323/9 (2006)

[1515] Kollman, P.A., Kuhn, B., Donini, O., Perakyla, M., Stanton, R. & Bakowies, D. Elucidating the nature of enzyme catalysis utilizing a new twist on an old methodology: Quantum mechanical free energy calculations on chemical reactions in enzymes and in aqueous solution. *Acc. Chem. Res.* **34**, 72-79 (2001)

[1516] van Speybroeck, V. & Meier, R.J. A recent development in computational chemistry: Chemical reactions from first principles molecular dynamics simulations. *Chem. Soc. Rev.* **32**, 151-157 (2003)

[1517] Cukier, R.I. Theory and simulation of proton-coupled electron transfer, hydrogen-atom transfer, and proton translocation in proteins. *Biochem. Biophys. Acta - Bioenergetics* **1655**, 37-44 (2004)

[1518] Noodleman, L., Lovell, T., Han, W.G., Li, J. & Himo, F. Quantum chemical studies of intermediates and reaction pathways in selected enzymes and catalytic synthetic systems. *Chem. Rev.* **104**, 459-508 (2004)

[1519] Amara, P., Volbeda, A., Fontecilla-Camps, J.C. & Field, M.J. A quantum chemical study of the reaction mechanism of acetyl-coenzyme a synthase. *J. Am. Chem. Soc.* **127**, 2776-2784 (2005)

[1520] Rod, T.H. & Ryde, U. Quantum mechanical free energy barrier for an enzymatic reaction. *Phys. Rev. Lett.* **94**, 138302/1-138302/4 (2005)

[1521] Blomberg, M.R.A. & Siegbahn, P.E.M. Different types of biological proton transfer reactions studied by quantum chemical methods. *Biochem. Biophys. Acta - Bioenergetics* **1757**, 969-980 (2006)

[1522] Eckert-Maksic, M., Vazdar, M., Barbatti, M., Lischka, H. & Maksic, Z.B. Automerization reaction of cyclobutadiene and its barrier height: An ab initio benchmark multireference average-quadratic coupled cluster study. *J. Chem. Phys.* **125**, 064310/1-064310/9 (2006)

[1523] Major, D.T. & Gao, J.L. A combined quantum mechanical and molecular mechanical study of the reaction mechanism and alpha-amino acidity in alanine racemase. *J. Am. Chem. Soc.* **128**, 16345-16357 (2006)

[1524] Bing, D., Zhao, Y.F., Hao, F.Y., Li, X.Y., Liu, F.L., Zhang, G.H. & Zhang, P.X. Ab initio study on the reaction mechanism of ozone with bromine atom. *Int. J. Quant. Chem.* **107**, 1085-1091 (2007)

[1525] Crim, F.F. Chemical reaction dynamics. *Proc. Natl. Acad. Sci. USA* **105**, 12647-12648 (2008)

[1526] Namazian, M. & Heidary, H. Ab initio calculations of pKa values of some organic acids in aqueous solution. *J. Mol. Struct. (Theochem)* **620**, 257-263 (2003)

[1527] Mrazek, J. & Burda, J.V. Can the pH value of water solutions be estimated by quantum chemical calculations of small water clusters? *J. Chem. Phys.* **125**, 194518/1-194518/14 (2006)

[1528] da Silva, G., Kennedy, E.M. & Dlugogorski, B.Z. Ab initio procedure for aqueous-phase pKa calculation: The acidity of nitrous acid. *J. Phys. Chem. A* **110**, 11371-11376 (2006)

[1529] Bucher, D., Guidoni, L. & Röthlisberger, U. The protonation state of the Glu-71/Asp-80 residues in the KcsA potassium channel: A first-principles QM/MM molecular dynamics study. *Biophys. J.* **93**, 2315-2324 (2007)

[1530] Dolg, M. Effective core potentials. In: *Modern methods and algorithms of quantum chemistry. Volume 3.* Grotendorst, J., Ed. John von Neumann Institute for Computing, Jülich, Germany; NIC Series, pp 507-540 (2000)

[1531] Reiher, M. & Hess, B. Relativistic electronic-structure calculations for atoms and molecules. In: *Modern methods and algorithms of quantum chemistry. Volume 3.* Grotendorst, J., Ed. John von Neumann Institute for Computing, Jülich, Germany; NIC Series, pp 479-505 (2000)

[1532] Reiher, M. & Wolf, A. Relativistic quantum chemistry. Wiley-VCH, Weinheim, Germany (2009)

[1533] Car, R. & Parrinello, M. Unified approach for molecular dynamics and density-functional theory. *Phys. Rev. Lett.* **55**, 2471-2474 (1985)

[1534] Feynmann, R.P. Space-time approach to non-relativistic quantum mechanics. *Rev. Mod. Phys.* **20**, 367-387 (1948)

[1535] Feynman, R.P. & Hibbs, A.R. Quantum mechanics and path integrals. McGraw-Hill Companies, New York, USA (1965)

[1536] Szabo, A. & Ostlund, N.S. Modern quantum chemistry: Introduction to advanced electronic structure theory. Dover, New York, USA (1996)

[1537] Knowles, P., Schütz, M. & Werner, H.-J. Ab initio methods for electron correlation in molecules. In: *Modern methods and algorithms of quantum chemistry. Volume 3.* Grotendorst, J., Ed. John von Neumann Institute for Computing, Jülich, Germany; NIC Series, pp 97-179 (2000)

[1538] Carsky, P. Recent progress in coupled cluster methods: Theory and applications. Edition 1. Springer-Verlag GmbH, Berlin, Germany (2010)

[1539] Pople, J.A. & Segal, G.A. Approximate self-consistent molecular orbital theory. 3. CNDO results for AB_2 and AB_3 systems. *J. Chem. Phys.* **44**, 3289-3296 (1966)

[1540] Pople, J.A., Beveridge, D.L. & Dobosh, P.A. Approximate self-consistent molecular-orbital theory. 5. Intermediate neglect of differential overlap. *J. Chem. Phys.* **47**, 2026-2033 (1967)

[1541] Delbene, J. & Jaffe, H.H. Use of CNDO method in spectroscopy. I. Benzene pyridine and diazines. *J. Chem. Phys.* **48**, 1807-1813 (1968)

[1542] Ridley, J. & Zerner, M. Intermediate neglect of differential overlap technique for spectroscopy - pyrrole and azines. *Theor. Chim. Acta* **32**, 111-134 (1973)

[1543] Bingham, R.C., Dewar, M.J.S. & Lo, D.H. Ground-states of molecules. 25. MINDO-3 - improved version of MINDO semiempirical SCF-MO method. *J. Am. Chem. Soc.* **97**, 1285-1293 (1975)

[1544] Dewar, M.J.S. & Thiel, W. Ground-states of molecules. 38. MNDO method - approximations and parameters. *J. Am. Chem. Soc.* **99**, 4899-4907 (1977)

[1545] Dewar, M.J.S. & Thiel, W. Ground-states of molecules. 39. MNDO results for molecules containing hydrogen, carbon, nitrogen, and oxygen. *J. Am. Chem. Soc.* **99**, 4907-4917 (1977)

[1546] Bacon, A.D. & Zerner, M.C. Intermediate neglect of differential overlap theory for transition-metal complexes - Fe, Co and Cu chlorides. *Theor. Chim. Acta* **53**, 21-54 (1979)

[1547] Nanda, D.N. & Jug, K. SINDO1 - a semi-empirical SCF-MO method for molecular-binding energy and geometry. 1. Approximations and parametrization. *Theor. Chim. Acta* **57**, 95-106 (1980)

[1548] Thiel, W. The MNDOC method, a correlated version of the MNDO model. *J. Am. Chem. Soc.* **103**, 1413-1420 (1981)

[1549] Dewar, M.J.S., Zoebisch, E.G., Healy, E.F. & Stewart, J.J.P. The development and use of quantum-mechanical molecular-models. 76. AM1 - a new general-purpose quantum-mechanical molecular-model. *J. Am. Chem. Soc.* **107**, 3902-3909 (1985)

[1550] Jug, K., Iffert, R. & Schulz, J. Development and parametrization of SINDO1 for 2nd-row elements. *Int. J. Quant. Chem.* **32**, 265-277 (1987)

[1551] Stewart, J.J.P. Optimization of parameters for semiempirical methods. 1. Method. *J. Comp. Chem.* **10**, 209-220 (1989)

[1552] Stewart, J.J.P. Optimization of parameters for semiempirical methods. 2. Applications. *J. Comp. Chem.* **10**, 221-264 (1989)

[1553] Thiel, W. & Voityuk, A.A. Extension of the MNDO formalism to d-orbitals - integral approximations and preliminary numerical results. *Theor. Chim. Acta* **81**, 391-404 (1992)

[1554] Dewar, M.J.S., Jie, C.X. & Yu, J.G. SAM1 - the 1st of a new series of general-purpose quantum-mechanical molecular-models. *Tetrahedron* **49**, 5003-5038 (1993)

[1555] Holder, A.J., Dennington II, R.D. & Jie, C. Addendum to SAM1 results previously published. *Tetrahedron* **50**, 627-638 (1994)

[1556] Thiel, W. & Voityuk, A.A. Extension of the MDNO formalism to d orbitals: Integral approximations and preliminary numerical results. *Theor. Chim. Acta* **93**, 315-315 (1996)

[1557] Thiel, W. & Voityuk, A.A. Extension of MNDO to d orbitals: Parameters and results for the second-row elements and for the zinc group. *J. Phys. Chem.* **100**, 616-626 (1996)

[1558] Ahlswede, B. & Jug, K. Consistent modifications of SINDO1: I. Approximations and parameters. *J. Comp. Chem.* **20**, 563-571 (1999)

[1559] Ahlswede, B. & Jug, K. Consistent modifications of SINDO1: II. Applications to first- and second-row elements. *J. Comp. Chem.* **20**, 572-578 (1999)

[1560] Voityuk, A.A., Zerner, M.C. & Rösch, N. Extension of the neglect of diatomic differential overlap method to spectroscopy. NDDO-G parametrization and results for organic molecules. *J. Phys. Chem. A* **103**, 4553-4559 (1999)

[1561] Thiel, W. Semiempirical methods. In: *Modern methods and algorithms of quantum chemistry. Volume 3.* Grotendorst, J., Ed. John von Neumann Institute for Computing, Jülich, Germany; NIC Series, pp 261-283 (2000)

[1562] Sherrill, C.D. & Schaefer, H.F. III. The configuration interaction method: Advances in highly correlated approaches. *Adv. Quantum Chem.* **34**, 143-269 (1999)

[1563] Bakowies, D. Accurate extrapolation of electron correlation energies from small basis sets. *J. Chem. Phys.* **127**, 164109/1-164109/12 (2007)

[1564] Bakowies, D. Extrapolation of electron correlation energies to finite and complete basis set targets. *J. Chem. Phys.* **127**, 084105/1-084105/23 (2007)

[1565] Hartree, D.R. The wave mechanics of an atom with a non-Coulomb central field. Part

III. Term values and intensities in series in optical spectra. *Proc. Cambridge Philos. Soc.* **24**, 426-437 (1928)
[1566] Hartree, D.R. The wave mechanics of an atom with a non-Coulomb central field. Part II. Some results and discussion. *Proc. Cambridge Philos. Soc.* **24**, 111-132 (1928)
[1567] Hartree, D.R. The wave mechanics of an atom with a non-Coulomb central field. Part I. Theory and methods. *Proc. Cambridge Philos. Soc.* **24**, 89-110 (1928)
[1568] Fock, V. Näherungsmethoden zur Lösung des quantenmechanischen Mehrkörperproblems. *Z. Phys.* **61**, 126-148 (1930)
[1569] Kohn, W. & Sham, L.J. Self-consistent equations including exchange and correlation effects. *Phys. Rev.* **140**, A1133-A1138 (1965)
[1570] Parr, R.G. & Weitao, Y. Density-functional theory of atoms and molecules. Oxford Univ. Press, New York, USA (1989)
[1571] Leung, K., Rempe, S.B. & von Lilienfeld, O.A. *Ab initio* molecular dynamics calculations of ion hydration free energies. *J. Chem. Phys.* **130**, 204507/1-204507/11 (2009)
[1572] Rempe, S.B. & Leung, K. Response to "Comment on '*Ab initio* molecular dynamics calculations of ion hydration free energies' " [J. Chem. Phys. 133, 047103 (2010)]. *J. Chem. Phys.* **133**, 047104/1-047104/2 (2010)
[1573] Seidel, R., Faubel, M., Winter, B. & Blumberger, J. Single-ion reorganization free energy of aqueous $Ru(bpy)_3^{2+/3+}$ and $Ru(H_2O)_6^{2+/3+}$ from photoemission spectroscopy and density functional molecular dynamics simulation. *J. Am. Chem. Soc.* **131**, 16127-16137 (2009)
[1574] Sulpizi, M. & Sprik, M. Acidity constants from vertical energy gaps: density functional theory based molecular dynamics implementation. *Phys. Chem. Chem. Phys.* **10**, 5238-5249 (2008)
[1575] Tuckerman, M.E., Laasonen, K., Sprik, M. & Parrinello, M. *Ab initio* molecular dynamics simulation of the solvation and transport of hydronium and hydroxyl ions in water. *J. Chem. Phys.* **103**, 150-161 (1995)
[1576] Tuckerman, M., Laasonen, K., Sprik, M. & Parrinello, M. Ab initio molecular dynamics simulation of the solvation and transport of H_3O^+ and OH^- ions in water. *J. Phys. Chem.* **99**, 5749-5752 (1995)
[1577] Marx, D., Sprik, M. & Parrinello, M. Ab initio molecular dynamics of ion solvation. The case of Be^{2+} in water. *Chem. Phys. Lett.* **273**, 360-366 (1997)
[1578] Ramaniah, L.M., Bernasconi, M. & Parrinello, M. *Ab initio* molecular-dynamics simulation of K^+ solvation in water. *J. Chem. Phys.* **111**, 1587-1591 (1999)
[1579] Lubin, M.I., Bylaska, E.J. & Weare, J.H. Ab initio molecular dynamics simulations of aluminum ion solvation in water clusters. *Chem. Phys. Lett.* **322**, 447-453 (2000)
[1580] Lyubartsev, A.P., Laasonen, K. & Laaksonen, A. Hydration of Li^+ ion. An ab initio molecular dynamics simulation. *J. Chem. Phys.* **114**, 3120-3126 (2001)
[1581] Tobias, D.J., Jungwirth, P. & Parrinello, M. Surface solvation of halogen anions in water clusters: An ab initio molecular dynamics study of the $Cl^-(H_2O)_6$ complex. *J. Chem. Phys.* **114**, 7036-7044 (2001)
[1582] Lightstone, F.C., Schwegler, E., Hood, R.Q., Gygi, F. & Galli, G. A first principles molecular dynamics simulation of the hydrated magnesium ion. *Chem. Phys. Lett.* **343**, 549-555 (2001)
[1583] Raugei, S. & Klein, M.L. Dynamics of water molecules in the Br^- solvation shell: An ab initio molecular dynamics study. *J. Am. Chem. Soc.* **123**, 9484-9485 (2001)
[1584] Raugei, S. & Klein, M.L. An *ab initio* study of water molecules in the bromide ion solvation shell. *J. Chem. Phys.* **116**, 196-202 (2002)
[1585] Bakó, I., Hutter, J. & Pálinkás, G. Car-Parrinello molecular dynamics simulation of the hydrated calcium ion. *J. Chem. Phys.* **117**, 9838-9843 (2002)
[1586] Chen, B., Ivanov, I., Park, J.M., Parrinello, M. & Klein, M.L. Solvation structure and mobility mechanism of OH^-: A Car-Parrinello molecular dynamics investigation of alkaline solutions. *J. Phys. Chem. B* **106**, 12006-12016 (2002)
[1587] Heuft, J.M. & Meijer, E.J. Density functional theory based molecular-dynamics study of aqueous chloride solvation. *J. Chem. Phys.* **119**, 11788-11791 (2003)
[1588] Ikeda, T., Hirata, M. & Kimura, T. *Ab initio* molecular dynamics study of polarization effects on ionic hydration in aqueous $AlCl_3$ solution. *J. Chem. Phys.* **119**, 12386-12392 (2003)
[1589] Leung, K. & Rempe, S.B. Ab initio molecular dynamics study of formate ion hydration. *J. Am. Chem. Soc.* **126**, 344-351 (2004)
[1590] Heuft, J.M. & Meijer, E.J. Density functional theory based molecular-dynamics study of aqueous fluoride solvation. *J. Chem. Phys.* **122**, 094501/1-094501/7 (2005)
[1591] Heuft, J.M. & Meijer, E.J. Density functional theory based molecular-dynamics study of aqueous iodide solvation. *J. Chem. Phys.* **123**, 094506/1-094506/7 (2005)
[1592] Ikeda, T., Hirata, M. & Kimura, T. Hydration of Y^{3+} ion: A Car-Parrinello molecular dynamics study. *J. Chem. Phys.* **122**, 024510/1-024510/5 (2005)
[1593] Izvekov, S. & Voth, G.A. Ab initio molecular-dynamics simulation of aqueous proton

solvation and transport revisited. *J. Chem. Phys.* **123**, 044505/1-044501/9 (2005)

[1594] Izvekov, S. & Voth, G.A. Erratum to "Ab initio molecular-dynamics simulation of aqueous proton solvation and transport revisited" [J. Chem. Phys. 123 044505/1-044501/9 (2005)]. *J. Chem. Phys.* **124**, 039901/1-039901/1 (2006)

[1595] Amira, S., Spångberg, D. & Hermansson, K. OD vibrations and hydration structure in an Al^{3+}(aq) solution from a Car-Parrinello molecular-dynamics simulation. *J. Chem. Phys.* **124**, 104501/1-104501/13 (2006)

[1596] Ikeda, T., Boero, M. & Terakura, K. Hydration of alkali ions from first principles molecular dynamics revisited. *J. Chem. Phys.* **126**, 034501/1-034501/9 (2007)

[1597] Ikeda, T., Boero, M. & Terakura, K. Hydration properties of magnesium and calcium ions from constrained first principles molecular dynamics. *J. Chem. Phys.* **127**, 074503/1-074503/8 (2007)

[1598] Sit, P.H.-L., Cococcioni, M. & Marzari, N. Car-Parrinello molecular dynamics in the DFT + U formalism: Structure and energetics of solvated ferrous and ferric ions. *J. Electroanal. Chem.* **607**, 107-112 (2007)

[1599] Krekeler, C. & delle Site, L. Solvation of positive ions in water: the dominant role of water-water interaction. *J. Phys.: Condens. Matter* **19**, 192101/1-192101/9 (2007)

[1600] Petit, L., Vuilleumier, R., Maldivi, P. & Adamo, C. Ab initio molecular dynamics study of a highly concentrated LiCl aqueous solution. *J. Chem. Theory Comput.* **4**, 1040-1048 (2008)

[1601] Petit, L., Vuilleumier, R., Maldivi, P. & Adamo, C. Molecular dynamics study of the coordination sphere of trivalent lanthanum in a highly concentrated LiCl aqueous solution: A combined classical and ab initio approach. *J. Phys. Chem. B* **112**, 10603-10607 (2008)

[1602] Beret, E.C., Martínez, J.M., Pappalardo, R.R., Marcos, E.S., Doltsinis, N.L. & Marx, D. Explaining asymmetric solvation of Pt(II) versus Pd(II) in aqueous solution revealed by ab initio molecular dynamics simulations. *J. Chem. Theory Comput.* **4**, 2108-2121 (2008)

[1603] Beret, E.C., Pappalardo, R.R., Doltsinis, N.L., Marx, D. & Marcos, E.S. Aqueous Pd^{II} and Pt^{II}: Anionic hydration revealed by Car-Parrinello simulations. *Chem. Phys. Chem.* **9**, 237-240 (2008)

[1604] Bucher, D. & Kuyucak, S. Polarization of water in the first hydration shell of K^+ and Ca^{2+} ions. *J. Phys. Chem. B* **112**, 10786-10790 (2008)

[1605] Costanzo, F. & Della Valle, R.G. Car-Parrinello MD simulations for the Na^+-phenylalanine complex in aqueous solution. *J. Phys. Chem. B* **112**, 12783-12789 (2008)

[1606] di Tommaso, D. & de Leeuw, N.H. The onset of calcium carbonate nucleation: A density functional theory molecular dynamics and hybrid microsolvation/continuum study. *J. Phys. Chem. B* **112**, 6965-6975 (2008)

[1607] Krekeler, C. & delle Site, L. Lone pair versus bonding pair electrons: The mechanism of electronic polarization of water in the presence of positive ions. *J. Chem. Phys.* **128**, 134515/1-134515/5 (2008)

[1608] Todorova, T., Hünenberger, P.H. & Hutter, J. Car-Parrinello molecular dynamics simulations of $CaCl_2$ aqueous solutions. *J. Chem. Theory Comput.* **4**, 779-789 (2008)

[1609] Rodríguez-Fortea, A., Nadal-Vilà, L. & Poblet, J.M. Hydration of hydrogentungstate anions at different pH conditions: A Car-Parrinello molecular dynamics study. *Inorg. Chem.* **47**, 7745-7750 (2008)

[1610] Kumar, P.P., Kalinichev, A.G. & Kirkpatrick, R.J. Hydrogen-bonding structure and dynamics of aqueous carbonate species from Car-Parrinello molecular dynamics simulations. *J. Phys. Chem. B* **113**, 794-802 (2009)

[1611] Beret, E.C., Galbis, E., Pappalardo, R.R. & Marcos, E.S. Opposite effects of successive hydration shells on the aqua ion structure of metal cations. *Mol. Simul.* **35**, 1007-1014 (2009)

[1612] Guàrdia, E. & Skarmoutsos, I. On ion and molecular polarization of halides in water *J. Chem. Theory Comput.* **5**, 1449-1453 (2009)

[1613] Webster, F.J., Schnitker, J., Friedrichs, M.S., Friesner, R.A. & Rossky, P.J. Solvation dynamics of the hydrated electron - a nonadiabatic quantum simulation. *Phys. Rev. Lett.* **66**, 3172-3175 (1991)

[1614] Gao, J.L. Absolute free-energy of solvation from Monte Carlo simulations using combined quantum and molecular mechanical potentials. *J. Phys. Chem.* **96**, 537-540 (1992)

[1615] Kerdcharoen, T., Liedl, K.R. & Rode, B.M. A QM/MM simulation method applied to the solution of Li^+ in liquid ammonia. *Chem. Phys.* **211**, 313-323 (1996)

[1616] Bryce, R.A., Vincent, M.A., Malcolm, N.O.J., Hillier, I.H. & Burton, N.A. Cooperative effects in the structuring of fluoride water clusters: *Ab initio* hybrid quantum mechanical/molecular mechanical model incorporating polarizable fluctuating charge solvent. *J. Chem. Phys.* **109**, 3077-3085 (1998)

[1617] Tongraar, A. & Rode, B.M. Preferential solvation of Li^+ in 18.45% aqueous ammonia: A Born-Oppenheimer ab initio quantum mechanics/molecular mechanics md simulation. *J. Phys. Chem. A* **103**, 8524-8527 (1999)

[1618] Cummins, P.L. & Gready, J.E. Coupled semiempirical quantum mechanics and molecular mechanics (QM/MM) calculations on the aqueous solvation free energies of ionized molecules. *J. Comput. Chem.* **20**, 1028-1038 (1999)

[1619] Marini, G.W., Liedl, K.R. & Rode, B.M. Investigation of Cu^{2+} hydration and the Jahn-Teller effect in solution by QM/MM Monte Carlo simulations. *J. Phys. Chem. A* **103**, 11387-11393 (1999)

[1620] Tongraar, A. & Rode, B.M. The role of non-additive contributions on the hydration shell structure of Mg^{2+} studied by Born-Oppenheimer ab initio quantum mechanical/-molecular mechanical molecular dynamics simulation. *Chem. Phys. Lett.* **346**, 485-491 (2001)

[1621] Tongraar, A. & Rode, B.M. A Born-Oppenheimer ab initio quantum mechanical/molecular mechanical molecular dynamics simulation on preferential solvation of Na^+ in aqueous ammonia solution. *J. Phys. Chem. A* **105**, 506-510 (2001)

[1622] Schwenk, C.F., Loeffler, H.H. & Rode, B.M. Molecular dynamics simulations of Ca^{2+} in water: Comparison of a classical simulation including three-body corrections and Born-Oppenheimer ab initio and density functional theory quantum mechanical/molecular mechanics simulations. *J. Chem. Phys.* **115**, 10808-10813 (2001)

[1623] Inada, Y., Mohammed, A.M., Loeffler, H.H. & Rode, B.M. Hydration structure and water exchange reaction of nickel(II)ion: Classical and QM/MM simulations. *J. Phys. Chem. A* **106**, 6783-6791 (2002)

[1624] Tongraar, A., Sagarik, K. & Rode, B.M. Preferential solvation of Ca^{2+} in aqueous ammonia solution: Classical and combined *ab initio* quantum mechanical/molecular mechanical molecular dynamics simulations. *Phys. Chem. Chem. Phys.* **4**, 628-634 (2002)

[1625] Inada, Y., Loeffler, H.H. & Rode, B.M. Librational, vibrational, and exchange motions of water molecules in aqueous Ni(II) solution: Classical and QM/MM molecular dynamics simulations. *Chem. Phys. Lett.* **358**, 449-458 (2002)

[1626] Loeffler, H.H. & Rode, B.M. The hydration structure of the lithium ion. *J. Chem. Phys.* **117**, 110-117 (2002)

[1627] Loeffler, H.H., Mohammed, A.M., Inada, Y. & Funahashi, S. Lithium(I) ion hydration: a QM/MM-MD study. *Chem. Phys. Lett.* **379**, 452-457 (2003)

[1628] Kerdcharoen, T. & Morokuma, K. Combined quantum mechanics and molecular mechanics simulation of Ca^{2+}/ammonia solution based on the ONIOM-XS method: Octahedral coordination and implication to biology. *J. Chem. Phys.* **118**, 8856-8862 (2003)

[1629] Kritayakornupong, C., Plankensteiner, K. & Rode, B.M. Structure and dynamics of the Cd^{2+} ion in aqueous solution: Ab initio QM/MM molecular dynamics simulation. *J. Phys. Chem. A* **107**, 10330-10334 (2003)

[1630] Kritayakornupong, C., Plankensteiner, K. & Rode, B.M. Structural and dynamical properties of Co(III) in aqueous solution: *Ab initio* quantum mechanical/molecular mechanical molecular dynamics simulation. *J. Chem. Phys.* **119**, 6068-6072 (2003)

[1631] Kritayakornupong, C., Plankensteiner, K. & Rode, B.M. Dynamics in the hydration shell of Hg^{2+} ion: Classical and ab initio QM/MM molecular dynamics simulations. *Chem. Phys. Lett.* **371**, 438-444 (2003)

[1632] Remsungnen, T. & Rode, B.M. Dynamical properties of the water molecules in the hydration shells of Fe(II) and Fe(III) ions: ab initio QM/MM molecular dynamics simulations. *Chem. Phys. Lett.* **367**, 586-592 (2003)

[1633] Schwenk, C.F., Loeffler, H.H. & Rode, B.M. Structure and dynamics of metal ions in solution: QM/MM molecular dynamics simulations of Mn^{2+} and V^{2+}. *J. Am. Chem. Soc.* **125**, 1618-1624 (2003)

[1634] Schwenk, C.F. & Rode, B.M. Extended *ab initio* quantum mechanical/molecular mechanical molecular dynamics simulations of hydrated Cu^{2+}. *J. Chem. Phys.* **119**, 9523-9531 (2003)

[1635] Schwenk, C.F. & Rode, B.M. New insights into the Jahn-Teller effect through ab initio quantum-mechanical/molecular-mechanical molecular dynamics simulations of Cu^{II} in water. *Chem. Phys. Chem.* **4**, 931-943 (2003)

[1636] Tongraar, A. & Rode, B.M. The hydration structures of F^- and Cl^- investigated by *ab initio* QM/MM molecular dynamics simulations. *Phys. Chem. Chem. Phys.* **5**, 357-362 (2003)

[1637] Tongraar, A. & Rode, B.M. Dynamical properties of water molecules in the hydration shells of Na^+ and K^+: ab initio QM/MM molecular dynamics simulations. *Chem. Phys. Lett.* **385**, 378-383 (2004)

[1638] Tongraar, A. & Rode, B.M. *Ab initio* QM/MM molecular dynamics simulation of preferential K^+ solvation in aqueous ammonia solution. *Phys. Chem. Chem. Phys.* **6**, 411-416 (2004)

[1639] Schwenk, C.F. & Rode, B.M. Ab initio QM/MM MD simulations of the hydrated Ca^{2+} ion. *Pure Appl. Chem.* **76**, 37-47 (2004)

[1640] Rode, B.M., Schwenk, C.F. & Tongraar, A. Structure and dynamics of hydrated ions -

New insights through quantum mechanical simulations. *J. Mol. Liq.* **110**, 105-122 (2004)
[1641] Kritayakornupong, C., Plankensteiner, K. & Rode, B.M. The Jahn-teller effect of the Ti^{III} ion in aqueous solution: Extended ab initio QM/MM molecular dynamics simulations. *Chem. Phys. Chem.* **5**, 1499-1506 (2004)
[1642] Armunanto, R., Schwenk, C.F., Tran, H.T. & Rode, B.M. Structure and dynamics of Au^+ ion in aqueous solution: Ab initio QM/MM md simulations. *J. Am. Chem. Soc.* **126**, 2582-2587 (2004)
[1643] Hermida-Ramón, J.M. & Karlström, G. Study of the hydronium ion in water. A combined quantum chemical and statistical mechanical treatment. *J. Mol. Struct. (Theochem)* **712**, 167-173 (2004)
[1644] Hofer, T.S. & Rode, B.M. The solvation structure of Pb(II) in dilute aqueous solution: An *ab initio* quantum mechanical/molecular mechanical molecular dynamics approach. *J. Chem. Phys.* **121**, 6406-6411 (2004)
[1645] Hofer, T.S., Pribil, A.B., Randolf, B.R. & Rode, B.M. Structure and dynamics of solvated Sn(II) in aqueous solution: An ab initio QM/MM MD approach. *J. Am. Chem. Soc.* **127**, 14231-14238 (2005)
[1646] Hofer, T.S., Rode, B.M. & Randolf, B.R. Structure and dynamics of solvated Ba(II) in dilute aqueous solution - an ab initio QM/MM MD approach. *Chem. Phys.* **312**, 81-88 (2005)
[1647] Armunanto, R., Schwenk, C.F. & Rode, B.M. Ab initio QM/MM simulation of Ag^+ in 18.6% aqueous ammonia solution: Structure and dynamics investigations. *J. Phys. Chem. A* **109**, 4437-4441 (2005)
[1648] Fatmi, M.Q., Hofer, T.S., Randolf, B.R. & Rode, B.M. An extended *ab initio* QM/MM MD approach to structure and dynamics of Zn(II) in aqueous solution. *J. Chem. Phys.* **123**, 054514/1-054514/8 (2005)
[1649] Hofer, T.S., Randolf, B.R. & Rode, B.M. Sr(II) in water: A labile hydrate with a highly mobile structure. *J. Phys. Chem. B* **110**, 20409-20417 (2006)
[1650] Morita, S. & Sakai, S. Theoretical studies on the Li^+ ion hydration system by the molecular dynamics simulations with ab initio IMiC MO method. *Bull. Chem. Soc. Jpn.* **79**, 397-405 (2006)
[1651] Intharathep, P., Tongraar, A. & Sagarik, K. *Ab initio* QM/MM dynamics of H_3O^+ in water. *J. Comput. Chem.* **27**, 1723-1732 (2006)
[1652] Hofer, T.S., Scharnagl, H., Randolf, B.R. & Rode, B.M. Structure and dynamics of La(III) in aqueous solution - An ab initio QM/MM MD approach. *Chem. Phys.* **327**, 31-42 (2006)
[1653] Hofer, T.S., Randolf, B.R. & Rode, B.M. The dynamics of the solvation of Pb(II) in aqueous solution obtained by an ab initio QM/MM MD approach. *Chem. Phys.* **323**, 473-478 (2006)
[1654] Fatmi, M.Q., Hofer, T.S., Randolf, B.R. & Rode, B.M. Quantum mechanical charge field molecular dynamics simulation of the TiO^{2+} ion in aqueous solution. *J. Comp. Chem.* **28**, 1704-1710 (2007)
[1655] Kritayakornupong, C. *Ab initio* QM/MM molecular dynamics simulations of Ru^{3+} in aqueous solution. *Chem. Phys. Lett.* **441**, 226-231 (2007)
[1656] Kritayakornupong, C. & Hannongbua, S. Structure and dynamics of high-spin Ru^{2+} in aqueous solution: *Ab initio* QM/MM molecular dynamics simulation. *Chem. Phys.* **332**, 95-101 (2007)
[1657] Senn, H.M. & Thiel, W. QM/MM methods for biological systems. *Top. Curr. Chem.* **268**, 173-290 (2007)
[1658] Vchirawongkwin, V. & Rode, B.M. Solvation energy and vibrational spectrum of sulfate in water - An *ab initio* quantum mechanical simulation. *Chem. Phys. Lett.* **443**, 152-157 (2007)
[1659] Vchirawongkwin, V., Rode, B.M. & Persson, I. Structure and dynamics of sulfate ion in aqueous solution - An *ab initio* QMCF MD simulation and large angle X-ray scattering study. *J. Phys. Chem. B* **111**, 4150-4155 (2007)
[1660] Takahashi, H., Ohno, H., Yamauchi, T., Kishi, R., Furukawa, S., Nakano, M. & Matubayasi, N. Investigation of the dominant hydration structures among the ionic species in aqueous solution: Novel quantum mechanics/molecular mechanics simulations combined with the theory of energy representation. *J. Chem. Phys.* **128**, 0645076/1-0645076/12 (2008)
[1661] Hofer, T.S., Randolf, B.R. & Rode, B.M. Al(III) hydration revisited. An ab initio quantum mechanical charge field molecular dynamics study. *J. Phys. Chem B* **112**, 11726-11733 (2008)
[1662] Hofer, T.S., Randolf, B.R. & Rode, B.M. The hydration of the mercury(I)-dimer - A quantum mechanical charge field molecular dynamics study. *Chem. Phys.* **349**, 210-218 (2008)
[1663] Brancato, G., Rega, N. & Barone, V. Microsolvation of the Zn(II) ion in aqueous solution: A hybrid QM/MM MD approach using non-periodic boundary conditions. *Chem.*

Phys. Lett. **451**, 53-57 (2008)

[1664] Haranczyk, M., Gutowski, M. & Warshel, A. Solvation free energies of molecules. The most stable anionic tautomers of uracil. *Phys. Chem. Chem. Phys.* **10**, 4442-4448 (2008)

[1665] Kritayakornupong, C., Vchirawongkwin, V., Hofer, T.S. & Rode, B.M. Structural and dynamical properties of hydrogen fluoride in aqueous solution: An ab initio quantum mechanical charge field molecular dynamics simulation. *J. Phys. Chem. B* **112**, 12032-12037 (2008)

[1666] Kritayakornupong, C. The Jahn-Teller effect of the Ag^{2+} ion in aqueous solution: A hybrid *ab initio* quantum mechanical/molecular mechanical molecular dynamics simulation. *Chem. Phys. Lett.* **455**, 207-212 (2008)

[1667] Kritayakornupong, C. The Jahn-Teller effect of the Cr^{2+} ion in aqueous solution: *Ab initio* QM/MM molecular dynamics simulations. *J. Comp. Chem.* **29**, 115-121 (2008)

[1668] Tongraar, A., Tangkawanwanit, P. & Rode, B.M. A combined QM/MM molecular dynamics simulations study of nitrate anion (NO_3^-) in aqueous solution. *J. Phys. Chem. A* **110**, 12918-12926 (2006)

[1669] Pribil, A.B., Hofer, T.S., Vchirawongkwin, V., Randolf, B.R. & Rode, B.M. Quantum mechanical simulation studies of molecular vibrations and dynamics of oxo-anions in water. *Chem. Phys.* **346**, 182-185 (2008)

[1670] Pribil, A.B., Hofer, T.S., Randolf, B.R. & Rode, B.M. Structure and dynamics of phosphate ion in aqueous solution: An *ab initio* QMCF MD study. *J. Comp. Chem.* **29**, 2330-2334 (2008)

[1671] Lim, L.H.V., Hofer, T.S., Pribil, A.B. & Rode, B.M. The hydration structure of Sn(II): An ab initio quantum mechanical charge field molecular dynamics study. *J. Phys. Chem. B* **113**, 4372-4378 (2009)

[1672] von Lilienfeld, O.A., Lins, R.D. & Röthlisberger, U. Variational particle number approach for rational compound design. *Phys. Rev. Lett.* **95**, 153002/1-153002/4 (2005)

[1673] Zeng, X.C., Hu, H., Hu, X.Q., Cohen, A.J. & Yang, W.T. Ab initio quantum mechanical/molecular mechanical simulation of electron transfer process: Fractional electron approach. *J. Chem. Phys.* **128**, 124510/1-124510/10 (2008)

[1674] Hess, G.H. Recherches thermodynamiques. *Bull. Sci. Acad. Imper. Sci.* **8**, 257-272 (1840)

[1675] Mayer, J.R. Bemerkungen über die Kräfte der unbelebten Natur. *Ann. Chemie Pharm.* **42**, 233-240 (1842)

[1676] Joule, J.P. On the calorific effects of magneto-electricity, and on the mechanical value of heat. *Phil. Mag. London* **23**, 435-443 (1843)

[1677] Joule, J.P. On the existence of an equivalent relation between heat and the ordinary forms of mechanical power. *Phil. Mag. London* **27**, 205-207 (1845)

[1678] von Helmholtz, H.L.F. Über die Erhaltung der Kraft. G. Reimer, Berlin, Germany (1847)

[1679] Clausius, R. Über die bewegende Kraft der Wärme, und die Gesetze, die sich daraus für die Wärmelehre ableiten lassen. *Ann. Phys. Chem.* **79**, 500-524 (1850)

[1680] Joule, J.P. On the mechanical equivalent of heat. *Phil. Trans. Roy. Soc. London* **140**, 61-82 (1850)

[1681] Thomson, W. On the dynamical theory of heat; with numerical results deduced from Mr. Joule's equivalent of a thermal unit and M. Regnault's observations on steam. *Phil. Mag.* **4**, 105-117 (1852)

[1682] Clausius, R. Über verschiedene für die Anwendung bequeme Formen der Hauptgleichungen der mechanischen Wärmetheorie. *Ann. Phys. Chem.* **125**, 353-400 (1865)

[1683] Carnot, S. Réflexions sur la puissance motrice du feu et sur les machines propres a développer cette puissance. Bachelier, Libraire, Paris, France (1824)

[1684] Clapeyron, E. Mémoire sur la puissance motrice de la chaleur. *J. de l'école polytech. (Paris)* **14**, 153-191 (1834)

[1685] Clausius, R. Über eine veränderte Form des zweiten Hauptsatzes der mechanischen Wärmetheorie. *Ann. Phys. Chem.* **93**, 481-506 (1854)

[1686] Kerker, M. Sadi Carnot and the steam engine engineers. *Isis* **51**, 257-270 (1960)

[1687] Nernst, W. Über die Berechnung chemischer Gleichgewichte aus thermischen Messungen. *Nachr. Kgl. Ges. Wiss. Gött.* **1**, 1-40 (1906)

[1688] Nernst, W. Über die Beziehung zwischen Wärmeentwicklung und maximaler Arbeit bei kondensierten Systemen. *Ber. Kgl. Preuss. Akad. Wiss.* **52**, 993-940 (1906)

[1689] Oganessian, Y.T., Utyonkov, V.K., Lobanov, Y.V., Abdullin, F.S., Polyakov, A.N., Sagaidak, R.N., Shirokovsky, I.V., Tsyganov, Y.S., Voinov, A.A., Gulbekian, G.G., Bogomolov, S.L, Gikal, B.N., Mezentsev, A.N., Iliev, S., Subbotin, V.G., Sukhov, A.M., Subotic, K., Zagrebaev, V.I., Vostokin, G.K., Itkis, M.G., Moody, K.J., Patin, J.B., Shaughnessy, D.A., Stoyer, M.A., Wilk, P.A., Kenneally, J.M., Landrum, J.H., Wild, J.F. & Lougheed, R.W. Synthesis of the isotopes of elements 118 and 116 in the ^{249}Cf and $^{245}Cm+^{48}Ca$ fusion reactions. *Phys. Rev. C* **74**, 044602/1-044602/9 (2006)

[1690] Gibbs, J.W. Graphical methods in the thermodynamics of fluids. *Trans. Connecticut Acad.* **2**, 309-342 (1873)

[1691] Gibbs, J.W. A method of geometrical representation of the thermodynamic properties of substances by means of surfaces. *Trans. Connecticut Acad.* **2**, 382-404 (1873)
[1692] Gibbs, J.W. On the equilibrium of heterogeneous substances. *Trans. Connecticut Acad.* **3**, 108-248 (1876)
[1693] Miller, W.L. The method of Willard Gibbs in chemical thermodynamics. *Chem. Rev.* **1**, 293-344 (1925)
[1694] Raman, V.V. The permeation of thermodynamics into nineteenth century chemistry. *Ind. J. Hist. Sci. Calcutta* **10**, 16-37 (1975)
[1695] Dolby, R.G.A. Thermochemistry versus thermodynamics: The nineteenth century controversy. *Hist. Sci.* **22**, 375-400 (1984)
[1696] Duhem, M.P. Sur la relation qui lie l'effet Peltier à la différence de deux métaux en contact. *Ann. Chim. Phys.* **12**, 433-471 (1887)
[1697] Duhem, M.P. Le potentiel thermodynamique et ses applications à la mécanique chimique et à l'étude des phénomènes électriques. Hermann, Paris, France (1886)
[1698] Kirchhoff, G. Über einen Satz der mechanischen Wärmetheorie, und einige Anwendungen desselben. *Ann. Phys. Chem.* **179**, 177-206 (1858)
[1699] Thomson, W. On the thermo-elastic and thermo-magnetic properties of matter. *Quart. J. Math.* **1**, 57-77 (1857)
[1700] Horstmann, A. Über den zweiten Hauptsatz der mechanischen Wärmetheorie und dessen Anwendung auf einige Zersetzungserscheinungen. *Ann. Chem. Pharm.* **S8**, 112-133 (1872)
[1701] Horstmann, A. Theorie der Dissociation. *Ann. Chem. Pharm.* **170**, 192-210 (1873)
[1702] von Helmholtz, H.L.F. Die Thermodynamik chemischer Vorgänge. *Ber. Kgl. Preuss. Akad. Wiss. Berlin* **I**, 22-39 (1882)
[1703] von Helmholtz, H.L.F. Die Thermodynamik chemischer Vorgänge. 2. Beitrag: Versuche an Chlorzink-Kalomel-Elementen. *Ber. Kgl. Preuss. Akad. Wiss. Berlin* **II**, 825-836 (1882)
[1704] von Helmholtz, H.L.F. Wissenschaftliche Abhandlungen. Volume 1. Barth, Leipzig, Germany (1882)
[1705] von Helmholtz, H.L.F. Die Thermodynamik chemischer Vorgänge. 3. Beitrag: Folgerungen die galvanische Polarisation betreffend. *Ber. Kgl. Preuss. Akad. Wiss. Berlin* **I**, 647-665 (1883)
[1706] von Helmholtz, H.L.F. Wissenschaftliche Abhandlungen. Volume 2. Barth, Leipzig, Germany (1883)
[1707] Czapski, S. Über die thermische Veränderlichkeit der electromotorischen Kraft galvanischer Elemente und ihre Beziehung zur freien Energie derselben. *Ann. Phys. Chem.* **21**, 209-243 (1884)
[1708] Gibbs, J.W. Electrochemical thermodynamics. (Letter from Professor Willard Gibbs to the secretary of the electrolysis committee of the British Association.) *Rep. Brit. Assoc. Adv. Sci.* **57**, 343-346 (1888)
[1709] Cahan, D. Hermann von Helmholtz and the foundations of nineteenth-century science. Univ. California Press , Berkeley, USA (1993)
[1710] Imai, T., Nomura, H., Kinoshita, M. & Hirata, F. Partial molar volume and compressibility of alkali-halide ions in aqueous solution: Hydration shell analysis with an integral equation theory of molecular liquids. *J. Phys. Chem. B* **106**, 7308-7314 (2002)
[1711] Dalton, J. Experimental essays on the constitution of mixed gases; on the force of steam or vapor from water and other liquids in different temperatures, both in a Torricellian vacuum and in air; on evaporation and on the expansion of gases by heat. *Mem. Manchester Lit. and Phil. Soc.* **5**, 535-602 (1802)
[1712] Ben-Naim, A. A new method of defining the activity functions of non-ideal gases and solutions. *J. Chem. Educ.* **39**, 242-245 (1962)
[1713] Trasatti, S. Relative and absolute electrochemical quantities. Components of the potential difference across the electrode/solution interface. *J. Chem. Soc. Farad. Trans. I* **70**, 1752-1768 (1974)
[1714] Trasatti, S. The concept of absolute electrode potential. An attempt at a calculation. *J. Electroanal. Chem.* **52**, 313-329 (1974)
[1715] Ben-Naim, A. Standard thermodynamics of transfer. Uses and misuses. *J. Phys. Chem.* **82**, 792-803 (1978)
[1716] Trasatti, S. The concept and physical meaning of absolute electrode potential. A reassessment. *J. Electroanal. Chem.* **139**, 1-13 (1982)
[1717] Trasatti, S. Components of the absolute electrode potential. Conceptions and misinterpretations. *Materials Chem. and Phys.* **15**, 427-438 (1986)
[1718] Trasatti, S. The absolute electrode potential: An explanatory note (Recommendations 1986). *Pure Appl. Chem.* **58**, 955-966 (1986)
[1719] Rockwood, A.L. Absolute half-cell thermodynamics: Electrode potentials. *Phys. Rev. A* **33**, 554-558 (1986)
[1720] Rockwood, A.L. Absolute half-cell entropy. *Phys. Rev. A* **36**, 1525-1526 (1987)
[1721] Trasatti, S. The "absolute" electrode potential - the end of the story. *Electrochim. Acta*

35, 269-271 (1990)
[1722] Marcus, Y. The thermodynamics of solvation of ions. Part 5. Gibbs free energy of hydration at 298.15 K. *J. Chem. Soc. Farad. Trans.* **87**, 2995-2999 (1991)
[1723] Bartmess, J.E. Thermodynamics of the electron and the proton. *J. Phys. Chem.* **98**, 6420-6424 (1994)
[1724] Ben-Naim, A. On the evolution of the concept of solvation thermodynamics. *J. Solut. Chem.* **5**, 475-487 (2001)
[1725] Pak, Y. & Wang, S. Accurate experimental values for the free energies of hydration of H^+, OH^- and H_3O^+. *J. Phys. Chem. A* **108**, 3692-3694 (2004)
[1726] Camaioni, D.M. & Schwerdtfeger, C.A. Comment on "Accurate experimental values for the free energies of hydration of H^+, OH^- and $H3O^+$" by Palascak M. W. and Shields G. C. (*J. Phys. Chem. A* 2004, 108, 3692). *J. Phys. Chem. A* **109**, 10795-10797 (2005)
[1727] Llano, J. & Eriksson, L.A. Response to "Comment on 'First principles electrochemistry: Electrons and protons reacting as independent ions' " [J. Chem. Phys. 122, 087103 (2005)]. *J. Chem. Phys.* **122**, 087104/1-087104/2 (2005)
[1728] Rockwood, A.L. Comment on "First principles electrochemistry: Electrons and protons reacting as independent ions" [J. Chem. Phys. 117, 10193 (2002)]. *J. Chem. Phys.* **122**, 087103/1-087103/2 (2005)
[1729] Bryantsev, V.S., Diallo, M.S. & Goddard III, W.A. Calculation of solvation free energies of charged solutes using mixed cluster/continuum models. *J. Phys. Chem. B* **112**, 9709-9719 (2008)
[1730] Rosenstock, H.M., Draxl, K., Steiner, B.W. & Herron, J.T. Energetics of gaseous ions. *J. Phys. Chem. Ref. Data 6 Suppl.* **1**, 1-783 (1977)
[1731] Cohen, E.R., Cvitas, T., Frey, J.G., Holmström, B., Kuchitsu, K., Marquardt, R., Mills, I., Pavese, F., Quack, M., Stohner, J., Strauss, H.L, Takami, M. & Thor, A.J. Quantities, units and symbols in physical chemistry. Edition 3. RSC Publishing, Cambridge, UK (2007)
[1732] Nernst, W. Die elektromotorische Wirksamkeit der Jonen. *Z. Phys. Chem. (Stöchiometrie u. Verwandtschaftslehre)* **4**, 129-181 (1889)
[1733] Lewis, G.N., Randall, M., Pitzer, K.S. & Brewer, L. Thermodynamics. McGraw-Hill Book Co., New York, USA (1961)
[1734] Debye, P. Zur Theorie der spezifischen Wärmen. *Ann. Phys.* **344**, 789-839 (1912)
[1735] Einstein, A. Die Plancksche Theorie der Strahlung und die Theorie der spezifischen Wärme. *Ann. Phys.* **327**, 180-190 (1906)
[1736] Einstein, A. Berichtigung zu meiner Arbeit: "Die Plancksche Theorie der Strahlung etc.". *Ann. Phys.* **327**, 800-800 (1907)
[1737] Mayer, J.E. & Mayer, M.G. Statistical mechanics. John Wiley and Sons, New York, USA (1940)
[1738] Evans, W.H. & Wagman, D.D. Thermodynamics of some simple sulfur containing molecules. *J. Res. Nat. Bur. Stand.* **49**, 141-148 (1952)
[1739] Tetrode, H. Berichtigung zu meiner Arbeit; "Die chemische Konstante der Gase und das elementare Wirkungsquantum". *Ann. Phys.* **344**, 255-256 (1912)
[1740] Tetrode, H. Die chemische Konstante der Gase und das elementare Wirkungsquantum. *Ann. Phys.* **343**, 434-442 (1912)
[1741] Sackur, O. Die universelle Bedeutung des sog. elementaren Wirkungsquantums. *Ann. Phys.* **345**, 67-86 (1913)
[1742] Mitchell, A.C.G. Entropie des Elektronengases auf Grund der Fermischen Statistik. *Z. Phys.* **50**, 570-576 (1928)
[1743] Sommerfeld, A. Zur Elektronentheorie der Metalle auf Grund der Fermischen Statistik. *Z. Phys.* **47**, 1-32 (1928)
[1744] Latimer, W.M. Oxidation potentials. Prentice-Hall, Inc., Englewood Cliffs, New York (1952)
[1745] Altshuller, A.P. Lattice entropies - Entropies of vaporization of ions from crystals. *J. Chem. Phys.* **26**, 404-406 (1957)
[1746] Morris, D.F.C. & Short, E.L. The Born-Fajans-Haber correlation. *Nature* **224**, 950-952 (1969)
[1747] Kelly, C.P., Cramer, C.J. & Truhlar, D.G. Aqueous solvation free energies of ions and ion-water clusters based on an accurate value for the absolute aqueous solvation free energy of the proton. *J. Phys. Chem. B* **110**, 16066-16081 (2006)
[1748] Latimer, W.M. The entropy of aqueous ions and the nature of the entropy of hydration. *Chem. Rev.* **18**, 349-358 (1936)
[1749] Breck, W.G. & Lin, J. Entropies of aqueous ions. *Trans. Farad. Soc.* **61**, 2223-2228 (1965)
[1750] Fawcett, W.R. Thermodynamic parameters for the solvation of monoatomic ions in water. *J. Phys. Chem. B* **103**, 11181-11185 (1999)
[1751] Lias, S.G., Bartmess, J.E., Liebman, J.F., Holmes, J.L., Levin, R.D. & Mallard, W.G. Gas-phase ion and neutral thermochemistry. *J. Phys. Chem. Ref. Data 16 Suppl.* **1**, 1-861 (1988)

[1752] Linstrom, P.J. & Mallard, W.G. NIST Chemistry WebBook, NIST standard reference database number 69 (http://webbook.nist.gov, retrieved November 2009). National institute of standards and technology, Gaithersburg, USA (2009)
[1753] Gokcen, N.A. & Reddy, R.G. Thermodynamics. Edition 2. Plenum Press, New York, USA (1996)
[1754] Bockris, J.O'M. & Reddy, A.K.N. The electrified interface. In: *Modern electrochemistry. Volume 2A*. Kluwern Academic/Plenum Publishers, New York, USA, pp 771-1033 (2000)
[1755] Conway, B.E. Electrochemical supercapacitors: Scientific fundamentals and technological applications. Edition 1. Springer, New York, USA (1999)
[1756] Grahame, D.C. The electrical double layer and the theory of electrocapillarity. *Chem. Rev.* **41**, 441-501 (1947)
[1757] Grahame, D.C. Electrode processes and the electrical double layer. *Ann. Rev. Phys. Chem.* **6**, 337-358 (1956)
[1758] Devanathan, M.A.V. & Tilak, B.V.K.S.R.A. Structure of electrical double layer at metal-solution interface. *Chem. Rev.* **65**, 635-684 (1965)
[1759] Parsons, R. Some problems of electrical double layer. *Rev. Pure Appl. Chem.* **18**, 91-95 (1968)
[1760] Payne, R. Electrical double layer: Problems and recent progress. *J. Electroanal. Chem.* **41**, 277-309 (1973)
[1761] Harrison, J.A., Randles, J.E.B. & Schiffrin, D.J. The entropy of formation of the mercury-aqueous solution interface and the structure of the inner layer. *J. Electroanal. Chem.* **48**, 359-381 (1973)
[1762] Trasatti, S. Solvent adsorption and double layer potential drop at electrodes. *Mod. Aspects Electrochem.* **13**, 81-206 (1979)
[1763] Carnie, S.L. & Torrie, G.M. The statistical mechanics of the electrical double layer. *Adv. Chem. Phys.* **56**, 141-253 (1984)
[1764] Martynov, G.A. & Salem, R.R. The dense part of the electrical double layer: Molecular or electronic capacitor. *Adv. Colloid. Interface Sci.* **22**, 229-296 (1985)
[1765] Borkowska, Z. The electrical double layer at the mercury solution interface in organic solvents. *J. Electroanal. Chem.* **244**, 1-13 (1988)
[1766] Parsons, R. Electrical double layer: Recent experimental and theoretical developments. *Chem. Rev.* **90**, 813-826 (1990)
[1767] Shubin, V.E. & Kekicheff, P. Electrical double layer structure revisited via a surface force apparatus. Mica interfaces in lithium-nitrate solutions. *J. Colloid. Interface Sci.* **155**, 108-123 (1993)
[1768] Gouy, G. Sur la constitution de la charge électrique à la surface d'un électrolyte. *J. Phys.* **9**, 457-468 (1910)
[1769] Chapman, D.L. A contribution to the theory of electrocapillarity. *Philos. Mag.* **25**, 475-481 (1913)
[1770] Stern, O. Theorie der elektrischen Doppelschicht. *Z. Elektrochem.* **30**, 508-516 (1924)
[1771] Quincke, G. Über eine neue Art elektrischer Ströme. *Ann. Phys.* **183**, 1-47 (1859)
[1772] Quincke, G. Über die Fortführung materieller Theilchen durch strömende Elektricität. *Ann. Phys.* **189**, 513-598 (1861)
[1773] von Helmholtz, H.L.F. Studien über electrische Grenzschichten. *Ann. Phys.* **243**, 337-382 (1879)
[1774] Rybkin, Y.R. Stabilisation of the surface potential and determination of the activities of individual ions in solution. *Russ. Chem. Rev.* **44**, 625-636 (1975)
[1775] Langmuir, I. The relation between contact potentials and electrochemical action. *Trans. Am. Electrochem. Soc.* **29**, 125-182 (1916)
[1776] Ampère, A.-M. Théorie mathématique des phénomènes électro-dynamiques uniquement déduits de l'expérience. *Mémoires de l'Académie Royale des Science de l'Institut de France* **VI**, 175-388 (1823)
[1777] Faraday, M. Experimental researches in electricity. Volume 1. Edition 2. Richard and John Edward Taylor, printers and publishers to the University of London, reprinted from the Philosophical Transactions of 1831-1838, London, UK (1839)
[1778] Lange, E. & Mishchenko, K.P. Zur Thermodynamik der Ionensolvatation. *Z. Phys. Chem. A* **149**, 1-41 (1930)
[1779] Trasatti, S. & Parsons, R. Interphases in systems of conducting phases. *Pure Appl. Chem.* **55**, 1251-1268 (1983)
[1780] Bockris, J.O'M. & Reddy, A.K.N. Electrodics. In: *Modern electrochemistry. Volume 2A*. Kluwern Academic/Plenum Publishers, New York, USA, pp 1035-1400 (2000)
[1781] Koczorowski, Z. Voltaic cells in electrochemistry and surface chemistry of liquids. In: *Modern aspects of electrochemistry. Volume 34*. Bockris, J.O'M., Ed. Kluwer Academic/Plenum Publishers, New York, USA, pp 13-52 (2001)
[1782] Frumkin, A.N. & Damaskin, B. Remark on the paper of S. Trasatti: The concept of absolute electrode potential. An attempt at a calculation. *J. Electroanal. Chem.* **66**, 150-154 (1975)
[1783] Trasatti, S. On the concept and the possibility of experimental determination of absolute

electrode potentials. *J. Chem. Phys.* **69**, 2938-2939 (1978)

[1784] Gomer, R. & Tryson, G. Reply to Comment "On the concept and the possibility of experimental determination of absolute electrode potentials" by S. Trasatti. *J. Chem. Phys.* **69**, 2939-2941 (1978)

[1785] Reiss, H. The Fermi level and the redox potential. *J. Phys. Chem.* **89**, 3783-3791 (1985)

[1786] Reiss, H. The absolute electrode potential. Tying the loose ends. *J. Electrochem. Soc.* **135**, 247C-258C (1988)

[1787] Tsiplakides, D. & Vayenas, C.G. Electrode work function and absolute potential scale in solid-state electrochemistry. *J. Electrochem. Soc.* **148**, E189-E202 (2001)

[1788] Tsiplakides, D. & Vayenas, C.G. The absolute potential scale in solid state electrochemistry. *Solid State Ionics* **152-153**, 625-639 (2002)

[1789] Hansen, W.N. & Kolb, D.M. The work function of emersed electrodes. *J. Electroanal. Chem.* **100**, 493-500 (1979)

[1790] Kötz, E.R., Neff, H. & Müller, K. A UPS, XPS and work function study of emersed silver, platinum and gold electrodes. *J. Electroanal. Chem.* **215**, 331-344 (1986)

[1791] Trasatti, S. Physical, chemical and structural aspects of the electrode/solution interface. *Electrochim. Acta* **28**, 1083-1093 (1983)

[1792] Samec, Z., Johnson, B.W. & Doblhofer, K. The absolute electrode potential of metal electrodes emersed from liquid electrolytes. *Surf. Sci.* **264**, 440-448 (1992)

[1793] Donald, W.A., Leib, R.D., O'Brien J.T., Bush, M.F., Williams, Absolute standard hydrogen electrode potential measured by reduction of aqueous nanodrops in the gas phase. *J. Am. Chem. Soc.* **130**, 3371-3381 (2008)

[1794] Donald, W.A., Leib, R.D., O'Brien J.T., Williams, Directly relating gas-phase cluster measurements to solution-phase hydrolysis, the absolute standard hydrogen electrode potential, and the absolute proton solvation energy. *Chem. Eur. J.* **15**, 5926-5934 (2009)

[1795] Donald, W.A., Leib, R.D., Demireva, M., O'Brien J.T., Prell, J.S., Williams, Directly relating reduction energies of gaseous $Eu(H_2O)_n^{3+}$, $n = 55 - 140$, to aqueous solution: the absolute SHE potential and real proton solvation energy. *J. Am. Chem. Soc.* **131**, 13328-13337 (2009)

[1796] Pleskov, Y.V. Comments on "The Absolute Potential of a Standard Hydrogen Electrode: A new Estimate", by H. Reiss and A. Heller. *J. Phys. Chem.* **91**, 1691-1692 (1987)

[1797] Hansen, W.F. & Hansen, G.J. Absolute half-cell potential: A simple direct measurement. *Phys. Rev. A* **36**, 1396-1402 (1987)

[1798] Heyrovska, R. Absolute potentials of the hydrogen electrode and of aqueous redox couples. *Electrochem. Solid-State Lett.* **12**, F29-F30 (2009)

[1799] Slater, J.C. Atomic shielding constants. *Phys. Rev.* **36**, 57-64 (1930)

[1800] Hehre, W.J., Radom, L., v. R. Schleyer, P. & Pople, J.A. Ab initio molecular orbital theory. Wiley, New York, USA (1986)

[1801] Nagata, T. Noble gas clusters doped with a metal ion I: Ab initio studies of Na^+Ar_n. *J. Phys. Chem. A* **108**, 683-690 (2004)

[1802] Jortner, J. & Noyes, R.M. Some thermodynamic properties of the hydrated electron. *J. Phys. Chem.* **70**, 770-774 (1966)

[1803] Han, P. & Bartels, P.M. Reevaluation of arrhenius parameters for $H + OH^- = (e^-)_{aq} + H_2O$ and the enthalpy and entropy of hydrated electrons. *J. Phys. Chem.* **94**, 7294-7299 (1990)

[1804] Boero, M., Parrinello, M., Terakura, K., Ikeshoji, T. & Liew, C.C. First-principles molecular-dynamics simulations of a hydrated electron in normal and supercritical water. *Phys. Rev. Lett.* **90**, 226403/1-226403/4 (2003)

[1805] Bartels, D.M., Takahashi, K., Cline, J.A., Marin, T.W. & Jonah, C.D. Pulse radiolysis of supercritical water. 3. Spectrum and thermodynamics of the hydrated electron. *J. Phys. Chem. A* **109**, 1299-1307 (2005)

[1806] Novakovskaya, Y.V. Electron hydration energy: Nonempirical estimate. *Protect. Metals* **43**, 129-140 (2007)

[1807] Wang, X.-J., Zhu, Q. & Xiang-Yuan, L. Hydration free energy and absorption spectrum of an extra electron in water by quantum-continuum model. *J. Theor. Comp. Chem.* **7**, 767-775 (2008)

[1808] Donald, W.A., Leib, R.D., O'Brien, J.T., Holm, A.I.S. & Williams, E.R. Nanocalorimetry in mass spectrometry: A route to understanding ion and electron solvation. *Proc. Natl. Acad. Sci. USA* **105**, 18102-18107 (2008)

[1809] Donald, W.A., Demireva, M., Leib, R.D., Aiken, M.J. & Williams, E.R. Electron hydration and ion-electron pairs in water clusters containing trivalent metal ions. *J. Am. Chem. Soc.* **132**, 4633-4640 (2010)

[1810] Lang, N.D. & Kohn, W. Theory of metal surfaces: Work function. *Phys. Rev. B* **3**, 1215-1223 (1971)

[1811] Fonda, L. Emission electron diffraction and holography: A theoretical survey. *Surf. Rev. Lett.* **3**, 1603-1626 (1996)

[1812] Völkl, E., Allard, L.F. & Joy, D.C. Introduction to electron holography. Kluwer Academic/Plenum Publishers, New York, USA (1999)

[1813] Lichte, H. & Lehmann, M. Electron holography - basics and applications. *Reports Prog. Phys.* **71**, 016102/1-016102/46 (2008)
[1814] Midgley, P.A. & Dunin-Borkowski, R.E. Electron tomography and holography in materials science. *Nature Materials* **8**, 271-280 (2009)
[1815] Guinier, A. X-ray diffraction: In crystals, imperfect crystals, and amorphous bodies. Dover Publications, N.Y., USA (1994)
[1816] Als-Nielsen, J. Elements of modern x-ray physics. John Wiley & Sons, New York, USA (2001)
[1817] Peng, L.M., Dudarev, S.L. & Whelan, M.J. High-energy electron diffraction and microscopy. Oxford University Press, Oxford, UK (2004)
[1818] Leung, K. & Marsman, M. Energies of ions in water and nanopores within density functional theory. *J. Chem. Phys.* **127**, 154722/1-154722/10 (2007)
[1819] Bard, A.J. & Faulkner, L.R. Electrochemical methods: Fundamentals and applications. Edition 2. Wiley, New York, USA (2000)
[1820] Lewis, G.N. & Kraus, C.A. The potential of the sodium electrode. *J. Am. Chem. Soc.* **32**, 1459-1468 (1910)
[1821] Lewis, G.N. & Keyes, F.G. The potential of the potassium electrode. *J. Am. Chem. Soc.* **34**, 119-122 (1912)
[1822] Lewis, G.N. & Keyes, F.G. The potential of the lithium electrode. *J. Am. Chem. Soc.* **35**, 340-344 (1913)
[1823] Lewis, G.N. & Argo, W.L. The potential of the rubidium electrode. *J. Am. Chem. Soc.* **37**, 1983-1990 (1915)
[1824] Bent, H.E., Forbes, G.S. & Forziati, A.F. The normal electrode potential of cesium. *J. Am. Chem. Soc.* **61**, 709-715 (1939)
[1825] Robinson, R.A. & Stokes, R.H. Electrolyte solutions. Edition 2. Butterworths, London, UK (1970)
[1826] Harned, H.S. & Owen, B.B. The physical chemistry of electrolytic solutions. Edition 3. Reinhold Publishing Corp., New York, USA (1958)
[1827] Henderson, P. Zur Thermodynamik der Flüssigkeitsketten. *Z. Phys. Chem. (Stöchiometrie u. Verwandtschaftslehre)* **59**, 118-127 (1907)
[1828] Henderson, P. Zur Thermodynamik der Flüssigkeitsketten. *Z. Phys. Chem. (Stöchiometrie u. Verwandtschaftslehre)* **63**, 325-345 (1908)
[1829] de Bethune, A.J., Licht, T.S. & Swendeman, N. The temperature coefficients of electrode potentials. The isothermal and thermal coefficients - The standard ionic entropy of electrochemical transport of the hydrogen ion. *J. Electrochem. Soc.* **106**, 616-625 (1959)
[1830] Agar, J.N. & Turner, J.C.R. Thermal diffusion in solutions of electrolytes. *Proc. Roy. Soc. London A* **255**, 307-330 (1960)
[1831] Tyrrell, H.J.C. Diffusion and heat flow in liquids. Butterworths, London, UK (1961)
[1832] Agar, J.N. Thermogalvanic cells. In: *Advances in electrochemistry and electrochemical engineering. Volume 3.* Delahay, P., Ed. Interscience Publishers, Div. of John Wiley & Sons, New York, USA, pp 31-121 (1963)
[1833] Guyot, M.J. Effet Volta et couches monomoléculaires. *Comptes rendus hebdomadaires des séances de l'académie des sciences* **159**, 307-311 (1914)
[1834] Guyot, M.J. Effet Volta métal-électrolyte et couches monomoléculaires. *Ann. Physique* **2**, 506-519 (1924)
[1835] Bichat, E. & Blondlot, R. Mesure de la différence de potentiel des couches électriques qui recouvrent deux liquides au contact. *J. Physique* **2**, 533-551 (1883)
[1836] Parsons, R. Manual of symbols and terminology for physicochemical quantities and units. *Pure. Appl. Chem.* **37**, 501-516 (1974)
[1837] Hamelin, A. & Lecoeur, J. The orientation dependence of zero charge potentials and surface energies of gold crystal faces. *Surf. Sci.* **57**, 771-774 (1976)
[1838] Frumkin, A.N., Petrii, O.A. & Damaskin, B.B. Potentials of zero charge. In: *Comprehensive treatise of electrochemistry. Volume 1.* Bockris, J.O'M., Conway, B.E. & Yeager, E., Eds. Plenum, New York, USA, pp 201-289 (1980)
[1839] Trasatti, S. & Lust, E. The potential of zero charge. In: *Modern aspects of electrochemistry. Volume 33.* White, R.E., Bockris, J.O. & Conway, B.E., Eds. Kluwer Academic/-Plenum Publ., New York, USA, pp 1-215 (1999)
[1840] Lippmann, G. Relations entre les phénomènes électriques et capillaires. *Ann. Chim. Phys.* **5**, 494-549 (1875)
[1841] Lippmann, G. Beziehungen zwischen den capillaren und elektrischen Erscheinungen. *Ann. Phys.* **225**, 546-561 (1873)
[1842] Lippmann, G. Bemerkungen über einige neuere electrocapillare Versuche. *Ann. Phys.* **247**, 316-324 (1880)
[1843] Frumkin, A.N. Die Elektrokapillarkurve. *Ergeb. Exakt. Naturw.* **7**, 235-275 (1928)
[1844] Morcos, I. The electrocapillary phenomena of solid electrodes. *J. Electroanal. Chem.* **62**, 313-340 (1975)
[1845] Bichat, E. & Blondlot, R. Sur les différences électriques entre les liquides et sur le rôle de

l'air dans la mesure électrométrique de ces différences. *Comptes rendus hebdomadaires des séances de l'académie des sciences* **100**, 791-793 (1885)
[1846] Meyer, G. Zur Theorie des Capillarelectrometers. *Ann. Phys.* **281**, 508-522 (1892)
[1847] Wiedeburg, O. Ueber die Potentialdifferenzen zwischen Metallen und Elektrolyten. *Ann. Phys.* **295**, 742-749 (1896)
[1848] Schreber, K. Zur Theorie des Capillarelectrometers. *Ann. Phys.* **289**, 109-134 (1894)
[1849] Meyer, G. Capillarelectrometer und Tropfelectroden. *Ann. Phys.* **289**, 845-873 (1894)
[1850] Meyer, G. Ueber die Potentialdifferenzen zwischen Metallen und Flüssigkeiten. *Ann. Phys.* **292**, 680-699 (1895)
[1851] Palmaer, W. Über das absolute Potential der Kalomelelektrode. *Z. Elektrochem.* **9**, 754-757 (1903)
[1852] Billitzer, J. Zu den kapillarelektrischen Bewegungen und über einen Strom im offenen Element. *Ann. Phys.* **318**, 827-835 (1904)
[1853] Nernst, W, Ueber Berührungselectricität. *Ann. Phys.* **S294**, 1-13 (1896)
[1854] Braun, F. Ueber Tropfelectroden. *Ann. Phys.* **277**, 448-462 (1890)
[1855] Palmaer, W. Ueber die Wirkungsart der Tropfelektrode. *Z. Phys. Chem. (Stöchiometrie u. Verwandtschaftslehre)* **25**, 265-283 (1898)
[1856] Meyer, G. Ueber Tropfelektroden. *Ann. Phys.* **303**, 433-438 (1899)
[1857] Billitzer, J. Nachtrag zu meiner Abhandlung: Versuche mit Tropfelektroden usw. *Z. Phys. Chem. (Stöchiometrie u. Verwandtschaftslehre)* **49**, 709-710 (1904)
[1858] Billitzer, J. Elektrische Doppelschicht und absolutes Potential. Kontaktelektrische Studien I. *Ann. Phys.* **316**, 902-913 (1903)
[1859] Billitzer, J. Über die Elektrizitätserregung durch die Bewegung fester Körper in Flüssigkeiten. Kontaktelektrische Studien II. *Ann. Phys.* **316**, 937-956 (1903)
[1860] Whitney, W.R. & Blake, J.C. The migration of colloids. *J. Am. Chem. Soc.* **26**, 1339-1387 (1904)
[1861] Goodwin, H.M. & Sosman, R.B. Billitzers Methode zur Bestimmung absoluter Potentialdifferenzen. *Z. Elektrochem.* **12**, 192-193 (1906)
[1862] Billitzer, J. Zur Bestimmung absoluter Potentialdifferenzen. *Z. Elektrochem.* **12**, 281-316 (1906)
[1863] Freundlich, H. & Mäkelt, E. Über den absoluten Nullpunkt des Potentials. *Z. Elektrochem.* **15**, 161-165 (1909)
[1864] Gordon, C.M. Eine neue Methode für die Bestimmung der Polarisationskapazität. *Z. Elektrochem.* **3**, 163-164 (1896)
[1865] Gordon, C.M. Über Messung der Polarisationskapazität. *Wied. Ann.* **61**, 1-29 (1897)
[1866] Scott, A.M. Studien über Polarisationskapazität. *Wied. Ann.* **67**, 388-420 (1899)
[1867] Krüger, F. Über Polarisationskapazität. *Z. Phys. Chem. (Stöchiometrie u. Verwandtschaftslehre)* **45**, 1-74 (1903)
[1868] Krüger, F. Theorie der Polarisationskapazität. *Nachr. Kgl. Ges. Wiss. Gött.* **(?)**, 59-74 (1903)
[1869] Fajans, K. Löslichkeit und Ionisation vom Standpunkte der Atomstruktur. *Naturwiss.* **9**, 729-738 (1921)
[1870] Baughan, E.C. The heat of hydration of the proton. *J. Chem. Soc.* **(Oct.)**, 1403-1403 (1940)
[1871] Benjamin, L. & Gold, V. A table of thermodynamic functions of ionic hydration. *Trans. Faraday Soc.* **50**, 797-799 (1954)
[1872] Morris, D.F.C. Ionic radii and enthalpies of hydration of ions. *Struct. Bond.* **4**, 63-82 (1968)
[1873] Salomon, M. The thermodynamics of ion solvation in water and propylene carbonate. *J. Phys. Chem.* **74**, 2519-2524 (1970)
[1874] Morris, D.F.C Estimation of thermodynamic properties for hydration of individual alkali metal and halide ions. *Electrochim. Acta* **27**, 1481-1486 (1982)
[1875] Tissandier, M.D., Cowen, K.A., Feng, W.Y., Gundlach, E., Cohen, M.H., Earhart, A.D., Coe, J.V. & Tuttle Jr., T.R. The proton's absolute aqueous enthalpy and Gibbs free energy of solvation from cluster-ion solvation data. *J. Phys. Chem. A* **102**, 7787-7794 (1998)
[1876] Tissandier, M.D., Cowen, K.A., Feng, W.Y., Gundlach, E., Cohen, M.H., Earhart, A.D., Coe, J.V. & Tuttle Jr., T.R. Correction to "The proton's absolute aqueous enthalpy and Gibbs free energy of solvation from cluster-ion solvation data.", J. Phys. Chem. A 102, 7787-7794 (1998). *J. Phys. Chem. A* **102**, 9308-9308 (1998)
[1877] Marcus, Y. Ion properties. Marcel Dekker, Inc., New York, USA (1997)
[1878] Grunwald, E., Baugham, G. & Kohnstam, G. The solvation of electrolytes in dioxane-water mixtures, as deduced from the effect of solvent change on the standard partial molar free energy. *J. Am. Chem. Soc.* **82**, 5801-5811 (1960)
[1879] Koepp, H.-M., Wendt, H. & Strehlow, H. Der Vergleich der Spannungsreihen in verschiedenen Solventien. II. *Z. Elektrochem.* **64**, 483-491 (1960)
[1880] Somsen, G. Enthalpies of solvation of alkali halides in formamide. III. Structural con-

siderations. *Rec. Trav. Chim.* **85**, 526-537 (1966)
[1881] Criss, C.M., Held, R.P. & Luksha, E. Thermodynamic properties of nonaqueous solutions. V. Ionic entropies: Their estimation and relationship to the structure of electrolyte solutions. *J. Phys. Chem.* **72**, 2970-2975 (1968)
[1882] Somsen, G. & Weeda, L. The evaluation of ionic enthalpies of solvation. *J. Electroanal. Chem.* **29**, 375-382 (1971)
[1883] Krishnan, K.S. & Friedman, H.L. Solvation enthalpies of electrolytes in methanol and dimethylformamide. *J. Phys. Chem.* **75**, 3606-3612 (1971)
[1884] Kim, J.I. Preferential solvation of single ions. A critical study of the Ph4AsPh4B assumption for single ion thermodynamics in amphiprotic and dipolar-aprotic solvents. *J. Phys. Chem.* **82**, 191-199 (1978)
[1885] Chakravarty, S.K. & Lahiri, S.C. The thermodynamics of ionisation of glycine in methanol+water mixtures and the determination of single ion thermodynamics. *Thermochimica Acta* **99**, 243-251 (1986)
[1886] Chakravorty, S.K., Sarkar, S.K. & Lahiri, S.C. The thermodynamics of ionization of α-alanine in methanol+water mixtures and the determination of single ion thermodynamics. *Thermochimica Acta* **114**, 245-256 (1987)
[1887] Woldan, M. Standard enthalpies of transfer of alkali-metal and halide ions from water to water-urea mixtures. *Thermochimica Acta* **120**, 97-106 (1987)
[1888] Kelly, C.P., Cramer, C.J. & Truhlar, D.G. Single-ion solvation free energies and the normal hydrogen electrode potential in methanol, acetonitrile, and dimethyl sulfoxide. *J. Phys. Chem. B.* **111**, 408-422 (2007)
[1889] Hefter, G.T., Grolier, J.-P. E., Roux, A.H. & Roux-Desgranges, G. Apparent molar heat capacities and volumes of electrolytes and ions in acetonitrile-water mixtures. *J. Solut. Chem.* **19**, 207-223 (1990)
[1890] Marcus, Y. & Hefter, G. The standard partial molar volumes of electrolytes and ions in non-aqueous solvents. *Chem. Rev.* **104**, 3405-3452 (2004)
[1891] Izutsu, K. Electrochemistry in nonaqueous solutions. Edition 2. Wiley-VCH, Weinheim, Germany (2009)
[1892] Feakins, D. & Watson, P. Studies in ion solvation in non-aqueous solvents and their aqueous mixtures. Part II. Properties of ion constituents. *J. Chem. Soc.* **(Oct.)**, 4734-4741 (1963)
[1893] Feakins, D. & Watson, P. Studies in ion solvation in non-aqueous solvents and their aqueous mixtures. Part I. The cell H_2|HX|AgX-Ag (X=Br,I) in 10% and 43·12% mixtures of methanol and water. *J. Chem. Soc.* **(Oct.)**, 4686-4691 (1963)
[1894] Arnett, E.M. & McKelvey, D.R. Enthalpies of transfer from water to dimethyl sulfoxide for some ions and molecules. *J. Am. Chem. Soc.* **88**, 2598-2599 (1966)
[1895] Feakins, D., Smith, B.C. & Thakur, L. Studies in ion solvation in non-aqueous solvents and their aqueous mixtures. 4. Enthalpies of transfer of alkali-metal halides from water to a 20% mixture of dioxan and water. *J. Chem. Soc. A* **(6)**, 714-718 (1966)
[1896] Alexander, R. & Parker, A.J. Solvation of ions. XII. Changes in the standard chemical potential of anions on transfer from protic to dipolar aprotic solvents. *J. Am. Chem. Soc.* **89**, 5549-5551 (1967)
[1897] Friedman, H.L. Regularities and specific effects in enthalpies of transfer of ions from water to aprotic solvents. *J. Phys. Chem.* **71**, 1723-1726 (1967)
[1898] Coetzee, J.F. & Campion, J.J. Solute-solvent interaction. II. Relative activities of anions in acetonitrile and water. *J. Am. Chem. Soc.* **89**, 2517-2521 (1967)
[1899] Coetzee, J.F. & Campion, J.J. Solute-solvent interaction. I. Evaluations of relative activities of reference cations in acetonitrile and water. *J. Am. Chem. Soc.* **89**, 2513-2517 (1967)
[1900] Abraham, M.H. Entropies of transfer of tetra-alkylammonium ions from water to methanol, dimethylformamide, and acetonitrile. *J. Chem. Soc. - Chem. Comm.* **15**, 888-889 (1972)
[1901] Feakins, D. & Voice, P.J. Studies in ion solvation in non-aqueous solvents and their aqueous mixtures. 14. Free energies of transfer of alkali-metal chlorides from water to 10-99% (w/w) methanol-water mixtures at 25°C. *J. Chem. Soc. Faraday Trans. I* **68**, 1390-1405 (1972)
[1902] Cox, B.G. & Parker, A.J. Solvation of ions. XVII. Free energies, heats and entropies of transfer of single ions from protic to dipolar aprotic solvents. *J. Am. Chem. Soc.* **95**, 402-407 (1973)
[1903] Diggle, J.W. & Parker, A.J. Solvation of ions - XX. The ferrocene-ferricinium couple and its role in the estimation of free energies of transfer of single ions. *Electrochim. Acta* **18**, 975-979 (1973)
[1904] Feakins, D. & Voice, P.J. Studies in ion solvation in non-aqueous solvents and their aqueous mixtures. 16. Free energies of transfer of sodium bromide and iodide from water to 10-99% (w/w) methanol+water mixtures at 25°C. *J. Chem. Soc. Faraday Trans.* **I**, 1711-1720 (1973)
[1905] Fuchs, R. & Hagan, C.P. Single-ion enthalpies of transfer from water to aqueous dimethyl sulfoxide solutions. *J. Phys. Chem.* **77**, 1797-1800 (1973)

[1906] Wells, C.F. Ionic solvation in methanol+water mixtures. Free energies of transfer from water. *J. Chem. Soc. Faraday Trans. 1* **69**, 984-992 (1973)

[1907] Cox, B.G. & Parker, A.J. Entropies of solution of ions in water. *J. Am. Chem. Soc.* **95**, 6879-6884 (1973)

[1908] Feakins, D., Willmott, A.S. & Willmott, A.R. Studies in ion solvation in non-aqueous solvents and their aqueous mixtures. 15. Free-energies of transfer of cadmium chloride from water to 10-40% (w/w) methanol-water mixtures at 25°C; dissociation of $CdCl^+$; use of dilute cadmium amalgam electrodes. The group II cations. *J. Chem. Soc. Faraday Trans. I* **69**, 122-131 (1973)

[1909] Cox, B.G., Hedwig, G.R., Parker, A.J. & Watts, D.W. Solvation of ions. 19. Thermodynamic properties for transfer of single ions between protic and dipolar aprotic-solvents. *Aust. J. Chem.* **27**, 477-501 (1974)

[1910] Hedwig, G.R. & Parker, A.J. Solvation of ions. XXIII. Enthalpies of transfer of some divalent metal ions from water to nonaqueous solvents. *J. Am. Chem. Soc.* **96**, 6589-6593 (1974)

[1911] Wells, C.F. Ionic solvation in water+co-solvent mixtures. Part 2. Free energies of transfer of single ions from water into mixtures of water with acetone, isopropanol, glycerol or methanol. *J. Chem. Soc. Faraday Trans. 1* **70**, 694-704 (1974)

[1912] Feakins, D., Hickey, B.E., Lorimer, J.P. & Voice, P.J. Studies in ion solvation in non-aqueous solvents and their aqueous mixtures. 17. Free energies of transfer of alkali-metal chlorides from water to 10-40% (w-w) dioxan + water mixtures, and of potassium bromide and iodide to 20% mixture, at 25°C. *J. Chem. Soc. Faraday Trans. I* **71**, 780-783 (1975)

[1913] Hedwig, G.R., Owensy, D.A. & Parker, A.J. Solvation of ions. XXIV. Entropies of transfer of some divalent metal ions from water to nonaqueous solvents. *J. Am. Chem. Soc.* **97**, 3888-3894 (1975)

[1914] Wells, C.F. Ionic solvation in water+co-solvent mixtures. Part 3. Free energies of transfer of single ions from water into water+ethylene glycol mixtures. *J. Chem. Soc. Faraday Trans. 1* **71**, 1868-1875 (1975)

[1915] Feakins, D. & Allan, C.T. Studies in ion solvation in non-aqueous solvents and their aqueous mixtures. Part 18. Enthalpies of transfer of alkali-metal halides from water to dioxan+water mixtures; structural effects and comparison with large ions. *J. Chem. Soc. Faraday Trans. 1* **72**, 314-322 (1976)

[1916] Hopkins Jr., H.P. & Alexander, C.J. Transfer Gibbs free energies from H_2O to CH_3OH for a series of substituted phenyltropylium ions, malachite green, and crystal violet. *J. Solut. Chem.* **5**, 249-255 (1976)

[1917] Wells, C.F. Ionic solvation in water+co-solvent mixtures. Part 4. Free energies of transfer of single ions from water into water+t-butyl alcohol mixtures. *J. Chem. Soc. Faraday Trans. 1* **72**, 601-609 (1976)

[1918] Parker, A.J. Solvation of ions - Enthalpies, entropies and free energies of transfer. *Electrochim. Acta* **21**, 671-679 (1976)

[1919] Clune, T.A., Feakins, D. & McCarthy, P.J. Free-energies of transfer of alkali-metal chlorides from water to tert-butanol-water mixtures: Comparison between amalgam and glass-electrode results. *J. Electroanal. Chem.* **84**, 199-201 (1977)

[1920] Nedermeijer-Denessen, H.J.M., de Ligny, C.L. & Remijnse, A.G. The significance of large negative ions in the estimation of standard free enthalpies of transfer of single ions. *J. Electroanal. Chem* **77**, 153-161 (1977)

[1921] Wells, C.F. Ionic solvation in water+co-solvent mixtures. Part 5. Free energies of transfer of large single ions from water into water+methanol with the "neutral" component removed. *J. Chem. Soc. Faraday Trans. 1* **74**, 636-643 (1977)

[1922] Mayer, U. Zusammenhang zwischen freien Standard-Transferenthalpien von Ionen und empirischen Lösungsmittelparametern. *Monatshefte f. Chemie* **108**, 1479-1495 (1977)

[1923] Parker, A.J. & Waghorne, W.E. Solvation of ions. XXVI. Free energies of transfer of ions to multi-site solvents and solvent mixtures. *Aust. J. Chem.* **31**, 1181-1187 (1978)

[1924] Wells, C.F. Ionic solvation in water+co-solvent mixtures. Part 6. Free energies of transfer of single ions from water into water+dioxan mixtures. *J. Chem. Soc. Faraday Trans. 1* **74**, 1569-1582 (1978)

[1925] Wells, C.F. Ionic solvation in water+co-solvent mixtures. Part 5. Free energies of transfer of large single ions from water into water+methanol with "neutral" component removed. *J. Chem. Soc. Faraday Trans. I* **74**, 636-643 (1978)

[1926] Abraham, M.H. & de Namor, A.F.D. Free energies and entropies of transfer of ions from water to methanol, ethanol and 1-propanol. *J. Chem. Soc. Faraday Trans. 1* **74**, 2101-2110 (1978)

[1927] Feakins, D., Hickey, B.E. & Voice, P.J. Studies in ion solvation in non-aqueous solvents and their aqueous mixtures. Part 19. Free-energies of transfer of alkali-metal chlorides from water to DMSO + water mixtures up to 40% (w-w) DMSO, of sodium-chloride to 60% (w-w) DMSO and of potassium-bromide and iodide to 10% (w-w) DMSO. Comparison with silver-chloride. *J. Chem. Soc. Faraday Trans. I* **75**, 907-913 (1979)

[1928] Kundu, K.K. & Das, A.K. Transfer free energies of some ions from water to dimethylsulfoxide-water and urea-water mixtures. *J. Solut. Chem.* **8**, 259-265 (1979)

[1929] Lahiri, S.C. & Aditya, S. Solvation and free energies of transfer of single ions. *J. Ind. Chem. Soc.* **56**, 1112-1124 (1979)

[1930] Basumullick, I.N. & Kundu, K.K. Transfer free energies of alkali metal chlorides and of some individual ions in glycerol and water mixtures. *Can. J. Chem.* **58**, 79-85 (1980)

[1931] de Valera, E., Feakins, D. & Waghorne, W.E. Studies in ion solvation in non-aqueous solvents and their aqueous mixtures. Part 20. Enthalpies of transfer of alkali-metal halides in the methanol water-system from enthalpies of dilution. *J. Chem. Soc. Faraday Trans. I* **76**, 560-569 (1980)

[1932] Kundu, K.K. & Parker, A.J. Solvation of ions. XXVII. Comparison of methods to calculate single ion free energies of transfer in mixed solvents. *J. Solut. Chem.* **10**, 847-861 (1981)

[1933] Sidahmed, I.M. & Wells, C.F. Solubility of salts of hexachlorothenate(IV) ions with complex cations in water and in water and alcohol mixtures: Free energies of transfer of the complex ions. *Dalton Trans.* **10**, 2034-2038 (1981)

[1934] Wells, C.F. Ionic solvation in water+co-solvent mixtures. Part 7. Free energies of transfer of single ions from water into water+dimethylsulphoxide mixtures. *J. Chem. Soc. Faraday Trans. 1* **77**, 1515-1528 (1981)

[1935] Burgess, J., Peacock, R.D. & Rodgers, J.H. Enthalpies of transfer of ions from water into aqueous hydrogen fluoride. *J. Fluor. Chem.* **19**, 333-347 (1982)

[1936] Singh, P., Macleod, I.D. & Parker, A.J. Solvation of ions. 30. Thermodynamics of transfer of copper ions from water to solvent mixtures. *J. Solut. Chem.* **11**, 495-508 (1982)

[1937] Wells, C.F. Ionic solvation in water + co-solvent mixtures. Part 8. Total free energies of transfer and free energies of transfer with the "neutral" component removed of single ions from water into water + acetone. *Thermochimica Acta* **53**, 67-87 (1982)

[1938] de Namor, A.F.D., Hill, T. & Sigstad, E. Free energies of transfer of 1:1 electrolytes from water to nitrobenzene. Partition of ions in the water+nitrobenzene system. *J. Chem. Soc. Faraday Trans. 1* **79**, 2713-2722 (1983)

[1939] Marcus, Y. Thermodynamic functions of transfer of single ions from water to non-aqueous and mixed solvents: Part 1 - Gibbs free energies of transfer to nonaqueous solvents. *Pure Appl. Chem.* **55**, 977-1021 (1983)

[1940] Wells, C.F. The spectrophotometric solvent sorting method for the determination of free energies of transfer of individual ions - A critical appraisal. *Aust. J. Chem.* **36**, 1739-1752 (1983)

[1941] de Namor, A F.D., Contreras, E. & Sigstad, E. Free energies of transfer of 1:1 electrolytes from water to butan-1-ol. *J. Chem. Soc. Faraday Trans. 1* **79**, 1001-1007 (1983)

[1942] Miyaji, K. & Morinaga, K. Enthalpies of transfer of monovalent ions from water to water acetonitrile mixtures. *Bull. Chem. Soc. Jpn.* **56**, 1861-1862 (1983)

[1943] Feakins, D., Knox, M. & Hickey, B.E. Studies in ion solvation in non-aqueous solvents and their aqueous mixtures. Part 21. Free energies of transfer of alkali-metal halides in the acetone+water system to 60% (w/w) acetone with cation-selective electrodes. *J. Chem. Soc. Faraday Trans. I* **80**, 961-968 (1984)

[1944] Ishiguro, S. & Ohtaki, H. Enthalpies of transfer of single ions and metal-complexes from water to an aqueous dioxane solution. *Bull. Chem. Soc. Jpn.* **57**, 2622-2627 (1984)

[1945] Wells, C.F. Ionic solvation in water+cosolvent mixtures. Part 9. Free energies of transfer of single ions from water into water+ethanol mixtures. *J. Chem. Soc. Faraday Trans. 1* **80**, 2445-2458 (1984)

[1946] Chakraborty, S.K. & Lahiri, S.C. Enthalpy and entropy of transfer of hydrogen ion from water to mixed solvents. *J. Therm. Anal.* **29**, 815-820 (1984)

[1947] Groves, G.S. & Wells, C.F. Ionic solvation in water-cosolvent mixtures. Part 11. Free energies of transfer of single ions from water into water-urea mixtures. *J. Chem. Soc. Faraday Trans. 1* **81**, 3091-3102 (1985)

[1948] Groves, G.S. & Wells, C.F. Ionic solvation in water-cosolvent mixtures. Part 10. Free energies of transfer of single ions from water into water+ethanonitrile mixtures. *J. Chem. Soc. Faraday Trans. 1* **81**, 1985-1997 (1985)

[1949] Groves, G.S. & Wells, C.F. Ionic solvation in water cosolvent mixtures. Part 2. Free energies of transfer of single ions from water into water urea mixtures. *J. Chem. Soc. Faraday Trans. I* **81**, 3091-3102 (1985)

[1950] Marcus, Y. Thermodynamic functions of transfer of single ions from water to non-aqueous and mixed solvents. Part 3: Standard potentials of selected electrodes. *Pure Appl. Chem.* **57**, 1129-1132 (1985)

[1951] Marcus, Y. Thermodynamic functions of transfer of single ions from water to non-aqueous and mixed solvents. Part 2: Enthalpies and entropies of transfer to nonaqueous solvents. *Pure Appl. Chem.* **57**, 1103-1128 (1985)

[1952] Elsemongy, M.M. & Reicha, F.M. Absolute electrode potentials in dimethyl sulphoxide-water mixtures and transfer free energies of individual ions. *Thermochimica Acta* **108**, 115-131 (1986)

[1953] Groves, G.S. & Wells, C.F. Free energy of transfer of cobalt(III) complexes from water into water-t-butyl alcohol mixtures. *J. Solut. Chem.* **15**, 211-219 (1986)

[1954] Sidahmed, I.M. & Wells, C.F. Ionic solvation in water-cosolvent mixtures. Part 12. Free energies of transfer of single ions from water into water-propan-1-ol mixtures. *J. Chem. Soc. Faraday Trans. 1* **82**, 2577-2588 (1986)

[1955] Hefter, G.T. & McLay, P.J. Solvation of fluoride ions. II. Enthalpies and entropies of transfer from water to aqueous methanol. *Aust. J. Chem.* **41**, 1971-1975 (1986)

[1956] Carthy, G., Feakins, D. & Waghorne, W.E. Enthalpies of transfer of tetra-alkylammonium halides from water to water propan-1-ol mixtures at 25°C. *J. Chem. Soc. Faraday Trans. I* **83**, 2585-2592 (1987)

[1957] Elsemongy, M.M. & Abu Elnader, H.M. Absolute electrode potentials in dioxane-water solvent mixtures and transfer free energies of individual ions. *Thermochimica Acta* **120**, 261-278 (1987)

[1958] Sidahmed, I.M. & Wells, C.F. Ionic solvation in water cosolvent mixtures. Part 13. Free energies of transfer of single ions from water into water tetrahydrofuran mixtures. *J. Chem. Soc. Faraday Trans. I* **83**, 439-449 (1987)

[1959] Groves, G.S., Halawani, K.H. & Wells, C.F. Ionic solvation in water-cosolvent mixtures. Part 14. Free energies of transfer of single ions from water into water-ethylene carbonate and water-propylene carbonate mixtures. *J. Chem. Soc. Faraday Trans. 1* **83**, 1281-1291 (1987)

[1960] Johnsson, M. & Persson, I. Determination of Gibbs free energy of transfer for some univalent ions from water to methanol, acetonitrile, dimethylsulfoxide, pyridine, tetrahydrothiophene and liquid ammonia; standard electrode potentials of some couples in these solvents. *Inorg. Chim. Acta* **127**, 15-24 (1987)

[1961] Johnsson, M. & Persson, I. Determination of heats and entropies of transfer for some univalent ions from water to methanol, acetonitrile, dimethylsulfoxide, pyridine and tetrahydrothiophene. *Inorg. Chim. Acta* **127**, 25-34 (1987)

[1962] Feakins, D., McCarthy, P.J. & Clune, T.A. Thermodynamics of ion solvation in mixed aqueous solvents. Part 1. Some free-energies of transfer of hydrochloric-acid, alkali-metal and alkaline-earth-metal chlorides and potassium halides in the trans-butyl alcohol water-system at 25

[1963] Hefter, G.T. & McLay, P.J. The solvation of fluoride ions. I. Free energies for transfer from water to aqueous alcohol and acetonitrile mixtures. *J. Solut. Chem.* **17**, 535-546 (1988)

[1964] Kondo, Y., Uematsu, R., Nakamura, Y. & Kusabayashi, S. Empirical analysis on the constituent terms of transfer enthalpies. *J. Chem. Soc. Faraday Trans. 1* **84**, 111-116 (1988)

[1965] Sidahmed, I.M. & Wells, C.F. Ionic solvation in water-cosolvent mixtures. Part 15. Free energies of transfer of single ions from water into water-dimethylformamide mixtures. *J. Chem. Soc. Faraday Trans. 1* **84**, 1153-1162 (1988)

[1966] Tissier, C. Alkaline earth ions in methanol and water+methanol mixtures. 2. Single-ion Gibbs free energies of transfer. *Bull. Soc. Chim. Franc.* **5**, 787-792 (1988)

[1967] Wells, C.F. Ionic solvation in water + co-solvent mixtures. Part 17. The "neutral" component of free energies of transfer of single ions from water into water + ethanol mixtures. *Thermochimica Acta* **132**, 141-154 (1988)

[1968] Bhattacharyya, A.K. & Lahiri, S.C. Determination of the free energy of solvation of ferrous ion in water and free energies of transfer of ferrous ion from water to ethanol-water mixtures. *Thermochimica Acta* **127**, 119-124 (1988)

[1969] Feakins, D., Hickey, B.E., Knox, M., McCarthy, P.J., Waghorne, E. & Clune, T.A. Thermodynamics of ion solvation in mixed aqueous solvents. Part 2. Effect of steric hindrance on free energies of transfer of cations: Correlations with structural determinations. *J. Chem. Soc. Faraday Trans. I* **84**, 4219-4233 (1988)

[1970] Wells, C.F. Ionic solvation in water + co-solvent mixtures. Part 16. Free energies of transfer of large single ions with the "neutral" component removed from water into water + ethanol mixtures. *Thermochimica Acta* **130**, 127-139 (1988)

[1971] Gomaa, E.A. Free energies of transfer for some monovalent ions and Ph_4SbBPh_4 from water to acetonitrile and acetonitrile-water mixtures using the asymmetric Ph_4AsBPh_4 assumption. *Thermochimica Acta* **152**, 371-379 (1989)

[1972] Miyaji, K., Nozawa, K. & Morinaga, K. Preferential solvation of Ni(II) and Mg(II) ions in water-acetonitrile mixtures. Enthalpies of transfer and the electronic spectra. *Bull. Chem. Soc. Jpn.* **62**, 1472-1476 (1989)

[1973] Feakins, D., Johnson, K.F. & Waghorne, W.E. Thermodynamics of ion solvation in mixed aqueous solvents. 3. Effect of steric hindrance on the free-energies of transfer of cations to 5% and 20% (w/w) propan-1-ol-water mixtures. *Proc. Roy. Irish Acad. B* **89**, 321-327 (1989)

[1974] Gomaa, E.A. Transfer free energies of ions from water to N,N-dimethylformamide and its aqueous mixtures, based on Ph_4AsBPh_4 and Ph_4SbBPh_4 assumptions. *Thermochimica Acta* **142**, 19-27 (1989)

[1975] Halawani, K.H.M. & Wells, C.F. Ionic solvation in water + co-solvent mixtures. Part 19. Free energies of transfer of single ions from water into mixtures of water with polyhydroxy compounds: An examination of the assumptions used in determining $\Delta G^\circ(i)$. *Thermochimica Acta* **155**, 57-76 (1989)

[1976] Halawani, K.H. & Wells, C.F. Ionic solvation in water-co-solvent mixtures. Part 18. Free energies of transfer of single ions from water into water-2-methoxyethanol mixtures. *J. Chem. Soc. Faraday Trans. I* **85**, 2185-2197 (1989)

[1977] Piekarska, A. & Taniewska-Osińska, S. Single-ion transfer enthalpies for $Ph_4P^+{=}BPh_4^-$, Na^+ and I^- ions in methanol-N,N-dimethylformamide mixtures. *Thermochimica Acta* **170**, 189-195 (1990)

[1978] Saxton, J. & Wells, C.F. Ionic solvation in water-co-solvent mixtures. Part 21. - Free energies of transfer of single ions from water into water-2-ethoxyethanol mixtures. *J. Chem. Soc. Faraday Trans.* **86**, 1471-1475 (1990)

[1979] Wells, C.F. Ionic solvation in water + co-solvent mixtures. Part 20. The "neutral" component of free energies of transfer of single ions from water into mixtures of water + hydroxylic co-solvents. *Thermochimica Acta* **161**, 223-237 (1990)

[1980] Marcus, Y. Thermodynamic functions of transfer of single ions from water to non-aqueous and mixed solvents (based on an extrathermodynamic assumption): Part 5. Gibbs energies of transfer into aqueous alcohols. *Pure Appl. Chem.* **62**, 899-940 (1990)

[1981] Coetzee, J.F, Dollard, W.J. & Istone, W.K. Measurement of real free energies of transfer of individual ions from water to other solvents with the jet cell. *J. Solut. Chem.* **20**, 957-975 (1991)

[1982] Elsemongy, M.M. & Abdel-Khalek, A.A. Gibbs free energies of transfer of individual ions from water into water-urea mixtures. *Thermochimica Acta* **181**, 95-107 (1991)

[1983] Feakins, D., Mullally, J. & Waghorne, W.E. Enthalpies of transfer of tetrabutylammonium bromide as indicators of the structure of aqueous solvents: Aqueous methanol, ethanol, propan-1-ol, 2-methylpropan-2-ol and 1,4-dioxane systems. *J. Chem. Soc. Faraday Trans.* **87**, 87-91 (1991)

[1984] Jóźwiak, M., Nowicka, B. & Taniewska-Osińska, S. Enthalpies of transfer of several ions from water to water-n-propanol mixtures at 298.15 K. *Thermochimica Acta* **190**, 319-323 (1991)

[1985] El-Subruiti, G.M.A., Wells, C.F. & Sidahmed, I.M. Solubility of dichlorotetraalkylpyridinecobalt III hexachlororhenate(IV) in water + methanol mixtures: Free energies of transfer of the complex cation from water into the mixtures. *J. Solut. Chem.* **20**, 403-415 (1991)

[1986] Halawani, K.H.M. & Wells, C.F. Ionic solvation in water + co-solvent mixtures. Part 23. Free energies of transfer of single ions from water into water + diethylene glycol mixtures. *Thermochimica Acta* **191**, 121-130 (1991)

[1987] Wells, C.F. Ionic solvation in water + co-solvent mixtures. Part 22. Free energies of transfer of complex ions from water into water + methanol mixtures. *Thermochimica Acta* **185**, 183-203 (1991)

[1988] Shao, Y., Stewart, A.A. & Girault, H.H. Determination of the half-wave potential of the species limiting the potential window. *J. Chem. Soc. Faraday Trans.* **87**, 2593-2597 (1991)

[1989] Gritzner, G. & Lewandowski, A. Temperature coefficients of half-wave potentials and entropies of transfer of cations in aprotic solvents. *J. Chem. Soc. Faraday Trans.* **87**, 2599-2602 (1991)

[1990] Gritzner, G. & Hörzenberger, F. Gibbs energies, enthalpies and entropies of transfer of cations from acetonitrile and N,N-dimethylformamide into water. *J. Chem. Soc. Faraday Trans.* **88**, 3013-3017 (1992)

[1991] Pietrzak, A. & Taniewska-Osińska, S. Single-ion transfer enthalpies in methanol-acetonitrile mixtures at 298.15 K based on the $Ph_4P^+{=}BPh_4^-$ assumption. *Thermochimica Acta* **194**, 109-116 (1992)

[1992] Taniewska-Osińska, S., Nowicka, B. & Pietrzak, A. Correlation of transfer enthalpies of Ph_4PCl from water to water-organic mixtures with parameters describing organic cosolvent properties. *Thermochimica Acta* **196**, 125-129 (1992)

[1993] Wells, C.F. Ionic solvation in water + co-solvent mixtures. Part 24. Free energies of transfer of single ions from water into water + 1,2-dimethoxyethane mixtures. *Thermochimica Acta* **208**, 323-339 (1992)

[1994] Hakin, A.W. & Beswick, C.L. Single-ion enthalpies and entropies of transfer from water to aqueous urea solutions at 298.15 K. *Can. J. Chem.* **70**, 1666-1670 (1992)

[1995] Hörzenberger, F. & Gritzner, G. Temperature coefficients of half-wave potentials and entropies of transfer of cations in aprotic solvents. *J. Chem. Soc. Faraday Trans.* **88**, 695-697 (1992)

[1996] Lewandowski, A. & Gritzner, G. Temperature coefficients of $Cu|Cu^{2+}$ and $Hg|Hg^{2+}$ electrode potentials and Gibbs energies, entropies and enthalpies of transfer for Cu^{2+} and Hg^{2+}. *J. Chem. Soc. Faraday Trans.* **89**, 3553-3556 (1993)

[1997] Feakins, D., Mullally, J. & Waghorne, W.E. Enthalpies of transfer of sodium and potassium-chlorides from water to aqueous 1,4-dioxane mixtures at 25°C. *J. Solut. Chem.* **22**, 311-319 (1993)

[1998] Hörzenberger, F. & Gritzner, G. Gibbs energies, entropies and enthalpies of transfer for monovalent cations from acetonitrile to several ions. *J. Chem. Soc. Faraday Trans.* **89**,

3557-3564 (1993)

[1999] Taniewska-Osińska, S., Nowicka, B., Pietrzak, A., Romanowski, S. & Pietrzak, T.M. Enthalpies of transfer of selected ions from water to water-propan-2-ol mixtures. Some quantum chemical aspects of the ionic solvation. *Thermochimica Acta* **225**, 9-16 (1993)

[2000] Gritzner, G. & Lewandowski, A. Temperature coefficients of electrode potentials and Gibbs energies, entropies and enthalpies of transfer of Ag^+, Na^+ and Tl^+ from N,N-dimethylformamide to its mixtures with N,N-dimethylthioformamide and with water. *J. Chem. Soc. Faraday Trans.* **89**, 2007-2010 (1993)

[2001] Majumdar, K., Lahiri, S.C. & Mukherjee, D.C. Thermodynamics of transfer of hydrogen ion from water to dioxane-water mixtures. *J. Ind. Chem. Soc.* **70**, 365-374 (1993)

[2002] Kondo, Y., Nonaka, O., Iwasaki, K., Kuwamoto, T. & Takagi, T. Single-ion enthalpies of transfer for conjugate-base anions of imides in acetonitrile-methanol mixtures. *J. Chem. Soc. Faraday Trans.* **90**, 121-125 (1994)

[2003] Mount, L.A. & Wells, C.F. Ionic solvation in water + cosolvent mixtures. Part 25. Gibbs free energies of transfer of single ions from water into water + sulpholane mixtures. *Thermochimica Acta* **237**, 55-71 (1994)

[2004] Piekarska, A. Ion solvation in methanol-organic cosolvent mixtures. Part 5. Enthalpies of transfer of inorganic ions in mixtures of methanol and propylene carbonate at 298.15 K. *Thermochimica Acta* **244**, 61-67 (1994)

[2005] Kundu, K.K. Transfer entropies and structuredness of solvents. *Pure Appl. Chem.* **66**, 411-417 (1994)

[2006] Gritzner, G. & Hörzenberger, F. Gibbs energies, enthalpies and entropies of transfer for divalent cations to several solvents. *J. Chem. Soc. Faraday Trans.* **91**, 3843-3850 (1995)

[2007] Koidel, M. & Ishiguro, S.-I. Molar enthalpies of transfer of divalent transition metal ions and their chloro complexes from N,N-dimethylformamide to N,N-dimethylacetamide. *J. Solut. Chem.* **24**, 511-522 (1995)

[2008] Marcus, Y. Transfer Gibbs free energies of divalent anions from water to organic and mixed aqueous-organic solvents. *Z. Naturforsch. A* **50**, 51-58 (1995)

[2009] Wells, C.F. Equilibria involving protons in mixtures of water with propane-1,2-diol. Gibbs energies of transfer of protons and other ions from water into the mixture. *J. Chem. Soc. Faraday Trans.* **93**, 273-277 (1997)

[2010] Katsuta, S., Takagi, C., Tanaka, M., Fukada, N. & Takeda, Y. Stabilities in water and Gibbs free energies of transfer from water to nonaqueous solvents of alkali metal ion complexes with 1,2-bis[2-(2-methoxyethoxy)ethoxy]benzene(acyclic polyether). *J. Chem. Soc. Faraday Trans.* **94**, 365-368 (1998)

[2011] Kalidas, C., Hefter, G. & Marcus, Y. Gibbs energy transfer of cations from water to mixed aqueous organic solvents. *Chem. Rev.* **100**, 819-852 (2000)

[2012] Wells, C.F. Ions in aqueous acetic acid mixtures: Solvent reorganization around protons and Gibbs energies of transfer from water. *J. Solut. Chem.* **29**, 271-287 (2000)

[2013] Mandal, R. & Lahiri, S.C. Gibbs energies of transfer of hydrogen ion from water to tetrahydrofuran + water mixtures and basicity and structuredness of aquo-ethers. *Z. Phys. Chem.* **214**, 1-13 (2000)

[2014] Hefter, G., Marcus, Y. & Waghorne, W.E. Enthalpies and entropies of transfer of electrolytes and ions from water to mixed aqueous organic solvents. *Chem. Rev.* **102**, 2773-2836 (2002)

[2015] Goldie, R., Luck, D. & Wells, C.F. Gibbs energies of transfer of the proton from water into mixtures of water with chain multiple ethers: Equilibria involving changes of solvation of the proton. *Phys. Chem. Chem. Phys.* **5**, 896-901 (2003)

[2016] Hora, J., McCarthy, R. & Waghorne, W.E. Enthalpies of transfer of the CH_2 moiety into aqueous acetonitrile mixtures; comparison of values from n-alcohols and tetraalkylammonium ions; effect of temperature variation. *J. Chem. Thermodyn.* **37**, 83-88 (2004)

[2017] Kamieńska-Piotrowicz, E. Analysis of the enthalpies of transfer of Co(II) ion in mixed solvents by means of the theory of preferential solvation. *Thermochimica Acta* **427**, 1-7 (2005)

[2018] Sokolov, V.N., Kobenin, V.A. & Gorelov, V.N. The entropies of solvation and transfer of ions in the water-ethanol binary system at 298.15 K. *Russ. J. Phys. Chem.* **79**, 168-172 (2005)

[2019] Marcus, Y. Gibbs energies of transfer of anions from water to mixed aqueous organic solvents. *Chem. Rev.* **107**, 3880-3897 (2007)

[2020] Gschneidner, K.A. Physical properties and interrelationships of metallic and semimetallic elements. *Solid State Phys. - Adv. Res. Appl.* **16**, 275-426 (1964)

[2021] Ho, C.Y. & Taylor, R.E. Thermal expansion of solids. ASM International, USA (1998)

[2022] Kaye, G.W.C. & Laby, T.H. Tables of Physical, Chemical Constants. 3.1.2 Properties of the elements. Kaye, Laby Online. Version 1.0. Available at: *www.kayelaby.npl.co.uk* (2005)

[2023] Kaye, G.W.C. & Laby, T.H. Tables of Physical, Chemical Constants. 2.3.5 Thermal Expansion. Kaye, Laby Online. Version 1.0. Available at: *www.kayelaby.npl.co.uk* (2005)

[2024] Richards, T.W. The compressibilities of the elements and their relations to other properties. *Proc. Natl. Acad. Sci. USA* **1**, 411-415 (1915)

[2025] Petit, A.-T. & Dulong, P.-L. Recherches sur quelques points importants de la théorie de la chaleur. *Ann. Chim. Phys.* **10**, 395-413 (1819)

[2026] Fox, R. The background to the discovery of Dulong and Petit's law. *Brit. J. Hist. Sci.* **4**, 1-22 (1968)

[2027] Simon, J.D. & McQuarrie, D.A. Molecular thermodynamics. Edition 1. University Science Books, Sausalito, California, USA (1999)

[2028] Rossiter, B.W. & Baetzold, R.C. Physical methods of chemistry. Determination of elastic and mechanical properties. Volume 7. Edition 2. Wiley-Interscience, New York, USA (1991)

[2029] Håkansson, B. & Ross, R.G. Thermal conductivity and heat capacity of solid LiBr and RbF under pressure. *J. Phys.: Condens. Matter* **1**, 3977-3985 (1989)

[2030] Roberts, R.W., Ultrasonic parameters in the Born model of the rubidium halides. *J. Phys. Chem. Solids* **31**, 2397-2400 (1970)

[2031] Kelley, K.K. Contributions to the data on theoretical metallurgy. X. High-temperature heat content, heat capacity, and entropy. Data for inorganic compounds. Volume 476. US Bureau of Mines Bulletin (1949)

[2032] Dworkin, A.S. & Bredig, M.A. The heat of fusion of the alkali metal halides. *J. Phys. Chem.* **64**, 269-272 (1960)

[2033] Luce, R.G. & Trischka, J.W. Molecular constants of cesium chloride by the molecular beam electric resonance method. *J. Chem. Phys.* **21**, 105-109 (1952)

[2034] Rusk, J.R. & Gordy, W. Millimeter wave molecular beam spectroscopy: Alkali bromides and iodides. *Phys. Rev.* **127**, 817-831 (1962)

[2035] Veazey, S.E. & Gordy, W. Millimeter-wave molecular beam spectroscopy: Alkali fluorides. *Phys. Rev.* **138**, 1303-1313 (1965)

[2036] Pearson, E.F. & Gordy, W. Millimeter- and submillimeter-wave spectra and molecular constants of LiF and LiCl. *Phys. Rev.* **177**, 52-179 (1969)

[2037] Dunham, J.L. The energy levels of a rotating vibrator. *Phys. Rev.* **41**, 721-731 (1932)

[2038] Maslen, V.W. Crystal ionic radii. *Proc. Phys. Soc. London* **91**, 259-260 (1967)

[2039] Jensen, H. Zur physikalischen Deutung der kristallographischen Ionenradien. *Angew. Chem.* **52**, 583-586 (1939)

[2040] Kucharczyk, W. Ionic radii and optical susceptibilities in the halite-type alkali halides. *Acta Crystallogr. B* **43**, 454-456 (1987)

[2041] Pauling, L. The influence of relative ionic sizes on the properties of ionic compounds. *J. Am. Chem. Soc.* **50**, 1036-1045 (1928)

[2042] Pauling, L. The sizes of ions and their influence on the properties of salt-like compounds. *Z. Kristallographie* **67**, 377-404 (1928)

[2043] Holbrook, J.B., Khaled, F.M. & Smith, B.C. Soft-sphere ionic-radii for Group 1 and Group 2 metal halides and ammonium halides. *Dalton Trans.* **12**, 1631-1634 (1978)

[2044] Collin, R.J. & Smith, B.C. Ionic radii for Group 1 halide crystals and ion-pairs. *Dalton Trans.* **4**, 702-705 (2005)

[2045] Landé, A. Über die Grösse der Atome. *Z. Physik* **1**, 191-197 (1920)

[2046] Ahrens, L.H. The use of ionization potentials. Part 1. Ionic radii of the elements. *Geochim. Cosmochim. Acta* **2**, 155-169 (1952)

[2047] Pauling, L. The sizes of ions and the structure of ionic crystals. *J. Am. Chem. Soc.* **49**, 765-790 (1927)

[2048] Wasastjerna, J.A. On the radii of ions. *Soc. Sci. Fenn. Comm. Phys. Math.* **38**, 1-25 (1923)

[2049] Zachariasen, W.H. A set of empirical crystal radii for ions with inert gas configuration. *Z. Kristallographie* **80**, 137-153 (1931)

[2050] Thomas, L.H. The calculation of atomic fields. *Proc. Cambridge Philos. Soc.* **33**, 542-548 (1927)

[2051] Fermi, E. Statistical methods of investigating electrons in atoms. *Z. Phys.* **48**, 73-79 (1928)

[2052] Waddington, T.C. Ionic radii and the method of the undetermined parameter. *Trans. Faraday Soc.* **62**, 1482-1492 (1966)

[2053] Johnson, O. Ionic radii for spherical potential ions. I. *Inorg. Chem.* **12**, 780-785 (1973)

[2054] Damm, J.Z. & Chvoj, Z. Optimization of Tosi-Fumi ionic radii for F.C.C. alkali halide crystals. *Phys. Stat. Sol. B* **114**, 413-418 (1982)

[2055] Yildiran, H., Ayata, S. & Tunçgenç, M. A theoretical study on calculation of ionic radii using limiting equivalent conductivities. *Ionics* **13**, 83-86 (2007)

[2056] Goldschmidt, V.M. Geochemische Verteilungsgesetze der Elemente VII: Die Gesetze der Krystallchemie. *Skrifter Norske Vidensk. Akad. Oslo 1 Mat.-Nat. Kl. No.* **2**, 1-117 (1926)

[2057] Herz, W. Beziehungen der Schmelzpunktsmolvolumen zu den Ionenradien bei Alkalihaloiden. *Z. anorg. allgem. Chem.* **184**, 303-304 (1929)

[2058] Shannon, R.D. & Prewitt, C.T. Effective ionic radii in oxides and fluorides. *Acta Crystallogr. B* **25**, 925-946 (1969)

[2059] Shannon, R.D. & Prewitt, C.T. Revised values of effective ionic radii. *Acta. Crystallogr.*

B **26**, 1046-1048 (1970)

[2060] Shannon, R.D. Revised values of effective ionic radii and systematic studies of interatomic distances in halides and chalcogenides. *Acta. Crystallogr. A* **32**, 751-767 (1976)

[2061] Schoknecht, G. Röntgen-Kristallstrukturanalyse mit Faltungsintegralen. 2. Messverfahren und Bestimmung der Elektronendichte in NaCl. *Z. Naturforsch. A* **12**, 983-996 (1957)

[2062] Witte, H. & Wölfel, E. Electron distributions in NaCl, LiF, CaF_2, and Al. *Rev. Mod. Phys.* **30**, 51-55 (1958)

[2063] Gourary, B.S. & Adrian, F.J. Wave functions for electron-excess color centers in alkali halide crystals. *Solid State Phys.* **10**, 127-247 (1960)

[2064] Stockar, K. Zur Kenntnis der Radien positiver Atomionen. *Helv. Chim. Acta* **33**, 1409-1420 (1950)

[2065] Rosseinsky, D.R. Ionic radii from a continuum model for alkali-metal halide lattices. *J. Chem. Soc. A* **(4)**, 608-610 (1971)

[2066] Batsanov, S.S. Determination of ionic radii from metal compressibilities. *J. Struct. Chem.* **45**, 896-899 (2004)

[2067] Philipps, J.C. Dielectric definition of electronegativity. *Phys. Rev. Lett.* **20**, 550-553 (1968)

[2068] Philipps, J.C. & van Vechten, J.A. Dielectric classification of crystal structures, ionization potentials, and band structures. *Phys. Rev. Lett.* **22**, 705-708 (1969)

[2069] van Vechten, J.A. Quantum dielectric theory of electronegativity in covalent systems. I. Electronic dielectric constant. *Phys. Rev.* **182**, 891-905 (1969)

[2070] Stokes, G.G. On the effect of the internal friction of fluids on the motion of pendulums. *Philos. Mag.* **1**, 337-339 (1851)

[2071] Stokes, G.G. On the effect of the internal friction of fluids on the motion of pendulums. *Trans. Cambridge Philos. Soc.* **9**, 8-106 (1856)

[2072] Sutherland, W. A dynamical theory of diffusion for non-electrolytes and the molecular mass of albumin. *Philos. Mag.* **9**, 781-785 (1905)

[2073] Einstein, A. Über die von der molekularkinetischen Theorie der Wärme geförderte Bewegung von in ruhenden Flüssigkeiten suspendierten Teilchen. *Ann. Phys.* **322**, 549-560 (1905)

[2074] Nightingale Jr., E.R. Phenomenological theory of ion solvation. Effective radii of hydrated ions. *J. Phys. Chem.* **63**, 1381- (1959)

[2075] Vieillard, P. Une nouvelle echelle des rayons ioniques de Pauling. *Acta Crystallogr. B* **43**, 513-517 (1987)

[2076] Ignatiev, V.D. Sizes of atoms and ions and covalency of bonding in molecules and crystals. *J. Struct. Chem.* **46**, 744-751 (2005)

[2077] Ignatiev, V. Relation between interatomic distances and sizes of ions in molecules and crystals. *Acta Crystallogr. B* **58**, 770-779 (2002)

[2078] Ignatiev, V. Interpermeable atoms in homonuclear diatomic molecules. *J. Mol. Struct. (Theochem)* **819**, 102-108 (2007)

[2079] Waber, J.T. & Cromer, D.T. Orbital radii of atoms and ions. *J. Chem. Phys.* **42**, 4116-4123 (1965)

[2080] Sen, K.D. & Politzer, P. Characteristic features of the electrostatic potentials of singly negative monoatomic ions. *J. Chem. Phys.* **90**, 4370-4372 (1989)

[2081] Ghosh, D.C. & Biswas, R. Theoretical calculation of absolute radii of atoms and ions. Part 2. The ionic radii. *Int. J. Mol. Sci.* **4**, 379-407 (2003)

[2082] Barrera, M. & Zuloagat, F. Determination of the ionic radii by means of the Kohn-Sham potential: Identification of the chemical potential. *Int. J. Quant. Chem.* **106**, 2044-2053 (2006)

[2083] Hirschfelder, J.O., Curtiss, C.F. & Bird, R.B. Molecular theory of gases and liquids. John Wiley & Sons, Inc., New York, USA (1954)

[2084] Grant, I.P. Relativistic calculation of atomic structures. *Adv. Phys.* **19**, 747-811 (1970)

[2085] Slater, J.C. Atomic radii in crystals. *J. Chem. Phys.* **41**, 3199-3204 (1964)

[2086] Liberman, D., Waber, J.T. & Cromer, D.T. Self-consistent-field Dirac-Slater wave functions for atoms and ions. I. Comparison with previous calculations. *Phys. Rev.* **137**, A27-A34 (1965)

[2087] Gilbert, T.L. Soft-sphere model for closed-shell atoms and ions. *J. Chem. Phys.* **49**, 2640-2642 (1968)

[2088] Barrera, M. & Zuloaga, F. Electronic properties of atoms and covalent radius determined by means of an exchange potential new model containing self interaction and gradient corrections. *J. Chilean Chem. Soc.* **48**, 31-37 (2003)

[2089] van Leeuwen, R. & Baerends, E.J. Exchange-correlation potential with correct asymptotic behaviour. *Phys. Rev. A* **49**, 2421-2431 (1994)

[2090] Perdew, J.P. & Zunger, A. Self-interaction correction to density functional approximations for many-electron systems. *Phys. Rev. B* **23**, 5048-5079 (1981)

[2091] Rittner, E.S. Binding energy and dipole moment of alkali halide molecules. *J. Chem. Phys.* **19**, 1030-1035 (1951)

[2092] Brumer, P. & Karplus, M. Perturbation theory and ionic models for alkali-halide systems. 1. Diatomics. *J. Chem. Phys.* **58**, 3903-3918 (1973)

[2093] Jia, Y.Q. Crystal radii and effective ionic radii of the rare earth ions. *J. Solid State Chem.* **95**, 184-187 (1991)

[2094] Johnson, O. Ionic radii for spherical potential ions. 2. Radii for rare-earth, actinide, transition metal and D10 cations. *Chemica Scripta* **7**, 5-10 (1975)

[2095] Goyal, S.C. & Shanker, J. Calculation of ionic radii in silver-halide and alkaline-earth oxide crystals. *Current Science* **41**, 891-891 (1972)

[2096] Whittaker, J.W. & Muntus, R. Ionic radii for use in geochemistry. *Geochim. Cosmochim. Acta* **34**, 945-956 (1970)

[2097] Boswarva, I.M. Ionic radii in alkaline earth chalcogenides. *J. Phys. C: Solid State Physics* **1**, 582-585 (1968)

[2098] Geller, S. & Mitchell, D.W. Rare earth ion radii in the iron garnets. *Acta Crystallogr.* **12**, 936-936 (1959)

[2099] Salvi, G.R. & de Bethune, A.J. The temperature coefficients of electrode potentials. 2. The second isothermal temperature coefficient. *J. Electrochem. Soc.* **108**, 672-676 (1961)

[2100] Giovanelli, D., Lawrence, N.S. & Compton, R.G. Electrochemistry at high pressures: A review. *Electroanalysis* **16**, 789-810 (2004)

[2101] Randall, M. & Rossini, F.D. Heat capacities in aqueous salt solutions. *J. Am. Chem. Soc.* **51**, 323-345 (1929)

[2102] Gucker, F.T. & Schminke, K.H. A study of the heat capacity and related thermodynamic properties of aqueous solutions of lithium chloride, hydrochloric acid and potassium hydroxide at 25°C. *J. Am. Chem. Soc.* **54**, 1358-1373 (1932)

[2103] Fortier, J.-L., Leduc, P.-A. & Desnoyers, J.E. Thermodynamic properties of alkali halides. II. Enthalpies of dilution and heat capacities in water at 25°C. *J. Solut. Chem.* **3**, 323-349 (1974)

[2104] Desnoyers, J.E., de Visser, C., Perron, G. & Picker, P. Reexamination of the heat capacities obtained by flow micrometry. Recommendation for the use of a chemical standard. *J. Solut. Chem.* **5**, 605-616 (1976)

[2105] Hedwig, G.R. & Hakin, A.W. Partial molar volumes and heat capacities of single ions in aqueous solution over the temperature range 288.15 to 328.15 K. *Phys. Chem. Chem. Phys.* **6**, 4690-4700 (2004)

[2106] Geffcken, W. & Price, D. Zur Frage der Konzentrationsabhängigkeit des scheinbaren Molvolumens und der scheinbaren Molrefraktion in verdünnten Lösungen. *Z. Phys. Chem. B* **26**, 81-99 (1934)

[2107] Gibson, R.E. & Loeffler, O.H. Pressure-volume-temperature relations in solutions. IV. The apparent volumes and thermal expansibilities of sodium chloride and sodium bromide in aqueous solutions between 25 and 95°C. *J. Am. Chem. Soc.* **63**, 443-449 (1941)

[2108] Owen, B.B. & Brinkley Jr., S.R. Calculation of the effect of pressure upon ionic equilibria in pure water and in salt solutions. *Chem. Rev.* **29**, 461-474 (1941)

[2109] MacInnes, D.A. & Dayhoff, M.O. The partial molal volumes of potassium chloride, potassium and sodium iodides and of iodine in aqueous solution at 25°C. *J. Am. Chem. Soc.* **74**, 1017-1020 (1952)

[2110] Zen, E.-A. Partial molar volumes of some salts in aqueous solution. *Geochim. Cosmochim. Acta* **12**, 103-122 (1956)

[2111] Robertson, R.E., Sugamori, S.E., Tse, R. & Wu, C.-Y. The solvent isotope effect and the partial molar volume of ions. *Can. J. Chem.* **44**, 487-494 (1966)

[2112] Vaslow, F. The apparent molal volumes of the alkali metal chlorides in aqueous solution and evidence for salt-induced structure transitions. *J. Phys. Chem.* **70**, 2286-2294 (1966)

[2113] Millero, F.J. Apparent molal volumes of sodium fluoride in aqueous solutions at 25°C *J. Phys. Chem.* **71**, 4567-4569 (1967)

[2114] Dunn, L.A. Apparent molar volumes of electrolytes. Part 2. Some 1-1 electrolytes in aqueous solution at 25°C. *Trans. Farad. Soc.* **64**, 1898-1903 (1968)

[2115] Millero, F.J. The partial molal volumes of tetraphenylarsonium tetraphenylboron in water at infinite dilution. Ionic partial molal volumes. *J. Phys. Chem.* **75**, 280-282 (1971)

[2116] Perron, G., Fortier, J.-L. & Desnoyers, J.E. The apparent molar heat capacities and volumes of aqueous NaCl from 0.01 to 3 mol kg^{-1} in the temperature range 274.65 to 318.15 K. *J. Chem. Thermodyn.* **7**, 1177-1184 (1975)

[2117] Desnoyers, J.E. & Arel, M. Apparent molal volumes of n-alkylamine hydrobromides in water at 25°C: Hydrophobic hydration and volume changes. *Can. J. Chem.* **45**, 359-366 (1967)

[2118] Millero, F.J. High precision magnetic float densimeter. *Rev. Sci. Instrum.* **38**, 1441-1444 (1967)

[2119] Redlich, O. & Rosenfeld, P. Das partielle molare Volumen von gelösten Elektrolyten. I. *Z. Phys. Chem. A* **155**, 65-74 (1931)

[2120] Redlich, O. & Rosenfeld, P. Zur Theorie des Molvolumens gelöster Elektrolyte. II. *Z.*

Elektrochem. **37**, 705-711 (1931)
[2121] Zana, R. & Yeager, E. Determination of ionic partial molal volumes from ionic vibration potentials. *J. Phys. Chem.* **70**, 954-955 (1966)
[2122] Zana, R. & Yeager, E. Ultrasonic vibration potentials and their use in the determination of ionic partial molal volumes. *J. Phys. Chem.* **71**, 521-536 (1967)
[2123] Garnsey, R., Boe, R.J., Mahoney, R. & Litovitz, T.A. Determination of electrolyte apparent molal compressibilities at infinite dilution using a high-precision ultrasonic velocimeter. *J. Chem. Phys.* **50**, 5222-5228 (1969)
[2124] Gucker, F.T. The apparent molal heat capacity, volume, and compressibility of electrolytes. *Chem. Rev.* **13**, 111-130 (1933)
[2125] Owen, B.B. & Simons, L. Standard partial molal compressibilities by ultrasonics. I. Sodium chloride and potassium chloride at 25°C. *J. Phys. Chem.* **61**, 479-482 (1957)
[2126] Allam, D.S. & Lee, W.H. Ultrasonic studies of electrolyte solutions. Part II. Compressibilities of electrolytes. *J. Chem. Soc. A* **0**, 5-9 (1966)
[2127] Mathieson, J.G. & Conway, B.E. Partial molal compressibilities of salts in aqueous solution and assignment of ionic contributions. *J. Solut. Chem.* **3**, 455-477 (1974)
[2128] Millero, F.J., Ward, G.K., Lepple, F.K. & Hoff, E.V. Isothermal compressibility of aqueous sodium chloride, magnesium chloride, sodium sulfate, and magnesium sulfate solutions from 0 to 45°C at 1 atm. *J. Phys. Chem.* **78**, 1636-1643 (1974)
[2129] Millero, F.J. & Drost-Hansen, W. Apparent molal volumes of aqueous monovalent salt solutions at various temperatures. *J. Chem. Engin. Data* **13**, 330-333 (1968)
[2130] Sidgwick, N.V. The chemical elements and their compounds. Volume 1. Oxford Univ. Press, Oxford, UK (1950)
[2131] Ralchenko, Y., Kramida, A.E., Reader, J., NIST atomic spectra database version 3.15. Available at: *http://physics.nist.gov/asd3* (2008)
[2132] Radziemski, L.J. & Kaufman, V. Wavelengths, energy levels, and analysis of neutral atomic chlorine (Cl I). *J. Opt. Soc. Am.* **59**, 424-442 (1969)
[2133] Moore, C.E. Atomic energy levels as derived from the analyses of optical spectra. 1H to 23V (Vol I). *National Standard Reference Data Series* **35**, 1-358 (1971)
[2134] Moore, C.E. Atomic energy levels as derived from the analyses of optical spectra. 24Cr to 41Nb (Vol II). *National Standard Reference Data Series* **35**, 1-263 (1971)
[2135] Moore, C.E. Atomic energy levels as derived from the analyses of optical spectra. 42Mo to 57La, 72Hf to 89Ac (Vol III). *National Standard Reference Data Series* **35**, 1-289 (1971)
[2136] Laguna, G.A. & Beattie, W.H. Direct absorption measurement of the spin-orbit splitting of the ground state in atomic fluorine. *Chem. Phys. Lett.* **88**, 439-440 (1982)
[2137] Uehara, H. & Horiai, K. Infrared-microwave double resonance of atomic chlorine on laser-magnetic-resonance fine-structure transitions. *J. Opt. Soc. Am. B* **4**, 1217-1221 (1987)
[2138] Langmuir, I. Vapor pressure of metallic tungsten. *Phys. Rev.* **2**, 329-342 (1913)
[2139] Knudsen, M. Kinetic theory of gases. Methuen, London, UK (1934)
[2140] Hertz, H. Uber die Verdunstung der Flüssigkeiten, insbesondere des Quecksilbers, im luftleeren Raume. *Ann. Phys.* **253**, 177-193 (1882)
[2141] Langmuir, I. Chemical reactions at very low pressures: II. *J. Am. Chem. Soc.* **35**, 931-945 (1913)
[2142] Knudsen, M. Die maximale Verdampfungsgeschwindigkeit des Quecksilbers. *Ann. Phys.* **352**, 697-708 (1915)
[2143] Smirnov, B.M. Physics of atoms and ions. Edition 1. Springer, New York, USA (2003)
[2144] Martin, W.C., Musgrove, A., Kotochigova, S. & Sansonetti, J.E. NIST Ground levels and ionization energies for the neutral atoms. Version 1.3. Available at: *http://physics.nist.gov/IonEnergy* (2003)
[2145] Born, M. & Landé, A. Über die absolute Berechnung der Kristalleigenschaften mit Hilfe bohrscher Atommodelle. *Ber. Kgl. Preuss. Akad. Wiss. Berlin* **45**, 1048-1068 (1918)
[2146] Born, M. & Landé, A. Die Abstände der Atome im Molekül und im Kristalle. *Naturwiss.* **6**, 496-496 (1918)
[2147] Madelung, E. Das elektrische Feld in Systemen von regelmässig angeordneten Punktladungen. *Phys. Z.* **19**, 524-533 (1919)
[2148] Jenkins, H.D.B. Thermodynamics of the relationship between lattice energy and lattice enthalpy. *J. Chem. Edu.* **82**, 950-952 (2005)
[2149] Barron, T.H.K., Berg, W.T. & Morrison, J.A. Specific heat of LiCl at low temperatures. *Proc. Roy. Soc. London Ser. A* **242**, 478-492 (1957)
[2150] Scales, W.W. Specific heat of LiF and KI at low temperatures. *Phys. Rev.* **112**, 49-54 (1958)
[2151] Moyer, D.F. Specific heat of LiCl at low temperatures. *J. Phys. Chem. Solids* **26**, 1459-1462 (1965)
[2152] Sharko, A.V. & Botaki, A.A. Debye temperature of alkali halide crystals. *Russ. J. Phys.* **14**, 360-364 (1971)
[2153] Jalenti, R. & Caramazza, R. Thermodynamic functions of hydration of alkali metal and

halide ions. *J. Chem. Soc. Farad. Trans. I* **72**, 715-722 (1976)
[2154] Mayer. J.E., Helmholz, Die Gitterenergie der Alkalihalogenide und die Elektronenaffinität der Halogene. *Z. Phys. A* **75**, 19-29 (1932)
[2155] Verwey, E.J.W. Absolute grootte van thermodynamische grootheden voor ionen in waterige oplossing en van den elektrischen potentialsprong aan het vrije oppervlak van water. *Chem. Weekblad* **36**, 530-535 (1940)
[2156] Eucken, A. Ionenhydrate in wässriger Lösung. *Z. Elektrochem.* **52**, 6-24 (1948)
[2157] Scott, A.F. The apparent volumes of salts in solutions. I. The test of the empirical rule of Masson. *J. Phys. Chem.* **35**, 2315-2329 (1931)
[2158] Millero, F.J. The apparent and partial molal volume of aqueous sodium chloride solutions at various temperatures. *J. Phys. Chem.* **74**, 356-362 (1970)
[2159] Dack, M.R.J., Bird, K.J. & Parker, A.J. Solvation of ions. XXV. Partial molal volumes of single ions in protic and dipolar aprotic solvents. *Aust. J. Chem.* **28**, 955-963 (1975)
[2160] Singh, P.P. Partial molar volumes of some aqueous electrolytes and a transition model. *J. Am. Chem. Soc.* **100**, 681-684 (1978)
[2161] Monnin, C. An ion interaction model for the volumetric properties of natural waters: Density of the solution and partial molal volumes of electrolytes to high concentrations at 25°C. *Geochim. Cosmochim. Acta* **53**, 1177-1188 (1989)
[2162] Jauch, K. Die spezifische Wärme wässeriger Salzlösungen. *Z. Physik* **4**, 441-447 (1921)
[2163] Millero, F.J. The molal volumes of electrolytes. *Chem. Rev.* **71**, 147-176 (1971)
[2164] Eastman, D.E. Photoelectric work functions of transition, rare-earth, and noble metals. *Phys. Rev. B* **2**, 1-2 (1970)
[2165] Shan, B. & Cho, K.J. First principles study of work functions of single wall carbon nanotubes. *Phys. Rev. Lett.* **94**, 236602/1-236602/4 (2005)
[2166] Su, W.S., Leung, T.C. & Chan, C.T. Work function of single-walled and multiwalled carbon nanotubes: First-principles study. *Phys. Rev. B* **76**, 235413/1-235413/8 (2007)
[2167] Lohmann, Z. Fermi-Niveau und Flachbandpotential von Molekülkristallen aromatischer Kohlenwasserstoffe. *Z. Naturforsch. A* **22**, 843-844 (1967)
[2168] Gurevich, Y.Y., Determination of the electrochemical potential and reorganization energy in a solution containing a redox system. *Elektrokhimiya* **18**, 1315-1320 (1982)
[2169] Frumkin, A.N., Petrii, O.A. & Damaskin, B. Notion of electrode charge and Lippmann equation. *J. Electroanal. Chem.* **27**, 81-100 (1970)
[2170] Frumkin, A.N. & Petrii, O.A. Potentials of zero total and zero free charge of platinum group metals. *Electrochim. Acta* **20**, 347-359 (1975)
[2171] Frumkin, A.N., Petrii, O.A. & Kolotyrkinasafonova, T.Y. Effect of solution pH on state of platinized platinum-electrode. *Dokl. Akad. Nauk. SSSR* **222**, 1159-1162 (1975)
[2172] Latimer, W.M. The energy of solution of gaseous ions in relation to the effect of a charge upon the dielectric. *J. Chem. Phys.* **48**, 1234-1239 (1926)
[2173] Garrison, A. A determination of absolute single electrode potentials. *J. Am. Chem. Soc.* **45**, 37-44 (1923)
[2174] Randles, J.E.B. & Schiffrin, D.J. The temperature-dependence of the surface potential of aqueous electrolytes. *J. Electroanal. Chem.* **10**, 480-484 (1965)
[2175] Rossini, F.D., Wagman, D.D., Evans, S.L. & Jaffe, I. Selected values of chemical thermodynamic properties. Circular of the National Bureau of Standards 500. US Government Printing Office, Washington D.C., USA (1952)
[2176] Moore, C.E. Atomic energy levels (hydrogen through vanadium). Circular of the National Bureau of Standards 467. Volume 1. US Government Printing Office, Washington D.C., USA (1949)
[2177] Moore, C.E. Atomic energy levels (chromium through niobium). Circular of the National Bureau of Standards 467. Volume 2. US Government Printing Office, Washington D.C., USA (1952)
[2178] Moore, C.E. Atomic energy levels (molybdenum through lanthanum and hafnium through actinium). Circular of the National Bureau of Standards 467. Volume 3. US Government Printing Office, Washington D.C., USA (1958)
[2179] Stull, D.R. & Prophet, H. JANAF thermochemical tables (2nd ed.; Supplements in 1974, 1975, 1978). *National Standard Reference Data Series* **37**, 1-1141 (1971)
[2180] CODATA recommended key values for thermodynamics 1977. *CODATA Bulletin* **28**, 1-17 (1978)
[2181] Lide, D.R. CRC Handbook of Chemistry and Physics. Edition 80. CRC Press, Boca Raton, Florida (1999)
[2182] Blandamer, M.J. & Symons, M.C. Significance of new values for ionic radii to solvation phenomena in aqueous solution. *J. Phys. Chem.* **67**, 1304-1306 (1963)
[2183] Jenkins, H.D.B. & Pritchett, M.S.F. A new approach to the analysis of absolute free energies, enthalpies and entropies of hydration of individual gaseous ions and absolute single-ion viscosity *B*-coefficients. *J. Chem. Soc. Faraday Trans. 1* **80**, 721-737 (1984)
[2184] Fawcett, W.R. The solvent dependence of ionic properties in solution in the limit of infinite dilution. *Mol. Phys.* **95**, 507-514 (1998)
[2185] Marcus, Y. The thermodynamics of solvation of ions. Part 2. The enthalpy of hydration

at 298.15 K. *J. Chem. Soc. Farad. Trans. I* **83**, 339-349 (1987)
[2186] Marcus, Y. Thermodynamics of ion hydration and its interpretation in terms of a common model. *Pure Appl. Chem.* **59**, 1093-1101 (1987)
[2187] Marcus, Y. The thermodynamics of solvation of ions. Part 4. Application of the tetraphenylarsonium tetraphenylborate (TATB) assumption to the hydration of ions and to properties of hydrated ions. *J. Chem. Soc. Farad. Trans. I* **83**, 2985-2992 (1987)
[2188] Coulter, L.V. The thermochemistry of the alkali and alkaline earth metals and halides in liquid ammonia at -33°. *J. Phys. Chem.* **57**, 553-558 (1953)
[2189] Frank, H.S. & Evans, M.W. Free volume and entropy in condensed systems. III. Entropy in binary liquid mixtures; partial molar entropy in dilute solutions; structure and thermodynamics in aqueous electrolytes. *J. Chem. Phys.* **13**, 507-532 (1945)
[2190] Bhattacharyya, M.M. Partitioning of entropy of solvation: Single-ion entropy of solvation in aqueous and non-aqueous solvents and the correlation of electrostatic entropy of solvation with electrostriction viscosity B and NMR B' coefficients. *J. Chem. Soc. Farad. Trans.* **91**, 3373-3378 (1995)
[2191] Verwey, E.J.W. & de Boer, J.H. Molecular energy of alkali halides. *Rec. Trav. Chim. Pays-Bas* **55**, 431-443 (1936)
[2192] Latimer, W.M. & Buffington, R.M. The entropy of aqueous ions. *J. Am. Chem. Soc.* **48**, 2297-2305 (1926)
[2193] Ulich, H. Ionenentropie und Solvatation. *Z. Elektrochem.* **36**, 497-506 (1930)
[2194] Latimer, W.M., Schutz, P.W. & Hicks Jr., J.F.G. A summary of the entropies of aqueous ions. *J. Chem. Phys.* **2**, 82-84 (1934)
[2195] Latimer, W.M., Pitzer, K.S. & Smith, W.V. The entropies of aqueous ions. *J. Am. Chem. Soc.* **60**, 1829-1831 (1938)
[2196] Criss, C.M. & Cobble, J.W. The thermodynamic properties of high temperature aqueous solutions. V. The calculation of ionic heat capacities up to 200°. Entropies and heat capacities above 200°. *J. Am. Chem. Soc.* **86**, 5390-5393 (1964)
[2197] Franks, F. & Reid, D.S. Ionic solvation entropies in mixed aqueous solvents. *J. Phys. Chem.* **73**, 3152-3154 (1969)
[2198] Eastman, E.D. Electromotive force of the electrolytic thermocouples and thermocells and the entropy of transfer and absolute entropy of ions. *J. Am. Chem. Soc.* **50**, 292-297 (1928)
[2199] Bernhardt, H.A. & Crockford, H.D. The determination of the entropy of the chloride ion. *J. Phys. Chem.* **46**, 473-476 (1942)
[2200] Crockford, H.D. & Hall, J.L. The entropy of the chloride ion at 12.5°C. *J. Phys. Chem.* **54**, 731-734 (1950)
[2201] Jolicoeur, C. & Mercier, J.C. An ionic scale for the partial molal heat capacities of aqueous electrolytes from chemical models. *J. Phys. Chem.* **81**, 1119-1121 (1977)
[2202] Abraham, M.H. & Marcus, Y. The thermodynamics of solvation of ions. Part 1. - The heat capacity of hydration at 298.15 K. *J. Chem. Soc. Faraday Trans. I* **82**, 3255-3274 (1986)
[2203] Shin, C., Worley, I. & Criss, C.M. Partial molal heat capacities of aqueous tetraalkylammonium bromides as functions of temperature. *J. Solut. Chem.* **5**, 867-879 (1980)
[2204] French, R.N. & Criss, C.M. Effect of charge on the standard partial molar volumes and heat capacities of organic electrolytes in methanol and water. *J. Solut. Chem.* **11**, 625-648 (1982)
[2205] Fajans, K. & Johnson, O. Apparent volumes of individual ions in aqueous solution. *J. Am. Chem. Soc.* **64**, 668-678 (1942)
[2206] Couture, A.M. & Laidler, K.J. The partial molal volumes of ions in aqueous solution. I. Dependence on charge and radius. *Can. J. Chem.* **34**, 1209-1216 (1956)
[2207] Stokes, R.H. & Robinson, R.A. The application of volume fraction statistics to the calculation of activities of hydrated electrolytes. *Trans. Faraday Soc.* **53**, 301-304 (1957)
[2208] Mukerjee, P. On ion-solvent interactions. Part II. Internal pressure and electrostriction of aqueous solutions of electrolytes. *J. Phys. Chem.* **65**, 744-746 (1961)
[2209] Padova, J. Solvation approach to ion-solvent interaction. *J. Chem. Phys.* **40**, 691-694 (1964)
[2210] Millero, F.J. & Drost-Hansen, W. Apparent molal volumes of ammonium chloride and some symmetrical tetraalkylammonium chlorides at various temperatures. *J. Phys. Chem.* **72**, 1758-1763 (1968)
[2211] Jákli, G. Evaluation of individual ionic partial molar volumes in aqueous solutions. *J. Chem. Thermodyn.* **40**, 770-776 (2008)
[2212] Glueckauf, E. Molar volumes of ions. *Trans. Faraday Soc.* **61**, 914-921 (1965)
[2213] Conway, B.E., Verrall, R.E. & Desnoyers, J.E. Partial molal volumes of tetraalkylammonium halides and assignment of individual ionic contributions. *Trans. Faraday Soc.* **62**, 2738-2749 (1966)
[2214] Glueckauf, E. Molar volumes of ions. *Trans. Faraday Soc.* **64**, 2423-2432 (1968)
[2215] Krumgalz, B.S. Ionic limiting partial molal volumes in various solvents. *J. Chem. Soc. Faraday I* **76**, 1887-1904 (1980)

[2216] Laliberté, L.H. & Conway, B.E. Solute and solvent structure effects in volumes and compressibilities of organic ions in solution. *J. Phys. Chem.* **74**, 4116-4125 (1970)

[2217] Millero, F.J. Apparent molal expansibilities of some divalent chlorides in aqueous solution at 25°C. *J. Phys. Chem.* **72**, 4589-4593 (1968)

[2218] Millero, F.J. The partial molal volumes of electrolytes in aqueous solutions. In: *Water and aqueous solutions: Structure, thermodynamics, and transport processes.* Horne, R.A., Ed. Wiley-Interscience, New York, USA, pp 519-595 (1972)

[2219] de Ligny, C.L., Alfenaar, M. & van der Veen, N.G. The standard chemical free enthalpy, enthalpy, entropy and heat capacity of hydration of the hydrogen ion, and the surface potential of water at 25°*C*. *Rec. Trav. Chim. Pays-Bas* **87**, 585-598 (1968)

[2220] Zhou, Y., Stell, G. & Friedman, H.L. Note on the standard free energy of transfer and partitioning of ionic species between two fluid phases. *J. Chem. Phys.* **89**, 3836-3839 (1988)

[2221] Pratt, L.R. Contact potentials of solution interfaces: Phase equilibrium and interfacial electric fields. *J. Phys. Chem.* **96**, 25-33 (1992)

[2222] Frumkin, A.N. Note on B. Kamienski's paper "The nature of the electric potential at the free surface of aqueous solutions". *Electrochim. Acta* **2**, 351-354 (1960)

[2223] Jakuszewski, B., Partyka, S. & Przasnyski, M. Zero charge potentials of mercury in ethanol-water mixtures. *Rocz. Chém.* **46**, 921-927 (1972)

[2224] Kochurova, N.N. & Rusanov, A.I. Dynamic surface properties of water: Surface tension and surface potential. *J. Colloid. Interface Sci.* **81**, 297-303 (1981)

[2225] Chalmers, M.A. & Pasquille, F. The potential difference at an air-water interface. *Philos. Mag.* **23**, 88-96 (1937)

[2226] Frumkin, A.N. Calculation of the absolute potential of the normal calomel electrode from the free energy of hydration of gaseous ions. *J. Chem. Phys.* **7**, 552-553 (1939)

[2227] Croxton, C.A. Molecular orientation and interfacial properties of liquid water. *Physica A* **106**, 239-259 (1981)

[2228] Izmailov, N.A. A new method of determining energy and solvation heat of individual ions. *Dokl. Akad. Nauk. SSSR* **149**, 884-887 (1963)

[2229] Conway, B.E. & Barradas R.G., Eds. Chemical physics of ionic solutions. Wiley, New York, USA (1966)

[2230] Conway, B.E.; ed. Bockris, J.O'M., Conway, Modern aspects of electrochemistry. Volume 3. Butterworths Scientific Publications Ltd., London, UK (1964)

[2231] Parfenyuk, V.I. & Chankina, T.I. Estimation of ion contribution to the value of the surface potential of organic solvents. *Zh. Fiz. Khim.* **71**, 547-550 (1997)

[2232] Buch, V., Milet, A., Vácha, R., Jungwirth, P. & Devlin, J.P. Water surface is acidic. *Proc. Natl. Acad. Sci. USA* **104**, 7342-7347 (2007)

[2233] Vácha, R., Buch, V., Milet, A. & Jungwirth, P. Autoionization at the surface of neat water: is the top layer pH neutral, basic, or acidic? *Phys. Chem. Chem. Phys.* **9**, 4736-4747 (2007)

[2234] Iuchi, S., Chen, H., Paesani, F. & Voth, G.A. Hydrated excess proton at water-hydrophobic interfaces. *J. Phys. Chem. B* **113**, 4017-4030 (2009)

[2235] Mundy, C.J., Kuo, I.-F.W., Tuckerman, M.E., Lee, H.-S. & Tobias, D.J. Hydroxide anion at the air-water interface. *Chem. Phys. Lett.* **481**, 2-8 (2009)

[2236] Petersen, P.B. & Saykally, R.J. Is the liquid water surface basic or acidic? Macroscopic *vs.* molecular-scale investigations. *Chem. Phys. Lett.* **458**, 255-261 (2008)

[2237] Zimmermann, R., Freudenberg, U., Schweiss, R., Küttner, D. & Werner, C. Hydroxide and hydronium ion adsorption. A survey. *Curr. Opin. Coll. Interface Sci.* **15**, 196-202 (2010)

[2238] Winter, B., Faubel, M., Vácha, R. & Jungwirth, P. Reply to comments on "Behavior of hydroxide at the water/vapor interface" [Chem. Phys. Lett. 474 (2009) 241]. *Chem. Phys. Lett.* **481**, 19-21 (2009)

[2239] Winter, B., Faubel, M., Vácha, R. & Jungwirth, P. Behavior of hydroxide at the water/vapor interface. *Chem. Phys. Lett.* **481**, 241-247 (2009)

[2240] Wick, C.D. & Dang, L.X. The behavior of NaOH at the air-water interface: A computational study. *J. Chem. Phys.* **133**, 024705/1-024705/8 (2010)

[2241] Beattie, J.K. Comment on "Behavior of hydroxide at the water/vapor interface" [Chem. Phys. Lett. 474 (2009) 241]. *Chem. Phys. Lett.* **481**, 17-18 (2009)

[2242] Beattie, J.K., Djerdjev, A.M. & Warr, G.G. The surface of neat water is basic. *Faraday Discuss.* **141**, 31-39 (2009)

[2243] Gray-Weale, A. Comment on "Behavior of hydroxide at the water/vapor interface" [Chem. Phys. Lett. 474 (2009) 241]. *Chem. Phys. Lett.* **481**, 22-24 (2009)

[2244] Gray-Weale, A. & Beattie, J.K. An explanation for the charge on water's surface. *Phys. Chem. Chem. Phys.* **11**, 10994-11005 (2009)

[2245] Enami, S., Hoffmann, M.R. & Colussi, A.J. Proton availability at the air/water interface. *J. Phys. Chem. Lett.* **1**, 1599-1604 (2010)

[2246] Fawcett, W.R. Acidity and basicity scales for polar solvents. *J. Phys. Chem.* **97**, 9540-9546 (1993)

[2247] Marcus, Y., Hefter, G. & Chen, T. Application of the tetraphenylarsonium tetraphenylborate (TATB) assumption to the hydration entropies of ions. *J. Chem. Thermodyn.* **32**, 639-649 (2000)

[2248] Pliego, J.R. & Riveros, J.M. New values for the absolute solvation free energy of univalent ions in aqueous solution. *Chem. Phys. Lett.* **332**, 597-602 (2000)

[2249] Fawcett, W.R. & Blum, L. The role of dipole-dipole interactions in the solvation of monoatomic monovalent ions in water on the basis of the mean spherical approximation. *J. Electroanal. Chem.* **355**, 253-263 (1993)

[2250] Florián, J. & Warshel, A. Calculations of hydration entropies of hydrophobic, polar, and ionic solutes in the framework of the Langevin dipoles solvation model. *J. Phys. Chem. B* **103**, 10282-10288 (1999)

[2251] Blum, L. & Fawcett, W.R. Application of the mean spherical approximation to describe the Gibbs solvation energies of monovalent monoatomic ions in polar solvents. *J. Phys. Chem.* **96**, 408-414 (1992)

[2252] Fawcett, W.R. & Blum, L. Application of the mean spherical approximation to describe the entropy of solvation of spherical ions in polar solvents. *J. Chem. Soc. Farad. Trans.* **88**, 3339-3344 (1992)

[2253] Azzam, A.M. Studies on ionic solvation. Part III. New theory for calculating heats of hydration of monovalent ions at 25°C. *Can. J. Chem.* **38**, 2203-2216 (1960)

[2254] Azzam, A.M. Über eine neue Theorie zur Berechnung der Ionensolvatation. *Z. Elektrochem.* **58**, 889-899 (1954)

[2255] Pauling, L. The nature of the chemical bond and the structure of molecules and crystals. Edition 1. Cornell University Press, Ithaca NY, USA (1940)

[2256] Saluja, P.P.S Environment of ions in aqueous solutions. In: *International review of science. Electrochemistry. Volume 6.* Butterworths, London, UK, pp 1-51 (1976)

[2257] Parsons, R. The single electrode potential: Its significance and calculation. In: *Standard potentials in aqueous solution.* Bard, A.J., Parsons, R. & Jordan, J., Eds. Dekker, New York, USA, pp 13-38 (1985)

[2258] Lister, M.W., Nyburg, S.C. & Poyntz, R.B. Absolute enthalpy of hydration of the proton using data for doubly-charged complex ions. *J. Chem. Soc. Farad. Trans. I* **70**, 685-693 (1974)

[2259] Tuttle Jr., T.R., Malaxos, S. & Coe, J.V. A new cluster pair method of determining absolute single ion solvation free energies demonstrated in water and applied to ammonia. *J. Phys. Chem. A* **106**, 925-932 (2002)

[2260] Lee, N., Keesee, R.G. & Castelman Jr., A.W. On the correlation of the total and partial enthalpies of ion solvation and the relationship to the energy barrier of nucleation. *J. Colloid. Interface Sci.* **75**, 555-565 (1980)

[2261] Krestov, G.A. Thermodynamics of solvation: Solution and dissolution, ions and solvents, structure and energetics. Ellis Horwood, New York, US (1991)

[2262] Lin, J. & Breck, W.G. Entropy differences for some related pairs of complex ions. *Can. J. Chem.* **43**, 766-771 (1965)

[2263] Jones, G. & Dole, M. The viscosity of aqueous solutions of strong electrolytes with special reference to barium chloride. *J. Am. Chem. Soc.* **51**, 2950-2964 (1929)

[2264] Cox, W.M. & Wolfenden, J.H. The viscosity of strong electrolytes measured by a differential method. *Proc. Roy. Soc. Lond. A* **145**, 475-488 (1934)

[2265] Lange, E. & Hesse, T. Experimenteller Nachweis von Überführungswärmen in elektrolytischen Peltier-Wärmen. *Z. Elektrochem.* **39**, 374-384 (1933)

[2266] Young, M.B. Thesis. University of California. (1935)

[2267] Goodrich, J.C., Goyan, F.M., Morse, E.E., Preston, R.C. & Young, M.B. Applications of the Eastman thermocell equation. I. Certain absolute ionic entropies and entropies of transfer of alkali metal and tetraalkylammonium bromides and halides. *J. Am. Chem. Soc.* **72**, 4411-4418 (1950)

[2268] Ikeda, T. Absolute value of the partial molar standard entropy of the hydrogen ion in aqueous solutions. *J. Chem. Phys.* **43**, 3412-3413 (1965)

[2269] Tyrrell, H.J.V. & Hollis, G.L. Thermal diffusion potentials in non-isothermal electrolytic systems. *Trans. Faraday Soc.* **45**, 411-423 (1949)

[2270] Eastman, E.D. Theory of the Soret effect. *J. Am. Chem. Soc.* **50**, 283-291 (1928)

[2271] Kirkwood, J.G. & Buff, F.P. The statistical mechanical theory of solutions. I. *J. Chem. Phys.* **19**, 774-777 (1951)

[2272] Chandler, D. & Andersen, H.C. Optimized cluster expansions for classical fluids. II. Theory of molecular liquids. *J. Chem. Phys.* **57**, 1930-1937 (1972)

[2273] Imai, T., Kinoshita, M. & Hirata, F. Theoretical study for partial molar volume of amino acids in aqueous solution: Implication of ideal fluctuation volume. *J. Chem. Phys.* **112**, 9469-9478 (2000)

[2274] Ohba, M., Kawaizumi, F. & Nomura, H. Effects of molecular conformation on the packing density in the liquid state. 3. Partial molar volume differences of cis and trans conformers at infinite dilution. *J. Phys. Chem.* **96**, 5129-5133 (1992)

[2275] King, E.J. Absolute partial molar ionic volumes. *J. Phys. Chem.* **74**, 4590-4592 (1970)

[2276] Zana, R. & Yeager, E. Ultrasonic vibration potentials in tetraalkylammonium halide solutions. *J. Phys. Chem.* **71**, 4241-4244 (1967)

[2277] Millero, F.J. Addendum to chapter 13. Compilation of the partial molal volumes of electrolytes at infinite dilution, $\overline{V}^{\circ}$, and the apparent molal volume concentration dependence constants, S_V^* and b_V, at various temperatures. In: *Water and aqueous solutions: Structure, thermodynamics, and transport processes.* Horne, R.A., Ed. Wiley-Interscience, New York, USA, pp 565-595 (1972)

[2278] Das, K. Single ionic partial molar volumes and solvation of ions. *J. Solut. Chem.* **18**, 1085-1093 (1989)

[2279] Tawa, G.J., Topol, I.A., Burt, S.K., Caldwell, R.A. & Rashin, A.A. Calculation of the aqueous solvation free energy of the proton. *J. Chem. Phys.* **109**, 4852-4863 (1998)

[2280] Pliego, J.R. & Riveros, J.M. Ab initio study of the hydroxide ion-water clusters: An accurate determination of the thermodynamic properties for the processes $n\mathrm{H_2O} + \mathrm{OH^-} \rightarrow \mathrm{HO^-(H2O)}_n$ (n = 1-4). *J. Chem. Phys.* **112**, 4045-4052 (2000)

[2281] Gurney, R.W. Ions in solution. University Press, Cambridge, UK (1936)

[2282] Baldwin, R.L. How Hofmeister ion interactions affect protein stability. *Biophys. J.* **71**, 2056-2063 (1996)

[2283] Collins, K.D. Ions from the Hofmeister series and osmolytes: effects on proteins in solution and in the crystallization process. *Methods* **34**, 300-311 (2004)

[2284] Kunz, W., Henle, J. & Ninham, B.W. "Zur Lehre von der Wirkung der Salze" (about the science of the effect of salts): Franz Hofmeister's historical papers. *Curr. Opin. Chem. Biol.* **10**, 658-663 (2004)

[2285] Zhang, Y. & Cremer, P.S. Interactions between macromolecules and ions: the Hofmeister series. *Curr. Opin. Chem. Biol.* **10**, 658-663 (2006)

[2286] Vlachy, N., Jagoda-Cwiklik, B., Vácha, R., Touraud, D., Jungwirth, P. & Kunz, W. Hofmeister series and specific interactions of charged headgroups with aqueous ions. *Adv. Colloid. Interface Sci.* **146**, 42-47 (2008)

[2287] Hofmeister, F. Zur Lehre von der Wirkung der Salze. Zweite Mittheilung. Ueber Regelmässigkeiten in der eiweissfällenden Wirkung der Salze und ihre Beziehung zum physiologischen Verhalten derselben. *Arch. Exp. Pathol. Pharmakol.* **24**, 247-260 (1887)

[2288] Hofmeister, F. Zur Lehre von der Wirkung der Salze. Dritte Mittheilung. Ueber die wasserentziehende Wirkung der Salze. *Arch. Exp. Pathol. Pharmakol.* **25**, 1-30 (1888)

[2289] Hofmeister, F. Zur Lehre von der Wirkung der Salze. Fünfte Mittheilung. Untersuchungen über den Quellungsvorgang. *Arch. Exp. Pathol. Pharmakol.* **27**, 395-413 (1890)

[2290] Hofmeister, F. Zur Lehre von der Wirkung der Salze. Sechste Mittheilung. Die Betheiligung gelöster Stoffe an Quellungsvorgängen. *Arch. Exp. Pathol. Pharmakol.* **28**, 210-238 (1891)

[2291] Lewith, S. Zur Lehre von der Wirkung der Salze. Erste Mittheilung. Das Verhalten der Eiweisskörper des Blutserums gegen Salze. *Arch. Exp. Pathol. Pharmakol.* **24**, 1-16 (1887)

[2292] Münzer, E. Zur Lehre von der Wirkung der Salze. Siebte Mittheilung. Die Allgemeinwirkung der Salze. *Arch. Exp. Pathol. Pharmakol.* **41**, 71-96 (1898)

[2293] dos Santos, A.P., Diehl, A. & Levin, Y. Surface tensions, surface potentials, and the Hofmeister series of electrolyte solutions. *Langmuir* **26**, 10778-10783 (2010)

[2294] v. Limbeck, R. Zur Lehre von der Wirkung der Salze. Vierte Mittheilung. Ueber die diuretische Wirkung der Salze. *Arch. Exp. Pathol. Pharmakol.* **25**, 69-86 (1888)

[2295] Omta, A.W., Kropman, M.F., Woutersen, S. & Bakker, H.J. Negligible effect of ions on the hydrogen-bond structure in liquid water. *Science* **301**, 347-349 (2003)

[2296] Omta, A.W., Kropman, M.F., Woutersen, S. & Bakker, H.J. Influence of ions on the hydrogen-bond structure in liquid water. *J. Chem. Phys.* **119**, 12457-12461 (2003)

[2297] Kropman, M.F. & Bakker, H.J. Vibrational relaxation of liquid water in ionic solvation shells. *Chem. Phys. Lett.* **370**, 741-746 (2003)

[2298] Kropman, M.F. & Bakker, H.J. Effect of ions on the vibrational relaxation of liquid water. *J. Am. Chem. Soc.* **126**, 9135-9141 (2004)

[2299] Wachter, W., Kunz, W., Buchner, R. & Hefter, G. Is there an anionic Hofmeister effect on water dynamics? Dielectric spectroscopy of aqueous solutions of NaBr, NaI, NaNO3, NaClO4, and NaSCN. *J. Phys. Chem. A* **109**, 8675-8683 (2005)

[2300] Hart, E.J. & Boag, J.W. Absorption spectrum of the hydrated electron in water and in aqueous solutions. *J. Am. Chem. Soc.* **84**, 4090-4095 (1962)

[2301] Buxton, G.V., Greenstock, C.L., Helman, W.P. & Ross, A.B. Critical review of rate constants for reactions of hydrated electrons, hydrogen atoms and hydroxyl radicals ($\cdot\mathrm{OH}/\cdot\mathrm{O}^{-1}$ in aqueous solution). *J. Phys. Chem. Ref. Data* **17**, 513-886 (1988)

[2302] Larsen, R.E. Does the hydrated electron occupy a cavity? *Science* **329**, 65-69 (2010)

[2303] Schnitker, J. & Rossky, P.J. An electron-water pseudopotential for condensed phase simulation. *J. Chem. Phys.* **86**, 3462-3470 (1987)

[2304] Schnitker, J. & Rossky, P.J. Quantum simulation study of the hydrated electron. *J. Chem. Phys.* **86**, 3471-3485 (1987)

[2305] Rossky, P.J. & Schnitker, J. The hydrated electron: Quantum simulation of structure, spectroscopy, and dynamics. *J. Phys. Chem.* **92**, 4277-4285 (1988)

[2306] Larsen, R.E., Glover, W.J. & Schwartz, B.J. Comment on "An electron-water pseudopotential for condensed phase simulation" [J. Chem. Phys. 86, 3462 (1987)]. *J. Chem. Phys.* **131**, 037101/1-037101/2 (2009)

[2307] Schnitker, J. & Rossky, P.J. Response to "Comment on 'An electron-water pseudopotential for condensed phase simulation' " [J. Chem. Phys. 131, 037101 (2009)]. *J. Chem. Phys.* **131**, 037102/1-037102/1 (2009)

[2308] Tuttle Jr., T.R. & Golden, S. Solvated electrons: what is solvated? *J. Phys. Chem.* **95**, 5725-5736 (1991)

[2309] Sagar, D.M., Bain, C.D. & Verlet, J.R.R. Hydrated electrons at the water/air interface. *J. Am. Chem. Soc.* **132**, 6917-6919 (2010)

[2310] Riniker, S., Kunz, A.-P.E. & van Gunsteren, W.F. On the calculation of the dielectric permittivity of molecular models in the liquid phase. *J. Chem. Theory Comput., submitted* (2011)

[2311] Rashin, A.A. & Bukatin, M.A. Calculation of hydration entropies of alkali and halide ions based on the continuum approach. *J. Phys. Chem.* **97**, 1974-1979 (1993)

[2312] Tolman, R.C. Consideration of the Gibbs theory of surface tension. *J. Chem. Phys.* **16**, 758-774 (1948)

[2313] Tolman, R.C. The superficial density of matter at a liquid-vapor boundary. *J. Chem. Phys.* **17**, 118-127 (1949)

[2314] Buff, F.P. & Kirkwood, J.G. Remarks on the surface tension of small droplets. *J. Chem. Phys.* **18**, 991-992 (1950)

[2315] Buff, F.P. The spherical interface. I. Thermodynamics. *J. Chem. Phys.* **19**, 1591-1594 (1951)

[2316] Buff, F.P. The spherical interface. II. Molecular theory. *J. Chem. Phys.* **23**, 419-427 (1955)

[2317] Kirkwood, J.G. & Buff, F.P. The statistical mechanical theory of surface tension. *J. Chem. Phys.* **17**, 338-343 (1948)

[2318] Kondo, S. Thermodynamical fundamental equation for spherical interface. *J. Chem. Phys.* **25**, 662-669 (1956)

[2319] Iwamatsu, M. The surface tension and Tolman's length of a drop. *J. Phys. Condens. Mat.* **6**, L173-L177 (1994)

[2320] Granasy, L. Semiempirical van der Waals/Cahn-Hilliard theory: Size dependence of the Tolman length. *J. Chem. Phys.* **109**, 9660-9663 (1998)

[2321] Koga, K., Zeng, X.C. & Shchekin, A.K. Validity of Tolman's equation: How large should a droplet be? *J. Chem. Phys.* **109**, 4063-4070 (1998)

[2322] Bykov, T.V. & Zeng, X.C. A patching model for surface tension and the Tolman length. *J. Chem. Phys.* **111**, 3705-3713 (1999)

[2323] Bykov, T.V. & Zeng, X.C. Statistical mechanics of surface tension and Tolman length of dipolar fluids. *J. Phys. Chem. B* **105**, 11586-11594 (2001)

[2324] Bartell, L.S. Tolman's delta, surface curvature, compressibility effects, and the free energy of drops. *J. Phys. Chem. B* **105**, 11615-11618 (2001)

[2325] Lei, Y.A., Bykov, T., Yoo, S. & Zeng, X.C. The Tolman length: Is it positive or negative? *J. Am. Chem. Soc.* **127**, 15346-15347 (2005)

[2326] Blokhuis, E.M. & Kuipers, J. Thermodynamic expressions for the Tolman length. *J. Chem. Phys.* **124**, 074701/1-074701/8 (2006)

[2327] Floris, F.M., Selmi, M., Tani, A. & Tomasi, J. Free energy and entropy for inserting cavities in water: Comparison of Monte Carlo simulation and scaled particle theory results. *J. Chem. Phys.* **107**, 6353-6365 (1997)

[2328] Marcus, Y. Ionic radii in aqueous solutions. *Chem. Rev.* **88**, 1475-1498 (1988)

[2329] Wertheim, M.S. Exact solution of the mean spherical model for fluids of hard spheres with permanent electric dipole moments. *J. Chem. Phys.* **55**, 4291-4298 (1971)

[2330] Smith, P.E. & van Gunsteren, W.F. Consistent dielectric properties of the simple point charge and extended simple point charge water models at 277 and 300 K. *J. Chem. Phys.* **100**, 3169-3174 (1994)

[2331] Placzek, G., Nijboer, B.R.A. & van Hove, L. Effect of short wavelength interference on neutron scattering by dense systems of heavy nuclei. *Phys. Rev.* **82**, 392-403 (1951)

[2332] Cichocki, B, Felderhof, B.U. & Hinsen, K. Electrostatic interactions in periodic Coulomb and dipolar systems. *Phys. Rev. A* **39**, 5350-5358 (1989)

[2333] Peter, C., van Gunsteren, W.F. & Hünenberger, P.H. Solving the Poisson equation for solute-solvent systems using fast Fourier transforms. *J. Chem. Phys.* **116**, 7434-7451 (2002)

[2334] Fukushima, N., Tamura, T. & Ohtaki, H. Dissolution of alkali fluoride and chloride crystals in water studied by molecular dynamics simulations. *Z. Naturforsch. A* **46**, 193-202 (1991)

[2335] Degrève, L. & da Silva, F.L.B. Structure of concentrated aqueous NaCl solution: A Monte Carlo study. *J. Chem. Phys.* **110**, 3070-3078 (1999)

[2336] Degrève, L. & da Silva, F.L.B. Large ionic clusters in concentrated aqueous NaCl solution. *J. Chem. Phys.* **111**, 5150-5156 (1999)

[2337] Koneshan, S. & Rasaiah, J.C. Computer simulation studies of aqueous sodium chloride solutions at 298 K and 683 K. *J. Chem. Phys.* **113**, 8125-8137 (2000)

[2338] Chitra, R. & Smith, P.E. Molecular association in solution: A Kirkwood-Buff analysis of sodium chloride, ammonium sulfate, guanidinium chloride, urea, and 2,2,2-trifluoroethanol in water. *J. Phys. Chem. B* **106**, 1491-1500 (2002)

[2339] Villa, A. & Mark, A.E. Calculation of the free energy of solvation for neutral analogues of amino acid side chains *J. Comput. Chem.* **23**, 548-553 (2002)

[2340] Kunz, W., Belloni, L., Bernard, O. & Ninham, B.W. Osmotic coefficients and surface tensions of aqueous electrolyte solutions: Role of dispersion forces. *J. Phys. Chem. B* **108**, 2398-2404 (2004)

[2341] Yang, Y., Meng, S. & Wang, E.G. A molecular dynamics study of hydration and dissolution of NaCl nanocrystal in liquid water. *J. Phys. Condens. Matter* **18**, 10165-10177 (2006)

[2342] Sanz, E. & Vega, C. Solubility of KF and NaCl in water by molecular simulation. *J. Chem. Phys.* **126**, 014507/1-014507/13 (2007)

[2343] Gu, B., Zhang, F.S., Wang, Z.P. & Zhou, H.Y. The solvation of NaCl in model water with different hydrogen bond strength. *J. Chem. Phys.* **129**, 184505/1-184505/7 (2008)

[2344] Gu, B., Zhang, F.S., Wang, Z.P. & Zhou, H.Y. The non-ideal solvation of NaCl in solvent: a simulation study. *Mol. Phys.* **106**, 1047-1054 (2008)

[2345] del Popolo, M.G., Kohanoff, J., Lynden-Bell, R.M. & Pinilla, C. Clusters, liquids, and crystals of dialkyimidazolium salts. A combined perspective from ab initio and classical computer simulations. *Acc. Chem. Res.* **40**, 1156-1164 (2007)

[2346] Lantelme, F., Turq, P., Quentrec, B. & Lewis, J.W.E. Application of the molecular dynamics method to a liquid system with long range forces (Molten NaCl). *Mol. Phys.* **28**, 1537-1549 (1974)

[2347] Lewis, J.W.E., Singer, K. & Woodcock, L.V. Thermodynamic and structural properties of liquid ionic salts obtained by Monte Carlo computation. *J. Chem. Soc. Farad. Trans. II* **71**, 301-312 (1975)

[2348] Amini, M., Fincham, D. & Hockney, R.W. Molecular dynamics study of the melting of alkali halide crystals. *J. Phys. C* **12**, 4707-4720 (1979)

[2349] Baranyai, A., Ruff, I. & McGreevy, R.L. Monte Carlo simulation of the complete set of molten alkali halides. *J. Phys. C* **19**, 453-465 (1986)

[2350] Tissen, J.T.W.M. & Janssen, G.J.M. Molecular dynamics simulation of molten alkali carbonates. *Mol. Phys.* **71**, 413-426 (1990)

[2351] Tissen, J.T.W.M., Janssen, G.J.M. & van der Eerden, J.P. Molecular dynamics simulation of binary mixtures of molten alkali carbonates. *Mol. Phys.* **82**, 101-111 (1994)

[2352] Guissani, Y. & Guillot, B. Coexisting phases and criticality in NaCl by computer simulation. *J. Chem. Phys.* **101**, 490-509 (1994)

[2353] Ciccotti, G., Jacucci, G. & McDonald, I.R. Transport properties of molten alkali halides. *Phys. Rev. A* **13**, 426-436 (1976)

[2354] Okazaki, S., Ohtori, N. & Okada, I. Molecular dynamics studies on molten alkali hydroxides. 2. Rotational and translational motions of ions in molten LiOH. *J. Chem. Phys.* **93**, 5954-5960 (1990)

[2355] Okazaki, S., Ohtori, N. & Okada, I. Molecular dynamics studies on molten alkali hydroxides. 1. Static properties of molten LiOH. *J. Chem. Phys.* **92**, 7505-7514 (1990)

[2356] Vohringer, G. & Richter, J. Molecular dynamics simulations of molten alkali nitrates. *Z. Naturforsch. A* **56**, 337-341 (2001)

[2357] Galamba, N., de Castro, C.A.N & Ely, J.F. Thermal conductivity of molten alkali halides from equilibrium molecular dynamics simulations. *J. Chem. Phys.* **120**, 8676-8692 (2004)

[2358] Galamba, N., de Castro, C.A.N & Ely, J.F. Molecular dynamics simulation of the shear viscosity of molten alkali halides. *J. Phys. Chem. B* **108**, 3658-3662 (2004)

[2359] Galamba, N., de Castro, C.A.N & Ely, J.F. Shear viscosity of molten alkali halides from equilibrium and nonequilibrium molecular-dynamics simulations. *J. Chem. Phys.* **122**, 224501/1-224501/9 (2005)

[2360] Galamba, N., de Castro, C.A.N & Ely, J.F. Equilibrium and nonequilibrium molecular dynamics simulations of the thermal conductivity of molten alkali halides. *J. Chem. Phys.* **126**, 204511/1-204511/10 (2007)

[2361] Galamba, N. & Costa Cabral, B.J. First principles molecular dynamics of molten NaCl. *J. Chem. Phys.* **126**, 124504/1-124504/8 (2007)

[2362] Rodrigues, P.C.R. & Fernandes, F.M.S.S. Phase diagrams of alkali halides using two interaction models: A molecular dynamics and free energy study. *J. Chem. Phys.* **126**, 024503/1-024503/10 (2007)

[2363] Jacucci, G., Klein, M.L. & McDonald, I.R. Molecular-dynamics study of lattice-vibrations of sodium-chloride. *J. Phys. Lett. (Paris)* **36**, L97-L100 (1975)

[2364] Michielsen, J., Woerlee, P., Graaf, F.V.D. & Ketelaar, J.A.A. Pair potential for alkali metal halides with rock salt crystal structure. *J. Chem. Soc. Farad. Trans. II* **71**, 1730-1741 (1975)

[2365] Rodrigues, P.C.R. & Fernandes, F.M.S.S. Molecular dynamics of phase transitions in clusters of alkali halides. *Int. J. Quant. Chem.* **84**, 169-180 (2001)

[2366] Brunsteiner, M. & Price, S.L. Surface structure of a complex inorganic crystal in aqueous solution from classical molecular simulation. *J. Phys. Chem. B* **108**, 12537-12546 (2004)

[2367] Lim, I.S., Laerdahl, J.K. & Schwerdtfeger, P. Fully relativistic coupled-cluster static dipole polarizabilities of the positively charged alkali ions from Li^+ to 119^+. *J. Chem. Phys.* **116**, 172-178 (2002)

[2368] Pyper, N.C., Pike, C.G. & Edwards, P.P. The polarizabilities of species present in ionic-solutions. *Mol. Phys.* **76**, 353-372 (1992)

[2369] Marzari, N. & Vanderbilt, D. Maximally localized generalized Wannier functions for composite energy bands. *Phys. Rev. B* **56**, 12847-12865 (1997)

[2370] Warshel, A., Kato, M. & Pisliakov, A.V. Polarizable force fields: History, test cases, and prospects. *J. Chem. Theory Comput.* **3**, 2034-2045 (2007)

[2371] Guillot, B. A reappraisal of what we have learnt during three decades of computer simulations on water. *J. Mol. Liq.* **101**, 219-260 (2002)

[2372] Thole, B.T. Molecular polarizabilities calculated with a modified dipole interaction. *Chem. Phys.* **59**, 341-350 (1981)

[2373] Baranyai, A. & Bartok, A. Classical interaction model for the water molecule. *J. Chem. Phys.* **126**, 184508/1-184508/7 (2007)

[2374] Dang, L.X. Solvation of ammonium ion. A molecular dynamics simulation with nonadditive potentials. *Chem. Phys. Lett.* **213**, 541-546 (1993)

[2375] Dang, L.X. & Smith, D.E. Molecular dynamics simulations of aqueous ionic clusters using polarizable water. *J. Chem. Phys.* **99**, 6950-6956 (1993)

[2376] Caleman, C. & van der Spoel, D. Evaporation from water clusters containing singly charged ions. *Phys. Chem. Chem. Phys.* **9**, 5105-5111 (2007)

[2377] Miller, Y., Thomas, J.L., Kemp, D.D., Finlayson-Pitts, B.J., Gordon, M.S., Tobias, D.J. & Gerber, R.B. Structure of large nitrate-water clusters at ambient temperatures: Simulations with effective fragment potentials and force fields with implications for atmospheric chemistry. *J. Phys. Chem. A* **113**, 12805-12814 (2009)

[2378] Wick, C.D. & Xantheas, S.S. Computational investigation of the first solvation shell structure of interfacial and bulk aqueous chloride and iodide ions. *J. Phys. Chem. B* **113**, 4141-4146 (2009)

[2379] Babu, C.S. & Lim, C. Empirical force fields for biologically active divalent metal cations in water. *J. Phys. Chem. A* **110**, 691-699 (2006)

[2380] Wheatley, R.J. The solvation of sodium ions in water clusters: intermolecular potentials for Na^+-H_2O and H_2O-H_2O. *Mol. Phys.* **87**, 1083-1116 (1996)

[2381] dos Santos, D.J.V.A., Müller-Plathe, F. & Weiss, V. Consistency of ion adsorption and excess surface tension in molecular dynamics simulations of aqueous solutions. *J. Phys. Chem.* **112**, 19431-19442 (2008)

[2382] Rogers, D.M. & Beck, T.L. Quasichemical and structural analysis of polarizable anion hydration. *J. Chem. Phys.* **132**, 014505/1-014505/12 (2010)

[2383] Wick, C.D., Kuo, I.-F.W., Mundy, C.J. & Dang, L.X. The effect of polarizability for understanding the molecular structure of aqueous interfaces. *J. Chem. Theory Comput.* **3**, 2002-2010 (2007)

[2384] Roux, B. Non-additivity in cation-peptide interactions. A molecular dynamics and ab initio study of Na^+ in the gramicidin channel. *Chem. Phys. Lett.* **212**, 231-240 (1993)

[2385] Allen, T.W., Andersen, O.S. & Roux, B. Ion permeation through a narrow channel: Using gramicidin to ascertain all-atom molecular dynamics potential of mean force methodology and biomolecular force fields. *Biophys. J.* **90**, 3447-3468 (2006)

[2386] Baker, C.M., Lopes, P.E.M., Zhu, X., Roux, B. & McKerell, A.D. Accurate calculation of hydration free energies using pair-specific Lennard-Jones parameters in the CHARMM Drude polarizable force field. *J. Chem. Theory Comput.* **6**, 1181-1198 (2010)

[2387] Hansen, H.S. & Hünenberger, P.H. A reoptimized GROMOS force field for hexopyranose-based carbohydrates accounting for the relative free energies of ring conformers, anomers, epimers, hydroxymethyl rotamers and glycosidic linkage conformers. *J. Comput. Chem., in press* (2011)

[2388] Li, G., Zhang, X. & Cui, Q. Free energy perturbation calculations with combined QM/MM potentials. Complications, simplifications, and applications to redox potential calculations. *J. Phys. Chem. B* **107**, 8643-8653 (2003)

[2389] Kamerlin, S.C.L., Haranczyk, M. & Warshel, A. Progress in *ab initio* QM/MM free-energy simulations of electrostatic energies in proteins: Accelerated QM/MM studies of pK, redox reactions and solvation free energies. *J. Phys. Chem. B* **113**, 1253-1272 (2009)

[2390] Zeng, X., Hu, H., Hu, X. & Yang, W. Calculating solution redox free energies with ab initio quantum mechanical/molecular mechanical minimum free energy path method. *J. Chem. Phys.* **130**, 164111/1-164111/8 (2009)

[2391] Asthagiri, D., Pratt, L.R. & Ashbaugh, H.S. Absolute hydration free energies of ions, ion-water clusters and quasichemical theory. *J. Chem. Phys.* **119**, 2702-2708 (2003)

[2392] Rempe, S.B., Asthagiri, D. & Pratt, L.R. Inner shell definition and absolute hydration free energy of K^+(aq) on the basis of quasi-chemical theory and *ab initio* molecular dynamics. *Phys. Chem. Chem. Phys.* **6**, 1966-1969 (2004)
[2393] Asthagiri, D., Pratt, L.R. & Kress, J.D. *Ab initio* molecular dynamics and quasichemical study of H^+(aq). *Proc. Natl. Acad. Sci. USA* **102**, 6704-7608 (2005)
[2394] Pliego, J.R. & Riveros, J.M. The cluster-continuum model for the calculation of the solvation free energy of ionic species. *J. Phys. Chem. A* **105**, 7241-7247 (2001)
[2395] Topol, I.A., Tawa, G.J., Burt, S.K. & Rashin, A.A. On the structure and thermodynamics of solvated monoatomic ions using a hybrid solvation model. *J. Chem. Phys.* **111**, 10998-11014 (1999)
[2396] Mejías. J.A., Lago, Calculation of the absolute hydration enthalpy and free energy of H^+ and OH^-. *J. Chem. Phys.* **113**, 7306-7316 (2000)
[2397] Zhan, C.-G. & Dixon, D.A. First-principles determination of the absolute hydration free energy of the hydroxide ion. *J. Phys. Chem. A* **106**, 9737-9744 (2002)
[2398] Zhan, C.-G. & Dixon, D.A. Hydration of the fluoride anion: Structures and absolute hydration free energy from first-principles electronic structure calculations. *J. Phys. Chem. A* **108**, 2020-2029 (2004)
[2399] Rosta, E., Haranczyk, M., Chu, Z.T. & Warshel, A. Accelerating QM/MM free energy calculations: Representing the surroundings by an updated mean charge distribution. *J. Phys. Chem. B* **112**, 5680-5692 (2008)
[2400] Min, D., Zheng, L., Harris, W., Chen, M., Lv, C. & Yang, W. Practically efficient QM/MM alchemical free energy simulations: The orthogonal space random walk strategy. *J. Chem. Theory Comput.* **6**, 2253-2266 (2010)
[2401] Chen, E.S. & Chen, E.C.M. Comment on "Ab initio molecular dynamics calculation of ion hydration free energies" [J. Chem. Phys. 130, 204507 (2010)]. *J. Chem. Phys.* **133**, 047103/1-047103/2 (2010)
[2402] Sulpizi, M. & Sprik, M. Acidity constants from DFT-based molecular dynamics simulations. *J. Phys.: Condens. Matter* **22**, 284116/1-284116/8 (2010)
[2403] Cheng, J., Sulpizi, M. & Sprik, M. Redox potentials and pK_a for benzoquinone from density functional theory based molecular dynamics. *J. Chem. Phys.* **131**, 154504/1-154504/20 (2009)
[2404] Blumberger, J. & Sprik, M. Redox free energies from vertical energy gaps: Ab initio molecular dynamics implementation. *Lect. Notes Phys.* **704**, 481-506 (2006)
[2405] Oberhofer, H. & Blumberger, J. Charge constrained density functional molecular dynamics for simulation of condensed phase electron transfer reactions. *J. Chem. Phys.* **131**, 06401/1-06401/11 (2009)
[2406] Costanzo, F., Sulpizi, M., Della Valle, R.G. & Sprik, M. First principles study of alkali-tyrosine complexes: Alkali solvation and redox properties. *Phys. Chem. Chem. Phys.* **4**, 1049-1056 (2008)

Table index

FIGURE INDEX

Keyword index

A

B

C

D

E

G

H

I

Q

R

S

T

U

V

W

X

Z